To Edward John Sm

in recognition of his great achievements and contribution to Science.

with best wishes,

Dr. Nikolaos Solomos
Director of the Eudoxos' Observatories
Mount Ainos, Kefalonia, Greece.

May 7, 2008.

RECENT ADVANCES IN ASTRONOMY AND ASTROPHYSICS

RECENT ADVANCES IN ASTRONOMY AND ASTROPHYSICS

7th International Conference of the Hellenic Astronomical Society

Lixourion, Kefallinia Island, Greece *8 – 11 September 2005*

EDITOR

Nikolaos H. Solomos

Physical Sciences Sector, Hellenic Naval Academy
Piraeus, Greece
and
EUDOXOS National Astronomy Center
Ainos Mountain, Kefallinia Island, Greece

SPONSORING ORGANIZATIONS

Municipality of Lixourion Pallis
Alexander S. Onassis Foundation
Hellenic Ministry of Tourism
EUDOXOS National Observatory of Education
Prefecture of Kefallinia and Ithaki
Hellenic National Committee for Astronomy
Hellenic Naval Academy

AIP
75 Years of Service

Melville, New York, 2006
AIP CONFERENCE PROCEEDINGS ■ 848

Editor

Nikolaos H. Solomos
Pure & Applied Physics Laboratory
Physical Sciences Sector
Hellenic Naval Academy
Hatzikyriakeion, Piraeus 18503
Greece
and
EUDOXOS National Astronomy Center
Ainos Mountain, Kefallinia Island, Greece

E-mail: nsolom@snd.edu.gr

L.C. Catalog Card No. 2006928750
ISBN 0-7354-0343-0
ISSN 0094-243X

Printed in the United States of America

Hellenic Astronomical Society

"RECENT ADVANCES IN ASTRONOMY & ASTROPHYSICS"

(7th International Conference)

Lixourion, Kefallinia Island, Greece
8-11 September 2005

SCIENTIFIC ORGANIZING COMMITTEE

- P.G. Laskarides (University of Athens) [Chairman]
- K. Tsinganos (University of Athens)
- E. Theodossiou (University of Athens)
- X. Moussas (University of Athens)
- D. Hadjidimitriou (University of Crete)
- K. Kokkotas (University of Thessaloniki)
- V. Geroyiannis (University of Patras)
- H. Dara (Academy of Athens)
- N. Kylafis (University of Crete)
- P. Niarchos (University of Athens)
- E. Harlaftis+ (National Observatory of Athens)
- S. Papadakis (University of Crete)
- J. Georgantopoulos (National Observatory of Athens)
- J.D. Hadjidemetriou (University of Thessaloniki)
- S. Persides (University of Thessaloniki)
- J. Seiradakis (University of Thessaloniki)
- N. H. Solomos (Hellenic Naval Academy & 'Eudoxos' National Observatory)
- Ch. Tomboulides (Ministry of Education & Religious Affairs)

LOCAL ORGANIZING COMMITTEE

- Nikolaos Solomos [Chairman] (Physical Sciences Sector, Hellenic Naval Academy, Piraeus & 'EUDOXOS' National Observatory of Education, Mount Ainos, Kefallinia)
- Alexandros Heilaris (Hellenic Navy & Physical Sciences Sector, Hellenic Naval Academy, Piraeus)
- Aikaterini Zachariadou (Institute of Nuclear Physics, NRCPS 'DEMOKRITOS', Athens & NOE 'EUDOXOS' Kefallinia)
- Olga Malandraki (National Observatory of Athens)
- Georgios Fanourakis (Institute of Nuclear Physics, NRCPS 'DEMOKRITOS', Athens & NOE 'EUDOXOS' Kefallinia)
- Dionysios Patrikios (NOE 'EUDOXOS' Kefallinia)
- Vasilios Kokalis(NOE 'EUDOXOS' Kefallinia)

Editorial Supervision by Sabine Kessler (American Institute of Physics)
Photographs by Maria Solomou & Haralampos Solomos

Ευχήν Οδυσσεί

ΑΣΤΗΡ ΠΡΙΝ ΜΕΝ ΕΛΑΜΠΕΣ ΕΝΙ ΖΩΟΙΣΙΝ
[ΕΩΟΣ`
ΝΥΝ ΔΕ ΘΑΝΩΝ ΛΑΜΠΕΙΣ ΕΣΠΕΡΟΣ ΕΝ
[ΦΘΙΜΕΝΟΙΣ.

Πλάτων

CONTENTS

PART 1

SUN, PLANETS, AND THE INTERPLANETARY MEDIUM

(Conveners: X. Moussas and H. Dara)

PART 2

OUR GALAXY: STARS, CLUSTERS, AND THE INTERSTELLAR MEDIUM

(Conveners: N. Kylafis, P. Niarchos, and E. T. Harlaftis (deceased))

PART 3

EXTRAGALACTIC ASTROPHYSICS

(Conveners: S. Papadakis and J. Georgantopoulos)

PART 4

THEORETICAL INSIGHTS IN DYNAMICAL ASTRONOMY, RELATIVITY, AND COSMOLOGY

(Conveners: J. D. Hadjidemetriou, K. Kokkotas, and S. Persides)

PART 5

INSTRUMENTATION AND METHODS OF ASTRONOMICAL OBSERVATIONS

(Conveners: J. Seiradakis and N. H. Solomos)

PART 6

HISTORY AND EDUCATION IN ASTRONOMY

(Conveners: E. Theodossiou and H. Tomboulides)

INVITED REVIEW LECTURERS

1. Professor **Keith Horne**, Head of Astronomy, School of Physics & Astronomy, University of St. Andrews, St. Andrews, Scotland, UK (*Emilios Harlaftis Lecture*: *"ExtraSolar Planets")*.

2. Professor **Eric Priest**, Fellow of the Royal Society, Bishop Wardlaw Professor at University of St.Andrews, St.Andrews, Scotland, UK (*"Our Enigmatic Sun"*)

3. Professor **Attilio Ferrari**, Dipartimento di Fisica Generale, Consorzio Interuniversitario Universita di Torino, Fisica Spaziale, Italy *("Astrophysical Jets")*

4. Professor **James Truran Jr.**, Department of Astronomy & Astrophysics, the Enrico Fermi Institute, and the College, ASCI Flash Center, University of Chicago, USA (*"Nucleosynthesis, Challenges and New Developments")*

5. Professor **Mike Edmunds,** Head of School of Physics and Astronomy, University of Wales, Cardiff, Wales, UK *("The Antikythera Mechanism")*.

6. Dr **Athena Coustenis,** LESIA, Observatoire de Paris, Meudon, France *("The Cassini/Huygens mission to Saturn/Titan").*

List of Participants

Alissandrakis, Costas	University of Ioannina, Greece
Anastasiadis, Anastasios	National Observatory of Athens, Institute for Space Applications and Remote Sensing
Angelakis, Emmanouil	Max Planck Institute for Radio Astronomy
Antoniadou, Irine	University of Athens
Antoniou, Antonios	University of Athens Faculty of Physics
Antonopoulou, Evgenia	National And Kapodistrian University of Athens
Apostolatos, Theocharis	University of Athens
Athanasiadis, Dimitris	National Observatory of Athens
Athanassoula, Lia	Observatoire de Marseille Provence, France
Avgoloupis, Stavros	Aristotle University Thessaloniki
Baan, Willem	ASTRON, Netherlands
Berti, Emanuele	Washington University in Saint Louis, USA
Bloomfield, Shaun D.	Queens University of Belfast
Bosma, Albert	Observatoire de Marseille, France
Bouratzis, Costas	University of Athens
Caranicolas, Nikolaos	University of Thessaloniki
Cavus, Huseyin	Canakkale Onsekiz Mart Uni. Arts & Science Fac. Physics Dep.
Chartas, George	Penn State University, USA
Chatzichristou, Eleni	National Observatory of Athens
Contadakis, Michael E.	Department of Surveying and Geodesy, Aristotle University of Thessaloniki
Contopoulos, George	Academy of Athens, Greece
Contopoulos, Ioannis	Research Centre for Astronomy & Applied Mathematics, Academy of Athens
Coustenis, Athena	LESIA, Paris-Meudon Observatory, France
Dallas, T.G.	University of Thessaly
Danezis, Emmanuel	University of Athens
Dara, H.	Research Centre for Astronomy Academy of Athens, Greece
Dialynas, Kostas	University of Athens
Dimitrakoudis, Stavros	National and Kapodistrian University of Athens
Dionatos, Odysseas	University of Athens
Douskos, Christos	Department of Engineering Sciences, University of Patras
Edmunds, Mike	School of Physics and Astronomy, Cardiff University
Efthymiopoulos, Christos	Research Centre for Astronomy and Applied Mathematics Academy of Athens
Emmanoulopoulos, Dimitrios	Landessternwarte Heidelberg, Germany
Exarhos, Georgios	University of Athens - Physics Department
Fanourakis, Giorgos	Eudoxos National Observatory & Institute of Nuclear Physics, NRCPS "Demokritos"
Ferrari, Attilio	Università di Torino, Italy
Freeth, Tony	Cardiff University, Wales
Gavras, Panagiotis	National & Kapodestrian University of Athens, Physics Department, Section of Astronomy ,Astrophysics and Mechanics
Gazeas, Kosmas	National and Kapodistrian University of Athens

Georgantopoulos, Ioannis	Institute of Astronomy & Astrophysics, National Observatory of Athens
Georgiou, Marina	University of Athens
Georgopoulos, Petros	Eudoxos - National Observatory of Education, Greece
Geranios, Thanassis	Nuclear Physics and Elementary Particles
Geronikolou, Stella	Academy of Athens ,RCAAM
Gerontidou, Maria	University of Athens
Geroyannis, Vassilis	Department of Physics University of Patras
Giannikakis, John	National and Kapodistrian University of Athens
Giannios, Dimitrios	Max Planck Institute for Astrophysics, Garching, Germany
Gini, Evgenia	University of Athens
Giouvanellis, Georgios	National and Kapodistrian University of Athens
Gizani, Nectaria	GAUC, Universidade de Coimbra, Portugal
Gonidakis, Ioannis	National and Kapodistrian University of Athens
Gontikakis, Costis	Research Centre for Astronomy and Applied Mathematics Academy of Athens
Goudis, Christos	Institute of Astronomy & Astrophysics, National Observatory of Athens
Gracia, Jose	IASA, University of Athens
Gradoula, Georgia-Peristera	National and Kapodistrian University of Athens
Griv, Evgeny	Ben-Gurion University, Israel
Hadjidemetriou, John	University of Thessaloniki
Hatzidimitriou, Despina	University of Crete
Hillaris, Alexander	Hellenic Navy & Physical Sciences Sector, Hellenic Naval Academy, Piraeus
Horne, Keith	SUPA , St.Andrews University, Scotland
Hussain, Sadiq	National Institute of Sciences and Technology
Isliker, Heinz	Section of Astrophysics, Astronomy, and Mechanics, Dept. of Physics, Aristotle University of Thessaloniki
Ivanova, Aleksandra	Main Astronomical Observatory National Academy of Sciences of Ukraine
Kalantonis, Vassilis	Department of Engineering Sciences, University of Patras
Kalas, Paul	University of California, Berkeley, CA, USA.
Kalimeris, Anastasios	National and Kapodistrian University of Athens
Kalvouridis, Tilemahos	National Technical University of Athens
Karalidi, Theodora	Department of Astrophysics, Astronomy and Mechanics, University of Athens,
Karastergiou, Aris	Institut de Radio Astronomie Millimétrique, France
Katsanikas, Manthos	University of Athens
Katsikas, Valadis	National & Kapodistrian University of Athens
Katsiyannis, Thanassis	Royal Observatory of Belgium
Kazanas, Demosthenes	NASA/GSFC, USA
Kitsionas, Spyridon	Astrophysikalisches Institut Potsdam, Germany
Klimchuk, James	U.S. Naval Research Laboratory
Kokkotas, Kostas	Department of Physics, Aristotle University of Thessaloniki
Kontogeorgos, Athanasios	Technological Educational Institute, Lamia
Kontogiannis, Giannis	University of Athens, Greece
Kotoulas, Thomas	Aristotle University of Thessaloniki
Koukouli, Mariliza	Laboratory of Atmospheric Physics, Physics Department, Aristotle University of

	Thessaloniki
Koulouridis, Elias	Physics Department, University of Patras.
Krimigis, Stamatios	Applied Physics Laboratory, Johns Hopkins University, Laurel, MD, USA & Academy of Athens, Chair of Science of Space, Athens, Greece
Kylafis, Nikolaos	University of Crete
Labropoulos, Panagiotis	Division of Astronomy, Astrophysics and Mechanics, Department of Physics University of Athens
Laguna, Pablo	Penn State University
Laskarides, Paul	Faculty of Physics, University of Athens, Greece
Lazaridis, Kosmas	Aristotle University of Thessaloniki
Livadiotis, George	University of Athens
Livanou, Evdokia	University of Athens
Loukes Gerakopoulos, George	University of Athens
Lykou, Foteini	University of Crete, Greece
Lyratzi, Evaggelia	University of Athens, Department of Physics, Greece
Magkanari, Maria	University of Athens, Greece
Malandraki, Olga	National Observatory of Athens, Institute for Space Applications and Remote Sensing & Space Research Lab., Democritus University of Thrace
Manimanis, Vassilios	National & Kapodistrian University of Athens
Manos, Thanos (Athanasios)	OAMP (Observartoire de Marseille, France) CRANS (Department of Mathematics, University of Patras, Greece)
Manousakis, Antonis	University of Crete, Greece
Marhavilas, Panagiotis	Demokritos University of Thrace
Mariatos, George	University of Athens
Markellos, Vassilis	Department of Engineering Sciences, University of Patras
Marsh, Thomas	Department of Physics University of Warwick, UK
Mastichiadis, Apostolos	University of Athens, Department of Physics, Greece
Mathioudakis, Mihalis	Queens University of Belfast, UK
Mavromichalaki, Helen	University of Athens Physics Department , Nuclear and Particle Physics Section
Mcateer, James	Queens University, Belfast
Meli, Athina	Institute of Physics University of Dortmund
Melnikov, Victor	Radiophysical Research Institute, Russia
Metallinou, Fiori - Anastasia	Aristotle University of Thessaloniki
Metheniti, Kiriaki	University of Athens
Misiriotis, Angelos	Physics Department, University of Crete
Mislis, Dimitris	Section of Astronomy & Astrophysics, Department Of Physics, Aristotle University of Thessaloniki, Greece
Mitsakou, Eleftheria	University of Athens, Greece
Moraitis, Kostas	University of Athens, Physics Department, Section of Astrophysics, Astronomy and Mechanics
Moussas, Xenophon	National And Kapodistrian University of Athens
Nandra, Kirpal	Astrophysics Group, Imperial College, London
Nanouris, Nikolaos	National and Kapodistrian University of Athens
Nestoras, John	Aristotle University of Thessaloniki - Department of Physics - Section of Astronomy, Astrophysics & Mechanics

Niarchos, Panagiotis	University of Athens, Greece
Niarchou, Anastasia	Imperial College, London
Nikolaidis, Dimitris	University of Athens, Greece
Nindos, Alexander	University of Ioannina, Physics Department, Greece
Nykytyuk, Tetyana	Main Astronomical Observatory of National Academy of Sciences of Ukraine
Paouris, Evaggelos	Astronomy, Astrophysics and Mechanism Section, Physics Department, University of Athens
Papadakis, Iossif	Physics Department University of Crete
Papadimitriou, Christos	National Observatory of Athens, University of Athens
Papadopoulos, Nikolaos	Department of Physics, Section of Astrophysics, Astronomy and Mechanics, University of Thessaloniki
Papaspirou, Panagiotis	National and Kapodistrian University of Athens
Papathanassoglou, Dimitrios	National & Kapodistrian University of Athens
Patsourakos, Spiros	U.S. Naval Research Lab & George Mason University
Perdios, Efstathios	Department of Engineering Sciences, University of Patras
Peres, Giovanni	Dipartimento Di Scienze Fisiche ed Astronomiche - Universita` Di Palermo
Perivolaropoulos, Leandros	University of Ioannina, Physics Department
Persides, Sotirios	University of Thessaloniki
Petoussis, Vlassis	Athens University & Technological Education Institute Lamia
Petropoulos, Vassilis	Academy of Athens, Research Center for Astronomy and Applied Mathematics
Petropoulos, Georgios	Aristotle University of Thessaloniki
Plainaki, Christina	University of Athens
Polatidis, Antonios	Max Planck Institute for Radio Astronomy
Pothitakis, George	University of Athens
Pourbaix, Dimitri	Universite Libre de Bruxelles
Preka-Papadema, Panagiota	University of Athens
Priest, Eric	St Andrews University, Scotland
Pyrlis, Orestis	University of Athens
Pyrzas, Stylianos	Section of Astronomy, Astrophysics & Mechanics, Department of Physics, Aristotle University of Thessaloniki
Reznikova, Veronika	Radiophysical Research Institute, Russia
Rigos, Alexandros	University of Athens
Romeou, Zaharenia	Dipartimento di Astronomia e Scienza Dello Spazio,Universita Degli Studi di Firenze, Italy
Roumelioti, Eleni	University of Athens
Sakelliou, Irini	Max-Planck-Institute for Astronomy (MPIA)
Sarlanis, Christos	University of Athens
Sarris, Emmanuel	Democritus University of Thrace, Space Research Laboratory
Sarris, Theodoros	Democritus University of Thrace
Seiradakis, John	University of Thessaloniki Greece
Serber, Alexander	Institute of Applied Physics of The Russian Academy of Sciences, Russia
Sergis, Nick	National and Kapodistrian University of Athens
Shoemaker, Deirdre	Penn State University, USA
Siopis, Christos	University of Michigan, USA

Sipahi, Esin	Ege University Observatory, Bornova, Izmir, Turkey
Skokos, Charalampos	Academy of Athens & University of Patras
Smponias, Theodoros	
Solomos, Nikolaos	Physics Sector, Hellenic Naval Academy, & Eudoxos National Observatory
Souvatzoglou, George	University of Athens
Stamatellos, Dimitris	School of Physics & Astronomy, Cardiff University, Wales
Stavrinos, Panayiotis	Division of Algebra and Geometry, Department Of Mathematics University of Athens
Stergioulas, Nikolaos	Department of Physics Aristotle University of Thessaloniki
Theodossiou, Efstratios	University of Athens, School of Physics
Tomboulides, Hariton	1st Lyceum of Naousa
Tremou, Evagelia	University of Thessaloniki
Trichas, Markos	Imperial College, London
Tritakis, Vasilis	Research Centre for Astronomy, Academy of Athens
Truran, James	University of Chicago
Tsalmantza, Paraskevi	National and Kapodistrian University of Athens
Tsamis, Yiannis	Observatoire de Paris-Meudon
Tsinganos, Kanaris	University of Athens
Tsitsipis, Panagiotis	Technological Educational Institute of Lamia
Tzanavaris, Panayiotis	Institute of Astronomy & Astrophysics National Observatory of Athens
Tzatzakis, Vassilis	University of Ioannina, Greece
Tzioumis, Anastasios	Australia Telescope National Facility (ATNF), CSIRO
Vamvatira-Nakou, Chloe	National & Kapodistrian University of Athens
Varvoglis, Charalampos	Aristotle University of Thessaloniki
Vavoulidis, Miltiadis	Aristotle University of Thessaloniki
Ventura, Joseph	Department of Physics, University of Crete, Greece
Vlahakis, Nektarios	University of Athens
Vlahos, Loukas	Department of Physics University of Thessaloniki
Voglis, Nikos	Research Centre for Astronomy, Academy of Athens
Voitsekhovska, Anna	Main Astronomical Observatory of National Academy of Sciences of Ukraine
Voyatzi, Despina	Department of Physics University of Thessaloniki
Vozikis, Christos	Aristotle University of Thessaloniki
Xanthopoulos, Emily	University of California Davis & Lawrence Livermore National Laboratory
Zacharopoulou, Olga	University of Athens
Zavros, Panayiotis	National & Kapodistrian University of Athens, Department of Physics
Zoulias, Manolis	

PREFACE

In 1905, the Annus Mirabilis of Physics, Einstein published three ground-breaking papers in which he proposed that light consists of energy quanta, demonstrated his ability to extend Maxwell's electromagnetism to the moving bodies introducing the postulates of the special theory of relativity and, finally, examined the dependence of the inertia of a body upon its energy content. It is exactly this trilogy of insights that laid the foundations of Quantum Mechanics and Relativity which, in turn, influenced the transformation of 'Astronomy' to 'Astrophysics' through the ability to analyze starlight, to infer about the properties of matter and to apply a theory of gravity to the universe as a whole. Whether or not Dark Matter and Dark Energy will eventually be proved Einstein's greatest triumph, relativistic astrophysics and cosmology represent his main astronomical legacy: nowadays several new ground-based observatories, instruments and space missions are joining efforts to create progress in the fields of Gravitational-Wave Astrophysics, Dark Matter-Energy, Large Scale Structure of the Universe, AGNs and Microquasars, Neutron Stars, Supermassive Black Holes in Galaxies and the possible association of black hole creation with Gamma Ray Bursts.

Just at the 100th birthday of the quantum, an International Conference on 'Recent Advances In Astronomy & Astrophysics', marking the 7th biannual conference of the Hellenic Astronomical Society, took place in the beautiful Mediterranean town of Lixourion in Kefallinia Island, Greece.

The purpose of the conference was to summarize recent progress in the field and to create an environment to disseminate the results of the current theoretical and observational studies in Astronomy, Astrophysics and Space Physics. Also, to provide a forum for exchange of views and best practices in those fields and stress both the diversity and interdisciplinary character of astronomical enquiry. Its aims were also extended to the fostering of the cooperation of astrophysicists from Greece or Greek Diaspora with their colleagues worldwide and to the creation of a source of inspiration for the younger generation.

The scientific agenda of the conference contained plenary, parallel and poster sessions as well as two special meetings-moderated discussions- open to all interested participants. The Sessions of the 7th Conference were devoted to: Sun, Planets & Interplanetary Medium, Our Galaxy: Stars, Clusters, Interstellar Medium, Extragalactic Astrophysics, Theoretical Insights in Dynamical Astronomy, Relativity and Cosmology, Instrumentation & Methods of Astronomical Observations, History and Education in Astronomy. The special meetings were focused on "Future of Astronomy in Greece" and "Greece and the European Space Agency".

It was obvious, that the outstanding academic and cultural tradition of the small town of Lixourion for more than four centuries, its natural beauty, and its idiosyncratic lively atmosphere, should be matched by a meeting with the broadest appeal possible to the astronomical community.

It was for these reasons that it was decided that younger scientists -especially postgraduate students and post-docs- were to play an important role in the conference, both as attendants and speakers and, in parallel, established astronomers should be given a chance to pause for a moment from their own fields of research and get an insight into the activities in other fields of astronomy, possibly related to their own.

All the above considerations reflect in the present Volume: Not only is it important to include the contributions of junior scientists for their own benefit, but it is our belief that the larger the number of young people involved , the more chances to choose the most fertile fields for active research, and the greatest the influence the conference will have long after the return to their home institutions. We offered all participants the possibility to publish work after having their contribution accepted by the committee of conveners of each session. No arbitrary choices had to be made by the editor to the results of those committees. The poster papers in particular –mostly from the young scientists- were limited to 5 pages in extent and, therefore, must be viewed not just as a hint or pointer towards a more detailed presentation elsewhere but as a quite detailed exposition of work.

Given the fact that we live in the Golden Age of Astrophysics as there has been a tremendous growth in the number and depth of astrophysical investigations over the past 30 years, (..and that many important contributors were unable to have their manuscript ready for publication in time), one should not expect to find here the latest developments in every subfield of Astronomy. Thus, the title of the conference and consequently the present Volume should be considered rather as an idealistic assumption than as the kind of achievement implied by its title. Nevertheless, there is a good grasp of excellent invited reviews and latest multidisciplinary observational and theoretical results covering a broad range of topics within this volume, and, -as expected-, a clue for an appreciation of the trends and current interests of the Greek astronomical community.

The present volume contains the majority of the contributions made. It is divided in Parts corresponding to the conference sessions. Although the Invited Review Lectures were plenary, they have been distributed to the Parts by subject and are placed first. As a rule, they are followed by the invited presentations, the 'oral' papers and the 'poster' ones . However, in a few cases for the sake of homogeneity and integrity in the course of a particular topic, the editor has changed the above sequence by grouping papers by subject. The invited presentations are explicitly indicated in the Table of Contents.

The conference was hosted in the luxurious Conference Center of the town of Lixourion, from September 8 to 11, 2005. Opening remarks were made by the HEL.A.S President Prof. P. Laskarides, and Academician Prof. G. Contopoulos.

Some 170 astronomers from Greece and other countries all over the world participated to the conference and were given a very warm welcome from the Prefect of Kefallinia & Ithaki Mr. D. Lefkaditis and the Mayor of Lixourion Mr. V. Rouhotas.

The Municipality of Lixourion-Pallis offered the participants a reception in the gardens of the Iakovatios Library, a magnificent 18th century building a few hundred meters from the town centre. Also, in the evening of the last day, participants enjoyed a musical concert with lyrical opera pieces and songs by F. Metallinou, a violin concert by N. Pylarinos, works of greek composers of classical music and composers from Kefallinia. Among them the string quartet 'Opera Quattro' played the contemporary composition 'Of The Stars' written especially for the conference by the composer Gerasimos Pylarinos.

The conference was sponsored by the Municipality of Lixourion Pallis, the Alexander S. Onassis Foundation, the National Astronomy Centre 'Eudoxos', the Greek Ministry of Tourism, the Prefecture of Kefallinia & Ithaki, the Greek National Committee for Astronomy and the Hellenic Naval Academy. We are grateful for this support which allowed us to welcome a large number of participants in excellent conditions.

As usual, many individuals whose primary goal was a successful conference rather than any kind of personal recognition, helped decisively. I would like to thank the Chairman and the other members of the SOC for their efforts in the preparation of the program and the members of the LOC for their close collaboration. Particular thanks are due to M. Kambrianis, F. Georgatos, D. Apostolou, Apollon Oikonomopoulos, I. Roussos, C. Zotos all affiliated with the EUDOXOS National Observatory, who devised many informatic and networking tools that made possible not only to handle the enormous administrative tasks of the organization but to establish stable wireless networking connections with the Observatory over the Bay of Livadion. This fact offered the participants the possibility to enjoy broadband internet facilities for the whole duration of the conference. I also extend my thanks to Mr. Nestor Ventouras and Sokrates Papadatos, also fellows in the municipality council of Lixourion, for helping arrange numerous issues of preparation and transportation, to the members of the Small Choir of Lixourion for the excellent short concert they gave us during the reception in Iakovatios Library gardens, to Mrs. Grekousi for the superb guidance in the Iakovatios Museum and her demonstrated knowledge of the historic subjects that impressed all the participants, as well as to Mrs. M. Solomou for her photographical coverage and Mr. M. Zoulias for his excellent artistic work in implementing the conference logo and poster according to the guidelines given.

The conference could not have taken place without the generous help of Mrs. M. Moschopoulou in solving the financial problems so effectively and Mr. D. Voros and Othon Koutavas who solved organizational and audiovisual problems, respectively. In addition, many citizens of Lixourion and even more, are to be thanked for helping in various ways to achieve the smooth running of the Conference.

Last but not least, we are deeply indebted to the American Institute of Physics (AIP) and particularly to Mrs.Sabine Kessler, Mrs. Maya Flikop and Mrs. Tina Choy for their patience, interest and effort put in this volume; otherwise this collection of papers would not now rest in your hands.

Nikolaos H. Solomos, Editor

Pure and Applied Physics Laboratory, Hellenic Naval Academy, Piraeus, 18503
National Astronomy Centre "*Eudoxos*", Mount Ainos, Kefallinia, Greece,

PART 1

SUN, PLANETS, AND THE INTERPLANETARY MEDIUM

(Conveners: X. Moussas and H. Dara)

Professor Eric Priest, St. Andrews University, Scotland

Our Enigmatic Sun

E.R. Priest

Mathematics Institute, St Andrews University, St Andrews KY16 9SS, UK

Abstract. Many major puzzles in Solar Physics have not yet been answered and are of wide importance for astrophysics in general, such as: How is the magnetic field generated? How is the corona heated? How do solar flares occur? What is the effect of solar variations on the Earth's climate? How is the solar wind accelerated? In each case, the questions have multiple parts, some of which have been recently answered with the help of spectacular space observations, while others are being refined in ways that will be described.

Keywords: Sun, MHD

INTRODUCTION

I am from St Andrews University in Scotland, where there is a very strong philosophy department with one of the world experts on Plato whom I greatly enjoy hearing. The university was founded in 1410, a little earlier than Athens Univiersity but of course well after The Academy! It is a very great pleasure for me to come to such a wonderful country with such a civilised approach to life. It has also been marvellous to meet up with old friends again such as Kanaris Tsinganos and Loukas Vlahos, whom I first met when they were research students in Chicago and Maryland and whose work I have long admired.

Worship of the Sun has taken place for centuries, and indeed many people come today to the beautiful beaches of Greece to continue that tradition, but why should we study the Sun today? Firstly, because it is of very great interest in its own right. Secondly, because it has profound influences on the Earth and its climate. Thirdly, because it is of key importance for astronomy as a whole, since we can study fundamental cosmic processes at work on the Sun.

However, many of the basic properties of the Sun are still a mystery. For example, we do not know: how the Sun's magnetic field is generated; how the solar wind is accelerated; how the corona is heated; how ejections of mass occur or how particles are accelerated in solar flares. Nevertheless, great progress on each of these questions has been made over the past five years and they have all been greatly refined and unpacked. So, the Sun is at present one of the liveliest branches of astronomy.

Traditionally, there has always been a close link between mathematics and astronomy, which was certainly true in St Andrews in the seventeenth century. I am fortunate to hold the Gregory chair of mathematics, named after James Gregory (1638-1675), who invented the gregorian telescope and was also one of the co-founders of calculus (with Newton and Leibniz). He was elected a Fellow of the Royal Society (an academician) in 1668 at the age of 30, and the same year was appointed the first regius professor of mathematics in St Andrews (Figure 1).

CP848, *Recent Advances in Astronomy and Astrophysics,*
7th *International Conference of the Hellenic Astronomical Society,* edited by N. H. Solomos

FIGURE 1. James Gregory

Gregory was given what is now known as Upper Parliament Hall (Figure 2) as his laboratory. He left St Andrews in 1674 because of "a prejudice the masters of the University did take at the mathematics", and went to Edinburgh, but tragically he died a year later – at the age of only 37. Before that, however, he had been the first to discover many basic elements of calculus that we take for granted today, such as: the general binomial theorem, Taylor expansions (40 years before Taylor), the ratio test for convergence of a series, the series expansions for sin x and tan x, the integrals of sec x and log x, the fact that differentiation is the inverse of integration, and how to use a change of variable in integration.

The role of theory should not be, in my view, to reproduce images or to explain every observation, but rather to understand the basic processes, and to do so in a step-by-step manner, starting with a simple model and gradually making it more and more sophisticated and hopefully more realistic. In doing this, it is crucial to listen carefully to what observers are saying – indeed, I have learnt much more about the Sun from listening attentively to observers than from books. In addition, the different types of theory complement one another, namely, analytical theory, computational experiments and data analysis.

My title today is "Our Enigmatic Sun" in the sense that the Sun is full of surprises. After introducing you to the structure of the Sun (Section 2) and sunspots and the corona, I shall describe briefly MHD and reconnection (Section 3) and will go on to discuss three topics on which there has been progress recently, namely, the solar interior, eruptive solar flares and coronal heating (Section 4). Many of the Sun's surprises are caused by

FIGURE 2. A picture of James Gregory in his laboratory, with the meridian line on the floor (along which he lined up his telescope) and an early pendulum clock on the wall, both of which you can see today.

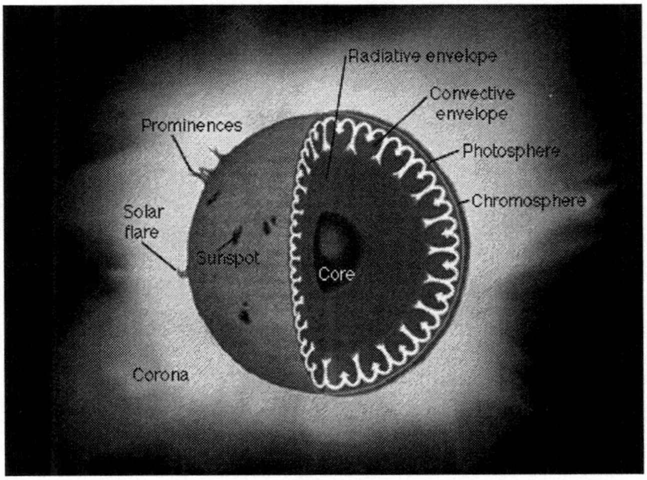

FIGURE 3. The interior of the Sun

the magnetic field. This is because the Sun is of course a plasma and so is coupled in an intimate, subtle way to the magnetic field, which exerts a force on a plasma and can store substantial amounts of energy.

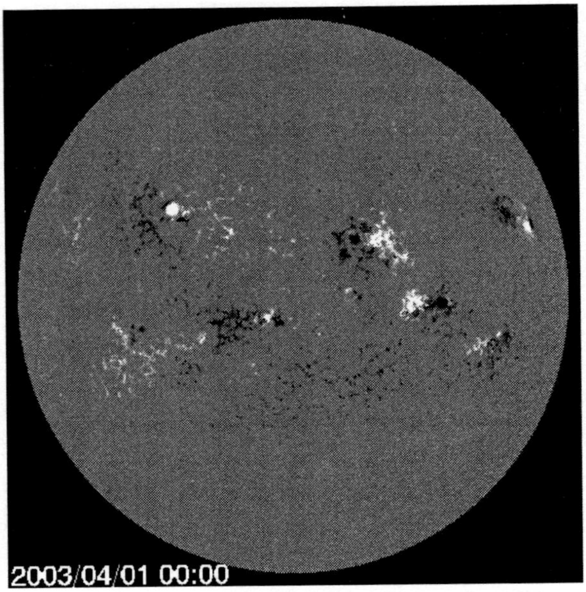

FIGURE 4. A magnetogram of the photospheric magnetic field, showing the line-of-sight magnetic field, with white being directed towards you and black away from you (from the MDI instrument on SoHO)

THE SUN'S GLOBAL STRUCTURE

Inside the Sun (Figure 3) there is a core (extending out to about $0.25R_0$, where the radius (R_0) of the Sun is about 700 Mm), a radiative zone, and a convective zone (extending between $0.7R_0$ and the surface). The atmosphere consists of the photosphere (the surface layer at about 6000 K), the rarer chromosphere, and the corona (which reaches to the Earth and beyond).

The photosphere

The latest models of the photosphere and corona (Carlsson and Stein, 2002) show them to be highly time-dependent and inhomogenous. The surface of the Sun (photosphere) is covered with turbulent convection cells, namely, granulation (with typical diameters of 1 Mm) and supergranulation (with diameters of 15 Mm). It rotates differentially, with the equator rotating more rapidly than the polar regions.

A map of the photospheric magnetic field (Figure 4) reveals that active regions around sunspots are often large-scale bipolar regions. They form a global pattern, with black polarity to the left in the northern hemisphere in Figure 4 and to the right in the southern. This pattern reverses with the start of a new sunspot cycle. It occurs because differential rotation in the interior shears up poloidal flux and creates toroidal flux of one polarity

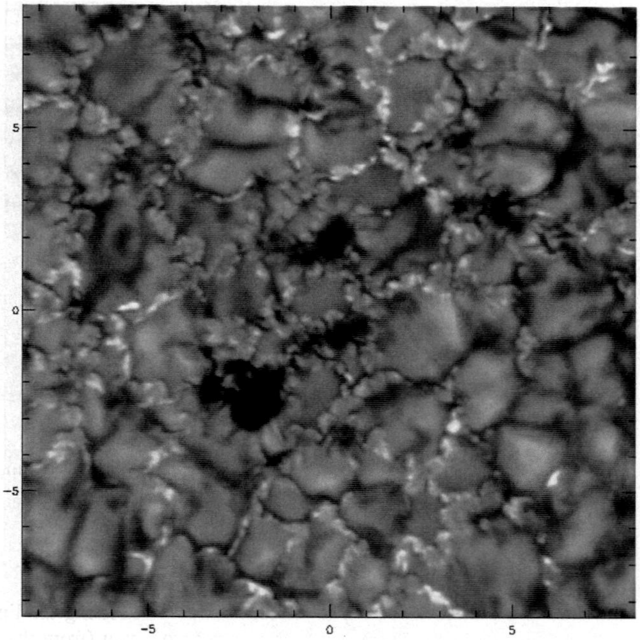

FIGURE 5. A close-up of a few granule cells in the photosphere (Swedish Solar Telescope, La Palma, courtesy of Luc Rouppe van der Voort)

in the north and the opposite polarity in the south. It is when such toroidal flux rises by magnetic buoyancy that it creates a pair of sunspots where it breaks through the surface. Another surprise is that outside active regions intense magnetic fields cover the whole Sun, concentrated at the boundaries of supergranules, whence they are carried by the supergranule flow.

In white light at 0.1 arcsec amazing images have been obtained from the Swedish Solar Telescope which reveal incredible detail in and between the granulation (Scharmer *et al*, 2002; van der Voort *et al*, 2004). Tiny bright points are probable locations of intense magnetic flux tubes at the edges of supergranules, where it was thought until this year that 90% of the quiet-Sun flux is located. However, close-ups of a few granules (Figure 5) reveal for the first time bright points, "flowers" and ribbons in the intergranular lanes around granules and suggest the presence of surprisingly many intense magnetic tubes throughout the centres of supergranules. Indeed, Trujillo Bueno *et al* (2004) have suggested that there may be five or six times as much magnetic flux there than we thought previously.

Sunspots are dark because they are cool (at about 3,700 K), which in turn is because the magnetic field stops granulation. They vary with the 11-year cycle. The dark umbra of a spot is the central part where the magnetic field is close to vertical. The striated penumbra surrounds the umbra and possesses a spreading magnetic field that is far from vertical. However, the penumbra is certainly not simple: recent observations from the

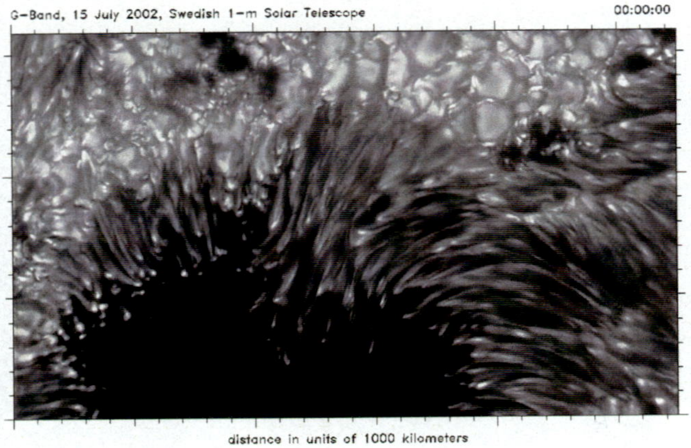

FIGURE 6. High-resolution observation of a sunspot penumbra by G Scharmer and L van der Voort

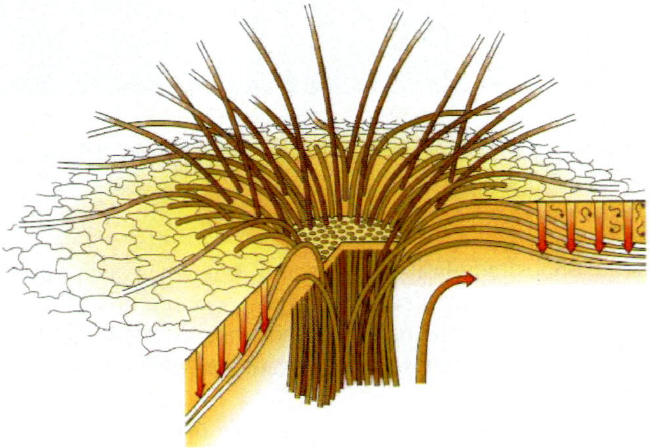

FIGURE 7. New model of sunspot penumbra due to Thomas *et al*, 2002 and Weiss *et al*, 2004

Swedish Solar Telescope (Scharmer *et al*, 2002) have revealed bright flows moving both inward and outward, together with strange dark cores (Figure 6). Also, the bright penumbral filaments are thought to be at intermediate angles to the vertical and to represent magnetic field lines that go far from the sunspot, whereas the dark filaments are lower-lying and so return to the solar surface close to the spot (Figure 7), probably held down by granular pumping (Thomas *et al*, 2002; Weiss *et al*, 2004).

Nonmagnetic simulations of the chromosphere show it to be extremely dynamic, but more realistically Hansteen and Carlsson have undertaken a simulation of a magnetic chromosphere in two dimensions. Figure 8 shows a couple of supergranule cells with a simple arched magnetic structure shown by black curves: the lowest layer is a convecting photosphere (with red representing dense sinking plasma and blue light rising

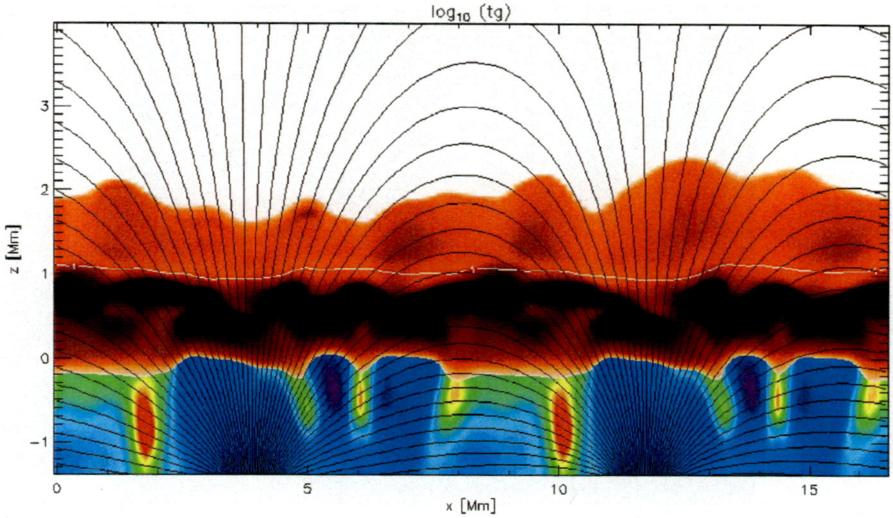

FIGURE 8. Simulation of magnetic photosphere, chromosphere and corona (courtesy V Hansteen and M Carlsson).

plasma) and above it lies the (brown) chromosphere and (white) corona. The curve $\beta = 1$ separates the upper chromosphere where the magnetic pressure exceeds the plasma pressure from the lower chromosphere where the plasma pressure dominates. The transition region is the curve separating the brown chromosphere from the white corona. The response to the slowly convecting photosphere is the generation of waves which become highly nonlinear and pervade the chromosphere so that it and the overlying corona are highly dynamic. Mode conversions from one type to another occur where the sound speed equals the Alfven speed near the $\beta = 1$ surface.

The Corona

The corona can be viewed during a solar eclipse, when the glare of the photosphere is cut out by the moon or by a disc in a coronagraph. An early surprise was the discovery by Edlen (1940) that the coronal temperature is a million degrees or so. In the corona, the magnetic field dominates the plasma, creating its beautiful structure in the way elegantly described by Kanaris Tsinganos. The magnetic field also heats the corona, but the mechanism has not yet been identified, although there are now some promising ideas. As well as viewing the corona during an eclipse, it can also be observed direct with an x-ray or euv telescope (Figure 9), and indeed Yohkoh has revealed it to be a magnetic world with an amazingly rich variety of MHD phenomena.

Earlier rocket images and images from Skylab showed that the corona consists of dots called x-ray bright points, together with coronal holes (dark, magnetically open regions from which the fast solar wind escapes), coronal loops and active regions which

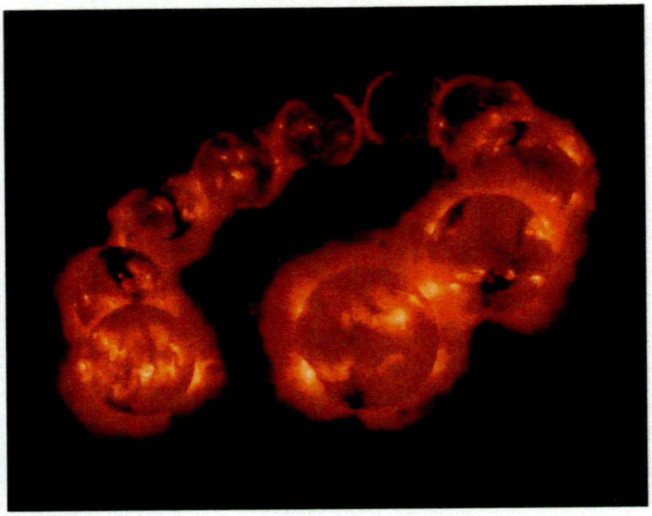

FIGURE 9. The corona over a solar cycle from Yohkoh (S Tsuneta)

FIGURE 10. Coronal loops imaged by TRACE (courtesy Alan Title)

are rather fuzzy. However, the TRACE mission has shown active regions in incredible detail (Figure 10). It has revealed that the corona is made up of intricate loops of plasma aligned along the magnetic field. This highly intricate structure is caused by a combination of the dominance of the magnetic field by the plasma, the small size of the heating sources and the small thermal conduction across the magnetic field.

MAGNETOHYDRODYNAMICS (MHD)

MHD Equations

The equations of magnetohydrodynamics (MHD) may be used to model the interaction of a magnetic field with a plasma, treated as a continuous medium. They are a unification of the equations of fluid mechanics and of Maxwell and are a set of partial differential equations for the behaviour in space and time of the magnetic field ($\mathbf{B}$), plasma velocity ($\mathbf{v}$), density (ρ), pressure (p) and temperature (T) as functions of (x, y, z, t). They include equations of induction, continuity and motion, namely

$$\frac{\partial \mathbf{B}}{\partial t} = \nabla \times (\mathbf{v} \times \mathbf{B}) + \eta \nabla^2 \mathbf{B}, \qquad (1)$$

$$\frac{d\rho}{dt} = -\rho \nabla \cdot \mathbf{v}, \qquad (2)$$

$$\rho \frac{d\mathbf{v}}{dt} = -\nabla p + \mathbf{j} \times \mathbf{B} + \rho \mathbf{g}, \qquad (3)$$

where

$$\nabla \cdot \mathbf{B} = 0 \qquad \text{and} \qquad p = R\rho T,$$

together with an energy equation and an equation

$$\mathbf{j} = \nabla \times \mathbf{B}/\mu \qquad (4)$$

for the electric current ($\mathbf{j}$). The equations are nonlinear and to me as beautiful as the Sun itself. They are impossible to solve completely, and so we approximate them in imaginative ways and use them to construct models for many complex phenomena on the Sun.

The induction equation (1) describes how the magnetic field at a fixed point in space changes in time due to two terms on the right-hand side, namely, transport of the magnetic field with the plasma and diffusion of the magnetic field through the plasma. The first term on the right is very much larger than the second in most of the Universe and so the magnetic field is frozen to the plasma and hangs on to its energy.

The exception is in singularities where the gradient of the magnetic field, and therefore by (4), the electric current is extremely large. These singularities form at null points where the magnetic field vanishes and near which the magnetic field lines have an X-type topology. In the singularities the magnetic field lines break and reconnect. In the process, the magnetic energy is converted into other forms: in particular, the large electric currents imply strong ohmic heating (j^2/σ) and the large electric field is able to accelerate particles to high energy. Such magnetic reconnection is thought to operate at the core of a solar flare and in myriads of coronal heating events.

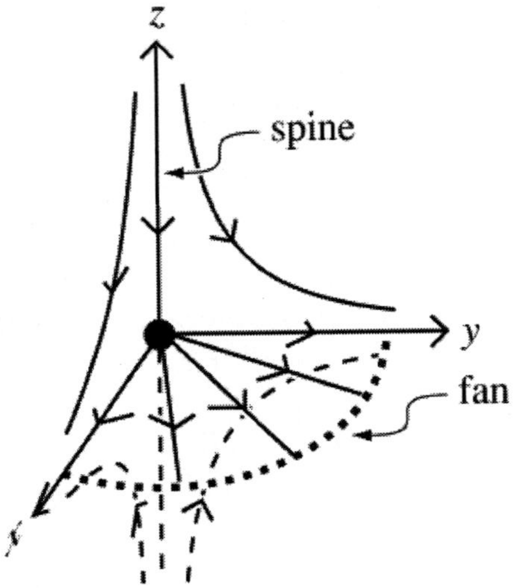

FIGURE 11. A three-dimensional null point

Three-Dimensional Reconnection

The theory for magnetic reconnection in two dimensions is well understood, but recently we have started trying to understand how it works in three dimensions (e.g., Priest and Forbes, 2000). However, so far, the theory is in a rudimentary state since many features of 3D reconnection are completely different from 2D. First of all, the structure of a null point is quite different. In two dimensions, you find either X-type null points or 0-type nulls, but in three dimensions there are two families of field lines through the null, namely, an isolated *spine* field line, which approaches (or recedes from) the null from two sides, and a *fan* surface of field lines which recede from (or approach) the null, as shown in Figure 11.

A second difference between 2D and 3D concerns the topology of the magnetic field. In 2D the field due to four sources of alternating sign in a line possesses an X-type null point (Figure 12a), from which emanate four separatrix curves: these special field lines divide the plane up into topologically distinct regions, in the sense that in each region all the field lines start from the same positive source and end at the same negative source. In 3D, in contrast, the field due to four sources (such as sunspots) on a plane possesses two separatrix surfaces (or *separatrices*) in the form of domes which separate the volume into topologically distinct regions, in the sense that all the field lines in one region have the same sources as their start-points and end-points (Figure 12b). The separatrix surfaces intersect in a special field line called a *separator*, which joins one 3D null point to another.

Three-dimensional reconnection can occur either at a null point or in the absence of a null point. At a null, spine reconnection has the current concentrate along the spine,

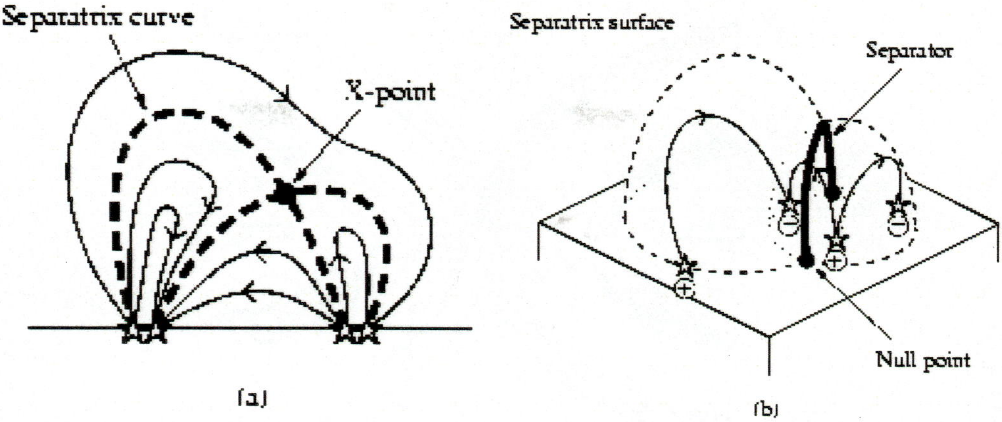

FIGURE 12. Complex topology in (a) 2D and (b) 3D.

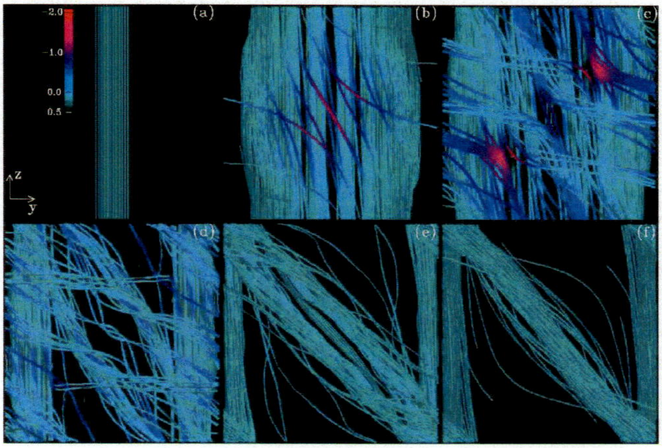

FIGURE 13. Field lines reconnecting as two flux tubes approach one another (Linton and Priest, 2003)

whereas fan reconnection has it concentrate along the fan and separator reconnection along the separator. An example of non-null reconnection has been studied by Linton and Priest, in which they started with two flux tubes at right angles, forced towards each other by a stagnation-point flow. The field lines for one tube during the reconnection process reveal several features (Figure 13). The reconnection fragments and occurs in several places. Complex twisting and braiding is produced, with the conservation of magnetic helicity giving a final flux tube having a twist (Φ) of π: thus, equating the initial mutual helicity and final self-helicity implies that

$$F^2 = 2 \times \frac{\Phi}{2\pi} F^2,$$

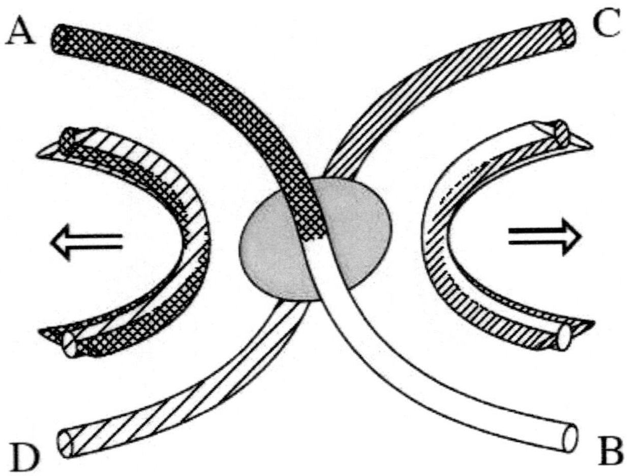

FIGURE 14. Breaking and partial rejoining of two flux tubes in 3D to form four new flux tubes

giving $\Phi = \pi$. Furthermore, the higher the magnetic Reynolds number, the more reconnection locations and the more braiding is produced.

Another difference between 2D and 3D concerns the nature of field line velocities ($\mathbf{w}$) obeying

$$\mathbf{E} + \mathbf{w} \times \mathbf{B} = \mathbf{0}, \tag{5}$$

so that outside a diffusion region, $\mathbf{w}$ and the plasma velocity ($\mathbf{v}$) are equal. In 2D, inside the diffusion region, $\mathbf{w}$ does not exist for an isolated region (Priest, Hornig and Pontin, 2003) i.e., there is no solution for $\mathbf{w}$ to Equation (5). The field lines continually change their connections in the diffusion region, while flux tubes split, flip and in general do not rejoin perfectly (Hornig and Priest, 2003; Pontin, Priest and Hornig, 2003), as indicated in Figure 14.

KEY ADVANCES FROM SOHO

In 1995 SoHO (the Solar and Heliospheric Observatory) was launched as a joint ESA-NASA mission. It is orbiting the Sun at the L1 point in phase with the Earth and is observing the Sun continuously for the first time. It has transformed our understanding of the Sun and has produced many surprises, only three of which I have time to describe today.

The Interior of the Sun

An important question is: what is the structure of the solar interior? Just as a bouzouki or mandalin string can oscillate beautifully in different normal modes, so can a three-

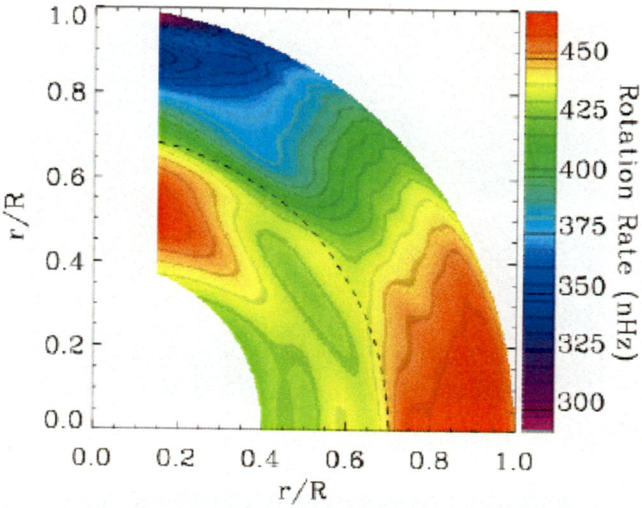

FIGURE 15. The Sun's internal rotation deduced from the MDI instrument on SoHO (A Kosovichev)

dimensional object such as the Sun. Indeed, several million different normal modes of vibration of the Sun have now been discovered, and with the MDI instrument on SoHO the line-of-sight velocity at a million points on the solar surface per minute is being measured. By summing such observations for several months and using techniques of solar seismology, the temperature inside the Sun as a function of radius has been measured and found to agree with the standard model to within 1%.

The MDI observations have, more interestingly, also been used to deduce the Sun's internal rotation. At the solar surface we know the equator rotates faster than the polar regions, but how is this differential rotation continued into the solar interior? The expectation was that the rotation would be constant on cylinders around the Sun's axis of rotation and that the magnetic field would be generated by dynamo action throughout the convection zone. The surprise from MDI (Figure 15) is that the rotation tends to be constant along cones in the convection zone and is then fairly uniform in the radiative zone, which implies the existence of a strong shear layer at the base of the convection zone (known as the tacocline). In addition, it is now thought that the magnetic field of sunspots and active regions is generated in and near the tacocline, so that theorists are now attempting to build tacocline dynamo models.

Another important development is the use of methods of local helioseismology to probe the structure below sunspots. It has been discovered with them that there is a region of reduced wave speed (by 10%) below a sunspot for a distance of about a sunspot radius, presumably because of the reduced temperature there. Also, there is a second region of enhanced wave speed for a distance of about three sunspot radii, presumably because the temperature is no longer cooler but the magnetic field is enhanced.

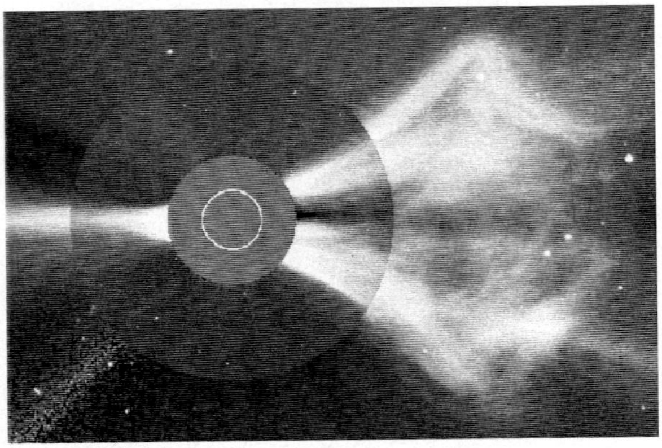

FIGURE 16. Coronal mass ejection observed by LASCO instrument on SoHO

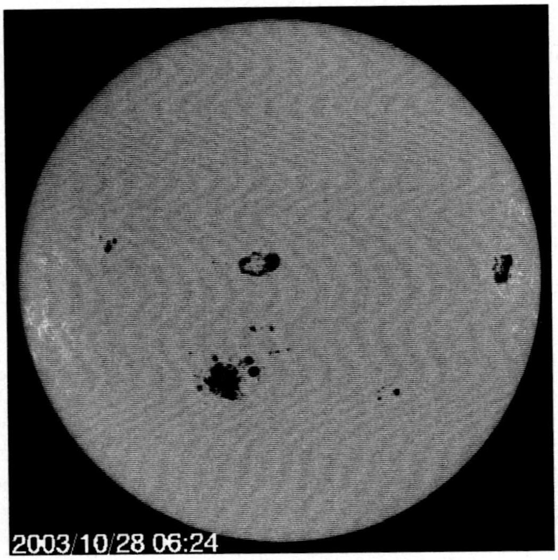

FIGURE 17. Solar photosphere from MDI on SoHO on October 28, 2003

Eruptive Behaviour (Flares and CME's)

How do eruptive solar flares and coronal mass ejections occur? This is an important question that has not yet been answered. The LASCO instrument on SoHO is a coronagraph which has discovered huge ejections of mass and magnetic flux called *Coronal Mass Ejections (CME's)*, which can sometimes reach the Earth and disrupt communications and space satellites (Figure 16).

An incredibly large and complicated group of sunspots (Figure 17 was crossing the solar disc in October, 2003, and was eagerly followed around the world as it spawned a series of giant flares. On October 28, for example, the 3rd largest solar flare ever recorded

FIGURE 18. CME on October 28, 2003, viewed by LASCO instrument on SoHO

took place. It produced a halo CME, namely, one that forms a halo round the Sun (Figure 18) since it is either coming right towards the Earth or is moving away from it. This one was travelling at 2000 km s^{-1}, five times faster than normal. High-energy particles taking only an hour to reach SoHO (by comparison with the CME itself, which takes a couple of days) produced "snow" as they bombarded the CCD detector plates (Figure 18), and, when the CME did reach Earth in a record 19 hours, it produced beautiful aurora that we viewed eagerly in St Andrews for a couple of nights. One week later, when the sunspot group had reached the limb of the Sun, the fireworks continued as it gave birth to the largest solar flare ever recorded.

What happens in outline during an eruptive flare (Figure /refoverallflare) is that a sheared and twisted magnetic tube with an overlying arcade either loses equilibrium (Priest and Forbes, 1990; Forbes and Priest, 1995) or goes kink unstable (Török and Kliem, Fan) or breaks out (Antiochos et al, 1999; Maclean et al, 2004). As the tube erupts, it drives reconnection in the arcade under the erupting tube. The reconnection heats a loop to high temperatures, which then cools down and drains as new loops are heated and the reconnection location rises. The result is the appearance of a rising arcade of hot loops with cool loops beneath them. A particularly fine example was caught by the TRACE satellite and the RHESSI flare satellite on April 21, 2002 (Figure 20). The RHESSI contours of hard x-ray flux at 12-25 keV show emission from the reconnecting current sheet above the 1.5 MK TRACE loops, while the 50-100 keV contours show emission from high-energy electrons accelerated in the reconnection and having travelled down to the feet of the loops. Particle acceleration is thought to be partly by DC acceleration in the current sheet and partly by Fermi and betatron acceleration in the field lines that are springing downwards after reconnection.

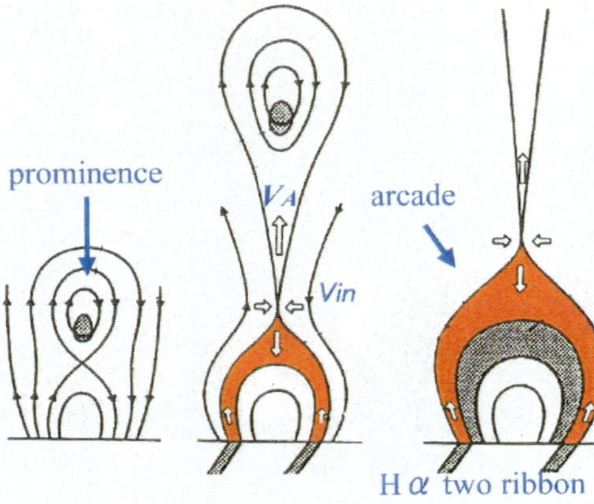

FIGURE 19. Overall picture of eruption driving reconnection in a flare

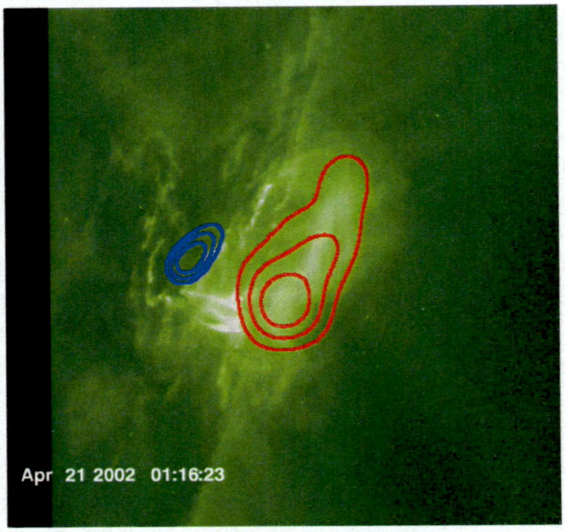

FIGURE 20. Overlay of RHESSI contours of hard x-ray flux and TRACE image in 195 A (P Gallagher).

Heating the Corona

Another question is: how is the solar corona heated to several million degrees by comparison with the photospheric temperature of only 6000 K? We know that the magnetic field is responsible and the mechanism is likely to be magnetic reconnection, but the exact process is still uncertain. A key discovery from SoHO is, however, the existence of the *magnetic carpet* (Schrijver *et al*, 1997), the fact that the photospheric sources of the coronal magnetic field are highly fragmentary and concentrated into intense flux tubes threading the solar surface (Figure 21). These sources are also highly

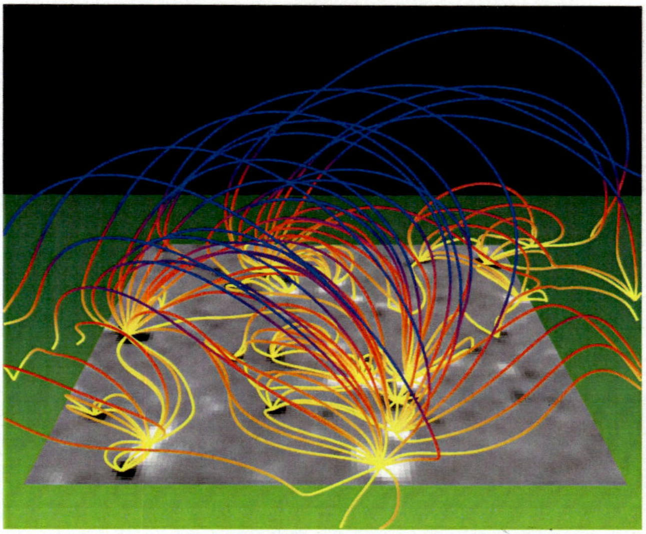

FIGURE 21. The magnetic carpet – coronal field lines calculated from an observed photospheric magnetogram (Close et al, 2004)

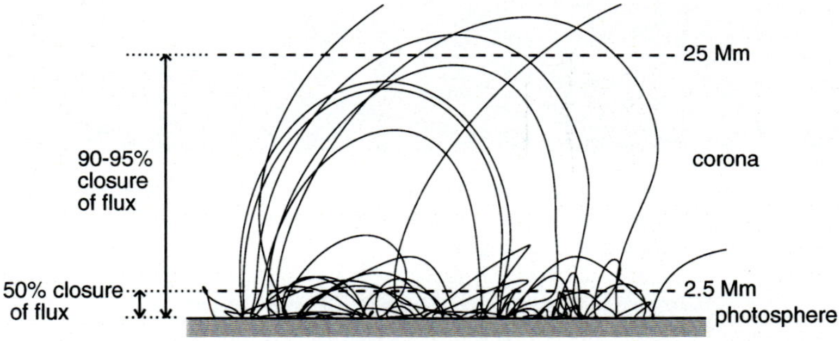

FIGURE 22. Magnetic field lines in the quiet Sun (Close et al, 2004)

dynamic, magnetic flux emerging continually in the quiet Sun and then undergoing processes of fragmentation, merging and cancellation, in such a way that the quiet Sun flux is reprocessed very quickly, in only 14 hours (Hagenaar, 2001).

Observed quiet-Sun magnetograms have recently been used from the MDI instrument on SoHO by Close *et al* (2004), who constructed the coronal field lines and studied their statistical properties. For the region they considered, 50% of the flux closed down within 2.5 Mm of the photosphere and 95% within 25 Mm, the remaining 5% extending to larger distances or being open (Figure 22). They then tracked the motion of individual magnetic fragments in the magnetogram and recalculated the coronal field lines and their connectivity. In doing so, they discovered the startling fact that the time for all the field lines in the quiet Sun to change their connections is only 1.5 hours. In other words,

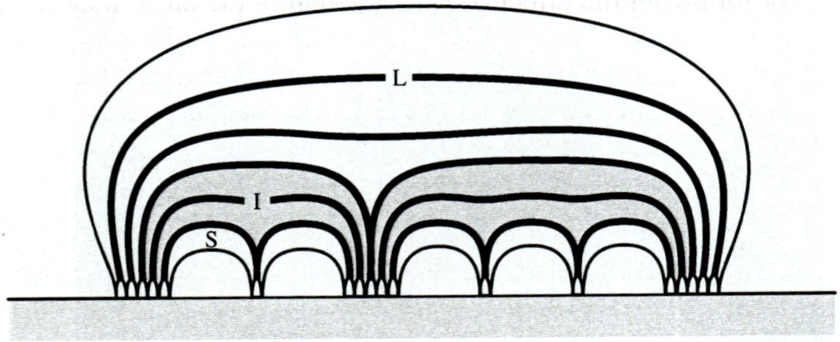

FIGURE 23. Coronal Tectonics model

FIGURE 24. A simulation of coronal tectonics showing (a) the initial state of a single flux tube whose flux has two discrete patches at both ends of the tube and (b) the electric current structure after the tube has been twisted (by I de Moortel and K Galsgaard).

an incredible amount of reconnection is continually taking place – indeed, enough to provide the required heating of the corona.

A *Coronal Tectonics Model* has been proposed (Priest *et al*, 2002), which seeks to determine the effect of the magnetic carpet on coronal heating. Each observed coronal loop reaches down to the surface in many sources, so the flux from each of these tiny sources is separated by separatrix surfaces (*separatrices*). As the sources move around, they generate current sheets on the separatrices and separators, where reconnection and heating takes place (Figure 23). In other words, the idea is that the corona is filled with myriads of separatrix and separator current sheets continually heating impulsively.

The *Coronal Tectonics Model* is essentially a refinement of Parker braiding which produces current sheets much more effectively. Ineke de Moortel and Klaus Galsgaard have recently conducted a very interesting experiment on coronal tectonics. If a uniform magnetic flux is slowly twisted, a weak current will form along the tube with negligible heating. However, if instead the flux of the tube is anchored in two discrete flux patches at one end of a cylinder and in two other patches at the other end, they found quite a different result. By following the effect of slowly twisting the ends, they found that current sheets do indeed form along separators or, if there is a weak background field, along quasi-separators, as expected (Figure 24). Indeed, the subsequent current sheet structure is highly fragmented and an extremely efficient coronal heater.

CONCLUSIONS

Currently, solar physics is in a golden age of discovery, driven by a wealth of new observations from space and ground-based telescopes. In addition, the development of MHD theory is playing an important role. Magnetic reconnection, in particular, is likely to be at work in solar flares and coronal heating.

In future, the current sense of vitality is certain to continue. The present missions are still working well, such as SoHO (launched in 1995), TRACE (in 1998) and RHESSI (in 2002). However, a fleet of new missions is planned to build on this success and answer new questions: these include STEREO (a NASA mission to be launched in 2006 consisting of two satellites, one moving ahead of the Earth and the other lagging behind, which will provide stereoscopic images of the corona), SolarB (a Japan-US-UK mission, to be launched in 2006, which will focus on understanding the subtle connections between the solar surface and the corona), Solar Dynamics Observatory (to be launched in 2008 as a successor to SoHO with a super-MDI and a super-TRACE), and Solar Orbiter (an ESA mission scheduled for launch in 2013, which will go three times closer to the Sun than Mercury).

ACKNOWLEDGMENTS

I am most grateful to PPARC for financial support and to Eleni Dara for making my trip to Greece so memorable.

REFERENCES

1. Antiochos, S., DeVore, R. & Klimchuk, J. (1999) *Astrophys. J.* 510, 485-493.
2. Carlsson, M. & Stein, R. (2002), *Astrophys. J.* 572, 626-635
3. Close, R., Parnell, C.E. & Priest, E.R. (2004) *Astrophys. J.*, 612, L81–L84.
4. Edlen, B. (1945) *Mon. Not. Roy. Astron. Soc.* 105, 323-333.
5. Forbes, T.G. and Priest, E.R. (1995) *Astrophys. J.* 446, 377-389.
6. Hagenaar, H.J. (2001) *Astrophys. J.* 555, 448.
7. Hornig, G. & Priest, E.R. (2003), *Phys. Plasmas* 10, 2712-2721
8. Linton, M. & Priest, E.R. (2003), *Astophys. J.* 595, 1259-1276
9. Longcope, D. (2001), *Phys. Plasmas* 8, 5277-5289
10. Maclean, R., Beveridge, C., Longcope, D., Brown, D. & Priest, E.R. (2004) *Proc. Roy. Soc.*, 461, 2099–2120.
11. Parker, E.N. (1972), *Astrophys. J.* 174, 499-510
12. Parnell, C.E., Priest, E.R. & Golub, L. (1994), *Solar Phys.* 151, 57-74
13. Pontin, D.I., Priest, E.R. & Hornig, G. (2003), *Solar Phys* 212, 319-342
14. Priest, E.R. and Forbes, T.G. (1990) *Solar Phys.* 126, 319-350.
15. Priest, E.R., Heyvaerts, J. & Title, A. (2002) *Astrophys. J.* 576, 533-551.

16. Priest, E.R. and Forbes, T.G. (2000) *Magnetic Reconnection* Cambridge University Press, Cambridge, UK.
17. Priest, E.R., Hornig, G. & Pontin, D.I. (2003), *J. Geophys. Res.* 108, SSH 6.1-6.8.
18. Priest, E.R., Longcope, D. & Heyvaerts, J. (2004), *Astrophys. J.* 624, 1057 – 1071.
19. Priest, E.R., Longcope, D. & Titov, V.S. (2004), *Astrophys. J.* 598, 667 – 677.
20. Rouppe vander Voort, L.H.M., Lofdahl, M.G., Kiselman, D. & Scharmer, G.B. (2004) *Astron. Astrophys.* 414, 717-726.
21. Sanchez Almeida, J. and Lites, B. (2000) *Astrophys. J.* 532, 1215-1229.
22. Scharmer, G.B., Gudiksen, B.V., Kiselman, D., Lofdahl, M.G. & Rouppe vander Voort, L.H.M. (2002) *Nature* 420, 151.
23. Schrijver, C.J., Title, A.M., van Ballegooijen, A.A., Hagenaar, H.J. & Shine, R.A. (1997) *Astrophys. J.* 597, L165-L168.
24. Schrijver, C.J. and Title (2003) *Astrophys. J.* 487, 424-436.
25. Thomas, J.H., Weiss, N.O., Tobias, S.M. & Brummel, N.H. (2002) *Nature* 420, 390.
26. Trujillo-Bueno, J., Shchukina, N. & Asensio Ramos, A. (2004) it Nature 430, 326-329.
27. Weiss, N.O., Thomas, J.H., Tobias, S.M. & Brummel, N.H. (2004) *Astrophys. J.* 600, 1073-1090.

Titan and the Cassini-Huygens mission

Athena Coustenis

LESIA, Paris-Meudon Observatory
5, place Jules Jannsen, 92195 Meudon Cedex, FRANCE

Abstract. The Cassini-Huygens mission has initiated its long exploration of the Saturnian system in July 2004, after a 7,5 year trek through our solar system. Since the Saturn Orbit Insertion, we have witnessed the great success of the Huygens mission, the probe descent through Titan's atmosphere, on January 14, 2005. One of the main targets of the Cassini-Huygens mission was Titan. The combined orbiter and probe data have been a precious tool in the description of Titan's atmosphere and surface returning wonderful new data whose analysis have revealed an amazing new world, 10 time further from our Sun and yet so close to our own planet. Indeed, Titan is currently the only exobiological system that we can study in reference to conditions which may have prevailed on the primitive Earth.

Keywords: Titan, Cassini-Huygens mission, atmosphere, surface, satellites of Saturn
PACS: 96.12.Jt, 96.12.Kz, 96.12.Ma, 96.25.Vt

INTRODUCTION

Titan, Saturn's largest satellite, has attracted the interest of the scientific community ever since its discovery by C. Huygens in 1655, and in particular since the realization in the early 20th century that it has an extended atmosphere. This atmosphere is composed essentially of molecular nitrogen (1.5 bar), like the Earth, and is the host of an active organic chemistry due to the presence of methane. CH_4 and N_2 combine to produce a large variety of hydrocarbons and nitriles. Some of the latter (such as HCN, called "prebiotic") are considered to be the precursors of life on our own planet. An exterior source brings traces of oxygen in Titan, in the form of H_2O, CO and CO_2. The temperature on Titan is very low (varies from 70 to180 K in the atmosphere) due to the distance to the Sun (10 AU). At the surface, the temperature reaches 94 K thanks to a greenhouse effect, as on our planet.

Titan was mainly explored in the 80s by the Voyager 1 mission which during one flyby returned its atmospheric composition and other characteristics. However, such things as the nature of the surface or the aerosol distribution, or the structure in the troposphere remained unknown.

The Cassini-Huygens mission to the Saturnian system was launched in 1997 and arrived in the vicinity of the Saturnian system in 2004. The instruments aboard the mission have since then been returning wonderful new information on the satellite during each flyby. 44 of them are expected until the end of the mission in 2008. In what follows I discuss some of this information that has been revealed to us by the

CP848, *Recent Advances in Astronomy and Astrophysics,*
7th *International Conference of the Hellenic Astronomical Society,* edited by N. H. Solomos

Cassini-Huygens mission and how our perception of Titan physics (Figure 1) has evolved over the past year or so.

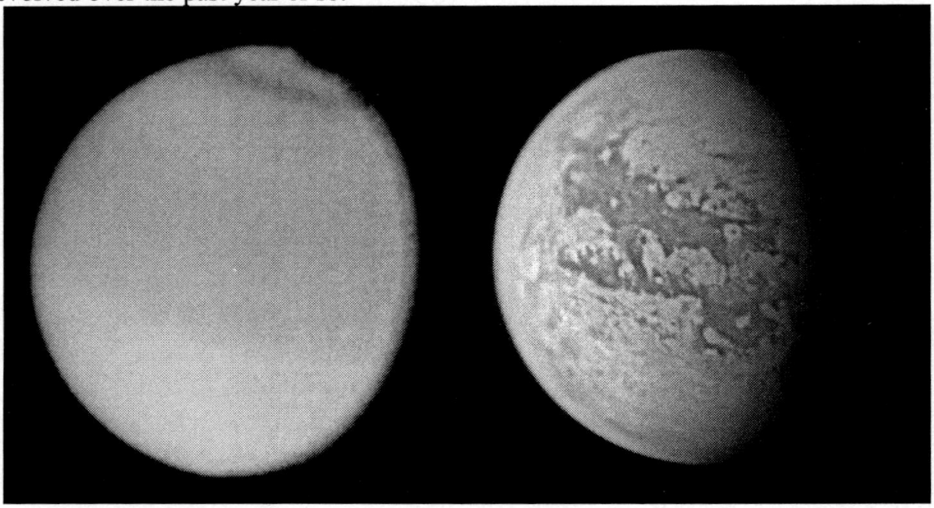

FIGURE 1. Titan observed in 1980 with the cameras of Voyager 1 in the visible and in 2004 with the Cassini/ISS camera (Porco et al., 2005) at 0.94 micron. In the first case, the bland appearance of the satellite belies a complex world. The only features apparent in the images taken by Voyager were the detached haze layers, the dark polar hood and generally a difference in brightness between the two hemispheres. In contrast one of the most recent images by Cassini shows Titan's surface features. *Image Credit: NASA/JPL.*

THE CASSINI-HUYGENS MISSION

Cassini-Huygens is a very ambitious mission, an extremely successful collaboration between ESA and NASA (with contribution from 17 countries), composed of an orbiter and a probe (Huygens). Although the mission's objectives span the entire Saturnian system, for Cassini (as for Voyager before it) Titan is a privileged target and the mission is designed to address our principal questions about the satellite. The spacecraft is equipped with 18 science instruments (12 on the orbiter and 6 carried by the probe), gathering both remote sensing and in situ data (Figure 2). It communicates through one high-gain and two low-gain antennas. Power is provided through three Radioisotope Thermoelectric Generators (or RTGs).

The 5,650-kilogram (6-ton) Cassini-Huygens spacecraft was launched successfully on the 15th of October 1997 from the Kennedy Space Center at Cape Canaveral at 4 :43 a.m. EDT. Because of its massive weight, Cassini could not be sent directly to Saturn but used the « gravity assist » technique to gain the energy required by looping twice around the Sun. This allowed it to perform flybys by Venus (April 26, 1998 and June 24, 1999), Earth (August 18, 1999) and Jupiter (December 30, 2000). Cassini-Huygens reached Saturn in July 2004 and performed a flawless Saturn Orbit Insertion (SOI), becoming trapped forever in orbit like one of Saturn's moons.

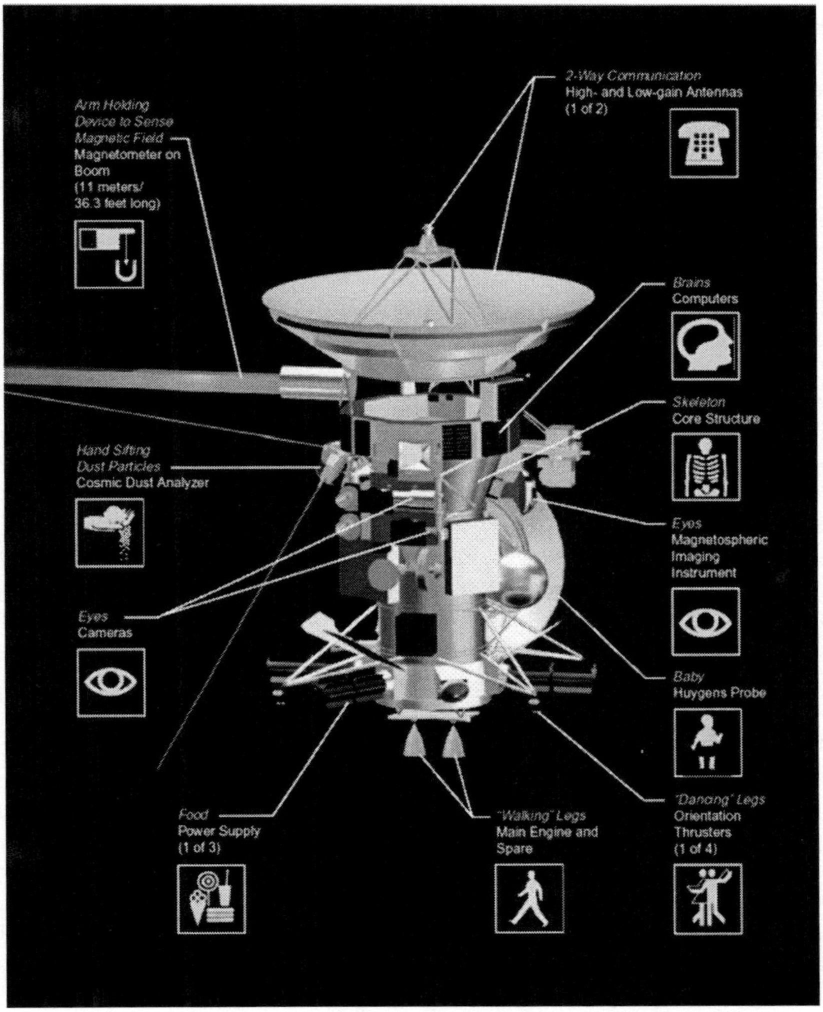

FIGURE 2. The Cassini-Huygens mission and its functions. *Image Credit: NASA/JPL.*

The Cassini instruments (Figure 2) have since then returned a great amount of data concerning the Saturnian system. During its four-year nominal mission, the Cassini Orbiter will make about 40 flybys of Titan, some as close as 1000 km (Voyager 1 flew by at 4400 km) from the surface and perform direct measurements with the visible, infrared, and radar instruments.

Additionally, the mission saw the deployment of the European-built Huygens probe. After release from the Cassini orbiter, on December 25, 2004, this 300 kg probe plunged into Titan's atmosphere on January 14, 2005 at 11:04 UTC and descended through it by means of several parachute breaks which slowed the probe from super sonic speeds of 6 km/s during entry and down to 5 m/s at impact. The five batteries onboard the probe lasted much longer than expected, allowing Huygens to collect descent data for 2hrs and 27 mns and surface data for 1hr and 12 mns. During its descent, Huygens' camera returned more than 750 images, while the Probe's other five instruments sampled Titan's atmosphere to help determine its composition and structure. The telemetry data from Huygens was stored onboard Cassini's Solid State recorders (SSRs) at a rate of 8 kilobits per second, while the spacecraft was at an altitude of 60,000 km from Titan. Although some data from Huygens was lost during its transmission to Cassini through a stream called « Channel A », in the end all of the measurements were recovered because Titan's weak signal was captured by Earth-based radio telescopes !

As well as imaging the atmosphere and surface, the probe took samples of the haze and atmosphere. These *in-situ* measurements complement the remote-sensing measurements made from the orbiter. The latter carries a host of instruments designed to perform in-situ (on-site) studies of elements of Saturn, its atmosphere, moons, rings and magnetosphere. The instruments study the temperatures in various locations, the plasma levels, the neutral and charged particles, the surface composition, the atmospheres and rings, the solar wind, and even the dust grains in the Saturn system. Other instruments perform spectral mapping for high-quality images of the ringed planet, its moons and rings.

Among them is a multimode radar, which completely penetrates the hazy atmosphere of Titan. This Radio Detection and Ranging (RADAR) imaging system operates as a radiometer (to measure surface temperature or emissivity), a scatterometer and altimeter (to measure the reflectivity and topography along the orbiter groundtrack) and as a synthetic aperture imager. This latter mode, nearest closest approach, images Titan's surface at 0.5 to 2 km resolution (i.e. several times poorer resolution than Magellan) over about 1% of Titan's surface for each flyby devoted to radar measurements. Thus perhaps 20% of Titan's surface will be mapped - in long thin strips - during the mission. The imager on the orbiter carries filters tuned to the windows (e.g. 940 nm) in between the methane bands, and so – like the HST - should be able to measure surface contrasts. Additionally, polarizers are carried which should be able to remove most of the light scattered by the haze at near 90° phase angle, so these measurements too will study the surface. The exact resolution achievable will depend on the scene contrast and the haze optical depth at the time of the mission, as well as the image motion compensation that can be achieved, but is likely to be better than 100 m. Other filters are able to probe different altitudes in the atmosphere.

The Visual and Infrared Mapping Spectrometer (VIMS) instrument spans other spectral windows between 0.6 and 5 microns. This will allow spectral identification of surface materials with high (~500m resolution) as well as resolved composition measurements. Looking at Titan's night-side, the instrument may be able to spot lightning, or thermal emission from active cryovolcanism.

The Composite Infrared Spectrometer (CIRS) instrument allows the temperature to be profiled at different locations in the atmosphere, as well as spatially-resolved composition measurements. CIRS operates in three different focal planes to cover the whole infrared range from 10 to 1500 cm^{-1}, with spectral resolutions varying from 0.5 to 15 cm^{-1}. These data are invaluable for verifying and refining models of chemistry, photochemistry and atmospheric circulation.

The interaction of Titan with the Saturnian magnetosphere is being studied by the Magnetospheric Imaging Instrument (MIMI), the Cassini Plasma Spectrometer (CAPS) and the Planetary Radio Astronomy (PRA). The latter instrument searches also for radio emissions from lightning on Titan, although a similar search by Voyager failed to indicate any such emission.

The Cosmic Dust Analyzer (CDA) measures the mass, velocity and composition of particles in Titan's vicinity. These data will be valuable in understanding the origin of oxygen compounds in Titan's reducing atmosphere, believed to be due to photochemical reactions involving meteoric water.

The radio system on the orbiter is used to study Titan in two ways - first by tracking the spacecraft from Earth, to determine Titan's gravity field. This in turn constrains its internal structure (e.g. the size of a rock core, and perhaps the rigidity of the crust), and second by multiple radio occultations. These will measure a temperature profile, and indicate the extent of Titan's ionosphere.

Direct measurements of the composition of Titan's atmosphere versus altitude were made above the Huygens landing site by the Gas Chromatograph Mass Spectrometer (GCMS) on the probe - as Titan's atmosphere has so many components, separation in two dimensions (by chromatography as well as mass spectroscopy) is required. The GCMS also analyzed the pyrolysis products from the Aerosol Collector and Pyrolyser (ACP), which sucked haze particles into the probe and trapped them in a filter which was subsequently baked in an oven to break down the haze macromolecules into smaller fragments that can be studied in the GCMS. The GCMS also had a heated inlet, so that the volatile component of the surface material at the landing site can be determined.

The atmospheric composition at high altitude was also sampled directly, during the closest flybys by the orbiter - the Ion and Neutral Mass Spectrometer analyzes atomic and molecular composition at ~1000 km altitude. These, and spacecraft dynamics measurements, allow direct comparison with the density profile measured by the entry deceleration of the Huygens probe.

The entry deceleration was measured by the Huygens Atmospheric Structure Instrument (HASI) which is the only probe instrument to operate prior to parachute deployment on the probe: the deceleration is proportional to density, and from the density profile and hydrostatic equilibrium, a temperature profile of the upper atmosphere was derived. The temperature, and pressure were measured directly from 170km down as the probe descended by parachute. HASI also includes a Plasma Wave Analyzer (PWA) which measured the electrical properties of the atmosphere (important in determining haze charging and coagulation physics), searched for thunder and lightning, and the dielectric properties of the surface material. The probe also carried a radar altimeter, which estimated radar reflectivity and surface topography. The radar altimeter, part of the probe system itself, passed its signal to the PWA for science data processing.

The Descent Imager/Spectral Radiometer (DISR) is an optical instrument that makes measurements at solar wavelengths from the Probe to look at the scattering properties of the aerosols and to find out at which height levels most of the energy from the Sun is deposited in the atmosphere. DISR also operates in the near-infrared range up to 1.6 micron. It also took pictures of the surface on the way down and allowed for the construction of large panoramas and 3-D images. The DISR was capable of retrieving the data with a resolution orders of magnitude better than has been available from the orbit, albeit of a much smaller region of Titan's surface. Observations in the methane bands determined the methane mole fraction.

The surface material was directly investigated by the Surface Science Package (SSP). This is a suite of sensors on the probe destined to measure in the case of a liquid surface (which was not found at the Huygens landing site) refractive index, density, thermal and other properties, allowing a coarse identification. An acoustic sounder measured the speed of sound in the atmosphere during descent (constraining temperature and relative molecular mass with high altitude resolution) and placed bounds on the depth of any liquid reservoir. Prior to impact, the sounder also estimated the surface roughness at the landing site. Having landed on a solid surface, SSP's accelerometer and penetrometer measured the mechanical properties (e.g. particle size, stickiness) of the surface material. Tilt sensors measured the probe's attitude during descent.

The Doppler Wind Experiment (DWE) expected data on the Titan wind field were finally recovered by measuring the probe signal on Earth by radiotelescopes.

The Cassini-Huygens mission has already provided a wealth of data. The analysis is in the first stages and the Cassini orbiter promises to unveil yet more of Titan's secrets in the years to come.

THE ATMOSPHERE OF TITAN

One of the most interesting aspects of Titan is the presence of this thick nitrogen atmosphere and the cloud deck surrounding the satellite.

Thermal Structure

Titan's temperature profile was measured in situ by the Cassini-Huygens HASI instrument at the probe's landing site (15 °S, 192 °W) from 1400 km down to the surface, where 93.65 ± 0.25 K were measured for a surface pressure of 1467 ±1 hPa. As Voyager did before, HASI found Titan's atmosphere to exhibit the features that characterize the Earth's thermal structure: an exosphere, a mesosphere, a stratosphere and a troposphere, with two major temperature inversions at 40 and 250 km, corresponding to the tropopause and stratopause, with temperatures of 70.43 K (min) and 186 K (max) respectively (Fig. 3). Another inversion region, less contrasted than the previous ones and than what was inferred from Voyager 1 data, and corresponding to the mesopause can be found at 490 km (for 152 K).

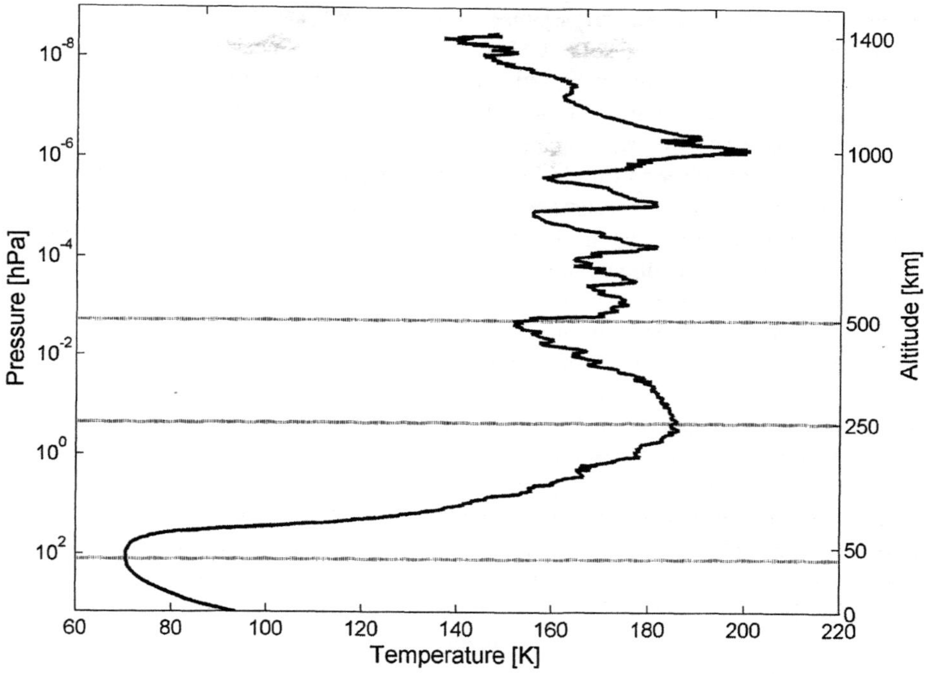

FIGURE 3. Titan's temperature profile as derived from Huygens/HASI measurements. (Fulchignoni et al., 2005).

Prior to these measurements, Flasar et al. (2005) working with the CIRS data provided information on the lower and upper stratosphere from roughly 70 to 400 km in altitude, indicating the presence of a stratopause around 310 km of altitude for a maximum temperature of 186 K.

The HASI data taken on January, 14, 2005 yield more precise and new information on the upper part of the Titan atmosphere, the thermosphere, where several temperature fluctuations are observed due to dynamical (gravity and tidal) phenomena. Indeed, gravity waves signatures of 10-20 K in amplitude were recorded above 500 km around an average temperature of 170 K (Figure 3). These temperatures are higher than predicted by the models. The absence of a marked mesosphere and the wave signatures observed in the upper regions indicate a strong regime of gravity waves as compared with radiative processes. HASI furthermore reported a lower ionospheric layer between 140 km and 40 km, with electrical conductivity peaking near 60 km. A tentative detection of lightning is under investigation (Fulchignoni et al., 2005).

In the lower atmosphere, below about 200 km, all current measurements on Titan agree with the Voyager 1 profile described in section. Analytical radiative transfer microphysical models (as those developed initially by C. McKay and updated by P.

Rannou and co-workers to include fractal particles) show that the temperature inversion observed at the tropopause (70 K at 40 km) is the result of strong stratospheric absorption of solar UV and penetration to near the surface of longer wavelength solar visible radiation. The temperature lapse rate measured by Huygens/HASI shows a cut-off from the dry adiabatic lapse rate. This means that convective instabilities may occur in Titan's lower atmosphere. The surface temperature as determined by HASI (12K warmer than the effective temperature of 82K) is maintained by a greenhouse warming (mostly due to methane and nitrogen) of 22K, offset by cooling by the haze. The models also place constraints on the properties of the haze, the surface albedo and the presence of clouds.

Chemical composition

Curiously, the bulk composition of Titan was more difficult to determine than the abundances of the trace constituents. Cassini finally allowed firm determinations for the major components, with a methane mole fraction of 1.41×10^{-2} in the stratosphere, increasing below the tropopause and reaching 4.9×10^{-2} near the surface. These Huygens GCMS measurements (Niemann et al., 2005) are in good agreement with the stratospheric CH_4 value inferred by CIRS on the Cassini orbiter ($1.6 \pm 0.5 \times 10^{-2}$) and the surface estimate given by the Huygens DISR spectra (also 5%). The GCMS also saw a rapid increase of the methane signal after landing suggests that liquid methane exists on the surface, together with other trace organic species, including cyanogen, benzene, ethane and carbon dioxide. The only noble gas detected to date is argon, found in the form of primordial ^{36}Ar (2.8×10^{-7}) and its radiogenic isotope ^{40}Ar (4.32×10^{-5}) by GCMS which also determined isotopic ratios such as $^{12}C/^{13}C$ (82.3±1, terrestrial-like), $^{14}N/^{15}N$ (measured in N_2, 183±5, that is 1.5 times less than on Earth) and D/H (measured in HD, $2.3 \pm 0.5 \times 10^{-4}$, found then to be enhanced). The low abundance of primordial noble gases on Titan implies that nitrogen was originally captured as NH_3 rather than N_2. Subsequent photolysis may have created the N_2 atmosphere we see today.

Ground-based observations of the $3\nu_2$ monodeuterated methane band at 1.6 μm confirm the value of $\sim 1.5 \times 10^{-4}$ found from Voyager data analyses of the ν_6 CH_3D band at 8.6 μm and support the evidence for a deuterium enrichment in Titan's atmosphere with respect to the protosolar value and that of the giant planets (D/H ~ 2-3.4×10^{-5}). However, the enrichment factor is not known with enough precision to allow a firm determination of its origin. Several hypotheses have been suggested. A primitive nebula model was proposed in which two distinct deuterium reservoirs co-existed before the formation of the solar system: the main one in gaseous hydrogen (HD) and the second smaller one in deuterated isotopes of CNO compounds (such as methane, ammonia, and water) trapped in ices or clathrates or adsorbed on grains. Or, Titan's atmosphere could result from volatile degassing of grains originating from the second reservoir and having accreted to form the satellite. The fractionation mechanisms acting during or after the formation of the satellite could have contributed to the deuterium enrichment observed on Titan. Several processes have been suggested, including exchanges of deuterium between methane gas and cloud particles or the putative ethane-methane ocean, or even the icy crust underneath. An additional possibility could involve isotopic exchange catalyzed in the presence of metallic grains in the Saturnian nebula.

The infrared spectra taken by the Cassini CIRS instrument in the 400-1500 cm^{-1} range at resolutions of 0.5 or 2.5 cm^{-1} depending on the objective (Figure 4), confirmed the presence of hydrocarbons such as methane, acetylene, ethylene, ethane, diacetylene, methylacetylene, propane and monodeuterated methane. Also, the signatures of nitriles, such as cyanoacetylene, cyanogen and dicyanoacetylene (the latter is observed as ice in emission at 478 cm^{-1} and has no equivalent gaseous band), as well as CO_2 were reported in CIRS spectra (Flasar et al., 2005). These species were first discovered in a special Voyager IRIS observational sequence, consisting of ~ 30 spectra, recorded at grazing incidence over Titan's north polar region in 1980.

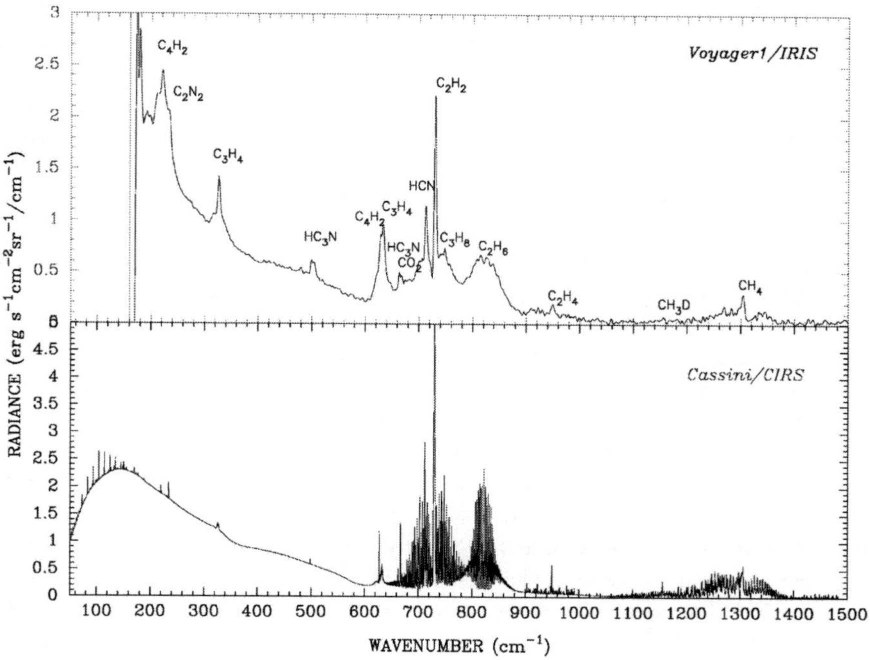

FIGURE 4. A complete CIRS spectrum (lower panel) taken during TB at high resolution and mid-latitudes compared to a 1980 full V1/IRIS observation of Titan in the 200-1500 cm-1 range (Sample E, 50° N), at original resolutions (0.53 vs 4.3 cm^{-1} respectively). The 10-200 cm^{-1} spectral region was observed with CIRS for the first time from space (from Coustenis et al., 2006).

This data yielded vertical distributions for most of the hydrocarbons and nitriles. The vertical distributions generally increase with altitude, confirming the prediction of photochemical models that these species form in the upper atmosphere and then diffuse downwards in the stratosphere. Below the condensation level of each gas, the distributions are assumed to decrease following the respective vapor saturation law.

Ground-based high-resolution heterodyne millimeter observations of Titan led by French and American teams offered the opportunity to determine vertical profiles and partial mapping in some cases of HCN, CO, HC_3N, and CH_3CN which showed that the nitrile abundances increase with altitude, in agreement with predictions by photochemical models which place the production zone above 300 km in the mesosphere. Subsidence causes the abundance of these species to decrease in the lower atmosphere. CH_3CN and CO were not observed on Titan with Voyager or ISO.

In 1997 the Infrared Space Observatory (ISO) made new detections of Titan atmospheric constituents. ISO was put into orbit around the Earth in November 1995 and lasted for 28 months. It carried a 60-cm diameter He-cooled telescope and operated in the 2-200 micron spectral region through 4 instruments: 2 spectrometers, a photometer and a camera. Titan spectra, acquired by Coustenis and colleagues at high-resolution (about 10 times higher than Voyager/IRIS) acquired on January 10 and December 27, 1997, were disk-averages. Those recorded by the Short Wavelength Spectrometer in the grating mode cover the range from roughly 2 to 50 micron and show emission signatures of all of the expected minor constituents in Titan's stratosphere (hydrocarbons, nitriles and CO_2), albeit with a higher resolution allowing us to resolve the bands and distinguish the various contributions. As a consequence, a better determination of the abundances and vertical distributions of these components is achieved on a disk-average scale. The ISO/SWS spectra also provided the first detection of water vapor in Titan's atmosphere from 2 emission lines around 40 micron (the associated mole fraction derived at 400 km of altitude is about 10^{-8}), as well as the first hint at the presence of benzene (C_6H_6) at 674 cm^{-1} for a mole fraction on the order of a few 10^{-10} (Coustenis et al., 1998; 2003). Since then, the benzene detection has been confirmed by Cassini (Flasar et al., 2005). The water vapor abundance although seemingly small, implies a water influx on Titan significantly superior to what might be expected based on local and interplanetary sources alone (rather in favor of Saturn).

With the detection of water vapor the source for the atmospheric carbon monoxide and carbon dioxide observed in Titan was at last found. Water quickly dissociates into OH which combines with methane photolysis products (such as CH_2, CH_3, etc..) and produces CO and CO_2. The source for water on Titan can be found in the rings of Saturn, the meteorites or comets. The Cassini/CIRS FP1 data have not so far allowed a confirmation of the water vapor detection, but they are compatible with the H_2O profile derived by ISO. Finally, the 3-micron region (only partly attainable from Earth due to the terrestrial H_2O interference but containing another methane window near 2.75 micron) was observed with ISO, providing new information on Titan's lower atmosphere and surface content.

Carbon monoxide (CO) and carbon dioxide (CO_2) are the only two oxygen compounds found in Titan's atmosphere to date. Their abundances are coupled and related to the presence of traces of water vapor in Titan's atmosphere. With this independent source provided in the form of hydroxyl radicals (OH, from an external water influx contained in chondritic or icy meteorites, consequently photodissociated to produce OH radicals in the high atmosphere), CO can then be formed from reactions of OH and CH_2 or CH_3. Carbon monoxide then moves downwards and is destroyed near 500 km by action of OH with subsequent formation of carbon dioxide.

On the other hand, through photolysis, various reactions and mainly condensation, CO_2 can be removed to restore CO. The photochemical lifetime of carbon monoxide is very long (comparable to the age of Titan), and as a consequence, it is expected to be uniformly mixed with the background N_2 atmosphere.

Carbon dioxide was identified in the Voyager 1/IRIS spectra from the emission feature around 667 cm-1, at an average mole fraction in Titan's stratosphere of about 15 ppb, representative of the 8 mbar level and assuming a constant mixing ratio above the condensation level.

Carbon monoxide (CO) was first detected in the near-infrared from ground-based observations around 6350 cm^{-1} and a constant mixing ratio of $5x10^{-5}$ was derived in the troposphere with a factor of 3 uncertainty (therefore in accordance with the Voyager value). Microwave measurements confirmed this value and it seemed that CO was, as expected, uniformly mixed throughout the whole Titan atmosphere. Observations using heterodyne techniques suggest however that the carbon monoxide mixing ratio is more than an order of magnitude lower than previous estimates. When compared to the tropospheric value, this abundance suggests that CO is depleted in the stratosphere. We are therefore faced with an enigma regarding the carbon monoxide abundance on Titan. On the one hand, if CO is indeed much less abundant in the stratosphere than in the troposphere, some mechanism - other than the classical photochemical concepts - must be found to produce a satisfactory explanation. It has been suggested that aerosols might alter the vertical distribution of carbon monoxide by adsorbing CO molecules and transporting them in the troposphere in sufficient quantities so as to generate the enrichment observed. This hypothesis requires massive condensation of methane and therefore, if all the observations are correct, there may be implication for large amounts of methane ice in the Titanian atmosphere. On the other hand, either or both sets of observations may be incorrect.

Haze and clouds on Titan

Starting from the upper atmosphere, the Cassini ISS camera showed a faint thin haze layer that encircles the denser stratospheric haze and could be the equivalent of the "detached haze layer" observed by Voyager 25 years ago, except for the difference in altitudes : the thin current haze layer is indeed located 150-200 km higher than the one seen by Voyager. Current models are still unable to render the complexity of seasonal phenomena or circulation patterns on Titan which could be responsible for such an upward shift.

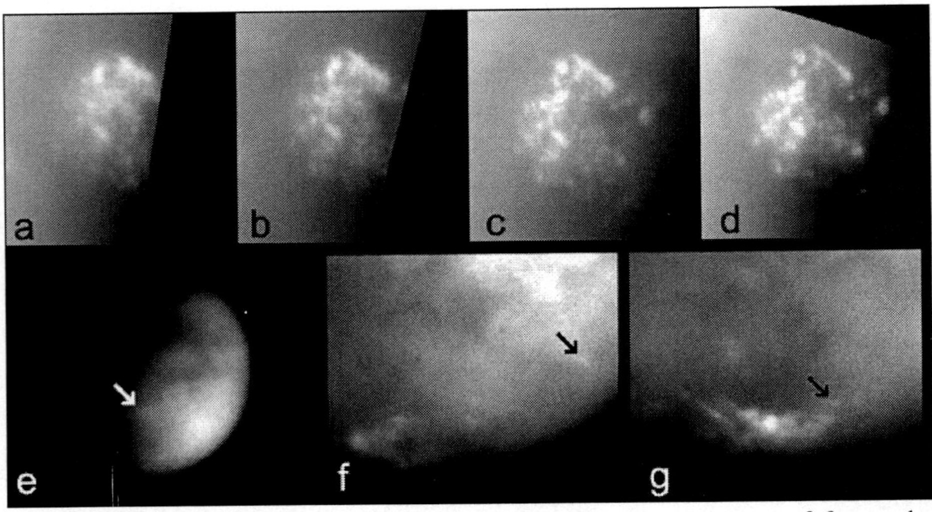

FIGURE 5. Titan's meteorology observed with Cassini/ISS: a-d, A sequence of four methane continuum (IRP0-IR3, 928 nm) images showing the temporal evolution over the period 05:05-09:38 of the Titan south polar cloud field on 2 July 2004. e-g, Three examples of discrete mid-latitude clouds (arrows) for which motions have been tracked in CB3 images. e, 38° S, 81° W (29 May 2004); this image was also viewed through an infrared polarizing filter. f, 43° S, 67° W (23 October 2004); g, 65° S, 110° W (25 October 2004). From Porco et al. (2005).

Cassini images also show a multi-layer structure in the North polar hood region and at lower latitudes in some cases (Porco et al., 2005). These features could be due to gravity waves that have been detected on Titan at lower altitudes. Some of these layers may be related to the two global inversion layers observed in stellar occultations of Titan above 400 km in altitude.

The nature of the haze aerosols measured by Huygens/DISR during the descent through Titan's lower atmosphere came as a surprise to scientists recalling the results from Pioneer and Voyager, as well as predictions by cloud physics models with sedimentation and coagulation. The new observations estimate the monomer radius to be 0.05 mm, in good agreement with previous values. However, contrary to previous assumptions, the DISR data show that the haze optical depth varies from about 2 at 935nm to only about 4.5 at 531 nm, and the number of monomers in a haze particle is therefore probably several hundred (instead of 45). A value of 256 for N gives a projected area equal to that of a sphere of radius 0.65 mm, about twice as large as previously assumed. With N = 512, the equivalent sphere with the same projected area has a radius of 0.9 mm, nearly three times the size previously used. In any case, it seems that the size of the aggregate particles is several times as large as in some of the older models (Tomasko et al., 2005).

In addition, measurements by the DISR violet photometer will extend the optical measurements of the haze to wavelengths as short as the band from 350 to 480 nm, also helping to constrain the size of the haze particles. The number density of the haze particles does not increase with depth nearly as dramatically as predicted by the older cloud physics models. In fact, the number density increases by only a factor of a few over the altitude range from 150 km to the surface. This implies that vertical mixing is much less than had been assumed in the older models where the particles were distributed approximately as the gas is with altitude. In any event, the clear space at low altitudes was suggested earlier, was not observed.

The methane mole fraction of 1.6% measured in the stratosphere by the CIRS and the GCMS is consistent with the DISR spectral measurements. At very low altitudes (20 m) DISR measured 5 ± 1% for the methane mole fraction (Tomasko et al., 2005).

Cassini-Huygens has provided new information on the role of methane and the methane cycle in Titan's atmosphere. The relative humidity of methane (about 50%) at the surface found by DISR and the evaporation witnessed by the GCMS show that there has been and will probably again exist fluid flows on the surface, implying precipitation of methane through the atmosphere.

Although some argument took place for a considerable time as to whether Titan's lower atmosphere could support convection or not and as to whether methane was supersaturated or not, there is clear evidence today that clouds exist in Titan's troposphere, although in general they tend to appear higher than expected.

Methane clouds in Titan's troposphere were first suspected from variability in the methane spectrum observed from the ground by Griffith et al. (2003). Direct imaging of clouds on Titan has been achieved from Earth-based observatories since the turning of the century (e. g. Roe et al., 2002; Hirtzig et al., 2006). Most of the currently-detected clouds are located in Titan's southern hemisphere, as expected given the season on Titan (summer in the South) which means that solar heating is concentrated there as well as rising motions. Other than the large, bright South Pole system observed for the past 5 years or so, discrete clouds detected at mid-latitudes are infrequent, small and short-lived (Cassini/VIMS observations tend to indicate that they rise quickly to the upper troposphere and dissipate through rain within an hour). Keck and Gemini data by Roe and colleagues indicate that they tend to cluster near 350 °W and 40 °S. They may be related to some surface-atmosphere exchange (such as geysering or cryovolcanism) because they don't seem to be easily explained by a shift in global circulation. A dozen or so of large-scale zonal streaks have also been observed by Cassini preferentially at low southern latitudes and mostly between 50-200 °W.

The large south polar system has been visible consistently essentially in the near-infrared (at 2.12 micron for instance) since 1999, while no previous indication of it was aver reported. It was extremely bright in 2001-2002 and recent Cassini images have shown that it is disappearing (indeed it was visible only during the few first Titan flybys and not afterwards, see Figure 5 and Porco et al., 2005). Its shape is irregular and changing with time, recently resembling more a cluster of smaller-scale clouds than a large compact field. Should it prove that this system's life was indeed on the order of 5-6 years (fairly close to a Titan season), stringent constraints can be retrieved on seasonal and circulation patterns on Titan.

Note that DISR reported no definite detection of clouds during its descent through Titan's atmosphere. However, the data are compatible with a thin haze layer at around 21 km of altitude which could be due to methane condensation.

THE SURFACE OF TITAN

Perhaps the most intriguing part of the Titan science will be to determine the nature of Titan's surface. This feature of Titan's was investigated in situ by the Huygens probe on January 14, 2005.

The surface from the orbiter

As mentioned previously, Titan's surface has been known to be covered with bright regions separated by darker material from ground-based and HST observations (Coustenis et al., 2005). These variations are more readily attributed to the presence on the surface of constituents with different albedos rather than topography, although contribution from the latter is also expected. The reason is that the Cassini camera (Porco et al., 2005) observing at 0.94 micron cannot see shadows and also Titan's icy bulk does not plead for high topographic structures on the surface (mountains should not exceed 3 km or so).

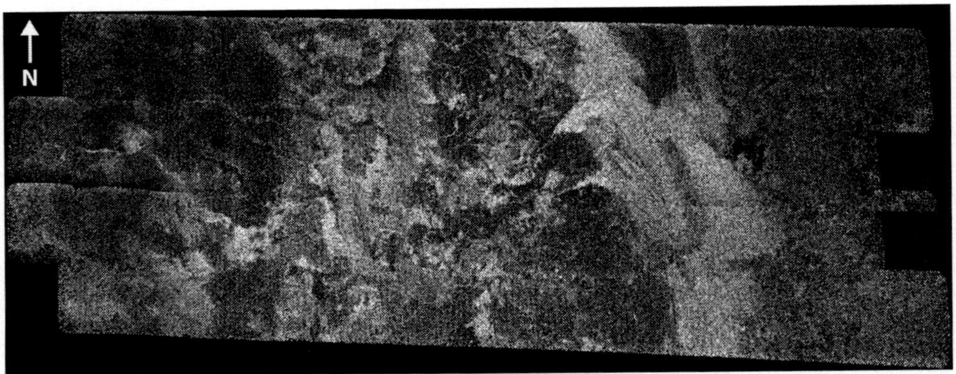

FIGURE 6. Cassini RADAR image (P.I. Elachi) of Titan's surface showing the highly contrasted terrain with a variety of geological features like the round one on the upper left. This radar image taken by the NASA/ESA/ASI Cassini-Huygens spacecraft, is a part of a larger strip created from data taken on 26 October 2004, when Cassini-Huygens flew approximately 1200 kilometres above Titan's surface. The area shown is in the northern hemisphere of Titan and is about 150 kilometres wide by 300 kilometres long. *Image Credit: NASA/JPL/Space Science Institute.*

The ISS and VIMS cameras confirmed these results and showed that the borders of these regions were linear but not smooth and that dramatic changes in surface albedo could be noted in the maps produced by these measurements. The best resolution achieved by ISS was of a few kilometers on Titan's surface. The large bright area around the equator first observed by the HST and the adaptive optics in 1994 was resolved and finely observed by Cassini instruments. It is centered at 10 °S and 100 °W and officially named "Xanadu Regio". The mid-latitude regions around the equator on Titan were found to be rather uniformly bright, while the Southern pole is relatively dark. What exactly is causing the albedo variations is still uncertain. A plausible candidate for the darker regions could be accumulations of hydrocarbons (in liquid or solid form), precipitating down from the atmosphere.

For the brighter regions the task of interpreting the data is more difficult. It has been hypothesized that they could be associated with some topography and more exposed dirty ice content and this tends to be in agreement with findings by the Huygens/DISR instrument whose stereoscopic imaging revealed that the brighter terrain was also more elevated than the darker, smother and lower-ice regions. The exact ice which can satisfy the constraints imposed by all the observations is not easy to determine, hydrocarbon ice has been invoked on the basis of Xanadu appearing bright at all the near-infrared wavelengths observed to date (Coustenis et al., 2005).

Titan may well exhibit cryovolcanic activity, in view of various recent reports (Sotin et al., 2005) from Cassini/ISS. A bright circular structure (about 30 km in diameter) found in the VIMS hyperspectral images is interpreted as a cryovolcanic dome in an area dominated by extension. The VIMS team hypothesized that is the dry channels observed on Titan are related to upwelling "hot ice" and contaminated by hydrocarbons that vaporize as they get close to the surface (to account for the methane gas in the atmosphere), mechanisms similar to those operating for silicate volcanism on Earth (using tidal heating as an energy source) may lead to flows of non-H_2O ices on Titan's surface. Following such eruptions, methane rain could produce the dendritic dark structures seen by Cassini-Huygens. If these structures are indeed channels, they could have dried out due to the short timescale for methane dissociation in the atmosphere. Studying volcanism on Titan (if Cassini definitely yields evidence for it) is important, not only to understand the thermal history of Titan (which since it differs in its incorporation of volatiles from the Galilean satellites, must surely have evolved differently) but also how volatiles - in particular, methane - were delivered to the surface.

Titan's present environment is very placid - tidal currents are weak, rainfall - if it occurs - is soft, the diurnal temperature contrasts small (and therefore winds are gentle). The solubility of ice in hydrocarbons is smaller than that of most rocks in water. Thus, except where the surface is more susceptible to erosion, due to organic deposits or perhaps water-ammonia ice, Titan's topography should not be significantly modified by erosion.

The Cassini instruments have found no obvious evidence for a heavy craterization on the bright or the dark areas of Titan so far. A few features interpreted as impact craters have been announced to date: Cassini's Radar and VIMS saw a 440-km large in diameter impact crater on Titan during two separate flybys in early 2005 (Figure 6). The coloring of the feature indicates that its terrain is rough, with different material for the crater floor and the ejecta and tilted towards the radar during the observations. The multi-ringed impact basin was named "Circus Maximus" by the science team. A smaller crater of about 40 km was also observed, exhibiting a parabolic-shaped ejecta blanket. Nevertheless, craters identified by the RADAR, VIMS or the ISS are rare. This may mean that the surface of Titan is young (less than a billion years) or highly eroded/modified.

Other features observed by the Cassini orbiter include "tiger scratches" or "cat scratches", a set of linear dark features visible across a large part of the RADAR swath to the west of the large crater. Their similarity to dunes as observed on Earth has led some scientists to hypothesize that they could be due to fine-grained material on a stringer, perhaps connected to fluid flows across the surface.

The surface from the probe

On January 14, 2005 the Huygens probe manufactured by ESA landed at 10.3 °S and 192.3 °W on Titan, providing the "ground truth" for the orbital measurements in terms of composition, structure and geomorphology. The probe flew over an icy surface, floated down and drifted eastwards for about 160 km. Several of the instruments on board contributed to our knowledge of Titan's surface conditions: DISR, SSP, HASI, GCMS.

The HASI instrument measured surface temperature and pressure at the landing site to be: 93.65 ± 0.25 K for 1.467 ± 1 atmospheres (Fulchignoni et al., 2005). The fact that the surface is solid but unconsolidated was verified by all the data. The first part of the probe to touch the surface was the SSP penetrometer whose data are now interpreted as indicative of the probe first hitting one of the icy pebbles littering the landing area before sinking into the softer darker ground material. The SSP detected the ground from 88 km in altitude by acoustic sounding, revealing a relatively smooth, but not flat surface for which our best current hypothesis is gravel wet sand, wet clay, or lightly packed snow. With a landing speed of about 5 m/s the front of the probe followed and penetrated the surface, then slid slightly before settling to allow the DISR camera to take several pictures of a Mars-like landscape, complete with a dark riverbed and brighter pebbles. No evidence for liquid was found, but the surface is expected to be very humid since methane evaporation (a 40% increase of the abundance) was measured by the GCMS after landing. Huygens landed on an organic-rich surface, with trace organic species such as cyanogens and ethane detected on the ground.

In spite of some misadventures (loss of the sun sensor measurements, of about half the images from Channel B and the probe's erratic motion), the DISR imager and spectrometer gathered a precious set of data both in spectroscopy and imaging. Starting from the first surface image at 49 km, down to the unprecedented-quality snapshots of the Huygens landing site, and through the lamp-on data recorded below 700 km in altitude, this instrument played a decisive part in untangling the enigma of Titan's surface and lower atmosphere. Panoramic mosaics constructed from a set of images taken at different altitudes show brighter regions separated by lanes or lineaments of darker material, interpreted as channels, which come in short stubby features or more complex ones with many branches (Figure 7). This latter dendritic network can be caused by rainfall creating drainage channels, implying a liquid source somewhere or at some times on Titan's surface. The former stubby channels are wider and rectilinear. They often start or end in dark circular areas suggesting dried lakes or pits. No obvious crater features were observed.

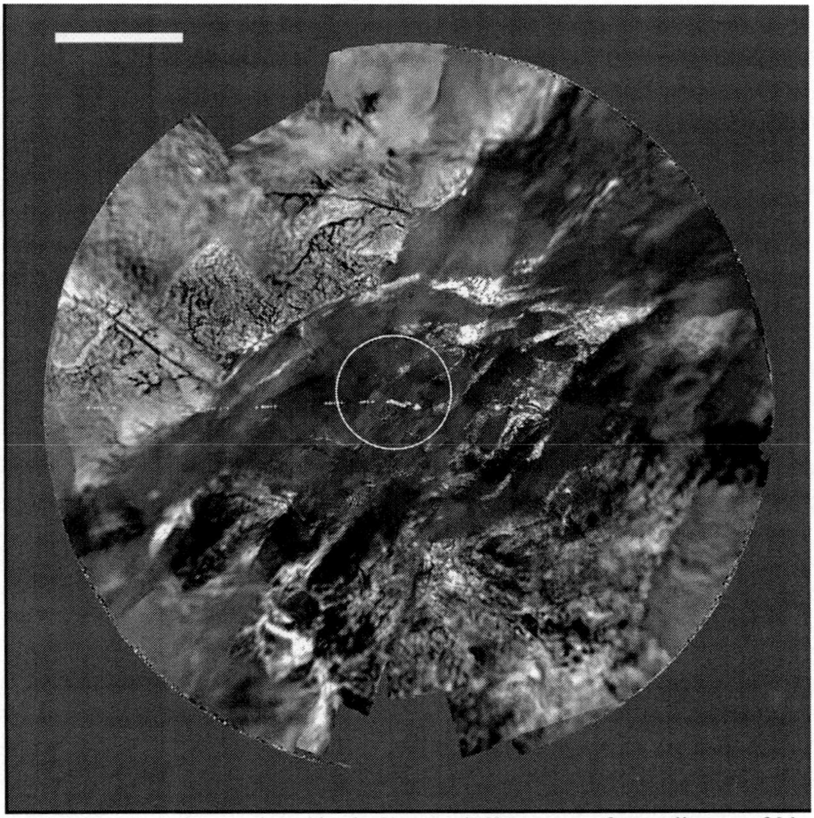

FIGURE 7 Titan's surface as viewed by the Huygens/DISR cameras from a distance of 8 km in altitude. (Tomasko et al., 2005).

Stereoscopic analysis was performed on the DISR images indicating that the bright area cut with the dendritic systems is 50-200 km higher than the large darker plane to the South (Tomasko et al, 2005). If the latter feature is a dried lakebed, it seems too large by Earth standards to have been created by the creeks and channels seen on the images and could be due to larger rivers or a catastrophic event in the past. The dark channels visible in Fig. 7 could be due to liquid methane irrigating the bright elevated terrains before being carried through the channels to the region offshore in south-easterly flows. This migration towards the lower regions leads to water (?) ice being exposed along the upstream faces of the ridges. The slopes are generally on the order of 30°. Some of the bright linear streaks seen on the images could be due to icy flows from the interior of Titan emerging through fissures.

The images taken after the probe had landed on Titan's surface show a dark riverbed strewn with brighter round rocks. 15-cm in diameter at most, more of these "stones" could possibly be hydrocarbons-coated water ice pebbles (Tomasko et al., 2005).

The spectra acquired during the descent gave information on the atmospheric properties, but also on the surface properties. Indeed, it was shown from spectral reflectance data of the region seen from the probe that the differences in albedo were related to differences in topography which in turn can be connected to the spectral behavior of the ground constituents. Thus, the higher brighter regions were also found to be redder than the lowland lakebeds. The regions near the mouths of the rivers are also redder than the lake regions. The spectra taken by DISR are compatible with the presence of water ice on Titan's surface (Tomasko et al., 2005), something that had already been suggested from ground-based observations. The most intriguing feature found in the spectra was however the featureless quasi-linear unidentified blue slope observed between 830 and 1,420 nm. No combination of any ice and organic material from laboratory measurements has been adequate in reproducing this characteristic. The jury is still out on the constituent(s) that create(s) this signature.

Although many questions still remain about the sequence of flooding and the formation of all the complex structures observed by DISR, this data tends to clear the picture we have of Titan today and at the same time enhance the impression that by studying Saturn's satellite we're looking at an environment resembling the Earth more closely than any other place in our Solar System.

No biotic signatures were found on Titan. One of the elements in the negative response (at least so far as the present or past life is concerned) was found by the GCMS in the $^{13}C/^{14}C$ isotopic ratio which showed that no active biota exist on Titan and that the methane on Titan is of non-biologic source.

The reality pictured by the Cassini-Huygens instruments went beyond anything that has been speculated about Titan's surface. The diversity of the terrain includes impact craters, dark plains with some brighter flows, mysterious linear black features possibly related to winds, sand dunes, snow dunes and a host of possible actors: solids, winds, liquids, ices, volcanism, etc... Titan has proven to be a much more complex world than originally thought and much tougher to unveil.

ACKNOWLEDGMENTS

The author is part of the CIRS Team (PI M. Flasar), the HASI Team (PI: M. Fulchignoni) and the DISR Team (PI: M. Tomasko).

REFERENCES

1. A. Coustenis etal., *Icarus* **177**,89-105, (2006).
2. A. Coustenis etal., *submitted for publication,* (2006).
3. F. M. Flasar etal., *Science* **308**, 975-978 (2005).
4. M. Fulchignoni etal., *Nature* **438**, 785-791 (2005).
5. C. Griffith etal., *Science* **300**, 628-630 (2003).
6. M. Hirtzig etal., *submitted for publication* (2006).
7. H. Niemann etal., *Nature* **438**, 785-791 (2005).
8. C. Porco et al. *Nature* **434**, 159-168.
9. H. Roe et al. *Icarus* **157**, 254.
10. C. Sotin etal., *Nature* **435**, 786-789 (2005).
11. M. Tomasko etal., *Nature* **438**, 785-791 (2005).

The Corona of the Sun as a Star

Giovanni Peres *, Salvatore Orlando† and Fabio Reale**

* *Dipartimento di Scienze Fisiche ed Astronomiche - Università di Palermo - Palermo; Italy*
† *INAF - Osservatorio Astronomico di Palermo - Palermo; Italy*
** *Dipartimento di Scienze Fisiche ed Astronomiche - Università di Palermo - Palermo; Italy*

Abstract. We study the physics of the solar corona as a whole, i.e. of the Sun as a Star, in order to understand its global features and to provide a template for stellar coronae. In this process we strive to understand the features of various structures which compose the solar corona. This process in not straightforward given the problems of observing the Sun as a whole: e.g., no recent X-ray wide-band, medium-resolution, spectrum of the Sun is avaible, unlike stars and no X-ray spectral monitoring of the Sun at various activity phases is available. The presentation will discuss our work in this field; we present the method we have devised, based on Yohkoh/SXT data, to derive the Differential Emission Measure vs. Temperature, and the X-ray spectrum, of the solar corona, and the related main results. Finally we present some new results on flares and microflares and the implications on solar-stellar coronal heating.

Keywords: X-ray, Solar physics, Corona, Flares, Solar Activity, Stellar Coronae, Stellar Activity, Main-sequence: late-type stars (G, K, and M)
PACS: 95.85.Nv, 96.60.-j, 96.60.P-, 96.60.Q-, 96.60.qe, 97.10.Ex, 97.10.Jb, 97.20.Jg

INTRODUCTION

The Sun is a typical, albeit old, late-type star. On the Sun we can observe directly many phenomena that we typically cannot resolve on other stars. Its proximity makes it an important target of observation and an excellent test-bed for various problems of stellar physics and, more in general, of Astrophysics; the Sun is an excellent laboratory for plasma physics too. We have been able to understand much of the solar inner structure, thanks to helioseismology, we can study the solar dynamo, chemical abundances, and the physics of magnetized plasma; on the Sun we can also identify individual sources of activity (spots, plages, active regions, loops, flares etc.) and study their features and evolution; instead on stars we just see the emission integrated over the whole disk. The set of problems to be addressed within the solar-stellar connection, i.e. on using the Sun as a template for stellar physics, is rather vast; the scope of our work is limited to coronal activity and related features.

In principle we should easily find an answer to stellar (and astrophysical) problems just using solar observations; in practice it is not so. Observing the Sun with the same methods and tools used for stars is very difficult. Furthermore, solar coronal observations are quite different from stellar ones in many respects: for instance the emphasis of solar observations is toward higher and higher resolution, in order to reveal the sub-arcsecond fine structuring of the magnetized plasma, and to understand the fundamental features of coronal heating; looking for finer and finer structures, some solar instruments observe just a portion of the solar disk.

As a consequence most solar instruments and their data are very different from those

CP848, *Recent Advances in Astronomy and Astrophysics,*
7th International Conference of the Hellenic Astronomical Society, edited by N. H. Solomos

of other astrophysical observations, certainly not simplifying the comparison. As an example of the difference between solar and stellar coronal physics, nobody measures the entire X-ray solar spectrum from 0.1 to 10 keV (neither of the whole solar corona nor of its composing structures) with a spectral resolution similar to what now is routinely achieved for stars with Chandra and XMM-Newton.

However solar and stellar coronal physics could complement nicely because the Sun offers the physical and structural detail while the stars allow us to explore a large range of physical conditions and of stellar parameters, like age, rotation rate, chemical abundance, formation conditions etc. The stellar observations allow us to trade the advantages of spatial resolution for the ability to constrain physical parameters which the Sun does not allow us to change

Various groups work on problems related to the activity of the Sun (as a star) and to the solar-stellar connection. A complete list would push this paper beyond its limits; here we will briefly mention just a few.

Judge and collaborators [1] studied the chromospheric and coronal emission, mostly focusing on soft X-ray and EUV emission. Schrijver and collaborators have worked on many aspect of the solar-stellar connection; for instance they have built the whole solar corona as a collection of coronal loops extrapolated from photospheric magnetic fields and constrained the model to resemble X-ray coronal images [2]. Mandrini et al. [3] have studied the heating of whole corona and inferred possible mechanisms at work. The group of the late Rodonò, Lanza and Pagano, studied the solar-stellar connection mostly in the optical band [4], [5]. Acton and collaborators, [6], used calibrated Yohkoh/SXT images to derive X-ray solar irradiance and coronal temperature, as global coronal characteristics. Stern [7] draws an important analogy between the solar and the stellar coronae.

We concentrate on the solar coronal features. With the scope of studying the Sun as a star, we transform Yohkoh/SXT data into other data virtually identical to those of X-ray stellar observations made with instruments like, for instance, ROSAT/PSPC, Chandra/ACIS and XMM-Newton/EPIC. We then apply to them standard methods of analysis used for stellar observations and compare the results with those of stellar X-ray studies. We aim at studying several aspects among which: the X-ray emission, including X-ray spectra, of the whole corona but also to identify the emission of various structures resolved on the Sun, from which X-rays originate, so as to characterize their role in determining the whole solar emission and the X-ray spectra. The scope is determining whether analogous phenomena (or structures) are also present in stellar coronae. A basic assumption is that the Sun and the late-type stars share the same physics i.e. the solar-stellar analogy, but many authors, facing the estreme stellar X-ray luminosity and activity wonder on "how far does the solar model apply to other stars"? and "how do the underlying processes differ"? So we also indirectly test the solar-stellar analogy through the comparison of solar and stellar X-ray emission.

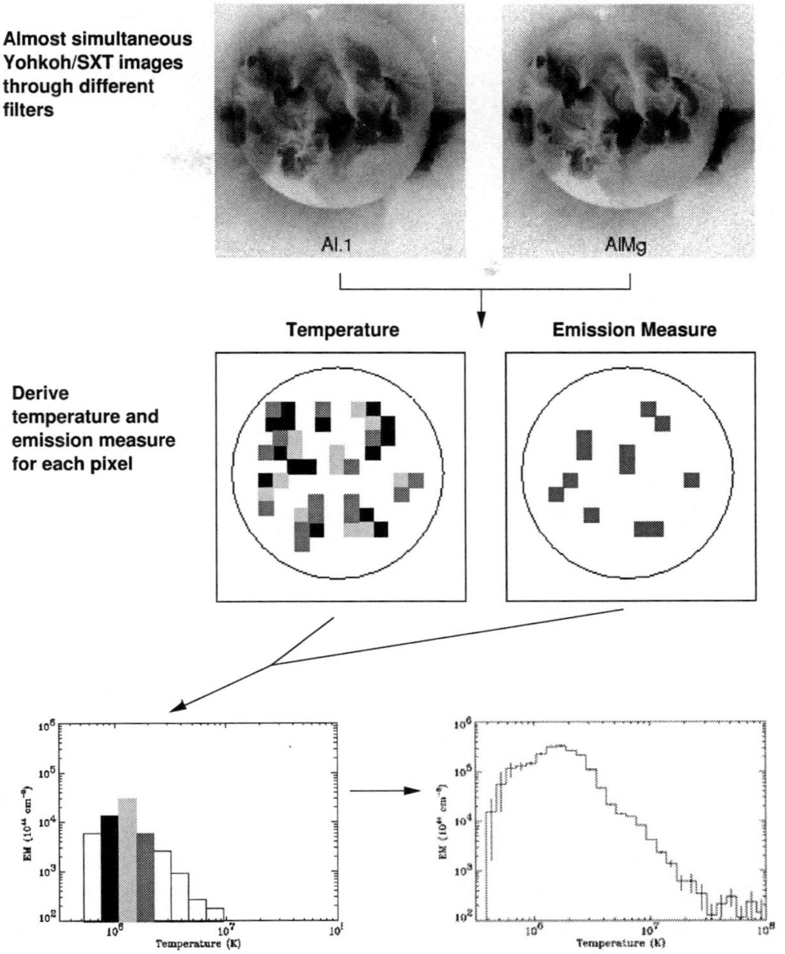

FIGURE 1. An outline of the first steps of our method: from two simultaneous Yohkoh/SXT images taken in different bands (top panels) we derive the maps of temperature and emission measure (middle panels), plotting the amount of emission measure at any temperature (derived from the maps) vs. the relevant temperature we obtain the EM(T) distribution (bottom panels).

THE METHOD

Deriving the emission measure of the solar corona

The first step of our process is deriving the emission measure distribution vs. temperature of the solar corona; it applies to the whole corona but can also be easily applied to any of its composing structures or parts. Fig. 1 shows a scheme of our method. From

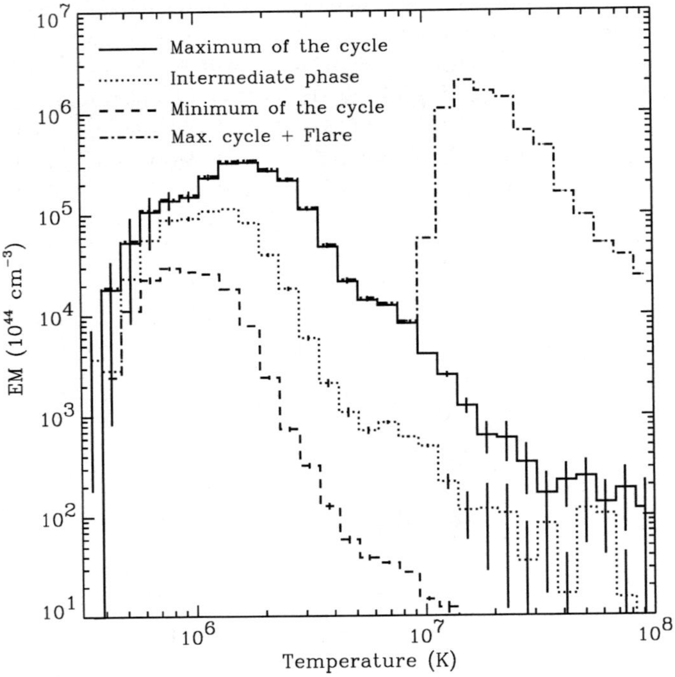

FIGURE 2. Emission measure distribution at three phases of the solar cycle: maximum, intermediate phase and minimum; the figure also shows the **total** coronal emission measure at the maximum of an X9 solar flare. The flare effect consist in adding the large peak above 10^7 K. (Adapted from [9]).

two virtually simultaneous and co-aligned X-ray images taken with Yohkoh/SXT in two different filters we derive the temperature and the emission measure in each pixel, using the standard Yohkoh data analysis software. We end up having two matching data sets of the corona, namely the the maps of temperature and the emission measure, with values in each pixel. We then identify all the pixels at a given temperature and sum up all their emission measure. Repeating this process for each temperature to which Yohkoh/SXT is nominally sensitive ($5.5 < log(T) < 8$), binning the emission measure values within temperarure ranges (dictated by the temperature resolution of the band-ratio-method of Yohkoh/SXT), we generate the distribution of emission measure vs. temperature of the corona. This method has been defined, refined and tested through a large set of simulations as illustrated in [8], in order to ascertain the advantages and limits of the method.

The features of the emission measure distribution are interesting by themselves for solar coronal physics. Fig. 2, adapted from [9] shows the typical emission measure distribution of the non-flaring corona at three phases of the cycle: near the maximum, at an intermediate phase and at a very deep minimum of a cycle. The emission measure distribution at the maximum of the cycle increases with temperature, reaches a maximum around two million degrees and then decreases approximately as a power law. As the cycle proceeds and the Sun is less active, the whole distribution decreases at any temperature while maintaining, approximately, the same shape but the maximum moves progressively toward lower temperatures, and the descent at higher temperatures becomes steeper and steeper. This evolution is particularly evident comparing the distributions at the maximum and at the minimum of the cycle: the peak of the second distribution is lower by more than a factor of ten and the high temperature descent is evidently steeper. The scope of the work in [9] included determining the changes of the solar quiescent luminosity in the soft X-ray band along the solar cycle, amounting to slightly more than a factor 10, and comparing the results of two-thermal-components fittings (typical of stellar X-ray spectral analysis) with the features actually present in the emission measure distribution.

Fig. 2 also illustrates the whole coronal distribution during a large GOES-class X9 flare. The bump in the distribution occurring at high temperature and with a maximum above 10^7 K is a snapshot of the evolving emission measure of the flare, near its maximum. The large bump due to the flare is important for the recent developments of our work, illustrated in the final part of this paper.

Deriving the spectrum and stellar-like features of the corona

The next step of our analysis involves deriving the X-ray coronal spectrum, folding it through the instrumental response, taking into account other observational features like the stellar distance and the exposure time, so as to obtain a realistic focal plane X-ray spectrum similar to those of other stars.

Using a code of spectral synthesis we generate the spectrum emitted by the plasma at each temperature; specifically we use the MEKAL code [10], since it is one of the most reliable and updated codes. Summing all the spectra at any temperature we derive the X- ray spectrum of the whole solar corona. Then folding the total spectrum through the spectral response and the response matrix of the X-ray observatory of interest we obtain the focal plane X-ray spectrum of the Sun as if we were observing it with that observatory.

The example shown in Fig. 3 uses the Chandra-ACIS instrumental features for the synthesis; we have also assumed that the solar-clone star is at a distance of 1 parsec and an exposure time of fifty thousand seconds (a typical value for stellar observations) for easy comparison with stellar observations. Since the final results of our syntheses are in a format and with a content identical to those of stellar coronal observations, we can then apply the standard stellar analysis to these "solar X-ray spectra" and compare them with stellar observations. We remark that equivalent solar X-ray spectra are not available.

AN OVERVIEW OF OUR RESULTS

The project, based on the method outlined above, addressed several problems of coronal physics, and several aspects of the solar-stellar connection in the context of coronal

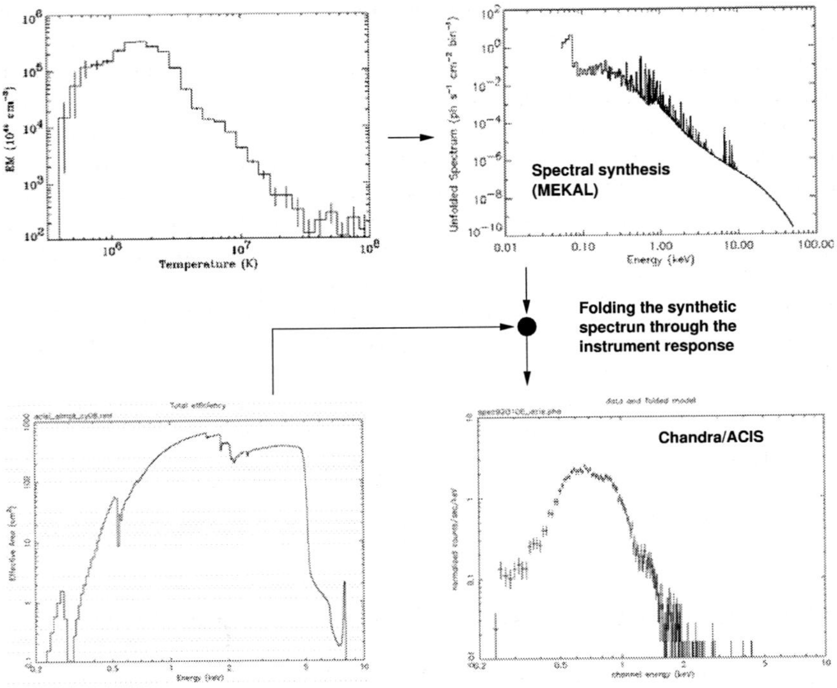

FIGURE 3. The second part of our method: translating solar data into a stellar context and format. From the EM(T) (upper left) we synthesize the spectrum of plasma at any temperature and sum all the contributions (upper right); folding through the instrumental response (lower left) and considering an appropriate exposure time and stellar distance, we obtain the spectrum in the focal plane of the instrument of interest (lower right).

physics.

Drake et al. [11] aimed at comparing some features of the EM(T) with those of observed stellar coronal emission measure distributions derived for the intermediate-activity stars ε Eri (K2 V) and ξ Boo A (G8 V) from Extreme Ultraviolet Explorer spectroscopic observations. It was found that the temperatures of the peaks of the observed stellar distributions EM(T), as well as their slopes in the temperature range $6.0 < logT < 6.5$, are very similar to those obtained for the brightest parts of the solar active regions in the 1992 January 6 SXT images. The observed slopes correspond approximately to $EM \approx T^{\beta}$ with $\beta \approx 4$, which is much steeper than predicted by static, uniformly heated loop models. The finding was the basis of a speculation on the coronal heating by flares and microflares.

Reale et al. [12] was dedicated to apply extensively the method to solar flares as a template for comparison with stellar flares and hot stellar thermal components. Eight flares, from weak (GOES class C5.8) to very intense ones (X9) were selected as representative of the flaring Sun. The EM(T) of the flaring regions was derived from Yohkoh/SXT observations in the rise, peak, and decay of the flares: it turned out to be, typically rather peaked and centered around $T \approx 10^7$ K for most of the time. It usually, grows during the rise phase of the flare, and then decreases and shifts toward lower temperatures during the decay. The most intense flare had significant emission measure even at very high temperatures ($T \approx 10^8$ K).

Orlando et al. [13] identified three classes of coronal structures of different brightness, low (background quiet corona), medium (active regions), and high (active region cores), and studied their contribution, analysing a large sample of selected full-disk Yohkoh/SXT observations taken between the maximum and the minimum of solar cycle 22. Using the diagram of X-ray surface flux in the ROSAT/PSPC band vs. spectral hardness ratio in the same band (F_X vs. HR) it was found that the Sun during the cycle fits well in the stellar scenario as a low-activity star. HR is, in practice, a spectral figure describing the hardness of the spectrum and closely related to the average plasma temperature. Both F_X and HR increase for increasing activity, from the background corona to the active regions and the cores of the active regions, and range between the values of low and intermediate activity stars.

Peres et al. [14] aimed at connecting the emission measure of individual solar coronal loops to the integrated emission measure of the entire solar corona, to deduce the energetic properties of the corona, and to "disentangle" (to some extent) the contribution of loops at each temperature from the global EM(T); the scope is also to apply the method to observations of solar-like stellar coronae.

Orlando et al. [15] focused on a single active region evolution, from its emergence to its almost disappearance, studying the changes along four solar rotations, and the rotational modulation due to this active region, the only one present at that time. The presence of the isolated active region caused significant variability of the average X-ray flux, with only moderate spectral variation, in the pass-bands of Rosat/PSPC, XMM-Newton/EPIC and Chandra/ACIS. Comparing the variability due to the rotational modulation, to the solar cycle, and to flares, they [15] found that in the $F_X$vs.HR diagram, both the solar cycle and the rotational modulation lead to the same correlation while flares also produce a correlation but with a much flatter slope than in the other two cases. Analogous results have been found putting data in the format XMM-Newton/EPIC and Chandra/ACIS.

THE F_X VS. HR DISTRIBUTION AND THE SCALING LAWS

Fig. 4, adapted from [16] and also based on previous work, shows a work to compare solar Yohkoh observations of the solar corona, of solar coronal active region and of the components of the solar corona, with a large set of ROSAT observations of stellar coronae.

Schmitt [17] studied all the late-type stars observed within 13 pc with ROSAT/PSPC. His work included their distribution of F_X vs. HR. Stars of different activity levels form an evident, single-parameter, strip ranging from low activity (also characterized by low flux and low temperature) to high activity (high surface flux and high temperature).

Fig. 4 also shows the average evolution of the Sun along the whole solar cycle (from [13] and [15]): evidently its characteristics are typical of a low activity star, in good agreement with the features of the Sun. We have also determined the loci occupied in this

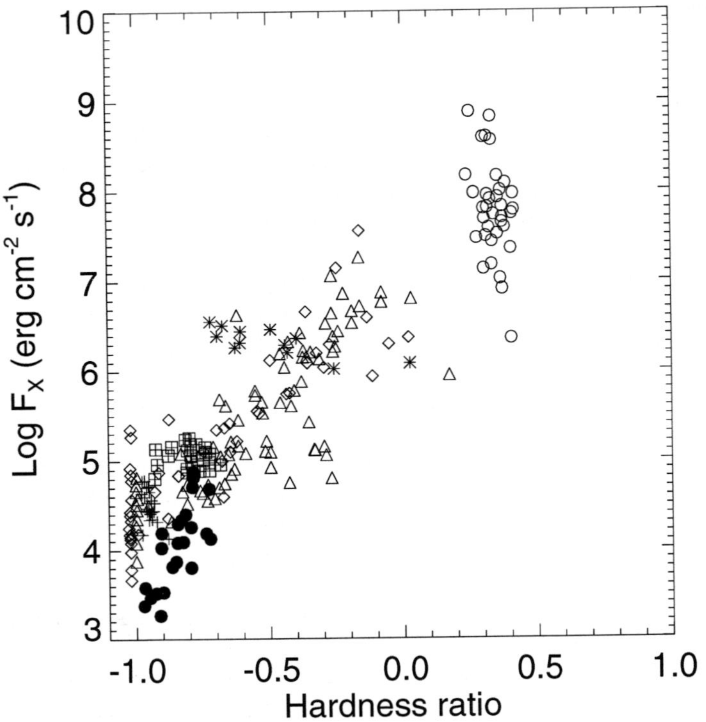

FIGURE 4. Surface X-ray flux vs. Hardness Ratio plot for late-type stars of [17] (triangles and diamonds), the whole solar corona along the cycle (filled dots), quiet coronal regions (crosses), active regions (squares), cores of active regions(stars) and flares (circles). (Adapted from [16])

graph by the various components of the solar corona: the active regions, the cores of the active regions, and the parts of the corona outside active regions (i.e. the quiet regions): they fall in characteristic, well identified and different regions of the strip. All the flares cover an additional, well characterized, region which occurs at higher temperature and surface flux than any star.

Analyzing this graph, it appears that stars of higher activity level can be explained with an higher fraction of corona filled with active regions. However, it is equally evident from the graph that even if covered with active regions or even with their cores, a stellar corona cannot reach the surface flux and temperature of the most active stars: in order to achieve their features, a significant fraction of the stellar surface has to be covered with structures having the features of flares.

There is, evidently, a common behaviour linking the less active stars to the most active ones. In order to model the common behaviour of the various components of the solar corona, of the solar corona as a whole (i.e. the Sun as a star) and of the stellar coronae, along the lines shown in the previous figures, we first translate hardness ratio into temperature using the conversion laws typical of ROSAT/PSPC, we then apply the physics of the coronal loops.

Adopting the reasonable assumption that the X-ray emission from stars is from plasma magnetically confined into loops, that loops have almost constant cross section and almost uniform heating, that roughly one half of the heating goes into radiation, and considering the scaling laws in [18] relating pressure, temperature, loop length and volumetric heating, we translate the previous graph into another form, correlating basic parameters of the coronal loops with the temperature.

Fig. 5 illustrates the results reformulating the F_X vs. HR correlation, showing analogous correlations in each panel of, respectively, pressure, density, loop length, and volumetric heating with temperature. The relevant distributions have been fitted with power laws, shown in each panel; these power laws illustrate other, somehow related, correlations among the variables of interest. We may also assume physical conditions slightly different from the above, for the coronal loops, amd the above correlations may change a little but will still be present; these curves illustrate a unified behavior of coronal structures from very active stars and active components of the solar corona, to faint stars and faint solar structures. It is worth noting that we have been able to remove free parameters from these correlations, with the only exception of the surface filling factor, thus linking the various parameters of coronal confined plasma.

Peres et al. [16] claim that the strong correlation between surface flux and temperature as well as the power laws shown above are just the effect of more fundamental physical mechanisms driving the coronal structures of all the late-type stars (including the Sun) from the emergence of new magnetic structures to their dispersal and dissipation.

FLARES, MICROFLARES AND THE PRESENCE OF VERY HOT PLASMA

For the following discussion let us just remind that in the previous section we inferred the continuous presence of flare-like structures present in coronae of very active stars in order to explain their loci in the graph, at surface flux and hardness ratio significantly higher than the cores of active regions.

Analogously, many spectroscopic observations of very active stars typically show an excess of the emission measure distribution vs. temperature at high temperature, often a "bump", around 10^7 K. Fig. 6 shows just an example, taken from [19]; also, the more active the star, the hotter and bigger is this hot bump. Many authors claim that this excess is due to the simultaneous presence of many flares, at any time, on the star. The light curve of the flares overlap yielding a practically continuous hot component. We remark the analogy with the big bump in the emission measure due to the maximum of the big flare shown in Fig. 2, but that was just one moment of the flare; the stellar excess we refer to is, instead, continuous.

On the Sun it is hard to observe these features because the flares are observed separately and only some of them are analysed systematically. The GOES satellites observes all of them but their integrated effect over time is not studied. Also the flares on the Sun are relatively less frequent than on active stars and certainly there are not many flare simultaneously present, even during the most active phases of the cycle.

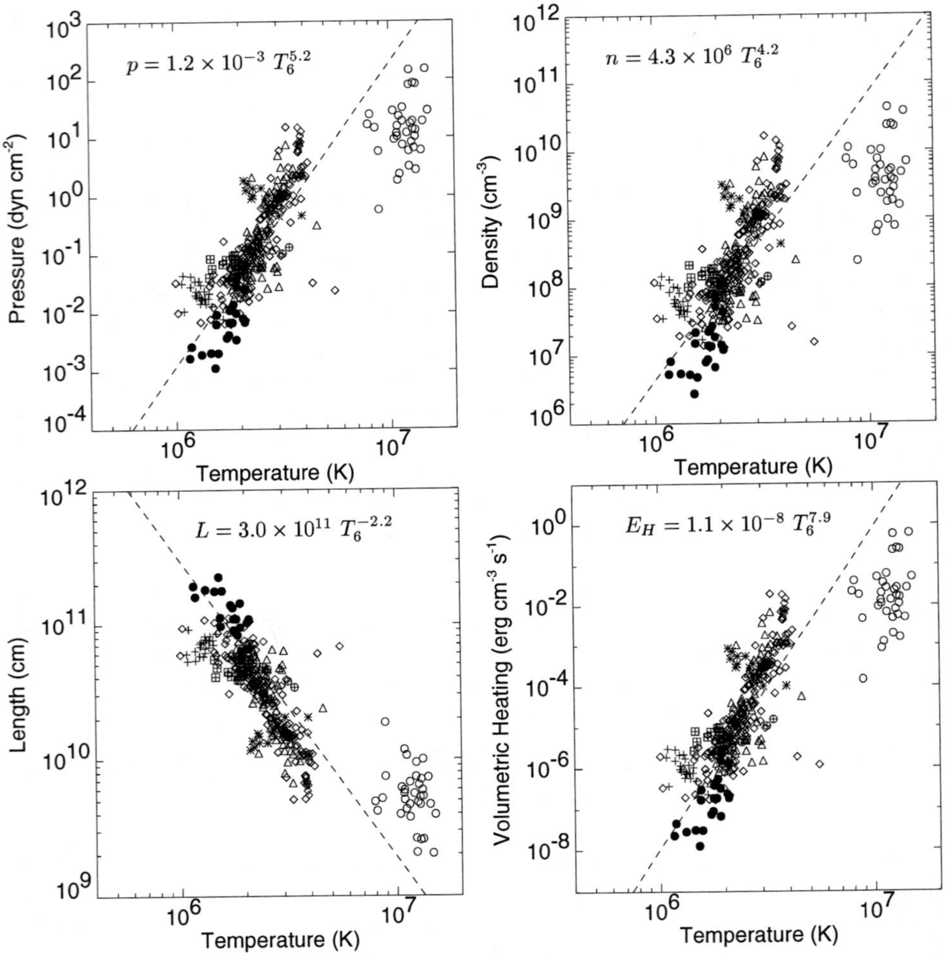

FIGURE 5. Distribution of plasma pressure (upper left), density (upper right), loops length (lower left), and heating per unit volume (lower right), vs. plasma temperature, of all the stars and coronal structures shown in figure 4, the dashed lines mark the power law fitting to all the data points also illustrated by the inset formula. (Adapted from [16])

However one wonders on the relative role of flares and not flaring corona in determining the time-averaged EM(T) of the Sun and how the ratio of the emission measures of the flaring and the quiet solar corona compares to what we see in stars of intermediate or very high activity.

So a question we have in the context of the solar stellar connection is: if we average the solar emission, including flares, over a time lapse long enough (say over a month) to include a large enough number of flares, do we get an emission measure with a

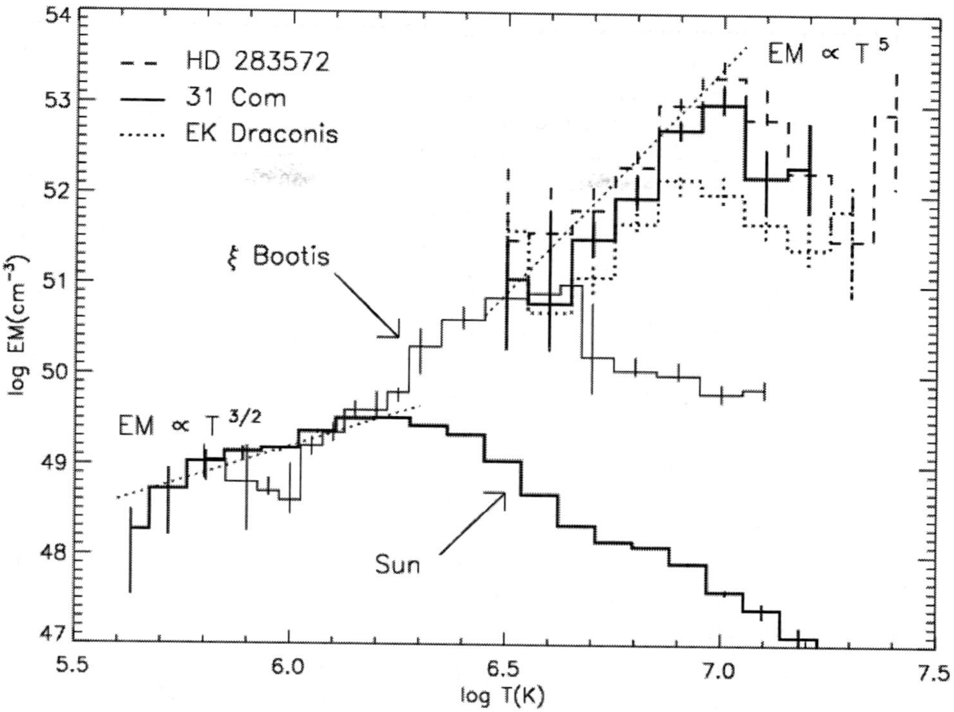

FIGURE 6. Emission measure distribution of five stars of very different X-ray luminosity: the Sun, χ Bootis, EK Draco, 31 Com, and HD 283572, in order of increasing luminosity. The figure illustrates the high temperature bump in the distribution and how it increases with luminosity, and activity. From [19].

significant bump at high temperature also for the Sun? (i.e. "Do the time averaged flares show up prominently with a bump or not?") and if the answer is yes, "How does the contribution of flares compare with that of the quiet corona?" Vice-versa, the time-averaged EM(T) of solar flares could turn out to be irrelevant and merge with that of non-flaring corona.

To do that, we cannot simply use the tecnique we have already developed applying it extensively to Yohkoh/SXT data: while Yohkoh observes routinely the corona on a daily basis (and the whole corona EM(T) varies quite slowly) it does not observe all the flares but only a subset, for different reasons, e.g. because some are not considered important, or big enough, or because the satellite is in eclipse, or because it gets saturated etc. This is crucial for our analysis since we have to include all the flares in a selected time interval. We thus turned to GOES satellite because it observes all the flares; however we need to cross calibrate Yohkoh/SXT and GOES as for temperature and emission measure determination.

We have used the data from nine flares of rather different characteristics for the cross calibration, to be sure of testing the calibration in very different conditions. So we took the instantaneous values of effective temperature and of total emission measure determined with both instruments and generated these two calibration curves. We had to sum up, for any time, the emission measure of all the flare pixels in Yohkoh/SXT and to determine the average temperature, in order to obtain a figure analogous to the one obtained with GOES observations. The results of the cross calibration are shown in Fig. 7 and 8, along with the relevant fitting functions of the cross-calibration.

Then we selected a month of quite high solar activity, namely December '91, and considered all the flares that GOES observed during that month, and generated this time-averaged whole-Sun emission measure distribution plus the emission measure of all the flares, rescaled, i.e. multiplied by the fraction of time each flares was active. Fig. 9 shows the cumulative distribution. The time averaged contribution of flares to the emission measure is evident; it is not as big bump as we see on active stars but that is not surprising. This is a preliminary work and these are the first results of the project. We still need to analize more cases and different phases of the solar cycle to see how things change along the cycle.

CONCLUSION

We have devised a method to translate the Yohkoh/SXT data into a context and format alike stellar X-ray observations, in order to allow a direct comparison of solar coronal observations with stellar ones. As for the results of the project, we have found:

- that the EM(T) evolves significantly along the solar cycle;
- that different kinds of coronal structures have very different features, and emission measure;
- that the more active a star the more it is likely to be covered with active regions, cores of active regions and flares (or structures as hot as flares);
- that in order to explain the most active stars we require the steady presence of very hot flares or flare-like structures;
- there is a universal scaling of coronal structures from the less active stars and quiet Sun to the most active stars and solar flares;
- as part of an on-going work, that the time-averaged effect of flares yields a bump in the EM(T) similar to the excess of EM(T) at high temperature found on active stars (but significantly smaller)

But then we have several questions concerning this last result.

If the steady corona is heated by micro-flares i.e. just a downsized version of flares, why the flare-time-averaged EM(T) does not merge with the steady corona EM(T)? Which is the difference between flares and steady coronal heating?
How does the "hot solar bump" due to flares change with phases of cycle and how to scale it to extreme stellar activity?
These questions open perspectives for our future work.

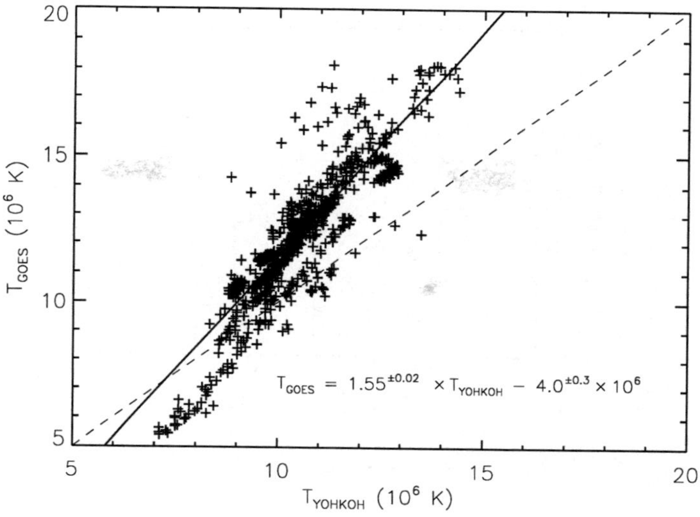

FIGURE 7. Cross calibration of flare instantaneous temperature derived from GOES data with the analogous figure derived from Yohkoh/SXT data, for a selected sample of flares of rather different characteristics.

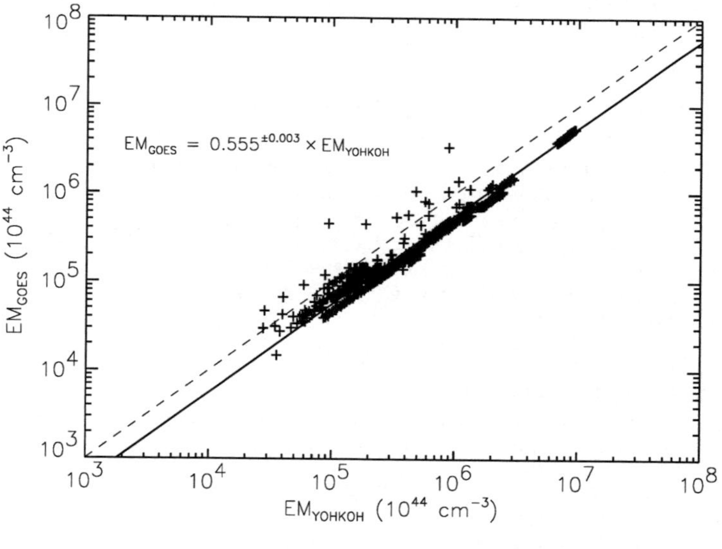

FIGURE 8. The same as Fig. 7 for the emission measure

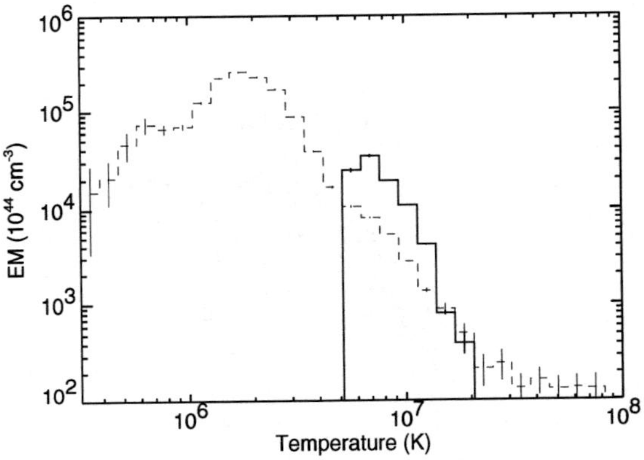

FIGURE 9. Emission measure distribution of the non flaring corona (dashed) and the time-averaged contribution of flares, during January 1991, as derived from Yohkoh/SXT data (corona) and GOES (flares).

ACKNOWLEDGMENTS

Our work had partial support of grants from the Italian Space Agency (ASI) and from the Italian Ministry for University and Research (MIUR). GP wishes to thank the warm hospitality of the Hellenic Astronomical Society, of the organizing committees, and, in particular, of Heleni Dara and Kanaris Tsinganos.

REFERENCES

1. Judge, P. G., Saar, S. H., Carlsson, M., & Ayres, T. R. 2004, The ApJ, 609, 392
2. Schrijver, C. J., Sandman, A. W., Aschwanden, M. J., & DeRosa, M. L. 2004, The ApJ, 615, 512
3. Mandrini, C. H., Démoulin, P., & Klimchuk, J. A. 2000, The ApJ, 530, 999
4. Lanza, A. F., Rodonò, M., Pagano, I., Barge, P., & Llebaria, A. 2003, A&A, 403, 1135
5. Lanza, A. F., Rodonò, M., & Pagano, I. 2004, A&A, 425, 707
6. Acton, L. W., Weston, D. C., & Bruner, M. E. 1999, JGR, 104, 14827
7. Stern, R. A. 1987, Sol.Ph., 113, 291
8. Orlando, S., Peres, G., & Reale, F. 2000, The ApJ, 528, 524
9. Peres, G., Orlando, S., Reale, F., Rosner, R., & Hudson, H. 2000, The ApJ, 528, 537
10. Mewe, R., Kaastra, J. S., & Liedhal, D. A. 1995, Legacy, 6, 16
11. Drake, J. J., Peres, G., Orlando, S., Laming, J. M., & Maggio, A. 2000, The ApJ, 545, 1074
12. Reale, F., Peres, G., & Orlando, S. 2001, The ApJ, 557, 906
13. Orlando, S., Peres, G., & Reale, F. 2001, The ApJ, 560, 499
14. Peres, G., Orlando, S., Reale, F., & Rosner, R. 2001, The ApJ, 563, 1045
15. Orlando, S., Peres, G., & Reale, F. 2004, A&A, 424, 677
16. Peres, G., Orlando, S., & Reale, F. 2004, The ApJ, 612, 472
17. Schmitt, J.H.M.M. 1997, A&A, 318, 215
18. Rosner, R., Tucker, W. H., & Vaiana, G. S. 1978, The ApJ, 220, 643
19. Scelsi, L., Maggio, A., Peres, G., & Pallavicini, R. 2005, A&A, 432, 671

Coronal Heating

J. A. Klimchuk and M. C. López Fuentes

Space Science Division, Naval Research Lab, Washington, DC 20375, USA

Abstract. Coronal heating is one of the most challenging and important problems in astrophysics. We here review some of the critical aspects of the problem and suggest that the most likely explanation is Parker's long-standing idea of nanoflares occurring in magnetic fields that become tangled by photospheric convection. An exciting new development is the identification of the "secondary instability" as the likely mechanism of energy release. We also present some preliminary new results indicating that the rise, main, and decay phases of the soft X-ray light curves of observed coronal loops might be associated with the development, maintenance, and destruction of self-organized critical systems.

Keywords: hydrodynamics; magnetohydrodynamics and plasmas; solar physics; solar magnetism; corona; coronal loops; solar electromagnetic emission; transition region
PACS: 95.30.Lz; 95.30.Qd; 96.60.-j; 96.60.Hv; 96.60.P-; 96.60.pf; 96.60.Tf; 96.60.Xy

1. GENERAL COMMENTS

Simply stated, the coronal heating problem is to identify and understand the physical mechanism responsible for heating the corona to multi-million degree temperatures, several hundred times hotter than the underlying photosphere. It has been and remains one of the most important problems in astrophysics. This short article outlines a number of concepts and recommendations that are described fully in Klimchuk (2006). Numerous references are cited therein and are not repeated here. Some preliminary new results are presented at the end.

A definitive solution to the coronal heating problem involves several critical steps, as outlined in the coronal heating flowchart of Figure 1. We must identify a source of energy and a mechanism for converting that energy into heat. We must determine how the plasma responds to the heating. And finally, we must predict the spectrum of emitted radiation and its manifestation as observable quantities. Only by accomplishing *all* of these steps can a truly rigorous and meaningful comparison with actual observations be made. Each of the steps is extremely challenging, and a variety of MHD and hydrodynamic modelling tools have been employed to make significant progress. Much additional work is left to be done, however.

Several themes cut across the different steps listed above. First, the coronal heating problem involves *highly disparate and coupled spatial scales*. Structuring is present both along and across the magnetic field. Because classical dissipation coefficients are extremely small in the corona, significant heating generally requires the formation of very steep gradients and very small spatial scales. Magnetic gradients lead to heating by reconnection and Ohmic dissipation, while velocity gradients lead to heating by viscous dissipation. Small-scale features of particular significance include

CP848, *Recent Advances in Astronomy and Astrophysics,*
7th *International Conference of the Hellenic Astronomical Society,* edited by N. H. Solomos

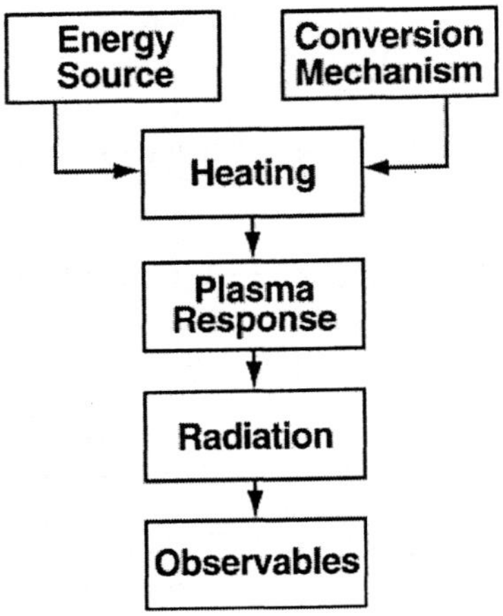

FIGURE 1. Flowchart showing the different steps that are involved in solving the coronal heating problem.

electric current sheets, resonance layers, shocks, and transition regions where the temperature and density change abruptly.

Second, physical *connections between the corona and lower atmosphere* are fundamentally important, both for the source and conversion of energy and for the response of the plasma to heating. The properties of the coronal magnetic field, including its energy content, current sheets, and instability, are determined by evolving photospheric boundary conditions. The properties of the coronal plasma, including its temperature, density, and velocity, are greatly influenced by the heat flux to the base and the associated transfer of mass via chromospheric evaporation and condensation. To understand the radiative signatures of coronal heating, it is imperative to treat the corona and transition region together as a coupled system.

Third, *microphysics* can be extremely important in all steps of the coronal heating problem. Collisionless reconnection may play a role in the conversion of magnetic energy into heat. Heat flux saturation can affect how the plasma responds to the heating. And ionization nonequilibrium impacts the radiation spectrum that is ultimately used to diagnose the heating.

Fourth, both theoretical and observational considerations lead us to conclude that coronal heating involves a high degree of *variability and dynamics*, as we now discuss.

2. ELEMENTAL FLUX STRANDS AND IMPULSIVE HEATING

To understand how coronal plasmas respond to heating, it is necessary to adopt the perspective of elemental magnetic flux strands. We define an elemental strand to be a flux tube where the heating rate and plasma properties (temperature, density, and velocity) are approximately uniform over a cross section. Elemental strands are likely to be very thin and well below the resolving capabilities of current instrumentation.

Nearly every realistic theory of coronal heating predicts that the energy release within elemental strands is very impulsive. Even steady Petschek-like reconnection has this property. In Petschek reconnection, individual strands are heated only while they are in contact with standing slow shocks that emanate from the reconnection X-point. A newly reconnected strand snaps away from the X-point at the Alfvén speed, so the heat pulse in the strand lasts a few tens of seconds at most (e.g., a $2x10^9$ cm long shock and $2x10^8$ cm s^{-1} Alfvén speed give a pulse of 10 s). Resonant wave absorption is another example of heating that is usually thought of as steady, but in fact is highly impulsive when looked at from the proper perspective. Because of chromospheric evaporation and condensation, the resonance layers drift back and forth throughout the loop, heating each strand only briefly as they pass by.

An independent indication that strands are heated impulsively comes from observations of the density and temperature relationships of coronal loops. Warm ($T \sim$ 1 MK) loops observed by *TRACE* and EIT tend to be over dense compared to the predictions of static equilibrium theory, while hot ($T > 2$ MK) loops observed by *Yohkoh* tend to be under dense compared to the predictions. This pattern is highly suggestive of plasmas that are cooling after having been impulsively heated. However, loops seen by *Yohkoh* and *TRACE* live much longer than a cooling time, so we must invoke a bundle of unresolved strands. Each strand is evolves rapidly, but the entire unresolved bundle appears to evolve slowly. *Yohkoh* preferentially detects the hot strands in the early stage of cooling, which is dominated by thermal conduction and characterized by low densities. *TRACE* preferentially detects the warm strands in the late stage of cooling, which is dominated by radiation and characterized by high densities.

3. TANGLED FIELDS AND SECONDARY INSTABILITY

The picture of loops as bundles of impulsively heated elemental strands fits nicely with Parker's idea of coronal magnetic fields that are tangled by turbulent photospheric convection. High-resolution observations of the photosphere reveal that the field is concentrated into small flux tubes of kilogauss strength. These tubes are so small that many tens of them must be contained within a single *TRACE* loop. Convective motions randomly displace the footpoints and cause the tubes to become intertwined and braided. Current sheets form at the coronal interfaces where the tubes are misaligned. Recent results suggest that a process called the secondary instability produces an explosive release of magnetic energy whenever the misalignment angle reaches a critical value of about 40°. A misalignment angle of this magnitude corresponds to a typical tilt from vertical at the coronal base of approximately 20°, which is the same angle that has been inferred from global energy balance considerations. This and other reasons give us growing confidence that the secondary instability is the primary mechanism of energy release in the corona.

Whatever the mechanism of energy release, it is clear that some form of magnetic reconnection (i.e., some process that changes the magnetic connectivity, including the secondary instability) must be occurring at significant altitudes in the corona. Without it, the degree of tangling produced photospheric convection would increase

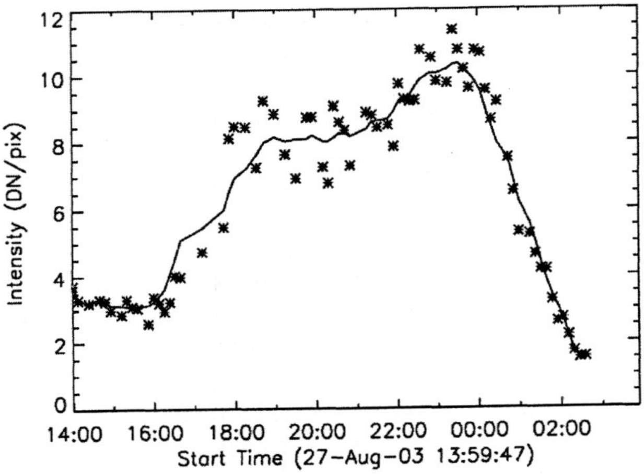

FIGURE 2. Soft X-ray light curve showing intensity versus time for a coronal loop observed by the SXI instrument on *GOES-12* using the open filter position. Solid line is a 10-point running average.

indefinitely. Magnetic stresses would build to ridiculous levels, and no coherent structures such as coronal loops would exist. The reconnection must occur above the magnetic carpet (above about 5×10^8 cm), since reconnection between a long coronal flux tube and a short carpet loop actually causes the tangling to increase rather than decrease.

4. LOOP EVOLUTION AND SELF-ORGANIZED CRITICALITY

We now present some preliminary results from López Fuentes, Mandrini, & Klimchuk (2004) and López Fuentes & Klimchuk (2005). Coronal loops observed by the Soft X-ray Imager (SXI) on the *GOES-12* spacecraft generally have light curves with three distinct phases: rise, main, and decay. A typical example is shown in Figure 2. We have begun to investigate the hypothesis these light curves indicate the development, maintenance, and destruction of self-organized critical systems. The basic idea is that loops begin as bundles of minimally tangled kG tubes. Photospheric convection shuffles the footpoints and progressively increases the tangling. In the early stages, relatively few pairs of adjacent tubes reach the secondary instability threshold, and the soft X-ray emission is relatively faint. As time proceeds, the emission brightens as the number of tube pairs reaching the threshold steadily increases. This is the rise phase. Eventually, a self-organized critical state is achieved in which the stressing of the field by convection is balanced by the destressing of the field by impulsive energy releases (nanoflares). This is the main phase, when the emission is roughly steady. Finally, something causes the effectiveness of the footpoint shuffling to diminish, and the loop enters a decay phase when the critical state can no longer be maintained. The cause of this loss of effectiveness is not entirely clear. One possibility is that the concentration of kG tubes decreases as the

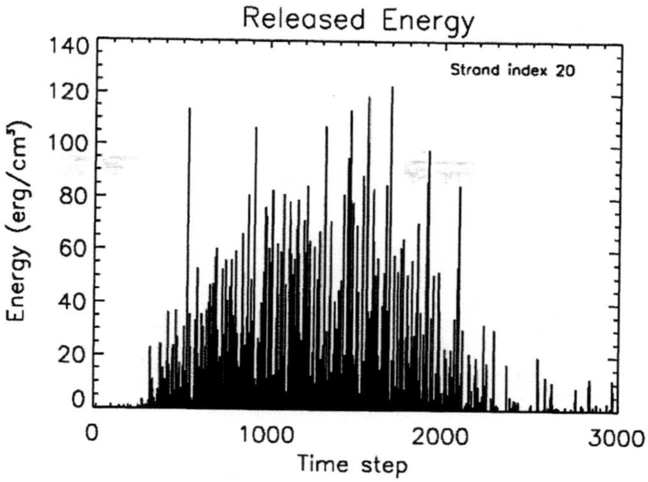

FIGURE 3. Magnetic energy release versus time step for one of the flux tubes in a self-organized criticality simulation (see text).

tubes diffuse across the photosphere. Footpoint shuffling is efficient at tangling the field only when the random walk step size is longer than the characteristic footpoint separation. When this condition is not met, a random step will not necessarily wrap a flux tube around its neighbor and there is a reduced likelihood that the level of tangling will increase.

We have investigated this basic scenario with a simple model. We consider a linear array of tubes with vertical field B_v. We mimic footpoint shuffling by randomly modifying the horizontal stress component B_h, initially taken to be zero. Whenever the difference ΔB_h between adjacent tubes exceeds a critical value (corresponding to a critical misalignment angle), we redistribute the B_h between the tubes. B_h is decreased in one tube by the same amount that it is increased in the other. There is a resulting decrease in the total free magnetic energy ($B_h^2/8\pi$) of the two tubes, which we assume deposits heat equally in both. We then use an approximate hydro model, discussed below, to follow the evolution of the plasma temperature and density. Note that B_h must be fixed to zero for the end tubes of the array in order for self-organized criticality to be established. The justification for this is a matter of discussion.

Figure 3 shows the nanoflare energy release as a function of time step for one of the tubes in a 20-tube simulation. Each spike corresponds to a separate nanoflare. Figure 4 shows the evolution of the heating rate, temperature, and density in the tube for a 5000 s sub-interval during the main phase. Eleven nanoflares occur in the tube during this interval, and they are assumed to produce triangular heating profiles lasting 100 s. This nanoflare duration is similar to what we find in our MHD simulations of the secondary instability. Finally, Figure 5 shows the evolution of the soft X-ray intensity that would be observed by the SXI instrument if all 20 tubes were observed together as an unresolved bundle. The full 25 hours of the simulation are shown, as in Figure 3.

We have used reasonable parameters for the model and are encouraged by the overall similarity between the simulation light curve of Figure 5 and the observed light

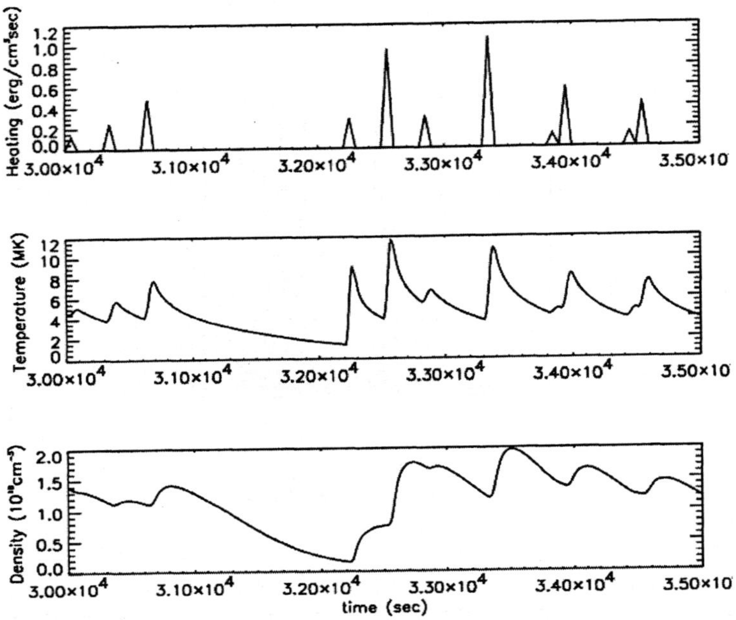

FIGURE 4. Heating rate (top), temperature (middle), and density (bottom) versus time in the flux tube for a 5000 s sub-interval during the main phase.

curve of Figure 2. The model assumes a vertical field $B_v = 40$ G at the base of the corona where the kG flux tubes have flared out to become space filling, a loop length of 40 Mm, a random walk step length of 1 Mm, and a random walk time step of 10^2 s. This corresponds to a somewhat rapid footpoint velocity of 10 km s^{-1}, but is necessary to obtain a rise time similar to that observed. We also include a reconnection efficiency factor of 0.02 to account for the fact that only a fraction of the tube volume reconnects during a single nanoflare event. Finally, we use two probability distribution functions to randomly modify B_h. The first corresponds to a high density of footpoints and highly efficient tangling, and the second corresponds to a low density of footpoints and less efficient tangling. We transition from the first to the second at $t = 15$ hr, and this causes the system to lose criticality and enter the decay phase.

Much more work needs to be done before we can say anything definitive, but we feel that the basic scenario of development, maintenance, and destruction of a self-organized critical system is a promising explanation for the rise, main, and decay phases of observed coronal loops.

5. HYDRO MODEL

The simulation above uses a new model for describing the plasma evolution in a loop strand that is subjected to variable heating (Klimchuk, Patsourakos, & Cargill 2005). It is a 0D model in the sense that it describes the average temperature and

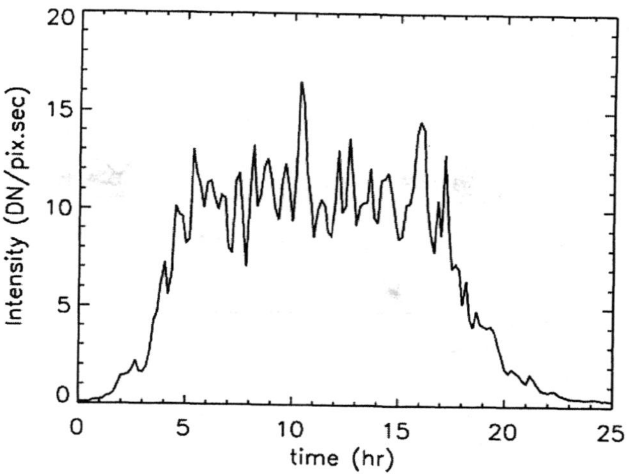

FIGURE 5. Soft X-ray intensity versus time for the unresolved bundle of 20 flux tubes, as would be observed by the *GOES-12* SXI instrument using the open filter position.

density along the strand. The basic idea behind the model is that the ratio of the radiative losses in the transition region to the radiative losses in the corona is at all times the value it would have if the loop were in static equilibrium. This value is easily computed. The transition region is heated by a downward heat flux from the corona, and any differences between the heat flux and the radiative losses are equated with an enthalpy flux of flowing material. If there is an excess heat flux, it drives a chromospheric evaporation upflow. If there is a deficiency in the heat flux, plasma drains from the corona and condenses back onto the chromosphere.

This simple model works exceptionally well. Figure 6 shows a comparison between a 0D simulation (dashed) and a sophisticated simulation that accurately solves the full 1D hydro equations (solid). The volumetric heating rate is spatially uniform along the strand and evolves in the manner shown in the bottom panel. It decreases slowly, increases abruptly, remains constant, decreases abruptly, then remains constant at a low level. The strand-averaged temperature and density are shown in the top two panels. The 0D solution tracks the 1D solution very closely, even though it requires 4 orders of magnitude less computing time! The 0D model is thus a powerful new tool for studying coronal heating. It differs from previous 0D models in that it can accommodate any heating profile and it accounts for radiative and conductive losses at all times during the simulation.

6. CONCLUSIONS

A number of different lines of reasoning point to the conclusion that the corona consists of sub-resolution strands that are impulsively heated. The photospheric field is observed to be concentrated in small kG flux tubes that are randomly displaced by turbulent convection. The tangled tubes *must* "reconnect" in the corona in order to avoid a catastrophic build-up of complexity and stress. The secondary instability is

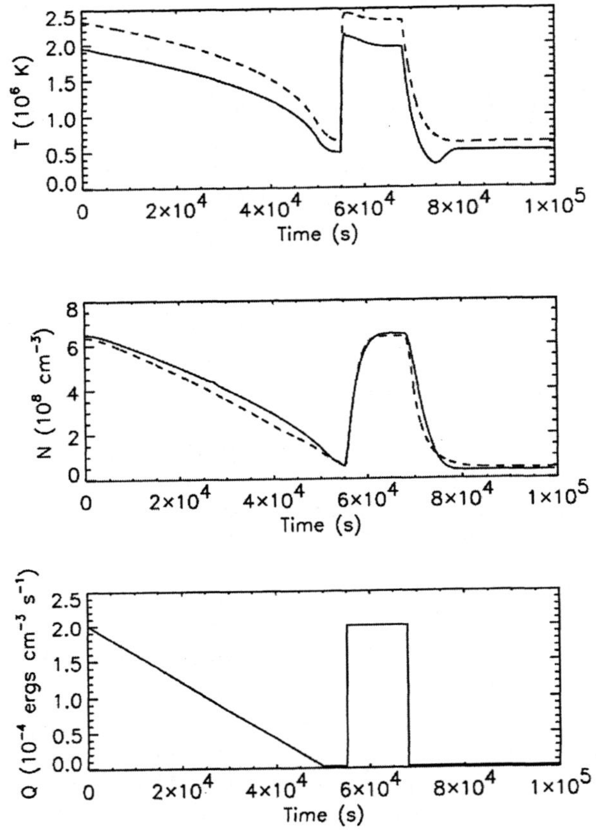

FIGURE 6. Temperature (top), density (middle), and heating rate (bottom) for loop strand simulations using the simplified 0D model (dashed) and the sophisticated 1D model (solid).

expected to produce explosive energy release whenever adjacent tubes reach a critical misalignment angle. The observed under and over densities of hot and warm loops are best explained by impulsive heating and subsequent cooling. The three-part light curves of soft X-ray loops can be reproduced by a self-organized criticality model based on tangled, reconnecting tubes. Finally, the observed symmetry of loops (i.e., the tendency for loops to have legs of similar width) might be also be explained by tangled, reconnected tubes, as explained in Klimchuk (2006).

A definitive test of this or any other theory of coronal heating requires a quantitative prediction of observable quantities that is based on a detailed, first principles treatment. This treatment must explain the source of the energy, the mechanism of converting that energy into heat, the response of the plasma to the heating, and the production of a radiation spectrum. This is monumental task. Many researchers have made outstanding contributions to the coronal heating problem in the past, as cited in Klimchuk (2006), and no doubt they will continue to do so until we arrive at our final objective.

ACKNOWLEDGMENTS

I thank the meeting organizers and especially Eleni Dara for their kind hospitality. This work was supported by NASA and the U.S. Office of Naval Research.

REFERENCES

1. J.A. Klimchuk, *Solar Phys.* **234**, in press (2006), http://arxiv.org/abs/astro-ph/0511841.
2. J.A. Klimchuk, S. Patsourakos, and P.J. Cargill, AGU Spring Mtg., *SP14A-06* (2005).
3. M.C. López Fuentes, C.H. Mandrini, and J.A. Klimchuk, *Bull. A. A. S.* **36**, 761 (2004).
4. M.C. López Fuentes and J.A. Klimchuk, AGU Spring Mtg., *SP14A-03* (2005).

The Role of Magnetic Helicity in Solar Eruptions

A. Nindos

Section of Astrogeophysics, Physics Department, University of Ioannina, Ioannina 45110, Greece, email: anindos@cc.uoi.gr

Abstract. Magnetic helicity is a quantity that descibes the chiral properties of magnetic structures. It has the unique feature that it is probably the only physical quantity which is approximately conserved even in resistive MHD. This makes magnetic helicity an ideal tool for the exploration of the physics of coronal mass ejections (CMEs). CMEs carry away from the Sun twisted magnetic fields and the concept of helicity can be used to monitor the whole history of a CME event from the emergence of twisted magnetic flux from the convective zone to the eruption and propagation of the CME into interplanetary space. I discuss the sources of the helicity shed by CMEs and the role of magnetic helicity in the initiation of CMEs.

Keywords: Sun:magnetic fields, sun:coronal mass ejections, sun:photosphere, sun:corona
PACS: 96. 60. ph

1. INTRODUCTION

Coronal mass ejections (CMEs) are large-scale expulsions of magnetized plasma from the Sun. They are observed with a coronagraph above the occulting disk as projections on the plane of the sky. CMEs have attracted significant attention lately because they are the primary cause of the largest space weather disturbances (e.g. Gosling 1993). In an average event, 10^{14} - 10^{16} gr of plasma is ejected into the heliosphere with speeds ranging from 100 to 2000 km/s (e.g. Hundhausen et al. 1994).

The occurrence of CMEs shows a strong solar cycle dependence: during solar minimum there is one CME every 2-3 days on average, while during solar maximum they are more frequently observed with several to more than 10 every day. CMEs sometimes, but not always, are associated with flares. Overall the occurrence rate of flares is larger than the occurrence rate of CMEs (averaged over a solar cycle, there are 2-3 CMEs per day while the corresponding occurence rate of GOES X-ray flares is 5-6 events per day).

Observations with a coronagraph do not reveal the correlation of CMEs with the configuration and evolution of the underlying solar structures. However, high resolution soft X-ray (SXR) and EUV observations show that the low corona CME counterparts (see the article by Hudson & Cliver 2001 for a review) may include coronal waves, EUV and/or SXR "dimmings", and long-duration SXR flares that occur after the CME eruption. Also SXR data reveal a relationship between coronal X-ray sigmoids and eruptivity. A flare may also destabilize an adjacent transequatorial loop structure, thus launching a CME. Note, however, that not every CME is accompanied by all the noncoronagraphic signatures listed here.

The essential physics of CMEs probably resides in the magnetic field. This argument is based on the fact that the magnetic field dominates the plasma throughout the corona because the coronal plasma $\beta \ll 1$. Therefore the most important forces which determine

CP848, *Recent Advances in Astronomy and Astrophysics,*
7th International Conference of the Hellenic Astronomical Society, edited by N. H. Solomos

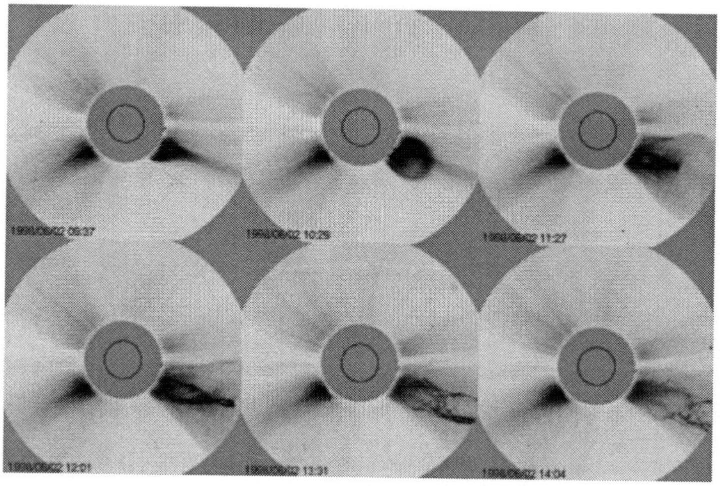

FIGURE 1. The evolution of a CME observed on June 2, 1998 with LASCO's C2 coronagraph.

the eruption initiation and dynamics are magnetic. For the same reason, the energy required to lift the mass of a CME against gravity and accelerate it to the observed velocities is believed to be magnetic (e.g. Vourlidas et al. 2000). The fraction of magnetic energy which is converted to other forms yielding the CME initiation and motion comes from non-potential magnetic field because energy cannot be extracted from a current-free potential field. The magnetic field carried away by CMEs can be obviously helical at times. An example is given in Figure 1: the CME's stressed helical structure shows clearly in the bottom row of the figure. Fig. 1 also demonstrates another interesting property of the magnetic field evolution during CMEs: a part of the pre-CME magnetic configuration may contain closed field lines that open up as a result of the eruption (compare for example, the top left and the bottom right frames of fig. 1).

In this article, I focus on the physics of CMEs as a magnetic phenomenon. For such treatment the concept of magnetic helicity is of great importance. Magnetic helicity quantifies the chiral properties of magnetic structures and has some unique features which can help us understand several aspects of the CME physics. Note that I do not attempt to provide an exhaustive review of either CMEs or helicity; the interested reader is referred to the articles in Crooker et al. (1997) and to the articles in Brown et al. (1999), respectively. All I will try to do is to review the aspects of the CME phenomenon that are relevant to the concept of magnetic helicity. Also, for reasons that will become clear below, confined flares (i.e. flares that are not associated with CMEs) are not discussed in detail.

The article is organized as follows. In §2, the definition of magnetic helicity is given. In §3, I discuss the sources of helicity carried away by CMEs and in §4 I discuss what magnetic helicity can tell us about CME initiation. I present conclusions and suggestions for future work in §5.

2. THE CONCEPT OF MAGNETIC HELICITY

2.1. Definition

Magnetic helicity is a quantity which describes nonpotential magnetic fields. For a magnetic field **B** within a volume V, it is defined as:

$$H_m = \int_V \mathbf{A}\cdot\mathbf{B}dV \tag{1}$$

where **A** is the magnetic vector potential ($\mathbf{B} = \nabla\times\mathbf{A}$). Eq. (1) is physically meaninful only when the magnetic field is fully contained inside V (i.e. at any point of the surface S surrounding V, the field's normal component B_n vanishes); this is so because the vector potential is defined through a gauge transformation ($\mathbf{A}' = \mathbf{A} + \nabla\Phi$), then H_m is gauge-invariant only when $B_n = 0$.

In the case of the solar atmosphere, magnetic flux passes through S and therefore the above condition is not satisfied. However, Berger & Field (1984) and Finn & Antonsen (1985) have shown that when $B_n \neq 0$ on S, we can define a gauge-invariant relative magnetic helicity H (hereafter referred to as helicity) of **B** with respect to the helicity of a reference field $\mathbf{B_p}$ having the same distribution of normal magnetic flux on the surface S surrounding V:

$$H = \int_V \mathbf{A}\cdot\mathbf{B}dV - \int_V \mathbf{A_p}\cdot\mathbf{B_p}dV \tag{2}$$

where $\mathbf{A_p}$ is the vector potential of $\mathbf{B_p}$. The quantity H does not depend on the common extension of **B** and $\mathbf{B_p}$ outside V. Being a potential field it is a convinient choice for $\mathbf{B_p}$. If in addition $\nabla\cdot\mathbf{A_p} = 0$ and $(A_p)_n = 0$ on S then the term $\int_V \mathbf{A_p}\cdot\mathbf{B_p}dV$ of eq. (2) vanishes (Berger 1988).

In ideal MHD, helicity is a conserved quantity (e.g. Berger 1984), and the frozen-in condition can be stated equivalently as the absolute conservation of helicity in all sub-volumes as well as of the total helicity in the whole domain (Zhang & Low 2005).

It is worth noting that the "natural" unit of helicity is the square of magnetic flux (Mx^2) and therefore the helicity of a twisted flux tube with N turns and magnetic flux equal to unity is simply N. For more complex magnetic topologies, helicity can be regarded as a measure of the topological complexity of the magnetic field (e.g. linkage and twistedness in the field).

2.2. Flux and flux density of Magnetic Helicity

In the solar corona, the electric resistivity is very low but not zero. Therefore in the most general case in the corona, the amount of helicity within V can change either due to helicity flux crossing S or/and due to dissipation within V. Berger (1984) has demonstrated that the helicity dissipation rate is negligible in all processes taking place in the corona, including reconnection and all non-ideal processes. Helicity's dissipation time scale is the global diffusion time scale. Consequently, helicity can be regarded as an (almost) conserved quantity even in resistive MHD. The difference is that while in ideal MHD helicity is conserved in any sub-volume closed by a flux surface, in resistive MHD only the total helicity within the whole volume is approximately conserved.

The fact that helicity is one of the few (probably the only) quantities that it is preserved in the absence or near absence of resistivity makes it a powerful tool that can help us trace the transport of magnetic field from the sub-photospheric layers to the corona, and then its ejection into the interplanetary medium.

The temporal evolution of helicity across the photospheric boundary S_p is given by

$$\frac{dH}{dt} = 2\int_{S_p} [(\mathbf{A_p}\cdot\mathbf{B_t})v_n - (\mathbf{v_t}\cdot\mathbf{A_p})B_n]dS \tag{3}$$

where B_t and B_n are the tangential and normal components of the magnetic field on the photosphere and $\mathbf{v_t}$ and $\mathbf{v_n}$ the tangential and normal components of the photospheric plasma velocity (Berger 1999). The first term gives the change of helicity due to the emergence of twisted field lines that cross the photosphere while the second term gives the change of helicity due to horizontal motions on the photospheric surface. Such motions may come either from differential rotation and/or from transient photospheric shearing flows.

When series of photospheric longitudinal magnetograms are available, the horizontal velocity appearing in eq. (3) can be computed using the local correlation tracking (LCT) technique. Several authors have computed the corresponding helicity injection rate (Chae 2001, Nindos & Zhang 2002, Moon et al. 2002a,b, Nindos et al. 2003, Chae et al. 2004). However, as Démoulin & Berger (2003) have pointed out, with the magnetograms one follows the photospheric intersection of the magnetic flux tubes but not the evolution of the plasma (generally the two velocities are different): the LCT method detects the velocity of the footpoints of the flux tube which is

$$\mathbf{u} = \mathbf{v_t} - \frac{v_n}{B_n}\mathbf{B_t} \tag{4}$$

If we insert this expression for the velocity in eq. (3) we get

$$\frac{dH}{dt} = -2\int_{S_p} (\mathbf{u}\cdot\mathbf{A_p})B_n dS \tag{5}$$

and consequently, one may use the quantity $G_A(\mathbf{x}) = -2(\mathbf{u}\cdot\mathbf{A_p})B_n$ as a proxy to the helicity flux density. Pariat et al. (2005) have proposed a different proxy to the helicity flux density. Its concept is based on the fact that the helicity injection rate can be understood as the summation of the relative rotation rate $d\theta(\mathbf{r})/dt$ of all pairs of elementary fluxes weighted with their magnetic flux (Berger 1984). The resulting quantity

$$G_\theta(\mathbf{x}) = -\frac{B_n}{2\pi}\int_{S'} \frac{\mathrm{d}\theta(\mathbf{r})}{\mathrm{d}t} B'_n dS' \tag{6}$$

can be derived from observations. Pariat et al. (2005) have shown that G_A can exhibit spurious bipolar signals of comparable magnitude while G_θ reduces the fake polarities by more than an order of magnitude and therefore it is a much better proxy of the magnetic helicity flux density. In fig. 2 (taken from Pariat et al. 2006) we show two

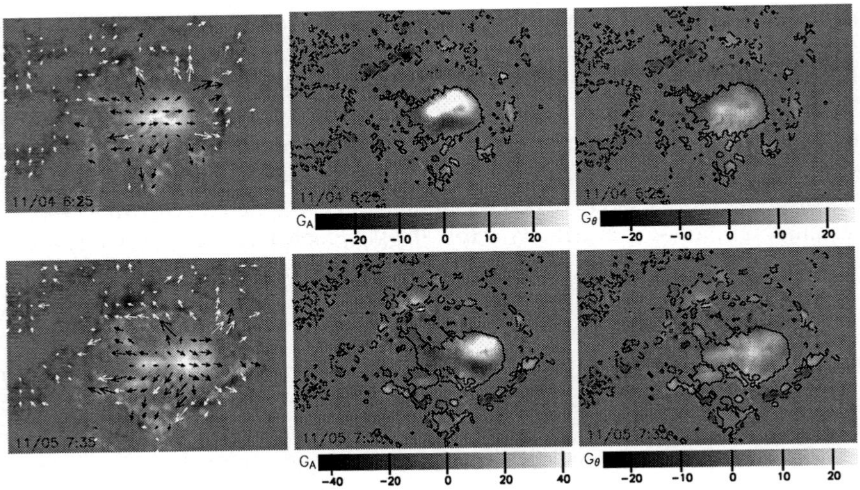

FIGURE 2. AR8375 at 06:25 UT on November 4, 1998 (top) and at 07:35 UT on November 5, 1998 (bottom). Left panels: B_n magnetograms with velocity fields (arrows). Middle panels: G_A maps. Right panels: G_θ maps. G_A and G_θ maps are in units of 10^6 Wb^2 m^{-2} s^{-1}, and on each one B_n isocontours of ± 300 G have been overlaid (from Pariat et al. 2006).

examples of G_A and G_θ computations for active region AR8375: note the bipolar pattern above the sunspot in the G_A images and the much more uniform appearance of the corresponding G_θ images.

Since both G_A and G_θ are derived from eq. (3), theoretically they should yield strictly equal helicity injection rates, dH/dt. The magnetogram image cubes analyzed by Pariat et al. (2006) indicate that indeed the total helicity fluxes derived from G_A and G_θ show very small differences on time scales larger than their transient temporal variations.

3. THE SOURCES OF HELICITY CARRIED AWAY BY CMES

3.1. CMEs as a Way of Removing Helicity from the Corona

The solar cycle dependence of the CME occurrence rate may imply that CMEs are somehow connected to the solar dynamo. Over the years, it has been realized that in several occasions magnetic fields emerging from the solar interior to the photosphere are twisted (e.g. Leka et al. 1996). This indicates that a significant fraction of active region's helicity is created by the dynamo and then transported into the corona through the photosphere with the emerging magnetic flux. Since helicity is not destroyed under reconnection this process accumulates helicity in the corona. Furthermore, on the global scale, helicity emerges predominantly negative in the northern hemisphere and predominantly positive in the southern hemisphere (e.g. Pevtsov et al. 1995). And also this hemispheric helicity sign pattern does not change from solar cycle to solar cycle (Pevtsov et al. 2001). Consequently, on the global scale, mutual cancellation of helicity of opposite signs cannot relieve the Sun from excess accumulated helicity. It has been suggested (e.g. Low 1996) that CMEs, as expulsions of twisted magnetic fields, consist an important process through which accumulated helicity is removed from the corona.

The above statement is supported from solar wind observations. In situ magnetic field measurements show that there are sporadic intervals when magnetic fields are twisted in either a right-handed or left-handed sense, corresponding to positive and negative helicity, respectively. Such disturbances are called magnetic clouds (MCs) when they show enhanced magnetic field whose vector direction rotates gradually through 180°. MCs have been interpreted as magnetic flux ropes ejected from the corona as CMEs (e.g. Rust 1999; note, however, that not all CMEs give MCs at 1 AU). The observations show (Rust 1994, Bothmer & Schwenn 1994) that the number of MCs with positive chirality (i.e. handedness) is roughly equal to the number of MCs with negative chirality. But CMEs that originate from the northern hemisphere tend strongly to have negative chirality in the corresponding MCs, and those from the southern hemisphere have positive chirality in their MCs.

3.2. The Helicity Budget of CME-Productive Active Regions

Once it was realized that CMEs remove helicity from the corona, the obvious question was where that helicity comes from. For the computation of the helicity budget of an active region, one needs to compute the helicity injection rate (see §2.2) and also the helicity stored in the corona and the helicity ejected into interplanetary space.

The first process which was studied was differential rotation (DeVore 2000). Démoulin et al. (2002a) and Green et al. (2002) studied the long-term evolution of the helicity injected by differential rotation into the coronal part of two active regions. The coronal magnetic field was modeled under the force-free field assumption ($\nabla \times \mathbf{B} = \alpha \mathbf{B}$, e.g. see Alissandrakis 1981). The best value of α, α_{best}, is determined by comparing the computed field lines with the observed SXR or EUV coronal structures. Then the computation of the coronal helicity is relatively straightforward (Berger 1985). The helicity carried away by CMEs cannot be directly computed. In order to overcome this problem one assumes that each CME ejected from the active region under study produces a MC, and that the helicity carried away by each CME is equal to the helicity in the corresponding MC. The helicity per unit of length in a MC can be calculated if we know the axial magnetic field B_0 and the radius R of the cloud's flux rope (DeVore 2000). These parameters can be calculated using a linear force-free magnetic field model (Lepping et al. 1990). For the calculation of the total MC helicity, the cloud's flux tube axis ℓ is needed which cannot be obtained observationally; usually a lower limit of $\ell = 0.5$ AU and an upper limit of $\ell = 2$ AU are used. Démoulin et al. (2002a) and Green et al. (2002) found that the helicity injected by differential rotation is at least about an order of magnitude smaller than the helicity stored in the corona and the helicity carried away by CMEs.

Nindos et al. (2003) were able to partially reconcile the amount of helicity injected into the corona with the helicity carried away by the CMEs in the 6 active regions they studied. When they assumed that the length of the MC's flux tube ℓ is determined by the condition for the initiation of the kink instability in the coronal flux rope or when $\ell = 0.5$

AU then the total CME helicities and the total helicities injected into the corona were broadly consistent. However, for $\ell = 2$ AU, the total helicities injected into the corona were a factor of 2.9-4 lower than the total CME helicities.

It is important to note that the LCT method has serious limitations that lead to underestimation of the computed helicities (Démoulin & Berger 2003). Furthermore one cannot separate the contribution of the shearing term of eq. (3) from the contribution of the advection term of eq. (3) to the helicity injected into the corona. The theoretical work by Démoulin et al. (2002b) who separated the helicity injection into two terms, twist and writhe, indicates that shearing motion is an inefficient way of providing helicity on the active region scale which will be subsequently removed by CMEs. Recently, alternative approaches have been developed which attempt to compute separately both the shearing and advection term using photospheric vector magnetograms. Kusano et al. (2002) proposed a method which uses the vertical component of the induction equation. In fact the velocity of flux tubes cannot be deduced fully from the induction equation and part of the velocity is still computed from the LCT method (Welsch et al. 2004). When the transverse component of the magnetic field is available, Kusano et al. (2004) developed a method which minimizes the input from the LCT technique. Longcope (2004) introduced a technique which demands that the photospheric flow agree with the observed photospheric field evolution according to the magnetic induction equation. It selects, from all consistent flows, that with the smallest overall flow speed by demanding that it minimize an energy functional. Georgoulis & LaBonte (2006) introduce a minimum structure reconstruction technique to infer the velocity field vector. Their analysis simultaneously determines the field-aligned flows and enforces a unique cross-field velocity solution of the induction equation. All these methods have not been tested extensively with data.

4. MAGNETIC HELICITY AND CME INITIATION

CMEs carry away coronal plasma and open up magnetic field lines which were closed before the eruption. Therefore, the pre-eruption state must contain enough magnetic energy to account for the observed gravitational potential and kinetic energies of the CMEs and also the energy required to open up the magnetic field. The latter imposes a difficulty to all CME models that comes from the Aly-Sturrock theorem (Aly 1991, Sturrock 1991). This theorem establishes that in the case of an open field in which all of the field lines extend to infinity, the magnetic energy increases during the eruption. Consequently, one needs a process capable of opening the field and at the same time decrease its energy by the amount required to power the mass motion. However, as Forbes (2000) has summarized, there are several ways to bypass this constraint: (1) the CME involves flux from several flux systems so most of the field involved is not opened (Antiochos et al. 1999), (2) the CME involves a detached flux rope (Low 1996), (3) an ideal MHD process may extend field lines such that they do not open all the way to infinity, (4) the pre-CME corona is not force-free and cross-field currents are present (Wolfson & Dlamini 1997), (5) non-ideal processes, specifically reconnection, might be important.

The processes leading to CMEs require helicity accumulation to the magnetic structure that will subsequently erupt. This can be done by shearing the magnetic field or by assuming a pre-existing flux rope. Several mechanisms are capable of shearing the pre-eruption field: emergence of twisted flux, shearing, twisting, and convering motions. Then the eruption can be regarded as a failure of field confinement. Overall, the current status of our understanding of CME initiation gives a very important role to magnetic helicity and reconnection. In order to appreciate fully the role of helicity in the initiation of CMEs we discuss in the next subsection two important theoretical arguments.

4.1. Woltjer Theorem and Taylor Relaxation

Woltjer (1958) showed that for a perfectly conducting plasma, the total magnetic helicity remains invariant during the evolution of any closed flux system and the minimum energy state of this system corresponds to a linear force-free field configuration. This statement has been proved mathematically and is known as Woltjer theorem. Taylor (1974) applied that theorem to the subject of plasma relaxation by proposing that turbulent magnetic reconnection will occur in a plasma of small but finite resistivity to change field topology and transport helicity from one part of the plasma to another, until the field reaches its minimum energy state which is the linear force-free configuration, according to Woltjer theorem. Therefore, while in the limiting case of ideal MHD the helicity of each field line will be an invariant of motion, Taylor's conjecture suggests that only the total helicity of the flux system will be approximately invariant during its evolution to the minimum energy state. A way to understand Taylor's theory is that the magnetic energy should be selectively dissipated faster than the magnetic helicity and that the final state of the linear force-free field might be self-organized.

The relevance of Taylor's theory to solar plasmas is still under debate. While some 3D numerical MHD simulations of relaxation processes in the corona indeed show such a relaxation toward a constant α-field (Kusano et al. 1994), other simulations (Amari & Luciani 2000, Antiochos & DeVore 1999) show a more complicated behavior at variance with Taylor's theory. Antiochos & DeVore (1999) have argued that Taylor's theory requirement that complete reconnection occurs, i.e. reconnection continues until magnetic energy reaches its lowest possible state, is wrong because complete reconnection requires the formation of numerous current sheets that do not form easily in the corona due to photospheric line-tying. However, Nandy et al. (2003) report the detection of a Taylor-like plasma relaxation process in the corona: their statistical study of highly flare-productive active regions implies that the magnetic field relaxes toward a constant-α configuration.

4.2. The Hydromagnetic Origin of CMEs

Low and Zhang in a series of papers (Low 1996, 1999, Low & Zhang 2002, Zhang & Low 2001, 2003; see also Zhang & Low 2005 for an extensive review) provided a unified theoretical view of CMEs as the last physical mechanism in the chain of

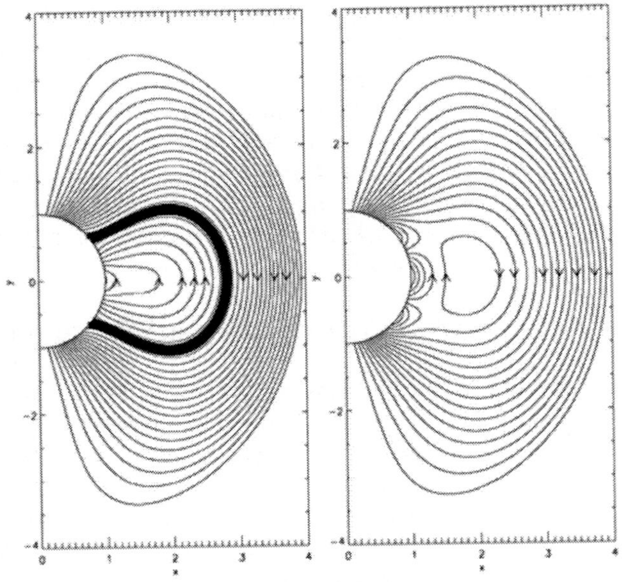

FIGURE 3. Contours of the stream function of an initial current-sheet magnetic field (left panel) and the final minimum-energy state of the field (right panel). The thick solid line in the current-sheet field shows the location of the current sheet (from Zhang & Low 2003).

processes that transfer magnetic flux and helicity from the base of the convective zone into the interplanetary medium. In this section I summarize their results with emphasis on the role of helicity in the initiation of CMEs. New active regions emerge with twisted magnetic field and magnetic polarities opposite to that of the pre-existing field. When the new field enters the corona, repeated reconnections between the new and pre-existing field take place. This process simplifies the magnetic topology and the dissipated magnetic energy produces flares. Finally the field achieves its minimum-energy state which is the constant-α force-free configuration required by the Woltjer-Taylor theories. An example of such process is given in Figure 3 taken by Zhang & Low (2003). The left panel of fig. 3 shows the initial state of a current-sheet field whose interface (i.e. the current sheet denoted by the thick solid line) separates two axisymmetric force-free fields. The final minimum-energy state of the field resulting from Taylor relaxation is given in the right panel of fig. 3. Note that the relaxation resulted in the formation of a flux rope: the redistribution of helicity, whose total amount is conserved, yields a minimum-energy state where a significant part of the total helicity of the system is contained within the flux rope.

The fate of the flux rope is determined by the efficiency of its confinement by its surrounding anchored field. Flux rope ejection occurs when the magnetic energy it contains is sufficient to drive an outward expansion against the confining field.

We may generalize the above findings as follows. Reconnection liberates magnetic energy by changing the field topology, but not all the magnetic free energy can be so released due to the conservation of the total helicity. Therefore, so long as a magnetic

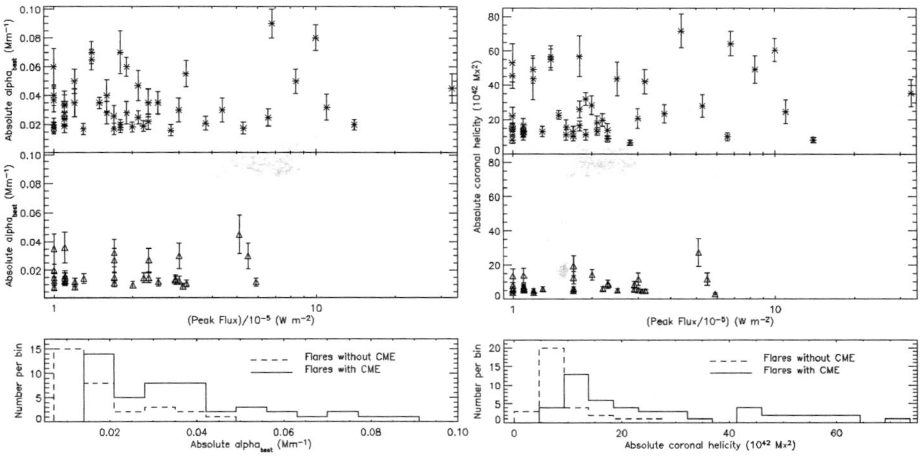

FIGURE 4. *Left column, top*: Scatter plot of the pre-flare absolute values of α_{best} as a function of the flare's peak X-ray flux for ARs producing CME-associated flares. *Left column, middle*: Same as top panel, but for ARs producing flares that do not have associated CMEs. *Left column, bottom*: Histograms of the values of α_{best} appearing in the top and middle panels. The solid line represents the histogram of α_{best} of the ARs which give CME-associated flares while the dashed line is the histogram of α_{best} of the ARs which produce flares that do not have CMEs. *Right column*: The absolute coronal helicity of the ARs appearing in the left panel. The format is identical to the format of the left column (modified from Nindos & Andrews 2004).

structure is confined to the corona during its relaxation through reconnection and its total helicity is significant, its relaxed state is a twisted field that conserves that total helicity. Trapped with that total helicity is a fraction of the magnetic free energy that cannot be released by reconnection. The removal of the trapped helicity and magnetic energy is achieved by the bodily expulsion of the twisted structure, i.e. the CME (Low 1994, 1996).

The above physical view is supported by the work by Nindos & Andrews (2004). The starting point of their work was the study by Andrews (2003) who found that approximately 40% of M-class flares between 1996 and 1999 are not associated with CMEs. Nindos & Andrews (2004) modeled under the constant α force-free field approximation, the pre-flare coronal magnetic field of 78 active regions from Andrews's (2003) data set. Then from the derived values of α_{best} they computed the corresponding coronal helicites H_c. Their results appear in Figure 4: both the scatter plots and histograms show clearly that several ARs that give big flares without CMEs tend to have smaller absolute values of α_{best} and H_c than those producing CME-associated flares. They found that this result is statistically significant: their analysis demonstrated that in a statistical sense, both the preflare absolute value of α and the corresponding coronal helicity of the active regions producing CME-associated big flares are larger than the absolute α and coronal helicity of those that do not have associated CMEs. Their study indicates that the amount of the preflare stored coronal helicity may determine whether a big flare will be associated with a CME or not.

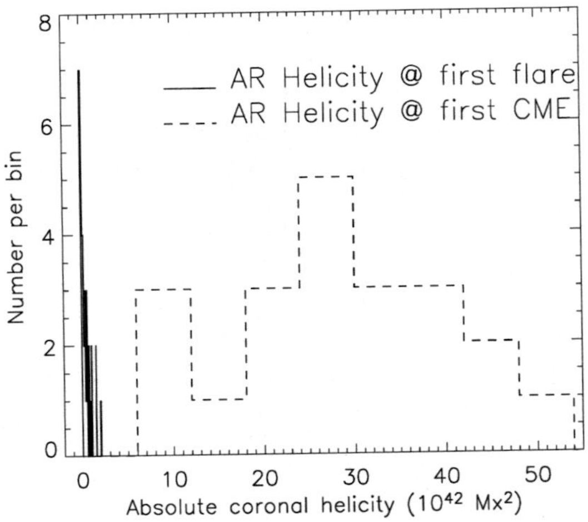

FIGURE 5. Solid-line histogram: histogram of the helicity of emerging ARs just before their first flare (above C1.0). Dashed-line histogram: histogram of the helicity of the same ARs just before the occurrence of their first CME.

Additional support for this paradigm is provided by our recent study that determines when emerging active regions produce their first CME. The details of this work will be reported elsewhere. Here I will only summarize the main result. My work does not focus on the well-known fact that emerging flux regions are sometimes associated with eruption of nearby filaments; instead I have studied whether *the AR itself* that forms after the bipole emergence erupts or not and what helicity has to do with the eruption. I selected ARs that did not emerge with significant twist and I computed their coronal helicity just before their first confined flare and just before their first CME using techniques described in §2.2 and §3.2. The results appear in Fig. 5 and show that the two helicity histograms do not overlap. Our work implies that a confined flare does not necessarily need significant amount of helicity in order to occur, whereas a CME typically needs it.

Our two studies presented briefly here yield direct observational support for the paradigm that helicity accumulation is a necessary condition for CME initiation: CMEs may be regarded as helicity expulsion events while confined flares may be regarded as reconnection events. Note that the above results are independent from the AR's magnetic flux or the strength of the flare.

4.3. Other approaches

Several other approaches to the problem of CME initiation have appeared in the literature. The purpose of this article is not to present an exhaustive review of all CME models; the interested reader may refer to the excellent articles by Forbes (2000) and Klimchuk (2001). Here I focus on the role of helicity in selected representative CME models. Overall, in most models the initiation of a CME comes from the interplay between the accumulation of helicity into the corona and reconnection. In most models the pre-eruption topology is either a sheared arcade (bipolar or quadrupolar configuration) or a flux rope. It has been suggested that in CMEs showing the "three-part structure" (i.e. a bright frontal loop, a dark cavity underneath, and an embedded bright core), cavities correspond to flux ropes seen edge-on. However, not all CMEs have a clear three-part structure. Additional evidence of the flux rope topology in some events is the in situ observation of a rotating magnetic field pattern in magnetic clouds. In §4.2, we discussed that in the framework of Taylor relaxation, flux ropes may be regarded as the inevitable outcome in a structure with sufficient accumulation of helicity. But the observation of a flux rope in a CME does not necessarily suggest that the flux rope was part of the initial configuration because reconnection may cause flux ropes to form from erupting sheared arcades (Gosling 1990).

Antiochos et al. (1999) developed the "breakout model" which follows the evolution of a quadrupolar magnetic field in which the inner part of the central arcade is sheared by antiparallel footpoint motions near the neutral line. As a result, reconnection between the sheared field and its neighboring field triggers an eruption. Reconnection removes the overlying unsheared field to allow the low-lying sheared field to open up. MacNeice et al. (2004) have studied the evolution of helicity under the breakout model. Their results show that although CMEs remove the bulk of the coronal helicity, some fraction remains behind. Phillips et al. (2005) presented simulations of the breakout model where eruption occurs at a fixed magnitude of free energy in the corona, independent of the value of helicity.

Amari et al. (2003a, 2003b) have constructed a set of force-free fields having different magnetic flux and helicity contents and used them as initial conditions by applying converging motions or a diffusion-driven evolution. These processes can trigger eruptive events that may be either confined or unconfined, depending on the value of the initial helicity.

Other CME models include the catastrophe model developed by Forbes and colleagues (e.g. Forbes 1990, Lin, Forbes, & Isenberg 2001) where the CME comes from a catastrophic loss of equilibrium. In the model by Wu et al. (1999) the CME is produced from the interaction between a pre-existing flux rope and the overlying helmet streamer. In the model by Linker & Mikic (1995), the CME initiation requires reconnection at the base of a sheared arcade. The model by Chen (1996) involves an equilibrium flux rope which contains a low-density hot component and a denser cold component. The eruption of the entire flux rope is triggered by an increase in the azimuthal magnetic flux of the structure.

5. CONCLUSIONS AND FUTURE WORK

Magnetic helicity provides an important tool for the study of CME physics. Its use is justified by two key properties of CMEs: (1) the pre-eruption magnetic topology is non-potential and (2) CMEs carry away twisted magnetic fields. Of course, there are also other physical quantities that describe non-potential fields, for example the α parameter of the force-free field approximation. But helicity is superior because of its unique feature of being (approximately) conserved even in resistive MHD. This makes helicity probably the only physical quantity which can monitor the entire history of the event: from the transfer of magnetic field from the convective zone all the way to the eruption and the escape of the CME into the interplanetary medium. On the other hand, the calculations of helicity are difficult and only recently attempts have been made to measure helicity using solar observations.

Once the importance of helicity was realized, a lot of effort was put on the helicity budget of CME-productive active regions and on the mechanisms which provide the helicity shed by CMEs. Theoretical considerations have demonstrated rigorously that shearing motions (either differential rotation and/or transient flows) on the photospheric surface is an inefficient way of providing helicity on the active region scale. However, computations using high-cadence longitudinal magnetograms give the total helicity flux and cannot separate the shearing term from the advection term. Furthermore, the computation of velocities using the LCT method has serious limitations. Attempts for the computation of the shearing and advection term separately have been made using vector magnetograms. But the algorithms that have been developed have not been applied extensively to observations. Even more serious uncertainties are associated with the computation of the helicity carried away by the CMEs. The use of MCs as proxies for the calculation of CME helicity is only a zero-order approximation. All the above problems contribute to the discrepancies concerning the helicity budget of active regions. At this point, these uncertainties have been cleared up only partially and much work needs to be done on this issue.

The new generation of ground-based and space-born vector magnetograms will provide important new data that should trigger the development of improved algoriths for the computation of the helicity injected into the solar atmosphere. Concerning the helicity of magnetic clouds, we need to constrain more accurately the length of magnetic cloud flux tubes and investigate whether the twist along the interplanetary flux tube is uniform. Hopefully, the coming STEREO mission will provide important observational input to such tasks.

As for the role of helicity in the initiation of CMEs, there is a general consensus that for the CME initiation, helicity must be accumulated to the pre-eruption topology. However, diverse opinions exist about the importance of helicity in the overall coronal evolution which leads to a CME. It has been argued that conservation of global helicity plays a minor role in determining the nature and consequences of reconnection in the Sun. The basic argument that supports this point of view is that the coronal magnetic field violates Taylor-theory's requirement that reconnection continues until the magnetic energy reaches its lowest possible state.

Helicity is a crucial element in the other approach which views CMEs as the last chain of a unified process that starts with emergence of twisted magnetic flux from the sub-photospheric layers. This point of view exploits the conservation of total helicity and accepts that magnetic energy should be selectively dissipated faster than magnetic helicity; it takes these arguments to their full logical conclusions and implications. Flux emergence charges the corona with helicity and the excess magnetic energy is removed by reconnection events (i.e. flares). But not all the magnetic energy can be so released due to the conservation of total helicity. The relaxed state is a twisted structure that conserves the total helicity. The CME is the removal of the trapped helicity, enabling the field to reach its minimum-energy state. Recent results from observations give support to that picture. On the theoretical front, the picture should be refined using more realistic magnetic topologies. Observationally, it will be desirable to test whether there is a helicity threshold above which the accumulated helicity results to CME initiation.

Acknowledgments I thank the organizers of the 7th HELAS Conference for inviting me to give this review. I also thank C.E. Alissandrakis, P. Demoulin, M. Georgoulis, S. Antiochos, D. Rust and N. Gopalswamy for useful discussions. The research on magnetic helicity and eruptive phenomena at the University of Ioannina is partially supported by Greek Ministry of Education's grant "Pythagoras II".

REFERENCES

1. Alissandrakis, C.E., A&A, 100, 197-200, 1981
2. Aly, J.J., ApJ (Lett), 375, L61-64, 1991.
3. Amari, T., and J.F. Luciani, Phys. Rev. Lett., 84, 1196-1199, 2000.
4. Amari, T., J.F. Luciani, J.J. Aly, Z. Mikic, and J. Linker, ApJ, 585, 1073-1086, 2003a.
5. Amari, T., J.F. Luciani, J.J. Aly, Z. Mikic, and J. Linker, ApJ, 595, 1231-1250, 2003b.
6. Andrews, M.D., Sol. Phys., 218, 261-279, 2003.
7. Antiochos, S.K., and C.R. DeVore, in *Physics of Magnetic Flux Ropes, Geophys. Monogr. Ser.*, vol. 111, edited by M.R. Brown, R.C. Canfield, and A.A. Pevtsov, pp. 187-196, AGU, Washington DC, 1999.
8. Antiochos, S.K., C.R. DeVore, and J.A. Klimchuk, ApJ, 510, 485-493, 1999.
9. Berger, M.A., PhD thesis, Harvard University, Cambridge MA., 1984.
10. Berger, M.A., ApJ (Suppl.), 59, 433-444, 1985.
11. Berger, M.A., A&A, 201, 355-361, 1988.
12. Berger, M.A., in *Physics of Magnetic Flux Ropes, Geophys. Monogr. Ser.*, vol. 111, edited by M.R. Brown, R.C. Canfield, and A.A. Pevtsov, pp. 1-9, AGU, Washington DC, 1999.
13. Berger, M.A., and G.B. Field, Journal of Fluid Mech., 147, 133-148, 1984.
14. Bothmer, V., and Schwenn, R., Space Sci. Rev., 70, 215, 1994.
15. Brown, M.R., R.C. Canfield, A.A. Pevtsov (Eds.), *Helicity in Space and Laboratory Plasmas,*, 304 pp., Geophys. Monogr. Ser., vol. 111, AGU, Washington DC, 1999
16. Chae, J., ApJ, 560, L95-98, 2001.
17. Chae, J., Y.-J. Moon, and Y.-D. Park, Sol. Phys., 223, 39-55, 2004.
18. Chen, J., JGR, 101, 27499-27520, 1996.
19. Crooker, N., J.A. Joselyn, and J. Feynman (Eds.), *Coronal Mass Ejections,* 299 pp., Geophys. Monogr. Ser. vol. 99, AGU, Washington DC, 1997.
20. Démoulin, P., and M.A. Berger, Sol. Phys., 215, 203-215, 2003.
21. Démoulin, P., C.H. Mandrini, L. van Driel-Gesztelyi, B.J. Thompson, S. Plunkett, Zs. Kovari, G. Aulanier, and A. Young, A&A, 382, 650-665, 2002a
22. Démoulin, P., C.H. Mandrini, L. van Driel-Gesztelyi, M.C. Lopez Fuentes, and G. Aulanier, Sol. Phys., 207, 87-110, 2002b
23. DeVore, C.R., ApJ, 539, 944-953, 2000.
24. Forbes, T.G., JGR, 105, 23153-23165, 2000.
25. Georgoulis, M.K., and B.J. LaBonte, ApJ, 636, 475-495, 2006.
26. Gosling, J.T., in *Physics of Magnetic Flux Ropes, Geophys. Monogr. Ser.*, vol. 58, edited by C.T. Russell, E.R. Priest, and L.C. Lee, pp. 343-364, AGU, Washington DC, 1990.
27. Gosling, J.T., JGR, 98, 18, 18937-18949, 1993.
28. Green, L.M., M.C. Lopez-Fuentes, C.H. Mandrini, P. Démoulin, L. van Driel-Gesztelyi, and J.L. Culhane, Sol. Phys., 208, 43-68, 2002.
29. Hudson, H.S., and E.W. Cliver, JGR, 106, 25199-25213, 2001.

30. Hundhausen, A.J., J.T. Burkepile, and O.C. St. Cyr, JGR, 99,6543-6552
31. Klimchuk, J.A., in *Space Weather, Geophys. Monogr. Ser.*, vol. 125, edited by P. Song, H.J. Singer, and G.L. Siscoe, pp. 143-158, AGU, Washington DC, 2001.
32. Kusano, K., Y. Suzuki, H. Kubo, T. Miyoshi, and K. Nishikawa, ApJ, 433, 361-378 1994.
33. Kusano, K., T. Maeshiro, T. Yokoyama, and T. Sakurai, ApJ, 577, 501-512, 2002.
34. Kusano, K., T. Maeshiro, T. Yokoyama, and T. Sakurai, in ASP Conference Series, 175, 2004
35. Leka, K.D., R.C. Canfield, A.N. McClymont, and L. van Driel-Gesztelyi, ApJ, 462, 547-560, 1996.
36. Lepping, R.P., L.F. Burlaga, J.A. Jones, JGR, 95, 11957-11965, 1990.
37. Lin, J., T.G. Forbes, P.A. Isenberg, JGR, 106, 25053-25074, 2001.
38. Linker, J.A., and Z. Mikic, ApJ, 438, L45-L48, 1995.
39. Longcope, D.W., ApJ, 612, 1181-1192, 2004
40. Low, B.C., Phys. Plasmas, 1, 1684-1690, 1994.
41. Low, B.C., Sol. Phys., 167, 217-265, 1996.
42. Low, B.C., and M. Zhang, ApJ, 564, L53-56, 2002.
43. MacNeice, P., S.K. Antiochos, A. Phililips, D.S. Spicer, C.R. DeVore, and K. Olson, ApJ, 614, 1028-1041.
44. Moon, Y.-J., J. Chae, H. Wang, G.S. Choe, Y.D. Park, ApJ, 580, 528-537, 2002a.
45. Moon, Y.-J., et al., ApJ, 574, 1066-1073, 2002b.
46. Nandy,D., M. Hahn, R.C. Canfield, and D.W. Longcope, ApJ, 597, L73-L76, 2003.
47. Nindos, A., and H. Zhang, ApJ, 573, L133-136, 2002.
48. Nindos, A., and M.D. Andrews, ApJ, 616, L175-178, 2004.
49. Nindos, A., J. Zhang, and H. Zhang, ApJ, 594, 1033-1048, 2003.
50. Pariat, E., P. Démoulin, and M.A. Berger, A&A, 439, 1191, 2005
51. Pariat, E., A. Nindos, P. Démoulin, and M.A. Berger, A&A, submitted, 2006
52. Pevtsov, A.A., R.C. Canfield, and T.R. Metcalf, ApJ, 440, L109-111, 1995.
53. Pevtsov, A.A., R.C. Canfield, and S.M. Latushko, ApJ, 549, L261-263, 2001.
54. Phillips, A.D., P.J. MacNeice, and S.K. Antiochos, ApJ, 624, L129-132, 2005.
55. Rust, D.M., GRL, 21, 241-244, 1994.
56. Rust, D.M., in *Physics of Magnetic Flux Ropes, Geophys. Monogr. Ser.*, vol. 111, edited by M.R. Brown, R.C. Canfield, and A.A. Pevtsov, pp. 221-227, AGU, Washington DC, 1999.
57. Sturrock, P.A., ApJ, 380, 655-659, 1991.
58. Taylor, J.B., Phys. Rev. Lett., 33, 1139-1141, 1974.
59. Vourlidas, A., P. Subramanian, K.P. Dere, and R.A. Howard, ApJ, 534, 456-467, 2000.
60. Welsch, B.T., G.H. Fisher, W.P. Abbett, and S. Regnier, ApJ, 610, 1148-1156, 2004
61. Wolfson, R., and B. Dlamini, ApJ, 483, 961-971, 1997
62. Woltjer, L.A., in *Proc. Natl. Acad. Sci. USA*,, 44, 489, 1958.
63. Wu, S.T., W.P. Guo, D.J. Michels, L.F. Burlaga, JGR, 104, 14789-14802, 1999.
64. Zhang, M., and B.C. ApJ, 561, 406-419, 2001.
65. Zhang, M., and B.C. Low, ApJ, 584, 479-496, 2003.
66. Zhang, M., and B.C. Low, ARA&A, 43, 103-137, 2005.

Acceleration and Radiation Model of Particles in Solar Active Regions

Anastasios Anastasiadis*, Cyril Dauphin† and Nicole Vilmer†

*National Observatory of Athens, Institute for Space Applications and Remote Sensing, Penteli, GR-15236 GREECE
†LESIA, Observatoire de Paris-Meudon, 92195 Meudon Cedex, FRANCE

Abstract. Cellular Automata (CA) models have successfully reproduced several statistical properties of solar flares such as the peak flux or the total flux distribution. We are using a CA model based on the concept of self organized criticality (SOC) to model the evolution of the magnetic energy released in a solar flare. Each burst of magnetic energy released is assumed to be the consequence of a magnetic reconnection process, where the particles are accelerated by a direct electric field. We relate the difference of energy gain of particles (alpha particles, protons and electrons) to the magnetic energy released and we calculate the resulting kinetic energy distributions and the emitted radiation.

Keywords: Sun: Solar flares; particle acceleration; radiation; self-organized criticality
PACS: 96.60.-j; 96.6.qe; 96.50.Pw; 96.60.tk

INTRODUCTION

During the last two decades, due to the existence of several space-born solar instruments and of a number of ground based solar telescopes, a number of statistical studies of the solar flaring activity has been performed (see e.g. Vilmer 1993; Vilmer & Mackinnon 2003, for reviews). These observations (Dennis 1985; Benz 1985; Benz & Aschwanden 1992; Crosby et al. 1993; Aschwanden et al. 1995; Crosby et al. 1998; Krucker & Benz 1998; Aschwanden et al. 2000) established that the frequency distributions of impulsive events (in particular solar flares) as a function of total energy, peak luminosity and duration are well defined power laws, extending over several orders of magnitude. Several qualitative models have been developed in order to model the dynamic evolution of solar flares (for reviews see van den Oord 1994; Vlahos 1996; Bastian & Vlahos 1997) following the above observational evidence. These models revealed the necessity to study and understand the global behavior of the evolution of the complex active regions and particle acceleration in such a complex environment.

In order to study the evolution and the dynamics of complex active regions two different approaches can be used: a) MHD numerical simulations (e.g. Galsgaard & Nordlund 1996; Einaudi et al. 1996; Dmitruk et al. 1998; Georgoulis et al. 1998; Galtier & Pouquet 1998; Walsh & Galtier 2000). b) Cellular Automata (CA) models either based on Self Organized Criticality (SOC), (e.g. Lu & Hamilton 1991; Lu et al. 1993; Vlahos et al. 1995; Georgoulis & Vlahos 1996; 1998) or in percolation theory (i.e. probabilistic CA models) (e.g. MacKinnon et al. 1996; MacKinnon & Macpherson 1999; Vlahos et al. 2002).

Both, MHD simulations and CA models, have advantages and disadvantages, but they

CP848, *Recent Advances in Astronomy and Astrophysics,*
7th *International Conference of the Hellenic Astronomical Society,* edited by N. H. Solomos

are complementary approaches. Several efforts have been done to find a connection between them. Isliker et al. (1998; 2000) revealed the role of several components of CA models, such as the physical interpretations of the grid-variable, the nature of the energy release process and the role of diffusivity. These efforts leaded to a construction of a new type of CA models for solar flares (Isliker et al. 2001) which are compatible to MHD theory and produce statistical results in agreement with the observations. In addition to the above efforts, hybrid models, which are intermediate between CA models and full MHD and reduced MHD models, have been constructed mainly to account for the coronal heating problem (e.g. Einaudi & Velli 1999; Buchlin et al. 2003).

The approach used for particle acceleration models proposed for solar flares (for review on acceleration models see Miller et al. 1997; Anastasiadis 2002) usually is based on the decoupling of the different processes (i.e. energy release, acceleration, transport and radiation). It is clear that in order to construct global models for solar flares, one must consider the coupling between these different processes. The different time and spatial scales of the processes and the fact that usually are acting simultaneously, make the coupling not an easy task to do (see Cargill 2002 for discussion of coupling models).

In previous studies, Anastasiadis and Vlahos (1991; 1994) proposed a model for the acceleration of particles (electrons and ions) by an ensemble of shock waves, assuming that the energy is released by means of many localized, small - scale explosive phenomena which were the drivers of a number of shock fronts (small - scale, short - lived discontinuities). In these early models there was still a vague and only qualitative association between the acceleration mechanism and the energy release process.

Anastasiadis et al. (1997; 2004) tried to connect the energy release process with the acceleration of electrons in solar flares, using a CA model for the energy release. The acceleration was based on a random number of localized electric fields (super-Dreicer electric fields) closely associated with the energy release process. The goal of the present work, is to improve and extend the above models, calculating in a more consistent way the electric fields and incorporating the radiation process.

THE MODEL

We assume that the observed complexity of a flaring active region and the turbulent driver, which represents the photoshpere, can give birth to multiple Reconnecting Current Sheets (RCS) as a result of magnetic reconnection processes, occurring during the energy release process (for details on magnetic reconnection see Priest & Forbes (2000)). For the study of the energy release process in the active region we use a CA model, and for modelling the acceleration of particles we consider the super-Dreicer electric fields associated with the RCS.

The CA model for energy release

We use a 3-D Cellular Automaton (CA) model based on the Self - Organized Criticality (SOC). The basic idea is that the evolution of active regions can be simulated by the continuous addition or change of new magnetic flux on the pre-existing magnetic

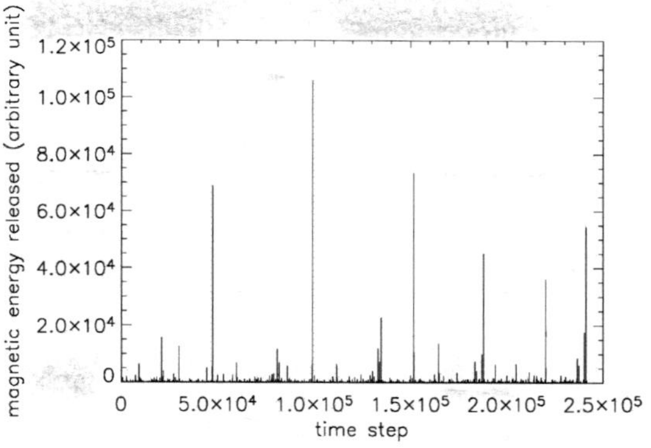

FIGURE 1. The energy release time series for the cellular automaton model. Note the intermittent nature of the released energy.

topology, until the local magnetic gradient reaches a threshold. At this point the local unstable magnetic topology is relaxing by rearranging the excess magnetic field to its nearest neighbors. This redistribution of the magnetic field can cause the lack of stability in the neighbors and the appearance of flares in form of avalanches in all scales of our system. In the following we only outline the basic rules of the CA model, for a detailed description see Vlahos et al. (1995), Isliker et al. (1998) and Georgoulis & Vlahos (1998).

The basic rules of the CA model are:(1) Initial loading (2) Ongoing random loading with increment δB given by the equation:

$$prob(\delta B) \approx (\delta B)^{-5/3} \tag{1}$$

(3) Relaxation process due to reconnection of magnetic field, leading to the generation of Reconnecting Current Sheets (RCS), according to the equation:

$$\vec{\nabla} \times \vec{B} \approx \vec{J} \tag{2}$$

(4) The energy release is calculated using:

$$\varepsilon \approx \left(B_i - \frac{6}{7} B_{\mathrm{cr}} \right)^2 \tag{3}$$

where B_i is the value of the magnetic field of given grid point i, which is becoming unstable when $B_i \geq B_{\mathrm{cr}}$, with B_{cr} being a critical value of the magnetic field.

The most important result from the CA model, for our purpose, is that an energy release time series ($\varepsilon(t)$) can be constructed, using Eq. 3. This time series (see Fig. 1) is highly intermittent, obeys a double power-law frequency distribution, and also exhibits a scale-invariant behavior, enclosing a self - similar nature.

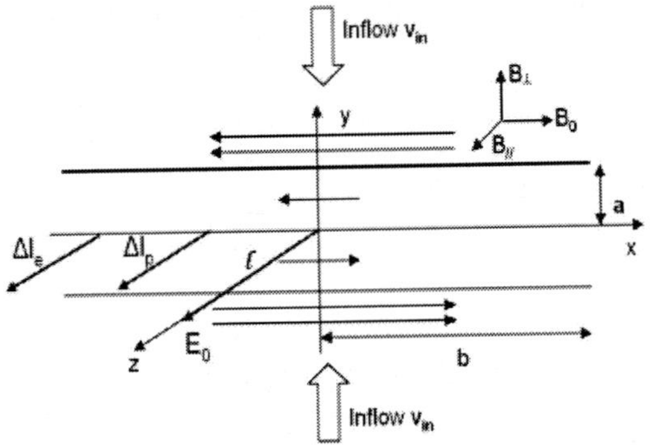

FIGURE 2. The assumed geometry of each Reconnecting Current Sheet (RCS).

The acceleration model

The simplest way to accelerate particles is by a large scale quasi-static electric field. A number of acceleration models with sub-Dreicer electric fields have been proposed for the case of solar flares (see e.g. Tsuneta 1985; Holman & Benka 1992). These models can explain the bulk energization of electrons up to 100 keV, but they are facing several problems such as: the requirement of a very long electric field parallel to magnetic field or the existence of highly filamented current channels, causing current closure and particle escaping problems. Application of super-Dreicer electric field models to solar flares have been also carried out in the past (e.g. Martens 1988; Litvinenko 1996; 2000). These models invoke a RCS with a significant magnetic field, which, together with the assumed inflow plasma velocity, can produce a convective electric field of a super-Dreicer value.

In our model we are following the above approach of the super-Dreicer electric field models. We assume that each energy release process is a reconnection site where the associated current sheet has the general geometry given by Fig. 2. An important parameter in our study is the value of the longitudinal reconnected magnetic field component.We must emphasize that recently, Efthymiopoulos et al. (2005) examined the particle dynamics in a 3-D RCS and found regular and chaotic behavior for particle orbits, depending on the value of the longitudinal magnetic field in a Harris type configuration. We distinguish here, three cases for the $B_{\|}$ component: (1) the component is zero, (2) the component is large, and (3) the component has intermediate value. The case (1) was studied by Speicer (1965) and the case (2) was explored by Litvinenko (1996).

We can calculate the corresponding electric field seen by the particles in each elementary RCS by equating the flux of the magnetic energy into the sheet to the energy gained by the accelerated particles per unit time. Following this assumption we find that the electric field is given by the relation:

$$E = \frac{B_o^2}{4\pi e(<\Delta l>_e n_e + <\Delta l>_p n_p)} \tag{4}$$

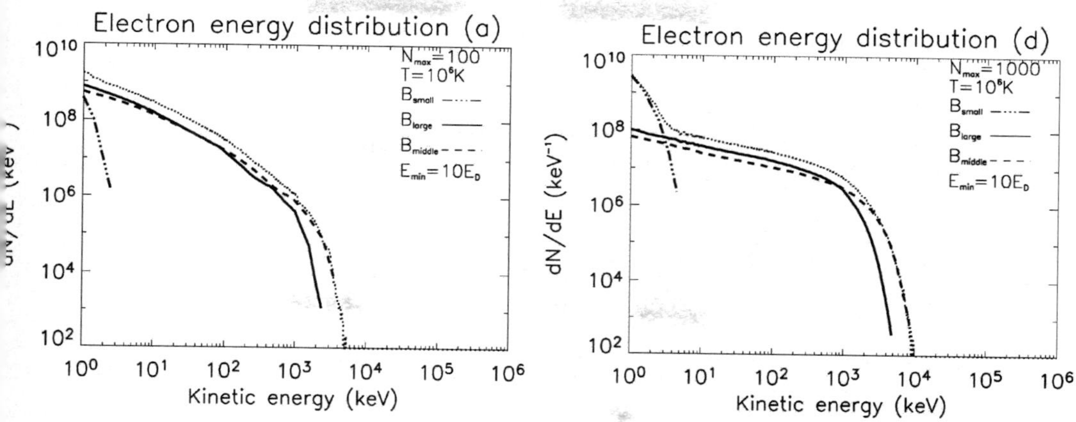

FIGURE 3. The resulting energy distribution of electrons for maximum number of interactions $N = 100$ and $N = 1000$. The normalized electric filed is $E_{min} = 10 \times E_D$. In each panel, the dashed dotted line corresponds to the case B_{small}, the dashed line corresponds to the case B_{middle}, the black line corresponds to the case B_{large} and the dotted line corresponds to the sum of these three distributions.

where $< \Delta l >_{\{e,p\}}$ is the average length of the RCS over which the particles are accelerated and $n_{\{e,p\}}$ is the particle density.

As it was state earlier, we assume that the energy release time series, produced by the CA model through the Eq. 3, is closely associated with the presence of a number of RCS in our system. This assumption is based on the fact that the energy release is produced by magnetic reconnection processes, simulated in the CA model by its redistribution rules. As the released energy calculated by the CA model is $\varepsilon(t) \sim B^2{}_o(t)$ (i.e. Eq. 3), we can construct a virtual electric field time series ($E(t)$) from the energy release time series using Eq. 4 for the electric field. Finally we can normalized this electric field time series to the Dreicer electric field value E_D ($= 5.5 \times 10^{-4}$ V/cm for the parameters used in our study).

Each injected particle enters into the acceleration volume and interacts successively with randomly selected, from the interval $[1, N]$, elements of the electric field time series by performing a free flight between each interaction. At each particle - RCS interaction, the kinetic energy change is given by the relation:

$$\Delta E \; = \pm \; Ze \, E(t) \, \Delta l \qquad (5)$$

where the plus (minus) sign corresponds to in (out of) phase interaction, Ze is the charge, $E(t)$ is the selected element of the virtual electric field time series given by the CA model and $\Delta l = \alpha < \Delta l >_{\{e,p\}}$. The parameter α is selected randomly to vary between zero and one at each particle - RCS interaction and plays the role of the acceleration coefficient, controlling the efficiency of the process.

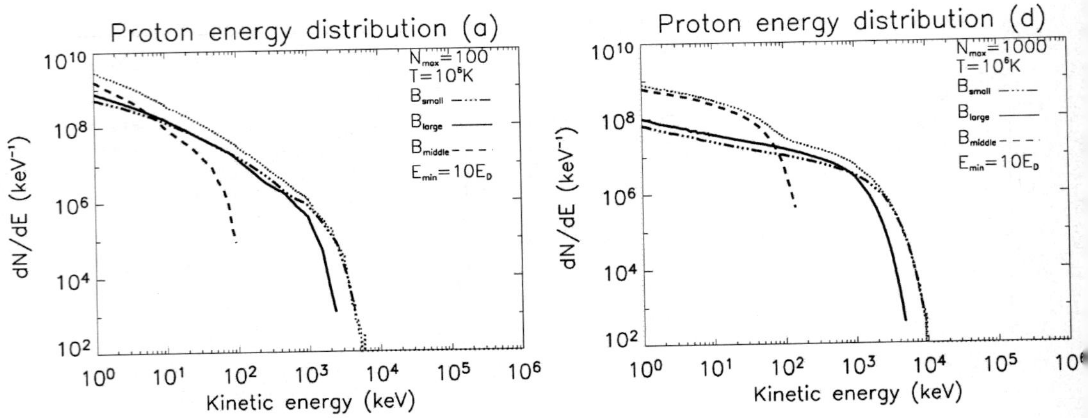

FIGURE 4. Same as Fig. 3 but for the case of protons.

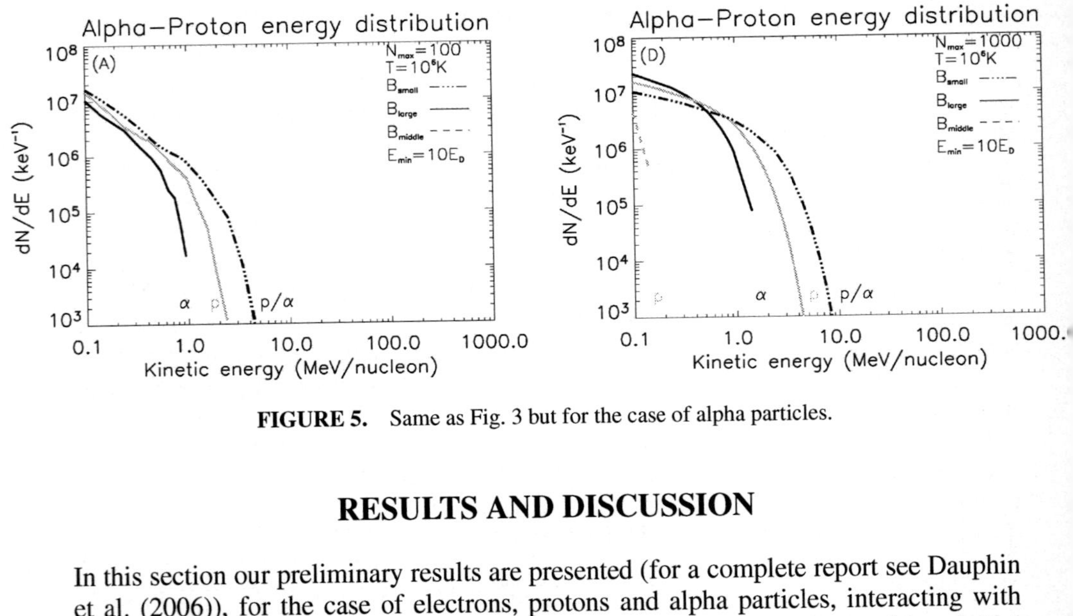

FIGURE 5. Same as Fig. 3 but for the case of alpha particles.

RESULTS AND DISCUSSION

In this section our preliminary results are presented (for a complete report see Dauphin et al. (2006)), for the case of electrons, protons and alpha particles, interacting with N RCS. Each particle population is injected into the acceleration volume with an initial thermal distribution of $T = 10^6$ K. We can perform a parametric study with respect to the number of interacting RCS (N), to the value of the longitudinal reconnecting magnetic field ($B_{\|}$) and to the normalization of the electric field with respect to the Dreicer value (E_{min}/E_D). In Figs. 3, 4, 5, we present the resulting particle (electrons, protons and alpha particles) distributions, for the case of $N = 100$ and $N = 1000$ with normalized electric field $E_{min} = 10 \times E_D$.

In addition to the acceleration process we can compute the resulting radiation flux of the accelerated particle distributions. In order to do so we choose to consider as a first

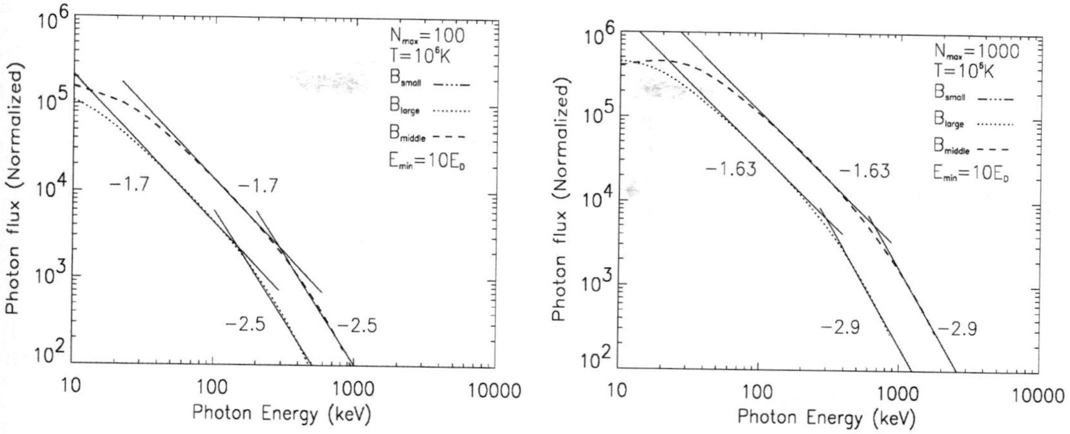

FIGURE 6. The emitted X-ray radiation calculated by the thick target approach for the case of electron distributions presented in Fig. 3.

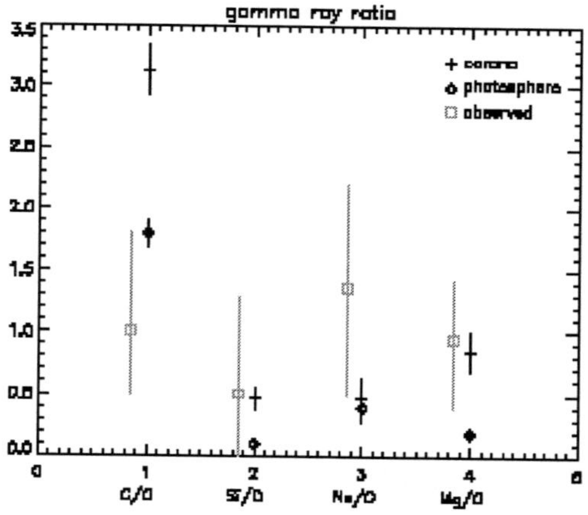

FIGURE 7. Ratio of the direct gamma ray lines calculated in the thick target approach from the proton and alpha energy distributions.The squares are deduced from the observations of Share and Murphy (1995), the crosses correspond to solar corona environment and the diamonds to photosphere.

approach, that energetic particles produce thick target radiation (for details see Brown 1971; Vilmer et al 1982). In other words, we assume that the accelerated particles after their escape from the acceleration volume are injected in a dense medium where they radiate all their energy instantaneously with no modification due to transport.

In Fig. 6 the emitted X-ray radiation of the electrons distributions presented in Fig. 3 is shown. The fitting was done by a power law in the energy ranges 10 -100 keV (considered as the low energy part) and 100 - 1000 keV (considered as the high energy part).

Finally, in Fig. 7 the ratio of the direct gamma ray lines calculated for the thick target approach from the proton and alpha particle energy distributions, for different realizations of our model parameters, are presented. The read marks in Fig. 7 corresponds to the observations reported by Share & Murphy (1995).

Summarizing, in this work we consider a global model for the acceleration and radiation process of particles (electrons, protons and alpha particles) in the complex evolving active regions (for details see Dauphin et al. 2006). Considering that the evolution of active regions can be simulated by the evolution of a Cellular Automaton (CA) model based on Self - Organized Criticality (SOC), we tried to connect the energy release process occurring in solar flares with the acceleration and radiation processes of energetic particles. We presented preliminary results for the accelerated particle distributions in respect to our model parameters (i.e. the number of reconnecting current sheets N, the value of the longitudinal reconnecting magnetic field $B_{||}$ and the normalization of the electric field with respect to the Dreicer value E_{min}/E_D). Finally we considered the emitted radiation of the accelerated particle distributions, assuming the thick target approximation. More work is clearly needed in the future if we want to incorporate the transport of the particles and to include the radiation losses due to collisions inside the acceleration volume (i.e. use of the thin target emission) and to compere our numerical results with the existing observations.

ACKNOWLEDGMENTS

We would like to thank Loukas Vlahos, Heinz Isliker and Manolis Georgoulis for many stimulating discussions on cellular automata models and their applications to solar flares.This work was partially supported by the Greek General Secretariat for Research and Technology.

REFERENCES

1. Anastasiadis, A., 2002, J. Atm. Sollar - Terr. Phys., 64(5-6), 481
2. Anastasiadis, A., & Vlahos, L. 1991, A&A, 245, 271
3. Anastasiadis, A., & Vlahos, L. 1994, ApJ, 428, 819
4. Anastasiadis, A., Vlahos, L., & Georgoulis, M. K. 1997, ApJ, 489, 367
5. Anastasiadis, A., Gontikakis, C., Vlimer, N., & Vlahos, L. 2004, A&A, 422, 323
6. Aschwanden, M. J., Tarbell, T. D., Nightingle, R. W., et al. 2000, ApJ, 535, 1047
7. Aschwanden, M. J, Montello, M., Dennis, B. R., & Benz, A. O. 1995, ApJ, 440, 394
8. Bastian, T. S., & Vlahos, L. 1997, in Lecture Notes in Physics, Vol. 483, ed. G. Trottet, (Springer-Verlag), 68
9. Benz, A. O. 1985, Sol. Phys., 96, 357
10. Benz, A. O., & Aschwanden, M. J. 1992, in Lecture Notes in Physics, Vol. 399, ed. Z. Svestka et al., (Springer-Verlag), 106
11. Brown, J. C. 1971, Sol. Phys., 18, 489
12. Buchlin, E., Aletti, V., Galtier, S., Velli, M., et al. 2003, A&A, 406, 1061
13. Cargill, P. J. 2002, in SOLMAG: Magnetic Coupling of the Solar Atmosphere, ESA SP-505, 245
14. Crosby, N. B., Aschwanden, M. J. & Dennis, R. B. 1993, Sol. Phys., 143, 275
15. Crosby, N. B., Vilmer, N., Lund, N., & Sunyaev, R. 1998, A&A, 334, 299
16. Dauphin, C., Vilmer, N., & Anastasiadis, A. 2006, A&A, to be submitted.
17. Dennis, B. R. 1985, Sol. Phys., 100, 465

18. Dmitruk, P., Gomez, D. O., & DeLuca, E. E. 1998, ApJ, 505, 974
19. Efthymiopoulos, C., Gontikakis, C., & Anastasiadis, A., 2005, A&A, 443, 663
20. Einaudi, G., & Velli, M. 1999, Phys. Plasmas, 6, 4146
21. Einaudi, G., Velli, M., Politano, H., & Pouquet, A. 1996, ApJ, 457, L13
22. Galsgaard, K., & Nordlund, A. 1996, J. Geophys. Res., 101, 13445
23. Galtier, S., & Pouquet, A. 1998, Sol. Phys., 179, 141
24. Georgoulis, M., & Vlahos, L. 1996, ApJ, 469, L135
25. Georgoulis, M., & Vlahos, L. 1998, A&A, 336, 721
26. Georgoulis, M., Velli, M., & Einaudi, G. 1998, ApJ, 497, 957
27. Holman, G. D., & Benka, S. G. 1992, ApJ, 400, L79
28. Isliker, H., Anastasiadis, A., & Vlahos, L. 2000, A&A, 361, 1134
29. Isliker, H., Anastasiadis, A., & Vlahos, L. 2001, A&A, 377, 1068
30. Isliker, H., Anastasiadis, A., Vassiliadis, D., & Vlahos, L. 1998, A&A, 335, 1085
31. Krucker, S., & Benz, A. O. 1998, ApJ, 501, L213
32. Lu, E. T., & Hamilton, R. J. 1991, ApJ, 380, L89
33. Lu, E. T., Hamilton, R. J., McTierman, J. M., & Bromund, K. R. 1993, ApJ, 412, 841
34. Litvinenko, Y. E. 1996, ApJ, 462, 997
35. Litvinenko, Y. E. 2000, Sol. Phys., 194, 324
36. MacKinnon, A. L., Macpherson, K. P., & Vlahos, L., 1996, A&A, 310, L9
37. Macpherson, K. P., & MacKinnon, A. L., 1999, A&A, 350, 1040
38. Martens, P. C. 1988, ApJ, 330, L131
39. Miller, J. A., et al. 1997, J. Geophys. Res., 102, 14659
40. Priest, E. R., & Forbes, T. 2000, Magnetic Reconnection, MHD Theory and Applications, (Cambridge Univercity Press)
41. Share, G. H., & Murphy, R. J., 1995, ApJ, 452, 933
42. Speicer, T. W. 1965, J. Geophys. Res., 70, 4219
43. Tsuneta, G. D. 1985, ApJ, 290, L353
44. van den Oord, G. H. J., ed. 1994, Fragmented Energy Release in Sun and Stars (Kluwer)
45. Vestrand, W. T. 1988, Sol. Phys., 118, 95
46. Vilmer, N., & MacKinnon, A. L. 2003, in Energy Conversion and Particle Acceleration, ed. K.-L. Klein, (Springer-Verlag), 127
47. Vilmer, N. 1993, Adv. Space Res., 13(9), 221
48. Vilmer, N., Kane, S. R, & Trottet, G., 1982, A&A 108, 306
49. Vlahos, L. 1996, in Radio Emission from the Stars and the Sun, ASP Confer. Ser., 93, ed. A. R. Taylor and J. M. Paredes (ASP Press), 355
50. Vlahos, L., Fragos, T., Isliker, H., & Georgoulis, M. 2002, ApJ, 575, L87
51. Vlahos, L., Georgoulis, M. K., Kluiving, R., & Paschos, P. 1995, A&A, 299, 897
52. Walsh, R. W., & Galtier, S. 2000, Sol. Phys., 197, 57

Electron Orbits in Solar Reconnecting Current Sheets

C. Gontikakis*, C. Efthymiopoulos* and A. Anastasiadis†

* *Academy of Athens, Research Center of Astronomy and Applied Mathematics, Soranou Efessiou 4, GR-11527 Athens Greece.*
†*National Observatory of Athens, Institute for Space Applications and Remote sensing 15236 Penteli Greece*

Abstract. We investigate the orbits of charged particles (electrons) that interact with a 3D-solar type reconnecting current sheet (RCS). The magnetic field used is a linearized Harris type model. Particles are accelerated by a super-Dreicer electric field. Taking advantage of the translational symmetries of the fields, the orbits computation is simplified to integration of the equations of motion in a Hamiltonian of two degrees of freedom. The application of canonical transformations, constructed with Lie series, transforms the Hamiltonian to a normal form. The conditions for which electrons are trapped in the RCS are derived analytically. For escaping particles, the amount of final kinetic energy gain or loss is described with analytical expressions that are in good agreement with the numerical results.

Keywords: Solar flares; non-linear dynamics; particles acceleration; Chaotic motion
PACS: 95.10.Fh;96.60.ql;96.50.Pw

INTRODUCTION

Charged particle's acceleration occurs in various space environments. One case of particular interest is the eruption of solar flares. Evidence that particle acceleration takes place during these events is given by the spectrum of hard X-rays, produced by *supra thermal* electron distributions. Such distributions include more high energy electrons than expected by a Maxwellian distribution of 10^7 K, which is the temperature of solar flares plasma [19].

The mechanism that accelerates particles should involve reconnecting current sheets (RCS). A RCS is a magnetic discontinuty where oppositely oriented magnetic fields come close in very small length scales. Current sheets are formed spontaneously in the solar magnetic field [16] in sites of magnetic reconnection [17]. Most magnetic reconnection topologies involve strong electric fields that are crucial to the effect of particle acceleration.

The study of particles motion in such magnetic topologies started soon after the discovery of the Earth magnetotail current sheet, which is responsible for the particle precipitation in the Earth atmosphere during substorms [17]. As the magnetic field reversal takes place in length scales comparable to the Larmor radius, one has to solve the equations of motion in detail in order to study the particle orbits. In the work of [18] a familly of orbits is found based on an analytical study. Several studies have been performed, based on detailed analysis of particles interacting with the Earth Magnetotail (see [3] for a review).

CP848, *Recent Advances in Astronomy and Astrophysics, 7th International Conference of the Hellenic Astronomical Society,* edited by N. H. Solomos

In the case of solar flares, the same formalism can be applied despite the difference in the values of magnetic and electric fields, which are much larger in the solar atmosphere than in the Earth magnetotail. In the solar plasma, particle acceleration can also be treated as a collisionless process when strong electric fields are present. This is because the electric field accelerates the particles to large velocities compared to the mean thermal velocity ([1], [14]).

One the other hand, measurements at the level of the photosphere using vector magnetographs [8] have shown that, during the first phase of flares, the magnetic field is strongly sheared. This suggests the importance of a *longitudinal* magnetic component, parallel to the electric field. As in the case of the Earth magnetotail, the perpendicular magnetic component introduces a perturbation to the particle's motions [18]. Particle acceleration in a RCS with non zero shear was investigated in [10], [11], [12]. In these papers, scaling formulas are derived for the final kinetic energy gain of particles as a function of the values of the various components of the magnetic field. In [5] and [6], we studied the behavior of particle orbits by dynamical systems' methods.

In the following section, numerical simulations are presented in order to show some characteristics of the electron motion in RCSs. This includes examples of electrons trapped inside current sheets or of escaping electrons gaining or losing kinetic energy. Poincaré sections are used to show the nature of the orbits inside the RCS. A detailed analysis of these phenomena is given elsewhere ([5],[6]).

THE MODEL

The topology of the magnetic and electric fields is given by Eqs (1). The B_x component represents a first order truncation of the Harris-type configuration, as the variation across the current sheet is linear. The components along the y and z axes are constant. The component along the z axis is called longitudinal component and represents the shear of the magnetic field.

$$\begin{aligned} E &= (0,0,E), \quad B=(-y,\xi_\perp,\xi_\parallel)B_0 \qquad \text{for } |y| \le 1 \\ E &= (0,0,0), \quad B=(-y/|y|,\xi_\perp,\xi_\parallel)B_0 \quad \text{for } |y| > 1 \end{aligned} \tag{1}$$

The RCS extents up to $|y| = 1$. Beyond this limit, the magnetic field is considered constant. In our numerical simulations we consider that when a particle reaches this limit it escapes from the current sheet. The equations of motion of a charged particle moving in these fields are given by

$$\frac{d^2x}{dt^2} = \xi_\parallel \frac{dy}{dt} - \xi_\perp \frac{dz}{dt} \tag{2}$$

$$\frac{d^2y}{dt^2} = -\xi_\parallel \frac{dx}{dt} - y\frac{dz}{dt} \tag{3}$$

$$\frac{d^2z}{dt^2} = \varepsilon + \xi_\perp \frac{dx}{dt} + y\frac{dy}{dt} \tag{4}$$

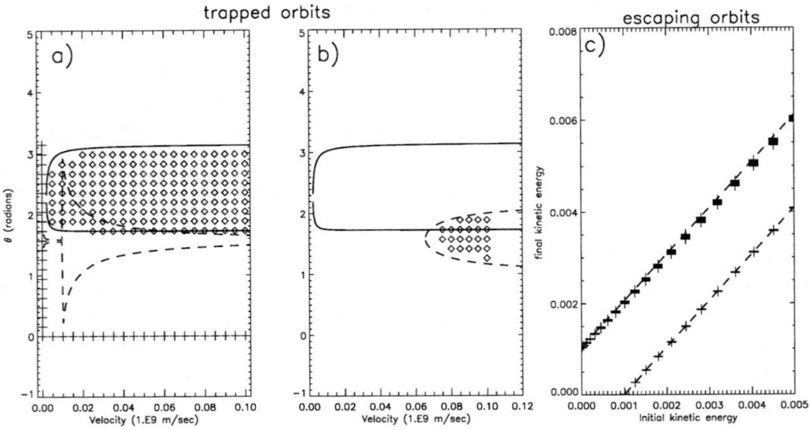

FIGURE 1. Initial conditions of trapped orbits in an RCS with $\xi_\perp = 10^{-2}, \varepsilon = 10^{-5}$ and $\xi_\parallel = 0$ (no-shear). Diamonds correspond to the initial conditions of injection (υ, θ) (see text) for which a particle remains trapped in the current sheet. The solid curve represents an analytical approximation of the trapping domain developed in [6], which is valid in the no-shear limit ($\xi_\parallel \to 0$). The dashed curve represents a different analytical approximation of the trapping domains based on a low order adiabatic invariant [5] [6], which is not succesful. b) Same as a) when the shear is strong ($\xi_\parallel = 1$). The second analytical approximation is now succesful. c) Final kinetic energy of escaping particles against their initial kinetic energy. The RCS parameters are as in (a). The energy unit is 1.9×10^3 keV.

Equations 2,3,4 are written for a positive charge and are scaled to dimensionless units so that the electric field is given by $\varepsilon = mE/(qB_0^2)$ whereas the unit of time is the inverse of the Larmor frequency $t = \omega^{-1} = m/(qB_0)$. In this study the magnetic field is $B_0 = 100$ G and the electric field $E = 100$ V/m, which corresponds to $\varepsilon = 10^{-5}$. These are typical values for the fields measured in solar flares (see [14]). Furthermore, except of the total energy E, a second constant of motion $I_2 = \dot{x} - \xi_\parallel y + \xi_\perp z$, can be derived by integrating Eq. 2. Using I_2 one can derive a Hamiltonian of two degrees of freedom to describe the particles motion instead of three as in Eqs. (2,3,4).

$$H = \frac{1}{2}p_y^2 + \frac{1}{2}(c_4 + \frac{1}{2}y^2)^2 + \frac{1}{2}(I_2 - \xi_\perp z + \xi_\parallel y)^2 - \varepsilon z \tag{5}$$

where $c_4 = \xi_\perp x + p_z$. The new canonically conjugate pairs are (y, p_y) and (z, c_4), where: $p_y = \dot{y}$ and $p_z = \dot{z} - (\xi_\perp x + 0.5y^2)$. The canonical momentum p_z is the sum of the velocity and the magnetic vector potential component (see [5], [6]). Integrating the Hamiltonian equations using he function $H(y, z, p_y, c_4)$ described by Eq. 5, we performed numerical simulations presented in the following section.

NUMERICAL SIMULATIONS

Trapped and escaping particles

We investigated two examples of RCS models with zero shear ($\xi_{\parallel} = 0$) or strong shear ($\xi_{\parallel} = 1$). In our scaled units, the electric field is $\varepsilon = 10^{-5}$ and the perpendicular component $\xi_{\perp} = 10^{-2}$. Particles are initially injected at the center of the current sheet (y=z=0). The initial velocity orientation is chosen in a range from $\theta = 0$ to π where θ is the angle of the velocity with the z axis. The dimensionless velocity range is 0 - 0.1, corresponding to 0-10^8 m/sec. For several orbits starting with these initial conditions, particles remain trapped into the sheet ($|y| < 1$) or eventually escape from it. The computation of the final kinetic energy provides a measure of the effectiveness of the RCS as an accelerator.

In Figure 1, the axes refer to a particle's initial velocity (orientation θ, and magnitude υ_0). Each diamond corresponds to a particle that remained trapped in the sheet until the end of the numerical integration. In the case of non-shear (Fig 1a) we notice that orbits are trapped for any value of υ_0 as long as $1.7 < \theta < \pi$. In the case of large shear, (Fig 1b), the trapped domain shrinks considerably so that for $\upsilon_0 < 0.07$ particles always escape from the sheet.

The domain of trapped orbits, defined in the (υ_0, θ) plane can be described by analytical approximations. This was done in two limits: 1) $\xi_{\parallel} \to 0$ and 2) $\xi_{\parallel} \to 1$ (see [5], [6]). Both theoretical limits are represented in Fig. 1, the first with a solid curve and the second with a dashed curve. Each approximation is succesfull in the case where the corresponding limit holds (case 1 for Fig 1a and case 2 for Fig 1b respectively).

In Fig. 1c, the kinetic energy gain of particles escaping the RCS with strong shear is plotted against their initial kinetic energy. Two groups of particles are distinguished: particles gainning or losing kinetic energy. The maximum kinetic energy gain of a particle can be described with good accuracy from our analytic model (see [5], [6]) shown as the dashed lines in Fig 1c. In particular, for particles escaping along chaotic orbits, the minimum and maximum acceleration lengths are provided by the zero velocity curves, found from Eq. (5) by setting $\dot{y} = \dot{z} = 0$:

$$\xi_{\parallel}^2 y^2 + \xi_{\perp}^2 z^2 - 2\xi_{\parallel}\xi_{\perp} yz - 2\varepsilon z = 2E \qquad (6)$$

When $\xi_{\parallel} \neq 0$ Eq. (6) is a parabola open in the half plane $y > 0$. For $y = 1$, Eq. (6) yields two solutions: $z = z_{min}$ or $z = z_{max}$. A particle leaving the sheet from z_{max} has the maximum possible kinetic energy gain. If the particle initial kinetic energy is strong enough, the particle can leave the current sheet from $y = -1$. In this case the particle loses kinetic energy (particles in Fig 1c below the diagonal).

Poincaré sections

A useful diagnostic tool for studying orbits is to compute Poincaré sections. In the present study, Poincaré sections are computed by tracing the crossing orbits on the $(\dot{z}, z)$

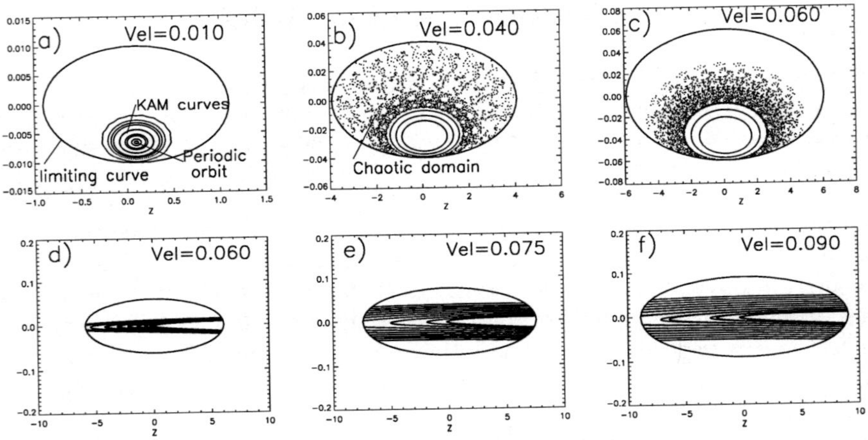

FIGURE 2. Poincaré sections $(\dot{z},z)$ for several initial kinetic energies of particles. The first row of panels is computed using $\xi_\perp = 10^{-2}, \xi_\parallel = 0, \varepsilon = 10^{-5}$ (as panel a of Fig 1). In the second row $\xi_\perp = 10^{-2}, \xi_\parallel = 1, \varepsilon = 10^{-5}$. The initial particle velocity is indicated in each panel.

plane when $y = 0, \dot{y} > 0$. Fig. 2 shows several examples of Poincaré sections computed for orbits with different initial energies for the same model parameters as in the two examples of Fig 1.

On the Poincaré section, orbits cannot give intersections outside a limiting curve. The limiting curve, from Eq. (5), is $\dot{z}^2 + \xi_\perp^2 (z - \varepsilon/\xi_\perp^2)^2 = 2E + (\varepsilon/\xi_\perp)^2$ where E is the energy of the particle. One of the trapped orbits is periodic and intersects the Poincaré section at a single point. Most other trapped orbits intersect the Poincaré section in a way forming KAM curves ([13],[2], [15]) which are either closed, around the periodic orbit (Figs 2a,b,c) or elliptic arcs that intersect the limiting curve (Figs 2d,e,f). The KAM curves determine the region of non-chaotic motion where particles follow either periodic or quasi-periodic orbits. Between the last KAM curve and the limiting curve lies a large chaotic region. Chaotic orbits explore all the available phase space between the last KAM torus and the zero velocity surface. The KAM curves cross the $z = 0$ axis in Fig 2, so that trapped orbits of Fig 1 starting at $y = z = 0$ correspond to KAM curves. In the case of no shear (Figs. 2a,b,c) the Poincaré section shows a region of regular motion organized around periodic orbits, surrounded by a chaotic domain. In this case, only chaotic particles escape from the sheet. The last KAM curve, which separates the two domains, is tangent to the limiting curve. As we increase the initial particle velocity, the area included in the limiting curve increases. For strong magnetic shear, (Figs. 2d,e,f) the morphology is very different. KAM curves cover all the available area, and they form elongated elliptic arcs that intersect the limiting curve. This results in that particles can escape form the sheet along regular orbits, when the corresponding KAM torus intersects the planes $y \pm 1$.

THEORETICAL ANALYSIS

The existence of regular orbits forming KAM curves on the Poincaré surfaces is a manifastation of the existence of a third constant of motion (named J_1') besides E, I_2. J_1' is preserved along regular orbits but not along chaotic orbits. J_1' corresponds to an adiabatic invariant of the particle bouncing inside the magnetic trap (see [9] Chapter 2.3). In [5] J_1' is computed as a series in the parameters $\xi_\perp, \xi_\|, \varepsilon$. The initial Hamiltonian is transformed into a normal form using Lie series. For $\xi_\| = 1$, the expression for J_1' is simplified and one can compute the KAM curves as well as the shape of the trapping domain for (see [5], [6]).

For $\xi_\| = 0$, the last KAM is an ellipse which is almost tangent to the limiting curve (see Figs. 2a,b,c). The shape of KAM as well as the position of the periodic orbit on the Poincaré sections, are obtained [6] using the averaging technique known in celestrial mechanics as 'Hadjidimitriou mapping' [7].

CONCLUSIONS

In this work the motion of electrons inside a RCS is analysed. Two types of RCS are studied, with zero or strong magnetic shear.

Numerical simulations show that, in both cases, electrons for particular initial conditions can be trapped in the RCS. The trapping domain is more extended in the case of zero shear.

With the help of Poincaré sections, we find that in the case of strong shear, both trapped and escaping electrons mostly follow regular orbits. On the other hand, in the case of zero shear electrons escape from the sheet following chaotic orbits while most regular orbits are trapped.

Regular orbits can be analysed by calculating a third integral of motion. This integral is calculated by the method of Lie-series (see [5] for the mathematical details.)

Trapped particles must be important for the self-consistency of solar RCS, (see [4] section 5.4). They could also affect the total number of accelerated particles during solar flares.

ACKNOWLEDGMENTS

This work was supported by the Research Committee of the Academy of Athens.

REFERENCES

1. A. Anastasiadis, C. Gontikakis, N. Vilmer and L. Vlahos, *Astron. & Astroph.*, **422** 323–330, (2004).
2. V. I. Arnold, *Russ. Math. Surv.*, 18 (5), 9; (6), 91 1963
3. J. Chen, *J. Geophys. Res.*, **97**, 15011–15050, (1992).
4. S. W. H. Cowley, *Solar Systems Magnetic Fields, Geophysics and Astrophysics Monographs*, Ed. E. Priest, Reidel Publishing Company, 121–155 (1985).
5. C. Efthymiopoulos, C. Gontikakis, and A. Anastasiadis, *Astron. & Astroph.*, **443** 663–678, (2005).

6. C. Gontikakis, C. Efthymiopoulos, and A. Anastasiadis, *M.N.R.A.S.*, in press (2006).
7. J.D. Hadjidemetriou, *Predictability, Stability and Chaos in N-Body Dynamical Systems*, Ed. A.E. Roy, Plenum Press, New York, 157-175 (1991).
8. M. J. Hagyard *Mem. S.A. It.* **61**, 337. (1990).
9. A. J. Lichtenberg, and M. A. Lieberman *Regular and Chaotic Dynamics*, Springer Verlag, New York, (1991).
10. Y. E Litvinenko, *Sol. Phys.*, **147**, 337. (1993).
11. Y. E. Litvinenko, and B. V. Somov, *Sol. Phys.*, **146**, 127 (1993).
12. Y. E. Litvinenko, *Astrophys. J.*, **462**, 997 (1996).
13. A. N. Kolmogorov, *Dokl. Akad. Nauk. SSSR* **119**, 527 (1954).
14. P. C. H. Martens and A. Young *Astrophys. J. Suppl.*, **73**, 333–342 (1990).
15. J. Moser, *Nachr. Acad. Wiss. Göttingen II. Math. Phys. Kl.*, 1, 1962.
16. G. J. D. Petrie, submitted to *Astroph. J. Suppl.* (2006)
17. E. Priest and T. Forbes, *Magnetic Reconnection, MHD theory and applications*, Cambridge University Press, Cambridge UK, (2000).
18. T. W. Speiser, *J. Geophys. Res.*, **70** 4129–4226, (1965).
19. N. Vilmer and A. L. MacKinnon in *Energy conversion and particle acceleration in the Solar Coronna* Ed. L. Klein, Lecture Notes in Physics, **612**, 127–160, (2003).

First results from SECIS observations of the 2001 total Solar eclipse

A.C. Katsiyannis[*,†], D.R. Williams[**,†], F. Murtagh[‡,§], R.T.J. McAteer[¶,†] and F.P. Keenan[†]

[*]*Department of Solar Physics, Royal Observatory of Belgium, Avenue Circulaire -3-Ringlaan,B-1180, Belgium. T.Katsiyannis<AT>oma.be*
[†]*Department of Pure and Applied Physics, Queen's University Belfast, Belfast, BT7 1NN, U.K.*
[**]*Mullard Space Science Laboratory, University College London, Holmbury St. Mary, Dorking, Surrey, RH5 6NT, U.K.*
[‡]*Department of Computer Science, Royal Holloway, University of London, Egham, Surrey TW20 0EX, England*
[§]*Observatoire Astronomique de Strasbourg, 11 rue de l'Université, 67000 Strasbourg, France.*
[¶]*National Research Council Research Associate, NASA Goddard Space Flight Center, Greenbelt, MD 20771, U.S.A.*

Abstract. SECIS observations of the June 2001 total solar eclipse were taken using an Fe XIV 5303 Å filter. Existing software was modified and new code was developed for the reduction and analysis of these data. The observations, data reduction, study of the atmospheric and instrumental effects, together with some preliminary results are discussed. Emphasis is given to the techniques used for the automated alignment of the 8000 images, the software developed for the automated detection of intensity oscillations using wavelet analysis and the application of the Monte Carlo randomisation test as a means of checking the reliability of those detections. In line with findings from the 1999 SECIS total eclipse observations, intensity oscillations with periods in the range of 7-8 s, lying outside coronal loops were also detected.

Keywords: MHD – waves – eclipses – Sun: activity – Sun: corona – Sun: oscillations
PACS: 96.60.-j, 95.10.Gi, 96.60.P-, 96.60.pf, 96.60.th

INTRODUCTION

The magnetohydrodynamic (MHD) waves in corona loops have been investigated as a possible cause of coronal heating by numerous authors. Hollweg [1] first proposed a mechanism that involved MHD waves dumping their energy into the solar corona through ion viscosity and electrical resistivity, while other authors favoured magnetic reconnection followed by current dissipation (for example see Parker [2]). Priest and Schrijver [3] have published a review article that contains all the theoretical attempts and their limitations.

MHD oscillations are usually divided in two main categories depending on the presence or not of density perturbations. Magnetoacoustic MHD waves are defined as the density, pressure and temperature perturbations (which are in turn divided into slow and fast modes), while the incompressible waves are called Alfvén and are also divided into those with movements perpendicular to the magnetic field and torsional. Roberts et al. [4] studied a low-β plasma, using reasonable values for the solar corona approximations and predicted that fast magnetoacoustic (sausage-mode) oscillations with frequencies

CP848, *Recent Advances in Astronomy and Astrophysics,*
7th International Conference of the Hellenic Astronomical Society, edited by N. H. Solomos

around 1 Hz can be excited in coronal loops. They also study the signature of a wave train, created by a pulsation at $x = 0$ observed at a point $x = x'$. Much later, [5] used numerical analysis to confirm the initial study and provide a more accurate simulation.

The Solar Eclipse Coronal Imaging System (SECIS) is an eclipse imager that was developed specifically to detect corona oscillations, following a long history of similar attempts [see 6, 7, 8, 9, 10, 11]. An Fe XIV 5303 Å filter was used by the SECIS project to observe the August 1999 and June 2001 total solar eclipses and the results of the 1999 observations were reported in a sequence of papers (Williams et al. [12, 13], Katsiyannis et al. [14]), while a detailed description of the instrument is provided by Phillips et al. [15]. [12, 13, 14] consistently detected several periodicities in the range of 4-7 s, while [13] reported a propagating wave train travelling through the corona loop with a phase-speed of $\sim$2100 km s^{-1}, making a fast-mode MHD wave as the most likely explanation of the observed intensity perturbation. By working on a different loop of the same active region [14] found intensity oscillations just outside but fully aligned with visual corona loops. Their detections were in the same frequency range and have similar amplitudes as with [12, 13]. After calculating the physical characteristics of the loop (physical dimensions, density, etc) and with reasonable assumptions regarding the strength of the magnetic field, Roberts (private communication) confirmed that the frequencies reported by the above authors [12, 13, 14] are well within the range predicted by [4].

Following the observations, reduction and data analysis of the 2001 eclipse were reported by Katsiyannis et al. [16]. In this paper we will present further progress in the analysis of the data, a detailed statistical analysis of the effects of the earth's atmosphere and some preliminary detections of oscillations in the lower solar corona.

An analytical description of the observations taken by SECIS during the August 1999 total solar eclipse is presented by [15]. Detail description of the 2001 observations can be found in [16], while similar procedures were followed in the 1999 observations as described by [15, 12].

DATA REDUCTION AND ANALYSIS

The data reduction software developed for the analysis of the 2001 eclipse data is described in detail by [16]. Since then various improvements were made mostly to the way the images were aligned. A brief description of the whole reduction and analysis is below and the differences with [16] will be explicitly mentioned.

In a first stage dark current and sky flat subtraction was performed in the standard way for astronomical images.

The second stage was the 8000 images to be aligned with an accuracy of 1 pixel. The first step of the alignment procedure was to use the Sobel filter to calculate the edge of the moon disk and then, assuming a constant lunar radius, a fit of the disk was produced and compared with the edge fitted by the Sobel function. Those points that were lying outside the fit by more than 3σ were rejected. In the next step we assumed that the centre of the moon was moving in respect to the Sun with a constant velocity and the images were corrected for this effect. In the final stage further accuracy was achieved by using an area of the 1000th image containing featureless parts of corona as well as a sharp transition from the moon disk as a reference. This area was correlated

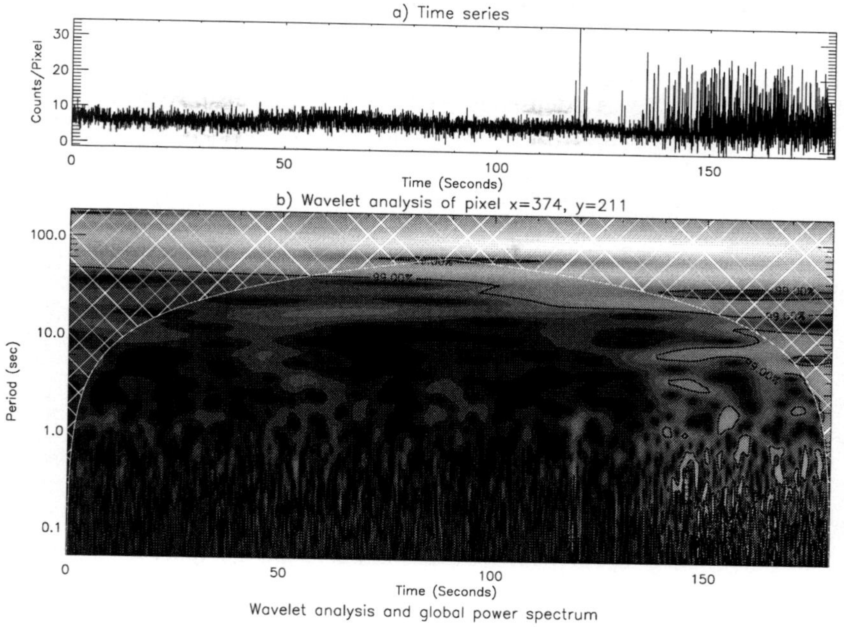

FIGURE 1. Wavelet transform analysis of point x=374, y=211 of the aligned data set. Panel (a) contains the time series while (b) is the wavelet transform. The contours in panel (b) highlight the areas where the detected power is at the 99% confidence level and the hatched area is the cone-of-influence (COI).

with the equivalent areas of the rest of the 8000 images and the pixel shifts for which this correlation was maximum were considered the final shifts for the alignment of the images.

In line with previous work on SECIS [12, 13, 14, 16] continuous wavelet analysis was used to detect intensity oscillations. Torrence and Compo [17] described in detail the algorithm used and provide a discussion of its benefits and application on various scientific problems.

The results of the wavelet analysis described above applied to pixel x=374, y=211 of the aligned 2001 data set are shown on Figure 1. Panel (a) is a plot of the time series of pixel x=374, y=211 of the aligned data. Panel (b) displays the power density wavelet transform of the above time series with the lighter areas representing the higher values. The hatched region marks the cone-of-influence (COI) and represents the areas that suffer from edge effects. Everything inside the COI should be discarded. More details on the edge effects introduced by a finite time-series and how the hatched area is affected by this can be found in [17] and the references their in. The contours of panel (a) surround the area where the detected power exceeds the 99% confidence level.

AUTOMATED DETECTION OF OSCILLATIONS

Software for the automated detection of intensity oscillations on the SECIS data set was developed by [16] and was extensively used for this study also. As with [16] this was done due to the large number of pixels included in the data set. In total 5400 pixels were passed through the scanning program for the determination of atmospheric and instrumental effect and the detection of lower corona loop oscillation (for more details see sections 5 and 6).

The following criteria were applied for the detections to be considered valid. Both the choice of the mother wavelet and the selection criteria were chosen for consistency with previous work by [12, 13] and [14]. See [14] for a detailed discussion on the establishment of these criteria.

- All values of coefficients falling within the COI are discarded.
- Only areas with 99% confidence level or higher are taken into account.
- All oscillations lasting less than the time length of three periods were discarded.

A brief description of the software, written in IDL, follows:

1. For a given pixel of the data set the continuous wavelet coefficients of the time series was produced.
2. For the lowest periodicity the number of the first sample in time that is unaffected by the COI is determined. This sample is called t.
3. For the same periodicity the last sample that is unaffected by the COI, t', was calculated.
4. The number of the sample that predates t' by three periodicities is determined. This sample is called t_{max}
5. The confidence level of the sample t is calculated
6. If the confidence level of sample t is 99% or higher, then the confidence level of the sample that is three periods later, say t_{+3}, is also extracted. Otherwise we go to step 8.
7. If the confidence level of t_{+3} is also 99% or higher the co-ordinates of the pixel, the current periodicity and t are recorded. In this case the algorithm goes to step 9. In this point, it is assumed that if t and t_{+3} have both confidence level of 99% or higher, all samples between them will have confidence level of 99% or higher in the same periodicity. This assumption is rarely wrong. All automated detections we have ever inspected manually fulfil this hypothesis.
8. Go to the next t and repeat steps 5-7 until t becomes t_{max}.
9. Go to next periodicity and repeat steps 2-8 until the periodicity reaches the limit of 70.9 *s*.
10. Move to the next pixel of the array and start again the entire procedure.

70.9 *s* is the upper limit of periodicities that can be detected in the SECIS 2001 data set. This is because any oscillations with longer periods do not have a long enough part of the time series outside the COI to satisfy the criteria mentioned above.

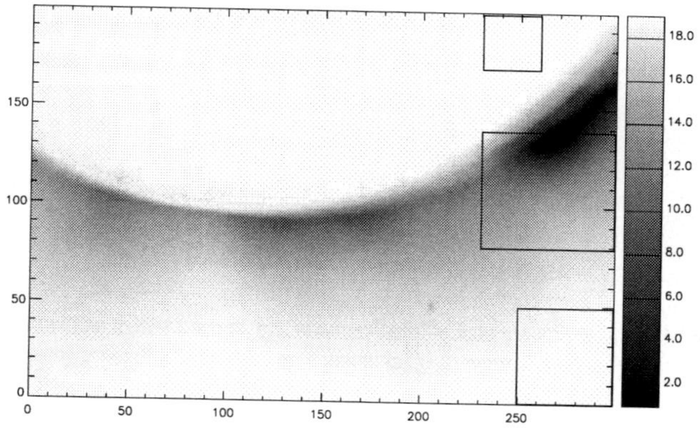

FIGURE 2. The 2001 total solar eclipse as observed by SECIS. This image is produced by averaging over the time axis the three-dimensional aligned array. After alignment the Sun remains in the same pixel area and the Moon crescent moves in respect to the solar corona. Areas of the Moon, outer corona and lower corona close to NOAA Active Region 9513 (AR 9513) are highlighted. These areas were used to test the instrumental and atmospheric effects on the data.

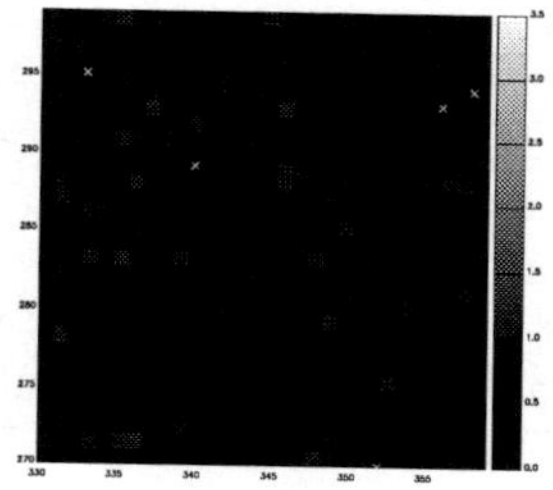

FIGURE 3. A 30×30 pixels area covered by the Moon. Out of the 900 pixels, 5 were found to oscillate with a periodicity of 7-8 *s*. The pixels where those oscillations were found, are marked with 'x'. On average, the chance of a pixel of this area to be found to oscillate is 0.56%.

ATMOSPHERIC AND INSTRUMENTAL EFFECTS

Each frame of the SECIS eclipse observations can be divided into three parts. The first is the area covered by the moon disk and contains only signal from scattered line from the Earth's atmosphere. The second part has signal mainly from the solar corona and the third part is dominated by light coming from the outer corona. Figure 2 contains a time-average image of the whole eclipse observations with samples of the three areas mentioned above highlighted. Those three samples were chosen to be close or containing NOAA Active Region 9513 (AR 9513), as it is expected that AR 9513 has the highest probability to produce coronal light intensity oscillations.

The different properties of these three areas are used in order to estimate the atmosphere and instrumental effects to the detection of intensity oscillations. The automated

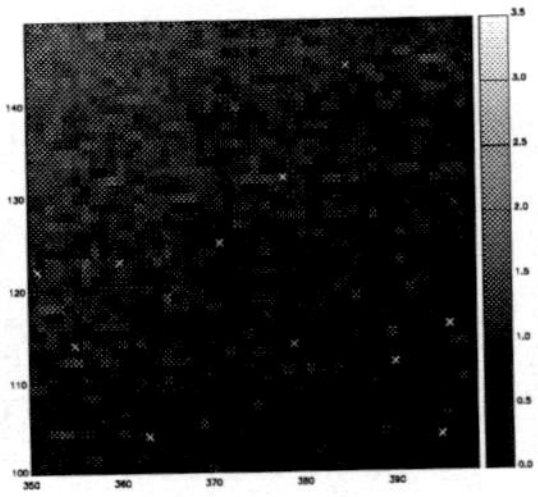

FIGURE 4. A 50×50 pixels area of the upper corona. Out of the 2500 pixels, 11 were found to oscillate with a periodicity of 7-8 *s*. On average, the chance of a pixel of this area to be found to oscillate is 0.44%.

method and the criteria for detecting oscillations, as defined in the previous sections, were applied to a sample area of the moon and the outer corona. Figures 3 and 4 contain the averaged, over the whole time sequence, images of an area of the moon and outer corona respectively. The two axes of both images are the pixel coordinates of the aligned data set and the green scale are the average pixel values. Both areas were chosen to be close to AR 9513 but significantly distant to the edge of the moon (for the disk sample) and the edges of the useful area (for the outer corona sample).

When applied to the sample moon area, for the detection of oscillations with 7-8*s* periodicity, the automated software produced 5 detections of oscillations out of 900 pixels. Since no signal from the lower corona is detected in this area directly, all counts generated will be either CCD read-out noise, or scattered atmospheric light. When different ranges of periodicity are scanned by the same software in the same area different pixels in random locations are found to oscillate. This led us to believe that the moon detections behave randomly and normal statistical procedures can be used to determine the atmospheric and instrumental effects. Further verification of this assumption is provided by the 50×50 area of the outer corona displayed in Figure 4. Using the same software, 11 oscillations were found in the same periodicity range, making the average possibility of a pixel of this coronal area to found with intensity oscillations 0.44%. This is very close to 0.56% for the moon area reported above and confirms our assumption that the intensity oscillation in the outer corona are too weak for SECIS to detect them. Therefore any oscillations found in the upper are mostly due to atmospheric and instrumental effects. To provide further confirmation to this assumption more sample areas will be used from both the lunar disk and outer corona areas and the possibility of a pixel to oscillate will be calculated for each sample area separately. The standard deviation of this measurements will provide us with an estimation of how consistent the atmospheric and instrumental effects are throughout the data set.

DETECTIONS OF OSCILLATIONS AROUND AR 9513

Figure 5 presents an area of the lower corona close to AR 9513 as observed by SECIS. Pixels with intensity oscillations that passes the established criteria are marked with 'x'. The detections included in this Figure are, in their majority, due to plasma intensity

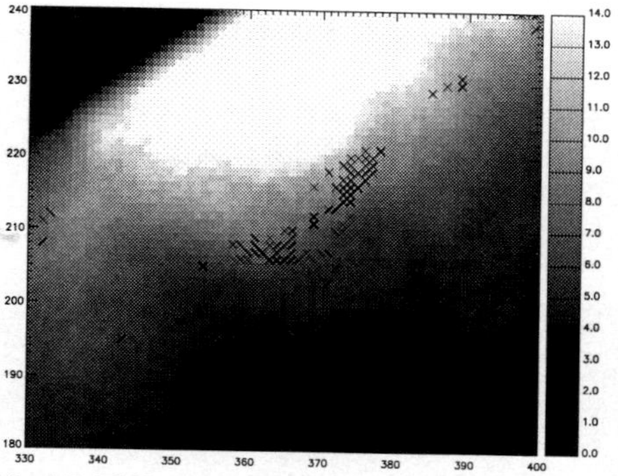

FIGURE 5. The area outside AR 9513 where the intensity oscillations with periodicity of 7-8s were detected. All pixels with intensity oscillations are marked with 'x'.

oscillation in the solar corona and not due to atmospheric or instrumental effects for mainly two reasons:

1. The spatial distribution of the detected oscillations, unlike that of Figures 3 and 4, is not random, but highly concentrated in one area.
2. All 66 detections of interest are gathered in a very compact area. Figure 6 contains all oscillations in a 25×19 pixels frame of 475 pixels in total. Based on the previous statistics an area of the moon or outer corona of the same size should have 2.4 detections due to atmospheric or instrumental effects.

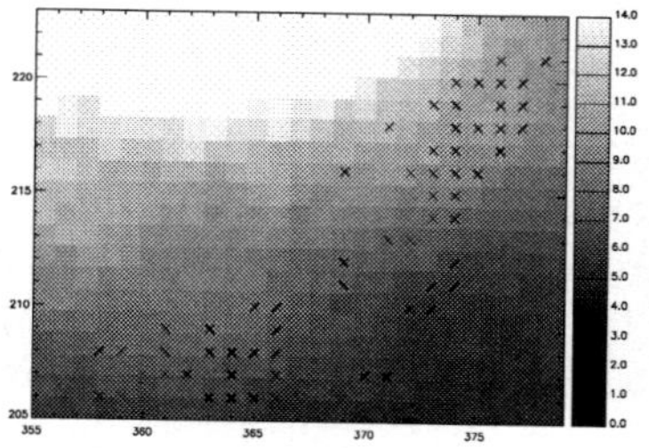

FIGURE 6. The area of AR 9513 with the highest population of oscillations with 7-8s periodicity. This area is 25×19 pixels in size (475 pixels in total). Based on the statistics calculated by the sample Moon and upper corona areas (see Figures 3 and 4), on average, there should be 2.4 detections due to atmospheric or instrumental effects.

Figure 1 is a display the wavelet analysis of the example point x=374, y=211 while Figure 7 is a zoom-in of the area of Figure 1 that contains the oscillations that pass the criteria established by [14].

Figure 7 has an interesting feature that appears in some of the 66 detections reported here and has not been seen in the 1999 observations yet. Although not detected by the automated software, as it does not satisfy the duration criteria established by [14], there is an oscillation contemporary to the main detection at exactly half the periodicity. This detection lasts for $\sim$12s and starts almost simultaneously with the main oscillation,

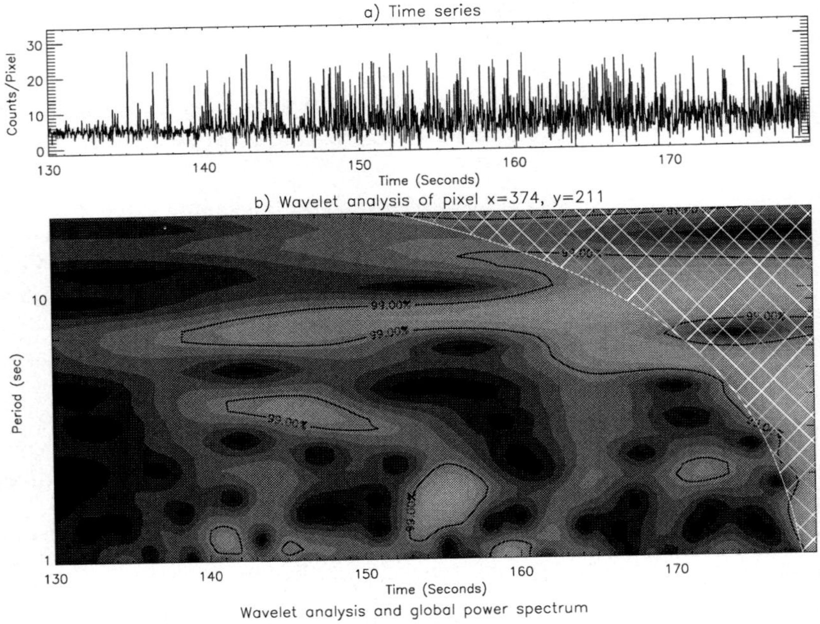

FIGURE 7. Wavelet analysis of pixel x=374, y=211. This is a typical sample of a detection of intensity oscillations in the area of Figure 6. The rest of the detections seen on Figure 1 and this image do not pass the criteria established by [14].

7.5 *s*. As this secondary oscillation does not satisfy the duration criteria of [14], we would normally ignore it, but there is a number of coincidences that make it worth reporting. Namely, the oscillation starts practically simultaneously with the main one (and definitely within the limits of the mother wavelet time resolution) and it has exactly half of the periodicity of the main detection. Those two properties make it a perfect candidate for identification as a second order sausage-mode oscillation.

MONTE CARLO RANDOMISATION TEST

The scanning of large areas of the images covered by the Moon and the outer corona revealed how atmospheric and possibly other instrumental effects can produce false detections of intensity oscillations. This comes as no surprise as it is well known that atmospheric seeing can produced intensity variations due to changes in atmospheric absorption, in the periodicity range of 5 to 19 s.

Although the effect of those "false" detections is too small to alter the significance of the detections reported here (only $\approx$2 detections are expected to appear in the area of Figure 6 based on the statistics produced by the Moon and outer coronal areas), a Monte Carlo randomisation test was applied by Katsiyannis et al. [18] as a means to establish a further criterion to distinguish the "false" detections. For each individual pixel of these areas the maximum power of the wavelet transformation was recorded and then compared to the maximum value of the 1000 shuffled time series produced from the original. The percentage p of those shuffled time series that had maximum power of their wavelet transformation larger than that of the original time series was then recorded. When the original time series is random noise of any given type (as this test is distribution free) we expect that 50% of the shuffled time series will have wavelet transformations with higher maximum power than the original data. For the purposes of this analysis we will consider any value p smaller than 1% (ie less than 1% of the shuffled time series has wavelet transformation with power more than that of the original time series) is indicating that the original time series has a strong signal comparing to the noise level.

60%, 65% and 5% of the pixels of the Moon, outer and lower corona areas were found to to have $p \geq 1\%$. This is an important result with various implications: First of all the areas of the Moon and outer corona have more that the half of their pixels having a relatively low S/N ratio, while the equivalent percentage of the lower corona pixels is much lower (5 %). Second, the percentage of pixels that have $p \geq 1\%$ is approximately the same in the Moon area (where we know that the signal of all pixels is due to not direct observation of the Sun) and the outer corona (where we know the signal is partially directly from the solar corona and partially scattered light). This is important as it indicates that the scattered light on this image area is a significant portion of the signal and it will be very difficult to distinguish which detections are "real" in this area by applying the existing criteria.

The third and by far the most significant result from the Monte Carlo analysis came when p was calculated for the pixels that were found to oscillate (i.e. the 5 pixels of the Moon area, 11 pixels of the outer corona area and 66 pixels of the lower corona). All pixels from lower corona were found to have $p < 0.1\%$ (i.e. none of the 1000 shuffled time series was found to have a wavelet transformation with higher power than that of the original time series), while 4 out of 5 of the detections from the Moon area and 10 out of 11 detections from the outer corona, were found with $p > 0.1\%$. Therefor, a criterion can be establish that will use the randomisation test described here to reject those pixels with $p > 0.1\%$.

ACKNOWLEDGMENTS

ACK acknowledges the Leverhume Trust for funding via grant F00203/A. PPARC has funded part of this work. DRW and JMA thank DEL and QUB for studentships. FPK is grateful to AWE Aldermaston for the award of the William Penney fellowship.

REFERENCES

1. J. Hollweg, *Sol. Phys.* **70**, 25 (1981).
2. E. Parker, *AJ* **330**, 474 (1988).
3. E. Priest, and C. Schrijver, *Sol. Phys.* **190**, 1 (1999).
4. B. Roberts, P. Edwin, and B. A.O., *Astrophisical Journal* **279**, 857 (1984).

5. V. Nakariakov, T. Arber, C. Ault, A. Katsiyannis, D. Williams, and F. Keenan, *MNRAS* **349**, 705 (2004).
6. S. Koutchmy, Y. Žugžda, and V. Locǎns, *A&A* **120**, 185 (1983).
7. J. Pasachoff, and D. Landman, *Sol. Phys.* **90**, 325 (1984).
8. J. Singh, R. Cowsik, A. Raveendran, S. Bagare, A. K. Saxena, K. Sundararaman, V. Krishan, N. Naidu, J. Samson, and F. Gabriel, *Sol. Phys.* **170**, 235 (1997).
9. R. Cowsik, J. Singh, A. Saxena, R. Srinivasan, and A. Raveendran, *Sol. Phys.* **188**, 89 (1999).
10. J. Pasachoff, B. Babcock, K. Russell, T. McConnochie, and S. Diaz, *Sol. Phys.* **195**, 281 (2000).
11. J. Pasachoff, B. Babcock, K. Russell, and D. Seaton, *Sol. Phys.* **207**, 241 (2002).
12. D. Williams, K. Phillips, P. Rudawy, M. Mathioudakis, P. Gallagher, E. O'Shea, F. Keenan, P. Read, and B. Rompolt, *MNRAS* **326**, 428 (2001).
13. D. Williams, M. Mathioudakis, P. Gallagher, K. Phillips, R. McAteer, F. Keenan, P. Rudawy, and A. Katsiyannis, *MNRAS* **336**, 747 (2002).
14. A. Katsiyannis, D. Williams, R. McAteer, P. Gallagher, F. Keenan, and M. F., *A&A* **406**, 709 (2003).
15. K. Phillips, P. Read, G. P.T., F. Keenan, P. Rudawy, B. Rompolt, A. Berlicki, A. Buczylko, F. Diego, R. Barnsley, R. Smartt, J. Pasachoff, and B. Babcock, *Sol. Phys.* **193**, 259 (2000).
16. A. Katsiyannis, R. McAteer, D. Williams, P. Gallagher, and K. F.P., "Initial results from SECIS observations of the 2001 eclipse," in *Wave, oscillations and small scale transient events in the solar atmosphere: A joint view of SOHO & TRACE*, edited by H. Lacoste, ESA, 2004, vol. ESA SP-547, p. 459.
17. C. Torrence, and G. Compo, *Bull. Amer. Meteor.* **79**, 61 (1998).
18. A. Katsiyannis, F. Murtagh, and F. Keenan, *Sol. Phys.* **228**, 325 (2005).

Forced MHD Turbulence Simulations for Coronal Loop Heating

Z.Romeou*,†, M.Velli* and G. Einaudi**

*Department of Astronomy and Space Sciences, University of Florence, Largo E.Fermi 2, Florence, FI 50125, Italy
†Department of Physics, University of Patras, 26500 Patras, Greece
**Dipartimento di Fisica, Universita di Pisa, Largo B. Pontecorvo, Pisa, PI 56100, Italy

Abstract. In this work we revisit the question of whether the assumption of a turbulent photosphere provides an efficient mechanism for the disposition of energy in the solar corona. Through a two-dimensional incompressible MHD spectral code and appropriate analysis we investigate the long time statistical behavior of a two-dimensional cross section of a coronal loop. In particular we study the transition to turbulence from a large scale quasi-stationary coherent forcing analyzing the effects of the finite Reynolds and Lundquist numbers and the role of noise in triggering resistive instabilities and subsequent cascades. Simulations of the average energy dissipation and the spectral and spatial distribution at a given time show the self-organization of the loop at large scales via an inverse MHD cascade. To quantify the nonlinearity of the response in the case of constant time forcing, we derive scaling laws against resistivity of the difference between the numerical solution and the linear approximation as well as of the time it takes the system to reach the peak after exceeding the linear approximation solution. We finally comment on the response of the loop also on the most general case of time dependent random forcing comparing with the first case.

Keywords: Magnetohydrodynamics (MHD) -Turbulence - Sun:Activity - Sun: Corona - Sun: Magnetic fields
PACS: 96.60.Q-, 96.50.Tf, 96.60.Iv, 96.60.-j, 96.60.P-, 96.60.pf

INTRODUCTION

Magnetic turbulence has been recognized over the latest years as a very powerful tool in our attempt to obtain a better understanding of the so far unresolved problem of coronal heating. Different MHD turbulent mechanisms effective in the development of the fi ne scales in the solar corona has been discussed by many authors ([6] [16] [13] [24] [2] [27]).

Numerical modelling in two and three dimensions ([7], [8], [17], [20]) has been developed to validate or disprove the ideas of topological dissipation by Parker [23] , further developed into the so-called nanoflare heating theory. [10], and [12] were the fi rst to show how the dissipation time-series of the two dimensional magnetically forced simulations of MHD turbulence displayed intermittent behaviour with events following a power-low behaviour. [9], [14], [15] also showed how this model of turbulence of a coronal loop could be relevant to the coronal heating problem. See also related work in [1], [10], [21].

Here we are making a step forward by studying how the dissipation end evolution of the 2D system depends on the time-scales associated with the photospheric forcing. To begin, we will study the case of a constant forcing, resembling a photospheric shearing

CP848, *Recent Advances in Astronomy and Astrophysics,*
7th *International Conference of the Hellenic Astronomical Society,* edited by N. H. Solomos

motion. We then carry out simulations with various random generated but spatially large scale-forcing profi les and variable in time over a range of time-scales compared with the internal dynamical time-scale of the turbulence.

The paper is constructed as follows. In Section 2 we present the assumptions of the model we have used, Section 3 goes further in the details of the initial conditions and the numerical setup, in Section 4 we present and discuss the numerical results. We summarise the results and conclude in Section 5.

THE NUMERICAL MODEL

We consider a cross section of a coronal loop threaded by a strong axial magnetic fi eld B_O with footpoints rooted in the photosphere. Boundary disturbances propagate along the axial direction z with the associated Alfvénic velocity giving rise to perpedicular magnetic and velocity fi elds $b_\perp$ and $u_\perp$. In the limit of a large loop aspect ratio, one may follow the evolution by using the reduced MHD equations [26]:

$$\rho(\frac{\partial \vec{u_\perp}}{\partial t} + \vec{u_\perp} \cdot \vec{\nabla} \vec{u_\perp}) = -\vec{\nabla}_\perp (p + \frac{1}{2} b_\perp^2) + \vec{b_\perp} \cdot \vec{\nabla} \vec{b_\perp} + B_0 \frac{\partial \vec{b_\perp}}{\partial z} + \nu \nabla^2 \vec{u_\perp} \tag{1}$$

$$\frac{\partial \vec{b_\perp}}{\partial t} = \vec{b_\perp} \cdot \vec{\nabla} \vec{u_\perp} - \vec{u_\perp} \cdot \vec{\nabla} \vec{b_\perp} + B_0 \frac{\partial \vec{u_\perp}}{\partial z} + \eta \nabla^2 \vec{b_\perp} \tag{2}$$

where ρ is the mass density, p is the plasma pressure and ν and η are the collisional dissipation coeffi cients, the kinematic viscosity and the resistivity respectively.

These equations are valid for a plasma with small ratio of kinetic to magnetic pressures, in the limit of a large loop-aspect ratio ($\varepsilon \equiv l/L \ll 1$, L being the length of the loop and l being the minor radius of the loop) and of a small ratio of poloidal to axial magnetic fi eld ($\vec{b_\perp}/B_0 \leq \varepsilon$). Incompressibility of plasma motion on the transverse plane follows from the latter condition [3]. Consequently, the density, initiallly considered to be uniform remains so, allowing use of the same units for velocities and magnetic fi elds via the normalization $b \to b/\rho^{1/2}$. We can alternatively write the above set of equations as follows:

$$\partial_t \alpha = B_0 \partial_z \psi + [\psi, \alpha] + \eta \nabla^2 \alpha \tag{3}$$

$$\partial_t \omega = B_0 \partial_z j + [\psi, \omega] - [\alpha, j] + \nu \nabla^2 \psi, \tag{4}$$

where the brackets $[c,d] = \vec{e}_z \cdot (\nabla c \times \nabla d)$ are the standard Poisson brackets and where the functions α and ψ are the transverse components of the magnetic and velocity fi elds given by $\vec{u} = \nabla \times (\psi \vec{e}_z)$ and $\vec{b} = \nabla \times (\alpha \vec{e}_z)$ respectively.

Eq. (3) describes the advection of the potential α and Eq. (4) corresponds to the evolution of vorticity ω.

The nonlinear terms represented by the Poisson brackets couple the normal modes in such a way that energy and other ideal invariants can be transferred between them.

The terms ∇^2 are standard resistive and viscous dissipation terms.

Of special interest are the terms $B_0\partial_z$ which represent the coupling between neighbouring z=const. planes transfering energy from the footpoints into the coronal part of of the loop thus enabling communication across planes at different z.

If as in our case we are interested in only following the evolution of the transverse fields on the planes instead of the fully three dimensional propagation along z, we can consider them as forcing terms that can be modeled consistently with our understanding of the photospheric driver.

Thus we write

$$B_0\frac{\partial \vec{u}_\perp}{\partial z} = \vec{F}_m(x,y,t) \tag{5}$$

$$B_0\frac{\partial \vec{b}_\perp}{\partial z} = \vec{F}_u(x,y,t) \tag{6}$$

where $\vec{F}_m$ and $\vec{F}_u$ are unknown external forcing functions.

Following [10] and [12], we proceed further by imposing $\vec{F}_u = 0$ and by rewriting $\vec{F}_m$ in the more convenient form

$$\vec{F}_m = \nabla \times f_m \vec{e}_z \tag{7}$$

The first assumption can be justified by the argument that in the more realistic three dimensional case the plasma pressure is much smaller than the magnetic pressure. Consequently the rms velocity fields remain consistenly much smaller than the magnetic fields.

The assumption about $\vec{F}_m$ facilitates our numerical setup and will be discussed further in the following section. Note that in this case we need to impose a functional profile for f_m in contrast with the three dimensional case where forcing is imposed by the boundary conditions for the velocity field.

INITIAL CONDITIONS AND NUMERICAL SETUP

First we investigate the effect of a time constant forcing. Therefore we substitute f_m with a non time dependent functional profile. Thus we assume

$$f_m = \sum_{nm} a_{nm} \sin(k_n x + k_m y + \zeta_{nm}) \tag{8}$$

The coefficients a_{nm} satisfy the condition $< f^2 >= 1$ which fixes our physical units in terms of the large scale field B_0 (in velocity units), the typical photospheric velocity u_{ph}, the loop length L and the aspect ratio $1/\varepsilon$. Denoting by b_0 , $l_\perp$ and τ our magnetic field, length and time units respectively we have $\tau = \frac{l_\perp}{B_0}$ and

$$1 \sim < f_m > \sim \frac{B_0}{b_0} u_{ph} \frac{\tau}{L} \tag{9}$$

which leads to

$$b_0 \sim B_0 \left(\frac{l_\perp u_{ph}}{B_0 L} \right)^{1/2} \tag{10}$$

On the assumption of an aspect ratio of 10 and a ratio of photospheric velocity to coronal Alfvén speed of about 1/1000, it follows that $b_0/B_0 \simeq 0.01$ and $\tau \sim 10L/B_0 s$. An estimation of $B_0 \sim 1000 km s^{-1}$ and $L = 10^4 km$ leads to $\tau \sim 100 s$. The unit for the energy dissipation follows directly from the parameters given above and an average value of $10^9 cm^{-3}$ for the coronal density:

$$[\eta j^2] = 4\pi\rho b_0^2/\tau \tag{11}$$

which for a density of $10^9 cm^{-3}$ yields a dissipation unit of $2.1 \times 10^{-4} erg/cm^3/s$.

Photospheric boundary motion is confi ned wihin the ring $3 \leq \sqrt{k_n^2 + k_m^2} \leq 4$ being zero elsewhere, to simulate a large scale forcing, compared to the scales at which dissipation occurs.

The explicit profi le we assumed for f_m in our calculations was $f_m = \sqrt{2}\sin(4x+1)$. In this time-constant case the fi eld we chose corresponds to a shearing motion with a scale equal to 4.

The initial profi les we used for the fi elds ψ and α were of the well studied Orszag-Tang type [22] as follows

$$\psi(x,y) = 2scv(\cos x + \cos y) \tag{12}$$

and

$$\alpha(x,y) = scb(2\cos x + \cos 2y) \tag{13}$$

where scv and scb denote the initial amplitudes of the fi elds.

To facilitate our calculations, we have further assumed $scv = scb$ and $\eta = \nu$.

The numerical calculations have been carried out with a code based on a pseudospectral method [5] by use of an FFT technique where periodic boundary conditions are applied. Full de-aliasing by the 2/3 rule mode truncation treats the aliasing effect.

Linear terms are treated in a full implicit way by use of a Crank-Nicolson method, while nonlinear terms are evolved using a Runge-Kutta scheme of third order. To provide for the stability of time integration the CFL condition is satisfi ed at every time step.

The divergence-free restriction is satisfi ed by use of the magnetic potential function.

NUMERICAL RESULTS

We have calculated time-series of the spatial average of the current dissipation ηj^2 with varying resolutions of 64×64, 128×128 and 256×256. We must note here that the values of η used are far larger than the ones expected for the corona, but more than the values themselves we are interested in deriving the trend as we use successively smaller values (see also [19]).

We have shown,(see [25] for a detailed discussion) that for smaller values of η, which are closer to the turbulent regime, the changes are signifi cant both qualitatively and quantitively passing from 64 $\times$ 64 to 128 $\times$ 128 but not from 128 $\times$ 128 to 256 $\times$ 256. Furthermore, for higher values of η, the change is only quantitive and from passing from 64 $\times$ 64 to 128 $\times$ 128. Passing from 128 $\times$ 128 to 256 $\times$ 256 in these cases does not cause a considerable change. In the following runs we kept a resolution of 128×128 grid points with uniform spacing in both directions.

We next investigated how the evolution at long times depends on the initial noise. For that reason we plotted timeseries of the mean dissipation of each η keeping the amplitudes of the initial profi les of the fi elds, namely scb, scv as parameters, and succesively reducing them. Fig. 1 shows clearly a strong dependance on these two parameters which can be regarded as the seeds of the noise necessary for reconnection to start up. The smaller they get the later the peak appears and the higher a value it attains until it eventually reaches a value either asymptotically, or in a slightly oscillating manner, see case (a) in Fig. 1. Also, as scb, scv become smaller although the peak is reached at later times, the time from the moment that the linear solution is overriden until the peak is reached, namely dt_{peak}, also reaches an asymptotic value and therefore is not dependant on the initial noise.

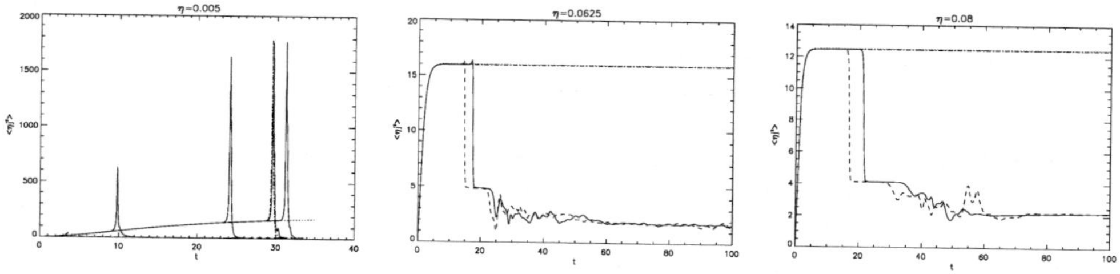

FIGURE 1. Dependance of dissipation $< \eta j^2 >$ on scb, scv, namely the amplitudes of the initial profiles given for α and ψ, for a range of η's from 0.005 to 0.08. In each graph timeseries of averaged current dissipation are plotted for successively smaller values of scb, scv from left to right. It is shown that for a fixed η and for successively smaller values of scb, scv, the peak is reached later in time and the dissipation peak gets bigger until it reaches an asymptotic value. For comparison, the linear approximation solution is overplotted, dotted line in all graphs. Solutions with scb, scv= 0.00 fall into the linear approximation profile. Note that for $\eta > 0.0625$ the solution differs in two significant ways: does not overrun the linear approximation profile and keeps longer the higher value attained before it eventually falls off.

It is worth noticing that after a certain value of η the peak does not exceed the linear approximation profi le any more but it reaches it asymptotically. This behavior is shown for $\eta > 0,065$, the value for which $\eta k^2 > 1$ in the analytical expression for the dissipation in the linear approximation. It can be easily shown that the later is given by Eq. (14)

$$< \eta j^2 > \sim (1/\eta)(1 - \exp(-\eta k^2 t))^2 \tag{14}$$

Also it keeps the higher value attained for longer times unlike the previous cases that the peak appeared and disappeared very sharply. Decreasing η seems to suppress the contribution of the linear terms and strengthen nonlinearities. These conclusions only regard the initial phase of the evolution. The long time evolution is not very much affected as long as scb and scv are not too small to allow instabilities to develop. What changes signifi cantly though with resistivity is the value of the peak and the time it is reached.

TABLE 1. The values used to plot dis_{peak}, Δ_{peak} and Δ_t as a function of η

η	dis_{peak}	Δ_{peak}	Δ_t
0.01	305.053	205.053	1.05
0.02	91.567	41.567	0.925
0.04	36.648	11.648	0.920
0.05	23.316	3.316	0.862
0.0625	16.346	0.346	0.55

The time dissipation leaves off the linear approximation solution is indicative of the time it takes for the nonlinearities to evolve and for the energy to redistribute and be released via resistive modes.

The appearance of a strong sharp peak therefore reveals the tendency of the system to respond rapidly and intensively to what in the absence of resistivity would have been an ideal MHD energy release.

In order to further investigate how dissipation changes with η we plotted this asymptotic value of the maximum of mean dissipation against η. Whether the asymptotic value was not obvious, we used the mean value of the closest to the lowest scb, scv runs. The results are listed in Table 1.

We used these values to plot the dissipation peak against resistivity, the difference of this peak from the linear approximation asymptotic value again as a function of η, and fi nally the time it takes to reach this peak after it leaves off the linear approximation solution. From the best fi t for the lines in the corresponding log-log graphs we obtained the following scaling laws:

- For the dissipation peak, $dis_{max} = 0.23\eta^{-1.56}$.
- For the difference of this peak from the linear approximation asymptotic value as a function of η, $d(dis_{max}) = 0.04\eta^{-2.4}$.
- Finally for the time it takes to reach this peak after it leaves off the linear approximation solution, $dt_{peak} = 0.64\eta^{-0.1}$.

It is evident from these results that with lower resistivity the dissipation peak increases rapidly from the point of view of the order of the magnitude. Nonlinear effects become more effective also.

As the time dt_{peak} is concerned it seems that with decreasing resistivity it also increases.

We further investigated the structure of the fi elds and the currents. In all cases there were three distinct phases developing.

An initial phase before dissipation reaches its peak where the system still resembles the initial confi guration and where magnetic fi eldlines and currents still follow the linear approximation profi le indicating that the nonlinearity has not yet come into effect.

During this time the magnetic energy spectra emit predominately in the $k = 4$ which is imposed by the forcing term, see Fig. 4.

A second phase appears later, where a sharp peak is reached and dissipation provides a mechanism for releasing and redistributing the energy by forming magnetic fi eld islands.

It is during this reconnection phase that nonlinearities also come into play, indicated by the surpass of the linear approximation profi le by the numerical solution (except for the cases where $\eta > 0.0625$, see previous discussion).

A fi nal stage follows where the islands deformate giving way to larger structures. This response is typical for the whole range of η's we tried, see Fig. 2 and Fig. 3.

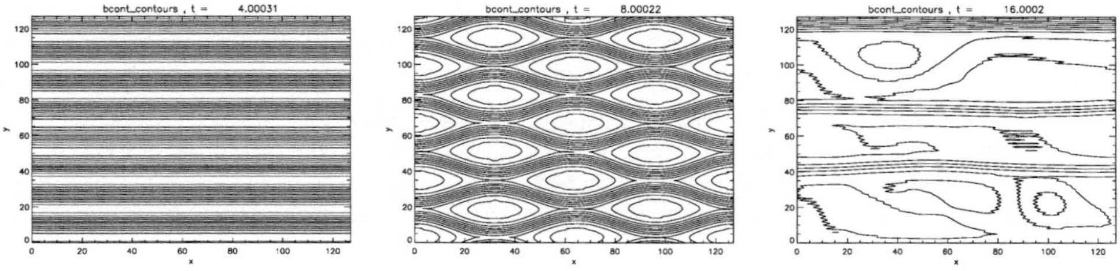

FIGURE 2. Magnetic fieldlines contours for $\eta = 0.01$.Three dinstinct phases are demonstrated clearly: An initial linear phase followed by the formation of a compact islands structure which at later times disentagles giving way to a more diffused magnetic fieldlines structure.

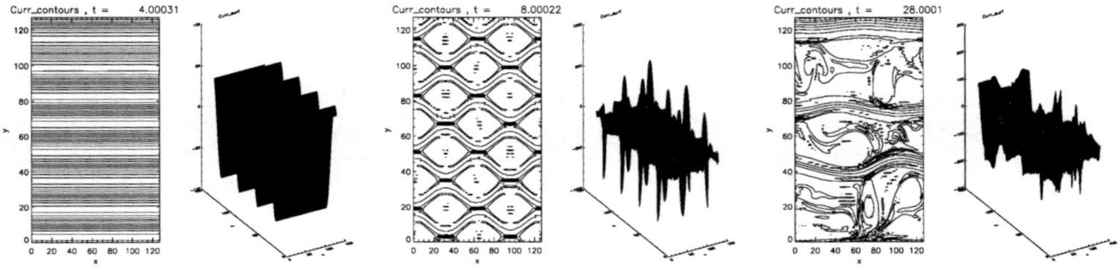

FIGURE 3. Contours and surface plots of the current density for $\eta = 0.01$. The three dinstinct phases are demonstrated clearly also here, compare with Fig. 2

Another interesting point comes from the study of the spectra. Initially, before the peak is reached the dominant mode for the emission of magnetic energy is the one imposed by the forcing profi le, namely $k = 4$. Right after, energy is allowed to redistribute in other modes as well. In the long run, the system seems to develop a secondary inverse cascade activity as it is expected for 2D MHD magnetic invariants.

This is more clearly seen if we look at the spectra of the magnetic square potential at different time instants, in particular before, after and long after the peak, see Fig. 4,left. The shown profi le is typical for a range of η going down to 0.001.

Keeping this tendency in mind we investigated further how magnetic energy spectra scale with η. For this reason we calculated the magnetic energy spectra for each particular value of η at a time long after the system has clearly gone through the peak phase.

At that time, different for each particular η but of the order of 40 time units for the range of η we have used (see Fig. 4, centre), we have parametrized the magnetic energy spectra $Em(k)$ through an expression of the form

$$Em(k) = Ck^{\Theta}. \tag{15}$$

For each η we derive Θ by the corresponding $\ln Em(k)_{,\eta} - \ln k_{,\eta}$ plot, see Fig. 4, centre. In these graphs Θ represents the slope $d\ln Em(k)/d\ln k$ of the calculated line.

This is of cource only a rough estimation in our attempt to quantify the scaling of $Em(k)$ with η.

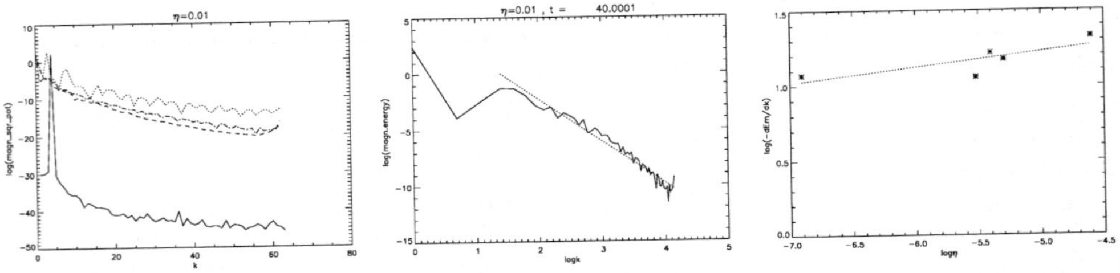

FIGURE 4. Left Graph: Snapshots of spectra for $\eta = 0.01$. The graph shows spectra of the magnetic square potential at a time before the peak (solid line),at a time right after the peak (dotted line), and for times long after the peak (dashed and dashed-dotted lines). It is clearly shown the redistribution of the magnetic energy before, during, after, and long after the peak in modes other than the $k = 4$ imposed by the forcing. It is also evident the expected appearance of a secondary inverse cascade at longer times where the system seems to relax to larger scales. Centre Graph: Snapshots of spectra for $\eta = 0.01$. The graph shows spectra of the magnetic energy for times long after the peak. From the best approximation for the line drawn, we derive the value of the slope and replace for Θ in Table 2. The right graph shows the calculated points and the best fit of the corresponding line.

TABLE 2. Dependance of $dEm(k)/dk$ on η

η	0.001	0.004	0.0045	0.005	0.01
Θ	-2.92	-2.88	-3.38	-3.23	-3.76

In Table 2. we list the results and in Fig. 4, right, we show the calculated points. From the best fi t of the corresponding line for the calculated points we obtain $dEm/dk = -5.630\eta^{0.101}$.

The random forcing case

To compare this results with a random time dependent forcing profi le, we followed [12], [10], [11] and assumed for the forcing profi le the form:

$$f_m = A_1(x,y)sin^2(\frac{\pi t}{2t_{kinck}}) + A_2(x,y)sin^2(\frac{\pi t}{2t_{kinck}} + \frac{\pi}{2}) \quad (16)$$

where the eddy turnover time t_{kinck} is introduced and where

$$A_i = \sum_{nm} a^i_{nm} \sin(k_n x + k_m y + \phi^i_{nm}) \quad (17)$$

and

$$3 \leq (k_n^2 + k_m^2)^{\frac{1}{2}} \leq 4 \quad (18)$$

as before. Every t_{kinck}, the a^i_{nm}, and the ϕ^i_{nm} are randomly changed alternately for eddy 1 and eddy 2.

For a range of different values of η and for $t_{kinck} = 16, 32, 64$ we calculated the timeseries of mean dissipation $< \eta j^2 >$ from which one can derive the dependance of $< \eta j^2 >_{mean}$ either with η or with t_{kinck}.

We first investigated the dependance with resolution and concluded that seems not to be important when resolution is high enough, for a discussion see[25].

We confirmed that initial conditions are soon forgotten and the profiles of the time series resemble the later (after the dissipation peak) stages of the constant forcing case.

Finally we calculated long time timeseries for different t_{kinck} from which we can clearly see how the relationship between photospheric and internal time scale makes a difference, Fig. 5.

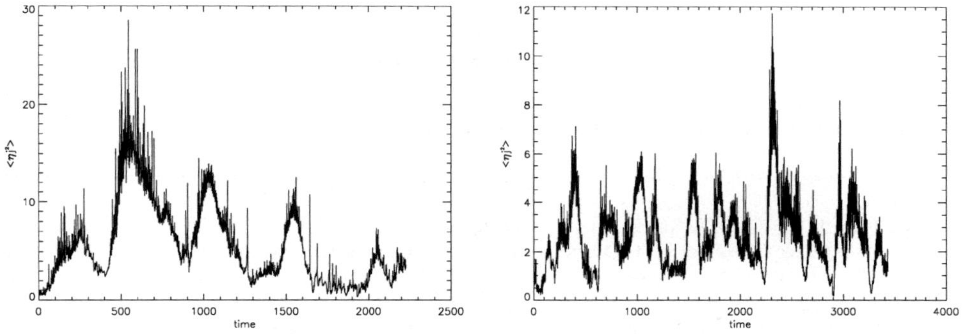

FIGURE 5. Typical long time timeseries of $< \eta j^2 >$ with random forcing, for $t_{kinck} = 128$, 128 X 128 resolution, and $\eta = 0.004$ (left) and $\eta = 0.01$(right). The role of the time scaling is demonstrated, notice the appearance of dissipation events within intervals of the order of the t_{kinck}.

A systematical study for quantifying this dependance is ongoing work.

CONCLUSIONS

Under the assumptions of the Reduced MHD approximation, we have investigated the spatial and temporal response of a coronal loop system in two different cases. Firstly by assuming constant time forcing and secondly by applying random time dependent forcing. Using a numerical pseudospectral method for the calculation of timeseries of ηj^2 we have showed:

- the dependance with η,
- the effects of resolution,
- the effects of the initial conditions,
- the role of the time scaling in the mechanisms involved.

We have shown that there are three phases appearing, indicated by the development of a strong and fast dissipation peak which falls off rapidly. The fields are organised into a compact magnetic island structure, which gradually disentagle before a secondary inverse cascade takes over. Study of the spectral response corroborates this behaviour. The strength and the duration of the dissipation peak have been scaled with η in order to quantify the non linear response of the system. Simulations of long time timeseries for the second case where we set the system subject to random forcing for a range of different t_{kinck} and η's revealed the appearance of dissipation events within intervals of the order of the corresponding t_{kinck} demonstrating the signifi cance of the scaling between the photospheric and the internal times of the system.

ACKNOWLEDGMENTS

This work was supported by EU grant Contract No HRRN-CT-00100310. We thank R.Betta, M.Ghiglione and Eric Buchlin for numerical help and useful discussions.

REFERENCES

1. V. Aletti et al., *The Astrophysical Journal*, **544**, pp. 550–557, (2000).
2. A. A. van Ballegooijen, *The Astrophysical Journal*, **311**, pp. 1001–1014, (1986).
3. D. Biscamp, in *N*onlinear Magnetoydrodynamics, edited by Cambridge University Press, Cambridge, 1993.
4. G. Boffetta, A. Celani A. and R. Prandi, *Phys. Rev. E, 61*, **4**, pp. 4329–4335, (2000).
5. V. Canuto., M. Y. Hussaini, A. Quarteroni and T. A. Zang,in *S*pectral methods in fluid Mechanics, edited by Springer Verlag, 1988.
6. P. Cargill, *The Astrophysical Journal*, **422**, pp. 381–393, (1994).
7. R. B. Dahlburg and G. Einaudi, *Phys. Let. A*, **294**, pp. 101–107, (2002).
8. R. B. Dahlburg et al., *Adv.Space Res.*, **32**, 6, pp. 1029–1034, (2003).
9. P. Dmitruk, D. O. Gomez and E. E. DeLuca, *The Astrophysical Journal*, **505**, pp. 974–983, (1998).
10. G. Einaudi, M. Velli , H. Politano and A. Pouquet, *The Astrophysical Journal*, **457**, L113–116, (1996).
11. G. Einaudi and M. Velli, *Phys. of Plasmas*, **6**, 11, pp. 4146–4153, (1999).
12. M. Georgoulis, M. Velli and G. Einaudi, *The Astrophysical Journal*, **497**,pp. 957–966, (1998).
13. V. Glukhov, *The Astrophysical Journal*, **469**, pp. 936–943, (1996).
14. T. Gomez , H. Politano and A. Pouquet, *Phys.Fluids*, **11**, 8, pp. 2298–2306, (1999).
15. D. O. Gomez, P. A. Dmitruk and L. J. Milano,*Solar Physics*, **195**, pp. 299-388, (2000).
16. A. B. Hasam, *The Astrophysical Journal*, **348**, pp. 778–780, (1990).
17. D. L. Hendrix and G. Van Hoven, *The Astrophysical Journal*, **467**, pp. 887–893, (1996).
18. J. Heyvaerts and E. R. Priest, *The Astrophysical Journal*, **390**, pp. 297–308, (1992).
19. D. W. Longcope and R. N. Sudan, *The Astrophysical Journal*, **437**, pp. 491–504, (1994).
20. Z. Mikic et al., *The Astrophysical Journal*, **338**, pp. 1148–1157, (1989).
21. J. M. Milano and D. O. Gomez, *The Astrophysical Journal*, **490**, pp. 442–451, (1997).
22. S. A. Orszag and Tsa–Mei Tang, *J. Fluid Mech.*, **90**, pp. 129–143, (1979).
23. E. N. Parker, *The Astrophysical Journal*, **330**, pp.474–479, (1988).
24. E. R. Priest et al., *The Astrophysical Journal*, **539**, pp.1002–1022, (2000).
25. Z. Romeou, M. Velli and G. Einaudi, in *Proccedings of the SOHO 15 Workshop- Coronal Heating*, edited by R. W. Walsh, J. Ireland, D. Danesy, B. Fleck, Paris: European Space Agency, ESA SP-575, 2004, pp.573–579.
26. H. R. Strauss, *Phys. Fluids*, **19**, pp. 134–140, (1976).
27. G. Vekstein and Y. Katsukawa , *The Astrophysical Journal*, **541**, pp. 1096–1103, (2000).

Line Broadening of EUV Lines at the Solar Limb Observed with SUMER/SoHO. Relation to Spicules.

Giannikakis J.[1,2], Tsiropoula G.[2], Tziotziou K.[2], Antonopoulou E.[1] and Doyle J.G.[3]

[1] *Department of Astrophysics, Astronomy and Mechanics, Faculty of Physics, National University of Athens, Panepistimiopolis, GR 15783, Zographos, Athens, Greece.*
[2] *National Observatory of Athens, Institute for Space Applications and Remote Sensing, Lofos Koufos, 15236 Palea Penteli, Greece*
[3] *Armagh Observatory, College Hill, Armagh BT61 9DG, N. Ireland*

Abstract. We study the broadening of EUV line widths as we move from the disk through and off the solar limb. This broadening maximizes at approximately 10 Mm above the limb. We use one raster scan and two time series obtained by SUMER/SoHO to examine a possible correlation between this phenomenon and spicule activity. It appears that spicule presence leads to even wider profiles whereas the appearance of macrospicules extends the broadening to even greater distances above the limb. We suggest that this correlation might imply that spicules are responsible for this disk-to-limb behavior even in regions where they cannot be observed.

Keywords: Sun, spicules, line broadening, solar atmosphere, transition region
PACS: 96.60.-j, 95.55.Ev, 96.60.Xy.

INTRODUCTION

Limb broadening is a solar center-to-limb effect referred to the increase of the profile width of spectral lines above the limb. Broadening of line profiles has been investigated by numerous workers, since it can provide important information concerning ion temperatures, sub-resolution turbulent motions, velocity fluctuations associated with MHD waves and, if coronal lines are used, information on coronal heating may be obtained. Line broadening when observed in solar transition region lines peaks at about 10″-15″ above the limb [1], [2]. The physical origin of this excessive broadening still remains controversial. A possible correlation between this behavior and solar spicules is given by Doyle et al. [3].

Solar spicules are spike-like features crossing the limb. They have a typical length of 10 Mm and width of 1Mm and their lifetime is about 10 mins. Initially a spicule moves upwards but after its expansion phase, where it reaches a maximum height, it either disappears or falls back, not always along the same path. EUV spicules have been studied by several authors [4], [5], [6], [7]. Structures in the polar coronal holes which look like spicules but are significantly longer are called macrospicules [7].

CP848, *Recent Advances in Astronomy and Astrophysics, 7th International Conference of the Hellenic Astronomical Society,* edited by N. H. Solomos

OBSERVATIONAL DATA

The data sets used in this study were obtained by the SUMER spectrometer on-board the SoHO satellite. The specifications and observational capabilities of the SUMER instrument were described by Wilhelm et al. [8], [9] and Lemaire et al. [10]. The first data set was a raster sequence taken on August 10, 1996 from 00:03 to 16:09 UT. While targeting at a constant Solar Y at 950″, the slit moved from Solar X -699″ to +721″ with a step size of 1.5″. Slit 2 (i.e. 1″ x 300″ centered) and detector A were used for taking the 60s exposures. The lines of interest in this dataset were the 1st order N v 1238.82 Å and the 2nd order O v 629.73 Å.

The second and third datasets were time series obtained at the south and north polar coronal holes (PCH) respectively. The time series at the south PCH was taken on February 21, 1997 from 01:20 to 01:36 UT. The targeting of the instrument was at a fixed position with Solar X at 0″ and Solar Y at -984.75″. Slit 3 (i.e. 1″ x 300″ top) and detector B were used for taking the 15s exposures of this 15 min series in the C III 977 Å. The time series at the north PCH was taken on December 13, 1996 and it lasted from 16:32 to 19:22 UT. The observing position for the north PCH was at fixed Solar X = 0″ and Solar Y = 985.12″. The spectral line observed was S VI 933.38 Å, the exposure time 30s, while slit 2 and detector B were used. Details on the data sets are given in Table 1.

TABLE 1. Description of data sets obtained by SUMER.

Item	Date	Time	Solar X , Y	Lines	Exp. Time
Raster scan	10/08/96	00:03 - 16:09	(-699 - 721) , 950	O v 629.73 Å , N v 1238.82 Å	60s
PCH south	21/02/97	01:20 - 01:36	0 , -984.75	C III 977.02 Å	15s
PCH north	13/12/96	16:32 -19:22	0 , 985.12	S vi 933.38 Å	30s

DATA ANALYSIS

Various corrections such as decompression, reversion and also flat-field, dead-time, local-gain and geometrical distortion corrections had to be applied to the data. For these corrections the standard SUMER reduction software was used. Most of the UV and EUV lines originated from the transition region are optically thin so that the Gaussian fitting method to the line profiles is quite efficient. The Gaussian fitting yields parameters such as total intensity, continuum intensity and line width. An automated one line Gaussian fit was applied (Solarsoft routine XCFIT_BLOCK) giving a set of parameters (i.e. line intensity, line center position and FWHM) for each pixel (according to the binning of the data) within a dataset. The behavior of the parameters was analysed in different areas along vertical stripes from the disk towards and off the limb. These stripes are parallel to the slit (i.e. the Solar Y-coordinate).

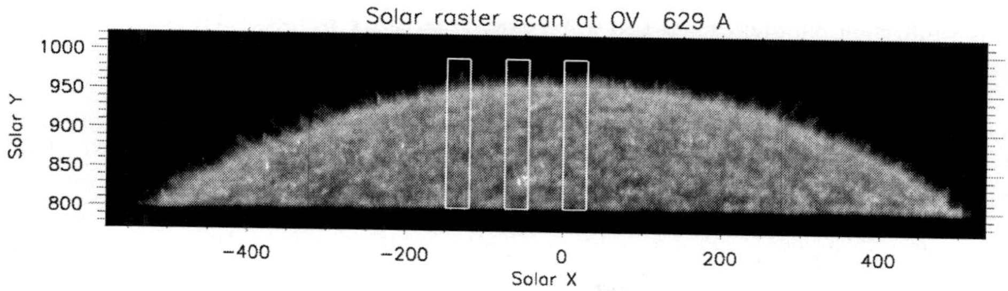

FIGURE 1. A sub-set of the O v 629 Å image obtained on August 10, 1996. The vertical stripes from where the three data plots given in Fig. 2. were produced are also shown. Axes are scaled in arcsec.

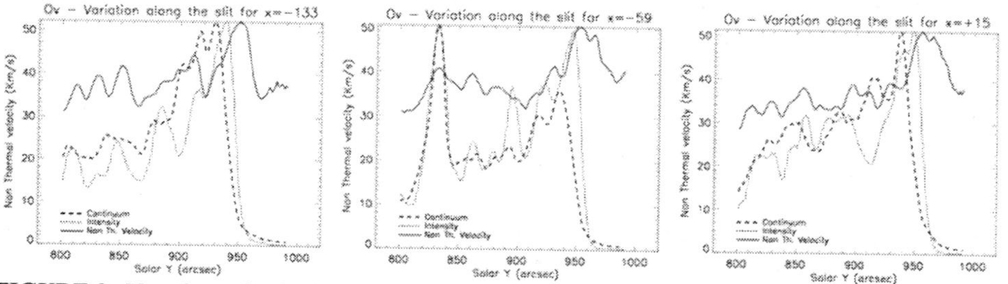

FIGURE 2. Non-thermal velocities derived from O v 629 Å line profile along the three vertical stripes shown in Fig. 1. Line intensity and continuum normalized to the non-thermal velocity range of values are also included.

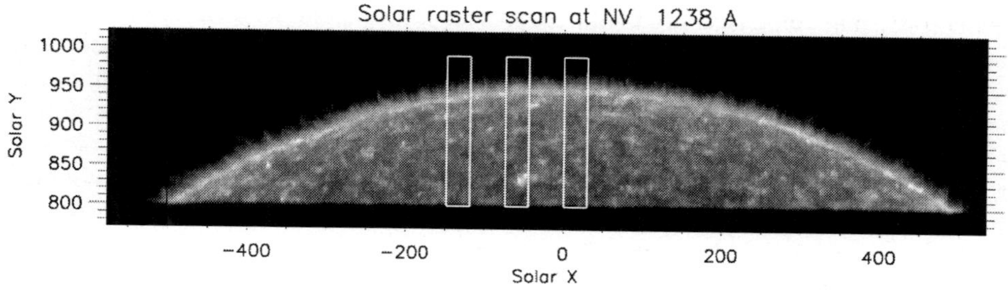

FIGURE 3. A sub-set of the N v 1238 Å image obtained on August 10, 1996. The vertical stripes from where the three data plots given in Fig. 4 were produced are also shown. Axes are scaled in arcsec.

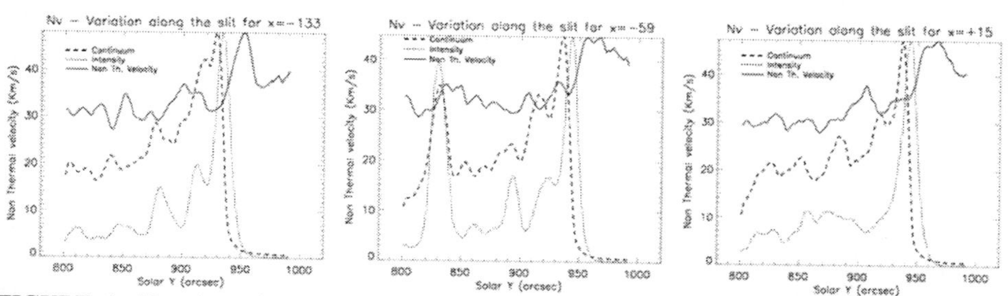

FIGURE 4. Non-thermal velocities derived from N v 1238 Å line profile along the three vertical stripes shown in Fig. 3. Line intensity and continuum normalized to the non-thermal velocity range of values are also included.

A significant parameter that had to be calculated is the non-thermal velocity which was derived from the line width given by:

$$FWHM = \sqrt{(\Delta\lambda_{inst})^2 + 4\ln 2 \frac{\lambda_0^2}{c^2}\left(\frac{2kT}{M} + \xi^2\right)} \qquad (1)$$

where $\Delta\lambda_{inst}$ is the instrumental width, λ_0 is the unshifted wavelength of the line, c is the speed of light, k is the Boltzmann constant, T is the ion temperature, M is the atomic mass and ξ is the non-thermal velocity. The instrumental broadening for SUMER was corrected by using the Solarsoft routine CON_WIDTH_FUNCT_3. Finally a correction for the spectral line shift caused by thermal deformations of the optical bench of SUMER [11] was applied.

Raster Sequence

From the raster sequence three regions were selected The analysis was performed along the vertical stripes at positions -133, -59 and +15 with respect to the central Solar X-coordinate and for both spectral lines (O v and N v) (Fig. 1 and Fig. 3). In order to increase the S/N ratio 21 pixels in the X-direction and 9 pixels in the Y-direction were averaged. In Fig. 2 and Fig. 4 the variations of the intensity, continuum and non-thermal velocity along the three vertical stripes from the disk through and off limb are shown. In order to infer the height variation of the non-thermal velocity and since the true solar limb cannot be determined from the data a reference height has to be defined as a zero point. In this work we defined as a zero point the position at which the continuum of each individual line attains its maximum value. It is obvious that the non-thermal velocity peaks at $\approx 15''$ above the zero point. It is worth noticing that although variations of the non-thermal velocities are observed on the disk the substantial increase is observed off limb and, furthermore, that in most cases the profile is getting narrower before the final broadening. A possible explanation for this increase is that the broadening reflects the summed profile of many spicules found along the line-of-sight. The randomness of the spicules' moving directions keeps the line position unchanged, but it has as a result a wider profile. This is also consistent with the fact that the peak appears at an off-limb point, where we expect to have the maximum number of spicules along the line-of-sight.

Time Series Data at the South PCH

In Fig. 5 (left) the time evolution of the intensity of a region found at the southern coronal hole and observed in the C III 977 Å line is displayed. Although this time series lasts for only $\approx$15 min, an intense spicule activity can be detected, which is more clearly seen in the dopplergram in Fig. 5 (right) as alternating red and blue shifts with respect to time. The Doppler shifts at four locations shown in white dashed lines in the

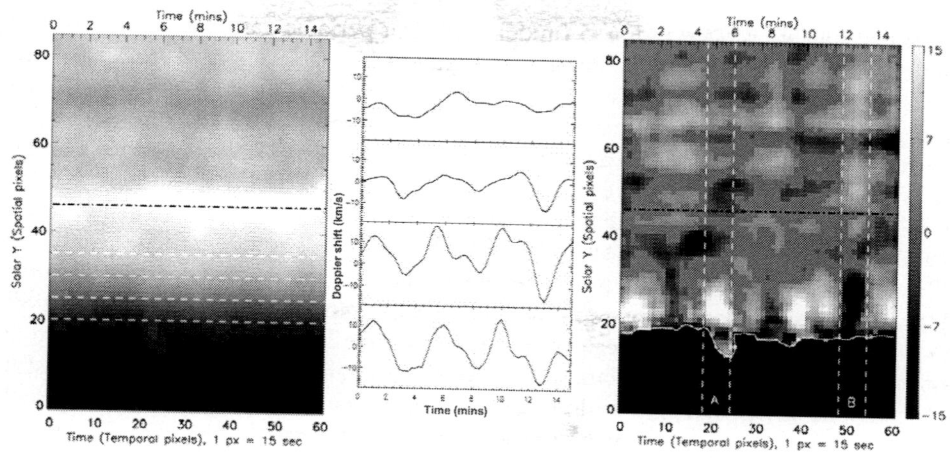

FIGURE 5. LEFT: The C III 977 Å time series set obtained on February 21, 1997 showing the limb (black dashed/dotted line) as seen in the continuum short-ward of the spectral line. Y-axis represents slit spatial pixels and it is scaled in arcsec. The four lines (white dashed) show the positions in which the Doppler shift plots on the middle were obtained. The evolution of these Doppler shifts indicates an intense spicule activity, which can also be seen in the dopplergram (right).
MIDDLE : Time variability of Doppler shift at the four locations shown in the intensity plot (left).
RIGHT: Dopplergram of the 21 February 1997 time series in the C III 977 Å line. The bar on the right represents Doppler shifts in Km/s showing that redshifts are drawn in tones of white and blushifts in tones of black while gray areas represent areas having very small or zero value. The white solid line separates the useful data area from the area beneath which contains very poor count rate and thus it cannot be processed. The two regions of interest are marked using the vertical white dashed lines.

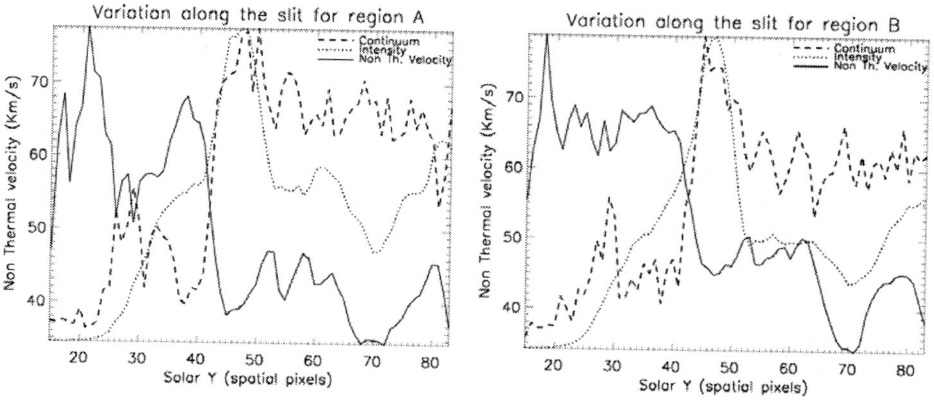

FIGURE 6. Non-thermal velocities for regions A & B shown in Fig. 5 (right). Line intensity and continuum normalized to the non-thermal velocity range of values are also included.

intensity map are plotted in Fig. 5 (middle). A clear periodic variation of the Doppler shift is seen. The period is present at all heights and is ≈ 5 min. This value has been recently reported for Hα spicules observed on the disk [12]. Two regions were selected here, shown in the dopplergram image of Fig. 5: region A from 4.5 to 6 mins corresponding to a red-shifted area and region B from 12 to 13.5 mins corresponding to a blue-shifted area. As we can see, in both regions the non-thermal velocity also peaks a few arcsecs above the line continuum limb (≈10″). In this case, where we have an obvious spicule activity, the non-thermal velocity keeps it's high values for about 20″ arcsecs. Especially in region A, where material is present in a greater distance from the limb, a second increase in the non-thermal velocity can be observed. Notice, in both regions, that even though we are far above the limb at heights where there is no evidence for a significant plasma presence (very small intensity values) we can still observe line shifts and line width broadenings. This is consistent with the typical characteristics of spicules, which cannot contribute significantly to the total intensity but they present a behavior which can contribute to the broadening of a line's width.

Time Series Data at the North PCH

In Fig 7 and 8 the intensity and Doppler shift variation of a long duration data set obtained in the S VI 933 Å line in the northern coronal hole is displayed. The height reference point, as defined in the raster scan paragraph, is marked by a dashed-dotted line in Fig. 7. In Fig. 8 time periods with less intense (regions A and B) and more intense (region C) spicule activity, as well as a time period with an elongated spicule-like feature (region D), maybe a macrospicule, are marked by vertical lines. The non-thermal velocity values (Fig. 9) show the same behavior: they increase above the continuum limb peaking at ≈ 15″ above it and then decrease after approximately 7″-8″. This behavior is more evident in the non-thermal velocity plot for region C, where we have a clear spicule activity but also in the plots for regions A & B

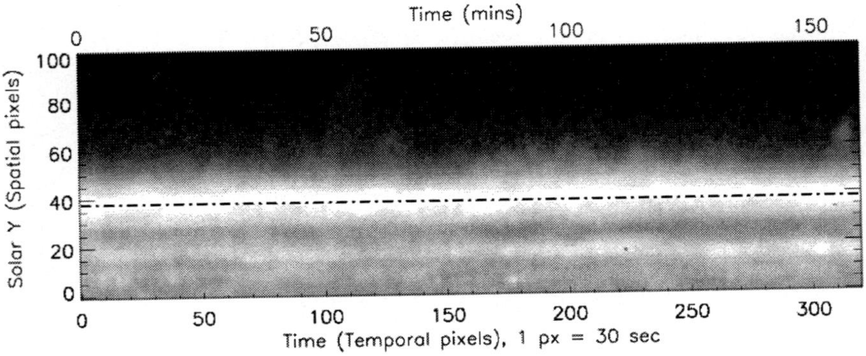

FIGURE 7. The S VI 933 Å time series set obtained on 13 December 1996 showing the limb (black dashed/dotted line) as seen in the continuum short-ward of the spectral line. Y-axis represents slit spatial pixels and it is scaled in arcsec.

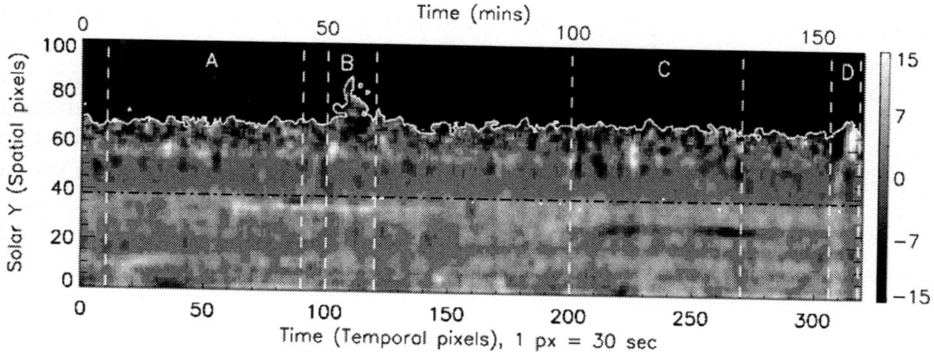

FIGURE 8. Dopplergram of the 13 December 1996 time series in the S VI 933 Å line. The bar on the right represents Doppler shifts in Km/s showing that redshifts are drawn in tones of white and blushifts in tones of black while gray areas represent areas having very small or zero value. The white solid line separates the useful data area from the above area which contains very poor count rate and thus it cannot be processed. The four regions of interest are marked using the vertical white dashed lines.

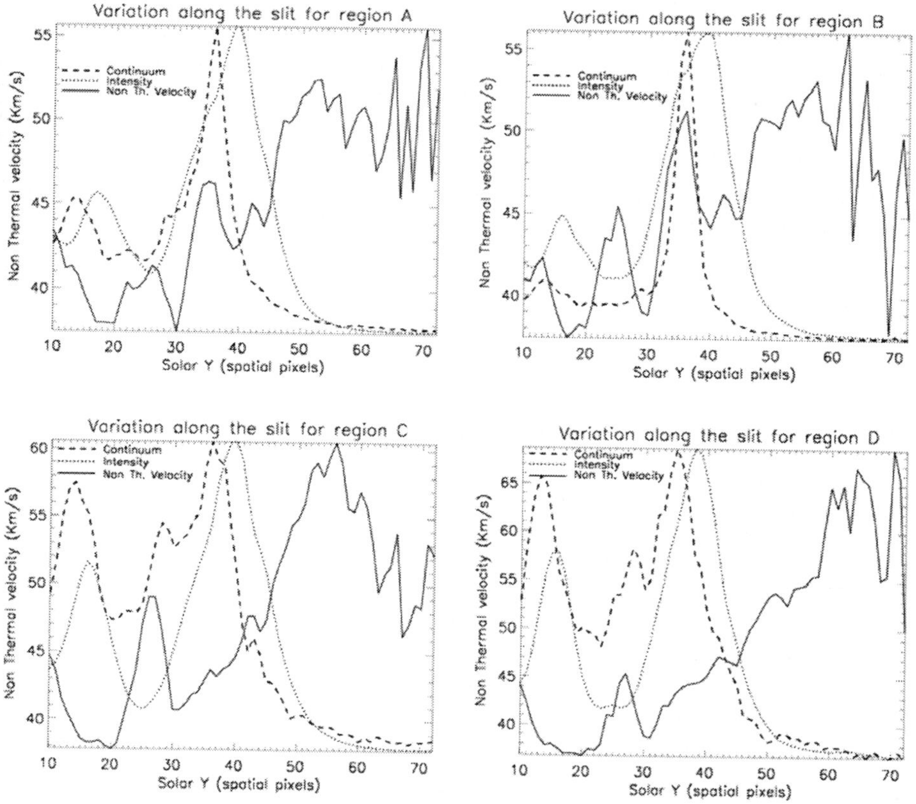

Figure 9. Non-thermal velocities derived from N V 1238 Å line for regions A, B, C & D shown in Fig. 8. Line intensity and continuum normalized to the non-thermal velocity range of values are also included.

where after the decrease point we can see only noise, thus we have no significant presence of a spectral line to be fitted properly. Notice that in region C the values of the non-thermal velocity attain a larger peak value giving an indication that the broadening is connected with spicules. Region D presents the most interesting behavior. The non-thermal velocity increases similarly to the other three regions up until the usual peak at 15″ above the limb. It also takes approximately the same values of 55 – 60 Km/s, depending on the spicule activity as we've discussed, that appear in the other three regions. But here it shows an additional increase right where the macrospicule appears.

CONCLUSIONS

We have seen that the limb broadening effect is clearly related to the presence of spicules and macrospicules. The phenomenon seems to be more intense when the spicule activity increases. Also it seems to follow the spatial existence of spicules and macrospicules with respect to the distance from the limb.

The possible explanations of the limb broadening in terms of spicules could be (a) the random directions of many spicules and/or (b) the non-thermal motions within spicules. According to (a), we have many spicules along the line-of-sight aligned in random directions. As we've already discussed the LOS projections of these plasma velocities can give a wide profile as a result. On the other hand, there are non-thermal motions within each spicule and macrospicule. Studies exist even for rotational motions within macrospicules [13]. Region D in Fig. 8 could present such a behavior. Assumption (b) might explain the limb broadening effect as a sum of these non-thermal motions.

REFERENCES

1. J. T. Mariska, U. Feldman, G. A. Doschek, *Astronomy and Astrophysics* **73**, 361-363 (1979).
2. H. Peter, C. Vocks, *Astronomy and Astrophysics* **411**, L481-L485 (2003).
3. J. G. Doyle, J. Giannikakis, L. D. Xia, M. S. Madjarska, *Astronomy and Astrophysics* **431**, L17-L20 (2005).
4. K. P. Dere, J.-D. F. Bartoe, G. E. Brueckner, *ApJ* **267**, L65-L68 (1983).
5. K. P. Dere, J.-D. F. Bartoe, G. E. Brueckner, F. Recely, *ApJ* **345**, L95-L97 (1989).
6. K. Wilhelm, I. E. Dammasch, E. Marsch, D. M. Hassler, *Astronomy and Astrophysics* **353**, 749-756 (2000).
7. K. Wilhelm, *Astronomy and Astrophysics* **360**, 351-362 (2000).
8. K. Wilhelm, W. Curdt, E. Marsch, U. Schuhle, P. Lemaire, A. Gabriel, J.-C. Vial, M. Grewing, M. C. E. Huber, S. D. Jordan, A. I. Poland, R. J. Thomas, M. Kuhne, J. G. Timothy, D. M. Hassler, O. H. W. Siegmund, *Sol. Phys.* **162**, 189-231 (1995).
9. K. Wilhelm, P. Lemaire, W. Curdt, U. Schuhle, E. Marsch, A. I. Poland, S. D. Jordan, R. J. Thomas, D. M. Hassler, M. C. E. Huber, J.-C. Vial, M. Kuhne, O. H. W. Siegmund, A. Gabriel, J. G. Timothy, M. Grewing, U. Feldman, J. Hollandt, P. Brekke, *Sol. Phys.* **170**, 75-104 (1997).
10. P. Lemaire, K. Wilhelm, W. Curdt, U. Schule, E. Marsch, A. I. Poland, S. D. Jordan, R. J. Thomas, D. M. Hassler, J. C. Vial, M. Kuhne, M. C. E. Huber, O. H. W. Siegmund, A. Gabriel, J. G. Timothy, M. Grewing, *Sol. Phys.* **170**, 105-122 (1997).
11. I. E. Dammasch, K. Wilhelm, W. Curdt, D. M. Hassler, *Astronomy and Astrophysics* **346**, 285-294 (1999).
12. K. Tziotziou, G. Tsiropoula, P. Mein, *Astronomy and Astrophysics* **423**, 1133 (2004).
13. C. D. Pike and H. E. Mason, *Sol. Phys.* **182**, 333-348 (1998)

Electron Spatial Distribution in Microwave Flaring Loops

Victor F. Melnikov*, Veronika E. Reznikova*, Sergei P. Gorbikov* and Kiyoto Shibasaki†

*Radiophysical Research Institute (NIRFI), Nizhny Novgorod, 603950, Russia
†National Astronomical Observatory of Japan, Minamimaki, Minamisaku, Nagano 384-1305, Japan

Abstract. In this paper we review recent studies of Nobeyama Radioheliograph observations concerning 1) spatial distribution of microwave brightness along flaring loops; 2) peculiarities ofits temporal dynamics in different parts of a loop; 3) consequences of the obtained findings on spatial and pitch-angle distributions of high energy electrons.

Keywords: the Sun: flares, microwave emission - particle acceleration and transport
PACS: 96.60.qe

1. INTRODUCTION

Spatial distribution of nonthermal emission along a single flare loop can bring us important information on the particle acceleration and transport. Studies of the microwave spatial distribution began in 80s using the WSRT and VLA facilities. The discovery of two kinds of microwave sources, single compact loop-top sources and double sources with their peaks located close to the conjugate magnetic footpoints [1-4] stimulated the development of theoretical simulations of gyrosynchrotron brightness distribution along model magnetic loops (see [5-7]). Later the similar studies were carried out using the Nobeyama Radioheliograph (NoRH) observations at single frequency 17 GHz [8-10]. They showed the existence of the double and triple radio sources associated with the footpoints of large and smaller flaring loops.

In recent years the new NoRH facilities have allowed to obtain images of solar flares at two high frequencies, 17 and 34 GHz, simultaneously with high angular ($5'' - 10''$) and temporal (0.1 s) resolution. This gives a chance to study the brightness distribution having the information on microwave spectrum slope, and, therefore, on the optical thickness in different parts of a flaring loop. Such high capabilities provide us with unique opportunity to get constraints on properties of mildly relativistic electrons accelerated and propagating in flaring magnetic loops. Already the first reports on these observations [11, 12] gave quite unexpected results showing evident disagreement between the observable and simulated brightness distributions of optically thin microwave emission. Several new studies were published during last years developing further understanding of the problem [13-20].

This paper is restricted to properties of single well resolved loop-like microwave sources. We review mainly our recent studies of Nobeyama observations concerning 1) spatial distribution of microwave brightness along flaring loops; 2) peculiarities of its

CP848, *Recent Advances in Astronomy and Astrophysics,*
7th International Conference of the Hellenic Astronomical Society, edited by N. H. Solomos

temporal dynamics in different parts of a loop; 3) consequences of the obtained findings on spatial and pitch-angle distributions of high energy electrons. We show that detailed comparative analysis of the microwave brightness distribution along single magnetic loops using the new NoRH observations and model simulations gives us some new constraints on the acceleration/injection mechanisms and kinetics of mildly relativistic electrons in solar flares.

2. BRIGHTNESS DISTRIBUTION ALONG EXTENDED FLARING LOOPS

The existence of the double and triple radio sources associated with the footpoints of flaring loops, as well as asymmetric sources, shown in [8-10], agrees well with the existing theoretical models of radio emission [5-7]. The obvious reason of such the location is strong dependence of gyrosynchrotron intensity on magnetic field strength. On the other hand, the location of the brightness peak of optically thin microwave emission in the looptop is a challenge to our understanding of processes in flaring loops. We will consider this problem in detail.

For the detailed study of the spatial properties of microwave emission from a flare loop we selected only flares, which are characterized by the presence of well resolved extended single microwave sources and relatively simple time profiles both at 17 and 34 GHz. We have chosen the flares occured on 1999 August 28, 2000 January 12, March 13, 2001 October 23, and 2002 August 24 [14]. Their sizes ($25'' - 90''$) exceed considerably the beam size of the Nobeyama Radioheliograph (NoRH) ($\Delta\varphi \sim 5''$ and $\sim 10''$ at 34 GHz and 17 GHz, respectively). The 1st, 3d and 4th flares occurred on the solar disk, the 2nd one - near the East limb. The projections of the radio sources on the disk display a well developed loop-like structure at both frequencies.

All the radio sources are optically thin at least at 34 GHz. Indeed, as analysis shows, the flux density spectral slope in the range 17-34 GHz is negative during the impulsive phase of the microwave bursts.

The spatial profiles of the brightness along the loop-like radio sources for each event are shown in the Fig.1. We can clearly see the brightness maxima to be located close to the central part of the sources. This is especially well pronounced for the highest frequency, 34 GHz, where the sources must be optically thin.

Our analysis of MDI magnetographic data has shown that this loop-top source is not associated with any photospheric features (spots, magnetic field enhancements etc.) under the loop top. At the same time the MDI maps display the magnetic field enhancements of opposite polarities near two ends of the loop-like structures (supposed to be the indications of the footpoints). Note that the similar brightness distributions at the peak time of three limb flares observed with NoRH were obtained by Kundu et al. [13]. This proves that such kind of the brightness distribution for extended flaring loops is quite common and, therefore, should be studied in detail.

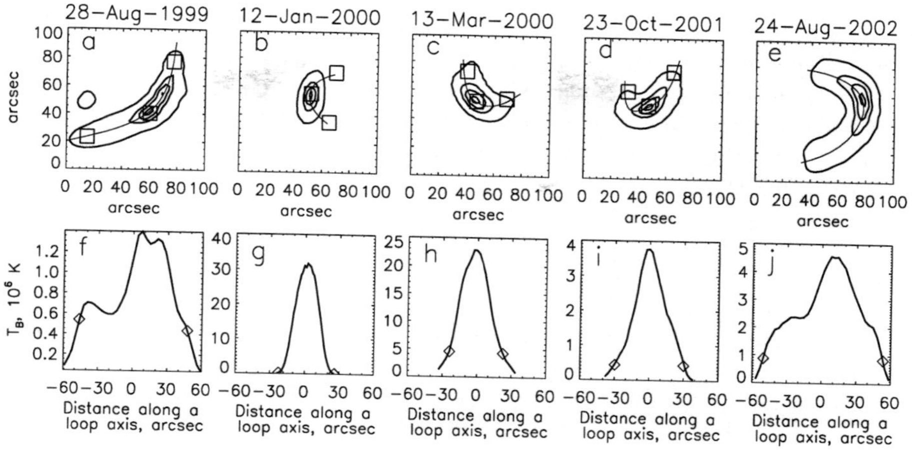

FIGURE 1. *Top panel*: 34 GHz contour images of the sources at the moment of the bursts peaks (except e, this image is for the rise phase) for five events under study. Contours show 0.1, 0.5, 0.75 and 0.95 levels of the maximum brightness temperature. Thin curve shows the visible flaring loop axis. Three boxes of the size $10'' \times 10''$, two near footpoints and one near the loop-top show the areas which the flux time profiles were calculated from. *Bottom panel*: spatial distributions of radio brightness temperature at 34 GHz along a visible flaring loop axes shown on top panel for each event by thin curve. Zero position on the x-axes of bottom panel corresponds to the middle of the loop axis, calculated as a half of the axis length. Negative values correspond to left footpoint on fig. a, c, d and lower footpoint on fig. b, e.

3. DYNAMICS OF BRIGHTNESS DISTRIBUTION AND TIME DELAYS

Further analysis of the microwave brightness distribution along the five well resolved flaring loops shows that it does not remain constant during a single peak of a burst [15]. In almost all the analyzed events there is the same tendency (see, for example, Fig.2, left panel). In the beginning of a burst there is always a time interval when brightness peaks (or one peak) are situated near the footpoints of a loop (both or only one). And close to the burst maximum or on the decay phase, an area near the looptop becomes most bright. A similar dynamics of brightness distribution was observed and studied in detail in [16] for the event 1999 May 29.

To show this tendency more quantitatively, in Fig. 2, right panel, we compare the time profiles of fluxes and their ratios from the 10"x10" boxes, located near the left footpoint (*a*), right footpoint (*c*) and in the loop top (*b*) of the flaring loop in the event 2000 March 13. The location of the boxes are shown on Fig.1(*c*). The *dotted* and *solid* lines correspond to the 17 and 34 GHz emission, respectively.

It is well seen that the burst of emission from the loop top is delayed against the burst from the footpoint source by several seconds. This delay is more pronounced at 34 GHz than at 17 GHz. Furthermore, it is clearly seen that the time profiles of emission from the loop top are wider and their decay is slower than those from the region near the footpoint.

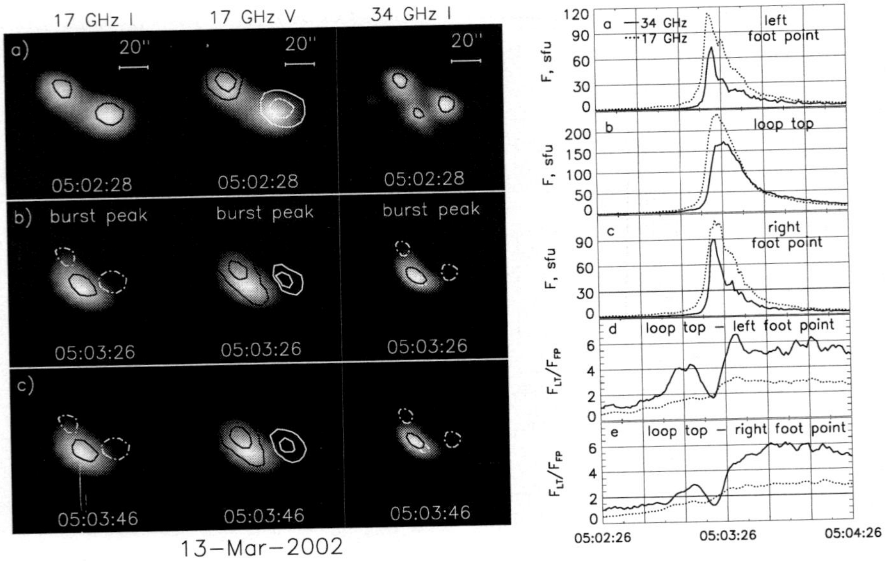

FIGURE 2. *Left*: NoRH images at the rise and peak phases of the flare 2000 March 13. Left and right panels: intensity at 17 and 34 GHz with the contours at the level 95% of the maximum at the current time (black solid line) and at the moment 05:02:28 UT (white dashed-dotted line). Middle panel: polarization V with the contours at the levels 40% and 80% of the maximum, black for positive V, white for negative V. *Right*: NoRH time profiles of the flux densities from the 10"x10" boxes, located at the left footpoint (*a*), right footpoint (*c*) and in the loop top (*b*) of the flare 2000 March 13. The flux densities ratio F_{LT}/F_{FP} from the loop-top and the footpoint regions is calculated for the left footpoint (*d*) and for the right one (*e*).

The similar delays and differences in the characteristic decay time are observed for all the events from our sample. Note that for all the events the flux peaks from the footpoint sources are coincident in time with the flux peaks of hard X-ray emission [15].

A comparison of the time profiles at the frequencies 17 and 34 GHz from the same parts of a loop-like radio source shows that the emission at higher frequency, 34 GHz, is delayed against that at 17 GHz. This is well pronounced for the loop-top part of the sources. However, there are no such delays for the footpoint parts.

The flux densities ratio F_{LT}/F_{FP} from the loop-top and the footpoint regions is calculated in Fig. 2, right panel, for the left footpoint (*d*) and for the right one (*e*). It is well seen that the ratios increase during the burst, especially on the decay phase where maximum values of ratios are 3-4 in the early decay phase.

Quite different evolution of the brightness distribution is realized during the limb event 2002 August 24. During the main peak of the burst (see Fig. 3), on its rising phase from 00:59:00, the southern footpoint of the loop becomes most bright at 34 GHz and remains the brightest part of the loop till the maximum at 01:00:30. At the peak time the brightness temperature reaches extremely high value at 34 GHz: $T_{Bmax} \approx 250$ MK. In Fig. 3*a*, *b* we can also see two other brightness peaks: one near the opposite (nothern) footpoint and one near the looptop. But they are much weaker. Only on the decay phase the looptop becomes relatively brighter then the footpoint sources which almost disappear to the moment of the valley before the second peak on the time profile

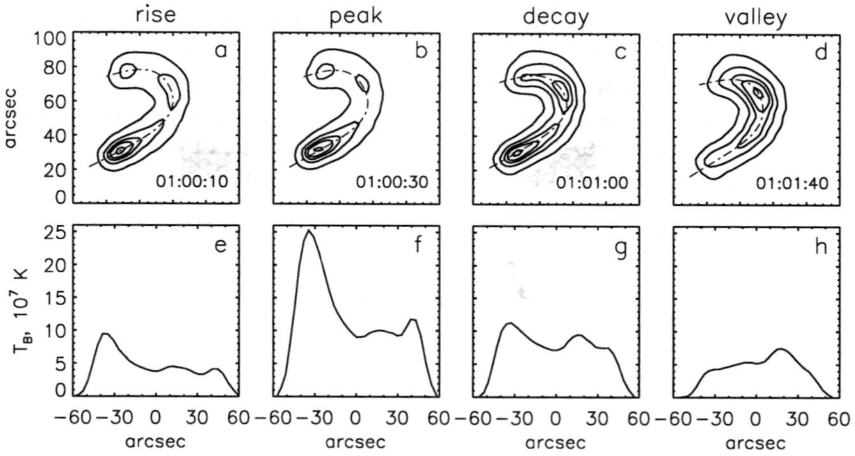

FIGURE 3. Evolution of the brightness distribution during the limb event 2002 August 24.

(Fig. 3*c*, *d*). During the decay the brightness temperature in the southern footpoint source decreases by 5 times, and in the looptop only by 20%.

It is worth to note that during the main peak, when the total flux changes by 20-30 times, the visible size of the microwave flaring loop has almost not changed. It means that the emission flux changes entirely due to changes in number of mildly relativistic electrons in the loop but not in the area or volume of the source. More or less similar situation was observed for other flares from our set.

4. COMPARISON WITH MODEL PREDICTIONS

The most interesting feature of all the flares is that the brightness peak of optically thin microwave emission at the burst maxima is located near the loop top. This fact disagrees with the well known model simulations [5], and [6] that predict the brightness peaks of optically thin emission to be near the footpoints of extended loops with a nonuniform magnetic field. The increase of the brightness temperature occurs because of strong dependence of gyrosynchrotron intensity, I_f, on the magnetic field strength, B, that is expected to be larger near the feet of an extended loop. For example, if the electron power law spectral index $\delta = 4$, then according to [21]

$$I_f \propto NB^{3.4}(\sin\theta)^{2.2}, \qquad (1)$$

where N is the number of nonthermal electrons, θ is the angle between the magnetic vector and the line of sight.

The striking fact is that the brightness peak at the looptop is also observed for limb flares. For a limb loop which plane is almost perpendicular to the line of sight, $\theta = 80 - 90^o$, like for the events 2000 January 12 and 2002 August 24 from our set, the ratio of brightness temperatures in the loop top, I_{fLT}, and at the peak near footpoints,

I_{fFP}, is expected to be $I_{fLT}/I_{fFP} \sim 0.1-0.2$. As can be seen in Fig.1 (and also from the analysis of the limb events [13, 16]) the observable ratios at the highest frequency 34 GHz are just opposite: $I_{fLT}/I_{fFP} \sim 3-20$.

In the frame of the existing models such a strong increase of the brightness near the loop top is not possible to explain with any reasonable set of parameters. The only opportunity to get the loop top brighter than the footpoints in the framework of the mentioned models is the assumption of almost constant magnetic field along the flaring loops [22]. However, this may work only for loops near the solar disk center where the angle between magnetic field and line of sight is small near the footpoints. For loops close to the limb, even under this assumption, quite unlikely for extended loops, the increase of brightness in the loop top has been shown to be too small (< 2) to explain the observable ratios I_{fLT}/I_{fFP} and the shape of spatial profile along a loop [13, 16].

We have to notice that the existing model simulations were made for the homogeneous distribution of energetic electrons along a loop, which can strongly deviate from the reality. Indeed, the existence of delays between intensity time profiles at the higher and lower frequencies from the loop top source and between time profiles from the loop-top and the footpoints with no noticeable delays between the emissions from the footpoints is a strong evidence for the importance of trapping and accumulation of nonthermal electrons in the looptop part of flaring loops.

A strong relative enhancement of energetic electron density near the loop top can eliminate the contradiction between the observations and the model predictions [11, 14]. Under this assumption we can easily derive the magnitude of the electron number density enhancement from Eq.(1) and the observed ratios I_{fLT}/I_{fFP}. The corresponding difference in the number densities in the loop top and footpoints falls in the range $N_{LT}/N_{FP} \sim 10-100$.

To illustrate this idea, Fig.4 displays three types of normalized spatial distributions along a loop [14]. On the left hand the number density (dashed line) and column number density (solid line) electron distributions, and magnetic field distribution (dotted line) are shown. These types of electron spatial distributions are obtained in the collisionless case when the first adiabatic invariant constancy along a magnetic field line with nonuniform strength $B(s)$ is fulfilled: $[1-\mu^2(s)]/B(s) = const$, and the corresponding electron pitch-angle distributions in the loop center are taken to be (a) beam-like $\phi(\mu) = exp[-(1-\mu^2)/\mu_{\|}^2]$, with the beam width along magnetic field $\mu_{\|} = 0.3$, (b) isotropic (with taking into account the loss cone), and (c) pancake-like $\phi(\mu) = exp[-\mu^2/\mu_{\perp}^2]$, with the beam width transverse to magnetic field $\mu_{\perp} = 0.4$, where $\mu = cos\alpha_0$, α_0 - pitch-angle in the loop center.

The corresponding gyrosynchrotron intensity distributions along the model loop in optically thin regime (at 34 GHz) are shown on the right-hand panels of Fig. 4. The emission is calculated using the exact formalism by [23, 18]. It is supposed that electrons have a power law energy dependence: $g(E) = g_0 E^{-\delta}$; the model loop is located close to the limb and the angle between the magnetic field vector and line of sight $\theta = 80^o$. Two strong peaks of radio emission are observed even for the isotropic injection while the well defined peak of the number density exists in the loop center (see Fig.4b,e). Only in the case of the pancake distribution we can get a pronounced radio brightness peak at the center of a limb loop (Fig.4c,f). A degree of brightness concentration to the loop

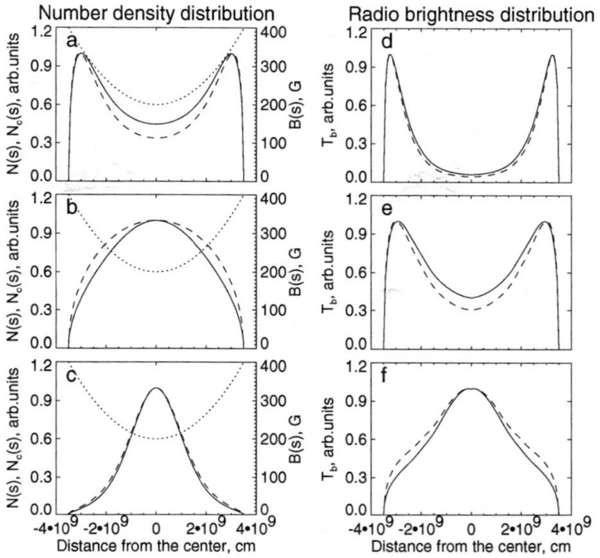

FIGURE 4. Normalized spatial distributions along a loop [14]. Left column: number density (dashed line) and column number density (solid line) for the beam-like (a), isotropic (b), and pancake (c) electron pitch-angle distributions; magnetic field distribution is shown by dotted line. Right column: corresponding gyrosynchrotron intensity in optically thin regime (at 34 GHz).

center (the ratio I_{fLT}/I_{fFP}) can vary in wide range depending on the parameter $\mu_\perp$ of the Gaussian distribution function.

To understand the origin of peculiarities of microwave brightness distribution and its dynamics along flaring loops we can do modeling of the time evolution of electron spatial distribution along a magnetic loop by solving the non-stationary Fokker-Planck equation under different assumptions on the physical conditions in the loop and for different positions of the injection site (loop top, loop foot).

We consider the non-stationary Fokker-Planck equation in the form [24], which takes into account Coulomb collisions and magnetic mirroring:

$$\frac{\partial f}{\partial t} = -c\beta\mu\frac{\partial f}{\partial s} + c\beta\frac{d\,lnB}{ds}\frac{\partial}{\partial\mu}\left[\frac{1-\mu^2}{2}f\right]$$

$$+\frac{c}{\lambda_0}\frac{\partial}{\partial E}\left(\frac{f}{\beta}\right) + \frac{c}{\lambda_0\beta^3\gamma^2}\frac{\partial}{\partial\mu}\left[\left(1-\mu^2\right)\frac{\partial f}{\partial\mu}\right] + S, \qquad (2)$$

where $f = f(E,\mu,s,t)$ is electron distribution function of kinetic energy $E = \gamma - 1$ (in units of mc^2), pitch-angle cosine $\mu = cos\ \alpha$, distance from the flaring loop center s, and time t, $S = S(E,\mu,s,t)$ is injection rate, $\beta = v/c$, v and c are electron velocity and speed of light, $\gamma = 1/\sqrt{1-\beta^2}$ is Lorentz factor, $B = B(s)$ is magnetic field distribution along the loop, $\lambda_0 = 10^{24}/n(s)ln\Lambda$, $n(s)$ is plasma density distribution, $ln\Lambda$ is Coulomb logarithm.

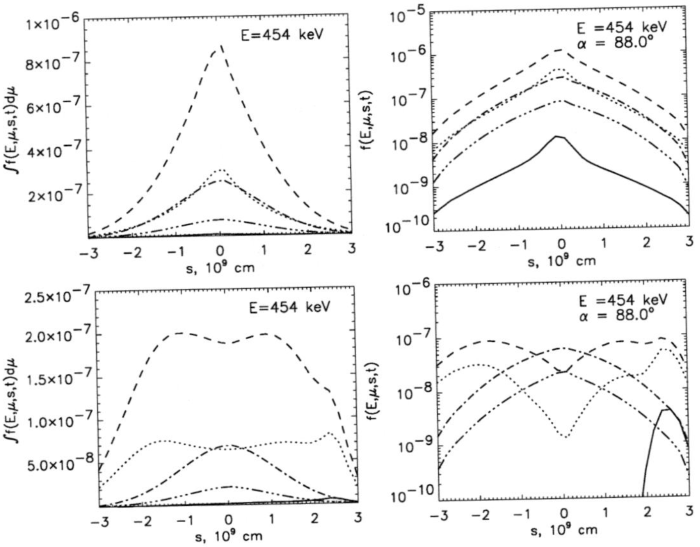

FIGURE 5. Results of the model simulations using the Fokker-Planck equation [25]. The distribution functions for the rising phase of injection are shown by solid ($t = 0.2$ s), dotted ($t = 1.7$ s) and dashed ($t = 3.25$ s) lines, and for the decay phase by dot-dashed ($t = 7.1$ s) and dot-dot-dot-dashed ($t = 10$ s) lines. Top panel represents the time evolution for Model 1 (the maximum of injection function is located in the center of the trap, $s = 0$), and bottom panel for Model 2 (the maximum of injection function is located near the foot, $s = 2.4 \times 10^9$ cm). On the left: the electron distribution function is integrated over all μ and displayed in linear scale. On the right: the distribution function is shown only for electrons with pitch-angle $\alpha = 88^o$.

We conducted our numerical calculations for two models using the method developed by [25]. In the first model the source of high energy electrons is located in the magnetic trap center $s = 0$, and in the second one near the trap foot $s = 2.4 \times 10^9$ cm. In both models the trap (loop) is symmetrical and its half-length is 3×10^9 cm and magnetic mirror ratio $B_{max}/B_{min} = 5$. Plasma density is homogeneous along the loop with $n(s) = 2.5 \times 10^{10}$ cm^{-3}. The injection function $S(E, \mu, s, t)$ is supposed to be a product of functions dependent only on one variable (energy E, cosine of pitch-angle μ, position s, and time t):

$$S(E, \mu, s, t) = S_1(E) S_2(\mu) S_3(s) S_4(t), \tag{3}$$

where the energy dependence is a power law $S_1(E) = (E/E_{min})^{-\delta}$, $E_{min} = 30$ keV, with spectral index $\delta = 5$; pitch-angle distribution is isotropic $S_2(\mu) = 1$; time dependence is Gaussian $S_4(t) = exp[-(t - t_m)^2/t_0^2]$, $t_m = 2.5$ s, $t_0 = 1.4$ s; spatial distribution is also Gaussian. For the first model: $S_3(s) = exp(-s^2/s_0^2)$, and for the second model: $S_3(s) = exp[-(s - s_1)^2/s_0^2]$, where $s_0 = 3 \times 10^8$ cm, $s_1 = 2.4 \times 10^9$ cm.

The results of the model simulations are presented in Fig. 5 where one can see the time evolution of electron number density along the loop for electrons with energy $E = 454$ kev.

It follows from Fig. 5 that for the case of Model 1 (top panel) the shape of the spatial distribution does not change much with time. During all the injection phases the distribution remains similar, with the peak located at the looptop. The simulation also reveals strong perpendicular anisotropy of the electron pitch-angle distribution in the loop. The degree of electron concentration near the looptop is strong enough to produce a peak of the microwave brightness distribution (see Fig. 4).

In the case of Model 2 (lower panel), the shape of the spatial distribution changes with time dramatically. Such distributions should produce two well pronounced radio brightness peaks near the footpoints in the beginning and maximum phases of a burst, and the brightness peak at the loop top in the decay phase of the burst.

5. DISCUSSION AND CONCLUSION

Recent Nobeyama Radioheliograph observations reveal some properties of microwave flaring loops which put new constraints on the particle acceleration/injection and kinetics of high energy electrons in flaring magnetic loops.

It is found that electron distributions along flaring loops may considerably differ from flare to flare and during a single flare. The type of the distributions strongly depends on the location where accelerated electrons are injected in a loop.

It follows from our analysis that the looptop peak of the electron distribution on the maximum phase of a burst may be produced if the injection region is located near the loop top. The injected pitch-angle distribution should be isotropic or anisotropic of the loss-cone ("pancake") type. Such kind of injection is considered in some acceleration/injection models [27-29]. Recently some authors [19, 29, 30] developed models which consider a magnetic trap collapsing after reconnection and providing conditions for the betatron acceleration of electrons transverse to magnetic field lines. On the other hand, there are models suggesting the DC-electric field acceleration along the axes of a twisted loop or the stochastic MHD-turbulent cascade acceleration, in which the acceleration occurs along the magnetic field lines (see for a review [31, 32]). Our simulations show that this type of injection can not produce microwave brightness peak in the looptop during the main phase of injection. However, we can not exclude the chance of the looptop source formation due to pitch-angle scattering and accumulation of trapped electrons in the top part of a loop on the decay phase.

The results of section 4 show that microwave brightness peaks near the footpoints of a loop is easy to get if the injection region is located near one of the footpoints. This conclusion is quite general, simulations reveal that it works for any pitch-angle distribution of injected electrons. It means that the acceleration mechanism itself is not very important, the main thing here is the position of the acceleration/injection region. We can conclude that in cases when microwave flaring loops show brightness maxima near footpoints (like for the event 24 August 2002 or the events observed by [8-10], the acceleration may take place in the loop leg close to a footpoint. We have to note, however, that this explanation is not unique. The footpoint brightenings could be observed even if the acceleration takes place in the looptop. But in this case the injection should occur preferably along the loop axes ("beam-like" injection). To distinguish between the two possibilities we have to involve some other observational properties of the microwave flaring loop. For this some further detailed investigations are needed. A more rigid set of constraints can be obtained from a study of spatial/spectral dynamics of microwave/HXR-emissions.

ACKNOWLEDGMENTS

The work was partly supported by the RFBR grants No.04-02-39029, 04-02-16753. V.M. and V.R. thank the National Astronomical Observatory of Japan for the support of they visit to NRO. Authors are also grateful to Drs. H.Nakajima and Y.Hanaoka for providing the corrected microwave data on the event of 2002, August 24 for use in this publication.

REFERENCES

1. Marsh, K.A., and Hurford, G.J.*ApJ*, **240**, p. L111 (1980)
2. Kundu, M., Schmall, E.J., & Velusamy, T. *ApJ*, **253**, p.963 (1982)
3. Kawabata, K, et al. , *"Proc. Hinotori Symp. on Solar Flares", (Tokyo: ISAS)*, p.168 (1982)
4. Nakajima, H. , *Sol. Phys.*, **86**, p.427 (1983)
5. Alissandrakis, C.E., and Preka-Papadema, P., *A&A*, **139**, 507 (1984)
6. Klein, K.-L., and Trottet, G., *A&A*, **141**, 67 (1984)
7. Bastian, T.S., Benz, A.O, & Gary, D.E., *Ann. Rev. Astron. Astrophys.*, **36**, 131 (1998)
8. Hanaoka, Y., *Sol. Phys.*, **165**, 275 (1996)
9. Nishio, M., Yaji, K., Kosugi, T., Nakajima, H., and Sakurai, T., *ApJ*, **489**, 976 (1997)
10. Hanaoka, Y. , *Pabl. Astron. Soc. Japan*, **51**, 483 (1999)
11. Melnikov, V.F., Shibasaki, K., Yokoyama, T., Nakajima, H., and Reznikova, V.E., *Abstracts of the CESRA workshop, July 2-6, 2001, Munich, Germany* (2001)
12. Nindos, A., Kundu, M. and White, S., *Abstracts of the CESRA workshop, July 2-6, 2001, Munich, Germany* (2001)
13. Kundu, M.R., Nindos, A., White, S.M., and Grechnev, V.V., *ApJ*, **557**, 880 (2001)
14. Melnikov V.F., Shibasaki K. & Reznikova V.E., *ApJ*, **580**, L185 (2002)
15. Melnikov, V.F., Yokoyama, T., Shibasaki, K., and Reznikova, V.E., *ESA SP-506*, 339 (2002)
16. White, S., Kundu M., Garaimov V., Yokoyama, T., & Sato J., *ApJ*, **576**, 505 (2002)
17. Yokoyama, T. , Nakajima, H., Shibasaki, K., Melnikov, V.F., & Stepanov, A.V., *ApJ*, **576**, L87 (2002)
18. Fleishman, G. D., & Melnikov, V. F., *ApJ*, **587**, 823 (2003)
19. Karlicky, M. & Kosugi, T., *A&A*, **419**, 1159 (2004)
20. Su, Y.N. & Huang, G.L., *Sol.Phys.*, **219**, 159 (2004)
21. Dulk, G. A., *Ann. Rev. Astron. Astrophys.*, **23**, 169 (1985)
22. Petrosian, V., *ApJ*, **255**, L85 (1982)
23. Ramaty, R., *ApJ*, **158**, 753 (1969)
24. Hamilton R.J., Lu E.T. & Petrosian V., *ApJ*, **354**, 726 (1990)
25. Gorbikov G.D. & Melnikov V.F., *Mathematical Modeling*, in press, (2006)
26. Somov, B.V., and Kosugi, T., *ApJ*, **485**, 859 (1997)
27. Fletcher, L., *ESA SP-448*, **2**, 693 (1999)
28. Petrosian, V., Donaghy, T. Q., & McTierman, J. M., *ApJ*, **569**, 459 (2002)
29. Jakimiec, J., *Proc. 10th European Solar Physics Meeting, Solar Variability: From Core to Outer Frontiers, Prague, Czech Republic*, **645** (2002)
30. Somov B.V.& Bogachev S.A., *Astronomy Letters*, **29**, 621 (2003)
31. Miller, J.A., Cargill, P.J., Emsly, A.G. et al., *GRL 102*, No.A7, **14**, 631 (1997)
32. Aschwanden, M.J., *Space Science Reviews*, **101**, 1 (2002)

MHD-Oscillation Modes of a Flaring Loop Using Microwave Observations With High Spatial Resolution

Veronika E. Reznikova*, Victor F. Melnikov*, Valery M. Nakariakov† and Kiyoto Shibasaki**

*Radiophysical Research Institute (NIRFI), Nizhny Novgorod, 603950, Russia
† Physics Department, University of Warwick, Coventry, CV4 7AL, UK
**National Astronomical Observatory of Japan, Minamimaki, Minamisaku, Nagano 384-1305, Japan

Abstract. Study of an oscillating loop observed with high spatial resolution (NoRH) shows the presence of two Fourier spectral peaks: P_1 = 14-17 s, which is more pronounced at the loop apex and P_2 = 8-11 s, that is stronger at the loop legs. We found the phase shift, $P_2/4$, between the pulsations at the northern leg and at the loop top for the P_2 spectral component and a lack of the phase shift between different parts of the flare loop for the P_1 component. The computed phase speeds and periods of MHD modes of a coronal loop confirm identification of the first periodicity P_1 as the global (fundamental) sausage mode. The second periodicity can be associated with several modes: the second and the third longitudinal harmonics of the kink mode, the third harmonics of the sausage mode and the second harmonics of ballooning mode.

Keywords: Magnetohydrodynamics (MHD)– waves – Sun: activity – Sun: corona – Sun: oscillations – Sun: radio pulsations
PACS: 96.50.Tf

INTRODUCTION

High spatial and temporal resolution EUV imagers onboard SOHO and TRACE spacecraft have brought a major breakthrough in the observational study of coronal MHD wave activity, revealing the presence of kink and slow magnetoacoustic oscillations of coronal loops [1]. Obviously, this progress was connected with the possibility to resolve the wavelength of the oscillation, the opportunity that has been traditionally neglected in radio observations. However, some radio observational facilities, e.g. NoRH, have quite a reasonable spatial resolution ($10''$ in 17 GHz and $5''$ in 34 GHz), providing us with a tool for the study of the spatial structure of the oscillations, similarly to the EUV band. This, together with the excellent time resolution, can allow for confident identification of observed MHD modes.

First attempts to analyse the microwave pulsations of flaring loops using NoRH images were undertaken by Asai et al. [2], Melnikov et al. [3], Grechnev et al. [4], Nakariakov et al. [5]. In the latter paper, it was demonstrated that the spatial distribution of an observed oscillation is crucial for the mode identification.

Numerous radio observations of quasiperiodic pulsations of the emission during the flares made without spatial resolution [6, 7] could not allow for the confident identifica-

CP848, *Recent Advances in Astronomy and Astrophysics,*
7th *International Conference of the Hellenic Astronomical Society,* edited by N. H. Solomos

tion of specific oscillation modes because of too many free parameters. The identification of the oscillation mode should be based upon the determination of the observational period, the longitudinal and transverse sizes of the magnetic loop, and the spatial distribution of the oscillation amplitude and phase along the loop, as well as the plasma density and magnetic field strength inside and outside the loop.

OBSERVATIONS

The analysed flare occurred on the 12th of January, 2000, at the Eastern limb (N13E80) and was observed with both the Nobeyama Radioheliograph (NoRH) and the Radiopolarimeter (NoRP). Also, the flare was detected with SXT and HXT instruments onboard Yohkoh.

The spectral analysis of the NoRP burst light curves reveals 8-15 s oscillations superposed on the smooth variations of the microwave flux. During the main (second) peak the mean amplitude of the observed quasi-periodical pulsations is about 100 s.f.u.

The existence of the pulsations is confirmed by more sensitive observations with NoRH. Figure 1a,c shows the average correlation amplitude $\rho(t)$ of the signals recorded by NoRH at 17 and 34 GHz, obtained by averaging the amplitude of observed correlation coefficients over the higher spatial frequencies. Consequently, it eliminates background emission and reflects integrated radio fluxes only from fine spatial structures smaller than about 24" (see [8]). For both correlation curves, we have analysed the relative change of the signal, $\Delta\rho/\rho = (\rho(t) - \rho_0)/\rho_0$, corresponding to the modulation depth (Figure 1b and d). The slowly varying mean signal ρ_0 is obtained by the 10-s smoothing of the signal. The average modulation depth is about 15%. The Fourier spectral analysis shows that the analysed signal contains well-pronounced periodicities: the strongest of 14-17 s, and another one of about 8-10 s (Figure 1, right plots).

The identification of observed modes requires the estimation of physical parameters in the oscillating loop. Figure 2 shows the radio images taken at the flare maximum at 01:36:15 UT at 34 GHz (filled pixels) and 17 GHz (the dotted contour curves). The contour levels are 0.7, 0.85 and 0.98 of the maximum brightness temperature T_{B17}. The brightness temperatures reached values 5.7×10^7 K at 17 GHz and 3.7×10^7 K at 34 GHz. The black contours show the positioning of the elongated HXR source, obtained in the channel L (13.9-22.7 keV) of Yohkoh/HXT, and correspond to the levels 0.2, 0.4, 0.6 and 0.9 of the maximum value. The HXR signal had been integrated over the time interval 01:35:52 - 01:36:15 UT. The source's double peak structure shows the places of strong electron precipitation.

The radio source projections are seen to be of elliptical shape at both frequencies. We can clearly see the microwave brightness maxima to be located close to the midpoint of the two HXR brightness peaks. The observations in the SXR band, with Yohkoh/SXT, were delayed and started 9 min after the peak of the radio burst. The analysis of SXR images confirms the presence of a magnetic loop that links two HXR brightness peaks corresponding to the loop footpoints.

According to the NoRH 34 GHz radio image, the loop length is about 2.5×10^9 cm (about 34"), and the average width at half intensity is about 6×10^8 cm (about 8").

Diagnostics of physical conditions in the loop based on microwave and SXR data

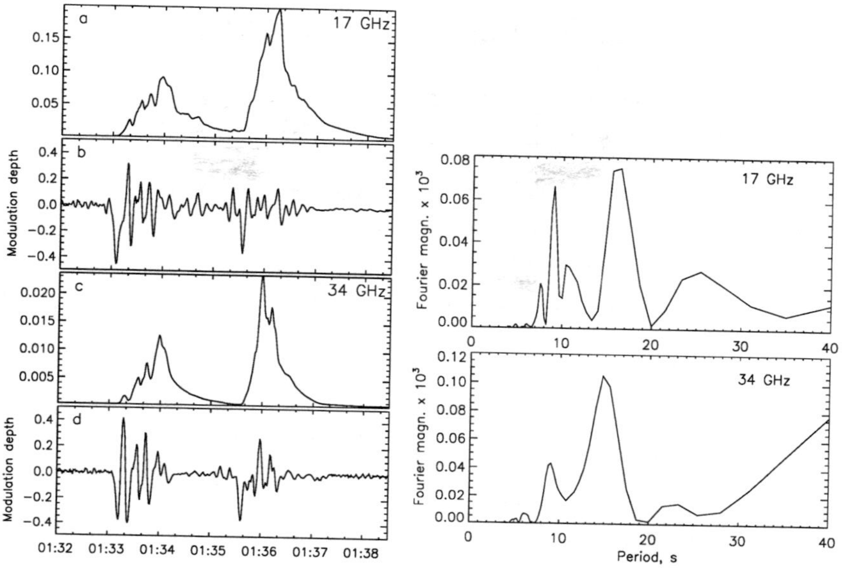

FIGURE 1. Left: The averaged correlation amplitude $\rho(t)$ of the signals recorded by NoRH at 17 (a) and 34 GHz (c) and relative variations of the signal, $\Delta\rho/\rho = (\rho(t) - \rho_0)/\rho_0$, corresponding to the modulation depth at each frequency (b, d). Right: Fourier spectra of the relative variations $\Delta\rho/\rho = (\rho(t) - \rho_0)/\rho_0$ of the NoRH correlation amplitude signal in the interval ± 40 s about 01:36:20 UT at 17 and 34 GHz. The spectra show the presence of at least two well pronounced spectral components, with the periods 14-17 s and 8-10 s.

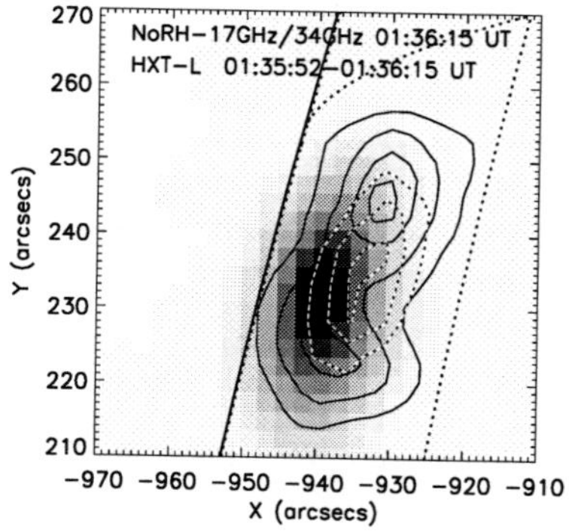

FIGURE 2. Radio images made at the flare maximum at 01:36:15 UT at 34 GHz (filled pixels) and 17 GHz (dotted contour curves, the contour levels are 0.7, 0.85 and 0.98 of the maximum brightness temperature T_{B17}). The black contour shows the positioning of the HXR sources, obtained in the channel L (13.9-22.7 keV) of Yohkoh/HXT. The HXR signal had been integrated over the time interval 01:35:52 - 01:36:15 UT, the contours correspond to the levels 0.2, 0.4, 0.6 and 0.9 of the maximum value.

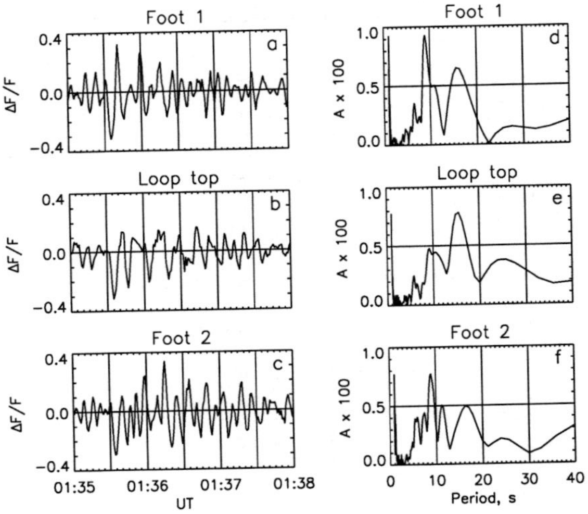

FIGURE 3. Left column: Radio flux variations $\Delta F/F = [F(t) - F_0]/F_0$, normalised on the slowly varying component of the emission, F_0, obtained by smoothing the observed flux $F(t)$ over 10 s. Right column: The corresponding Fourier power spectra of the pulsations shown in panels a,b,c. Two dominant spectral components with periods $P_1 = 14$-17 s and $P_2 = 8$-11 s are clearly seen for the pulsations detected at the different parts of the loop.

(described in [11]) shows the loop to be filled by a dense plasma with the concentration $n_0 \approx 10^{17}$ m^{-3}, penetrated by the magnetic field changing from $B_0 \approx 100$ G near the loop top to $B_0 \approx 200$ G near the north footpoint.

A search for a phase shift between oscillations in different parts of coronal loops is very important for the discrimination between different MHD modes and their harmonics. In Fig. 4 we compare phases of oscillations for the peak periods of the two spectral components, P_1 and P_2. The phases in the interval ± 70 s around the burst maximum 01:36:00 UT were calculated using imaginary and real parts of the Fast Fourier Transform of the oscillating signal in the running window of Gaussian shape

$$g(t,t_0) = \exp\left[-(t-t_0)^2/\tau^2\right]$$

with the width $\tau = 90$ s that is comparable with the burst half amplitude duration. The phase of the oscillations for the loop top source is shown by the solid line, and for the northern (Foot 2) and southern (Foot 1) leg sources - by the dotted and dashed lines, respectively. Phases of both spectral components are quite stable in the time interval ± 70 s around 01:36:00 UT. As one can see, the phase shift between oscillations from the loop top and legs sources is very small for the P_1 component, less than 0.5 rad or $1/12 \times P_1 \approx 1.3$ s, and two times less for the opposite leg sources. However it is well pronounced for the P_2 spectral component. The value of the phase shift between P_2-oscillations from the loop top and the northern leg sources is about 1.5 rad or $1/4 \times P_2 = 2.2$ s. For the southern leg source the phase difference has the same sign, but the value is three times smaller, only ≈ 0.5 rad or $1/12 \times P_2 \approx 0.7$ s. The small

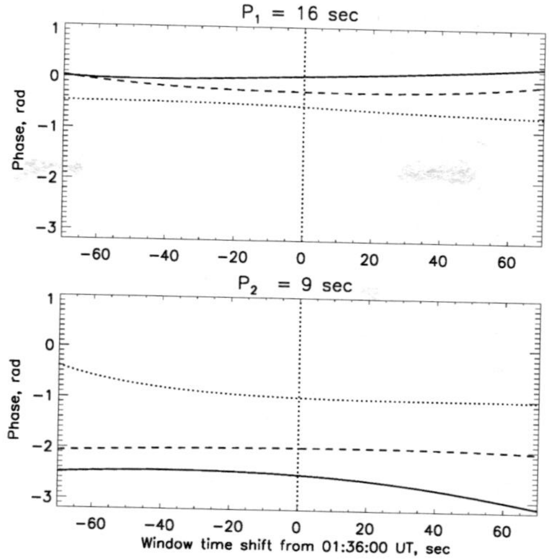

FIGURE 4. The phase evolution near the burst maximum at 01:36:00 UT for the two spectral peaks of oscillations: $P_1 = 14 - 17$ s and $P_2 = 8 - 11$ s. The phase of oscillations for the loop top source is shown by the solid line, and for the northern (Foot 2) and southern (Foot 1) leg sources by the dotted and dashed lines, respectively. The phase shift between oscillations from the loop top and foot sources is well pronounced for the P_2 spectral component and is very small for the P_1 component.

phase difference for this source is possibly connected with its proximity to the loop top source.

MECHANISMS FOR THE MICROWAVE PULSATIONS

The most probable mechanisms for the quasi-periodic modulation of the emission intensity could be a) variations of the magnetic field strength in the loop, b) variations of the angle between the line-of-sight and the magnetic field vector, c) variations of the mirror ratio in the loop, modulating the loss cone and, therefore, the electron precipitation, and d) a quasi-periodic regime of the energetic electron acceleration/injection in the loop. In any case, the period of the observed oscillations (of order of 10 s) strongly indicates the MHD nature of the pulsations. Indeed, MHD modes of the flaring loop can modulate the efficiency of the gyrosynchrotron emission of the electrons in the loop or affect the electron acceleration, injection or accumulation/precipitation in the loop.

Assuming that the pulsations are produced by variation of the value and direction of the magnetic field, and taking into account the observed modulation depth of the radio pulsations, 15-20%, we can estimate the amplitude of the magnetic field perturbations. For the estimations we use Dulk & Marsh's approximated formula for the intensity, I_f, of the optically thin gyrosynchrotron emission at a frequency f [9]. As it is shown in [11], the applicability of the optically thin model is justified for all parts of microwave source at both frequencies.

Under the assumption that the direction of the magnetic field in the loop does not significantly change (e.g., sausage modes) and that the pulsations are connected with periodic expansions and contractions of the loop tube, we express the relationship of the intensity variation with the magnetic field strength B,

$$I_f \propto B^{0.90\delta - 0.22}, \tag{1}$$

which for the electron spectral index $\delta = 5$ gives the relative perturbation of the magnetic field $\Delta B/B \approx 3 - 4\%$.

On the other hand, if we assume that the pulsations are associated with the variations of the angle θ between the magnetic field and the line-of-sight and that the absolute value of the field does not change (e.g., kink modes), the expression for the intensity reduces to

$$I_f \propto (\sin\theta)^{-0.43+0.65\delta}. \tag{2}$$

Taking $\theta \approx 80°$, as the flare takes place at the limb, we obtain that $\Delta\theta \approx 12° - 15°$ or, for the relative variation of the angle, $\Delta\theta/\theta \approx 15\% - 19\%$. This value seems to be rather too high, especially if we take into account that these variations can affect the emission only for the source located near footpoints. For the loop top source the viewing angle practically does not change, as the loop is on the limb.

MODES OF MHD-OSCILLATION OF CORONAL LOOPS

According to the theory developed by Zaitsev & Stepanov [7] and Edwin & Roberts [10] linear MHD perturbations of a straight plasma cylinder with the radius $r = a$ are described by dispersion relation

$$\rho_e(\omega^2 - k_z^2 C_{Ae}^2) m_0 \frac{I'_m(m_0 a)}{I_m(m_0 a)} + \rho_0(k_z^2 C_{A0}^2 - \omega^2) m_e \frac{K'_m(m_e a)}{K_m(m_e a)} = 0, \tag{3}$$

where $I_m(x)$ and $K_m(x)$ are modified Bessel functions of order m, and the prime denotes the derivative with respect to argument x. The functions m_0 and m_e which may be considered as radial wave numbers of the perturbations inside and outside the cylinder, respectively, are defined through

$$m_i^2 = \frac{(k_z^2 C_{si}^2 - \omega^2)(k_z^2 C_{Ai}^2 - \omega^2)}{(C_{si}^2 + C_{Ai}^2)(k_z^2 C_{Ti}^2 - \omega^2)}, \tag{4}$$

where $i = 0, e$. In the internal and external media, the sound speeds are C_{s0} and C_{se}, the Alfvén speeds are C_{A0} and C_{Ae}, and the tube speeds are C_{T0} and C_{Te}, respectively. (The definitions of the speeds are standard, see, e.g., [1]). Relations between those characteristic speeds determine properties of MHD modes guided by the cylinder. For modes that are confined to the tube (evanescent outside, in $r > a$), the condition $m_e^2 > 0$ has to be fulfilled. The integer m is the azimuthal mode number. It determines the azimuthal modal structure: waves with $m = 0$ are sausage modes, waves with $m = 1$ are kink modes, waves with higher m are referred to as flute or ballooning modes. The existence and properties of the modes are determined by the equilibrium physical

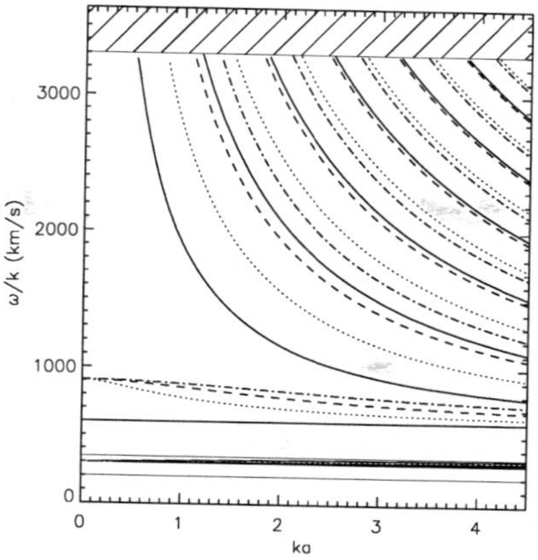

FIGURE 5. Phase speeds of MHD modes of a straight magnetic cylinder as function of the longitudinal wave number normalised by the cylinder radius a, for $C_{A0} = 600$ km/s, $C_{Ae} = 3,300$ km/s, $C_{s0} = 340$ km/s and $C_{se} = 200$ km/s. Solid curves correspond to sausage ($m = 0$) modes of different radial numbers ($l = 1, 2, 3, ...$), dotted curves to kink ($m = 1$) modes, dashed and dashed-dotted curves - to the ballooning modes $m = 2$ and $m = 3$, respectively. For fast modes, the lowest phase speed curves of each line style correspond to the $l = 1$ mode. Horizontal straight lines show the characteristic speeds mentioned above. The hatched region shows where no trapped mode can exist.

quantities. In particular, for $m_0^2 < 0$, the internal radial structure of the modes is described by the Bessel functions $J_m(x)$, and the radial dependence of the oscillation is quasi-periodic.

Figure 5 shows dispersive curves of MHD modes guided by a magnetic cylinder with the characteristic speeds $C_{A0} = 600$ km/s, $C_{Ae} = 3,300$ km/s, $C_{s0} = 340$ km/s and $C_{se} = 200$ km/s that are obtained using the loop parameters (see Section Observations) derived from microwave and X-ray diagnostics (it is described in [11]).

There is an infinite number of modes with a given azimuthal number m, corresponding to different radial numbers, l. The modes with higher radial numbers l have higher phase speeds along the loop and exist for higher values of the longitudinal wave number k.

The horizontal axis of Fig. 5 shows the normalised longitudinal wave number, which may be rewritten through the longitudinal wave length $ka = 2\pi a/\lambda$. As fast modes of a coronal loop should have nodes of the transverse velocity perturbations at the loop footpoints, the wave lengths of the modes are quantized, $\lambda = 2L/n$, where n is an integer corresponding to the mode number (or the number of maxima of the transverse velocity perturbations along the loop) and L is the length of the loop. Identification of a particular mode should be based upon the analysis of the mode period and the mode radial and longitudinal numbers. The radial mode number l is perhaps the most difficult to measure, as the loop width is usually either unresolved or resolved very poorly.

The lowest $n = 1$ mode is a fundamental or global mode. Global kink ($n = 1$, $m = 1$) modes have been discovered with TRACE EUV imaging telescope [12, 13].

Identification of MHD-modes

The length of the analysed loop is estimated as $L = 25$ Mm. The observationally determined distribution of the oscillation power along the loop has one maximum, near the loop apex, suggesting that the observed mode is global. In the loop of the observed length, the global (n=1) mode with the period of 16 s should have the phase speed about 3,130 km/s. According to Fig. 5 the longest wave length mode which can have such a speed is the sausage mode. The corresponding normalised longitudinal wave number $ka \approx 0.54$, which gives us $a \approx L/6 \approx 4.3$ Mm. As the observed loop is sufficiently thick, this estimation is realistic. The next longest wave length global mode, the kink l=2 mode, requires $ka = 0.85$, which does not seem to be realistic as the loop would be too thick in this case. All other global modes require the loop to be even thicker and should be ruled out on those grounds. Similarly, the longitudinal modes are excluded because they cannot reach the required phase speed. In addition, according to Fig. 4 (upper panel) all parts of the loop analysed are observed to oscillate in phase with each other. This confirms the identification of the observed mode as the global sausage $l = 1$ mode, as suggested in [5].

It is more difficult to interpret the shorter period mode observed with the period of 8-11 s. To have such a period, the mode should have much shorter longitudinal wave length than the global mode. Because of the very steep dependence of the phase speed of the sausage mode upon the longitudinal wave number, it is not likely that the observed 8-11 s periodicity is associated with the second longitudinal harmonics of this mode. Table 1 shows phase speeds and periods of the modes which exist for the longitudinal wave numbers $ka \approx 1$ and $ka \approx 1.5$, corresponding to the second ($n = 2$) and third ($n = 3$) harmonics of the global mode determined above. As one can see, the second and the third longitudinal harmonics of the kink mode $m = 1, l = 2$ have periods in the required range (9.8 s and 9.2 s, respectively). However, other modes, e.g. the third harmonics of the sausage mode (with the period 12.6 s) and the second harmonics of $m = 2, l = 2$–ballooning mode (7.6 s) should not be ruled out, because the dispersion relation is very sensitive to the uncertainties in characteristic speeds. As the phase shift observed along the loop (see Fig. 4, lower panel) is by a factor of 2 smaller than the expected phase shift between maxima in a mode with $n > 1$, we interpret it in favour of a mode with a more complicated radial structure ($l > 1$). In this case, there can be maxima and minima in the radial direction at almost the same position along the loop and the observed phase shift can be smaller than half of the period because of the integration effect. The suitable modes are, e.g., ($n = 2$, $m = 1$, $l = 2$), ($n = 2$, $m = 2$, $l = 2$) and ($n = 3$, $m = 1$, $l = 2$).

The theory discussed above does not take into account another MHD mode, the torsional one. Torsional waves are twistings of the tube, propagating at the local Alfvén speed. They can easily be observed via modulation of the gyrosynchrotron emission produced by the local variations of the angle between the magnetic field and the LOS. In a coronal loop, the resonant periods of the torsional mode are given by the expression

$$P = 2L/nC_{A0}. \tag{5}$$

A first possible identification of a global torsional mode has been discussed by Zaqarashvili [14]. In the considered loop, the resonant periods of lower torsional harmonics

TABLE 1. Phase speeds V_p and periods P of MHD modes of a coronal loop. Modes with lowest azimuthal m and radial l numbers and with longest longitudinal wave lengths λ are considered only.

n	m	l	ka	λ/Mm	V_p/[km/s]	P/s
1	0	1	0.54	50	3128	16.0
1	1	1	0.54	50	836	59.8
1	2	1	0.54	50	886	56.4
1	3	1	0.54	50	897	55.7
2	0	1	1.09	25	1775	14.1
2	1	1	1.09	25	768	32.5
2	1	2	1.09	25	2549	9.8
2	2	1	1.09	25	847	29.5
2	2	2	1.09	25	3299	7.6
2	3	1	1.09	25	876	28.5
3	0	1	1.64	16.7	1323	12.6
3	0	2	1.64	16.7	2460	6.8
3	1	1	1.64	16.7	723	23
3	1	2	1.64	16.7	1816	9.2
3	1	3	1.64	16.7	3033	5.5
3	2	1	1.64	16.7	807	20.6
3	2	2	1.64	16.7	2292	7.3
3	3	1	1.64	16.7	850	19.6
3	3	2	1.64	16.7	2771	6.0

are 83 s, 42 s and 27.8 s for $n = 1$, 2 and 3, respectively. To give the observed periodicities, the suitable mode numbers are $n = 5$ ($P = 16.8$ s) and $n = 6$ ($P = 13.9$ s). The oscillations with the period in the range 8-11 s could be associated with $n = 8 - 11$ ($P = 10.4 - 7.6$ s). However, it is difficult to explain why a mode with a particular high number is excited only. Also, the wavelength of higher-n modes is much smaller than the pixel size of NoRH, and the positive and negative semi-periods would cancel each other. Consequently, it is not likely that torsional modes are responsible for the observed periodicities.

CONCLUSIONS

Measured phase speeds and periods of MHD modes and the observed properties of pulsations indicate the presence of at least two MHD modes in the loop: the global sausage mode with the period $P_1 = 14 - 17$ s, and another one with the period $P_2 = 8 - 11$ s. The global sausage mode has a maximum magnetic field perturbation at the loop apex and nodes at the footpoints. Indeed, all parts of the loop analysed are observed to oscillate in phase in this mode. This interpretation is in agreement with our previous conclusion [5]. We exclude the interpretation of the longest period component (P_1) in terms of other MHD modes (kink, ballooning or torsional).

The second periodicity $P_2 = 8 - 11$ s can be associated with several modes: the second and the third longitudinal harmonics of the kink mode $m = 1$, $l = 2$, the third harmonics of the sausage mode and the second harmonics of $m = 2, l = 2$ - ballooning mode. The observed distribution of the oscillation phase along the loop suggests that this mode is likely to have a complicated radial structure, $l > 1$. The likely candidates are the modes with ($n = 2$, $m = 1$, $l = 2$), ($n = 2$, $m = 2$, $l = 2$) and ($n = 3$, $m = 1$, $l = 2$).

ACKNOWLEDGMENTS

The work was partly supported by the RFBR grants No.04-02-39029, 04-02-16753 and by the Royal Society British-Russian Research Collaboration grant. V.M.N. acknowledges the support of a Royal Society Leverhulme Trust Senior Research Fellowship. We acknowledge the MSSL SURF (http://surfwww.mssl.ucl.ac.uk/surf/) for providing SXR data for use in this publication. The authors are grateful to Erwin Verwichte for the help with Figure 5.

REFERENCES

1. Nakariakov V.M., In: *Dynamic Sun*, Ed. B. Dwivedi, Cambridge University Press, Cambridge, 2003, pp.314-334
2. Asai A. Shimojio M., Isobe H. et al. *ApJ*, **562**, L103-L106 (2001)
3. Melnikov V.F, Reznikova V.E., & Shibasaki K., *Proc. of Intern. Conf. "Active processes on the Sun and stars"*, Eds. Zaitsev V.V. & Yasnov L.V., 2002, pp.225-228
4. Grechnev V.V., White S.M., Kundu M.R., *ApJ*, **588**, pp.1163-1175 (2003)
5. Nakariakov V. M., Melnikov V.F. and Reznikova V.E., *A&A*, **412**, pp.L7-L10 (2003)
6. Rosenberg H., *A&A*, **9**, pp.159-162 (1970)
7. Zaitsev V.V., Stepanov A.V., *Issled. Geomagn. Aeron. Fiz. Solntsa*, **37**, p.3 (1975)
8. Nobeyama Radioheliograph Catalog of Events. NRO, Minamimaki, Minamisaku, Nagano 384-1305, Japan, 2002
9. Dulk G.A., Marsh K.A., *ApJ*, **259**, pp.350-358 (1982)
10. Edwin, P. M., Roberts, B., 1983, *Sol. Phys.*, **88**, pp.179-191 (1983)
11. Melnikov V. F., Reznikova V. E., Shibasaki K., Nakariakov V.M., *A&A*, **439**, pp.727-736 (2005)
12. Aschwanden M.J., Fletcher L., Schrijver C.J. ey al. *ApJ.*, **520**, No 2., pp. 880-894 (1999).
13. Nakariakov V.M., Ofman L., DeLuka E.E., Roberts B., and Davila J.M. *Science*, **285**, pp. 862-864 (1999).
14. Zaqarashvili, T.V., *A&A*, **399**, pp.L15-L18 (2003)

The Edge of The Heliosphere: Voyager Beyond the Termination Shock (Abstract)

Stamatios Krimigis

Applied Physics Laboratory, Johns Hopkins University, Laurel, MD, 20723, USA
Academy of Athens, Chair of Science of Space, Athens, Greece

A simple, MHD-based, model has shaped our expectations on the boundary between the solar wind and interstellar space. Predictions of the location of the putative termination shock (TS) in particular have ranged over the years from ~ 5 to > 100 AU, depending on the choice of parameters inserted into the balance equation. It is now clear that Voyager has reached the vicinity of the TS at ~85-95 AU, and may have crossed (Krimigis et al, 2003) or still upstream but close (McDonald et al, 2003) and definitively crossed (Decker et al, 2005) on December 16, 2004. In either case, the observations are at substantial variance with many, if not all, the expected characteristics. (a) Solar wind convection is no longer dominating flows in this region; instead, intense anisotropic beams of ions, apparently field-aligned, are flowing tagentially to the radial direction and dominantly outward away from the sun. (b) Relativistic electrons are present concurently with the ions, and exhibit similar intensity-time profiles. (c) Ion composition is characteristic of anomalous cosmic rays (ACR) but spectra are power laws at lower (< 50 Mev total energy) energies and still exhibit a "hump" at ~100 Mev, an ACR characteristic. (d) Galactic cosmic rays (GCR) no longer show the expected solar cycle variation at V1, and recent (May 05) V2 GCR intensities are nearly at the same level as those at Voyager 1. (e) The interplanetary magnetic field (IMF), contrary to previous examples in the vicinity of interplanetary shocks, appears to be insensitive to the low energy particle activity (Burlaga et al, 2003), but is viewed as weakly corelated to GCR variations. (f) Beam anisotropies preclude a diffusive TS, contrary to expectations, although the power law spectrum is consistent with shock predictions. The totality of observations suggest that there is a belt of energetic (> 40keV) ions surrounding the high (34° N) latitude heliosphere, a picture drastically different from what MHD models have predicted.

REFERENCES

Krimigis, S. M., et al, *Nature*, **426**,45-48 (2003)
McDonald, F. D., et al, *Nature*, **426**, 48-51 (2003)
Burlaga, L. F., et al, *Geophys. Res. Lett.*, **30** (20), doi.1029/2003GL018291, (2003)
Decker, R. B., S. M. Krimigis, and E.C. Roelof, EGU, Vienna, April 2005 (Abstract)

CP848, *Recent Advances in Astronomy and Astrophysics,*
7th *International Conference of the Hellenic Astronomical Society,* edited by N. H. Solomos

Determining the Location of the Dispersionless Injection Boundary During Substorms

Theodore E. Sarris[1,2], Xinlin Li[2] and Nikolaos Tsaggas[1]

Demokritus University of Thrace, Vas. Sofias 1, Xanthi, 67100, Greece[1], Laboratory for Atmospheric and Space Physics, 1234 Innovation Drive, Boulder, CO 80303, USA[2]

Abstract. Particle acceleration in the inner magnetosphere during magnetospheric substorms is usually manifested with the appearance of particle injections, i.e. with sudden enhancements in the fluxes of energetic electrons and ions. Injections that show simultaneous flux increases in all energy ranges of a detector are called "dispersionless injections"; whether particle injections will appear to be dispersionless depends on the location of the spacecraft relative to the injection region. How particles of different energies can be transported to inside geosynchronous orbit without dispersion and where these particles come from are long standing issues in magnetospheric physics. We present a modeling study of the extent and temporal evolution of the region where particle acceleration and Dispersionless injections occur. The study is based on simulation results from a model of the expansion of the electric and magnetic fields associated with a substorm. The model that we use is able to reproduce simultaneous particle flux measurements from various spacecraft locations with considerable success.

Keywords: Magnetospheric Substorms, Dispersionless Injections, Substorm Injections, Pulse Fields

PACS: 94.30.Lr, 94.30Xy

INTRODUCTION

Magnetospheric substorms play a key role in the study of the geospace environment, particularly in the way the energy that originates from the solar wind is dissipated into the magnetosphere [Baker et al., 1997]. The energy that is stored in the Earth's magnetotail during a substorms' growth phase, is subsequently released in what is called the "expansion phase" in association with many phenomena, most notably earthward high speed plasma flows [Angelopoulos et al., 1992], plasmoid detection in the far tail [Hones et al., 1984], near-Earth cross-tail current disruption [Fairfield, et al., 1970], current-wedge formation [Atkinson, 1967], increase of O+ concentration in the magnetosphere [Daglis et al., 1994] and dispersionless injections of accelerated particles to and inside geosynchronous orbit [Baker et al., 1982, Li et al., 1998]. As a consequence of the current disruption, the magnetic field returns to a more dipolar configuration [Opgenoorth et al., 1994].

One of perhaps the most difficult questions involves the timing of the various processes described above. We constructed a model to simulate the transition from expansion phase to growth phase; in the model the transient electric field associated with the dipolarization process is simulated by a time-dependent Gaussian-shaped

CP848, *Recent Advances in Astronomy and Astrophysics,*
7th *International Conference of the Hellenic Astronomical Society,* edited by N. H. Solomos

pulse whereas the consistent magnetic field is obtained from Faraday's law. The fields are initialized in the magnetotail, are earthward-propagating and slow down as they approach the inner magnetosphere, in a fashion similar to what has often been observed in data provided by two radially displaced satellites [Russell and McPherron, 1973, Moore et al., 1981, Reeves et al., 1996]. As they propagate towards the Earth, the pulse-fields interact with pre-existing particles. The particles are traced using guiding center equations and are recorded in the simulation whenever they cross a virtual satellite's location, enabling us to reconstruct the fluxes actually measured by spacecraft during substorms. The recordings of the model fields at the satellite locations simulate the actual electric and magnetic field measurements. This interaction simulates, in a manner that will be discussed, the appearance of high-speed flows, cross-tail current disruption and field dipolarization, dispersionless injections and subsequent particle drift-echoes.

In particular, the model simulates the substorm-associated particle injections that accompanied the expansion phase of the substorms of August 27th 2001. This substorm was fortuitously observed by a multiple of spacecraft at various locations of the magnetosphere (CLUSTER at ~$19R_E$ tailward, LANL at the morning sector of geosynchronous orbit, GOES-8 at $6.6R_E$ pre-midnight and POLAR at $7R_E$ post-midnight) and offered an excellent understanding of the extent and the succession of the events [Baker et al., 2002], as well as providing constraints about the magnitude, direction and propagation of the electric and magnetic fields involved.

MODEL

A. Dynamic Pulse

The transient field in the simulation is associated with the dipolarization process in the magnetotail and is modelled as an electromagnetic pulse of localized radial and longitudinal extent which propagates earthward. Pulse fields have been used in the past in simulations of storm-time and substorm processes by Li et al. [1993, 1998]; the difference from the previous models of Li et al. [1993, 1998], is the introduction of pulse propagation velocity variation with radial distance, in a fashion similar to Sarris et al. [2002]. The propagation velocity of the electromagnetic pulse in this model drops with radial distance as the pulse propagates earthward, which is consistent with observations where the propagation speed of substorm injections was calculated based on measurements from two radially displaced spacecraft [Russell and McPherron, 1973; Moore et al., 1981; Reeves et al., 1996].

The electric field is modelled as a time-dependent Gaussian pulse with a purely azimuthal electric field component that decreases away from midnight [Li et al., 1998] and propagates radially inward at a decreasing speed. In the usual spherical coordinate system (r,θ,ϕ), where $r = 0$ at the centre of the Earth, $\theta = 0^o$ defines the equatorial plane and $\varphi = 0^o$ at local noon, positive eastward, the electric field is given by:

$$\mathbf{E}_\phi = -\hat{e}_\phi E_0 \left(1 + c_1 \cos(\phi - \phi_0)\right)^p \exp\left(-\xi^2\right) \quad (1)$$

where $\xi = [r - r_i + v(r)(t - t_0)]/d$ determines the location of the maximum value of the pulse; $v(r) = a + b \cdot r$ is the pulse-front velocity as a function of radial distance r; d is the width of the pulse; c_1 (> 0) and p (> 0) describe the local time dependence of the electric field amplitude, which is largest at φ_0; $t_a = (c_2 R_E / v_a)(1-\cos(\varphi - \varphi_0))$ represents the delay of the pulse from φ_0 to other local times; c_2 determines the magnitude of the delay, v_a is the longitudinal propagation speed of the pulse (assumed constant) and r_i is a parameter in the simulation that determines the arrival time of the pulse. The consistent magnetic field of the dipolarization process for a given electric field is obtained from Faraday's law:

$$\frac{\partial \mathbf{B}}{\partial t} = -\nabla \times \mathbf{E} \tag{2}$$

From (1) and (2), after performing the curl calculation in spherical coordinates and integrating, we obtain the magnetic field:

$$\vec{B}_\theta = \hat{e}_\vartheta E_0 (1 + c_1 \cos(\varphi - \varphi_0))^p \times \left[\frac{1 + v'(r)(t - t_{ph})}{v(r)} e^{-\xi^2} - \frac{d v'(r)\sqrt{\pi}}{2v(r)^2}(1 + \mathrm{erf}(\xi)) + \frac{d\sqrt{\pi}}{2rv(r)}(1 + \mathrm{erf}(\xi)) \right] \tag{3}$$

where $erf(x) = 2/\sqrt{\pi} \int_0^\infty e^{-x^2} dx$ is the error function.

The magnetic field of the pulse is superimposed on a background magnetic field, B_E, which is modelled as an asymmetrical dipole field with a compressed dayside and a weakened night side as in Li et al. [1998], which has a magnitude of 105 nT at local noon, 75 nT at midnight, and 90 nT at dawn and dusk at geosynchronous orbit. The pulse field and background field satisfy $\mathbf{E}_\phi \cdot (\mathbf{B}_\theta + \mathbf{B}_E) = 0$ and $\nabla \cdot (\mathbf{B}_\theta + \mathbf{B}_E) = 0$. In the simulation we consider only equatorially mirroring particles, which move on average according to the relativistic guiding center equation described in [Northrop, 1963].

B. Particle Motion in the Model Fields

Under the effect of the Earth's model magnetic field B_E, which is constant in time, electrons will drift eastward and protons will drift westward following constant-B contours. In Figure 1, the transient convective field that is modelled as a Gaussian pulse is superimposed on the constant Earth magnetic field B_E for two time instances of the pulse's earthward propagation. On the vertical axis, the magnetic field strength is plotted in the X-Y plane. The Earth's background magnetic field is asymmetric in the X direction, as it is compressed in the dayside. In the same figure, an electron's drift path is shown, as it is following a contour of constant magnetic field strength. Upon arrival of the convective pulse fields, the normal inward gradient of the constant magnetic field is reversed (region marked with circle), causing a local reversal of the particles' gradient-B drift. Particles "trapped" in this region will be transported inwards, following the **E** x **B** drift that is caused by the azimuthal pulse electric field and the northward magnetic fields. In Figure 2, the orbits of four 25-keV test-electrons are traced in the model fields following the motion that was described above.

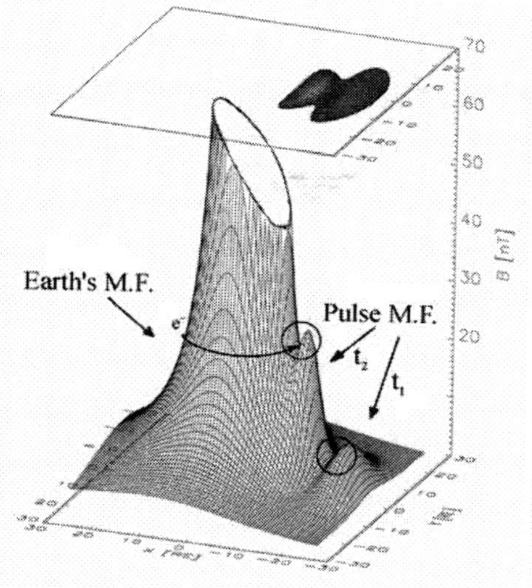

FIGURE 1. The pulse magnetic field is superimposed on the Earth's asymmetric dipole magnetic field. Two time-instances of a pulse are plotted on the X-Y equatorial plane The normal inward gradient of the magnetic field, which is responsible for grad-B drifting of the particles, can be canceled and even reversed in the areas in front of the pulse fields marked by the circles in the figure. An electron's gradient-B drift-path is drawn on a contour of constant magnetic field strength. The electron's drift is canceled by the pulse at the gradient-B reversal areas.

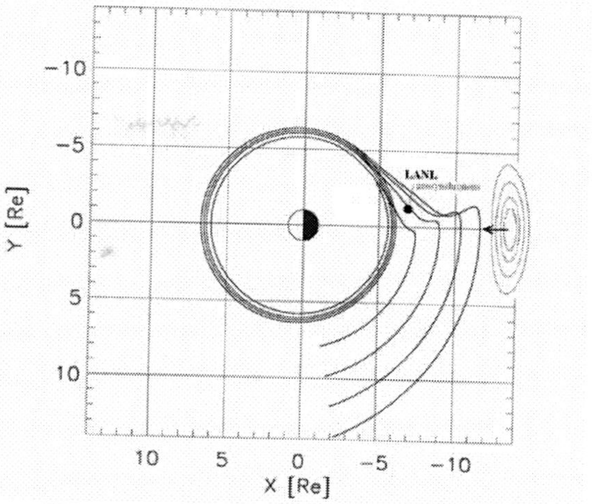

FIGURE 2. Four 25 keV electrons drift eastward in the equatorial plane; at the regions where the local magnetic field gradient is canceled, particles are transported earthward, following an **ExB** surf-like drift in the region in front of the pulse. As the pulse reaches inner magnetosphere, the Earth's magnetic field dominates again, and particles are once again left to grad-B drift in a stronger magnetic field, having been energized.

C. Simulating Particle Flux

In the model particles are initially distributed randomly in radial distance and azimuth. The initial energies start at 6 keV with a 5% increment, up to 361 keV for electrons and 418 keV for protons. In the post-processing stage each particle is given a weight which depends on the assumed initial distribution in energy and in radial distance. The weight factor is of the form:

$$Q = \sigma_j(W_f)\cdot Q_w(W_i)\cdot Q_L(L_i)\cdot (L_i/L_0)^2\cdot (v/v_d) \qquad (4)$$

where W_i and W_f are the initial and final particle energy respectively, L_i is the initial particle radial distance, L_0 is an arbitrary radial distance, Q_w is the weight related to the initial particle distribution in energy, Q_L is the weight related to the initial radial dependence, v is the particle's velocity and v_d the particle's guiding center velocity. The initial particle energy distribution Q_w is assumed to be a kappa distribution [Vasyliunas, 1968], and the initial radial dependence Q_L is set as by Li et al. [1998], who used an analytic model to assign a differential flux to each particle. The factor

$(L_i/L_0)^2$ implies that an electron initially at larger radial distance represents a larger phase-space since particles are distributed randomly in L_i. The factor υ/υ_d corrects for the fact that flux is proportional to the actual particle velocity, υ, and not υ_d. Finally, the factor $\sigma_i(W)$ is used to account for the energy dependent geometric factor for the energy channels of the detectors. Here $i = 1,2,...,N$, where N is the number of energy channels for each of the spacecraft.

OBSERVATIONS VS SIMULATION

The event modelled is a relatively isolated magnetospheric substorm event that was detected on August 27, 2001 by many Earth-orbiting spacecraft [Baker et al., 2002]. The locations of the spacecraft that observed the event are shown in Figure 3.

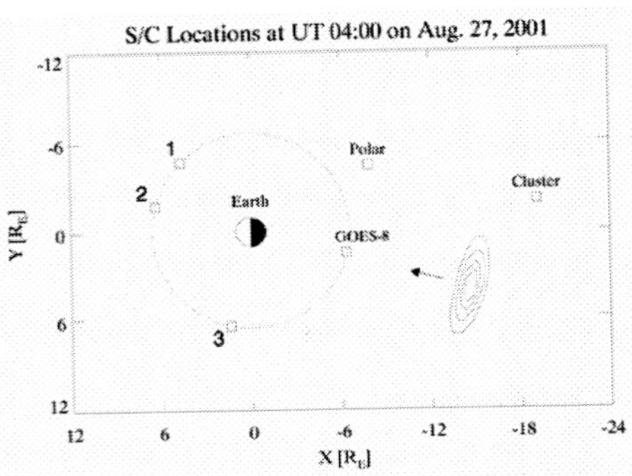

FIGURE 3. The locations of the spacecraft that observed the substorm of August 27, 2001 are plotted: Numbers 1, 2 and 3 correspond to geosynchronous spacecraft LANL-97A, 1994-084 and 1991-080, respectively. The locations of GOES-8 geosynchronous, POLAR, and the CLUSTER constellation are also plotted. The pulse-field that best reproduced the field signatures and the particle measurements is located in the pre-midnight (φ_0=160°) region, and is plotted in this figure with respect to the satellite locations.

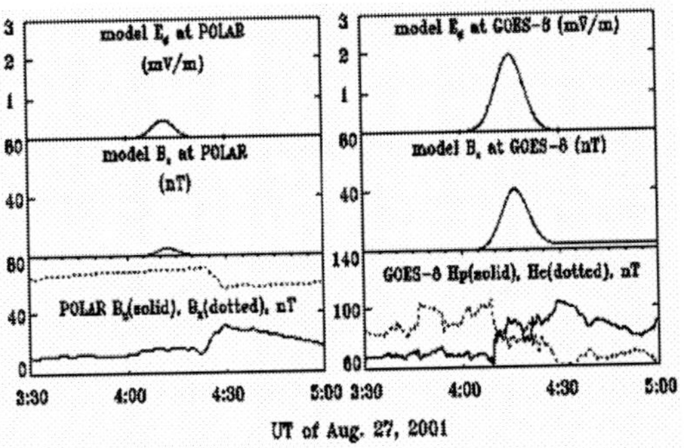

FIGURE 4. The pulse electric and magnetic field signatures are plotted in the upper panels for the simulated POLAR (left) and GOES-8 (right) locations. On the lower left panel, the actual POLAR measurements are plotted: With a dotted line is the POLAR Bz (northward) component; with a solid line we plot the Bx component. On the lower right panel, GOES-8 measurements are plotted: with a dotted line is the Hp component (parallel); with a solid line we plot the He (eastward) component.

The CLUSTER constellation of spacecraft were near apogee (~19R_E) and slightly post-midnight and were near the Z_{GSM} = 0 plane. At ~0401 *UT*, all of the four CLUSTER spacecraft detected negative $\mathbf{B}_z$ (southward magnetic field) and strong tailward flows in the plasmasheet that persisted for several minutes [Baker et al., 2002]. About 8 minutes later, GOES-8 spacecraft at geosynchronous orbit near local midnight and at Z_{GSM} ~ 1R_E measured a clear magnetic field dipolarization, shown in the lower right panel of Figure 4. Concurrent measurements from POLAR between 0400–0410 UT, at X_{GSM} ~ 9R_E and near 0200 LT and about 2R_E above Z_{GSM} = 0 plane, showed little change in the magnetic field, shown in the lower left panel of Figure 4, and in the particle flux (particle data not shown here). GOES-10 at 1900 LT showed no obvious sign of dipolarization (data not shown here).

Energetic electron injections were first detected at satellite LANL-97A (satellite 1 in Figure 5A), which was closer to the region where dipolarization signatures were observed, and subsequently at LANL 1994-084 (satellite 2) and LANL 1991-080 (satellite 3), as the electrons drifted eastward. Energetic ion injections were first observed at satellite LANL 1991-080 (satellite 3 in Figure 5B), which was close to dawn at 0411 UT, and were subsequently observed at LANL 1994-084 (satellite 2) and LANL-97A (satellite 1), as the ions drifted westward. Both the injected electron and ion populations were detected more than once by the same satellites. This is shown as multiple peaks in the flux measurements. Such features are referred to as drift echoes. These observations are consistent with magnetic reconnection occurring on the earthward side of the CLUSTER constellation and resulting in magnetic field dipolarization within a limited local time sector [Baker et al., 1996], which propagates toward the Earth.

Many simulations of different pulse-field intensities, azimuthal locations and speeds were performed, and the values of the parameters of the time dependent pulse fields that best reproduced the event are the following: $\varphi_0 = 160°$, $E_0 = 2$mV/m, $c_1 = 1$, $c_2 = 0.5$ R$_E$, $a = 86.0$ km/s, $b = 0.0145$ sec^{-1}, $p = 8$, $v_a = 20$ km/s, $r_i = 100$ R$_E$, $d = 2.8\text{x}10^8$m. For the purpose of comparison, another simulation was carried out for a post-midnight pulse, with $\varphi_0 = 200°$. The comparison of the pulse field signatures with the actual measurements at POLAR and GOES-8 are shown in the upper panels of Figure 4.

In this simulation 500,000 electrons and 750,000 protons were followed as they drifted on the equatorial plane in the combined pulse and background fields, and their energy, arrival time, and radial distance as well as initial conditions were recorded as they passed various local times of interest. Both electrons and protons were initially distributed randomly in radial distance between 4 and 14 R$_E$ and at all local times in the equatorial plane. The initial energies start at 6 keV with a 5% increment, up to 361 keV for electrons and up to 418 keV for protons. In the post-processing stage, each particle is given a weight based on its initial position and energy to represent its contribution to the initial distribution. The initial energy distribution was a kappa distribution [Vasyliunas, 1968] with $k = 1.8$ and $E_0 = 0.5$ keV for electrons and $k = 2.7$ and $E_0 = 2.5$ keV for protons. These parameters are typical for a moderately active plasma sheet [Christon et al., 1991] and similar to the ones used by Birn et al. [1997, 1998]. The initial radial dependence parameters of (7) were assigned values $a_0 = 3$, $nl = 4$, $ml = 10$, $a_{0d} = 6$ for both electrons and protons, $r_i = 12$ R$_E$ for electrons and 9.5 R$_E$ for protons. Thus, given an initial particle distribution, one can obtain particle fluxes and distributions at any location and time and can compare the simulation results with the LANL observations. The particle fluxes were summed over ± 0.4 R$_E$ at $r = 6.6$ R$_E$ and plotted with a time resolution of 60 seconds for protons and 90 seconds for electrons. There are no new injections in the simulation.

Figure 5A displays a comparison of energetic electron injections observed by LANL sensors on three satellites at geosynchronous orbit, shown in the left panels, and the simulation results from traced particles, shown in the right panels, that were recorded at three virtual satellite locations corresponding to the LANL satellites. The measured electron fluxes are omni-directional, with a time resolution of 10 seconds [Belian et al., 1992]; however a 60-second resolution was used in Figure 5A, as it is sufficient to yield the dispersive characteristics of the injection. A similar 60-second time resolution was also used in the simulation, in which the detectors' response has been incorporated. In the right-hand side panels, the simulation results for a pre-midnight pulse with $\varphi_0 = 160^\circ$ are plotted with a solid line. For comparison, simulation results for a post-midnight pulse with $\varphi_0 = 200^\circ$ are plotted with a dashed line.

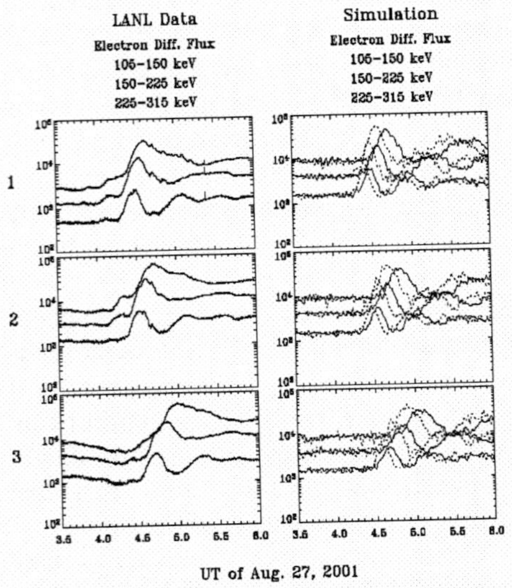

FIGURE 5A. LANL electron observations (left) are compared to simulation results (right) for a pre-midnight pulse (solid line) and a post-midnight pulse (dashed line). Numbers 1, 2 and 3 correspond to spacecraft LANL-97A, 1994-084 and 1991-080 respectively.

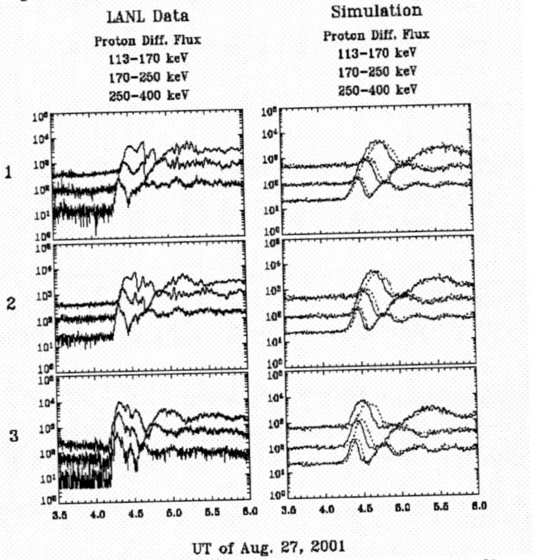

FIGURE 5B. LANL proton observations (left) are compared to simulation results (right) for a pre-midnight pulse (solid) and a post-midnight pulse (dashed line). Numbers 1, 2 and 3 correspond to spacecraft LANL-97A, 1994-084 and 1991-080 respectively.

The electron injections plotted with a dashed line precede the ones in solid line, as electrons drift for a shorter distance after being injected by the post-midnight pulse. The dispersion feature, drift echo, and width and shape of the fluxes are more or less reproduced. The electrons in the simulation seem to drift faster and more coherently than the measurements since the model only traces 90° pitch angle particles at the magnetic equator. The simulated electron enhancements fall off more rapidly than those in the observations. We attribute this firstly to the lack of scattering processes, such as pitch-angle scattering, which are not included in the simulation, and secondly to the rapid fall-off of B_z in the model as compared to the observations, which is also the reason for the more rapid falloff of the simulated proton enhancements to be discussed below. Figure 5B shows a similar comparison but for protons (most ions are protons for this event [Baker et al., 2002]). The post-midnight pulse injections, plotted with a dashed line in Figure 5B, are lagging by a few minutes, as protons are drifting for a larger distance after being injected by the post-midnight pulse to or inside of geosynchronous orbit. In the simulation, the identical model field is used but the initial particle distribution is different from the electrons, as described earlier.

DISCUSSIONS - SUMMARY

One of the main questions regarding dispersionless injections involves the azimuthal drift of the particles, which is energy dependent. The injection was usually assumed to be fast so that azimuthal drifts were considered negligible during the injection; a slow injection would allow higher-energy particles to drift over a larger range of local times than lower-energy particles, resulting in dispersion. However, dispersionless injections were still observed for very slow injections [Reeves et al., 1996]. Under the field configuration discussed earlier, a decelerating transient magnetic field can cancel and/or reverse the magnetic field gradient, thus stopping and/or reversing the electrons' gradient-B drift. Meanwhile, the transient electric field results in an **E**x**B** drift which is energy independent, and moves particles toward the Earth, following the pulse propagation front. During this motion, betatron acceleration by the transient fields leads to the energization of charged particles. On the basis of these ideas it has been shown that under this model injections can remain dispersionless at surprisingly low propagation speeds.

The timing of the injection signatures of particles has been used before to mark the location of the injection region through numerical back-tracing of eastward drifting electrons and westward drifting ions. Herein we use instead the timing of the injections at various satellites as well as field signatures to assign the azimuthal location, the extent and the propagation characteristics of a pulse-field model that simulates the dipolarization process during a substorm. The physical equivalent of the pulse-field model is a fast flow in the form of a plasma bubble that slows down as it propagates earthward. According to this model, at any given radius a dispersionless injection will be observed when the front of the pulse passes that radius, bringing energized particles with it.

It was demonstrated in *Sarris et al.* [2002] that a slow-moving, faster-decelerating pulse will exhibit sharper gradients and will thus be more effective in accelerating and transporting energetic particles inward. If the physical equivalent of the pulse-field model is an inward-propagating fast flow in the form of a plasma bubble, such as described in the above, then it would brake due to the balance between an Earthward magnetic tension force and a tailward pressure force, as described by *Shiokawa et al.* [1997]. The tailward pressure force, P_T, is given by: $P_T = B^2/2\mu_0 + n_i k T_i$ where B is the magnetic field intensity, k is the Boltzmann constant, T_i is the ion temperature and n_i is the ion density. Thus a sharp increase in the ion density and/or the magnetic field strength at a certain radial distances would cause a sudden compression in the plasma bubble, which would correspond to a sudden increase in the magnitude and the gradients of the pulse electric and magnetic fields. This increase would result in increased occurrence rates and intensity of the dispersionless injections over that particular transition region. This concept has been suggested before by *Moore* [1986], who proposed that the magnetosphere is typically configured so as to produce wave breaking near synchronous orbit, without explicit reference to the cause of the braking. In their description, *Shiokawa et al.* [1997] suggest that these flows are stopped at a clear boundary between the regions of dipolar field (high-pressure gradient) and tail-like field (low-pressure gradient) in the plasma sheet.

In the simulation that we performed, the pulse-fields were placed in the pre-midnight region, so as to better reproduce the injection signatures at the various satellites. Thus the braking of the pulse fields and the associated dispersionless injections are also occurring in the pre-midnight sector. This is actually typical of substorm-related injections in the inner magnetosphere: *Nagai* [1982], *Arnoldy and Moore* [1983] and *Lopez et al.* [1988] have shown through statistical studies that the current wedge is initially centered in the pre-midnight region and gradually expands towards midnight and post-midnight. We suggest herein that the increased occurrence of dispersionless injections in the pre-midnight sector and the cause and location of the braking of fast flows might be closely associated with the observations of the plasmaspheric behaviour: it is well established that the plasmasphere is asymmetric, having a clear dusk-side (pre-midnight) bulge [*Carpender,* 1970; *Chappel et al.,* 1970; *Carpender et al.*, 1993]. The classic picture of plasmaspheric dynamics following a storm is that the plasmapause moves in, while a plume develops in the dusk side [e.g., *Chen et al.*, 1975; *Elphic et al.*, 1996]. However, new results by *Moldwin et al.* [2002] using plasmapause crossings of CRRES, have demonstrated an outward, in general, motion of the dusk side of the plasmapause during active times of the magnetosphere, and a systematic inward motion by 1 R_E of the dawn side. If the dense cold plasma of the plasmasphere is the cause for the braking of the fast flows, through the increase of the tailward pressure force, $n_i kT_i$, then the earthward flows following a substorm would first brake at the pre-midnight bulge, causing dispersionless injections; the earthward flow breaking region would subsequently expand in the azimuthal directions, primarily towards the midnight and post-midnight sectors, as is shown in the schematic of Figure 6.

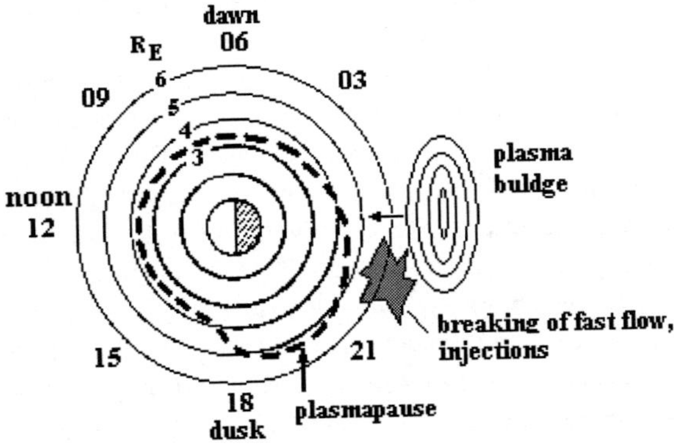

FIGURE 6. Schematic of the breaking of a fast flow at the plasmapause. The plasmaspheric plume is asymmetric towards the dusk side, thus causing a pre-midnight initial breaking of the fast flow.

REFERENCES

1. G. Atkinson, "An approximate flow equation for geomagnetic flux tubes and its application to polar substorms", in *J. Geophys. Res.*, **72**, 5373-5382 (1967).
2. V. Angelopoulos, et al., "Bursty bulk flows in the inner central plasma sheet", *J. Geophys. Res.*, **97**, 4027-4039 (1992).
3. R. L. Arnoldy and T. E. Moore, The longitudinal structure of substorm injections at synchronous orbit, *J. Geophys. Res.*, **88**, 6213-6220 (1983).

4. D. N. Baker, et al., "Observations and modelling of energetic particles at synchronous orbit on July 29, 1977", *J. Geophys. Res.*, **87**, 5917-5932 (1982).
5. D. N. Baker, J. B. Blake, L. B. Callis, R. D. Belian, and T. E. Cayton, "Relativistic electrons near geostationary orbit: Evidence for internal magnetospheric acceleration", *Geophys. Res. Lett.*, **16**, 559 (1989).
6. D. N. Baker, A. J. Klimas, D. Vassiliadis, T. I. Pulkkinen, and R. L. McPherron, "Re-examination of driven and unloading aspects of magnetospheric substorms", *J. Geophys. Res.*, **102**, 7169-7177 (1997).
7. D. N. Baker, et al., "Timing of magnetic reconnection initiation during a global magnetospheric substorm onset", *Geophys. Res. Lett.*, **29**, 24, 2190 (2002).
8. R. D. Belian, et al., High-Z energetic particles at geosynchronous orbit during the great solar proton event series of October 1989, *J. Geophys. Res.*, **97**, 16,897 (1992).
9. J. Birn et al., "Substorm ion injections: Geosynchronous observations and test particle orbits in three-dimensional dynamic MHD fields", *J. Geophys. Res.*, **102**, 2325 (1997).
10. J. Birn et al., "Substorm electron injections: Geosynchronous observations and test particle stimulations", in *J. Geophys. Res.*, **103**, 9235 (1998).
11. S. P. Christon, et al., "Spectral characteristics of plasma sheet ion and electron populations during disturbed geomagnetic conditions", *J. Geophys. Res.*, **96**, 1 (1991).
12. I. A. Daglis, S. Livi, E. T. Sarris, and B. Wilken, "Energy density of ionospheric and solar-wind-origin ions in the near-earth magnetotail during substorms", *J. Geophys. Res.*, **99**, 5691-5703 (1994).
13. D. H. Fairfield, and N. F. Ness, "Configuration of the geomagnetic tail during substorms", *J. Geophys. Res.*, **75**, 7032 (1970).
14. E. W. Hones, et al., "Structure of the magnetotail at 220 Re", *Geophys. Res. Lett.*, **11**, 5-7 (1984).
15. X. Li, I. Roth, M. Temerin, J. Wygant, M. K. Hudson, and J. B. Blake, "Simulation of the prompt energization and transport of radiation particles during the March 23, 1991", SSC, *Geophys. Res. Lett.*, **20**, 2423 (1993).
16. X. Li, D. N. Baker, M. Temerin, G. D. Reeves, and R. D. Belian, "Simulation of dispersionless injections and drift echoes of energetic electrons associated with substorms", *Geophys. Res. Lett.*, **25**, 3763 (1998).
17. X. Li, D. N. Baker, M. Temerin, G. D. Reeves, and R. D. Belian, "Dispersionless injection simulations explore auroral substorm origins", *Eos Trans. AGU*, **80**, 405 (Sep.7th, 1999).
18. X. Li, T. E. Sarris, D. N. Baker, W. K. Peterson, H. J. Singer, "Simulation of energetic particle injections associated with a substorm on August 27, 2001", *Geophys. Res. Lett.*, **30**, 1004 (2003).
19. R. E. Lopez, D. N. Baker, A. T. Y. Lui, D. G. Sibeck, R. D. Belian, R. W. McEntire, T. A. Potemra, and S. M. Krimigis, "The radial and longitudinal propagation characteristics of substorm injections", *Adv. Space Res.*, **8**, (9)91-(9)95 (1988).
20. T. E. Moore et al., "Propagating substorm injection fronts", *J. Geophys. Res.*, **86**, 6713 (1981).
21. T. Nagai, "Observed Magnetic substorm signatures at synchronous altitude", *J. Geophys. Res.*, **87**, 4405-4417 (1982).
22. T. G. Northrop, *The Adiabatic Motion of Charged Particles*, Wiley-Interscience, New York, 1963, pp 29.
23. S. Ohtani, et al., "AMPTE/CCE-SCATHA simultaneous observations of substorm-associated magnetic fluctuations", *J. Geophys. Res.*, **103**, 4671-4682 (1998).
24. H. J. Opgenoorth, M. A. L. Persson, T. I. Pulkkinen, and R. J. Pellinen, "The recovery phase of magnetospheric substorms and its association with morning-sector aurora", *J. Geophys. Res.*, **99**, 4115- (1994).
25. G. D. Reeves, "Radial propagation of substorm injections", *International Conference on Substorms-3*, Eur. Space Agency Spec. Publ., ESA SP-339, 1996. pp. 579.
26. C. T. Russell, and R. L. McPherron, "The magnetotail and substorms", *Space Sci. Rev.*, **15**, 205 (1973).
27. T. E. Sarris, X. Li, N. Tsagas, and N. Paschalidis, "Modelling energetic particle injections in dynamic pulse fields with varying propagation speeds", *J. Geophys. Res.*, **107**, 10.1029/2001JA900166 (2002).
28. D.J. Vasyliunas, "A survey of low-energy electrons in the evening sector of the magnetosphere with OGO 1 and OGO 3", *J. Geophys. Res.*, **73**, 2839 (1968).

Prediction of basic elements of the forthcoming solar cycles 24 and 25 (years 2005-2027)

Vasilis Tritakis*, Helen Mavromichalaki†, George Giouvanellis †

* *Research Center of Astronomy and Applied Mathematics, Academy of Athens,*

† *Nuclear and Particle Physics Section, Physics Department, University of Athens,*

vas@academyofathens.gr, emavromi@cc.uoa.gr, giogiouvas@yahoo.gr

Abstract. Fundamental parameters of a solar cycle could be predicted if we take in mind an existed interdependence between successive 11-year and 22-year cycles. Elements which could be predicted are, the time of rise, the total duration and the yearly values of the activity on the maximum and minimum epochs of the cycles as well as the time of maximum and the second minimum. Predicted values of these parameters for the cycles No. 24 (2005-2016) and 25 (2016-2027) are tabulated in this contribution. The small value of standard error (high coincidence between observed and calculated values) which has been succeeded during the application of the present method in the past solar cycles implies success of the proposed method of prediction for the forthcoming cycles.

Keywords: Sunspot number, prediction, solar cycle, odd-even cycles, space weather
PACS: 94.20.wq

INTRODUCTION

The impressive progress in the solar physics research during the last decades has not succeeded to describe the exact mechanism of the solar activity development, yet. This is a very serious difficulty to define a confident solar cycle prediction based on pure physics. However, the rapid evolution of new and dynamic branches of solar and space physics like, Space Weather, Solar Variability, Global Change have made the necessity of accurate short-term and medium – term solar activity predictions very urgent.

Since the prediction of the solar activity by physical methods is still impossible but the demand for predictions is very intense, it is reasonable to turn to the only way left which is the development and the improvement of empirical methods which can approach an approximated profile of a forthcoming solar cycle.

This is a very long and old way along which many methods of this category have been developed and a large number of papers have been published on this subject. The common point of these methods is the use of statistical characteristics of solar cycles or various optical, geomagnetic and interplanetary precursors. All these methods are separated in three major categories concerning to the range of the prediction they can possibly succeed. The short–range category comprises methods they attempt to predict solar activity for the next few days or weeks, the middle-range methods offer forecasting for the next 11-year solar cycle, while the long range methods risk

CP848, *Recent Advances in Astronomy and Astrophysics,*
7th *International Conference of the Hellenic Astronomical Society,* edited by N. H. Solomos

predictions for several solar cycles. Methods of the last category are rather hopeless or they contain only philological interest because the relatively short duration of systematic observations of the Sun (120 years or 10-12 solar cycles) does not allow the formulation of a significant prediction way.

On the other hand, short-term methods are mainly based on the continuously recording of the solar disk and the near interplanetary medium seeking for some indication of solar activity change. The medium –range methods are based on empirico-statistical characteristics of the previous solar cycles which can be extrapolated to the forthcoming cycles. There are tenths of methods which can be classified in the last category the oldest of which were introduced by legends of the solar activity subject like [1], [2], [3], [4]. In the present study we attempt to forecast the main parameters of the next couple of 11-year solar cycles using some very close relations they exist between two successive solar cycles especially a couple of an even-odd cycles they form a 22-year magnetic cycle [5].

DATA ANALYSIS

A substantial point of the following method is that uses alternative data from the typical Wolf's numbers. These data come from the relation,

$$I_\alpha = 1/2\,(\sqrt{A} + \sqrt{f}\,) \qquad (1)$$

where I_α is the area index, A and f the areas of sunspots and faculae, measured by the Greenwich Observatory, respectively [6], [7]. The reason we prefer to use this index instead to the well and wide accepted Wolf Numbers (W.N) is the presence of faculae which add physical meaning in the final formula. It is widely known that the faculae are good indicators of the small scale solar magnetic fields they surround active centers and make solar activity more effective. In addition, this index varies smoothly within a rather narrow range of values that makes it very predictable, in contrast to the Wolf's numbers they vary abruptly within a wide range of values. The variations of both the Wolf's numbers (upper panel) and the area index (bottom panel) for the time span 1884-2004 are depicted in Fig. 1. It is obvious that sunspot number values are ranged between 0 and 200 and Area index varies in the range from 7 and 55.

Fundamental data of both indexes can be easily found and downloaded by the sites:
http://www.ngdc.noaa.gov/stp/SOLAR/ftpwhitelightfaculae.html
and http://www.ngdc.noaa.gov/stp/SOLAR/ftpsunspotnumber.html.

A serious problem we faced during processing was the limited record of faculae which exceeded until the year 1969. After that year, the Greenwich Observatory which compiled this record abolished the areas and faculae observations patrol. Fortunately, the correlation between the I_α and the $\sqrt{R}$ is extremely high, almost 98% (Figure 2), so we were able to complete our faculae time series until nowadays, in a high confidence level 97% by extrapolating the relation,

$$I_\alpha = 3.98\sqrt{R} + 4.43 \qquad (2)$$

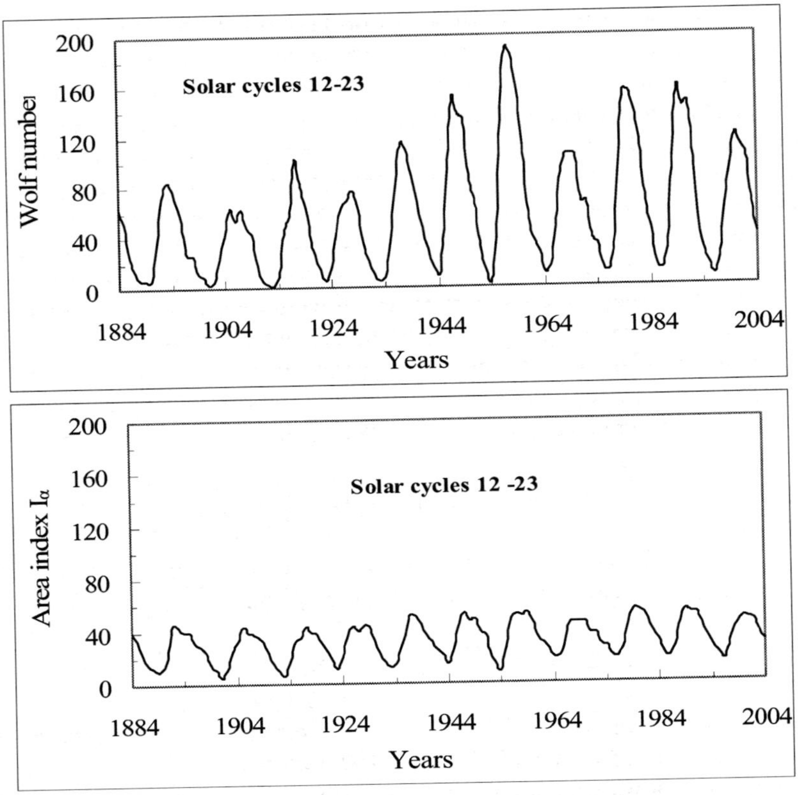

FIGURE 1. Time distribution of the Wolf's numbers (upper panel) and the area index (lower panel) for the solar cycles 12-23 (1884-2004).

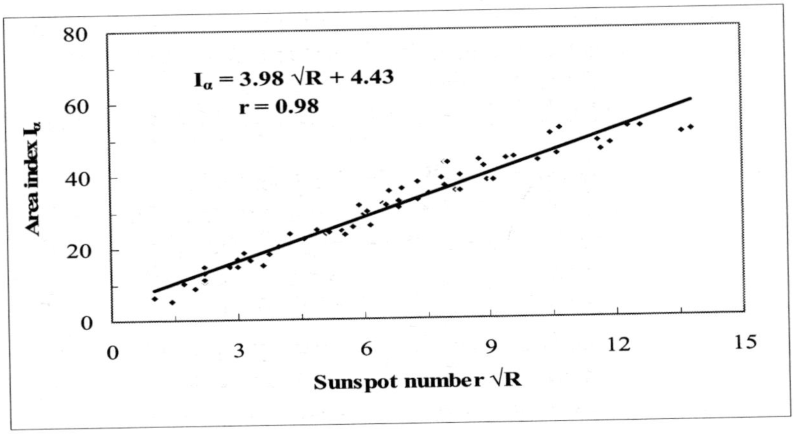

FIGURE 2. Correlation diagram between the area index I_α and the sunspot number $\sqrt{R}$.

THE METHOD

According to our way of prediction, the most important point of a solar cycle is the quasi-triangle area which is limited by the ascending branch of a solar cycle and the vertical from the maximum to the time axis, called "ascending triangle" (Figure 3).

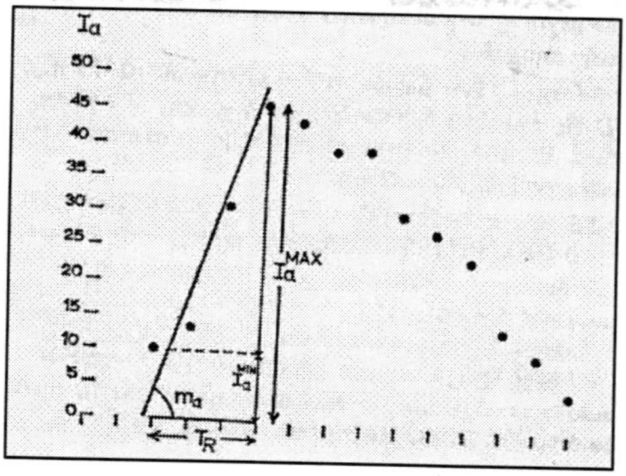

Figure 3: The ascending triangle of a typical solar cycle.
Parameters T_R , m_α , I_α^{max} and I_α^{min} are defined.

Two very important parameters of this triangle is the slope of the ascending branch m_α that could be calculated by the least square method and the time of rise T_R that is, the time interval from the appearance of the preliminary indications of a new solar cycle to the maximum of it.

It is known that there are differences in solar activity from cycle to cycle. There are series of cycles with very high activity level (odd cycles) as well as series of cycles with quite low activity (even cycles). A different behavior between even and odd solar cycles is presented in solar activity, where even sunspot cycles are characterized by two well defined 'still stands' in the level of activity during the declining phases of such cycles. It is known that there are systematic differences in the overall shapes of successive 11-year modulation cycles and similarities in the shapes of alternative 11-year cycles that are related to the 22-year magnetic cycle and to the polarity reversals of the polar magnetic field of the Sun [8]. So, considering a couple of an even and an odd cycle according the Waldmeier's numbering system of solar cycles (the cycle that starts in the year 1610 is cycle No.1) we verify that there are strong correlations among basic parameters of the cycles they belong to the same couple. At first, there is a strong correlation between the time of rise and the slope of the ascending branch of the even cycle. In addition, the inverse slope of the ascending branch of the all studied here even cycles correlate very well with the rising time of the corresponding next odd cycles given a correlation coefficient 90%. The regression line of both even and odd cycles is given in Fig. 4, separately. From this figure it is clear that the rising time of the odd cycles is smaller than that of the even cycles, a result that was expected due to a previous work [9]. The regression in both panels of the figure 4 gives a very good formula for predicting the time of rise of a forthcoming cycle by the slope of

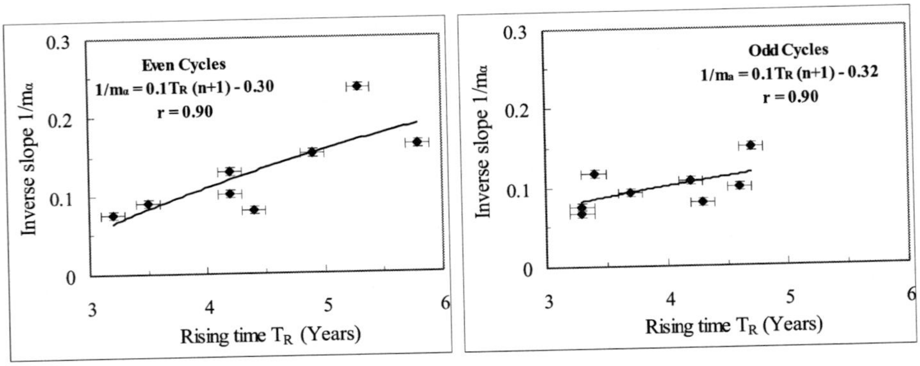

FIGURE 4. Correlation diagrams of the inverse slope of an even cycle to the time of rise of the next odd cycle (left panel) and of the same parameters of a couple odd -even cycle (right panel).

a cycle which has competed the ascending branch.

The relations which could help in the prediction of T_R are defined ,

$$
\begin{aligned}
1/m_\alpha &= 0.1T_{R(n+1)} - 0.30 \quad \text{(for even cycles)} \\
1/m_\alpha &= 0.1T_{R(n+1)} - 0.32 \quad \text{(for odd cycles)}
\end{aligned}
\qquad (3)
$$

where m_α is the slope of a certain cycle and $T_{R(n+1)}$ is the rising time of the next cycle.

The application of the above relations to past solar cycles gave accuracy, that is, a coincidence between real and predicted values of the slope, 91% which is a very high criterion of confidence.

The accuracy for odd cycles is about 91% and for even cycles is also high and about 92%. The choice of two relations and not only one had been guided by the differences of odd and even cycles, that are described by many researchers [10], [11], [8]. Generally, the odd cycles are characterized by one maximum, hard rising time, "saddle-like" shape and "mesa-type" maximum, whereas even cycles are characterized by two maxima, soft rising time, "peak-like" shape and "point-type" maximum [12].

When we succeed to predict the time of rise of a forthcoming cycle we can proceed to the next step which is to predict the total time of a forthcoming solar cycle that is, the time span between the first and the second minimum of the cycle. It could be done by dividing all the solar cycles in two groups. Group1 contains the cycles ({1, 2}, {5, 6}, {9, 10}, {13, 14}, {17,18}, {21, 22}, {25, 26}...) and Group 2 contains ({3,4}, {7,

8}, {11,12}, {15,16}, {19,20}, {23, 24}...). The way we divide solar cycles in the above two groups implies a quasi periodicity of 22 years which coincides to the solar magnetic field inversion [8]. In the figure 5, the correlation between the total time and the time of rise of each cycle for the groups 1 and 2 is depicted, while the equations (5) and (6) below express the relevant relations. If T_R, for each group are the predicted from the relations (3) and (4) values, we can predict the total time with accuracy 95% and 92% for the groups 1 and 2 respectively (Fig. 5)

$$T_{tot} = 0.85\ T_R + 7.63 \quad \text{(group 1)} \quad (5)$$
$$T_{tot} = 0.55\ T_R + 7.62 \quad \text{(group 2)} \quad (6)$$

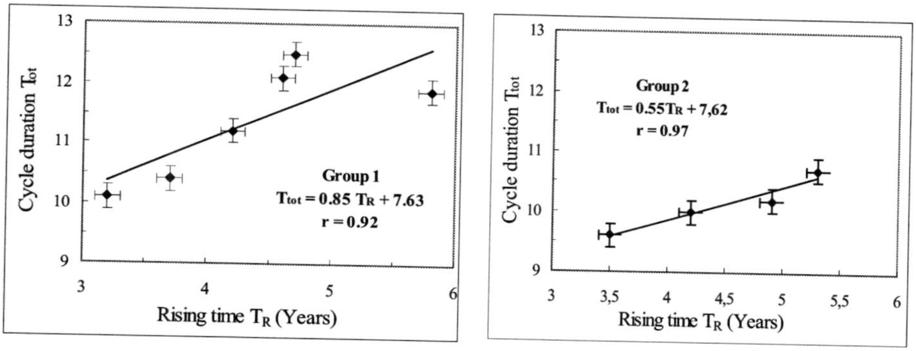

FIGURE 5. Correlation diagrams between the total time and the rising time for both group 1 and group 2 cycles.

The maximum area index during a certain cycle could be predicted by the following relations (7) and (8), where T_R is the rising time predicted by the relations (3) with accuracy 96% for group 1 and 95% for group 2 presented in Fig. 6.

$$I_\alpha^{max} = -5.30\ T_R + 70.22 \quad \text{(group 1)} \quad (7)$$
$$I_\alpha^{max} = -6.20\ T_R + 72.16 \quad \text{(group 2)} \quad (8)$$

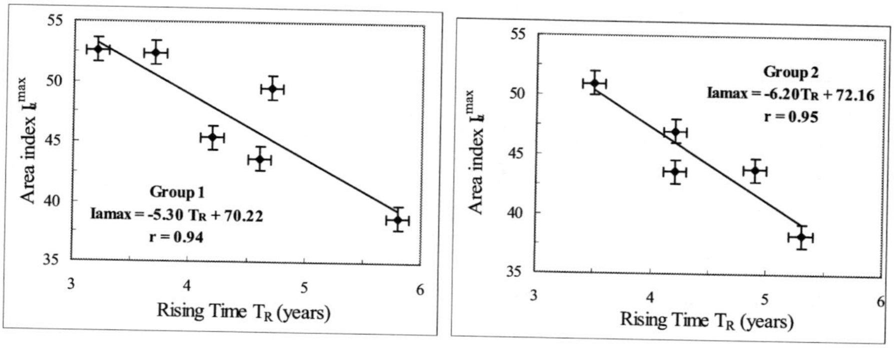

FIGURE 6. The regression lines between the area index and the rising time for both group 1 and group 2 cycles.

RESULTS

Application to Past Cycles: As an example we present an application of the proposed method to the past cycles 21 and 22. The time evolution of these solar cycles (1975-1995) expressed by the area index I_α are depicted in Figure 7, while a comparison between observed and predicted by this method values of some important elements of the above cycles are tabulated in the Tables I(a) and I(b). It is obvious that the coincidence between observed and predicted values is very high, something which makes us very optimists concerning the prediction of the same elements for the forthcoming cycles No. 24-25 (2005-2027). It is important to be noticed that the predicted values of the basic elements of cycle 21 are very close to the observed ones. It is noteworthy that the epoch I_α^{max} deviates only four months from the observed value.

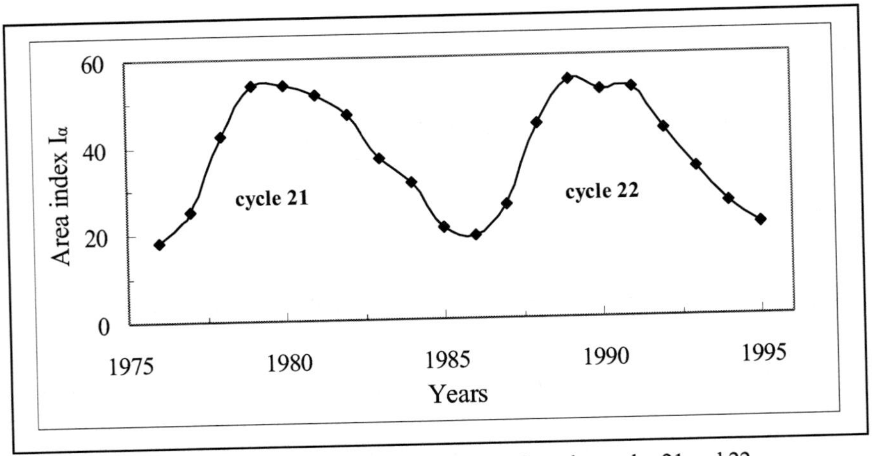

FIGURE 7. Time distribution of area index I_α for solar cycles 21 and 22

Parameters	Used data cycle 19	Predicted data cycle 21	Observed data cycle 21	Residuals Obs-Cal
m_α	15.02	-	-	-
T_R		3.8 years	4.3 years	**0.5** years
T_{tot}	-	11 years	10.6 years	**0.4** years
I_α^{max}	-	47.43	50.52	**3.09** units
I_α^{max} epoch	-	Jun 1979	Oct 1979	**4** months

TABLE I(a). Results of the application of the proposed model to the cycle 21

Parameters	Used data cycle 20	Predicted data cycle 22	Observed data cycle 22	Residuals Obs-Cal
m_{α}	7.63	-	-	-
T_R		4.4 years	4 years	**0.4** years
T_{tot}	-	10.7 years	11.2 years	**0.5** years
I_{α}^{max}	-	47.6	46.9	**0.7** units
I_{α}^{max} epoch	-	Jul 1989	Mar 1989	**4** months

TABLE I(b). Results of the application of the proposed model to the cycle 22

Application to Forthcoming Cycles: In the Table II purely predicted values of basic elements of the forthcoming solar cycles No. 24 and 25 (2005-2027) are tabulated. In the last column the accuracies between observed and predicted values of these elements for the past solar cycles No. 12-23 are bolded. The high values of coincidence between observed and calculated elements imply optimistic results for our prediction.

Parameters	Used data cycle 22	Predicted data cycle 24	Accuracy
m_{α}	12.51	-	-
T_R		3.7 years	91%
T_{tot}	-	10.7 years	92%
I_{α}^{max}	-	50.26	92%
Rmax	-	132.59	92%
I_{α}^{max} epoch	-	Jun 2009?!!	90%

Parameters	Used data cycle23	Predicted data cycle 25	Accuracy
m_{α}	8,46	-	-
T_R		4.4 years	90%
T_{tot}	-	11.3 years	95%
I_{α}^{max}	-	46.9	95%
Rmax	-	113.86	92%
I_{α}^{max} epoch	-	Dec 2020?!!	90%

TABLE II. Predicted values of the basic elements of the forthcoming cycles No. 24 and 25.

CONCLUSIONS

The application of the above prediction method to all the confident past solar cycles supports the aspect that there is a great interdependence between two successive 11-year cycles they form a couple of even-odd cycle, according to the Waldmeier's numbering system. It is obvious that the evolution of an odd cycle depends very much on the evolutionary behaviour of the preceding even cycle. A reasonable exploitation of this interdependence can lead to a high accuracy prediction of the same basic parameters of a forthcoming 11-year cycle. The important point of this method is that the use of the area index values instead of the widely known Wolf's numbers or relative sunspot numbers as solar activity indicator. The reason for this change is the high predictability of the area index in contrast to the Wolf's numbers. The predicted values tabulated above in the Table I(a) give high grade of coincidence between observed and predicted values which in all cases exceed 90%. Concerning to the special characteristics of the predicted cycles we could underline that the prediction for the forthcoming solar cycle No. 24 gives a high solar activity cycle with maximum on June 2009 (fast rising and high maximum). On the other hand, the prediction for solar cycle 25 forms the profile of a moderate solar activity cycle with maximum on December 2020 (slow rising and medium maximum).

ACKNOWLEDGMENTS

Thanks are due to the Hellenic Astronomical Society supporting the student G. Giouvanellis to present this work to the 7th Astronomy Conference 2005.

RERERENCES

1. M. Waldmeier, *Astron. Mitt. Zurich*, **14**, 439 (1939)
2. W. Gleissberg, *Solar Phys.*, **21**, 240 (1971)
3. A. I. Ohl, Solnechnaya Dannyye 9 73 (1976),
4. A. McNish and J.V. Lincoln, *Trans. Amer. Geoph. Union*, **30**, 5 (1949)
5. V. Tritakis, Astrophys. Space Sci. 82, 463 (1982)
6. J. Xanthakis, Nature, 210, 1942 (1966)
7. C. Poulakos and V. Tritakis, *Solar Phys.*, 229 (1973)
8. H. Mavromichalaki, A. Belehaki and X. Rafios, *Astron. and Astrophys.* **330**, 764 (1998)
9. H. Mavromichalaki, E. Marmatsouri and A. Vassilaki, *Earth, Moon and Planets*, **42**, 233, 1988
10. H. W.Dodson and E. R. Hedeman, Solar Phys. 42, 121 (1975)
11. M. Storini, *Adv. Space Res.* **16**, 53 (1995)
12. H. Mavromichalaki, A. Belehaki, X. Rafios and I. Tsagouri, *Astrophysics and Space Science* **246**, 7 (1997)

Mars: Determination of the most appropriate electron energy for the bow shock identification, using MGS data

N. Sergis[1], K. Dialynas[1], G. Babasides[1], E. Roussos[2] and X. Moussas[1]

[1]*Department of Astrophysics, Astronomy and Mechanics, Faculty of Physics, National University of Athens, Panepistimiopolis, GR 15783, Zographos, Athens, Greece.*
[2]*Max Planck Institute fuer Sonnensystemforschung, 37191, Katlenburg Lindau, Germany.*

Abstract. The bow shock of Mars is re-examined using combined magnetic field and electron flux data obtained by the Mars Global Surveyor during 1998. The location of the bow shock, seems to depend on the actual energy chosen when the electron flux is used as a diagnostic tool. We compare the electron data from 4 different energy channels to the corresponding magnetic field data, in order to determine the most accurate electron energy for the identification of the exact bow shock location, in the lack of magnetic field measurements (eg. Mars Express mission). The results show that the location of the bow shock can be accurately detected by using electron energy channels from 50 up to 200 eV. The planetocentric difference between the magnetic and electron flux configuration of the bow shock was found to be small, compared to the scale of the Martian magnetosphere. This difference in most cases was found to be comparable to the actual thickness of the bow shock (~70 km). A systematic delay at the detection time between the two methods was also found, where the magnetic field disturbance was the first to indicate the bow shock's outer limit (foot of the bow shock).

Keywords: Mars, Martian bow shock, bow shock location
PACS: 96.50.Fm, 96.30.Gc, 96.25.Qr

INTRODUCTION

During the MGS Mission, Mars was found to posses a unique kind of an intrinsic magnetic field. Its main characteristic is that it consists of several highly variable magnetic dipoles or multipoles, of crustal origin [1]. These localized magnetic sources seem to be more densely distributed at the equatorial zone and the southern hemisphere of the planet in accordance to the Martian dichotomy. The crustal magnetic environment, in combination with the ionospheric magnetic field of the planet, provides an efficient obstacle in the solar wind flow, which is revealed through a bow shock configuration. The bow shock is the first dynamical surface that intimates the planet's presence through the interplanetary plasma environment.

From the magnetic field measurements obtained by the MAG/ER experiment onboard MGS [2], during 353 complete orbits around Mars in the time period from March to September of the year 1998, we selected 41 of the 148 bow shock crossings [3]. These crossings were characterized by a sharp and major increase of the magnetic

CP848, *Recent Advances in Astronomy and Astrophysics,*
7th International Conference of the Hellenic Astronomical Society, edited by N. H. Solomos

field magnitude with decreasing planetocentric distance. Using electron flux data, an obvious and sharp increase at the bow shock crossing was observed almost simultaneously in all energies from 50 eV up to 10 KeV. Although the electron flux behaviour also indicates the presence of the bow shock, it is not certain whether these data are an equally precise diagnostic tool for the bow shock detection, when magnetic field measurements are not available, like in the case of the Mars Express mission. In addition, we must address the question of which would be the most suitable electron energy range for such an analysis.

ANALYSIS AND RESULTS

A thorough examination of the 41 selected MGS passages revealed a slight delay in the detection of the bow shock surface when different electron energy channels were used, as seen in Fig. 1 (left). Although both flux signals seem to diminish simultaneously, it is clearly observed that the ratio of the two fluxes present a distinct peak, implying that the location of the bow shock depends on the actual energy channel chosen for the analysis (Fig. 1, right).

In order to ascertain the consequences of these observations, we used four different energy channels from the ER data (11eV, 36eV, 79eV, 314eV), which seems to constitute the most appropriate electron energy range for the BS detection (eg. Fig. 3, right), combined with the magnetic field data provided by the MAG experiment onboard MGS.

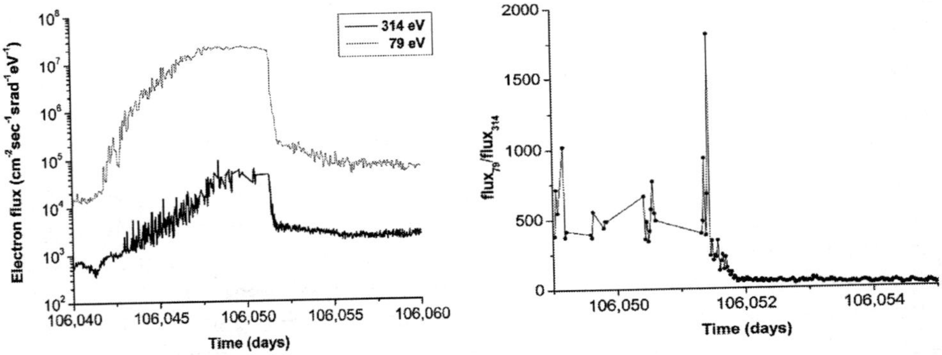

FIGURE 1. *Left:* Characteristic example of the electron flux behaviour during an MGS passage through the Martian bow shock, for two different energy channels (314 eV and 79 eV) on day 106 of 1998. *Right:* The ratio of the fluxes recorded within the two electron channels in respect to the recording time. Notice the obvious peak, which indicates the time difference in the bow shock detection when different energy channels are used.

As the spacecraft moves towards the planet, the MGS magnetometer is the first to detect the presence of the bow shock, as a sharp and major increase of the magnetic field. At the same time, the Electron Reflectometer, continues measuring solar wind electrons, while, only a few minutes later, the bow shock is also detected as a major increase in the electron flux.

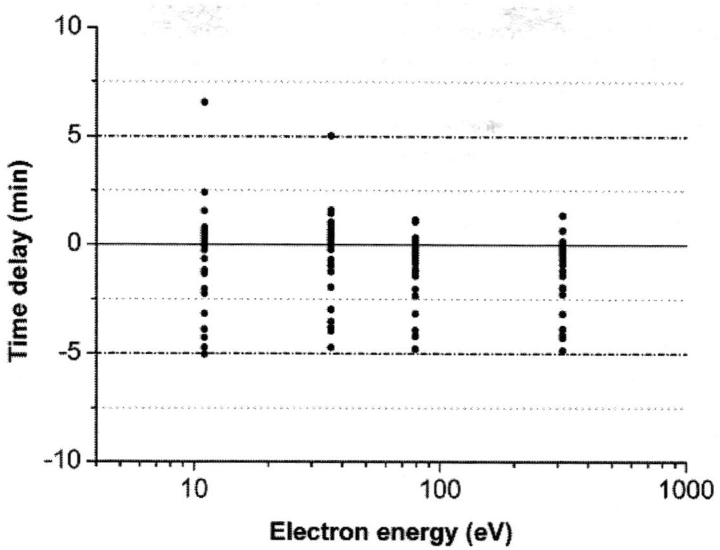

FIGURE 2. The time difference observed in the Martian bow shock detection for the 4 energy channels used in our analysis, in respect to the actual detection time as resulted from the magnetic field data.

The systematic presence of negative time delay values in Fig. 2, indicates that the Martian bow shock is detected slightly further from the planet when the magnetic field is used as a diagnostic tool, verifying that the plasma compression takes place right behind the foot of the bow shock, as generally expected in planetary shocks [4].

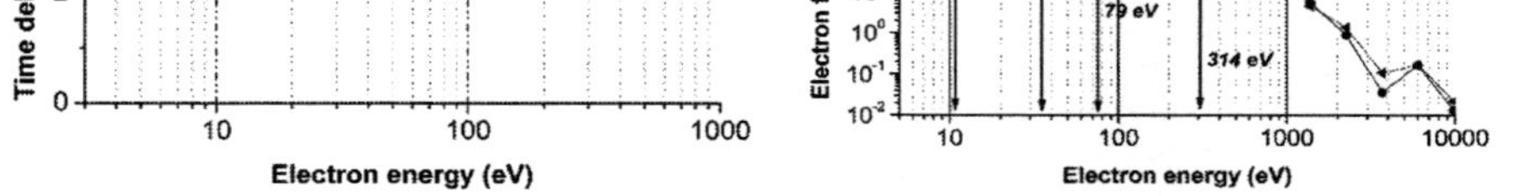

FIGURE 3. *Left:* Standard deviation of the time difference between the two methods of detection (electron flux data-magnetic field data), vs. electron energy for the 4 energy channels used. Notice the minimum values around 100 eV. *Right:* Characteristic electron energy spectra for two distinct spacecraft positions, one inside the undisturbed solar wind near Mars and one inside the Martian bow shock. The vertical arrows correspond to the energy channels used in this study.

Although the electron flux data seem to systematically exhibit a time delay in respect to the magnetic field data in all energy channels, they do not denote the bow shock crossing simultaneously. For energies close to 100 eV (Fig. 3 left panel) the deviation of the time difference between the two methods of detection reaches minimum values, indicating the proper energy range for the most precise bow shock detection using electron flux data. It seems, however, that for a large energy range, the accuracy of the detection is almost equally satisfying.

By constructing the energy spectra (Fig. 3, right), we were able to notice the obvious flux increase for energies below 300 eV, while the spacecraft is moving within the bow shock, marking out the energy range for which the Martian bow shock can be observed by the electron flux measurements. Although this flux increase is apparent even for low energies, it should be noted that these channels below 20 eV should always be handled quite carefully, mainly due to the fact that the plasma population at these energies includes scattered electrons.

Treating each MGS passage from the Martian bow shock separately, we compare the position vectors with respect to the center of the planet for both the magnetically identified bow shock and the bow shock resulted from the 314 eV electron energy channel. Assuming that the angle among them is practically insignificant (not exceeding 2.2 degrees) we were able to quite accurately calculate the planetocenrtic difference between the two independently detected BS surfaces. At 67% of the cases, this planetocentric difference was limited below 100 km, indicating that the difference imposed in the location of the Martian bow shock by the selection of the electron flux as a diagnostic tool, is generally insignificant in global scale (R_M=3397 km). However, it could become important when the actual width and the inner structure of the bow shock is examined, considering the typical 1-15 km radius of the electron helices in the local magnetic field.

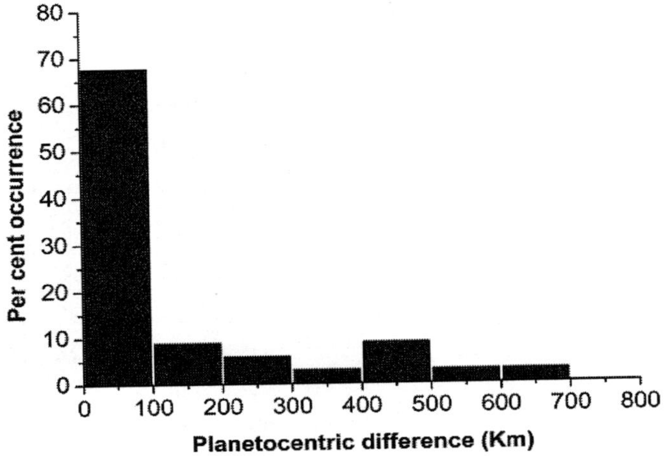

FIGURE 4. Statistical illustration of the absolute difference between the planetocentric distance in which the bow shock is magnetically detected over the corresponding distance in which the bow shock is identified using the 314 eV electron flux data.

CONCLUSIONS

The Martian bow shock was re-examined using combined electron flux and magnetic field data. Its location was found to be slightly different and systematically closer to the planet, compared to the magnetically identified bow shock, verifying the expected plasma compression within the bow shock.

The modification imposed in the bow shock location by the use of electron flux data, has been proved insignificant in global scale, with the difference in the planetocentric distances being almost always below 100 km, only a few times the local electron gyroradius. However, the detected asymphony could be important when the structure of the bow shock is examined in more detail.

The most appropriate energy range for the detection of the Martian bow shock was also determined, using both the electron flux time series and selected energy spectra inside and outside the bow shock. Energies in the range between 50 and 200 eV, seem to provide almost equally precise detection of the Martian bow shock, under normal solar wind conditions. For high electron energies (>350 eV), the Martian bow shock is practically invisible, while for low electron energies (<20 eV), the electron population is usually influenced by several electron propagating mechanisms.

A similar investigation is already being carried out for the electron flux based detection of the Magnetic Pileup Boundary (MPB) of Mars.

ACKNOWLEDGMENTS

We thank Dr. M. Acuña, Principal Investigator of the MAG/ER team, Dr. J. Connerney and Dr. S. Joy, PDS/PPI operation manager, for the MGS MAG/ER data.

REFERENCES

1. M. H. Acuña, J. E. P. Connerney, P. Wasilewski, R. P. Lin, K. A. Anderson, C. W. Carlson, J. McFadden, D. W. Curtis, D. Mitchell, H. Reme, C. Mazelle, J. A. Sauvaud, C. D'Uston, A. Cros, J. L. Medale, S. J. Bauer, P. Cloutier, M. Mayhew, D. Winterhalter, N. F. Ness, *Science* **279**, p. 1676 (1998)
2. M. H. Acuña, J. E. P. Connerney, P. Wasilewski, R. P. Lin, K. A. Anderson, C. W. Carlson, J. McFadden, D. W. Curtis, H. Réme, A. Cros, J. L. Médale, J. A. Sauvaud, C. d'Uston, S. J. Bauer, P. Cloutier, Mayhew, Michael; N. F. Ness, *J. Geophys. Res.* **97**, pp. 7799-7814 (1992).
3. N. Sergis and X. Moussas, *Solar Phys.* **202**, pp.191-200 (2001).
4. D. Burges, "Collisionless Shocks", in *Introduction to space physics,* edited by M. G. Kivelson and C. T. Russell, Cambridge University Press, Cambridge, 1995, pp. 129-163.

The Influence of Substorm-Induced Electric Fields on the Build-up of Particle Radiation during Geospace Magnetic Storms

Fiori-Anastasia Metallinou, Ioannis A. Daglis, Dominique C. Delcourt

University of Thessaloniki, Department of Physics, GR-541 24 Thessaloniki, Greece
National Observatory of Athens, Institute for Space Applications, GR-152 36 Penteli, Greece
Centre d'étude des Environnements Terrestre et Planétaires, Saint-Maur des Fossés, France

Abstract. The enhanced magnetospheric convection electric field and the substorm-induced electric fields are the drivers of plasma acceleration during geospace magnetic storms. The relative influence of the two drivers on the development of the storm-time ring current is the topic of this study. We use a three-dimensional dynamic ion-tracing model, to construct maps of the temporal and spatial variations of ion¬ number densities in the inner magnetosphere. Outflowing ionospheric oxygen ions and protons are traced under two distinct scenarios: in the first case, we follow their transport and acceleration under the influence of a large convection electric field only. In the second case, we add an impulsive electric field due to magnetic field dipolarization, as observed by spacecrafts during substorm expansion. Our aim is to evaluate the contribution of substorms to the build-up of particle radiation in the inner magnetosphere.

Keywords: Magnetic storm, magnetospheric substorm, ring current.
PACS: 94.30.Lr, 94.30.Kq

INTRODUCTION

We call 'convection' the concurrent drift of the plasma and the field lines as a whole. The momentum of the solar wind flow is the major source that influences magnetospheric convection. During magnetic storms the large scale magnetospheric convection is able to inject plasma to the inner magnetosphere and generate a realistic ring current around the earth. [e.g., Kamide et al., 1998]. In several recent studies convection has been regarded to be the dominant driver of ring current enhancement during geomagnetic storms.

Liemohn et al. [2001] suggested that the main phase ring current is mostly partial, where ions drift past the duskside magnetosphere on open trajectories and are lost through the dayside magnetopause. According to this scenario, the simple global enhancement of the magnetospheric convection electric field should be sufficient for the storm-time ring current formation.

On the other hand, during the expansion phase of substorms magnetic field lines are rapidly relaxed from a tail-like to a more dipolar configuration (dipolarization). Consequently, large induced electric fields develop and give rise to injections of

CP848, *Recent Advances in Astronomy and Astrophysics,*
7th *International Conference of the Hellenic Astronomical Society,* edited by N. H. Solomos

energetic particles to the inner magnetosphere [eg. Aggson et al., 1983, Lopez et al., 1989].

Even though magnetospheric substorms can occur at periods during which no significant magnetic storm is in progress, observations have shown that during intense magnetic storms, a succession of intense substorms also takes place [e.g., Akasofu et al., 2003].

The induced electric field which develops during magnetospheric substorms is not global, but localized to the midnight/dusk sector, and presumably plays an important role in the storm-time ring current formation [Daglis and Kamide, 2003].

Hence, both convection and substorm-induced electric fields can inject energetic particles into the inner magnetosphere and thus contribute to the ring current enhancement. The question that rises is how we can clarify the contribution of the substorm injections to the ring current build up, compared to the contribution of the large-scale convection during magnetic storms. Numerous studies have been undertaken in order to identify the relative contribution of the two processes to the ring current build-up.

In order to clarify the relative influence of magnetospheric convection and substorm injections on ring current development, we need to examine temporal and spatial variations of ion¬ energy densities in the inner magnetosphere during storms with and without substorm occurrence.

MODELING A STORM-TIME SUBSTORM

In order to calculate particle motion during storm time substorms we have to model a substorm event under storm-time conditions.

The geomagnetic field used is simulated through the Tsyganenko model [Tsyganenko, 1989]. The Tsyganenko model gives a description of the average magnetic field configuration for 6 different levels of geomagnetic activity ($K_p = 0, 0^+, K_p = 1^-, 1, 1^+, K_p = 2^-, 2, 2^+, K_p = 3^-, 3, 3^+, K_p = 4^-, 4, 4^+, K_p \geq 5^-$). It is a semi-empirical best-fit representation for the magnetic field, based on a large number of observations of the IMP and HEOS satellites. It includes contributions from external sources such as the ring current, the magnetotail current system, the magnetopause currents and the large-scale system of field-aligned currents.

The large-scale steady convection electric field in the magnetosphere is calculated by the Volland-Stern model [Volland, 1973; Stern, 1975]. It is a widely used steady-state e-field model, based on a quasi-static electric potential field. Semi-empirical analytical expressions of the electric field at all latitudes and all times have been considered. It has been arranged to fit most general features of electric fields observed by polar orbiting satellites. The free parameters of the model are the convection revisal boundary (CRB) and the polar-cap potential drop. CRB is adjusted to the position of the last closed field line at the noon. In the code used, the convection model is tilted toward the nightside (5 degrees offset on the midnight meridian) for a better reproducing of the auroral oval which is not centered on the magnetic pole. The last closed field line at noon is anchored at 78 deg invariant latitude, equivalently 12 deg colatitude. CRB in the code is placed at 17 deg (12 deg colatitude + 5 deg offset toward nightside). This value of course varies with the Kp level considered, which leads to different positions of the open-closed field line boundary.

The cross polar-cap potential drop (Φpc) characterizes the strength of the convection and as a result the intensity of the magnetic storm activity. It is calculated from the electric fields measured by polar-orbiting spacecraft on a pass over the high-latitude ionosphere. It typically ranges from 20kV, during geomagnetically quite periods, to 150 kV, during highly disturbed conditions.

The electric field induced by a transition of the geomagnetic field, from an initial level to a final one, more or less disturbed, is derived by the vector potential technique of Delcourt [Delcourt et al., 1990a, Delcourt and Sauvand, 1994]. The 3D particle code of Delcourt computes the guiding center (GC) equation in the near-Earth region and the full equation of the particle motion (FP) at distances larger than 3RE.

We model a storm-time substorm according to the following pattern:

1. We primarily set convection conditions in the magnetosphere by the cross-polar cap potential drop. Φpc is set to the value of 40 kV and at some time before the onset of the substorm it changes to the value of 80kV, and remains on that level for about one hour.

2. During the growth phase of a magnetospheric substorm the magnetospheric magnetic field is stretched to a less dipolar configuration, forming an elongated magnetotail [e.g., Baker et al., 1996]. The growth phase of a substorm can be simulated by the evolution of a given magnetic field level-a, which is represented by a given value of Kp geomagnetic index, to a more disturbed one, level-b (a higher value of Kp). The time scale of the magnetic transition is of the order of 30 minutes, which is regarded to be a typical value for a substorm growth phase [e.g., Kennel, 1995, Fok et al., 1999].

3. As the expansion phase of a substorm is characterized by a collapse of the geomagnetic tail, this dipolarization can be modeled by an evolution of the magnetic field level-b to the initial level-a disturbance. The time scale of this second transition is in the order of a few minutes.

4. The magnetic field remains at the ground level for the next few hours. Under the above sequence of transitions the magnetic field is calculated by the Tsyganenko model. The convection and induced electric field are derived from the Volland-Stern and Delcourt models respectively.

The sum of the total electric field should be oriented westward, and always cause a drift of the magnetospheric plasma towards the Earth. So, the calculation of the total electric field plays an important role in our study. The convection electric field is oriented westward at any time, but it is not the same with the induced electric field. The induced electric field is oriented eastward during the stretching of the magnetic field (growth phase of the substorm) and westward during the dipolarization of the substorm (expansion phase).

$$\nabla \times \mathbf{E} = \frac{-\partial B}{\partial t} \qquad (1)$$

According to Faraday's law eq. (1), reduction of ΔB, which is controlled by the Kp levels, or increase of Δt (duration of magnetic field transition) the induced electric field can be controlled in order to lead to magnetospheric plasma drift towards the Earth.

Figure (1) depicts the time history of the electric field level during the magnetospheric substorm scenario described above. The electric field starts from a negative level and within a few minutes it drops to a lower level and remains there for about one hour. In that way we introduce conditions of steady convection, and thus storm-time conditions to the magnetosphere. At t = 0s the first transition takes place, Kp reaches a higher level and the total electric field increases. At the end of the first transition the field remains at its final level for a short time (of the order of a few minutes). At that point we introduce the second transition. Within one or two minutes Kp returns to it's initial value and the total electric field drops to negative values. At the end of the second transition the field remains at the ground level.

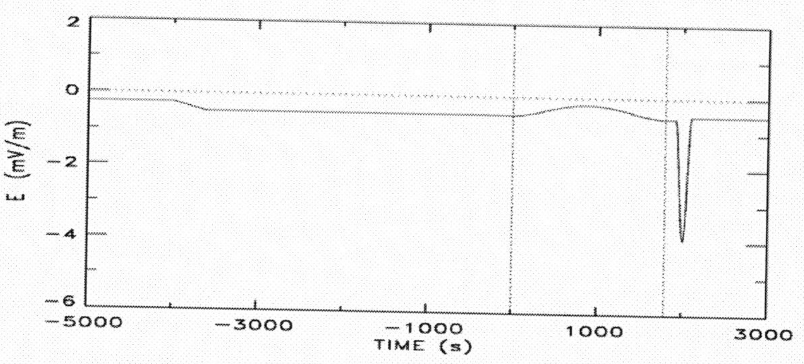

FIGURE 1. The total electric field calculated during the evolution of a storm-time substorm.

As seen from Fig. (1), the shortest time that a transition of the magnetic field takes place, the largest the induced electric field.

ANALYSIS AND DISCUSSION

Ionosphere originating ions are regarded to be the major source for the storm-time ring current in several studies [e.g., Daglis, 1997; Daglis et. al, 1999].

During the first step of our study we set the magnetosphere in a storm state, by choosing a cross-polar cap potential drop (Φpc) of 80 kV. Ionospheric particles are released at the altitude of R=5.103 km, at Magnetic Local Time (MLT) near midnight and magnetic latitude between 60-70 degrees. The particles' pitch angle is 170o and their initial energy 100eV, which represents a typical energy value of outflowing ionospheric particles at that altitude. The ionospheric particles (oxygen ions and protons) start drifting under the influence of the large convection electric field for 12,760 sec (~3,5 hours). During frequent time intervals the number density and energy density of the particles are calculated. The results are used to produce an informative video that show the diffusion of the particles in space and time. Figures (2) represents characteristic snapshots of the temporal and spatial variations of energy densities of ionosphere-originating H+, in the equatorial plane, under the influence of a large convection electric field only.

As a second step in our study, we introduce to the above storm-state condition a disturbance on the Earth's magnetic field which represents a magnetospheric substorm. 600 sec after launching the ionospheric particles, the magnetic field stretches during a time interval of 2,100 sec (35 minutes), from an initial Kp level (a) to a final level (b). Within the following 60 seconds the magnetic field relaxes to its initial state, where it remains for the next 104sec (~2,7 hours). In this case the ionospheric particles move under conditions of a storm-time substorm. Figures (3) and (4) represent the temporal and spatial variations of energy densities of ionosphere-Considering that ionospheric ions are a significant part of the storm-time ring current and can in fact dominate the ring current during intense storms [Daglis, 1997], we conclude that the results of our simulations are a strong indication of the importance of substorm occurrence for the efficient growth of the storm-time ring current.

In the near future we will investigate the features of ionospheric ion transport and energization for a wider range of initial conditions. Furthermore, we will investigate the efficiency of convection-only and convection-plus-substorms energization of plasma sheet O+ and H+ ions. This way we will gain a more comprehensive view of the storm-time ring current growth and of the critical parameters behind it.

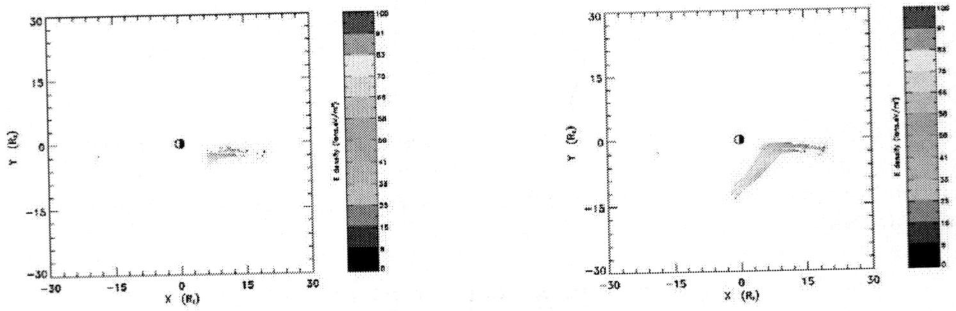

FIGURE 2. Temporal and spatial variations of energy densities of ionosphere-originating H+ particles, in the equatorial plane under the influence of a large convection el. field.

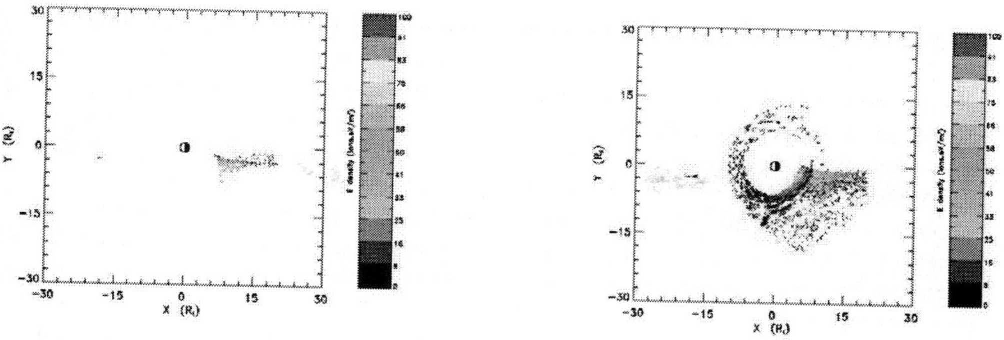

FIGURE 3. Temporal and spatial variations of energy densities of ionosphere-originating H+ particles, in the equatorial plane under the influence of a storm-time substorm. (a) during the growth phase, (b) during the expansion phase.

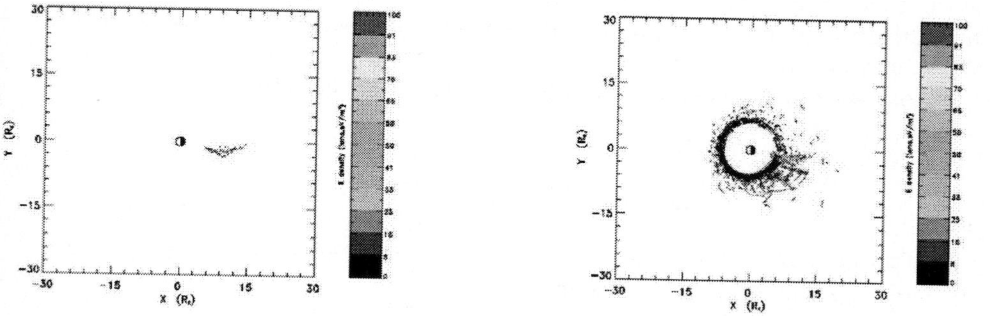

FIGURE 4. Temporal and spatial variations of energy densities of ionosphere-originating O+ particles, in the equatorial plane under the influence of a storm-time substorm. (a) during the growth phase, (b) during the expansion phase.

CONCLUSIONS

Contrary to the classic Chapman-Akasofu paradigm according to which magnetic storms are the accumulative result of successive substorms, a number of studies have recently concluded that storms are the sole result of the solar wind driven magnetospheric convection [e.g., McPherron, 1997].

This paper addresses the importance of substorm-induced electric fields in the build-up of particle radiation during geospace magnetic storms. We do this through a test-particle approach of ion transport and energization, simulating a convection-only storm and a storm-with-substorms case.

In this first-step investigation we consider the case of outflowing ionospheric H+ and O+ ions. The results of the simulation show that for both ion species the inclusion of substorm-induced electric fields renders ion acceleration much more efficient. The difference in energization is much more prominent for O+ ions, which have been know to be preferentially accelerated by substorm-induced electric fields [e.g., Daglis and Axford, 1996].

Our study suggests that substorm-induced electric fields are instrumental in energizing outflowing ionospheric ions during magnetic storms. Large-scale magnetospheric convection alone is not sufficient to accelerate these particles to ring current energies.

Considering that ionospheric ions are a significant part of the storm-time ring current and can in fact dominate the ring current during intense storms [Daglis, 1997], we conclude that the results of our simulations are a strong indication of the importance of substorm occurrence for the efficient growth of the storm-time ring current.

In the near future we will investigate the features of ionospheric ion transport and energization for a wider range of initial conditions. Furthermore, we will investigate the efficiency of convection-only and convection-plus-substorms energization of plasma sheet O+ and H+ ions. This way we will gain a more comprehensive view of the storm-time ring current growth and of the critical parameters behind it.

ACKNOWLEDGMENTS

One of us (F-A. M) is supported through a State Scholarship under the Greek Ministry of Education project “Heraklitos”.

REFERENCES

1. T. L. Aggson, J.P. Heppner and N.C. Maynard, “Observations of large magnetospheric electric fields during the onset phase of a substorm”, *Journal of Geophysical Research*, vol. 88, pp. 3981-3990 (1983).
2. S. -I. Akasofu, W. Sun, and B.-H. Ahn, “Comments on Some Long-standing Problems in Storm/Substorm Studies, in Disturbances in Geospace: The Storm-Substorm Relationship”, edited by A. S. Sharma, Y. Kamide, and G. S. Lakhina, pp. 243-254, DOI 10.1029/142GM20 (2003)
3. D. N. Baker, T. I. Pulkkinen, V. Angelopoulos, W. Baumjohann, and R. L. McPherron, “Neutral line model of substorms: Past results and present view”, *Journal of Geophysical Research*, 101, 12,975-13,010, 1996.
4. I. A. Daglis, “The role of magnetosphere-ionosphere coupling in magnetic storm dynamics, in Magnetic Storms”, edited by B. T. Tsurutani, W. D. Gonzalez, Y. Kamide, and J. K. Arballo, American Geophysical Union, Washington, DC, 1997.
5. I. A. Daglis, and W. I. Axford, “Fast ionospheric response to enhanced activity in geospace: Ion feeding of the inner magnetotail”, *Journal of Geophysical Research*, 101, pp. 5047-5065 (1996)
6. I. A. Daglis, and Y. Kamide, “The role of substorms in storm-time particle acceleration, in Disturbances in Geospace: The Storm-Substorm Relationship”, edited by A. S. Sharma, Y. Kamide, and G. S. Lakhina, pp. 119-129, Geophysical Monograph Series 142, DOI 10.1029/142GM11, American Geophysical Union, Washington, DC, 2003.
7. I. A. Daglis, R. M. Thorne, W. Baumjohann, and S. Orsini, “The terrestrial ring current: Origin, formation, evolution and decay”, *Reviews of Geophysics*, 37, 407-438, 1999.
8. D. C. Delcourt, J.-A. Sauvaud, and A. Pedersen: “Dynamics of single-particle orbits during substorm expansion phase”, *Journal of Geophysical Research*, vol. 95, pp. 20,853-20,865 (1990).

9. D. C. Delcourt, "Particle acceleration by inductive electric fields in the inner magnetosphere", *Journal of Atmospheric and Solar-Terrestrial Physics*, vol.64, pp. 551-559 (2002).
10. M. -C Fok, T. E. Moore and, D. C. Delcourt: "Modeling of inner plasma sheet and ring current during substorms", *Journal of Geophysical Research*, vol. 104, no. A7, pp. 14,557-14,569, (1999).
11. Y. Kamide, et al., "Current understanding of magnetic storms: Storm/ substorm relationships", *J. Geophys. Res.*, 103, 17,705-17,728, 1998.
12. C. F. Kennel, "Convection and Substorms - Paradigms of magnetospheric phenomenology", Oxford University Press, New York, 1995.
13. M. W. Liemohn, J.U. Kozyra, M.F. Thomsen, J.L. Roeder, G. Lu, J.E. Borovsky, and T.E. Cayton: "Dominant of the asymmetric ring current in producing the stormtime Dst*", *Journal of Geophysical Research*, vol. 106, no. A6, pp. 10,883-10,904 (2001).
14. R. E. Lopez, A.T.E. Lui, D.G. Sibeck, K. Takahashi, R.W. McEntire, L.J. Zanetti, S.M. Krimigis: "On the relationship between the energetic particle flux morphology and the change in the magnetic field magnitude during substorms", *Journal of Geophysical Research*, vol. 94, pp. 17,105-17,119 (1989).
15. R. L. McPherron, "The role of substorms in the generation of magnetic storms", in Magnetic Storms, Geophys. Monogr. Ser., vol. 98, edited by B. T. Tsurutani, W. D. Gonzalez, Y. Kamide, and J. K. Arballo, pp. 131-147, AGU, Washington, DC, 1997.
16. D. P. Stern, "The motion of a proton in the equatorial magnetosphere", *Journal of Geophysical Research*, vol. 80, pp. 595-599, 1975.
17. N. A. Tsyganenko, "A magnetospheric magnetic field model with a warped tail current sheet", *Planet. Space Sci.*, vol. 37, no. 1, pp. 5-20, 1989.
18. H. Volland, "A semi-empirical model of large-scale magnetospheric electric field", *Journal of Geophysical Research*, vol. 78, pp. 171-180 (1973).

ACE/EPAM Energetic Particle Signatures during the October/November 2003 Interplanetary Coronal Mass Ejection Events

O.E. Malandraki[*,†], D. Lario[¥], L.J. Lanzerotti[‡,¶], E.T. Sarris[ǂ], and A. Geranios[¤]

[*]*Research and Scientific Support Department of ESA, ESTEC, Noordwijk, The Netherlands*

[†]*National Observatory of Athens, Institute for Space Applications and Remote Sensing, Athens, Greece*

[¥]*Applied Physics Laboratory, Johns Hopkins University, Laurel, Maryland, USA*

[‡]*NJ Institute of Technology,Center of Solar-Terrestrial Research, Newark, New Jersey,USA*

[¶]*Bell Laboratories, Lucent Technologies, Murray Hill, New Jersey, USA*

Space Research Laboratory, Democritus University of Thrace, Xanthi, Greece

[¤]*Nuclear and Particle Physics Department, University of Athens, Athens, Greece*

Abstract. Solar energetic particles are utilized as diagnostic tracers of the large-scale structure and topology of the Interplanetary Magnetic Field embedded within two Interplanetary Coronal Mass Ejections detected in late October and early November 2003 at 1 AU by the *ACE* spacecraft. Two candidate scenarios to account for the observations in terms of open and closed magnetic field configurations are examined. We also use *ACE*/EPAM observations to reassess the boundaries of the ICMEs with respect to those previously proposed.

INTRODUCTION

A series of intense solar flares and fast Coronal Mass Ejections (CMEs) which have generated much interest in the space physics community were observed in late October and early November 2003, during the declining phase of solar cycle 23. The associated SEP events were among the largest in solar cycle 23 as observed in the ecliptic plane and at a heliocentric distance of 1 AU [1]. We provide a detailed analysis of the energetic particle signatures detected at 1 AU by the EPAM (Electron, Proton and Alpha Monitor) experiment [3] (the flight spare of the *Ulysses* HI-SCALE instrument) onboard ACE during the passage of two Interplanetary Coronal Mass Ejections (ICMEs) preceded by extremely fast solar wind observed on 29 and 30 October, 2003 [2]. The ICME leading and trailing edges are also investigated in the context of energetic particle observations with respect to those previously identified [2].

CP848, *Recent Advances in Astronomy and Astrophysics,*
7th *International Conference of the Hellenic Astronomical Society,* edited by N. H. Solomos

OBSERVATIONS AND DATA ANALYSIS

Figure 1 shows an overview of the 1-min averaged differential intensities of 38-315 keV electrons (e) in 4 energy channels and 1.9-4.8 MeV ions measured by the ACE/EPAM experiment from 0000 UT on 28 Oct to 1200 UT on 3 Nov, 2003. The vertical arrows indicate the time of observation of CMEs (short thick arrows, http://cdaw.gsfc.nasa.gov) and the onset of the temporally associated soft X-ray flare emission (long thin arrows, http://sgd.ngdc.noaa.gov/sgd/jsp/solarindex.jsp). Solid vertical lines S1 and S2 indicate two shocks which were followed by the passage of two ICMEs.

Because of the uncertainties in the SWEPAM measurements [2] energetic particle observations provide the required piece of information to determine the boundaries of the 1st ICME. We argue that the two vertical dashed lines in Fig. 2 denote a more accurate determination of its boundaries. At the time of the 1st vertical dashed line smooth rotations of the magnetic field direction began to be observed, a change in B from high to low variance in both magnitude and direction as well as a decrease in the ~MeV ion intensity. The e fluxes were observed to rise simultaneously at all energies at 2145 UT on 29 Oct. No velocity dispersion was observed due to the high pre-event ambient intensities that mask the onset of this e event.

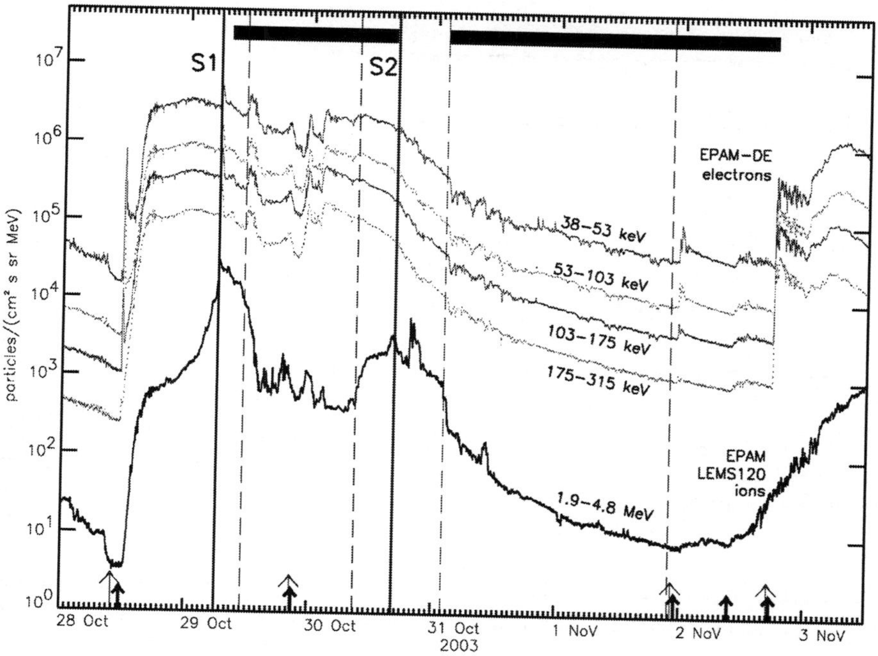

FIGURE 1. The black horizontal bars indicate the passage of ICMEs as identified by [2]. The dashed vertical lines indicate the passage of the ICME boundaries based upon our energetic particle observations (see text).

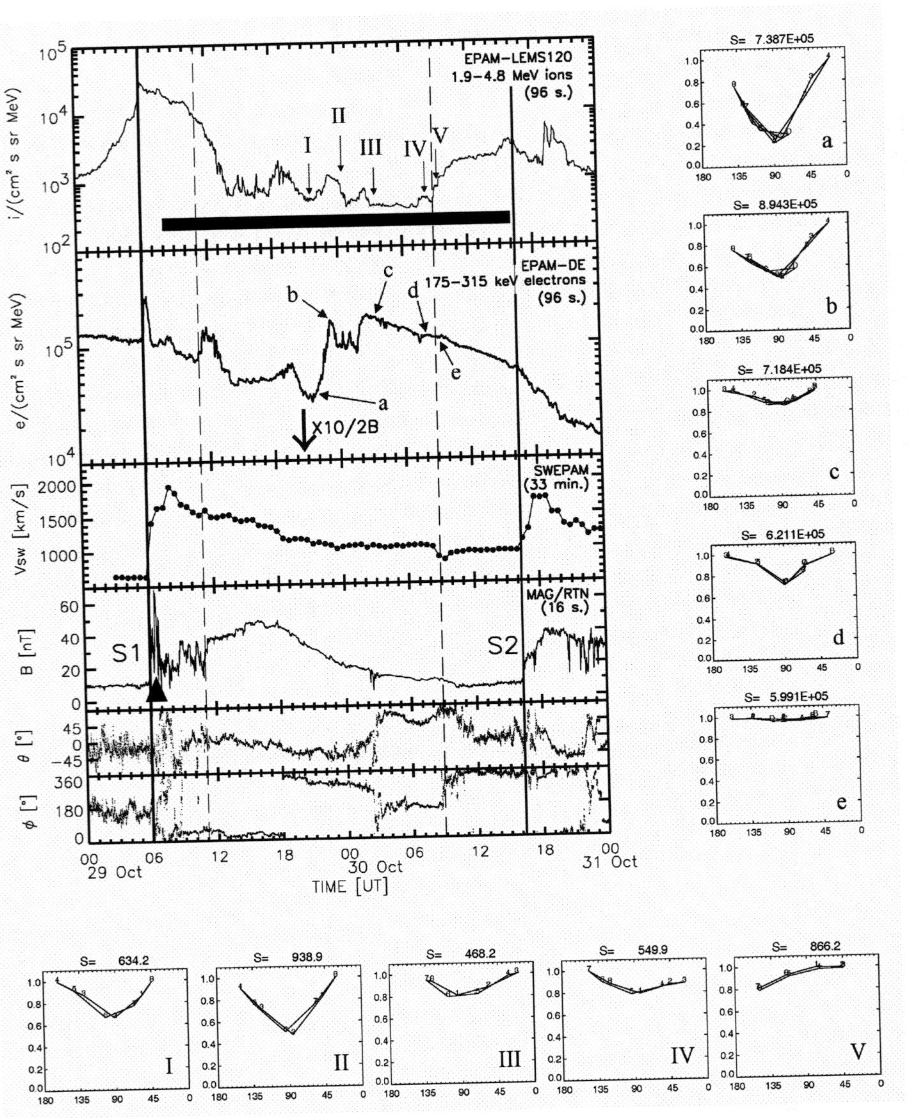

FIGURE 2. Ion, electron, solar wind and magnetic field observations by the EPAM, SWEPAM and MAG experiments onboard ACE during the passage of the 1st ICME. 64-112 keV electron (panels a-e) and 1.9-4.8 MeV ion (panels I-V) PAD snapshots.

Pitch Angle Distributions (PADs) with a much stronger bi-directional (BD) character start to be detected at the time of the e enhancement. Bi-directional PADs with variable anisotropy magnitudes (panels a-d) were observed till 0855 UT on 30 October (2nd dashed line), when the e PADs switched to isotropic also coinciding with a change in the B orientation. Furthermore, BD ~ MeV ion flows were observed until this time (panels I-IV), when both ion intensities started to increase and antisunward weakly unidirectional flows started to be observed (panel V).

After the shock S2 crossed the spacecraft the e intensities started decaying faster (see also Fig. 1). The entry into the ICME was characterized by an abrupt decrease in ion intensities, a decrease in e intensities and a change of the decay rate of the e and ion intensities (Fig. 3). A slower particle intensity decay rate compared to that observed outside the ICME was established. Consistent with our previous studies of near-relativistic e observations during the passage of ICMEs (e.g. [4], [5]), isotropic PADs were observed before the entry into the ICME (Fig. 4, panels a, b) and abruptly switched to strongly BD inside (beginning with panel c) persistently observed till 2150 UT on 1 Nov. BD ~MeV ion flows were clearly observed till 0200 on 1 Nov (panels I-III). No other periods of BD PADs of either e or ions were observed throughout the rest of the time interval considered in Fig. 3. Based on these observations, our identification of the leading edge of this ICME agrees with [2]. However, the different signatures used by [2] indicate different times for the trailing edge of this ICME. We argue that the ICME trailing edge passed over the spacecraft at 2150 UT on 1 Nov (2nd dashed vertical line, Fig. 3), i.e. ~1.5 h before the onset of the new prompt SEP event (panel k, Fig. 4), when the EPAM BD e flows ceased to be observed, close to the time when the magnetic field orientation changed from a smooth low-variance evolution to an oscillating changing orientation.

DISCUSSION

The prompt detection of an SEP event within the 1st ICME implies that the field lines threading through this structure are still rooted at the Sun, allowing direct access of solar e to the interior of the ICME. Assuming the ICME is an open structure, when ACE is inside the ICME it may establish magnetic connection directly with the downstream region of the shock S1 (now beyond the observer) and also connect to the shock S2 that is beginning to propagate away from the Sun. The enhanced magnetic field region behind the shock S1, reaching 68 nT in intensity moved past ACE from ~0600 UT to 1100 UT on 29 Oct and traveled radially away from the Sun with a solar wind speed of ~1900 km s^{-1}. At the time of the X10 flare at ~2037 UT on 29 Oct when new energetic e were injected from the Sun, this compressed B region was located at a ~0.6 AU radial distance upstream from ACE. In such an open B configuration, energetic e streaming in an antisunward direction can magnetically mirror at this magnetic constriction in space, reverse direction, and propagate in the sunward direction. Subsequent reflection by the increasing B close to the Sun or by the following shock S2 resulted in an efficient confinement of particles and hence the BD particle flows. However, we do not know whether the increase in the B magnitude behind the shock S1 at ~1.6 AU was still strong enough to mirror these near-relativistic e. In order to estimate the earliest arrival time of the e injected at the time of the X10 flare at 2037 UT on 29 Oct and the formation of BD e PADs observed by ACE, we can assume that the IMF from the Sun to the shock S1 was radial and that particles propagated scatter-free with a pitch angle cosine close to 1. The lower limit

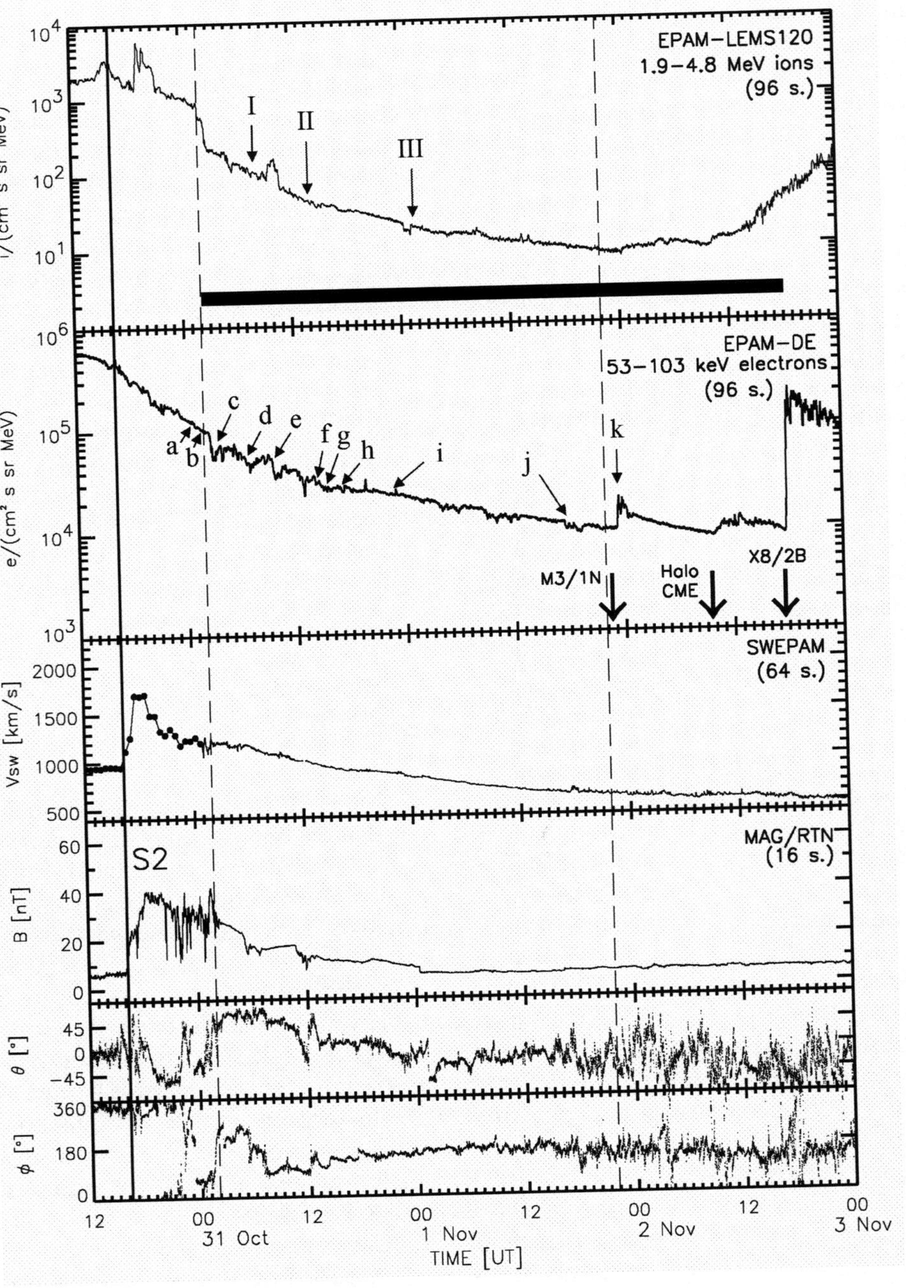

FIGURE 3. Ion, near-relativistic electron, solar wind and magnetic field observations by the EPAM, SWEPAM and MAG experiments onboard ACE during the passage of the 2nd ICME. The black horizontal bar indicates the time interval identified by [2] as the 2nd ICME. The dashed vertical lines indicate our identification of the ICME boundaries.

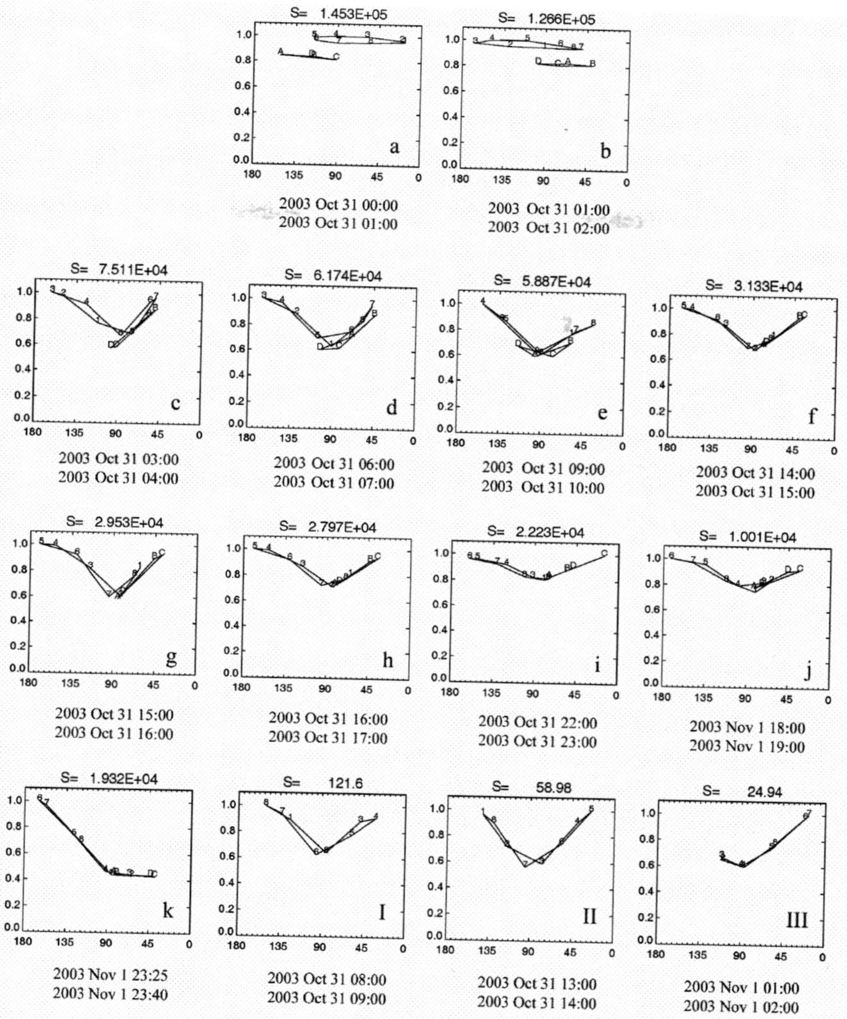

FIGURE 4. 64-112 keV electron and 1.9-4.8 MeV ion PADs measured by the LEFS60 and LEFS150 telescopes of the EPAM instrument [3] at different times during the passage and in the vicinity of the 2nd ICME.

deduced gives an arrival time for the 53-103 keV e at 1 AU at ~2054 UT on 29 Oct, assuming that they were injected at the Sun at ~2037 UT. After mirroring, the e would be expected to be observed at ACE ~2113 UT on 29 Oct, consistent with the observations.

Another configuration consistent with the observations is a closed loop B topology. When the observer is inside the ICME there is no direct connection between the observer and S1 or its downstream region (beyond the observer). Energetic ion intensities at the leading edge of the ICME would be depressed with respect to those measured at the time of the shock. The observed strong BD PADs in this scenario could result from the injection of e at both legs of looped field lines by the 29 Oct solar event. Particles may remain trapped within this configuration by reflection and re-acceleration at S2 or by reflection at the increasing B close to the Sun, bouncing back and forth between the legs of the looped B lines. Very early in the event, an asymmetry in BD flows (or even a unidirectional flow) might be expected, but if present, this was most likely masked by the high particle intensities existing prior to the onset of the event.

The observation of BD PADs within the 2nd ICME also supports the presence of an interplanetary structure able to confine energetic particles propagating between reflecting points. Assuming an open B structure, it is not definite whether the enhanced turbulent B formed downstream of S2 is efficient enough to reflect ions and e propagating between the Sun and this enhanced B region. A configuration consisting of looped B lines rooted at the Sun is consistent with the existence of BD PADs within the ICME, as well as the decrease in the ion intensity observed at the entry of ACE into the ICME.

CONCLUSIONS

The solar wind signatures (including counterstreaming suprathermal e flows, [2]) together with the characteristics of the time-intensity profiles argue in favor of closed looped B lines (see also [6], [7]). The low-energy ion intensity depressions at the entry into and the intensity increase at the exit from the 1st ICME, together with the onset of the e event within this ICME, suggest a closed looped field structure connected to the Sun at both ends. Similarly, the low-energy ion intensity depression at the entry of ACE into the 2nd ICME, the slower intra-ICME decay intensity rate and the presence of BD ion, near-relativistic and suprathermal e flows argue in favor of a closed looped B configuration.

ACKNOWLEDGEMENTS

We are thankful to our HI-SCALE team colleagues for their support and encouragement. We also thank the ACE SWEPAM and ACE MAG instrument teams and the ACE Science Center for providing the ACE data. We thank R.M. Skoug and T.H. Zurbuchen for providing ACE solar wind data collected by the SWEPAM instrument in the "search" mode during the intense events analyzed in this paper. This paper is prepared in the frame of the PYTHAGORAS II project, granted by the Greek Ministry of Education. NASA, through the Jet Propulsion Laboratory, supported a portion of the research at both APL/JHU and at NJIT.

REFERENCES

1. C. M. S. Cohen et al., *J. Geophys. Res.* **110,** A09S16, doi:10.1029/2005JA011004 (2005).
2. R.M. Skoug et al., *J. Geophys. Res.* **109**, A09102, doi:10.1029/2004JA010494 (2004).
3. R. E. Gold et al., *Space Sci. Rev.* **86**, 541 (1998).
4. O. E. Malandraki et al., *Space Sci. Rev.* **97,** 263 (2001).
5. O. E. Malandraki et al., *Ann. Geophys.* **21,** 1249 (2003).
6. J. T. Gosling, "Coronal Mass Ejections" in *26th International Cosmic Ray Conference: Invited, Rapporteur and Highlight Papers,* edited by B. L. Dingus et al., AIP Conference Proceedings 516, Melville, New York, 2000, pp. 59-79.
7. I. G. Richardson, "Using energetic particles to probe the magnetic topology of ejecta" in *Coronal Mass Ejections",* edited by N. Crooker et al., Geophysical Monograph Series 99, AGU, Washington, D. C., 1997, pp. 189-198.

Long-term Cosmic-ray Modulation during Solar Cycle 23

Helen Mavromichalaki*, Evagelos Paouris† and Theothora Karalidi†

National and Kapodistrian University of Athens, Physics Department,
Panepistimiopolis GR 15783, Athens, Greece emavromi@phys.uoa.gr; epevan@yahoo.com
** Section of Nuclear and Particle Physics*
† Section of Astrophysics, Astronomy and Mechanics

Abstract. The long-term modulation of the current solar cycle 23 is of great interest, as this cycle is characterized by many peculiarities with double peaks. Many quiet periods (Gnevyshev gaps [1]) are interrupted by extreme solar activity, as for example in April 2001, October-November 2003 and January 2005. Previous works proposed an empirical model to describe the long-term cosmic-ray intensity modulation during solar cycles 20, 21 and 22. In this work an attempt to apply this model to the solar cycle 23 is made analyzing monthly cosmic ray data from the Neutron Monitor Stations of Oulu (cut-off rigidity 0.81 GV) and Moscow (cut-off rigidity 2.42 GV). An empirical relation obtained from the linear combination of the sunspot number R_z, the solar flares N_f and the geomagnetic index A_p, is outlined incorporating the time-lag of these parameters against cosmic ray intensity. An extended study of the hysteresis phenomenon of cosmic ray intensity against different solar, interplanetary and terrestrial parameters during the time period 1996-2005 is also presented. This effect was found to have a high value, in average about 14 months, related to the size of the heliosphere and characterizing an odd solar cycle. Analytical modeling and numerical simulations are qualitatively consistent with experimental data. In our days a satisfied long-term cosmic ray model would be very useful for Space Weather studies given the possibility for cosmic-ray intensity prediction, especially now where these and satellite data are provided in real time to the Internet. A first evidence for the important role of coronal mass ejections to the long-term cosmic ray modulation is also discussed.

Keywords: Cosmic rays, hysteresis effect, long-term modulation, solar cycle, coronal mass ejections
PACS: 94.20.wq

INTRODUCTION

The cosmic-ray (CR) intensity as is well known exhibits an approximate 11-year variation anticorrelated with solar activity with perhaps some time lag [2], [3], [4]. A great effort is carried out in order to express this long-term variation of galactic cosmic-ray intensity by appropriate solar indices. Some of these indices have been used in previous works, such as the sunspot number by Nagashima and Morishita [5], solar flares by Hatton [6] and geomagnetic index by Chirkov and Kuzmin [7]. Other authors like Xanthakis, Mavromichalaki and Petropoulos [8] and Nagashima and Morishita [9] have taken into account the contribution of more than one solar and/or

CP848, *Recent Advances in Astronomy and Astrophysics,*
7th International Conference of the Hellenic Astronomical Society, edited by N. H. Solomos

geophysical parameters to the modulation process. A relation between the modulated cosmic-ray intensity during the 20th solar cycle, and a combination of the relative sunspot number, the number of proton events and the geomagnetic index A_p has been found by [10]. In a later work [11], this empirical relation was improved by including the number of corotating solar wind streams. On the other hand, a close relationship between the magnitude and frequency of Forbush decreases and the 11-year cosmic-ray variation was found by [12]; they concluded that the effect of Forbush and other transient decreases is a dominant factor in the long-term intensity modulation. Particular consideration of the cosmic ray modulation is given to the correlation of long-term cosmic ray variations with different solar-heliospheric parameters and to empirical models of cosmic ray intensity, as it is described in the review paper [13]. Recently, Lantos [14] proposed a method to predict cosmic ray intensity and solar modulation parameters. This method gives satisfactory results applied to prediction of the dose received on-board commercial aeroplane flights. He notes that prediction of the galactic cosmic ray intensity observed at a given station is preferable than prediction of the different potentials like as the modulation potential in terms of sunspot numbers [15]. The importance of this choice is that the cosmic ray intensity is the only variable directly observed. Records of cosmic ray intensity are available, and homogeneous, over a long period that is not the case for the data obtained from space observations.

It is known that there are differences in solar activity from cycle to cycle. There are series of cycles with very high activity level (odd cycles) as well as series of cycles with quite low activity (even cycles). Solar cycle 23rd is a cycle with a lot of extra violent phenomena even we are not in the maximum of solar activity. Some of those periods are April 2001, October-November 2003, January 2005, July 2005, August-September 2005 when the neutron monitors detected some of the strongest Forbush Decreases. Many quiet periods are interrupted by extreme bursts of solar activity, as in March 2001 and October 2003 [1],[16]. A result of this activity is the observation of aurora in Athens (Latitude: 37° 58' N) in November 20, 2003[17].

In this contribution a simulation of the long-term cosmic-ray modulation for the 23rd solar cycle being very close to its end, has been attempted taking into account the influence of the sunspot number, the solar flares (≥ 1 B) and the geomagnetic index A_p. The monthly cosmic-ray intensity using the above mentioned parameters has been reproduced taking into account the effect of time-lag of them against the cosmic ray intensity. A study of the hysteresis effect of some other solar and heliospheric parameters confirm once again the different characteristics of even and odd solar cycles presenting a high value of the time lag during the odd cycles [18]. As it is easy to understand, the study of a solar cycle is difficult, so the study of a solar cycle with much more violent activity than others make the whole study far more difficult. If the empirical modulation could give us a good value of standard deviation between the observed and calculated from the model values, it would be a very trustful result and this study would be used in the future in order to study the solar cycles which are follow.

DATA COLLECTION

In order to study the long-term modulation in cycle 23, monthly values of cosmic-ray intensity data from two neutron monitor stations (super NM-64) with different cut-off rigidities, Moscow (2.42 GV) and Oulu (0.81 GV) have been used. The pressure-corrected data for each station were normalized with the intensity taken equal to 1.00 at cosmic-ray intensity minimum (August 1997) and equal to 0.00 at maximum (October 2003). In this study we have used also monthly values of the following parameters: the sunspot number Rz, the number of solar flares with importance ≥ 1 B N_f and the geomagnetic index A_p taken from the National Geophysical Data Center: ftp://ftp.ngdc.noaa.gov/STP/SOLAR_DATA. Moreover, monthly values of the interplanetary magnetic field IMF from Ulysses mission http://helio.esa.int/ulysses/archive/ and flare index FI from: http://www.koeri.boun.edu.tr/astronomy/findex.html are also obtained. Data for coronal mass ejections CMEs are taken from SOHO/LASCO CME Catalog in website http://lasco-www.nrl.navy.mil/cmelist.html. This CME catalog is generated and maintained at the CDAW Data Center by NASA and the Catholic University of America in cooperation with the Naval Research Laboratory. In this point we must mentioned the fact that there are not data for CMEs for the months July, August and September of 1998 and January of 1999. It is interesting to note that a new index P_i based on the number of CMEs per month (*Nc*) and the mean plasma velocity (*Vp*) of CMEs during the examined monthly period, is defined in this work according to the following relation

$$P_i = 0.65 \cdot Nc + 0.35 \cdot Vp \qquad (1)$$

The factors 0.65 and 0.35 have been calculated by the best correlation coefficient values in linear fit. This index can well explain the cosmic ray intensity fluctuations due to the solar activity, as the main cause of Forbush decreases of cosmic ray intensity at the Earth are the coronal mass ejections traveling in the interplanetary space [19].

Time profiles of all parameters used in this work, as a function of time from 1996 to 2005 (23rd solar cycle), are given in Fig. 1. It is of great interest that this cycle presents the main features of an old solar cycle, as there are described in previous works [20], [21]. It appears a ‘saddle-like’ shape and a ‘mesa-type’ maximum. The recovery phase is of long duration is about 6-8 years. As it is seen in this figure the solar parameters sunspot number, solar flares and flare index present one maximum in the year 2000 together with the first maximum of cosmic ray intensity. The interplanetary parameter IMF presents also one maximum shifted to the year 2001, as it was expected, whereas the geomagnetic index A_p, the cosmic ray intensity and the coronal mass ejections index present a maximum in the year 2003 consistent with the second great burst of solar activity in the declining phase of the current solar cycle. The first one was in April 2001 [16]. It is remarkable that the time behavior of the defined coronal mass ejections index follows the cosmic ray intensity indicating the close relationship of these two parameters. As coronal mass ejections are recorded at the Earth orbit, it means that it is a very important index to the cosmic ray modulation recorded at ground based neutron monitors.

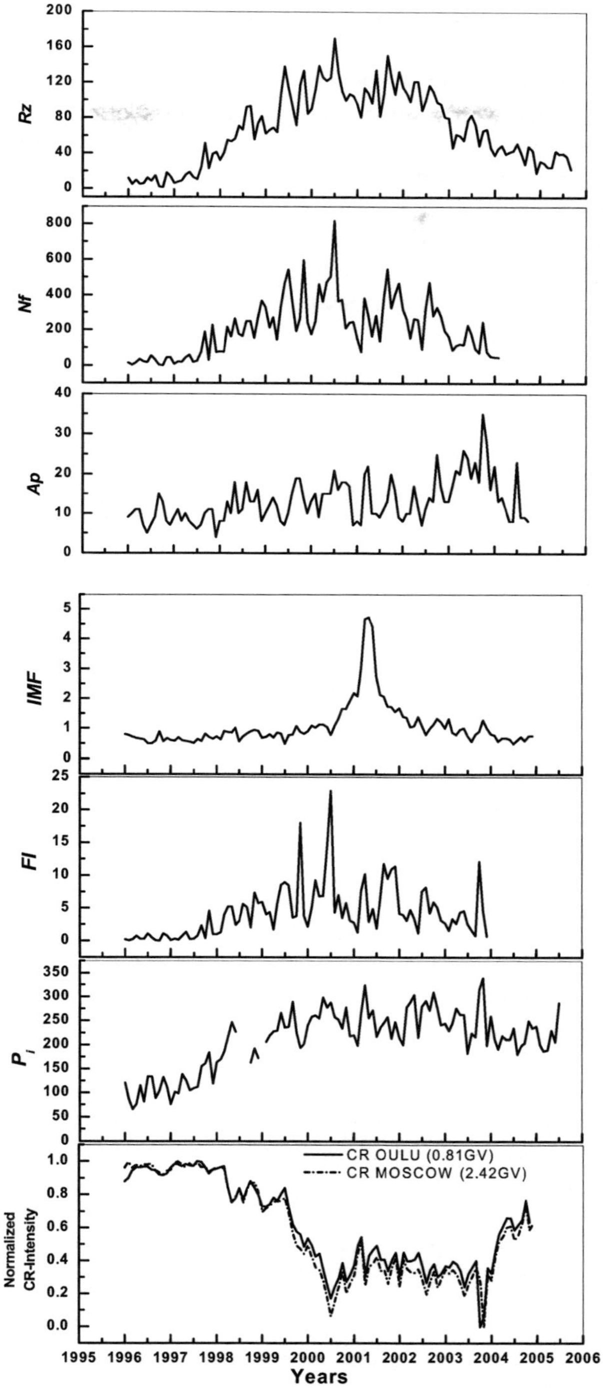

FIGURE 1. Time profiles of all parameters used in this work for the 23rd solar cycle.

HYSTERESIS EFFECT

The 11-year modulation of the cosmic ray intensity shows some time lag behind the solar activity, in other words some kind of hysteresis effect against the activity [22],[23]. A correlation analysis between the monthly values of the cosmic-ray intensity at Neutron Monitor Energies for the 23rd solar cycle and the solar activity indicated by the sunspot number Rz, the grouped solar flares Nf and the geomagnetic index Ap for the time period 1995-2005 as a function of the cosmic-ray intensity lag with respect to these parameters is performed [6], [11]. The same analysis has also been done for the interplanetary magnetic field IMF, the flare index FI and the

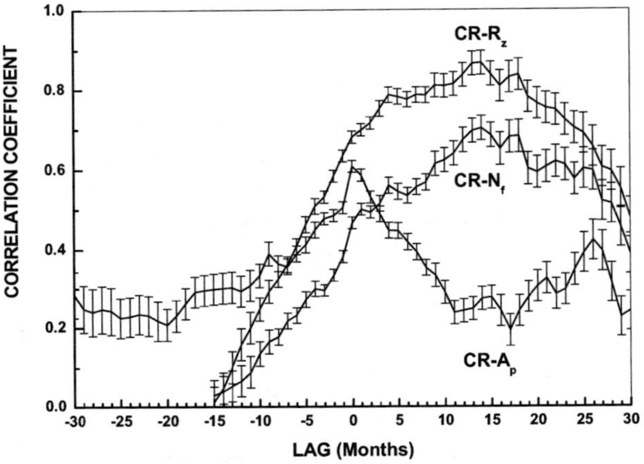

FIGURE 2. Correlation coefficients between monthly cosmic-ray intensity and sunspot number, grouped solar flares and Ap-index as functions of cosmic-ray intensity time-lag with respect to these indices for the 23rd solar cycle.

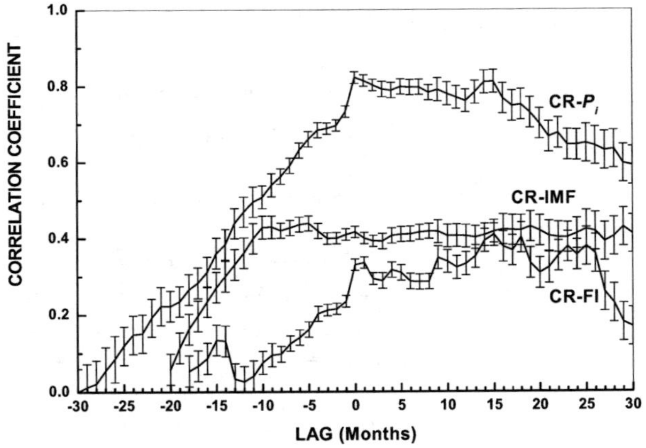

FIGURE 3. Correlation coefficients between monthly cosmic-ray intensity and the coronal mass ejections index (Pi), the interplanetary magnetic field (IMF) and the flare index (FI) for different cosmic ray time-lag with respect to these indices for the 23rd solar cycle.

TABLE I
Cross-correlation coefficients and the corresponding time lags for the 23rd solar cycle

Indices	Correlation Coefficient (r)	Time-Lag (months)
Sunspot number: *Rz*	-0.87 ± 0.01	14
Grouped Solar Flares: *Nf*	-0.70 ± 0.01	14
Geomagnetic index: *Ap*	-0.61 ± 0.02	0
Interplanetary magnetic field: *IMF*	-0.44 ± 0.02	-5
Flare index: *FI*	-0.41 ± 0.02	15
Coronal mass ejections index: P_i	-0.82 ± 0.01	0

coronal mass ejections index P_i. The correlation coefficients with their errors for different time lags calculating over the 23rd solar cycle are presented in Figs 2 and 3. The best correlation coefficients with their errors and the time lags of cosmic ray intensity corresponding to the cross-correlation coefficient of each parameter for the 23rd cycle are given in Table I. The high correlation values between cosmic rays and sunspot numbers (r = 0.87) and coronal mass ejections index (r = 0.82) as well, are indicated. The same value concerning the sunspot number was also calculated for the 21st cycle in a previous work [18]. This high correlation value of coronal mass ejections is given for first time, as the measurements of them have begun since 1996 covering only one solar cycle. An example of the correlation diagrams for the sunspot numbers and the Ap index is given in Fig. 4. The interrupted lines include points with confidence level 95%.

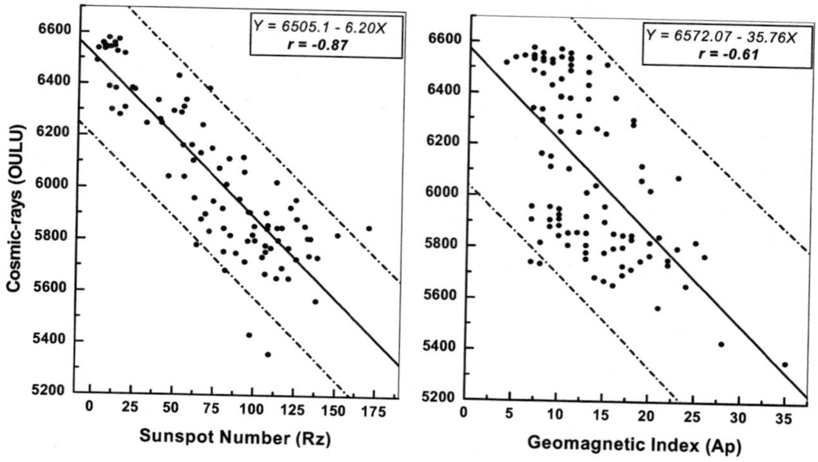

FIGURE 4. Cross-correlation coefficients between monthly cosmic-ray intensity and sunspot number Rz (left panel) and geomagnetic index Ap (right panel)

On the other hand, it is noteworthy that the sunspot number phase lag for the 23rd solar cycle is remarkably large (13.5 ± 0.6 months), whereas it was small in previous 20th and 22nd cycles (2 months and 4 months respectively) which are even cycles. This gives us more evidence that there is a distinction between even and odd solar cycles concerning the hysteresis phenomenon. To clarify this distinction, we present the time-lag of sunspot numbers with respect to cosmic-ray intensity for the last seven solar cycles in Table II.

TABLE II
Solar cycle dependence of the cosmic-ray intensity time-lag behind the sunspot number

Solar cycle	17	18	19	20	21	22	23
Time-lag (in months)	9	1	10-11	2	16	4	14

This result for the first three solar cycles has been adapted from Nagashima and Morishita [9], while the hysteresis for the last three cycles has been computed for the purposes of a previous work [18]. Inspecting the whole set of results, we can clearly distinguish between even and odd solar cycles as far as the sunspot number time-lag in concerned. This is due to the 22-year variation in the time-lag already found [9], [24], [18]. Indeed particles reach the Earth more easily when their access route is by the heliospheric polar regions than when they gain access along the recurrent sheet. In this case, as the route of access becomes longer due to the waviness of the neutral sheet [25], the time lag is also longer as one would expect from theoretical considerations. This model can't explain, however, the double-maximum structure of the even cycles.

EMPIRICAL MODULATION

An empirical model to describe the long-term cosmic-ray modulation during solar cycles 20, 21 and 22 was presented in previous works [8],[23],[25]. In the work [18], a generalized model applied over the three solar cycles mentioned before was proposed, as data are not available for earlier cycles. This model is derived by a generalization of Simpson's solar wind model using the diffusion-convection-drift model [5]. In this work we attempt to produce the monthly cosmic-ray intensity values as a function of solar and interplanetary indices for 23rd solar cycle. According to this, the modulated cosmic-ray intensity expressed by a constant *C* and the sum of some source functions appropriately selected from the solar and interplanetary indices that affect the cosmic-ray modulation. The empirical relation, used in the previous work, is given by the following expression:

$$I = C - 10^3 \times (a_1\, Rz + a_2\, Nf - a_3\, Ap) \tag{2}$$

Where the constant *C* depends linearly on the cut-off rigidity of each station, Rz, Nf and Ap are the solar-terrestrial parameters incorporating the time-lag and i (i=1 to 3) are factors calculated by the RMS-minimization (5.1, 0.5 and 0.1 respectively).
In a previous work [23] found that the constant *C* is linearly correlated to the cut-off rigidity of each station, as:

$$C = 0.95 + 0.005 \cdot P \tag{3}$$

where P is the cut-off rigidity for each station.

The observed and the calculated by the equation (2) values of the cosmic-ray intensity for Oulu Neutron Monitor station for 23rd solar cycle are presented in the upper panel of Fig. 5. The residuals are also indicated in the lower panel of the same figure. The standard deviation of this empirical modulation for the ascending and descending part of the cycle is about 12% that suggest a good approximation. It is noteworthy that this formula simulates fairly well the cosmic-ray intensity observed at the Earth during the onset and the declining phase of the solar cycles, whereas results are not satisfactory during the maximum phase of solar activity that is extreme long in this cycle. This is expected, because during the maximum phase of the cycle the solar magnetic polarity usually changes configuration. It is known that this change takes place over a period of several months. In our case the period from 1999.84 to 2001.99 years are characterized by magnetic field reversals according to [14]. Indeed, during this time interval the differences between observed and calculated values of the cosmic ray intensity seem to be high. It is also remarkable the fact that this solar cycle gave a lot of extra violent activity after the maximum of the cycle, so the need for better understanding leads to further improvements of this simulation. In the future this factor will be used for the improved empirical modulation.

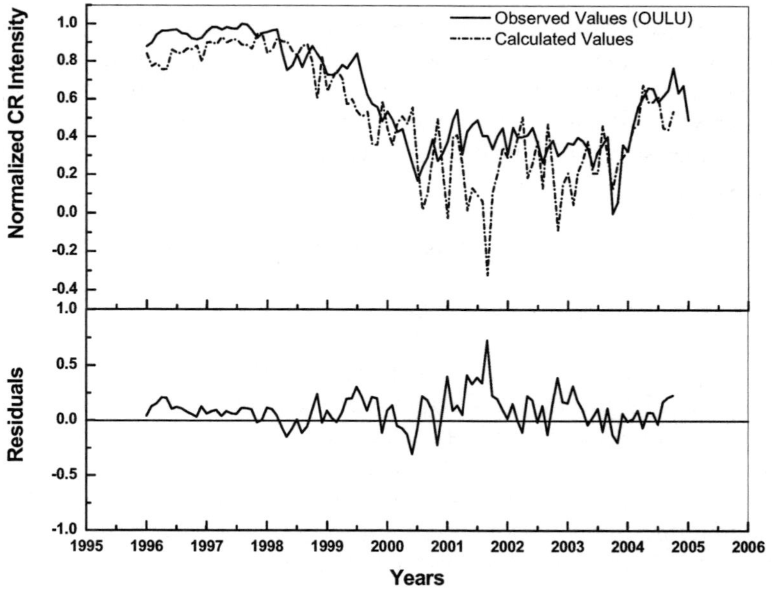

FIGURE 5. The observed and the calculated by the equation (2) values of the cosmic-ray intensity for Oulu Neutron Monitor station for 23rd solar cycle are presented (upper panel). The differences between these values are also depicted (lower panel)

DISCUSSION

The cosmic-ray modulation depends upon various factors namely: the magnitude and direction of regular magnetic fields, the level of magnetic disturbances, the solar wind speed, the size and shape of the heliosphere. In a review paper [13], it was noted that the current knowledge of cosmic ray modulation depends on observations of the cosmic ray modulation at the Earth and main characteristics of the accumulated experimental data, manifestations of the solar magnetic cycle in cosmic rays; the effect of hysteresis and its relation to the size of the heliosphere; the rigidity spectrum of long-term cosmic ray variations; the influence of the sporadic effects on long-term modulation; long-term variations of cosmic ray anisotropy and gradients; the place of ground level observations in current studies of cosmic ray modulation and their future prospects.

In this work we have invoked the relation

$$I(t) = I - \int f(r)S(t-r)dr \tag{4}$$

Where I and $I(t)$ are, respectively, the galactic (unmodulated) and modulated cosmic-ray intensities, $S(t\text{-}r)$ is the source function representing some proper solar activity index at a time $t\text{-}r$ ($r \geq 0$) and $f(r)$ is the characteristic function which expresses the time dependence of solar disturbances represented by $S(t\text{-}r)$.

The modulation of cosmic-ray intensity is described on a monthly basis empirically by the source function of Equation (4) which is an arbitrary linear combination of the three indices: the sunspot number Rz, the solar flares Nf and the geomagnetic index Ap. The characteristic function $f(r)$ of all these indices has a constant value during this solar cycle calculated by the RMS-minimization method. By this way the modulated cosmic-ray intensity is equal to galactic cosmic-ray intensity (unmodulated) at a finite distance, corrected by a few appropriate solar, interplanetary and terrestrial activity indices, which cause the disturbances in interplanetary space and thus modulate the CR intensity.

CONCLUSIONS

From the above we can summarize the following:

It is noteworthy the fact that the 23rd solar cycle, which is an odd cycle, has a remarkable large time lag as it was expected due to the magnetic cycle [18]. This cycle shows a time-lag phase of about 14 months for the solar and heliospheric parameters consistent with the characteristics of an odd cycle.

On the other hand, the empirical modulation showed that there are very good results for the ascending and descending part of the 23rd cycle, but it is also obvious the need for further improvements for the period of the solar maximum. Studying better the connection between the solar activity, specially the violent activity of the sun and the cosmic-ray intensity, we expect that the use of the new index of coronal mass ejections will improve the attempted cosmic ray simulation throwing some more light to the unexplained yet mechanism of long term modulation[19],[27]. Theoretical and statistical analysis of this index and its key role to the cosmic-ray intensity modulation will be published elsewhere very soon. This is another characteristic of the fact of the violence of this solar cycle.

ACKNOWLEDGMENTS

Thanks are due to our colleagues from Moscow and Oulu Neutron Monitor stations who have kindly provided us with their cosmic ray data used in this analysis. Thanks are also expressed to the Hellenic Astronomical Society supporting the undergraduate students E. Paouris and T. Karalidi to participate the 7th Hellenic Astronomy Conference.

REFERENCES

1. Gnevyshev M.N., *Solar Phys.* **51**, 175 (1967)
2. Forbush, S.E., *J. Geophys. Res.* **63**, 651 (1958)
3. Pomerantz M. A. and Dugal S. R., *Rev. Geophys. Space Phys.* 12, 343 (1974)
4. Perko J. S. and Fisk L. A., J. Geophys. Res. 12, 343 (1983)
5. Nagashima K. and Morishita I., *Plan. Space Sci.* **28**, 117 (1980a)
6. Hatton G.J., *Solar Phys.* **66**, 159 (1980)
7. Chirkov N. P. and Kuzmin A.I , *Proc. 16th ICRC 4, 360* (1979)
8. Xanthakis J., Mavromichalaki H. and Petropoulos B., *Astrophys. Space Sci.* **74**, 303 (1981)
9. Nagashima K. and Morishita I., *Plan. Space Sci.* **28**, 177 (1980b)
10. Mavromichalaki H. and Petropoulos B., *Astrophys. Space Sci.,* **106**, 61 (1984)
11. Mavromichalaki H. and Petropoulos B., *Earth, Moon and Planets.* **37**, 79 (1987)
12. Lockwood J.A. and W.R. Webber, *J. Geophys. Res.* **83**, 17 (1984)
13. Belov A., *Space Science Reviews*, **93**, 79 (2000)
14. Lantos P., *Solar Physics*, **229**, 373 (2005)
15. Badhwar G. D. and O'Neil P. M., *Proc. 23rd ICRC* **3**, 535 (1993)
16. Eroshenko E., Belov A., Mavromichalaki H., Mariatos G., Oleneva V., Plainaki C., Yanke V., *Solar Physics*, **224**, 345 (2004)
17. Belov A., Baisultanova L., Eroshenko E., Mavromichalaki H., Yanke V, Pchelkin V, Plainaki C., Mariatos G., *J Geophys. Research-Space Physics* **110**, 9, A09S20, doi:10.1029/2005JA011067, 2005
18. Mavromichalaki H., Belehaki A. and Rafios X., *Astron. and Astrophys.* **330**, 764 (1998)
19. Cane H. V., *Space Science Reviews*, **93**, 55 (2000)
20. H. Mavromichalaki, E. Marmatsouri and A. Vassilaki, *Earth, Moon and Planets*, **42**, 233 (1988)
21. Mavromichalaki H., A. Belehaki, X. Rafios and I. Tsagouri, *Astrophysics and Space Science* **246**, 7 (1997)
22. Moraal H., *Space Sci. Rev.* **19**, 845 (1976)
23. Mavromichalaki H., Marmatsouri E. and Vassilaki A., *Solar Phys.* **125**, 409 (1990)
24. Otaola J., Perez-Enriquez R. and Valdes-Galicia J.F., *Proc. 19th ICRC (La Jolla)* 4, 93 (1985)
25. Kota J. and Jokipii J.R., *Geophys. Res. Let.* **8**, 1979 (1991)
26. Marmatsouri L., Vassilaki A. and Mavromichalaki H., *Adv. Space Res.* **16** (9), 245 (1995)
27. Mavromichalaki H., A. Papaioannou, A. Petrides, B. Assimakopoulos, C. Sarlanis and G. Souvatzoglou, *Modern Journal of Physics A*, **20**, 6714 (2005)

Spatial & Temporal Characteristics of Ha flares during the period 1975–2002 (comparison with SXR flares)

E. Gini*, J. Polygiannakis*,†, A. Hillaris*,**, P. Preka–Papadema* and X. Moussas*

**Section of Astrophysics, Astronomy and Mechanics, Department of Physics, University of Athens, 15784 Panepistimiopolis Zografos, Athens, Greece*
†Deceased
***Hellenic Naval Academy, Piraeus 18503, GREECE*

Abstract. Although the energetic phenomena of the Sun (flares, coronal mass injections etc.) exhibit intermittent stochastic behavior in their rate of occurrence, they are well correlated to the variations of the solar cycle. In this work we study the spatial and temporal characteristics of transient solar activity in an attempt to statistically interpret the evolution of these phenomena through the solar cycle, in terms of the self-organized criticality theory.

Keywords: Solar activity; Corona; Sun spots, solar cycles; Flares
PACS: 96.60.Q; 96.60.P; 96.60.qd; 96.60.qe

INTRODUCTION

A statistics description of solar energetic phenomena, furnishes much information despite the inherently complex and, hence, incompetely undestood processes of magnetic energy storage and release which drives them.

The *Self Organised Criticality (SOC)* constitutes such a promising approach (cf. [2] for a review). In the SOC paradigm of sand pile model periods of stasis are interrupted by intermittent sand slides, or avalanches (a domino effect) and the number of avalanches $N(s)$ of size, s, is given by power law distribution $N(s) \approx s^a$, where, a, is the power-law exponent, characteristic of the particular physical process. Power–law size distributions can be reproduced computationally as the output of cellular automata models, in which energy is added, cascaded and released as a collective phenomenon on a grid.

Solar flares exemplify *Self Organised Criticality (SOC)* because they represent an almost explosive release of previously stored magnetic energy ([10], [11]); this has been duly demonstrated in [7], [8], [6], [1], [9], [14] etc. They have simulated SOC by means of *cellular automata (CA)* models, which reproduced the power law behavior in the distribution functions of the total energy, the peak luminosity and the duration of the flares. The results were observationally confirmed as regards the distribution of the HXR peak flux, total energy associated, flare duration (cf. [13], [3], [4], [5]) etc.

The flaring process, has an intermittent and, apparently, random character modelled well by the SOC. The rate of magnetic energy emergence, on the other hand, follows in general, the solar cyle evolution and exhibits recuring patterns as regards the appearance

CP848, *Recent Advances in Astronomy and Astrophysics,*
7th International Conference of the Hellenic Astronomical Society, edited by N. H. Solomos

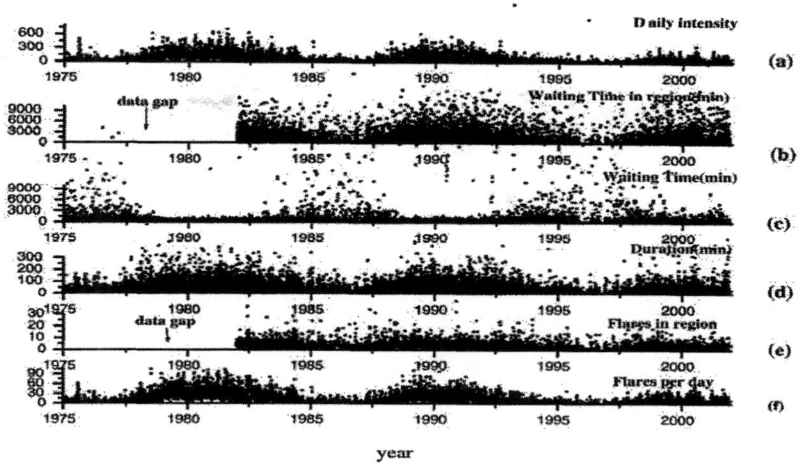

FIGURE 1. Variation of Flare Parameters with the Solar Cycle: (a) Daily Intensity (SXR); it has been included for comparison as it marks the Solar Cycle in exactly the same way as the Sunspot Number. (b) Waiting Time between Successive Ha flares in the same active region. The data gap is due to the fact that AR number fro Ha flare has not been included in SGR since 1982. (c)Waiting Time between Successive Ha flares (d)Ha Flare Duration (e) Number of flares per day in the same active region. There is a data gap from 1975 to 1982 (f). Number of Flares Daily.

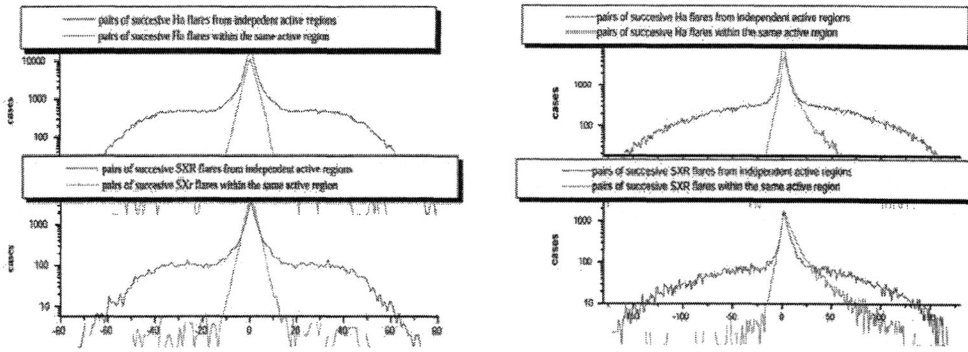

FIGURE 2. Angular Separation of Successive Flares; Upper panels refer to Ha flares lower to SXR: Left: Separation in Lattitude Right: Separation in Longitude. The peak corresponds to events within the same active region, the plateau to flares on the disk in general.

on the disk (butterfly); it provides, thus, a quasi–periodic component in the flaring activity as is represents the rate of accumulation of the energy which, in turn, is explosively released in flares.

In this report we expand on a previous work ([12]) on the spatial and temporal

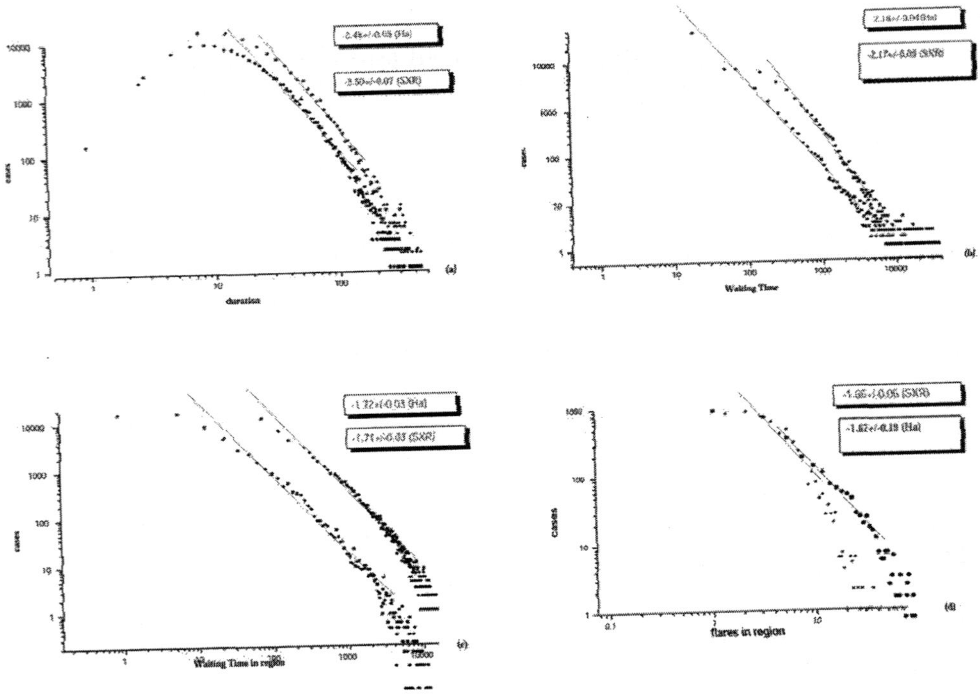

FIGURE 3. Ha Flare Statistics Compared to SXR: Top Left: Flare Duration (D); Bottom Left: Interval Between Successive Flares in the same Active Region (ΔT_{AR}) Top Right:Interval Between Successive Flares ΔT; Bottom Right: Number of Flares per Active Region (NF).

distribution of energetic phenomena on the sun. This work was based on GOES SXR data in the 1975–2002 period; in this report we extend our analysis, including the SGD Ha flares in the same period.

DATA ANALYSIS AND RESULTS

From the *Solar Geophysical Data (SGD)* we have analysed and compared SXR and Ha flare parameters in the 1975–2002 period. We only used events with recorded time of start. Since the data cover more than one solar cycle (in particular cycles 21, 22 and partly 23) we were able to study their dependence on the to the cycleŠs phase .

In figure 1 we note that the total intensity[1], (1(a)), is positively correlated with the daily number of events, (1(f)), the number of events in an active region (1(e)), the event duration, (1(d)), and the waiting time between events within the same region, (1(b)), while is anticorrelated with the waiting time between events regardless of region, (1(c)).

[1] Which follows the Sunspot Number

The distributions of event duration (D) (figure 3(a)), waiting times between successive events on disk (ΔT), (figure 3(b)) and in the same active region (ΔT_{AR}), (figure 3(c)), and number of flares in the same region (NF_{AR}), (figure 3(d)) exhibit power–law tails. This reflects the wide range of energy release time scales in flares. Althougth the non–Poisson, power–law distributions of the intervals between successive events have been challenged at times (cf. [5] and references within) they appear to underline that the individual flares are correlated. Long range time correlations are expected since the active regions are magnetically connected and the flares represent a redistribution of the magnetic energy and alteration of the magnetic topology by reconnection. This affects the lifetime of neighbouring structures and often triggers new flares, thus the trend for flares to appear clustered in time.

The spatial distribution of successive events on the solar disk was also studied; we derived the distribution of angular separation of pairs of successive events. Most occur in the same active region, with exponentially decaying probability distribution. The distribution of separation in longitude is asymmetric because of the solar differential rotation (cf. figure 2).

CONCLUSIONS

The conclusions support and reaffirm previous results of [12]. In particular:

1. The daily number of SXR and Ha flares in same active region, event duration and waiting time in same active region follow the general solar cycle. The same solar cycle dependence was derived for the flares on the whole disk regardless of active region. What constitutes a marked exception to this trend is the waiting time between successive flares on disk; this was found anticorrelated to the solar cycle.
2. The distributions N, of event duration (D),waiting time between successive events[2] (ΔT, ΔT_{AR} and number of flares occurring in the same active region NF_{AR} exhibit power–law parts over a broad range of parameters. The exponents calculated are:
    - $N \propto D^{-2.48\pm0.05}$ for Ha compared to $N \propto D^{-2.50\pm0.07}$ for SXR[3]
    - $N \propto \Delta T^{-2.18\pm0.04}$ for the waiting time between Ha flares compared to $N \propto \Delta T^{-2.17\pm0.05}$ for SXR
    - $N \propto \Delta T_{AR}{}^{-1.72\pm0.03}$ for the waiting time between Ha flares in the same AR compared to $N \propto \Delta T_{AR}{}^{-1.71\pm0.03}$ for SXR
    - $N \propto NF_{AR}{}^{-1.67\pm0.19}$ for the number of Ha flares within the same active region compared to $N \propto NF_{AR}{}^{-1.66\pm0.06}$ for SXR.

   The power–law exponents are, within the margin of statistical error, the same for Ha and SXR. They confirm thus the previous conclusions by [12] in support of the hypothesis of meta-stable, SOC state of magnetic structures of the Sun. The power–law distributions of waiting times indicate that each event triggers others due to the spatial interconnection of the erupting magnetic structures.

---

[2] on disk or within each active region; the latter has been subscripted with AR

[3] In [13] the SXR duration power–law differs a little, being $N \propto D^{-2.93\pm0.12}$

3. The study of the angular separation between successive events shows that mostly, energetic events tend to cluster within the same active region.

ACKNOWLEDGMENTS

This work is dedicated to the departed member of the ARTEMIS–IV Group, Dr. John Polygiannakis.

REFERENCES

1. A. Anastasiadis, *First School on Physics and Technology of Fusion Proceedings, (Eds. A. Grekos)* 115–129, (2003).
2. P., Bak, *How nature works, the science of self–organized criticality, Copernicus, Springer–Verlag, New York*, (1996).
3. N. B., Crosby, M. J., Aschwanden, B. R., Dennis, *Sol.Phys.*, **143**, 275–299, (1993).
4. N. B., Crosby, M. J., Aschwanden, B. R., Dennis, *Advances in Space Research*, **13**, 179–182, (1993).
5. N. B., Crosby, M. K., Georgoulis, , N., Vilmer, *ESA SP* **446**, 247–250, (1999).
6. M. K., Georgoulis, L., Vlahos, *Astron. Astrophys.*, **336**, 721–734, (1998).
7. E. T., Lu and R. J, Hamilton, *Astrophysical Journal*, **380**, L89–L92 (1991).
8. E. T., Lu and R. J, Hamilton, J. M., Mc Tiernan and K. R., Bromund, *Astrophysical Journal*, **412** 841-852 (1993).
9. A. L., MacKinnon, K. P., Macpherson, L., Vlahos, *Astron. Astrophys.*, **310**, L9–L12, (1996).
10. E. N., Parker, *Astrophys. J*, **330**, 474–479, (1988).
11. E. N., Parker, *Sol. Phys.*, **121**, 271-288, (1989).
12. J. M., Polygiannakis, A., Nikolopoulou, P., Preka-Papadema, X., Moussas, A., Hillaris, *ESA SP* **505**, 541–544, 2002.
13. A., Veronig, M., Temmer, A., Hanslmeier, W., Otruba, and M., Messerotti, *Astron. Astrophys.*, **382**, 1070–1080 (2002)
14. L., Vlahos, M. K., Georgoulis, R., Kluiving and P., Paschos, *Astron. Astrophys.*, **299**, 897–911 (1995).

Type II and IV radio bursts in the active period October-November 2003

V. Petoussis*, P. Tsitsipis*, A. Kontogeorgos*, X. Moussas†, P. Preka–Papadema†, A. Hillaris†,**, C. Caroubalos†, C. E. Alissandrakis‡, J.-L. Bougeret§ and G. Dumas§

*Technological Education Institute of Lamia, Lamia, Greece
†University of Athens, 15783 Athens, Greece
**Hellenic Naval Academy, Piraeus 18503, GREECE
‡University of Ioannina, 45110 Ioannina, Greece
§Observatoire de Paris, CNRS UA 264, 92195 Meudon Cedex, France

Abstract. In this report we present the Type II and IV radio bursts observed and analyzed by the radio spectrograph ARTEMIS IV[1], in the 650-20MHz frequency range, during the active period October-November 2003. These bursts exhibit very rich fine structures such fibers, pulsations and zebra patterns which is associated with certain characteristics of the associated solar flares and CMEs.

Keywords: Solar activity; Corona; Radio; Radiation and spectra; Solar electromagnetic emission
PACS: 96.60.Q; 96.60.P; 95.85.Bh; 96.25.Tg; 96.60.Tf

INTRODUCTION

The active period from 20 October to 5 November 2003 has attracted considerable attention ([5], [8], [10], [11], [6], [12], [13]) because of its abundance in of powerful flares and large coronal mass ejections (CMEs) as well as by strong space weather disturbances. This activity was associated with the appearance of a global complex consisting of three large, remote but connected active regions (AR): AR 484 (Carrington coordinates, N04; L = 354), AR 486 (S15; L = 283) and AR 488 (N08; L = 291) (cf. [5]); while in the visible hemisphere of the sun, this complex, produced 8 X–class and 12 M–class flares.

The Artemis IV solar radio-spectrograph operating at Thermopylae since 1996 ([2]) consists of a 7–m parabolic antenna covering the metric range, to which a dipole antenna was added recently in order to cover the decametric range ([3], [9]). Two receivers operate in parallel, a sweep frequency analyser (ASG) covering the 650–20 MHz range in 630 data channels with a cadence of 10 samples/sec and a high sensitivity multichannel acousto–optical analyser (SAO), which covers the 270–450 MHz range in 128 channels with a high time resolution of 100 samples/sec.

In October–November 2003 ARTEMIS-IV observed six major events on October 23 (08:24 UT), October 25 (05:20 UT), October 26 (06:16 UT), October 28 (11:02 UT),

---

[1] Appareil de Routine pour le Traitement et l' Enregistrement Magnetique de l' Information Spectral

CP848, *Recent Advances in Astronomy and Astrophysics,*
7th International Conference of the Hellenic Astronomical Society, edited by N. H. Solomos

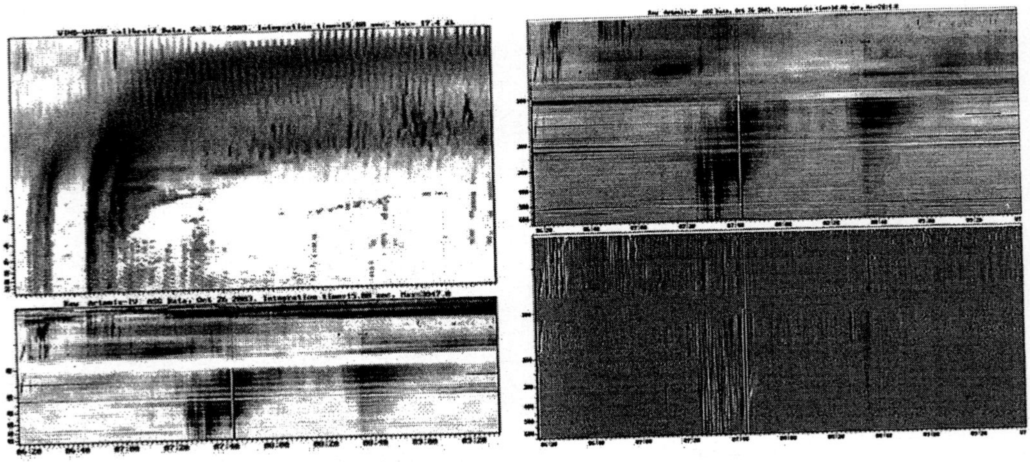

FIGURE 1. LEFT:Combined ARTEMIS-IV and WIND–WAVES spectra of the October 26, 2003 event. RIGHT: ARTEMIS–IV pectra in the frequency range 20–650 MHz; Intensity (top)and its time derivative (bottom), enhancing fine temporal details

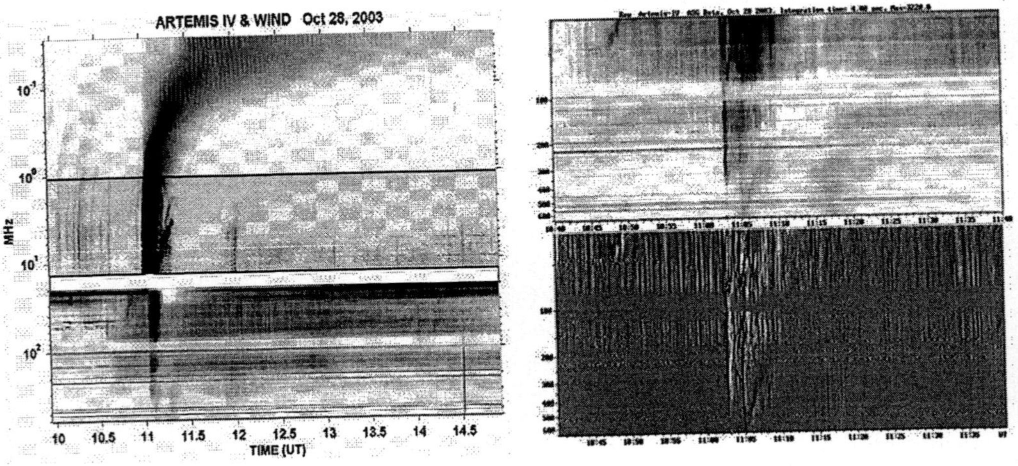

FIGURE 2. LEFT: Combined ARTEMIS-IV and WIND–WAVES spectra of the October 28, 2003 event. RIGHT: ARTEMIS–IV spectra in the frequency range 20–650 MHz; Intensity (top)and its time derivative (bottom), enhancing fine temporal details

November 2 (a behind–the–limb event, 09:20 UT) and November 3 (09:49); Combined with data from WIND–WAVES ([1]), these observations provide a complete view of the radio emission induced by shock waves and electron beams from the low corona to about 1 A.U. Out of those, we selected three events to give an overview here.

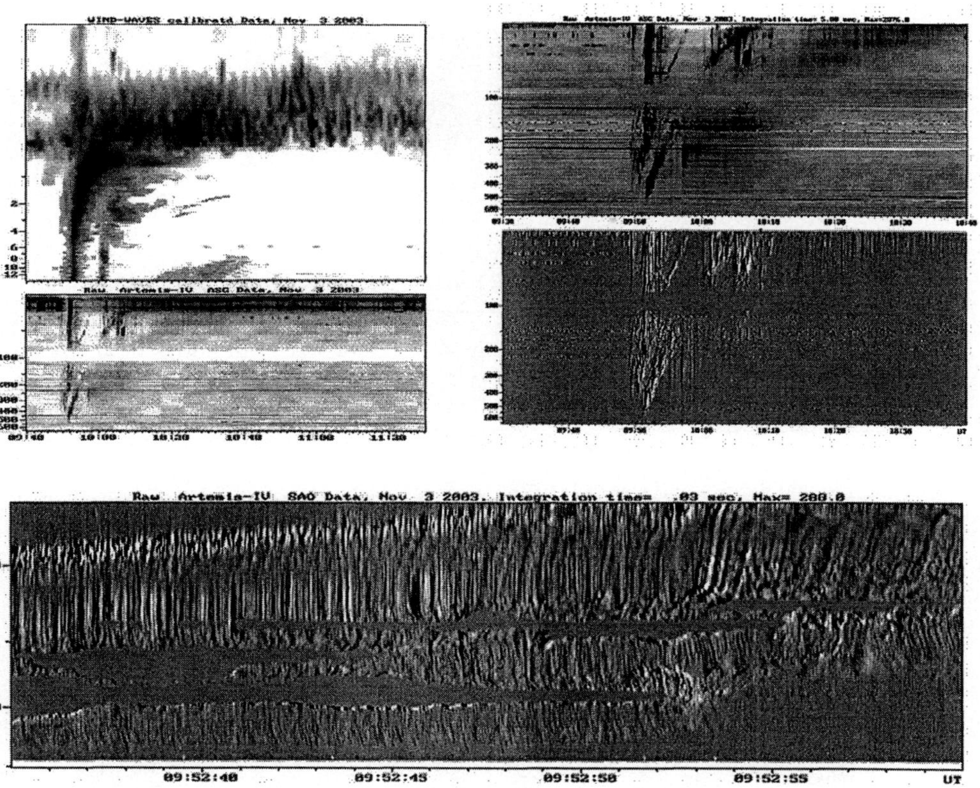

FIGURE 3. UPPER LEFT: Combined ARTEMIS-IV and WIND–WAVES spectra of the November 3, 2003 event. UPPER RIGHT: ARTEMIS–IV ASG Spectra in the frequency range 20–650 MHz; Intensity (top)and its time derivative (bottom), enhancing fine temporal details. BOTTOM PANEL: Differential SAO Spectrum with high (0.03 sec) resolution of a part Type II lane, showing U–shaped *herringbone* structures.

SELECTED EVENTS DESCRIPTION

Event Overview

The event of October 26, 2003

A type II with fundamental–harmonic emission, starts at 06:16 followed by high frequency type IV emission starting at 06:20 (high frequencies) and 06:44 (low frequencies); follows high frequency continuum emission at 07:20 and at 08:35 UT, with considerable fine structure (cf. figure 1).

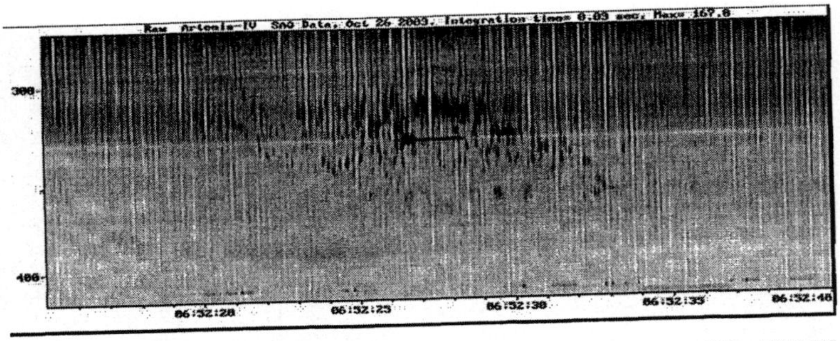

FIGURE 4. ARTEMIS–IV SAO differential spectrum, in the frequency range 270–450 MHz, of Narrowband Spikes (26 October 2003)

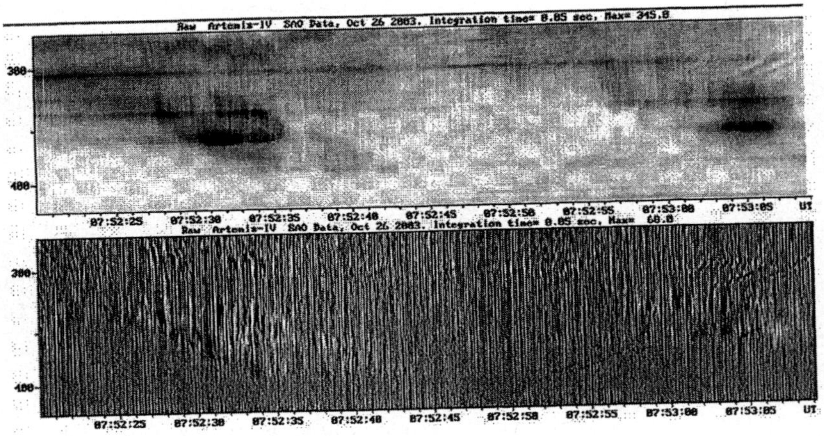

FIGURE 5. ARTEMIS–IV SAO spectra, in the frequency range 350–450 MHz, of Slowly Drifting Bursts (cf. [7] their figure 7) and fibers overlayed on a pulsating structure (26 October 2003). Top: Intensity Spectrum, Bottom: Differential Spectrum.

The event of October 28, 2003

The events start with a type II burst starting at 10:47 in ARTEMIS–IV lower frequencies, and extending into the WIND–WAVES range. Follows a complex event, comprising an intense type III group, a type II with fundamental–harmonic structure followed by a type IV continuum; all extend into the WIND–WAVES range. (cf. figure 2)

The event of November 3, 2003

The events starts at 09:49 UT with weak type III bursts, followed by a very strong type III group extending down to 150 kHz. Follows a type II, displaying fundamental–harmonic structure. There is an indication of multiple shocks since three lanes appear

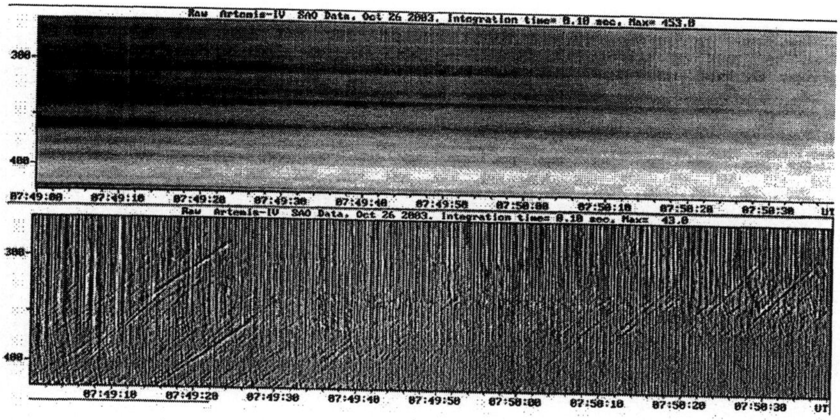

FIGURE 6. ARTEMIS–IV SAO spectra, in the frequency range 270–450 MHz, of Pulsations and Fibers (26 October 2003). Top: Intensity Spectrum, Bottom: Differential Spectrum.

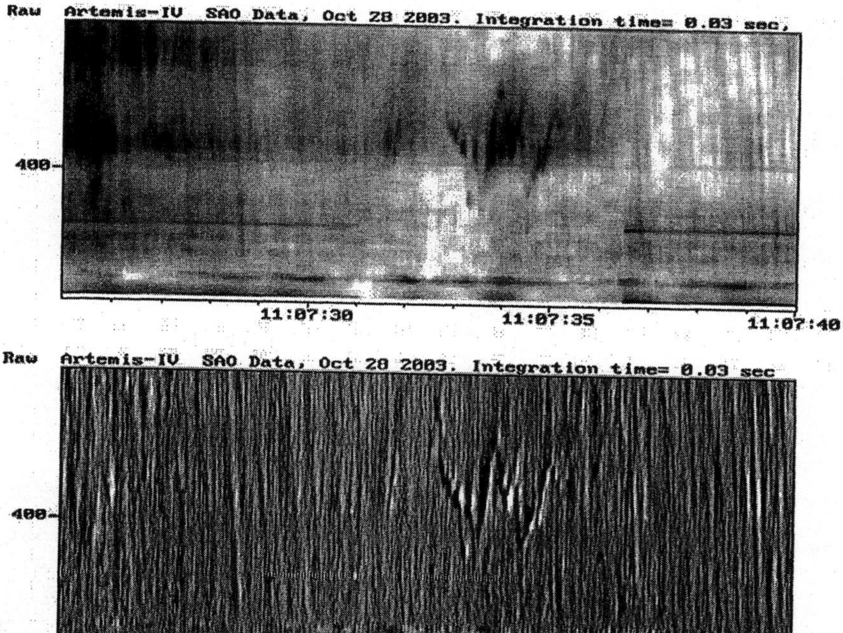

FIGURE 7. ARTEMIS–IV SAO spectra (28 October 2003), in the frequency range 270–450 MHz, of a lace type structure (cf. [7] their figure 11) overlayed with spikes. Top: Intensity Spectrum, Bottom: Differential Spectrum.

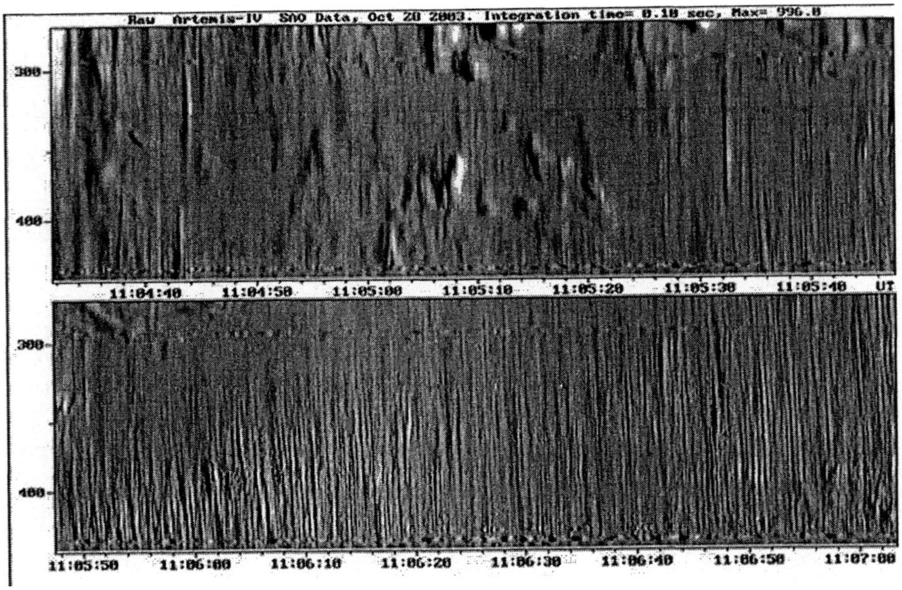

FIGURE 8. ARTEMIS–IV SAO differential spectrum, in the frequency range 270–450 MHz, of Narrowband Type III Bursts ([7] their figure 6) Pulsations and Fibers (28 October 2003). Zebra patterns are superposed on the fibers and the pulsating structure.

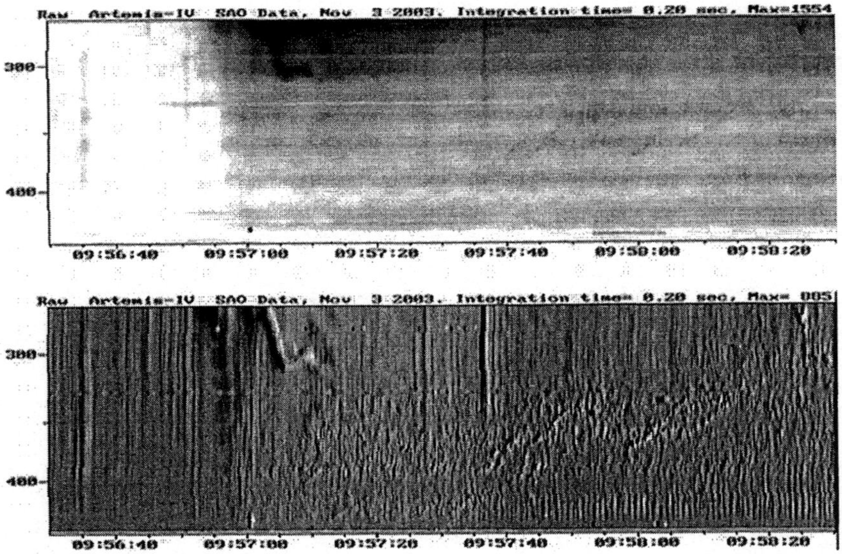

FIGURE 9. ARTEMIS–IV SAO spectra (03 November 2003), in the frequency range 270–450 MHz, of Isolated Broadband Pulse ([7] their figure 4); in the lower frequency part of the spectrum the lower part of a pulsating structure with an EEL–type form([4] their figure 1) appears. Top: Intensity Spectrum, Bottom: Differential Spectrum.

on the spectra; firstly a fast type II, probably fundamental, followed by the fundamental-harmonic pair; these shocks were associated with a coronal wave by [12]. The event extends into the high frequency part of the WIND–WAVES spectrum, down to about 9 MHz. A type IV continuum follows the type II, also extending into the WIND–WAVES band range (cf. figure 3 upper panels).

The SAO recordings of the type II harmonic, with high resolution, exhibit *herringbone* structure, resulting from *type III–like* emissions due to electron populations accelerated by the MHD shock. In this recordings however the spectral shape of the *herringbone* is *type U–like*, thus indicating the presence of some type of magnetic confinement for the exciter (cf. figure 3 lower panels).

Fine Structure

The high sensitivity and time resolution of the SAO facilitated an examination on fine structure within the three studied periods; all exhibited rich fine structure embedded in the Type–IV continua. In our analysis, the continuum background is removed by the use of high–pass filtering on the dynamic spectra (differential spectra in this case).

We present certain examples, which are divided according to a published morphological classification scheme ([7]) based on Ondrejov Radiospectrograph recordings in the 0.8–2.0 GHz range. In our recordings we have detected narrow band spikes (figure 4), Slowly Drifting Bursts (figure 5), fibers (figures 5,6), Narrowband Type III Bursts (figure 8), laces (figure 7), zebra patterns (figure 8), and isolated broadband pulses (figure 9). An EEL type emission ([4]) was found to border a pulsating structure (figure 9).

DISCUSSION & CONCLUSIONS

The ARTEMIS–IV radio–spectrograph, operating in the range of 650–20 MHz, observed 6 complex events during the super–active period of October–November 2003; three of them were presented here. WIND-WAVES data complement nicely the ARTEMIS data and trace the radio emission from the middle corona all the way to almost 1 A.U. The high resolution SAO recordings on the other hand reveal a variety of fine structure which, almost, matches the comprehensive Ondrejov Catalogue ([7]). This last, although it refers to the spectral range 0.8–2 GHz, seems to produce similar fine structure with the metric range.

Further analysis of the metric radio bursts fine structure is in progress.

ACKNOWLEDGMENTS

This work was supported in part by the Greek Secretariat for Research and Technology through the program Pythagoras II. The LASCO CME catalogue is generated and maintained by the Center for Solar Physics and Space Weather, The Catholic University of America in cooperation with the Naval Research Laboratory and NASA. SOHO is a project of international cooperation between ESA and NASA. The Solar Geophysical Data Catalogue is compiled and maintained by the US Department of Commerce.

REFERENCES

1. J. L. Bougeret, M. L., Kaiser, P. J., Kellogg, R., Manning, K., Goetz, S. J., Monson, N., Monge, L., Friel, C. A., Meetre, C., Perche, L., Sitruk, S., Hoang, *Space Science Reviews*, **71**, 231–263, (1995).
2. C., Caroubalos, D., Maroulis, N., Patavalis, J. L., Bougeret, G., Dumas, C., Perche, C., Alissandrakis, A., Hillaris, X., Moussas, P., Preka–Papadema, A., Kontogeorgos, P., Tsitsipis, P., and G., Kanellakis, *Experimental Astronomy*, **11**,23–32, (2001).
3. C., Caroubalos, C. E., Alissandrakis, A., Hillaris, P., Preka–Papadema, J., Polygiannakis, X., Moussas, P., Tsitsipis, A., Kontogeorgos, V., Petoussis, C., Bouratzis, J. L., Bougeret, G., Dumas and A., Nindos, *These Proceedings* (2006).
4. G.P. Chernov, A.K. Markeev, M. Poquerusse, J.L. Bougeret, K.-L. Klein, G. Mann, H. Aurass, and M.J. Aschwanden, *Astron. Astrophys.*, **334**, 314Ű324, (1998)
5. I. M., Chertok, V. V., Fomichev, A. A., Gnezdilov, R. V. Gorgutsa, A. K., Markeev & D. E., Sobolev, *Astronomical and Astrophysical Transactions*, **24**, 45-52, (2005).
6. N., Gopalswamy, L., Barbieri, E. W., Cliver, G., Lu, S. P., Plunkett, R. M., Skoug, *Solar Physics*, **226**, 337–357, (2005).
7. K., Jiricka, M., Karlicky, H., Meszarosova and V., Snizek, *Astronomy and Astrophysics*, **375**, 243-250, (2001)
8. J., Kiener, M., Gros, V., Tatischeff, and G. Weidenspointner, *Astronomy and Astrophysics*, **445**, 725–733 (2006)
9. Á. Kontogeorgos, P. Tsitsipis, C. Caroubalos, X. Moussas, P. Preka-Papadema, A. Hillaris, V. Petousis, J-L Bougeret, C.E. Alissandrakis, G. Dumas, J. Polygiannakis, *13th International Symposium on Measurements for Research and Industry Applications, Athens 29 Sept.-1 Oct.*, 39–44, 2004.
10. E. Mitsakou, M. Thanasa , P. Preka–Papadema X. Moussas, A. Hillaris, C. Caroubalos, C. E. Alissandrakis P. Tsitsipis, A. Kontogeorgos, J. L. Bougeret, G. Dumas, *These Proceedings* (2006).
11. Cheng-Ming Tan, Qi-Jun Fu, Yi-Hua Yan and Yu-Ying Liu, *Chin. J. Astron. Astrophys*, **4**, 205-208, (2004).
12. B., Vrsnak, A., Warmuth, M., Temmer, A., Veronig, J., Magdalenic, A., Hillaris, M., Karlicky., *Astronomy and Astrophysics*, (accepted) (2006).
13. Wang Yuming, Ye Pinzhong, Zhou Guiping, Wang Shujuan, S., Wang, Yan Yihua, Wang Jingxiu, *Solar Physics*, **226**, 337-357, (2005).
14. S., Yashiro, N., Gopalswamy, O. C., St. Cyr, G., Lawrence, G., Michalek, C. A., Young, S. P., Plunkett, & R. A., Howard, *AGU Spring Meeting* Abstract **SH31C-10**, 2001

Nonlinear Magneto-Hydrodynamical Modeling Of Solar Envelope

Huseyin Cavus[1] and Aysel I. Karafistan[2]

Canakkale Onsekiz Mart University, Arts&Science Faculty, Physics Department Canakkale-TURKEY
[1] h_cavus@comu.edu.tr , [2] akarafistan@comu.edu.tr

Abstract. In this study, our objective was to investigate the variation of the physical parameters in the 30 % outermost convective solar layer with the use equations of magneto-hydrodynamics (MHD) for non-linear case. The anelastic approximation was used in our calculations. The variations in physical parameters were performed by Maple. Hydrostatics equilibrium state was taken as a Standard Solar Model (SSM) excluding both rotation and magnetic field.

Keywords: Sun: magnetic fields: magneto-hydrodynamics, anelastic approximation.

PACS: 96.60.-j, 95.30.Qd, 96.12.Hg

INTRODUCTION

Magnetic field in the Sun has been the most attractive subject of solar physics since about fifty years (eg Chitre et. al, 1973 and Parker, 1955). More recent arguments support the hypothesis that the magnetic field, which is responsible for the solar activity, is stored in the base of the convective zone (Schüssler et. al, 1994). A solar envelope model taking into account the magnetic field was constructed in this work. In this study we used the definition of anelastic approximation with the help of meridional circulation. This approximation is a filtering approximation for the equations of motion that eliminates sound waves by assuming that the flow has velocities and phase speeds much smaller than the speed of sound. In its purest form, it requires that the reference state be isentropic as well as hydrostatic, although in practice the reference state is often taken to be nonisentropic which can have deleterious effects on the energy conservation properties of the full set of equations.

FORMULATION

The ideal MHD equations are the equation of continuity (1), the momentum transfer (2), the magnetic induction (3), and the absence of magnetic monopoles (4).

$$\nabla\cdot(\rho\mathbf{v}) = 0 \tag{1}$$

CP848, *Recent Advances in Astronomy and Astrophysics,*
7th International Conference of the Hellenic Astronomical Society, edited by N. H. Solomos

$$\rho\frac{d\mathbf{v}}{dt}=-\nabla p+\rho\mathbf{g}-\nabla.(\rho\mathbf{uu})+\frac{1}{\mu_0}(\nabla\times\mathbf{B})\times\mathbf{B}\ (2)$$

$$\frac{\partial\mathbf{B}}{\partial t}=-\nabla.(\mathbf{vB}-\mathbf{Bv})=(\mathbf{B}.\nabla)\mathbf{v}-(\nabla.\mathbf{v})\mathbf{B}-(\mathbf{v}.\nabla)\mathbf{B} \quad (3)$$

$$\nabla.\mathbf{B}=0\,. \quad (4)$$

Here ρ, **v**, p and **B** denote the mean values of density, velocity, pressure and magnetic field respectively. Details about the averaging procedure for the turbulent velocity **u** can be found in the work of Gabriel (1996). Equations (1)-(4) given above and describing the motion of such a mean flow are in fact highly non-linear and difficult to solve. It is not easy to solve them even impossible them analytically or numerically without any approximation (e.g. linearization, incompressibility…).

The velocity can be written as

$$\mathbf{v}=\Omega r\sin\theta\,\mathbf{e}_\phi+\mathbf{v}_m \quad (5)$$

where Ω is the angular velocity and $\mathbf{v}_m=v_r\mathbf{e}_r+v_\theta\mathbf{e}_r$ is the meridional circulation (Forgacs-Dajka and Petrovay, 2002). The value of Ω depends on the latitude and radius since there is differential rotation in the Sun. Wilson et al. (1996) suggested a constant on cylinders model with an analytical formula

$$\frac{\Omega(r,\theta)}{2\pi}=462-77[1-(X\sin\theta)^2]-42[1-(X\sin\theta)^2]^2 \text{ in nHz.} \quad (6)$$

X=(r/R⊙) gives the relative radius. X=0.713 is the base of the convection zone. Figure 1 shows the changes of $\Omega/2\pi$ with respect to relative radius at 30° and 45° latitudes.

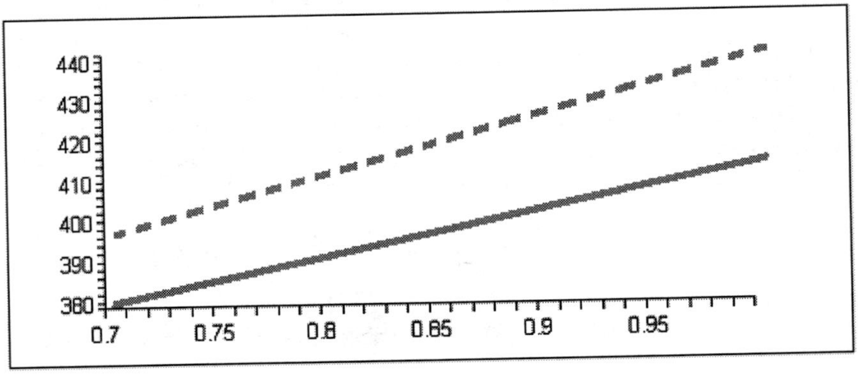

FIGURE 1. Changes of angular velocity at 30° and 45° latitudes in nHz (dashed: 30°, line: 45°).

The morphology and the profile of magnetic field of the Sun is one of the unknown of the solar physics. Certain studies (e.g. Moreno-Insertis et al., 1992) concerning the magnetic flux tubes in the Sun show that significant properties of the sunspots require the existence of an azimuthal flux system ranging between 10 T and 100 T at the base of the convective layer. According to Schüssler et al. (1994) the reason for such high azimuthal values for the stored magnetic flux, prior to eruption, is due to the fact that flux tubes with weaker field would vanish at the top of the convection zone because of the Coriolis force.

Using the above approximations together with the relation $B_\phi \rangle\rangle B_r$, B_θ among the radial (B_r), the polar (B_θ) and the azimuthal (B_ϕ) components of the magnetic field, given by Kleeorin et al., (1996) in spherical coordinates, we can consider the magnetic to be in the toroidal direction. Magnetic field in the solar convection zone of the present Sun is given with use of helioseismic methods (Antia et al., 2000):

$$\mathbf{B} = B\mathbf{e}_\phi \tag{11}$$

$$B = a(r)\frac{dP_l}{d\theta}(\cos\theta) \tag{12}$$

with the amplitude

$$a(r) = \begin{cases} \sqrt{2\mu_0 p\beta}(1-(\frac{r-r_0}{d})^2) & |r-r_0| \le d \\ 0 & \text{otherwise} \end{cases}. \tag{13}$$

Here P_l is the Legendre polynomial of degree *l*, *p* the gas pressure, β a constant giving the ratio of magnetic pressure to the gas pressure. r_0 and *d* are constants defining the mean position and thickness of the layer where the field is supposed to be mostly concentrated. The values of r_0, d and l are taken as 0.713R⊙, 0.287 R⊙ and 2 respectively in our work. Figure 2 shows the isolines of the magnetic field with respect to relative radius and polar angle θ in radians (left part), the changes of the magnetic field at 45° latitudes with the use of equations 11-13.

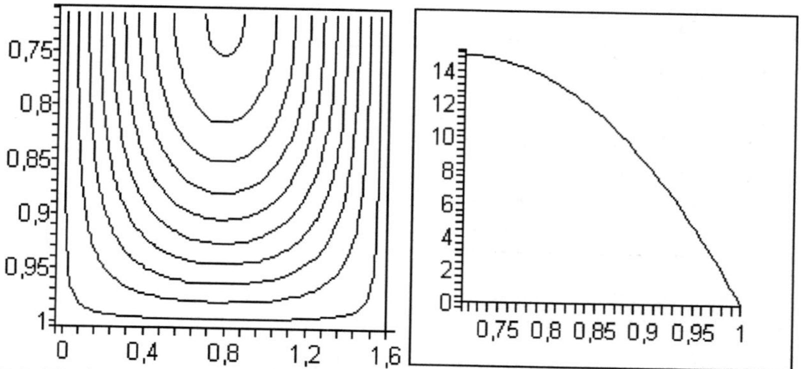

FIGURE 2. The isolines of the magnetic field shown in rectangular fashion in order to be understood (left part) and the change of magnetic field at 45° latitude with respect to relative radius (right part).

The maximum value of magnetic field takes place at the base of the convection zone for 45° latitude shown in left part. Ossendrijver (2003) is also taken into consideration while choosing this profile of the magnetic field. It has its maximum value at the base of the convection zone with an approximate value of 14 Tesla. Ossendrijver states that the maximum value of the magnetic field takes pace at the base convection zone with β of 10^5.

RESULTS

In our study reference model is the model of Christensen-Dalsgaard et al. (1996). Figure 3 and 4 show the changes of components of meridional velocity with the use equation 7 –10 at 30° and 45° latitude respectively. As it can be seen from the figures there are big jumps at 0.78 and 0.95. The reason for the jump at 0.78 may be the overshooting mechanism. The jump at 0.95 can be predicted from the sharp change of adiabatic index taken by reference model. Its theoretical value is 5/3 for fully ionised monatomic gas. In order to check our results with apriority of anelastic approximation, we look at the changes of the components of the meridional circulation between 0.79 and 0.95 we see that, the maximum value of the radial component is approximately equal to -6 m/s. It is 25 m/s for polar component at 30° latitude as shown in Figure 5.

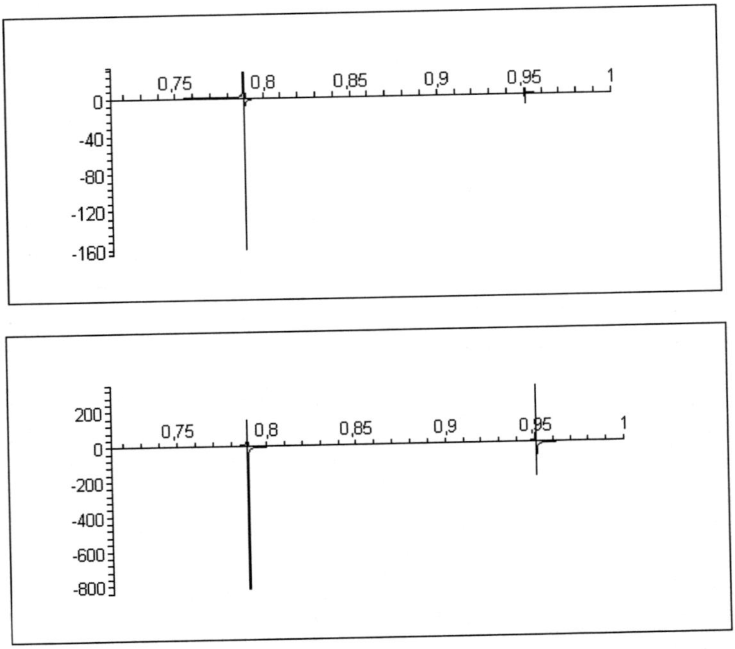

FIGURE 3. Variations of the components of the meridional circulation at 30° latitude with respect to relative radius. Radial (upper), polar (lower).

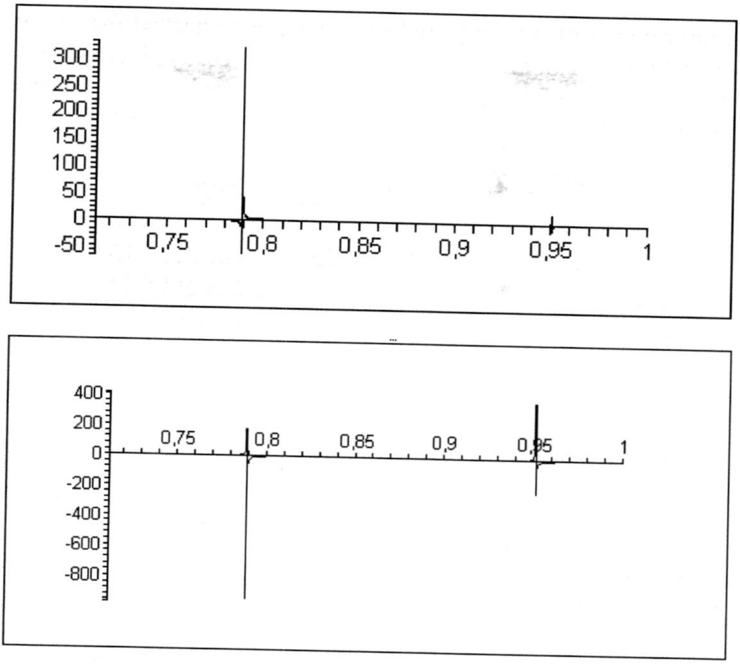

FIGURE 4. Same as Figure 2 but for 45° latitude.

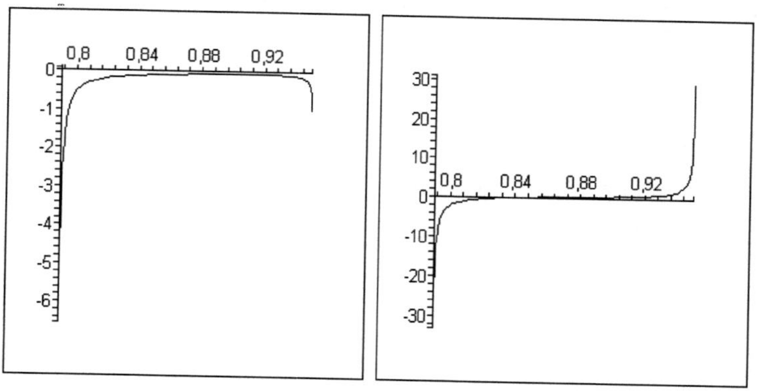

FIGURE 5. Variations of the components of the meridional circulation at 30° latitude with respect to relative radius between 0.79 and 0.95. Radial (left), polar (right).

DISCUSSION AND CONCLUSION

In this study, the investigation for the variation of the physical parameters in the 30 % outermost convective solar layer was done with the use equations of MHD. Model of Christensen-Dalsgaard et al. (1996) was taken as a reference model. We saw important jumps in the components of meridional circulation, when we look the figure 3 and 4. The reasons of these jumps are still under study. We think that these jumps are because of nonlinearity and chaotic behaviours. We checked the apriority of the anelastic approximation with the use of the figure 5. Our study gives consistent results when compare them with reference model that is speeds are less than the sound speed.

ACKNOWLEDGEMENTS

This work is being supported by Canakkale Onsekiz Mart University-Scientific Research Projects Commission (COMU-BAP) (Project No: 2005/113)

REFERENCES

1. H.M. Antia, S.M. Chitre and Thompson, M.J., *J. Astrophys. & Astron.* **21**, 343-347 (2000).
2. S.M. Chitre, D. Ezer and R. Stothers, *Astrophys. Letters* **14**, 37-40 (1973).
3. J. Christensen-Dalsgaard, W. Däppen, S.V. Ajukov, E.R. Anderson, H.M. Antia, S. Basu, V.A. Baturin, G., G. Berthomieu, B. Chaboyer, S.M. Chitre, A.N. Cox, P. Demarque, J. Donatowicz, W.A. Dziembowski, M. Gabriel, D.O. Gough, D.B. Guenther, J.A. Guzik, J.W. Harvey, F. Hill, G. Houdek, C.A. Iglesias, A.G. Kosovichev, J.W. Leibacher, P. Morel, C.R. Proffitt, J. Provost, J. Reiter, E.J. Rhodes Jr., F.J. Rogers, I.W. Roxburgh, M.J. Thompson and R.K. Ulrich, *Science* **272**, 1286-1292 (1996).
4. E. Forgacs-Dajka and K. Petrovay, *Astron. & Astrophys.* **389**, 629-640 (2002).
5. M. Gabriel, *Bull. Astr. Soc. India* **24**, 233-243 (1996).
6. N. Kleeorin, M. Mond and I. Rogachevskii, *Astron. & Astrophys.* **307**, 293-309 (1996).
7. F. Moreno-Insertis, M. Schüssler and A. Ferriz-Mass, *Astron. & Astrophys.* **264**, 686-700 (1992).
8. M. Ossendrijver, *The Astron. Astrophys. Rev.***11**, 287-367 (2003).
9. E.N. Parker, *Astrophys. J.* **122**, 293-314 (1955).
10. M. Schüssler, P. Caligari, A. Ferriz-Mas and F. Moreno-Insertis, *Astron. & Astrophys.* **281**, L69-L72 (1994).
11. P.R. Wilson, D.Burtonclay and Y. Li, *Astrophys. J.* **470**, 621-628 (1996).

Radio Bursts in the Active Period January 2005

K. Bouratzis*, P. Preka–Papadema*, A. Hillaris*,†, X. Moussas*, C. Caroubalos*, V. Petoussis**, P. Tsitsipis** and A. Kontogeorgos**

*University of Athens, 15783 Athens, Greece
†Hellenic Naval Academy, Piraeus 18503, GREECE
**Technological Education Institute of Lamia, Lamia, Greece

Abstract. We present complex radio bursts recorded by the radiospectrograph ARTEMIS-IV in the active period of January 2005. The wide spectral coverage of this recorder, in the 650-20 MHz range, permits an analysis of the radio bursts from the base of the Solar Corona to 2 Solar Radii; it thus facilitates the association of radio activity with other types of solar energetic phenomena. Furthermore the ARTEMIS-IV[1], high time resolution (1/100 sec) in the 450–270 MHz range, makes possible the detection and analysis of the fine structure which most of the major radio events exhibit.

Keywords: Solar activity; Corona; Radio; Radiation and spectra; Solar electromagnetic emission
PACS: 96.60.Q; 96.60.P; 95.85.Bh; 96.25.Tg; 96.60.Tf

INSTRUMENTATION & OBSERVATIONS

The Artemis–IV solar radio–spectrograph ([2], [3], [5]) covers the metric–decametric range with two receivers operating in parallel: A sweep frequency analyser (ASG) covering the 650–20 MHz range at 10 samples/sec and a high sensitivity acousto–optical analyser (SAO), which covers the 270–450 MHz range with a time resolution of 100 samples/sec.

In the active period 14–20 January 2005 the ARTEMIS–IV observed five major events January 14 (12:36 UT cf. figure 1 Top Two panels), January 15 (06:06 UT cf. figure 1 Bottom Two panels), January 17 (09:00 UT cf. figure 2 Top Two panels), January 19 (08:05 UT cf. figure 2 Bottom Two panels), January 20 (06:39 UT cf. figure 3) Combined with data from WIND–WAVES ([1]), these observations provide a complete view of the radio emission induced by shock waves and electron beams from the low corona to about 1 A.U. These recordings were supplemented with CME data from the LASCO lists on line[2] ([6]) and SXR (GOES) reports from SGD[3]. A brief overview of the five events is presented in the Table.

The high sensitivity and time resolution of the SAO facilitated an examination on fine structure within the three studied periods; all exhibited rich fine structure embedded in the Type–IV continua. In our analysis, the continuum background is removed by the use of high–pass filtering on the dynamic spectra (differential spectra in this case).

---

1 Appareil de Routine pour le Traitement et l' Enregistrement Magnetique de l' Information Spectral
2 http://cdaw.gsfc.nasa.gov/CME list
3 http//www.sel.noaa.gov/ftpmenu/indices

CP848, *Recent Advances in Astronomy and Astrophysics,*
7th International Conference of the Hellenic Astronomical Society, edited by N. H. Solomos

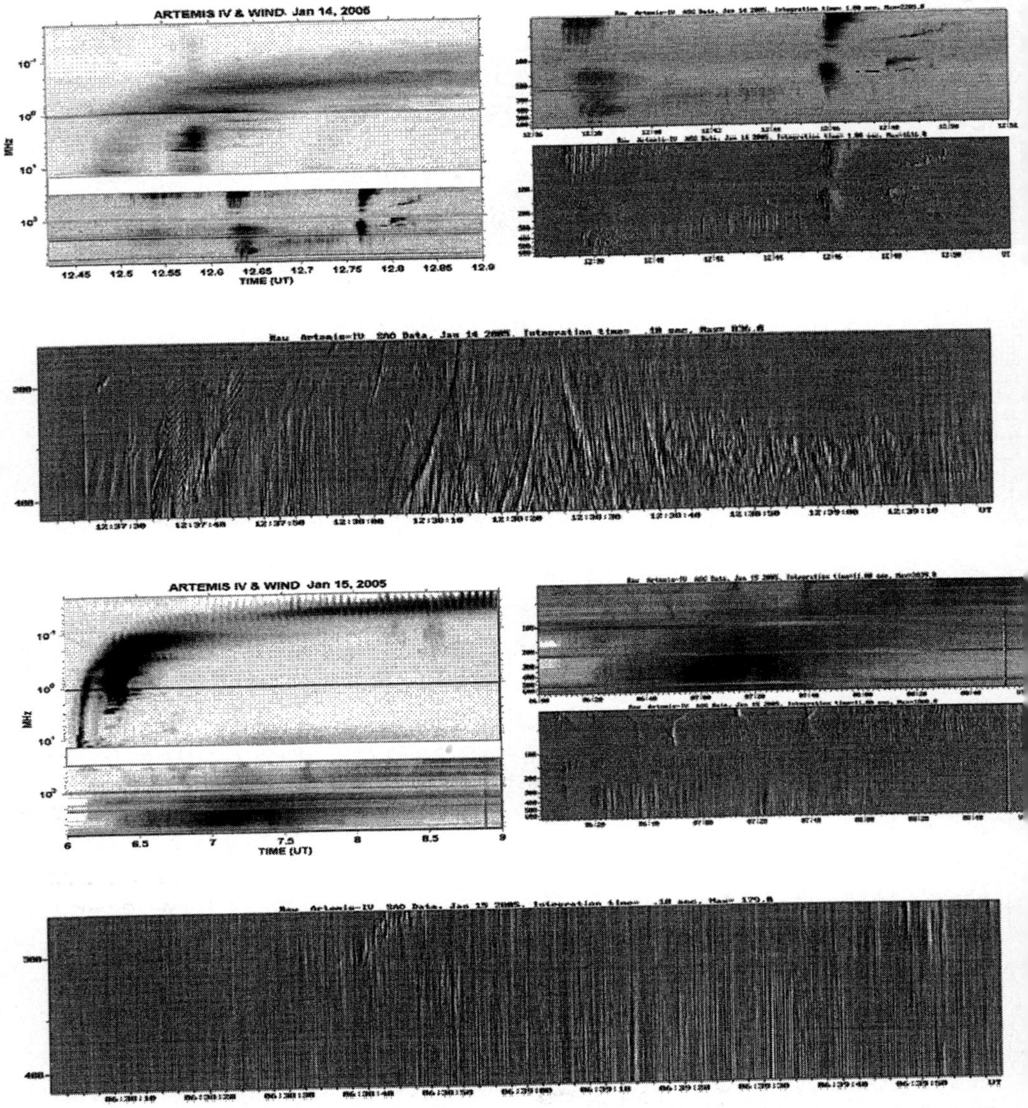

FIGURE 1. UPPER PANEL: LEFT: Combined ARTEMIS-IV and WIND–WAVES spectra of the January 14, 2005 event. RIGHT: ARTEMIS–IV ASG Spectra in the frequency range 20–650 MHz; Intensity (top)and its time derivative (bottom), enhancing fine temporal details. SECOND FROM TOP: Fibers, Zebra, Pulsations and *Lace Bursts* (cf. [4] from the same event (Differential Spectrum).) THIRD PANEL: Same as UPPER PANEL but for the event of January 15, 2005. LAST PANEL:Narrow Band Spikes & Pulsations from the same event (Differential Spectrum).

We present certain examples, which are divided according to a published morphological classification scheme ([4]) based on Ondrejov Radiospectrograph recordings in the

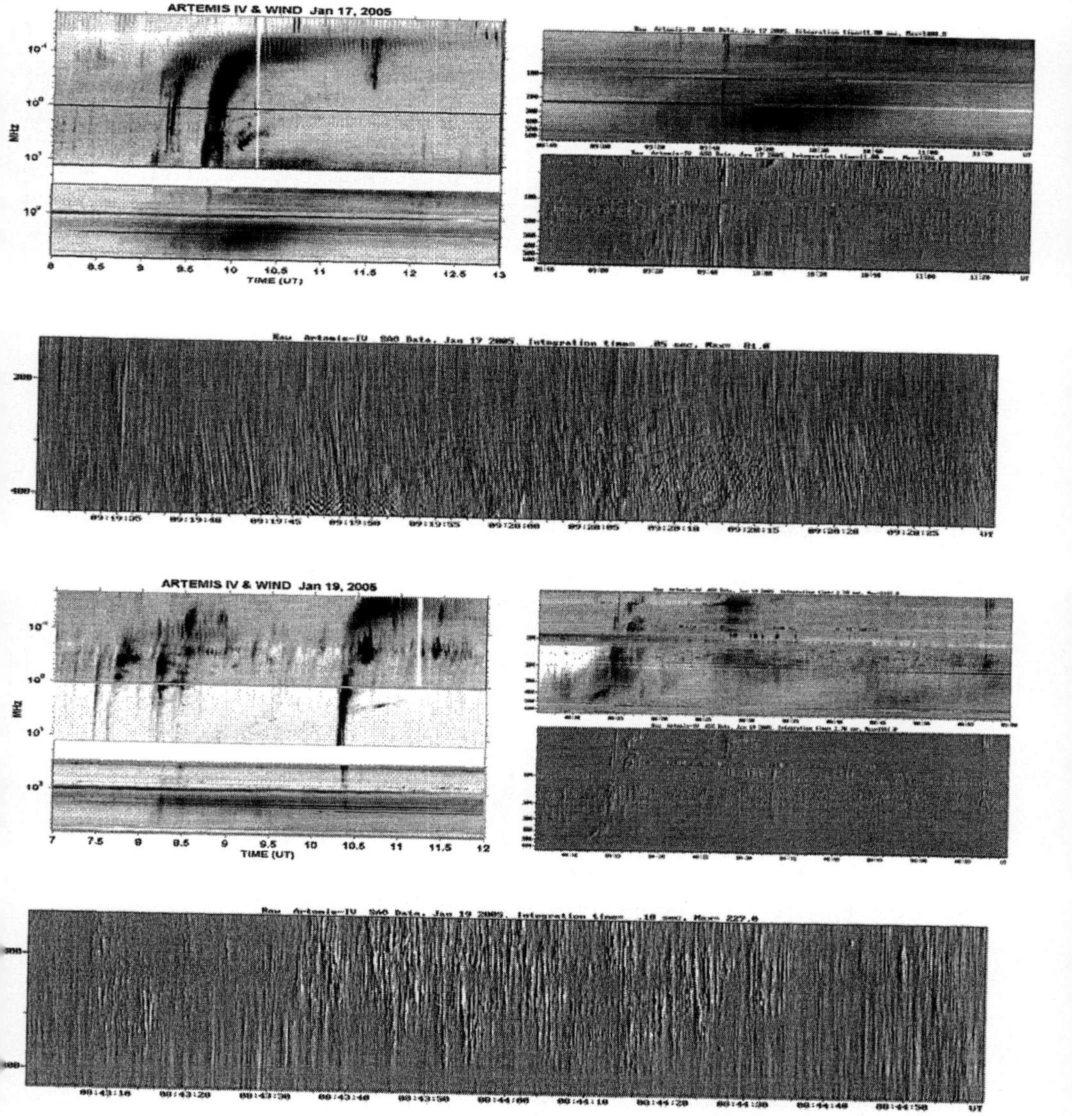

FIGURE 2. UPPER PANEL: LEFT: Same as figure 1 but for the events of January 17, 2005. SECOND FROM TOP: Fibers, Zebra, Pulsations and a Narrow band spike group from the same event (Differential Spectrum).) THIRD PANEL: Same as UPPER PANEL but for the event of of January 19, 2005. LAST PANEL:Narrow Band Spikes & Pulsations from the same event (Differential Spectrum).

0.8–2.0 GHz range. In our recordings we have detected:

- Narrow Band Spikes (figures 1, 2, 3)
- Fibers (figures 2,1)

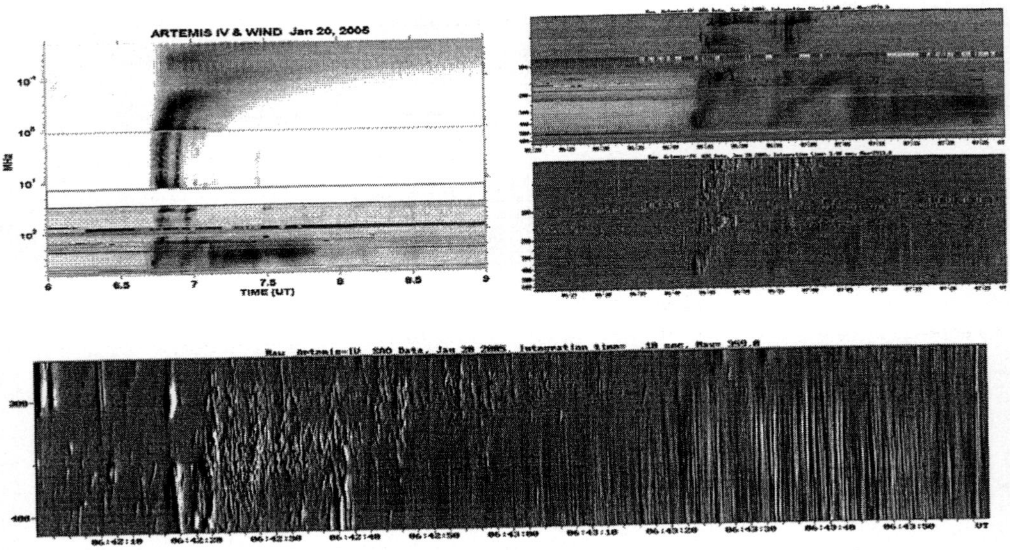

FIGURE 3. UPPER PANEL: Same as figures 1 & 2 but for the event of January 20, 2005. LOWER PANEL: Narrow Band Type III, Spikes & Pulsations from the same event (Differential Spectrum).

- Narrowband Type III Bursts (figure 3)
- Laces (figure 1)
- Zebra patterns (figure 2)

DISCUSSION & CONCLUSIONS

The ARTEMIS–IV radio–spectrograph, operating in the range of 650–20 MHz, observed 5 complex events during the super–active period 14–20 January 2005. WIND-WAVES data complement nicely the ARTEMIS data and trace the radio emission from the middle corona all the way to almost 1 A.U. The high resolution SAO recordings on the other hand reveal a variety of fine structure which, almost, matches the comprehensive Ondrejov Catalogue ([4]). This last, although it refers to the spectral range 0.8–2 GHz, seems to produce similar fine structure with the metric range.

ACKNOWLEDGMENTS

The LASCO CME catalogue is generated and maintained by the Center for Solar Physics and Space Weather, The Catholic University of America in cooperation with the Naval Research Laboratory and NASA. SOHO is a project of international cooperation between ESA and NASA. The Solar Geophysical Data Catalogue is compiled and maintained by the US Department of Commerce.

TABLE 1. Overview of the Events 14–20 January 2005

Event	Start UT	AR	Freq. Range MHz	Class	Speed Km/sec	Accel Km/sec^2
14 Jan 05						
Type II	12:48	718	30–70		1190	
SXR	12:33	718		C4.6		
15 Jan 05						
Type II	06:08	720	100–400		690	
Type IV	06:15	720	20–650			
SXR	05:54	720		M8.6		
CME (lift off)	06:00				2049	-30.7
17 Jan 05						
Type II	09:43	720	150–500		3256	
Type IV	09:16	720	20–400			
SXR	06:59	720		X3.8		
CME (lift off)	09:15				2094	-118.0
19 Jan 05						
Type II	08:12	720	20–500		1624	
Type IV	08:06	720	150–400			
SXR	08:03	720		X1.3		
CME (lift off)	08:15				2020	-43.8
20 Jan 05						
Type II	06:43	720	20–500		2294	
Type IV	06:39	720	150–450			
SXR	06:36	720		X7.1		
CME (lift off)	05:55				882	16

REFERENCES

1. J. L. Bougeret, M. L., Kaiser, P. J., Kellogg, R., Manning, K., Goetz, S. J., Monson, N., Monge, L., Friel, C. A., Meetre, C., Perche, L., Sitruk, S., Hoang, *Space Science Reviews*, **71**, 231–263, (1995).
2. C., Caroubalos, D., Maroulis, N., Patavalis, J. L., Bougeret, G., Dumas, C., Perche, C., Alissandrakis, A., Hillaris, X., Moussas, P., Preka–Papadema, A., Kontogeorgos, P., Tsitsipis, P., and G., Kanellakis, *Experimental Astronomy*, **11**,23–32, (2001).
3. C., Caroubalos, C. E., Alissandrakis, A., Hillaris, P., Preka–Papadema, J., Polygiannakis, X., Moussas, P., Tsitsipis, A., Kontogeorgos, V., Petoussis, C., Bouratzis, J. L., Bougeret, G., Dumas and A., Nindos, *These Proceedings* (2006).
4. K., Jiricka, M., Karlicky, H., Meszarosova and V., Snizek, *Astronomy and Astrophysics*, **375**, 243-250, (2001)
5. Á. Kontogeorgos, P. Tsitsipis, C. Caroubalos, X. Moussas, P. Preka-Papadema, A. Hillaris, V. Petousis, J-L Bougeret, C.E. Alissandrakis, G. Dumas, J. Polygiannakis, *13th International Symposium on Measurements for Research and Industry Applications, Athens 29 Sept.-1 Oct.*, 39–44, 2004.
6. S., Yashiro, N., Gopalswamy, O. C., St. Cyr, G., Lawrence, G., Michalek, C. A., Young, S. P., Plunkett, & R. A., Howard, *AGU Spring Meeting* Abstract **SH31C-10**, 2001

A Study of Halo Coronal Mass Ejections and Related Flare and Radio Burst Observations in Solar Cycle 23

M. Georgiou*, E. Mitsakou*, G. Pothitakis*, A. Hillaris*,
P. Preka-Papadema* and X. Moussas*

*University of Athens, 15784, Panepistimiopolis Zografos, Athens, Greece

Abstract. We present a statistical study of dynamical and kinetic characteristics of CMEs which show temporal and spatial association with flares and type II radio bursts or complex radio events of type II bursts and type IV continua. This study is based on a set of earth-directed full halo CMEs occurring during the present solar cycle, with data from the Large Angle Spectrometric Coronagraphs (LASCO) and Extreme-Ultraviolet Imaging Telescope (EIT) aboard the Solar and Heliospheric Observatory (SOHO) mission and the Magnetic Fields Investigation (MFI) and 3-D Plasma and Energetic Particle Analyzer Investigation experiment on board the WIND spacecraft.

Keywords: Solar activity; Corona; Radio; Radiation and spectra; Coronal Mass Ejections
PACS: 96.60.Q; 96.60.P;95.85.Bh; 96.25.Tg

INTRODUCTION

Coronal Mass Ejections (CMEs) are solar transients characterised by the expulsion of large quantities of plasma and frozen in magnetic field from the corona to the interplanetary space. The response of the heliosphere to the CMEs is manifested as magnetic clouds, IP shocks and energetic particle events (SEPs), which in turn, constitute the *Space Weather* (cf. for example [1], [2]). The influence on *Space Weather* of Earth–Bound halo CMEs is significant, as they are moving along the Sun–Earth line; meanwhile, these constitute only a small fraction $\approx 3,5\%$ of all CMEs observed. Solar radio bursts, on the other hand, represent an efficient diagnostic of energetic phenomena on the Sun. Type II bursts result either from a flare blast wave or are driven by a CME ([3]), while the type IV continua originate from energetic electrons trapped within plasmoids, magnetic structures or substructures of CMEs ([4]).

In this paper we present a statistical analysis with emphasis on the correlation between the earth-directed halo CMEs, metric type II and type IV radio bursts and flares.

DATA ANALYSIS & RESULTS

The occurence of a halo CME in the time interval from January 1996 to December 2003 was identified from the LASCO catalogues online ([5]); the CMEs parameters,which included the CME's linear speed, acceleration and take off time, were obtained from the same catalogues. Secondly, we used five minutes averaged data of the MFI and 3-D Plasma and Energetic Particle Analyzer Investigation experiment onboard the WIND

CP848, *Recent Advances in Astronomy and Astrophysics,*
7th *International Conference of the Hellenic Astronomical Society,* edited by N. H. Solomos

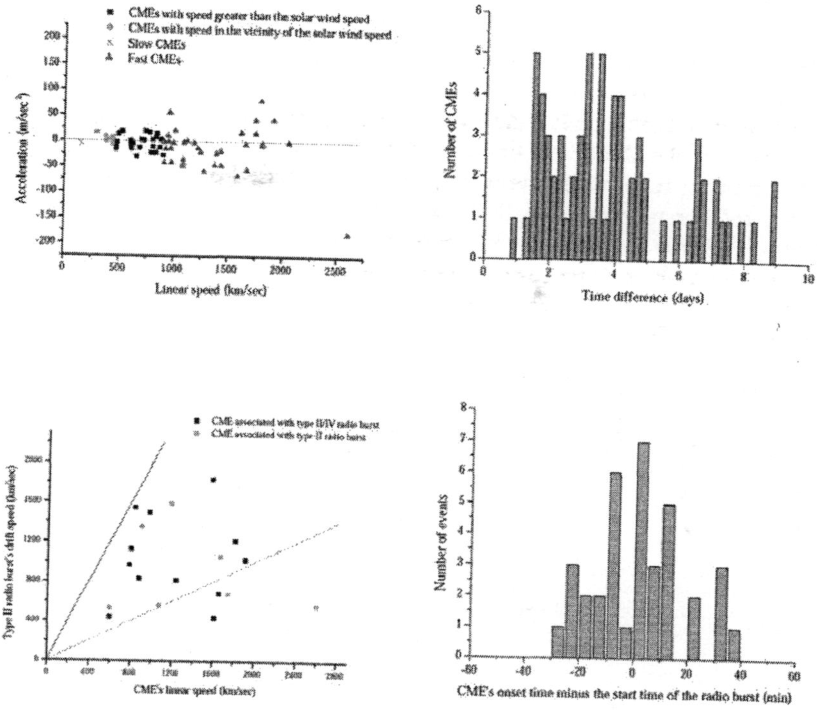

FIGURE 1. *Upper left panel*: CME acceleration as a function of the CME's linear speed. *Upper right panel*:Histogram of time elapsed between the observation of the CME by LASCO coronagraphs and the detection of the material expelled by WIND instruments. The time difference ranged from less than one day to more than four days. *Lower left panel*: Type II radio burst's drift speed versus the CME's linear speed. *Lower right panel*: Histogram (with bin size of 5 minutes) of time difference between the time when the CME was at 1 Ro and the radio burst's start time.

spacecraft to identify the passage of the internal material of the CME through the following distinctive signatures: a transient shock marked by an abrupt increase in the strength of the interplanetary magnetic field, which was complemented by a depression of the density, temperature and speed of the particles populations within the post-shock flow. As a result, 67 earth-directed full halo CMEs were identified for this study.

The movies of EUV images obtained by the SOHO/EIT, Yohkoh SXT images and Ha filtergrams were examined for associated activity on the visible solar disk within a time window of 60 minutes before and after the extrapolated CME take off time. The associated solar flares properties were derived using data from the Soft X-rays Sensor (SXR) aboard the GOES satellite. All of the associated flares were simple events without multiple peaks, or truncated by the occurrence of another flare. The duration of each flare was measured using the GOES 1-8A channel and it was defined as the time between the point were the gradient of the X-rays flux turned positive, to the point where it returned to the pre–flare value or leveled off. The peak colour temperature for each flare event was estimated from the ratio of fluxes in the two GOES channels, following the standard calibration of *Garcia* [6]. In this study, we focused our attention on two forms of activity, the flares and the metric type II and IV radio bursts. A 60 minutes time window between the CME's extrapolated take off time and the radio event's onset time was introduced as the criterion for establishing associations. Ground–based observations of the II and IV metric radio bursts recorded in SGD, were used for the estimation of the radio bursts parameters. The type II frequency drift and, in turn, the radial speed, was derived using a density model of the solar corona by *Newkirk* ([8] and [7]).

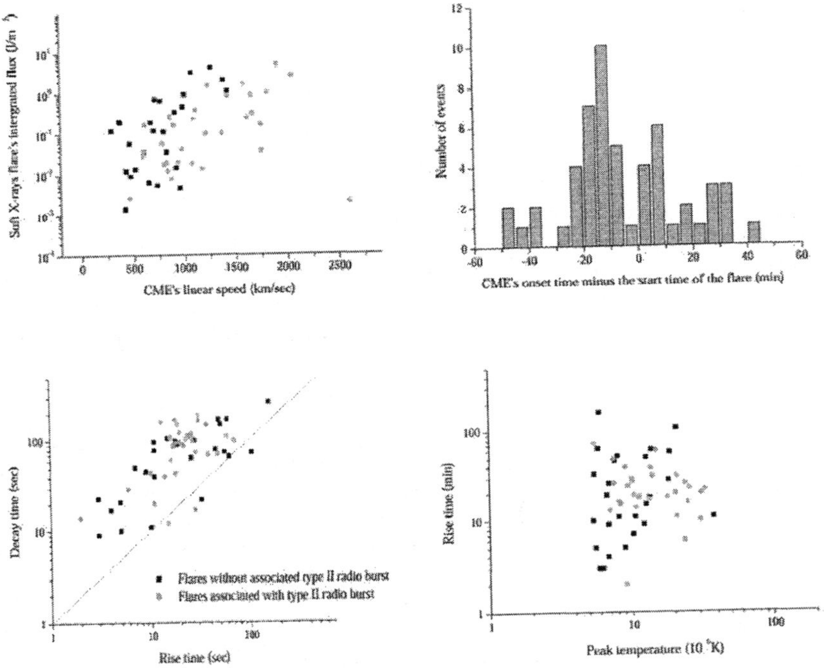

FIGURE 2. *Upper right panel*: Histogram (with bin size 5 minutes) of the time elapsed between the onset of the CME and the start time of the soft X-rays flare. *Upper left panel*: Soft X-rays flare's intergrated flux versus the CME's linear speed. *Lower left panel*: The time from the start of the flare to the time of flare's peak intensity versus the length of the the subsequent decrease in the flare's intensity. *Lower right panel:* The length of the initial increase in the flare's intensity versus the flare's peak temperature.

The CME's linear speed provides an estimation of the mean speed within the coronagraph's field of view and in our list it ranged from 150 km/s to 2600 km/s. With an average value of $\approx$ 1004 km/s, twice the average speed ($\approx$ 489 km/s) of all the CMEs observed in the same period ([9]), the CMEs in our list constitute a fraction of faster, hence, more energetic than average CMEs. CMEs with speed greater than the typically slow solar wind ($450\ km/s < Vcme < 900\ km/s$) show predominant deceleration ($a \approx -3.944 m/s^2$), as well as, fast CMEs ($900 km/s < Vcme$) with an average deceleration of $a \approx -13.487 m/s^2$, indicating that the retarding exceed the propelling forces than drive the CME outward in the heliosphere. On the other hand slower CMEs with speed in the vicinity of the solar wind speed ($250km/s < Vcme < 450km/s$) were found to have small acceleration ($a \approx 4.87\ m/s^2$)(cf. figure 1, left panel).

Observations of metric radio bursts were compared with the observations of the halo CMEs and 42 of these events ($\approx 63\%$) showed a close relation to type II radio bursts with respect to speed and extrapolated onset time (cf.figure 1). Type II frequency drift rates appeared to be all in excess of 400 km/s and the mean speed was found to be equal to $\approx 1009\ km/s$. Furthermore, a CME related group of 32 events ($\approx 48\%$) were accompanied by a type IV continuum. Considering the time difference between the time when the CME was at 1 Ro and the radio burst's start time, 27 of CMEs take offs preceded and 19 followed their corresponding radio burst onset; the mean value of the time difference was found within 0 to 5 minutes from the CMEs take off .

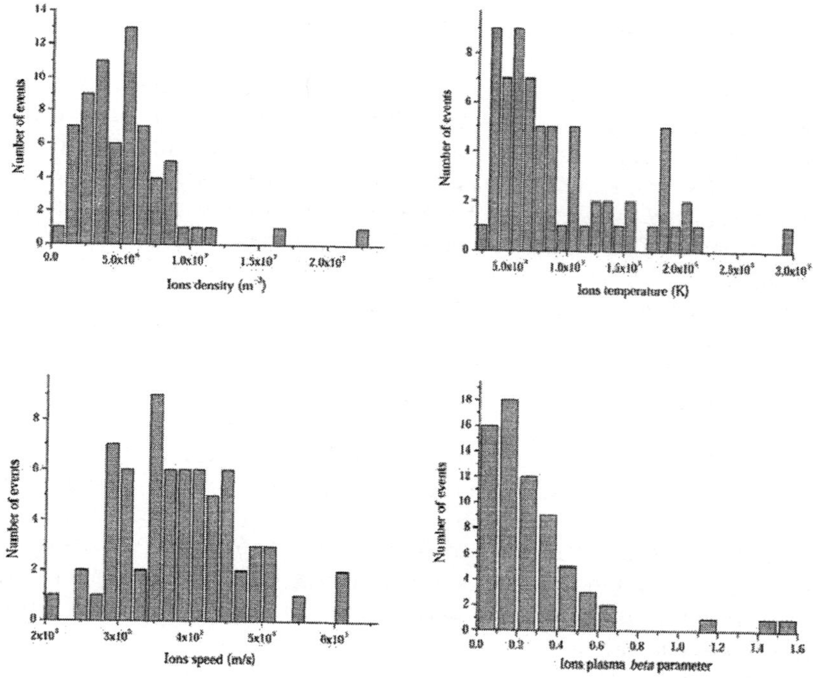

FIGURE 3. The average plasma properties within the CMEs observed by the WIND. From top left to bottom right are the distribution of the ions mean density, the ions mean temperature, the average speed and the average *beta* parameter within the ions population of the CMEs.

The majority of the sampled CMEs were associated with surface activity. In this study, the Ha and soft X–rays flares were the only form of activity taken into account.Flares of all sizes (the X-rays importance ranged from B to X), occurring at all longitudes (19 flares originated within 25 degrees from the central meridian and 5 flares within 30 degrees from the solar limb) were found to be associated with the CMEs. 57 of the CMEs ($\approx 85\%$) were associated with soft X-rays flares and 28 ($\approx 41\%$) were associated with a flare and a type II radio burst. 45 of the CMEs were associated with Ha flares ($\approx 67\%$). For all of the events appears a general correlation between the CME's linear speed and the soft X-rays integrated flux. The events associated with type II radio burst are clustered toward the greater velocity and flux ranges, while the CME speed exceeds a threshold of about 500km/s. A comparison between the relative times of the initial increase and the subsequent decrease in the flare's intensity indicates a clear corellation, meanwhile it appears to be a systematic difference between the two times, with the decrease lasting longer than the initial increase. 18 flares lasted longer than 180 minutes, while 18 events as short as 60 minutes and less also show to be associated with CMEs. As regards the peak temperature of the flares, events without associated radio emission tend to reach peak temperature faster than those associated with a radio burst. Considering the time elapsed between the onset of the CME and the start time of the flare, 23 of the CMEs lift offs preceded and 34 followed their corresponding flare onset. The mean of the distribution was found within 10 to 5 minutes before the CMEs take off (cf. figure 2).

The size of the interplanetary CMEs in the radial direction observed near the Earth's orbit ranged from 0.078A.U. to 0.58A.U. with a mean value of 0.26A.U., while the size of the core of the interplanetary post-shock flow ranged from 0.055A.U. to 0.36A.U. with a mean value of 0.17A.U.. The enormous radial size of the interplanetary CMEs cannot be considered to be typical of those observed close to the Sun, indicating an expansion of the CME in the heliosphere, which causes their radial size to increase with distance and may affect the average physical conditions within the CMEs.

The average density of ions inside the CMEs is about $5.3 \times 10^6 ions/cm^3$, slightly lower than the average background solar wind density, $9 \times 10^6 ions/cm^3$, which is in accordance with the CMEs expansion. The average density of ions ranged from $0.93 \times 10^6 ions/cm^3$ to $2.2 \times 10^7 ions/cm^3$. The average value of the ions temperature within the CMEs was estimated of about $10^6 K$, slightly lower than the average solar wind temperature, $1.210^6 K$, which is in accordance with the low temperature criterion used for the identification of the CMEs in the interplanetary medium. The average ions temperature ranged from $2 \times 10^4 K$ to $3 \times 10^5 K$. The average speed within the ions population of the CMEs lies in the range from 220 km/s to 620 km/s with a symmetrical distribution around their mean value of $Vi = 384 km/s$. Most CMEs were found to have a propagation speed similar to that of the average solar wind speed in the same period, $Vsw = 340 km/s$, and only a few had speed above the speed of the slow flow of the solar wind (cf. figure 3). These results are similar to the calculations of CMEs observed by WIND and ACE in the same time interval by *Liu et al.*([10]) and the results obtained from Helios 1 and 2 in the inner heliosphere([11]). The interplanetary post-shock flow is characterised by low plasma *beta* values between 0.025 and 1.6 with a mean value of 0.28 , as these were calculated from the ions parameters only. The plasma's magnetic pressure dominated the thermal pressure within the CMEs of our data set, due to their higher than average solar wind magnetic field strength and their unusually low plasma temperature and density, thus, contriduting to the radial expansion of the CMEs.

CONCLUSIONS

We have studied a fraction of full halo CMEs with distinctive multiple signature in the interplanetary medium, which constitute a population of fast and wide CMEs. The CMEs associated with coronal type II radio bursts are faster than the CMEs without any associated metric radio emission (cf. figure 2). The average speed of the CMEs associated with type II radio bursts was estimated of about 1200 km/s, greater than the average speed of the CMEs without associated radio emission ($\approx 810\ km/s$). Depending on the CME parameters and the interplanetary medium, a shock can be formed in the inner corona, which produces metric type II bursts. It must be noted that almost all of the CMEs with associated type II radio emission (28 out of 32 events) were also associated with an X-rays flare. The close coincidence and other correlations between flares, CMEs and type II radio bursts can be considered to suggest that the influence of flares in the production of type II bursts cannot be ruled out.

REFERENCES

1. M., Messerotti, P., Zlobec, I., Coretti, M., Jurcev and S., Padovan, *Mem. S.A.It.*, **74**, 770–773, (2003).
2. C., Lathuillere, M., Menvielle, J., Lilensten, T., Amari and S. M., Radicella, *Annales Geophysicae*, **20**, 1081-1104, (2002).
3. H. T., Classen, & H., Aurass, *A&A*, **384**, 1098–1106, (2002).
4. T. S., Bastian, M., Pick, A., Kedraon, D., Maia & A., Vourlidas, *ApJ.*, **558**, L65–L69, (2001).
5. S. Yashiro, N. Gopalswamy, O. C. St. Cyr & al, *AGU Spring Meeting*, Abstract SH31C-10, (2001).
6. H. A. Garcia, *Solar Physics*, **154**, 275–308, (1994).
7. G., Newkirk, *ApJ*, **133**, 983–1013, (1961).
8. G., Newkirk, *ARA&A*, **5**, 213–266, (1967).
9. N. Gopalswamy, *AdSpR*, **31**, 869–880, (2003).
10. Y., Liu, J. D., Richardson and J. W., Belcher, *Planetary and Space Science*, **53**, 3-17, (2005).
11. V., Bothmer, and R., Schwenn, *Annales Geophysicae*, **20**, 1-24, (1998).

Millisecond solar radio bursts in the metric wavelength range

J. Magdalenić*, A. Hillaris†, P. Zlobec** and B. Vršnak*

*Hvar Observatory, Faculty of Geodesy, University of Zagreb
†University of Athens, Greece
**INAF-Trieste Astronomical Observatory, Italy

Abstract. A study and classification of super-short structures (SSSs) recorded during metric type IV bursts is presented. The most important property of SSSs is their duration, at half power ranging from 4-50 ms, what is up to 10 times shorter than spikes at corresponding frequencies. The solar origin of the SSSs is confirmed by one-to-one correspondence between spectral recordings of Artemis-IV[1] and high time resolution single frequency measurements of the TSRS[2].

We have divided the SSSs in the following categories:

1. Broad-Band SSSs: They were partitioned in two subcategories, the *SSS-Pulses* and *Drifting SSSs*;
2. Narrow-band: They appear either as *Spike-Like SSSs* or as *Patch-Like SSSs*;
3. Complex SSS: They consist of the absorption-emission segments and were morphologically subdivided into *Rain-drop Bursts* (narrow-band emission *head* and a broad-band absorption *tail*) and *Blinkers*.

Keywords: Solar activity;Corona;Radio;Radiation and spectra
PACS: "96.60.Q; 96.60.P;95.85.Bh; 96.25.Tg

THE *SHORTEST*, AS YET, OBSERVED SOLAR RADIO EMISSIONS

The decimetric spikes (cf. [1] for a review, also [6]) have been long considered as the solar radio bursts with the shortest duration recorded; this has only been occasionally challenged ([3], [9], [4]). We present observational evidence and characteristics of a variety of solar radio bursts with durations shorter than spikes up to an order of magnitude. These are, hence, named *Super Short Structures*, or *SSS*. Our observations consist of radio data with 1-10 ms time resolution, which enable the detection of, as yet unobserved, fine structure embedded in the type IV continuum; they comprise in particular :

- Single frequency measurements at 237, 327, 408 610, 1420 and 2695 MHz, recorded with time resolution of 1 ms by the solar multichannel radiopolarimetric system of the INAF-Trieste Astronomical Observatory (TSRS-Trieste Solar Radio System).

---

[1] Thermopylae–Greece
[2] Trieste–Italy

CP848, *Recent Advances in Astronomy and Astrophysics,*
7th *International Conference of the Hellenic Astronomical Society,* edited by N. H. Solomos

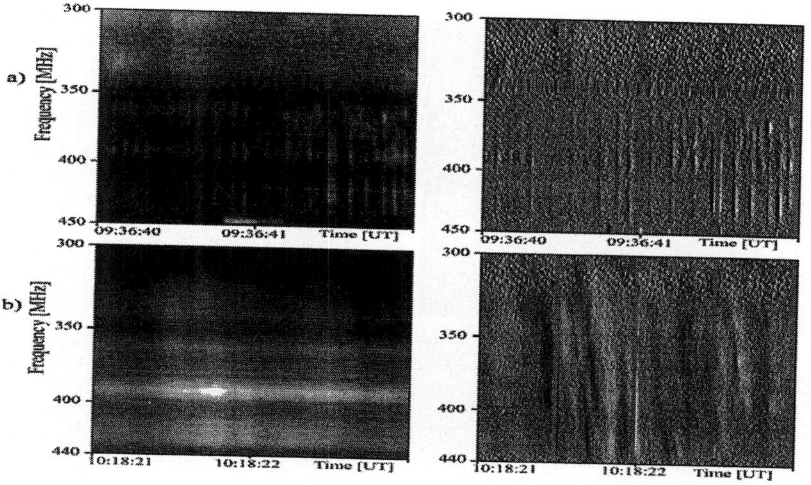

FIGURE 1. ARTEMIS-IV Dynamic Spectra (Intensity on the left and differential on the right) of broad-band SSS: (a) SSS pulses, (b) Drifting SSSs.

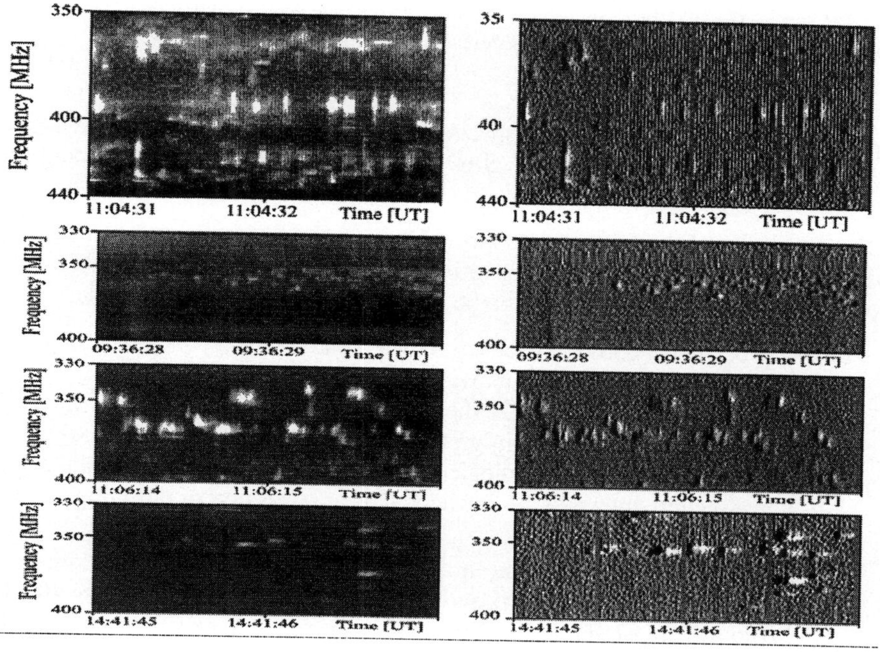

FIGURE 2. ARTEMIS-IV Dynamic Spectra of narrow band SSS. Top to Bottom: Spike–like SSSs, dot–like, sail–like, flag–like. Intensity spectra are on the left and differential on the right.

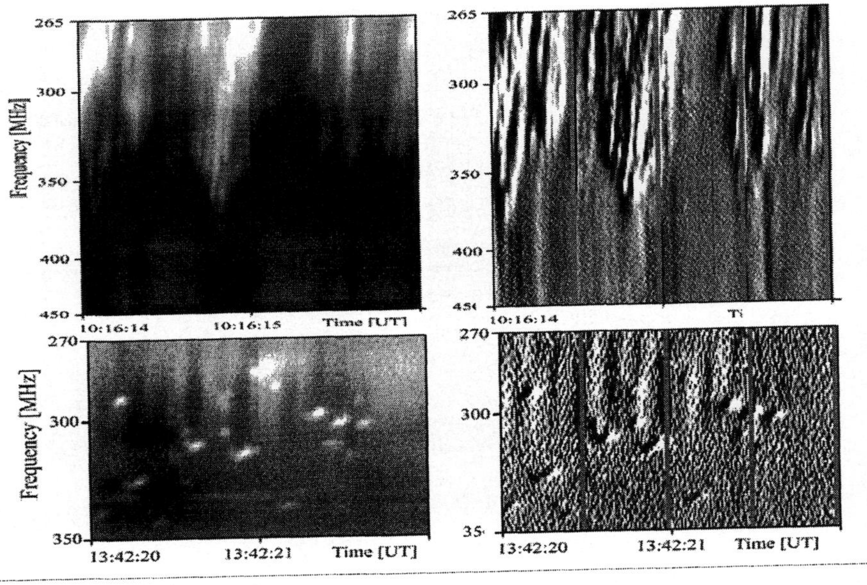

FIGURE 3. ARTEMIS-IV Dynamic Spectrum of Blinker (Upper Panel) and Raindrop-like SSS (Lower Panel). Intensity spectra are on the left and differential on the right.

- ARTEMIS-IV, cf. [2], dynamic spectra were obtained by the high sensitivity multichannel acoustooptical analyser (SAO) which covers the 265-450 MHz range, with time resolution of 10 ms.

The credibility of the *SSSs* was confirmed by one-to-one identification of individual SSS bursts in the single frequency recordings and in the corresponding ARTEMIS-IV spectra. We have classified the *SSSs* in the following morphological categories:

- *Simple broad-band SSSs*; characterized by a broad frequency bandwidth $\Delta f \geq 100$ MHz (Figure 1). They can be subdivided in two subcategories:
  - *SSS-pulses* have duration in the range 10-20 ms, and frequency bandwidth $\approx$100 MHz. They appear in groups and, occasionally, exhibit quasi-periodic behaviour. Due to morphological similarity with pulsations of the type IV fine structure, they may be considered as an extension of pulsating structures towards higher time scales.
  - *Drifting SSSs* have duration in the range 30-70 ms. Their bandwidth in general exceeds the 100 MHz, and frequency drift rates ($|\Delta f/\Delta t| \approx$ 400-1000 MHz/sec) are similar to the drift rates of the metric type III bursts. ([8], [7]).
- *Simple narrow-band SSSs* are distinguished by their narrow frequency bandwidth $\Delta f \leq$ 20MHz. They are, also, subdivided into two subcategories:
  - *Spike-like SSSs* are the shortest *SSSs* with a duration in the 4-30 ms range (Figure 2 top panel). Their frequency bandwidth is mostly $\Delta f \leq 20$ MHz. If

measurable, frequency drifts are $|\Delta f/\Delta t| \geq 800$ MHz/sec. On dynamic spectra they very much resemble spikes, ([1]) as their name implies.
  - *Patch–like SSSs* exhibit, due to their morphological diversity, a rather broad range of duration which varies between 4 and 50 ms (Figure 2, three lower panels). The frequency bandwidth is $\Delta f \leq 15$ MHz, and it can be as low as a few MHz. This qualifies *patch–like SSS* as the *SSSs* of the narrowest bandwidth. Their spectral appearance varies; they can resemble dots, sails or flags and were further subdivided accordingly to: *dot-like SSS*, *sail-like SSS* and *flag-like SSS*. *Dot-Like* structures in the 1000-2000 MHz range, recorded by the Brazilian Solar Spectroscope with 50 ms resolution, have been reported by [10].
- *Complex Super Short Structures* are characterised by an emission and an absorption element. Two subcategories could be distinguished:
  - *Rain-drop bursts* (Figure 3, lower panel) consist of a narrow-band emission *HEAD* ($\Delta f \approx 5MHz$) and a broad-band absorption *TAIL* ($\Delta f \approx 40MHz$). The durations are approximately 50 ms for the *HEAD* and 30 ms for the *TAIL*. Both the *HEAD* and the *TAIL* exhibit frequency drift which is $|\Delta f/\Delta t| \approx 60 \pm 10$ MHz/sec and $|\Delta f/\Delta t| \approx 1000 \pm 400$ MHz/sec, respectively. Morphologically these *SSSs* resemble, somehow, to the tadpole bursts ([11]).
  - *Blinkers* (Figure 3, upper panel) are drifting bursts ($|\Delta f/\Delta t| \approx 650$ MHz/sec) consisting of absorption element that is switching abruptly to emission element. Opposite cases have also been found (emission in the high-frequency part, and absorption in low-frequency part of the burst). The duration $d_{1/2}$=30–40 ms, is approximately the same along the whole burst. They are the *SSSs* with the largest frequency bandwidth $\Delta f > 150$ MHz.

DISCUSSION & CONCLUSIONS

The analysis of high time resolution spectral recordings (ARTEMIS-IV) and single frequency measurements (TSRS-data) reveals a number of different classes of *Super Short Structures*. The basic characteristics of the described *SSS* classes are summarized in Table 1.

It is stressed that all of the presented features have duration considerably shorter than spikes which in the frequency range 250-450 MHz have duration to 100–50 ms. This does not necessarily establishes *SSSs* as the shortest in existence, but only as the shortest recorded until now. It is to be expected that the improvement of instrumental resolution may reveal the existence of even shorter radio bursts.

The duration of the *SSSs* varies with frequency, and from event to event. Therefore, at present it is not possible to establish a systematical dependence of the *SSSs* duration on the observation frequency; this would be in favour of plasma radiation mechanisms which cannot be yet confirmed. For more specific results a detailed case study is under way.

Lastly, although some morphological similarities with already known burst types (pulsations, type IIIs, tadpoles) exist, there is no as yet conclusive evidence of a common

TABLE 1. Classification and Basic Characteristics of Super Short Structures (SSS)

Class	Subclass	Characteristics	Morphologically Similar Bursts
Simple Broad Band SSS	SSS Pulses	$\Delta f \approx$ 100MHz, Duration=10-20 ms Some Groups Exhibit Quasi Periodicity	Pulsations
	Drifting SSS	$\lvert\Delta f/\Delta t\rvert \approx$ 400-1000 MHz/sec Duration=30-70 ms, Δ f $\geq$ 100MHz	Type III like Bursts [8]
Simple Narrow Band SSS	Spike-like SSS	$\Delta f \leq$ 20 MHz, $\lvert\Delta f/\Delta t\rvert \geq$ 800 MHz/sec Duration=4-30 ms	Spikes [1]
	Patch-like SSS	$\Delta f \leq$ 15 MHz, Duration=4-50 ms Further Subdivision: Dot-like SSS Sail-like SSS and Flag-like SSS	 Dot like Structures [10] -
Complex SSS with emission and absorbtion element	Blinkers	$\lvert\Delta f/\Delta t\rvert \approx$ 650 MHz/sec Δf > 150 MHz, Duration=30-40 ms Absorption switches to emission	-
	Raindrop bursts	Emission *HEAD*, Duration$\approx$50 ms $\Delta f \approx 5MHz$, $\lvert\Delta f/\Delta t\rvert \approx 60 \pm 10$ MHz/sec Absorption *TAIL*, Duration$\approx$30 ms $\Delta f \approx 40MHz$, $\lvert\Delta f/\Delta t\rvert \approx 1000 \pm 400$ MHz/sec	Tadpole Bursts [11]

radiation mechanism.

REFERENCES

1. A. O. Benz, *Solar Physics*, **104**, 99–110, (1986).
2. C. Caroubalos, D. Maroulis, N. Patavalis, J.-L. Bougeret, G. Dumas, C. Perche , C. Alissandrakis, A. Hillaris, X. Moussas, P. Preka-Papadema, A. Kontogeorgos, P. Tsitsipis and G. Kanelakis, *Experimental Astronomy*, **11**, 23–32, (2001).
3. F., Droege, *Astronomy & Astrophysics*, **57**, 285–290, (1977).
4. O., Elgaroy, *Astronomy & Astrophysics*, **82**, 308–313, (1980).
5. O.Elgaroy & O. P. Sveen, *Nature*, **278**, 626–627, (1979).
6. G. D. Fleishman, V. F. Melnikov,*Magnetic Fields and Solar Processes. The 9th European Meeting on Solar Physics, held 12-18 September, 1999, in Florence, Italy.* Edited by A. Wilson. European Space Agency, ESA SP-448, 1999, pp. 1247–1251.
7. M., Gudel & A. O., Benz, A. O., *Astronomy & Astrophysics*, **231**, 202–212, (1990).
8. H., Isliker, A. O, Benz, *Astronomy & Astrophysics Supp.*, **104**, 145–160, (1994).
9. D., McConnell, *Astronomical Society of Australia, Proceedings*, **4**, 64–67, (1980).
10. H. S., Sawant, F. C. R., Fernandes, J. R., Cecatto, H. O., Vats, J. A. C. F., Neri, V. A., Portezani, A. R. F., Martinon, M., Karlicky, K., Jirika & H., Meszarosova, *Adv. Space Res.*, **29**, 349–354, (2002)
11. C., Slottje, *Solar Phys.*, **25**, 210–231, 1972.

Traveling Waves In Network Bright Points

G.Kontogiannis[1], S.D.Bloomfield[2], J.R.T.McAteer[2], M.Mathioudakis[2], E.Antonopoulou[1]

1. *Department of Astrophysics, Astronomy and Mechanics, University of Athens, Panepistimiopolis Zografou GR-15784 Athens, Greece*
2. *Physics and Astronomy, Queens University Belfast, Belfast BT7 1NN, N.Ireland*

Abstract. One of the main features of the quiet solar chromosphere is the Network Bright Points (NBP), formed by the emerging magnetic flux, at the boundaries of supergranular cells. Triggered by the motions of magnetic loop foot-points, at the top of the convection zone, MHD waves propagate inside the NBP's. Using SSW IDL routines and wavelet analysis of series of images in four bandpasses (CaII K_3, Mgb1-0.4, Mgb2 and Ha core) we detected these MHD wave modes. The observations have been analyzed using cross-correlation techniques and we have drawn conclusions on wave propagation and mode coupling.

Keywords: Chromosphere; Network Bright Points; MHD; waves;
PACS: 96.60.Na

INTRODUCTION

The main feature of the quiet solar chromosphere is the chromospheric network and it is outlined by the Network Bright Points (NBP's), formed by the emerging magnetic flux that the convective flow pushes at the supergranular cell boundaries. Considering the thin flux tube approximation and applying the MHD equations for the NBP's, wave propagation is possible, the excitation mechanism being the displacement of the magnetic footpoints due to the granular motion. The three dominant modes in non-homogeneous plasma are the *Alfven*, the *longitudinal* and *transverse* modes and only the last two of them are dispersive and could be responsible for energy deposition at the upper chromosphere.

It has been found that NBP's exhibit low frequency oscillations and the existence of many of these modes has been disputed [1] but the methods that have been used vary. By means of the method presented here, modes in the range of 1 – 4 mHz have been previously detected [3].

Due to the stratification of the solar chromosphere, the physical parameters, and consequently the propagation speeds, change with height and mode coupling might occur transforming a slow transverse wave into a fast longitudinal one [5]. The latter then might shock depositing their energy and therefore heating the upper chromosphere.

The aim of this study is the identification of propagating waves in the NBP's and the detection of mode coupling.

CP848, *Recent Advances in Astronomy and Astrophysics,*
7th *International Conference of the Hellenic Astronomical Society,* edited by N. H. Solomos

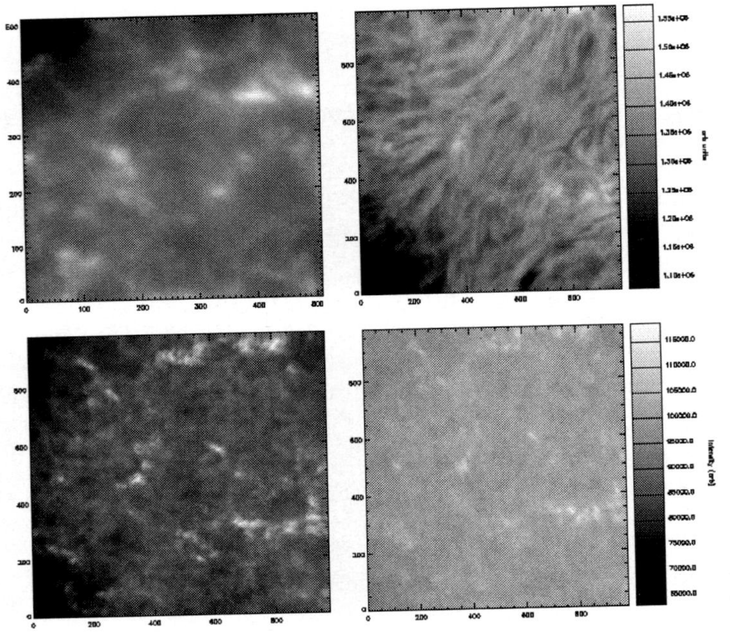

FIGURE 1. Images of a region of the chromospheric network in four bandpasses (top left, Ca II K3, top right, Hα, bottom left, MgI b1-0.4 A and bottom right Mg I b2)

DATA

Our data consist of sets of 1419 images of the quiet sun, taken from R.B.Dunn Solar Telescope at National Solar Observatory, Sacramento Peak, USA. The observations were carried out on October 14th 2002 and the images depict the same area of the sun in four bandpasses, that is, CaII K3 (3933.7 A), Mg I b1-0.4 (5172.7 A), Mg I b2 (5183.2 A), and Ha core (6562.8 A), representing four different heights in the solar atmosphere [6] (Fig,1).

ANALYSIS

Each image in the dataset was aligned with a reference image and then the clearer and better spatially localised NBP's were chosen (Fig.2). An area of 100 x 100 pixel around the NBP which was isolated using a contour method by McAteer et al. (2002).

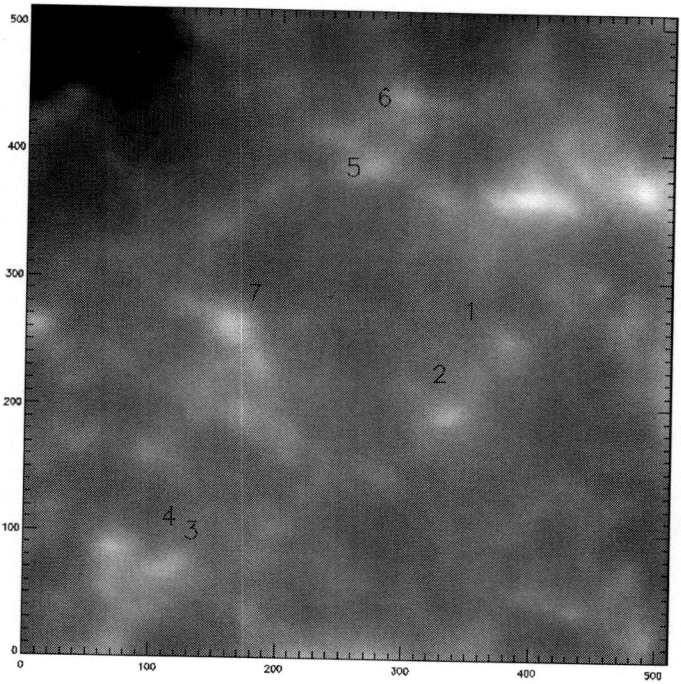

FIGURE 2. Time averaged image and the chosen NBP's

For each image, the background and threshold intensity were determined and the contour containing the NBP was drawn. A value of the intensity of the NBP was extracted and then a contrast value was calculated:

Thus a time series of contrast values was created, for each NBP and bandpass. Each light curve was filtered, to cut off very low frequencies (<0.5 mHz) by means of time series reconstruction, using wavelet analysis [4].

Wavelet transformation of the time series (Fig.3) reveals a number of frequencies between 0.5 – 6.0 mHz that represent oscillations at four different heights that correspond to each bandpass.

Cross correlation between the power curves of different bandpasses leads to detection of traveling waves. The output of this procedure is the value of the correlation coefficient and the time lag Δt that represents the time interval between the moment that the oscillation occurred at a height h1 and when the same frequency was detected at h2 (h2 > h1). Acceptable waves must have a correlation coefficient greater than 0.7.

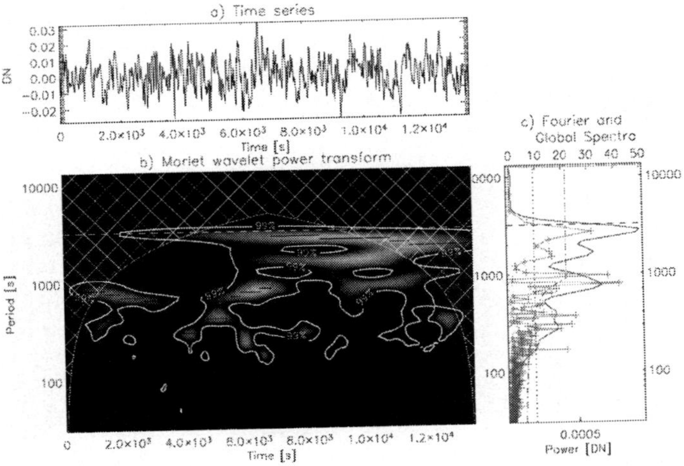

FIGURE 3. Wavelet analysis for the CaII K_3 light curve of the NBP 2. The top graph represents the filtered time series and at the bottom row are the wavelet and Fourier spectra.

RESULTS – DISCUSSION

In Table 1 we present the frequency range of the detected upward waves and their propagation height and velocity range. The latter has been calculated dividing the height range by the time lag Δt.

It appears that some of these waves propagate at all heights while others are confined in a small area of the NBP. These are excited by processes like granular overshoot, which are irrelevant to the flux tube. We note that oscillations at ~3mHz are due to the 5 min oscillation of the photosphere.

In order to detect mode coupling, a number of tests have been applied [3]: A transverse wave with frequency ν, in lower chromosphere should appear with much reduced power, in bigger heights while a longitudinal wave with frequency 2ν should be detected. Both modes should appear in the wavelet spectrum with a duration of at least twice their period and the speed of the transverse wave must be near the local sound speed.

The most likely mode coupling cases are: For NBP 1, between *0.75 – 1.50 mHz* which then shocks, for NBP 2, between *2.14 – 4.44* mHz which probably shocks higher in the chromosphere, for NBP 3, between *0.68 – 1.28 mHz* and for NBP 6, between *0.62 – 1.32* mHz and *1.8 – 3.15* mHz.

TABLE 1. NBP 6

Frequency (mHz)	Height Range (Km)	Speed (Km/s)
	MgI b_1-0.4 – MgI b_2	
1.310 – 1.510	500 – 550	2 – 4
1.755 – 1.955	500 – 550	5 – 7
	MgI b_1 – Hα	
0.710 – 0.910	1000 – 1500	> 5
3.470 – 3.670	1000 – 1500	3 – 9
	MgI b_2 – Hα	
0.535 – 0.640	450 – 1000	0.3 – 1.3
1.180 – 1.380	450 – 1000	0.6 – 1.5
3.850 – 4.050	450 – 1000	> 30
	MgI b_2 – CaII K_3	
0.510 – 0.710	950 – 1300	0.6 – 29
0.740 – 0.940	950 – 1300	6 – 29
	Hα – CaII K_3	
-	-	-

REFERENCES

1. Kalkofen, W. 1997, ApJ, 486, L145.
2. McAteer, R.T.J.,Gallagher, P.T.,Williams, D.R.,Mathioudakis, M,Phillips, K.J.H.,Keenan, F.P., 2002 ApJ, 567, L165.
3. McAteer, R.T.J.,Gallagher, P.T.,Williams, D.R.,Mathioudakis, M.,Bloomfield, D.S.,Phillips, K.J.H., Keenan, F.P., 2003, ApJ, 587, 806.
4. Torence,C., & Compo,G.P. 1998, Bull.Am.Meteor.Soc., 79, 61.
5. Ulmschneider, P., Zahringer, K., Musielak, Z.E. 1991, A&A, 241, 625.
6. Vernazza, J.E., Avrett, E.H., & Loeser, R., 1981, ApJS, 45, 635.

Type III events, flares and CMEs, in the extremely active period October-November 2003

E. Mitsakou*, M. Thanasa*, P. Preka–Papadema*, X. Moussas*, A. Hillaris*, C. Caroubalos†, C. E. Alissandrakis**, P. Tsitsipis‡, A. Kontogeorgos‡, J.-L. Bougeret§ and G. Dumas§

*Section of Astrophysics, Astronomy and Mechanics, Department of Physics, University of Athens, 15784 Panepistimiopolis Zografos, Athens, Greece
†Department of Informatics, University of Athens, 15783 Athens, Greece
**Section of Astro-Geophysics, Department of Physics, University of Ioannina, 45110 Ioannina, Greece
‡Department of Electronics, Technological Education Institute of Lamia, Lamia, Greece
§Observatoire de Paris, CNRS UA 264, 92195 Meudon Cedex, France

Abstract. The type III observations trace the propagation of energetic electron populations through the Solar Corona which, more often than not, precede or are associated with energy release on the Sun. A sample of Type III bursts in the range 20–650 MHz during the period of extraordinary solar activity (20 October to 4 November 2003) recorded by the ARTEMIS–IV[1] radio spectrograph is analysed; its parameters are compared with characteristics of associated flares (Ha and GOES SXR) and CMEs, observed in the same period and reported in the SGR and the LASCO archives respectively. In this report we attempt to establish a correlation between energetic particles and major manifestations of solar activity such as flares and CMEs.

Keywords: Solar activity; Corona; Radio; Radiation and spectra; Solar electromagnetic emission; Flares
PACS: 96.60.Q; 96.60.P; 95.85.Bh; 96.25.Tg; 96.60.Tf; 96.60.qe

INTRODUCTION

The active period from 20 October to 5 November 2003 has attracted considerable attention ([3], [5], [6], [4], [7], [8]) because of its abundance in of powerful flares and large coronal mass ejections (CMEs) as well as by strong space weather disturbances. Although within the descending phase of the solar cycle the intensity of the solar activity has been dramatic, culminating with the X17.2 flare of October 28, 2003 and the X28/3B flare of the 4 November 2003, one of the more powerful flares ever detected. This activity was associated with the appearance of a global complex consisting of three large, remote but connected active regions (AR): AR 484 (Carrington coordinates, N04; L = 354), AR 486 (S15; L = 283) and AR 488 (N08; L = 291) (cf. [3]).

The large number of flares, CMEs and radio bursts during this period, has provided a data set adequate for an initial study on the association of CMEs and flares with type III radio bursts. The importance of the radio bursts of the type III family in the

---

[1] Appareil de Routine pour le Traitement et l' Enregistrement Magnetique de l' Information Spectral

CP848, *Recent Advances in Astronomy and Astrophysics,*
7th *International Conference of the Hellenic Astronomical Society,* edited by N. H. Solomos

study of energy release processes is two fold: Firstly, the electron beams exciting the type III radio emission, stream along the coronal magnetic lines, thus tracing magnetic structures; secondly the study of these suprathermal electron populations is of interest because it is directly associated with particle acceleration mechanisms in the lower corona. In this paper we present a study with emphasis on the relationship between the

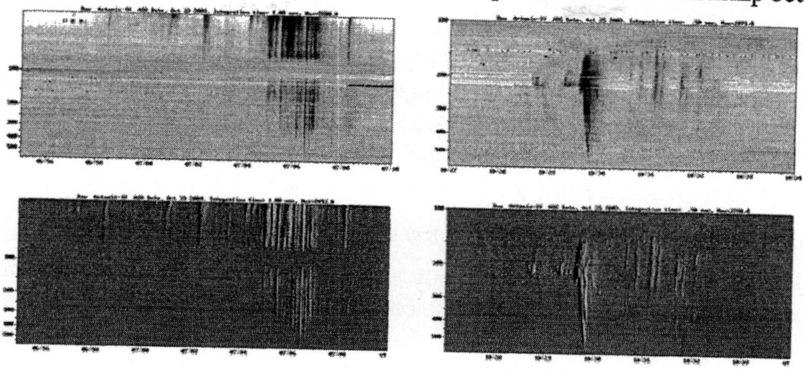

FIGURE 1. ARTEMIS–IV recordings of type III groups; The upper panels are intensity dynamic spectra, the lower are differential spectra. Left: A type III group 23 October 2003 06:55–07:10 UT; it was found associated with an M class SXR flare and a CME. Right: A type III group 25 October 2003 10:27–10:34 UT; it was found associated with an M class SXR flare without accompanying CME.

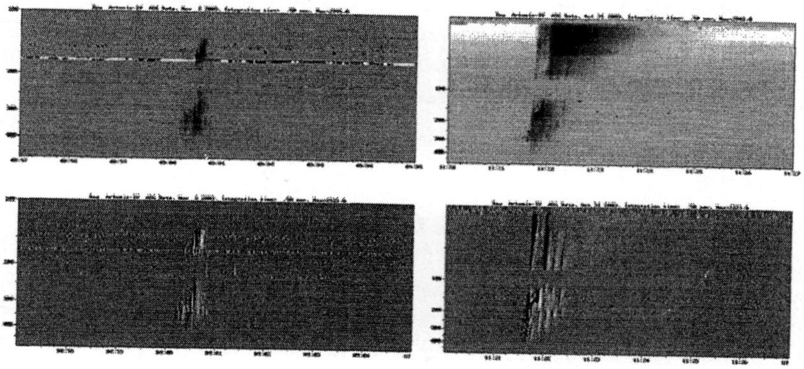

FIGURE 2. Same as 1 but for the Type III Groups: Left: Event 2 November 2003 08:55–09:05 UT with pronounced fundamental–harmonic structure; it was asociated with a halo CME without SXR flare. Right: 31 October 2003 11:20–11:30 UT; it was neither CME nor SXR flare associated.

metric solar type III radio bursts and coronal mass ejections (CMEs), and SXR flares.

DATA ANALYSIS & RESULTS

Our data set, covers the the period of intense solar activity from 20 October to 5 November 2003 and includes:

- Observations of the III metric radio bursts tabulated recorded by ARTEMIS–IV radiospectrograph ([1], [2]); it provides dynamic spectra in the frequency range

20–650 MHz, ie. from the base of the corona to two Solar Radii, at 10 samples per second. This is our primary data list and comprises 206 type III bursts and groups. Although from the dynamic spectra we cannot deduce the position of the bursts on the disk, we have assumed that the emission orginated near the position of the associated flare.

- SXR (GOES) and Ha flare observations, from SGD[2]. In our study, we establish their temporal association with type III activity, using a time window of one minute.
- CMEs selected from the LASCO lists on line[3] ([9]). The CMEs were characterised as type III associated when recorded half an hour before and after the burst. They were, furthermore, examined for spatial associtㅤion with the type III burst; this association was established if both were on the same quadrant of the disk.

The recorded type III bursts were separated into four categories, as follows:

1. *Type III with flare and CME*: Five only events were found temporally associated with an SXR flare and a CME; of these only two are spatially correlated.
2. *Type III with flare and without CME*: Twenty two events were closely associated with a flare, but not with a CME
3. *Type III without flare but with CME*: Five only, type III radio events were closely associated with a CME, but not with a flare
4. *Type III without flare and without CME*: For most of the type III radio events (173) neither a flare, nor a CME was recorded in the time window introduced for the association in our study. We note, however, that,in 111 of these bursts, considerable SXR enhancements have been observed, in the GOES recordings, which were not included as flares in the NGDC archives.

DISCUSSION & CONCLUSIONS

Our analysis is affected by the two time windows for Type III–flare & Type III–CME association respectively; it is tacitly affected, further more, by the threshold for the inclusion of an SXR enhancement in the NGDC archive or not. We are, hence, to expect, some uncertainty due to the compromise between rejection of significant correlations and inclusion of spurious ones, as it is the case with threshold based processes.

The results of this study point out that the number of type III events exceeds significantly the number of CMEs and flares recorded. This implies that even minor energy releases, and small flares may result in *runaway* suprathermal electron populations which excite type III bursts. Therefore the statistics of this report are deemed inconclusive. The importance, however, of the analysis of the energetic electron populations streaming within magnetic structures during the *liftoff* of CMEs or the onset of flares remains since they may provide indication about the magnetic structure, and the commencement

---

2 http//www.sel.noaa.gov/ftpmenu/indices

3 http://cdaw.gsfc.nasa.gov/CME list

of the magnetic reconnection. In the latter case, the examination of type III activity may complement the study of the type IV fine structure.

ACKNOWLEDGMENTS

The LASCO CME catalogue is generated and maintained by the Center for Solar Physics and Space Weather, The Catholic University of America in cooperation with the Naval Research Laboratory and NASA. SOHO is a project of international cooperation between ESA and NASA. The Solar Geophysical Data Catalogue is compiled and maintained by the US Department of Commerce.

REFERENCES

1. C., Caroubalos, D., Maroulis, N., Patavalis, J. L., Bougeret, G., Dumas, C., Perche, C., Alissandrakis, A., Hillaris, X., Moussas, P., Preka–Papadema, A., Kontogeorgos, P., Tsitsipis, P., and G., Kanellakis, *Experimental Astronomy*, **11**,23–32, (2001).
2. C., Caroubalos, C. E., Alissandrakis, A., Hillaris, P., Preka–Papadema, J., Polygiannakis, X., Moussas, P., Tsitsipis, A., Kontogeorgos, V., Petoussis, C., Bouratzis, J. L., Bougeret, G., Dumas and A., Nindos, *These Proceedings* (2006).
3. I. M., Chertok, V. V., Fomichev, A. A., Gnezdilov, R. V. Gorgutsa, A. K., Markeev & D. E., Sobolev, *Astronomical and Astrophysical Transactions*, **24**, 45-52, (2005).
4. N., Gopalswamy, L., Barbieri, E. W., Cliver, G., Lu, S. P., Plunkett, R. M., Skoug, *Solar Physics*, **226**, 337–357, (2005).
5. J., Kiener, M., Gros, V., Tatischeff, and G. Weidenspointner, *Astronomy and Astrophysics*, **445**, 725–733 (2006)
6. Cheng-Ming Tan, Qi-Jun Fu, Yi-Hua Yan and Yu-Ying Liu, *Chin. J. Astron. Astrophys*, **4**, 205-208, (2004).
7. B., Vrsnak, A., Warmuth, M., Temmer, A., Veronig, J., Magdalenic, A., Hillaris, M., Karlicky., *Astronomy and Astrophysics*, (accepted) (2006).
8. Wang Yuming, Ye Pinzhong, Zhou Guiping, Wang Shujuan, S., Wang, Yan Yihua, Wang Jingxiu, *Solar Physics*, **226**, 337-357, (2005).
9. S., Yashiro, N., Gopalswamy, O. C., St. Cyr, G., Lawrence, G., Michalek, C. A., Young, S. P., Plunkett, & R. A., Howard, *AGU Spring Meeting* Abstract **SH31C-10**, 2001

Type II Shocks Characteristics: Comparison with associated CMEs and Flares

G. Pothitakis*, E. Mitsakou*, P. Preka–Papadema*, X. Moussas*, C. Caroubalos*, C. E. Alissandrakis†, A. Hillaris*, P. Tsitsipis**, A. Kontogeorgos**, J.-L. Bougeret‡ and G. Dumas‡

*University of Athens, 15783 Athens, Greece
†University of Ioannina, 45110 Ioannina, Greece
**Technological Education Institute of Lamia, Lamia, Greece
‡Observatoire de Paris, CNRS UA 264, 92195 Meudon Cedex, France

Abstract. A number of metric (100-650 MHz) typeII bursts was recorded by the ARTEMIS-IV radiospectrograph in the 1998-2000 period; the sample includes both CME driven shocks and shocks originating from flare blasts. We study their characteristics in comparison with characteristics of associated CMEs and flares.

Keywords: Solar activity;Corona;Radio;Radiation and spectra
PACS: "96.60.Q; 96.60.P;95.85.Bh; 96.25.Tg

THE RELATIONSHIP OF TYPE II BURSTS WITH CMES AND FLARES.

Type II bursts are radio signatures of MHD shock waves in the solar corona; they either orginate from flare blasts or are driven by CMEs. Although the CME driven type II scenario has been widely accepted for the Interplanetary Shocks the issue is still open for their coronal counterparts (cf. for example [6]). The CMEs, on the other hand, are energetic phenomena consisting of mass and *frozen in* magnetic field expulsions from the Sun. They, more often than not, coincide with flares, although the cause and effect relationship is, as yet, obscure (cf. [5]). A thorough study of coronal bursts (in the metric wavelengths) with associated activity such as flares and CMEs may contribute to the resolution of these questions.

To this end, in this report, we examine a relatively small sample of type II bursts and compare their characteristics with these of associated CMEs and flares; it is, in fact, a three fold comparison of type II–SXR flare, type II–CME and flare–CME.

Our data set consists of eleven type II bursts accompanied by SXR flares; seven of them were associated with a CME while the remaining four were characterised as Non–CME events.

The type II radio bursts were recorded in the 100–650 MHz range by the ARTEMIS–IV radiospectrograph[1] ([1]) in the 1998-2000 period. A cataloque has been published in [2] where the characteristics of these bursts, such as type II radial velocity, frequency

[1] www.uoa.gr/ artemis

CP848, *Recent Advances in Astronomy and Astrophysics, 7th International Conference of the Hellenic Astronomical Society,* edited by N. H. Solomos

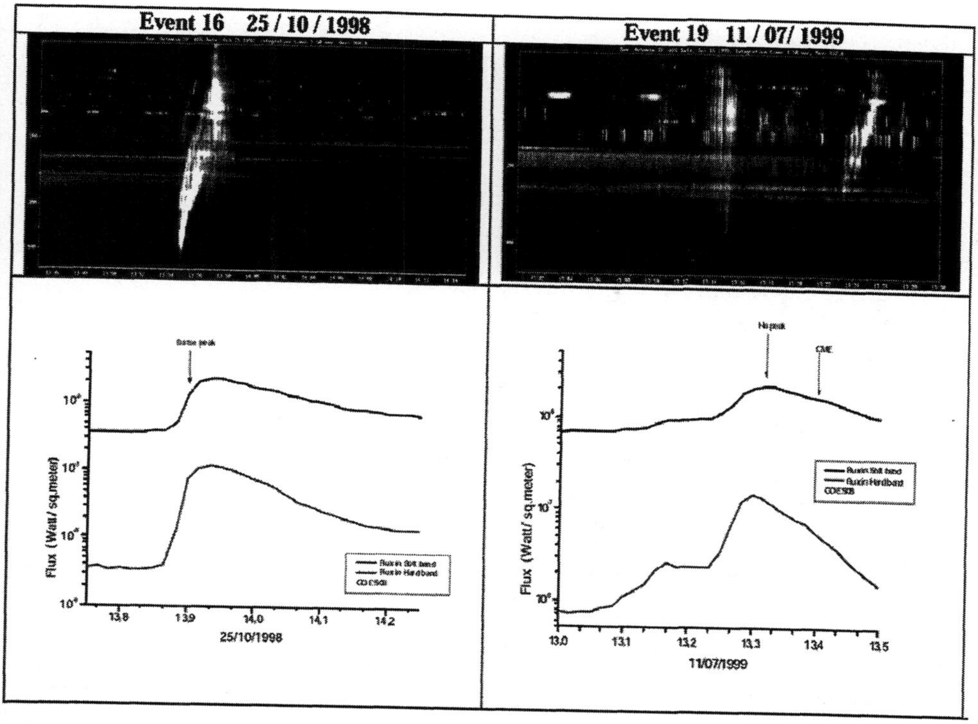

FIGURE 1. ARTEMIS–IV Dynamic Spectra (Top) and light curves of the associated GOES 08 SXR Flares (Bottom). Left Panels: The Non–CME type II burst of the 25th October 1998. right Panes: The CME–associated type II burst of the 11th June 1999. The latter is pereceded by a type III group.

range, duration, associated activity, etc. are reported. The associated CMEs were obtained from SOHO/LASCO lists[2] (cf. [11]); GOES SXR and Ha flare data were found in the NGDC[3]. Two examples of type II dynamic spectra and the associated GOES SXR light curves are demonstrated in figure 1.

The relationship of the following flare–type II–CME parameters have been studied:

- Flare average temperature: In [10] and [4] a method of calculation of the flare average temperature (or colour temperature) using the two channel fluxes from the GOES SXR detectors is demonstrated. We have obtained the flare temperature for each of the 11 SXR events of out data set following this method. In figure 2 we plot the average temperature (Ta) as a function of SXR integrated flux for these events.
- Rise, decay and Duration of SXR flares: These were obtained from the GOES profiles; in figure 3, left panel, we have plotted the flare duration versus the GOES SXR integrated flux, in the right panel of the same figure we present a comparison

[2] cdaw.gsfc.nasa.gov/CME_list
[3] www.ngdc.noaa.gov/stp/SOLAR/

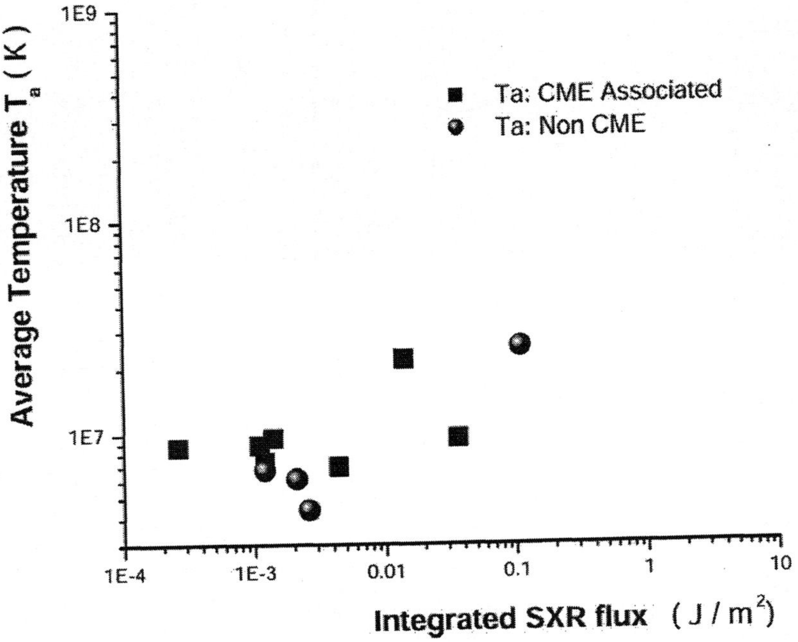

FIGURE 2. Average temperature (Ta) versus SXR integrated flux for the eleven events of our data set. CME associated events are squares while non–CME events are dots.

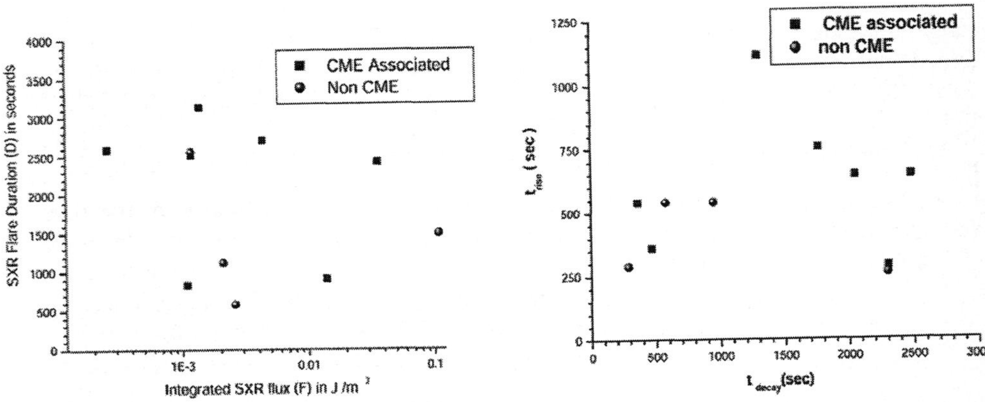

FIGURE 3. Comparison of parameter relationship between CME and Non–CME events; CME associated events are squares while non–CME events are dots.Left panel:Flare duration (D) versus SXR integrated flux (F). Right panel:Flare rise time versus decay time.

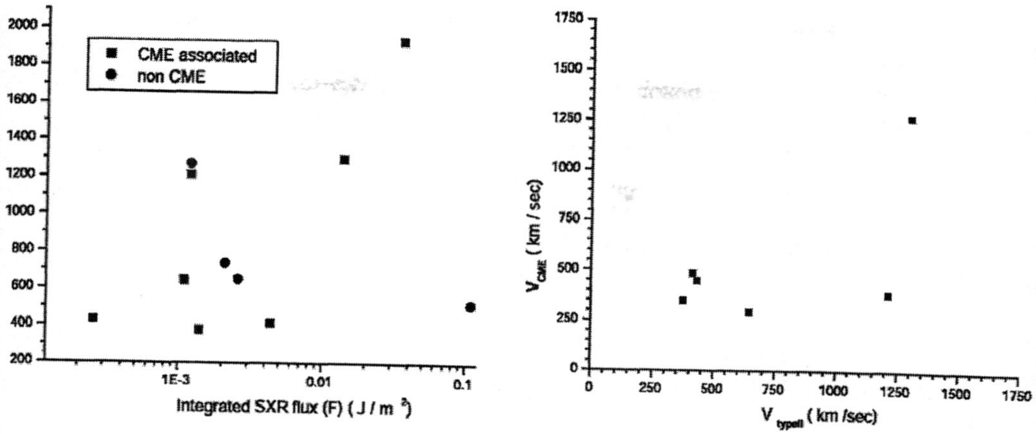

FIGURE 4. Left panel: Type II radial velocity versus SXR integrated Flux. The CME driven type II have radial velocities approximately equal (within a factor of order 2 [3], cf. also next panel) with the associated CME and appear to increase with SXR integrated Flux; the flare blast initiated type II do not present any systematic dependence on the SXR flux. Right Panel: CME Velocity versus Type II Velocity; the two velocities are within a factor of order 2 to each other.

of flare rise time versus decay time.

- Velocities of type II events and CMEs: The type II radial velocities were obtained from [2], the CMEs linear speeds from the SOHO/LASCO catalogues on line. In figure 4 left panel we exhibit the type II radial velocity versus SXR integrated flux; in the right panel the CME Velocity as a function of the type II Velocity.

REASULTS, DISCUSSION & CONCLUSIONS

We have analysed a relatively small group of complex events including type II shocks and SXR flares. A subgroup of these events was CME associated while the remaining were Non–CME events. We have studied the association of a number of characteristic parameters for each of the two subgroups, in an attempt to isolate features, or combinations thereoff, which might clearly distinguish them. The results are summarised as follows:

- Flare average temperature: Our results in figure 2 do not reveal a clear trend in the relationship of flare average temperature (Ta) versus SXR integrated flux; in [7] the CME–events have systematically lower tempreratures than Non–CME events for the same peak intensity.
- Rise, decay and Duration of SXR flares: With respect to flare duration (D) versus SXR integrated flux (F) the CME–events tend to have longer duration than Non–CME flares (cf. figure 3) in accordance with previous results in [5] and [7] although the latter derive these results comparing duration to peak intensity. As regards Flare rise time versus decay time no systematic difference appears for the two subgroups in accordance with similar results in [7].

- Velocities of type II events and CMEs: As is demonstrated in figure 4 (right) the type II and the CME velocities are within a factor of order 2 to each other, as expected from CME piston–driven shocks for the CME–associated subgroup ([3]). This justifies the type II velocity relationship to the SXR integrated flux in figure 4 (left). In this panel the CME driven type II have radial velocities approximately equal (within a factor of order 2 [3], with the associated CME and appear thus to increase with SXR integrated Flux as is expected for the CME speeds ([9], [2]); the flare blast initiated type II do not present any systematic dependence on the SXR flux as they are expected to be affected, mostly, by the ambient corona conditions ([8]).

An original comparison of the CME–associted and the Non–CME subgroups does not demonstrate a substantial difference in the characteristics between them but for a tendency of the latter group to be shorter in duration than the former. Further work, with the use of a much larger sample is in progress.

REFERENCES

1. C. Caroubalos, D. Maroulis, N. Patavalis, J.-L. Bougeret, G. Dumas, C. Perche , C. Alissandrakis, A. Hillaris, X. Moussas, P. Preka-Papadema, A. Kontogeorgos, P. Tsitsipis and G. Kanelakis, *Experimental Astronomy*, **11**, 23–32, (2001).
2. C. Caroubalos, A. Hillaris, C. Bouratzis, C. E. Alissandrakis , P. Preka–Papadema, J. Polygiannakis,P. Tsitsipis, A. Kontogeorgos, X. Moussas, J.-L. Bougeret, G. Dumas & C. Perche, *A&A* **413**, 1125Ű-1133 (2004).
3. H. T., Classen, & H., Aurass, *A&A* **384** 1098–1106, (2002).
4. H. A. Garcia, *Solar Physics* **154**, 275–308 (1994).
5. R. A., Harrison, *A&A*,**304**, 585–594, (1995).
6. S. W., Kahler, N., Sheeley, R. A., Howard et al, *Solar Physics* **93**, 133–141, (1984).
7. H. R. M. Kay, L. K. Harra, S. A. Matthews et al, *A&A*, **400**,779–784, (2003).
8. G., Mann, A., Klassen, H., Aurass, and H. T. Classen, *A&A*, **400**, 329Ű-336 (2003)
9. Y. J., Moon, G. S., Choe, Haimin Wang, Y. D., Park, C. Z., Cheng, *JKAS*, **36**, 61–66, (2003).
10. R. J. Thomas, R. Starr, & C. J. Crannell, *Solar Physics* **95**, 323–329 (1985).
11. S., Yashiro, N. Gopalswamy, O., C., St. Cyr, et al., *AGU Spring Meeting, Abstract SH31C-10*, (2001).

Periodicities in Fine Structure of Type IV Radio Solar Bursts

P. Tsitsipis*, A. Kontogeorgos*, X. Moussas†, P. Preka–Papadema†, A. Hillaris†,**, V. Petoussis*, C. Caroubalos†, C. E. Alissandrakis‡, J.-L. Bougeret§ and G. Dumas§

*Technological Education Institute of Lamia, Lamia, Greece
†University of Athens, 15783 Athens, Greece
**Hellenic Naval Academy, Piraeus 18503, GREECE
‡University of Ioannina, 45110 Ioannina, Greece
§Observatoire de Paris, CNRS UA 264, 92195 Meudon Cedex, France

Abstract. A study of periodicities appearing in the type–IV bursts fine structure (fibers and pulsations in particular) is presented. Our data set includes recordings by ARTEMIS IV with high time resolution; this facilitates the detection of lower than 1 sec periodicities. We introduce a method which employs directional filtering for the separation of intertwined fibers and pulsations for independent analysis of different events within complex radio bursts. For further analysis we use Short Time Fourier and Wavelet transforms which enable the study of the evolution of periodicities with time over specified periods of solar radio activity. Since, according to existing theoretical models, periodicities depend on basic parameters of active regions, this methodology provides the appropriate diagnostic tools.

Keywords: Solar activity; Corona; Radio; Radiation and spectra; Solar electromagnetic emission
PACS: 96.60.Q; 96.60.P; 95.85.Bh; 96.25.Tg; 96.60.Tf

INTRODUCTION

The type–IV solar radio bursts originate from energetic electron populations confined within magnetic structures such as coronal loops etc. On the dynamic spectra they have the form of continua yet, they are, often, rich in *fine structure* (fibers or intermediate drift bursts, pulsations, etc.). The fibers and pulsations, frequently exhibit quasi–periodic behaviour which is thought to be the effect of the magnetic structure oscillations, and of the corresponding variations of the magnetic field on the emission mechanism.
A study of periodicities appearing in the type–IV bursts fine structure (fibers and pulsations in particular) appears promising in revealing the variations of the basic parameters of active regions during this type of activity.
A study of this type of periodic behavior and the appropriate methodology is presented in this report.

DATA ANALYSIS

For the analysis of periodicities in the type–IV fine structure Short Time Fourier Transform wavelet transform are applied on radiospectrograph recordings. These have been

CP848, *Recent Advances in Astronomy and Astrophysics,*
7th *International Conference of the Hellenic Astronomical Society,* edited by N. H. Solomos

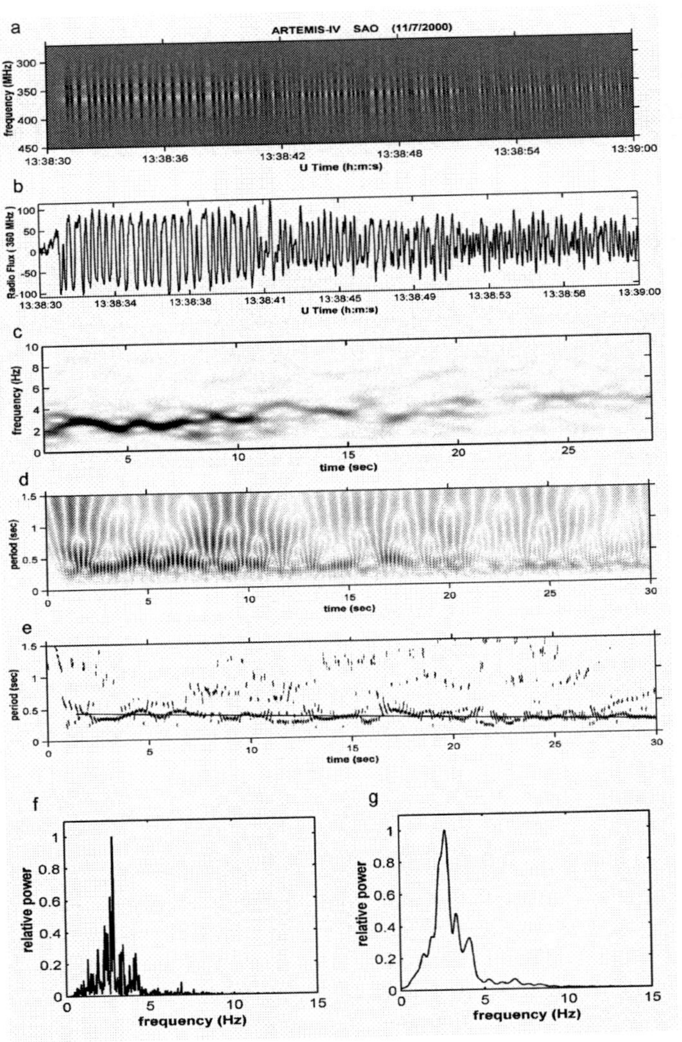

FIGURE 1. Periodicity Detection on a Pulsating Structure (2000 July 07 13:38:30–13:39:00); it is part of a type–IV event from the 1998–2000 ARTEMIS–IV Catalogue [3]. Top: The ARTEMIS–IV Dynamic Spectrum; the type–IV continuum background has been suppressed by means of a high pass filter. Second panel from top: Time profile of a single ARTEMIS–IV channel recording (360 MHz). Third panel: Short Fourier Transform of the time profile in the second panel. Fourth panel: Wavelet Transform of the same time profile and (Fifth panel) skeleton of the corresponding scalogram. Bottom: Comparison of the total Fourier Spectrum (left) and average wavelet spectrum (right).

obtained from the Solar Radio-spectrograph ARTEMIS-IV[1] (cf. [1], [4]) with a fre-

[1] Appareil de Routine pour le Traitement et l' Enregistrement Magnetique de l' Information Spectral

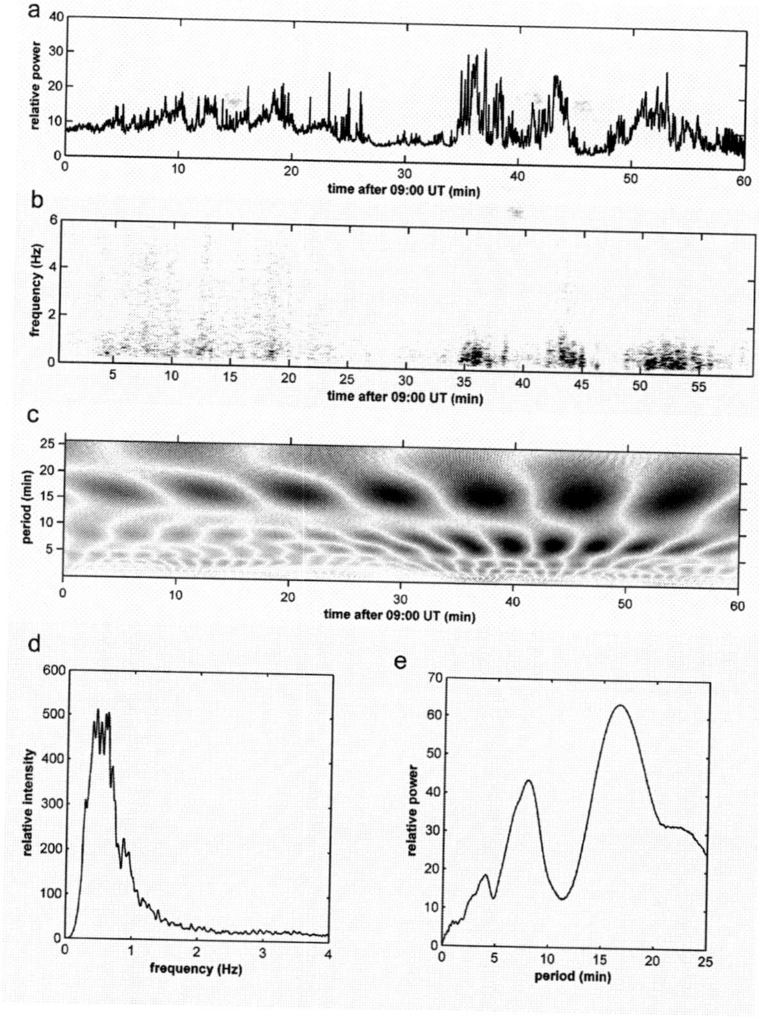

FIGURE 2. Periodicity Detection on fiber groups (2002 April 17 09:00:00–10:00:00 UT) recorded by ARTEMIS–IV; (a) Time evolution of the power of fibers. A single channel is used (channel 75, 376 MHz). (b) Time - frequency analysis with STFT of the signal of channel 75 (376 MHz). (c) Wavelet Transform of the same time profile. (d) Averaging of spectra of the second panel. Quasi-periodic character of fibers is obvious with a mean frequency of about 0.5Hz. (e) Averaging spectra of the third panel. Three dominant periods 4.1 min, 8.0min and 16.3min are detected.

quency coverage 650–20 MHz and time resolution 1/10-1/100 sec.

This type of analysis presents a particular difficulty, since the spectra of interest are quite complex comprising different groups of fibers and pulsating structures within the same period.

The separation of intertwined structures for individual analysis is attained by means of

TABLE 1. Periodicities of Fibers and Pulsations Within Selected Type–IV radio events

Date	Time (UT)	Pulsation Freq. (Hz)	Pulsation Power (minutes)	Fiber Freq. (Hz)	Fiber Power (minutes)
14–07–2000	10:30–11:25	0.31; 0.45	1.4; 2.7; 4.7; 14.4	0.32; 0.45 0.80; 1.2	4.7; 10.4 14.7
11–07–2000	13:00–14:00	0.35; 0.38; 0.47; 0.65; 0.73; 1.22 2.5; 2.8; 3.3	2.8; 8.6		2.7; 3.4; 4.4; 6.4; 11.6; 15.5
17–04–2002	08:40–09:10		3.6; 5.7; 7.9	0.44	4.12; 8.3; 16.3
17–04–2002	09:00–10:00		13.5; 18		
17–01–2005	09:30–10:30		2.0; 4.1; 6.6; 13.8	0.4	14.2
15–01–2005	07:00–08:00		2.4; 3.3; 4.8; 6.4; 22		1.7; 4.3 6.7; 14.2

an algorithm outlined in [7] and [8]; this algorithm has been used for the detection of different groups of fine structure based on their *frequency drift rate*, these have been separated, in turn, with directional filtering. The groups of fibers or pulsations thus isolated have been further analysed with Short Time Fourier and wavelet methods. The parameters of interest in this study were firstly, the characteristic frequencies of the recorded time–series and the characteristic periodicities of the power of the recorded signal. Examples of this analysis for a pulsating structure and for fibers are in figures 1 and 2.

RESULTS

For this work, a number of selected periods of type–IV activity recorded by ARTEMIS–IV were used; characteristic frequencies and periodicities were calculated and the results are summarized in the table. The period of the pulsations power varies in the range 1.4–22.0 minutes, while for the fibers was in the range 1.7–16.3 minutes. As regards the characteristic frequencies detected, they are in the range of 0.3–3.5 Hz for the pulsating structures and 0.3–0.8 Hz for fibers.
We note, at this point, that certain characteristic periodicities appear to be of particular interest. A periodicity of $\approx$ 3 minutes may be identified with the 3 min sunspot oscillation and another of $\approx$ 5 minutes with the global modes of solar oscillations (cf. [5], [6]).

ACKNOWLEDGMENTS

This work was partly supported by the Greek Ministry of Education with and *Archimedes II fund. The ARTEMIS–IV group thanks the Hellenic Telecommunication Organization for hosting the ARTEMIS–IV in the the Thermopylae Satellite Communication Station*

REFERENCES

1. C. Caroubalos, D. Maroulis, N. Patavalis, J. L. Bougeret, G. Dumas, C. Perche, C. Alissandrakis, A. Hillaris, X. Moussas, P. Preka–Papadema, A. Kontogeorgos, P. Tsitsipis and G. Kanellakis, *Experimental Astronomy* **11**, 23–32, (2001).
2. C. Caroubalos, C. E. Alissandrakis, A. Hillaris, A. Nindos, P. Tsitsipis, X. Moussas, J. L. Bougeret, C. Bouratzis, G. Dumas, G. Kanellakis, A. Kontogeorgos, D. Maroulis, N. Patavalis, C. Perche, J. Polygiannakis and P. Preka–Papadema, *Solar Physics* **204**, 167–179, (2001).
3. C. Caroubalos, A. Hillaris, C. Bouratzis, C. E. Alissandrakis, P. Preka-Papadema, J. Polygiannakis, P. Tsitsipis, A. Kontogeorgos, X. Moussas, J. L. Bougeret, G. Dumas, C. Perche, *A&A* **413** 1125-1133 (2004).
4. C., Caroubalos, C. E., Alissandrakis, A., Hillaris, P., Preka–Papadema, J., Polygiannakis, X., Moussas, P., Tsitsipis, A., Kontogeorgos, V., Petoussis, C., Bouratzis, J. L., Bougeret, G., Dumas and A., Nindos, *These Proceedings* (2006).
5. I. De Moortel, A. W. Hood, and J. Ireland, *A&A* **381**, 311-323 (2002).
6. I. De Moortel, A. W. Hood, J. Ireland and R.W. Walsh, *Solar Physics* **209**, 89-108 (2002).
7. P. Tsitsipis, A. Kontogeorgos, A. Hillaris, X. Moussas, C. Caroubalos, P. Preka–Papadema, *Pattern Recognition* (submitted)(2005).
8. P. Tsitsipis, A. Kontogeorgos, X. Moussas, P. Preka–Papadema, A. Hillaris, V. Petoussis, C. Caroubalos, C. E. Alissandrakis, J.-L. Bougeret and G. Dumas , *These Proceedings* (2006).

A Statistical Study of Microwave Flare Morphologies

V. Tzatzakis*, A. Nindos*, C. E. Alissandrakis* and K. Shibasaki†

*Section of Astrogeophysics, Physics Department, University of Ioannina, Ioannina 45110, Greece, email: me01621,anindos, callissan@cc.uoi.gr

†Nobeyama Radio Observatory, National Astronomical Observatory of Japan, Minamimaki, Minamisaku, Nagano 384-1305, Japan.

Abstract. Using Nobeyama Radioheliograph (NoRH) high spatial resolution images at 17 and 34 GHz, we study the morphology of several flare events that occured relatively close to the limb. Our study has been motivated by the recent detection of a small number of optically thin flares whose maximum emission peaks close to the loop top. These events may show significant anisotropy of the density and pitch angle distribution of the nonthermal electrons. However, we do not know how often they occur. Our sample consists of 104 flare events. Using data from the Nobeyama Polarimeter we were able to determine whether the 17 and 34 GHz emissions are optically thin or thick. Almost half of our events appear unresolved in the NoRH images. Among the resolved events, special attention is paid to those with circular polarization (*V*) showing two distinct sources. Such *V* sources are considered proxies to the flaring loop footpoints if the total intensity (*I*) morphology is consistent with a single flaring loop configuration. A small number of events, as expected at such high frequencies, are optically thick with *I* maximum between the loop footpoints. The number of optically thin events showing *I* maxima co-spatial with the *V* sources is approximately the same as the number of those optically thin resolved events showing emission peak between the *V* sources. Our future plans concerning the study of our database are also outlined.

Keywords: Sun:corona, Sun:flares, Sun:radio radiation
PACS: Flares 96. 60. qe

1. INTRODUCTION

Solar flare microwave emission comes from accelerated electrons: through the gyrosynchrotron mechanism, mildly relativistic electrons (energies of tens to several hundreds of KeV) emit at microwaves as they spiral along magnetic field lines in the corona. The gyrosynchrotron emission mechanism depends strongly on the coronal magnetic field strength and orientation.

At microwaves, for a given frequency the morphology of the radio source depends on the geometry of the flaring region and its location on the disk, as well as the properties of the nonthermal electrons. Assuming an isotropic distribution of nonthermal electrons, along the loop, typically, at high frequencies the emission is presumably optically thin and the loop footpoints are seen while at low frequencies the radio emission becomes optically thick and there is a tendency for more extended sources, indicating emission from the entire loop (e.g. Alissandrakis & Preka-Papadema 1984 and also the reviews by Alissandrakis 1986; Bastian, Benz & Gary 1998; Nindos 2005). At a fixed frequency higher energy electrons emit at the loop top (weak magnetic field) and lower energy electrons emit at the footpoints (strong magnetic field). A decrease of the frequency of

CP848, *Recent Advances in Astronomy and Astrophysics*, 7th *International Conference of the Hellenic Astronomical Society*, edited by N. H. Solomos observation has approximately the same effect as an increase of the magnetic field.

Recent observations have revealed the existence of limb flaring loops whose microwave emission peaks close to the loop top despite the fact that the emission is optically thin (e.g. Kundu, et al.2001a). In order to have emission from the top of an optically thin loop, the magnetic field should decrease slowly from the footpoints to the top of the loop. But very small variation in magnetic field between the loop top and the footpoints will not be efficient in trapping the electrons.

It has been suggested (Melnikov et al.2002) that loop top sources in optically thin events are due to highly inhomogeneous distribution of energetic electrons along the flaring loop: accelared electrons are concentrated in the upper part of the loop. Note that the properties of the emitted radiation (intensity, optically thin spectral index, degree of polarization) may depend sensitively on the type and degree of electron anisotropy (Fleishman & Melnikov 2003a,2003b). Melnikov et al. (2002) considered the question whether the anisotropic distribution of electrons along the loop is due to accelaration /injection process or due to transport effects. Their simulations seem to support the idea that the high concetration of electrons at the loop top is due to transverse pitch angle anisotropy of the injected electrons. But they also pointed out that transport effects (enhanced pitch angle scattering and energy losses near the footpoints due to either higher plasma density in the lower part of the loop or enhanced level of whistler wave turbulance) are able to create a similar morphology. However, we do not know how often such events occur. In order to check whether events with significant anisotropies of the nonthermal electron distributions occur rarely or not, we have undertaken a statistical study of the morphological properties of several limb flares observed with Nobeyama Radioheliograph (NoRH) at 17 and 34 GHz.

2. DATA ANALYSIS

We have used Nobeyama Radioheliograph (NoRH) high spatial resolution images at 17 and 34 GHz which have been proccessed by Fujiki's (17 GHz) and Hanaoka's (34 GHz) CLEAN programs. The NoRH observes the full solar disk in total intensity (*I*) and circular polarization (*V*) at 17 GHz and in (*I*) only, at 34 GHz. The spatial resolution of the images is about 10″-15″. We have used data of 104 flare events from 1996 to 2005 that occured relatively close to the limb. Our data have been taken from the Nobeyama's website (http://solar.nro.nao.ac.jp). We have studied events relatively close to the limb in order to make sure that possible optically thin loop-top sources do not result from projection effects. We have selected flare events that satisfy the following criterion. Their minimum heliographic longtitude should be 65 degrees to the East/West with respect to the sun's center. Also we have used total flux data from the Nobeyama Polarimeter to determine whether the 17 and 34 GHz emissions are optically thin or optically thick. The NoRP records total power fluxes at 7 frequencies between 1-80 GHz. The peak of the flux-frequency curve roughly divides the spectrum into two domains: the optically thin part at frequencies higher than the frequency corresponding to the peak flux, f_{peak}, and the optically thick part at frequencies lower than f_{peak}. For those with optically thin emissions, their spatial morphologies are studied.

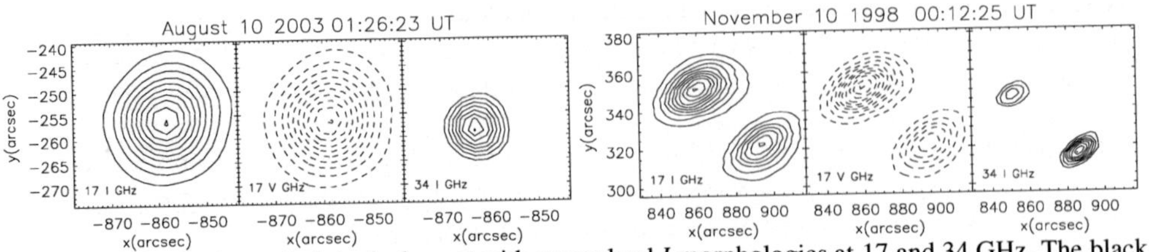

FIGURE 1. Left panel: A typical event with unresolved I morphologies at 17 and 34 GHz. The black solid contours show the 17 and 34 I emissions at the time of maximum. In this and subsequent figures the contour levels are 10, 20, 30, 40, 50, 60, 70, 80, 90, 99% for 17 I and 30, 40, 50, 60, 70, 80, 90, 99% for 34 I of the maximum brightness temperature, which is 26.7 and 2.4 MK at 17 and 34 GHz, respectively. At 17 GHz V, the solid and dashed contours show positive and negative emissions, respectively. The V contours are at ± 10, ± 20, ± 30, ± 40, ± 50, ± 60, ± 70, ± 80, ± 90, $\pm 99\%$ of the absolute maximum brightness temperature (which is 5.3 MK in this event). North is up and West to the right. The axes labels denote seconds of arc from disk center. Right panel: A typical optically thin event showing I emission associated with the two V sources (presumably the loop's footpoints). The maximum brightness temperatures are 10.8 and 0.93 MK for 17 and 34 GHz I, respectively, and 4.8 MK for 17 V.

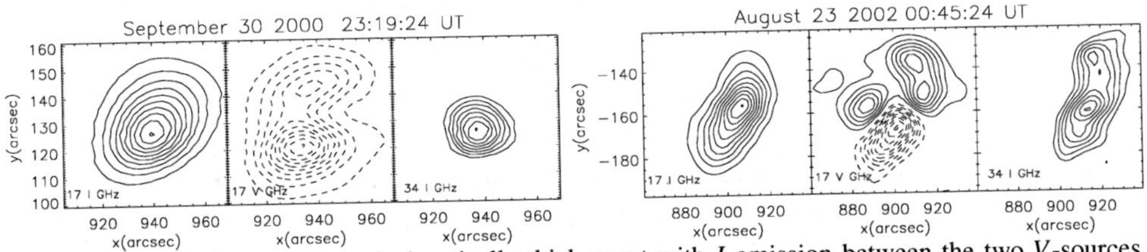

FIGURE 2. Left panel: A typical optically thick event with I emission between the two V-sources. The maximum brightness temperatures are 174 and 88 MK for 17 and 34 GHz I, respectively, and 5.2 MK for 17 V. Right panel: Resolved event with complex morphology (more than 1 loop). The maximum brightness temperatures are 4.77 and 0.13 MK for 17 and 34 GHz I, respectively, and 5.8 MK for 17 V.

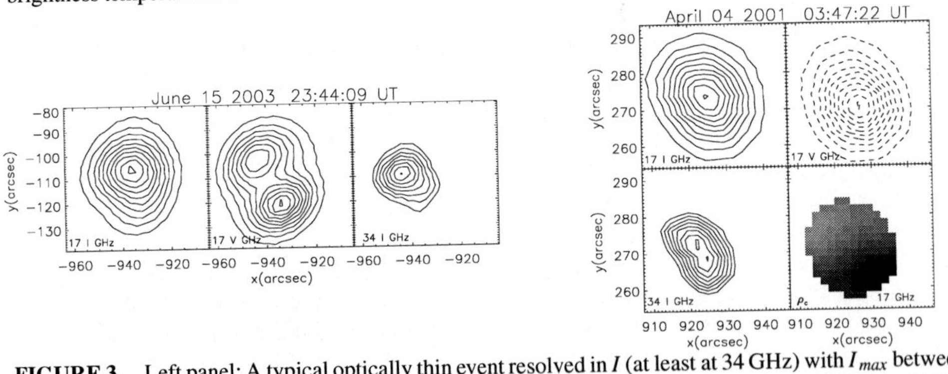

FIGURE 3. Left panel: A typical optically thin event resolved in I (at least at 34 GHz) with I_{max} between the two V-sources (presumably the loop's footpoints). The maximum temperatures are 227.8 and 36.6 MK for 17 and 34 GHz I, respectively, and 15.1 MK for 17 V. Right panel: An optically thin event that is slightly resolved at 34 GHz. The location of the footpoints is ambiguous. The map of the degree of circular polarization, ρ_c, is also presented. The maximum brightness temperatures are 13.3 and 2.35 MK for 17 and 34 GHz I, respectively, and 2.9 MK for 17 V.

TABLE 1. PROPERTIES AND MORPHOLOGIES OF FLARE EVENTS

Event Properties	**Unresolved**			**Resolved in *I* (1 loop)**				**More than 1 loop**	
Optical Depth	thin	thick	Not clear	thin	thin	thin	thick	thin	thick
Morphology *I* emission				between ftps†	unclear*	Above ftps	Between ftps		
Number (%)	31 29.8	10 9.6	8 7.7	10 9.6	11 10.6	14 13.5	6 5.8	8 7.7	6 5.8
Total		47.1%			33.6%		5.8%	13.5%	

* Footpoint location is unclear
† footpoints

3. RESULTS AND FUTURE WORK

Our study considers images at the peak of the events only. We grouped the events into 6 categories:

According to table 1, almost half of the events appeared unresolved in I (Figure 1 left panel). The unresolved events are divided according to their optical depth in 3 categories. The optically thin (31 events), the optically thick (10 events) and the events for which we could not tell whether they were optically thin or thick (8 events in a total of 104 flare events). Some of them could have been resolved if "super-resolution" imaging reconstruction techniques had been used. But even then, one expects that some of these events would still appear unresolved (Kundu et al.2001b).

We have studied the morphology of optically thin events (one loop configuration) showing resolved morphology in I at least at one frequency and clear two-component structure in V. The V sources are considered as proxies of the loop footpoints. The events showing such a morphology can be divided in two categories. The optically thin events with maximum intensity between the two V sources (footpoints of the loop) and the events with I emissions nearly co-spatial with the two V sources (emission above the footpoints). According to table 1 the number of the events that consist these two categories are approximately the same (10 events instead 11 among 104 events in total). The morphology of a characteristic optically thin event with I emission between the 2 V sources is presented in figure 3 (left panel) while a characteristic optically thin event with two I components co-spatial with the V sources is given in Figure 1 (right panel).

Some optically thin events show resolved I morphologies but the V maps do not show clear two-component structure. In this category there are 11 events among the 104. Studying and modeling the morphology of the corresponding circular polarization, ρ_c, maps (e.g. as in Hanaoka 1997) may help us constraint the flare configuration at least in some of them. In figure 3 (Right panel) we show such an event. The degree of polarization of our source is sloping. Such a slope can not be produced by a single source. As is demonstrated by Hanaoka (1997) such ρ_c morphology is consistent only with two oppositely polarized V sources.

As one expects at such high frequencies, the number of optically thick events is small (6 events among 104) according to table 1. The morphology of an optically thick event is shown at Figure 2 (Left panel). Table 1 shows that the number of the events with complex morphologies (showing more than one loop) is small (8 optically thin events and 6 optically thick among 104). Such morphologies are shown at Figure 2 (Right panel).

We have studied the frequency of appearance of the different morphologies as a function of the flare's heliographic longtitude. The statistics appearing in table 1 does not depend on the location of the flare events.

Concerning our future plans, it will be important to remove the effect of the clean beam from the images in order to derive more accurate conclusions about the true dimensions of the *I* sources. The true radio flare source sizes can be determined by considering the source in each radio map as the convolution of the true source (assume it as a two dimensional Gaussian in shape) with the Gaussian used for the restoring beam in the map. Also we need to investigate whether some of the events which show two distant *I* sources are caused by two interacting loops of different sizes: a compact unresolved loop associated with one of the *I* sources and a longer loop connecting the two *I* sources (as in Hanaoka 1996, 1997).

REFERENCES

1. Alissandrakis, C. E. 1986, *Sol.Phys.*, 104, 207
2. Alissandrakis, C. E., Preka-Papadema, P. 1984, A&A, 139, 507
3. Bastian, T. S., Benz, A. O., Gary, D. E. 1998, ARA&A, 36, 131
4. Dulk, G. A., Marsh, K. A. 1982, Apj, 259, 350
5. Fleishman, G.D., Melnikov, V.F. 2003a, Apj, 584, 1071
6. Fleishman, G.D., Melnikov, V.F. 2003b, Apj, 587, 823
7. Hanaoka, Y. 1996, *Sol. Phys.*, 165, 275
8. Hanaoka, Y. 1997, *Sol. Phys.*, 173,319
9. Kundu, M. R., White, S. M., Shibasaki, K., Sakurai., T., Grechnev, V. V. 2001b, ApJ, 547, 1090
10. Kundu, M. R., Nindos, A., White, S. M., Grechnev V. V, 2001a, ApJ, 557, 880
11. Melnikov, V. F., Shibasaki, K., Reznikova, V. E. 2002, Apj, 580,L185
12. Nindos, A. 2005, *in proceedings of the Nobeyama Symposium 2004, in press*

Prospects of Space Weather Prediction Based on Solar Proton Events

Maria V. Gerontidou*, Anatoly V. Belov†, Howard A. Garcia¶*, Helen A. Mavromichalaki* and Victoria G. Kurt¶

*Nuclear and Particle Physics Section Physics Department, University of Athens, Pan/polis, Zografos 15771 Athens Greece emavromi@cc.uoa.gr, mgeront@cc.uoa.gr

†IZMIRAN, Russian Academy of Science, Moscow, Russia abelov@izmiran.rssi.ru

¶*SEC/NOAA, BOULDER, Co 80303, USA

¶Institute of Nuclear Physics, Moscow State University, 119899, Leninskie Gory, Moscow, Russia

Abstract. Progress in Space environment analysis is expected from extensive collaborative efforts with near real –time data from spacecraft and ground-based data with the aim of establishing predictive systems relating the cause (solar activity) to the effect on technological systems and human activity. On the basis of solar energetic particle prediction a unique database of all proton enhancements and their sources covering the period 1970-2002 has been created and statistically analyzed by the team of Athens Cosmic Ray Group in collaboration with the Institute IZMIRAN of Russian Academy Sciences , Institute of Nuclear Physics of Lemonosov and Space Environment Center NOAA. Using this database and following Garcia's method, that relates thermal parameters to spatial flare dimension, a main parameter that controls proton escaping from the Sun to the interplanetary space, is determined. It is found out that the flare's loop dimension is a crucial factor for the accelerated protons to run away from the Sun. This statistical approach allows us to derive ratios between the flares with obvious evidence for proton production and flares with protons escaping from the Sun vicinity into interplanetary space. This is an important result for the predictive studies of Space Weather phenomena.

Keywords: solar proton event, space weather, solar energetic particles, particle emission.
PACS: 96.20.wq

INTRODUCTION

Solar energetic particles, high-energy neutral emissions, coronal mass ejections (CMEs) and shock waves associated with fast CMEs determine the Space Weather at the Earth's orbit. The most powerful sources of solar energetic particle (SEP) fluxes observed at 1 AU are flares and interplanetary shocks waves. Due to that the main charged component of SEP are protons, their appearance in the interplanetary space and in the Earth's environment is usually called as "solar proton event" (SPE). A more specific determination of SPE is that it is the enhancement of the proton fluxes >10 particles/cm^2 s sr (10 pfu) over the background level with energies >10 MeV in the interplanetary space related with the sporadic activity of the Sun. Protons with energies > 500 MeV are known as Ground Level Enhancements (GLEs) and are registered by Neutron Monitors stations when they are arrived at the Earth. An example of a recent SPE at January 20, 2005 is presented in Fig.1.

CP848, *Recent Advances in Astronomy and Astrophysics,*
7th *International Conference of the Hellenic Astronomical Society,* edited by N. H. Solomos

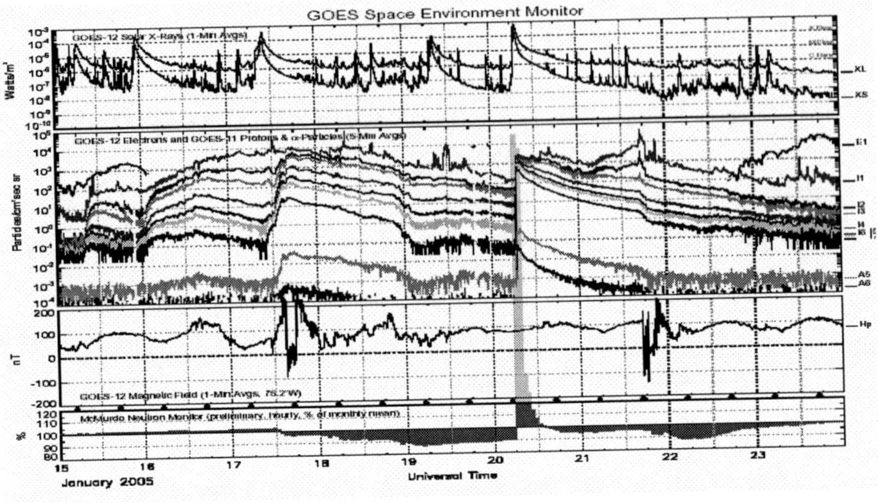

FIGURE 1. A paradigm of a recent SPE of 20 Jan 2005.After a big flare X7.1 (upper panel) a typical SPE was recorded by a) GOES satelites at 1 AU(midle panel) and b) Neutron Monitor in Earth (GLE) (lower panel)

The results of the present work and other recent publications on the relation of proton events and flares [1,2] attests to the physicality of these phenomena. Clearly, the correlation between certain intrinsic flare characteristics and proton events exists, and this relationship can be used as an aid in the radiation hazard forecast process.

STATISTICAL PROPERTIES

Based on an update database including 1144 proton enhancements with energy >10 MeV and flux I_p>0.01 pfu and their sources covering the period 1975-2003 [3] and a recent catalogue of 253 SPE with energies >10 MeV and I_p>10pfu at the Earth's orbit during the time interval 1970-2002 [4], a detail statistical analysis is carried out. In this analysis the integral proton fluxes (fluences) are obtained from the GOES and IMP-8 satellites as well as from ground based neutron monitors.

Longitudinal - Latitudinal distribution

The longitudinal distribution of the associated with SPEs flares as well as of the mean flux of corresponded SPEs demonstrates that are centered within the longitude interval 40^0E÷80^0W, that is in accordance with previous works [5,6,]. On the other hand, the great majority of flares, which are associated with proton enhancements, were located in a more wide longitude interval 30^0 E- 120^0 W [3].

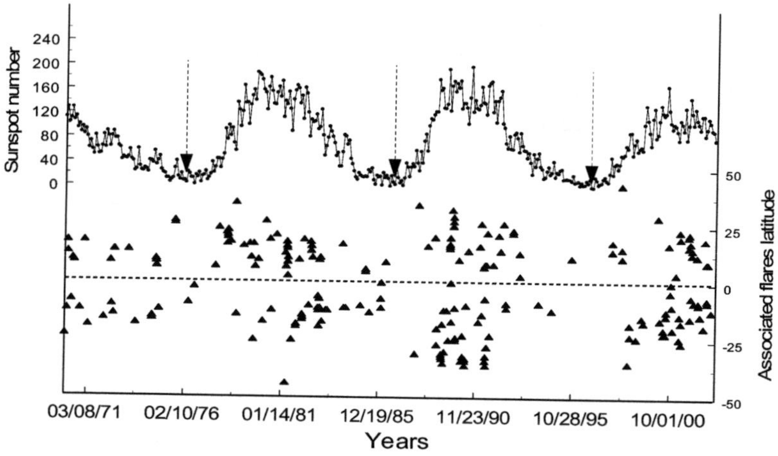

FIGURE 2. "Butterfly diagram" of the parent flares is presented in the lower curve. Time profile of the sunspot number is also given in the upper curve.

A latitudinal distribution of the "associated with SPEs flares" for the interval 1970-2002 is presented in Fig.2. Monthly values of sunspot number are also given in this figure, and the arrows indicate the start of each cycle. It is known that solar flares follow the "butterfly" diagram, by which every cycle starts with flares at high latitudes (greater than 35^0), while it ends with flares at low latitudes. It is interesting to note, that in our case the latitudinal distribution of the associated flares with SPEs as well as proton enhancements seems to have similar distribution as total flares and sunspot number.

Flare importance and SPE probability

The SPE probabilities (percent) of associated SXR flares (N= 18383) for different importance thresholds are calculated and presented in Fig. 3. To avoid the longitudinal dependence, only central and western flares (i.e., located to west of 20°E) have been used. It is evident from this figure that the SPE probability first becomes noticeably significant in flare importance range of ≥M1. One out of four X-ray flares of ≥M3 importance results in a measurable level of >10 MeV protons. At importance ≥X1, more than half of flares have measurable ≥10 MeV protons and GLE incidence becomes significant. Proton incidence and flux increases sharply above this level of importance.

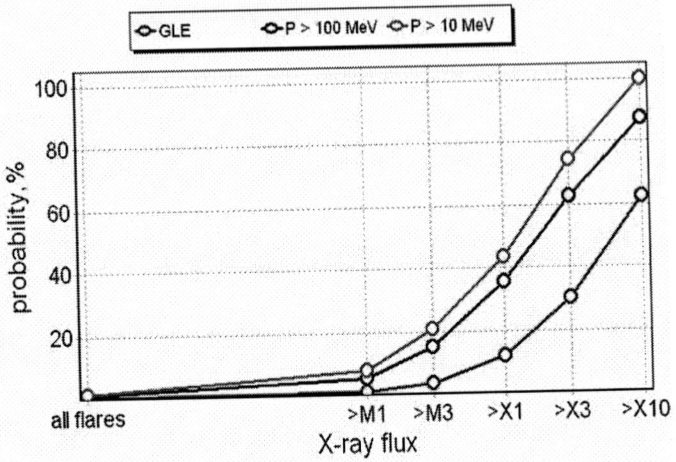

FIGURE 3. The SPEs probability after X-ray flares with different importance is presented. The botom line is referred to GLE , the middle line to SPEs with energy >100MeV and the upper line to SPEs with energy >10MeV.

Empirical method derived the loop length from the SXR characteristics

Based on theoretical studies [7] and utilizing Garcia's method, which has worked with a large population [8] of flares occurring during solar cycles 21, 22 and 23 and has computed the above empirical fits to the respective light curve

$$2L = 1.26 \cdot 10^3 \cdot \tau_r^x \cdot \tau_d^y \cdot T_m^z \qquad (1)$$

where L is the semi-loop lenght in km, τ_r is the rise time in sec, τ_d is the decay time in sec, T_m is the maximum temperature (K) and the empirical exponents x=0.026±0.14, y=0.752±0.18 and z=-0.111±0.05 . The standard deviation of residuals is equal to σ=1.7 10^4 km .

These calculations are based on the available GOES X-ray light curves of 1 min time resolution during 1986-1989 years in the "preferable" longitude interval 20°E-90°W. The result of our calculations demonstrates the linear relationship between τ_d and emitting SXR loop length (however it can be forecast from (1)). It gives the minumum values of τ_d =1.5*10^3 s for the SXR flares with escaping protons. The results of our calculations is presented in Fig.4

A first approximation for a prediction of SPEs gives that the main parameter that controls proton escaping from the Sun to the interplanetary space is the lop length. The result of our calculations gives a minimum loop length value for the flares associated with protons.

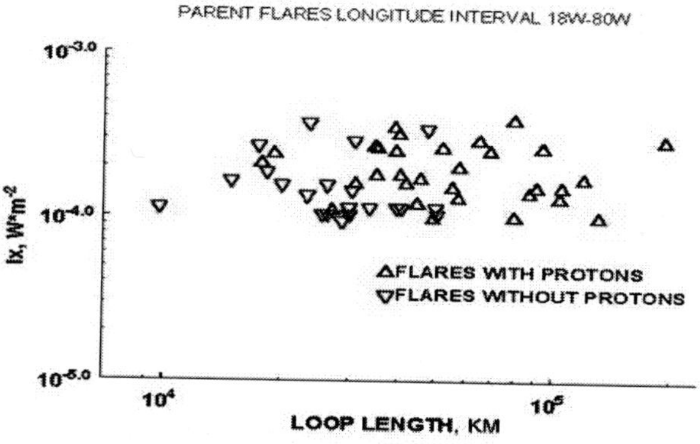

FIGURE 4. Calculated loop length versus importance of all parent flares located in a more narrow interval o f the "preferable connection region".

CONCLUSIONS

1) The longitudinal distribution of parent flare of solar proton events gives an existence of preference connection region.

2) Dependence of solar latitude distribution of parent flares of SPEs on solar cycle seems to have the same behavior as the sunspot number.

3) All X-Ray flares with importance ≥X10 are followed by protons of E >10 MeV and 2/3 of flares ≥X10 are followed by ultra relativistic protons (GLE).

4) It is found out that the flare's loop dimension is a crucial factor for the accelerated protons to run away from the Sun. The accelerated protons can be escaped if the loop length is ≥(5-6) 10^4 km.

5) This statistical approach allows us to derive ratios between the flares with obvious evidence for proton production and flares with protons escaping from the Sun vicinity into interplanetary space. This is an important result for the predictive studies of Space Weather phenomena.

ACKNOWLEDGMENTS

This work is supported by PENED project of the General Secretariat of Research and Technology and by Special Research Account of Athens University.

REFERENCES

1. Garcia, H.A., *Space Weather*, **2**, SW000001 (2004a)
2. Garcia, H.A, *Space Weather*, **2**, SW000035R. (2004b)
3. A. Belov , H.A. Garcia, V. Kurt, H. Mavromichalaki and M. Gerontidou, *Solar Phys*. **229,** 135-159 (2005)
4. Kurt, V., Belov, A., Mavromichalaki, H. and M. Gerontidou, *Annales Geophys*.**22**, 2255 -2271 (2004)
5. Sladkova, A.I. and Bazilevskaya G.A., *Adv. Space Res.*, **4**, 559-562 (1998)
6. Van Hollebeke M.A., MaSung L.S.and McDonald F.M., *Solar Phys.*, **41**, 189-223 (1975).
7. S.L. Hawley, G.H. Fisher, T.Simmon , S.L.Cully, S.E. Deustua, M.Jablonski, C.M. Johns-Krull, B.R. Pettersen,V. Smith, W.J. Spiesman and J.Valenti, *Astoph. J.* **453,** 464-479 (1995)
8. Garcia, H.A., *Ap. J. Suppl. Series*, **27**, 189-210 (2000)

Modeling of the solar energetic particles recorded at Neutron Monitors

Christina G. Plainaki [*], Helen A. Mavromichalaki [*], Anatoly V. Belov [†], Eugenia A. Eroshenko[†] and Victor A. Yanke [†]

[*]*Nuclear and Particle Physics Section, Physics Department, Athens University Pan/polis-Zografos 15771 Athens, Greece* cplainak@phys.uoa.gr, *emavromi@phys.uoa.gr*
[†]*Institute of Terrestrial Magnetism, Ionosphere and Radio Wave Propagation (IZMIRAN), 42092, Troitsk, Moscow Region, Russia* abelov@izmiran.rssi.ru

Abstract. Last year the worldwide network of neutron monitors recorded a new ground level enhancement (GLE) of cosmic ray intensity (CR) on 20th of January 2005, during the recovery phase of a series of Forbush effects taking place at a time-period very close to the minimum of the current cycle of solar activity. This enhancement seems to be the greatest GLE of the current solar cycle, reaching almost 5000% in some polar stations. A joint analysis of data from ground level stations (neutron monitors) and satellite measurements has been performed in order to calculate the amplitude and the anisotropy of the event as well as the energy of the more fast particles arriving at the Earth. A new GLE – model has been created in order to couple the primary solar cosmic ray flux on the top of the magnetosphere with the flux recorded by neutron monitors and define this way the primary solar cosmic ray spectrum during the event. Moreover, the absorption length of solar energetic particles propagating through the Earth's atmosphere has been calculated applying the Wilson et al. method (1967). These results have been compared with the respective results calculated for other ground level enhancements. Finally, a comparison between this enhancement and the big GLE of 1956 has been performed.

Keywords: GLE, solar energetic particles, neutron monitor, solar cosmic rays
PACS: 94.20.wq

INTRODUCTION

Solar cosmic rays can effectively be used for studying the processes of particle acceleration in the solar atmosphere and their propagation in interplanetary space, as well as for understanding the electromagnetic conditions at the Sun. When a high flux of solar nucleons with energy greater than a few hundred MeV strikes Earth's atmosphere, the nuclear products cascade to Earth's surface resulting in a "ground level enhancement" known as GLE [1].

A new ground level enhancement (GLE) was recorded on January 20, 2005 by the worldwide network of neutron monitors. It took place during the recovery phase of a Forbush decrease, on the background of a relatively quiet geomagnetic activity

CP848, *Recent Advances in Astronomy and Astrophysics,*
7th *International Conference of the Hellenic Astronomical Society,* edited by N. H. Solomos

(Kp changed from 2 to 4). This GLE was associated with an X7.1 flare starting at 6:36 UT of the same day, in AR720 (N12 W58). It is remarkable that this solar particle event, as well as the three ground level enhancements of October-November 2003 period [2], occurred in the descending phase of the 23rd cycle, at a time-period very close to the minimum of solar activity. Nevertheless it exceeds all previous events of this solar cycle and may be comparable by order of magnitude to the greatest GLEs over the history of observations. The peak of the cosmic ray variations on 20 January 2005 reached several thousands of percentages at southern polar stations. Time profiles of three of the most significant ever recorded GLEs using data from those neutron monitors that recorded maximum effect, are given in Figure 1.

In this work we analyzed the ground level observations on 20 January 2005 in order to define a model for the behavior of solar cosmic rays. Data from 37 neutron monitors were incorporated so that the main characteristics of this extraordinary event are revealed. Moreover, the absorption length of solar energetic particles propagating through the Earth's atmosphere has been calculated on the basis of Wilson et al. method [3], using data of two conventional neutron monitor stations (Erevan and Erevan3). A more complete analysis will be presented elsewhere.

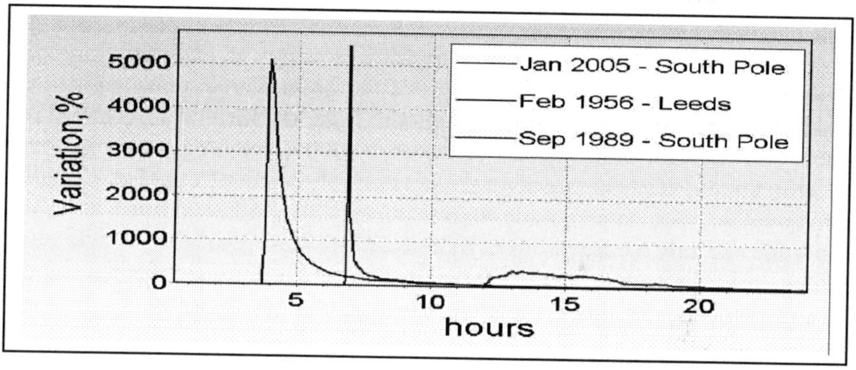

FIGURE 1. Time profiles of the greatest GLEs derived from 5-minute neutron monitor data.

MODEL FOR COSMIC-RAY VARIATIONS

Cosmic ray variations recorded by a ground level detector during a GLE of cosmic rays may be written as follows [4], [5], [6].

$$\frac{\Delta N}{N_0}(t,t_0) = \frac{\int_{E_c}^{E_u} W(E,t_0,h)\frac{\Delta I}{I_0}(t,E)dE}{\int_{E_c}^{E_u} W(E,t_0,h)dE} \quad (1)$$

where N_0 is the basic background counting rate resulted by galactic cosmic ray flux I_0 and measured at the moment t_0 , h is the atmospheric depth at the point of observation in b, Ec is the threshold kinetic energy of the primary protons that cause the secondary flux recorded by a specific detector and Eu is the upper energy limit for the solar particle registered during the event. Coupling coefficients $W(E, t_0, h)$ were calculated analytically for each station separately according to Clem and Dorman [7], taking into consideration, however, the different neutron monitor response in the lower energy range [6],[8]. A power law spectra of the form E^{γ} was considered for the primary CR intensity variations $I(t, E)$ which were supposed to consist of an isotropic part and an anisotropic one, characterized by an angular distribution function taken as:

$$\Psi = \exp[-n_a^{\ 2}\sin^2(x - x_0)] \tag{2}$$

where x_0 corresponds to the direction in which Ψ equals to maximal value 1.

Five-minute data from 37 neutron monitors have been processed and analyzed in order to fit the main equation of the GLE model, in the basis of a least square method.

RESULTS – DISCUSSION

Our analysis showed that the event of 20 January 2005 was extremely anisotropic. This fact is testified by the time profiles of cosmic ray variations registered in two neutron monitors of the same cut-off rigidity, located at different longitudes and hemispheres (Figure 2). The earliest onset of the ground level enhancement event, as well as the biggest effect was recorded at South Pole and McMurdo. Analysis of the 1-min data showed that enhancement of almost 5000% was registered in the neutron monitor of South Pole. On the contrary, a relatively weak effect (for a small rigidity station) together with late particle arrival rigidity was observed at Thule. All other high latitude stations recorded effect in between of these two: the particle response at Nain and Forth Smith was more similar to that at McMurdo, whereas Cape Shmidt demonstrates a behavior that is more common the other polar stations.

The primary spectrum was very hard in the beginning of the event, with an index of -0.7±0.2. At the same time the anisotropy index n_a, in equation (2), was calculated as 4.7±0.2, meaning that the first particles have come from the Sun in the form of a narrow beam. Moreover, our model placed the anisotropy source at longitude = 69±7° and latitude -60±3°. The big enhancements observed by the polar stations of South Pole and McMurdo, already in a very early phase of the event, occurred as a consequence of the specific beam direction of the anisotropic primary proton flux. The relative positions of the asymptotic cones of these two stations and the direction of the solar particle beam favored the record of big secondary fluxes, since they were placed very close to the location of the anisotropy source. In the same time the effect was observed quite ordinary at the other high latitude stations (Figure 3). All parameters changed essentially during the evolution of the event. In particular, the spectral index γ became more negative, the position of the anisotropy source moved southwest, and the beam of solar particles arriving in an anisotropic way widened.

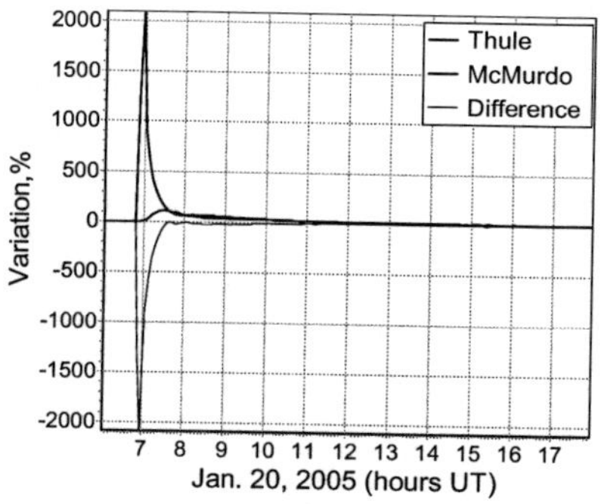

FIGURE 2. Anisotropic arrival of solar cosmic ray particles

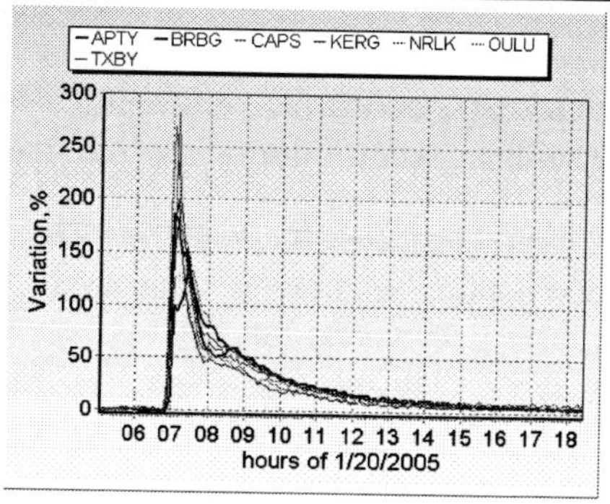

FIGURE 3. The GLE of 20 January 2005 observed by high latitude neutron monitors

In order to determine the absorption length of solar energetic particles propagating through the atmosphere of the earth, we applied a method proposed by Wilson et al. [3]. According to this method, during a solar particle event, any difference in the recorded cosmic ray variation between two conventional neutrons monitor stations (i.e. stations of the same cut-off rigidity and coordinates, differing in altitude) can be attributed to the atmospheric particle absorption. Taking into account the different fluxes recorded by the conventional stations of Erevan and Erevan-3 neutron monitors we found that the absorption length was $\lambda = (105 \pm 11) gr/cm^2$. This value coincides with the value given in literature [9] and it is also very close to the values calculated for other GLE events. For example Wilson et al. [3] have found an absorption length $\lambda = (103 \pm 3) gr/cm^2$ for the GLE of 28 January, 1967.

There are a lot of characteristics in common between the GLE of 20 January 2005 and the GLE of 23 February 1956. Both enhancements were extremely big: increases of ~5000% recorded by polar stations in both cases. Moreover, the fluxes of the first relativistic protons reached the Earth by very narrow beams and had a very hard spectrum. It seems that the recent event of 2005 has got even higher amplitude than that of the famous GLE05 in 1956. However, during the event of 2005, it seems that the solar particles were of smaller energy compared with those in 1956 or in 1989. It should also be noted that the effect either was recorded faint by the majority of mid-latitude stations with cut off rigidity more than 6GV or it was completely absent. The most outstanding feature of these both proton enhancements seems to be an extremely high anisotropy narrow and intensive beam of ultra relativistic particles arriving at Earth during the first minutes after the onset.

CONCLUSIONS

Our conclusions derived from the January 2005 GLE analysis presented above, can be summarized as follows:

a) The recorded cosmic ray enhancement in the counting rate at some southern polar neutron monitors on January 20, 2005 places this event among the greatest GLEs in history. Nevertheless, high energy particles (>3 GeV) turned out to be less than in February 1956 and September 1989 events.

b) The first particles came from the Sun by a narrow beam and had very hard spectrum. Some minutes after the onset the spectrum became soft keeping its form for several hours (spectral index was ranging from -3.0 to -4.0).

c) The absorption length of solar particles propagating in the atmosphere during this GLE is calculated as~105 gr/cm^2. This is consistent with the results from other events.

d) Anisotropy dominates during the first minutes of the event. The asymptotic cones of South Pole and McMurdo fall into the narrow beam of the solar particles arrival.

ACKNOWLEDGMENTS

Thanks are due to our colleagues from the Neutron Monitor stations who have kindly provided us with their cosmic ray data used in this analysis. This work is partly supported by the IRAKLITOS project (70/4/7218).

REFERENCES

1. J.W. Bieber, J. Clem, P. Evenson,, R. Pyle, D. Ruffolo, A. Saiz, *Geophys. Res. Letters,* 32, 3, CiteID L03S02 (2005)
2. C. Plainaki, A. Belov, E. Eroshenko, V. Kurt, H. Mavromichalaki, V. Yanke, *Adv. Space Res.*, 35, 4, 691-696 (2005)
3. B.G. Wilson, T. Nethews, and R.H. Johnson, *Phys. Rev. Lett.* 18 (126), 675-676 (1967)
4. L.I. Dorman, *Progress in Elementary Particle and Cosmic Ray Physics* Amsterdam: J. G. Wilson and S.A. Wouthuysen, North-Holland Publ. Co., 1963
5. A.V. Belov, E.A. Eroshenko , M.A. Livshits, *Proc. Eighth Int. Symp. on Solar Terrestrial Physics*, Sendai, Japan, 26 (1994).
6. A. Belov, E. Eroshenko, H. Mavromichalaki, C. Plainaki, V. Yanke, *Anal. Geophys.* 23, 1-11 (2005)
7. J. Clem and L. I. Dorman, *Space Sci. Rev.*, 93, 1, 335-359 (2000).
8. A.V. Belov and A.B. Struminsky, *Proc. 25-th ICRC*, 1, 201-204 (1997).
9. S.P. Duggal, *Rev. of Geophysics and Space Physics*, 17, 1021-1058 (1979)

Magnetospheric Electron Flows In The Martian Ionosphere. Detection And Implications.

N. Sergis[1], K. Dialynas[1], E. Roussos[2] and X. Moussas[1]

[1]*Department of Astrophysics, Astronomy and Mechanics, Faculty of Physics, National University of Athens, Panepistimiopolis, GR 15783, Zographos, Athens, Greece.*
[2]*Max Planck Institute fuer Sonnensystemforschung, 37191, Katlenburg Lindau, Germany.*

Abstract. A number of distinct hot plasma flux tubes are detected in the Martian ionosphere, primarily formed when the magnetospheric plasma is magnetically connected to the upper ionosphere. We examine the physical characteristics and topology of the flux tubes, which are observed during the highly elliptical orbits of the Mars Global Surveyor. Most of the flux tubes were detected at high solar zenith angles, while the inclination of the flow direction is almost entirely determined by the magnetic sources of the crust. Magnetospheric proton and electron precipitation is expected to produce energetic neutral atoms and cause multiple ionizations, possibly followed by electromagnetic emission (even soft X-rays). An extend network of magnetospheric plasma tubes could also result the uneven heating of the upper ionosphere. We also provide a first order estimation of the typical impact ionization rate within a flux tube and evaluate electron precipitation as an ion production mechanism.

Keywords: Mars, electron flow, flux tubes, Martian ionosphere, electron impact ionization.
PACS: 96.30.Gc, 96.60.Iv, 96.25.Ln

INTRODUCTION

During the last 7 years, the Mars Global Surveyor spacecraft (MGS) offered the opportunity of an in situ study of the Martian ionosphere. The planetary ionopause has also been mapped in great detail from a large number of crossings during the aerobraking and science phasing orbits of MGS [1]. The low altitude periapsis (160 km altitude) of the MGS during its science phasing orbits (March to September 1998), allowed the spacecraft to move deep inside the Martian ionosphere, collecting detailed magnetic field and electron flux measurements (MAG/ER experiment). Both the inbound and outbound ionopause crossings are easily identified in an electron flux vs. time diagram (Fig. 1), as an abrupt decrease or increase, respectively, of magnetospheric electrons (E>50 eV).

In addition, a number of narrow, well defined characteristic zones where magnetospheric electrons dominate within the Martian ionospheric environment are detected in the electron flux diagrams. These zones, which we will refer to as *flux tubes*, are the focus of this study. Based on electron flux and magnetic field data, we will try to investigate the characteristics of the flux tubes, propose a physical mechanism for their formation and further examine possible implications in local plasma environment.

CP848, *Recent Advances in Astronomy and Astrophysics, 7th International Conference of the Hellenic Astronomical Society*, edited by N. H. Solomos

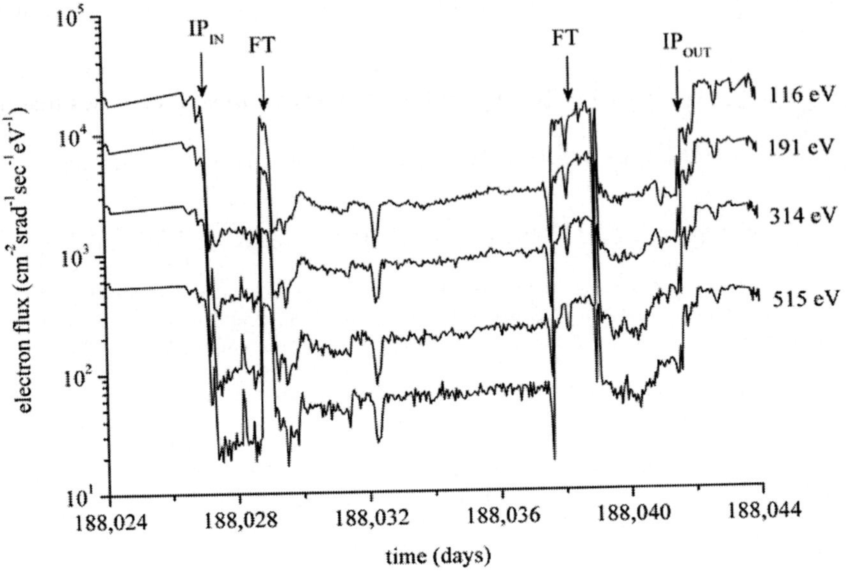

FIGURE 1. Characteristic example of the electron flux behaviour during a Mars Global Surveyor passage through the Martian ionopause (IP_{IN} and IP_{OUT}) and two electron flux tubes (FT), on day 188 of 1998, for central electron energies of 116, 191, 314 and 515 eV. Notice the obvious change in the plasma composition within the flux tubes.

DETECTION OF THE FLUX TUBES

The electron flux measurements provided by the Electron Reflectometer (ER), is the basic diagnostic tool considering the flux tube detection. As the electrons move in tight helices (less than 10 Km in our case) following the magnetic field lines, they offer very accurate information on both the location and size of each plasma formation, illustrating very clearly the abrupt change of the plasma environment whenever the ionopause or a flux tube is penetrated.

Each flux tube observation has a duration varying between 1 and 4 minutes, which is adequate for a satisfactory data collection. These distinct plasma formations were detected for about 25% of the orbits examined, as an obvious change in the electron plasma regime. As seen in Fig. 1, where a representative case of such flux tube detection is presented, the electron plasma inside the tube is characterized by considerably higher energies compared to the surrounding ionospheric plasma. Magnetospheric electrons seem to disappear inside the ionosphere (as expected) to appear back again while the spacecraft moves within the electron flux tubes, reaching their previous magnetosheath flux values.

The most probable origin of the hot flux tube electron plasma is expected to be the Martian magnetosphere, where electrons with energies of a few hundreds of eV dominate. In order to examine the accuracy of this assumption, several electron energy spectra taken inside and outside each flux tube and were constructed and compared to similar ones taken inside the magnetosphere, right above the Martian ionopause (Fig. 2). The energy distribution resulted for these regions, verifies in all cases that the hot plasma observed inside the flux tubes is indeed of magnetospheric origin.

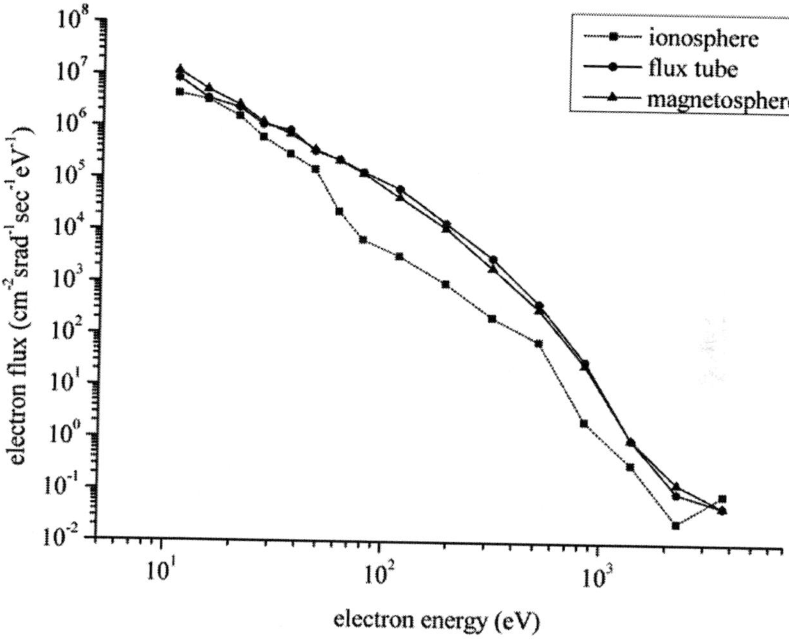

FIGURE 2. Electron energy spectra for three distinct positions of the Mars Global Surveyor, inside the ionosphere, within an electron flux tube and in the magnetosphere, right above the ionopause. A simple comparison of the spectra indicates that the electron plasma population of the flux tube has magnetospheric origin.

Some of the flux tube detections, such as those presented in Fig. 1, could as well be interpreted as multiple ionopause crossings, caused by variations of the ionopause altitude. This possibility was also examined and finally discarded, mainly because of two reasons: a) most of the flux tubes were detected at altitudes considerably lower than the 400 Km mean ionopause altitude, b) a dynamic boundary surface like the Martian ionopause, is not expected to locally curve as much as a multiple crossing situation would require. For example, in the case seen in Fig. 3, a multiple ionopause crossing interpretation would require the Martian ionopause to actually move from above 400 Km, down to 150 Km and back again to 400 Km, within a short time (less than 15 min) and in a limited spatial range (less than 1200 Km total horizontal distance covered).

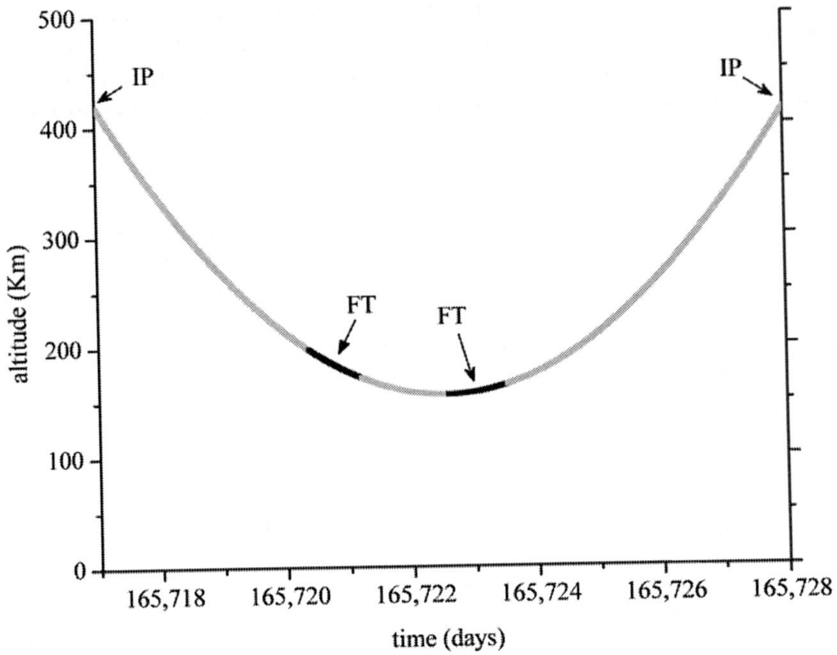

FIGURE 3. Time vs. spacecraft altitude diagram of double flux tube detection, during day 165 of 1998. The left and right ends of the curve are taken to be at the exact inbound and outbound ionopause crossings (IP). The black parts of the curve indicate the part of the orbit that the spacecraft is inside the flux tubes (FT).

FORMATION FOR THE FLUX TUBES

The definite identification of the electron plasma inside the detected flux tubes indicates that the most probable source has to be the magnetoseath just above the upper ionosphere. Both ionospheric and magnetospheric plasma flows are normally expected to be mostly horizontal and uniformly magnetized. However, according to Brain et al. [2] and Crider [3, 4], the surface magnetic fields are locally connected to the solar wind-induced ionospheric field, favouring non horizontal plasma flows. The typical magnetic field values in the case of Mars allow reconnection to occur at a wide range of altitudes, ensuring that the two plasma populations often interact.

Consequently, intense electron precipitation of magnetospheric electrons within the ionosphere could be the result of local magnetic reconnection, as the connected field lines provide an actual path, allowing the magnetospheric plasma to flow downwards. Crider [4] has also verified local reconnection through the detection of detached ionospheric plasma clouds inside the Martian magnetosphere.

As a result, the geometrical structure of each flux tube would be primarily determined by the configuration of the local magnetic field. Furthermore, one would expect the electron flows to be spatially correlated with the magnetic regions of the surface. However, the MGS spacecraft often detected flux tubes quite far from any distinct crustal magnetic sources. Magnetic reconnection is expected to provide the initial conditions needed for the setup of a non horizontal flow, while as we move away from the reconnection area, the connected field lines tend to become horizontal [2], indicating that the plasma flow structures must be quite extended, rather than strictly limited, forming an entire network in the upper Martian ionosphere.

Apart from the electrons, magnetospheric ions are also expected to locally precipitate inside the upper ionosphere, still, their large gyroradii prevents them from forming observable, well defined flows.

BASIC PROPERTIES OF THE FLUX TUBES

Even though the spacecraft intersects with a flux tube just twice during each passage, providing only two boundary points, we were able to approximately calculate some of the geometrical properties of the flux tubes. Unfortunately, the data collected by the MGS are not sufficient for a detailed reconstruction of the plasma tubes. Furthermore, the spacecraft's low altitude passages are concentrated on high latitude regions of the planet, limiting the part of the ionosphere that was examined. Still, through a total sample of 36 flux tube observations, we were able to extract some preliminary information regarding some characteristics of the plasma flows.

In Fig. 4 and 5 we present the flux tube distribution over their altitude of detection (Fig. 4) and over their solar zenith angle (Fig. 5).

Most of the flux tubes were observed below the mean ionopause altitude of 400 Km, well inside the Martian ionosphere. Some high altitude flux tube crossings correspond to a few ionopause positions of increased altitude, mostly due to the variability of the solar wind conditions.

During the passages used in this study, the MGS spacecraft maintained a minimum altitude of about 150 Km above the surface. The distribution shown in Fig. 4 indicates that the flux tube detections were not significantly affected by the minimum altitude of the passages.

The distribution of the detected flux tubes over solar zenith angle shows that all of the flux tubes were observed at SZA over 50°. However, this is practically imposed by the MGS orbit. During the period of data collection, the MGS spacecraft was moving in highly elliptical orbits, with an eastwards shifting periapsis, still always limited in high latitudes, close to the northern polar region. We should mention that these regions are not characterized by strong surface magnetic fields, with the exception of two distinct magnetic sources.

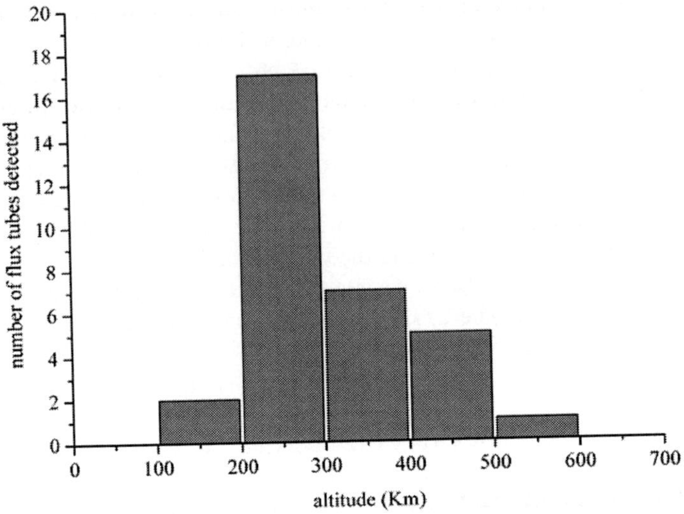

FIGURE 4. Statistical distribution of the flux tube detection altitude (mean of inbound and outbound values). The crossing altitudes are concentrated, as expected, between 200 and 400 Km, where the upper ionosphere is systematically observed.

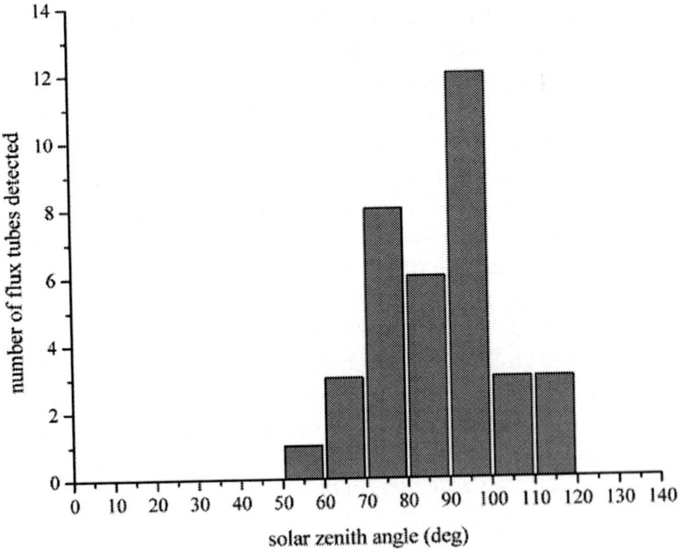

FIGURE 5. Statistical distribution of the flux tube solar zenith angle (mean of inbound and outbound values). The distribution is affected by the spacecraft's orbit (near polar periapsis) and therefore can not be safely used for understanding the global characteristics of the flux tube network.

The concentration of the detected flux tubes close to the ionospheric flanks, may not offer a representative sample for a global study of the non horizontal plasma flows in the upper ionosphere, however, the systematic detection of magnetospheric plasma tubes so close to the nightside of the planed, indicates that these flows could sometimes extend to the night ionosphere, especially where and when the magnetic field configuration becomes closer to horizontal. In this case, the possibility of magnetospheric electron precipitation partially supporting the nightside ionosphere should be further investigated.

THE CASE FOR ELECTRON IMPACT IONIZATION

The electron reflectometer (ER) measurements have shown that magnetospheric electrons are sometimes observed not only close to the ionopause, but at low altitudes (below 200 Km) as well. Such low altitude electron flows are able to interact with neutrals from the upper atmosphere of the planet, leading to intense electron impact ionization. Through this mechanism, magnetospheric electron flows are expected to deposit their energy and eventually decay and deform, as the flux tube comes up against increasingly denser atmospheric layers. The presence of electron impact ionization in the magnetic environment of Mars, has been reported and extensively studied by Haider et al. [6] and Crider et al. [3].

The vertical composition of both the dayside and nightside Martian atmosphere is quite well known, with CO_2, O, CO, N_2 and O_2 being the dominating neutrals for altitudes between 100 and 220 Km [6, 7].

In Fig. 6 we present the electron impact ionization cross sections for the most common neutrals and for several electron energies [8, 9, 10, 11, 12], along with the actual electron energy window of the flux tubes that the MGS observed. Although present in the Martian atmosphere, hydrogen was not included in our study due to its comparatively small electron impact ionization cross section.

It can be easily understood that the flux tube electrons are perfectly capable of supporting electron impact ionization of the upper Martian atmosphere. In a further try to determine whether the impact ionization due to precipitating magnetospheric electrons is significant compared to other ion production mechanisms, we thoroughly investigated four selected flux tubes, two of them detected at 160 Km and the other two at 180 Km. Using the corresponding density for each neutral specie [6], we were able carry out an upper limit estimation of the ion production rate imposed by the electron flows, for these two specific altitudes. The four flux tubes used for the calculation were observed at solar zenith angles between 91° and 98°, which correspond to the edge of the dayside. However, in order to obtain the maximum impact ion production result, the dayside neutral density profiles were adopted.

It should also be noticed that several other factors, like solar wind conditions or the seasonal phase, might have a local or global effect on the density and temperature profiles and in turn, on the impact ionization rate, still more important, on the magnetosphere-ionosphere coupling and the flux tube formation.

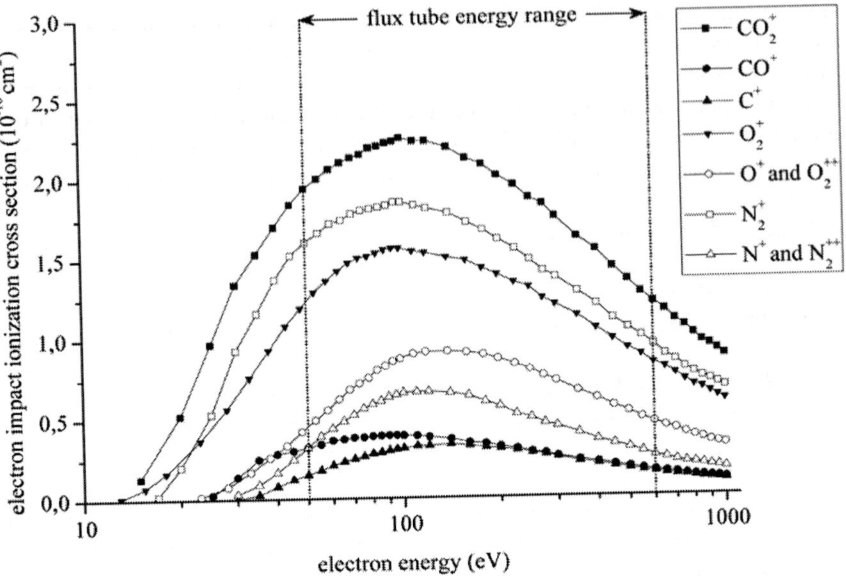

FIGURE 6. The electron impact ionization cross section vs. electron energy for the dominating atoms and molecules of the Martian atmosphere. The parallel vertical lines indicate the energy range of the flux tube electrons.

An approximate estimation of the electron impact ionization rate contribution R_i of each i-atmospheric component, can be made by integrating the product of the electron flux j_e, the cross section for each specific electron impact ionization reaction σ_{i-e} and the numerical density of the corresponding neutral population n_i, over the energy range E and the solid angle Ω covered by the ER instrument:

$$R_i = n_i \iint j_e(E)\sigma_{i-e}(E)dEd\Omega \tag{1}$$

The total electron impact ionization rate (ions produced per cm^3 and sec), will be given as the sum over all contributing atmospheric constituents:

$$R_{tot.} = \sum_i R_i \tag{2}$$

The resulted total ion production rate due to the electron impact ionization was found to be 1.8 ions cm^{-3} sec for the 160 Km and 0.4 ions cm^{-3} sec for the 180 Km altitude.

Figure 7 summarizes our results. The significance of the electron impact ionization effect as an ion production process can be observed through a direct comparison with the other mechanisms. The results indicate that the electron flux tubes can locally contribute to the total ion production, becoming comparable to the H^+-H ionization and dominating in the nightside of the planet, in the absence of EUV and X-ray radiation.

The actual influence that a single electron flux tube locally has on the ionosphere, further depends on its size (cross section) and its flow direction, which in the case of smooth flow, once the flux tube is stabilized, is expected to be mostly imposed by the local magnetic field configuration.

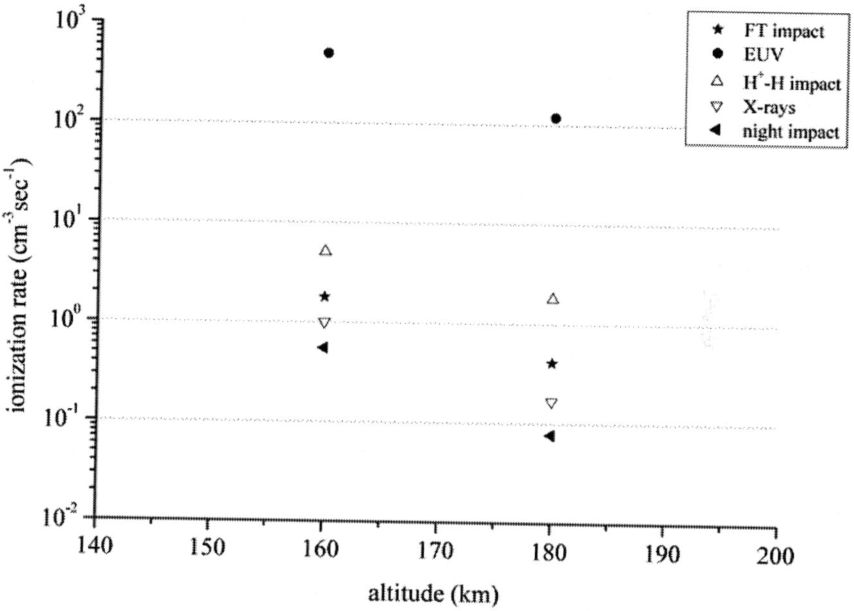

FIGURE 7. Comparative diagram of the ion production rates for the most important mechanisms active in the Martian ionosphere. The values for the EUV and X-ray ionization rates are calculated for SZA 75° (Haider et al., 2002). The contribution of the magnetospheric electron flows (stars) can become important, especially in the nightside of the planet, where the EUV and the X-ray radiations are practically absent and the H^+-H collisions are less frequent.

SUMMARY AND CONCLUSIONS

An extended network of low-altitude magnetospheric electron flows has been detected well inside the Martian ionosphere, using electron flux data collected by the MAG/ER experiment onboard the Mars Global Surveyor.

The magnetospheric origin of the plasma population inside the flux tubes was verified using electron energy spectra. Local reconnection between the magnetic fields of the surface and the mainly horizontal ionospheric field can provide the conditions needed for "hot" magnetospheric electrons to precipitate deep within the Martian ionosphere.

However, the flux tubes do not seem to be strictly correlated with the weak surface magnetic sources of the northern hemisphere, where the low altitude MGS passages were concentrated. The most probable explanation (apart from the limited amount of flux tube crossings) could be that these organized electron flows systematically extend quite far from the reconnection region, often becoming close to horizontal as the magnetic field configuration indicates.

The intense impact ionization of atmospheric neutrals is an immediate result of such low-altitude (as low as 150 km) magnetospheric plasma flows. The typical energies of the precipitating electrons (few hundreds of eV), makes them perfectly capable of impact-ionizing the upper atmospheric neutrals, (mostly CO_2, O, CO, N_2 and O_2).

A first order estimation of the impact ionization rate locally imposed by the flux tubes was carried out for two different altitudes (160 and 180 Km). The final outcome (1.8 and 0.4 ions cm^{-3} sec respectively) indicates that such a wide flow network could be a significant ion supplier for the dark side, possibly adequate to support a night Martian ionosphere.

ACKNOWLEDGMENTS

We thank Dr. M.A. Acuña, Principal Investigator of the MAG/ER team, Dr. J.E.P. Connerney and Dr. S. Joy, PDS/PPI operation manager, for the MGS MAG/ER data. We would also like to acknowledge useful discussions with Dr. A. Alevizos.

REFERENCES

1. M. H. Acuña, J. E. P. Connerney, P. Wasilewski, R. P. Lin, D. L. Mitchell, K. A. Anderson, C.W. Carlson, J. McFadden, H. Rème, C. Mazelle, D. Vignes, S. J. Bauer, P. Cloutier, N. F. Ness, *J. Geophys. Res.* **106**, 2001.
2. D. A. Brain, "The influence of crustal magnetic sources on the topology of the Martian magnetic environment", Ph.D. Thesis, University of Colorado, 2002.
3. D. H. Crider, *Adv. Space Res.*, **33**, 2004, pp. 152-160.
4. D. H. Crider, M. H. Acuña, D. L. Mitchell, "Mars Global Surveyor Observations of High Altitude Ionospheric Clouds", *American Geophysical Union Spring Meeting*, abstract #SM44A-01, 2004.
5. D. H. Crider, P. Cloutier, C. Law, P. Walker, Y. Chen, M. Acuña, J. Connerney, D. Mitchell, R. Lin, K. Anderson, C. Carlson, J. McFadden, H. Reme, C. Mazelle, C. d'Uston, J. Sauvaud, D. Vignes, D. Brain and N. Ness, *Geophys. Res. Let.*, **27**, 2000, pp. 45-48.
6. S. A. Haider, S. P Seth, E. Kallio and K. I. Oyama, *Icarus* **159**, 18-30 (2002).
7. J. T. Schofield, J. R. Barnes, D. Crisp, R. M. Haberle, S. Larsen, J. A. Magalhaes, J. R. Murphy, A. Seiff, G. Wilson, *Science*, **278**, (1997).
8. H. C. Straub, B. G. Lindsay, K. A. Smith and R.F. Stebbings, *Chem. Phys.*, **105**, 4015-4022, (1996).
9. H. C. Straub, P. Renault, B.G. Lindsay, K.A. Smith and R.F. Stebbings, *J. Phys. Rev.* A, **54**, 2146-2153, (1996).
10. B. G. Lindsay, M. A. Mangan, H. C. Straub, and R. F. Stebbings, *J. Chem. Phys.*, **112**, 2000, pp. 9404-9410.
11. M. A. Mangan, B. G. Lindsay, and R. F. Stebbings, *J. Phys. B: At. Mol. Opt. Phys.*, **33**, 2000, pp. 3225-3234.
12. H. Deutsch, K. Becker, P. Defrance, U. Onthong, R. Parajuli, M. Probst, S. Matt and T. D. Märk, *J. Phys. B: At. Mol. Opt. Phys.*, **35**, L65–L69, (2002).

Simulating the Effects of ULF Waves on Energetic Electron Populations

Theodore Sarris[1,2], Xinlin Li[2], and Mike Temerin[3]

Demokritus University of Thrace, Vas. Sofias 1, Xanthi, 67100, Greece[1],
Laboratory for Atmospheric and Space Physics, 1234 Innovation Drive, Boulder, CO 80303, USA[2]
Space Sciences Lab, Univ. of California, Berkeley, Grisley Peak Rd, Berkeley, CA 80303, USA[3]

Abstract. A model of magnetic and electric perturbations in the ULF regime is used to simulate magnetospheric variability. The model is based on the assumption that compressional oscillations propagate from the magnetopause into the inner magnetosphere, and are reflected at an inner boundary. Electrons are traced under the effect of the modeled ULF fluctuations and the radial diffusion coefficient is calculated through the electrons' radial displacement. The numerically calculated diffusion coefficient is compared to theoretical estimates, and is found to show a similar L-dependence.

Keywords: Radial Diffusion, ULF Pulsations, Particle Tracing
PACS: 92.60.Hk, 94.30.Bg, 94.30.Hn, 94.30.Ms, 94.30.Xy

INTRODUCTION

In the study of the Earth's radiation belts it has been observed that often, during periods of intense magnetic activity, and in particular during storm times, electron acceleration occurs, evidenced by MeV electron flux increases by orders of magnitude [1], within timescales of a couple of days from the onset of magnetic activity or high solar wind velocity. Further studies showed that southward interplanetary magnetic field, in addition to magnetic activity, can enhance trapped electron populations [2; 3]. Instead of refilling the radiation belts with relativistic electrons, magnetic storms can also empty them [4]. Several acceleration mechanisms have been proposed to explain such flux changes, and in particular their time scales. One theory that has been proposed early on is that of radial diffusion [5], according to which irregular fluctuations of the electromagnetic fields in the magnetosphere with spectral power at the frequency of an electron's drift orbit around the Earth can cause a random inward and outward radial motion of the electron, by violating its third adiabatic invariant. Energetic electrons in the radiation belts have drift periods of 150-600 sec, and the geomagnetic pulsations with corresponding frequency ranges are Ultra Low Frequency, ULF, waves in the Pc-5 range (for a classification of ULF waves see, [6]). As an electron moves to lower L-shells and stronger magnetic field under the effect of the pulsations, it gains energy. Such stochastic diffusion in the electrons' L-shell will result in net gain or loss depending on the initial distribution of the electrons, i.e. depending on the existence of sources and sinks in the magnetosphere.

CP848, *Recent Advances in Astronomy and Astrophysics,*
7th International Conference of the Hellenic Astronomical Society, edited by N. H. Solomos

In theory, the diffusion of particles is quantitatively described by the diffusion coefficient, D_{LL}, in which the effects of electric and magnetic field fluctuations are included. In the theoretical descriptions of [5], a distinction was made between electrostatic (D_{LL}^E) and electromagnetic (D_{LL}^B) contributions to the total diffusion coefficient D_{LL}. These expressions were found to be dependent on the power of the fluctuating fields.

In the present work, we calculate the diffusion coefficient D_{LL}^B by numerically tracing electrons in a model of ULF electromagnetic perturbations. Fluctuations in the magnetic field are simulated by superimposing a large number of propagating electric and consistent magnetic pulses. The simulated perturbations in this model represent compressional, poloidal ULF perturbations. In a theoretical qualitative description, compressional perturbations are caused by magnetopause instabilities or by solar wind pressure pulses that propagate inwards in the magnetosphere and are reflected at the plasmapause, due to the large gradient in the Alfven velocity. The model of superimposed pulses mimics magnetic field measurements made at geosynchronous orbit, as well as frequency spectra of these measurements. We subsequently attempt to determine D_{LL}^B from the average behavior of the particles in the simulation, as they gradually diffuse toward a relaxed configuration.

THEORY

The diffusion coefficients D_{LL}^E and D_{LL}^B have long been thought to be dependent on magnetospheric activity: it has been pointed out by [9] that temporal variations in the radial diffusion coefficients are a consequence of time-varying intensity of geomagnetic activity. In particular, the power of magnetospheric pulsations is enhanced characteristically during the main phase of magnetic storms [10, 11], establishing the link between the solar wind and magnetospheric fluctuations. Measurements of magnetic and electric fields have also shown a *kp* dependence on the power spectra and the D_{LL} derived from this spectra [e.g., 12]. Also, Brautigam and Albert [13] have solved the radial diffusion equation for both a *kp*-dependent and a *kp*-independent diffusion coefficient D_{LL}, and have shown that their model manages to capture the dynamics of the main phase dropout of a geomagnetic storm only for a *kp*-dependent diffusion coefficient. Radial diffusion calculations carried out by Selesnick and Blake [14] indicate that the radial diffusion coefficient can vary by a factor of 5 during disturbed geomagnetic conditions, suggesting that observed diffusive time scales at $L = 6$ can be as low as a few hours during storms.

The concepts of radial diffusion and theoretical calculations of the diffusion coefficients based on observations of the power spectra of field perturbations are contained in the monograph by Schulz and Lanzerotti [15]. In general, the diffusion coefficient, D_{LL}, corresponding to particle diffusion in L-shell, is defined as:

$$D_{LL} \equiv \frac{\left\langle (\Delta L)^2 \right\rangle}{2\tau} \tag{1}$$

The magnetic diffusion coefficient, $D_{LL}^{B,Sym}$, produced by electromagnetic fluctuations on particles of a single μ-value that are drifting in a symmetric background magnetic field, has been derived by Falthamar [5]. The expression of Falthamar [5] is non-relativistic and includes only fluctuations of mode number m=1 terms. Recently, Fei et al. [16] generalized Falthamar's expression to include relativistic electrons and multiple mode numbers of fluctuation:

$$D_{LL}^{B,Sym} = \frac{\mu^2}{8q^2 B_E^{\ 2} R_E^{\ 4}} \left(\frac{L^4}{\gamma^2} \right) \sum_{m=1}^{\infty} m^2 P_m^{\ B}(m\omega_d) \tag{2}$$

where μ is the value of the first adiabatic invariant of the electrons considered, q is the electron charge, γ is the Lorentz relativistic factor, B_E is the magnetic field strength at the surface of the Earth, R_E is one Earth radius, m is the azimuthal mode number of the fluctuation and $P_m^{\ B}(m\omega_d)$ is the power spectral density of the compressional wave magnetic field at frequency m-times the drift frequency ω_d of the electrons considered. In the non-relativistic case, γ is equal to one at all L; in the ultra-relativistic limit, γ is proportional to $L^{-1.5}$ and the factor L^4/γ^2 in equation (2) is proportional to L^7. For the electrons traced in the simulation, which have a μ-value of 1865 *MeV/G*, the L-dependence of the Lorentz factor can be approximately fitted as: $\gamma = 36 \cdot L^{-1.32}$.

MODEL

In the model that was constructed, a large number of radialy propagating Electric and consistent Magnetic Gaussian pulses are used to simulate magnetospheric compressional variability. In the spherical coordinate system(r,θ,ϕ), the electric field of a single pulse is given as:

$$\mathbf{E}_\phi = -\hat{e}_\phi E_0 \left(1 + c_1 \cos(\phi - \phi_0)\right)^p \exp\left((-\xi^2) - c_2 \exp(-\eta^2)\right) \tag{3}$$

where $\xi = (r - r_i + v_0 t)/d$ determines the location of the maximum value of the incoming pulse and $\eta = (r - 2r_i + r_d - v_0 t)/d$ determines the location of the reflecting pulse; d is the radial width of the pulse; v_0 is the radial speed of the pulse; $\hat{e}_\phi$ is the azimuthal direction; E_0 is the electric field amplitude; c_1 (= 1 in the simulation presented) and p (= 1 in the simulation presented) describe the local time dependence of the electric field amplitude, which is largest at ϕ_0; c_2 describes the percentage of the reflection (= 1 in the simulations presented, indicating 100% reflection) and r_i is a parameter in the simulation that determines the arrival time of the pulse. From (3) the pulse electric field is positive (westward) for incoming and negative (eastward) for reflecting pulses. The consistent magnetic field of the propagating electric pulse of equation (3) is obtained from Faraday's law, after performing the curl calculation of (3) in spherical coordinates and integrating.

The analytic expressions of the model fluctuating fields make possible the numeric calculation of the power of the fluctuations as a function of frequency and time at various L. In the left panel of **Figure 1**, the average power that an electron drifting in a dipole field would experience at different L is plotted as solid lines, one for each L, from L=2 to L=8, for the model fluctuating fields with the parameters that were described above. Subsequently, from the energy-versus-L relationship for this particular μ-value we can obtain the particle drift frequency at a particular L. In the left panel of **Figure 1** we plot vertical dotted lines at the drift frequencies ω_d of electrons at various L, from L=2 to L=8, for μ=1860 MeV/G. The power at each frequency ω_d is marked as a diamond. The power at the particular L is then plotted in the right panel of **Figure 1** as a function of L. Thus there is a one-to-one relation between the diamonds of the left and right panels. We then perform a fit through the points in the right panel, and obtain the power-versus-L relationship for the model fluctuating fields. This relation is then used in equation (2) to calculate the theoretical diffusion coefficient $D_{LL}^{B,Sym}$, which is plotted in the right panel of **Figure 2** as a dashed line.

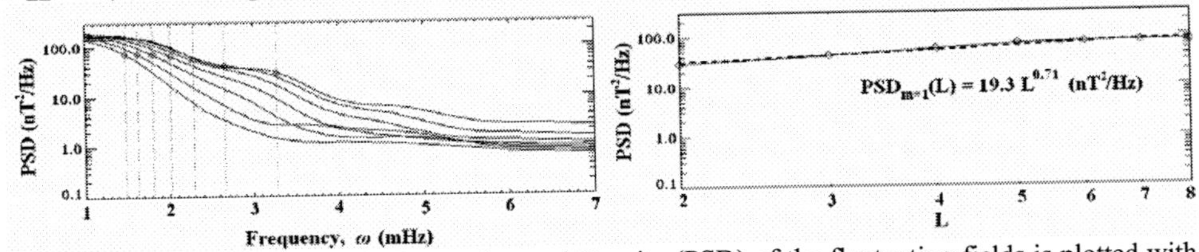

FIGURE 1. In the left panel, the Power Spectral Density (PSD) of the fluctuating fields is plotted with solid lines for various L from L=2 to L=8 as a function of frequency. The top-most line corresponds to L=2. The diamonds in the plot correspond to the power of fluctuations at frequency ω_d (drift frequency of electrons of μ=1865 MeV/G) at the particular L. The power at these L is plotted as a function of L in the right panel; a fit through these points gives the L-dependence function of the PSD.

SIMULATION

Using the model fluctuating fields that were described above, the diffusion coefficient was calculated numerically by the actual radial displacement in the orbits of relativistic 90° pitch-angle traced particles. The diffusion coefficient calculated under these field parameters corresponds to the electromagnetic symmetric radial diffusion coefficient, $D_{LL}^{B,Sym}$. In order to calculate $D_{LL}^{B,Sym}$ as a function of L, rings of particles were initialized at various L and their average squared displacement, $<(\Delta L)^2>$, was monitored in time. The results after monitoring the particles at each ring for 2 hours of simulation time are given in the left panel of **Figure 2**.

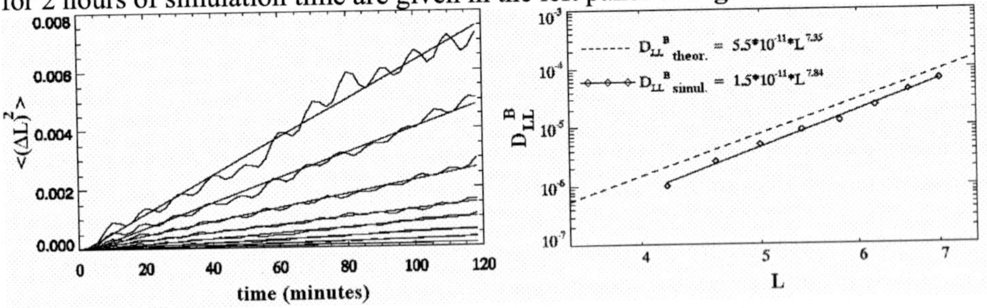

FIGURE 2. In the left panel, each line represents the average squared displacement, $<(\Delta L)^2>$, of particles evenly distributed in rings at L from 4.6 to 7.0 (largest slope). From the fit to each line the radial diffusion coefficient at L is calculated. In the right panel, $D_{LL}^{B,Sym}$ at various L is plotted as diamonds. The fit through these points (solid line) gives the L-dependence of $D_{LL}^{B,Sym}$, in units of 1/sec. The theoretical estimate for $D_{LL}^{B,Sym}$ is drawn with a dashed line.

In this figure, $<(\Delta L)^2>$ at each L is linearly fitted by a line, the slope of which is the diffusion coefficient of electromagnetic symmetric perturbations $D_{LL}^{B,Sym}$ at that L. On the right panel of **Figure 2** the slopes of the rate of change of $<(\Delta L)^2>$ are plotted for various rings of particles as a function of L, to yield the L-dependence of the diffusion coefficient. A fit through these points gives the numerically calculated diffusion coefficient, $D_{LL}^{B,Sym}$, as a function of L, which is plotted as a solid line.

CONCLUSIONS

A model of field fluctuations was presented that simulates compressional, poloidal ULF pulsations. The model is constructed by a superposition of a large number of electric and consistent magnetic pulses that are initialized at random initial radial distances with random amplitudes. This electromagnetic variability reproduces a realistic broad-band fluctuation. Electrons traced under the effect of this magnetospheric variability exhibit radial diffusion, which can be quantified by the radial diffusion coefficient, $D_{LL}^{B,Sym}$. We calculated $D_{LL}^{B,Sym}$ by the average squared displacement of electrons in the simulation and also by the theoretical expression of *Falthamar* [5], as it was generalized by Fei et al. [16]. The results show an agreement in their L-dependence.

ACKNOWLEDGMENTS

This work has been supported by the "Pythagoras" grant 2.2.3.στ, Ministry of Education, Greece.

REFERENCES

1. Paulikas, G.A., and Blake, J.B., "Effects of the solar wind on magnetospheric dynamics: Energetic electrons at the synchronous orbit" in *Quantitative Modeling of Magnetospheric Processes*, edited by W. P. Olson, Geophys. Monogr. Ser., 21, AGU, Washington, D. C., 1979, pp. 180.
2. Baker, D.N., Li, X., Turner, N., et al., *J. Geophys. Res.*, **102**, 14141 (1997)
3. Obara, T., Nagatsuma, T., and Onsager, T.G., "Effects of the interplanetary magnetic field (IMF) on the rapid enhancement of relativistic electrons in the outer radiation belt during the storm recovery phase" in *Fourth International Conference on Substorms*, Kluwer Acad, Norwell, Mass., 1998, pp. 215.
4. Friedel, R.H.W., Reeves, G.D., and Obara, T., *J. Atmos. Solar Terr. Phys.*, **64**, 265 (2002).
5. Falthammar, C.-G., *J. Geophys. Res.*, **70**, 2503 (1965)
6. Jacobs, J.A., Kato, Y., Matsushita, S., and Troitskaya, V.A., *J. Geophys. Res.*, **69**, 180 (1964)
7. Radoski, H.R., Hydromagnetic waves: Temporal development of coupled modes, Environ. Res. Pap. 559, Air Force Geophys. Lab., Hanscom AFB, 1976.
8. Wright, A.N., *J. Geophys. Res.*, **99**, 159 (1994)
9. Lanzerotti, L.J., Maclennan, C.G., and Shculz, M., *J. Geophys. Res.*, **75**, 5351-5371 (1970)
10. Baker, D.N., Pulkkinen, T.I., Li, X., Kanekal, S.G., Blake, J.B., Selesnick, R.S., Henderson, M.G., Reeves, G.D., Spence, H.D., Rostoker, G., *J. Geophys. Res.*, **103**, 17279 (1998)
11. O'Brien, T.P., McPherron, R.L., Somette, D., Reeves, G.D., Friedel, R., and Singer, H., *J. Geophys. Res.*, **106**, 15533 (2001)
12. Lanzerotti, L.J., Webb, D.C., and Arthur, C.W., *J. Geophys. Res.*, **83**, 3866 (1978)
13. Brautigam, D.H. and Albert, J.M., *J. Geophys. Res.*, **105**, 291 (2000)
14. Selesnick, R. S., and Blake, J. B., *J. Geophys. Res.*, **105**, 2607-2624, (2000)
15. Schulz, M., and Lanzerotti, L.J., *Particle diffusion in the radiation belts*, New York: Springer, 1974
16. Fei, Y., Chan, A.A., Elkington, S.R., and Wiltberger, M.J., submitted, *J. Geophys. Res.* (2006)

PART 2

OUR GALAXY:

STARS, CLUSTERS, AND THE INTERSTELLAR MEDIUM

(Conveners: N. Kylafis, P. Niarchos, and E. T. Harlaftis (deceased))

Supernova Explosions, Nucleosynthesis, and Cosmic Chemical Evolution

James W. Truran

Department of Astronomy and Astrophysics, Enrico Fermi Institute, University of Chicago, 5640 S. Ellis Avenue, Chicago, Illinois 60637 and Argonne National Laboratory, 9700 South Cass Avenue, Argonne, IL 60439

Abstract. The Universe emerged from its first three minutes with a composition consisting of hydrogen, deuterium, ^3He, ^4He, and ^7Li. These isotopes constitute the primordial compositions of galaxies. Within galaxies, the synthesis of heavier elements from carbon through uranium is understood to occur during the normal evolution of stars and in supernova explosions of Types I and II. This history is written in the compositions of the stars and gas in our Milky Way Galaxy and other galaxies. The contributions both from massive stars ($M>10\ M_{\odot}$) and associated Type II supernovae and from Type Ia (thermonuclear) supernovae are particularly noteworthy. We review both the nuclear processes by which this occurs and the compositions of the stellar components of our Galaxy as a function of time which reflect these nucleosynthesis processes. We then discuss how such observations inform us of the nature of the earliest stellar populations and of the abundance history of the Cosmos.

Keywords: supernovae, nucleosynthesis, chemical evolution
PACS: 26.30.+k, 95.30.-k

INTRODUCTION

Observations have taught us that the concentrations of the two most abundant elements in the Universe – hydrogen and helium – are substantially the same in galaxies of all types, all sizes, and all ages (e.g. at all red shifts). This a consequence of the fact that they are cosmological in origin, representing the main constituents emerging from the first three minutes of the cosmological "big bang." In contrast, observations of the diverse stellar and gas components of galaxies reveal that the heavy element abundance in Galactic matter increases monotonically with time. The life cycles of stars, from their births to the final stages of their evolution as planetary nebulae or supernovae, provide the settings for the diverse thermonuclear processes by which the synthesis of the heavy elements occurs. The heavy element enriched matter ejected in the formation of planetary nebulae and in supernova explosions serves to enrich the interstellar media of galaxies from which subsequent generations of stars are formed.

The processes by which heavy element synthesis occurs in stellar and supernova environments were first identified and defined in the classic papers by Burbidge et al. (1957) and Cameron (1957a,b). The elements from carbon through the iron peak (in the mass range A = 12-60) are formed in a sequence of stable nuclear burning phases which power the lifetimes of stars. The extent to which such burning proceeds is a

CP848, *Recent Advances in Astronomy and Astrophysics,*
7th International Conference of the Hellenic Astronomical Society, edited by N. H. Solomos

function of the stellar mass: more massive stars achieve higher core temperatures and therefore can burn heavier fuels. This leads to the identification of two distinct regimes of stellar mass with very different evolutionary behaviors. Low and intermediate mass stars M~1-10 $M_{\odot}$ (including the Sun) can ignite and burn only hydrogen and helium fuels and ultimately end their lives as planetary nebulae, leaving white dwarf remnants. In contrast, stars of mass M≥10 $M_{\odot}$ continue to burn neon, oxygen, and silicon fuels through the realization of an iron core.

The bulk of the nuclei past iron are synthesized by processes of neutron capture. The abundance features in the heavy element region (Figure 1) which are correlated

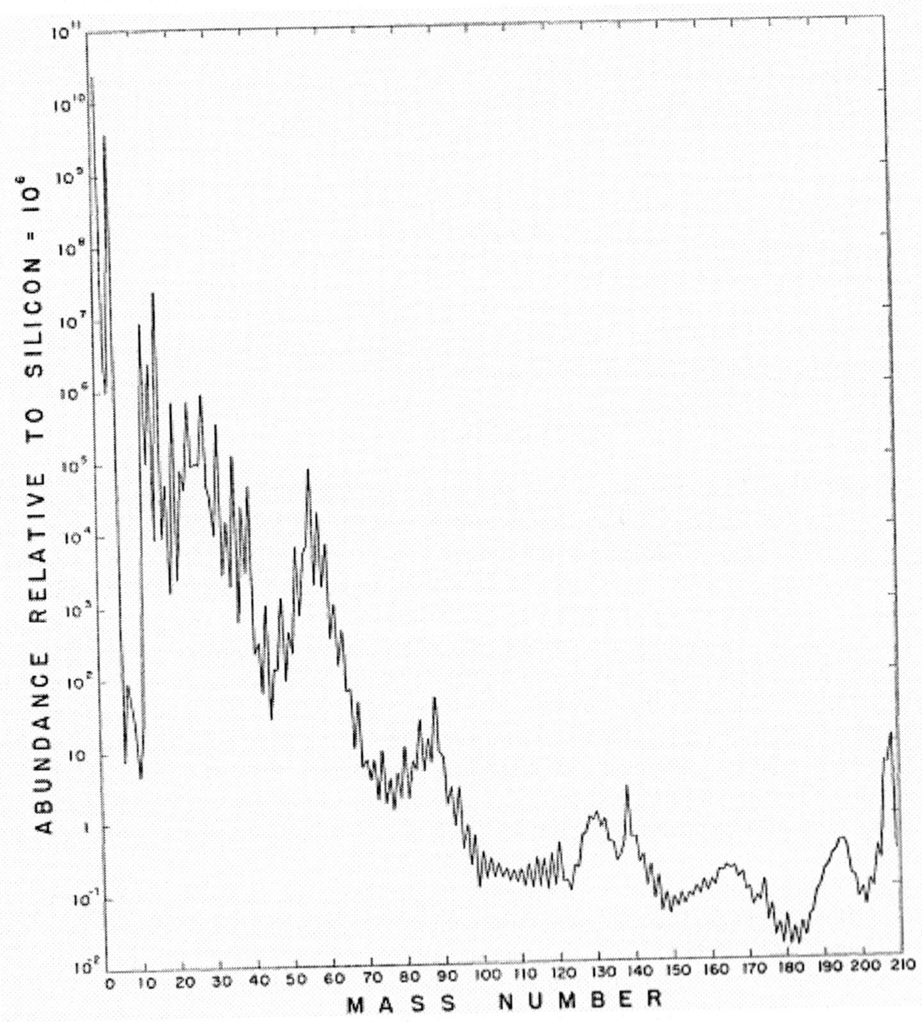

FIGURE 1. Abundances of the elements in the Solar System after Cameron (1982).

with neutron shell closures at neutron numbers N = 50, 82, and 126, strongly support this view. Scrutiny of the abundance patterns indeed reveals signature of two distinct neutron fluxes. This has led historically to the definition of two nucleosynthesis processes: a slow (s-)process occurring in red giant stars and a rapid (r-)process associated with supernovae.

STELLAR AND SUPERNOVA NUCLEOSYNTHESIS

We will confine our attention to three broad categories of stellar and supernova site with which specific nucleosynthesis products are understood to be associated: (1) the mass range $1 \leq M/M_{\odot} \leq 10$ of low and ``intermediate'' mass stars, for which substantial element production occurs during the asymptotic giant branch (AGB) phase of their evolution; (2) the mass range $M \geq 10\ M_{\odot}$, corresponding to the massive star progenitors of Type II (`core collapse') supernovae; and (3) Type Ia supernovae, which are generally believed to arise as a consequence of the evolution of intermediate mass stars in close binary systems. In the following, we review expectations for nucleosynthesis associated with each of these specific nucleosynthesis sites.

Low and Intermediate Mass Stars

Intermediate mass stars ($1 \leq M/M_{\odot} \leq 10$) provide important contributions to galactic abundances primarily as a consequence of the occurrence of thermal pulses in their helium burning shells on the ``asymptotic giant branch'' (see, e.g. the review by Busso *et al.* 1999). The principle products are ^{12}C, ^{14}N, and the **s**-process neutron capture elements. Discussions of nucleosynthesis in low and intermediate mass stars have also been presented by Iben & Truran (1978) and Renzine & Voli (1981). Carbon production results from the incomplete burning of helium in the thermally pulsing shells of AGB stars, followed by the dredge-up of this shell matter into the envelope. ``Carbon stars'' are a product of such dredge-up. This environment also produces the bulk of the heavy s-process (``slow neutron capture'') isotopes in the mass range $A \geq 90$, with the neutrons being provided by the $^{13}C(\alpha,n)^{16}O$ reaction. The main production of **s**-process isotopes occurs at the lower end of the mass range of intermediate mass stars ($M \approx 1$-$2\ M_{\odot}$), on a timescale of order 1-2 x 10^9 years.

Massive Stars and Type II Supernovae

Massive stars $M \geq 10\ M_{\odot}$ represent the progenitors of Type II (or `core collapse') supernovae. The combination of stellar burning stages and explosive nucleosynthesis in the supernova phase is understood to synthesize both nuclei from oxygen to nickel and the **r**-process heavy neutron capture products. Calculations of charged-particle nucleosynthesis in massive stars and Type II supernovae over the past decade (Woosley & Weaver 1995; Thielemann *et al.* 1996; Nomoto *et al.* 1997; Limongi *et al.* 2000; Woosley *et al.* 2002) reveal one particularly significant distinguishing feature of the emerging abundance patterns: the elements from oxygen through calcium (and titanium) are overproduced relative to iron by a factor ≈2-3. This means that SNe II produce only approximately 1/3 to 1/2 of the iron in Galactic matter. (We note that, while all such models necessarily make use of an artificially induced shock wave via thermal energy deposition (Thielemann *et al.* 1996) or a piston (Woosley & Weaver 1995), the general trends obtained from such nucleosynthesis studies are expected to be valid.) As we shall see in the next section, these trends are reflected in the abundance patterns of metal deficient stars, for which typically [O-Ca/Fe] ~+0.3 to +0.5.

It now also seems most likely that the r-process synthesis of the heavy neutron capture elements in the mass regime A ≥ 130-140 occurs in an environment associated with massive stars. The association of the **r**-process site with massive star environments results from two factors: (1) the two most promising mechanisms for **r**-process synthesis - a neutrino heated "hot bubble" (Woosley *et al.* 1994; Takahashi *et al.* 1994) and neutron star mergers (Lattimer *et al.* 1977; Freiburghaus & Rosswog 1999) - are both tied to environments associated with core collapse supernovae; and (2) observations of old stars (elaborated in the next section) confirm the early entry of **r**-process isotopes into Galactic matter.

Type Ia Supernovae

Type Ia supernovae are understood to represent the source of the 1/2 to 2/3rds of the iron-peak nuclei that are not produced in SNe II. The standard model for SNe Ia involves the growth of a white dwarf to the Chandrasekhar limiting mass in a close binary system (see e.g. the review of Type Ia progenitor models in Livio (2000)) and its subsequent incineration. Calculations of explosive nucleosynthesis associated with carbon deflagration models for Type Ia events (Thielemann *et al.* 1986; Iwamoto *et al.* 1999; Dominguez, Hoeflich, & Straniero 2001) predict that sufficient iron-peak nuclei (~0.6-0.8 $M_\odot$ of ^{56}Fe in the form of ^{56}Ni) are synthesized to explain both the powering of the light curves due to the decays of ^{56}Ni and ^{56}Co and the observed mass fraction of ^{56}Fe in Galactic matter. Estimates of the timescale for first entry of the ejecta of SNe Ia into the interstellar medium of our Galaxy yield ≈ 2 x 10^9 years, at a metallicity [Fe/H] ≈ -1 (Kobayashi *et al.* 2000; Goswami & Prantzos 2000).

Evolutionary Timescale Considerations and SNe Frequencies

A knowledge of the production timescales - the effective timescales for the return of a star's nucleosynthesis yields to the interstellar gas - is also crucial to our use of abundance patterns to constrain star formation histories. The lifetime of a 10 $M_\odot$ star is approximately 5x10^7 years. We can thus expect all massive stars $M \geq 10\ M_\odot$ to evolve on timescales $\tau_{SNeII} < \sim 10^8$ years, and to represent the first source of heavy element enrichment of stellar populations. In contrast, intermediate mass stars (IMS) evolve on timescales $\tau_{IMS} \geq 10^8$-10^9 years. Finally, the timescale for SNe Ia product enrichment is a complicated function of the binary history of Type Ia progenitors. Observations and theory suggest a timescale $\tau_{SNeIa} \geq \sim 1.5$-2 x 10^9 K. With these timescales in mind, we will now review the fossil record – abundances in stars over our Galaxy's history – and its implications for nucleosynthesis in early stars (the first stars?).

THE FOSSIL RECORD: HALO STAR ABUNDANCES

Spectroscopic abundance studies of the oldest stars in our Galaxy, down to metallicities [Fe/H] ≈ -4 to -3, have been most recently reviewed by Wheeler, Sneden, & Truran (1989) and McWilliam (1995). Such studies typically reveal abundance trends which can best be understood as reflecting the nucleosynthesis products of the massive stars and associated Type II supernovae that can be expected to evolve and to enrich the interstellar media of galaxies on rapid timescales ($\leq 10^8$ years). It is these trends in abundance patterns which we will review in this section.

α-Element Trends with Respect to Iron

A strong and unambiguous signature of these massive stars is a ratio of the abundances of the so called α-elements (O, Ne, Mg, Si, S, Ar, and Ca) to that of iron-peak nuclei which is approximately 2-3 times the Solar System value (Wheeler *et al.*

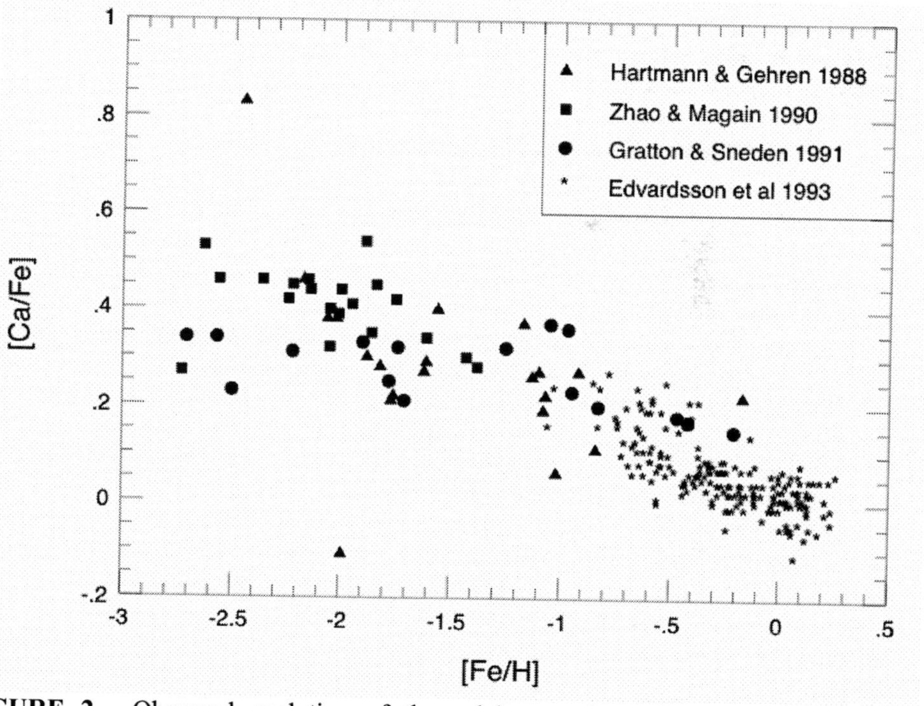

FIGURE 2. Observed evolution of the calcium to iron abundance ratio with metallicity (Truran & Timmes 1995).

1989; McWilliam 1995). The observed O/Fe and (Si-Ti)/Fe ratios fall gradually to the Solar value on longer timescales, as the iron-peak rich products of Type Ia supernovae are introduced, at metallicities [Fe/H] ≈ -1. High (Si-Ca)/Fe ratios are also characteristic of globular cluster stars in our Galaxy (Wheeler *et al.* 1989) and of some nearby dwarf spheroidal galaxies (Shetrone *et al.* 2001).

For purposes of illustration, the trends in [Ca/Fe] as a function of [Fe/H] is shown in Figure 2. Following our earlier discussion of the expected products of the evolution of massive stars and associated Type II supernovae, the observed trends with increasing metallicity can be understood in a straightforward manner. The ejecta of massive stars of short lifetimes dominate the nucleosynthyesis contributions until such time (after approximately 1.5 to 2 billion years of Galactic evolution) as the iron-peak products of Type Ia are first introduced. This is seen to occur at a metallicity [Fe/H] ≈ -1, approxmately the point at which the halo to disk `transition' occurred. Subsequently, the [O/Fe] and [Ca/Fe} ratios fall monotonically to the Solar value. (Note here that we use [Ca/Fe] as representative of the intermediate mass alpha-particle elements - Ne, Mg, Si, S, Ar, Ca, and Ti).

In contrast to the α-chain elements, the even-Z elements in the iron peak region - specifically, Cr and Ni - are found to have been produced in approximately Solar proportions, relative to iron, at least down to metallicities [Fe/H] ≈ -3. This is consistent with the formation of these iron ``equilibrium" peak nuclei in nuclear statistical equilibrium at high temperatures. The general consistency of the even-Z elemental abundance rations with Solar system abundance ratios (for stars in this metalicity range) extends to zinc.

The Heavy Element Neutron Capture Products

The nuclei in the mass region extending from just beyond the iron peak through the actinides (60 ≤A ≤ 238) are understood to have been formed by one of two processes of neutron capture: the s-process of (slow) neutron addition and the r-process of (rapid) neutron addition. As noted previously, intermediate mass stars are understood to be the site of formation of the heaviest s-process isotopes/elements while r-process synthesis is tied rather to an environment associated with massive stars, Type II supernovae, and their remnants. In the heavy element region, both ground based and space based studies (see, e.g., the review by Truran *et al.* 2002) have confirmed that the abundances of the heavy neutron-capture products (A ≥ 130-140) in the most extremely metal deficient halo fields stars ([Fe/H] ≤ -2.5) and globular cluster stars were formed in a robust r-process event (Truran 1981; Truran, Cowan, & Fields 2001). This is reflected, for example, in the extraordinary agreement of the heavy element patterns in the halo stars CS 22892-052 (Sneden *et al.* 2000) and HD 115444 (Westin *et al.* 2000) with solar system r-process abundances (Truran *et al.* 2002). This striking behavior is evident in Figure 3. The site for such production is associated with massive stars, and is thus consistent with nucleosynthetic enrichment in the early Galaxy.

In contrast, the evolution of the [Ba/Eu] ratio (a measure of the ratio of s-process/r-process contributions) with metallicity [Fe/H] in halo stars (Burris *et al.* 2000) reveals that the first significant s-process contributions appear at metallicities

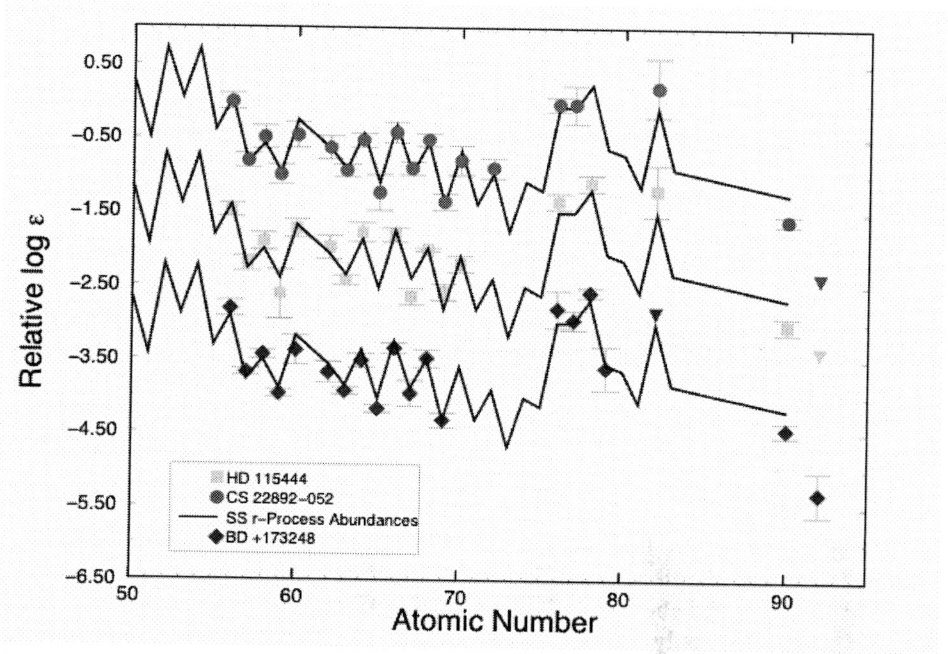

FIGURE 3. r-Process abundance patterns in halo stars. The robustness of the r-process mechanism for the production of elements past barium is reflected in the agreement between Solar abundances and those revealed in three extremely metal deficient halo stars (Truran *et al.* 2002).

[Fe/H] ≈-2.5 to -2.0. This is a consequence of the longer lifetimes of the 1-2 $M_\odot$ stars in which the synthesis of the heavy s-process elements is known to occur.

Abundance Trends at the Lowest Metallicities: Constraints on the Earliest Stellar Populations

The availability of data at increasingly low metallicities now provides further clues to early nucleosynthesis. Figure 4 displays two particularly interesting and seemingly contrasting features. (1) Trends are observed in iron-peak elements (Cayrel et al 2004) as a function of [Fe/H] which are quite well defined with a very low level of scatter. These reflect variations of elemental abundance ratios relative to iron with respect to Solar System abundances, but the smooth and relatively scatter free trends would seem to imply that the massive stars contributing to these trends were extremely robust in their synthesis of iron-peak nuclei.

This behavior is even more notable when one considers: (2) the stark contrast in behavior of the total r-process abundance levels in metal deficient stars. The scatter in the ratio of r-process abundances to iron, as reflected in [Eu/Fe] ranges broadly, reaching values as high as ≈ 30 or more times the Solar system abundance level. The

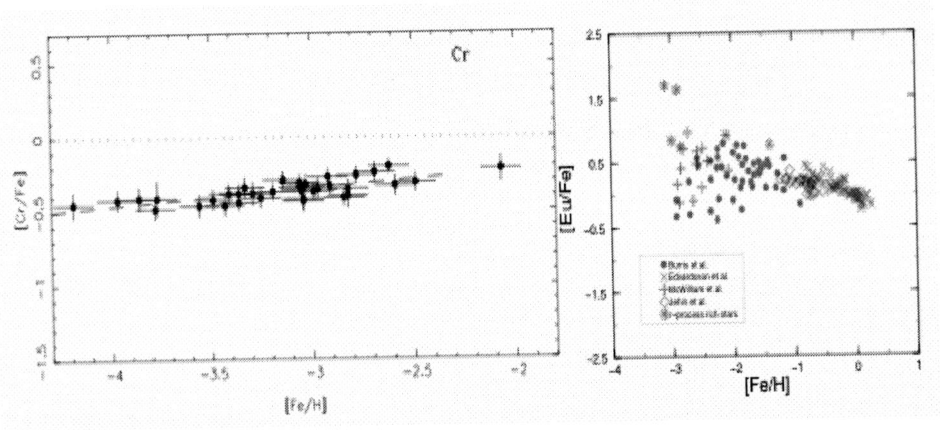

FIGURE 4. Trends in iron peak (Cayrel *et al.* 2005) and r-process (Truran *et al.* 2002) abundances at low metallicities. Note the great differences in the levels of scatter for the two cases.

star to star scatter in this figure can be understood on the basis of inhomogeneities resulting from contributions from individual nucleosynthesis events in an early, relatively unmixed, and chemically inhomogeneous Galaxy. The high achieved ratios of Eu/Fe suggest that only a small fraction of massive star environments give rise to the production of r-process nuclei.

In light of the levels of inhomogeneity reflected in the r-process/Fe ratios, the regularity in the trends in iron peak elements would seem to imply that the process of formation of iron-peak nuclei is quite robust for massive low metallicity stars with, perhaps, a dependence on the individual stellar mass.

SUMMARY AND DISCUSSION: THE FIRST STARS

In the wake of the cosmological Big Bang, the synthesis of heavy elements awaited the formation and evolution of the first stars, some 10^8 to 10^9 years later. The characteristics of the first stellar contributions to nucleosynthesis, whether associated with Population II or a Population III, reflect (as might be expected) the nucleosynthesis products of the evolution of massive stars ($M \geq 10M_\odot$) of short lifetimes ($\tau \leq 10^8$ years). Metal deficient stars ([Fe/H] $\leq$ -1) in our own Galaxy's halo show two significant variations with respect to solar abundances: the elements in the mass range from oxygen to calcium are overabundant - relative to iron-peak nuclei - by a factor $\approx$ 2-3, and the abundance pattern in the heavy element regime $A \geq$ 60-70 closely reflects the r-process abundance distribution that is characteristic of solar system matter, with no evidence for an s-process nucleosynthesis contribution. Both of these features are entirely consistent with nucleosynthesis expectations for massive stars and associated Type II supernovae.

The introduction of both the s-process products of the evolution of low mass stars ($M \approx$ 1-2 $M_\odot$) and of iron nuclei from Type Ia supernovae occurs on a significantly longer timescale. In our Galaxy, evidence for low mass star input is first seen at a metallicity $-2.5 \leq$ [Fe/H] ≤ -2, while the products of Type Ia supernovae seem first to have appeared at [Fe/H] ≈ -1.

These trends are consistent as well with abundance determinations for both galactic globular clusters and (some) nearby dwarf galaxies. The trends in α-elements to iron are also consistent with observations of the abundances and metallicity evolution in the high redshift Universe, as reflected in the damped Lyman-α systems (see, e.g.: Pettini 2003, 2005; Wolfe *et al.* 2005).

We can thus see how we can learn much concerning the nucleosynthesis contributions associated with the earliest stars formed in our Galaxy. We note that the abundance histories as a function of metallicity, time, or redshift *should* reflect, as well, the star formation histories of individual populations (or even of the Universe in the large). The abundance trends we have identified – at least down to metallicities [Fe/H] $\approx$ -3, can generally be understood on the basis of the evolution of 'normal stars' of a range of metallicities over the course of Galactic evolution. The elevated α-nuclei/Fe ratios and r-process dominance over s-process contributions at metallicities below [Fe/H] ≈ -1 clearly indicate that contributions from massive stars and associated Type II supernovae dominate at these epochs, but it is difficult to come to a firm conclusion as to whether this is a consequence simply of their shorter production timescales (lifetimes) or rather of an altered stellar initial mass function (IMF). The increasing scatter in the (r-process)/Fe ratio provides a measure of the degree of inhomogeneity of the Galactic halo gas, increasingly noticeable for metallicities below [Fe/H] ≈ -2. The trends in the abundances of "iron-peak" nuclei to iron (e.g. Cr/Fe. Mn/Fe. Co/Fe. and Zn/Fe) from the Cayrel et al. (2004) study indicate a new behavior/dependence of explosive (supernova) nucleosynthesis below [Fe/H] ≈ -2.5 (see e.g. the discussion by Frohlich *et al.* 2005). Perhaps more disconcerting is the fact that these trends do not seem to continue monotonically down to metallicities below [Fe/H] ≤ -5. It is clear that our understanding of the source of the explosive nuclear burning contributions from the very earliest stellar populations remains incomplete and thus that nucleosynthesis theorists continue to encounter interesting new challenges.

ACKNOWLEDGMENTS

This work has been supported in part by the National Science Foundation under Grant PHYS02-16783 for the Physic Frontier Center "Joint institute for Nuclear Astrophysics" (JINA), by the Department of Energy under Grant B523820 to the ASC Center for Astrophysical Thermonuclear Flashes at the University of Chicago, and by the DOE under Grant DE-FG 02-91ER 40606 in Nuclear Physics and Astrophysics.

REFERENCES

Burbidge, E. M., Burbidge, G. R., Fowler, W.A., and Hoyle, F., 1957, Rev. Mod. Phys. 29, 547

Burris, D.L., Pilachowski, C.A., Armandroff, T., Sneden, C., Cowan, J.J., & Roe, H., 2000, ApJ 544, 302
Busso, M., Gallino, R., & Wasserburg, G.J., 1999, AAR&A 37, 239
Cameron, A.G.W., 1957, PASP 69, 201
Cameron, A.G.W., 1957, Atomic Energy of Canada Limited, CRL-41
Cayrel, R. *et al.*, 2004, A&A 416, 1117
Dominguez, I., Hoeflich, P., & Straniero, O., 2001, ApJ 557, 229
Fields, B.D., Truran, J.W., & Cowan, J,J,, 2002, ApJ 575, 84
Freiburghaus, C., Rosswog, S., 1999, ApJ525, 121
Frohlich, C. et al., 2005, astro-ph/0511376
Goswami, A., & Prantzos, N., 2000, A&A 359, 191
Iben, I. Jr., & Truran, J.W., 1978, ApJ 220, 980
Iwamoto, K., Brachwitz, F., Nomoto, K., Kishimoto, N., Hix, R., & Thielemann, F.-K., 1999, ApJS 125, 439
Kobayashi, C., Tsujimoto, T., & Nomoto, K., 2000, ApJ 539, 26
Kulkarni, V.P., Fall, S.M., & Truran, J.W., 1997, ApJ 484, L7
Lattimer, J.M., Mackie, F., Ravenhall, D.G., & Schramm, D.N., 1977, ApJ 213, 225
Lauroesch, J.T., Truran, J.W., Welty, D.E., & York, D.G., 1996, PASP 108, 641
Limongi, M., Straniero, O., & Chieffi, A., 2000, ApJS 129, 625
Lu, L., Sargent, W.L.W., Barlow, T.A., Churchill, C.W., & Vogt, S.S., 1996, ApJS 107, 475
Madau, P., Della Valle, M., & Panagia, N., 1998, MNRAS 297, L17
McWilliam, A., 1995, ARAA 35, 503
Nomoto, K., Hashimoto, M., & Tsujimoto, T., 1997, Nucl. Phys. A161, 79
Pettini, M., 2003, astro-ph/0303272
Pettini, M., 2005, preprint
Goswami, A., & Prantzos, N., 2000, A&A 359, 191
Renzini, A., & Voli, M., 1981, A\&A 94, 175
Shetrone, M.D., Cote, P., & Sargent, W.L.W., 2001, ApJ 548, 592
Sneden, C., Cowan, J.J., Burris, D.L., & Truran, J.W., 1998, ApJ 496, 235
Sneden, C., Cowan, J.J., Ivans, I.I., Fuller, G.M., Burles, S., Beers, T.C., & Lawler, J.E., 2000, ApJ 533, L139
Takahashi, K., Witti, J., & Janka, H.-T., 1994, A&A 286, 857
Thielemann, F.-K., Nomoto, K., & Hashimoto, M., 1996, ApJ 460, 408
Thielemann, F.-K., Nomoto, K., & Yokoi, K., 1986, A&A 158, 17
Truran, J.W., & Cameron, A.G.W., 1971, A&SS 14, 179
Truran, J.W., 1981, A&A 97, 391
Truran, J.W., Cowan, J.J., & Fields, B.D., 2001, Nucl. Phys. A688, 330
Truran, J.W., Cowan, J.J., Pilacowski, C.A., & Sneden, C., 2002, PASP 114, 1293
Westin, J., Sneden, C., Gustafsson, B., & Cowan, J.J., 2000, ApJ 530, 783
Wheeler, J.C., Sneden, C., & Truran, J.W., 1989, ARAA 27, 279
Wolfe, A., Gawiser, E., & Prochaska, J.X., 2005, ARAA 43, 861
Woosley, S.A., \& Weaver, T.A., 1995, ApJS 101, 181
Woosley, S.A., Wilson, J.R., Mathews, G.J., Hoffman, R.D., & Meyer, B.S., 1994, ApJ 433, 229

Probing the Galactic Magnetic Field using Pulsar Rotation Measures

Athanasiadis, D.*, Kramer, M.*, Mitra, D.† and Lyne, A. G.*

*Jodrell Bank Observatory, Macclesfield, Cheshire, SK11 9DL
†NCRA, TIFR, Pune University Campus, Pune, 411 007

Abstract. Polarisation observations of millisecond pulsars taken with the Effelsberg telescope are presented. The aim of the observations is to determine the Rotation Measure of those pulsars and consequently to shed light on the magnetic field structure off the Galactic plane and in the vicinity of the Sun. Adding our sample to the already existing population of pulsar Rotation Measures we model the local magnetic field using two different types of models: (a) a field following the Galactic spiral arms in conjunction with an exponential drop off the Galactic Plane and (b) a combination of a dipolar and a toroidal field both of which are placed on the Galactic centre.

Keywords: pulsar, magnetic field, ISM, Milky Way
PACS: 97.60.Gb, 98.38.-j, 91.25.Cw, 97.10.Ld

1. INTRODUCTION

The linearly polarized emission from pulsars suffers Faraday rotation due to the birefringent magneto-ionized plasma lying between the pulsar and the observer. The amount of rotation of the polarisation plane is a function of the wavelength and the Rotation Measure (RM). The RM is the integrated sum of the magnetic field component along the line of sight, weighted by the local electron density, n_e:

$$\mathrm{RM} = 0.81 \int \left(\frac{n_e}{\mathrm{cm}^{-3}}\right) \left(\frac{\vec{B}}{\mu\mathrm{G}}\right) \left(\frac{d\vec{l}}{\mathrm{m}}\right) \ \mathrm{rad\,m}^{-2}. \quad (1)$$

A positive RM is defined as resulting from an (effective) magnetic field that points towards the Sun. Faraday rotation is very useful as it is the only tool that allows to determine the longitudal component of the magnetic field [1]. This can be achieved by combining knowledge of the RM with the Dispersion Measure (DM) which is the electron density in the line of sight to the object. This is a quantity determined for every pulsar upon detection and is given by:

$$\mathrm{DM} = \int \left(\frac{n_e}{\mathrm{cm}^{-3}}\right) \left(\frac{d\vec{l}}{\mathrm{m}}\right) \ \mathrm{pc\,cm}^{-3}. \quad (2)$$

As such, the RMs of pulsars can be employed to estimate the average magnetic field component parallel to the line of sight using:

$$< B_{||} >= 1.232 \left(\frac{RM}{\mathrm{rad\,m}^{-2}}\right) \left(\frac{DM}{\mathrm{pc\,cm}^{-3}}\right)^{-1} \mu\mathrm{G} \quad (3)$$

CP848, *Recent Advances in Astronomy and Astrophysics,*
7th *International Conference of the Hellenic Astronomical Society,* edited by N. H. Solomos

The theoretical approach of the formation and evolution of the Galactic magnetic field has resulted in the conclusion that the shape of the resulting magnetic field for any given galaxy is a factor of two terms in general: a) whether there is symmetry with regard to the Galactic plane and b) whether there is rotational axis symmetry. Evidence from higher latitudes is invaluable for the rejection or acceptance of such symmetries. The Galaxy seems to be antisymmetrical with respect to the Galactic plane and with symmetry about the rotation axis (a so-called A0 dynamo mode) in its co-rotating centre [2]. However, it is symmetrical both with respect to the Galactic plane and its rotation axis (S0 mode) away from the centre. Similar results are reported in [3] - indication is also provided there that the local field is quadrupolar but the RM structure towards the inner Galaxy is suggestive of (dipolar) asymmetry. In [4] the antisymmetrical structure found in the Galaxy is indicative of toroidal fields off the Galactic plane and poloidal fields at the Galactic centre perpendicular to the Galactic plane (A0 dynamo mode). However, such a structure is not confirmed in the Galactic quadrants where $90° < l < 180°$ and $270° < l < 360°$.

The exponential scale height for the local Galactic electron disk is 1.5 ± 0.1 kpc based on its full equivalent thickness of 3.0 ± 0.2 kpc [5], [3]. All-sky survey maps of the Galactic continuum at 408 MHz were used to model the synchrotron emission. This compares well with the NE2001 electron density model exponential scale height value of $\simeq 1$ kpc [6]. The exponential scale height of the magnetic disk in the Galactic plane is $h \simeq 1.2 \pm 0.4$ kpc [7] and this is considerably different than the value determined for an electron disk of constant density (approximately 170 pc). More detailed modelling sets the electron disk scale height to 0.7 kpc and the Galactic plane radius to 15 kpc. The magnetic field structure is largely parallel to the plane apart from a small component perpendicular to it which locally is $0.2 - 0.3$ μG and points from South to North [7]. This result was confirmed in [4] where a value of 0.37 ± 0.14 μG is found. The average strength of the magnetic field in the halo is $1 - 2$ μG.

As already mentioned the prominent large-scale feature in the Galactic magnetic field above and below the plane is believed to be antisymmetric: the Galactic plane and the plane defined by the meridian with l=8° are the planes of reference. In particular, each of the quadrants of the Galaxy defined by these two planes has an average magnetic field component parallel to the line of sight that points either towards us or away from us in a quadrupole configuration. This conclusion is derived from the RMs of pulsars and is confirmed by extragalactic sources - it is thought to be a result of the dynamo mode that the Galaxy evolved from. For a synopsis of the theoretical work in galaxy formation see [2].

Several authors have worked in determining the features of the Galactic magnetic field structure. The following sections will address determining the magnetic field off the Galactic Plane and in small ($d < 2$ kpc) and intermediate ($2 < d < 5$ kpc) distances from the Sun. Section 2 will briefly outline the observational strategy and the initial results while in Section 3 the two models studied will be presented along with the results of their fitting. Finally, in Section 4 the results are discussed.

2. OBSERVATIONS AND RESULTS

In order to study the local Galactic magnetic field in intermediate-to-high latitudes the already existing sample of RMs corresponding to that space needed to be expanded. As such, pulsars located off the Galactic plane ($|b| > 5°$) were selected for observation from all possible longitudes. The sample was reduced to include mostly millisecond pulsars (MSPs) i.e. pulsars with period $P < 50$ ms. MSPs are an old population of pulsars and thus have had time to travel away from the Galactic plane. They are therefore found at much higher Galactic latitudes than normal, young pulsars which are born in the Galactic plane. They are also found locally as they are subject to selection effects due to the difficulty in detecting them. The remaining sample was reduced down to those MSPs observable from Effelsberg telescope ($\delta > -28°$) within an integration time of less than 1.5 hours.

The observations took place from the 9th until the 11th of August 2002 and the 1.4 GHz HEMT cooled receiver of the 100m Effelsberg telescope installed at the prime focus was used. After the (native linear) horn the signal is fed into a hybrid which divides it into left and right hand circularly polarised signals. The receiver system temperature was approximately 30 K. The total bandwidth used was 28 MHz spread in four sub-bands (of 7 MHz each) that spanned 100 MHz in total. The two polarisations were down-converted to an intermediate frequency of 150 MHz and then input into the Effelsberg-Berkeley Pulsar Processor (EBPP) backend. The EBPP is ideal for this project as firstly it provides the capability for coherent, rather than incoherent, dedispersion as such promising high signal-to-noise profiles. Secondly, the sub-bands were so configured that they covered a wide frequency range of $\simeq$ 100 MHz with intermediate gaps (thus allowing for an accurate determination of the slope of the polarisation angle variation due to Faraday rotation). The data stream is folded to the topocentric pulsar period and recorded on hard disk for later reduction. For more details about the EBPP see [8] and [9]. In total, nine RM values of MSPs were robustly determined and they are presented in Table 1. The observations outlined in this work increased the number of published pulsar RMs to more than 320 values while an additional 200 RMs are to be published soon in (Han et al., in prep.).

3. MODELLING OF THE GALACTIC MAGNETIC FIELD

In order to investigate the local magnetic field all the pulsars with a known Rotation Measure including the MSPs observed as part of this work were fit with a model for the magnetic field structure. The only pulsar with a known RM not to be included was the Vela Pulsar which is known to be associated with the Gum Nebula which such requires special treatment. Also, in contrast to similar modelling attempts found in the literature, no extragalactic sources were included in the fit mainly because the focus of the models presented is on the local magnetic field.

The process involved numerically integrating Eq. 1 from the observer to the pulsar for each of the pulsar data points. The pulsar coordinates were retrieved from the pulsar online catalogues. The NE2001 model provides considerable detail about the pulsar distances and electron density in the local neighbourhood [6]. The parallel component of

TABLE 1. Determined RMs and the calculated $\langle B_{||} \rangle$. A main pulse is indicated by m, while an interpulse by i. The literature value for PSR B2111+46 is -224 ± 2 rad m^{-2} and is consistent with the value determined here.

Pulsar	Period (ms)	DM (pc cm^{-3})	b (deg)	z (kpc)	RM (rad m^{-2})	$\langle B_{\|\|} \rangle$ (μG)
1012+5307m	5.256	9.0207(4)	+50.86	0.401	−34±2.2	−4.6±0.3
1012+5307i	5.256	9.0207(4)	+50.86	0.401	−47±13	−6.4±1.8
1022+1001	16.45	10.250(10)	+51.10	0.466	−42±6	−5.0±0.7
1640+2224	3.163	18.415(2)	+38.27	0.734	−18±24	−1.2±1.6
1643−1224	4.622	62.41210(10)	+21.22	1.76	−300±40	−5.9±0.8
1713+0747	4.570	15.9920(10)	+25.22	0.38	−23±4	−1.8±0.3
1730−2304m	8.123	9.611(2)	+ 6.02	0.053	4.3±1.9	0.60±0.24
1730−2304i	8.123	9.611(2)	+ 6.02	0.053	8±7	1.0±0.9
1744−1134	4.075	3.1388(3)	+ 9.18	0.0265	−31±5	−12.2±2.0
1937+21m	1.558	71.0370(10)	− 0.29	−0.0488	−11.4±1.7	−0.20±0.03
1937+21i	1.558	71.0370(10)	− 0.29	−0.0488	−2±9	−0.03±0.16
2145−0750	16.05	9.00060(10)	−42.08	−0.335	−8±19	−1.1±2.6
2111+46	1014.7	141.50(4)	−1.27	−0.111	−238±17	−2.1±0.1

the magnetic field along the line of sight was derived from an assumed model and input in Eq. 1. The distance to which the model was applied was gradually increased so as to include a larger number of pulsars and to cover both the short and intermediate distances. The χ^2 was then calculated for each model by comparing the predicted RM values with the observationally determined values (and their errors). In order to determine the best fit parameters a SIMPLEX algorithm [10] was used to minimise the χ^2. All χ^2 values quoted hereafter refer to the normalised χ^2, i.e. divided by the appropriate number of degrees of freedom, unless otherwise noted.

3.1. Spiral Exponential Model

The first model consists of the bisymmetrical spiral model (suggested in [7]) and an exponential dependence with distance from the Galactic plane as mentioned in Sec. 1. The model is given in cylindrical coordinates by

$$B_r = B_0(r)\cos\left[\theta - \beta\, \ln\left(\frac{r}{r_0}\right)\right]\cos p \tag{4}$$

$$B_\theta = B_0(r)\cos\left[\theta - \beta\, \ln\left(\frac{r}{r_0}\right)\right]\sin p \tag{5}$$

$$B_z = 0.37(R_0/r)\exp(-z^2/h_z^2) \tag{6}$$

where p is the pitch angle of the spirals [in agreement with the [4] sign convention], r_0 is the distance from the Galactic Centre of the first magnetic field peak strength towards the anticentre, $\beta = \cot p$ and $R_0 = 8.5$ kpc. The pitch angle, is a measure of how 'tightly' the spirals wind around the Galactic Centre; in the case of $p = 0$ the spirals degenerate into concentric rings. Although in [4] a constant $B_0(r)$ function is assumed, here a radial dependence is adopted so that $B_0 = (3R_0/r)\ \mu$G and subsequently $B_{xy\odot} = 3\ \mu$G as suggested in [11]. This combines the more realistic radial dependence for the magnetic field with a Galactic plane strength at the location of the Sun which is similar to modern estimates. $B_{z\odot} = 0.37\ \mu$G and pointing towards the north Galactic pole as suggested in [4].

In this model there are three parameters to fit for: the pitch angle p, the distance r_0 and the scale-height h_z. The pitch angle p is thought to vary from $-4°$ to $-14°$[12]. Similarly, based on the literature presented in Sec. 1, reasonable ranges for r_0 and h_z were assumed to be [8,15] kpc and [0.5,3] kpc, respectively. The three parameters were allowed to vary in these ranges but were also extended to broader values. The χ^2 values were then calculated for a 3D grid covering all the parameters' combinations in order to investigate the full complexity of the χ^2 sphere around the minimum converged upon by the SIMPLEX algorithm. Grids of various sizes and coarseness were investigated and were all found to exhibit a similar structure as the one shown in Fig. 1.

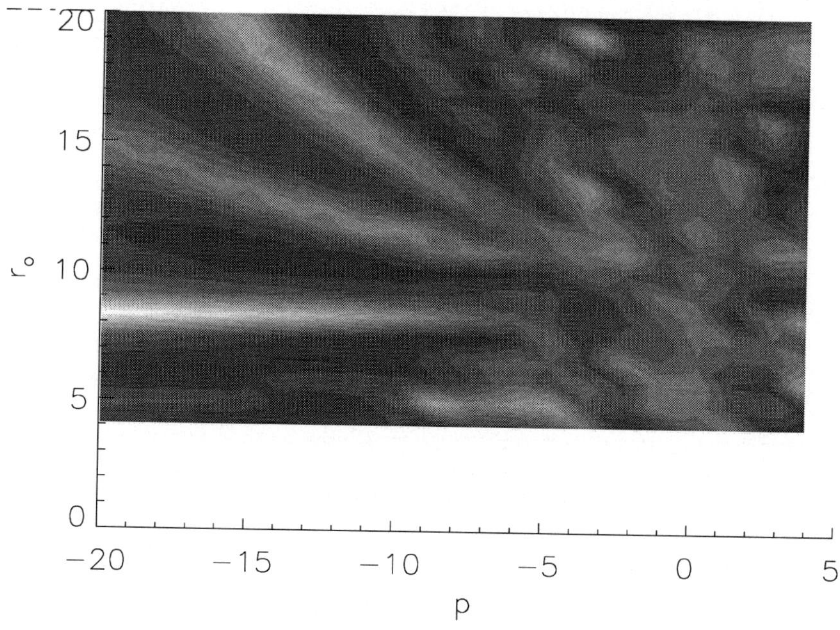

FIGURE 1. Projected χ^2-sphere for the spiral exponential model for a distance of 5 kpc. Parameter h_z is collapsed. See text for details.

In Fig. 1 the greyscale shows the χ^2 values corresponding to the combinations of parameters p and r_0 while the third parameter h_z is collapsed (and set to its value which corresponds to the minimum χ^2). The minimum χ^2 is found to be too large and at the same time Fig. 1 demonstrates the complexity of the χ^2-sphere for this model. Thus, even if the minimum χ^2 was reliable such a topology could not guarantee that a global minimum had been found. The irregularity of the parameter space also means that error contours cannot be drawn.

The process was applied to pulsars within 0.5, 1, 2 and 5 kpc from the Sun. The topology proved in all the cases to be similar to the 5 kpc shown here: in none of those distances could there be found a minimum $\chi^2 \approx 1$. Moreover, neither the χ^2 levels nor the χ^2 topology change significantly if an absolute pure exponential (i.e. a more conventional scale-height definition) is used in Eq. 4 instead of a squared exponential. Similar results are expected therefore to hold for that model variation too.

An alternate method to determine the accuracy of the model is to compare the number of RMs it predicts to have the same value as the observationally determined value within the errors. Similarly, the same comparison can be undertaken between the sign of those two values. The justification for the latter comparison is that it is the RM sign that determines the magnetic field direction. Thus, if the magnetic field model predicts the sign correctly, the prediction for the field direction will be robust, even though the magnitude prediction will be unreliable. That way at least the structure of the magnetic field can be deduced even if qualitatively. The two aforementioned comparisons are shown in the left side of Table 2 for all the four distances examined. The mean over the four distances is also included as an approximate measure of total model performance.

TABLE 2. Comparison of the two models predicted RM values and signs with the observationally determined values within the errors.

		Spiral Exponential Model		Dipolar Toroidal Model		
Distance (kpc)	no. of pulsars	% same RM sign	% same RM value	approx. χ^2	% same RM sign	% same RM value
0.5	16	69	25	370	81	25
1.0	43	47	12	780	53	11
2.0	108	49	09	4300	44	10
5.0	266	59	09	6000	62	08
Mean	-	56	13.75	-	60	13.75

3.2. Dipolar Toroidal Model

Perhaps the most plausible model currently for the Galactic magnetic field according to [13] consists of two components which are based on the formation theories about symmetry mentioned in Sec. 1. The first is a dipole centred on the Galactic Centre with the dipole symmetry axis aligned with the North-South axis. The second is a toroidal field with a sense that reverses on either side of the Galactic plane. The model was formulated in spherical coordinates in the following way:

$$\mathbf{B} = \mathbf{B}_{dipole} + \mathbf{B}_{toroidal} = \tag{7}$$

$$= m \cdot \frac{1}{r^3}(2\cos\theta\hat{r} + \sin\theta\hat{\theta}) + n \cdot \frac{sign(\phi)}{r}\hat{\theta}$$

The $sign(\phi)$ factor is introduced to show the sense reversal in the toroidal field on either side of the Galactic plane. Finally, a $1/r$ decrease in the toroidal field is also adopted which reflects the expected drop of magnetic field strength as distance from the Galactic centre increases.

The two parameters to fit for are m and n, which represent the relative strength of the two model components at a point in the Galaxy (weighted by the distance of that point from the Galactic Centre). A similar process to the one implemented in the previous section was employed to determine the optimal m and n values for this model. As no first estimate exists about them in the literature, χ^2 surfaces were created with very broad limits (initially on a logarithmic scale) so that the region of lowest values could be identified 'visually'. Again, the fit was gradually applied to 0.5, 1.0, 2.0 and 5 kpc.

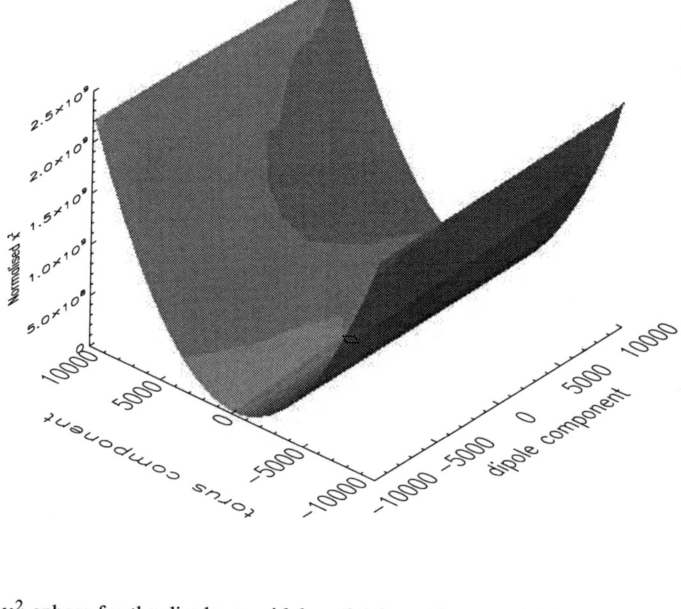

FIGURE 2. χ^2-sphere for the dipolar toroidal model for a distance of 5 kpc. The square indicates the minimum χ^2. See text for details.

Fig. 2 is an example representation of this model's parameter space at a distance of 5.0 kpc - the topology is similar for the other distances. The 3D surface shows the χ^2 values as a function of the fit parameters m and n. Again, the high χ^2 values found render the model unreliable. In this case however, the parameter space topology is considerably more regular and the χ^2 minimum can be positively identified to be the global one. Given the good behaviour of the χ^2 surface along with the lack of *a priori* knowledge of a model, the parameters found can be considered a first estimate despite the large χ^2 and the small-number statistics involved.

The dipolar toroidal model can be further examined with the plot of the model vs observational Rotation Measures in Fig. 3. In this plot, the more accurate the model is, the more points will concentrate in the quadrants with a same-sign abscissa and ordinate. The concentration of points however is only marginally larger in those quadrants which indicates that this model predicts the wrong magnetic field direction in a considerable fraction of the cases. This also can be seen in Table 2 where the agreement between the model and observational RM signs or values has a mean of 60% and 13.75%, respectively.

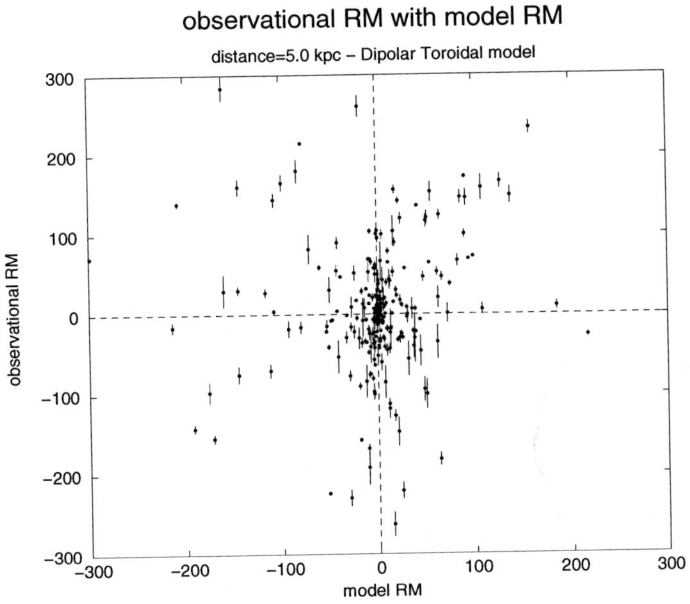

FIGURE 3. Observational RM with model RM for the dipolar toroidal model for a distance of 5.0 kpc.

4. DISCUSSION

As demonstrated in the previous sections the nature of the Galactic magnetic field remains highly non-analytical. There is no knowledge of a model that can fit the data fully and as such the models presented here are not expected to provide perfect fits. They will necessarily be at best an approximation and the basis for more accurate work to follow. In order to evaluate the success of the two models in describing the observations as a function of the amount of volume covered, it would be desirable to have a number of common criteria. The criteria and tests described here can be used as a set of tools to evaluate the success of any other magnetic field model, not just those described here.

The classic candidate for such an evaluation is the χ^2 value. The fact that no χ^2 value is close to unity is indicative that none of the models provides a good fit to the data. However, the complexity of the magnetic field under study, the small-number statistics involved in most of the cases and most importantly the fact that the actual underlying model and its parametrisation is not known *a priori* must be taken into consideration. Some models can be then thought of as being able to describe observations better than others.

In fact, χ^2 values for the fits are orders of magnitude larger than the ideal value of unity. Under this perspective, the χ^2 is best considered as a measure of the residual difference between the model and the observations - similarly to the interpretation assumed in [7] for their *RE* quantity. It is also worth noting that few, if any, authors quote actual χ^2 values in their modelling of the Galactic magnetic field.

The χ^2 topology is an alternate comparison tool between models. A simple inspection of Fig.1 is sufficient to demonstrate the complex structure of this spiral exponential model as mentioned in Sec. 3.1. The χ^2-sphere looks too irregular to allow for a reliable and unique solution - this is often overlooked by some authors. The complexity of this χ^2-sphere proves that such a detailed study of the topology should be undertaken whenever possible to ensure that the fit result is reliable.

Similarly, Table 2 provides an additional evaluation of the fit. The low to intermediate success of the model in predicting correct RM signs ($\approx 70\%$ at best) or RM values within the errors ($\approx 10\%$ on average) is evident. Especially in small distances the small-number statistics involved further degrade the model accuracy.

The spiral exponential model formalism of Eqs. 4 is more appropriate for intermediate distances (e.g. $\approx$ 5 kpc) rather than smaller ones as it traces the spiral arms. Thus, one would expect an increasingly better fit with distance (also enhanced by the better statistics at higher distances). However, the model fits for both the small and intermediate distances proved equally unsuccessful in converging to a reliable solution or even in showing any regularity in their χ^2-sphere topology. This is confirmed by Table 2 which shows no significant variation with distance in the agreement between observational and model RM values and signs.

The combination of the topology, the very high value of χ^2 for all the sensible parameter combinations and the RM values and signs comparisons most likely suggest that this particular model is not appropriate to describe the local Galactic magnetic field.

These results could perhaps be suggesting that the inclusion of the exponential decrease in the model, although intuitive and widely used in general, in this particular case fails to describe the observations. This conclusion is supported by the relative good agreement by which the spiral model seems to be describing the overall Galactic plane pulsar population with a known RM in [4].

In Sec. 3.2 the optimal values of (m,n) for the dipolar toroidal model are also associated with too high χ^2 values. However, given the good behaviour of the χ^2 surface (Fig. 2) along with the already mentioned lack of *a priori* knowledge of the model parametrisation, it can be considered a reliable estimate despite the large χ^2 and the small-number statistics involved. At the very least this is a more reliable solution than that identified in the irregular structure of the spiral exponential model.

However, the model and observational RM signs comparison given in Table 2 shows that only marginally is the dipolar toroidal model better on average than the spiral exponential. When comparing the RM values the two models prove equally unsuccesful both on average (mean $\approx 14\%$) and per specific volume size.

Taking into account all these methods and their shortcomings, it is the dipolar toroidal which seems to stand out as a marginally better model. It is also worth noting that the electron density model assumed plays a crucial role in the determination of model RM values. Replacement of NE2001 with its earlier version (TC1993; [14]) left the χ^2-spheres relatively unchanged qualitatively but resulted in different magnetic field fitted parameters and even the predicted RM signs were significantly different in some instances.

The work presented here can be extended in a number of ways. Inclusion of a larger pulsar sample in the models will be possible upon the publication of more RMs by (Han et al., in prep.). Extragalactic sources can also be used to augment the sample if care is exercised when weighing them in the existing population. The models themselves can be enhanced by adding a limited number of additional fit parameters (e.g. by assuming a hybrid spiral-dipolar model) or by allowing for specific special regions (SNR, HII, etc.) in a manner similar to the NE2001 model. The ultimate aim is an as detailed as possible magnetic field model of the Galaxy.

ACKNOWLEDGMENTS

DA would like to thank the staff of Jodrell Bank Observatory and in Effelsberg telescope for aiding in the retrieval and analysis of the data. DA would also like to thank Ioannis Georgantopoulos for allowing the use of the National Observatory of Athens facilities during most of the writeup of this paper. DA was funded by a PPARC scholarship during most of the work presented here.

REFERENCES

1. J. Tinbergen, *Astronomical Polarimetry*, Cambridge University Press, 1996, iSBN: 0521475317.
2. R. Wielebinski, and F. Krause, *A&AR* **4**, 449–485 (1993).
3. R. Beck, *Space Sci. Rev.* **99**, 243–260 (2001).
4. J. L. Han, R. N. Manchester, and G. J. Qiao, *MNRAS* **306**, 371–380 (1999).
5. K. Beuermann, G. Kanbach, and E. M. Berkhuijsen, *A&A* **153**, 17–34 (1985).
6. J. M. Cordes, and T. J. W. Lazio (2002), astro-ph/0207156.
7. J. L. Han, and G. J. Qiao, *A&A* **288**, 759–772 (1994).
8. D. C. Backer, M. R. Dexter, A. Zepka, N. D., D. J. Wertheimer, P. S. Ray, and R. S. Foster, *PASP* **109**, 61 (1997).
9. M. Kramer, K. M. Xilouris, F. Camilo, D. Nice, C. Lange, D. C. Backer, and O. Doroshenko, *ApJ* **520**, 324–334 (1999).
10. W. H. Press, S. A. Teukolsky, W. T. Vetterling, and B. P. Flannery, *Numerical Recipes: The Art of Scientific Computing, 2nd edition*, Cambridge University Press, Cambridge, 1992.
11. Y. Sofue, and M. Fujimoto, *ApJ* **265**, 722–729 (1983).
12. J. M. Reid, "The Magnetic Field of the Milky Way," in *Highlights of Astronomy, Vol. 12, 2002, IAU*, edited by H. Rickman, Astronomical Society of the Pacific, San Francisco, 2002, pp. 716–718.
13. J. L. Han, "Magnetic Fields in our Galaxy: How Much Do We Know? (II). Halo field and global structure," in *AIP Conf. Proc. 609: Astrophysical Polarized Backgrounds*, edited by S.Cecchini, S. Cortiglioni, R. Sault, and C. Sbarra, 2002.
14. J. H. Taylor, and J. M. Cordes, *ApJ* **411**, 674–684 (1993).

Modified Pulsar Spindown

I. Contopoulos[a] and A. Spitkovsky[b]

[a]*Research Center for Astronomy, Academy of Athens, Soranou Efessiou 4, Athens, GR-11527*
[b]*KIPAC, Stanford University, P.O. Box 20450, Stanford, CA 94309*

Abstract. We propose a new electromagnetic pulsar spindown law which takes into account the misalignment of the magnetic and rotation axes, and the magnetospheric particle acceleration gaps. We show that near the death line aligned rotators spin down much slower than orthogonal ones. We predict that the older pulsar population has preferentially smaller magnetic inclication angles, and we discuss a possible signature of such alignment in the existing pulsar data.

Keywords: Pulsars–MHD

INTRODUCTION

The current canonical pulsar paradigm is that of a magnetized rotating neutron star (see Mestel 1999 for a review). A spinning down neutron star with mass $M_* \sim 1.4 M_\odot$, radius $r_* \sim 10$ km, and angular velocity Ω loses rotational kinetic energy through electromagnetic torques in the magnetosphere at a rate

$$L = \frac{2}{5} M_* r_*^2 \Omega \dot{\Omega} \,. \tag{1}$$

To a first approximation, the stellar magnetic field may be considered as that of a rotating magnetic dipole (see Lyutikov & Thompson 2005 for recent observational confirmation). The general description of the field structure is a formidable three dimensional problem, since, in general, the magnetic and rotation axes do not coincide. Awaiting the results of time-dependent three dimensional numerical force-free ideal MHD simulations (Spitkovsky, under preparation), we can still derive important conclusions based on two idealized limiting cases which have been studied in great extent: the case of an aligned magnetic dipole rotating in an atmosphere with freely available electric charges, and that of a misaligned magnetic dipole rotating in vacuum.

The neutron star is not surrounded by vacuum, and one needs to take into consideration the electric fields that develop and the electric currents that flow in the rotating charged magnetosphere (Goldreich & Julian 1969). In that picture, the magnetosphere consists of a corotating region of closed fieldlines which extends up to a distance from the rotation axis equal to c/Ω (the 'light cylinder'), and an open fieldline region outside (see Contopoulos, Kazanas & Fendt 1999; Goodwin *et al.* 2004; Gruzinov 2005a; Contopoulos 2005; Timokhin 2005; Komisarov 2005). The neutron star spins down because of the establishment of a large scale poloidal electric current circuit flowing along open field lines, and returning along the edge of the open field line region. The electromagnetic energy loss in the simplest possible case, that of an aligned rotator, is equal

CP848, *Recent Advances in Astronomy and Astrophysics,*
7th *International Conference of the Hellenic Astronomical Society,* edited by N. H. Solomos

to

$$L_{\text{aligned}} = \frac{\Omega^3 \Omega_F B_*^2 r_*^6}{4c^3} \tag{2}$$

where B_* is the polar value of the magnetic field. Ω_F, the angular frequency of rotation of the open field lines, is set by the electric potential drop that develops accross open field lines, between the magnetic axis and the edge of the open field line region (Beskin 1997). This potential in the magnetosphere is in general smaller than the corresponding electric potential drop on the surface of the star. The difference between the two potential drops is just the particle acceleration gap potential which develops *along* open magnetic field lines in the vicinity of the polar cap. Consequently, the angular velocity of open field lines will in general be smaller than Ω. Models of particle acceleration and pair creation of rotation-powered pulsars yield values of the gap potential V_{gap} of the order of 10^{12} Volts (e.g. Hibschmann & Arons 2001). One can directly show (Contopoulos 2005) that

$$\Omega_F = \Omega - \Omega_{\text{death}} , \tag{3}$$

where,

$$\Omega_{\text{death}} \sim \frac{2V_{\text{gap}} c^2}{\Omega B_* r_*^3} . \tag{4}$$

This describes the so-called pulsar 'death', i.e. the stopping of pulsar emission. As the neutron star slows down and Ω drops below Ω_{death}, the gap potential cannot attain the value required for particle acceleration and consequent pulse generation, and the pulsar stops generating radio emission.

On the other hand, a misaligned dipole rotating in vacuum loses energy at a rate

$$L_{vacuum} = \frac{B_*^2 \Omega^4 r_*^6}{6c^3} \sin^2\theta , \tag{5}$$

where θ is the misalignment angle between the magnetic and rotation axes. Eqs. 2 & 5 being so similar led most researchers to ignore the dependence on θ and Ω_F in estimates of stellar magnetic fields, and in most studies of the $P - \dot{P}$ diagram. The aim of the present work is to show that the dependence on θ and Ω_F is important and should not be ignored, especially in old pulsars approaching their death.

ELECTROMAGNETIC ENERGY LOSSES

In real life, pulsars are neither aligned nor perpendicular rotators. It is natural, therefore, to expect that the two electromagnetic energy loss terms described in eqs. 2 & 5 contribute together in the total energy loss. A possible combination is:

$$L \sim L_{\text{vacuum}} + L_{\text{aligned}} \cos^2\theta$$

$$= \frac{B_*^2 \Omega^2 r_*^6}{4cr_c^2} (\alpha \sin^2\theta + \left(1 - \frac{\Omega_{\text{death}}}{\Omega}\right) \cos^2\theta) , \tag{6}$$

where α is of order unity. We stress that this combination, although inspired by physical limiting cases, is not a rigorous derivation. An exact expression will require the development of a detailed non-axisymmetric MHD theory of the pulsar magnetosphere[1].

Eq. 6 brings a new element to the discussion: *we directly introduce pulsar death and magnetic-rotation axes misalignment in the electromagnetic pulsar spindown expression.* Pulsar death expresses the inability of the magnetosphere to generate the particles required in the poloidal electric currents which generate spindown torques in the aligned rotator case. They do not affect the orthogonal spindown component though. Beyond the death line, the misaligned neutron star will continue to spin down without emitting pulsar radiation[2].

In order to study how pulsars spin down, we equate the values of L in eqs. 1 & 6 and thus obtain

$$\dot{P} = 3.3 \times 10^{-16} \left(\frac{P}{P_\circ}\right)^{-1} \left(\frac{B_*}{10^{12}\ \mathrm{G}}\right)^2 \left(\frac{P_\circ}{\mathrm{sec}}\right)^{-1} \times \left(\sin^2\theta + \left(1 - \frac{P}{P_{\mathrm{death}}}\right)\cos^2\theta\right) \tag{7}$$

where from (4)

$$P_{\mathrm{death}} = 3\ \mathrm{sec} \left(\frac{B_*}{10^{12}\mathrm{G}}\right)^{\frac{1}{2}} \left(\frac{V_{\mathrm{gap}}}{10^{12}\mathrm{Volt}}\right)^{-\frac{1}{2}} . \tag{8}$$

$P_\circ$ is the initial period at pulsar birth. Note that we have introduced an angular dependence in the definition of the death line $\dot{P}(P_{\mathrm{death}})$. As we will see below, this has interesting observational consequences. Note also that, in the limit $\theta \sim 90^\circ$, we obtain

$$\log \dot{P}_{\mathrm{death}} = 3 \log P_{\mathrm{death}} - 17.3 + 2(\log V_{\mathrm{gap}} - 12) \tag{9}$$

(compare with Zhang, Harding & Muslimov 2000). When $P \ll P_{\mathrm{death}}$, $B_* \propto (\dot{P}P)^{\frac{1}{2}}$, and pulsars evolve in the logarithmic $P - \dot{P}$ diagram along lines of constant magnetic field $\dot{P} \propto P^{-1}$. As P approaches P_{death}, however, the evolution curves down away from the above straight lines and ends at a point $P = P_{\mathrm{death}}$ and $\dot{P} = \dot{P}(P_{\mathrm{death}})$ (see figure 1). This evolution is similar to what one would expect if we had assumed magnetic field decay in the standard dipole spindown model (e.g. Gonthier *et al.* 2002, 2004).

As a test of the validity of our results, we tried to reproduce the gross features of the observed distribution of non-millisecond pulsars in the $P - \dot{P}$ diagram, through a simple Monte Carlo numerical experiment that follows Gonthier *et al.* 2002, 2004 (see Contopoulos & Spitkovsky 2006 for details). We decided to keep the number of initial magnetic field components to a minimum. In the standard spindown picture it is very difficult to have pulsars in the lower corner of the $P - \dot{P}$ triangle without polluting the low P – low $\dot{P}$ region of the diagram with low-field pulsars, which are not observed. In our spindown model the lower corner is filled with pulsars dropping from the higher $\dot{P}$

---

[1] A recent independent estimation yielded $\alpha \approx 1.8$ (Gruzinov 2005b).

[2] See, however, Gurevich, Beskin & Istomin 1993; Mestel, Panagi & Shibata 1999; Mestel 1999 for a different point of view where an orthogonal rotator does not spin down.

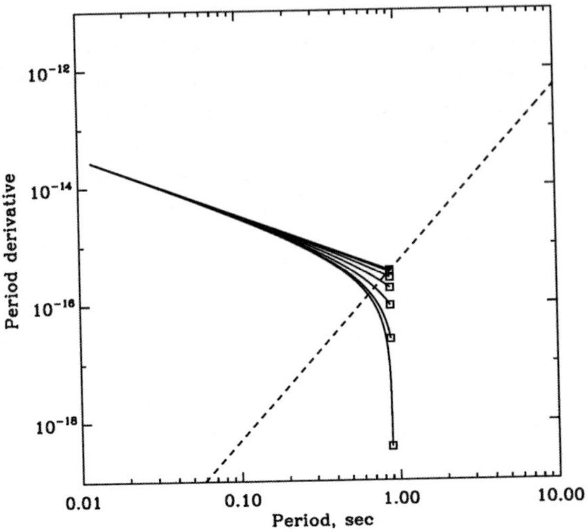

FIGURE 1. $P-\dot{P}$ evolutionary diagram that shows the effect of the misalignment angle θ. We show spindown trajectories for pulsars with initial period $P_\circ = 10$ msec, magnetic field $B_* = 10^{12}$ G, and $V_{gap} = 10^{13}$ Volt. Trajectories from bottom to top correspond to pulsars with increasing magnetic inclination angle from $\sim 0°$ to $90°$ with $15°$ increments. Rectangular dots indicate when pulsars reach P_{death} and turn off. The dashed line represents the theoretical death line as given in eq. 9 which does not take into account the misalignment angle dependence introduced in eq. 7.

(higher B_*) region, rather than the ones moving on straight tracks as in the standard case. This contributes to the triangular shape of the diagram.

Our results are shown in figure 2. In figure 3, we show the present day observed radio pulsar distribution from the ATNF pulsar catalog[3], and the similarity is quite good. Note that the introduction of the θ dependence that we have obtained in eq. 7 with random misalignment angles, generated a 'soft' death line boundary without the introduction of a 'death valey' (Gonthier *et al.* 2002) or field decay in our simulation.

Furthermore, from eqs. 1 & 6 we derive an expression for the pulsar braking index (defined as $n \equiv \ddot{\Omega}\Omega/\dot{\Omega}^2$), namely

$$n = 3 + \frac{1}{\Omega/(\Omega_{death}\cos^2\theta) - 1} . \qquad (10)$$

Around the pulsar death line $\Omega \approx \Omega_{death}$, this equation gives

$$n \approx 3 + \frac{\cos^2\theta}{1-\cos^2\theta} , \qquad (11)$$

[3] Available at http://www.atnf.csiro.au/research/pulsar/psrcat/

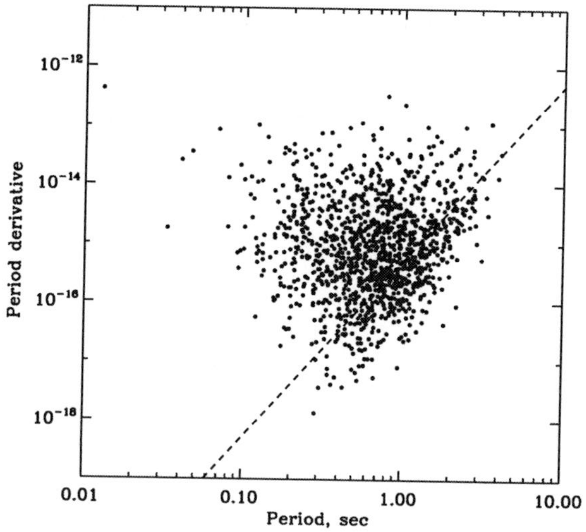

FIGURE 2. Monte Carlo simulation of the observed $P-\dot{P}$ diagram. Pulsars are injected with random misalignement angles, and a lognormal distribution of polar magnetic field values around $B_* = 10^{12}$ G (as in Gonthier *et al.* 2002). $V_{\rm gap} = 10^{13}$ Volt. $P_\circ$, the initial period at pulsar birth, is uniformly randomly distributed between 10 msec and 0.2 sec. We have also introduced an observational selection effect due to finite instrument detectability limit (see Contopoulos & Sptikovsky 2006 for details). The dotted line represents the theoretical death line as given in eq. 9. The introduction of the angle dependence results in the softening of the death line. An observational prediction of the present work is that pulsars lying near or below the theoretical death line should have preferentially smaller misalignment angles.

which, for aligned pulsars with $\cos\theta \approx 1$ may be much greater than 3. Such high braking index values may be observable (Johnston & Galloway 1999).

In figure 4 we plot the distribution of $\cos\theta$ in our Monte Carlo simulation that was presented in figure 2. In this simulation pulsars are injected (born) with random magnetic axis orientation with respect to the rotation axis as shown by the dashed line histogram. The pulsar population at the end of the simulation, however, shows a different distribution of misalignment angles as shown by the thick solid line. The selection bias towards smaller misalignment angles may be attributed to the fact that orthogonal rotators spin down faster (on the average), and therefore are more likely to evolve beyond the death line within the observation time interval. Finally, the thin solid line represents the distribution of misalignment angles in the pulsars observed lying below the theoretical death line given by eq. 9. One sees clearly that *pulsars observed near or below the death line should have preferentially smaller misalignment angles* θ *than the rest of the pulsar population*. This is, we believe, an important observational prediction of the present work.

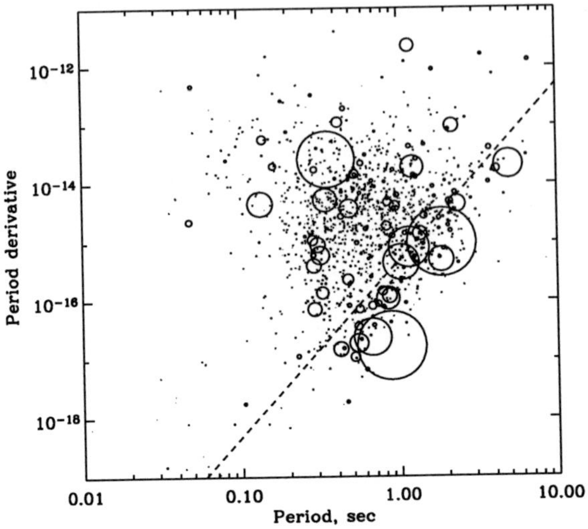

FIGURE 3. The present day observed radio pulsar distribution. We also show the distribution of alignment measure. Each pulsar is plotted as a circle with radius proportional to the quantity $WP^{-1/2}$ using the pulse width W at 50% of the peak. Data from the ATNF pulsar database.

DISCUSSION

Awaiting the development of a detailed three dimensional MHD theory for the rotating neutron star magnetosphere, let us summarize the main conclusions of the present work:

1. The electromagnetic energy loss of a pulsar depends not only on the surface magnetic field B_* and on the rotational angular velocity Ω, but also on the misalignment angle θ and on the angular velocity of the open magnetic field lines Ω_F (which is in general smaller than Ω).
2. The approach to pulsar death modifies the rate of energy loss. The death occurs when the pulsar slows down sufficiently so that the angular velocity of the open fieldlines $\Omega_F = 0$.
3. The energy loss close to the pulsar death is smaller than what is given by the standard dipolar spindown formula. This effect gives a good fit between the theoretical and observed distributions of pulsars near the death line, without invoking a magnetic field decay.
4. Pulsars near the death line have braking index values $n > 3$. Such high braking index values may be observable.
5. Pulsars near the death line may have preferentially smaller inclination angles.

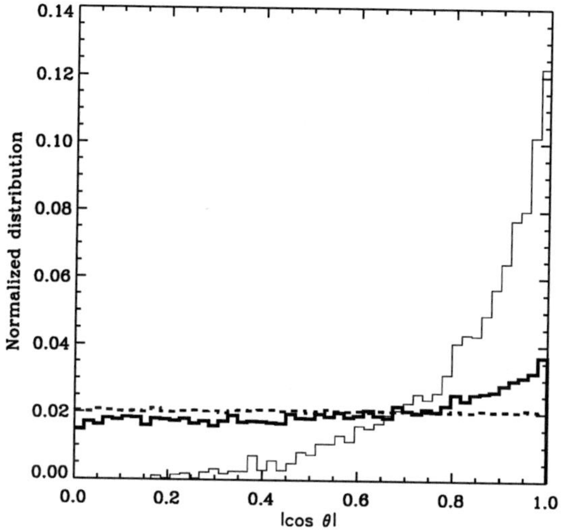

FIGURE 4. The distribution of $|\cos(\theta)|$ in our Monte Carlo experiment in figure 3. The dashed line represents the original distribution injected with uniform random misalignment angle. The thick solid line represents the distribution of misalignment angles in the simulated pulsar distribution as a whole. The thin solid line represents the distribution of misalignment angles for the pulsars observed lying below the theoretical death line given by eq. 9. An interesting observational prediction of the present work is that pulsars lying near or below the theoretical death line should have preferentially smaller misalignment angles.

A preliminary look at the ATNF catalog data suggests that the last point may have some observational support. One possible measure of the inclination angle of a pulsar is its fractional pulse width, or the ratio of the width of the pulse to the period of the pulsar. If the radio beam size is independent of inclination, then we would expect pulsars with smaller inclinations to be seen for a larger fraction of the period than pulsars with large inclinations. In order to test this hypothesis we took the available data for the pulse width at 50% of the pulse peak, $W_{50\%}$, from the ATNF catalog (1375 pulsars with $W_{50\%} \neq 0$). In order to be able to compare the fractional pulse width for pulsars of different periods, we have to correct for the intrinsic size of the pulsar beam. If the beam roughly follows the angular size of the open fieldlines, then the beam size falls as $P^{-1/2}$ with increasing period. Therefore, the quantity that should relate to the degree of alignment is $F_{\rm align} \equiv (W \times P^{1/2})/P = WP^{-1/2}$, where W is the measured pulse width. In figure 3 we plot the observed pulsars as circles with the radius of the circle linearly proportional to $F_{\rm align}$. A visual inspection of the plot shows that there is an excess of larger pulse fractions for older pulsars, and in particular for pulsars near the right edge of the $P-\dot{P}$ triangle. This region is near the pulsar death line (dashed line), and therefore, this feature is particularly interesting. Obviously, a much better analysis needs to be performed, but these results are quite encouraging.

In an effort to simplify our Monte Carlo fit we have assumed that the misalignment angle stays the same during the evolution, and there is no field decay. The reality may of course be more complicated, and there could be some amount of field decay, and potential alignment (and consequent precession) during the lifetime of a pulsar, due to electromagnetic torques on the star. These effects would introduce extra degrees of freedom to fitting the pulsar distribution. We hope, however, that the physically motivated spindown law of the type introduced in this work would find its way into detailed population synthesis models.

REFERENCES

1. Beskin, V.S. 1997, Physics-Uspekhi, 40, 659
2. Contopoulos, I., Kazanas, D. & Fendt, C. 1999, ApJ, 511, 351
3. Contopoulos, I. 2005, A&A, 442, 579
4. Contopoulos, I & Spitkovsky, A., 2006, ApJ, in press (astro-ph/0512002)
5. Goldreich, P. & Julian, W. H. 1969, ApJ, 157, 869
6. Goodwin, S. P., Mestel, J., Mestel, L. & Wright, G. A. E. 2004, MNRAS, 349, 213
7. Gonthier, P. L., Ouellette, M. S., Berrier, J., O'Brien, S. & Harding, A. K. 2002, ApJ, 565, 482
8. Gonthier, P. L., Van Guilder, R. & Harding, A. K. 2004, ApJ, 604, 775
10. Gruzinov A. 2005a, Phys. Rev. Lett., 94, 021101
10. Gruzinov A. 2005b, astro-ph/0510751
11. Gurevich, A., Beskin, V. & Istomin, Y., 1993, "Physics of the Pulsar Magnetosphere", Cambridge University Press
12. Hibschman, J. A. & Arons, J. 2001, ApJ, 554, 624
13. Johnston S. & Galloway D. 1999, MNRAS, 306L, 50
14. Komissarov, S., 2005, astro-ph/0510310
15. Lyutikov, M. & Thompson, C., 2005, astro-ph/0502333
16. Mestel, L. 1999, "Stellar Magnetism", Oxford University Press
17. Mestel, L., Panagi, P. & Shibata, S., 1999, MNRAS, 309, 388
18. Timokhin, A. 2005, astro-ph/0511817
19. Zhang B., Harding A. K., Muslimov A. G., 2000 ApJL, 531, 135

Pulsar Nulling Quantitative Analysis

Kosmas Lazaridis and John H. Seiradakis

Aristotle University of Thessaloniki, Department of Physics, Section of Astrophysics, Astronomy and Mechanics, GR-541 21, Thessaloniki, Greece

Abstract. Using long sequences of single pulses we analyzed the nulling behaviour of several pulsars. Each pulsar has been characterized by a "nulling parameter", which represents the average length of consecutive null pulses. The "nulling parameter" has been compared to other pulsar parameters. Some interesting correlations have been derived.

Keywords: neutron stars – pulsars – pulse nulling
PACS: 97.60.Gb, 97.60.Jd

INTRODUCTION

Pulsars are highly magnetized, fast rotating neutron stars, emitting narrow radio pulses along the magnetic axis, often highly polarized. Their magnetic field exceeds 1012 Gauss and their rotating period extends between 1.397 ms (716 Hz) and 8.5 sec. The surface temperature of pulsars is $T = 10^6 - 10^7$ K and the stellar diameter is about 20 km. Observations have revealed that high frequencies are emitted close to the polar surface and lower frequencies higher up. A few pulsars have been observed in X-rays, γ-rays and optical wavelengths. Already about 1700 pulsars have been detected, mostly in our Galaxy along the galactic plane and a few in the large Magellanic cloud.

The integrated profile of every pulsar is generally very stable and has a characteristic shape (Figure 1). On the other side when single (individual) pulses (Figure 2) are investigated with high time resolution a variety of shapes are observed from one period to another reflecting the magnetic field structure and a complex discharge pattern, that is assumed to take place above the magnetic poles.

The phenomenon of pulse nulling is a complex process during which the pulsar ceases to radiate for a few up to a few hundred pulses and it was first discovered in PSR 1929+10 by Backer (1970). Each pulsar has completely different nulling characteristics, which change from component to component. Some very important studies have been published that attempt to explain the nulling mechanism and to correlate the pulse nulling with some other pulsar characteristics as the age, period, period derivative and magnetic field, notably Ritchings (1976), Filippenlo and Radhakrishnan (1982), Rankin (1986), Biggs (1992), Kazbegi et al. (1996) and van Leeuwen et al. (2003). In this work, we describe the methods and techniques we used for the quantitative analysis of pulsar data and we present some interesting results and correlations for the phenomenon of nulling

CP848, *Recent Advances in Astronomy and Astrophysics,*
7th *International Conference of the Hellenic Astronomical Society,* edited by N. H. Solomos

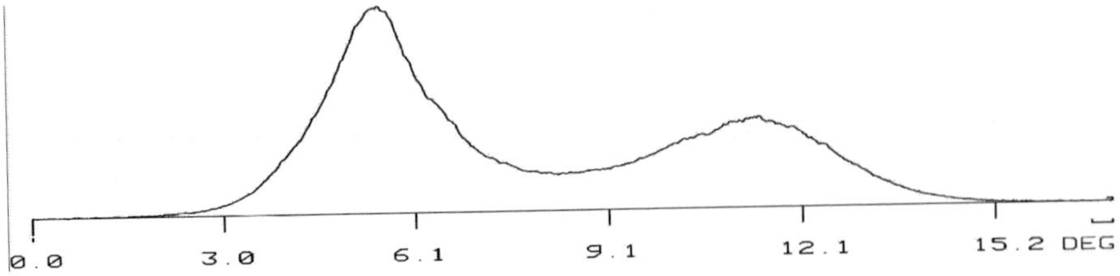

FIGURE 1. Integrated double profile (with two components) of PSR 1133+16.

FIGURE 2. Sequence of single pulses of PSR 1133+16.

OBSERVATIONS

Single pulses from 24 strong pulsars (Table 1) were observed with the 100m Effelsberg radio telescope of the Max-Planck-Institute for Radio Astronomy, Bonn in 1991-1994. The resolution extended from a few tens to a few hundreds of μs and we analyzed sequences of a few hundred to a few thousand single pulses at 21cm..

TABLE 1. Pulsars studied for their nulling properties

PSR 0301+19	PSR 0950+08	PSR 1944+17
PSR 0329+54	PSR 1133+16	PSR 1952+29
PSR 0450+55	PSR 1237+25	PSR 2016+28
PSR 0525+21	PSR 1642–03	PSR 2020+28
PSR 0628–28	PSR 1706–16	PSR 2021+51
PSR 0823+26	PSR 1749–28	PSR 2217+44
PSR 0301+19	PSR 1822–09	PSR 2310+60
PSR 0834+06	PSR 1933+16	PSR 2351+61

ANALYSIS

The program (jhspuls) that we used for computing and analyzing the observations from the 24 pulsars was created and developed by W. Sieber, A. Sievers, A. Jessner, J.H. Seiradakis and M. Kramer for studying and calculating the statistics of pulsar emission. We modified this program in order to make it capable of calculating pulse nulling. In Figure 3 the integrated profile of PSR 2319+60 and a sequence of 50 single pulses from the same pulsar are shown. There are some obvious nulling pulses in this figure. The spikes observed in the sequence of single pulses are due to interference. Their power is very small compared with the power of the single pulses. Therefore, they play insignificant role in our analysis.

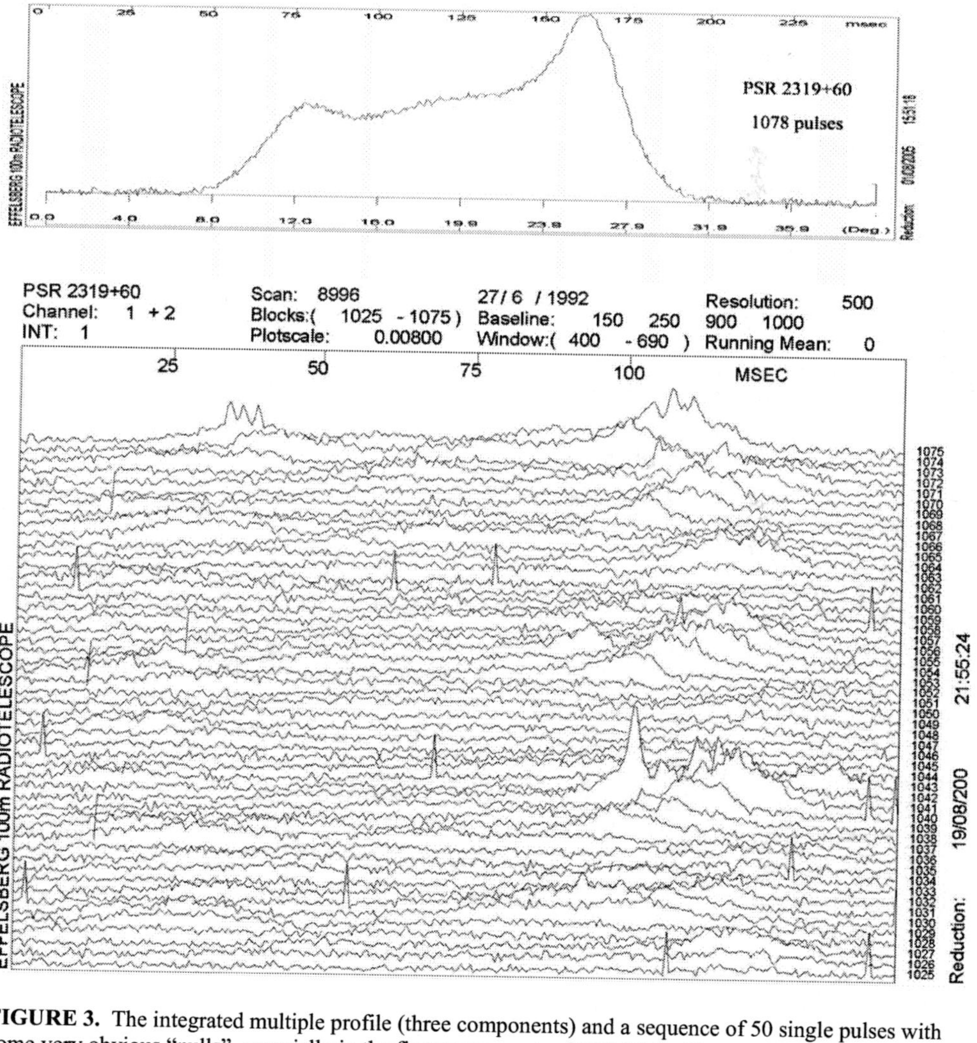

FIGURE 3. The integrated multiple profile (three components) and a sequence of 50 single pulses with some very obvious "nulls", especially in the first component, of PSR 2319+60.

The main concern in this analysis is to make sure that the number of nulls that our program calculates is equal or very close to the number of nulls one can visually obtain in long sequences of single pulses with good signal to noise. This problem was tackled by determining the average power of every single pulse within the integrated pulse profile. If this power was below a certain threshold, then the single pulse was considered to be a "null". Simple pulses are known to be much narrower than the integrated profile. Therefore, while calculating the average power, we all data points across the window of the integrated pulse profile which did not exceed the height of 3σ were excluded. The (standard deviation) σ was suitably calculated outside the integrated pulse window.

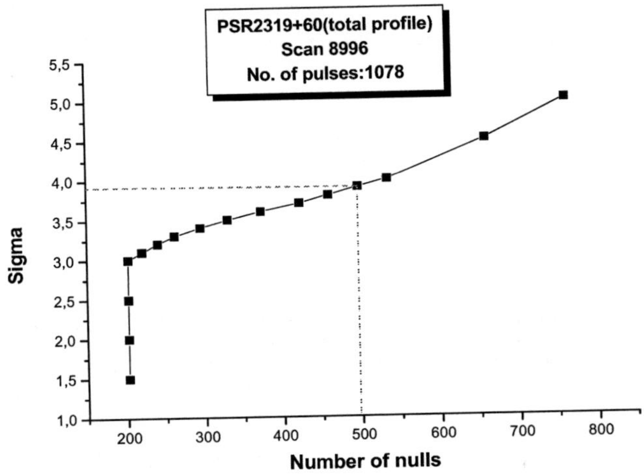

FIGURE 4. The graph of the threshold standard deviation, σ and the number of nulls for a sequence of 1078 pulses for PSR 2319+60.

Figure 4 shows how the threshold was obtained for a sequence of 1078 pulses with good signal-to-noise, in which we had visually counted 498 nulls. We run our program for a large number of threshold values. For each value the number of nulls was calculated. It is obvious in this figure that for the threshold value of 3.8σ, the program calculated 498 nulls. This procedure was repeated for a large number of different pulse sequences. All sequences with good signal-to-noise gave a threshold of equal or very close to 3.8σ. This threshold value was eventually adopted in our investigation.

The final steps of the analysis was to calculate all the nulls within about ten thousands single pulses, for the 24 pulsars and to clear out the sequences of weak pulses, which were due to the overall decreasing of the signal strength due to interstellar scintillation. So, we finally worked with 19 pulsars with excellent data and we came to the following preliminary conclusions concerning pulsar nulling.

RESULTS

In Table 2 we give the percentage of nulling for the individual components (C1, C2, C3, C4, and C5) of each of the 19 pulsars and the percentage of nulling for their total profile.

TABLE 2. Nulling properties of 19 pulsars

PSR	Logτ (yr)	Nulls%	C1%	C2%	C3%	C4%	C5%
0329+54	6,74	0,004995	0,26	0,06	0,0225	0,08	0,0675
0525+21	6,17	0,515135	0,596	0,592	--	--	--
0628-28	6,44	0,177686	0,24	0,425	--	--	--
0823+26	6,69	0,307692	--	--	--	--	--
0834+06	6,47	0,559322	0,56	0,81	--	--	--
0950+08	7,24	0,414141	--	--	--	--	--
1133+16	6,70	0,121517	0,17	0,247	--	--	--
1237+25	7,36	0,077589	0,283	0,213	0,49	0,217	0,163
1642-03	6,54	0,013619	--	--	--	--	--
1706-16	6,21	0,126087	--	--	--	--	--
1822-09	5,37	0,117647	0,29	0,18	--	--	--
1933+16	5,98	0	0	0	--	--	--
1944+17	8,46	0,658591	0,66	0,745	--	--	--
1952+29	9,62	0,916865	0,915	0,98	0,99	0,98	--
2016+28	7,77	0,003730	--	--	--	--	--
2020+28	6,46	0,288571	0,68	0,45	0,61	0,59	--

In Figure 5 a graph of the nulling percentage versus the pulsar age is shown, where it is obvious that there is a correlation between these two pulsar parameters. In Figure 6 the longest duration of nulling in each of the 19 pulsars versus their age is shown. It is obvious that there is quite a good correlation again.

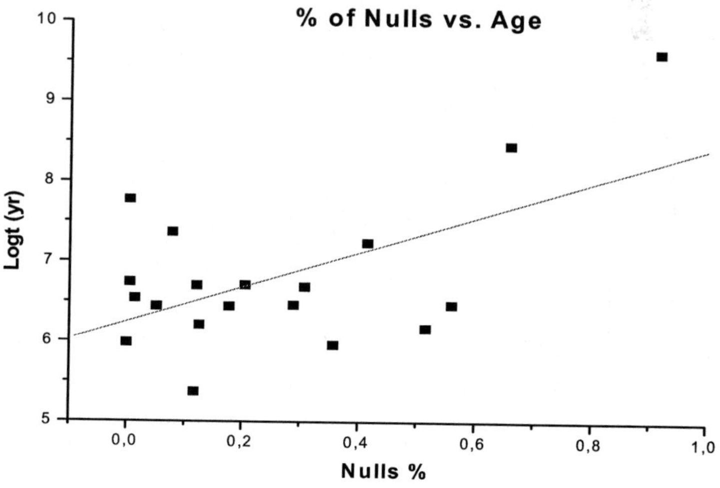

FIGURE 5. Graph for 19 pulsars that shows the correlation between pulsar age and pulse nulling..

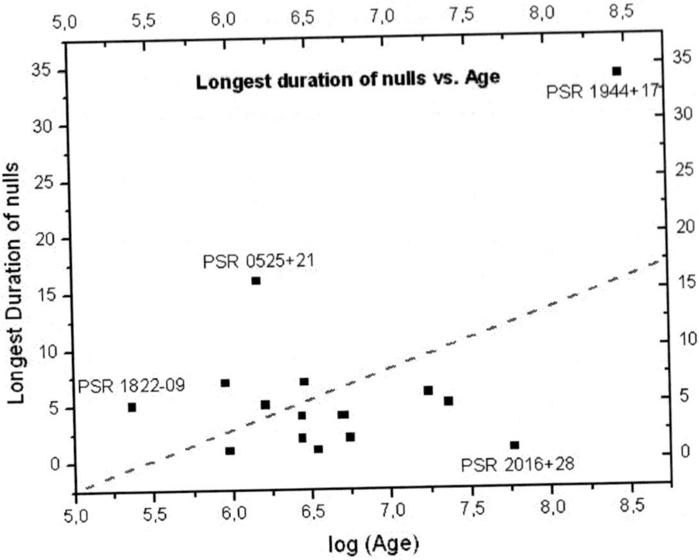

FIGURE 6. Graph for 19 pulsars that shows the correlation between pulsar age and the longest sequence of nulling

Finally we used our data in order to examine if there is a correlation between the ratio of the height of the components and the ratio of the nulls in these components. We tested this idea with some double profile pulsars as it is shown in Figure 7 and we concluded that the ratio of the heights is inversely proportional to the ratio of the nulls. Therefore single pulses have more or less equal flux density in all components, but weak components have more nulls.

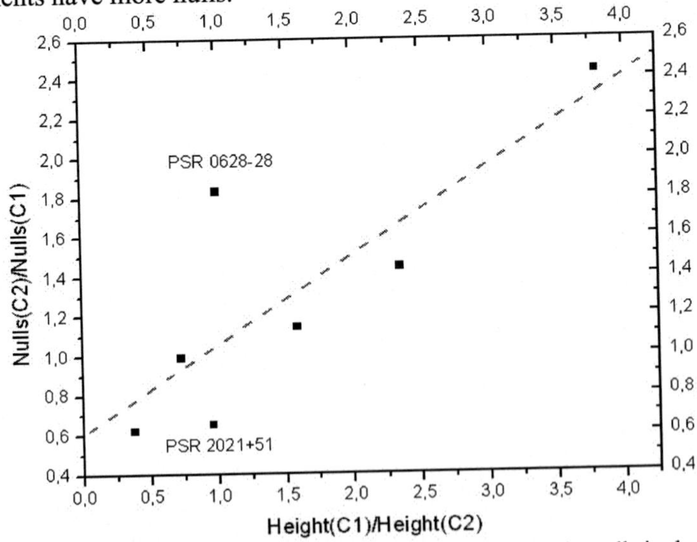

FIGURE 7. The ratio of the height of the components versus the ratio of the nulls in the components for some double profiled pulsars.

REFERENCES

1. Backer D.C., 1970. Nature, 228, 42
2. Biggs J.D., 1992. The Astrophysical Journal, 394, 574.
3. Filippenko V. & Radhakrishnan V., 1982. The Astrophysical Journal, 263, 828.
4. Kazbegi A., Machabeli G., Melikidze G. & Shukre C., 1996. Astron. Astrophys., 309, 515.
5. van Leeuwen A.G.J., B.W. Stappers, R. Ramachandran & Rankin J.M., 2003. A&A, 399, 223.
6. Rankin J.M., 1986. The Astrophysical Journal, 301, 901.
7. Rankin J.M., 1993. The Astrophysical Journal, 405, 285.
8. Redman S.L., Wright G.A.E. & Rankin J.M., 2005. Mon. Not. R. astr. Soc., 357, 859.
9. Ritchings R.T., 1976. Mon. Not. R. astr. Soc., 176, 249.
10. Seiradakis J.H., Gil J.A., Graham D.A., Jessner A., Kramer M., Malofeev V.M., Sieber W., Wielebinski R., 1995.Astron. Astrophys. Suppl. Ser., 111, 205.

A two-zone model for the emission from RX J1713.7-3946

K. Moraitis* and A. Mastichiadis*

**Department of Physics, University of Athens, Panepistimiopolis, 15783 Zografos, Athens*

Abstract. We study the acceleration and radiation of charged particles in the shock waves of supernova remnants using a recent version of the so-called "box model". Particles are accelerated in an energy–dependent region around the shock and lose energy through radiation. At the same time a fraction of the particles escapes behind the shock where it also radiates. We calculate the distribution function of both populations and obtain the photon spectra they produce. The application of the model to the supernova remnant RX J1713.7-3946, which was recently confirmed as a TeV source by H.E.S.S., shows that the X-ray emission can be attributed to electron synchrotron radiation and the γ-ray emission to inverse Compton scattering from the same electrons. The possible contribution from a hadronic component is not considered.

Keywords: Supernova remnants, Particle acceleration, Radiation mechanisms, Cosmic rays
PACS: 96.50.Pw, 98.38.Mz, 98.35.Eg, 98.70.Sa, 95.55.Ka, 95.85.Pw

INTRODUCTION

Supernova remnants (SNRs) have long thought to be the cites of galactic cosmic ray acceleration, at least up to energies close to that of the "knee" ($\sim 10^{15}$ eV) of the cosmic ray spectrum [1, 2, 3]. Direct evidence for this came only during the last decade with the observation of non-thermal X-ray emission from SNRs. This can be best explained as synchrotron radiation by electrons that have been accelerated to energies up to $\sim$ 100 TeV [4]. Despite this success, evidence for the presence of a relativistic hadronic component is less certain. Accelerated hadrons in SNRs are expected to emit high energy gamma rays from the decay of neutral pions (π^0) produced in proton-proton collisions [5]. These are best observed in the TeV regime [6], thus SNRs have been, during the last decade, one of the main targets for the ground based Čerenkov detectors [for a review see 7]. However, the above picture is complicated from the fact that the same high energy electrons which produce the synchrotron X-rays will Compton up-scatter the microwave photons to TeV energies [8, 9], contributing an extra component in this regime. Therefore one expects that any TeV detection of a SNR cannot be unequivocally attributed to the presence of a relativistic hadronic (or leptonic) component. Instead careful modeling is required to determine the origin of these gamma-rays.

In modeling the multi-wavelength spectrum of SNRs most authors assume for the particle energy distribution (electrons or protons) the form of a power-law with an exponential cut-off. The power-law is a natural consequence of Fermi acceleration while the exponential turnover is supposed to arise from energy losses or SNR age limitations. Despite the theoretical support for these assumptions, this approach is still

CP848, *Recent Advances in Astronomy and Astrophysics,*
7th *International Conference of the Hellenic Astronomical Society,* edited by N. H. Solomos

phenomenological. On the other hand, the use of extensive numerical codes which treat the acceleration of the particles has proved a valuable tool for modeling the data, especially as they allow more complex phenomena, such as shock modification, to be taken into account [10, 11]. Still it would have been interesting to apply an analytical model to the data as this offers a good opportunity to get an insight at some of the basic principles of particle acceleration.

A development in the direction of an analytical approach to the problem of particle acceleration has been made recently by Drury et al. [12]. This approach has expanded on the concept of the so-called "box" model for particle acceleration [13] by assuming an energy-dependent size for the acceleration region. This inserts some modifications on the particle spectra and is thus interesting to apply this model to fit the SNR multi-wavelength emission. A good starting point for the application of the aforementioned model is the first SNR with confirmed TeV emission, RX J1713.7-3946 [14].

RX J1713.7-3946 (G347.3-0.5) is the remnant of a type II supernova that was discovered during the ROSAT All Sky Survey [15] and is one of the brightest galactic X-ray sources. It is a young SNR with an age of $\sim$ 2000 years and lies at a distance $\sim$ 1 kpc. Its X-ray morphology, as revealed by Chandra, is quite complicated consisting of bright filaments and dark voids [16] indicating spatially inhomogeneous magnetic field and/or density of relativistic particles. There are also indications that a fraction of the remnant is interacting with a dense molecular cloud of hydrogen number density $\sim 300\ \mathrm{cm}^{-3}$ [17], complicating even further the picture. Finally, as we have already mentioned RX J1713.7-3946 emits also in the γ-ray regime.

In this paper we apply the Drury et al. [12] model to the supernova remnant RX J1713.7-3946 and draw some conclusions about its properties. In section 2 we mention the main characteristics of the model presented by Drury et al. [12] and obtain the distribution function of accelerated particles and also of particles that have escaped behind the shock. In section 3 we obtain the multi-wavelength photon spectrum from these distributions and apply it to RX J1713.7-3946. Finally, in section 4 we summarize and discuss our results.

THE ACCELERATION MODEL

Charged particles interacting with the shock wave of a supernova remnant can be accelerated to high energies through first order Fermi acceleration. Alfvén waves in the background medium scatter the charged particles, changing their direction but not their energy. Particles are thus confined to a region around the shock and are forced to cross it many times. The difference in the gas velocity upstream and downstream results in the statistical acceleration of particles after many shock crossings. While confined around the shock particles also lose energy because they emit radiation. At the same time a small fraction of particles escapes behind the shock where it also radiates. We will calculate the distribution function of both populations beginning with the ones at the shock. We denote quantities in the acceleration region with the subscript 1 and in the escape region with the subscript 2.

Although the shock in a SNR is spherical we can treat the acceleration in the plane shock geometry as long as the size of the acceleration region is much smaller than the

SNR radius. For the magnetic field we assume that it is uniform and normal to the shock. We also adopt the test particle approximation, i.e. we assume zero particle pressure. With these assumptions acceleration takes place in a region around the shock of size $L = L_u + L_d = K_u/U_u + K_d/U_d$ where $K_{u(d)}$, $U_{u(d)}$ are the upstream (downstream) diffusion coefficient and gas velocity respectively. Assuming that the diffusion coefficient is continuous across the shock, i.e. $K_u = K_d = K$, the size of the acceleration region can be further written $L = K(r+1)/U_u$ where $r = U_u/U_d$ is the compression ratio of the shock. Following Drury et al. [12] we let the diffusion coefficient to be momentum-dependent and adopt a general power-law dependence $K(p) = \kappa p^\delta$ with $\delta \geq 0$. This choice makes the mathematical analysis easier but also covers some physically interesting cases. For example, the case $\delta = 0$ corresponds to energy-independent diffusion, $\delta = 1$ to Bohm diffusion and $\delta = 1/3$ to diffusion from turbulent Alfvén waves with a Kolmogorov spectrum. The same power-law dependence holds for the size of the acceleration region $L(p) = \kappa(r+1)p^\delta/U_u$ as well as the acceleration timescale which is given by the relation

$$t_{\mathrm{acc}}(p) = \frac{3L(p)}{U_u - U_d}. \tag{1}$$

Particles also lose energy through synchrotron radiation and/or inverse Compton scattering. In the case where the latter is in the Thomson limit, we can write for the particle's total losses $\dot{p} = -\alpha_1 p^2$, where

$$\alpha_1 = \frac{4}{3}\frac{\sigma_T(U_{B_1} + U_{ph})}{m^2c^2} \tag{2}$$

is essentially the sum of the energy densities of the magnetic field and the target photon field, σ_T is the Thomson cross section, m the particle mass and c the speed of light. If there were no energy losses, particles would escape from the acceleration region with the downstream flow at the bulk velocity U_d. The presence of losses however generates a downward flux in momentum space and when combined with the increasing size of the acceleration region gives an additional escape process. The velocity of escape thus increases and the escape timescale is given by

$$t_{\mathrm{esc}}(p) = \frac{L(p)}{U_d + \alpha_1 p^2 \frac{dL_d}{dp}}. \tag{3}$$

The second term in the denominator is the addition of the Drury et al. [12] model to the standard box model and vanishes for constant diffusion coefficient.

The evolution of the spatially averaged isotropic distribution function of particles at the shock front, $N_1(p,t)$, is governed by the continuity equation

$$\frac{\partial N_1}{\partial t} + \frac{\partial}{\partial p}\left[N_1\left(\frac{p}{t_{\mathrm{acc}}(p)} - \alpha_1 p^2\right)\right] + \frac{N_1}{t_{\mathrm{esc}}(p)} = Q(p,t). \tag{4}$$

The source term is $Q(p,t) = Q_o\delta(p - p_o)\vartheta(t)$ representing constant injection of particles at low momenta ($\delta(x)$ and $\vartheta(x)$ are the Dirac delta and Heavyside functions respectively). From the equation describing the evolution of a particle's momentum

$$\frac{dp}{dt} = \frac{p}{t_{\mathrm{acc}}(p)} - \alpha_1 p^2 \tag{5}$$

it follows that particles are accelerated up to the momentum where the acceleration and cooling timescales are equal. This is true as long as SNR age limitations are not important and in that case the maximum momentum is

$$p_{\max} = \left[\frac{(r-1)U_{\rm u}^2}{3r(r+1)\alpha_1\kappa}\right]^{\frac{1}{\delta+1}}. \quad (6)$$

The solution of Eq. (4) for the distribution function is then

$$N_1(p,t) = \frac{Q_{\rm o}p_{\rm o}^{\delta-1}}{\alpha_1 p_{\max}^{\delta+1}}\left(\frac{p}{p_{\rm o}}\right)^{-s_1}\left[1-\left(\frac{p}{p_{\max}}\right)^{\delta+1}\right]^{-b}\left[1-\left(\frac{p_{\rm o}}{p_{\max}}\right)^{\delta+1}\right]^{b-1} \quad (7)$$

in the momentum range $p_{\rm o} < p < p_1(t)$ and zero otherwise. The accelerated particles form a power-law of index $s_1 = \frac{r+2}{r-1} - \delta$, smaller by δ from the index predicted by the standard box model. The power-law extends from the injection momentum $p_{\rm o}$ up to a maximum momentum $p_1(t)$ given at each instant by the solution of the characteristic equation, Eq. (5). This can be written in the implicit form

$$\alpha_1 p_{\max} t = \frac{x^\delta}{\delta}\,{}_2F_1\left(\frac{\delta}{\delta+1},1;1+\frac{\delta}{\delta+1};x^{\delta+1}\right)\Bigg|_{p_{\rm o}/p_{\max}}^{p_1(t)/p_{\max}}, \quad (8)$$

where ${}_2F_1$ is the hypergeometric function. This maximum momentum increases with time until energy losses become important and the acceleration saturates at $p_{\max}$ as can be seen in Fig. 1. The timescale that determines this is the minimum cooling timescale $(\alpha_1 p_{\max})^{-1}$. Thus eventually a steady-state is achieved given by Eq. (7) for $p_{\rm o} < p < p_{\max}$. In steady-state a pile-up or a cut-off of particles forms depending on whether the index $b = \frac{1}{\delta+1}\left(\frac{r-4}{r-1} + \frac{\delta}{r+1}\right)$ is positive or negative respectively [12].

Particles that escape towards the interior of the remnant play an important role in the emission [18, 19] and we calculate next their energy distribution. The distribution of escaping particles obeys a similar equation to Eq. (4) where now the source term is the escape rate of accelerated particles while there are no acceleration or escape terms

$$\frac{\partial N_2}{\partial t} - \frac{\partial}{\partial p}\left(\alpha_2 p^2 N_2\right) = \frac{N_1(p,t)}{t_{\rm esc}(p)}. \quad (9)$$

The energy losses have the same form as in the acceleration zone (see Eq. 2) but depend on the magnetic field of the escape zone. The solution of this equation can be easily obtained numerically but we prefer a simplified analytic approach. If the energy losses are negligible the solution of Eq. (9) is simply the escape rate integrated over time

$$\begin{aligned} N_2(p,t) &= \frac{Q_{\rm o}[t-t_1(p)]}{p_{\rm o}}\left(\frac{p}{p_{\rm o}}\right)^{-s_2}\left[1-\left(\frac{p}{p_{\max}}\right)^{\delta+1}\right]^{-b} \\ &\left[1-\left(\frac{p_{\rm o}}{p_{\max}}\right)^{\delta+1}\right]^{b-1}\left[\frac{3}{r-1}+\delta\frac{r}{r+1}\left(\frac{p}{p_{\max}}\right)^{\delta+1}\right], \end{aligned} \quad (10)$$

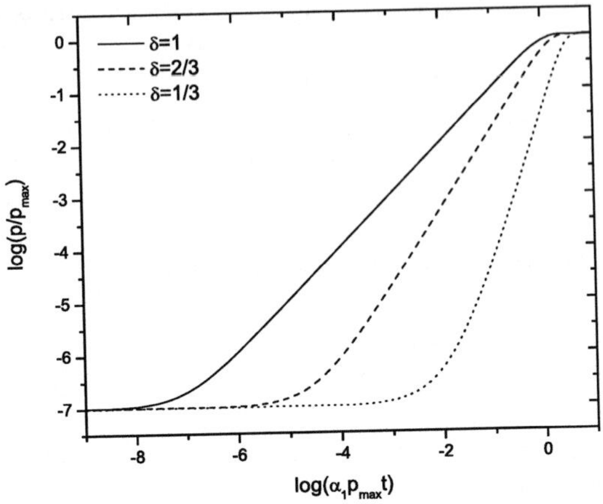

FIGURE 1. The evolution of a particle's momentum for various values of δ. The initial momentum is chosen $p_o = 10^{-7} p_{max}$.

where $s_2 = s_1 + \delta$ and $t_1(p)$ is the inverse function of $p_1(t)$ giving the time a particle needs to accelerate to momentum p. This result remains valid when losses are present but only for those particles that had not enough time to cool, i.e. those with momentum less than

$$p_{cool}(t) = \frac{p_1(t)}{1 + a_2 p_1(t) t}. \tag{11}$$

This equation describes the evolution of momentum of the highest energy particles as they cool in the escape zone. On the other hand, particles with momentum exceeding $p_{cool}(t)$ had the necessary time to cool and their distribution is steeper from the injected one by one power of p

$$N_{2,cool}(p,t) \simeq N_2(p,t) \frac{p_{cool}(t)}{p}. \tag{12}$$

The escaping particles thus have a broken power-law distribution, given by Eq. (10) for $p < p_{cool}(t)$ and by Eq. (12) for $p_{cool}(t) < p < p_1(t)$. The break occurs at the cooling momentum and depends on the magnetic field of the escape region. A low value of it will make the cooling unimportant and the distribution will then be the simple power-law, Eq. (10). In steady-state both distributions have the same maximum momentum p_{max} as this is defined by the energy losses in the acceleration region. We finally note that the escaping particles have steeper spectra from the accelerated ones, $s_2 = s_1 + \delta$.

APPLICATION TO THE SUPERNOVA REMNANT RX J1713.7-3946

In this section we apply the model developed so far to the case of the supernova remnant RX J1713.7-3946. For this we assume that the particles responsible for the emission are electrons, that is we neglect the possible contribution from protons. As we saw in the previous section there exist two populations of electrons, the ones in the acceleration zone and the ones that have escaped behind the shock. These electrons will produce synchrotron radiation in the magnetic field of the two regions B_1 and B_2 respectively. In the calculations we use the angle averaged synchrotron emissivity and assume that emission comes from a fraction f_B of the remnant's volume, that is we adopt the so called magnetic filling factor [e.g. 20]. Electrons will also Compton up-scatter the photons of the cosmic microwave and infrared background radiation fields which are the same in both zones. For the microwave photon field we take the standard values of temperature $T_{MW} = 2.7$ K and energy density $U_{MW} = 0.25$ eV cm^{-3} and for the infrared field $T_{IR} = 25$ K and $U_{IR} = 0.22$ eV cm^{-3} and use the full inverse Compton emissivity. Finally, the remnant may produce bremsstrahlung radiation in the region where it interacts with the cloud of enhanced hydrogen density $n \sim 300$ cm^{-3}, depending on the extent of interaction.

The photon spectrum derives by folding the two electron distribution functions with the single particle emissivity for the radiative processes considered. It depends on the same parameters that determine the electron distribution function in the acceleration and escape zones plus the factor f_B introduced above. These parameters are the compression ratio of the shock r, the index of the diffusion coefficient δ, the magnetic fields B_1 and B_2, the maximum electron energy $E_{e,max} = p_{max}c$ and the normalization constant Q_o. The injection momentum of electrons does not influence the spectrum significantly and so we take the arbitrary value $p_o = 4$ MeV/c. We should note that the calculation of the photon spectrum is for the present day $t = t_{SN} = 2000$ yr and that we do not assume that it is in steady-state since this depends on the unknown value of the magnetic field. The free parameters will be determined from the fit of our model to RX J1713.7-3946's spectrum.

The multi-wavelength spectrum of RX J1713.7-3946 is shown in Fig. 2. Radio data from the northwest part of the remnant taken with ATCA [20] are shown here multiplied by a factor of 2 to cover the whole remnant [21]. Likewise, the X-ray data of ASCA [22] are multiplied by a factor of ~ 4 so that the total X-ray flux from the remnant is obtained [23]. Finally, the TeV data are the recent observations of H.E.S.S. [21] of the entire remnant. The EGRET data of the nearby source 3EG J1714-3857 are also plotted, since as emphasized by Reimer and Pohl [24] they set an upper limit to the GeV emission of RX J1713.7-3946.

The best fit of the model to the multi-wavelength spectrum of RX J1713.7-3946 is shown in Fig. 2 together with the observational data. The fit is quite satisfactory and has a reduced chi square $\chi^2 = 1.5$. From the derived values of $r = 3.75$ and $\delta = 0.5$ we can calculate the two electron power-law indices $s_1 = 1.6$ and $s_2 = 2.1$. We note that the value of r implies a strong shock. The magnetic field turns out to have the same value in both regions $B_1 = B_2 = 2.7\ 10^{-5}$ G while the filling factor is $f_B = 0.15$.

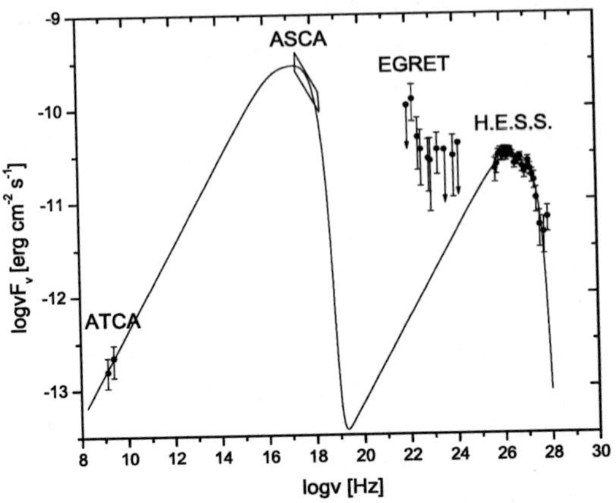

FIGURE 2. The multi-wavelength spectrum of RX J1713.7-3946 together with the best-fit leptonic model with parameters given in text.

This value indicates that synchrotron emitting electrons occupy $\sim 1/6$ of the volume of inverse Compton emitting electrons and is an order of magnitude larger from other estimates. The value of the magnetic field and the maximum electron energy $E_{e,max} =$ 50 TeV indicate that the energy losses in the escape zone are important since the cooling timescale is $t_{cool} = 300$ yr $< t_{SN}$. The escaping electrons thus have the broken power-law distribution given by Eqs. (10) and (12). The normalization constant $Q_o = 7\ 10^{40}$ s^{-1} determines the total energy content in electrons from both zones $\mathcal{E}_e = 9\ 10^{47}$ erg. The majority of this energy is carried by electrons in the escape region. For the typical value of supernova explosion energy $E_{SN} = 10^{51}$ erg, we find that a fraction $\sim 10^{-3}$ of it is converted to relativistic electrons. We also note that the size of the acceleration region for the maximum energy electrons is $L(p_{max}) = 0.35$ pc. This value is close to the size of the bright filaments of the remnant [16] and could imply that the filaments are the cites of particle acceleration. Bremsstrahlung emission for these parameters is not important unless the cloud is interacting with approximately half of the remnant, case not favored from the observations.

SUMMARY

We have shown that the multi-wavelength spectrum of the supernova remnant RX J1713.7-3946 can be explained with the acceleration mechanism proposed by Drury et al. [12]. In our approach we have considered the radiation from particles inside the acceleration region and also that of the particles which have escaped in the downstream region. We treat the energy losses self-consistently in both regions and calculate the resulting electron distribution functions. We find that the escaping particles produce most of the observed radiation, both in X-rays and in γ-rays. X-ray emission is attributed to synchrotron radiation, while γ-ray emission to inverse Compton scattering from the same electrons. In the γ-ray regime there could be a contribution from relativistic protons as well, but we have not considered this possibility.

ACKNOWLEDGMENTS

This research was supported in part by the "PYTHAGORAS" research funding program Grant No. 70/3/7396.

REFERENCES

1. G. F. Krymskii, *Dokl. Akad. Nauk SSSR [Sov. Phys. Dokl.]* **234**, 1306–1308 (1977).
2. A. R. Bell, *Mon. Not. R. Astron. Soc.* **182**, 147–156 (1978).
3. R. D. Blandford, and J. P. Ostriker, *Astrophys. J. Lett.* **221**, L29–L32 (1978).
4. S. P. Reynolds, and J. W. Keohane, *Astrophys. J.* **525**, 368–374 (1999).
5. E. A. Dorfi, *Astron. Astrophys.* **251**, 597–610 (1991).
6. L. O. Drury, F. A. Aharonian, and H. J. Voelk, *Astron. Astrophys.* **287**, 959–971 (1994).
7. T. C. Weekes et al., *Astroparticle Physics* **17**, 221–243 (2002).
8. A. Mastichiadis, *Astron. Astrophys.* **305**, L53 (1996).
9. M. Pohl, *Astron. Astrophys.* **307**, L57–L59 (1996).
10. D. C. Ellison, P. Slane, and B. M. Gaensler, *Astrophys. J.* **563**, 191–201 (2001).
11. E. G. Berezhko, L. T. Ksenofontov, and H. J. Völk, *Astron. Astrophys.* **395**, 943–953 (2002).
12. L. O. Drury, P. Duffy, D. Eichler, and A. Mastichiadis, *Astron. Astrophys.* **347**, 370–374 (1999).
13. T. J. Bogdan, and H. J. Volk, *Astron. Astrophys.* **122**, 129–136 (1983).
14. F. A. Aharonian et al., *Nature* **432**, 75–77 (2004).
15. E. Pfeffermann, and B. Aschenbach, "ROSAT Observation of a New Supernova Remnant in the Constellation Scorpius," in *Roentgenstrahlung from the Universe*, 1996, pp. 267–268.
16. Y. Uchiyama, F. A. Aharonian, and T. Takahashi, *Astron. Astrophys.* **400**, 567–574 (2003).
17. G. Cassam-Chenaï, A. Decourchelle, J. Ballet, U. Hwang, J. P. Hughes, R. Petre, and et al., *Astron. Astrophys.* **414**, 545–558 (2004).
18. L. Ball, and J. G. Kirk, *Astrophys. J. Lett.* **396**, L39–L42 (1992).
19. F. A. Aharonian, and A. M. Atoyan, *Astron. Astrophys.* **351**, 330–340 (1999).
20. J. S. Lazendic, P. O. Slane, B. M. Gaensler, S. P. Reynolds, P. P. Plucinsky, and J. P. Hughes, *Astrophys. J.* **602**, 271–285 (2004).
21. The HESS Collaboration: F. Aharonian, *astro-ph/0511678* (2005).
22. R. Enomoto et al., *Nature* **416**, 823–826 (2002).
23. H. Tomida, *Synchrotron Emission from the Shell-like Supernova Remnants and the Cosmic-Ray Origin*, Ph.D. thesis, Kyoto University (1999).
24. O. Reimer, and M. Pohl, *Astron. Astrophys.* **390**, L43–L46 (2002).

Further investigation of the high frequency optical oscillations during the flare phase of the red dwarf EV Lac

Contadakis M.E.*, Avgoloupis S.J. † and Seiradakis J.H . †

*Department of Surveying and Geodesy, Aristotle University of Thessaloniki, GR-54124, Thessaloniki Greece.
†Session of Astronomy, Astrophysics and Mechanics, Department of Physics, Scool of Sciences, Aristotle University of Thessaloniki, GR-54124, Thessaloniki Greece.

Abstract. In this paper we present the results of the analysis of the U-light curve for a flare of magnitude 1.2, which was observed on September 2004, with the help of the 30-inch Cassegrain telescope of the Stephanion Observatory. The results of both, DFT-analysis and wavelet analysis, indicate that (a) Oscillations with period 1 to 1.5min, 11s, 7.5s and 5.5s appear during the pre-flare state and persist during the whole flare state, (b) From the flare maximum phase on, a progressive increase of oscillations with periods 20s up to 4.0s is markedly indicated and (c) At the end of the flare only the oscillation of the pre-flare state do remain. This result is in favour of (or does not contradict) the suggested explanation, i.e. the evolution of a fast mode magneto-acoustic wave generated at the impulsive phase of the flare and traveling through the magnetic loop.

Keywords: Flare stars-DFT analysis- wavelet analysis
PACS: 97.30.Nr

INTRODUCTION

Apart of their spectacular brightness variation while flaring, it is well documented that flare stars exhibit a number of periodical and semi-periodical transient brightness variations. Long period brightness variations of EV Lac with period 5 years has been reported by Mavridis and Avgoloupis [7] and small period small amplitude brightness variations with period 4.378 days has been reported by Contadakis [2]. Similar behavior with periods 7.5 years and 3.5 days has been repeatedly reported for BY Drac by Mavridis [9]. Both types of variation were explained on the base of star's flare activity and spottedness in analogy of the solar flare activity (Alexeev and Gershberg [1]). High frequency optical oscillations during the quiescence phase of EV Lac with periods of 1 to 2 min, has been reported by Mavridis and Varvoglis [8]. High frequency, very small amplitude optical oscillations in dips with periods 1,95 and 4.25 min [11] 25 minutes before a flare on FF And, and periods ranging between 4.38 and 18.37 min before the pre-flare dip on V 1045 Oph., has been reported by Ventura et al.

CP848, *Recent Advances in Astronomy and Astrophysics,*
7th International Conference of the Hellenic Astronomical Society, edited by N. H. Solomos

[14]. The authors suggest that such brightness variations are common phenomena in active stars not connected with the flare phenomenon.

High-frequency small amplitude optical oscillations during a flare maximum of HII 2411 with a period of 14.4 sec has been reported by Rodono [13] while high-frequency optical oscillations of amplitude 10% in U and 2.5% in B and period 14.4 seconds on a flare of EV Lac around the flare maximum phase [17]. A multitude of high-frequency oscillations with amplitudes ranging between 1.4% and 2.6% and periods ranging between 6.9 sec and 20 sec around maximum phase and the subsequent stages of the flare evolution of a flare on EV Lac have been reported by Contadakis et al. [3]. The resolution of the data was 1.2 sec. It appears that the frequencies of the oscillations become higher and the amplitude decreases as the flare evolves. In addition Contadakis et al. [4] analyzing a flare of 1992 on EV Lac with a data resolution of 12 sec discover high-frequency oscillations on a flare of EV Lac with periods ranging between 30 sec and 125 sec. Finally, Zhilyaev et al. [18] reported high-frequency oscillations just after the flare maximum with periods ranging between 4 sec and 6 sec. These results are consistent with the phenomenology of the evolution of a fast mode magneto-acoustic wave generated at the impulsive phase of the flare and traveling through the loop. Nevertheless a better understanding of the high-frequency oscillations demand a unified analysis of the flare light-curve for a wider time window covering pre-flare, flare and post flare and a broader band of frequencies.

In this paper we present the results of Discrete Fourier Transform and Wavelet analysis of the light-curve in the U-Band of Johnson International System for a flare of EV Lac recorded with of 0.108 sec. The flare was observed on U.T $22^h\ 52^m\ 37^s.6$ of September 12, 2004 (Fig.1).

THE OBSERVATIONS

The observations were curried out with the help of the 30-inch Cassegrain tele-

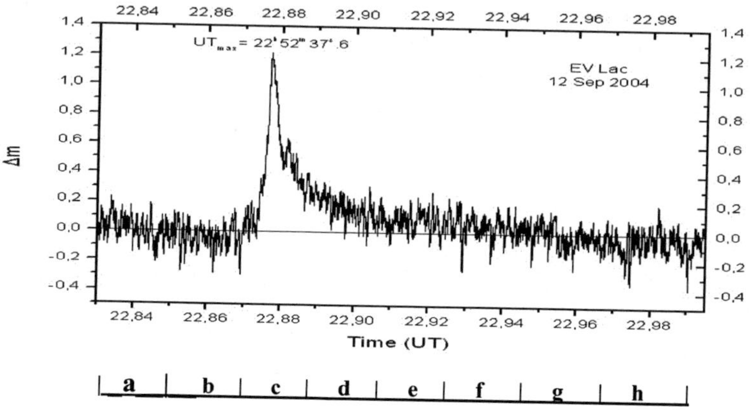

FIGURE 1. The light-curve of the flare in the U-band. The segments **a , b, c, d, e , f , g , h** are the parts of the light-curve for which DFT analysis were performed

scope of the Stephanion Observatory equipped with a Johnson photometer. The digitized recording system has a recording resolution of 0.108s. This facility allows the reliable analysis of our data for a wide range of frequencies (4.63Hz to $4.63x10^{-3}$ Hz i.e. periods 0.216s to 216s) and permits the reliable spectral analysis of small part of the light-curve. The accuracy of our photometric system is estimated from the standard deviation of the quite state star deflection a few minutes before the flare and is σ_m=0.05 in stellar magnitudes or σ_I=0.047 in stellar normalized brightness. However it appears that small semi-periodical fluctuations are present in the quiet state stellar deflection. In order to realize if these fluctuations are random or not we recall the notion of the fractal Brownian Walk. The quite star deflection is the sum of three components (a) the noisy stellar brightness (b) the inherent atmospheric noise which we assume to be random and (c) the white noise of the recording system. If all three deflection components were random, then the deflection should be a random Brownian Walk and should have fractal exponent of its Power Spectrum β=2.0 i.e. S~ $k^{-2.00}$, where k is the wave number. The Power Spectrum of the quiet state stellar deflection is $S \sim k^{0.75}$ i.e. strongly anti-persistent. This means that at least one of the components is not random. Having in mind the results of Mavridis and Varvoglis [8], Peres et al [11] and Ventura et al. [14] we reasonably assume that the star brightness induces the non random part. Next we note that for wave numbers k>0.05 the Power Spectrum is S~ $k^{-2.14}$ i.e. approaches that of the random Brownian Walk. So, we filter out frequencies lower than 0.050 and we get the random noise of the normalized deflection from which we determine $\sigma_{I(BW)}$=0.014. This is the estimation of the accuracy of our photometric system.

DATA ANALYSIS

The data consist of a sample of relative intensities i.e. $(I_f-I_o)/I_o$, where I_f is the flare intensity and I_o is the quiet star intensity. The extent as well as the resolution of our sample permit a reliable Discrete Fourier Transform analysis of sufficient light-curve parts for a better determination of possible transient oscillation in the frequency domain and Wavelet analysis of the whole light-curve for a better determination of the possible transient oscillation in the time domain.

Discrete Fourier Transform analysis

The sampling interval of the flare light-curve is B= 0.108 sec. This means that the Nyquist frequency is f_N=1/[2(0.108s)]= 4.626 Hz well above of the frequencies of our interest which is lower than 0.37 Hz. We separate the light-curve in eight samples of extent N=700 i.e. samples of length T= 700X0.108s=75.6s. The noise to signal ratio is $\sigma_S/S = (BT)^{-1/2}$ =0.0378 . For the estimation of the accuracy of a frequency determination for a real brightness variation two questions must be answered: (a) how well the spectral density frequency is been determined? and (b) does the observed spectral density correspond to a real brightness variation?

To this purpose we assume that the measured spectral density S, which is a sum of products of values of our sample x_i (i=1...n) (i.e. relative intensity residuals) with Normal distribution N(0, s), can be approximated by a random variable with χ_n^2 distribution of n statistical degrees of freedom [10]

$$S = \chi_n^2 = x_1^2 + x_2^2 + \ldots + x_n^2 \quad .$$

The mean value and the variance of the spectral density are $m_s = n\,s^2$ and $s_s^2 = 2\,ns^4$.

To answer the question (a) we consider that an approximate level of confidence of an estimation of a spectral density can be achieved by assuming the spectral density distribution for a particular frequency. Then using the above mentioned noise to signal ratio we find that the probability the observed spectral density S to be very close to the true mean value of the spectral density m is very high i.e.

$$P(0.8216 < S/m < 1.1959) = 1 - \alpha = 99.98\ \%$$

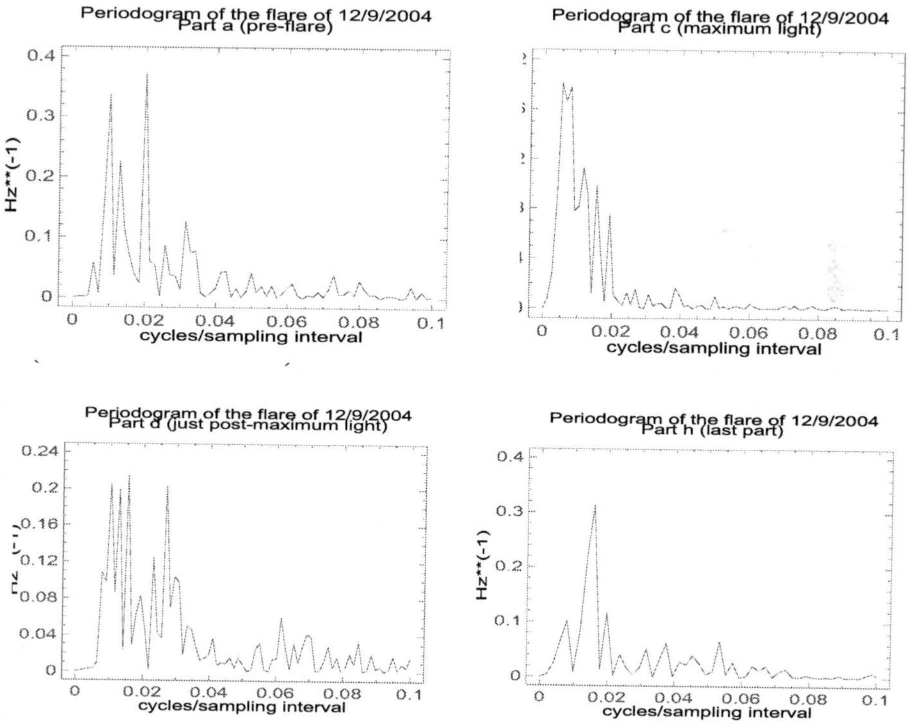

Figure 2. Periodograms for the pre-flare (**a** part), maximum (**c** part), post-maximum (**d** part) and end (**h** part) of the light-curve

Next we come to answer the question (b) i.e. to estimate the confidence level for a determination of a real brightness variation.

We set the hypothesis H_0 according to which the estimated spectral density corresponds to the random noise if $s^2 = s_q^2$ against the hypothesis H_1, that the spectral density corresponds to anintensity oscillation if $s^2 > s_q^2$ (s_q^2 is the variance of the stellar intensity in the quiet state). It is well known that the random variable $(n-1)\, s^2 / s_q^2$ follow the chi-square distribution (see for instance Fisz [5]). Then the hypothesis H_1 holds to a significant level α if $(n-1)\, s^2 / s_q^2 > \chi_\alpha^2$. Finally the estimated confidence

TABLE 1. Power Spectral Densities for different phases of flare evolution

Freq	Freq	Period		Spect.	Dens.	(Hz^{-1})				
(c/lag)	(Hz)	(sec)	**a**	b	**c**	d	**e**	**f**	**g**	**h**
0.04	0.37	2.7								
0.037	0.34	2.92		0.11		0.10				
0.033	0.3306	3.27						0.12		
0.030	0.278	3.58	0.12			0.205				
0.028	0.259	3.86					0.13			0.125
0.027	0.25	4.00		0.15		0.205				
0.026	0.247	4.15					0.22		0.122	
0.025	0.231	4.32						0.14		
0.024	0.222	4.5					0.095			
0.023	0.213	4.7		0.16		0.125			0.14	
0.021	0.194	5.14					0.210			
0.020	0.185	5.4	0.362							0.106
0.019	0.175	5.6				0.0805			0.140	
0.018	0.166	6.0		0.15			0.205			
0.017	0.157	6.35		0.14	0.09			0.14	0.14	
0.016	0.148	6.75								
0.015	0.139	7.2				0.215		0.320	0.088	0.315
0.014	0.129	7.7	0.221	0.418			0.315	0.18		
0.013	0.120	8.3			0.11	0.200				
0.012	0.111	9.0							0.16	0.145
0.011	0.102	9.8					0.18			
0.010	0.093	10.8	0.335			0.205				
0.009	0.083	12.0		0.275	0.17		0.26			
0.008	0.074	13.5				0.105				0.10
0.0075	0.069	14.4					0.10	0.10		
0.007	0.065	15.43								
0.006	0.056	18.0					0.102	0.110		
0.005	0.046	21.6			0.18			0.112		
0.004	0.037	27.6								
0.0038	0.035	28.0		0.22			0.27	0.52	0.69	
0.003	0.028	36.0								
0.002	0.019	54.0								0.400
0.0017	0.016	63.5								
0.0015	0.014	72.0	0.68	0.60	60	0.60	0.24	0.33	0.38	
Noise/	Signal					0.038				

level of a spectral density is the product of the above calculated confidence level with the confidence level of the basic hypothesis which is 66%.

In the frame of this test we find that Spectral Densities of magnitude $S > 0.15$ correspond to real brightness variation to a confidence level $P(1-\alpha) > 70\%$ while Spectral Densities of magnitude $0.1 < S < 0.15$ correspond to real brightness variation to a confidence level $44\% < P(1-\alpha) < 70\%$.

Figure 2 displays a sample of periodograms covering the whole flare periods and Table 1 displays in detail the presence of transient high frequency small amplitude oscillations through the whole flare. Transient oscillations with periods ranging from 72.0s and probably higher than 4s appear at the pre-flare period (**a** segment) increase in number (**b** segment) until the maximum (**c** segment) where the small period oscillations probably are been masked by the flare brightness. The post-maximum period is crowded with transient oscillations (**d, e, f, g, h** segments) of which the number and intensity decreases as we approach the end of the flare. It appears that the dumping of the oscillations with larger period is faster.

Wavelet analysis

Wavelet analysis will provide us a better time resolution for the time resolution. We use the Biorthogonal 6.8 for a Discrete as well as a Continuous Wavelet analysis [5] because of its resemblance with the assumed oscillations, it is symmetric, performs exact reconstruction and works on both Discrete and Continuous wavelet analysis. Figure 3 displays the characteristics of this wavelet.

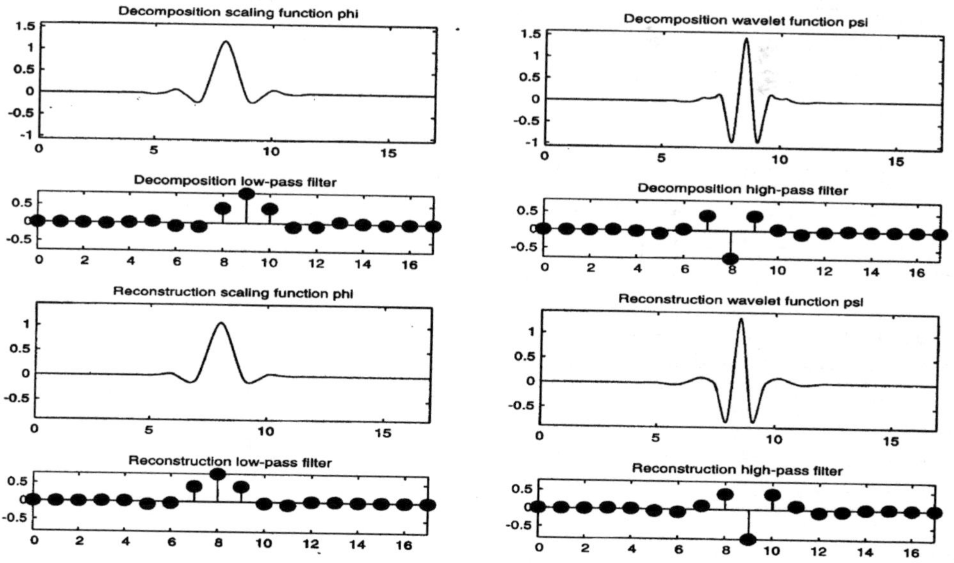

FIGURE 3. Biorthogonal 6.8 Wavelet.

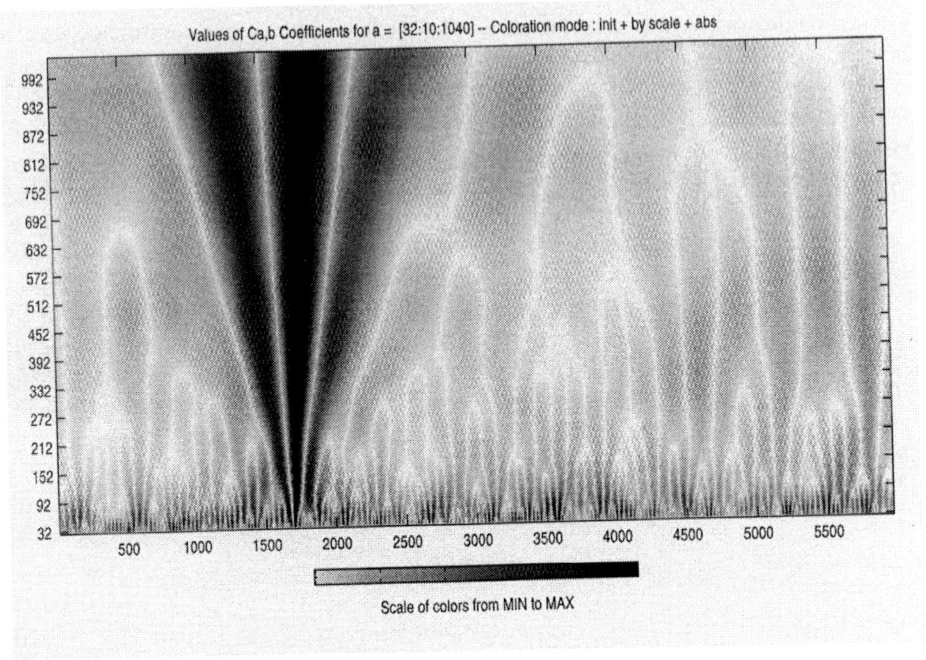

FIGURE 4. The colorated correlogram. The scales of both axis are multiples of the sampling interval 0.108sec

Figure 4 displays the colorate diagram of the correlation coefficients. This diagram depicts the way the various transient oscillations vary with time and it is in perfect agreement with the results of DFT analysis, revealing the detailed time variations of the transient oscillations. Figure 5 displays an example of the transient variation with period around 7sec as it was reconstructed by Discrete Wavelet analysis.

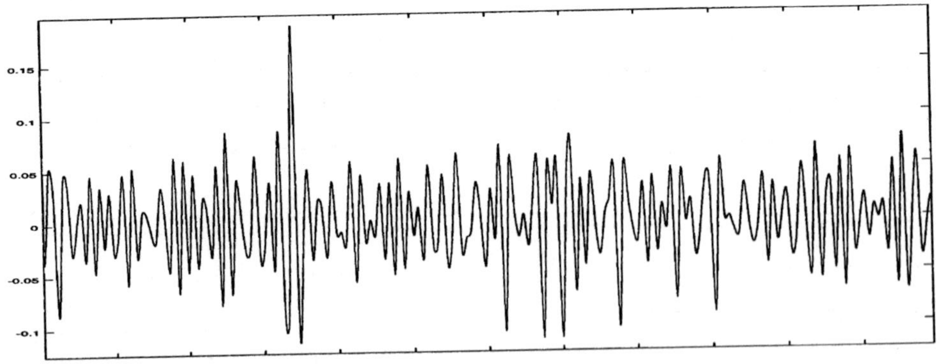

FIGURE 5. Variation of the transient oscillation with a period around 7 sec. The ordinate are the same with Figure 4

It is seen that the maximum oscillation amplitude corresponds to the time periods **b, e, f** and **h** in accordance with the results of DFT analysis.

CONCLUDING REMARKS

Both Discrete Fourier analysis and Wavelet analysis indicate that a multitude of high-frequency optical oscillations manifest their presence during the whole period of the flare phenomenon from the pre-flare state down to the end of the flare.There are evidences that high frequency low amplitude oscillations appear also during the quiescent state of the star in accordance with those reported by Mavridis and Varvoglis [8], Peres et al. [11] and Ventura et al. [14], but this phenomenon merits a better attention and will be the subject of a special paper.

As far as the flare event concerns, the frequencies of the transient oscillations range from 0.34Hz down to 0.014 Hz (i.e. periods from 3.5 sec to 72 sec) and even more. Their number increases from the pre-flare state to after the maximum and then decreases. Their intensities increase from the pre-flare state to after maximum and then decrease. As the flare developed the number of the higher frequencies constituent increases.

This phenomenology is consistent with the successive generation of fast mode magneto-acoustic waves generated at the base of the magnetic loop of the flare and traveling through the loop. The one, which appear at the impulsive phase, is the stronger. Evidences appear that similar activity, to a lesser degree, is also present during the quiet state of the star. According to Ventura et al. [14] this activity may be common phenomena in active stars not connected with the flare event.

Roberts et al. [12] describe the evolution of an impulsively generated fast wave as follows: (1) Initial periodic phase, beginning at a time $\approx 1/u_{Ae}$ after the initial impulse (Ae = external Alven velocity) and characterized by an increase in amplitude and frequency. (2) A quasi-periodic phase, showing an increase in amplitude at a time $\approx 1/u_A$. (3) A periodic decay phase, with a rapid decrease in amplitude (see also Williams et al. [15], [16] and references therein). It is quiet possible that observing a flare we observe the development of a succession of such waves.

REFERENCES

1. Alexeev, I.Y., Gershberg, R.E., "The activity of the red dwarf star EV Lac from UBVRI observations in Crimea in 1986-1995", in *The Earth and the Univers,* edited by G. Asteriadis, A. Bandelas, M.E. Contadakis et al. , Ziti Editions, Thessaloniki, Greece, 1997, pp. 43-57.
2. Contadakis, M.E., 1995, *A&A* 306, p. 819.
3. Contadakis, M. E., Avgoloupis, S. J., Seiradakis J. H., et al., 2004, *Astron. Nachr.*, 325, No. 5, pp. 428-432.
4. Contadakis, M.E., Avgoloupis, S. J., Seiradakis J. H., "High-frequency optical oscillation during the flare phase of the red dwarf EV Lac" in *The Small and the Large Scale of the Universe,* Proceedings of JENAM 2004, Granada, Spain, in preparation.
5. Daubechies, I, *Ten lectures on Wavelets*, SIAM, 1992.

6. Fisz, M., *Wahrscheinlichkeitsrechnung und mathematische Statistik,* Deutscher Verlag der Wissenschaften, Berlin, 1978.
7. Mavridis, L.N., Avgoloupis, S.J., 1986, *A&A* ,154/1, pp. 171-175.
8. Mavridis, L.N. and Varvoglis, P.P., "Very-short term photometric variability of the Flare star EV Lac" in *Proceedings of the First Panhellenic Astronomical Meeting*, edited by P. Lascaridis, National Astronomical Committee, Athens, Greece, 1993, pp. 203-208.
9. Mavridis, L.N., Asteriadis, G., Mahmoud, F.H., in *Compendium in Astronomy,* edited by Mariolo-poulos, E.G., Theocharis, B.S., Mavridis, L.N., Reidel Dordrecht, 1982 , p. 253.
10. Newland, D.E., *An introduction to random vibrations and spectral analysis*, 2nd ed., Longman, London, 1984.
11. Peres, G., Ventura, R., Pagano, I., Rodono, M., 1993, *A &A* ,278, p. 179.
12. Roberts, B., Edwin, P.M., Benz, A.O., 1984, *ApJ* , 279, p. 857.
13. Rodono, M., 1974, *A&A,* 32, 337.
14. Ventura, R., Peres, G., Pagano, I., Rodono, M.: 1995, *A & A* 303, 509
15. Williams, D.R., Phillips, K.J.H., Rudway, P., et al., 2001, *Mon. No. Astron. Soc.*, 326, pp. 428-436.
14. Williams, D.R., Mathioudakis, M., Gallager, P.T., et al., 2002, *Mon. No. Astron. Soc.*, 336, pp. 747-752.
17. Zhilyaev, B.E., Romanyuk, Ya.O., Verlyuk, I.A., et al. , 2000, *A & A*, 364, pp. 641-645.
18. Zhilyaev, B.E., Romanyuk, Ya.O., Verlyuk, I.A., et al. , 2005, *A & A*, in press

Kinematics of the $v=1$, $J=1\rightarrow 0$ SiO masers at 43 GHz around TX Cam - New 73 epoch movie

I. Gonidakis*, P. J. Diamond† and A. J. Kemball**

*Department of Physics Astronomy & Mechanics, Faculty of Physics, University of Athens, GR-15783, Greece
†Jodrell Bank Observatory, University of Manchester, Macclesfield, Cheshire, SK11 9DL, U.K.
**National Radio Astronomy Observatory, P. O. Box 0, Socorro, NM 87801, USA.

Abstract.

A new version of the movie of the SiO masers around the Mira variable TX Cam is presented. Observations were carried out using the Very Long Baseline Array (VLBA) in biweekly and later in weekly intervals, between the 24th of May 1997 and the 21st of May 2000 with data covering almost two complete stellar cycles. The result is a movie consisting of 73 epochs which is used to study the global properties of the emission in the extended atmosphere of the star, as well as the kinematics of individual components. The existence of shock waves and their contribution to the morphology and the kinematics of the gas is also examined.

Keywords: circumstellar matter – masers: SiO – shock waves – stars: individual (TX Camelopardalis) – stars: Miras.
PACS: 01.30.Cc, 95.75.Kk, 95.85.Bh, 97.10.Fy, 97.30.Jm

INTRODUCTION

Miras are evolved red giants with regular periodicity and fluctuations in the V-band $\Delta V \geq 2.5^m$, consisting of a degenerated core of O and C, covered by a helium shell and a hydrogen mantle. Their typical radius is more than 2 AU and their temperature about 2000 K, changing around 30% due to their pulsation. They have a high rate of expelling matter (10^{-7} to 10^{-4} $M_\odot$/yr) and the relatively low temperatures allow the formation of molecules in their circumstellar envelope.

The idea of the existence of shocks in the stellar atmospheres of Miras was firstly proposed in 1959 [10]. In order to explain the high mass loss rates, both shock waves and radiation pressure have to be taken into account. A shock is formed each time the star is in its expanding phase, levitating the gas into cooler areas where it condenses into dust. Radiation pressure then accelerates the dust resulting to a slow, cool wind [4].

Circumstellar SiO masers can be powerful tools in the study of circumstellar envelopes properties. The first excited level has an excitation temperature of 1800 K so, masers are located in a narrow area between the star's surface and the dust condensation region. Thus, SiO masers can serve as excellent tracers of the properties and dynamics of this area.

TX Cam is a cool Mira variable (M8-M10) with a period of 557.4 days [23]. It is located in the constellation of Camelopardalis (the Giraffe) in the northern hemisphere and its coordinates are RA: 05h 00m 51.1860s and DEC: 56° 10' 54.341" (J2000). Its distance can be estimated from the period-luminosity relationship, giving 317 pc [28]; it

CP848, *Recent Advances in Astronomy and Astrophysics,*
7th International Conference of the Hellenic Astronomical Society, edited by N. H. Solomos

still remains uncertain, as other studies have given values of 400 [27] and 1200 pc [5]. The mass loss rate for this star was estimated $\sim 1.1 \times 10^{-6}$ $M_{\odot}yr^{-1}$ [24] and its radius was calculated to bc between 1.3 [6] and 3.1 A.U [29].

The first report of SiO maser emission from this star came in 1977 [33]. Although many other chemical molecules are found in its atmosphere, silicon monoxide is the only one that exhibits maser emission. Transitions from several rotational levels for different vibrational states have been detected [19], as well as from different SiO isotopes [2]. The detection of several C- and S-molecular species indicates that TX Cam might be changing from an O- to a C-rich star [8].

SiO masers in TX Cam have previously found to be located in a well defined ring between 2 and 4 $R_{\star}$, in the extended atmosphere of the star. The amplification appeared to be tangential and the ring was dominated by a globally ordered outflow [12]. Recent results, based on a previous 44 epoch movie, revealed ballistic deceleration of the ring, governed mostly by outflowing material. The shape of the ring for most epochs was far from being symmetric, with a well defined inner border becoming more complicated in its outer part, with both diffuse emission and fine structure of filaments and arcs [11].

OBSERVATIONS AND DATA REDUCTION

The transition observed in this project was the v=1, J=1→0 at 43 GHz using the VLBA. The target source was observed from the 24th of May 1997 to the 21st of May 2000. Initially the interval between observations was almost biweekly, from September 9, 1999 approximately every month covering phases from 0.68 to 2.64, almost two complete cycles of the stellar pulsation. In order to ensure homogeneity in the data, both observations and data reduction were performed in the exact same way. Thus, for each epoch, data were recorded at a rest frequency of 43.11499 GHz in dual polarisation, in 128 channels over a 4 MHz bandwidth and centred at 9.0 km/sec. Typical system temperatures were 95 K for right and 105 K for left polarisation and sensitivities around 10.5 Jy/K. For each set, approximately 2.5 hours of data were obtained for TX Cam itself. Over the 6 hour scan, the rest of the time was dedicated in sources 3C 454.3, 0359+509 and 0609-157, which were used for residual group delay, delay, bandpass and polarisation calibration. The data were correlated at the VLBA correlator in Socorro giving all 4 Stokes parameters.

A pipeline based on the spectral-line polarimetry reduction technique ([22] and [21]) was developed by P.J. Diamond and A.J. Kemball as a POPS script using the AIPS package in order to ensure reduction consistency. The final images produced were 1028 × 1028 pixels, with pixel spacing of 0.1 mas and a restoring beam of 540×420 mas and position angle of 20 degrees. The images were concatenated into an animated movie, to provide a visual aid in monitoring the changes of the ring structure.

The main difficulty, raised by the lack of phase referencing, was the alignment of the subsequent frames, since the information concerning the absolute positions are lost. A cross correlation technique was applied, using task XYCOR in AIPS. The task searches for the brightest spots in the image and calculates their shift in both dimensions; in the occasions when the task failed, this correction was also estimated by eye.

In cases of epochs with no or bad data, where reduction was not possible, the frames

TABLE 1. List of epochs, observation dates and the corresponding phase of the star. $\surd$ in the status column indicates successfully reduced epochs, $\times$ interpolated epochs due to bad data and $\between$ interpolated epochs due to unscheduled observations. Missing epochs were interpolated in order to have a smooth, time coherent movie. The time when observations switched to monthly is indicated in the table.

Epoch	Date	Optical phase	
Observations every 2 weeks			
BD46A	24/05/97	0.68 ± 0.01	$\surd$
BD46B	07/06/97	0.70 ± 0.01	$\surd$
BD46C	22/06/97	0.73 ± 0.01	$\surd$
BD46D	06/07/97	0.75 ± 0.01	$\surd$
BD46E	19/07/97	0.78 ± 0.01	$\surd$
BD46F	02/08/97	0.80 ± 0.01	$\surd$
BD46G	16/08/97	0.83 ± 0.01	$\surd$
BD46H	28/08/97	0.85 ± 0.01	$\surd$
BD46I	12/09/97	0.87 ± 0.01	$\surd$
BD46J	26/09/97	0.90 ± 0.01	$\surd$
BD46K	09/10/97	0.92 ± 0.01	$\surd$
BD46L	25/10/97	0.95 ± 0.01	$\times$
BD46M	08/11/97	0.98 ± 0.01	$\surd$
BD46N	21/11/97	1.00 ± 0.01	$\surd$
BD46O	05/12/97	1.03 ± 0.01	$\surd$
BD46P	17/12/97	1.05 ± 0.01	$\surd$
BD46Q	30/12/97	1.07 ± 0.01	$\times$
BD46R	13/01/98	1.10 ± 0.01	$\surd$
BD46S	25/01/98	1.12 ± 0.01	$\between$
BD46T	05/02/98	1.14 ± 0.01	$\surd$
BD46U	22/02/98	1.17 ± 0.01	$\surd$
BD46V	05/03/98	1.19 ± 0.01	$\surd$
BD46W	21/03/98	1.22 ± 0.01	$\between$
BD46WX	04/04/98	1.24 ± 0.01	$\between$
BD46X	19/04/98	1.27 ± 0.01	$\surd$
BD46Z	10/05/98	1.30 ± 0.01	$\times$
BD46AA	22/05/98	1.33 ± 0.01	$\surd$
BD46AB	06/06/98	1.35 ± 0.01	$\surd$
BD46AC	18/06/98	1.37 ± 0.01	$\surd$
BD46AD	03/07/98	1.40 ± 0.01	$\surd$
BD46AE	23/07/98	1.44 ± 0.01	$\surd$
BD46AF	09/08/98	1.47 ± 0.01	$\surd$
BD46AG	23/08/98	1.49 ± 0.01	$\surd$
BD46AH	09/09/98	1.52 ± 0.01	$\surd$
BD46AI	25/09/98	1.55 ± 0.01	$\surd$
BD46AJ	14/10/98	1.59 ± 0.01	$\surd$
BD46AK	29/10/98	1.61 ± 0.01	$\surd$
BD46AL	17/10/98	1.65 ± 0.01	$\surd$
BD46AM	06/12/98	1.68 ± 0.01	$\surd$
BD46AN	23/12/98	1.71 ± 0.01	$\surd$
BD46AO	05/01/99	1.74 ± 0.01	$\times$
BD46AP	23/01/99	1.77 ± 0.01	$\surd$
BD46AQ	06/02/99	1.79 ± 0.01	$\surd$
BD46AR	19/02/99	1.82 ± 0.01	$\between$
BD57A	12/03/99	1.85 ± 0.01	$\surd$
BD57B	29/03/99	1.88 ± 0.01	$\times$
BD57C	12/04/99	1.91 ± 0.01	$\surd$
BD57D	24/04/99	1.93 ± 0.01	$\surd$
BD57E	14/05/99	1.97 ± 0.01	$\surd$
BD57F	31/05/99	2.00 ± 0.01	$\surd$
BD57G	12/06/99	2.02 ± 0.01	$\surd$
BD57H	27/06/99	2.05 ± 0.01	$\surd$
BD57I	12/07/99	2.07 ± 0.01	$\surd$
BD57J	30/07/99	2.11 ± 0.01	$\surd$
BD57K	11/08/99	2.13 ± 0.01	$\surd$
BD57L	24/08/99	2.15 ± 0.01	$\surd$
Observations every month			
BD62A	09/09/99	2.18 ± 0.01	$\surd$
BD62AB	27/09/99	2.21 ± 0.01	$\between$
BD62B	15/10/99	2.24 ± 0.01	$\times$
BD62BC	31/10/99	2.27 ± 0.01	$\between$
BD62C	14/11/99	2.30 ± 0.01	$\surd$
BD62CD	02/12/99	2.33 ± 0.01	$\between$
BD62D	19/12/99	2.36 ± 0.01	$\surd$
BD62DE	01/01/00	2.38 ± 0.01	$\between$
BD62E	15/01/00	2.41 ± 0.01	$\surd$
BD62EF	31/01/00	2.44 ± 0.01	$\between$
BD62F	14/02/00	2.46 ± 0.01	$\surd$
BD62FG	02/03/00	2.49 ± 0.01	$\between$
BD62G	17/03/00	2.52 ± 0.01	$\surd$
BD62GH	04/04/00	2.55 ± 0.01	$\between$
BD62H	21/04/00	2.58 ± 0.01	$\surd$
BD62HI	05/05/00	2.61 ± 0.01	$\between$
BD62I	21/05/00	2.64 ± 0.01	$\surd$

were interpolated as the average of the adjacent epochs, to ensure a regular interval between each frame. To enhance the visibility of each frame, the $\sqrt{I}$ was taken to minimise the amplitude range in the movie and a cutoff of $\sqrt{I} \lesssim 0.7$ was introduced to eliminate the noise. Then the images were converted into animated files with commercial software.

Table 1 includes all the epochs consisting the movie. Their code name and date of observation are stated and the optical phase is computed using the optical maximum at MJD [14] and uncertainty $\Delta\phi$ ~0.01. Epochs are marked to show whether they were reduced successfully.

ANALYSIS

The Movie

The movie starts at phase ϕ~0.68, near the stellar minimum. The projected ring appears to be almost circular, with a defined inner shell dominated by bright and compact maser spots and filaments. The movie can be seen at http://www.jb.man.ac.uk/research/masers/index.html. Diffuse emission is mainly observed in the outer parts of the ring. The structure reveals that the mechanism of amplification is tangential, since no emission is evident inside the ring. The overall motion of the spots seems to be that of expansion, only some features appear to have different kinematic behaviour. The overall circular shape is disrupted in the western part where bright emission is almost totally absent. During the first observed cycle, expansion appears to be dominant in the ring with emission reaching its maximum at ϕ~1.07 but the western part remains faded. The initial circular ring becomes somehow disrupted, getting an ellipsoidal shape with its major axis inclined in the NE-SW direction.

Despite the well defined inner shell of the ring, the outer parts seem to be more complicated, dominated by diffuse emission and peculiar kinematics. A distinctive part where emission disobeys the overall ordered pattern is the formation observed at the SE quadrant. This formation is eminent for almost a whole stellar cycle and its existence has been previously attributed to localised mass loss ([12] and [11]). Contrary to the dominant expansion, some features appear to infall. In this first cycle, there are two key features revealing such a behaviour. The first one, located in the northern part of the ring, is an individual filamentary maser spot that remains constant until ϕ~1.10 but then it starts to fall back toward the centre of the star and fades out at ϕ~1.65. The other, at the eastern part, is a group of very bright components that start to infall at phase ϕ~1.00. Around the end of the first cycle, these feature seem to bounce back, moving outward again, a behaviour that could be explained by interaction of the group with a shock wave. This peculiar motion results to a cavity in the ring and its transformation to a double ellipsoid, an 8-like structure.

In the second cycle observed (ϕ~1.65 to ϕ~2.64) the ring has the same properties (well defined inner shell, fragmented structure, diffuse emission in the outer parts etc.); the main difference is that contraction is also observed during this cycle. The ring starts as a double ellipsoid that becomes an ellipse, expanding until ϕ~2.05. For the first time the whole perimeter of the emission is obvious, even in the western part that was previously vacant. The contraction that follows is dominated by fewer but more compact and bright spots. By the end of the movie, the ring has changed its shape once again, resembling to an ellipse with its axis to the SE-NW direction.

Effects deviating from the overall motion are apparent in this cycle too. The group of masers that bounced in the previous cycle, are moving outwards reducing the cavity and giving the ring an elliptical shape again. At that time ($\phi \sim 1.96$) they start following tangential trajectories, moving along the perimeter of the ring from east to south before they fade at $\phi \sim 2.41$. At $\phi \sim 1.96$ the quadrant of emission from east to north splits up, creating a double arc; one of these is expanding, the other is contracting with the rest of the ring.

Variability

Frequent observations as well as the time span of the project give us the opportunity to study both the short and long-term variability, to observe how the maser intensity changes not only within a single cycle, but also from one cycle to the other and its correlation with the optical changes. The light curve provided by the AAVSO, was compared to the integrated maser flux calculated from the Stokes I data cubes using ISPEC in AIPS and the changes in maser intensity with respect to the phase, were used as tools to study variability.

Inner Shell

A program was developed in AIPS++ by A. Kemball to give a more robust estimation of the shell's radius, ignoring the structure of the ring. For this, the program divides the image into sectors (the number of them can be given by the user) and then estimates the maximum in the gradient of the radial intensity function

$$S(r,\theta) = \sum_{\Delta_{(r,\theta)}} I(\eta,\zeta) r^{-1} d\Delta_{(r,\theta)} \tag{1}$$

where $I(\eta,\zeta)$ is the flux distribution and $\Delta_{(r,\theta)}$ represents the triangular sector in which summation takes place, with the apex of the triangle at the centre of the image and its base being tangential to the part of the ring that it includes. The shell radius for each epoch is calculated by the mean of all values. The program generates an image of all fitted triangles for each epoch (Figure 1) and provides a list of values for the radius with respect to the stellar phase. In Figure 2 the inner radius is plotted with respect to the stellar phase. The co-efficient of the fitted polynomials were used to calculate the values of the gravitational acceleration or deceleration and the velocity of the ring.

An estimation of the change of the width of the ring with the stellar phase was made with AIPS task IRING. The total intensity maps were first modified to exclude the noise with task BLANK. Then with IRING we fitted circles with a common centre. The task averages the intensity in each ring and then plots it as a function of radius. The beginning and end of the intensity distribution curve were considered to be the borders of the ring giving an estimate of the ring width for every epoch.

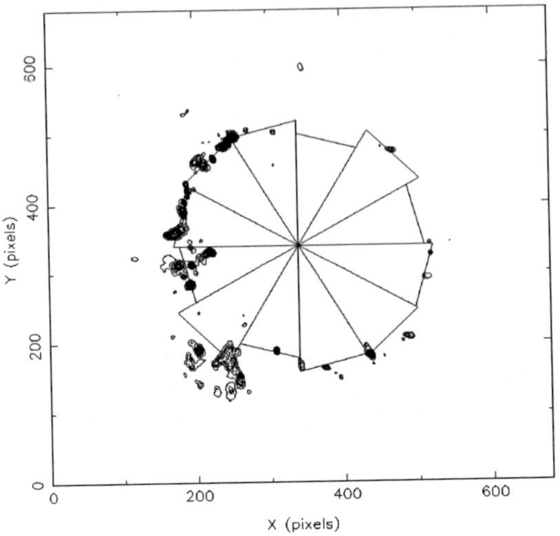

FIGURE 1. Plot generated by AIPS++ for the estimation of the mean shell radius.

Individual Components

Identification of individual maser components is essential in order to study the evolution of the ring and determine the kinematics in the area. The vast amount of data had to be handled in a simple and efficient way so an existing routine developed by P. J. Diamond was used, based on AIPS task SAD. For the component subtraction, a non constant Gaussian was used and features with flux more than 8 times the 1σ off source noise level were accepted. Candidates were adopted as components, if found in more than three consecutive channels within a Stokes I cube. The error estimation for R.A. and Dec. was based on the analysis by Kogan [25] and Condon [9] and the flux on the equation $\sigma_I = |I_{neg}|/\gamma$ where γ is 4.9021 for a 1024×1024 image [3].

Once all components for all epochs were identified, the evolution of individuals had to be determined. The trajectories for around 100 components were determined, using an interactive program written in IDL. The starting points were selected homogeneously in space and time to enable us to study different regions of the ring at different times. For each component the trajectory was fitted by a second degree polynomial to extract the values of velocity and acceleration.

CONCLUSIONS

The monitoring of the evolution of the SiO masers in the shell around TX Cam spans from $\phi \sim 0.68$ to $\phi \sim 2.64$, providing a movie of 73 epochs, covering two stellar pulsation cycles at millimetre spatial resolution. The properties of the v=*1*, $J-1 \rightarrow 0$ transition at 43 GHz are summarised in the rest of this paragraph.

- The masers are located in a ring-like structure at 2-4 $R_\star$ from the surface of the star

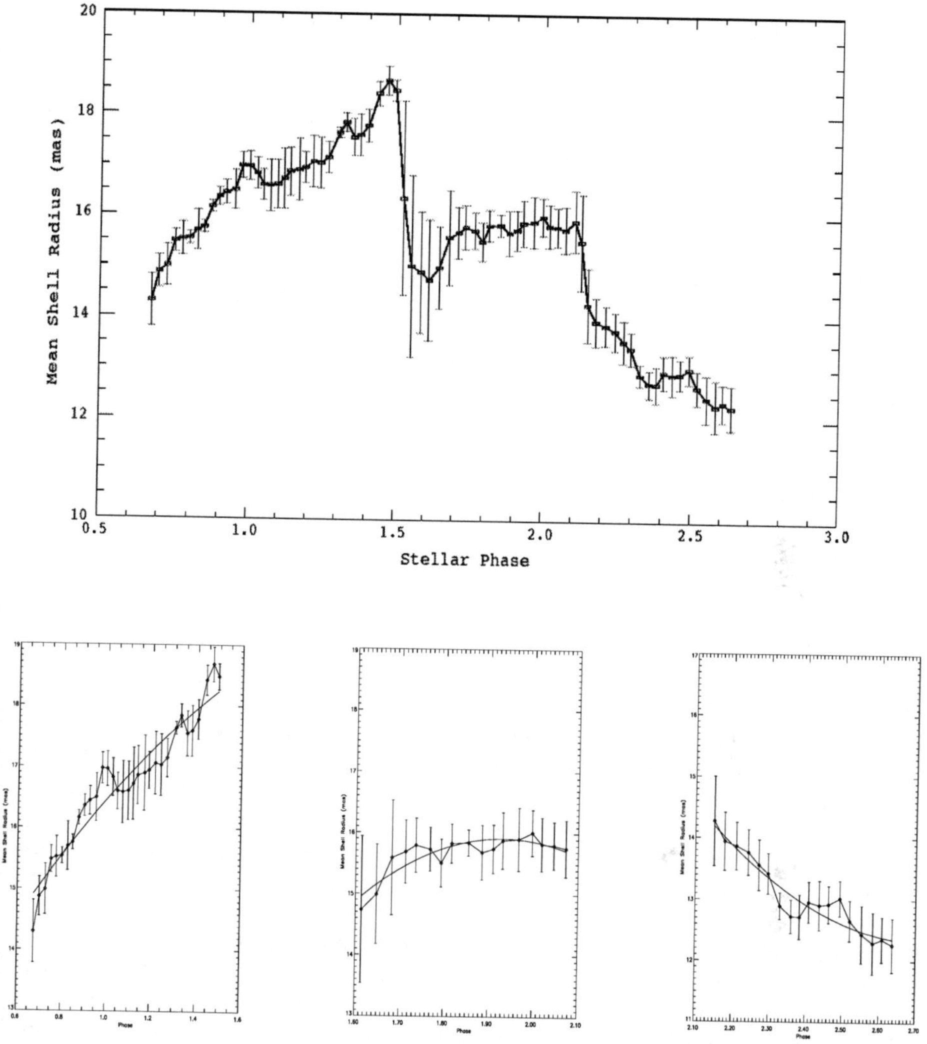

FIGURE 2. Graph of the mean shell radius with respect to the stellar phase.The three graphs below show the fitted parabolas

which can change from an almost perfect ring into a double ellipse. The ring is well defined at its inner part, having a fragmented filamentary structure, with well defined maser spots, but it is dominated by diffuse emission as the radius increases, making the detection of its outer boundary uncertain. The amplification is tangential, since no apparent emission is visible inside the ring.

• The overall structure is sometimes dominated by features that deviate from the dominant behaviour of ordered flow. Thus, there are regions of emission that are separated from the projected ring, an indicative behaviour of localised mass loss.

• The short and long term variability are in accordance with previous studies. The expected lag ([18], [7]) was observed during both cycles, being 12.5 % for the first and 10 % for the second. The variability of the masers follows the light curve of the optical (as already suggested [15]), but their intensities are not correlated [26]. The contrast is not constant, for the first cycle was estimated to be 3 and in the second 8, a property observed in other Miras [1].

• The number of spectral features, as well as their intensity, changes not only from one cycle to the other but also during each cycle. Maxima are dominated by two or three lines with long lifetimes, shifted with respect to the stellar velocity so that one of them is always blue and the other red shifted. The lines form at different times in the stellar cycle and have different lifetimes (in accordance with other studies [7]). The first to appear is the blue shifted at $\phi \sim 0.7$ and seems to live longer than the red shifted which forms at $\phi \sim 0.9$. The maximum of the blue shifted line is at $\phi \sim 0.1$ and the red at $\phi \sim 0.5$ The blue and red shifted lines in the spectra showed a dependence between their shift with respect to the LSR and their intensity.

• The ring undergoes expansion during the first cycle. A new ring forms at $\phi \sim 1.5$ and in the second cycle both expansion and contraction are observed in accordance to theoretical models. The masers seem to follow ballistical motions with gravitational deceleration $g_{SiO} \sim 1.51 \times 10^{-7}$ km/sec^2 for the first cycle and 1.53×10^{-7} km/sec^2 for the second, in accordance to what expected by theory. The expansion velocities for the inner part of the ring are 28.96 and 18.96 km/sec respectively. Given a radius for TX Cam 3.1 AU [29], the distance from the surface where the ring forms can be estimated to be between 1.5 to 4 AU.

• The width of the ring seems to follow the pulsation mode, delayed by the calculated lag. It ranges from 8 to 12 mas, corresponding to 3.1 and 4.7 AU.

• There is no correlation between the velocities and position angle along the ring. Velocity maps reveal that there is no coherence along the shell; an overall velocity gradient was not identified thus, the maser shell is not rotating. The radial velocities though are not randomly distributed along the ring but they tend to group together in arcs.

• The filamentary parts of the ring demonstrate a velocity gradient along their axes; this is due to a selection effect also found and explained by Yi et al. [36]. As we move away from the plane of sky, the angle between the spokes and the plane of sky increases, increasing at the same time the coherent path through the spokes. This path does not increase indefinitely; for a critical angle θ_c the velocity difference along the path equals the internal velocity dispersion and the maximum velocity coherent path is obtained. At this angle the beaming is enhanced along the path and the spokes appear bright.

• The study of individual components showed that generally they follow the overall behaviour of the ring. During the second cycle components are moving inwards following the contraction of the ring; although this is the behaviour predicted by models ([35], [4], [17], [16]) and also observed in other objects [3], contraction is not apparent in the first cycle observed. Proposed shock wave velocities in the photosphere are of 5-10 km/sec and do not exceed 25 km/sec [32]. Assuming a mass for TX Cam 1 $M_\odot$ and the distance where masers form $2R_*$, with R_* between 1.3 and 3.1 AU, then the escape velocity from that point would be between 16.9 and 26.1 km/sec. Component velocities in the first cycle are larger than the escape velocity, which could explain the lack of

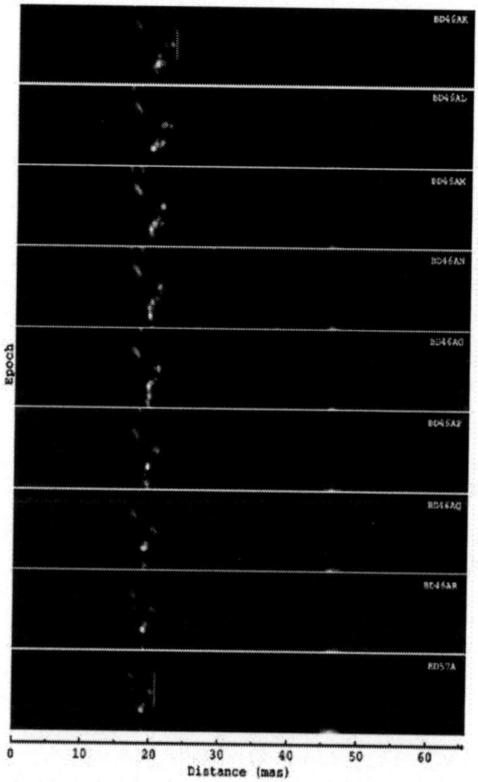

FIGURE 3. A collage of images from epochs BD46AK to BD57A showing the propagation of a shock front. At $\phi \sim 1.40$ a volume of gas is infalling toward the centre of the star. At $\phi \sim 1.49$ the infalling material is bouncing, changing direction and moving outward again, a behaviour that can be attributed to the encounter with a shock. The distance between the lines is 2.2 mas, indicating a velocity of 12.5 km/sec for a distance of 310 pc for TX Cam.

contraction since the shock can not reach the material that is moving faster to cause the expected contraction.

- Some components appear to be more complex kinematically. Not only they change their direction in the plane of the sky, but also along the line of sight, indicating a three dimensional kinematic behaviour.
- Component lifetimes showed that they follow a Gaussian distribution, with maximum value between 150 - 200 days. The lifetime distribution shows that components with very short (less than 50 days) or very long (more than 350 days) lifetimes are very unlikely to occur.
- Features that can be attributed to the interaction of the material with a shock wave were identified. These features enabled us to estimate the velocity of the shock, giving a

value of $U_{shock} \sim 12.5$ km/sec (Figure 3).

REFERENCES

1. Alcolea J., Pardo J. R., Bujarrabal V., Bachiller R., Barcia A., Colomer F., Gallego J. D., Gómez-González J., del Pino A., Planesas P., Rodríguez-Franco A., del Romero A., Tafalla M., de Vicente P., 1999, A&AS, 139, 461
2. Barcia A., Alcolea J., Bujarrabal V., 1989, A&A, 215L, 9
3. Boboltz D. A., Ph. D. Thesis, 1997, Virginia Polytechnic Institute
4. Bowen G. H., 1988, Ap.J, 329, 299
5. Bujarrabal V., Planesas P., del Romero A., 1987, A&A, 175, 164
6. Chan J. H., Elitzur M., 1979, ApJ, 231, 124
7. Cho S. H., Kaifu N., Ukita., 1996, AJ, 111, 1987
8. Cho S. H., Ukita N., 1995, PASJ, 47, L1
9. Condon J. J., 1997, Publications of the Astronomical Society of the Pacific, 109, 166
10. Deutsch A. J., Merrill P. W., 1959, ApJ,130, 570
11. Diamond P. J., Kemball A. J., 2003, ApJ, 599, 1372
12. Diamond P. J., Kemball A. J., Junor W., Zensus A., Benson J., Dhawan V., 1994, ApJ, 430, L61
13. Girard T., Willson L., 1987, A&A, 183, 247
14. Gray M. D., Humphreys E. M. L., Yates J. A., 1999, MNRAS, 304, 906
15. Hjalmarson Å., Olofsson H., 1979, ApJ, 234, L199
16. Humpreys E. M. L., Gray M. D., Yates J. A., Field D., Bowen G. and Diamond P. J., 2002, A&A, 386, 256
17. Humpreys E. M. L., Gray M. D., Yates J. A., Field D., Bowen G. and Diamond P. J., 1996, MNRAS, 282, 1359
18. Jewell P. R., Snyder L. E., Walmsley C. M., Wilson T. L., Gensheimer P. D., 1991, A&A, 242, 211
19. Jewell P. R., Dickinson D. F., Snyder L. E., Clemens D. P., 1987, ApJ, 323, 749
20. Josselin E., Mauron N., Planesas P., Bachiller R., 2000, A&A, 362, 255
21. Kemball A. J., Diamond P. J., 1997, ApJ, 481, L111-L114
22. Kemball A. J., Diamond P. J., Cotton, W. D., 1995, A&AS, 110, 383K
23. Kholopov P. N.,Samus N. N., Frolov M. S., Goranskij V. P., Gorynya N. A., Kireeva N. N., Kukarkina N. P., Kurochin N. E., Medvedeva G. I., Perova N. B., Shugarov S. Y., 1985, General Catalogue of Variable Stars, Moscow Publishing House, Moscow
24. Knapp G. R., Morris M., 1985, ApJ, 292, 640
25. Kogan L., 1996, NRAO AIPS Memo Series No. 92
26. Martinez A., Bujarrabal V., Alcolea J., 1988, A&AS, 74, 273
27. Olofsson H., Lindqvist M., Nyman L. A., Winnberg A., Rieu N. -Q., 1991, A&A, 254, 611
28. Patel N. A., Joseph A., Ganesan R., 1992, A&A, 13, 241
29. Pegourie B., 1987, Ap&SS, 136, 133
30. Schwarz H., Nyman L., Seaquist E., Ivison R., 1995, A&A, 303, 833
31. Seaquist E., Ivison R., Hall P., 1995, MNRAS, 276, 867
32. Slutz S., 1976, ApJ, 210, 750
33. Spencer J. H., Schwartz P. R., Waak J. A., Bologna J. M., 1977, AJ, 82, 706
34. Thomson R. R., Creech-Eakman M. J., Akeson R. L., 2002, ApJ, 570, 373
35. Wood P. R., 1979, Ap.J, 227, 220
36. Yi J., Booth R. S., Conway J. E., Diamond P. J., 2005, A&A, 432, 531

ACKNOWLEDGMENTS

The Very Long Baseline Array (VLBA) is operated by the National Radio Astronomy Observatory (NRAO), a facility of the National Science Foundation (NSF) under cooperative agreement by Associated Universities, Inc.

The dependence of the IMF on the density-temperature relation of prestellar gas

S. Kitsionas*,†, A.-K. Jappsen* and R. S. Klessen*

*Astrophysikalisches Institut Potsdam, An der Sternwarte 16, D-14482 Potsdam, Germany
†email: skitsionas@aip.de

Abstract. We review several theoretical calculations of the IMF resulting from simulations of gravoturbulent fragmentation. The power-law behaviour of the IMF at high masses is obtained in almost all cases. It is suggested to be associated (at least partly) with accretion onto a clump mass distribution imprinted on the turbulent velocity distribution. The peak of the IMF at lower masses is obtained irrespectively of the turbulent driving and it is related to the Jeans mass, hence the thermodynamic behaviour of the gas. Several experiments have shown that the peak of the IMF shifts to lower masses if the gas is allowed to cool slowly or to higher masses in case of gas in a starburst environment that gets progressively heated.

Keywords: stars: formation, methods: numerical, hydrodynamics, turbulence, equation of state, ISM: clouds, Mass Functions
PACS: 97.10.Bt, 97.10.Xq, 97.21.+a

IMF: A BRIEF OBSERVATIONAL OVERVIEW

The Initial Mass Function (IMF) represents the mass distribution at birth of all stars ever formed. It can also be interpreted as the mass probability function with which stars are formed in the universe. It is, therefore, one of the most important parameters that need to be constrained in order to achieve a better understanding of the process of star formation. In the context of a complete theory of star formation, one should be able to predict the probability of forming stars of a certain mass, or inversely, one should be able to explain the observed mass distribution of groups or clusters of young stars.

Much progress has been achieved since the seminal work by Salpeter [1] in which the first IMF for relatively massive stars was derived [2, 3]. In the last decade in particular, the IMF has been extended to lower masses, even below the hydrogen burning limit [4, 5]. The consensus to-date is that at high masses ($m > 1\ M_\odot$) the IMF can be approximated by a power-law with an exponent of ~ -1.3 (in logarithmic mass), whereas the distribution at lower masses is either a log-normal, or alternatively, it can be fitted by piecewise power-laws (Fig. 1). In either case, there is a definite peak in the IMF at a mass of $\sim 0.5\ M_\odot$, with the distribution steadily declining below the sub-stellar limit.

Despite the considerable recent progress in observational techniques and detector power, there are still great uncertainties in the derived observational IMF. Theoretical calculations using numerical simulations to overcome the complexity of the processes involved and of the systems studied have become popular over the last several years mainly due to the great increase in available computing power and computational techniques. In this paper, we review several studies of the IMF using numerical simulations

CP848, *Recent Advances in Astronomy and Astrophysics,*
7th International Conference of the Hellenic Astronomical Society, edited by N. H. Solomos

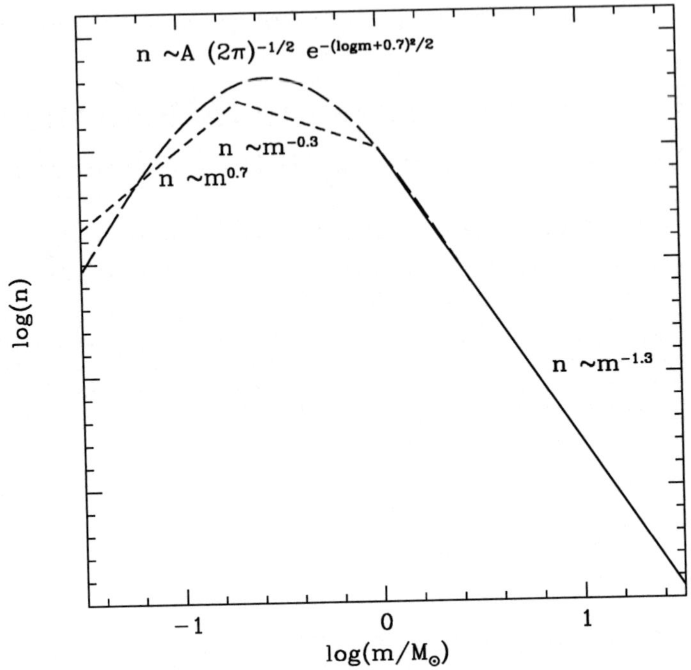

FIGURE 1. A graphical summary of the main characteristics of the observed IMF: a power-law at large masses ($m > 1$ $M_\odot$, solid line) and a peak at lower masses (~ 0.5 $M_\odot$) which can be fitted either by a log-normal function (long-dashed line) or piecewise power-laws (short-dashed lines).

of star formation. In particular, we summarise the results of theoretical investigations on the influence of the thermodynamic state of interstellar gas on the stellar mass distribution of groups of forming stars.

THEORETICAL IMF: NUMERICAL SIMULATIONS

Is the nature of turbulent support relevant to the IMF?

It is widely believed that star clusters form by gravoturbulent fragmentation of interstellar gas clouds (for details see the review by Mac Low & Klessen [6]). The supersonic turbulence ubiquitously observed in Galactic molecular gas generates strong density fluctuations with gravity taking over in the densest and most massive regions. Once such dense cloud regions become gravitationally unstable, collapse sets in to build up stars and star clusters.

Turbulence plays a dual role. On global scales it provides support, while at the same time it can promote local collapse. Stellar birth is thus intimately linked to the dynamical behaviour of the parental gas cloud, which determines when and where protostellar

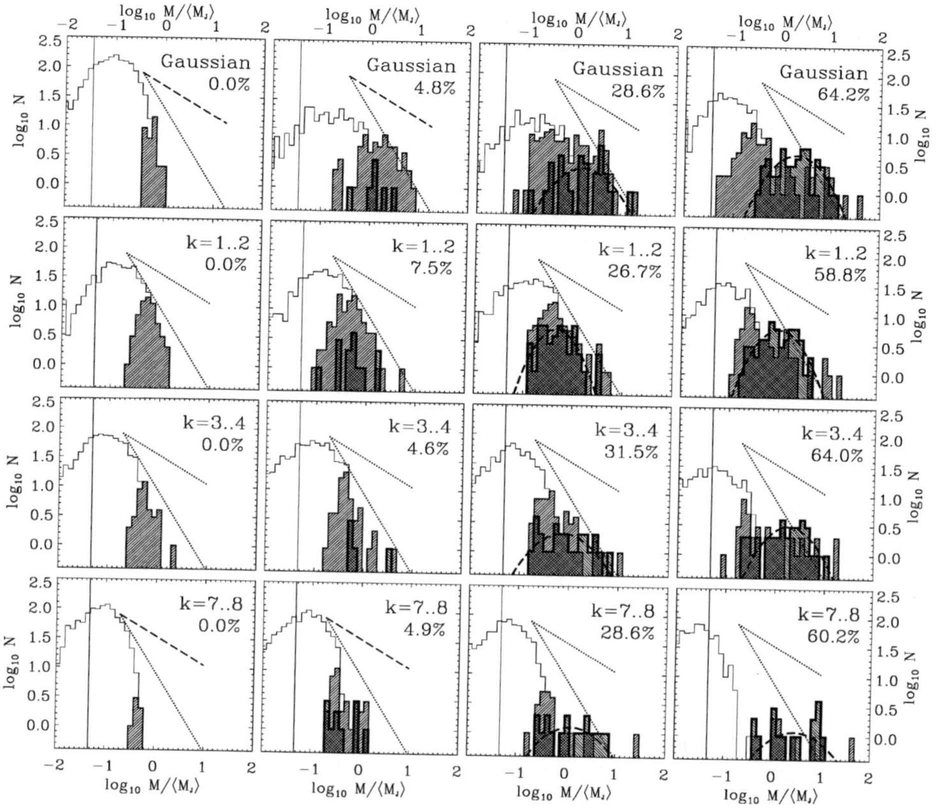

FIGURE 2. Mass spectra of all gas clumps (thin lines), of the subset of Jeans unstable clumps (thin lines, hatched distribution), and of collapsed cores (protostars: hatched thick-lined histograms). Different evolutionary phases are defined by the fraction of mass converted into protostars and are indicated in the upper right corner of each plot. Masses are binned logarithmically and normalised to the average Jeans mass $\langle M_J \rangle$. The vertical line shows the SPH resolution limit. Shown also are two power-law spectra with exponents -0.5 and -1.3 (figure courtesy of Klessen [7]).

cores form, and how they contract and grow in mass via accretion from the surrounding cloud material to build up stars. Slow, inefficient, isolated star formation is a hallmark of turbulent support, whereas fast, efficient, clustered star formation occurs in its absence.

Klessen [7] has studied the dependence of the final stellar mass on the gravoturbulent fragmentation process, and analysed four numerical models which span the full parameter range from strongly clustered to very isolated star formation. This study was based on numerical simulations using the smoothed particle hydrodynamics (SPH) method [8] and including sink (protostar) particles [9]. The models studied involved the isothermal evolution of turbulent clouds in the absence of further turbulent driving (turbulence decay, upper panel of Fig. 2), as well as of clouds in which turbulence was continuously driven between wavenumbers $k = k_{min}..k_{max}$, corresponding to length scales $l \propto L/k$, where L is the size of the computational volume (lower three panels of Fig. 2).

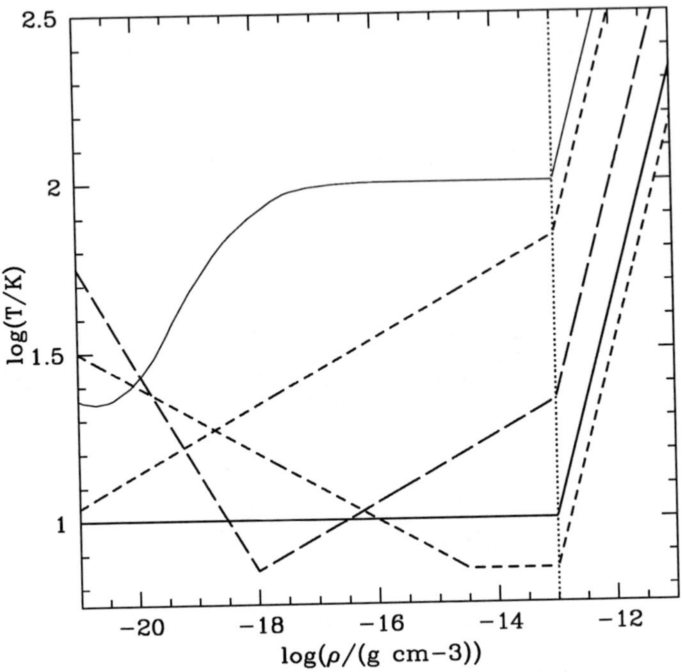

FIGURE 3. Summary of the equations of state adopted by different authors. The thick, solid, horizontal line describes an isothermal equation of state ($\gamma = 1.0$). The short-dashed lines give two examples of equations of state with constant γ (shown here are the equations of state with $\gamma = 0.9$ and 1.1) [10]. The long-dashed line is the piecewise polytropic equation of state used by Jappsen et al. [11]. The thin, solid line is the equation of state used by Spaans & Silk [12, 13]. The vertical line shows the characteristic density corresponding to the opacity limit for fragmentation. The effect of adiabatic heating at densities higher than this characteristic density is shown for all equations of state.

Fig. 2 plots the mass distribution of all gas clumps, of the subset of Jeans critical clumps, and of collapsed cores. We show four different evolutionary phases, initially just when gravity is 'switched on', and after turbulent fragmentation has lead to the accumulation of $M_* \approx 5\%$, $M_* \approx 30\%$ and $M_* \approx 60\%$ of the total mass in protostars. In the completely prestellar phase the clump mass spectrum is very steep (about the Salpeter slope [1], -1.3, or less) at the high-mass end. It has a break and gets shallower below $M \approx 0.4\langle M_J \rangle$ with slope -0.5. The spectrum strongly declines beyond the SPH resolution limit. Individual clumps are hardly more massive than a few $\langle M_J \rangle$.

Gravitational evolution modifies the distribution of clump masses considerably. As clumps merge and grow bigger, the spectrum becomes flatter and extends towards larger masses. Consequently the number of cores that exceed the Jeans limit increases. This is most evident in the Gaussian model of decayed turbulence, where the clump mass spectrum exhibits a slope -0.5.

The mass spectrum depends on the wavelength of the dominant velocity modes.

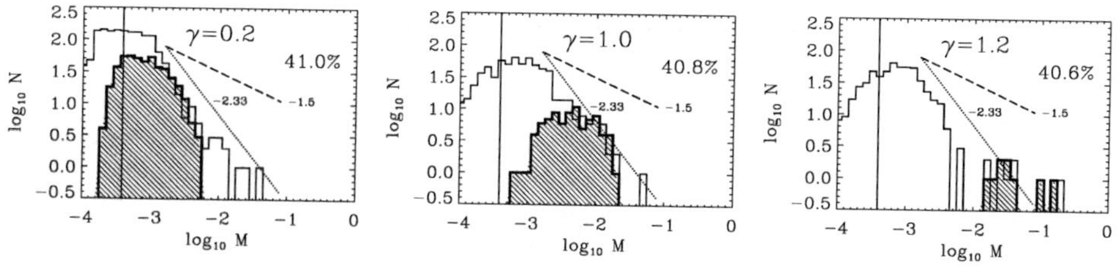

FIGURE 4. Mass spectra of gas clumps (*thin lines*) and of protostars (collapsed cores: *hatched thick-lined histograms*) for experiments with different γ. The percentage shows the fraction of total mass accreted onto protostars. The vertical line shows the SPH resolution limit. Shown also are two power-law spectra (in linear mass) with exponents -1.5 (dashed line) and -2.33 (dotted line). (figure courtesy of Li et al. [10].)

Small-scale turbulence does not allow for massive, coherent and strongly self-gravitating structures. Together with the short interval between shock passages, this prohibits efficient merging and the build up of a large number of massive clumps. Only few clumps become Jeans unstable and collapse to form stars. This occurs at random locations and times. The clump mass spectrum remains steep. Increasing the driving wavelength leads to more coherent and rapid core formation, resulting in a larger number of protostars.

Long-wavelength turbulence or turbulent decay produces a core mass spectrum that is well approximated by a *log-normal* distribution. It roughly peaks at the *average thermal Jeans mass* $\langle M_J \rangle$ of the system [14, 15] and is comparable in width with the observed IMF [4, 5]. However, this is still debated [16]. The log-normal shape of the mass distribution may be explained by invoking the central limit theorem [17], as protostellar cores form and evolve through a sequence of highly stochastic events (resulting from supersonic turbulence and/or competitive accretion [18, 19]).

In summary, the above results have shown that the power-law part of the IMF at high masses may be the consequence of accretion on a distribution of prestellar cores directly delivered by the initial gas turbulence [20] (a fact that has been recently debated by Ballesteros et al. [21]) and that the peak of the IMF at lower masses is a product of (turbulent) fragmentation with preference to objects with mass of the order of the local Jeans mass. However, the role of turbulence driving is not important as both decaying turbulence and turbulence driven on large scales can reproduce the log-normal mass distribution at low masses. On the other hand, it is the thermodynamic state of the gas that must play an important role in determining the peak of the IMF, as the Jeans mass $M_J \propto T^{3/2}\rho^{-1/2}$, where T and ρ are the gas temperature and density respectively.

Bate, Bonnell & Bromm [19] and Bate & Bonnell [22] have also performed isothermal simulations of decaying turbulence in which they have included adiabatic heating in the equation of state above a critical density dependent on dust opacity [23]. This has produced an additional dependence of the produced mass spectrum on the gas thermodynamics in terms of the minimum fragment mass corresponding to this opacity limit (*cf.* Fig. 3 for the isothermal case, where $M_J \propto \rho^{-1/2}$ until adiabatic heating switches on, after which M_J increases with increasing density).

Non-isothermal equation of state

From the above isothermal calculations [7, 19] it has become obvious that the thermodynamic state of turbulent interstellar gas is of particular importance for the IMF resulting from gravoturbulent fragmentation. In fact, it has been shown that the simple isothermal assumption on the gas thermodynamic behaviour is not to be taken at face value, despite the fact that gas temperature has been observed to be approximately constant over several orders of magnitude in density (thick, solid line in Fig. 3).

To depart from isothermality, Li, Klessen & Mac Low [10] have experimented by using a constant polytropic exponent γ, taking values between 0.2 and 1.4, for the equation of state of turbulent gas (short-dashed lines in Fig. 3). They have found that for $\gamma > 1$ the increasing temperature, and thus pressure support, gives rise to less but more massive objects than for $\gamma \leq 1$. In particular, increased cooling ($\gamma < 1$) was shown to shift the peak of the IMF to smaller masses (Fig. 4).

Larson [24] claims that interstellar gas is slowly cooling with increasing density due to molecular line cooling below densities of $\rho_{crit} \sim 10^{-18}$ g cm^{-3}, but then it slowly heats up for densities higher than ρ_{crit} and less than the density set by the opacity limit for fragmentation, due to the thermal coupling of the gas to the dust. Based on Larson [24], Jappsen et al. [11] have calculated the IMF resulting from turbulent gas that is subject to a piecewise equation of state (long-dashed line in Fig. 3). In particular, they have used a $\gamma_1 = 0.7$ for the cooling regime and a $\gamma_2 = 1.1$ for the heating regime, with the transition density ρ_{crit} ranging between $\sim 10^{-19}$ and $\sim 10^{-16}$ g cm^{-3}. Their results are summarised in Fig. 5. Their conclusion is similar to that of Li et al. [10] in the sense that for more cooling (i.e. higher ρ_{crit}) the peak of the IMF shifts to smaller masses.

Klessen, Spaans & Jappsen (in prep.) have followed the evolution of turbulent gas obeying an equation of state characteristic of starburst environments [12, 13]. Similarly to the above, they have found that for gas that gets progressively heated with increasing density (thin, solid line in Fig. 3) the mass spectrum peaks at considerably higher masses and that the distribution does not demonstrate a Salpeter [1] power-law at the high-mass end but it is rather flat. In fact their IMF is consistent with the flat IMF observed at the Arches cluster [25].

SUMMARY AND CONCLUSIONS

We have reviewed several theoretical calculations of the IMF resulting from simulations of gravoturbulent fragmentation. The power-law behaviour of the IMF at high masses is obtained in almost all cases. It is suggested to be associated (at least partly) with accretion onto a clump mass distribution imprinted on the turbulent velocity distribution. The peak of the IMF at lower masses is obtained irrespectively of the turbulent driving and it is related to the Jeans mass, hence the thermodynamic behaviour of the gas. Several experiments have shown that the peak of the IMF shifts to lower masses if the gas is allowed to cool slowly (as it is suggested by Larson [24]) or to higher masses in case of gas in a starburst environment that gets progressively heated.

Since the opacity limit for fragmentation sets the minimum fragment mass, it is interesting to extend the calculations of Jappsen et al. [11] to the adiabatic regime, in

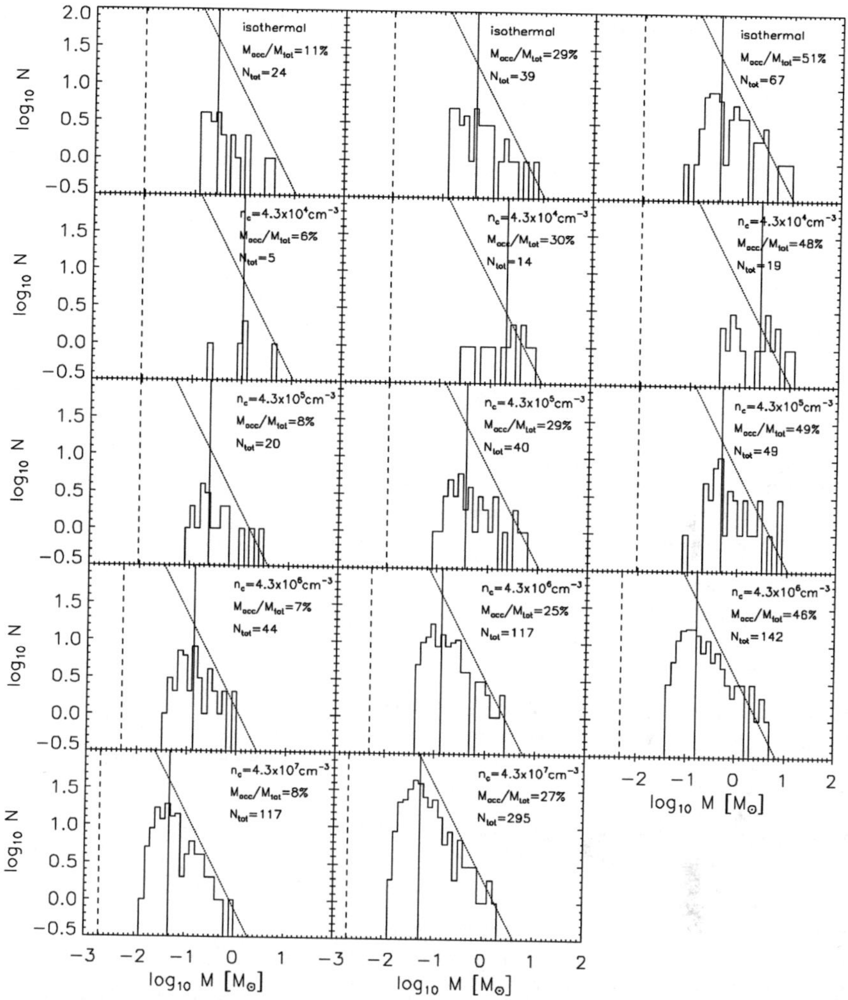

FIGURE 5. Mass spectra of protostellar objects for models with increasing ρ_{crit} at the stages when 10%, 30% and 50% of the total gas mass has been accreted on the protostars. For comparison we also show in the first row the mass spectra of an isothermal run. Critical density n_c (in particles per cm^3), ratio of accreted gas mass to total gas mass M_{acc}/M_{tot} and number of protostellar objects are given in the plots. The *vertical solid line* shows the position of the median mass. The *dotted line* has a slope of -1.3 and serves as a reference to the Salpeter value [1]. The *dashed line* indicates the mass resolution limit of the simulations (figure courtesy of Jappsen et al. [11]).

order to explore the effect of the minimum mass set by the opacity limit on the IMF peak.

The main conclusion we would like to draw here is that real progress on theoretical calculations of the IMF would be achieved only when a self-consistent treatment of the gas thermodynamics in terms of full radiation transport will be attainable. Unfortunately, this is not foreseen to be computationally feasible before five to ten years from now.

ACKNOWLEDGMENTS

We thank Javier Ballesteros-Paredes, Enrique Vázquez-Semadeni, Yuexing Li, Mordecai Mac Low, Anthony Whitworth, Richard Larson, Stefan Schmeja and Hans Zinnecker for continuing discussions and collaborations. SK acknowledges support by the EU Commission in terms of a Marie Curie Intra-European Fellowship (MEIF-CT-2004-011226). AKJ and RSK acknowledge support by the Emmy Noether Programme of the Deutsche Forschungsgemeinschaft (grant No. KL1358/1).

REFERENCES

1. E. E. Salpeter, *ApJ* **121**, 161–167 (1955).
2. G. E. Miller, and J. M. Scalo, *ApJ Supp. Ser.* **41**, 513–547 (1979).
3. J. Scalo, "The IMF revisited: A case for variations," in *The Stellar Initial Mass Function: Proceedings of the 38th Herstmonceux Conference*, edited by G. Gilmore, and D. Howell, ASP Conf. Series, 1998, p. 201.
4. P. Kroupa, *Science* **295**, 82–91 (2002).
5. G. Chabrier, *PASP* **115**, 763–795 (2003).
6. M.-M. Mac Low, and R. S. Klessen, *Rev. Mod. Phys.* **76**, 125 (2004).
7. R. S. Klessen, *ApJ* **556**, 837 (2001).
8. J. J. Monaghan, *Ann. Rev. A&A* **30**, 543–574 (1992).
9. M. R. Bate, I. A. Bonnell, and N. M. Price, *MNRAS* **277**, 362–376 (1995).
10. Y. Li, R. S. Klessen, and M.-M. Mac Low, *ApJ* **592**, 975 (2003).
11. A.-K. Jappsen, R. S. Klessen, R. B. Larson, Y. Li, and M.-M. Mac Low, *A&A* **435**, 611 (2005).
12. M. Spaans, and J. Silk, *ApJ* **538**, 115 (2000).
13. M. Spaans, and J. Silk, *ApJ* **626**, 644 (2005).
14. R. S. Klessen, and A. Burkert, *ApJ Supp. Ser.* **128**, 287 (2000).
15. R. S. Klessen, and A. Burkert, *ApJ* **549**, 386 (2001).
16. H. Martel, N. J. Evans II, and P. R. Shapiro, *ApJ Supp. Ser.* p. in press (2006).
17. H. Zinnecker, *MNRAS* **210**, 43 (1984).
18. B. Reipurth, and C. Clarke, *AJ* **122**, 432–439 (2001).
19. M. R. Bate, I. A. Bonnell, and V. Bromm, *MNRAS* **339**, 577–599 (2003).
20. P. Padoan, and A. Nordlund, *ApJ* **576**, 870–879 (2002).
21. J. Ballesteros-Paredes, A. Gazol, J. Kim, A.-K. Jappsen, R. S. Klessen, and E. Tejero, *ApJ* p. in press (2006).
22. M. R. Bate, and I. A. Bonnell, *MNRAS* **356**, 1201–1221 (2005).
23. J. E. Tohline, *Hydrodynamic collapse*, Gordon and Breach Science Publishers, 1982, vol. 8 of *Fundamentals of cosmic physics*, pp. 1–82.
24. R. B. Larson, *MNRAS* **359**, 211–222 (2005).
25. A. Stolte, W. Brandner, E. K. Grebel, R. Lenzer, and A.-M. Lagrange, *ApJ* **628**, L113–117 (2005).

The distribution of the ISM in the Milky Way

A. Misiriotis* and E. M. Xilouris†

*University of Crete, Physics Department, P.O. Box 2208, 710 03 Heraklion, Crete, Greece
†National Observatory of Athens, I. Metaxa & Vas. Pavlou str., Palaia Penteli, GR-15236, Athens, Greece

Abstract. We use the COBE/DIRBE (1.2, 2.2, 60, 100, 140 and 240 μm) maps and the COBE/FIRAS spectra (for the wavelength range 100 - 1000 μm) to constrain a model for the spatial distribution of the dust, the stars and the gas in the Milky Way. By assumming exponential axisymmetric distributions for the dust and the stars and by performing the corresponding radiative transfer calculations we reproduce the FIR and NIR maps of the Milky Way. Similar distributions for the atomic and molecular hydrogen in the disk are used (with an inner cut off radius for the atomic Hydrogen) to fit the gas data. The star formation rate surface density is ploted against the gas surface density and an "intrinsic" Galactic Schmidt law is derived with excellent agreement with the "external" Schmidt law found for spiral galaxies. The Milky Way is found to consume $\sim$ 1% and $\sim$ 10% of its gas in the outer and inner regions respectively (for a period of 0.1 Gyr) for making stars. The dust-induced B-V colour excess observed in various directions and distances (up to $\sim$ 6.5 kpc) to well studied Cepheid stars is compared with the model predictions showing a good agreement. The simple assumption of exponential distributions of stars and dust in the Galaxy is found to be quite instructive and adequate in modelling all the available data sets from 0.45 μm (B-band) to 1000 μm.

Keywords: dust, extinction; ISM: structure; Galaxy: structure
PACS: 98.35.-a

INTRODUCTION

Our Galaxy, the Milky Way, constitutes the best laboratory for studying the properties of the interstellar medium (ISM) in spiral galaxies. Being inside the Galaxy, the observer has the advantage of looking through different lines of sight passing both through dense environments (e.g. the Galactic center) and through almost free of dust and gas areas (e.g. the Galactic poles). The way that dust and gas (both in molecular and neutral phase) are distributed inside the Galaxy not only shapes the appearance of the Galaxy in different wavelengths, but also drives the star formation, a key process that determines the Galactic evolution.

There are several tracers for measuring the star formation activity in a galaxy (e.g., ultra-violet (UV) emission, far-infrared (FIR) emission, Ha and radio emission), yet it is far from trivial to get an accurate estimate (see Kennicutt, [11] for a review). Most of the star formation rate (SFR) tracers are affected by dust extinction thus rendering their use quite limited and uncertain (see however Kewley et al. [13] for a discussion on properly accounting for extinction). The far infrared (FIR) emission (if modelled properly in order to distinguish between the diffuse dust emission and the emission coming from the star formation complexes) has been proven to be one of the most promising SFR indicators (see Misiriotis et al. [16] and for a review Kylafis & Misiriotis [15]).

CP848, *Recent Advances in Astronomy and Astrophysics*, 7th *International Conference of the Hellenic Astronomical Society*, edited by N. H. Solomos

Some efforts have been made in modelling the interstellar medium (ISM) in the Milky Way. Kent et al. ([12]) used the K-band data from the SPACELAB infrared telescope and modelled the distribution of the light in the Galaxy by assuming an exponential distribution for the starlight in the disk and the bulge. Using the COBE 140 and 240 μm maps along with radio surveys of HI and H_2 Sodroski et al. ([23]) modelled the FIR emission of the Galaxy and derived the dust temperature profiles as well as gas-to-dust mass ratios and emissivities along different lines of sight. A composite of COBE/DIRBE and IRAS/ISSA maps made by Schlegel et al. ([20]) provides a unique tool for estimating the Galactic extinction along different directions. Davies et al. ([3]) modelled the FIR emission from the Galaxy by fitting exponential dust distributions to the COBE/DIRBE maps. This study claimed that the best fit could be achieved by assuming that the dust distribution is more extended than that of the stars, namely that the scalelength of the dust is 1.5 times that of the stars and the scaleheight of the dust is twice that of the stars. Drimmel & Spergel ([5]) presented a very detailed modelling of the Milky Way dust content. Based on a total of 48 parameters (26 parameters for the dust and 22 parameters for the stars) this model fitted quite nicely the NIR (J-band) and FIR (240 μm) radial profiles. In a subsequent paper (Drimmel et al. [4]) this model was used in order to calculate the extinction to any point within the Galactic disk.

In this paper we make use of a simple (with the minimal number of parameters required) but realistic, three-dimensional model, which includes the effects of absorption and scattering of the stellar light by the dust in the optical and near infrared (NIR) wavelengths. We also take into account the emission from the diffuse dust and the emission from star forming regions in the FIR. In order to constrain the model parameters, we fit the model to all the available data simultaneously. The data that we use consist of the COBE/DIRBE (1.2, 2.2, 60, 100, 140 and 240 μm) maps and the COBE/FIRAS spectra (for the wavelength coverage between 100 and 10000 μm). Using published maps for the atomic and molecular hydrogen we model the gas distribution. After calculating the SFR density throughout the Galactic disk and comparing it with the observed gas surface density we investigate the validity of the Schmidt ([21]) law as quantified by Kennicutt ([10]). Finally, using a catalogue of Cepheid stars with good estimates of their distance and B-V colour excess, we confirm the validity of our model in the optical region.

THE DATA

The Cosmic Background Explorer (COBE) satellite, launched on November 18, 1989, has been proved to be a very successful space mission (Boggess et al. [1]). With the Diffuse Infrared Background Experiment (DIRBE) instrument, COBE surveyed the entire sky in 10 photometric bands covering the wavelength region from 1 to 240 μm at an angular resolution of 0.7 square degrees (Sodroski et al. [23]). Using another instrument, the Far Infrared Absolute Spectrometer (FIRAS), COBE measured the spectrum of the dust emission of our Galaxy in the wavelength range from 120 μm to 10000 μm (Fixsen et al. [7]). These two instruments provide us with a unique set of data of the dust emission for a wide range of wavelengths, ideal for studying the properties and the distribution of dust in the Galaxy.

The Leiden/Argentine/Bonn survey of the Galactic HI (Kalberla et al. [9]) is a merging of the Argentino de Radioastronomía and the Leiden/Dwingeloo surveys. The angular resolution of the combined survey is 0.6 square degrees (comparable to that of the COBE maps) and is the most sensitive, to date, HI map of the Milky Way.

A composite CO survey of the Milky Way constructed from 37 individual surveys has been carried out by Dame et al. ([2]) [1]. The total area covered by this survey is 9,353 square degrees which accounts for nearly one half of the area within 30 degrees of the Galactic plane. Using a CO-to-H_2 mass conversion factor of $X = 1.8 \pm 0.3 \times 10^{20}$ cm^{-2} K^{-1} km^{-1} s (Dame et al. [2]), an H_2 map is constructed.

THE MODEL

The model that we use simulates quite accurately the physical processes that take place inside the Galaxy for the wavelength range between optical and submillimeter (submm). In the optical/NIR wavelength range, absorption and scattering of the stellar light by the dust grains occurs, while emission from the heated dust grains is taking place in the FIR/submm wavelengths. The different nature of the processes allow us to treat the two wavelength ranges separately as we describe below. Since the dust reveals itself by its FIR/submm emission, it is reasonable to begin with the modelling in these wavelengths in order to constrain the dust parameters. By fixing the dust parameters in the radiative transfer model we can then model the optical/NIR emission of the Galaxy.

The FIR/submm wavelength range

The emission model that is used for the FIR/submm wavelengths is along the lines of the model described in Popescu et al. ([18]; see also Misiriotis et al. [17]). Following these studies we assume that the Galactic dust can be described by two components. The first component is the warm dust associated with star-formation complexes, which is heated locally by the young stars. The second component is the cold diffuse dust, which is heated by the diffuse radiation filed in the Galaxy. Axisymetric exponential spatial distributions are adopted for both the warm and the cold components. For the dust grain emissivity in these wavelengths we use the values reported in Weingartner & Draine ([24]).

The Optical/NIR wavelength range

Given the dust distribution, as this is inferred by the FIR emission, one can calculate the extinction along any line of sight at any distance from the Earth. Using this information, we calculate the observed starlight intensity along any line of sight at 1.2 and $2.2\mu m$. (see Kylafis & Bahcall [14]). For the spatial distribution of the stars we follow

---

[1] http://cfa-www.harvard.edu/mmw/MilkyWayinMolClouds.html

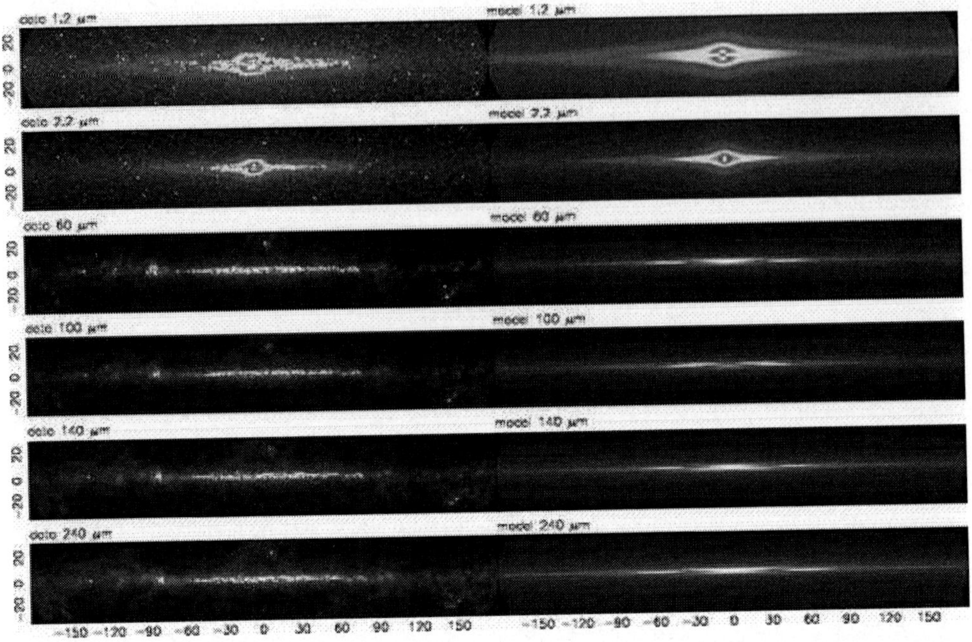

FIGURE 1. The COBE/DIRBE maps of the Galaxy (left panels) in direct comparison with the fitted model (right panels). From top to bottom the 1.2, 2.2, 60, 100, 140, and 240μm maps are presented for the Galactic latitude range of -30 to 30 degrees. The maps were made in Mollweide projection (Snyder [22]).

Xilouris et al. ([25]; [26]; [27]) thus assuming an exponential axisymmetric disk and an axisymmetric bulge.

The molecular and atomic Hydrogen

The two gas phases (HI and H_2), being totally independent from the dust and stars are treated separately from the rest of the components. The distribution of the molecular Hydrogen (H_2) follows the CO emission. Although it is rather clumpy it can be, to a first approximation, described by an axisymmetric exponential disk.

For the atomic Hydrogen (HI) we assume a similar distribution but we also introduce an inner truncation radius in order to account for the absence of HI in the central parts of the galaxy as indicated by the relevant map (see Kalberla et al. [9]).

RESULTS

Having determined all the parameters for the dust and the stellar distributions, we can now create the model images of the Galaxy at each wavelength in order to compare with

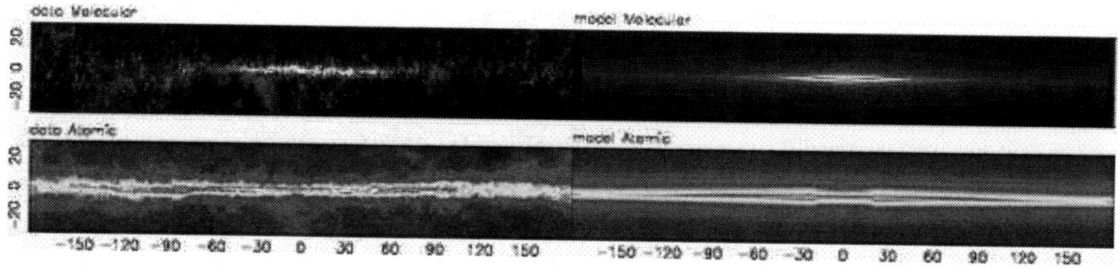

FIGURE 2. The gas maps of the Galaxy (left panels) in direct comparison with the fitted model (right panels). The upper left panel shows the molecular Hydrogen distribution (Dame et al. [2]), while the atomic Hydrogen distribution (Kalberla et al. [9]) is presented in the bottom left panel. Both maps are given for the Galactic latitude range of -30 to 30 degrees.

the observed images. This is what we do in Fig. 1, where the observations at different wavelengths (left panels) are compared with the model images (right panel). The 1.2, 2.2, 60, 100, 140 and 240 μm images are shown from top to bottom. In these images the x and y-axis are the Galactic longitude and latitude in degrees with the Galactic center being at (0,0) while positive and negative longitudes are the southern and northern regions of the sky respectively. Positive and negative latitudes are regions above and below the Galactic plane respectively.

As it is obvious from Fig. 1, our model and the observations compare quite well for all the different wavelengths. The NIR (1.2, 2.2 μm) maps (two top panels) show the characteristic dust lane crossing the bulge along the Galactic plane which, at these wavelengths, is poorly visible (compared to the optical wavelengths). The stellar disk is well described by the smooth exponential distribution used in the model. The FIR (60, 100, 140 and 240 μm) maps (third, fourth, fifth, and sixth panels from the top respectively) show the diffuse dust emission along the Galactic plane. Given the simplicity of the axisymmetric distributions that were used, the model reproduces quite accurately the observed emission at these wavelengths.

Using the values of the parameters of the model and adopting the Weingartner & Draine ([24]) values for the extinction cross section at 1.2μm and at 2.2μm we can calculate the optical depth (and subsequently the extinction) between any two points in the Galaxy. In particular, the central face-on optical depth $\tau_\lambda^f = 2\kappa_\lambda z_c$, is 0.33 at 1.2 μm and 0.17 at 2.2 μm.

In Fig. 2 we compare the maps of the hydrogen column density with the corresponding model. The top panel on the left shows the molecular hydrogen map as this is inferred from the CO observations (Dame et al. [2]). The respective model column density is shown in the top right panel. From the comparison between the model and the observations it is evident that, despite the large clumpiness of the data, the large scale structure is fairly well represented by the model. In the bottom left panel we show the map of the atomic hydrogen (Kalberla et al. [9]). The lack of this component in the central part of the galaxy shows up as a decrease of the coloumn density. The model column density is shown in the bottom right panel. In this image, the lack of atomic hydrogen in the center of the Galaxy is more prominent due to the sharp cut introduced by the model at the truncation radius R_t.

Given the parameters describing the distributions it is straightforward to compute the total mass of the cold dust M_c, the warm dust M_w, the mass of the molecular hydrogen M_{H_2} and the mass of the atomic hydrogen M_{HI}. Using the values for the fitted parameters we find $M_c = 7.0 \times 10^7 M_\odot$, $M_w = 2.2 \times 10^5 M_\odot$, $M_{H_2} = 1.3 \times 10^9 M_\odot$ and $M_{HI} = 8.2 \times 10^9 M_\odot$. The total gas-to-dust mass ratio becomes $(M_{H_2}+M_{HI})/(M_c+M_w) = 135$, which is in excellent agreement with the value of 160 ± 60 reported in Sodroski et al. ([23]).

DISCUSSION

Comparison of the model with the FIRAS data

To further establish the validity of the model, we compare it with the COBE/FIRAS data.

This is shown in Fig. 3 with the dots being the COBE/FIRAS data, the open circles the COBE/DIRBE 60 and 100 μm data, the dashed line the warm dust emission, the dot-dashed line the cold dust emission and the solid line the total FIR/submm emission predicted by the model. This astonishing agreement, which does not come from fitting the model to the data but from comparison of the data with our model, gives us the confidence that our model gives a consistent representation of the Galaxy in wavelengths extending up to 1000 μm.

In their analysis, Reach et al. ([19]) reported emission from a very cold component (4-7 K). This component is not evident in our analysis when looking at the averaged SED of the Galaxy (Fig. 3).

Star formation efficiency in the Milky Way disk and the Schmidt law

As already mentioned in the Introduction, a good indicator for the SFR is the FIR emission. In order to exploit the FIR emission we make use of the strong correlation between the SFR and the 100 μm flux found in Misiriotis et al. ([16]) for a large sample of spiral galaxies. Using this relation and calculating the total 100 μm luminosity we derive a SFR of $2.7 M_\odot yr^{-1}$ for the Milky Way.

The SFR surface density can then be calculated in the disk of the Milky Way, using the same relation presented in Misiriotis et al. ([16]), A simple power-law relation between SFR and gas content of external galaxies, introduced by Schmidt ([21]) and further explored by Kennicutt ([10]), is well established and tested for large samples of galaxies (Misiriotis et al. [16]). This relation is expressed in terms of the SFR surface density Σ_{SFR} and the gas surface density Σ_{gas} and has the form:

$$\Sigma_{SFR} = A\Sigma_{gas}^N, \qquad (1)$$

with A being the absolute SFR efficiency and N the slope of the power-law. Kennicutt ([10]) examined two samples of galaxies showing normal and starbursting behaviour and derived a Schmidt law for normal galaxies (with $N = 2.47 \pm 0.39$) and for starburst galaxies (with $N = 1.40 \pm 0.13$).

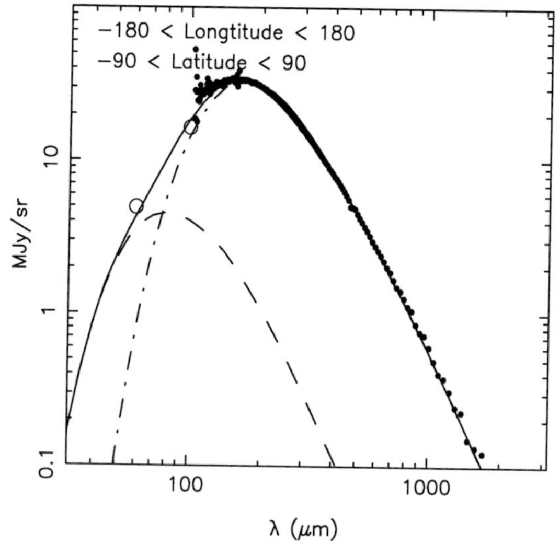

FIGURE 3. The global SED of the Milky Way averaged over the whole sky. The filled circles are the COBE/FIRAS data and the open circles are the COBE/DIRBE data while the dashed line is the contribution to the SED of the warm dust component, the dot-dashed line is the contribution of the cold dust component and the solid line is the total modelled SED.

In our study we have, for the first time, the opportunity to examine the validity of this law not for a sample of external galaxies, by measuring their global properties, but for different regions of the same galaxy (the Milky Way). This is shown in Fig. 4 with the SFR surface density (Σ_{SFR}) calculated for different radii along the Galactic disk plotted against the gas surface density $\Sigma_{\rm gas}$ of the same region. These data are shown with solid circles with the numbers close to the points indicating the distance to the Galactic center (in kpc). Furthermore it is evident that for regions within the Galaxy with small gas content (at the outer part of the disk) the SFR is low while it gets more intense for regions with larger gas density (close to the center of the galaxy). From this plot it is evident that the Galactic Schmidt law also follows a power-law. A best fit to these data (indicated with a solid line in the plot) yields

$$\frac{\Sigma_{\rm SFR}}{{\rm M_{\odot} yr^{-1} pc^{-2}}} = 4.45 \times 10^{-5} \left(\frac{\Sigma_{\rm gas}}{{\rm M_{\odot} kpc^{-2}}} \right)^{2.18} . \quad (2)$$

The power-law index of 2.2 ± 0.2 that we find is in a very good agreement with the value of 2.47 ± 0.39 quoted by Kennicutt ([10]) for normal galaxies. This is also evident by looking at the distribution of the normal galaxies (open circles) on this plot (Fig. 4). The points for the external galaxies (open circles) are well correlated with the measurements for different regions within the Milky Way (solid circles). This indicates the strong correlation between the gas content and the SFR activity. The striking similarity of the "internal" Schmidt law for an individual galaxy (in this case the Milky Way) and the "external" Schmidt law for the global properties of normal galaxies may indicate a

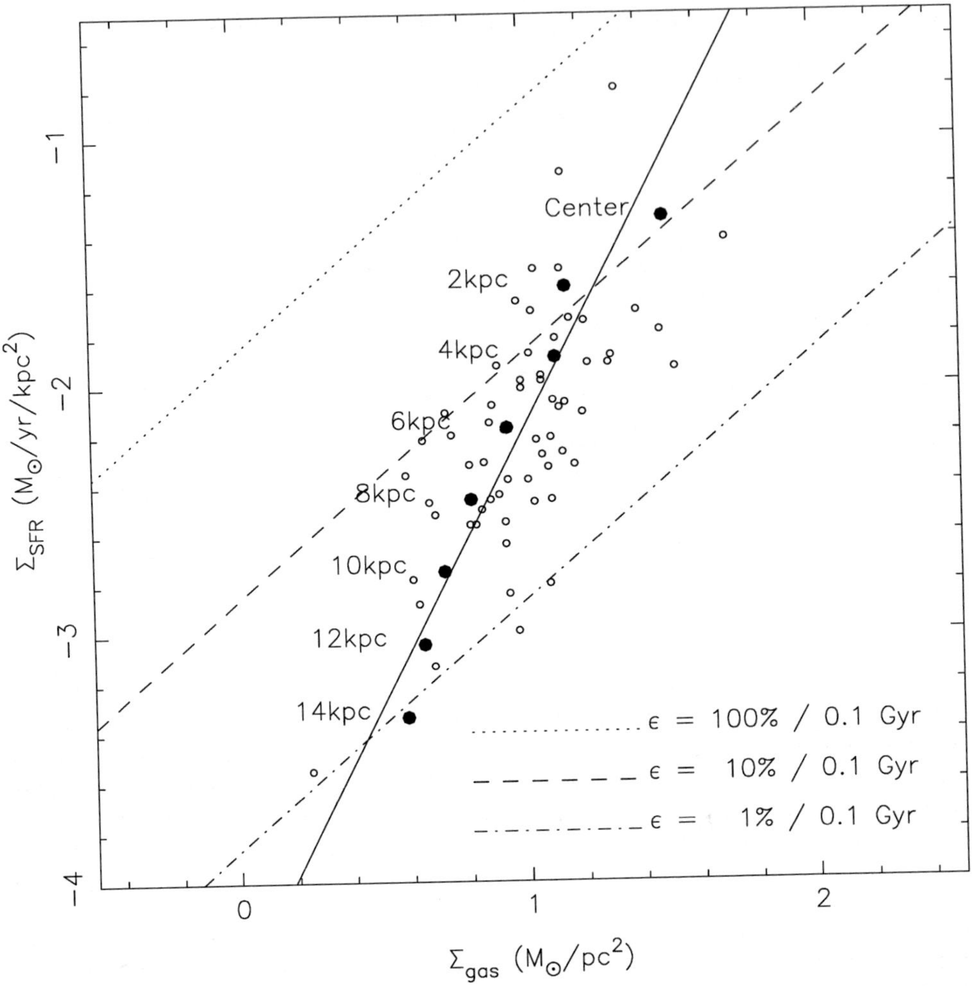

FIGURE 4. Star formation rate density as a function of gas surface density in the Milky Way disk. The solid circles indicate the measurements for the Milky Way (for distances along the disk as indicated by the numbers close to the solid circles) while the solid line is the best fit of a Schmidt law to these points (see the text for details). The open circles are the global measurements for a sample of external galaxies presented in Kennicutt ([10]). The three parallel dashed and dotted lines correspond to constant star formation efficiencies of 1%, 10% and 100% per 0.1 Gyr.

universal gas-SFR correlation not only on large scales (for galaxies as a whole) but also for small-scales (for different regions inside a galaxy).

Comparing the model with optical data

In order to check the validity of our model in the optical wavelengths, we use the Database of Galactic Classical Cepheids (Fernie et el. [6]). This database contains information of over 500 classical Cepheids in the Galaxy with determined distances ranging from a couple of hundred parsec up to ~ 6.5 kpc. The E(B-V) colour excess reported in this catalogue for each star was then checked against the E(B-V) determined with our model by calculating the extinction in the B and V bands along the directions of the stars and until the reported distance to each star. The comparison between the observed data and the model predictions for E(B-V) is shown in Fig. 5 with the x-axis being the B-V color excess observed and the y-axis the B-V color excess predicted by the model. From this plot it is evident that the correlation between the observed and the predicted measurements is extremely good with the average of the points (filled circles), binned every 0.2 magnitudes of E(B-V) in both axis, falling, within the error bars, along the diagonal of the plot. This is also evident by looking at the individual observations (open circles) which are scattered symmetrically with respect to the diagonal.

ACKNOWLEDGMENTS

We thank Dr. N. D. Kylafis for his significant contribution in the final stages of this work. We would also like to thank Dr. V. Charmandaris for useful discussions concerning the FIRAS data and Dr. P. M. W. Kalberla for kindly providing us with the map of the Leiden/Argentine/Bonn Survey of Galactic HI. We acknowledge the use of the Legacy Archive for Microwave Background Data Analysis (LAMBDA). Support for LAMBDA is provided by the NASA Office of Space Science. This work has been supported in part by a Pythagoras II research program of the Ministry of Education of Greece.

REFERENCES

1. Boggess, N. W., Mather, J. C., Weiss, R., et al., 1992, ApJ, 397, 420
2. Dame, T. M. Hartmann, Dap., Thaddeus, P., 2001, ApJ, 547, 792
3. Davies, J. I., Trewhella, M., Jones, H., Lisk, C., Madden, A., Moss, J., 1997, MNRAS, 288, 679
4. Drimmel, R., Cabrera-Lavers, A., LÃşpez-Corredoira, M., 2003, A&A, 409, 205
5. Drimmel, R., Spergel, D. N., 2001, ApJ, 556, 181
6. Fernie, J. D., Beattie, B., Evans, N. R., and Seager, S. 1995, IBVS No. 4148
7. Fixsen, D. J., Cheng, E. S., Gales, J. M., et al., 1996, ApJ, 473, 576
8. Hauser, M. G., Arendt, R. G., Kelsall, T., et al., 1998, ApJ, 508, 25
9. Kalberla, P. M. W., Burton, W. B., Hartmann, Dap, 2005, A&A, 440, 775
10. Kennicutt, R. C. Jr, 1998a, ApJ, 498, 541
11. Kennicutt, R.C. Jr, 1998b, ARA&A, 36, 189
12. Kent, S. M., Dame, T. M., Fazio, G., 1991, ApJ, 378, 131
13. Kewley, L. J., Geller, M. J., Jansen, R. A., Dopita, M. A., 2002, AJ, 124, 3135
14. Kylafis, N. D., Bahcall, J. N., 1987, ApJ, 317, 637
15. Kylafis, N. D., Misiriotis, A., 2005, astro-ph/0501373
16. Misiriotis, A., Papadakis, I. E., Kylafis, N. D., Papamastorakis, J., 2004, A&A, 417, 39
17. Misiriotis, A., Popescu, C. C., Tuffs, R. J., Kylafis, N. D., 2001, A&A, 372, 775
18. Popescu, C. C., Misiriotis, A., Kylafis, N. D., Tuffs, R. J., Fischera, J., 2000, A&A, 362, 138
19. Reach, W. T., Dwek, E., Fixsen, D. J., et al., 1995, ApJ, 451, 188

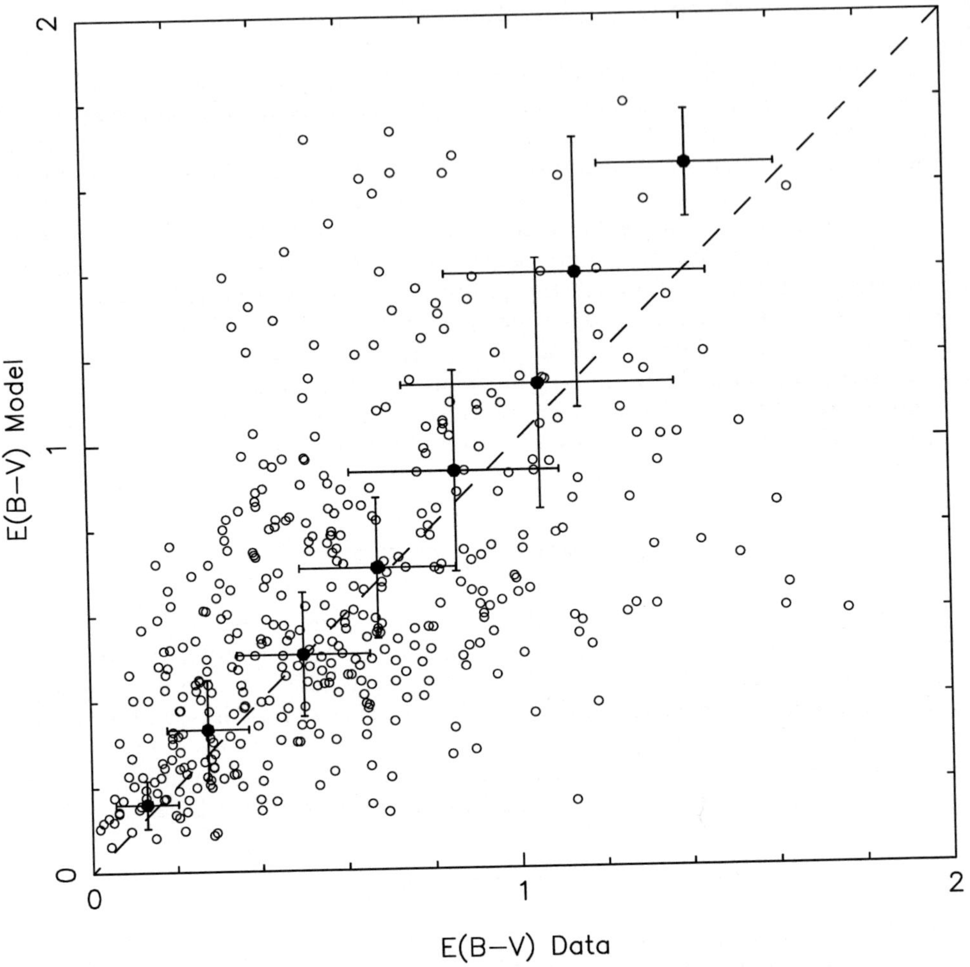

FIGURE 5. The B-V color excess for the sample of the Galactic Classical Cepheids (Fernie et el. 1995; x-axis) compared with the color excess along the same directions as derived from the model.

20. Schlegel, D. J., Finkbeiner, D. P., Davis, M., 1998, ApJ, 500, 525
21. Schmidt, M. 1959, ApJ, 129, 243
22. Snyder, J. P., 1987, Map Projections–A working manual. U. S. Geological Survey Professional Paper 1395. Washington, DC: U. S. Government Printing Office
23. Sodroski, T. J.; Bennett, C.; Boggess, N., et al., 1994, ApJ, 428, 638
24. Weingartner J. C., Draine, B. T., 2001, ApJ, 548, 296
25. Xilouris, E. M., Kylafis, N. D., Papamastorakis, J., Paleologou, E. V., Haerendel, G., 1997, A&A, 325, 135
26. Xilouris, E. M., Alton, P. B., Davies, J. I., et al., 1998, A&A, 331, 894
27. Xilouris, E. M., Byun, Y. I., Kylafis, N. D., Paleologou, E. V., Papamastorakis, J., 1999, A&A, 344, 868

The Impact of the GAIA Experiment on Eclipsing Binary Star Studies

P.G. Niarchos

Department of Astrophysics, Astronomy and Mechanics, National and Kapodistrian University of Athens, Panepistimiopolis, GR 157 84 Zografos, Athens, Greece

Abstract. GAIA, the approved cornerstone 6 ESA mission, will observe up to a billion stars in our Galaxy and obtain their astrometric positions on a micro-arcsec level, multi-band photometry as well as spectroscopic observations. The GAIA large-scale photometric survey will have significant intrinsic scientific value for the study of variable stars of nearly all types, including detached eclipsing binaries, near-contact or contact binaries and pulsating stars. It is expected that about 1×10^6 Eclipsing Binaries (EBs) (with $V \leq 16$ mag) will be discovered and the observing fashion will be quite similar to Hipparcos/Tycho mission operational mode. On the other hand, GAIA's spectrograph will be able to secure useful radial velocities for binaries down to $V \approx 17$ mag. The combined astrometric, photometric and spectroscopic data will be used to compute the physical parameters of the observed EBs. Even if for only 1% of the observed EBs reliable physical parameters could be derived, this would be a great contribution to stellar astrophysics and a giant leap in comparison with what has been obtained so far from ground-based observations. From a study of a small sample of EBs, it is shown that the agreement between the fundamental stellar parameters, derived from ground-based and Hipparcos (GAIA-like) observations, is more than satisfactory. The percentage difference between the absolute elements derived from GAIA-like observations and those from ground-based observations is of the order 2% or lower. These results are encouraging and demonstrate that GAIA data will be suitable to obtain accurate binary solutions.

Keywords: stars:binaries:eclipsing; stars:fundamental parameters; surveys:GAIA
PACS: 97.80.Fk, 97.80.Hn, 97.10.Nf, 97.10.Pg, 97.10.Ri, 95.55.-n

1. INTRODUCTION

Stellar astrophysics is a cornerstone for fundamental cosmogonic enquiry. In order to approach the broader basic questions properly, we must know, as confidently as possible, the physics of stars and the limits of this knowledge. GAIA is designed to obtain extremely precise astrometry (in the micro-arcsec regime), multi-band photometry and medium-resolution spectroscopy for a large sample of stars. This large amount of data of extreme accuracy will stimulate a revolution in the exploration of stellar and Galactic formation and evolution, and the determination of the cosmic distance scale.

The main scientific objectives of the GAIA mission and the expected benefits for astrophysical research are given by Gilmore et al. (1998), Perryman et al. (2001) and Mignard (2005). An overview of the GAIA mission is presented by Perryman (2005), the GAIA payload and spacecraft design is detailed by Pace (2005), and the goals of GAIA spectroscopy and photometry are discussed by Munari (1999a,b; 2002) and Høg (2002). The role of GAIA photometry and spectroscopy on eclipsing binaries has been reviewed in details by Munari et al. (2001) and Zwitter (2002, 2003). The GAIA large-scale photometric survey will have significant intrinsic scientific value for the study of

CP848, *Recent Advances in Astronomy and Astrophysics,*
7th *International Conference of the Hellenic Astronomical Society,* edited by N. H. Solomos

variable stars of nearly all types, including detached eclipsing binaries, near contact or contact binaries and pulsating stars. It is expected that about 10^6 Eclipsing Binaries (EBs), (with V$\leq$ 16 mag), will be discovered. In the five-year mission lifetime the number of photometric observations per target and for each one of the 4 broad bands will be about 100 and about 150 for each of the 11 intermediate bands, and the observing fashion will be quite similar to Hipparcos/Tycho mission operational mode. Even if reliable physical parameters could be derived for only 1% of the observed EBs, this would be a great contribution to stellar astrophysics and a giant leap in comparison with what has been obtained so far from ground based observations (Andersen 1991).

The study of EB systems provides insights into nearly all areas of astrophysics, including stellar interiors and atmospheres, stellar evolution, nucleosynthesis, plasma physics, magnetic dynamos (in cool stars), and relativistic physics to name a partial list. Recently, the study of EBs in other galaxies and clusters makes it possible to explore stellar evolution and establish mass-luminosity laws for galaxies with vastly different evolutionary and chemical histories from our Galaxy (such as LMC and SMC). Moreover, EBs are beginning to play an important role in cosmology as distant indicators to nearby galaxies. As more data are accumulated, the studies of these systems may lead to an improvement in the extragalactic distance scale.

2. SCIENTIFIC AND MISSION OBJECTIVES OF GAIA

Following the cut of the ESA Science budget, the system level design of GAIA has been reassessed, with the aim to drastically reduce the cost at completion by at least 20%, keeping the same scientific requirements. According to the new ESA Science Programme, approved by the ESA Science Programme Committee in May 2002, the GAIA spacecraft has been redesigned to be launched in 2011 with a Soyuz-Fregat launcher and it will be put in a Lissajous-type orbit around the Sun-Earth Lagrangian point L2. The new GAIA Technical Baseline at satellite level is summarized in detail by Pace (2005). That paper summarizes the current mission technical concept and introduces the technological developments required to make the GAIA mission feasible.

The objectives of the GAIA mission are to build a catalogue of 10^9 stars with accurate positions, parallaxes, proper motions, magnitudes and radial velocities. The catalogue will be complete up to V = 20 mag. The overall mission contains three parts: **(i)** astrometry, with the measurement of accurate stellar positions, parallaxes and proper motions; **(ii)** photometry, with measurements in different spectral bands; **(iii)** radial velocity measurements up to V = 17 mag.

2.1. Astrometry with GAIA

The main objective of the GAIA mission is to perform global or wide-field astrometry instead of local or narrow field astrometry (Pace 2005). While in local astrometry the star is measured with respect to a neighboring star in the same field, in global astrometry each star is connected to a large number of other stars in every direction. In the first case, even with an accurate instrument, the propagation of errors results in uncertain values when making a sky survey. In the case of global astrometry it is necessary to observe simultaneously two fields of view where the positions of the star are measured and compared. To obtain this, the payload will provide two lines of sight, obtained with two separate telescopes. Unlike Hipparcos, the two images will be focalized, slightly spaced, on a unique focal plane assembly. The angle between the instrument lines of sight defines what is called the basic angle. The scanning law is similar to that of Hipparcos and ensures the coverage of the sky to build up the star network. The lines of sight of both instruments are perpendicular to the spin axis, and the two instruments successively scan the same sky area, which can be viewed as a band whose height is equal to that of the field of view. From the instrument standpoint, the stars are crossing each field of view in a regular motion. A detailed description of the measurement principle is given by Pace (2005).

The expected accuracy for the astrometry mission is 10 μas in positional and parallax accuracy and 10 μas yr^{-1} for proper motion accuracy, for stars of V = 15 mag. The parallax accuracy is equivalent to an accuracy of 10% on the distance of a star at 10 kpc, while the proper-motion accuracy is equivalent to an accuracy of 1 km/s on the velocity of a star at 20 kpc. As for the astrometric data reduction is concerned, a more general global approach for the data reduction is being considered. Large amounts of information for the instruments are derived from the data reduction. The final catalogue will be built from the best estimate of all star positions and of the instrument parameters computed from the total amount of data collected during the mission. The final accuracy is only reached at the end of the mission, i.e., after 5 yr of operation, and after a complete processing of all the data.

2.2. Photometric observations of Eclipsing Binaries with GAIA

The photometry mission will be multi-colour and multi-epoch. Magnitudes will be obtained in 4 wide spectral bands for all stars up to V = 20 mag, complemented with measurements in 11 narrow bands for stars up to V = 17 mag. The multi-epoch aspect is performed by a regular observation of the stars throughout the mission. The photometric information obtained by the GAIA satellite will be far superior to that of Hipparcos. Considerable effort has been devoted to the design of an optimum filter system for GAIA, taking into account the spectral energy distribution of the main Galactic stellar populations, as derived from model atmospheres and spectrophotometric observations (Pickles 1998; Munari 1999b; Straizys 1999; Grenon et al. 1999a; Grenon et al. 1999b). The choice of filters for GAIA photometry is extremely important considering that GAIA multi-colour photometry will be the only astrophysical information available for the most of the brightest one billion stars on the sky during the next perhaps fifty years. The result of this effort is a baseline system, which provides the required accuracy for chromatic correction of the astrometric observations, and a good accuracy in the astrophysical information. The baseline system is considered to be a preliminary choice, and refinements with the same number of bands are being studied (Jordi & Høg 2005). The system features four broad wavelength bands (the Broad Band Photometer, BBP) and eleven bands of medium to narrow width (the Medium Band Photometer,

MBP). The bands span the complete spectral range from $\approx$ 280 to $\approx$ 1000 nm. The photometric filters are placed in two different locations: the four broad band filters are mainly dedicated to supplying the information necessary for the chromatic correction; they are thus located on top of the trailing CCD chips in the astrometric fields of view, thus ensuring strict simultaneity, as well as the use of identical CCD chips and same pixel resolution as the astrometric observations. The bands whose main purpose is to yield astrophysical information are located within the dedicated spectrometric viewing direction, as shown in Figure 2.13 of the GAIA manual (Concept and Technology Study Report, ESA-SCI(2000)4) where the different image scale allows for longer integration times and thus higher S/N.

The photometric data analysis is carried out on the ground. GAIA will discover huge numbers of spectroscopic and eclipsing binaries. The numbers are orders of magnitude larger than everything collected in the last century from ground. It is expected that about 1×10^6 eclipsing binaries (with $V \leq 16$ mag), will be discovered. In the five-year mission lifetime the number of photometric observations per target and for each one of the 4 broad bands will be about 100 and about 150 for each of the 11 intermediate bands, and the observing fashion will be quite similar to Hipparcos/Tycho mission operating mode. In many cases the observations obtained by GAIA will be good enough to determine system parameters at 1-2% accuracy level. They will have an immense impact on theories of stellar structure and evolution. Such a large amount of data requires an automatic process in all stages of reduction and interpretation. It is not possible by simple visual inspection to recognize photometric eclipses or winging radial velocity curves for the hundred thousands or even millions of the observed systems. All steps that were so far performed by astronomers, like the choice of initial fitting values, will have to be incorporated in an integrated reduction pipeline.

2.3. Spectroscopic observations of Eclipsing Binaries with GAIA

Eclipsing Binaries are a prime tool to derive fundamental stellar parameters like mass and radius, or the temperature scale. They can be used as accurate, geometric distance indicators (current best distances to LMC and globular clusters). Moreover, eclipsing binary stars represent the most astrophysically relevant type of binary stars that GAIA will observe spectroscopically during its mission. GAIA will survey a wide range of binary star separations and orbital periods. For the case of very long orbital periods (> 100 yrs) the components will be resolved on the sky and their binary nature could be judged from their equal parallax, proper motion and a similar radial velocity. Close binaries, down to orbital periods of days, will be discovered by the curved, non-linear proper motion path. Spectroscopy will be useful to discover even shorter period binaries. Therefore, GAIA will be able to study the binarity phenomenon for the whole range of orbital periods.

The spectrograph aboard GAIA will have a square primary mirror with the surface area of $0.5 \times 0.5\ m^2$ and a field of view of $1.65^\circ \times 1.6^\circ$. There is no slit on the spectrograph and the stars drift across the focal plane at a rate of $60''$/s. This gives a fixed exposure time of 99 seconds for each transit of the star across the focal plane. Because of a precessional motion of the spin axis of the satellite the whole celestial sphere will be covered, and each position on the sky will be observed on average 100-times during a 5-year mission. GAIA will record spectra covering the 8490-8750 Å wavelength range, at 0.75 Å/pix dispersion as currently baselined in the Concept and Technology Study Report (ESA SP(2000)4). The GAIA 8490-8750 Å interval is centered on the near-IR Ca II triplet and head of the Paschen series. It also includes numerous N I, Fe I, Ti I, Cr I, Si I, Ni I, S II, Mg I and Mn I lines (Munari 2002). GAIA wavelength range is the only spectral window in the near-IR, which is not crunched by telluric absorptions, allowing uncontaminated ground-based preparatory and follow-up observations.

The GAIA spectral targets (approaching 10^8 in current estimates) will require fully automatic data treatment and analysis. Although line intensities and ratios may be useful in pre-flight ground-based preparatory work and/or quick-look classification and re-direction along the reduction pipeline of actual GAIA data, it is clear that proper exploitation of GAIA spectra will require a smart spectral analysis approach. A review of GAIA spectroscopy has been presented by Munari (1999a, 2002), while the performance of the spectrograph and the expected radial velocity accuracy as a function of magnitude and spectral type was discussed by Zwitter (2002, 2003).

3. GAIA AND THE FUNDAMENTAL STELLAR PARAMETERS FROM ECLIPSING BINARIES

As mentioned above, the number of eclipsing binaries discovered by GAIA will reach hundreds of thousands. Therefore, suitable codes are needed to reduce and analyze the combined astrometric, photometric and spectroscopic observations of these systems in order reliable orbital and physical parameters are determined. How can these observations be compared with the state-of-the-art ground-based observations? What is the accuracy to which Eclipsing Binaries can be investigated using GAIA data alone? These are two crucial questions regarding the reliability of the derived stellar parameters from GAIA observations. Zwitter (2002, 2003) gives good estimates of the errors on single-star parameters (effective temperature, gravity, metallicity and distance) obtained from a combination of astrometry, photometry and spectroscopy. These error values could be checked by observations of some 4×10^5 eclipsing binaries with $V \leq 15$, of which some 10^5 will be double-lined. It is expected that for a few tens of thousands of such systems excellent orbital solutions with secure determination of stellar parameters will be obtained. On the other hand, for fainter systems ($V > 15$), although good quality light curves could be obtained, the spectroscopic observations will be of limited accuracy. In these cases only good estimates of stellar parameters can be secured.

The accuracy of the computed elements depends strongly on both the quality and number of the photometric and spectroscopic observations and the methods of light curve and spectral analysis. For ground-based observations these methods have reached a very high level, although there is space for further improvement. For GAIA observations, it was shown by Munari et al. (2001), Zwitter (2002, 2003), Marrese et al. (2004) and Milone et al. (2005) that stellar parameters could be determined (in many cases) at about 2% accuracy level. Given that GAIA observations will be superior to those of Hipparcos/Tycho mission, the results obtained by using Hipparcos observations should

TABLE 1. Light curve modelling of RZ Dra and V1010 Oph. (The absolute elements *M, R and L are in solar units)*

Parameter	RZ Dra		V1010 Oph	
	Ground obs.	Hipparcos	Ground obs.	Hipparcos
i (degrees)	87.21(27)	84.18(4)	85.86(68)	81.06(95)
T_1 (K)	8150	8150	8200	8200
T_2 (K)	4887(30)	5270(125)	5652(33)	5748(73)
Ω_1	2.824(4)	2.846(28)	2.775(27)	2.775(24)
Ω_2	2.767(5)	2.767(22)	2.772(28)	2.777(25)
$q = m_2/m_1$	0.444	0.444	0.448	0.448
$L_1/(L_1+L_2)$	0.909(3)	0.919(15)	0.896(4)	0.889(11)
r_1 (roche)	0.440(1)	0.435(6)	0.452(2)	0.452(4)
r_2 (roche)	0.311(1)	0.311(2)	0.312(12)	0.311(7)
M_1	1.40(4)	1.42(5)	1.87(12)	1.88(14)
M_2	0.62(3)	0.59(4)	0.90(5)	0.89(7)
R_1	1.62(1)	1.55(1)	2.08(6)	2.08(1)
R_2	1.12(1)	1.11(1)	1.48(8)	1.46(1)
L_1	10.1(4)	9.60(47)	10.7(7)	11.7(8)
L_2	1.01(6)	0.86(5)	3.47(35)	3.07(8)

present a lower limit to the GAIA expected accuracy. Another point to be stressed is that for double-lined detached and contact systems the results will be of the accuracy level mentioned before. The problem will be with the near-contact (semi-detached) systems, in which only one spectral line can be observed with medium size (2 m) ground-based telescopes. For bright such systems it is possible to observe with large telescopes both lines and disentagle them by modern techniques. The problem will again remain for GAIA spectroscopic observations of semi-detached systems, although the detection limit is increased.

In a recent study Niarchos & Manimanis (2003) made a first evaluation of the GAIA performance on the photometry of contact and near-contact eclipsing binaries by analyzing the light curves of selected systems using ground-based and Hipparcos (GAIA like) observations. In the present study we include systems of all basic spectral types (from A to K) and of both W and A type (in the case of contact systems). We used systems with very good observations taken from both the ground (in V filter) and the Hipparcos/Tycho mission in H_p filter (very close to V filter). For all systems the spectroscopic mass-ratio (derived by modern techniques) was used as fixed parameter in the light curve analysis. The temperature T_1 was also fixed at the value according to the spectral type of the system. The systems analysed are: the near contact systems RZ Dra (Kreiner et al. 1994) and V1010 Oph (Shaw et al. 1990, Manimanis 2002); and the contact systems: XZ Leo (Rucinski & Lu 1999, unpublished observations by K. Gazeas), ε CrA (Goecking & Duerbeck 1993, Twigg 1979), YY CrB (Rucinski et al. 2000, Gazeas et al. 2005), V839 Oph (Rucinski & Lu 1999, unpublished observations by K. Gazeas), AB And (Hrivnak 1988) and XY Leo (Hrivnak et al. 1984, unpublished observations by K. Gazeas). The Wilson-Devinney code was used for the light curve analysis and the results are given in Tables 1-3. In the same tables, together with the geometric and photometric elements, the fundamental parameters (absolute elements) of the analysed systems are

TABLE 2. Light curve modelling of XZ Leo, ε CrA and YY CrB. (The absolute elements *M, R and L are in solar units)*

Parameter	XZ Leo		ε CrA		YY CrB	
	Ground obs.	Hipparcos	Ground obs.	Hipparcos	Ground obs.	Hipparcos
i (degrees)	75.74(17)	76.35(66)	72.25(62)	73.60(15)	78.27(39)	81.82(100)
T_1 (K)	7250	7250	7100	7100	6100	6100
T_2 (K)	7045(18)	7151(57)	6639(40)	6549(74)	6221(11)	6236(27)
$\Omega_1=\Omega_2$	2.528(3)	2.530(11)	1.973(16)	1.983(8)	2.261(4)	2.288(7)
$q=m_2/m_1$	0.348	0.348	0.112	0.112	0.243	0.243
$L_1/(L_1+L_2)$	0.742(3)	0.731(9)	0.911(5)	0.905(7)	0.757(2)	0.760(5)
r_1 (roche)	0.486(1)	0.486(3)	0.581(2)	0.577(4)	0.531(1)	0.522(2)
r_2 (roche)	0.305(1)	0.304(3)	0.224(2)	0.219(3)	0.292(1)	0.282(2)
M_1	1.83(5)	1.82(13)	1.75(4)	1.69(7)	1.41(9)	1.37(16)
M_2	0.64(2)	0.63(6)	0.21(2)	0.22(5)	0.34(2)	0.33(4)
R_1	1.71(2)	1.71(2)	2.20(3)	2.12(12)	1.40(7)	1.36(7)
R_2	1.07(3)	1.07(3)	0.80(1)	0.80(4)	0.77(10)	0.74(10)
L_1	7.29(29)	7.26(29)	11.1(1)	10.3(20)	2.46(26)	2.33(25)
L_2	2.56(16)	2.69(19)	1.08(1)	1.07(14)	0.81(20)	0.74(20)

also given. From an inspection of Tables 1-3 it is obvious that the agreement between the derived elements and the fundamental stellar parameters, derived from ground-based and Hipparcos (GAIA-like) observations is more than satisfactory. The derived absolute elements from Hipparcos observations differ from those derived from ground-based observations within the limits of the combined errors. Since GAIA photometry will have excellent precision and much larger number of photometric bands than Hipparcos, it is expected that the resulting stellar parameters will be comparable to those obtained from high quality ground-based observations. This will not be the case for all observed eclipsing binaries, but only for the systems where a good phase coverage (especially during eclipses) is obtained. The eclipse coverage will be better in intermediate pass-band filters.

A good example of the phase coverage of spectroscopic, narrow- and broad-band photometric observations to be obtained by GAIA for a typical double-lined eclipsing system (the GK Dra) is given in Figure 4 of Zwitter (2003). In the same paper, the accuracies of the fundamental parameters for 6 systems studied so far by Munari et al. (2001) and Zwitter et al. (2003) are given. With some exceptions for individual radii, which are not well determined due to the incomplete phase coverage of eclipses, the relative errors in most parameters are 2% or lower. On the average, errors of the same order were found for the four systems studied in the work of Niarchos & Manimanis (2003).

4. SUMMARY AND CONCLUSIONS

There are some special advantages of the GAIA mission (with combined astrometric, spectroscopic and photometric observations) compared to the classical ground-based approach to eclipsing binary studies. GAIA will provide light curves for millions of faint

TABLE 3. Light curve modelling of V839 Oph, AB And and XY Leo. (The absolute elements *M, R and L are in solar units*)

Parameter	V839 Oph		AB And		XY Leo	
	Ground obs.	Hipparcos	Ground obs.	Hipparcos	Ground obs.	Hipparcos
i (degrees)	80.82(20)	81.93(100)	84.95(21)	86.89(179)	68.98(11)	68.60(61)
T_1 (K)	6200	6200	5470	5470	4850	4850
T_2 (K)	6406(5)	6302(32)	5305(6)	5194(31)	4435(8)	4664(40)
Ω_1=Ω_2	2.384(2)	2.387(10)	2.934(4)	2.891(20)	2.852(3)	2.877(11)
$q = m_2/m_1$	0.305	0.305	0.560	0.560	0.500	0.500
$L_1/(L_1+L_2)$	0.667(2)	0.671(4)	0.622(1)	0.629(6)	0.757(2)	0.702(13)
r_1 (roche)	0.514(1)	0.514(3)	0.444(1)	0.455(5)	0.448(1)	0.442(2)
r_2 (roche)	0.313(1)	0.312(3)	0.343(1)	0.354(4)	0.327(1)	0.320(2)
M_1	1.62(4)	1.61(17)	1.01(2)	1.00(9)	0.82(4)	0.83(21)
M_2	0.49(1)	0.49(5)	0.56(1)	0.55(5)	0.41(2)	0.41(10)
R_1	1.53(6)	1.52(6)	1.04(4)	1.06(4)	0.87(1)	0.86(1)
R_2	0.93(8)	0.92(8)	0.80(5)	0.83(5)	0.64(2)	0.62(2)
L_1	3.12(26)	3.10(26)	0.87(7)	0.91(7)	0.38(2)	0.37(2)
L_2	1.32(23)	1.22(21)	0.46(5)	0.45(5)	0.14(1)	0.17(1)

eclipsing binaries and also their absolute luminosities and temperatures. These data, combined with the radial velocity curves obtained from spectroscopic observations, will allow us to determine the fundamental parameters (radii, masses, luminosities, temperatures) of hundreds of thousands eclipsing systems. Although the expected accuracy will be moderate, the large amount of data will allow us to look for large deviations from the "normal" mass-radius-luminosity relations. An analysis of photometric observations for a small sample of eclipsing systems, obtained by ground-based and (GAIA-like) Hipparcos photometry, shows that the fundamental parameters of these systems can be determined at $\sim 2\%$ accuracy level.

The strength of the GAIA mission is in the numbers. As it was mentioned before, GAIA will observe 4×10^5 eclipsing binaries brighter than V =15 and 10^5 of these will be double-lined systems (Munari et al. 2001). According to Niarchos & Manimanis (2003) GAIA will observe about 4 hundred times more contact and near-contact binaries than those that are known from ground-based observations. Even if the stellar parameters will be determined at 1% accuracy for only 1% of them, this is still 25-times more than what has been obtained from all ground-based observations in the past (Andersen 1991). This will definitely have an immense impact on theories of stellar structure and evolution. Moreover most of the GAIA binaries will be of G-K spectral type (Zwitter & Henden 2003) for which accurate solutions exist only for a small number of systems.

REFERENCES

1. J. Andersen, *A&ARv*, **3**, No 2, 91-126 (1991).
2. *GAIA Consept and Technology Study Report*, ESA-SCI 4 (2000).
3. K. D. Gazeas, A. Baran, P. Niarchos, et al., *Acta Astron.*, **55**, 123-140 (2005).
4. G. Gilmore, M. Perryman, L. Lindegren, et al., "GAIA: origin and evolution of the Milky WAY", in *Astronomical Interferometry*, edited by R.D. Reasenberg, SPIE Conference Proceedings 3350, The International Society for Optical Engineering, Bellingham WA, USA, 1998, pp. 541-550.

5. K. -D. Goecking, and H. Duerbeck, *A&A*, **278**, 463-466 (1993).
6. M. Grenon, C. Jordi, F. Figueras, and J. Torra, "Stellar Classification from GAIA Intermediate Band System", Technical Report MGUB-PWG-003, Obs. De Geneve, 1999a.
7. M. Grenon, C. Jordi, F. Figueras, and J. Torra, "GAIA Broad Band Photometry", Technical Report MGUB-PWG-003, Obs. De Geneve, 1999b.
8. E. Høg, "Photometric and Imaging Performance", in *GAIA: A European Space Project*, edited by O. Bienayme and C. Turon, EAS Pub. Ser. 2, Les Houches, France, 2002, pp. 27-37.
9. B. J. Hrivnak, *ApJ*, **335**, 319-330 (1988).
10. B. J. Hrivnak, E. F. Milone, G. Hill, and W. A. Fisher, *ApJ*, **285**, 683-687 (1984).
11. C. Jordi, and E. Høg, "Photometric System Design and Performances", in *The Three-Dimensional Universe with Gaia*, edited by C. Turon, K.S. O'Flaherty and M.A.C. Perryman, ESA SP-576, Paris, 2005, pp. 43-50.
12. J. M. Kreiner, G. Pajdosz, J. Tremko, and S. Zola, *A&A*, **285**, 459-466 (1994).
13. V. Manimanis, *Ph.D. Dissertation*, University of Athens, Greece (2002).
14. P. M. Marrese, U. Munari, A. Siviero, et al., *A&A*, **413**, 635-642 (2004).
15. F. Mignard, "Overall Science Goals of the Gaia Mission", in *The Three-Dimensional Universe with Gaia*, edited by C. Turon, K.S. O'Flaherty and M.A.C. Perryman, ESA SP-576, Paris, 2005, pp. 5-14.
16. E. F. Milone, U. Munari, P. M. Marrese, et al., *A&A*, **441**, 605-613 (2005).
17. U. Munari, *Baltic Astronomy*, **8**, 73-96 (1999a).
18. U. Munari, *Baltic Astronomy*, **8**, 123-128 (1999b).
19. U. Munari, T. Tomov, T. Zwitter, et al., *A&A*, **378**, 477-486 (2001).
20. U. Munari, "GAIA Spectroscopy and Radial Velocities", in *GAIA: A European Space Project*, edited by O. Bienayme and C. Turon., EAS Pub. Ser. 2, Les Houches, France, 2002, pp. 39-54.
21. P. Niarchos, and V. Manimanis, *A&A*, **405**, 263-269 (2003).
22. O. Pace, "Gaia: The Satellite and Payload", in *The Three-Dimensional Universe with Gaia*, edited by C. Turon, K.S. O'Flaherty and M.A.C. Perryman, ESA SP-576, Paris, 2005, pp. 23-28.
23. M. Perryman, K. S. De Boer, G. Gilmore, et al., *A&A*, **369**, 339-363 (2001).
24. M. Perryman , "Overview of the Gaia Mission", in *The Three-Dimensional Universe with Gaia*, edited by C. Turon, K.S. O'Flaherty and M.A.C. Perryman, ESA SP-576, Paris, 2005, pp. 15-22.
25. A. J. Pickles, *PASP*, **110**, 863-878 (1998).
26. S. M. Rucinski, and W. -X. Lu, *AJ*, **118**, 2451-2459 (1999).
27. S. M. Rucinski, W. -X. Lu, and S. W. Mochnacki, *AJ*, **120**, 1133-1139 (2000).
28. J. S. Shaw, E. F. Guinan, and C. J. Garasi, *BAAS* **22**, 1296 (1990).
29. V. Straizys, *Baltic Astronomy*, **8**, 109-121 (1999).
30. L. W. Twigg, *MNRAS* **189**, 907-909 (1979).
31. T. Zwitter: 2002, "GAIA Survey of Galactic Eclipsing Binaries", in *Exotic Stars as Challenges to Evolution*, edited by C. A. Toot and W. Van Hamme, ASP Conf. Ser. 279, Astronomical Society of the Pacific, San Francisco, 2002, p.31.
32. T. Zwitter, "Observational tests of the GAIA expected harvest on eclipsing binaries", in *GAIA Spectroscopy, Science and Technology*, edited by U. Munari, ASP Conf. Ser. 298, Astronomical Society of the Pacific, San Francisco, 2003, p. 329.
33. T. Zwitter, and A. A. Henden, "Crowding in the GAIA spectrograph focal plane", in *GAIA Spectroscopy, Science and Technology*, edited by U. Munari, ASP Conf. Ser. 298, Astronomical Society of the Pacific, San Francisco, 2003, p. 489.
34. T. Zwitter, U. Munari, P. M. Marrese, et al., *A&A* , **404**, 333-340 (2003).

Long Term Variability of the Radial Velocities in the Coronal and Post-coronal Regions of the Oe Star HD 93521

Antoniou, A.[1], Danezis, E.[1], Lyratzi, E.[1], Nikolaidis, D.[1], Popović, L. Č.[2], Dimitrijević, M. S.[2] and Theodossiou, E.[1]

1. University of Athens, School of Physics, Department of Astrophysics, Astronomy and Mechanics, Panepistimiopolis, Zografos 157 84, Athens – Greece
2. Astronomical Observatory, Volgina 7, 11160 Belgrade, Serbia

Abstract. We examine the timescale changes of C IV, N IV and N V spectral lines of the Oe star HD 93521, during a period of 16 years, applying the model proposed by Danezis et al. (2005). We found that the spectral lines consist of one or more Satellite Absorption Components (SACs or DACs) which construct the whole spectral profile. In this paper we present the time scale variation of the radial velocities.

Keywords: Oe stars, SACs, radial velocity.
PACS: 97.10.Ex, 97.10.Fy, 97.20.Ec, 97.30.Eh

INTRODUCTION

HD 93521 is a relatively bright, very rapidly rotating O9.5V star (Hobbs et al. 1982). These characteristics, together with its exceptionally high Galactic latitude (b=+63.130, Galactic length l=183.30, Costero & Stalio 1984) have made it a favorite target for studies regarding stars out of the Galactic plane (e.g. Pettini & West 1982).

Since its adoption as a spectrophotometric standard for IUE (Bohlin et al. 1980), it has acquired an increasingly well – documented record of spectroscopic variability (Garmany et al. 1980). The ultraviolet spectrum shows wind signatures at C IV, N V, and Si IV. The presence of a strong Si IV wind line is exceptional for a luminosity class V star; indeed, all the wind profiles have unusual morphologies (Prinja & Howarth 1986), which have been interpreted as evidence for a cylindrically (as opposed to spherically) symmetric wind (Massa 1992). According to C IV resonance line profile of HD 93521, Massa (1992) and Howarth & Reid (1993) also suggested, that there is a high – speed component in the polar outflow from the star as well as a low – speed component in the equatorial regions. Howarth & Reid (1993) supported that the mean profiles of the resonance lines of C IV, NV and SiIV show that the morphology of the lines is very unusual, and it is possible to identify three separate components: very strong, low - velocity absorption in Si IV and C IV, which is saturated out about -500 km/s; weaker absorption which extends to about -1200 km/s in C IV and N V and emission in C IV and N V which is unusually strong for a late O main sequence star.

CP848, *Recent Advances in Astronomy and Astrophysics,*
7th *International Conference of the Hellenic Astronomical Society,* edited by N. H. Solomos

THE GAUSSIAN - ROTATIONAL MODEL (GR-MODEL)

With the following model we can calculate the apparent rotational and radial velocities, the Gaussian deviation of the random motions of the ions, the random velocities of these motions, as well as the optical depth, the Full Width at Half Maximum (FWHM), the absorbed and the emitted energy and the product of the Source function S and the optical depth ξ of the independent regions of matter which produce the main and the satellites components (SACs) of the studied spectral lines.

For our study the line broadening is caused by two reasons: The first one is the rotational velocity of the spherical region that produce the spectral line and the second one the random velocities of the ions, which make thermal random motions. In this model we present a new approach, which describes both of these factors.

We consider that the area of gas, which creates a specific spectral line consists of i independent absorbing shells followed by j independent shells that both absorbs and emits and an outer absorbing shell. Such a structure produces DACs or SACs (Danezis et al. 2003) and the final line function is:

$$I_\lambda = \left[I_{\lambda 0} \prod_i \exp\{-L_i \xi_i\} + \sum_j S_{\lambda ej} \left(1 - \exp\{-L_{ej}\xi_{ej}\}\right) \right] \exp\{-L_g \xi_g\} \qquad (1)$$

where: $I_{\lambda 0}$**:** the initial radiation intensity,

L_i, L_{ej}, L_g: are the distribution functions of the absorption coefficients $k_{\lambda i}$, $k_{\lambda ej}$, $k_{\lambda g}$ respectively. Each L depends on the values of the apparent rotational velocity as well as of the radial expansion/contraction velocity of the density shell, which forms the spectral line *(V_{rot}, V_{exp})*,

$\xi = \int_0^s \Omega \rho ds$ is an expression of the optical depth τ, where Ω is an expression of k_λ, $S_{\lambda ej}$: the source function, which, at the moment when the spectrum is taken, is constant.

The function (1) does not depend on the geometry of the regions which create the observed feature. This means that L may represent any distribution which considers certain geometry, without changing anything in I_λ.

One of the hypotheses when we constructed the rotational model was that the line's width $\Delta\lambda$ is only a rotational effect. This means that the random velocities were very low and they did not contribute to the line broadening. In a new approach of the problem we also consider the parameter of random velocities in the calculation of the distribution function *L* and we have:

$$L(\lambda) = \frac{\sqrt{\pi}}{2\lambda_0 z} \int_{-\frac{\pi}{2}}^{\frac{\pi}{2}} \left[erf\left(\frac{\lambda - \lambda_0}{\sigma\sqrt{2}} + \frac{\lambda_0 z}{\sigma\sqrt{2}} \cos\theta \right) - erf\left(\frac{\lambda - \lambda_0}{\sigma\sqrt{2}} - \frac{\lambda_0 z}{\sigma\sqrt{2}} \cos\theta \right) \right] \cos\theta d\theta \qquad (2)$$

where λ_0 is the laboratory wavelength of a spectral line that arises from a specific point A_i of the equator of a spherical shell, $z = \frac{V_{rot}}{c}$ (V_{rot} is the rotational velocity of the point A_i). This *L(λ)* of the equation (2) is the distribution that replaces the rotational distribution *L* that Danezis et al (2003) proposed (see Danezis et al. 2003 and Danezis et al. 2005a,b).

LONG TERM VARIABILITY OF THE RADIAL VELOCITIES

The Analysis Of The Shapes

This study is based on eleven different spectra of HD 93521 taken with the IUE – Data satellite. We study the structure of the spectral lines C IV λλ 1548.155 Å, 1550.774 Å, N IV λ 1718.80 Å and N V λλ 1238.821, 1242.804 Å.

We present some spectral lines and their best fits of the C IV, N IV and N V regions of three different dates. These are the IUE - data SWP04472 (4.3.79), SWP44900 (9.6.72) and SWP30086 (12.1.87) respectively.

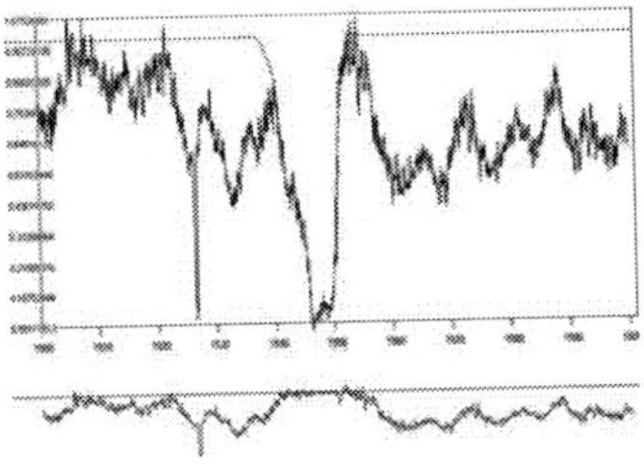

FIGURE 1. The C IV λλ 1548.155, 1550.774 Å resonance lines in the spectrum of HD 93521. Each of C IV spectral lines consists of five SACs. The graph bellow the fit indicates the difference between the observed spectrum and the fit.

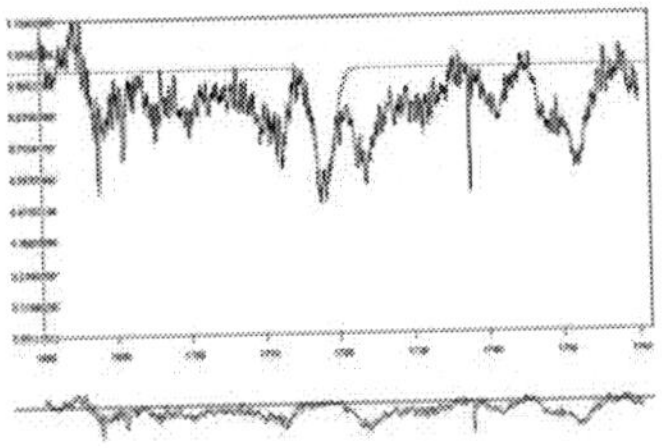

FIGURE 2. The N IV λ 1718.80 Å absorption line in the spectrum of HD 93521. The N IV spectral line consists of one SAC. The graph bellow the fit indicates the difference between the observed spectrum and the fit.

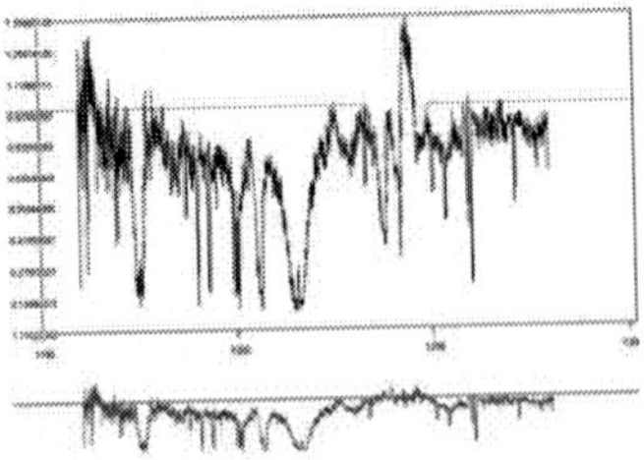

FIGURE 3. Each of N V λλ 1238.821, 1242.804 Å resonance lines in the spectrum of HD 93521 shows a characteristic P Cygni profile. Each of these spectral lines consists of one SAC. The graph bellow the fit indicates the difference between the observed spectrum and the fit.

Results

The following diagrams describe the time scale changes of the radial velocities of the coronal regions where are created the C IV, the N IV and the N V ions. We took into account that the shift of the interstellar lines corresponds to a mean value of the radial velocity about -387 km/s. We took also into account that the radial velocity of the star is about -11 km/s (Garmany et al. 1980).

The C IV Region

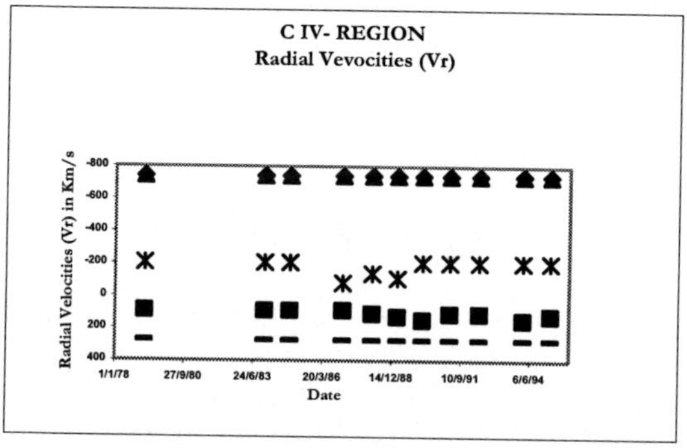

FIGURE 4. Timescale variations of the apparent radial velocities V_{rad} (km/s) of the λλ 1548.155, 1550.774 Å C IV resonance lines for the independent density regions of matter which create the 5 satellite components. We see that the radial velocity in each component remains constant.

The N IV Region

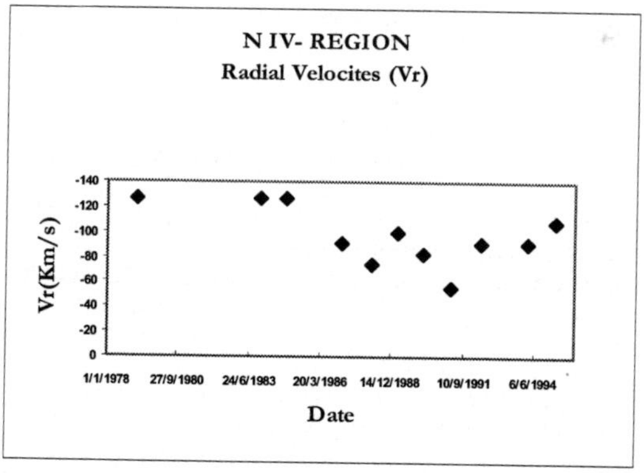

FIGURE 5. Time scale changes of the apparent radial velocities V_{rad} (km/s) of the density region which creates the N IV spectral line λ 1718.8 Å. The time scale variability is remarkable.

The N V Region

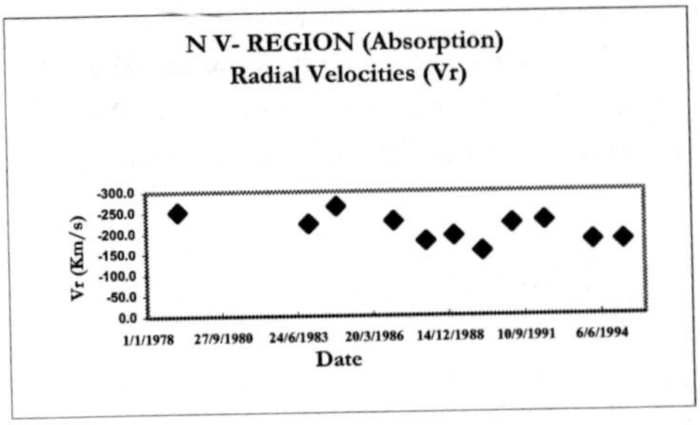

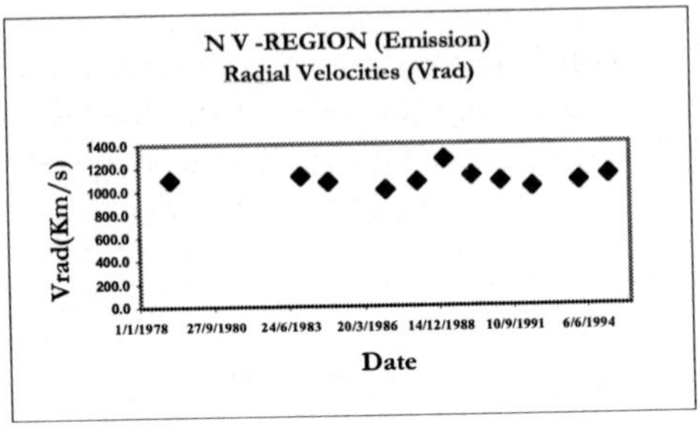

FIGURE 6. Timescale variations of the mean values of the apparent radial velocity V_{rad} (km/s) of the absorption and emission component of the N V resonance lines λλ 1238.821, 1242.804 Å. In each case we note also remarkable time scale variability.

ACKNOWLEDGMENTS

This research project is progressing at the University of Athens, Department of Astrophysics, Astronomy and Mechanics, under the financial support of the Special Account for Research Grants, which we thank very much. This work also was supported by Ministry of Science and Environment Protection of Serbia, through the projects "Influence of collisional processes on astrophysical plasma line shapes" and "Astrophysical spectroscopy of extragalactic objects".

REFERENCES

1. R. C. Bohlin, W. M. Sparks, A. V. Holm, B. D. Savage and M. A. Snijders, *Astronomy & Astrophysics*, **85**, 1 (1980).
2. R. Costero, and R. Stalio, *Astronomy & Astrophysics Supplement Series,* **58,** 95 (1984).
3. E. Danezis, D. Nikolaidis, V. Lyratzi, M. Stathopoulou, E. Theodossiou, A. Kosionidis, C. Drakopoulos, G. Christou and P. Koutsouris, A&SS, **284**, 1119 (2003).
4. E. Danezis, D. Nikolaidis, E. Lyratzi, L. Č. Popović, M. S. Dimitrijević, E. Theodossiou and A. Antoniou, Mem. S.A.It., in press (2005a).
5. E. Danezis, D. Nikolaidis, E. Lyratzi, L. Č. Popović, M. S. Dimitrijević, E. Theodossiou and A. Antoniou, Proceedings of 7th HEL.A.S. Conference, in press (2005b).
6. C. D. Garmany, P. S. Conti and P. Massey, *Astrophysical Journal*, **242**,1063 (1980).
7. L. M. Hobbs, G. Wallerstein and E. M. Huu, *Astrophysical Journal,*.**252**, L17 (1982).
8. I. D. Howarth and A. Reid, *Astronomy and Astrophysics*, **279**, 148H (1993).
9. D. Massa, *ASP Conference Series,* **22,** 84 (1992).
10. M. Pettini and K. A. West, *In ESA 3rd European IUE Conf.,* pp. 435-437 (1982).
11. R. Prinja and I. D. Howarth, *Astrophysical Journal Supplement Series* **61,** 357 (1986).

First Photometric Observations of Recently Discovered Binaries

E.Antonopoulou, G. Kontogiannis, N.Nanouris, G.Giannikakis, O.Dionatos

Department of Astrophysics, Astronomy and Mechanics, University of Athens, Panepistimiopolis Zografou GR-15784 Athens, Greece

Abstract. We present preliminary results of BVRI photometry of binary stars out of an observing list of short period binaries without any previous multifilter observations. The purpose of this project is to obtain photometric elements and the absolute parameters for further understanding of the evolutionary processes for these systems. The observations took place at the Kryonerion Astronomical Station (NOA). New times of minima are presented and a more accurate ephemeris is suggested.

Keywords: Eclipsing binary systems, Photometry, V1025 Her, FO Aur, V1051 Her.
PACS: 97.80.Hn.

INTRODUCTION

Binary stars V1025 Her, FO Aur and V1051 Her are eclipsing binaries out of a list of targets that have not been previously studied. Our target list includes 45 selected close-binary systems with period less than 1 day. These stars are rapid rotators and might exhibit interesting phenomena such as chromospheric activity. These systems have been registered in recently published catalogues, such as ROTSE-1 survey database, where their type and ephemeris is included. However, we find that these elements are not always accurate and need to be re-estimated.

DATA ACQUISITION

Our data were obtained using the 1.2m reflector of the Kryonerion Astronomical Station of the Athens National Observatory equipped with a *Photometrics* CCD camera (512x512 pixels, field of view 2′x2′) in four filters (B,V,R,I). Photometric results were obtained using IDL routines for the radiometric calibrations and GAIA software package (Starlink) for the aperture photometry. Observations took place between March and July, 2005.

ANALYSIS - DISCUSSION

V1025 Her *(GSC 2047:270)* is a relatively faint target ($V_{ROTSE-1}$ ~ 12.1 mag). Due to our limited field of view, only three fainter nearby objects could be simultaneously

CP848, *Recent Advances in Astronomy and Astrophysics,*
7th International Conference of the Hellenic Astronomical Society, edited by N. H. Solomos

observed. After thoroughly testing their stability in time, we selected the two most appropriate as comparison and check stars.

At first, the observations were made during two periods, May 16-19 and June 14-16, 2005. System's phases were calculated using the ephemeris given by Diethelm in the ROTSE-1 CCD data VI catalogue (Diethelm 2001):

$$MinI(HJD) = 2451286.881(2) + 0.5615(2) \cdot E.$$

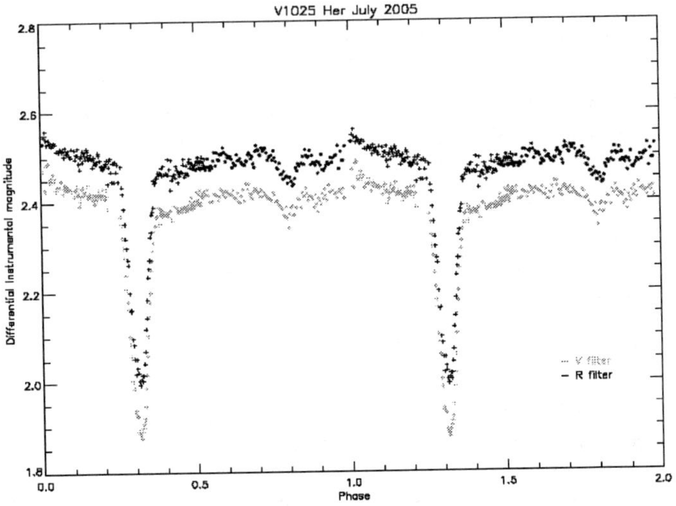

FIGURE 1. VR photometry of V1025 Her, during July, 2005 (from top to bottom are R and V light curves).

The light curve acquired in June 2005 revealed a considerable displacement in the phase of primary minimum. However, it appears that this displacement was not due to any physical mechanism but stems from the fact that the period of the binary has not been determined accurately. Indeed, after we had determined ten new times of minima, during a consecutive interval of four months (TABLE 1.), the ephemeris was corrected and eventually a new one, more precise, is suggested below:

$$MinI(HJD) = 2453559.45614(22) + 0.563358(29) \cdot E.$$

Higher resolution photometry in filters V and R, which was performed in July 2005, (FIGURE 1.) shows variations outside eclipses, indicating intrinsic variability of at least one of the members or mass transfer from the one component to the other. There are also indications of a distortion wave, probably due to chromospheric activity. Further observations and analysis need to be carried out in order to clarify the nature of this variability.

TABLE 1. The new CCD times of minima of V1025 Her.

Date	*Min (HJD)*	*Error*	*Type*	*Filter*
17 May 2005	2453508.4705	0.0005	II	R
17 May 2005	2453508.4661	0.0008	II	I
15 June 2005	2453537.4845	0.0012	I	B
15 June 2005	2453537.48263	0.00028	I	V
15 June 2005	2453537.4822	0.0003	I	R
15 June 2005	2453537.4806	0.0004	I	I
5 July 2005	2453557.47568	0.00028	II	V
5 July 2005	2453557.4784	0.0004	II	R
7 July 2005	2453559.4551	0.0003	I	V
7 July 2005	2453559.45614	0.00022	I	R

F0 Aur *(GSC 2409:1108)* is a faint target ($V_{GCVS} \sim 14.3$ mag). Due to our limited field of view, only five fainter nearby objects could be simultaneously observed. After thoroughly testing their stability in time, we selected the two most appropriate as comparison and check stars. The observations were made during March 14-17 and April 11-14, 2005. The phases were calculated using the ephemeris given by the GCVS v.4 catalogue (Kholopov et al., 1985-2004):

$$MinI(HJD) = 2426224.55 + 0.6100525 \cdot E.$$

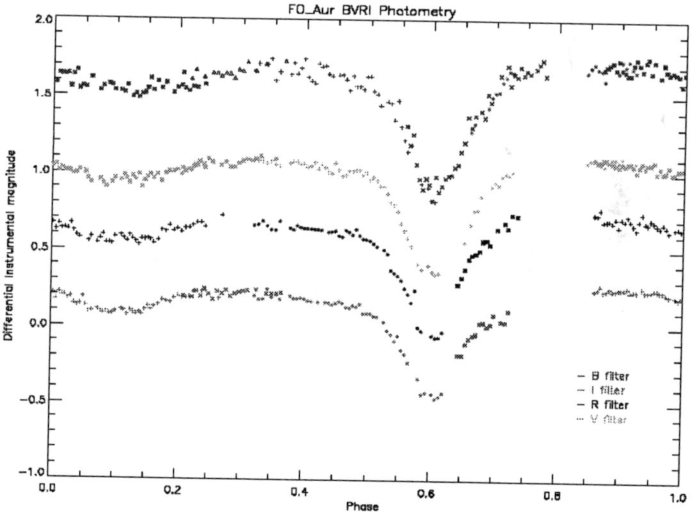

FIGURE 2. BVRI photometry of FO Aur, during March and April, 2005 (from top to bottom are B, V, R and I light curves).

Although it has been classified as EA-type binary in the GCVS catalogue, it turns out that it is an EB-type binary (FIGURE 2.). The displacement of the light curve could be either due to physical mechanisms active in the binary or because of the ephemeris being old and inaccurate. The asymmetry between the ascending and descending branch of the primary minimum may be an indication of spot activity. Six new times of minima during March were accurately determined and are presented below (TABLE 2.).

TABLE 2. The new CCD times of minima of FO Aur.

Date	*Min (HJD)*	*Error*	*Type*	*Filter*
15 March 2005	2453445.4761	0.0012	I	R
16 March 2005	2453446.3936	0.0010	II	B
16 March 2005	2453446.3978	0.0008	II	V
16 March 2005	2453446.3956	0.0005	II	R
16 March 2005	2453446.3921	0.0004	II	I
17 March 2005	2453447.3031	0.0010	I	B

***V1051 Her** (GSC 2069:150)* is a relatively faint target ($V_{ROTSE-1}$ ~ 12.5 mag). Only two fainter objects were observable within our field view, which were selected as comparison and check stars. The VRI observations took place in July 6-8, 2005. The phases were calculated using the ephemeris given by Diethelm in the ROTSE-1 CCD data VI catalogue (Diethelm 2001):

$$MinI(HJD) = 2451286.8856(17) + 0.5582(1) \cdot E.$$

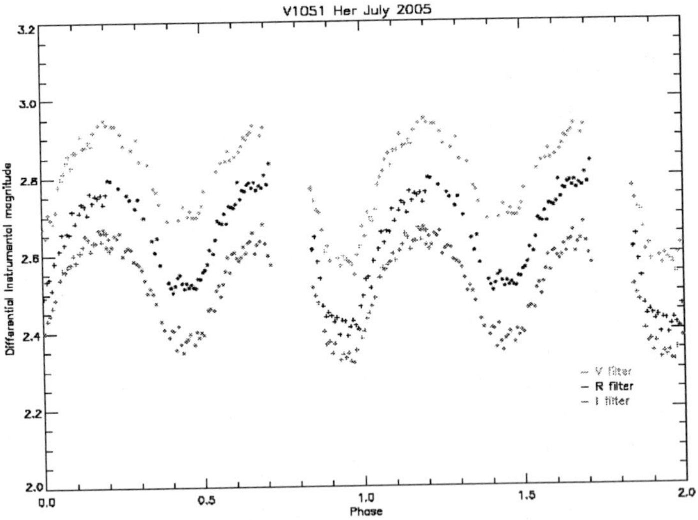

FIGURE 3. VRI photometry of V1051 Her, during July, 2005 (from top to bottom are R and V light curves).

V1051 Her is classified as an EB-type binary in the GCVS catalogue but it is undoubtedly an EW-type (FIGURE 3.). Once again there is a shift in the phase of minima possibly due to inaccurate period estimation.

We have to point out that the determination of all times of minima above was carried out using the K&W method (Kwee & van Woerden 1956). However, apart from the primary minimum of V1025 Her which is strictly symmetric, the shape of minimum profiles is characterised by obvious irregularities, reducing the validity of the previous method. Hence, a re-estimation of times of minima using least-squares approximation by polynomial fit (Breinhorst et al. 1973) is under consideration.

CONCLUSIONS

It appears that there are points to be reconsidered and defined, concerning the three binaries presented above, especially V1025 Her, which seems to be a very interesting case. New ephemeris' need to be constructed before a further analysis takes place.

Our attempt in near future focuses on the estimation of some colour indices of the examined systems, using standard stars, in order to determine approximately their temperature. Therefore, further spectroscopic observations and new photometric ones are needed to get the absolute parameters of their components and identify their nature.

ACKNOWLEDGMENTS

The authors would like to thank the National Observatory of Athens (NOA) and its stuff for providing with observation time, facilities and technical support. We also thank the University of Athens (Special Account for Research Grants) for providing funds for the computing facilities for data reduction and analysis.

REFERENCES

1. Breinhorst R., Pfleiderer J., Reinhardt M., Karimie M., 1973, A&A, 22, 239.
2. Diethelm R., 2001, IBVS 5060.
3. Kholopov P. et al., 1985-2004 (General Catalogue of Variable Stars, fourth edition, electronic version).
4. Kwee K. & van Woerden H., 1956, BIAN, 12, 327.

A New Modeling Approach For DACs And SACs Regions In The Atmospheres Of Hot Emissions Stars

Danezis, E.[1], Nikolaidis, D.[1], Lyratzi, E.[1], Popović, L. Č.[2], Dimitrijević, M. S.[2], Theodossiou, E.[1] and Antoniou, A.[1]

1. University of Athens, Faculty of Physics, Department of Astrophysics, Astronomy and Mechanics,Panepistimioupoli, GR 157 84 Zographou, Athens, Greece
2. Astronomical Observatory, Volgina 7, 11160, Belgrade, Serbia

Abstract. The presence of Discrete Absorption Components (DACs) or Satellite Absorption Components (SACs) is a very common phenomenon in the atmospheres of hot emission stars (Danezis et al. 2003, Lyratzi & Danezis 2004) and result to the complex line profiles of these stars. The shapes of these lines are interpreted by the existence of two or more independent layers of matter nearby a star. These structures are responsible for the formation of a series of satellite components for each spectral line. Here we will present a model reproducing the complex profile of the spectral lines of Oe and Be stars with DACs and SACs (Danezis et al. 2003, Lyratzi & Danezis 2004). In general, this model has a line function for the complex structure of the spectral lines with DACs or SACs and include a function L that considers the kinematic (geometry) of an independent region. In the calculation of the function L we have considered the rotational velocities of the independent regions, as well as the random velocities within them. This means that the new function of L is a synthesis of the rotational distribution and a physical Gaussian. Finally, we calculate the optical depth (ξ) and the column density (d) of each independent density region.

Keywords: Hot emission stars, models, DACs.
PACS: 97.10.Ex, 97.10.Fy, 97.20.Ec, 97.30.Eh

INTRODUCTION

One of the most important phenomena in the spectra of hot emission stars is the DACs (Discrete Absorption Components) phenomenon (Peton 1974, Underhill 1975, Lamers et al. 1982, Sahade et al. 1984, Sahade & Brandi 1985, Hutsemékers 1985, Danezis 1984, 1987, Danezis et al. 1991, 2003).

DACs are discrete but not unknown absorption spectral lines. They are spectral lines of the same ion and the same wavelength as a main spectral line, shifted at different Δλ, as they are created in different density regions which rotate and move radially with different velocities (Danezis et al. 2003, Lyratzi & Danezis 2004). DACs are lines, easily observed, when the regions that give rise to such lines, rotate with low velocities and move radially with high velocities. However, if the regions, that give rise to such lines, rotate with large velocities and move radially with small velocities,

CP848, *Recent Advances in Astronomy and Astrophysics,*
7th *International Conference of the Hellenic Astronomical Society,* edited by N. H. Solomos

the produced lines are very broadened but have small shifts. As a result they are blended among themselves as well as with the main spectral line and thus they are not discrete. In such a case the name Discrete Absorption Component is inappropriate and we use only the name SACs (Satellite Absorption Components).

DESCRIPTION OF THE MODEL

The Line Profile Function

Some years ago our group proposed a new model to explain the complex structure of the density regions of hot stars, where the spectral lines that present SACs or DACs are created (Danezis et al. 1991, 1998, 2000a,b,c, 2002a,b,c, 2003), Laskarides et al. 1992a,b).

The main hypothesis of this model is that the atmospherical region where a specific line is created is not continuous, but it is composed of a number of successive independent absorbing density regions, a number of emission regions and an external general absorption region.

By solving the equations of radiation transfer through a complex structure, as the one described, we conclude to a function for the line's profile, able to give the best fit for the main spectral line and its Satellite Components in the same time (Equation 1).

$$I_\lambda = \left[I_{\lambda 0} \prod_i e^{-x_{ai}} + \sum_j S_{\lambda ej}\left(1 - e^{-x_{ej}}\right) \right] e^{-x_g} \qquad (1)$$

where: $I_{\lambda 0}$: is the initial radiation intensity, $S_{\lambda ej}$ is the source function, which, at the moment when the spectrum is taken, is constant and $e^{-x_{ai}}$, $e^{-x_{ej}}$, e^{-x_g} are the distribution functions of the absorption, emission and general absorption lines respectively. This function I_λ does not depend on the geometry of the regions which create the observed feature.

The Rotation Distribution Function

One of the main hypotheses when we constructed the old version of the model (rotation model) was that the line's width $\Delta\lambda$ is only a rotational effect and we consider spherical symmetry for the independent density regions, which create the satellite components. This means that the random velocities were very low and they did not contribute to the line broadening. In such a case Eq. (1) becomes:

$$I_\lambda = \left[I_{\lambda 0} \prod_i e^{-L_{ai}\xi_{ai}} + \sum_j S_{\lambda ej}\left(1 - e^{-L_{ej}\xi_{ej}}\right) \right] e^{-L_g \xi_g} \qquad (2)$$

where: $I_{\lambda 0}$: is the initial radiation intensity, L_{ai}, L_{ej}, L_g: are the distribution functions (Rotation distribution) of the absorption coefficients $k_{\lambda ai}$, $k_{\lambda ej}$, $k_{\lambda g}$, respectively and ξ is the optical depth.

In the present work we propose a new approach of the problem, as we also consider the parameter of random velocities in the calculation of the distribution function L (See Danezis et al. 2005). This new L is a synthesis of the rotation distribution that we had presented in the old rotational model and a Gaussian. This means that the new L has two limits, the first one gives us a Gaussian and the other the old rotation distribution.

Calculation Of The New Distribution Function (Gauss-Rotation)

Let us consider a spherical shell and a point A_i in its equator (See Fig. 1a). If the laboratory wavelength of a spectral line that arises from A_i is λ_{lab}, the observed wavelength will be $\lambda_0=\lambda_{lab}+\Delta\lambda_{rad}$.

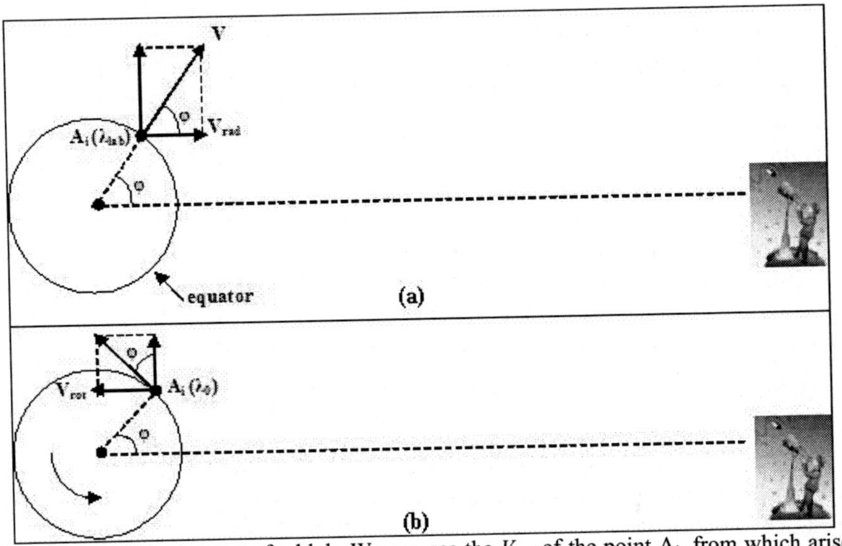

FIGURE 1. View of the equator of a blob. We can see the V_{rad} of the point A_i, from which arise the $\Delta\lambda_{rad}$ (a) and the V_{rot} from which arise the $\Delta\lambda_{rot}$ (b).

If the spherical density region rotates (See Fig. 1b), we will observe a displacement $\Delta\lambda_{rot}$ and the new wavelength of the center of the line λ_i is $\lambda_i = \lambda_0 \pm \Delta\lambda_{rot}$, where $\Delta\lambda_{rot} = \lambda_0 z \sin\varphi$ and $z = \frac{V_{rot}}{c} = \frac{\Delta\lambda_{rot}}{\lambda_0 \sin\varphi}$, where V_{rot} is the observed rotational velocity of the point A_i.

This means that $\lambda_i = \lambda_0 \pm \lambda_0 z \sin\varphi = \lambda_0(1 \pm z\sin\varphi)$ and if $-\frac{\pi}{2} < \varphi < \frac{\pi}{2}$ then $\lambda_i = \lambda_0(1 - z\sin\varphi)$.

If we consider that the spectral line profile is a Gaussian distribution we have: $P(\lambda) = \frac{1}{\sqrt{2\pi}\sigma} e^{-\left[\frac{\lambda-\kappa}{\sigma\sqrt{2}}\right]^2}$ where κ is the mean value of the distribution and in the case of the line profile it indicates the center of the spectral line that arises from A_i. This means that $P(\lambda) = \frac{1}{\sigma\sqrt{2\pi}} e^{-\left[\frac{\lambda-\lambda_0(1-z\sin\varphi)}{\sigma\sqrt{2}}\right]^2} = \frac{1}{\sigma\sqrt{2\pi}} e^{-\frac{[\lambda-\lambda_0(1-z\sin\varphi)]^2}{2\sigma^2}}$. For all the semi-equator we have $L(\lambda)=\int_{-\frac{\pi}{2}}^{\frac{\pi}{2}}\frac{1}{\sqrt{2\pi\sigma}}e^{-\frac{[\lambda-\lambda_0(1-z\sin\varphi)]^2}{2\sigma^2}}\cos\varphi d\varphi$. If we make the transformation $\sin\varphi=x$ and $u=\frac{\lambda-\lambda_0(1-zx)}{\sqrt{2}\sigma}$ then $L(\lambda)=\frac{1}{\lambda_0 z\sqrt{\pi}}\int_{\frac{\lambda-\lambda_0(1+z)}{\sigma\sqrt{2}}}^{\frac{\lambda-\lambda_0(1-z)}{\sigma\sqrt{2}}}e^{-u^2}du$ or

$$L(\lambda)=\frac{1}{\lambda_0 z\sqrt{\pi}}\left[\int_0^{\frac{\lambda-\lambda_0(1-z)}{\sigma\sqrt{2}}}e^{-u^2}du-\int_0^{\frac{\lambda-\lambda_0(1+z)}{\sigma\sqrt{2}}}e^{-u^2}du\right]$$

and $L(\lambda)=\frac{\sqrt{\pi}}{2\lambda_0 z}\left[erf\left(\frac{\lambda-\lambda_0(1-z)}{\sqrt{2}\sigma}\right)-erf\left(\frac{\lambda-\lambda_0(1+z)}{\sqrt{2}\sigma}\right)\right]$.

The distribution function from the semi-spherical region is

$$L_{final}(\lambda)=\frac{\sqrt{\pi}}{2\lambda_0 z}\int_{-\frac{\pi}{2}}^{\frac{\pi}{2}}\left[erf\left(\frac{\lambda-\lambda_0}{\sqrt{2}\sigma}+\frac{\lambda_0 z}{\sqrt{2}\sigma}\cos\theta\right)-erf\left(\frac{\lambda-\lambda_0}{\sqrt{2}\sigma}-\frac{\lambda_0 z}{\sqrt{2}\sigma}\cos\theta\right)\right]\cos\theta d\theta \quad (3)$$

(Method Simpson).

This $L_{final}(\lambda)$ is the distribution that replaces the old rotational distribution L in equation (2) that our group proposed some years ago (Danezis et al 2003).

ACKNOWLEDGMENTS

This research project is progressing at the University of Athens, Department of Astrophysics, Astronomy and Mechanics, under the financial support of the Special Account for Research Grants, which we thank very much. This work also was supported by Ministry of Science and Environment Protection of Serbia, through the projects "Influence of collisional processes on astrophysical plasma line shapes" and "Astrophysical spectroscopy of extragalactic objects".

REFERENCES

1. E. Danezis, "The nature of Be stars", PhD Thesis, University of Athens, 1984.
2. E. Danezis, in *Physics of Be Stars*, edited by A. Slettebak and T. P. Snow, Proceedings IAU Colloq. No 92, (Cambridge University Press), 1987, p.149.
3. E. Danezis, E. Theodossiou and P. G. Laskarides, A&SS, **179**, 111-139 (1991).
4. E. Danezis, D. Nikolaidis, E. Theodossiou, M. Stathopoulou and A. Kosionidis, "Satellite spectral components and moving shells in the atmospheres of Early Type Stars - The example of HD 175754", Proceedings of the 65th Czech and the 7th European Joint Conference, JENAM '98, Prague, Czech Republic (September 1998).
5. E. Danezis, D. Nikolaides, E. Theodossiou, A. Kosionidis, M. Stathopoulou, E. Lyratzi, C. Drakopoulos, and P. Bourma, "A simple model for the complex structure of moving atmospheric shells in Oe and Be stars. The example of HD 36861, some first conclusions", Proceedings of JENAM 2000, Moscow, Russia (29 May-3 June 2000a).
6. E. Danezis, D. Nikolaides, E. Theodossiou, A. Kosionidis, M. Stathopoulou, E. Lyratzi, C. Drakopoulos and P. Bourma, "A simple model for the complex structure of moving atmospheric shells in Oe and Be stars. The

example of HD 66811, some first conclusions", Proceedings of JENAM 2000, Moscow, Russia (29 May-3 June 2000b).
7. E. Danezis, D. Nikolaides, E. Theodossiou, A. Kosionidis, M. Stathopoulou, E. Lyratzi, C. Drakopoulos, and P. Bourma, "A simple model for the complex structure of moving atmospheric shells in Oe and Be stars. The example of HD 152408, some first conclusions", Proceedings of JENAM 2000, Moscow, Russia (29 May-3 June 2000c).
8. E. Danezis, D. Nikolaidis, E. Lyratzi, M. Stathopoulou, E. Theodossiou, A. Kosionidis, C. Drakopoulos, G. Christou and P. Koutsouris, "A new model of density layers of matter in the expanding gaseous envelope of Oe and Be stars", Proceedings of the IAU Symposium 210: Modeling of stellar atmospheres, Uppsala, Sweden (17-21 June 2002a).
9. E. Danezis, E. Lyratzi, D. Nikolaidis, E. Theodossiou, M. Stathopoulou, G. Christou, and A. Soulikias, "The complex structure of Mg II regions in the gaseous envelope of Be V stars", Proceedings of the IAU Symposium 210: Modeling of stellar atmospheres, Uppsala, Sweden (17-21 June 2002b).
10. E. Danezis, E. Lyratzi, G. Christou, E. Theodossiou, M. Stathopoulou, D. Nikolaides, and A. Kyriakopoulou, "Hyperionisation phenomena in the coronal and post-coronal regions of the moving atmosphere of the Oe star HD 66811 (ζ Puppis)", Proceedings of the IAU Symposium 210: Modeling of stellar atmospheres, Uppsala, Sweden (17-21 June 2002c).
11. E. Danezis, D. Nikolaidis, V. Lyratzi, M. Stathopoulou, E. Theodossiou, A. Kosionidis, C. Drakopoulos, G. Christou and P. Koutsouris, A&SS, **284**, 1119 (2003).
12. E. Danezis, D. Nikolaidis, E. Lyratzi, L. Č. Popović, M. S. Dimitrijević, E. Theodossiou and A. Antoniou, Mem. S.A.It., in press (2005).
13. D. Hutsemékers, A&AS, **60**, 373 (1985).
14. H. J. G. L. M. Lamers, R. Gathier, and T. P. Snow, ApJ, **258**, 186, (1982).
15. P. G. Laskarides, E. Danezis, and E. Theodossiou, A&SS, **179**, 13 (1992a)
16. P. G. Laskarides, E. Danezis, and E. Theodossiou, A&SS, **183**, 67 (1992b)
17. E. Lyratzi and E. Danezis, in *The Physics of Ionized Gases: 22nd Summer School and International Symposium on the Physics of Ionized Gases*, edited by Lj. Hadzijevski, T. Grozdanov and N. Bibić, AIP Conference Proceedings 740, Bajina Bašta, pp. 458-473 (23-27 August 2004).
18. A. Peton, Space Sci. Rev., **30**, 481 (1974).
19. J. Sahade, E. Brandi and J. M. Fontela, A&AS, **56**, 17 (1984).
20. J. Sahade and E. Brandi, RMxAA, **10**, 229 (1985).
21. A. B. Underhill, ApJ, **199**, 691 (1975).

Seeking Tidal Streams Emanating from the Globular Cluster NGC 5897

Odysseas Dionatos[*+], Carl J. Grillmair[+]

*Department of Physics, Section of Astrophysics, Astronomy & Mechanics, University of Athens 15784, Greece

+Spitzer Science Center – California Institute of Technology CA 91125 Mail Stop 220-6

Abstract. NGC 5897 is one of a number of halo globular clusters which, based on their inferred orbital parameters, have a high likelihood of being tidally distorted or disrupted by the Galactic potential field. Using MOSAIC images taken on the Blanco 4-meter telescope of a 4x4 degree field surrounding NGC 5897, we present deep, two color, stellar photometry to V of approximately 24. We use color selection and matched filter techniques to search for evidence of tidal tails extending from the cluster.

Keywords: Galaxy: Halo, Galaxy: Globular Clusters: General, globular clusters: individual (NGC5897)
PACS: 98.20.Gm

INTRODUCTION

Numerical simulations show that the tidal stresses imposed by our Galaxy on globular clusters leads to the development of tidal tails. Mapping such tidal tails will enable us to measure the orbit shapes of globular clusters, and the extended, three-dimensional potential field of our Galaxy (Grillmair 1997; Johnston et al. 1999). Tidal tails extending out as far as one or two degrees from their parent clusters have been discovered for a number of Galactic globular clusters (eg. Grillmair et al. 1995; Lehmann & Scholz 1997; Zaggia, Piotto, & Capaccioli 1997; Leon, Meylan, & Combes 1999) and in M31 (Grillmair et al. 1996; Holland, Fahlman & Richer 1997). Recently, tidal tails extending about ten degrees from Palomar 5 were discovered using data from the SDSS photometric catalogue (Odenkirchen et al. 2001, 2003). Here we attempt to detect tidal tails in NGC 5897 using deep two color (V, I) photometry in order to select targets for the upcoming Space Inteferometry Mission (SIM, estimated launch date 2011).

One particular SIM Key Project ("Taking Measure of the Milky Way", Principal Investigator: Prof. Steven Majewski) will measure proper motions for stars in the tidal streams generated by globular clusters and dwarf galaxies orbiting the Galaxy. Combined with positions and radial velocities, these measurements will enable us to model the potential of the Galaxy to an accuracy of order 1%. However, to reduce the

CP848, *Recent Advances in Astronomy and Astrophysics,*
7th International Conference of the Hellenic Astronomical Society, edited by N. H. Solomos

uncertainties to this level, tidal streams will need to be traced out to tens of degrees from their parent clusters.

DATA AND REDUCTION

Galactic tidal streams are by nature and proximity rather large features on the sky. Mapping of such features requires an imager with a wide field of view. Our data set was obtained using the 4-meter Blanco telescope (CTIO) equipped with the MOSAIC2 camera. An area of 4x4 sq degrees around the cluster is covered by 49 deep, two color (V,I) images. Every image overlaps with its neighbors by 3 arcmin, allowing us to use the stars in those areas in order to ensure photometric homogeneity across the entire field. Deep photometry is necessary in order to take advantage of the rise in the globular cluster luminosity function and to provide the photometric accuracy necessary to prevent large numbers of foreground stars from contaminating the star counts and overwhelming the tidal tails.

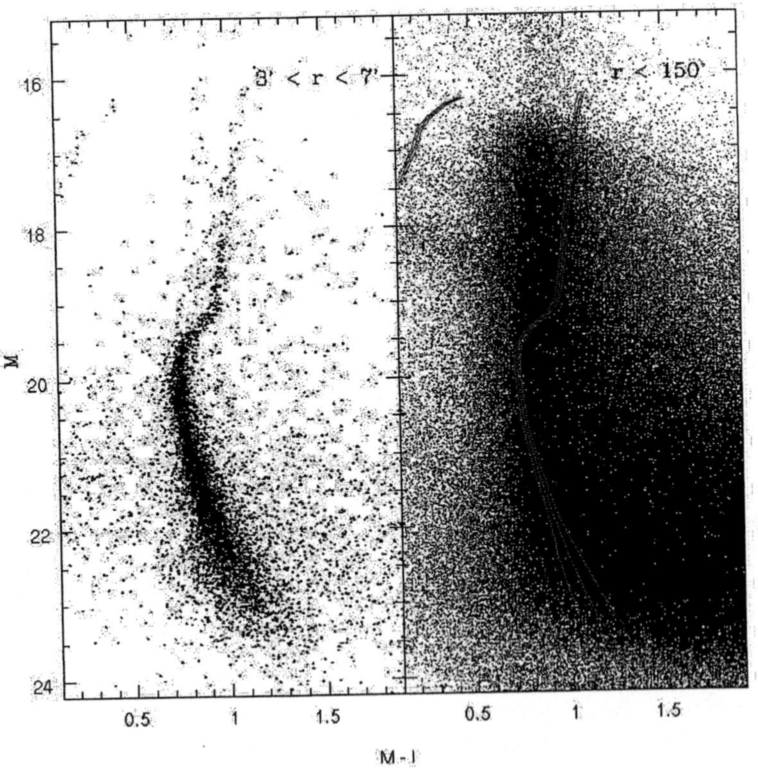

FIGURE 1. Color magnitude distribution of ~4000 stars within 7 arcminutes of the center of NGC 5897 (left) and of $\sim 10^6$ stars over a 4x4 degree region surrounding the cluster (right). Color selecting stars using the 1-σ photometric envelope about the cluster sequence (shown by the red dashed lines in the right hand panel) removes more than 97% of field stars from the sample.

For the data reduction we used PSF-fitting photometry routines, namely the IRAF implementations of DAOPHOT and ALLSTAR (P. Stetson, 1987). During the reduction each CCD is treated separately and the differences in response for each CCD of the imager are calibrated using V and I photometric standards. In Figure 1 we present the color magnitude diagram for stars surrounding NGC5897. On the left panel which represents an area of 7 arcminute radius centered to the cluster, the main sequence, turn off and the giant branch are immediately apparent and stand out well from the color - magnitude distribution of foreground field stars. Weighting stars based on their distance from the 1σ photometric RGB/main sequence envelope shown in the left part of figure 1 we arrive at the smoothed, weighted star count image shown in figure 2. By applying this color selection criterion to the observed field, we ensure optimal contrast between cluster and field stars (Rockosi et al. 2002).

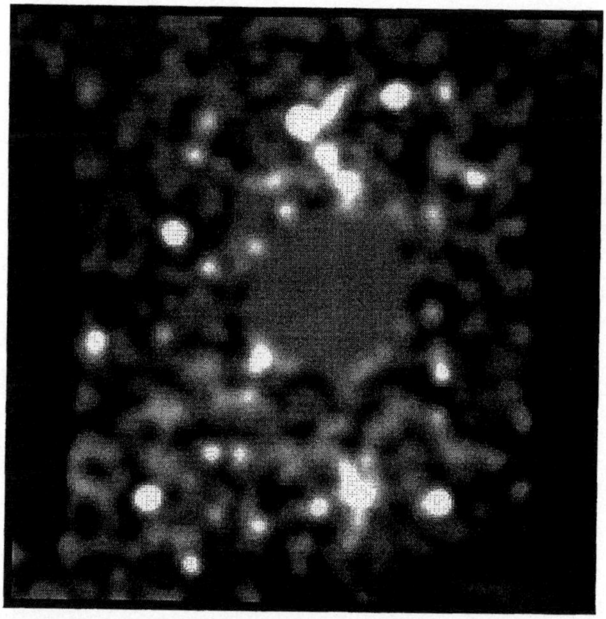

FIGURE 2. Smoothed, color weighted surface density distribution of stars in the 4x4 degree field surrounding NGC5897. The cluster itself has been masked out to avoid overwhelming the tidal tails extending to the North and South. The observed tidal tails are consistent with the proper motion measurements of the cluster

An excess of stars is clearly detected emanating to the North and South of the cluster, in agreement with the proper motion measurements (Dinescu et al. 1999). This represents the first detection of tidal tails in NGC5897 (Grillmair, C. J., Dionatos O., et al 2005). We have now acquired and are processing data for an additional 13 square degrees of coverage to the North and South of the cluster.

CONCLUSIONS

NGC 5897 has detectable tidal tails. The North-South orientation of the extensions in Figure 2 is consistent with the measured proper motion of the cluster. We continue to refine our analysis procedures to improve the signal-to-noise ratio, with the eventual goal of applying these procedures, in an automated fashion, to similar data for several other observed halo globulars. In parallel, we will kinematically identify the bona fide globular cluster stars in Figure 2 using moderate resolution spectroscopy. The brightest confirmed tidal tail stars will be included in the SIM target list for proper motion measurements.

REFERENCES

1. Cudworth, K., & Hanson, R. 1993, AJ, 105, 168
2. Grillmair, C. 1997, *Galactic Halos: A UC Santa Cruz Workshop,* ed. D. Zaritsky, ASP Conference series 136, 45
3. Grillmair, C Freeman, K. Irwin, M. & Quinn, P. 1995, AJ, 109, 2553
4. Grillmair et al. 1996, AJ, 111, 2293
5. Holland, S. Fahlman, G. & H. Richer 1997, AJ, 114, 1488
6. Johnston, K. Zhao, H., Spergel, D. & Hernquist, L. 1999, ApJ, 512, 109
7. Lehmann & Sholtz 1997, A&A 320, 776
8. Schlegel, D., Finkbeiner, D., & Davis, M. 1998 ApJ, 550, 525
9. Zaggia,S. Piotto,G., & Capaccioli, M. 1997 A&A 237, 1004
10. Dinescu, Dana I., van Altena, William F., Girard, Terrence M., López, Carlos E.1999 ,ApJ, 117:277-285
11. Grillmair, C. J., Dionatos O., et al 2005 (2005AAS, 205.2307G)
12. Rockosi et al 2002, AJ 124, 349

CCD Measurements of Hipparcos Wide Visual Double Stars*

D. Sinachopoulos (1), P. Gavras (1,2), Medupe Th. (3), Ducourant Ch. (4), O. Dionatos (2)

(1) National Observatory of Athens, Institute of Astronomy and Astrophysics
(2) National & Kapodistrian University of Athens, Dept. of Astronomy, Astrophysics and Mechanics
(3) South African Astronomical Observatory
(4) Observatory of the University of Bordeaux

Abstract. We present final astrometric and photometric results of CCD observations of HIPPARCOS visual double stars obtained with the Kryonerion 1.2m telescope. The new measurements together with several hundreds more obtained during the last two years in both hemispheres, will permit the determination of the physical pairs in the sample.

Keywords: stars: binaries: Hipparcos - visual - astrometry – photometry
PACS: 01.10.Fv

THE PROJECT AND THE OBSERVATIONS

The HIPPARCOS catalogue has been published for mean epoch 1991.25 (Perryman et al., 1997). Its proper motions were based on the three years of the satellite observations and have a mean precision of 25 mas per year (Lindegren et al., 1997). Therefore now, more than one decade later, the relative positions of the HIPPARCOS visual double star components are known with a precision of about 50 mas.

Modern astrometric CCD observations of these objects on the other hand, can achieve an angular separation accuracy of about fifteen mas per observation (Sinachopoulos et al., 1999), or factor three more accurate than the knowledge of the HIPPARCOS positions nowadays.

For this reason, we have obtained new CCD measurements during the last two years using the 1.2m telescope of the Kryonerion, the transit instrument of Bordeaux observatories in the north and the 1.m South African Astronomical Observatory SAAO telescope in the south, hemispheres. The CCD camera of the Kryonerion 1.2m telescope is described by Sinachopoulos et al. (1999). SAAO 1m telescope uses a 1K CCD chip of 24μX24μ pixel.

More than 400 HIPPARCOS double stars have been observed in the frame of the present project. We show here first final results of the observations obtained with the Kryonerion 1.2m telescope in March 2005. A detailed description of the telescope and its CCD camera can be found in Sinachopoulos et al. (1999).

---

* Based on observations obtained with the 1m South African Astronomical Observatory (SAAO), the transit instrument of the Bordeaux observatory and the 1.2m Kryonerion (National Observatory of Athens) telescopes

CP848, *Recent Advances in Astronomy and Astrophysics,*
7th International Conference of the Hellenic Astronomical Society, edited by N. H. Solomos

The new measurements will be used together with the corresponding HIPPARCOS ones and all old observations known, in order to determine the most accurate possible relative proper motion of the components of these double stars since Brosche and Sinachopoulos (1988, 1990). These relative proper motions will permit the extraction of binaries (physical pairs) and the double stars which have components of common origin from the sample. This study will contribute to the improvement our understanding of stellar formation and evolution rates of double stars in the solar neighborhood

CALIBRATION OF THE POSITION ANGLE

Stellar traces near the equator and meridian have been taken every night in order to determine the deviation (offset) of the CCD camera orientation from the East-West direction of the CCD chip. This offset is the correction of the position angle –in degrees- to be added to the instrumental value of the double star position angle.

In addition to these traces, about 200 CCD frames containing trellar traces have been taken with the 1.2m telescope of Kryonerion and 300 with the 1m SAAO one, grouped at different declinations and hour angles, in order to correct the observed position angle of the double stars for the misalignment between the telescope axis and the axis of the rotation of the Earth..

Figure 1 shows a stellar trace from SAAO telescopes taken with seeing of 1.5 arcseconds and Fig. 2 from Kryonerion (seeing 2.3 arcseconds).

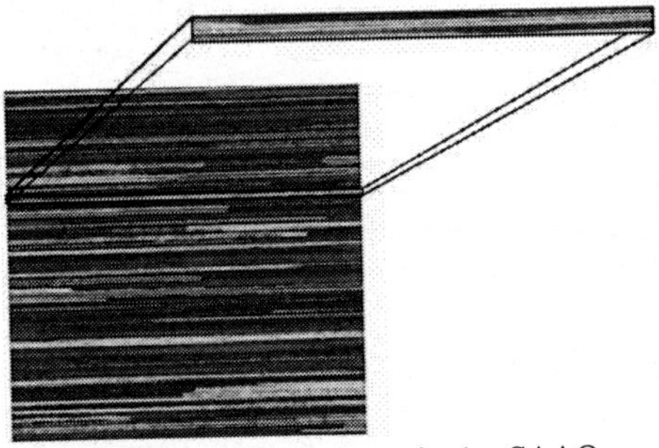

FIGURE 1. Stellar traces from the 1m SAAO (Seeing 1.5 arcseconds)

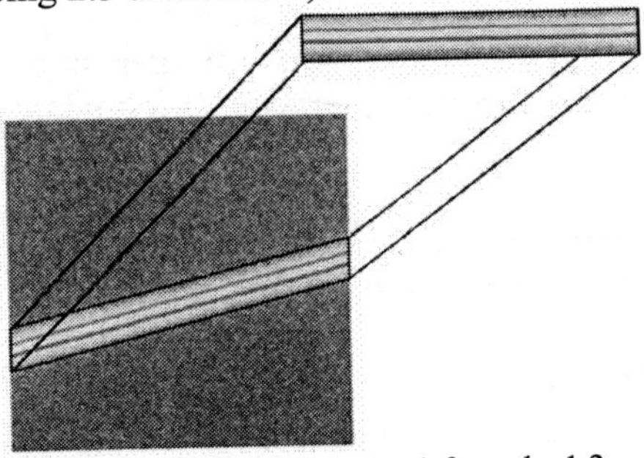

FIGURE 2. Stellar traces (two) from the 1.2m Kryonerion (Seeing 2.3 arcseconds)

ESO-MIDAS software has been used to compute the photocenter position for each column of these CCD frames of each stellar trace. A linear regression was applied on these photocenter positions and the inclination of the stellar trace was determined. Figure3 shows the residuals between the photocenter position and the fitted line of the stellar trace in Figure 1.

Table 1 shows the inclination of the stellar traces on the CCD of the Kryonerion 1.2m telescope given for different hour angles and declinations. Obviously, there is dependence between the inclination of the star trails and the declination. Especially in the declination range between +36 and +65 degrees corrections should be applied to the position angle measurement according to $\Delta\vartheta = i + S_i \cdot H.A. \cdot \cos(\delta)$.

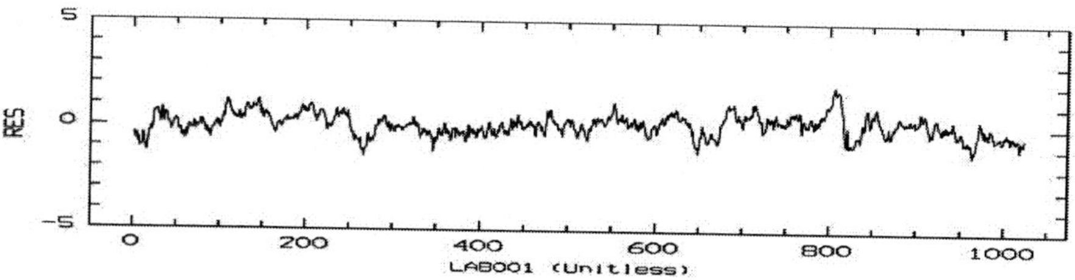

FIGURE 3. Residuals [O-C] of one SAAO trace (see Fig. 1)

In higher declinations between 66 and 90 degrees stellar traces have been taken for each double star observed.

TABLE 1. Inclination of the stellar traces on the CCD of the Kryonerion 1.2m telescope

Validity Range		Inclination		Slope of		No of
Declination	**H.A.*cos(δ)**	**of trails (i)**		**inclination (S_i)**		**trails**
		Degrees				
-05 - +35	-45 - +45	+0.0	±0.004	-0.0	±0.0	587
+36 - +50	-15 - +40	+0.008	±0.045	-0.0066	±0.0022	88
+51 - +65	-10 - +30	+0.140	±0.043	-0.0065	±0.0023	75
+66 - +78	-04 - +05	+0.17	±0.14			24
+83 - +85	-02 - +03	+0.58	±0.10			8

ble 2 shows corresponding results for the SAAO 1.m telescope in South Africa. It shows that lination of stellar traces is the same for all the range of declinations and hour angles tested. This ves that the polar axis of the 1.m SAAO telescope is practically parallel to the axis of the Earth.

TABLE 2. Inclination of the stellar traces on the CCD of the SAAO 1.m telescope

	- 3h 0m	**- 2h 0m**	**- 1h 30m**	**0h**	**+ 1h 30m**	**+ 2h 0m**	**+ 3h 0m**
0°	+0.03±0.03		+0.03±0.02	+0.04±0.02	+0.10±0.02		+0.06±0.03
-20°	+0.02±0.01		-0.01±0.02	+0.01±0.02	+0.01±0.01		+0.04±0.01
-40°	+0.08±0.07		+0.02±0.01	-0.02±0.01	-0.03±0.01		+0.3±0.03
-50°	-0.01±0.01		-0.01±0.01	-0.02±0.01	+0.02±0.02		-0.05±0.02
-60°	-0.04±0.02	-0.10±0.02	-0.03±0.03	-0.02±0.01	-0.02±0.02	-0.01±0.02	-0.01±0.02
-70°	-0.08±0.06		-0.06±0.03	+0.01±0.03	+0.03±0.01		-0.01±0.01
-80°	+0.00±0.01		-0.00±0.01	-0.03±0.03	+0.04±0.02		+0.02±0.02

CALIBRATION OF THE ANGULAR SEPARATION

HIPPARCOS data of both components in the four double stars listed in Table 3 have been used for the determination of the scale of the telescope and its CCD camera. The mean value of the scale was determined in this way, which was found 0.30068 ± 0.00067 arcseconds / pixel. This new value is statistically the same with the scale published by Sinachopoulos et al. (1999).

TABLE 3. CCD scale determination at the 1.2m Kryonerion telescope using HIPPARCOS stars

HIPPARCOS A	HIPPARCOS B	$\rho_{HIPPARCOS}$ ('')	ρ_{CCD} (pixels)	CCD scale (''/pixel)
40527	40532	207.646	69.166	0.300213
43270	43265	258.863	86.595	0.298935
56202	56205	122.910	40.842	0.300939
58693	58695	120.112	39.687	0.302647

The double WDS 08483+0033 (Mason et al. 2005 and Mason et al., 2001) star was excluded from these computations due to very low accuracy of the satellite results.

DATA REDUCTION AND RESULTS

These calibrations were used for creating the results of the 14 stars listed in table 4, which were observed at Kryonerion in the nights of 23 and 24 March 2005 (epoch 2005.22). Figure 4 shows the double star WDS 12016+7051, a double star with large components magnitude difference. For double stars like this having angular separation smaller than 7 arcseconds, a simultaneous fitting of a Moffat function to the components was used for the data reduction (Sinachopoulos et al., 1999).In this particular case, the big ΔV of the components combined with the poor seeing conditions during observations (2".2), with problems in telescope tracking and the relatively small components angular separation (5".3) resulted to poor photometric results (see table 4).

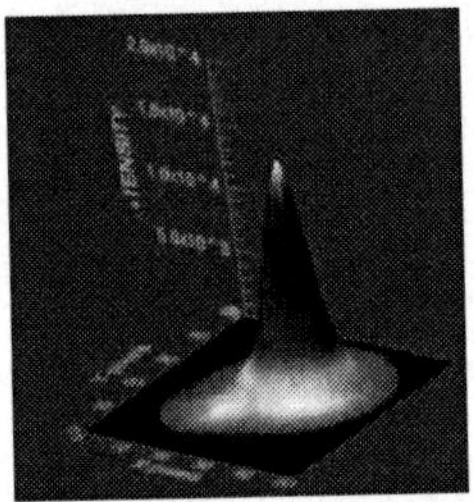

FIGURE 4. Double star WDS 12016+7051

FIGURE 5. The very wide double star WDS 08165+7930

Figure 5 shows the very wide double star WDS 08165+7930 (ρ = 20".8). For angular separation larger than 7 arcseconds, smaller windows -corresponding to 3"X3"- were cut around each component. Two dimensional Gaussians were fitted to them for the determination of the photocenter and DAOPHOT aperture photometry was performed to the double star CCD frames of such double stars.

We present our results in Table 4. Column two contains the WDS identifier of the visual double star, followed in the next columns by the exact 2000 coordinates of the double star taken from the WDS. The next four columns contain the relative coordinates angular separation and position angle **(ρ,θ)** with their r.m.s. error, followed by the magnitude difference ΔV of the components and its accuracy. Last column contains the number of the exposures used for obtaining these results.

TABLE 4. Results of 14 visual double stars observed in the nights of 23 and 24 March 2005 (epoch 2005.22).

No	WDS	RA	Dec	ρ	σ_ρ	θ	σ_θ	ΔV	$\sigma_{\Delta V}$	N_{exp}
1	08165+7930	08 16 31.2	+79 30 04.0	20.797	.047	14.79	.02	0.291	.001	16
2	08222+7449	08 22 09.5	+74 49 12.0	10.455	.025	195.24	.03	3.840	.023	11
3	08420+6201	08 41 59.1	+62 01 24.0	18.397	.041	209.38	.03	0.861	.001	16
4	08483+0033	08 48 18.3	+00 33 18.0	12.648	.028	56.38	.03	3.506	.004	16
5	08489+7201	08 48 54.0	+72 01 13.0	26.038	.058	245.98	.02	2.321	.001	16
6	09092+1514	09 09 14.4	+15 14 27.0	7.837	.018	62.51	.04	1.707	.003	12
7	09174+2339	09 17 19.2	+23 39 10.0	5.921	.014	28.43	.04	0.319	.001	16
8	09551+6854CD	09 55 04.0	+68 54 06.0	8.953	.020	272.76	.03	0.085	.001	16
9	11092+6230	11 09 13.0	+62 29 50.0	9.926	.022	51.18	.03	0.192	.001	16
10	11313+5942	11 31 18.5	+59 42 03.0	12.281	.027	90.73	.02	0.879	.000	16
11	11323+5606	11 32 20.5	+56 05 42.0	11.483	.026	348.86	.02	2.964	.003	16
12	11366+5608	11 36 35.8	+56 08 08.0	6.149	.016	165.83	.07	.509	.002	4
13	12016+7051	12 01 37.8	+70 50 59.0	5.312	.019	234.48	.09			11
14	12023+7222	12 02 15.9	+72 21 43.0	11.933	.027	35.89	.01	.675	.001	16

CONCLUSION

These first results show that an even much higher astrometric precision will be achieved in the angular separation when all 400 visual double stars observed in the frame of this project will have been homogeneously reduced. Table 4 shows again that CCD observations can achieve a precision at least twice as good as the corresponding HIPPARCOS relative coordinates calculated for the current epoch. A unique precision of relative coordinates and proper motion of these visual double stars will be achieved, once the results of this project will be combined will the 1991.25 HIPPARCOS coordinates and all known older astrometric observations of each double star of the sample. This accuracy will be smaller only from the precision expected by the ESA mission of the satellite GAIA (de Boer et al., 2000)

ACKNOWLEDGEMENTS

This research was partially carried out in the framework of the project "ATHENOGAIA", financed by the Greek General Secretary for Research and Development.

D. Sinachopoulos would like to thank the Universities of Bordeaux (France) and South-West (South Africa) for support and accommodation.

REFERENCES

1. De Boer, K.S., et al. (eds), 2000, ESA-SCI(2000)4, "GAIA, Composition, Formation and Evolution of the Galaxy"
2. Brosche, P., Sinachopoulos, D, 1988, BICDS, **34**, p 39
3. Brosche, P., Sinachopoulos, D, 1990, BICDS, **37**, p 109
4. Lindegren, L. et al., 1997, A&A **323**, p.L53-L56
5. Mason, B.D., Wycoff, G.L. and Hartkopf, W.I, 2005, http://ad.usno.navy.mil/proj/WDS/
8. Mason, B.D., et al., 2001, AJ **122**, 6, pp. 3466-3471
9. Perryman, M.A.C. et al., 1997, A&A **323**, p.L49-L52
10. Sinachopoulos et al., 1999, A&AS **136**, p 525

New results on the pulsating variable star GSC 4778:0324

K.D. Gazeas, V.N. Manimanis, P.G. Niarchos

Department of Astrophysics, Astronomy and Mechanics, Faculty of Physics, University of Athens

Abstract. We report on the results of a series of photometric observations of the low-amplitude *δ Scuti* star GSC 4778:0324, in the course of a multi-site campaign aiming at monitoring V1162 Ori. The two dominant frequencies, found in the previous analyses based on the 2000-2002 data, are confirmed and their amplitudes are found to be stable on a time-scale of a few years.

Keywords: stars: pulsating variables, photometry
PACS: 97.10.Sj, 95.75.-z

INTRODUCTION

The low-amplitude *δ-Scuti* variable GSC 4778:0324 was discovered by Van Cauteren & Lampens in 2000, in the frame of a multi-site campaign, aiming at monitoring V1162 Ori (Arentoft et al. 2001a,b).

The Strömgren colour indices reported by Handler (2000) place this star in the middle of the δ Scuti instability strip. GSC 4778:0324 shows light variations with total amplitude of about 0.01 mag with respect to GSC 4778:0019, the same comparison star used in the campaign for V1162 Ori.

Some results of the 1999-2002 campaign have already been presented in two publications (Lampens et al. 2001, 2002). Since the proposed frequency solutions, based on the available data set, are rather incomplete, we decided to continue the observations for the period 2004-2005, in an effort to improve the frequency determination and check the stability of the known frequencies.

OBSERVATIONS AND REDUCTION

The observations of GSC 4778:0324 were made on 11 nights between September 2004 – March 2005. The instruments used were the 0.40 m Cassegrain reflector at the University of Athens Observatory, Greece, and a ST-8 CCD camera. The CCD camera uses a Kodak KAF-1600 CCD chip, cooled by a two-stage Peltier element, which provides a CCD working temperature of -30 degrees C below ambient. The CCD chip has 1536×1024 useful pixels of 9×9 microns, covering an area of approximately 10×15 arcmin. The CCD camera is equipped with a set U, B, V, R and I Bessell filters. The scale on the chip is 1.24 arcsec/pixel in 2x2 binning mode.

CP848, *Recent Advances in Astronomy and Astrophysics*,
7th *International Conference of the Hellenic Astronomical Society*, edited by N. H. Solomos

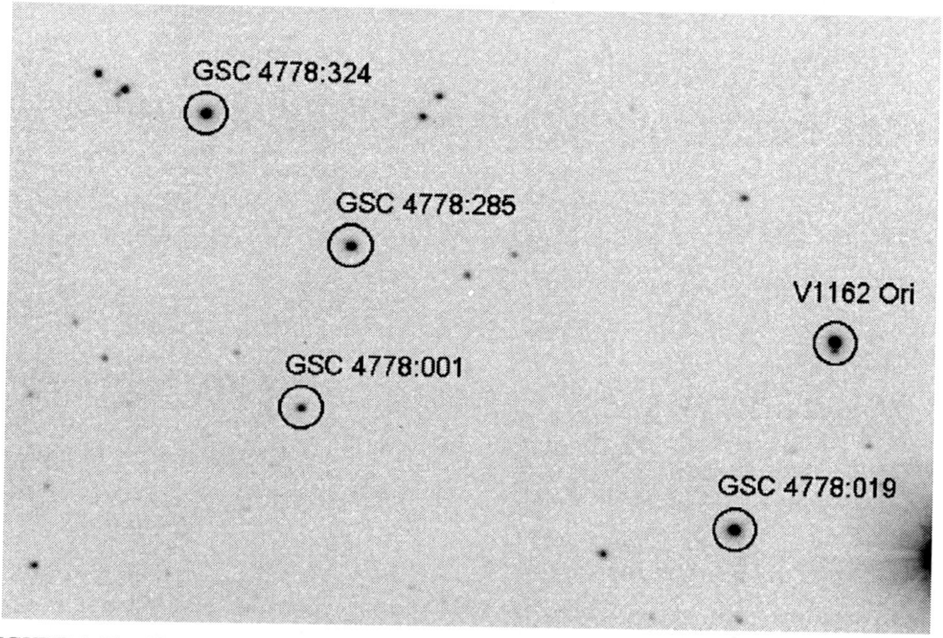

FIGURE 1. The 10×15 arcmin field around GSC 4778:0324. The stars used are the comparison (GSC 4778:0019), the check (GSC 4778:0285) and the variable (GSC 4778:0324).

All observations were made by using the V filter. The data reduction was made with the package AIP4WIN (Berry & Burnell 2000). In total 1753 observations were collected to study the variability of GSC 4778:0324 (V=10.26, b-y=0.20). The stars GSC 4778:0019 (V=9.73, B-V=1.55, b-y=0.91) and GSC 4778:0285 (with a photographic magnitude of 11.8 mag) were used as comparison and check star, respectively.

Although the comparison star is known to show nightly zero point shifts at the level of 0.01-0.02 mag (Arentoft et al. 2001a), this will not affect the frequency search in the range of interest (10-20 cycles/day). In addition the differential data have been rescaled as much as possible making use of the check star, so that all magnitudes were on the same level of brightness. The time base length of the complete data set is 163 days. This allows a theoretical frequency resolution of 1.5/T or 0.0092 cycles/day (cpd). The star field observed is shown in Figure 1. A typical sample of the light curves, obtained from the variables in this field is shown in Figure 2.

FREQUENCY ANALYSIS

We applied corrections to the differential V magnitudes due to night-to-night shifts of the mean differences between the check and the comparison star. The mean value of the nightly rms error (check - comparison differences) is 14 mmag.

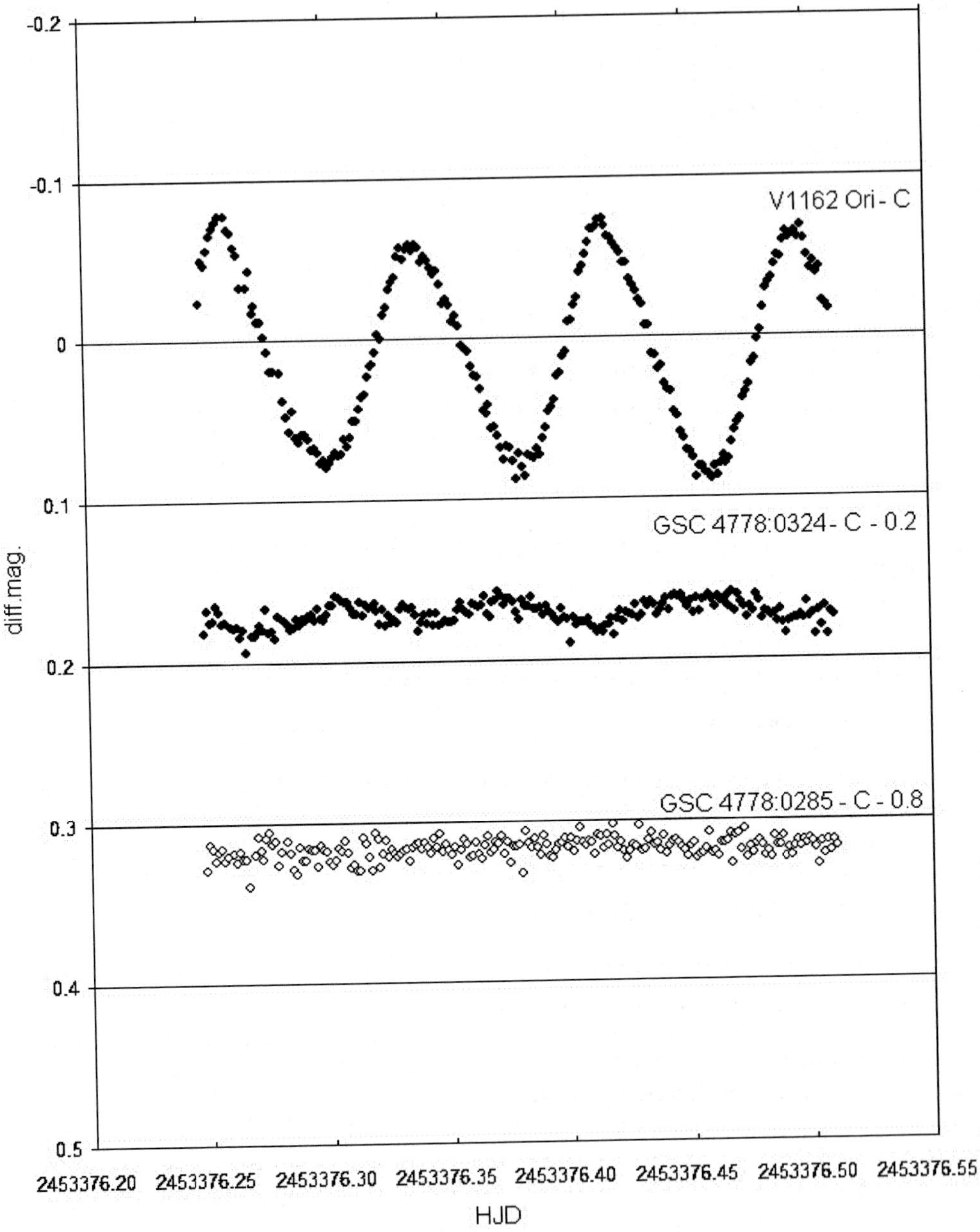

FIGURE 2. The variability of GSC 4778:0324 and V1162 Ori in comparison with the star GSC 4778:0019, which is marked as C. The comparison between C and K (GSC 4778:0285) is also displayed at the bottom.

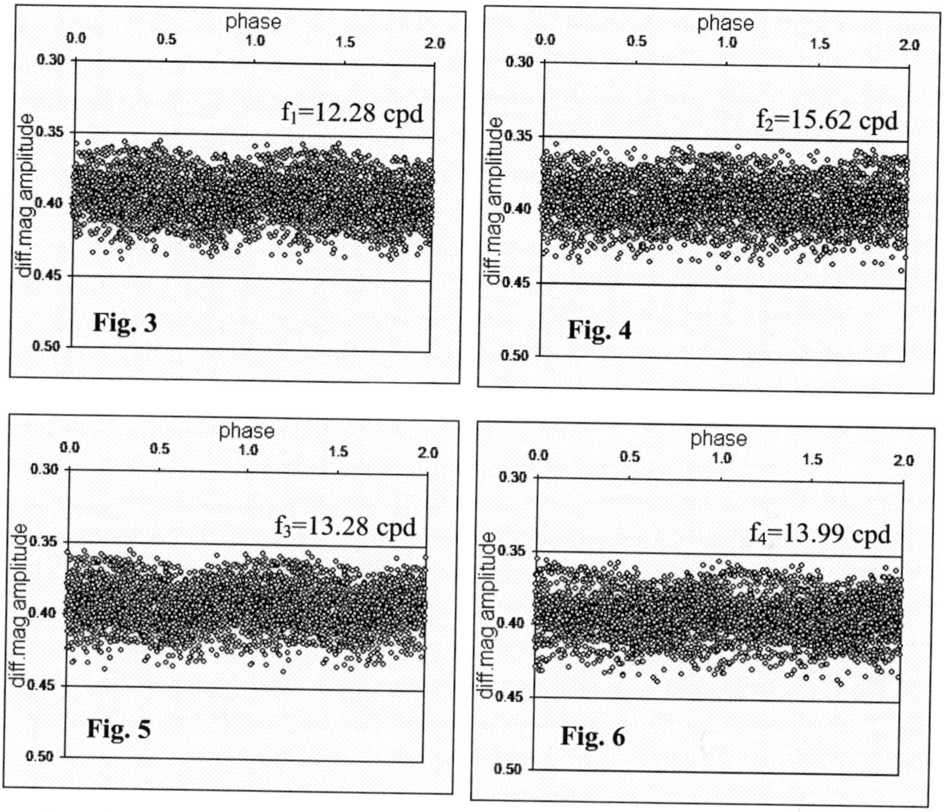

FIGURES 3-6. Phase-amplitude diagrams of the most dominant frequencies found in this study.

We performed a frequency analysis of this data set in the range 0-30 cpd using the program Period98 (Sperl 1998). The following three frequencies with the largest signal-to-noise ratio were found: 12.2814, 15.6166 and 13.2838 cpd.

The first two frequencies are obviously linked to intrinsic light variations of the δ Scuti star, because they are very nearly identical to those found in the frequency analysis of the previous campaigns (see Table 1). The third frequency is a new one. A similar frequency (13.2568 cpd) appeared in a previous analysis of the combined data of the 1999-2001 campaign. In this study it seems that the amplitude is relatively high. This may imply that the star GSC 4778:0324 may have a third pulsating mode. In the same analysis we were able to identify the frequency of 13.9892 cpd, which could be an alias of f_3.

Further frequency search results in low frequencies in the range 0-5 cpd. These are either related to atmospheric effects or to a possible variability of the comparison star. Figures 3-6 show the phase diagrams of the four most dominant frequencies found in this study.

Table 1. Frequency analysis of GSC 4778:0324 based on the data collected in the present and previous studies.

Season	Nr of data	Ident.	Frequency (cpd)	Ampl. (mag)	Phase (2π rad)	S/N
1999-2001	2949	f_1	12.2810	0.0041	0.136	8.1
		f_2	15.6169	0.0036	0.135	6.8
		f_3	10.5852	0.0021	0.421	4.7
2004-2005	1753	f_1	12.2814	0.0047	0.700	9.4
		f_2	15.6163	0.0024	0.789	4.8
		f_3	13.2839	0.0034	0.800	6.8
		f_4	13.9887	0.0023	0.830	4.6

SUMMARY AND CONCLUSIONS

New observations of GSC 4778:0324, obtained at the University of Athens Observatory, were used for a follow-up study of the pulsation behavior of this small-amplitude δ *Scuti* star. With the frequency analysis we were able to confirm the presence of the two most dominant frequencies identified in the previous campaigns of 1999-2001. We have also found that not only the frequencies themselves but also their amplitudes are stable on a time-scale of at least a few years. Again the presently available data set does not permit the identification of more frequencies, although an alias of f_3 was detected, but at a low σ level.

REFERENCES

1. Arentoft, T., C. Sterken, G. Handler, L.M. Freyhammer, A. Bruch, P. Niarchos, K. Gazeas, K. Manimanis, P. Van Cauteren, E. Poretti, D.W. Dawson, Z.L. Liu, A.Y. Zhou, B.T. Du, R.R. Shobbrook, R. Garrido, R. Fried, M.C. Akan, C. Ibanoglu, S. Evren, G. Tas, D. Johnson, C. Blake & D.W. Kurtz, *A&A*, **374**, 1056 (2001a)
2. Arentoft, T., Sterken, C., Knudsen, M.R., G. Handler, P. Niarchos, K. Gazeas, K. Manimanis, M.B. Moalusi, F.F Vuthela & P. Van Cauteren, *A&A*, **378**, 33 (2001b)
3. Berry, R. & Burnell, J., AIP4WIN: *Astronomical Image Processing for Windows*, Version 1.4, Published by Willmann-Bell, Inc. (2000)
4. Handler, G., *IBVS*, 4817 (2000)
5. Lampens, P., Van Cauteren, P., Niarchos, P. et al., *IAU Coll. 185, ASP Conf. Series* 259, p. 344 (2002a)
6. Lampens, P., Van Cauteren, P., Niarchos, P., Gazeas, K., Manimanis, V., Arentoft, T., Wils, P., Bruch, A., Garrido, R. & Shobbrook, R., *Communications in Asteroseismology*, **141**, 65 (Paper I) (2002b)
7. Sperl, M., *PERIOD98*: V 1.0.9. (http://www.astro.univie.ac.at/~dsn) (1998)
8. Van Cauteren, P. & Lampens, P., *IBVS*, 4849 (2000)

Correlation of *DARWIN* mission star catalogue with radio source surveys

F. Lykou* and H. J. A. Röttgering†

*University of Crete, P.O. Box 2208, 71003 Heraklio, Greece
†Leiden Sterrewacht, P.O. Box 9513, 2300RA Leiden, The Netherlands

Abstract. We present a search for radio source counterparts of 508 main sequence stars, that are part of the target list of the Infra Red Space Interferometer DARWIN. We searched all available online radio surveys to find counterparts for DARWIN stars, and we only found 2 candidate identifications out of 508 in two surveys made with VLA radio telescope in New Mexico, USA. We required that candidate identifications should have fluxes larger than 3σ times the noise in the radio maps and have a radio-optical positional difference less than 15″. Both stars had radio counterparts in NVSS (NRAO VLA Sky Survey) and one of them also in FIRST (Faint Images of the Radio Sky at Twenty-centimeters). Both surveys are 1.4 GHz continuum surveys. Our goal for this project is to show that there are main sequence stars whose radio emission is detectable and get a statistical limit for the radio flux of the DARWIN stars. The method we performed was to use the radio maps and coadd these maps at the location of the stars.

Keywords: Radio counterparts; exoplanets
PACS: 95.80.+p, 98.62.Tc

DARWIN

Six space telescopes of 1.5m in diameter in a hexagonal pattern, and all in the same distance from the beam combiner, compose DARWIN. The interferometer will be accompanied by a control and communications satellite, which will stand behind the hexagonal pattern. DARWIN's main goal is to search the Galaxy in the mid-infrared, for earthlike exoplanets orbiting stars at a distance less than 25 pc from our Solar system and to study them [4]. The aim is to launch the interferometer in 2015.

THE RADIO SURVEYS

NVSS (NRAO VLA Sky Survey) [2] surveyed the entire sky north of -40 Dec, and detected more than 1.8 million radio sources, with the rms uncertainties in RA and Dec vary from less than $1''$ for relatively strong (S > 15 mJy) point sources to $7''$ for the faintest (S = 2.3 mJy) detectable sources. The maps have a typical rms of 0.45 mJy/beam and an angular resolution of $45''$. The completeness limit is about 2.5 mJy and the sky density was estimated $\sim$ 50 per square degree.

FIRST (Faint Images of the Radio Sky at Twenty-centimeters) [1] surveyed the sky region defined by: $7h \leq RA \leq 17.5h$ and $-8 \leq Dec \leq +57.6$ North; $21.3h \leq RA \leq 3.3h$ and $-11.5 \leq Dec \leq +1.6$. The resulting maps have rms 0.15 mJy, and a resolution of $5''$. There are about 90 sources per square degree at the 1mJy detection threshold. The

CP848, *Recent Advances in Astronomy and Astrophysics,*
7th International Conference of the Hellenic Astronomical Society, edited by N. H. Solomos

astrometric reference frame of the maps is accurate to 0.05″, and individual sources have 90 percent confidence error circles of radius less than 0.5″ at the 3 mJy level and 1″ at the survey threshold.

POSSIBLE DETECTIONS

In Table 1 the results of this research project are given. As it can be seen, the flux at the center of FIRST radio map (where the DARWIN star is positioned) for the first match is smaller than that of the NVSS radio map. This is due to FIRST's better resolution; the radio map obtained by that survey is a 'zoom' of the radio map from NVSS. Although

TABLE 1. Possible matches

DARWIN*	Counterpart	d†	S**	rms	Source
16 13 18.452, +13 31 36.88	16 13 18.217, +13 31 34.46	4.2″	2.11	0.16	FIRST
	16 13 18.17, +13 31 33.0	5.68″	8.55	0.49	NVSS
18 45 39.725, +20 32 46.71	18 45 38.87, +20 32 48.6	12.2″	1.53	0.43	NVSS

* coordinates (J2000.0) : *left*, main sequence stars, *right*, radio source

† positional difference

** flux and rms noise in mJy

it is clear that both sources have fluxes above the 3σ limit, we can not be certain yet, whether those counterparts are true detections. FIRST sources have 1″ uncertainties and we found a positional difference of 4.2″ for the first source. That value exceeds the one given by FIRST survey. This indicates that this could not be a detection. It is clear that, in the end, we might have not found a real identification and that more research has to be done in this area, to help us clarify what kind of radio sources are close to the DARWIN stars or if these stars have any detectable radio emission.

In Figure 1 images and contour plots of the stars and counterparts are shown. The radio maps in the first row were taken by FIRST and NVSS. The images in the second row are the contour plots of the radio maps above, with the position of the DARWIN star marked by a cross in the middle of each plot.

CONCLUSIONS

The nature of identity of the possible radio counterparts is not known. The search for citations in NED, SIMBAD and VIZIER only gave the properties of the main sequence stars. Future work should reveal whether the identifications are correct and what is the origin of the radio emission. Since it is not clear that the sources found are counterparts, we could say that we did not find any peculiarities in our DARWIN sample in the radio waves, so the sample in that area is homogeneous.

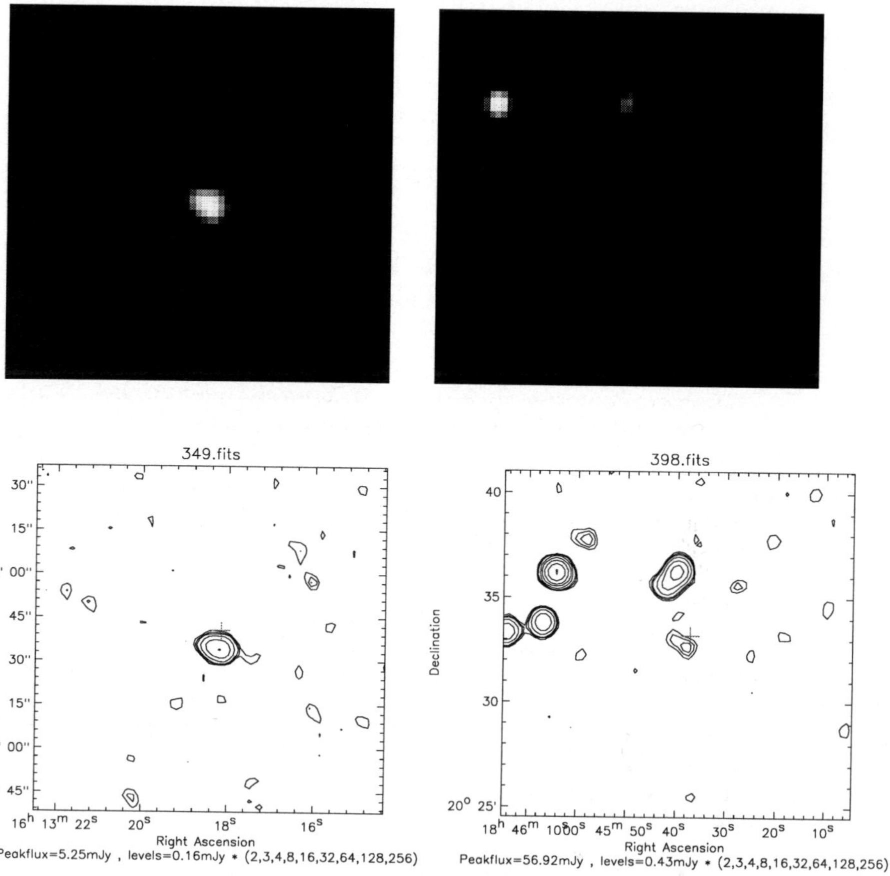

FIGURE 1. *First row:* $2' \times 2'$ 1.4 GHz radio map from *left* FIRST survey and *right* NVSS. *Second row:* contour plots of the radio maps above; red cross at the center defines the position of DARWIN star.

REFERENCES

1. Becker, R. H., et al., *ApJ*, **450**, 559 (1995)
2. Condon, J. J., et al., *AJ*, **115**, 1693 (1998)
3. De Ruiter, H. R., et al., *A&AS*, **28**, 211 (1997)
4. Fridlund, C. V. M., *AdSpR*, **34**, 613 (2003)
5. Katgert, P., et al., *A&A*, **195**, 21 (1988)
6. Longair, M. S., *MNRAS*, **133**, 421 (1966)

A New Approach For The Structure Of Ha Regions In 120 Be-type Stars

Lyratzi, E.[1], Danezis, E.[1], Nikolaidis, D.[1], Antoniou, A.[1], Popović, L. Č.[2], Dimitrijević, M. S.[2], Stathopoulou, M. and Theodossiou, E.[1]

1. University of Athens, Faculty of Physics, Department of Astrophysics, Astronomy and Mechanics,Panepistimioupoli, GR 157 84 Zographou, Athens, Greece
2. Astronomical Observatory, Volgina 7, 11160, Belgrade, Serbia

Abstract. In this paper we present a statistical study of the Hα line profiles of 120 Be-type stars using the model proposed by Danezis et al. (2003) and Lyratzi & Danezis (2004). This model proposes that the density layers which produce the Hα line lie in different regions in the stellar atmosphere. In the Be-type stellar atmospheres, there are two regions that can produce the Hα satellite components. The first one lies in the chromosphere and the second one in the cool extended envelope. By fitting the Hα line profiles with the line function of the proposed model we are able to calculate: a) For the chromospheric absorption components we calculated the rotational and radial velocities as well as the optical depth. b) For the emission and absorption components which are created in the cool extended envelope we calculated the radial velocities, the FWHM and the optical depth. Finally, we present the relation between these parameters with the spectral subtype and the luminosity class.

Keywords: Hot emission stars, models, DACs.
PACS: 97.10.Ex, 97.10.Fy, 97.20.Ec, 97.30.Eh

INTRODUCTION

In the spectra of many Oe and Be stars some spectral lines are accompanied by Discrete Absorption Components (DACs) (Bates & Halliwell 1986, Prinja 1988, Willis et al. 1989, Bates & Gilheany 1990, Gilheany et al. 1990, Waldron et al. 1994, Henrichs et al. 1994, Telting et al. 1993, Telting & Kaper 1994, Cranmer & Owocki 1996, Prinja et al. 1997, Fullerton et al. 1997, Kaper et al. 1996, 1997, 1999, Cranmer et al. 2000) or Satellite Absorption Components (SACs) (Peton 1974, Lamers et al. 1982, Sahade et al. 1984, Sahade & Brandi 1985, Hutsemékers 1985, Danezis 1987, Danezis et al. 1991, 2003, Laskarides et al. 1992a,b, Lyratzi and Danezis 2004).

The DACs were considered to be unknown spectral lines, which accompanied some spectral lines (Si IV, C IV, N IV, N V, Mg II) in the spectra of Oe and Be stars (Bates & Halliwell 1986, Prinja 1988, Willis et al. 1989, Bates & Gilheany 1990, Gilheany 1990, Kaper et al. 1990, Waldron et al. 1994, Henrichs et al. 1994, Telting et al. 1993, Telting & Kaper 1994, Cranmer & Owocki 1996, Prinja et al. 1997, Fullerton et al. 1997, Kaper et al. 1996, 1997, 1999, Cranmer et al. 2000).

CP848, *Recent Advances in Astronomy and Astrophysics,*
7th *International Conference of the Hellenic Astronomical Society,* edited by N. H. Solomos

DACs, now, are not unknown absorption spectral lines, but spectral lines of the same ion and the same wavelength as the main spectral line, shifted at different $\Delta\lambda$, as they are created in different density regions which rotate and move radially with different velocities.

If the regions, where such lines are created, rotate quickly and move radially slowly, the produced lines are quite broadened and with small shifts. So, they may not be discrete absorption spectral lines, but blended among themselves. In such a case, they are not observable, but we can detect them through the analysis of the profile. As Peton (1974) first pointed out, these components appear as "satellites" in the violet or in the red side of a main spectral line, as a function of the time or the phase, in the case of a binary system. For these two reasons and in order to include all these components, either they are discreet or not, to a unique name, we prefer to name them Satellite Absorption Components (SACs) and not Discrete Absorption Components (DACs).

MECHANISMS RESPONSIBLE FOR THE SACs' CREATION

The creation of SACs is due to mechanisms which allow the existence of structures which cover all or a significant part of the stellar disk, such as shells, blobs or puffs (Underhill 1975, Underhill & Fahey 1984, Bates & Halliwell 1986, Grady et al. 1987, Lamers et al. 1988, Cranmer & Owocki 1996, Kaper et al. 1996, 1997, 1999, Markova 2000) or interaction of fast and slow wind components, Corrotation Interaction Regions (CIRs), structures due to magnetic fields or spiral streams as a result of the star's rotation (Underhill & Fahey 1984, Mullan 1984a,b, 1986, Prinja & Howarth 1988, Cranmer & Owocki 1996, Fullerton et al. 1997, Kaper et al. 1996, 1997, 1999, Cranmer et al. 2000).

Though we do not know yet the mechanism responsible for the formation of such structures, it is positive that the SACs result from independent high density regions in the stars' environment (Fig. 1). These regions are formed by the specific ions which create a specific spectral line.

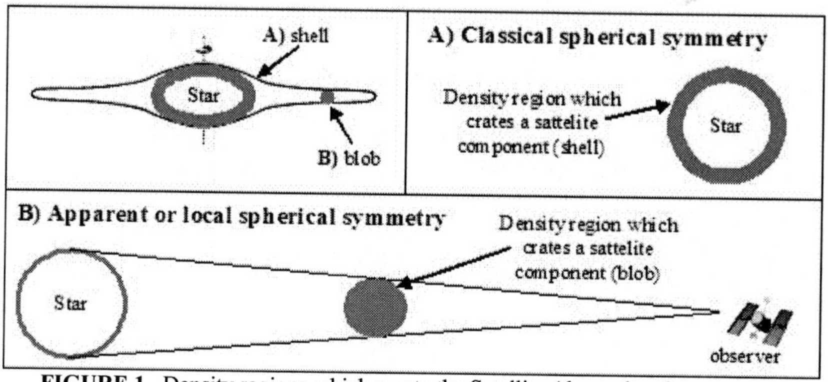

FIGURE 1. Density regions which create the Satellite Absorption Components.

THE USED MODEL

Danezis et al. (2003) constructed a mathematical model, in order to study the atmospheric regions that give rise to SACs.

By solving the equations of radiation transfer through a complex structure as the one described, we conclude to a function for the line's profile, able to give the best fit for the main spectral line and its Satellite Absorption Components in the same time. Such a best fit, through the function of the line's profile, enables us to calculate some parameters (rotational and radial velocities, FWHM, optical depth) of the independent layers of matter, which form the main spectral line and its satellite absorption components. (See Danezis et al. 2005a, b, Lyratzi et al. 2005).

$$I_\lambda = \left[I_{\lambda 0} \prod_i e^{-\tau_i} + \sum_j S_{\lambda ej}\left(1 - e^{-\tau_j}\right) \right] e^{-\tau_g} \tag{1}$$

where: $I_{\lambda 0}$: is the initial radiation intensity, $S_{\lambda ej}$ is the source function, which, at the moment when the spectrum is taken, is constant and $e^{-\tau_i}$, $e^{-\tau_j}$, $e^{-\tau_g}$ are the appropriate distribution functions (Gauss, Lorentz, Voigt, Rotation) of the absorption, emission and general absorption lines respectively. This function I_λ does not depend on the geometry of the regions which create the observed feature.

APPLICATION TO THE Hα LINE OF 120 Be STARS

In our study we use the stellar spectrographs which were taken by Andrillat & Fehrenbach (1982) and Andrillat (1983) (resolution 5,5 and 27 Å). We applied the model on the Ha line 6562.817 Å in the spectra of 120 Be stars of all the spectral subtypes and luminosity classes. In most of the Be stellar spectra the Ha line presents peculiar and complex profiles. Usually the Ha line's profile consists of: a very broad absorption feature, an emission feature and a narrow absorption feature (Fig. 2).

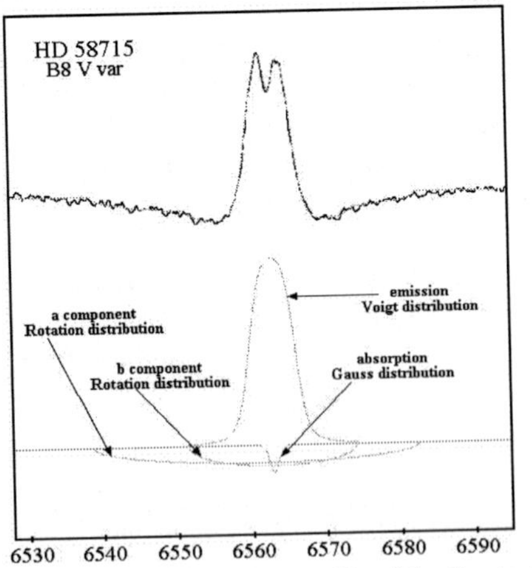

FIGURE 2. Analysis of the Ha line profile of the Be star HD 58715.

We applied the proposed model in order to reproduce these complex profiles. We tried to fit the observed profiles by applying all the classical distribution functions (Gauss, Lorentz, Voigt, Rotation). We concluded that the best fit is accomplished when we fit: a) the very broad absorption component with Rotation distribution, b) the emission component with Voigt distribution and c) the narrow absorption component with Gauss distribution.

CONCLUSIONS

The proposed line function $I_\lambda = \left[I_{\lambda 0} \prod_i e^{-\tau_i} + \sum_j S_{\lambda ej} \left(1 - e^{-\tau_j}\right) \right] e^{-\tau_g}$ is able to reproduce accurately the complex Hα profiles of all the 120 studied Be-type stars. This means that the regions where the Ha line is created are not continuous, but they consist of successive independent density regions. In the place of the exponential $e^{-\tau}$, which gives the profile of each component, we apply the appropriate distribution function. The choice of the appropriate distribution function depends on the physical conditions of the regions which create the SACs. The most important point is that, in any case, the proposed line function remains the same. The important advantage of this method is that we are able to accomplish the best fit of the observed spectral lines, by applying a line function, to which we conclude after the solution of the radiation transfer equations, through a complex atmospherical structure, and not by a graphical composition of mathematical distribution functions with no physical meaning.

The existence of SACs is a general phenomenon in the spectra of Be-type stars.

The absorption regions of the Hα line, lie in two different atmospherical regions: in the chromosphere and in the cool extended envelope. In the chromosphere we detected one to five successive, independent density regions, which rotate with 5200±1192 km/s, 990±170 km/s, 536±68 km/s, 352±37 km/s and 152±46 km/s and move radially with 15±121 km/s, 7±123 km/s, 19±62 km/s, 15±60 km/s and -2±42 km/s. Each region creates one Satellite Absorption Component (SAC). In the Cool Extended Envelope there are the density regions which create the emission components and the narrow absorption components. The emission regions move radially with 20 km/s and create SACs with Full Width at Half Maximum (FWHM) about 7.1 Å. In 7 of the 120 stars where we detected one more emission region, with the same radial velocity and FWHM about 2.0 Å. The narrow absorption components have FWHM about 2.0 Å and the regions which create them have radial velocity of 0 km/s.

The profiles of the studied Hα lines appear to be peculiar and complex, as they do not present only one spectral line, but a number of SACs, which are created in independent density regions. All the studied stars do not present the same number of independent density regions.

ACKNOWLEDGMENTS

This research project is progressing at the University of Athens, Department of Astrophysics, Astronomy and Mechanics, under the financial support of the Special

Account for Research Grants, which we thank very much. This work also was supported by Ministry of Science and Environment Protection of Serbia, through the projects "Influence of collisional processes on astrophysical plasma line shapes" and "Astrophysical spectroscopy of extragalactic objects".

REFERENCES

1. Y. Andrillat, A&AS, **53**, 319 (1983).
2. Y. Andrillat and Ch. Fehrenbach, A&AS, **48**, 93 (1982).
3. B. Bates and D. R. Halliwell, MNRAS, **223**, 673 (1986).
4. B. Bates and S. Gilheany, MNRAS, **243**, 320 (1990).
5. S. R. Cranmer and S. P. Owocki, ApJ, **462**, 469 (1996).
6. S. R. Cranmer, M. A. Smith and R. D. Robinson, ApJ, **537**, 433 (2000).
7. E. Danezis, in *Physics of Be Stars*, edited by A. Slettebak and T. P. Snow, Proceedings IAU Colloq. No 92, (Cambridge University Press), 1987, p.149.
8. E. Danezis, E. Theodossiou and P. G. Laskarides, A&SS, **179**, 111-139 (1991).
9. E. Danezis, D. Nikolaidis, V. Lyratzi, M. Stathopoulou, E. Theodossiou, A. Kosionidis, C. Drakopoulos, G. Christou and P. Koutsouris, A&SS, **284**, 1119 (2003).
10. E. Danezis, D. Nikolaidis, E. Lyratzi, L. Č. Popović, M. S. Dimitrijević, E. Theodossiou and A. Antoniou, Mem. S.A.It., in press (2005a).
11. E. Danezis, D. Nikolaidis, E. Lyratzi, L. Č. Popović, M. S. Dimitrijević, E. Theodossiou and A. Antoniou, Proceedings of 7th HEL.A.S. Conference, in press (2005b).
12. A. W. Fullerton, D. L. Massa, R. K. Prinja, S. P. Owocki and S. R. Cranmer, A&A, **327**, 699 (1997).
13. S. Gilheany, B. Bates, M. G. Catney and P. L. Dufton, A&SS, **169**, 85 (1990).
14. Grady, G. Sonneborn, Wu Chi-chao and H. F. Henrichs, ApJS, **65**, 673 (1987).
15. H. F. Henrichs, L. Kaper and J. S. Nichols, A&A, **285**, 565 (1994).
16. D. Hutsemékers, A&AS, **60**, 373 (1985).
17. L. Kaper, H. F. Henrichs, G. A. A. Zwarthoed and J. Nichols-Bohlin, in *Angular Momentum and Mass Loss for Hot Stars*, edited by L. A. Willson and R. Stalio (Dordrecht: Kluwer), 1990, p. 213.
18. L. Kaper, H. F. Henrichs, J. S. Nichols, L. C. Snoek, H. Volten and G. A. A. Zwarthoed, A&AS, **116**, 257 (1996).
19. L. Kaper, H. G. Henrichs, A. W. Fullerton, H. Ando, K. S. Bjorkman, D. R. Gies, R. Hirata, E. Dambe, D. McDavid and J. S. Nichols, A&A, **327**, 281 (1997).
20. L. Kaper, H. F. Henrichs, J. S. Nichols and J. H. Telting, A&A, **344**, 231 (1999).
21. H. J. G. L. M. Lamers, R. Gathier, and T. P. Snow, ApJ, **258**, 186, (1982).
22. H. J. G. L. M. Lamers, T. P. Snow, C. de Jager and A. Langerwerf, ApJ, **325**, 342 (1988).
23. P. G. Laskarides, E. Danezis, and E. Theodossiou, A&SS, **179**, 13 (1992a).
24. P. G. Laskarides, E. Danezis, and E. Theodossiou, A&SS, **183**, 67 (1992b).
25. E. Lyratzi and E. Danezis, in *The Physics of Ionized Gases: 22nd Summer School and International Symposium on the Physics of Ionized Gases*, edited by Lj. Hadzijevski, T. Grozdanov and N. Bibić, AIP Conference Proceedings 740, Bajina Bašta, pp. 458-473 (23-27 August 2004).
26. E. Lyratzi, E. Danezis, D. Nikolaidis, L. Č. Popović, M. S. Dimitrijević, E. Theodossiou and A. Antoniou, Mem. S.A.It., in press (2005).
27. N. Markova, A&AS, **144**, 391 (2000).
28. D. J. Mullan, ApJ, **283**, 303 (1984a).
29. D. J. Mullan, ApJ, **284**, 769 (1984b).
30. D. J. Mullan, A&A, **165**, 157 (1986).
31. A. Peton, Space Sci. Rev., **30**, 481 (1974).
32. R. K. Prinja, MNRAS, **231**, 21 (1988).
33. R. K. Prinja and I. D. Howarth, MNRAS, **233**, 123 (1988).
34. R. K. Prinja, D. Massa, A. W. Fullerton, I. D. Howarth and M. Pontefract, A&A, **318**, 157 (1997).
35. J. Sahade, E. Brandi and J. M. Fontela, A&AS, **56**, 17 (1984).
36. J. Sahade and E. Brandi, RMxAA, **10**, 229 (1985).
37. J. H. Telting, L. B. F. M. Waters, P. Persi and S. Dunlop, A&A, **270**, 355 (1993).
38. J. H. Telting, and L. Kaper, A&A, **284**, 515 (1994).
39. A. B. Underhill, ApJ, **199**, 691 (1975).
40. A. B. Underhill and R. P. Fahey, ApJ, **280**, 712 (1984).
41. W. L. Waldron, L. Klein and B. Altner, ApJ, **426**, 725 (1994).
42. A. J. Willis, I. D. Howarth, D. J. Stickland and S. R. Heap, ApJ, **347**, 413 (1989).

A photometric study of the near-contact binaries V1034 Cygni and AX Virginis

V.N. Manimanis*, P.G. Niarchos* and K.D. Gazeas*

*Department of Astrophysics, Astronomy and Mechanics, Faculty of Physics, National and Kapodistrian University of Athens, GR 15784 Zographou, Athens, Greece

Abstract. The first complete CCD light curves of the eclipsing binary systems AX Virginis and V1034 Cygni have been obtained in the B, V, R, I and V, R, I filters, respectively, during the spring and summer of 2004 with the 40-cm telescope of the University of Athens Observatory. The light curves were analyzed with the W-D program in order to determine the geometrical and physical parameters of the systems.

Keywords: binaries: eclipsing, stars: individual: AX Vir, V1034 Cyg, photometry
PACS: 97.80.Fk, 97.80.Hn

1. INTRODUCTION

AX Virginis (= BD+04°2748 = PPM 159769 = HIP 65660) with an A2 primary and a G3 secondary was discovered as a variable star by Hoffmeister (1935) and shortly after that Jensch (1935) identified its light curve as beta Lyrae (EB) type. Afterwards, AX Vir has been observed only to obtain times of minima, with the exception of Liu & Yang (1985), who obtained a complete photoelectric light curve in only one filter (V) in 1982, which was analyzed and solved by Leung et al. (1985) using the Wilson-Devinney method. They found the system to be a semi-detached one with the secondary filling its Roche lobe. However, when Barone et al. (1991) reanalyzed *the same* light curve, they came to a "remarkably different" solution: they found that the primary fills its Roche lobe, while the secondary is clearly detached. Brancewicz & Dworak (1980) include it in their catalogue of parameters as a contact system with an EB-type light curve. Giuricin et al. (1983) mention it in their statistical paper. Shaw (1994) includes it in his second catalogue of near-contact systems as a system of unknown state. The reason for the conflicting solutions is probably that one-filter photometry does not allow the correct determination of the origin of light in the intermediate phases in spite of the large temperature difference. Therefore, we obtained the first simultaneous complete light curves in 4 filters (BVRI) of this system in order to determine its state. The period of the system is constant within the limits of the history of observations; an (O–C) diagram with the published minima, containing also many visual minima times from the AAVSO database, is given by Robert Nelson at his website.

V1034 Cygni (= BD+30°3884 = HD 333491 = BV 114 Cygni) was discovered by W. Strohmeier and R. Knigge during the Bamberg program, and was described first by them (1961) and Sazonov (1961). Its period was determined by Strohmeier & Bauernfeind (1968). Koch (1974) gives for the system a (B–V) value of -0.01 and a slightly larger-than-average unreddened CN index of 0.226 ± 0.003. Times of minima have been

CP848, *Recent Advances in Astronomy and Astrophysics,*
7th *International Conference of the Hellenic Astronomical Society,* edited by N. H. Solomos

published by various authors. Brancewicz & Dworak (1980) include it in their catalogue of parameters as a detached system with an EB-type light curve. Shaw (1994) includes it in his second catalogue of near-contact systems as a possible Algol-type system. The spectral type of the primary is cited as A0. Ours are the first complete light curves of any kind for this star.

2. THE OBSERVATIONS

The CCD observations of both stars were conducted in 2004 with the 40-cm Cassegrain telescope of the University of Athens Observatory and an SBIG-8 CCD camera cooled about 30 degrees lower than the ambient temperature.

The Virgo star was observed on five nights: February 7, March 30 and April 3, 6 and 7. The stars GSC 00303:00289 ($V = 10.50$) and GSC 00303:00193 were used as comparison and check stars, respectively. In total, 272 blue, 256 yellow, 268 red and 288 infrared observations were obtained. The probable error of a single observation was estimated to be ± 0.010 mag in infrared, ± 0.012 mag in red, ± 0.015 mag in yellow and ± 0.022 mag in blue.

The Cygnus star was observed on June 15, June 22, July 29, August 27, 29 and 30, and September 9, 10, 16 and 30. The stars GSC 02670:01960 ($V = 10.8$) and GSC 02670:03186 were used as comparison and check stars, respectively. In total, 850 yellow, 821 red and 747 infrared observations were obtained in the ten nights. The probable error of a single observation was estimated to be ± 0.006 mag in infrared, ± 0.008 mag in red and ± 0.010 mag in yellow.

3. LIGHT CURVE SOLUTIONS

We solved the light curves assuming there are no spots on the components of the systems, since from a visual inspection of the curves we decided that, in first approximation, the observations are not affected by the presence of spots. All observational points for each color were used. The subscripts 1 and 2 refer to the star being eclipsed at primary and secondary minimum, respectively. The PHOEBE 0.27 version (Prša & Zwitter 2005) of the Wilson-Devinney program was used in the light curve analysis.

In the case of AX Vir, since no double-line spectroscopy was available, a "q-search" gave a starting value of $q = 0.4$. A preliminary set of the other input parameters was selected from the solution by Barone et al. (1991). The results converged with both Modes 4 and 5; the best fit (minimum χ^2 value averaged for all filters) was achieved with Mode 4, indicating that the primary star fills its Roche lobe. The adjustable parameters were q, the conjunction phase ϕ, the inclination i, the temperature T_2 of the secondary component, the non-dimensional surface potential Ω_1 (Mode 5), or Ω_2 (Mode 4), and the non-normalized monochromatic luminosity L_1. The fixed parameter was the temperature T_1 (from the spectral type of the primary). Assumed values for gravity and limb darkening, and bolometric albedos, according to the spectral types of the components, were used. Third light was assumed to be zero. The results are given in Table 1. The errors given are standard deviations. Theoretical light curves based on our Mode 4 solution

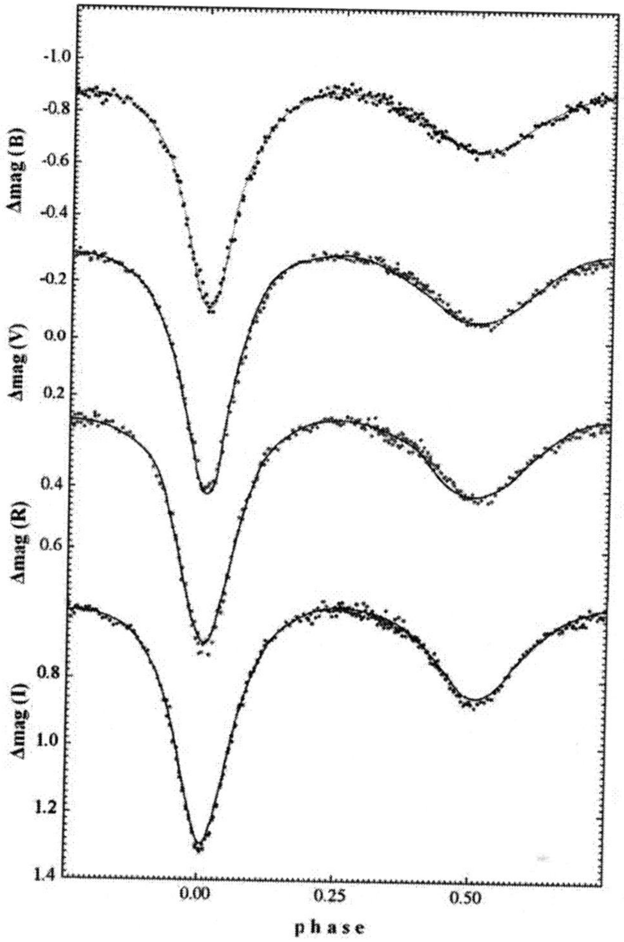

FIGURE 1. The observational points and theoretical light curves of AX Vir for our Mode 4 solution.

along with the observed ones are shown in Fig. 1.

In the case of V1034 Cyg, a "q-search" gave a starting value of $q = 0.2$. A preliminary set of the other input parameters was selected from Brancewicz & Dworak (1980). The results converged only with the Mode 5, indicating that the secondary star fills its Roche lobe. Modes 2 and 4 were also tried without achieving convergence. The adjustable parameters were the same as in the case of AX Vir. Third light was assumed to be zero. The results are given in Table 2. The errors given are standard deviations. Theoretical light curves based on our solution along with the observed ones are shown in Fig. 2.

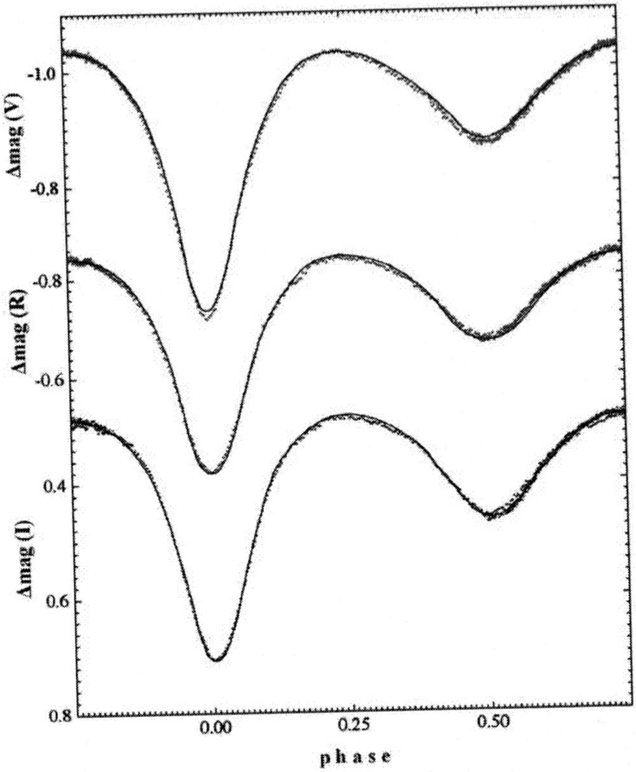

FIGURE 2. The observational points and theoretical light curves of V1034 Cyg for our Mode 5 solution.

4. SUMMARY AND CONCLUSIONS

From the results obtained using the WD program it seems almost certain that the system of AX Virginis is a semidetached one; furthermore, our solution supports the results by Barone et al. (1991), because the primary fills its Roche lobe, while the secondary is clearly detached, although certainly it is "near contact". However, the average χ^2 is almost the same for both Modes 4 and 5, which explains the discrepancy in the bibliography. Spectroscopic observations would be useful to solve this ambiguity, because a spectroscopic mass ratio would strongly favor the Mode 4 solution if close to 0.52 and the Mode 5 one if closer to 0.43. The V1034 Cygni system probably needs further analysis: The WD program converged with difficulty and in only one Mode, an unusual fact. This suggests that the presence of spots on the secondary (the primary is too hot to have magnetic spots in theory), or possibly circumstellar matter, have a small photometric effect, which should be taken into account in a more detailed future study. It could be in a contact state, although its period exceeds the period of almost all near-contact systems. Hence, we think that this is an evolved system with oversized components and all the special characteristics such a state presents.

TABLE 1. Light curve solutions of AX Virginis and V1034 Cygni.

Parameter	AX Virginis Mode 4	AX Virginis Mode 5	V1034 Cygni Mode 5
ϕ_0	0.0066(2)	0.0065(2)	0.0044(1)
i (°)	81.121(91)	80.821(203)	76.100(43)
T_1 (K)	9307*	9307*	9870*
T_2 (K)	6589(7)	6726(22)	6451(4)
g_1	1.0*	1.0*	1.0*
g_2	0.32*	0.32*	0.32*
A_1	1.0*	1.0*	1.0*
A_2	0.5*	0.5*	0.5*
Ω_1	2.917*	2.7626(92)	2.3055(15)
Ω_2	3.0600(142)	2.7451*	2.3006*
$q = M_2/M_1$	0.5217(39)	0.4332(43)	0.22782(40)
$L_1/(L_1+L_2)$ (B)	0.8877(7)	0.8892(15)	–
$L_1/(L_1+L_2)$ (V)	0.8594(6)	0.8483(16)	0.9239(2)
$L_1/(L_1+L_2)$ (R)	0.8548(7)	0.8441(15)	0.9151(2)
$L_1/(L_1+L_2)$ (I)	0.8557(11)	0.8450(18)	0.9121(5)
r_1 (pole)	0.410(2)	0.423(2)	0.480(2)
r_1 (side)	0.435(3)	0.450(3)	0.520(3)
r_1 (back)	0.464(2)	0.476(2)	0.545(3)
r_2 (pole)	0.281(5)	0.289(2)	0.242(1)
r_2 (side)	0.290(7)	0.301(8)	0.252(1)
r_2 (back)	0.312(10)	0.334(12)	0.284(2)
χ^2	0.01183	0.01188	0.01065

* assumed

ACKNOWLEDGMENTS

This research was included in the project for the support of research groups in the universities, co-funded by the European Social Fund (ESF) and National Resources (EPEAEK II) - *PYTHAGORAS*.

REFERENCES

1. F. Barone et al., *Ap&SS*, **183**, 117-127 (1991).
2. H.K. Brancewicz and T.Z. Dworak, *Acta Astronomica*, **30**, 501-524 (1980).
3. G. Giuricin, F. Mardirossian and M. Mezzetti, *A&A Suppl. Series*, **54**, 211-220 (1983).
4. C. Hoffmeister, *AN*, **255**, 401 (1935).
5. A. Jensch, *AN*, **256**, 279 (1935).
6. R.H. Koch, *AJ*, **79**, 34-41 (1974).
7. K.-C. Leung, D. Zhai, Q.-Y. Liu and Y.-L. Yang, *AJ*, **90**, 115-119 (1985).
8. Q.-Y. Liu and Y.-L. Yang, *Acta Astrophys. Sinica*, **5**, 124-131 (1985).
9. A. Prša and T. Zwitter, *ApJ*, **628**, 426-438 (2005).
10. V. Sazonov, *Astron. Tsirk.*, **219**, 25 (1961).
11. V. Sazonov, *Peremen. Zvezdy*, **13**, 440-442 (1961).
12. J.S. Shaw, *Mem. S. A. Ital.*, **65**, 95-103 (1994).
13. W. Strohmeier and H. Bauernfeind, *Bamberg Ver. Remeis-Sternw.*, **7**, 72 (1968).
14. W. Strohmeier and R. Knigge, *Bamberg Ver. Remeis-Sternw.*, **5**, 11 (1961).

CCD photometry and modelling of the ROTSE-1 variable GSC 2534:1121

K.D. Gazeas*, V.N. Manimanis*, P.G. Niarchos* and C. Vamvatira-Nakou*

*Department of Astrophysics, Astronomy and Mechanics, Faculty of Physics, National and Kapodistrian University of Athens, GR 15784 Zographou, Athens, Greece

Abstract. The first VRI CCD photometric observations of the newly discovered ROTSE-1 eclipsing variable GSC 2534:1121 are presented. Our new times of minima are used along with those of the last two years to calculate a new ephemeris of the system. The observed light curves are analyzed with the Wilson-Devinney program in order to derive the geometric and photometric elements of the system. The moderate O'Connell effect is explained by assuming a cool spot on the primary component.

Keywords: binaries: eclipsing, starspots, stars: individual: GSC 2534:1121, photometry
PACS: 97.80.Fk, 97.80.Hn

1. INTRODUCTION

The star GSC 2534:1121 was found to be an eclipsing variable of W UMa type during the ROTSE-1 program (Akerlof et al. 2000). Its catalogue number is ROTSE 1 J130625,38+342917,7 EW. The main goal of the ROTSE experiment (Robotic Optical Transient Search Experiment) is to identify Gamma Ray Bursts in the optical region of the electromagnetic spectrum. In the frame of this search many new objects were also detected. GSC 2534:1121 was also observed by Blättler & Diethelm (2004), who determined 11 times of minima from their unfiltered observations obtained during seven nights between JD 2453045 and 2453117. The system GSC 2534:1121 was included in our program of observations of selected W UMa systems with the aim to provide reliable system parameters for poorly observed or unobserved such systems.

2. OBSERVATIONS AND DATA REDUCTION

The observations of GSC 2534:1121 were made on 12 and 15 April 2005. The instruments used were the 0.40 m Cassegrain reflector at the University of Athens Observatory, Greece, and a CCD camera, whose chip has 1536×1024 useful pixels of 9×9 μm, covering an area of approximately 10×15 arcmin. A focal reducer f/6.3 was used, increasing the field to 15×22.5 arcmin. All observations were made using the V, R and I filters. The data reduction was made with the package AIP4WIN (Berry & Burnell 2000). In total, 1753 observations were collected in all bands used. The stars GSC 2534:1056 and GSC 2534:1091 were used as comparison and check star, respectively. Differential magnitudes were used to determine the mean standard deviation of our measurements, which is approximately 0.01 mag in all bands used. It was clear from the observations

CP848, *Recent Advances in Astronomy and Astrophysics,*
7th *International Conference of the Hellenic Astronomical Society,* edited by N. H. Solomos

that GSC 2534:1121 is a contact binary of the W UMa type.Three times of minima were computed from our observations with the method of Kwee & van Woerden (1956) and, along with the minima by Blättler & Diethelm (2004), were used to derive the following ephemeris:

$$\begin{aligned}\text{(HJD) Min I} &= 2453473.5077 + 0.3426870d \cdot E \\ &\qquad \pm 0.0010 \pm 0.0000004\end{aligned}$$

3. DETERMINATION OF THE EFFECTIVE TEMPERATURE

Transformation of the observed magnitudes of standard stars and the variable to the Standard System was done in order to find the color index and the temperature of the system, since the spectral types were unknown. Standard stars, selected from the Landolt (1983) catalogue, were observed in July 2005. The standard stars were observed simultaneously with the variable, in order to check the stability of the extinction coefficients. Extinction corrections of first and second order were applied. Since we are not interested in the standard magnitudes but only in the fluxes of the system, we did not transform all the magnitudes of the variable to the Standard System, but we converted them into flux units. Applying the first and second order corrections we found that the color index of the variable GSC 2534:1121 is: $B - V = 0.51 \pm 0.03$. Corrections for the color excess were applied to the calculated color index (Burstein & Heiles 1982). The corrections due to the reddening were almost negligible for the location defined by the coordinates of the variable. The coordinates where transformed from equatorial to galactic and it was found that GSC 2534:1121 is very close to the north galactic pole (galactic latitude 64°, galactic longitude 82°). Thus $E(B-V) \sim 0.0$ and $(B-V)_o = 0.51 \pm 0.03$. From the color index we found the temperature of the hotter component of the system to be $T_h \simeq 6250$ K.

4. LIGHT CURVE ANALYSIS

The light curves have almost equal minima, but the two maxima show a difference MaxII–MaxI = 0.02 mag. No other significant asymmetries are present. The shape of the light curves at the minima indicates that the eclipses are partial. Our light curves were analyzed with the Wilson-Devinney program (Wilson & Devinney 1971, Wilson 1998) running in Mode 3, under the form of the software PHOEBE (Prša & Zwitter, 2005). The pronounced O'Connell effect was explained by assuming the existence of at least one spot on one component of the system. There is no indication on which component the spot is located, since the temperatures are almost equal (the minima have almost the same depth). A "q-search" was made in order to find the most probable mass ratio for the binary; the diagram in Figure 1 shows the variation of the sum of the squares of the residuals with respect to different values of mass ratio. It is obvious that a minimum occurs at a value around $q = 2.5$. Thus, the mass ratio was set to an initial value of $q = 2.5$ and left as a free parameter. The free parameters to be determined by the program were the following: the phase of conjunction ϕ, the inclination i, the

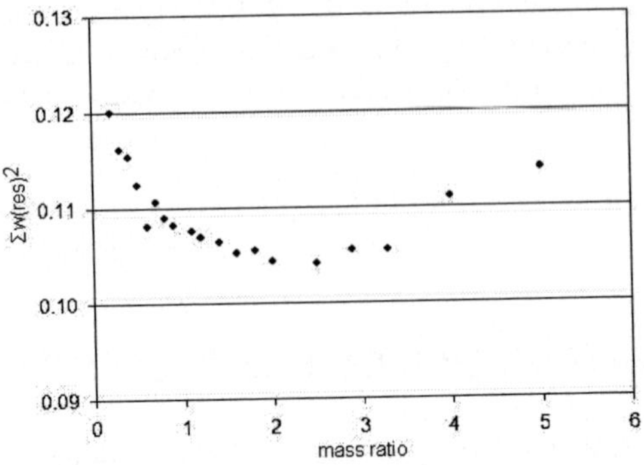

FIGURE 1. The "q-search" diagram for GSC 2534:1121.

TABLE 1. Light curve solutions of GSC 2534:1121.

Parameter	Mode 3, unspotted	Mode 3, spotted
ϕ_0	0.0031(5)	0.0030(5)
i (°)	61.7(2)	62.1(2)
T_1 (K)	6250*	6250*
T_2 (K)	6053(8)	6258(11)
$g_1(=g_2)$	0.32*	0.32*
$A_1(=A_2)$	0.5*	0.5*
$\Omega_1=\Omega_2$	5.855(1)	5.937(22)
% overcontact	15%	24%
$q=M_2/M_1$	2.501(6)	2.603(13)
$L_1/(L_1+L_2)$ (V)	0.3044(11)	0.2752(14)
$L_1/(L_1+L_2)$ (R)	0.3053(11)	0.2761(14)
$L_1/(L_1+L_2)$ (I)	0.3039(11)	0.2749(15)
r_1 (vol.)	0.3111(2)	0.3141(2)
r_1 (pole)	0.2896(2)	0.2909(2)
r_1 (side)	0.3029(2)	0.3049(2)
r_1 (back)	0.3408(2)	0.3466(2)
r_2 (vol.)	0.4701(2)	0.4793(2)
r_2 (pole)	0.4395(2)	0.4468(2)
r_2 (side)	0.4707(2)	0.4801(2)
r_2 (back)	0.5001(2)	0.5108(2)
spot co-latitude(°)	-	90*
spot longitude (°)	-	85(10)
spot radius (°)	-	20(3)
spot T factor	-	0.90(2)
Σw(res)^2	0.104	0.101

*assumed

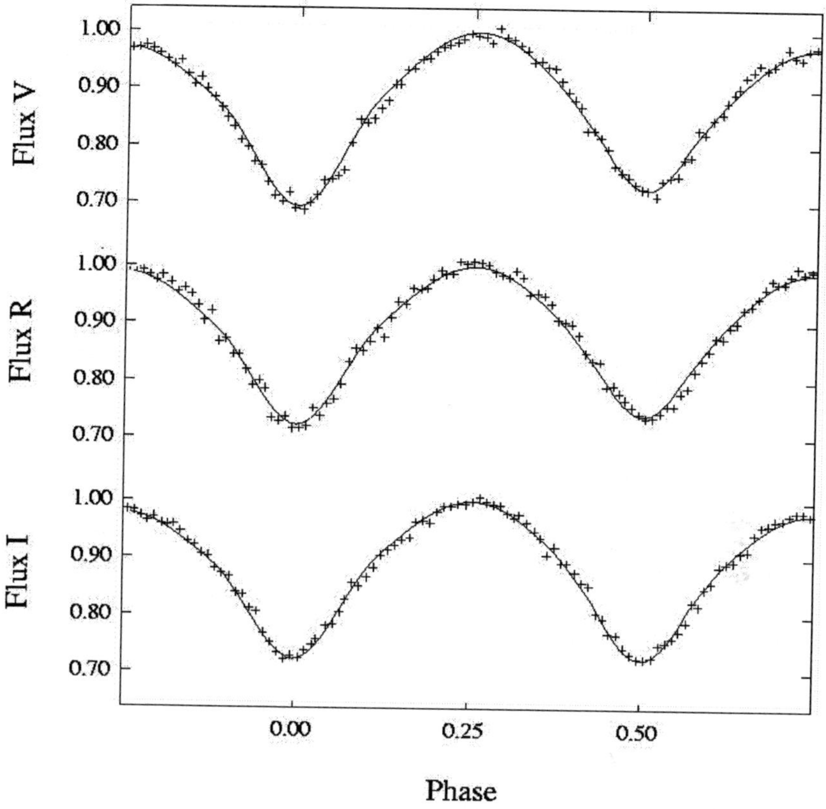

FIGURE 2. Observed (normal points) and theoretical light curves of GSC 2534:1121.

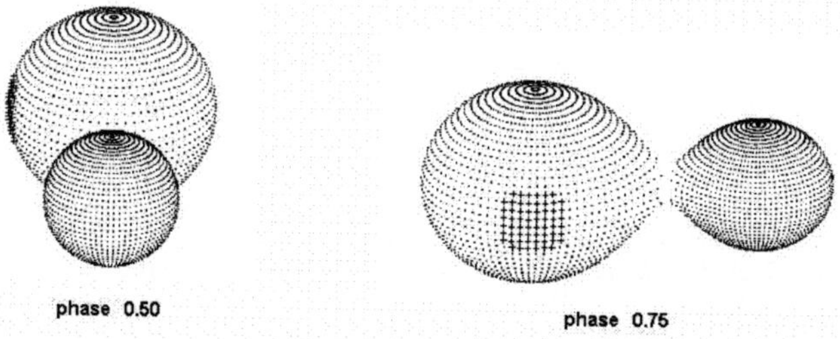

FIGURE 3. A three-dimensional model of GSC 2534:1121 at phases 0.50 and 0.75.

temperature of the secondary star T_2, the mass ratio q, the surface potentials Ω (both Ω_1 and Ω_2) and the non-normalized monochromatic luminosities of the primary (L_1) for the three filters. The temperature T_1 had a fixed value (from the calculated color index). The standard values from the literature were used for the coefficients of gravity darkening.

The limb darkening coefficients were taken from the tables of Claret (2000) and Claret, Díaz-Cordovés & Giménez (1995), according to the spectral type and wavelength of observation. Third light was assumed to be zero. The results are given in Table 1. The errors given are standard deviations. The derived parameters were used to construct theoretical light curves, which are shown along with the observed ones in Figure 2, while a 3-D model of the system is given in Figure 3.

5. SUMMARY AND CONCLUSIONS

The shape of the light curves of the variable GSC 2534:1121 indicates that the system is a W UMa type contact binary. The analysis of the light curves was very difficult because: (1) no spectroscopic mass ratio is available, (2) the light curves show partial eclipses with minima of almost the same depth, and (3) no spectral type is given. We tried to solve the light curves by applying advanced techniques of the Wilson-Devinney program. We made the "q-search" to locate the most probable value of the photometric mass ratio and we determined the effective temperature of the system by means of the color index, calculated from observations of standard stars and the variable. The results of the analysis show that the system is a W UMa type contact binary of W-type. The moderate O'Connell effect was explained by assuming a cool spot on the larger and more massive component. Spectroscopic observations are needed for the determination of the mass ratio, but the faintness of the system makes these observations quite difficult.

ACKNOWLEDGMENTS

This research was included in the project for the support of research groups in the universities, co-funded by the European Social Fund (ESF) and National Resources (EPEAEK II) - *PYTHAGORAS*.

REFERENCES

1. C. Akerlof et al., *AJ*, **119**, 1901-1913 (2000).
2. R. Berry and J. Burnell, *The Handbook of Astronomical Image Processing*, Willmann-Bell, Richmond, 2000.
3. E. Blättler and R. Diethelm, *IBVS*, No. 5541 (2004).
4. D. Burstein and C. Heiles, *AJ*, **87**, 1165-1189 (1982).
5. A. Claret, *A&A*, **363**, 1081-1190 (2000).
6. A. Claret, J. Díaz-Cordovés, A. Giménez, *A&A Suppl.Series*, **114**, 247-252 (1995).
7. K.K. Kwee and H. van Woerden, *Bull. Astron. Inst. of Netherlands*, **12**, 327-330 (1956).
8. A.U. Landolt, *AJ*, **104**, 340-371 and 436-491 (1992).
9. A. Prša and T. Zwitter, *ApJ*, **628**, 426-438 (2005).
10. R.E. Wilson, *Computing Binary Star Observables (Reference Manual to the Wilson-Devinney Programme)*, University of Florida, Gainesville, 1998.
11. R.E. Wilson and E.J. Devinney, *ApJ*, **166**, 605-619 (1971).

A spot model of the overcontact binary PY Lyrae

V.N. Manimanis*, P.G. Niarchos* and K.D. Gazeas*

*Department of Astrophysics, Astronomy and Mechanics, Faculty of Physics, National and Kapodistrian University of Athens, GR 15784 Zographou, Athens, Greece

Abstract. The first BVRI CCD photometric observations of PY Lyrae were obtained at the Kryoneri Observatory in the summer of 2004. Five new times of minima and a new ephemeris were calculated. The light curves were analyzed with the Wilson-Devinney program, in order to derive the geometric and photometric elements of the system. A spot model was assumed in order to explain the pronounced O'Connell effect.

Keywords: binaries: eclipsing, starspots, stars: individual: PY Lyr, photometry
PACS: 97.80.Fk, 97.80.Hn

1. INTRODUCTION

PY Lyr (= GSC 02136:03365) was classified as a variable by Wahmann (1961, 1963, 1964, 1966), who studied 310 objects in the region of the southern Cygnus star cloud. Skiff (1999) used a new method to identify these objects and improved their coordinates. The system is classified as a W UMa-type system in GCVS (Kholopov et al. 1985). Pribulla et al. (2003) included PY Lyr in their catalogue of 361 galactic contact binaries with a note of fast or complicated variation of the orbital period. No complete light curves of the system existed so far, but quite a few times of minima have been obtained by several authors (Hoffmann 1983; Safar & Zejda 2002; Zejda 2002; Demircan et al. 2003; Bakis et al. 2003a,b; Zejda 2004). Although no radial velocity observations are available for the system, we included the system in our W UMa program with the aim to provide the first photometric observations and present the first approximate model based on photometry alone.

2. THE OBSERVATIONS

The photometric observations of PY Lyr were made during three consecutive nights, on July 16, 17 and 18, 2004. The instruments used were the 1.22 m Cassegrain reflector at the Kryoneri Astronomical Station of the National Observatory of Athens, Greece, and a Photometrics CH250 CCD camera. The CCD chip has 512 × 512 useful pixels of 24 × 24 microns; the corresponding field of view is 2.5 × 2.5 arcmin.

The stars circled in Fig. 1 were used as comparison (C) and check (K) stars, respectively, while PY Lyr is labeled with V. A total number of 1035 frames were obtained in the B, V, R and I filters, covering five minima (3 primary and 2 secondary). The images were processed with the AIP4WIN program of Berry & Burnell (2000). Differential magnitudes were used to determine the mean standard deviation of our measurements, which is approximately 0.006 mag in all bands. It was clear from the observations that

CP848, *Recent Advances in Astronomy and Astrophysics,*
7th *International Conference of the Hellenic Astronomical Society,* edited by N. H. Solomos

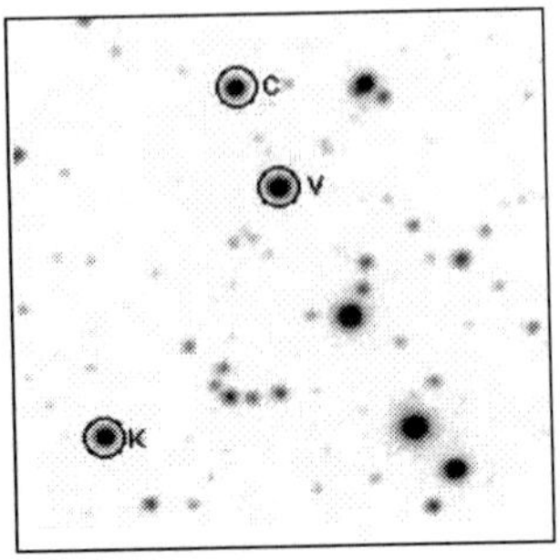

FIGURE 1. The 2.5 × 2.5 arcmin field around PY Lyrae. The stars used are the comparison (C), check (K) and the variable PY Lyr (V).

PY Lyr is a contact binary of the W UMa type. The times of minima were computed using the method of Kwee & van Woerden (1956) and were used to derive the following ephemeris:

$$\begin{aligned}(\mathrm{HJD})\ \mathrm{Min\,I} &= 2453203.1729 + 0.38593^{d} \cdot E \\ & \qquad \pm 0.0006 \pm 0.00016\end{aligned}$$

3. LIGHT CURVE ANALYSIS

The light curves have almost equal minima (MinI – MinII = 0.01 mag) but the two maxima show a difference MaxI – MaxII = 0.035 mag. No other significant asymmetries are present. There is no obvious indication of total eclipses and the two minima have almost equal depths, so that it is difficult to identify the primary (deeper) minimum. The pronounced O'Connell effect was explained by invoking a cool spot on one of the components of the system. It was found that the results of spot modeling were the same regardless of the component on which the spot is located. This is explained by the fact that the temperature of the two components should be almost identical. In order to obtain the physical parameters we used the Wilson-Devinney code (W-D) (Wilson & Devinney 1973, Wilson 1979). In the present case, where the solution is difficult due to the fact that the eclipses are partial and no spectroscopic mass ratio is available, we decided to apply a Monte Carlo search method (Kreiner et al. 2003). The main two advantages of this algorithm are: (a) it searches for a global minimum in a given range of parameters and (b) it is independent of correlations between parameters describing the system. The mass ratio was chosen as a free parameter. Two runs were done with ranges for q: $0.1 \rightarrow 1$ and $1 \rightarrow 10$, in order to search for A and/or W-type, respectively. The free parameters to be determined by the program were the following: the phase of conjunction ϕ, the inclination i, the temperature of the secondary star T_2, the mass ratio q, the surface potentials Ω (both Ω_1 and Ω_2) and the non-normalized monochromatic luminosities of the primary (L_1) for the four filters. The temperature of the primary component, T_1, was given a value according to its spectral type. The usual values from the literature were

TABLE 1. Spotted light curve solutions of PY Lyrae.

Parameter	Mode 3, $q < 1$	Mode 3, $q > 1$
ϕ_0	0.0004(1)	0.0004(1)
i (°)	74.66(25)	73.91(5)
T_1 (K)	6980*	6980*
T_2 (K)	6980(11)	6985(10)
$g_1(=g_2)$	0.32*	0.32*
$A_1(=A_2)$	0.5*	0.5*
$\Omega_1 = \Omega_2$	2.982(77)	4.025(15)
% overcontact	12%	10%
$q = M_2/M_1$	0.578(42)	1.207(10)
$L_1/(L_1+L_2)$ (B)	0.572(15)	0.418(2)
$L_1/(L_1+L_2)$ (V)	0.577(15)	0.422(2)
$L_1/(L_1+L_2)$ (R)	0.579(15)	0.423(2)
$L_1/(L_1+L_2)$ (I)	0.584(15)	0.427(2)
r_1 (vol.)	0.4362(18)	0.3701(13)
r_1 (pole)	0.4087(13)	0.3467(10)
r_1 (back)	0.4657(28)	0.3993(24)
r_2 (vol.)	0.3403(16)	0.4032(15)
r_2 (pole)	0.3180(14)	0.3780(11)
r_2 (back)	0.3699(20)	0.4324(19)
spot co-latitude(°)	90*	90*
spot longitude (°)	80(3)	260(3)
spot radius (°)	17(1)	17(1)
spot T factor	0.80(2)	0.80(2)
$\Sigma w(res)^2$	0.0395	0.0401

*assumed

used for the coefficients of gravity darkening. The limb darkening coefficients were taken from the tables of Claret (2000) (bolometric values) and Claret, Díaz-Cordovés & Giménez (1995), according to the spectral type and the wavelength of observation. The third light was assumed equal to zero. The results are given in Table 1. The errors given are standard deviations. The derived parameters were used to construct theoretical light curves, which are shown along with the observed ones in Fig. 2 for the $q < 1$ case. A 3-D model of the system is given in Fig. 3.

4. SUMMARY AND CONCLUSIONS

High quality photometric observations were used to derive the geometric and photometric elements of PY Lyr. The observed light curves are rather well behaved and symmetric, but a pronounced O'Connell effect is present (MaxII is brighter than MaxI). The two minima have almost the same depth, indicative of equal surface temperatures of the two components. Moreover the shapes of the two minima indicate that the eclipses are partial, a fact which makes the light curve analysis very difficult, especially in the absence of spectroscopic mass ratio. For this reason, two separate solutions were performed by applying the q-search method in the ranges of $0.1 < q < 1$ and $1 < q < 10$. The results

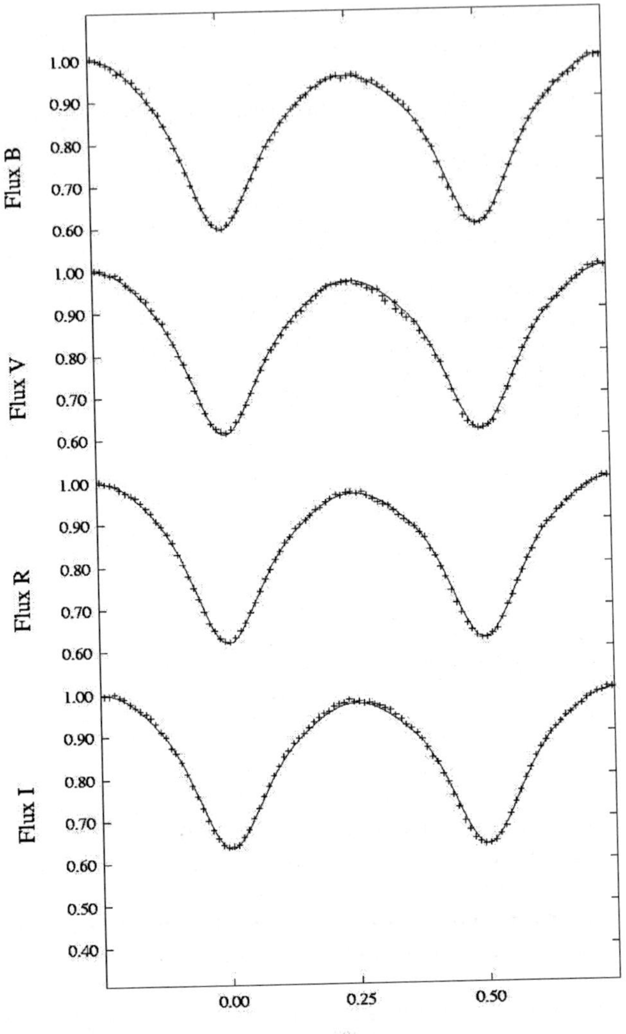

FIGURE 2. Observed (normal points) and theoretical light curves of PY Lyr for q<1.

from the two solutions describe extremely well the system and it is difficult to adopt one of them as a final solution. This is also the case for other W UMa systems with partial eclipses and components with equal surface temperatures. The problem can be solved if a spectroscopic mass ration of the system is determined. However, the faintness of the system makes its radial velocity determination very difficult and possible only with large (> 3m) ground-based telescopes.

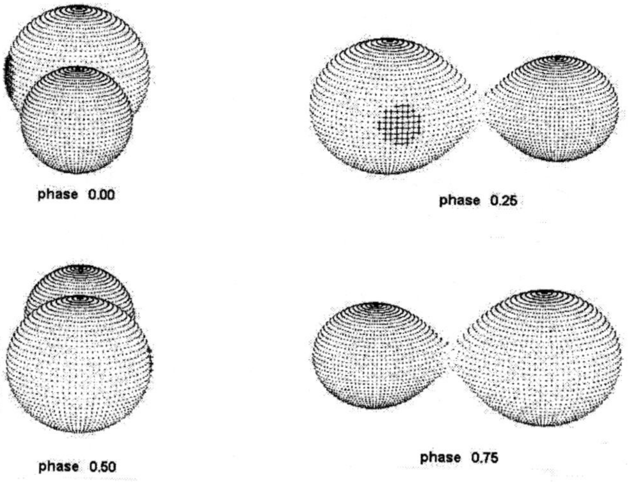

FIGURE 3. A three-dimensional model of PY Lyr (q<1) at phases 0.00, 0.25, 0.50 and 0.75.

ACKNOWLEDGMENTS

This research was included in the project for the support of research groups in the universities, co-funded by the European Social Fund (ESF) and National Resources (EPEAEK II) - *PYTHAGORAS*.

REFERENCES

1. V. Bakis et al., *IBVS*, No. 5399 (2003).
2. V. Bakis et al., *IBVS*, No. 5464 (2003).
3. R. Berry and J. Burnell, *The Handbook of Astronomical Image Processing*, Willmann-Bell, Richmond, 2000.
4. A. Claret, *A&A*, **363**, 1081-1190 (2000).
5. A. Claret, J. Díaz-Cordovés, A. Giménez, *A&A Suppl.Series*, **114**, 247-252 (1995).
6. O. Demircan et al., *IBVS*, No. 5364 (2003).
7. M. Hoffmann, *IBVS*, No. 2344 (1983).
8. P.N. Kholopov et al., *General Catalogue of Variable Stars*, vol. II, Moscow: Nauk Publ., 1985.
9. J.M. Kreiner et al., *A&A*, **412**, 465-471 (2003).
10. K.K. Kwee and H. van Woerden, *Bull. Astron. Inst. of Netherlands*, **12**, 327-330 (1956).
11. J. Safar and M. Zejda, *IBVS*, No. 5263 (2002).
12. B.A. Skiff, *IBVS*, No.4675 (1999).
13. A.A. Wachmann, *Astron.Abh. Hamburg. Sternw.*, **6**, 1-96 (1961).
14. A.A. Wachmann, *Astron.Abh. Hamburg. Sternw.*, **6**, 97-178 (1963).
15. A.A. Wachmann, *Astron.Abh. Hamburg. Sternw.*, **6**, 179-279 (1964).
16. A.A. Wachmann, *Astron.Abh. Hamburg. Sternw.*, **6**, 281-384 (1966).
17. R.E. Wilson, *ApJ*, **234**, 1054-1066 (1979).
18. R.E. Wilson and E.J. Devinney, *ApJ*, **182**, 539-547 (1973).
19. M. Zejda, *IBVS*, No. 5287 (2002).
20. M. Zejda, *IBVS*, No. 5583 (2004).

An Investigation of the Tidally-Driven Period Changes in Detached Binary Systems through O-C Time-Series Analysis

N.Nanouris[(1)], A.Kalimeris[(2)], E.Antonopoulou[(1)]

[(1)]*Department of Astrophysics, Astronomy and Mechanics, University of Athens, Panepistimiopolis Zografou GR-15784 Athens, Greece*

[(2)]*Technological & Educational Institute of the Ionian Islands, Kalvos Sq., 29 100 Zakynthos, Greece*

Abstract. Detached binary systems are subjected to a variety of physical mechanisms which are able to change their period in a continuous way and in broad band of timescales. Assuming that these changes should be detectable in O-C diagrams in a way specified by the physical theory of the underlying mechanisms, we carry out a preliminary study in a properly selected sample of detached binaries, where mass transfer is expected to lack. We primarily interested to find traces of tidal moments in observed orbital period changes (possibly coupled with stellar wind angular momentum losses). Findings of a preliminary study are presented here.

Keywords: Tidal evolution, O-C time-series, Short-period detached binaries, Main-sequence.
PACS: 97.80.Hn.

INTRODUCTION

Tidal interaction between the components of close binaries is one of the basic physical mechanisms driving their evolution. For instance, the interaction of tidal torques with the stellar wind seems to be the main mechanism driving the components of detached configurations into physical contact.

After the works of Claret & Cunha (1997) and Goodman & Oh (1997), a new era began on the study of tidal interaction. These authors find that Zahn's classical tidal theory anticipates a very low rate of orbital shrinkage or very long circularization time-scales (approximately by two orders of magnitude) in binaries with late-type main sequence (MS) components. Hence, new theories of dynamical tides developed, which take into account the complicated non-adiabatic interaction between the tidal flow raised by a close companion to a solar type star and the normal modes of oscillation of its interior. It was found that resonances between different tidal forcing frequencies and stellar normal *g*-modes could enhance a lot the tidal torque (possibly by a factor 10^3 to 10^4 ; Terquem et al. 1998). But, since these resonances found to be very narrow and widely spaced, each system is influenced by the enhanced tidal torques only for short time intervals. As a result, the discrepancy with the observed circularization cut-off periods remains, unless the binary parameters are such that a low orbital harmonic falls in the frequency regime of the very closely spaced *q*-mode resonances found by Savonije & Witte (2002a). The computed circularization cut-off periods of MS binaries fit much better the observed ones in this case, but not satisfactorily yet!

Given the aforementioned theoretical developments and uncertainties, we intend to check whether traces of tidal evolution could be detectable on the O-C time-series of detached binaries with MS components. We also interested to see whether "direct" tidal torque estimations, relied on the observed long time-scale variations of the orbital period, can be achieved. In particular, such estimations will be based on the generalized $\dot{J}-\dot{m}-\dot{P}$ type relation proposed by Kalimeris & Rovithis-Livaniou (2005):

$$\frac{\dot{J}}{J_{orb}}=\frac{1-q}{M_2}+\frac{2q+3}{3qM}(\dot{m}_{L2}+\dot{m}_W)-\frac{e}{1-e^2}\dot{e}+\frac{\dot{\Omega}}{\Omega}+\frac{4}{3}\frac{\dot{P}}{P}+\sum_{i=1}^{2}\frac{J_i}{J_{orb}}\left(5\frac{\dot{R}_i}{R_i}+\frac{\dot{\omega}_i}{\omega_i}+\frac{\dot{I}_{N,i}}{I_{N,i}}\right).$$

SELECTION CRITERIA

To isolate the effects of the classic thermally driven mass transfer from the orbital period variations, we are restricted to detached binaries. Among them, we initially focus to short period pairs, where O-C data files are finest, longest and the stellar parameters are more reliable. As a consequence, the tidal torques and the stellar wind angular momentum losses (AMLs) remain the most significant factors of binary evolution.

Although under conservative conditions tidal torques are expected to be very weak in short period binaries (due to earlier circularization and synchronization), AMLs due to a stellar wind perpetually impose *desynchronization trends to the components and in this way tidal interaction remains active*. In this case, as the stellar wind diffuses the components' spin angular momentum, the tidal torques replenish it from the orbital reservoir. Consequently, the two companions get closer, the tidal torques are enhanced, the components are enforced to spin faster, the stellar dynamo and the rate of AMLs increase, and finally, the orbit shrinks even faster. As this cyclic interaction continues, a feedback mechanism (often termed as *tidal or orbital instability*) is developed, and continuously shrinks the orbit. This instability can be either weak (when the rate the wind diffuses, the spin momentum is higher than the rate the tidal torques transfer momentum from the orbit to the spin) or strong (when the tidal moments transfer angular momentum from the orbit to the spin faster than the wind diffuses it). As a result, a tidally or wind driven orbital decay is expected to be observed in binaries with magnetically active components. On the other hand, if both components possess radiative envelopes (and presumably they are not surrounded by a magnetically-driven stellar wind) then only a weak tidal interaction is expected to drive a long-time scale orbital evolution.

CANDIDATE BINARIES

In this preliminary study, we have collected a large number of detached binaries with main sequence components and orbital periods less than about 2 days. Binaries with early-type components and very low (or no) wind AMLs possess an important role in our sample, since in such systems we expect to trace the effects of tidal torques (if any) isolated from the wind-tides cycling interaction.

We have used the "Atlas of O-C Diagrams of Eclipsing Binary Stars" (Kreiner et al. 2001) as the main guide of our research. The "Catalog of Chromospherically Active Binary Stars" (Strassmeier et al. 1993) and "Catalog and Atlas of Eclipsing Binaries" (Bradstreet et al., 2004) were also used as additional sources. Lack of knowledge of some parameters was completed/checked from the "General Catalog of Variable Stars" (Kholopov et al. 1985-2004). A list initially containing 124 systems formed in this way. Among them, 61 were *detached binaries with main sequence components* whose absolute parameters were available. They are listed on TABLE 1, sorted according to their period and their envelope type. When both components possess a radiative envelope, the system is classified as "R-R", those possessing convective envelopes as "C-C" systems, while "R-C" means that the primary component possesses a radiative envelope and the secondary a convective envelope. A mass criterion of 1.2 solar masses was used for this distinction.

TABLE 1. The list of candidate binaries sample.

No.	*System*	*Period (days)*	*Envelope Type*	*No.*	*System*	*Period (days)*	*Envelope Type*
1	BW3 V38	0.20	C-C	32	HS Hya	1.57	R-R
2	XY UMa	0.48	C-C	33	AO Vel	1.58	?
3	GU Boo	0.49	C-C	34	CM Lac	1.60	R-R
4	SV Cam	0.59	C-C	35	AY Vel	1.62	?
5	UV Leo	0.60	C-C	36	HS Her	1.64	R-R
6	RT And	0.63	R-C	37	PV Pup	1.67	R-R
7	CG Cyg	0.63	C-C	38	ζ Phe	1.67	R-R
8	ER Vul	0.70	C-C	39	U Oph	1.68	R-R
9	WY Hya	0.72	R-R	40	V885 Cyg	1.69	R-R
10	EU Hya	0.78	R-C	41	V760 Sco	1.73	R-R
11	YY Gem	0.81	C-C	42	DN UMa	1.73	R-R
12	BH Vir	0.82	C-C	43	IQ Per	1.74	R-R
13	WY Cnc	0.83	C-C	44	PV Cas	1.75	R-R
14	VZ CVn	0.84	R-R	45	AH Cep	1.77	R-R
15	UV Psc	0.86	C-C	46	RX Her	1.78	R-R
16	V772 Her	0.88	C-C	47	GG Lup	1.85	R-R
17	V541 Cas	0.91	?	48	V470 Cyg	1.87	R-R
18	AD And	0.99	R-R	49	AO Mon	1.88	R-R
19	RX Ari	1.03	R-C	50	RU Gru	1.89	?
20	TrES-Her0-07621	1.20	C-C	51	MN Cas	1.91	R-R
21	WX Cnc	1.22	R-C	52	FS Mon	1.91	R-R
22	ST Cen	1.22	?	53	V526 Sgr	1.92	R-R
23	NO Pup	1.26	R-R	54	BP Vul	1.94	R-R
24	FZ CMa	1.27	R-R	55	σ Aql	1.95	R-R
25	CM Dra	1.27	C-C	56	AG Per	2.03	R-R
26	AZ Cam	1.32	?	57	NN Cep	2.06	R-R
27	DU Leo	1.37	C-C	58	TX Her	2.06	R-R
28	AT Cam	1.40	?	59	AD Boo	2.07	R-R
29	DT Lup	1.45	R-C (?)	60	V478 Lyr	2.13	C-C
30	CO Lac	1.54	R-R	61	FL Lyr	2.18	R-C
31	V364 Cas	1.54	R-R				

We note that several studies have been drawn on a photometric determination of the mass ratio q (RX Ari, AT Cam, AZ Cam, MN Cas, V541 Cas, NN Cep, EU Hya, AO Vel, AY Vel, ST Cen, RU Gru, WY Hya and DT Lup). This fact has led to ambiguities about the exact values of Roche geometry parameters and hence about the evolutionary state of their components. Moreover, we point out the possibility of a magnetic A star in some of those candidate binaries (AO Vel and PV Cas).

TABLE 2. Number of systems in each O-C modulation and each physical mechanism.

O-C Diagram Modulation	***Percentage***	***Physical Mechanism***	***Percentage***
Constant	15 %	*Third Body (Suspected + Confirmed)*	28 %
Parabolic	7 %	*Third Body (Confirmed)*	11 %
Irregular	8 %	*Apsidal Motion*	21 %
Unclear	41 %	*Magnetic Cycles*	11 %

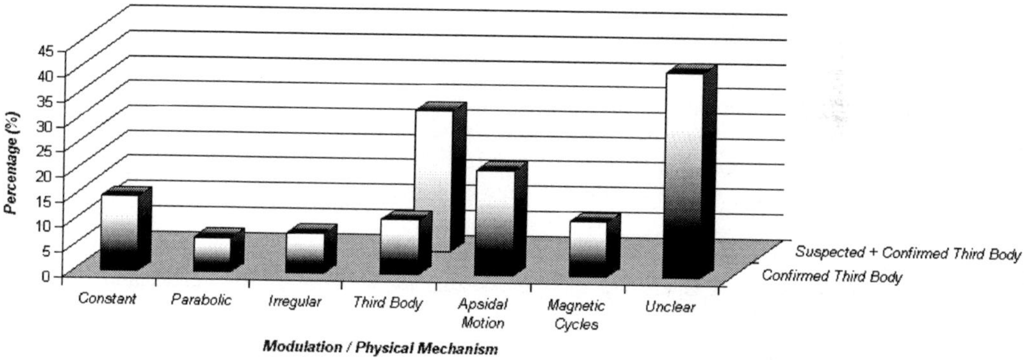

FIGURE 1. Plot of the distribution of systems in each O-C modulation and each physical mechanism.

RESULTS

According to a preliminary statistical analysis, shown in TABLE 2 in combination with FIGURE 1, a large portion of the examined systems (23 out of 61) reveals variations of a *sinusoidal type* owing to apsidal motion, to light-time effect due to the confirmed presence of a third body, or to possible cycles of intense magnetic activity. In particular, 13 systems reveal *apsidal motion* (NO Pup, CO Lac, AO Vel, HS Her, PV Pup, ζ Phe, U Oph, V760 Sco, IQ Per, PV Cas, GG Lup, V526 Sgr, AG Per), 4 reveal a light-time effect (FZ CMa, AO Vel, U Oph, AH Cep), while 7 systems (XY UMa, SV Cam, UV Leo, RT And, CG Cyg, ER Vul, UV Psc) reveal variations that are possibly connected with magnetic activity cycles. TX Her shows a considerable sinusoidal modulation, which is often ascribed from the hypothetical presence of a companion. For the most of these systems a sinusoidal term is dominant, and as a result, a long time-scale term (if any) cannot be traced through a preliminary inspection. We also neglect the 3 binaries for which the presence of the third body has been verified by means other than O-C diagram analysis (V772 Her, CM Dra, HS Hya). Hence our research is primarily focused on the remaining 43 systems.

Additionally, there were 25 systems whose times of minima were neither enough to give us the required information nor the modulation was clear (scatter is straightly comparable with the possible variations). Quite often, the presence of a vast number of visually determined times of minima increases the level of noise in the O-C diagrams.

In general, according to the aforementioned theoretical description, a long time-scale parabolic modulation of the O–C time series compatible with orbital period decrease (reflecting the expected existence of orbital instability), is expected to characterise all or most of the C-C or R-C binaries of the remaining 18 systems of our sample. Furthermore, very weak or no long time-scale period modulation is expected for R-R binaries. However, on inquiry, we came to the next three preliminary results:

- *No evidence of any kind of period changes was found to 9 systems.* Four of them (CM Lac, RX Her, MN Cas, BP Vul) belong to the R-R class, one (FL Lyr) belongs to the R-C class and remarkably three (ER Vul, BH Vir, UV Psc) belong to the C-C class. V541 Cas, has not yet been classified due to deficient knowledge of their components mass.

- *A nearly parabolic modulation was found to characterise only 4 systems.* Three of them reveal a decreasing period (RT And, YY Gem, WY Cnc) while UV Leo (perhaps SV Cam too) reveals an increasing period. Aside from RT And which is an R-C member, the rest belong to the C-C group.

- *Irregular changes were found to take place in the last 5 systems, four of them* (XY UMa, SV Cam, CG Cyg, EU Hya) belonging to C-C or R-C class and one (TX Her) to R-R class.

The preliminary results show a much more complicated scenery than was originally expected for the C-C or R-C class, as only (3 out of 18) systems of the C-C or R-C class (RT And, YY Gem, WY Cnc) show a consecutively systematic period reduction, while 3 more C-C type binaries (ER Vul, BH Vir, UV Psc) reveal a remarkably constant orbital period for over a century. On the other hand, the R-R class behaves in accordance to general expectations, as no one shows either a parabolic or an irregular modulation (apart from TX Her, highly suspected to contain an unseen companion).

REFERENCES

1. Bradstreet D., Steelman D., Sanders S., Hargis J., 2004, A&AS, 204, 501 (Eclipsing Binary On-Line Atlas).
2. Claret A. & Cunha N.C.S., 1997, A&A, 318, 187.
3. Goodman J. & Oh S.P., 1997, ApJ, 486, 403.
4. Kalimeris A. & Rovithis-Livaniou H., 2005, Proc. Conf. on Variable Stars in the 21st Century, Syros.
5. Kholopov P. et al., 1985-2004 (General Catalogue of Variable Stars , fourth edition, electronic version).
6. Kreiner J., Kim C.-H., Nha Il-S., 2001 (An Atlas of O-C Diagrams of Eclipsing Binary Stars).
7. Savonije G.J. & Witte M.G., 2002a, A&A, 386, 211.
8. Strassmeier K., Hall D., Fekel F., Scheck M., 1993, A&AS, 100, 173.
9. Terquem C. et al., 1998, ApJ 502, 788.
10. Witte M.G. & Savonije G.J., 2002b, A&A, 386, 222.
11. Zahn J-P., 1977, A&A, 57, 383.

Fourier Analysis of SS433 - High Speed Photometry

J.S. Nestoras*, J.H. Seiradakis*, E. Harlaftis†, O. Giannakis** and S. Kitsionas‡,**

*University of Thessaloniki, Department of Physics, Section of Astrophysics, Astronomy and Mechanics, GR-541 24 Thessaloniki, Greece
†Institute of Space Applications and Remote Sensing, Lofos Koufou, GR-152 36 Athens, Greece
**National Observatory of Athens, Institute of Astronomy and Astrophysics, Lofos Koufou, GR-152 36 Athens, Greece
‡Astrophysikalisches Institut Potsdam, An der Sternwarte 16, D-14482 Potsdam, Germany

Abstract. We present and discuss the periodograms of high-speed V-band and R-band optical photometry of the binary microquasar SS433, obtained with the 1.2m telescope of the National Observatory of Athens (NOA) in August 2004 and August 2005. We used power spectrum analysis with a time resolution adequate for the detection of oscillations longer than 30 sec. We report on the discovery of semi periodic oscillations between 20 - 50 min, that appear in both the 2004 and 2005 data. We also confirm the nutational period (6.28^d) from our photometric data in 2005 and we investigate the chaotic behaviour of the system on time scales smaller than a few hours. We finally compare the shape of the light curves observed in 2004 and 2005.

Keywords: SS433, Fourier Analysis, Accretion Disks, X-ray Binaries
PACS: 97.80.Jp

INTRODUCTION

SS433 or else V1343 Aql is famous as the first relativistic jet source in the galaxy and is considered to be a prototype microquasar. It is essentially a binary system of a compact object, probably a black hole, and probably an A4 supergiant donor star. It demonstrates precessing jets and a supercritical accretion disc around the black hole. SS433 is also in the center of the supernova remnant (SNR) W50. It is a highly variable system and shows different types of periodicities:

Precessional ($P_{prec} = 162.5^d$) which is observed by periodical Doppler shifts of H, He, I, Fe and other optical and X-ray emission lines also is clearly visible in optical and X-ray light curves. (A. M. Cherepashchuk et al., 2004)

Orbital ($P_{orb} = 13.082^d$) which is observed in optical, radio and X-ray bands. The shape of the optical light curve strongly depends on the precessional phase but its period remains stable for over 30 years. (A. M. Cherepashchuk et al., 2004)

Nutational ($P_{nut} = 6.28^d$) which is observed as a periodic deviation from the precessional Doppler motion of emission lines and can also be recovered from photometric data. Nutational radial velocity variations are delayed from the nutational photometric variations by 0.6 days. This delay corresponds to the travel time from the accretion disc centre to the region where the moving optical emission lines, downstream the relativistic jets are formed. This region is located at $l \approx 10^{14} - 10^{15}$ cm away from the disc center.

CP848, *Recent Advances in Astronomy and Astrophysics,*
7th International Conference of the Hellenic Astronomical Society, edited by N. H. Solomos

This periodicity remains stable for at least 16 years. (A. M. Cherepashchuk et al., 2004)

The radial velocity curves of the shifted lines are very well described by the Kinematic Model (Margon 1984), thereby revealing key parameters of the jets ($v = 0.26c$ $i = 79°$). These parameters combined with radio observations of the associated SNR W50 give a distance of 5.5kpc (K.M. Blundell et al., 2004). SS433 exhibits noticeable variability on shorter time scales as well. However at time scales smaller than hours this has been poorly studied. For example, erratic variability with amplitude of 5-10% on a time-scale of a few minutes was found in the optical band by Goranskij et al. (1987) and Zwitter et al. (1991). Recently, Kotani et al. (2002) detected X-ray variability of SS433 (RXTE, PCA) on a time scale of $\approx 50s$, when the source was in its active state. In this work we investigate the optical variability of SS433 on the order of one hour or less.

OBSERVATIONS AND METHODS OF ANALYSIS

We observed SS433 in the V and R bands in 2004 and 2005 respectively. In 2004 the observations were performed in V-band for the period between 26 and 31 August whereas in 2005 the system was observed in R from 1 to 10 August. The exposure times were 5 and 10 sec respectively. We used a 516x516 CCD camera mounted on the 1.2m telescope of NOA. The data were reduced using standard Starlink scripts including bias, dark and flat frames. Differential photometry was obtained with the automated optimal photometry package AUTOPHOTOM for SS433 using the comparison star *GSC 00471-01564* ($RA : 19^h 11^m 47.6^s Dec : +05^d 00^m 36.1^s$). Then fluxes were extracted with their corresponding errors. We used the Lomb-Scargle method in the PERIOD package to produce the power spectra of all the data. In order to make sure that the periodicities found were real, additional photometry of the comparison was performed with two other random stars in the field. These stars were *C1* ($RA : 19^h 11^m 49.9^s Dec : +05^d 00^m 05.5^s$) and *C2* ($RA : 19^h 11^m 47.7^s Dec : +05^d 00^m 04.0^s$). No similar periodicities were found in the comparison star data. Also the window function was constructed for every periodogram and the periodicities found were marked as false signals. Folding was performed in the 2005 data to confirm the nutational period (fig5). In all light curves (fig1 - fig4) presented here photometry of GSC 00471-01564 with respect to the two comparison stars (C1 and C2) is included in all figures. For clarity a constant was subtracted from the fluxes of the comparison stars. The upper curve depicts *C1* and the lower one *C2*.

RESULTS AND CONCLUSIONS

Below we present some representative examples of light curves and the corresponding power spectra of individual epochs (fig3-fig4) as well as the combined datasets of 2004 and 2005 (fig1-fig2). It is obvious from the light curves that SS433 is a highly variable system on time scales between a few minutes and a few hours. In almost all epochs a Quasi Periodic Oscillation (QPO) pattern is observed with periods of over 1100s, much longer than a usual QPO the period of which ranges from 60s to 1000s. This probably means that these variations are not actually due to a QPO behaviour, but rather due to **random** perturbations in the accretion disc. The fact that there is

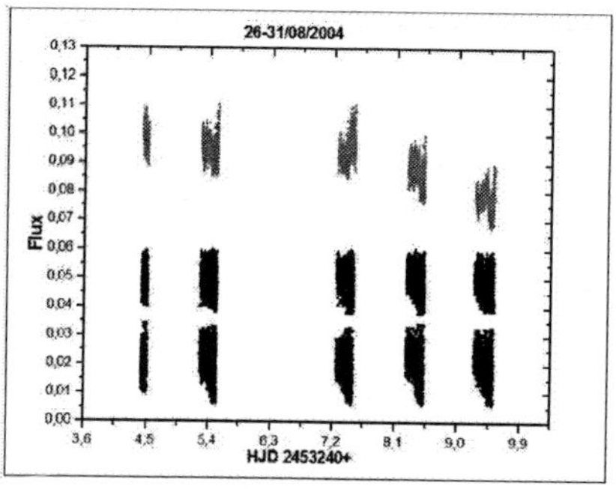

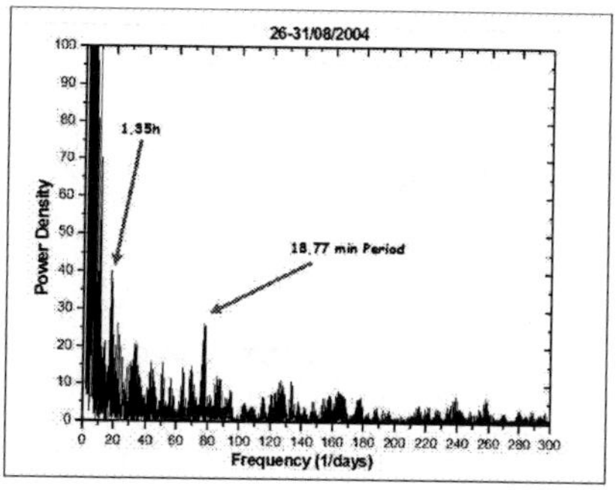

FIGURE 1. Light curve and the periodogram for the full 2004 observing period. We find a strong periodicity at 18.77min ± 0.7sec also we find a period at 1.35h ± 13.47sec with its two harmonics at 42,33min ± 3,62sec and 22.6min ± 1.04sec. Unfortunately the available six day observing period was not adequate to confirm the nutational period of the system.

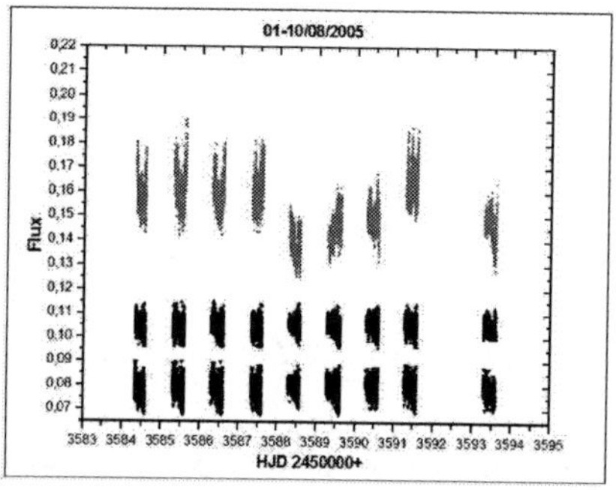

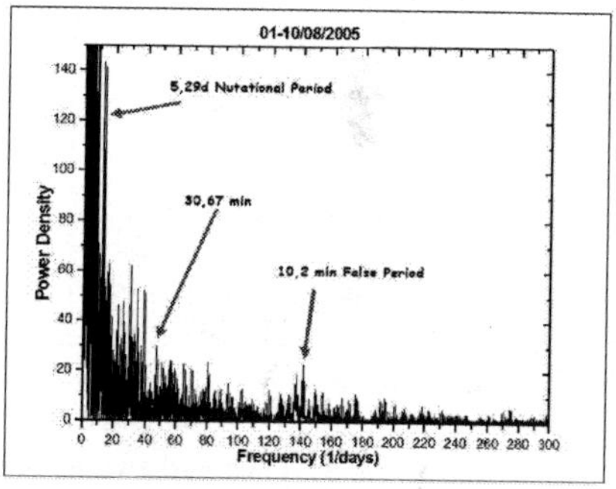

FIGURE 2. Light curve and periodogram for the full observing period in 2005. We report various periodicities. The chaotic behaviour of SS433 is clearly seen. We find a peak at 30.67min ± 0.26sec that could be associated with the individual periodicities that show up in each epoch. Its harmonics is also observed. There is also a possible confirmation of the nutational period (6.28^d) at 5.29^d ± 18.5h. The error here is quite large because the overall length of our observations is just about twice this period. The false period of 10.2min is of a known systematic error that was introduced only in August 03. The same periodicity appears in the comparison star's data and in the window function.

no periodicity stable enough to appear in two or more consecutive epochs could be explained by assuming that every periodicity is unique and short lived. The important fact is that there is in every epoch a periodicity on a scale of ten's of minutes showing that there is a physical mechanism capable for producing such a unique temporal behaviour. Further investigation is needed to find out whether is coherence between the periodicities reported here. If any relationship is found then the characterisation as "random" should be withdrawn. Another known characteristic of SS433 is that it has distinct active and passive states. During the active (flaring) states the signature of regular orbital and precessional variabilities in the light curves is destroyed. This fact is confirmed from our data by comparing the 2004 with the 2005 observations. Between 26 to 31 August 2004 (fig1) our data is in agreement with the published linear ephemeris for the orbital period: $MinI(HJD) = 2450023.62 + 13.08211^d E$ (V.P Goranskij et al., 1998), but in 2005, after 04/08, the flux instead of continuing to decrease, it increases until 08/08, after which it decreases again (fig2).

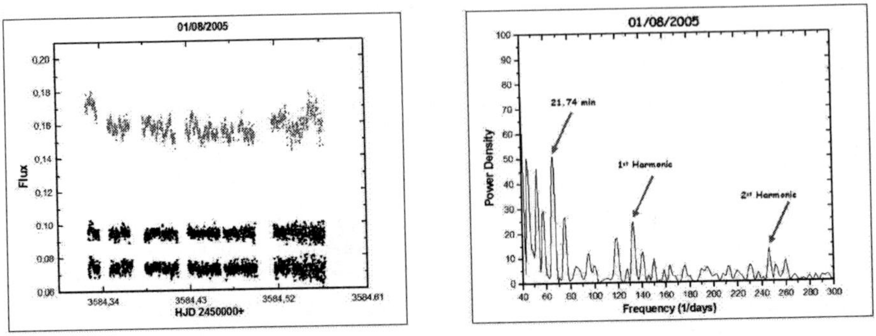

FIGURE 3. In the light curve and periodogram of SS433 for 01/08/2005 we see a conspicuous oscillation of 21.74min ± 20.3sec and its two harmonics at 10.87min ± 5.09sec and 5.84min ± 1.47sec. The light curve shows a QPO pattern but in much larger time scales.

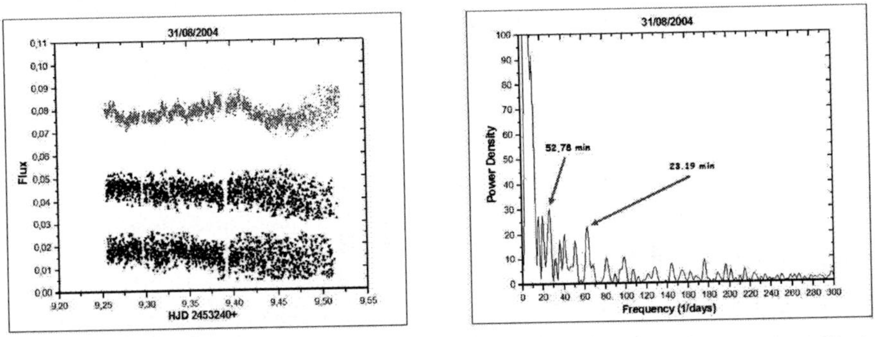

FIGURE 4. In 31/08/2004 we find two main periodicities, the first one at 52.78min ± 1.82min and probably its two harmonics at 28.33min ± 31.44sec and 14.65min ± 12.3sec. The second periodicity found is at 23.19min ± 21.08sec but without a strong harmonic. The light curve has the same chaotic pattern in small time scales.

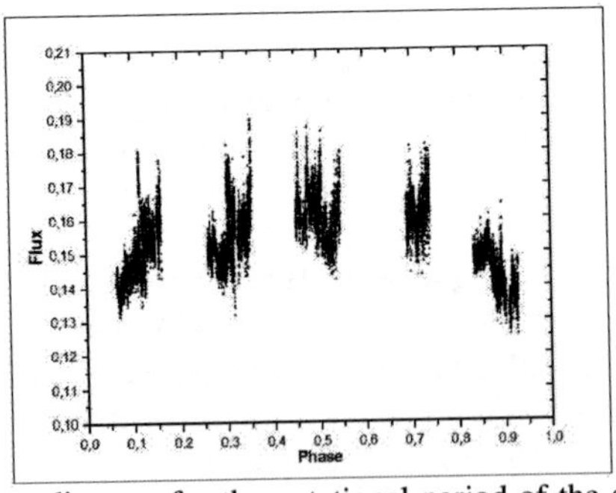

FIGURE 5. This is the phase diagram for the nutational period of the system. The folding was done in the 2005 data. The diagram was produced with the linear ephemeris $Max\Delta V = 2450000.94 + 5.2^d E$ Folding is very important for the confirmation of a periodicity. Here after folding, a half sine function appears indicating that the physical mechanism producing this kind of a periodicity follows a sinusoidal law. This fact is strong evidence that the signal we have detected is the nutational period of the system. The intermediate variations are probably caused by the chaotic nature of the system in relatively small time scales.

This behaviour could be attributed to a flaring event that occurred in SS433 (S. Fabrika et al., 2002), obscuring the orbital period of 13.08^d. It has been shown by Revnivtsev et al.(2004) that simultaneous observations show a clear correlation between the light curves patterns in X-rays and in the optical, with the X-rays being delayed by about 80s with respect to the optical. This confirms the fact that any variations discovered are produced by various physical phenomena in the accretion disc and not in the donor star.

FUTURE WORK

From the above conclusions it is obvious that further analysis to the data must be done. Because SS433 is a highly variable system with a chaotic behaviour, it produces light curves that are best analyzed with the wavelet algorithm and not by means of a simple Fourier transformation. The light curves of SS433 are stochastic signals, meaning that are multi-frequency and non-stationary in time. Thus the wavelet analysis will give us information on how the periodicities change with time in conjunction to the Fourier transform that only gives us the information of the existence or not of a periodicity. Thus the wavelet transform is the only way to see what really is happening to the time scales investigated in this paper.

ACKNOWLEDGMENTS

We would like to thank G.Dimou for his support during the observations and the National Observatory of Athens for the use of the 1.2m telescope at Kryonerion Observatory. Emilios Harlaftis passed away on February the 13th 2005. A big thanks from JSN to the late EH for initiating this project. This work is dedicated to his memory.

REFERENCES

1. V.P. Goranskij, "Analysis of Twenty-Year Multicolor Photometry of SS433(V1343 Aql), an Unique Eclipsing System with Relativistic Jets"
2. M. Revnivtsev, R. Burenin, S. Fabrika, K. Postnov, I. Bikmaev, M. Pavlinsky, R. Sunyaev, I. Khamitov, and Z. Aslan, "First simultanous X-ray and optical observations of rapid variability of supercritical accretor SS433", *A&A 424, L5-L8 (2004)*
3. A. M. Cherepashchuk, R. A. Sunyaev, S. N. Fabrika, S. V. Molkov, E. A. Barsukova, E. A. Antokhina, T. R. Irsmambetova, I. E. Panchenko, K. A. Postnov, E. V. Seifina, N. I. Shakura, A. N. Timokhin, I. F. Bikmaev, N. A. Sakhibullin, Yu. N. Gnedin, A. A. Arkharov, and V. M. Larionov, "Coordinated Integral and optical observations of SS433", *arXiv:astro-ph/0404479 v1 24 Apr 2004*
4. S. Fabrika, & T. Irsmambetova, "The central region in SS 433 supercritical disk and origin of flares", *arXiv:astro-ph/0207254 v1 11 Jul 2002*
5. A.A. Aslanov, A.M. Cherepashchuk, V.P Goranskij, V.Yu. Rakhimov and R.C. Vermeulen, "Multicolour photometry of SS433 duringthe monitoring campaign in May/June 1987", *A&A 270, 200-203 (1993)*
6. D.R. Gies, M.V. Mcswain, R.L. Riddle, Z. Wang, P.J. Wiita, and D.W. Wingert, "The spectral components of SS433", *The Astrophysical Journal, 566:10691083, 2002 February 20*
7. Katherine M. Blundell and Michael G. Bowler, "Symmetry in the changing jets of SS433 and its true distance from us", *The Astrophysical Journal, 616:L159 L162, 2004 December 1*
8. Fabrika, S. Abolmasov, P. Postnov K., Bikmaev I., Burenin R., Pavlinsky M., Sunyaev R., Khamitov I. , "Broad band variability of SS433: Accretion disk at work?", *eprint arXiv:astro-ph/0501516*
9. Nandi A., Pal S., Anandarao B. G., Mandal Soumen. , "Photometric Evidence of Bullets in SS433 Jet", *eprint arXiv:astro-ph/0501295*

RXJ0636 - A new intermediate polar

S. Pyrzas*, B. T. Gänsicke†, T. R. Marsh†, A. Edge**, D. Steeghs‡ and E. Harlaftis§

*Aristotle University of Thessaloniki, Department of Physics, Section of Astrophysics, Astronomy & Mechanics, 54124 Thessaloniki, Greece
†Department of Physics, University of Warwick, Coventry CV4 7AL, UK
**Departement of Physics, University of Durham, South Road, Durham DH1 3LE, UK
‡Harvard-Smithsonian Center for Astrophysics, 60 Garden Str, MS-67, Cambridge, MA 02138, US
§Institute of Space Applications & Remote Sensing, National Observatory of Athens, PO Box 20048, Athens 11810, Greece

Abstract. We report the identification of RXJ0636 as a new intermediate polar. The system's light curves and spectral analysis, together with a variety of its physical parameters are presented. A brief description of the observations leading to the above mentioned identification is also given.

Keywords: accretion, accretion discs - binaries: close - novae, cataclysmic variables
PACS: 97.80.Gm, 97.30.Qt, 97.10.Gz

1. INTRODUCTION

Cataclysmic variables (CVs) are interacting, semi-detached, short-period binary systems consisting of a white dwarf (primary star) and, usually, a low mass main sequence star, ie a red dwarf (secondary star). CVs are quite a diverse class of stellar objects. They can be classified in various ways, most common of which is the dependence on the primary's magnetic field strength. In this way two main categories, each one with two subcategories, can be created: (I) non-magnatic systems with (Ia) dwarf novae & (Ib) novalike variables and (II) magnetic systems with (IIa) intermediate polars & (IIb) polars.

In most CV cases, the secondary overfills its Roche lobe, thus enabling mass transfer through the inner Langrangian point L_1. A result of this procedure is the formation of an accretion disc around the primary (although that depends also on the magnetic field of the primary - see next section as well).

2. INTERMEDIATE POLARS

As seen in the CV classification, intermediate polars - or IPs - are a type of magnetic cataclysmic variables. In other words, the primary has a relative strong magnetic field, with magnetic field strength about $B \approx 10^6 G$. IPs are also known as DQ Her stars. As it is mentioned in the previous section, the accretion disc can interact with the primary's magnetic field. This is the case in IPs: the accretion disk is disrupted in the area near the white dwarf's magnetosphere, which causes accreting material to leave the disk and attach itself to the magnetic field lines, following them to the magnetic poles. At some

CP848, *Recent Advances in Astronomy and Astrophysics*,
7th *International Conference of the Hellenic Astronomical Society*, edited by N. H. Solomos

point this flow of material passes a strong shock, below which it hits the WD's surface, thus generating X-ray to optical radiation. This radiation is pulsated at the spin period of the white dwarf, as a result of the magnetic axis being offset to the white dwarf's spin axis.

3. TARGET SELECTION

RXJ0636+3535 was among the 174 targets selected for a search for new cataclysmic variables. The search is based on a combined ROSAT/2MASS (X-ray/infrared respectively) selection. To be more specific, the selection uses the ROSAT Bright Catalog and a 2MASS infrared colour-cut to discriminate against background AGN. Our search resulted in the discovery of twelve new cataclysmic variables, which constitutes a quite satisfactory hit rate of 7%, among which four (one third!) have been also identified as new Intermediate Polars. The following two tables show some parameters and features of RXJ0636:

TABLE 1. RXJ0636 parameters

PSPC [$10^{-2} counts^{-1}$]	**HR1**	**HR2**	**HeII** [$\dot{A}$]	**Hα** [$\dot{A}$]	**Hβ** [$\dot{A}$]
5.0 ± 1.4	0.71 ± 0.25	0.33 ± 0.29	8	16	7

TABLE 2. RXJ0636 magnitudes

B mag	**R mag**	**J mag**	**H mag**	**K mag**
15.7	15.9	15.4	15.2	15.1

4. OBSERVATIONS

Identification spectroscopy for all 174 targets was conducted with the Intermediate Dispersion Spectrograph on the Isaac Newton Telescope (IDS/INT), while time resolved follow-up spectroscopy for RXJ0636 was carried out using the ISIS double arm spectrograph on the William Herschel Telescope (WHT). Regarding photometric observations of RXJ0636, filterless CCD images were obtained with the 1.2m telescope and the PMIS SI-502 (516x516 pixel) camera at the Kryoneri Astronomical Station of the National Observatory of Athens (NOA).

5. ANALYSIS AND RESULTS

The physical processes that take place in intermediate polars have "characteristic signatures" in a lightcurve. This is how one can identify an IP. In general, the *IP criterion*, what is needed in order to characterize a CV as an IP, is the following: the detection of coherent short-period variability over a sufficiently long period of time. So, an intermediate polar lightcurve should demonstrate (if properly analyzed) a persistent and stable non-orbital (spin) periodicity.

Regarding RXJ0636, its spectrum (Figure 1) shows a relatively strong HeII emission line, a fact that indicates a possible magnetic nature.

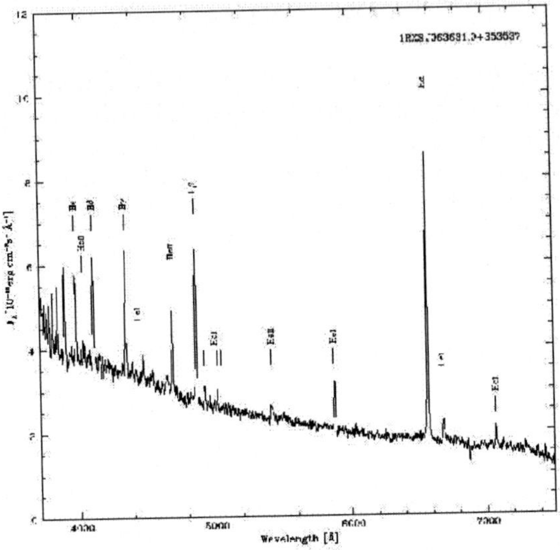

FIGURE 1. RXJ0636 spectrum

The CCD photometry clearly displays a variability of 15min superimposed on longer term modulations. Sample lightcurves can be seen in Figure 2.

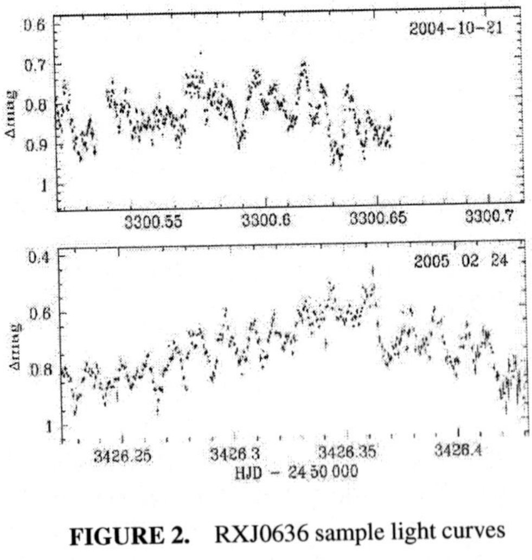

FIGURE 2. RXJ0636 sample light curves

The output of the scargle periodogram (Figure 3), which we applied on the data, shows two strong clusters of signals at $\approx 85.7d^{-1}$ and $\approx 92.9d^{-1}$. From the highest peaks of each cluster we can deduce the corresponding periods. Indeed, these are 1008.3408(19)s and 930.5829(40)s respectively.

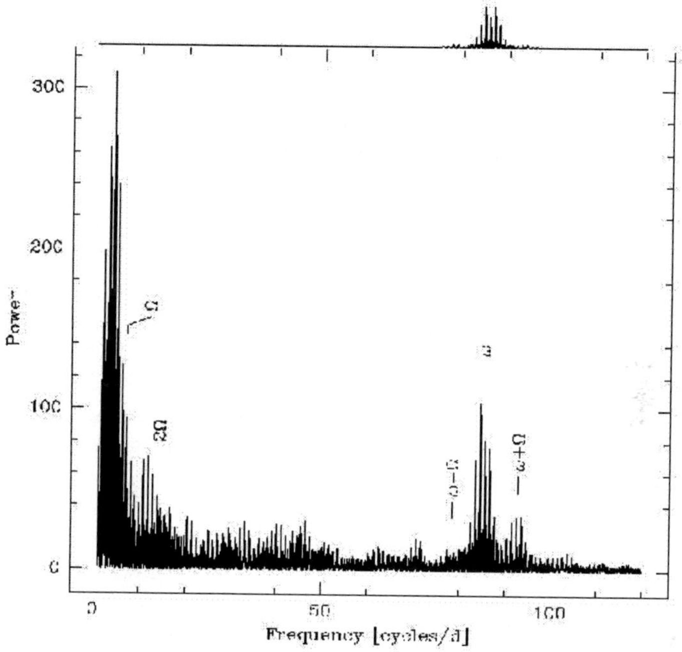

FIGURE 3. Scargle periodogram of RXJ0636

We tentatively interpret these two signals as the white dwarf spin frequency ω (or the corresponding P_{spin}) for the former and the side-band (beat) frequency $\omega \pm \Omega$ with Ω being the orbital frequency (or the corresponding P_{orb}) for the latter.

However, by using photometry only, we cannot identify the two signals unambiguously, which means that "in reality" the - true - spin and side-band frequencies could be swapped. Assuming these two signals are indeed ω and $\omega \pm \Omega$, would mean that RXJ0636 has an orbital period of $P_{orb} \approx 201 min$. As can be seen in Figure 3 no significant signal can be detected near the frequency of $\approx 7.15d^{-1}$, which corresponds to the orbital period. So, time resolved spectroscopy was used to test the orbital period hypothesis. This showed a broad peak centered at $7.3d^{-1}$ ($P_{orb} \approx 197 min$). Fitting a sine curve to the data resulted in an orbital period of 201(8)min, consistent to the photometric estimation.

Summing up, based on all current photometric and spectroscopic data we believe RXJ0636 to be an Intermediate Polar with an orbital period $P_{orb} = 201 min$ and a white dwarf spin period of either $P_{spin} = 1008.3498(19)s$ or $P_{spin} = 930.5829(40)s$.

6. DISCUSSION

Of the four subclasses of cataclysmic variables intermediate polars are probably the worst defined one. That is mainly because researchers apply the *IP criterion* with degrees of strictness, which vary substantially from one another. As a result, disagreements when it comes to IP head count rise. For example, K.Mukai in his IP webpage (`http://lheawww.gsfc.nasa.gov/users/mukai/iphome/iphome.html`) lists 25 systems, while Ritter & Kolb in their latest version (7.4) contain 37 and Downes et al. list 46 Intermediate Polars. Our group prefers Mukai's conservative classification. We have updated his list with 6 new IPs, bringing them to a total of 31 systems.

ACKNOWLEDGMENTS

Based in part on observations made with the William Herschel Telescope and the Isaac Newton Telescope operated on the island of La Palma by the Isaac Newton Group in the Spanish Observatorio del Roque de los Muchachos of the IAC; on observations made with the 1.2m telescope located at Kryoneri Korinthias and owned by the NOA, Greece.

Dedicated in the memory of Emilios Harlaftis, who passed away on Sunday February the 13th 2005, leaving a huge gap in the Greek astronomical community. We shall miss him dearly. May he rest in peace.

REFERENCES

1. Gänsicke B. T., Marsh T. R., Edge A., Rodríguez-Gil P., Steeghs D., Araujo-Betancor S., Harlaftis E., Giannakis O., Pyrzas S., Morales-Rueda L., Aungwerojwit A., "Cataclysmic variables from a ROSAT/2MASS selection - I. Four new intermediate polars", *MNRAS*, 361, 2005, pp. 141-154
2. Hellier C., "The accretion geometry of intermediate polars", *Cape Workshop on Magnetic Cataclysmic Variables*, ASP Conference Series, Vol. 85, 1995
3. Parker T. L., Norton A. J., Mukai, K. "X-ray orbital modulation in intermediate polars", *A&A*, 439, 2005, pp.213-225
4. Evans P. A. and Hellier C. "HT Camelopardalis: the simplest intermediate polar spin pulse", *MNRAS*, 359, 2005, pp. 1531-1536
5. Brian Warner, "Multiple orbital sidebands in intermediate polar", *MNRAS*, 219, 1986, pp. 347-356

KR Cyg Revisited: New Results from the Light Curve Analysis

E. Sipahi* and G. Taş

*esin.sipahi@mail.ege.edu.tr, gunay.tas@ege.edu.tr, Ege University Observatory, 35100 Bornova, Izmir-TURKEY

Abstract. KR Cyg is a β Lyr type eclipsing binary. It was classified by many different authors as the spectral types A0, A1, A2.5 and A4 V. Since the different spectral types mean different temperatures, we see the different light curve analyse results in the literature. In this study we attempted to make photometric investigations for KR Cyg. We simultaneously analysed U, B, V, R light curves of KR Cyg using latest version of Wilson-Devinney code (Wilson and van Hamme, 2003), and compared with the analyse results published by the other authors in the literature.

Keywords: eclipsing binary, light curve analyse

INTRODUCTION

KR Cyg was observed the first time by Schneller (1931). The visual light curves by Lassovsky (1936), Gaposchkin (1953), Cesevic (1954) and the photographic light curves by Nekrasova (1945) and Wachmann (1948) were obtained, and KR Cyg was classified as *"an Algol type binary"* based on these studies. The first photoelectric observation of the system was made by Vetesnik (1965) who classified the system as *"a β Lyr type eclipsing binary"*. The details on the history of KR Cyg can find in the study of Sipahi and Gülmen (2004).

In this work, we present the results of the light curve analyse obtained in U, B, V, and R filters. These contain the first U light curve analyse results. The light curves are analysed using the latest Wilson-Devinney code. The process using in this work is examined in section 2. The results are presented in section 3.

THE LIGHT CURVES AND THEIR ANALYSE

The light curves obtained in the year 2004 are taken from Sipahi (2005). As can be seen from the light curve of KR Cyg has the classic shape reflecting a β Lyr type eclipsing binary. The magnitudes at maxima are nearly the same. Neither 2004 light curve nor the others published in the literature show O'Connell effect. Therefore, we can use the magnitudes at phases 0.25 or 0.75 assuming that the maximum contribution to the total light of the system comes from primary component, although they contains the contribution of two components of the system. For this aim we determined the magnitudes and colours at phase 0.75 as follows;

$$V = 9^m.231, \quad U\text{-}B = 0^m.101, \quad B\text{-}V = 0^m.156, \quad V\text{-}R = 0^m.128$$

CP848, *Recent Advances in Astronomy and Astrophysics,*
7th *International Conference of the Hellenic Astronomical Society,* edited by N. H. Solomos

We used to determine $(B-V)_0$ value of the system the reddening free parameter Q. The standard relation to calculate Q parameter is given in Eq. (1).

$$Q = (U-B) - 0.72(B-V) \tag{1}$$

Hovhannessian (2004) suggests to use Eq. (2) for Cygnus region of the sky instead of Eq. (1),

$$Q = (U-B) - 0.81(B-V) \tag{2}$$

Replacing the observed (U-B) and (B-V) colour values in Eq. (2) we found the value of Q = -0.025 which corresponds to two de-reddening colours of $(B-V)_0$ = -$0^m.02$ and $0^m.15$, and the spectral types A0 IV and A5 V in Table 2 of Hovhannessian's work, respectively. If we plot $(B-V)_0$ and Q colours that we determined at phase 0.75, in Fig. 78 of Golay's book (1974), which it shows $(B-V)_0$ versus Q values, we can see that the Q value crosses both the spectral types A0 and A5. For the spectral type A0 V corresponding T_e given by different sources is 9850 K (Johnson 1966), 9520 K (Schmidt-Kaler 1982), 9727 K (Gray 1992) and 9570 K (Flower 1977). We meet with the same confusion in case the spectral type A5. Many authors give different temperature values of this spectral type like 8200 K (Flower 1977), 7880 K (Gray 1992). From these values, it is clear that, even for a well-determined spectral type, the uncertainty in the corresponding effective temperature of the star is rather large.

In this work we decided the effective temperatures from Flower (1996) corresponds to the values $(B-V)_0$, which it was determined for the spectral types A0 and A5 from the list of Hovhannessian (2004). These are 9840 K and 8016 K, respectively. These are the first adopted parameter for the light curve analyse. The limb darkening coefficients corresponds to the values of T_e and log g = 4.25 (Gray 1992) were taken from Claret et al. (1995) and Díaz-Cordovés et al. (1995). We assumed A_1=1, g_1=1 (for radiative stars) and A_1= 0.5, g_1= 0.32 (for convective stars) (Rucinski 1969, Lucy 1967). The subscripts 1 and 2 represent the primary and the secondary components, respectively. We used to analyse the U, B, V, R light curves simultaneously Mode 2 for the semi-detached configuration of the latest version of Wilson-Devinney code (Wilson and van Hamme 2003). The adjusted parameters are i, T_2, L_1, Ω_1, Ω_2. Then, we started to analyse the light curves for two temperatures values. At the beginning of analyse, we performed Wilson-Devinney code for the q values between 0.3 and 1.0, and we found that the analyse results for q = 0.51 and 0.40 have the minimum sums of squares of residuals for T_e = 9840 K and 8016 K, respectively. Then we reached the resulting parameters and listed in Table 1. In this Table, Solution I refer to the results of the effective temperature of 9840 K for the primary and the mass ratio of 0.51, while the Solution II corresponds to the results of T_e = 8016 K for primary and q = 0.40.

RESULTS

In this study we made the photometric investigation of the eclipsing binary KR Cyg and presented the results of the first U light curve analyse in the literature. Our work based on de-reddened parameter Q. Since there is no U observation of KR Cyg, Q parameter

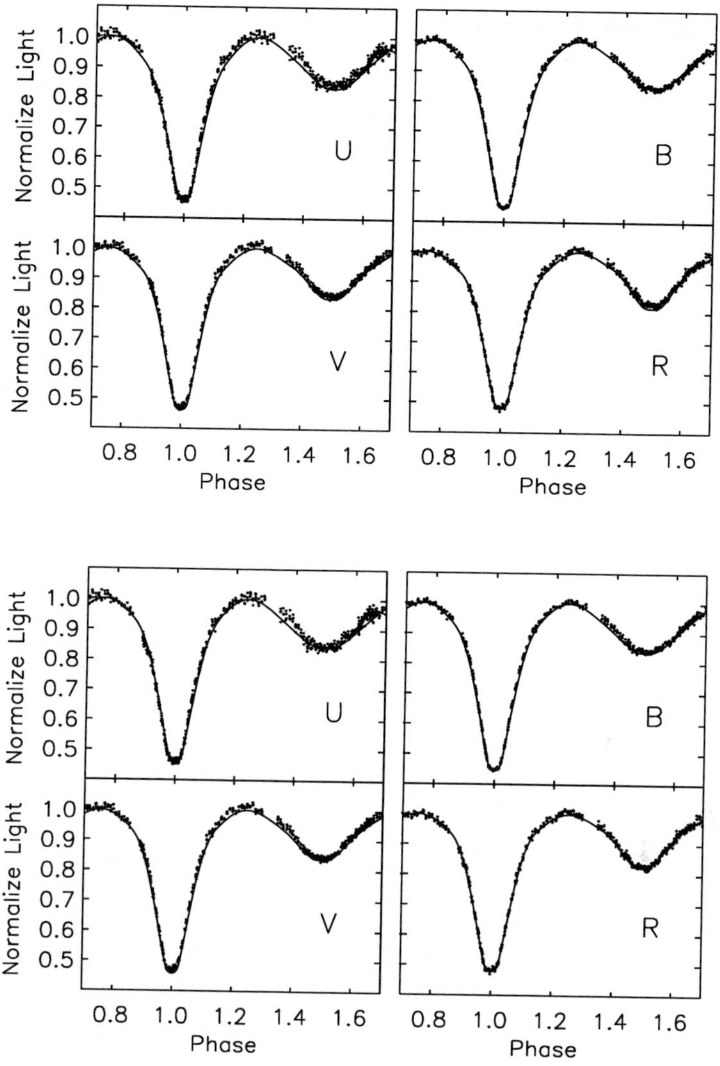

FIGURE 1. The light curve of KR Cyg obtained in U, B, V, R colours during 2004 observing season. The synthetic curves obtained from the results of the light curve analyse are also plotted in this figure with solid line: in the top panel shows the result of Solution I, and the bottom panel shows the result of Solution II.

TABLE 1. The light curve analyse results

Parameters	Solution I	Solution II	Parameters	Solution I	Solution II
$q = m_2 / m_1$	0.51	0.40	L_1 (U)	12.1599 (78)	12.2878 (72)
T_1 (K)	9840	8016	L_1 (B)	12.0582 (67)	12.2934 (74)
T_2 (K)	5385 (14)	4711 (11)	L_1 (V)	11.8230 (76)	12.0190 (84)
i°	85.81 (13)	86.12 (13)	L_1 (R)	11.5421 (83)	11.6731 (89)
A_1	1	1	L_2 (U)	0.1865	0.1094
A_2	0.5	0.5	L_2 (B)	0.2663	0.2449
g_1	1	1	L_2 (V)	0.5196	0.4828
g_2	0.32	0.32	L_2 (R)	0.8283	0.7856
Ω_1	2.9516 (16)	2.8008 (16)	Ω_2	3.0376 (22)	2.6820 (18)
χ_1 (U), χ_2 (U)	0.497, 0.905	0.600, 0.898	χ_1 (B), χ_2 (B)	0.569, 0.883	0.671, 0.880
χ_1 (V), χ_2 (V)	0.504, 0.796	0.604, 0.803	χ_1 (R), χ_2 (R)	0.404, 0.687	0.491, 0.700
r_1 (pole)	0.4034 (3)	0.4114 (3)	r_2 (pole)	0.2788 (3)	0.2818 (3)
r_1 (point)	0.5013 (9)	0.4885 (8)	r_2 (point)	0.3292 (7)	0.3886 (42)
r_1 (side)	0.4264 (3)	0.4344 (3)	r_2 (side)	0.2883 (4)	0.2936 (4)
r_1 (back)	0.4512 (4)	0.4547 (4)	r_2 (back)	0.3097 (5)	0.3258 (6)
$\Sigma W(O-C)^2$	0.143	0.142			

had not been use before this work. The results of light curve analyse show us that the maximum contribution to the total light of the system comes from primary. Therefore we can assume that the magnitudes and colours determined at phase 0.75 represent the primary. As can be see from Table 1, this assumption is not totally wrong. We used Eq. 3 (taken from Fitzpatrick 1999) to calculate E (U-B);

$$\frac{E(U-B)}{E(B-V)} = 0.70 + 0.05 \times E(B-V) \tag{3}$$

Using the reddening free B-V colour values taken from Hovhannessian (2004) we found that the colour excess E (B-V) is $0^m.006$ and $0^m.153$, for Solution 1 and 2, respectively. These values are possible because the galactic coordinates of KR Cyg (l = 68°.60 and b = -1°.32) correspond to one of the most transparent regions of our galaxy (Forbes 1985). Unfortunately KR Cyg is not the object measured its distance by Hipparcos. Therefore it can be rather roughly calculate the absolute physical parameters of the system. We listed the parameters of KR Cyg published in the literature together with those of this study in Table 2. We clearly see from Table 2 that the temperatures of the components and, of course, their spectral types change for each different study. It could be solved this confusion obtaining the radial velocity curve of the system. Besides the photometry it is necessary to obtain its radial velocity curve.

ACKNOWLEDGMENTS

One of the authors (E. Sipahi) thank very much to Prof. Dr. Semra Çalangu for the financial support.

TABLE 2. The solutions on KR Cyg published in the literature

Par.	I*	II†	III**	IV‡	Solution I §	Solution II ¶
q		0.478	0.51	0.425	0.51	0.40
T_1 (K)	12 000	9000	9210	8550	9840	8016
T_2 (K)	6500	3814	6370	4900	5385	4711
Sp. (pri)	B9 V				A0-A1 IV-V	A4-A5 IV-V
Sp. (sec)	F5 V				G8-G9 IV-V	K1-K2 IV-V
Method	Russell-Merrill	WD‖	Fourier	WD98	WD03	WD03

* Vetesnik (1965)
† Wilson and Rafert (1980)
** Al Naimiy et al. (1985)
‡ Sipahi and Gülmen (2004)
§ This study
¶ This study
‖ WD present the Wilson-Devinney code. The numbers as 98 and 03 refer to the version of code.

REFERENCES

1. Cesevic, V.P., 1954, Odessa Izv., 4, 263.
2. Claret, A., 1995, Díaz-Cordovés, J., Giménez, A., A&ASS, 114, 247.
3. Díaz-Cordovés, J., Claret, A., Giménez, A., 1995, A&ASS, 110, 329.
4. Fitzpatrick, E.L., 1999, PASP, 111, 63.
5. Flower, P.J., 1977, A&A, 54, 31.
6. Flower, P.J., 1996, ApJ, 469, 355.
7. Forbes, D., 1985, AJ, 90, 301.
8. Gaposchkin, S., 1953, 113, 104.
9. Golay, M., 1974, "Introduction to astronomical photometry", D. Reidel Publishing Company.
10. Gray, D., 1992, "The observation and analysis of stellar photospheres", Cambridge University Press.
11. Hovhannessian, R.Kh., 2004, Astrophysics, 47, 499.
12. Johnson, H.L., 1966, ARAA, 4, 193.
13. Lassovsky, K., 1936, AN, 258, 93.
14. Lucy, L.B., 1967, Zeitschrift für Astrophysik, 65, 89.
15. Nekrasova, C., 1945, Pulk. Bull., 134, 27.
16. Rucinski, S.M., 1969, AcA, 19, 245.
17. Schmidt-Kaler, Th., 1982, Bull. Inform. CDS, No. 23, 2.
18. Schneller, H., 1931, AN, 242, 180.
19. Sipahi, E., 2005, IBVS, No. 5635.
20. Sipahi, E., Gülmen, Ö., 2004, Ap&SS, 293, 307.
21. Vetesnik, M., 1965, VeBam, 27, 65.
22. Wachmann, A.A., 1948, Astr. Abh., Erg., AN, 11, No. 5.
23. Wilson, R.E. and van Hamme, W., 2003, "Computing binary star observables".

Modelling the equatorial activity in SS433

T. Smponias

Ioannina, Greece

Abstract. The jets in SS433 were modelled using a two-dimensional classical hydrocode and a dynamical mechanism was suggested to explain the observed equatorial radio features in SS433. The modelling provided constraints to a number of jet and stellar wind parameters for the jet-wind system. An ambient density falling off away from the jet base was found to offer the best results for enhancing the equatorial region activity.

Keywords: binaries:eclipsing –stars:winds, outflows–plasmas. radiation mechanisms: thermal – binaries: close – radio continuum: stars – stars: individual (SS433)

INTRODUCTION

In recent VLBI-scale maps (Paragi et al. 1998, 1999, Blundell et al. 2001, Rupen et al. 2004), two persistent components were detected along a direction roughly perpendicular to the main jets of SS433 in what has been called a 'ruff' structure. These regions have lower brightness than the main jet features, yet they appear to remain over several different epochs of observation.

Blundell et al. (2001) using MERLIN and VLBA data suggest that the equatorial emission is thermal in origin and comes from an equatorial outflow. On the other hand, Paragi et al. (2001) using EVN and global VLBI data interpret the equatorial emission as non-thermal, which suggests the presence of shocks in this region.

The reason for its existence may be because of a jet bow-shock-ambient medium interaction. Alternatively, it could be an excretion disk, formed from mass loss through a Langrangian point from the accretion disk environment (Fabrika 1993). A combination of the above effects is also possible, e.g. the jets' bow shocks compressing a stellar wind and interacting with an excretion disk. A preliminary study of ours showed that an interaction of flows at the equator may produce enhanced pressure and density, and this interaction is investigated further in this paper as a possible explanation of the 'ruff'.

MODELLING SETUP

The hydrocode

We used a hydrodynamical simulation computer program (hereafter hydrocode) in order to model the SS433 jets. The hydrocode was originally written by Lloyd et al. (1993). We used the two-dimensional axisymmetric version of the code, with cylindrical coordinates. The code models an ideal classical gas. Therefore, the jets' fluid flow has been taken to be purely hydrodynamical (hereafter HD). The HD simplification on the

CP848, *Recent Advances in Astronomy and Astrophysics*,
7th *International Conference of the Hellenic Astronomical Society*, edited by N. H. Solomos

one hand implies that the magnetic field contributes to the coupling of the jet matter that allows the fluid approximation. On the other hand, the field cannot affect the dynamics of the flow, as opposed to the MHD case (Rai-Choudhuri 1998). The method used in the hydrocode to advance the Euler equations in time is explicit 2nd order Godunov (Falle 1991).

The modelling

In thc model the resolution was around 20 cells across each jet, which allowed for the jet structure to be resolved. Various combinations of jet density, stellar wind, ambient density and jet Mach number were tried. This way, some insight was provided into how the changes in the above SS433 model parameters affect the conditions in the equatorial emission region of the SS433 jet model.

The equatorial activity was modelled using classical dynamics, as indicated from some observations (Stirling et al. 2002), since it occurs away from the relativistic jet, in a region where we can assume that the velocities are much smaller than the speed of light. Thus, in the equatorial region the matter is macroscopically classical, although it might still be microscopically relativistic.

Setup of the models

The jet densities varied from 10^4 to 10^6 particles per cm^3. The ambient density was either constant, or decreasing with $1/r^2$ away from the jet base (a stellar wind). The stellar wind densities refer to the middle of the first cell of the grid after the jet ejection nozzle, i.e to the middle of 4×10^{13}cm, or 2×10^{13}cm. The wind densities vary from 10 to 10^5. The jet densities are 10, 100 or 1000 times higher than the corresponding wind density of each run, justified, for being denser than the ambient medium, by the lack of deceleration of the knots in MERLIN scales (Spencer 1984). A proton jet was assumed for the jet flow, based on optical lines for SS433 jets (Margon 1984).

The values of the model parameters were selected to cover an estimated range of ambient and jet densities for SS433 (Begelman et al. 1980). The model jet is not continuous but it is made of a series of blobs ejected along the model jet axis, as seen in the radio images (Spencer 1984, Paragi et al. 1999).

The wind was assumed to originate from the jet base. The scale of the simulation was much larger than the distance of the binary orbit separation. Both the compact object and the companion star are located in the same computational cell in the scale of the simulation. The model system was assumed to have no extended disk (Fabrika 1993). The grid size was 1500 (parallel to the jet axis) by 500 (vertical to the jet ejection direction).

The jets of SS433 are modelled using classical dynamics and radiative transfer. One could argue that the 0.26c jets of SS433 are therefore modelled with limited accuracy, due to relativistic effects emerging at the above jet velocity. However, the main aim of the modelling presented here is to study the equatorial region of the system. And, as was mentioned earlier, the macroscopical velocities in the equatorial region are much less than the speed of light, c.

RESULTS OF THE JET SIMULATIONS

Morphology of model jets

The model jets consisted, in all the runs, of ejections of plasmoids (blobs) in partly random intervals. Thus, a group of blobs were ejected in relatively close succession, until a larger interval occured. The latter was of duration one order of magnitude higher than an interblob interval. Then another group of blobs, another large interval, and so on. The situation resembles a machine gun firing in bursts. The intrabullet interval is the equivalent of the interblob interval. The larger interburst interval of the machine gun then corresponds to the interval between groups of blobs.

A group of blobs corresponds to a series of VLBI-scale blob series, while being at the same time a MERLIN scale blob. Thus the blob ejection pattern was consistent with the different scales of observed features in SS433 jets near the core. Only the jet precession was left out, as it is a 3-dimensional effect. During a run, the jets might have completed more than half a precession. This could enhance the equatorial emission, as it brings the equatorial emission region closer to a jet each time. Hence if a result of equatorial emission is obtained here without precession, then in a precessing system the obtained equatorial emission should be of comparable strength.

In half the model runs, the ambient density was constant. Then, the activity in the equatorial region was much reduced, as the jet was well contained along its path.

Simulations of a jet in a wind

In the second half of model runs, the ambient density fell off as $1/r^2$ away from the jet base, as expected for a simple constant velocity mass conserving stellar wind. It is in these cases that the equatorial region appears more active, suggesting that the equatorial emission may be produced there (Figure 1). Regarding the pressure for the wind case, a prominent feature, apart from the pressurized entrained material in the equatorial region, is the low pressure interblob cavities that appear along the jet trail.

The jet head advances slower than the jet speed and this causes the blobs to strike it periodically from the back, creating a protrusion effect. The blobs are producing their individual bow shocks in the area of the jet cocoon. The latter is then a turbulent region where complex dynamical interactions take place, with oblique shocks and non-linear effects present. We should note here that the shock of each subsequent blob does not overtake the previous shocks. This has the obvious effect that the outer shock from the first blob does all the job on the equatorial region, while the rest of the blobs cannot affect the equatorial region directly. To investigate this further, we did also perform simulations with just one or two VLBI-scale blobs, the latter with a large interval between the two small blobs. These showed that the equatorial activity was again present, although less persistent after the blobs were gone.

The fall in the ambient density along the jet trail means that the outer jet's expansion shell is inflated so much faster than the inner, that some portion of it retreats and moves towards the equatorial plane. As a result, there is an interaction region near the equatorial plane, where meet a retreating shock from the outer jet, a sideshock from the inner jet and matter from the ambient medium. The above area of activity extends back to the equatorial plane where it meets its mirror area through a reflection boundary condition, representing the existence of a counterjet.

CONCLUSIONS

The jets in the SS433 system have been modelled with a two-dimensional ideal gas hydrocode. It was found in the modelling that an ambient medium density, falling with distance from the jet base, may cause the jet bow shock to create an active region at the equatorial plane.

Rupen (2004) suggested an excretion disk (see also Fabrika 1993) as a possible cause for the equatorial emission in SS433. Such an excretion flow would create an extended disk at the equatorial region and then the bow shocks from the jets, would have something to hit at the equatorial region, apart from the entrained stellar wind. Therefore, it is probable that a pre-existing disk is not only compatible with, but also in aid of our suggested mechanism for the equatorial emission.

It might be possible that the equatorial emission is a general phenomenon in microquasars, since its causes do not appear specific to the SS433 system but they are more fundamental properties of a microquasar system.

ACKNOWLEDGMENTS

The simulation was performed on the COBRA computer cluster at Jodrell Bank Observatory, The University of Manchester. The author thanks Dr Tim O'Brien and Dr Ralph Spencer for their advice and encouragement. The author acknowledges suppport from the Alexander S. Onassis public benefit foundation.

BIBLIOGRAPHY

Begelman M. C., Hatchett S. P., McKee C. F., Sarazin C. L., Arons J., 1980, ApJ, 238, 722

Blundell K. M., Mioduszewski A. J., Muxlow T. W. B., Podsiadlowski P., Rupen M. P., 2001, ApJ, 562, L79

Choudhuri A. R., 1998, The physics of fluids and plasmas, CUP, 1998

Fabrika S. N., 1993, MNRAS, 261, 241

Falle S. A. E. G., 1991, MNRAS, 250, 581

Lloyd H. M., Bode M. F., O'Brien T. J., Kahn F. D., 1993, MNRAS, 265, 457

Margon B., 1984, ARA&A, 22, 507

Paragi Z., Vermeulen R. C., Fejes I., Schilizzi R. T., Spencer R. E., Stirling A. M., 1998, NewAR, 42, 641

Paragi Z., Vermeulen R. C., Fejes I., Schilizzi R. T., Spencer R. E., Stirling A. M., 1999, A&A, 348, 910

Paragi Z., Fejes I., Vermeulen R. C., Schilizzi R. T., Spencer R. E., Stirling A. M., 2001, ApSSS, 276, 131

Rupen, M., Mioduszewski, A., Taylor, G., Walker, C., NRAO press releases, January 2004

Spencer R. E., 1984, MNRAS, 209, 869

Stirling A. M., Jowett F. H., Spencer R. E., Paragi Z., Ogley R. N., Cawthorne T. V., 2002, MNRAS, 337, 657

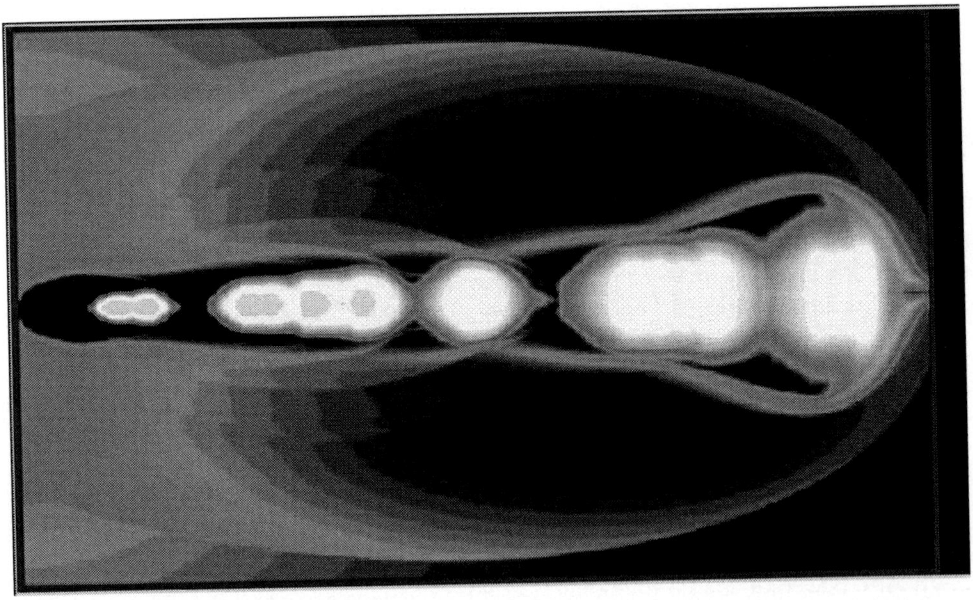

FIGURE 1. A snapshot of the density from a model run with decreasing ambient density (stellar wind). The jet base is at the middle of the left perpendicular side. The top half of the image is the computational space, while the bottom half is just a mirror image to enhance presentation. The ambient density is 10^5 cm^{-3} and the jet density 10^6 cm^{-3}. The jet is less constrained as it proceeds outwards, allowing the blobs to expand more. We can see the formation of activity at the equatorial region, which is a strong indicator of the origin of the observed equatorial emission.

CCD Observations of the Algol-type Binary AO Serpentis

Zavros P. and Rovithis-Livaniou H.

Section of Astrophysics, Astronomy & Mechanics, Dept. of Physics, Athens University, Panepistimiopolis, Zografos 157 84, Athens, Greece

Abstract. Our recent CCD observations of the short-period eclipsing binary AO Ser are presented. The observations were performed with the 40cm Cassegrain reflector at the Athens University Observatory, during June 2005, and confirm that the primary component of AO Ser is a pulsating star. Moreover,

Keywords: Eclipsing binaries - Pulsations, oscillations, and stellar seismology.
PACS: 97.80.Hn – 97.10.Sj

1. INTRODUCTION

AO Ser is a short-period (0.8793 days), Algol-type eclipsing binary, and was a rather neglected system till 2004. Then, it was proposed [1] to be included as a new member in a recently formed group of mass-accreting pulsating components in semi-detached, Algol-type, eclipsing binaries [2]. The members of this group are characterized by oEA, instead of simple EA, where "o" denotes the presence of oscillations; while, "EA" stands for an eclipsing binary of Algol-type, as usual.

The spectral type of its primary component is A2. Similar are the spectral types of the primaries of the rest members of this group. The detection of the oscillations in these systems is quit difficult, as their amplitude is very low in comparison to the large variations due to the eclipses. So, very accurate observational data can yield to their detection.

2. PHOTOMETRIC OBSERVATIONS OF AO SER

The photometric observations of *AO Ser*, presented here, were carried out in June 2005. They were obtained in 4 nights, between 5 & 10, using the 40cm Cassegrain reflector of the Athens University Observatory, and a ST8/SBIG CCD camera. The observations were made in three pass bands (VRI). The star BD +17 2940 was served as comparison, and the reduction of the observations was made with the AIP4WIN program. The following light elements were used for the calculation of the phases:

$$\text{Min I} = 2452334.2063 + 0^{d}.87934745 \cdot E$$

CP848, *Recent Advances in Astronomy and Astrophysics,*
7th International Conference of the Hellenic Astronomical Society, edited by N. H. Solomos

The corresponding light curves of *AO Ser* are presented in Figures 1, 2 & 3 for the three observational pass bands VRI, respectively.

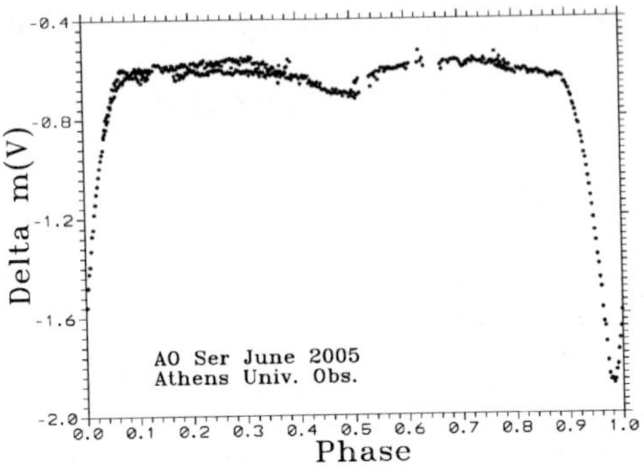

FIGURE 1. The V light curve of AO Ser as derived from our observations in June 2005.

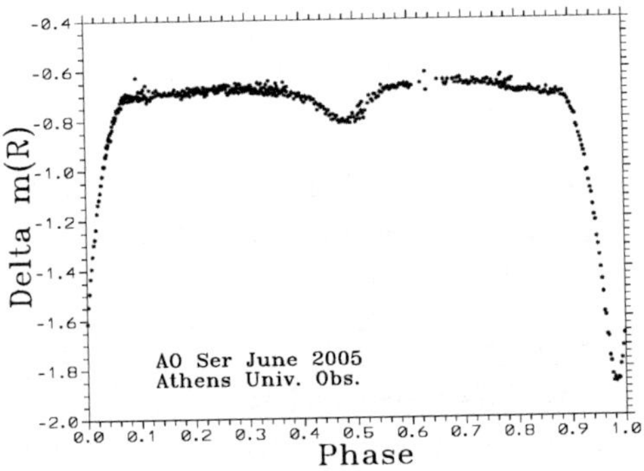

FIGURE 2. The R light curve of AO Ser as derived from our observations in June 2005.

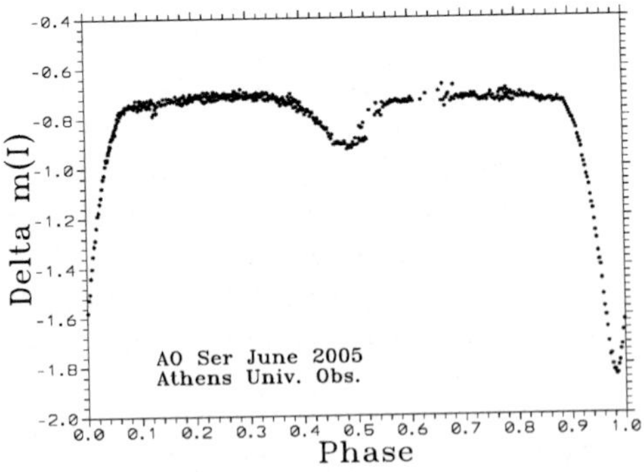

FIGURE 3. The light curve of AO Ser as derived from our observations in June 2005, in I.

Moreover, Figures 4, 5 & 6 present parts of our observations in various bands and in detail, for a better view. Actually, Figure 4 presents parts of our observations of June the 5th around primary minimum in I pass-band. Figure 5 presents our observations of June the 8th before and during secondary minimum in R pass-band. And Figure 6 presents part of our observations out-side eclipses in V filter, where the oscillations were more pronounced.

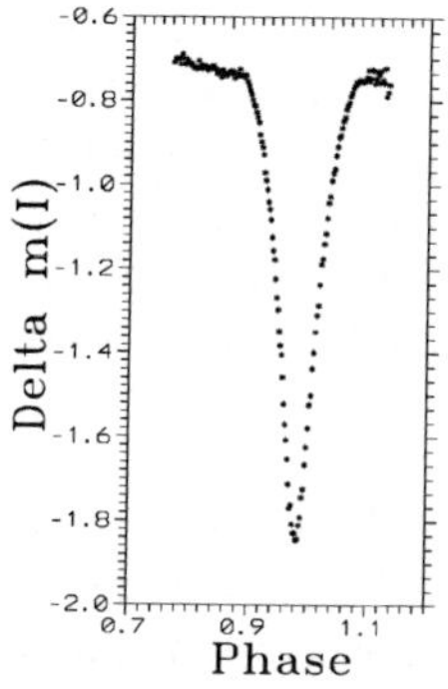

FIGURE 4. Part of our observations of AO Ser around primary minimum obtained in June the 5th 2005, in I pass-band.

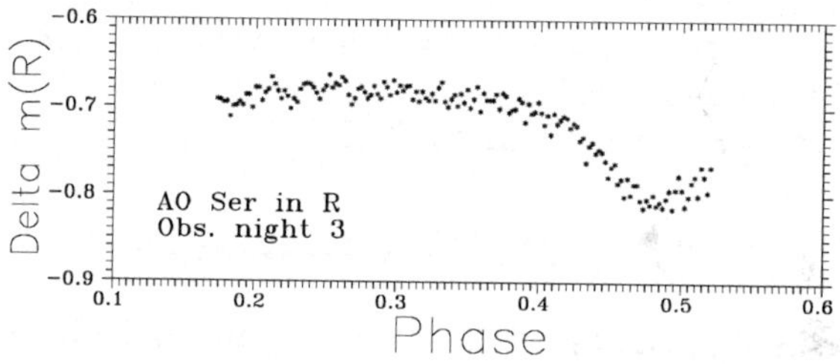

FIGURE 5. Our observations of AO Ser before and during the secondary minimum obtained in June the 8th 2005, in R pass-band.

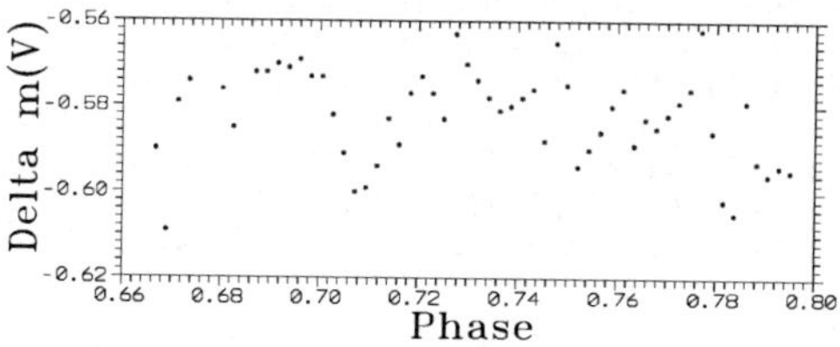

FIGURE 6. Part of our observations of AO Ser obtained in June the 10th 2005, shown an out-of-eclipses part in V pass-band, where the oscillations were more clearly seen.

3. DISCUSSION

Our photometric observations of *AO Ser* started in May, while they continued and completed in June 2005. The first, were performed with the 1.23cm Cassegrain telescope at the Kryonerion Astronomical Station of the Athens National Observatory. The second were carried out with the 40cm Cassegrain reflector of the Athens University Observatory, as was already mentioned. Here, we present our second set of observations, since they correspond to complete light curves of *AO Ser*.

As is obvious from the Figures, and especially from 5 & 6, our data undoubtedly show short-period and small amplitude oscillations, as had been firstly reported [1]. Actually, from Figure 6, maximum amplitude seems to be around $0.^{m}04$ with period of about $0.^{d}05$; while, there are smaller amplitudes of about $0.^{m}02$, and period of about $0.^{d}02$. Thus, it seems that *AO Ser* have multiple periods. This is in agreement with other findings, i.e. [1].

When *AO Ser* was proposed to be included in this newly formed group of Algol-type binaries, this group numbered only 11 such binaries. Within the one and a half year passed since then, their number has increased to 19 systems [3]. This is very promising, as these poorly studied stars are potentially attractive for asteroseismology. Their primary components are pulsators, sharing the pulsation properties of the classical δ Sct-type variables, although coming from different evolution [4]. In Figure 5 of paper [2] the positions of 5 such systems —with accurate known absolute parameters— are presented on $logT_e - logL/L_{\odot}$ diagram. And although we have not yet analyzed our light curves of *AO Ser* to get new elements, using some old ones [5], *AO Ser* is found to be just outside the blue edge (BE dashed line in the foregoing mentioned figure of [2]), very close to *AB Cas*.

ACKNOWLEDGMENTS

Athens University supported partly this work, (grant 70/4/3305).

REFERENCES

1. S-L Kim, Y-B Kang, J-R Koo , D.E. Mkrtichian and J.W. Lee, *I.B.V.S. No.* **5538** 2004.
2. D. E. Mkrtichian, A.V. Kusakin, E. Rodriguez, A. Yu. Gamarova, C. Kim, S-L Kim, J.W. Lee, J-H Youn, Y.W. Kang, E.C. Olson and K. Grankin, *Aston. & Astrophys.* **419**, 1015-1024 (2004).
3. S-L Kim, C-U Lee, J-R Koo, Y.B. Kang, J.W. Lee, D.E. Mkrtichian, *I.B.V.S. No.* **5669** (2005).
4. E. Rodríguez, J.M. Gaecía, A.Y. Garamova, V. Costa, J. Daszyńska-Daszkiewicz, M.J. López-Gonźalez, D.E. Mkrtichian and A. Rolland, *Mon. Not. R.A.S.* **353**, pp. 310-318 (2004).
5. H.K. Brancewicz and T.Z. Dworak, *Acta Aston.* **30**, pp. 501-524 (1980).

PART 3

EXTRAGALACTIC ASTROPHYSICS

(Conveners: S. Papadakis and J. Georgantopoulos)

Astrophysical Jets
Observations, Theory, Experiments

Attilio Ferrari[1,2]

[1]*Dipartimento di Fisica Generale, Università di Torino, Italy*
[2]*Department of Astronomy and Astrophysics, University of Chicago, USA*

Abstract. Collimated supersonic outflows, often relativistic, have become a common feature in many astrophysical settings, from star formation regions to extragalactic radio sources. In this review I shall illustrate the main observational characteristics of the various classes of jets and discuss the theoretical models used for their interpretation, In particular I shall outline the physical questions and challenges faced by these models, with specific reference to jets from Active Galactic Nuclei, and present experiments that are being used to test the validity of the theoretical approach. Finally I discuss in some detail the issue of heating of intracluster gas by jets from AGNs.

Keywords: jets, accretion, active galactic nuclei, stellar outflows.
PACS: 98.38.Fs, 98.58.CM, 98.54.Gr, 98.62.Mw, 97.21.+a.

OBSERVATIONS

In 1918 observations of the elliptical galaxy M87 (NGC 4486) carried out by Heber Curtis revealed a "curious straight jet ... apparently connected with the nucleus by a thin line of matter". These optical observations were not followed up by Curtis and it was not until after the development of radio astronomy that M87 was discovered in the 1953 to be associated with a powerful double-lobe radio galaxy with extended jets emanating from the optical nucleus (Fig. 1).

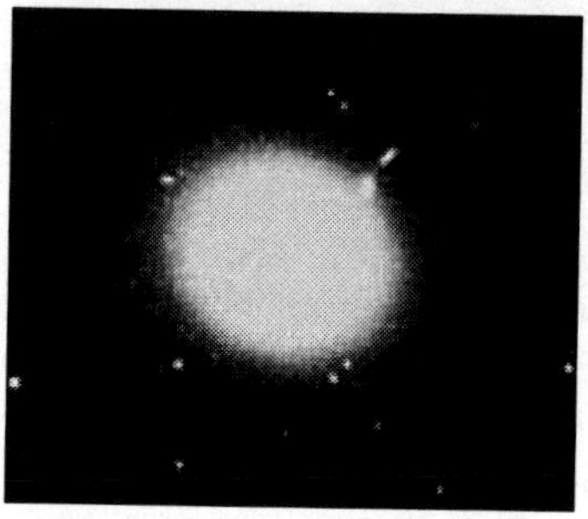

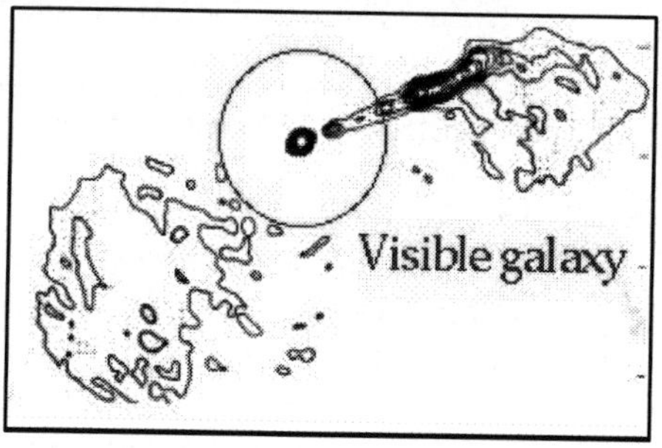

FIGURE 1. The elliptical galaxy M87 and the radio map

CP848, *Recent Advances in Astronomy and Astrophysics,*
7th *International Conference of the Hellenic Astronomical Society,* edited by N. H. Solomos

Many other similar collimated extended structures were discovered in the 1960's in elliptical galaxies and the study of jets became a major theme of research in astrophysics. Initially there was no way to tell whether jets were actually outflows, as radio emission had a non thermal continuum spectrum without any spectral lines. In the 1970's it was possible to measure proper motions of features, showing that jets flow in all cases at supersonic and often relativistic speeds [1]. Radio spectra were also used to show that radiation emission is synchrotron type, requiring relativistic electrons and magnetic fields all along the jet; and polarization maps showed organized morphologies.

From the theoretical point of view it was shown by Blandford & Rees [2] that a continuous outflow model could better explain the energy requirements and was more consistent with the morphologies than a single ejection event; in particular the double outflow ejection could be globally interpreted as a twin-exhaust generated in the potential well of rotating galactic nuclei.

Finally the advent of the Hubble Space Telescope and the XMM and Chandra satellites allowed to detect the optical and X-ray non thermal counterparts of powerful radio jets (Fig. 2). Jets were therefore recognized as the typical dynamical components of active galactic nuclei and they were also discovered in association with quasars.

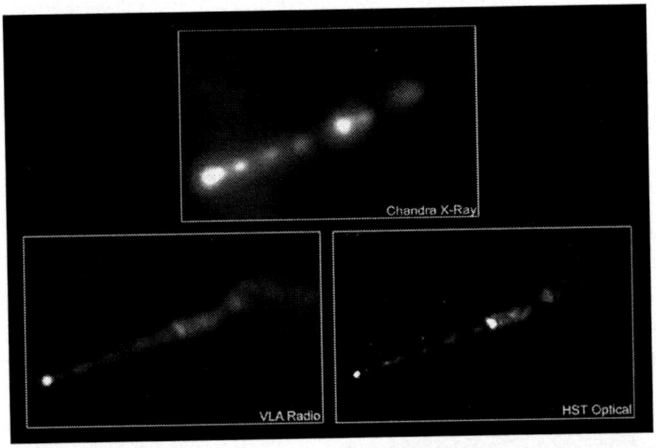

FIGURE 2. Multi-frequency maps of the radio jet of M87 (X-ray: NASA/CXC/MIT/H.Marshall et al.; Radio: F. Zhou, F.Owen (NRAO), J.Biretta (STScI); Optical: NASA/STScI/UMBC/E.Perlman et al.)

These observations posed a strong requirement for the global energetics: if extended radio galaxies are at cosmological distances their estimated power can reach up to 10^{47} erg s^{-1} for a source lifetime of at least 10^{8} yrs. Another question was how to accelerate jets to relativistic speeds and maintain those outflows collimated over enormous distances avoiding any sonic expansion. Finally, according to the interpretation of the continuum spectrum from radio to X-rays as synchrotron emission from relativistic electrons, the argument was made for reacceleration of electrons along the outflow given their relatively their short lifetimes.

Starting from the 1980's supersonic outflows were discovered in star formation regions [3] and relativistic jets in active stars, as GRS 1915+105 [4]. Therefore collimated outflows arising from dynamically active objects appear to be a common feature and we may also consider the effect they must have on environment.

Concerning their origin we must mention that high-resolution observations have now allowed to prove a non-circumstantial correlation of jets with accretion flows. A 1 $M_\odot$ yr^{-1} accretion rate onto a 10^8 $M_\odot$ black hole allows a release of gravitational energy that largely fulfill the energy requirement for the most powerful quasars. At the same time, as we shall discuss, the dynamics of accretion disks is well apt for launching outflows along their rotation axis.

PHYSICAL MODELS

Following the seminal twin-exhaust paper by Blandford & Rees [2] the physical study of jets, their production, stability and interaction with the ambient medium has adopted the scenario of the jet-disk connection. The physical ingredients of models are [5]:

1. a central super-massive object (most likely a black hole) pulling matter from the environment
2. a turbulent magnetized disk accreting into the super-massive object, emitting thermal radiation and blowing a wind out of a coronal structure
3. a two-component supersonic (relativistic) collimated magnetized outflow, including a thermal (electron-proton or electron-positron) plasma carrying inertia and shaping dynamical morphologies and a relativistic supra-thermal component responsible for non thermal radiation
4. a magnetic field connecting the disk with the supersonic outflow through magnetic stresses
5. an external medium into which the collimated outflow shocks an expanding supersonic bubble transferring its kinetic energy.

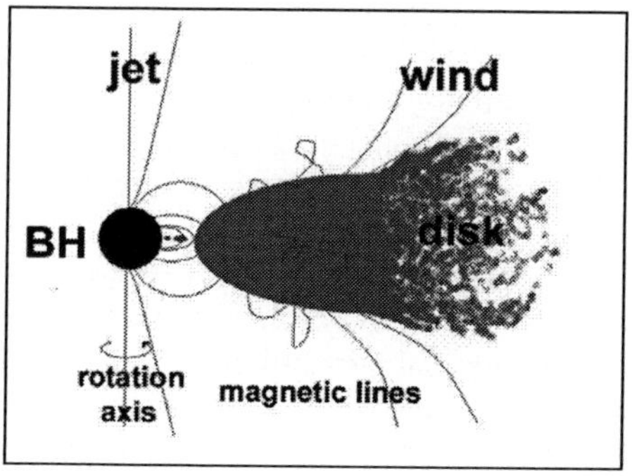

FIGURE 3. The disk-jet model

Clearly the interaction between all these components involves nonlinear plasma electrodynamic relativistic effects and cannot be treated in a completely self-consistent model. For this reason models have addressed specific aspects of the phenomenology separately and complex numerical simulations have been used to combine them. Nevertheless we do not have a global picture yet, especially because some of the physics involved is beyond reach of present models. Here I shall review the most relevant results obtained in this general scenario and discuss the research directions to follow to advance the subject making reference to jets connected to active galactic nuclei.

THE CENTRAL BLACK BOX: ACCRETION DISK

The origin of jets is related to a hot region where the combination of the inward strong gravitational pull and the anisotropic outward centrifugal and pressure forces originates and maintain a wind-type outflow. The above mentioned energetic requirements for active galactic nuclei can be fulfilled in the framework of a quasi-Keplerian accretion disk that steadily liberates gravitational energy feeding matters into a black hole. In active stars and star forming region the central gravitational pull is instead due to a star, but the physics for what concerns the generation of jets is substantially similar. The creation of such configuration requires a mass inflow from the ambient: in active stars mass exchange from a binary companion can be the source of inflow, while in galactic nuclei and star forming regions the ambient is rich enough of matter for a long evolutionary period.

From the modeling point of view the main question is how to maintain a steady quasi-Keplerian disk with an appropriate outward angular momentum transport that allows matter to fall regularly onto the central object. Shakura & Sunyaev [6] postulated the presence of a viscosity in the disk to transfer momentum from fast inner rings to slower outer rings ($\omega_K \propto r^{-3/2}$). However standard molecular viscosity is found too weak to produce the required transport; proposals to enhance "effective" viscosity are: hydrodynamic turbulence generated by nonlinear shear instabilities and spiral waves [7], magneto-rotational instabilities [8], ballooning-type instabilities [9]. The ensuing disk models depend also on the internal energy dissipation by radiation emission, that is directly connected with observations. Three classes of disks can exist: (1) thin disk, where radiation emission cools the plasma very rapidly and the gas pressure becomes not important; (2) thick disks, where radiation is not efficient and the plasma pressure inflates the structure; (3) accretion tori, where the interaction electrons-ions is weak because of the low density and correspondingly radiation emission from electrons is also weak and most of the internal energy is advected onto the central object by ions. Numerical simulations to include also nonlinear effects in this general scenario have been carried out in recent years [10,11,12].

However the physical picture is still not well refined, taking also into account that numerical codes cannot reach the range of physical parameters (viscosity, resistivity, conduction, etc.) of the astrophysical applications. Laboratory experiments are being developed in order to validate the numerical codes used in the simulations [14].

In particular the connection between disks and jet production requires a more correct representation: it looks, as we shall see in the next section, that relatively strong magnetic fields are needed to produce jets. In that case it would be the jet itself responsible for the angular momentum subtraction from the disk.

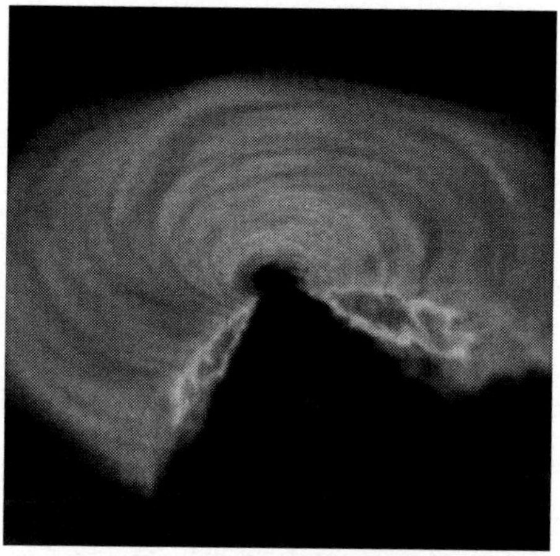

FIGURE 4. A 3-D model of a magnetized accretion torus [13]

JET LAUNCHING

The initial approach to the study of acceleration of jets followed the original twin-exhaust model. A hot central region in a galactic nucleus or a star forming region creates an overpressured gas bubble that can start to a wind-type outflow following the preferred direction of ejection is in two opposite directions along the rotation axis. Different geometric configurations have been explored with and without magnetic fields.

Hydrodynamic Winds

The main drawback of the twin-exhaust model was the transition to supersonic flow at a 100 parsec scale, too far from the central core, because observations indicate that jets are perfectly collimated and relativistic already at the 0.1 parsec scale. One possibility explored by Ferrari et al. [15] was to explicitly include the geometry of the accretion funnel and the radiation output from its walls which both act as momentum addition factors to the flow. In this way one can actually reach supersonic speeds well within the object core and also interpret large scale morphologies in terms of internal shocks arising from shear instabilities at the boundary layer with the ambient medium. However such model is intrinsically unfit to produce relativistic flows because of radiation drag effects; therefore it may be used to launch jets in star forming regions, but not jets from active galaxies.

Magneto-Hydrodynamic Winds

The presence of magnetic fields in AGN jets is well founded on synchrotron spectra, more uncertain in stellar jets. However, given the customary presence of magnetic fields in astrophysical plasmas, it looked immediately worthwhile to adopt electrodynamic effects to accelerate charged particle outflows. The first paper along this line was published by Blandford & Payne [16]; they proposed a self-similar solution of the ideal MHD equations yielding an outflow out of a magnetized Keplerian accretion disk by a magneto-centrifugal mechanism. The condition for starting the outflow is that magnetic lines are bent by a relatively large angle with respect to the spin axis of the disk (> 30° for a Keplerian disk); such bending can be consistently produced as the disk accretes towards the center of the gravitational well.

The accretion of a (slightly sub) Keplerian disk is a difficult problem by itself. As it was originally discussed in the framework of accretion onto X-ray binaries [6], the inflow requires a regular outward transport of angular momentum. Molecular viscosity in the disk is not efficient enough to this taks and turbulent viscosity by fluid/electromagnetic stresses is apparently excited only if some weak magnetic field is present leading to magneto-rotational instability [8].

Assuming that a steady state accretion can be reached and the magnetic field configuration has a force-free poloidal component plus a toroidal component generated by differential rotation, the Blandford & Payne solution has the following characteristics. Close to the accretion disk the jet is accelerated along the poloidal component of field lines and solutions scale with spherical radius along a given direction; centrifugal acceleration drives the gas out and then acceleration proceeds by magnetic pressure that also acts in collimating the outflow. Far away from the disk the acceleration is due to electrodynamic (Lorentz) forces and the predominantly toroidal field maintains collimation; the asymptotic velocity of the jet is of the order of the Keplerian velocity at the inner edge of the disk and in principle can be relativistic [17] two branches of solutions exist, one dominated by matter and another by Poynting flux. Several stability issues have been raised against this model. In addition it can be shown that the profile of the magnetic field is determined by the values of viscosity and resistivity, and the launch of a steady flow is allowed only for relatively large values of these parameters [18,19].

A different type of solutions has been discussed by Tsynganos & Trussoni [20], in which a condition of self-similarity is imposed on the poloidal angle fixing the shape of the magnetic field. In these solutions the energy equation is not solved and is assumed not adiabatic, but such to allow for the geometry configuration. Also the accretion disk is simply a boundary condition.

The original paper attempting a numerical simulation of the complex interaction between disks and jets is by Uchida & Shibata [21]; these authors were able to give hints of the initial stage of an outflow acceleration, although the physics of the disk was completely missing and the integration conditions very crude. Several numerical simulations have then been published [22,23]. Recent simulations based on the FLASH code [24] at high resolution with a self-consistent treatment of disk and jet have explored the conditions for stationary jet generation from disks at relatively high magnetic fields. In the case of relatively low resistivity hollow jets are formed, while at high resistivity jets are very fat (see Fig. 5).

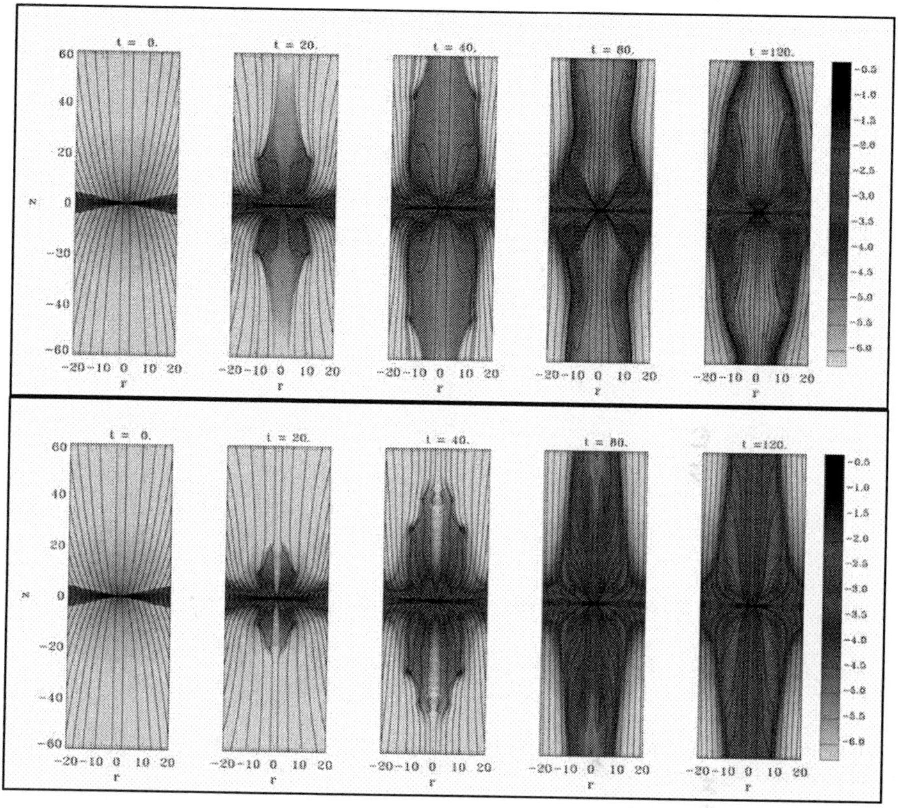

FIGURE 5. Numerical simulation of jet launch from accretion disk; density time evolution, lines trace the poloidal magnetic field; upper panel low resistivity case, lower panel large resistivity case [24].

Poynting-Flux Acceleration

An alternative approach to the launching mechanism, still in the framework of electrodynamic processes, was proposed by Lovelace [25] and Blandford [26] referring to the unipolar inductor mechanism that is likely to work in pulsars. They calculated that starting from an accretion disk threaded by a poloidal magnetic fields, differential rotation induces an electric field in the inertial frame that creates an open magnetosphere above the disk with a flow velocity along the spin axis. The equilibrium condition is $\rho\mathbf{E} + \mathbf{J} \times \mathbf{B} \approx 0$. This solution is not MHD as it allows charge separation and in particular an electron (+ positron ?) outflow to feed the large scale radio galaxy; the outflow is accelerated by the Poynting flux, is not a wind. Lovelace and collaborators have published a series of analytical papers deriving stationary solutions and testing them through non-linear numerical simulations using MHD codes at very low density. They find quasi-stationary outflows collimated along the jet plus an uncollimated wind from the outer part of the disk.

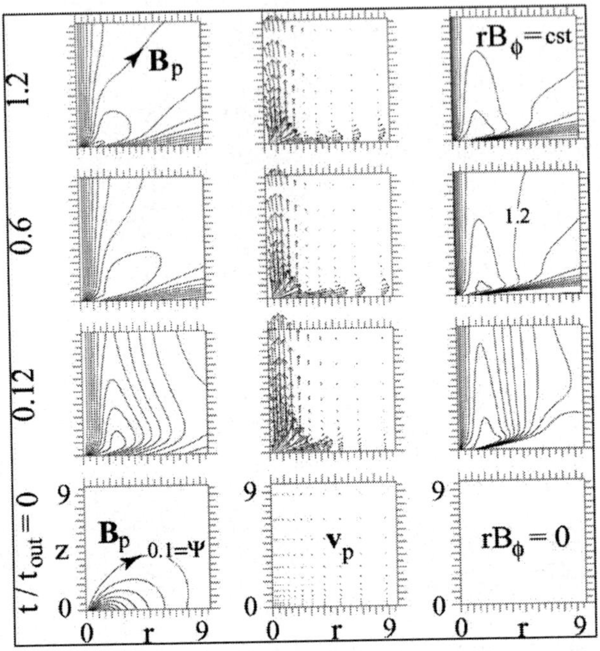

FIGURE 6. Numerical simulation of a Poynting jet accelerated from a disk with dipole magnetic field: $\mathbf{B}_p$ poloidal field, $\mathbf{B}_\varphi$ toroidal field, $\mathbf{v}_p$ poloidal outflow velocity, at four subsequent times [27].

This type of outflows has received less attention in the literature so far; in particular there is no self-consistent disk/jet solution. Instead Poynting jets are less subject to instabilities in the propagation, as we discuss below, and are typically cold, which means that they do not expand and remain collimated.

JET PROPAGATION

Propagation of jets is strictly related observable scales and therefore can be used to obtain direct information on the physical characteristics of these objects. For instance the issue of collimation and supersonic dynamics came originally from radio data of structures protruding from elliptical galaxies. High resolution VLBI observations show that jets are collimated well within the central black box, both in galactic and stellar jets, which is consistent with the acceleration/collimation models discussed in the previous section. In this region the bulk flow of active galactic nuclei is relativistic, as shown by the (apparent) superluminal velocities. Two problems arise at this point: how is collimation maintained over large scales (more than 8 orders of magnitude) and how is the jet slowed down to produce non-relativistic outflows and eventually the large radio lobe ?

The problem of large scale collimation has been partially relaxed as measurements of the ambient pressure from X-ray data in galaxies and IR data in stars have shown that the external pressure can provide confinement at least in some objects. On the other hand jets appear also to have structures evolving along their propagation that can be interpreted in terms of the flow shear instabilities, corresponding to an interaction with the external medium. Actually these instabilities (namely Kelvin-Helmholtz instabilities), in accordance with well-known laboratory results, were initially considered as capable of stopping the jets; an extended literature on this point with analytical results and numerical simulations exists (see, for instance [28]).

The breakthrough came when numerical simulations were performed taking into account the head of the jet and not simply the shear effects along an infinite boundary layer or a slab [29]. In this case a bow shock is formed in front of the jet's head that leaves behind an expanding overpressured cocoon where a backflow confines the jet. In this region mixing of the jet and cocoon matter slows down the flow and at the same time shear instabilities produce internal shocks and boundary effects that can be related to the observed morphologies (knots and wiggles) and to the heating and production of high-energy particles required by thermal and non-thermal radiation emission. An example of this process is given in Fig. 7 [30].

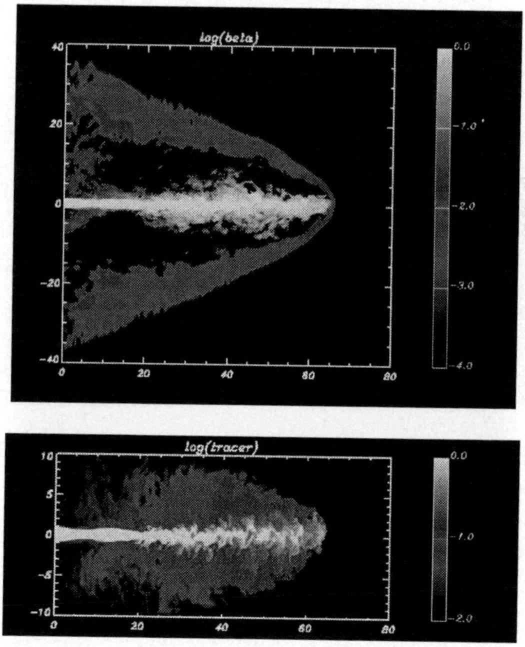

FIGURE 7. The cocoon around a relativistic jet with Mach number M = 3 and density ratio (ambient/jet) $\eta = 1000$. Upper panel: flow velocity $\beta = v/c$. Lower panel: passive tracer showing the mixing between the jet (yellow) and ambient (black) material.

These morphologies refer to supersonic jets with density lower than the ambient. Two different types of cocoons are suggested by simulations: fat and spearheaded. Fat cocoons are formed for high pressure jets or in general jets in which most of the energy is in internal energy and less in kinetic energy; this corresponds to very light jets with relatively low Mach number so that shear instabilities and turbulent mixing are very efficient. In the case of high Mach number and not so light jets the energy of the flow remains in kinetic form because shear instabilities and turbulent mixing are less efficient. In fact jets denser than the ambient medium do not develop mixing and their cocoons are very narrow: jets are ballistically confined. Referring to radio galaxies, FRI sources have extended mixing regions, while FRII release their energy in the final splash region.

An obvious fact emerging from these simulation is that entrainment, although important in shaping the cocoon morphology, cannot slow down the core of the jet that remains highly relativistic up to the bow shock. This result is at present not confirmed by data

Again the problem with numerical simulation is their applicability to astrophysical objects; in fact the effective viscosity and resistivity are much larger in simulations because of the finite spatial resolution of numerical codes, even using advanced techniques with adaptive mesh refinement on supercomputers. The validation of the codes is now being attempted experimentally in the laboratory. We mention here an experiment that is being carried by a collaboration between the University and Polytechnic School of Torino and the Polytechnic School of Milano [31].. It is based on a gun that injects a jet into a large vessel through a nozzle with a profile producing a jet with Mach number up to 20.

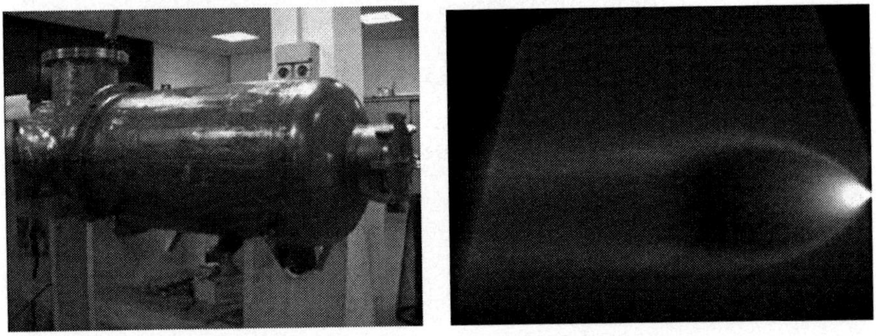

FIGURE 8. The Torino - Milano experiment on jets. Left panel: experimental setup; right panel: barrel shock and Mach disk of a jet with maximum Mach number before Mach disk $M = 17$.

A high-velocity video-camera allows to measure the time evolution of the jet in the ambient on scales comparable to the simulation setup. A sample of the stationary state of a supersonic jet is shown in Fig. 8.

JET TERMINATION

The global picture of a jet inflating a cocoon in the ambient medium is schematically represented in Fig. 9 with reference to a strong FRII source. Clearly the jet and the cocoon strongly modify the intergalactic gas transferring momentum and energy through shocks.

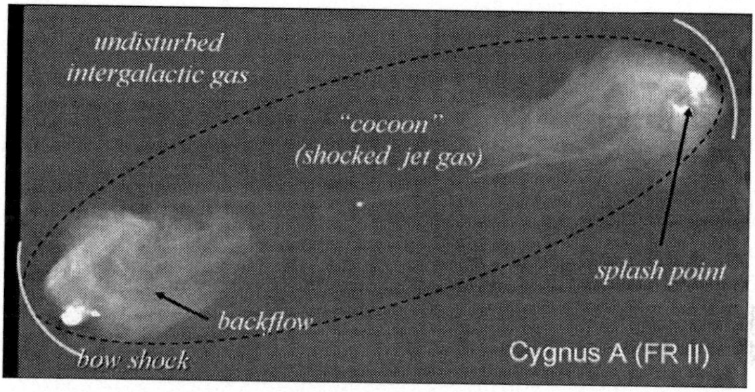

FIGURE 9. The physical ingredients of a large scale powerful jet

X-ray observations of extragalactic radio sources have recently provided strong information on the energy exchange between the jet plasma and the X-ray emitting gas in the ambient medium. In particular observation of Cygnus A with the ROSAT HRI [32] and with the Chandra X-ray Observatory [33] show deficits of X-ray emission in the cluster gas spatially coincident with the radio lobes. Observations of the Perseus cluster, also with ROSAT [34], and with the Chandra X-ray Observatory [35] showed that the cluster emitting gas is displaced by the radio lobes of the source NGC 1275; in addition there are hints of waves and ripples created in the cluster gas by the activity of the radio jets [36]. A similar behavior has been observed in the Hydra A cluster, hosting the radio source 3C 218 [37] and in Abell 2052 that displays regions devoid of X-ray emission coincident with the radio lobes of 3C 217 [38]. Disturbances by a radio source are also found in the gas halo of some giant elliptical galaxies such as M 87 and M 84 [39].

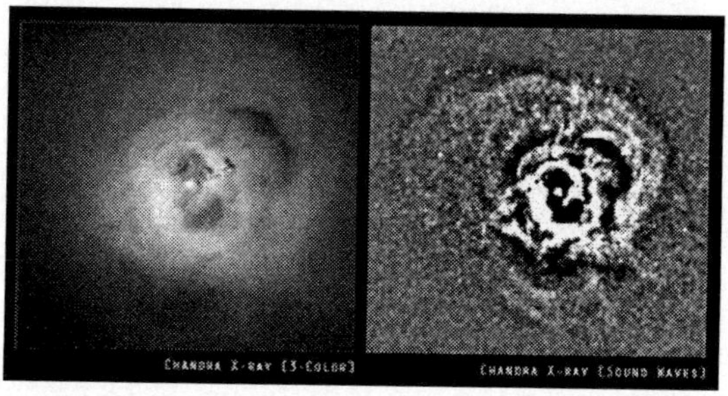

FIGURE 10. The X-ray structure of NGC 1275 in the Perseus cluster [36]

This scenario is suggestive of a situation consistent with the formation of overpressured cocoons around jets, capable of displacing and compressing the ambient gas: the end result of such interaction could indeed be the formation of cavities and shells in the X-ray emission. A simple one-dimensional model of this interaction was presented by Begelman & Cioffi [40].

Zanni et al. [41] have analyzed in detail, by using expanding numerical simulations, the evolution of the X-ray properties of a cocoon expanding in a stratified external medium and their dependence on the jet properties. The jets are characterized by their Mach number and their density ratio with the ambient medium density; the parameter domain has been widely covered in order to consider a wide range of jet powers. Two distinct and observationally different regimes of interaction exist. In the first phase the cocoon is highly overpressured with respect to the external medium and drives a strong shock: the ambient material is compressed in a thin and hot shell. Correspondingly the X-ray emission is dominated by a shell of enhanced luminosity surrounding a luminous cocoon. In the second phase, the cocoon pressure decays and tends to balance with the external value: the shock becomes weak and the shell widens and cools down. In this phase a bright X-ray shell surrounds a region of low luminosity coincident with the radio lobes corresponding to the cool cocoon. Zanni et al. show that the separation between the two regimes depends essentially on the jet kinetic power: the transition will occur at larger distances fro the galactic nucleus for high kinetic power jets and at lower distances for low kinetic power jets.

The global morphology of radio galaxies is related to the orientation of jets with respect to the ambient medium stratification as this affects the development of the cocoon. In an atmosphere with elliptical symmetry with the jet oriented along the major axis of stratification, the transverse steeper pressure gradient causes a lateral expansion of the cocoon at the same rate of the expansion along the major axis. This lateral expansion is collimated in an outflow that is similar to that observed in peculiar X-shaped radio sources [42]. If the jet propagates along the minor axis of the stratification a classical double radio source is formed.

HEATING OF INTERCLUSTER MEDIUM

The X-ray spectra of clusters and groups of galaxies is mainly thermal brems-strahlung, which indicates the presence of a hot Intra ClusterMedium (ICM) with temperatures ranging from 0.5 to more than 10 keV. A wealth of recent observations, coming in particular from the X-ray telescopes CHANDRA and XMM-Newton [43], have evidenced a complex ICM structure, as discussed in the previous section. While these aggregations can be understood in terms of the currently favoured hierarchical model of structure formation, there is an ongoing debate concerning the origin and this hot ICM. The most obvious way to explain the heating is to suppose an adiabatic heating of the gas accreting onto the potential well of clusters. Such heating is a natural outcome of the merging processes which form groups and clusters of galaxies in the hierarchical model picture. While Dark Matter halos merge, their gas is shocked and heats up, to form the groups and clusters atmospheres which are observed. The power spectrum of the initial density perturbations grow hierarchically to form groups and clusters and does not possess a characteristic length or mass scale, in the relevant range. The gravity itself has no characteristic scale. Thus, with the additional assumption that gas is in hydrostatic equilibrium with the cluster potential, it can be shown that this scenario predicts simple self-similar scaling relations among the various observable X-ray properties of groups and clusters. In particular, the relation between X-ray luminosity and the temperature is $L_X \propto T^2$, and entropy $S = T/n_e^{2/3}$ (where n_e is the electron number density of the gas) scales as $S \propto T$. Nevertheless, the observational situation appears to be quite different. The luminosity-temperature relation is steeper than this simple picture predicts, $L_X \propto T^\alpha$ with $\alpha = 2.5 - 3$ for clusters with temperature $T > 2$ keV [44]. There are indications of a steeper slope for groups with $T < 1$ keV [45]. The gas entropy in clusters is found to be in excess with respect to the model predictions, and its dependence on the temperature approximately follows the law $S \propto T^{2/3}$ [46].

The discrepancy between observations and theory suggests that the ICM is subject to a heating which is larger than what the simple self-similar model predicts. The ICM can be heated by two natural processes, which are present when the physics of the heavy baryonic component is considered. The first is the energy feedback from supernovae explosions, which directly heats cold gas and increases its entropy. The second is the radiative cooling of the gas itself, which subtracts low-entropy cold gas from the hot phase and turn it into stars, thus increasing the average temperature and entropy of the remaining gaseous medium. The efficiency of the two processes in heating the ICM have been extensively studied so far, both with semi-analytical techniques [47] and with numerical simulations [48]. Recent numerical studies also considered simultaneously the effect of radiative cooling, star formation and feedback from SN [49,50] showing how the fraction f_* of gas which is converted into a "cold" stellar phase results to be too high, in excess with respect to the value $f_* \leq 10\%$ which is obtained from measurements of the local luminosity density of stars. The overcooling problem which naturally descends from these results still calls for further energy contributions. A possible source for this energy is given by AGNs activity. The available energy budget is in principle more than adequate to heat the ICM to the desired temperature [51]. In fact, the accretion of gas onto supermassive black holes in galactic cores gives outputs of order 2×10^{62} $(M/10^9\ M_\odot)$ erg, where M is the accreted mass and a standard mass-energy conversion efficiency of 10^{-1} is used. We have already discussed in the previous section how observations with the ROSAT satellite suggested an interaction of the central AGN activity with the ICM in the Perseus cluster: the cluster gas is displaced by the radio lobes of the source 3C 84 which inflates cavities or "bubbles" in the X-ray emitting gas and creates ripples and sound waves propagating through the ICM.

Zanni et al. [52] have studied the energy exchange between AGN-powered jets and the surrounding ICM of group/cluster of galaxies, with 2D hydrodynamical numerical simulations. They extract the groups and clusters atmospheres from cosmological numerical simulations by Tornatore et al. [49], at redshift $z = 0$ and built a self-similar model for such atmospheres based on the NFW radial density profile for the mass distribution and the hydrostatic equilibrium with the consequent potential for the gas.

The model is not an *average* description of a typical group or cluster of galaxies, but was built to represent the output of a self-consistent numerical cosmological simulation. The jet is injected in the ICM for a time $t \approx 2 \times 10^7$ yr, then switched off. Varying the temperature scale and the kinetic energy of the jet one derives the energy balance of the ICM, the entropy evolution, and the resulting L_X - T_X relation.

The mechanism through which jets interact with the ambient medium can be summarized as follows: an underdense supersonic jet is slowed down by a terminal shock (Mach disk) where its kinetic energy is thermalized and inflates the overpressured cocoon; the expansion of the cocoon compresses the surrounding medium and, as long as the energy transferred to the ambient by the expansion work is higher than the background energy (even after the jet has ceased its activity), the compression of the medium will be mediated by an expanding shock (bow shock) where the energy transferred to the medium can be dissipated. A high fraction of the injected energy can be deposited into the ambient medium by expansion work (> 60%) while only a small fraction is retained in the cocoon and in the entrained ambient material (< 10%; see Fig.11). Once the jet has ceased its activity it quickly loses its collimation and disappears while the high entropy cocoon formed by the thermalized material of the jets buoyantly rises into the atmosphere entraining the ambient medium. The energy of the rising plumes (or bubbles) associated with the relics of the radio lobes can be dissipated thanks to the expansion work done as they rise in the decreasing pressure ambient, to the mixing between high and low entropy material or to the viscous dissipation of sound waves generated by the rising lobes, and generally is high enough to balance cooling losses and limit cooling flows in the cluster cores.

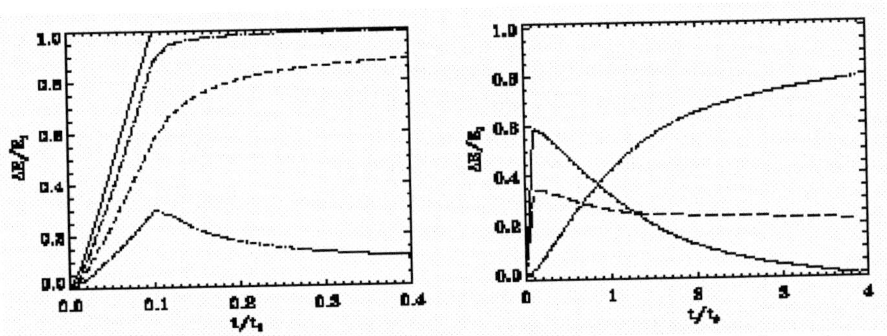

FIGURE 11. The temporal evolution of the variation of the ambient energy. *Left panel*: temporal behavior of the total ambient energy at the beginning of the cocoon evolution ($t/t_0 < 0.4$ where t_0 is the jet radius sound crossing time); plots of the total injected energy (*solid line*), total ambient energy (*dotted line*), total energy acquired by the shocked compressed material (*dashed line*) and by the entrained material (*dot-dashed line*). *Right panel*: plots of the ambient internal energy (*solid line*), the ambient kinetic energy (*long-dashed line*) and the ambient gravitational potential energy (*triple-dot-dashed line*).

The energy dissipated at the shock front has been evaluated looking at the entropy profiles of the shocked ambient medium: after the passage of the bow shock the entropy profile as a function of Lagrangian mass coordinates reaches a steady condition, confirming that the expanding bow shock is the main heating mechanism. The entropy of the ICM can be substantially raised as long as the energy associated with the expanding shock is greater than the energy of the unperturbed medium contained inside the shock radius: in fact only the higher power jets (5×10^{46} erg s^{-1}) can raise the entropy of the central core of clusters up to values > 125 $(h/0.7)^{-1/3}$ keV cm^2. For these sources the heating in the cluster cores around 1 keV per particle matches that required in cosmological simulations to satisfy observational constraints. Finally the impact of this heating on the L - T relation can be evaluated. First one must take into account that heated clusters tend to expand and cool down adiabatically: when they recover the gravitational equilibrium they have a less dense core and a temperature slightly higher than the initial one thus decreasing their X-ray luminosity. As it is shown in Fig. 12 the cases that are heated up to the observed values (Fig. 13) can expand to a new equilibrium to give X-ray luminosities which are consistent with the observed L - T relation.

AGN jets can explain the heating of the ICM. An extra heating of ≈ 1 keV per particle, is known to be sufficient for bringing the theoretical L_X - T relation in agreement with observations [50].We have shown how such a heating level can be provided by AGN jets having kinetic power in the range 10^{46} - 5×10^{46} erg s^{-1}. The situation of one single, powerful jet which heats the ICM to the observed level can be quite rare in reality. More probably, several, less powerful events contribute to the heating. We showed that the contribution of such events, while smaller, have a non negligible effect. Other non gravitational mechanism, like the "effervescent heating" or other interactions between AGN-originated bubbles or plumes of gas and the ICM, could also be simultaneously in action, and we regard them as alternative, and not exclusive.

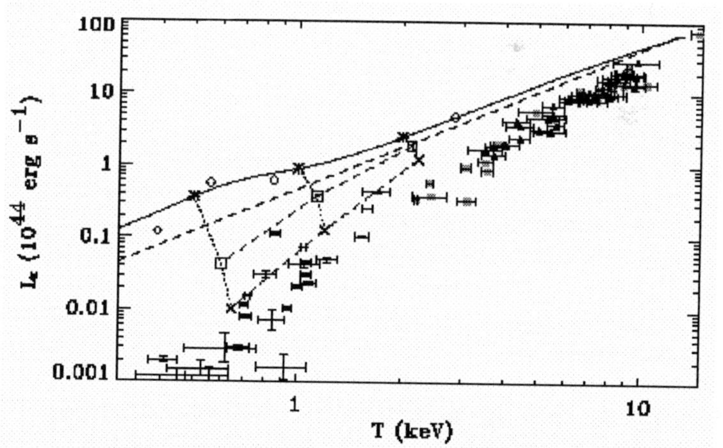

FIGURE 12. Bolometric luminosity-emission weighted temperature relation for simulations of jets with $Lj = 10^{46}$ erg s^{-1} (*empty squares*) and $L_j = 5 \times 10^{46}$ erg s^{-1} (*crosses*) in comparison with observational data (*pluses*: Helsdon & Ponman 2000; *filled squares*: Arnaud & Evrard 1999; *filled triangles* Markevitch 1999). Plotted are also the scaling of the bremsstrahlung luminosity (*dashed line*) and the X-ray luminosity (*solid line*) of simulation initial model. The bolometric luminosities of the three initial unperturbed clusters (*asterisks*) and of the four Tornatore et al. [49] simulations (*diamonds*) are shown for comparison.

Present calculations are based on the ICM physical characteristics *at a redshift* $z = 0$, while most probably a jet heating would occur at higher redshifts, in the environment given by proto-cluster structures. One could speculate that, depending on the actual number and power of heating events which happened in these proto-clusters, the final excess heating can significantly vary among different clusters, thus explaining not only the break in the self-similarity, but also the observed spread in the X-ray properties at a given temperature. For instance, from Fig. 12 one can see that, at a temperature $T \approx 0.7$ - 0.9 keV, the observed luminosity can vary by more than an order of magnitude. If the extra-heating of the clusters atmospheres is due to the AGN activity, and if such activity is somehow linked with the proto-cluster "cosmological" environment, the spread could be reminiscent of the formation history of the cluster. To put on firmer ground such speculations, the current analysis must be extended using several atmospheres of proto-cluster structures taken from *N*-Body/SPH simulations at high redshifts.

AN OUTLOOK TO THE FUTURE

The interrelated dynamics of accretion disks and jets is a fundamental element in different astrophysical phenomena ranging from star formation to active galactic nuclei. In this context several problems are still open and active subject of investigation. For accretion disk the transport of angular momentum is generally ascribed to the magneto-rotational instability, however its detailed development and efficiency are still under investigation; moreover there could be other mechanisms responsible for the angular momentum transport, particularly in cases when the magneto-rotational instability does not work, i.e. when magnetic fields are strong, almost at equipartition, or very weak. The acceleration and collimation of jets from accretion disks is another problem that is still far from being completely solved as it is the case for the propagation and stability of jets.

In order to investigate the physics of such objects numerical simulations have become an important tool. However if we compare the "typical" physical conditions that we find in astrophysical plasmas with the physical conditions that can be reproduced in numerical simulations, we realize that, at best, numerical simulations can only probe a very small portion of the astrophysical parameter space. One of the main problems is the large gap in Reynolds number between numerical simulations and the real astrophysical world. The vast range of spatial and temporal scales typical of astrophysical phenomena is well beyond the present capabilities of the most powerful supercomputers. Given this fact, we can then ask the question whether we can learn something useful from direct numerical simulations. Direct numerical simulations can answer questions about physical processes and phenomena in regimes characterized by dimensionless control parameters in a range that can be captured by the simulations. The study of nonlinear phenomena in these attainable parameter regimes, although they are outside the regimes encountered in astrophysics, can nevertheless teach our physical intuition for understanding and modeling the real astrophysical world. One important step in this process is the simulation code validation, as we want to answer the following questions: is the physical system being correctly described by our numerical code? what is the full domain of validity of the simulations? In this study a very important role can be played by laboratory experiments. Laboratory experiments can reproduce the regimes attainable by direct numerical simulation, whose results can then be directly compared with them. This comparison can answer to the questions outlined above and give confidence that our simulations capture correctly the important physical processes.

Moreover laboratory experiments can be, in general, relatively well-diagnosed allowing good comparisons with the results of direct numerical simulations. A third important ingredient in the process of modeling astrophysical phenomena through numerical simulations is the usage of sub-grid models, these models can allow to reach conditions closer to the astrophysical world however their validity has to be tested either against direct numerical simulations or against experimental results. In summary, it is the interplay between direct numerical simulations, laboratory experiments and sub-grid models that can give us tools for understanding astrophysical phenomena. Laboratory analogs of astrophysical systems and simulations of such analogs will allow us to validate our simulation codes and to build our physical intuition for such processes, even though their control parameters may be very different than their astrophysical counterparts. In addition sub-grid models, validated by the comparison both with direct numerical simulations and with laboratory experiments can allow the extend the validity of numerical simulations towards the regimes of the astrophysical world.

ACKNOWLEDGMENTS

I want to thank my collaborators Gianluigi Bodo, Andrea Mignone, Paola Rossi, Claudio Zanni and Elena Beltritti for discussions. I praise the Scientific Committee of the Hellenic Astronomical Society for organizing a very interesting meeting in a lively and fascinating environment. This work was supported by the FLASH Center at the University of Chicago and the Italian Ministry of University and Research at the University of Torino.

REFERENCES

1. K.I. Kellermann, D.L. Jauncey, M.H. Cohen, et al., *ApJ*, **169**, 1 (1971)
2. R.D. Blandford & M.J. Rees, *MNRAS* 169, 395 (1974)
3. C.J. Lada, *ARA&A*, **23**, 267 (1985)
4. I.F Mirabel & L.F. Rodriguez, *Nature*, **371**, 46 (1994)
5. A. Ferrari, *Astrophysics and Space Science* **293:** 15 (2004)
6. N.I. Shakura & R.A. Sunyaev, A&A **24**, 337 (1973)
7. G. Bodo, G. Murante, P. Rossi, G. Chagelishvili, A. Tevzadze, & A. Ferrari, *MHD Couette Flows: Experiments and Models*, Proceedings of the conference held 29 February-2 March 2004 in Acitrezza, Catania (Italy). Edited by Robert Rosner, Günther Rüdiger, and Alfio Bonanno. New York: American Institute of Physics, Vol. 733, 2004., p.177-190 (2004)

8. S.A. Balbus & J.F. Hawley, Ap. J. **392**, 662 (1992)
9. B. Coppi & E.A. Keyes, *ApJ*, **595**, 1000 (2003)
10. J.F. Hawley, S.A. Balbus, J.M. Stone, *ApJ*, **554**, L49 (2001)
11. G. Bodo, G. Chagelishvili, G. Murante, A. Tevzadze, P. Rossi, and A. Ferrari, *A&A* **437**, 9 (2005)
12. J.P. De Villiers, &J.F. Hawley, *ApJ*, **592**,1060 (2003)
13. J.F. Hawley, *ApJ*, **554**, 534 (2001)
14. H. Ji, J. Goodman, A. Kageyama, M. Burin, E. Schartman, & W. Liu, *MHD Couette Flows: Experiments and Models*, Proceedings of the conference held 29 February-2 March 2004 in Acitrezza, Catania (Italy). Edited by Robert Rosner, Günther Rüdiger, and Alfio Bonanno. New York: American Institute of Physics, Vol. 733, 2004., p.21-34 (2004)
15. A. Ferrari, E.Trussoni, R. Rosner & K. Tsinganos, *ApJ*, **294**, 397 (1985)
16. R.D.Blandford & D. Payne, *MNRAS* **199**, 883 (1982)
17. Z-Y. Li, T. Chiueh & M.C. Begelman, *ApJ*, **394**, 459 (1992)
18. G.I. Ogilvie & M. Livio, *ApJ,* **553**, 158 (2001)
19. F. Casse & J. Ferreira, *A&A* **361**, 1178 (2000)
20. K. Tsynganos & E. Trussoni, *A&A*, **231**, 270 (1990)
21. Y. Uchida & K. Shibata, *PASJ* **37**, 515 (1985)
22. T. Kuwabara, K. Shibata, T. Kudoh, & R. Matsumoto, *PASJ*, **52**, 1109 (2000)
23. F. Casse & R. Keppens, *ApJ* **581**, 988 (2002)
24. C. Zanni, A. Ferrari, S. Massaglia, G. Bodo & P. Rossi, *Mem. SAIt,* **76**, 372 (2005)
25. R.V.E. Lovelace, *Nature*, **262**, 649 (1976)
26. R.D. Blandford, *MNRAS*, **176**, 465 (1976)
27. G. V. Ustyugova, R. V. E. Lovelace, M. M. Romanova, H. Li & S. A. Colgate, *ApJ*, **541**, L21 (2000)
28. A. Ferrari, *ARA&A,* **36**, 539 (1998)
29. M.L. Norman, K.H. Winkler & L. Smarr, *Astrophysical Jets*; Proceedings of the International Workshop, Turin, Italy, October 7-9, 1982 (A84-22076 08-90). Eds. A. Ferrari & A.G. Pacholczyk, Dordrecht, D. Reidel Publishing Co., 1983, p. 227-250 (1983)
30. P. Rossi, G. Bodo, S. Massaglia, A. Ferrari, A. Mignone, et al., *A&SS*, **293**, 149 (2004)
31. M. Belan, S. de Ponte, S. Massaglia, & D. Tordella, *A&SS*, **293**, 225 (2004)
32. C.L. Carilli, R.A. Perley & D.E. Harris, *MNRAS*, **270**, 173 (1994)
33. D.A. Smith, A.S. Wilson, S. Andrew et al., *ApJ,* **565**, 195 (2002)
34. H. Böhringer, W. Voges, A.C. Fabian, A.C. Edge & D.M. Neumann, *MNRAS*, **264**, L25 (1993)
35. A.C. Fabian, J.S. Sanders, S. Ettori et al., *MNRAS* **318**, L65 (2000)
36. A.C. Fabian, J.S. Sanders, C.S. Crawford, & S. Ettori, MNRAS, 341, 729 (2003)
37. P.E.J. Nulsen, L.P. David, B.R. McNamara et al., *ApJ* **568**, 163 (2002)
38. E.L. Blanton, C.L. Sarazin, B.R. McNamara & M.W. Wise, *ApJ,* **558**, L15 (2001)
39. A. Finoguenov & C. Jones, *ApJ* **547**, L107 (2001)
40. M.C. Begelman & D.F. Cioffi, *ApJ* **345,** L21 (1989)
41. C. Zanni, G. Bodo, P. Rossi, S. Massaglia, A. Durbala & A. Ferrari, *A&A,* **402**, 949 (2003)
42. A. Capetti, S. Zamfir, P. Rossi, G. Bodo, C. Zanni, S. Massaglia, *A&A*, **394**, 39 (2002)
43. F.B.S. Paerels & S.M. Kahn, *ARA&A*, **41**, 291 (2003)
44. D.A. White, C. Jones & W. Forman, *MNRAS,* **292**, 419 (1997)
45. A.J.R. Sanderson, T.J. Ponman, A. Finoguenov, E.J. Lloyd-Davies & M. Markevitch, *MNRAS,* **340**, 989 (2003)
46. T.J. Ponman, A.J.R. Sanderson & A. Finoguenov, *MNRAS*, **343**, 331 (2003)
47. N. Menci & A. Cavaliere, *MNRAS*, **311**, 50 (2000)
48. A. Babul, M.L. Balogh, G.F. Lewis & G.B. Poole, *MNRAS*, **330**, 329 (2002)
49. L. Tornatore, S. Borgani, V. Springel, F. Matteucci, N. Menci, & G. Murante, *MNRAS*, **342**, 1025 (2003)
50. S. Borgani, G. Murante, V. Springel, et al., *MNRAS*, **348**, 1078 (2004)
51. A. Cavaliere, A. Lapi & N. Menci, *ApJ*, **581**, L1 (2002)
52. C. Zanni, G.Murante, G. Bodo, S. Massaglia, P. Rossi & A. Ferrari, *A&A,* **429**, 399 (2005)

X-ray Spectral Evolution of AGN

G. Chartas

Department of Astronomy & Astrophysics, Penn State University, University Park, PA 16802

Abstract. We review recent studies of X-ray spectral evolution of active galactic nuclei (AGN). One of the goals of these studies is to search for changes of the X-ray spectral properties of AGN that might have accompanied the significant evolution of the luminosity functions and the space densities of AGN recently found in optical and X-ray surveys of these objects. Determining the evolution of the spectra of AGN is commonly carried out by searching for correlations between the X-ray power-law photon indices, the optical-to-X-ray spectral slopes, and the accretion efficiencies of AGN with redshift. No significant changes in the X-ray emission properties of AGN are detected in most current studies, however, a recent survey of $z > 1.5$ lensed AGN indicates a possible change of accretion efficiency with redshift. Several observations of high redshift broad absorption line quasars and narrow line type-I quasars indicate the presence of massive relativistic outflows. It is proposed that these winds in combination with outflows produced by supernovae might be important in driving the evolution of the space density of AGN.

Keywords: <galaxies: active, galaxies: quasars, X-rays:general >
PACS: < PACS Category 90>

INTRODUCTION

Observations of luminous quasars [1] point towards a dramatic change of their comoving number density by a factor of about 20 between redshifts of 1 and 2. Proposed explanations of this change of quasar number density include the evolution of the environments of quasars which might impact the X-ray emission region, the evolution of accretion rates such that quasars at higher redshift are accreting close to or above their Eddington rates [2], quasar and supernovae outflows that can heat up and unbind the ISM gas from the host galaxy thus halting the growth of the black hole [3], and quasar and supernovae outflows on larger scales that eventually heat the IGM gas above a critical entropy level making cooling impossible, and inhibiting their own formation below $z \sim 2$ [4].

Several of the main questions pursued in current studies of AGN are : (a) what is the physics and structure of AGN? (b) how do the properties of the black hole, the accretion disk, the spectral energy distribution (SED), the surrounding environment and the AGN outflow evolve? (c) how do supermassive black holes form? (d) are AGN winds important in regulating structure formation and their own growth? This review will focus mainly on addressing questions (b) and (d) through studies of the X-ray spectral evolution of AGN. X-rays detected from AGN are thought to originate near the black hole region and the X-ray luminosity of an AGN is expected to change with accretion rate. Evolution in the environment or accretion rate of AGN with redshift may manifest itself as an observable change in the X-ray emission properties of AGN with redshift. Such an evolution has been observed in NLS1 galaxies [5]. Several surveys of high redshift quasars have been performed recently aimed at detecting such changes. The two main parameters that are commonly constrained by these surveys and are used

CP848, *Recent Advances in Astronomy and Astrophysics,*
7th *International Conference of the Hellenic Astronomical Society,* edited by N. H. Solomos

to characterize the emission properties of quasars are: (a) The photon index Γ of the power-law of the form $N(E) = N_0(E/E_0)^{-\Gamma}$ that is used to model the observed X-ray continuum. Extensive studies of AGN during the past decade indicate that this power-law component is produced by inverse Compton scattering of soft photons from the accretion disk by hot electrons in a corona [6], [7]. (b) the optical-to-X-ray spectral slope, quantified as $\alpha_{ox} = \log(f_{2\,\mathrm{keV}}/f_{2500\,\mathrm{\AA}})/\log(\nu_{2\mathrm{keV}}/\nu_{2500\mathrm{\AA}})$ [8], where f_{2keV} and $f_{2500\mathrm{\AA}}$ are the flux densities at 2 keV and 2500 Å in the quasar rest-frame respectively.

X-ray surveys have benefited greatly with the advent of the *Chandra* and the *XMM-Newton* observatories. *Chandra* with its sub-arcsec imaging capability provides accurate source-positions that are important for cross-correlating with surveys in other wavelengths. The deepest 2 Ms *Chandra* survey to date reached a 0.5-2 keV flux limit of about 2.5×10^{-17} erg cm^{-2} s^{-1}. *XMM-Newton* has a larger collecting area than *Chandra* resulting in deeper surveys at high energies ($\gtrsim$ 5 keV) and higher signal-to-noise ratio (S/N) spectra of resolved sources. We note that even with the larger effective areas of the current X-ray observatories there are several difficulties in accurately characterizing the properties of distant AGN and comparing them to those of nearby ones such as : (a) Most X-ray surveys of high redshift AGN are comprised mostly of the brightest objects resulting in a possible bias in these samples. (b) In many cases the available spectra have relatively low-to-moderate S/N. This problem occasionally leads to an over-simplification of the models used in fitting the X-ray spectra of these objects. (c) Present studies of AGN X-ray evolution sample objects over relatively large redshift ranges. The X-ray spectra of the AGN in these studies are usually fit in the observed-frame assuming that AGN spectra follow simple power laws. Due to possible departures of the AGN spectrum from a pure power law and the effect of cosmological redshift, comparison of AGN properties from analyses performed in the same observed-frame may lead to erroneous results. (d) A large fraction of heavily obscured AGN may be missed in current AGN surveys. For intrinsic column densities of $N_H \gtrsim 1.5 \times 10^{24}$ cm^{-2} an absorber becomes "Compton-thick" to X-ray radiation and most direct X-rays do not reach the observer due to Compton scattering and absorption. Compton-thick AGN can be detected indirectly from X-rays that are reflected or scattered from material in the vicinity of the AGN. Deeper observations with *Chandra* may be able to detect the indirect radiation from "Compton-thick" AGN and constrain better the completeness of X-ray AGN selection [9].

In §2 we briefly review recent optical and X-ray surveys that find a strong evolution of the luminosity function of AGN. In §3 and §4 we describe studies of X-ray spectral evolution of AGN that search for evolution of the optical-to-X-ray spectral slopes and X-ray photon indices of AGN, respectively. Preliminary results from a survey of gravitationally lensed quasars is presented in §4. Finally in §5 we discuss the importance of quasar outflows in driving quasar evolution.

Throughout this paper we adopt a Λ-dominated cosmology with $H_0 = 70$ km s^{-1} Mpc^{-1}, $\Omega_\Lambda = 0.7$, and $\Omega_M = 0.3$.

EVOLUTION OF THE LUMINOSITY FUNCTION OF AGN

Optical studies first showed that quasars evolved strongly with redshift [10], [11], [12], [13], [14]. The available data at the time could not distinguish between various proposed parameterizations for quasar evolution. For example, the early study by Schmidt [11], proposed pure density evolution, the work by Mathez [12], pointed towards pure luminosity evolution(PLE), an analysis of the Palomar bright quasar survey by Schmidt & Green [13], indicated luminosity-dependent density evolution(LDDE) and Boyle [14], reported pure luminosity evolution for $z < 2.2$ quasars. Current observations in the soft and hard X-ray bands [15], [16], [17] prefer luminosity dependent density evolution (i.e., the density evolution of quasars is luminosity dependent as indicated by the space density vs. z and luminosity studies).

Hasinger et al. [16], used a sample that contained *ROSAT*, *XMM-Newton* and *Chandra* observations of AGN. This study focused on the soft band (0.2-2 keV) in order to incorporate the soft *ROSAT* data and considered unabsorbed type-I AGN. Hereafter, the term unabsorbed will mean no detectable absorption in the X-ray band and the term type-I AGN will refer to AGN with broad Balmer emission lines with a small Balmer decrement. The analysis of the soft X-ray sample indicates that pure luminosity dependent evolution models fail completely at low luminosities. In Figure 1 we show the space density of AGN as a function of redshift in different luminosity classes as presented in [16]. The derived space density evolution from these studies are consistent with an anti-hierarchical black hole growth (also referred to as downsizing). The larger black holes formed first (the space density of quasars peaked first around $z \sim 2$) whereas the smaller black holes form more efficiently latter $z \sim 1$. Anti-hierarchical black hole growth is not predicted in most semi-analytic models based on Cold Dark Matter structure formation models.

In a related study Barger et al. [18], find that the X-ray luminosity function of the AGN in the deep and wide-field *Chandra* surveys also show strong evolution. Barger et al. [18], argue that the high luminosity quasars switch off at redshift of about 1 and that the large drop in luminosity below $z \sim 1$ is not due to an evolution in the accretion rate. Most of the detected AGN at low redshifts $z < 1$ have considerably smaller masses than AGN at $z > 1$.

Because of the known correlation between black hole mass and velocity dispersion of the stars in the host galaxy and the mass of the spheroidal component [19], [20], [21], [22], it is also possible to study the evolution of AGN by searching for changes of the properties of the host galaxies with redshift. Barger et al. [18], used this approach and studied the correlation of the luminosity of the host galaxy with redshift and luminosity of the central AGN. These studies showed that, (a) most luminous AGN reside in the most luminous galaxies as expected from the well known relation between bulge luminosity and black hole mass, (b) the luminosities of the host galaxies within a narrow range of X-ray luminosities of the central AGN do not vary much over redshift, and (c) the luminous host galaxies present at redshifts greater than $z \sim 1$ are not present at low redshift. Barger et al. [18], suggest that these findings are consistent with downsizing of the AGN population and are not a result of evolution in the AGN accretion rate.

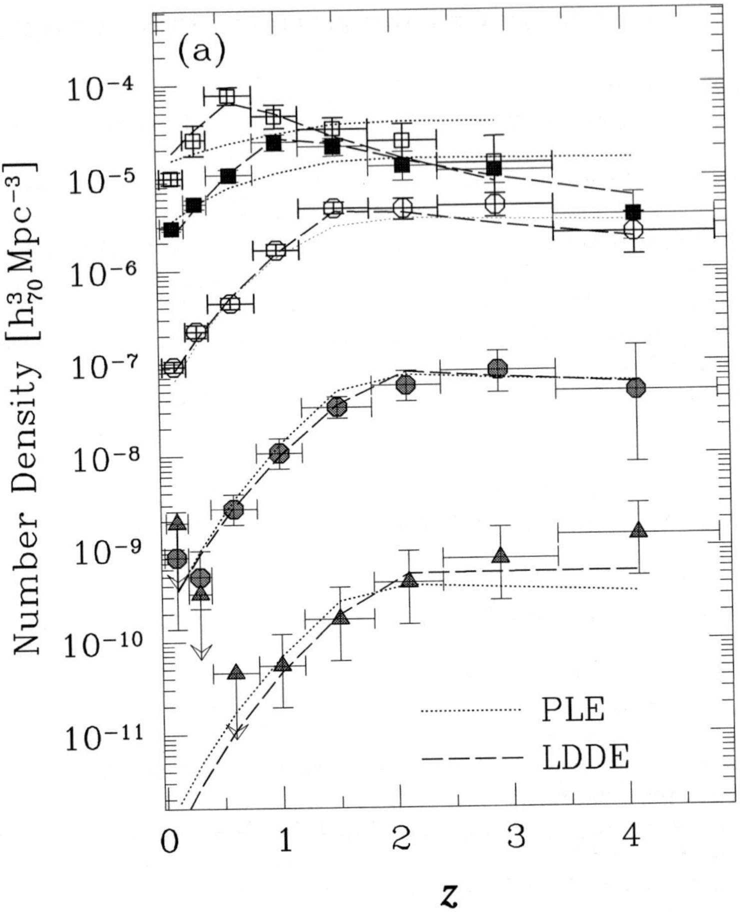

FIGURE 1. The space density of AGNs as a function of redshift in different luminosity classes and the sum over all luminosities with $\log L_X \geq 42$. Densities from the PLE and LDDE models are overplotted with solid lines. Figure courtesy of Hasinger et al. [16].

OPTICAL-TO-X-RAY SPECTRAL SLOPES OF AGN

The discovery of a luminosity dependent density evolution of the luminosity function of AGN raises several important questions; What causes the dramatic turning point in the space density of quasars near z $\sim$ 2? Would we expect any detectable changes in the accretion process with z for AGN especially near the turning point of their comoving number density. Is there any detectable increase in variability in the X-ray spectra of quasars near z $\sim$ 2 as the fueling supply is "choked off"? What is the effect of quasar outflows on quasar evolution and can they explain the different evolution of low and high-luminosity quasars?

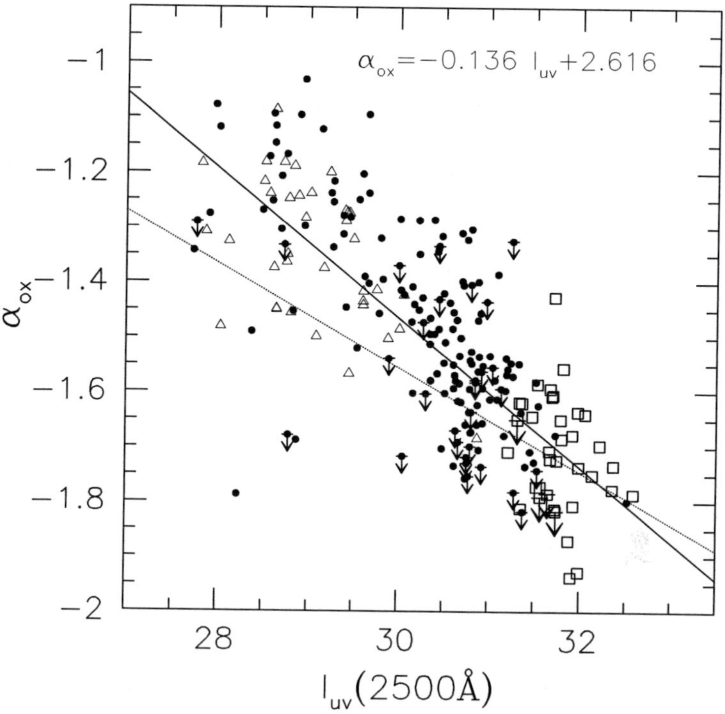

FIGURE 2. $\alpha_{\rm ox}$ dependence on the 2500 Å monochromatic luminosity. The main sample of AGN selected from the Sloan Digital Sky Survey is given by filled circles, the high-*z* sample by open squares, and the Sy 1 sample by open triangles; arrows in the left panel indicate 2 keV upper limits. The solid line is the linear relation found for the combined sample, and the dotted line is the corresponding relation from Vignali et al. [25], their eq. [4]). The anticorrelation is significant at the 10.6 σ level for the combined sample. Figure courtesy of Strateva et al. [23].

Several attempts have been made since the 80's to measure the optical-to-X-ray spectral slope and determine if there is any dependence of $\alpha_{\rm ox}$ with UV luminosity and/or redshift. A recent thorough study was reported by Strateva et al. [23].The Strateva et al. [23], sample contained 228 optically selected AGN in the 0.01–6.3 redshift range. 155 of these AGN were selected from the Sloan Digital Sky Survey (SDSS) in an unbiased way, and an additional 36 high redshift and 37 low redshift AGN were added to the SDSS sample. The main conclusions of the Strateva et al. [23], study were; (a) Partial correlation analysis confirms previous results [24] that α_{ox} is correlated to the UV luminosity through the relation $\alpha_{\rm ox} = -0.136 L_{\rm UV} + 2.616$, where, $L_{\rm UV}$ = $\log[L_\nu(2500A)$ erg s^{-1} Hz^{-1})]. The reason for this correlation is still not well understood. The significant correlation (at the 10.6 σ level) between α_{ox} and the 2500 Å monochromatic luminosity derived in the recent study by Strateva et al. [23] is shown in Figure 2, (b) Once the dependence of $\alpha_{\rm ox}$ with $L_{\rm UV}$ is taken into account no significant dependence of $\alpha_{\rm ox}$ with redshift is detected. To remove the UV luminosity dependence of $\alpha_{\rm ox}$ the parameter $\Delta\alpha_{ox}$ = α_{ox} - $\alpha_{ox}(L_{2500})$ is often calculated, where $\alpha_{ox}(L_{2500})$ is

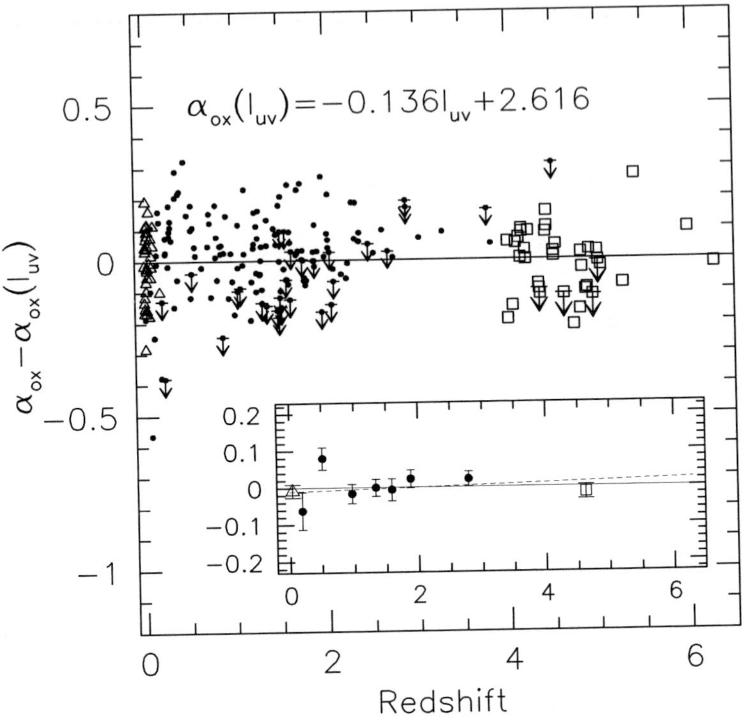

FIGURE 3. α_{ox} - $\alpha_{ox}(l_{UV})$ dependence on redshift. Single-parameter fit residuals for an α_{ox} dependence on l_{UV}. Symbols are as in Fig. 2. The small inset plot gives the Kaplan-Meier estimators of the means of the α_{ox} - $\alpha_{ox}(l_{UV})$ residuals. The dashed line in the inset plot corresponds to the weighted linear regression fit, $\langle\alpha_{ox}-\alpha_{ox}(l_{UV})\rangle=(0.005\pm0.012)z+(-0.010\pm0.023)$. Figure courtesy of Strateva et al. [23].

the expected α_{ox} for the monochromatic luminosity at 2500 Å. In Figure 3 we show the dependence of $\Delta\alpha_{ox}$ on redshift [23].No significant change of $\Delta\alpha_{ox}$ is detected for AGN with $z \lesssim 6$.

Strateva et al. indicated the following items that need to be taken into account for their study to yield reliable results :

- The sample needs to cover a large range in luminosity and redshift, in order to reveal possible correlations of α_{ox} with luminosity and z.
- Radio-loud objects should be removed from the sample. Part of the X-ray emission of radio-loud AGN may be dominated from synchrotron emission originating from a jet.
- Absorbed objects should be excluded from the sample.
- Appropriate statistical tools need to be implemented to search for correlations.
- Possible contamination from the host galaxy needs to be subtracted.
- Care must be taken in using data from different mission and processed with different reduction techniques.

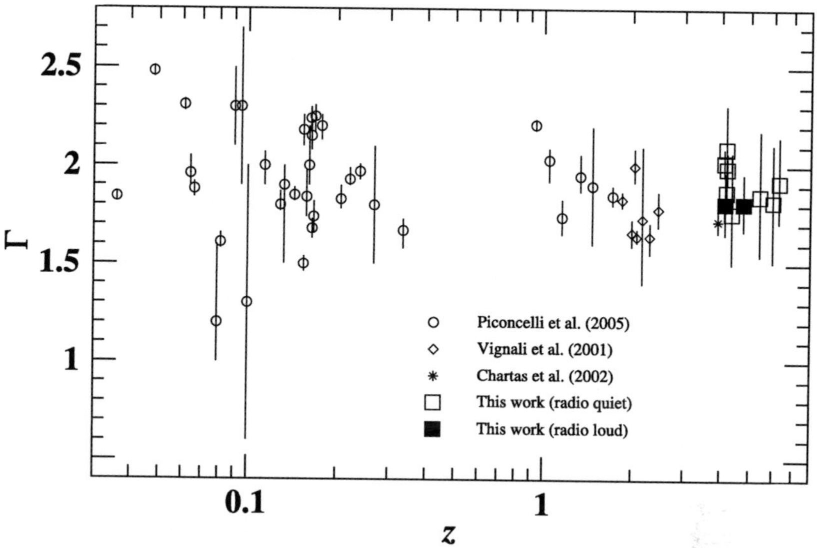

FIGURE 4. X-ray photon index vs. redshift, adapted from Shemmer et al. [29]. The eight (two) radio-quiet (moderately radio-loud) AGNs presented are marked with open (filled) squares. Note that the photon index does not show any clear dependence on redshift.

X-RAY PHOTON INDICES OF QUASARS

One approach of constraining possible evolution of quasars is to compare the X-ray spectra of the most distant ones to relatively nearby ones. One of the obvious difficulties of this approach is that most of the available spectra of high redshift quasars to date are of poor-to-moderate quality. Possible underlying complexity in the spectra of poor S/N spectra cannot be resolved and the inferred values of the spectral parameters may be incorrect. Several studies rely on either stacking many spectra of high redshift quasars to obtain moderate S/N spectra or performing deep X-ray observations of a few quasars. Conflicting results have appeared in the literature regarding the evolution of the power-law photon indices of quasars. A *Chandra* survey by Bechtold et al. [26] of 16 RQQs with redshifts between 3.70 and 6.28 finds that their mean power-law photon index is flatter than the average Γ of nearby quasars observed with ROSAT. Recently, Grupe et al. [27], reported a rather steep mean power-law photon index of $\Gamma = 2.21 \pm 0.52$ for a sample of 10 high-redshift RQQs. Vignali et al. [28], performed a joint fit to the *Chandra* spectra of 48 radio-quiet quasar in the range of z = 4–6.28 and found that the high redshift spectra of quasars look very similar and have similar X-ray power-law photon indices to those of lower redshift ones. They conclude that quasars feed and grow in the same way over the history of the Universe. Shemmer et al. [29], performed a detailed investigation of moderate quality X-ray spectra of 10 quasars in the redshift range of 4 – 6.28. Shemmer et al. [29], do not find any significant difference between the spectra of these high redshift quasars compared to ones at lower redshifts confirming the results of Vignali et al. [28]. In Figure 4 we show the observed power-law photon indices of $z > 4$

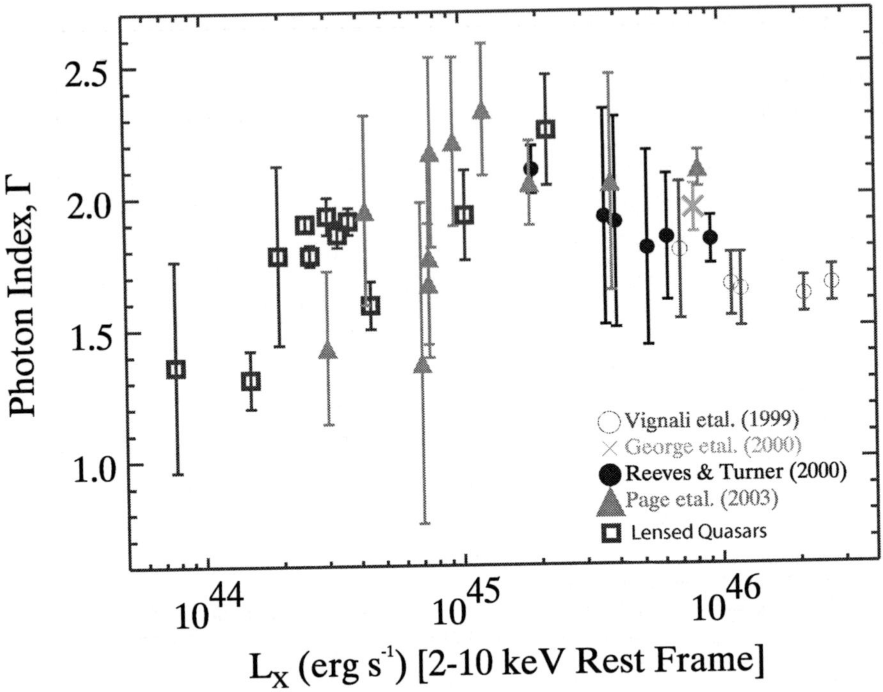

FIGURE 5. Γ–L_X diagram from our recent analyses of high redshift ($z > 1.5$) radio-quiet quasars. Adapted from Dai et al. [32].

quasars as a function redshift found by Shemmer et al. [29]. Even though significant scatter is found in Γ no systematic trend of Γ with z is detected. Recently, Mateos et al. [30], performed a spectral analysis of 123 AGN detected in the Lockman Hole with *XMM-Newton*. This sample has flux limits of 6×10^{-16} erg cm^{-2} s^{-1} in the 0.5 –2 and 2–10 keV bands. Mateos et al. [30], find no significant dependence of photon indices and intrinsic absorption with either redshift or luminosity. The cause of the significant scatter in the Γ vs. redshift and Γ vs. X-ray luminosity plots is not well understood. A possible contributor to this scatter is the evolution of the accretion efficiency of quasars.

GRAVITATIONAL LENSING AS A TOOL TO STUDY QUASAR EVOLUTION

A recent mini-survey of relatively high redshift $(1.5 < z < 4)$ gravitationally lensed radio-quiet quasars (RQQs) obtained with *Chandra* and *XMM-Newton* [31], [32] indicated a possible correlation between the X-ray power-law photon index and X-ray luminosity of the radio-quiet quasars of the sample. The power-law photon index becomes steeper as the X-ray luminosity increases (see Figure 5). Such a correlation is not found in nearby $z < 0.1$ quasars suggesting that quasars at redshifts near the peak of

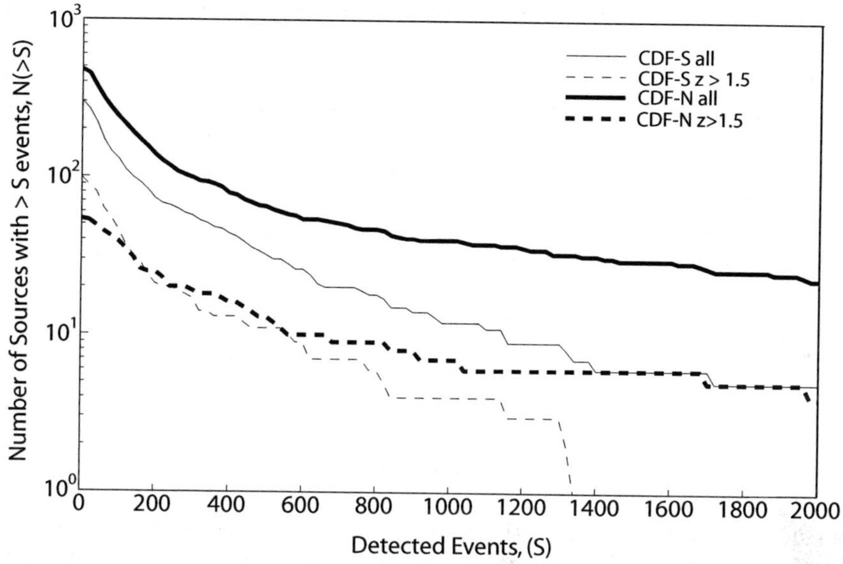

FIGURE 6. Number of sources with more than S detected photons versus the number of detected photons for the CDF-N (solid thick line), CDF-S (solid thin line), the CDF-N with $z > 1.5$ (dashed thick line) and CDF-S with $z > 1.5$ (dashed thin line).

their number density may have different accretion properties than low redshift quasars. The lensing magnification effect allows a search for changes in quasar spectroscopic and variability properties with redshift over three orders of magnitude in intrinsic X-ray luminosity.

The redshift dependence of the Γ–L_X correlation suggests that quasars at redshifts near the peak of their number density may have different accretion properties than low redshift quasars. In a recent study Dai et al. [32], provided interpretations to explain the Γ–L_X correlation in high-z quasars in the context of the hot corona model [6], [7].

The main interpretation was that the range of accreting black hole masses at high redshift is large (spanning two orders of magnitude), while the accretion rate relative to the Eddington rate is fairly constant and close to unity. In the context of this hypothesis and under the assumption that the optical depth of the corona is dominated by electron-positron pairs, Haardt, Maraschi, & Ghisellini [33], predict a correlation between the 2–10 keV photon index and luminosity just like the one reported [31], [32]. Moreover, as the optical depth reaches unity, the photon index reaches its maximum value and begins to decline for larger optical depths. This may explain the turnover observed in the Γ–L_X plot of Figure 5. The redshift dependence of the possible correlation implies that quasars near the peak of their comoving number density are accreting near Eddington and have different accretion properties than their low-z counterparts.

There is concern that the limited number of quasars in the present sample of lensed quasars, combined with the medium S/N of several of the observations and the relatively large fraction of BAL quasars, may have led to an unaccounted systematic effect. The quasars in our present sample with $L_X < 2.6 \times 10^{44}$erg s^{-1} are mini-BAL or BAL quasars. Unfortunately the number of available lensed RQQs suitable for this study is limited to a total of about 25 sources out of which the brightest 11 have been observed, analyzed and presented in Figure 5. In order to increase the size of our sample we are currently including high redshift quasars with moderate-to-high S/N spectra available from archived *Chandra* deep-field (CDF) surveys. Using the published data of [34], [35], [36], [37], we show in Figure 6 the number of X-ray sources with more than S detected photons as a function of the number of detected photons for the CDF-N and CDF-S. Based on the published X-ray classifications almost all sources in the CDF-N and CDF-S surveys with $z > 1.5$ and containing more than 400 counts are AGN. We find that the CDF-N and CDF-S surveys contain approximately 35 AGN with more than 400 counts and with $z > 1.5$.

THE IMPORTANCE OF QUASAR WINDS IN DRIVING QUASAR EVOLUTION

Recent studies indicate that the mass of the black holes are related to the velocity dispersion of the stars in their host and also related to the mass of the central bulge of the galaxy. These findings suggest some link between the growth of the black holes and their host galaxies. Several theoretical studies have proposed a connection between black hole and bulge growth in the host galaxy [38]. More recently, the potential importance of quasar outflows has been explicitly demonstrated in theoretical models of structure formation and galaxy mergers that incorporate the effects of quasar outflows [4], [3], [39]. Scannapieco & Oh [4], find that by choosing the fraction of the total bolometric energy released over a quasar's lifetime into the Interstellar Medium (ISM) and Intergalactic Medium (IGM) in the form of kinetic energy to be ε_k=0.05, they can successfully model the observed evolution of the B-band quasar luminosity function between redshifts of 0.25 and 6.25 (see Figure 7). Springel, Di Matteo, & Hernquist [39], simulated the growth of black holes in gas-rich galaxies with and without the presence of accretion feedback. The feedback in their model is thought to occur through energetic quasar outflows that interact with the gas of the host galaxy. It is assumed that about 5% of the radiated luminosity is thermally coupled via these outflows to the surrounding gas. They find that the growth of the black-hole mass is self-regulated and eventually saturates at a final value that depends on the initial amount of gas available for accretion.

With the advent of *XMM-Newton* and *Chandra* it has become possible to infer the kinematic and ionization properties of highly ionized X-ray absorbers in Seyfert 1 galaxies [40], [41], [42]. Approximately 60% of Seyfert 1 galaxies show outflowing UV and X-ray absorption by ionized gas with velocities up to $\approx$ 2,500 km s^{-1}. The absorbing outflow properties of more luminous quasars appear to differ in several ways from those of Seyfert 1s. The fraction of occurrence of UV and X-ray absorption has been reported to be lower in quasars [43], [44], however, these claims for a lower occurrence of X-ray absorption have been recently challenged by Porquet et al. [45], who find that about half of their sample of 21 low-redshift PG quasars harbor ionized absorbers. UV spectroscopic observations indicate that about 20% of quasars show broad absorption lines blueward of their resonant UV emission lines [46]. Outflow velocities of the UV

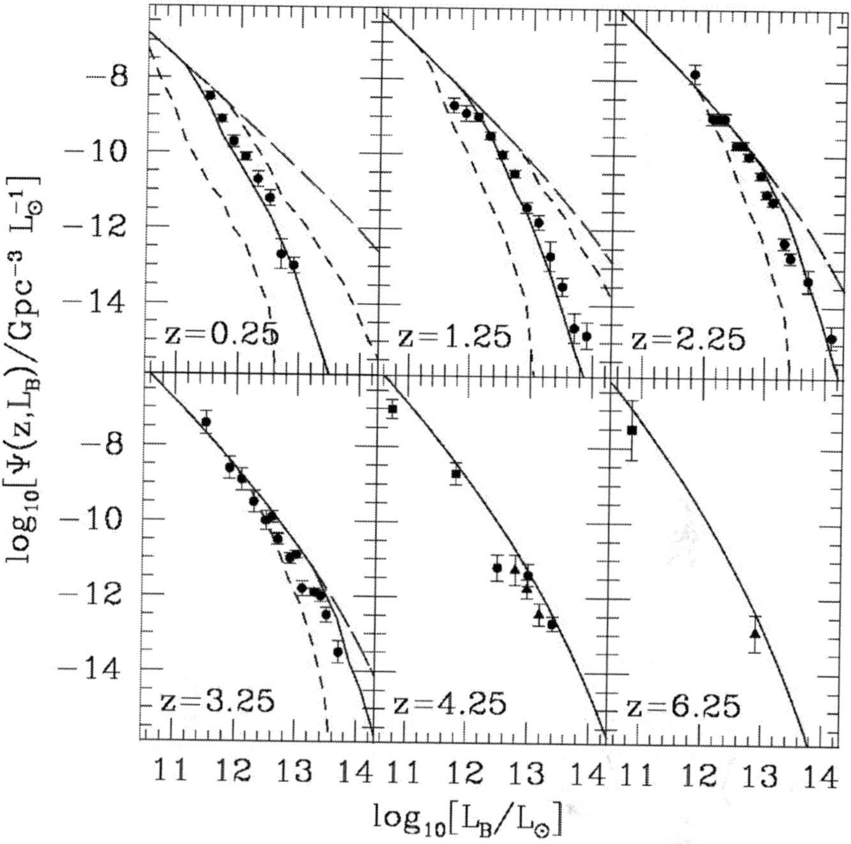

FIGURE 7. Evolution of *B*-band quasar luminosity function. Long dashed lines are from the WL03 model. Solid line is from Scannapieco & Oh [4], with $\varepsilon_k = 0.05$. The upper and lower short dashed lines are for $\varepsilon_k = 0.025$ and 0.1, respectively. Figure courtesy of Scannapieco & Oh [4].

absorbers in BAL quasars have been found to be as high as $\approx$ 60,000 km s^{-1}.

The importance of quasar winds driving quasar evolution will in part depend on the frequency of these outflows in quasars and the rate of kinetic energy injected into the IGM. The commonly accepted model suggested by Weymann et al. [47], posits that winds exist in most quasars; however because of the relatively small opening angles of the outflows only a small fraction of quasars have detectable BAL features.

Recent X-ray spectroscopic observations of BAL quasars have provided estimates of quasar outflow efficiencies. Specifically, X-ray observations of two BAL quasars [48], [49], have suggested the presence of relativistic outflows of X-ray absorbing material with velocities of up to ~0.4*c* [48], [49]. Figure 8 shows the $\Delta\chi$ residuals between a best-fit simple absorbed power-law model and the *XMM-Newton* PN and *Chandra* ACIS spectra of APM 08279+5255 , respectively. The detection of significant variability of the widths and energies of the high-energy absorption lines in APM 08279+5255 confirms

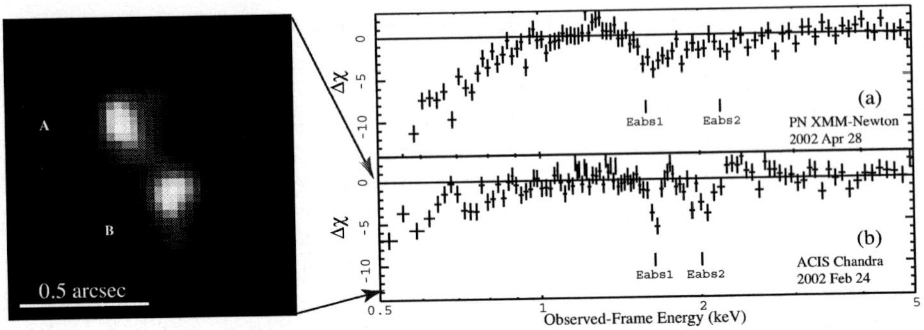

FIGURE 8. (left) : Deconvolved image of the 2002 Feb 24 89 ks *Chandra* observation of APM 08279+5255. (right): $\Delta\chi$ residuals between the best-fit Galactic absorption and power-law model and (a) the *XMM-Newton* pn spectrum and (b) the *Chandra* ACIS spectrum of APM 08279+5255.

the intrinsic origin of the high-energy absorption features in this BALQSO. The inferred hydrogen column densities ranging between 10^{22-23} cm^{-2} and relativistic velocities of these outflowing X-ray absorbers imply mass-outflow rates that are comparable to the estimated accretion rates of a few $M_\odot$ yr^{-1}. In a recent analysis Chartas et al. [50], constrained the fraction of the total bolometric energy released by quasars PG 1115+080 and APM 08279+5255 into the IGM in the form of kinetic energy to be $\varepsilon_k = 0.3^{+0.1}_{-0.2}$, and $\varepsilon_k = 0.09^{+0.03}_{-0.07}$ respectively. According to recent theoretical studies this range of efficiencies is large enough to influence significantly the formation of the host galaxy and regulate the growth of the central black hole. Additional observational evidence to support the presence of quasar feedback came with the detection of high-velocity blueshifted absorption-line features in the X-ray spectra of several quasars and Narrow-Line Seyfert 1 galaxies [51], [52], [53]. We note, however, that some of these claims have been disputed in recent re-analyses of the data [54], [55], [56].

The fraction of BAL quasars with massive relativistic outflows is not well constrained from available X-ray observations because of the relatively poor-to-moderate S/N X-ray spectra of BAL quasars observed to date. The two X-ray brightest BAL quasars APM 08279+5255 and PG 1115+080 show significant highly ionized and blueshifted Fe lines. The third brightest BAL quasar known to date CSO 755 shows a hint of an Fe absorption feature in the rest-frame energy range 7.1–9.3 keV [29]. Higher quality observations are required to confirm this result. The X-ray spectra of the remaining observed BAL quasars are of lower S/N and do not provide interesting constraints on the presence of high-energy Fe absorption features.

FUTURE PROSPECTS

To make significant advances in understanding the X-ray spectral evolution of AGN we will need to wait for upcoming X-ray missions such as *Constellation-X* and *XEUS*. These missions will provide: (a) A significant increase in the number of high-quality AGN spectra at high redshifts. The increase in the sample size will allow us to test for changes in AGN properties in narrower redshift bands and thus better constrain the epoch at which possible changes in the average emission properties of AGN occurred. (b) allow fits to the X-ray continuum spectra with more realistic models than the present simple power-law models used in most spectral analyses. (c) high resolution spectroscopy will be possible for quasars at high redshift as well. This will allow us to search for evolution in the physical and kinematic properties of the AGN outflows.

ACKNOWLEDGMENTS

This work was supported in part by NASA grants NNG04GK83G and NAG5-13543.

REFERENCES

1. Fan, X., Strauss, M. A., & Schneider, D. P., et al. 2001, AJ, 121, 54.
2. Kauffmann, G., & Haehnelt, M. 2000, MNRAS, 311, 576.
3. Granato, G. L., De Zotti, G., Silva, L., Bressan, A., & Danese, L. 2004, ApJ, 600, 580.
4. Scannapieco, E., & Oh, S. P. 2004, ApJ, 608, 62.
5. Brandt, W. N., Mathur, S., & Elvis, M. 1997, MNRAS, 285, L25.
6. Haardt & Maraschi 1993, ApJ, 413, 507.
7. Haardt, Maraschi, & Ghisellini 1994, ApJ, 432, 95.
8. Tananbaum, et al. 1979, ApJ, 234, L9.
9. Brandt, W. N., Alexander, D. M., Bauer, F. E., & Vignali, C. 2004, in Physics of Active Galactic Nuclei at All Scales, eds. Alloin D., Johnson R., Lira P. (Springer-Verlag, Berlin), astro-ph//0404543.
10. Schmidt, M. 1968, ApJ, 151, 193.
11. Schmidt, M. 1972, ApJ, 176, 273.
12. Mathez, G., 1976, A&A, 53, 15.
13. Schmidt, M., & Green, R. F. 1983, ApJ, 269, 352.
14. Boyle, B. J., Fong, R., Shanks, T., & Peterson, B. A. 1987, MNRAS, 227, 717.
15. Ueda, Y., Akiyama, M., Ohta, K., & Miyaji, T. 2003, ApJ, 598, 886.
16. Hasinger, G., Miyaji, T., & Schmidt, M. 2005, A&A, 441, 417.
17. Silverman, J. D., Green, P. J., Barkhouse, W. A. et al. 2005, ApJ, 624, 630.
18. Barger, A. J., Cowie, L. L., Mushotzky, R. F., Yang, Y., Wang, W.-H., Steffen, A. T., & Capak, P. 2005, AJ, 129, 578.
19. Magorrian et al. 1998, AJ, 115, 2285.
20. Ferrarese, L. & Merritt, D. 2000, ApJ, 539, L1.
21. Gebhardt, et al. 2000, 539, L13.
22. McLure R. J., & Dunlop, J. S. 2002, MNRAS, 331, 795.
23. Strateva, I. V., Brandt, W. N., Schneider, D. P., Vanden Berk, D. G., & Vignali, C. 2005, AJ, 130, 387.
24. Avni & Tananbaum, 1986, 305, 83.
25. Vignali, C., Brandt, W. N., & Schneider, D. P. 2003, AJ, 125, 433.
26. Bechtold, J., Siemiginowska, A., Shields, J.et al. 2003, ApJ, 588, 119.
27. Grupe, D., Pradhan, A. K., & Frank, S. 2005, AJ, 130, 355.
28. Vignali, C., Brandt, W. N., Schneider, D. P., & Kaspi, S. 2005, AJ, 129, 2519.
29. Shemmer, O., Brandt, W. N., Vignali, C., Schneider, D. P., Fan, X., Richards, G. T., & Strauss, M. A. 2005, ApJ, 630, 729.
30. Mateos, S., Barcons, X., & Carrera, F. J.et al. 2005, A&A, 433, 855.
31. Chartas, G., Brandt, W. N., Gallagher, S. C., & Garmire, G. P. 2003, Astronomische Nachrichten, 324, 163.
32. Dai, X., Chartas, G., Eracleous, M., & Garmire, G. P. 2004, ApJ, 605, 45.
33. Haardt, F., Maraschi, L., & Ghisellini, G. 1997, ApJ, 476, 620.
34. Alexander et al. 2003, AJ, 126, 539.
35. Barger et al. 2003, AJ, 126, 632.

36. Szokoly et al. 2004, ApJS, 155, 271.
37. Giacconi et al. 2002, ApJS, 139, 369.
38. Fabian, A. C. 1999, MNRAS, 308, L39.
39. Springel, V., Di Matteo, T., & Hernquist, L. 2005, ApJL, 620, L79.
40. Kaspi, S., Brandt, W. N., George, I. M., et al., 2002, ApJ, 574, 643.
41. Kaastra, J. S., Steenbrugge, K. C., Raassen, A. J. J., van der Meer, R. L. J., Brinkman, A. C., Liedahl, D. A., Behar, E., & de Rosa, A. 2002, A&A, 386, 427.
42. Netzer, H., Kaspi, S., Behar, E., et al., 2003, ApJ, 599, 933.
43. Ganguly, R., Bond, N. A., Charlton, J. C., Eracleous, M., Brandt, W. N., & Churchill, C. W. 2001, ApJ, 549, 133.
44. George, I. M., Turner, T. J., & Yaqoob, T. 2002, ASP Conf. Ser. 255: Mass Outflow in Active Galactic Nuclei: New Perspectives, 255, 49.
45. Porquet, D., Reeves, J. N., O'Brien, P., & Brinkmann, W. 2004, A&A, 422, 85.
46. Hewett, P. C., & Foltz, C. B. 2003, AJ, 125, 1784.
47. Weymann, R. J., Morris, S. L., Foltz, C. B., & Hewett, P. C. 1991, ApJ, 373, 23.
48. Chartas, G., Brandt, W. N., Gallagher, S. C., & Garmire, G. P. 2002, ApJ, 579, 169.
49. Chartas, G., Brandt, W. N., & Gallagher, S. C. 2003, ApJ, 595, 85.
50. Chartas, G., Brandt, W. N., Gallagher, S. C., & D. Proga, 2006, submitted to ApJ.
51. Reeves, J. N., O'Brien, P. T., & Ward, M. J. 2003, ApJL, 593, L65.
52. Pounds, K. A., King, A. R., Page, K. L., & O'Brien, P. T. 2003a, MNRAS, 346, 1025.
53. Pounds, K. A., Reeves, J. N., King, A. R., Page, K. L., O'Brien, P. T., & Turner, M. J. L. 2003b, MNRAS, 345, 705.
54. Kaspi, S., & Behar, E. 2006, ApJ, 636, 674.
55. McKernan, B., Yaqoob, T., & Reynolds, C. S. 2004, ApJ, 617, 232.
56. McKernan, B., Yaqoob, T., & Reynolds, C. S. 2005, MNRAS, 361, 1337.

Deep Surveys of Obscured Active Galactic Nuclei

Eleni T. Chatzichristou

Insitute for Astronomy and Astrophysics, National Observatory of Athens, GR-15236, Greece

Abstract. Several key goals require measuring the number of all Active Galactic Nuclei (AGN) in the Universe, and the evolution of the ratio of obscured to unobscured AGN with redshift. This reflects the structure of AGN and thus the development in the heart of all galaxies. Hard X-rays can penetrate most obscuring dust columns to reveal the AGN that remains hidden in all other wavelengths. Mid-IR surveys probe the thermal dust emission, that is, the continuum light from the central source after it is reprocessed by dust, and this emission dominates the bolometric luminosities of dusty high-redshift galaxies. Thus, combining deep mid-IR and hard X-ray surveys can provide us with accurate demographics of AGN especially at high redshifts. Multi-wavelength surveys aim to address these science goals by exploiting the unprecedented combination of great observatories such as HST, Chandra and SIRTF to survey the distant universe to the faintest flux limits across the broadest range of wavelengths. In this paper I discuss some of the results coming from multi-wavelength surveys placing particular focus on the systematic study of obscured AGN.

Keywords: AGN, Multiwavelength Surveys, Obscured AGN
PACS: 95.80.+p 98.54.Cm

MOTIVATION

Active Galactic Nuclei (AGN) are signposts for galaxy formation. It is thus very important to understand their nature and evolution vs. cosmic time, and the black hole-galaxy connection. Furthermore, AGN are of cosmological significance because they may play an important role in the re-ionization of the Universe and they probably comprise most of the X-ray background radiation ([31, 1, 2, 23, 34]). Moderate luminosity AGN near z$\sim$2, associated with the peak AGN activity, are thought to be the most representative sources for the entire AGN class throughout the Universe and to make up the majority of the X-ray background. AGN are almost certainly obscured by dust. According to the current AGN paradigm, dust in a torus or warped disk obscures for some lines of sight the optical, UV and soft X-ray continuum produced by the super massive black hole (SMBH) and the broad-line emission. At such orientations, AGN lack broad emission lines or a bright continuum and are called type 2 AGN, as opposed to type 1 AGN. Unification models imply that these objects have the same general structure, with the level of obscuration of the central source dependent upon the random orientation of the dusty torus surrounding it ([3]).

Obscured AGN most likely dominate the population of SMBH ([18]), however, very few obscured AGN are known beyond z$\sim$0.3 ([22, 36, 32, 12]) and it remains unclear whether this is due to an intrinsic rarity or is the result of biased surveys. Standard UV excess/color selection techniques or broad emission line survey methods usually miss the highly obscured AGN, being biased towards the highly luminous AGN at high z and the numerous weaker AGN observed in the local Universe. A key complicating factor is that usually the AGN coexist with a starburst, thus, contamination by the host galaxy

CP848, *Recent Advances in Astronomy and Astrophysics,*
7th *International Conference of the Hellenic Astronomical Society,* edited by N. H. Solomos

light and the star forming regions could be overwhelming the emission-line signatures of the AGN activity ([30]).

The obscured AGN can only be detected in the hard X-ray, mid and far-IR wavelengths, where the majority of their bolometric luminosity is observed. Hard X-rays (>2 keV) can penetrate most obscuring dust columns to reveal the AGN, thus such surveys (in particular when combined with optical and mid-IR) can provide us with accurate demographics of AGN at high z. Recent deep surveys with Chandra and XMM-Newton have resolved most of the X-ray background, containing high-redshift obscured AGN ([8, 34]). Population synthesis models involving large numbers of obscured AGN have been successful at matching the X-ray background intensity and spectrum ([27, 10, 17, 18, 19, 33]). The measurement of the X-ray spectrum or hardness ratio has typically been used as a barometer for the amount of obscuration in the AGN assuming a set dust-to-gas ratio, typically the Galactic value. However, it has been shown that many local AGN have non-galactic ratios ([28]). Mid-IR data is a far superior diagnostic of the warm dust associated with the AGN. However, many starbursts show strong far-IR emission too. Hence, the need for Multi-Wavelength Surveys.

Many deep multi-wavelength surveys, such as GOODS ([16]), are motivated by the possible discovery of a previously unknown high-z population of obscured AGN. A complementary approach, given the relatively low surface density of AGN (vs. normal galaxies), is to target high-luminosity AGN over a wider area of the sky. Examples of such surveys are the XMM-Newton/2dF Survey ([7]), CHAMP ([20]), CYDER ([9]), SEXSI ([21]), HELLAS2XMM ([4]).

BIASES

However, X-ray surveys are not devoid of biases. Indeed, we find that the observed redshift distribution of X-ray sources is peaked at lower redshift ($z \sim 0.7$) and the ratio of obscured to unobscured AGN is substantially lower (2-2.5) than the values (respectively, $z \sim 1.4$ and ratios of 4:1 locally to 10:1 at $z=1.3$) required by X-ray background models. The most obvious explanation is that X-ray flux-limited samples suffer incompleteness at high column densities, estimated to be (for the GOODS fields) 75% for $N_H=10^{23}$ cm^{-2} and 30% for $N_H=10^{24}$ cm^{-2} ([37]). Furthermore, as Treister et al. [37] showed, deriving a hard X-ray luminosity function (LF) or redshift distribution, imposes an effective optical cut at $R<24$ mag, since optical spectroscopy is required to obtain accurate redshifts. Thus, published redshift distributions can be missing optically-faint sources, which could be either low-luminosity unobscured AGN or obscured AGN, and this spectroscopic incompleteness is expected to increase with redshift. This selection effect explains the discrepancy shown between photometric and spectroscopic redshift distributions.

X-RAY-TO-OPTICAL FLUXES

The X-ray-to-optical flux ratio is known to be strongly indicative of high obscuration in high luminosity AGN, probably representing the high-rdshift obscured AGN population ([1, 25, 13, 29, 15]). Fiore et al. [13] found that 20% of hard X-ray selected sources have f_X/f_{opt} ~10 times higher than that of optically selected AGN. Among the sources with the highest ratio, the 2/3 are absorbed QSO the rest being type 1 QSO. Interestingly, Perola et al. [33] found that about 10% of the sources with log N_H >22 are optically classified as unabsorbed (type 1) AGN, and they all have log $L_{2-10keV}$ >44. There are two possible explanations for this: either the dust-to-gas ratio is substantially different than the assumed Galactic value, or the X-ray absorbing gas and the gas and dust obscuring the broad line region (BLR) are geometrically separated. One could add a third factor, the variability in the X-ray absorbing column or/and in the optical emission.

DIFFERENTIAL LUMINOSITY EVOLUTION

Hard X-ray LF have revealed that BROAD-LINE (BL) AGN dominate the number densities at the higher X-ray luminosities, while non-BL AGN dominate at the lower X-ray luminosities ([35]). Ueda et al. [38] confirmed the luminosity dependence of the fraction of absorbed sources (decreasing with luminosity), while its z dependence seems to not be significant. Fiore et al. [13] found that the low X-ray luminosity sources increase in number backward in time at a much slower rate than the high X-ray luminosity sources. The former reach a maximum at z~1 and then level off at z>2, similar to the evolution of actively star forming galaxies. The latter show strong evolution up to z=2-3, following the evolution of spheroids.

This differential evolution between obscured and unobscured AGN means indicates that, either the simple orientation unification model for AGN needs to be modified to include an X-ray dependent covering factor (e.g. the opening angle of the dust torus is larger in luminous AGN), or that the BL and non-BL AGN are intrinsically different. Perola et al. [33] suggested that modifying the shape and evolution of the LF and the NH distribution as a function of luminosity simultaneously, could help improve the models so that they reproduce the data. Ueda et al. [38] showed that the cosmological evolution of the hard X-ray LF could be best described by a luminosity-dependent density evolution model, where the cutoff redshift increases with luminosity. This would imply that the luminous AGN (quasars) formed in earlier epochs than the less luminous AGN (which is consistent with the claims of [11] and [24].

OPTICAL SPECTRAL TYPES AND HOST GALAXIES

Barger et al. [6] argued that, since the optical spectra of X-ray sources suffer varying degrees of AGN mixing with the host galaxy spectrum (this further depending on redshift), instead of using the conventional AGN classification scheme, they have devised four optical spectral classes for X-ray selected sources. They showed that one can separate BL AGN from non-BL AGN on the basis of X-ray colours/spectra, but for the narrow-line (NL) AGN there is no clear dependence of the absorbing column density of the optical spectral type or optical line widths. They do find a luminosity dependence in optical spectral type, both in the sense of BL AGN dominating at large $L_X > 10^{44}$ ergs s^{-1} noted earlier, as well as, between optically normal galaxies and AGN (the cross-over point lying at $L_X < 10^{43}$ ergs s^{-1} at z=0.25 and at $L_X < 10^{44}$ ergs s^{-1} near z=1). Their suggestion is pure luminosity evolution in both the LF and the spectral type mix, indicating, as before, a luminosity dependent unified model.

The rapid drop in the luminosities of AGN (both BL and NL) with decreasing redshift between z=1 and z=0, indicates a drop in the overall energy density production by the AGN. The authors put forward two possible explanations: mass starvation -declining accretion rate of all SMBH, regardless of mass-, or AGN downsizing -the more massive black holes are being preferentially starved, while the less massive continue to accrete-. They prefer the latter, since the most X-ray luminous AGN lie in the most optically luminous host galaxies, and the latter seems to have completely disappeared at low redshifts.

Furthermore, Barger et al. [6] discuss the fact that the accretion history of SMBH may be strongly linked to the star formation history in their host galaxies. The density evolution of luminous AGN resembles the formation history of early type galaxies ([14]), in accordance with the Magorrian relation ([26]) found in the local Universe. This put together with recent findings that the star formation of early type galaxies peaks earlier (z$\sim$3) than that of late type galaxies (z$\sim$1.5) ([5]), suggest that luminous AGN lived preferentially in galaxies with large spheroid components and their SMBH grew in parallel with strong starbursts at early epochs (possibly as a result of mergers), followed by a rapid decrease of their activities after z<$\sim$2. Late-type galaxies, on the other hand, made starbursts later or/and slower and are hosting less luminous AGN whose activity continued until recently (z=1). Furthermore, since there seems to be a difference in the peak star formation (earlier) and AGN (later) activity for both early and late type galaxies, as indicated by the above peak redshifts, this would imply that the AGN activity in a galaxy occurs in a later phase than a major starburst ([38]).

CONCLUSION

It is clear that only $\sim$ 1/2-1/4 of the SMBH mass density was fabricated in BL AGN, this leaving plenty of room for obscured accretion. All the new evidence outlined above, despite the problems and new issues that it raised, has helped enormously to improve our understanding of the elusive population of obscured AGN in the Universe, constraining the formation history of both the luminous unobscured quasars and the less luminous obscured AGN, which are the main contributors of the X-ray background.

REFERENCES

1. D.M. Alexander et al., *AJ*, **122**, 2156 (2001)
2. D.M. Alexander et al., *AJ*,**126**, 539 (2003)
3. R. Antonucci, *ARAA*, **31**, 473 (1993)
4. A. Baldi et al., *ApJ*, **564**, 190 (2002)
5. C. Balland et al., *MNRAS*, **343**, 107 (2003)
6. A.J. Barger et al., *AJ*, **129**, 578 (2005)
7. S. Basilakos et al., *ApJ*, **607**, L79 (2004)
8. W.N. Brandt et al., *AJ*, **122**, 2810 (2001)
9. F.J. Castander et al., *Astronomisch Nachrichten*, **324**, 40 (2003)
10. A. Comastri et al., *A&A*, **296**, 1 (1995)
11. L. L. Cowie et al., *ApJ*, **584**, L57 (2003)
12. S. Dawson et al., *AJ*, **125**, 1236 (2003)
13. F. Fiore et al., *A&A*, **409**, 79 (2003)
14. A. Franceschini et al., *MNRAS*, **310**, L5 (1999)
15. P. Gandhi et al., *MNRAS*, **348**, 529 (2004)
16. M. Giavalisco et al., *ApJL*, **600**, 93 (2004)
17. R. Gilli et al., *A&A*, **347**, 424 (1999)
18. R. Gilli et al., *A&A*, **366**, 407 (2001)
19. R. Gilli, *Astronomische Nachrichten*, **324**, 165 (2003)
20. P.J. Green, *ApJS*, **150**, 43 (2004)
21. F.A. Harrison et al., *ApJ*, **596**, 944 (2003)
22. G. Hasinger et al., *A&A*, **365**, L45 (2001)
23. G. Hasinger, in "New Visions of the X-ray Universe in the XMM-Newton and Chandra Era", Ed. F. Jansen (ESTEC:ESA SP-488), *astro-ph/0202430* (2002)
24. G. Hasinger, in "The Emergence of Cosmic Structure", Eds. S.S Holt & C. Reynolds, **666**, p.227 (2003)
25. A.M. Koekemoer et al., *ApJ*, **567**, 657 (2002)
26. J. Magorrian et al., *AJ*, **115**, 2285 (1998)
27. P. Madau et al., *MNRAS* (1994)
28. R. Maiolino, "The Central Kiloparsec pf Starbursts and AGN: The LaPalma Connection", *ASP Conf. Proc.*, **249**, 256 (2001)
29. M. Mignoli et al., *A&A*, **418**, 827 (2004)
30. E.C. Moran et al., *ApJ*, **579**, 71 (2002)
31. R.F. Mushotzky et al., *Nature*, **404**, 459 (2000)
32. C. Norman et al. 2002, *ApJ*, **571**, 218 (2002)
33. G.C. Perola et al., *A&A*, **421**, 491 (2004)
34. P. Rosati et al., *ApJ*, **566**, 667 (2002)
35. A.T. Steffen et al., *ApJ*, **596**, L23 (2003)
36. D. Stern et al., *ApJ*, **568**, 71 (2002)
37. E. Treister et al., *ApJ*, **616**, 123 (2004)
38. Y. Ueda et al., *ApJ*, **598**, 886 (2003)

Two Dimensional Velocity Fields of Low Surface Brightness Galaxies

Rachel Kuzio de Naray*, Stacy S. McGaugh*, W.J.G. de Blok† and A. Bosma**

*Astronomy Department, University of Maryland
†Research School of Astronomy and Astrophysics, Australian National University
**Observatoire Astronomique de Marseille Provence

Abstract. We have obtained high resolution two-dimensional velocity fields from integral field spectroscopy along with derived rotation curves for nine low surface brightness galaxies. This is a positive step forward in terms of both data quality and number of objects studied. We fit NFW and pseudo-isothermal halo models to the observations. We find that the pseudo-isothermal halo better represents the data in most cases than the NFW halo, as the resulting concentrations for the latter are lower than would be expected from ΛCDM. Cusps can clearly be detected by this method when they are present and the data are of high quality. Yet only one of our six best cases is consistent with a cusp.

Keywords: Kinematics, dynamics, and rotation; Galactic halos
PACS: 98.62.Dm; 98.62.Gq

INTRODUCTION

Though it is widely agreed that low surface brightness (LSB) galaxies are dark matter dominated at all radii (de Blok & McGaugh 1997) and, hence, are ideal test subjects for studying the dark matter distribution, a consensus has not yet been reached on the interpretation of their rotation curves. Are the dark matter halos of LSB galaxies "cuspy" NFW halos, as expected from cosmological numerical simulations (e.g., Navarro, Frenk, & White 1997; Moore et al. 1999; Klypin et al. 2001), or are they closer to "cored" pseudo-isothermal halos? NFW halos can be fit to the observations, but the cosmological parameters implied by the fits are inconsistent with the standard ΛCDM picture (McGaugh, Barker & de Blok 2003). Pseudo-isothermal halos provide much better fits, and in the case of a power-law fit to the density, $\rho \sim r^{\alpha}$, the optimum value of α seems to be -0.2 $\pm$ 0.2 (de Blok, Bosma & McGaugh 2003). However, such models have no cosmological dependence or theoretical basis. In hopes of salvaging the NFW halo, a number of systematic effects have been suggested to affect the data (Swaters et al. 2003a, 2003b).

In an effort to address these concerns, the quality of observations has been greatly improved. The first suggested effect which could "hide" a NFW-halo was beam smearing in two-dimensional 21-cm HI velocity fields. Subsequent long-slit Hα observations increased the resolution by an order of magnitude. When cusps did not appear along with the resolution increase, effects such as slit misplacement and non-circular motions came to the forefront. We address these issues with new high resolution, two-dimensional data.

CP848, *Recent Advances in Astronomy and Astrophysics,*
7th *International Conference of the Hellenic Astronomical Society,* edited by N. H. Solomos

OBSERVATIONS AND RESULTS

We observed twelve LSB galaxies using the DensePak Integrated Field Unit on the WIYN telescope at Kitt Peak. DensePak is comprised of 3" diameter fibers arranged in a 43" x 28" rectangle. The distances to the galaxies are such that a 3" fiber provides sub-kpc resolution. The fiber bundle orientation on the sky and number of pointings were tailored to each galaxy so that the critical central regions were covered by the fibers. Two 1800 sec exposures were taken at each pointing.

Velocities of the Hα, [NII]λ6584, [SII]λ6717 lines were measured for each fiber. The average of the measured velocities in each fiber was taken as the fiber velocity. Velocity fields were constructed by combining individual DensePak pointings and comparing an Hα flux image made from the DensePak data to an actual Hα image. Velocity fields were then transformed to rotation curves using the NEMO program ROTCUR. For three of the galaxies no useful data covering a large fraction of the galaxy has been obtained. The resulting rotation data agree very well with the previous long slit data obtained by McGaugh, Rubin & de Blok (2001) and de Blok & Bosma (2002).

We fit the combined rotation curves with pseudo-isothermal and NFW_{free} halos for the minimum disk case. We also fit an $NFW_{constrained}$ halo which was required to match the velocities at the outer radii of each galaxy, in effect, constraining the concentrations to agree with cosmology. Most of the galaxies, such as UGC 5750, are clearly well-fit by an pseudo-isothermal halo. We stress that the fit should extend to the whole rotation curve, and that the inner most points on their own do not constitute a good fit to a NFW halo. In our sample sofar, only NGC 4395 is fit by a NFW halo. The concentrations for the other galaxies do not agree with values expected for ΛCDM. That we did observe one halo consistent with the expected cusp shows that cusps can be detected when they are present and the data are of high quality.

Future work will include mass-modeling, reconstruction of slit placement, and a detailed assessment of the non-circular motions possibly indicated by the twisting of the kinematic axes.

Acknowledgement :This research was supported by NSF grant AST 0206078.

REFERENCES

1. de Blok, W.J.G., & Bosma, A. 2002, *A.&A.*, **385**, 816
2. de Blok, W.J.G., Bosma, A., & McGaugh, S.S. 2003, *MNRAS*, **340**, 567
3. de Blok, W.J.G., McGaugh, S.S., & Rubin, V.C. 2001, *A. J.*, **122**, 2396
4. de Blok, W.J.G., McGaugh, S.S., Bosma, A., & Rubin, V.C. 2001, *Ap. J.*, **552**, L23
5. de Blok, W.J.G., & McGaugh, S.S. 1997, *MNRAS*, **290**, 533
6. Klypin, A.A., Kravtsov, A.V., Bullock, J.S., & Primack, J.R. 2001, *Ap. J.*, **554**, 903
7. McGaugh, S.S., Barker, M.K. & de Blok, W.J.G. 2003, *Ap. J.*, **584**, 566
8. McGaugh, S.S., Rubin, V.C., & de Blok, W.J.G. 2001, *A. J.*, **122**, 2381
9. Moore, B., Quinn, T., Governato, F., Stadel, J., & Lake, G. 1999, *MNRAS*, **310**, 1147
10. Navarro, J.F., Frenk, C.S., & White, S.D.M. 1997, *Ap. J.*, **490**, 493
11. Swaters, R.A., Madore, B.F., van den Bosch, F.C., & Balcells, M. 2003, *Ap. J.*, **583**, 732
12. Swaters, R.A., Verheijen, M.A.W., Bershady, M.A., & Andersen, D.R. 2003, *Ap. J.*, **587**, L13

Improved constraints on the cosmological parameters using the VLA FIRST survey

E. Xanthopoulos[*,†], R. H. Becker [*,†], W. H. deVries [*,†] and R. L. White [**]

[*] *University of California Davis, Department of Physics, Davis, CA 95616*
[†] *IGPP/Lawrence Livermore National Laboratory, Livermore, CA 94550*
[**] *Space Telescope Science Institute, 3700 San Martin Drive, Baltimore, MD 21218*

Abstract. Using the final version of the VLA FIRST survey (Becker et al. 2003), the optical Sloan Digital Sky Survey Data Release 3 (DR3) quasar list and a series of carefully selected criteria, we have defined the largest homogeneous population of double-lobed sources. With the precise sample definition, the increased depth and sensitivity of the survey data, the large size of the dataset, and our self-consistent method of analysis, which addresses many of the problems associated with previous work in the area, we are able to: a) explore the correlations between the intrinsic properties of the double-lobed radio sources (the results are also confirmed by a non-parametric analysis) and study their evolution, b) place more interesting and tighter constraints on the cosmological parameters, distinguish among the different cosmology models, and determine the impact of the angular size-redshift studies in cosmology, c) further our understanding of the behavior of the intergalactic medium (IGM) density as a function of redshift and shed more light to the quasar-radio galaxy unification issue.

Keywords: Cosmological parameters, Radio galaxies
PACS: 98.80.-k, 98.80.Bp, 98.80.Es, 98.54.Gr

INTRODUCTION

The angular size-redshift (theta-z) relation for a cosmological population of standard rods is a powerful probe of the large-scale geometry of the Universe. However, previous attempts to measure the cosmological parameters from the angular size-redshift relation of double radio-lobed radio sources have been marred by a variety of selection effects, destroying the integrity of the data sets, and by inconsistencies in the analysis, undermining the results and leading to data consistent with a static Euclidean Universe rather than with standard Friedmann models (e.g. Wardle & Miley 1974, Barthel & Miley 1988, Nilsson et al. 1993). Interpretation of this observation has caused disagreements among various authors.

THE SAMPLE AND SELECTION CRITERIA

The VLA FIRST survey (Becker et al. 1995) has mapped $\sim$9000 deg^2 of the sky at 1.4 GHz to a sensitivity of $\sim$1 mJy with $5\overset{''}{.}4$ FWHM Gaussian beam and has cataloged 111,115 sources with subarcsecond positional accuracy. We cross-correlated this survey with a sample of 51,516 quasars compiled from the optical Sloan Digital Sky Survey Data Release 3 (DR3) quasar list and the list of quasars from the 2dF survey. By using a lower detection limit of 3σ for the Poisson probability distribution of finding

CP848, *Recent Advances in Astronomy and Astrophysics,*
7th *International Conference of the Hellenic Astronomical Society,* edited by N. H. Solomos

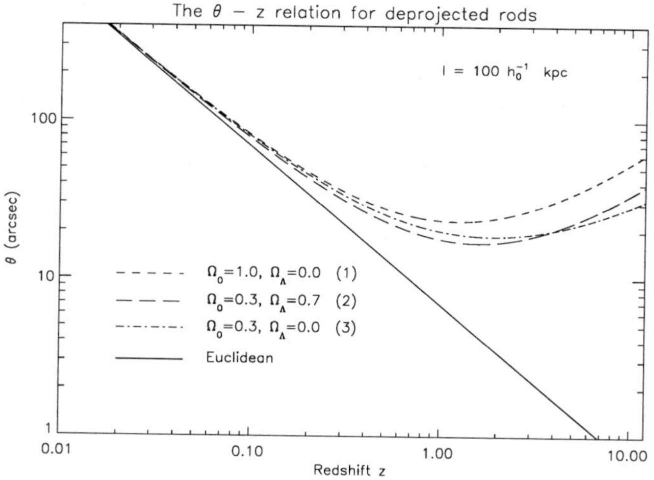

FIGURE 1. The angular size - redshift relation for deprojected rods of length 100 h_0^{-1} kpc for different cosmologies. The choice of the cosmological parameters for the three Friedmann models are listed on the figure. The curve for a static Euclidean universe is shown for comparison. In practice, the curves actually define upper limits to the observed angular sizes, since projection effects will scatter the observed sizes downward. Note the presence of the minimum near z$\sim$1.5.

companions of the quasars within an area of $5'$ radius from the quasar position, we defined an initial sample of double-lobed radio sources with their cores centered on the quasar positions. We then followed a series of carefully selected criteria, and specifically we: a) included only radio sources, FRIIs, with symmetric and collinear triple structure (core + two lobes), thereby minimizing asymmetrical effects that might distort the apparent angular size, such as relative motion with respect to the IGM, b) further restricted the sample to sources with redshift z>0.3 beyond which quasars begin to dominate and carry more info about the cosmology. This way we bypass the problem of different mean orientations or power-size correlations which produce non- cosmological effects in the theta-z plane. Our final clean sample consists of 389 FRIIs, which is the largest homogeneous population of double-lobed sources to date.

THE ANGULAR SIZE - REDSHIFT RELATIONSHIP

Fig. 1 illustrates the $\theta - z$ relationship for deprojected rods (ϕ = 90°) with an intrinsic size of l=100 h_0^{-1} kpc for three Friedmann cosmologies: a) an Einstein-de Sitter universe, b) a flat universe, and 3) a nonclosed, matter-dominated universe. The particular values of Ω_0 and Ω_Λ are listed in the figure. The curve for a static Euclidean universe, with $\theta \propto z^{-1}$ is also shown for comparison. The location of the minimum in the angular size (typically between z=1 and z=2) depends on Ω_0 and Ω_Λ.
Fig. 2 shows a scatter plot of the $\theta - z$ data. The errors in the measured values of θ are typically $\sim 1''$, far less that the scatter in the angular sizes at any redshift. With $5\farcs4$

FWHM beam, the FIRST survey can detect extended structure down to 2″. However, due to the survey resolution limit, uncertainties in the quasar optical positions, variations in the morphologies of double-lobed sources, and inspection of numerous FIRST radio maps, we also introduce an effective cutoff in the data of 12″, shown in Fig. 2, following Buchalter et al. (1998), below which an accurate morphological classification could not be assigned with certainty. Such a cutoff ensures that we do not include the so called core-jet, diffuse, cometary and other type of extended radio sources that may be mistaken for double-lobed objects at low resolution.

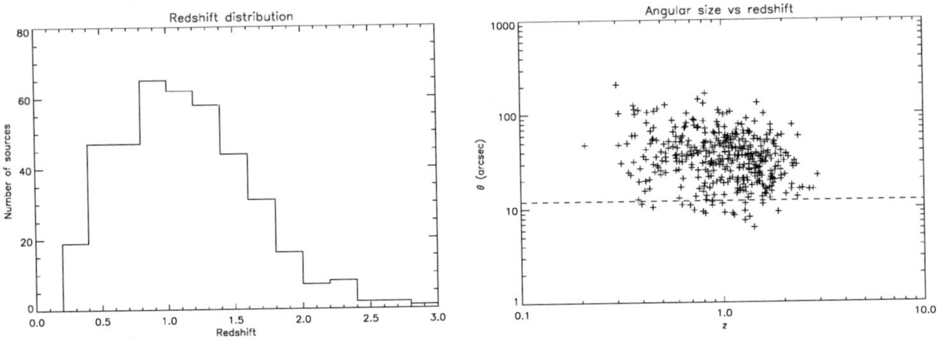

FIGURE 2. (Left:) The distribution of redshift in our sample. The sample includes sources out to a redshift of 2.94 significantly higher than the redshifts at which the minima in the theta-z curves typically occur for different Friedmann models, and in contrast to previous work which included samples that contained significant numbers of sources with z<1 where roughly Euclidean behavior is expected. The median redshift of our sample is 1.074 while that of the FIRST survey is 1, evidence that indicates that the two populations have similar distributions, and therefore our sample selected using the optical SDSS survey is fairly representative of double-lobed radio sources as a whole. (Right:) Scatter plot of the peak-to-peak angular sizes vs redshift. The dashed line represents the effective resolution limit at 12″, below which accurate morphological classifications could not be determined. The errors in the measured values of the angular size are typically $\sim 1''$, far less than the scatter in the angular sizes at any redshift.

STANDARD AND NONSTANDARD COSMOLOGICAL MODELS

For graphical purposes, we bin the data in redshift, in equal numbers per bin, and calculate the mean and median values of the theta, together we standard errors of the mean values and median absolute deviations for each bin. The results are shown in Fig. 3 and Fig. 4, for the median values of θ, along with the curves from Fig. 1, whose amplitudes (corresponding to the median intrinsic sizes), have been scaled to provide a rough visual fit. Apart from the Friedmann models we have also experimented with the Steady State and two nonstandard cosmology models, the Tired-Light and the Gauge models, the latter shown in Fig. 4. The most striking feature of the data is that, regardless of the binning details, the observed data seem to be more consistent with Friedmann models than with a Euclidean model. The Friedmann curves shown are not the best-fit results but are merely intended for qualitative reference.

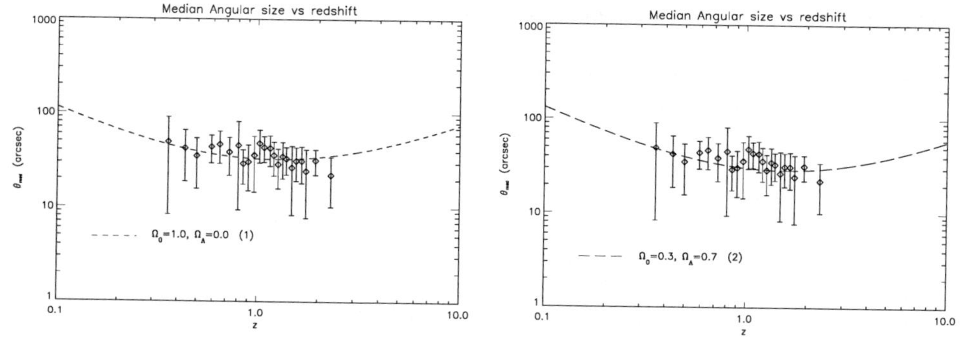

FIGURE 3. Median angular sizes, binned in redshift (to compensate for projection effects), with roughly equal numbers per bin (~square-root of N the number of sources in the sample). The Friedmann curves, corresponding to models 1 and 2 from Fig. 1, have been scaled to provide a rough visual fit. The error bars represent the median absolute deviation in each bin.

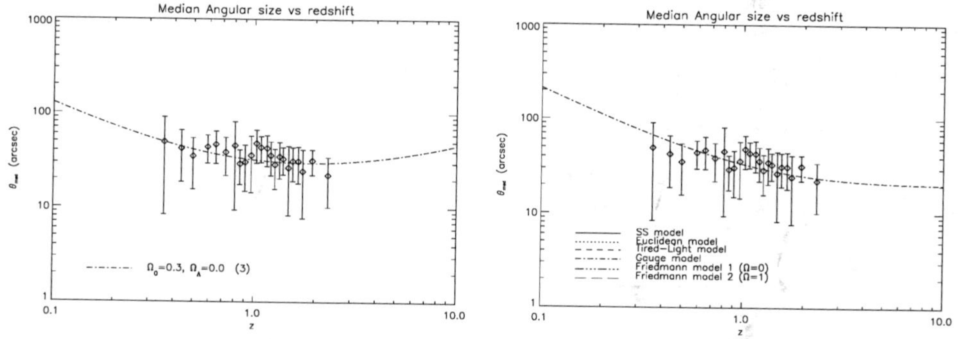

FIGURE 4. (Left:) Median angular sizes, binned in redshift (to compensate for projection effects), with roughly equal numbers per bin (~square-root of N the number of sources in the sample). The Friedmann curve, corresponding to model 3 from Fig. 1, has been scaled to provide a rough visual fit. The error bars represent the median absolute deviation in each bin. (Right:) The Gauge curve has been scaled to provide a qualitative reference. In the Gauge model, when z tends to infinity the angular size tends to a constant similar to the Friedmann model with qo=0 but more rapidly.

DISCUSSION AND FUTURE WORK

Before attempting to find the best-fit cosmological parameters, and determine whether we can distinguish with high significance between the different models with the present sample, we need to explore the relationships between the intrinsic properties (P intrinsic power, l projected linear size and redshift z) of the sources in our sample using both parametric and non-parametric analysis. Such correlations have important implications in determining the parameters. Therefore our next steps are to: a) Optimize the analysis by defining and exploring a parameter space and use an iterative procedure that will lead to the best model fit parameters, b) Apply a chi-square goodness-to-fit statistic to determine the best fit values for the free parameters in each model and, c) Examine the values of the Hubble constant implied by the data.

ACKNOWLEDGMENTS

EXs work was performed under the auspices of the U.S. Department of Energy, National Nuclear Security Administration by the University of California, Lawrence Livermore National Laboratory under contract No. W-7405-Eng-48 and she also acknowledges support from the National Science Foundation (grant AST 00-98355).

REFERENCES

1. Barthel, P. D, Miley, G. K., *Nature*, **333**, 319 (1988).
2. Becker, R. H., Helfand, D. J., White, R. L., Gregg, M. D., Laurent-Muehleisen, S. A., 2003, *VizieR On-line Data Catalog: VIII/71*.
3. Becker, R. H., White, R. L., Helfand, D. J., *Astrophys. J*, **450**, 559–577 (1995).
4. Buchalter, A., Helfand, D. J., Becker, R. H., White, R. L., *Astrophys. J*, **494**, 503–522 (1998).
5. Nilsson, K., Valtonen, M. J., Kotilainen, J., Jaakkola, T., *Astrophys. J*, **413**, 453–476 (1993).
6. Wardle, J. F. C., Miley, G. K., *Astron. & Astrophy.*, **30**, 305–315 (1974).

Elemental abundances in PNe and H II regions: New insights from observations and models

Yiannis G. Tsamis

Dept. of Physics & Astronomy, University College London, London WC1E 6BT, U.K.

Abstract.
The emission-line analysis of ionized gaseous nebulae such as planetary nebulae (PNe) and H II regions is one of the best established methods for obtaining the abundance of elements in our own and in external galaxies. Abundance studies of PNe reveal the composition of matter returned to the interstellar medium (ISM) after being processed by stellar nucleosynthesis, while those of H II regions provide a snapshot of the composition of the ISM in the latest episode of its evolution. We discuss current issues in these related fields, and show how concerted efforts from observational spectroscopy and detailed models have been instrumental in proposing solutions for outstanding problems in the astrophysics of photoionized nebulae, and could motivate developments on ISM chemical evolution topics. The potential of Greek astronomy in the pursuit of relevant research with modern local facilities such as the Aristarchos telescope is noted.

Keywords: ISM: abundances – planetary nebulae – H II regions – ISM: evolution – ISM: individual: 30 Dor

INTRODUCTION

Planetary nebulae (PNe) are the ionized ejected envelopes of evolved stars whose progenitors were quite similar to the Sun ($M \sim 1$–$8\,M_\odot$) and, as such, they are emblematic objects to study. Most stars in late-type galaxies (such as the Milky Way) will live through the PN phase in their final evolutionary stages. The study of PNe is therefore very important in a (extra-)galactic context. It is well known that before and during the PN phase the interstellar medium (ISM) is enriched with heavy elements from the evolving star, some of which (e.g. He, C, N) have been modified by nucleosynthetic processes so that they do not reflect the make-up of the stellar progenitor. These elements, along with freshly formed dust species and molecules, mix with the ambient ISM and alter its composition. The photoionized PN envelopes are ideal laboratories for the study of the exact chemical composition of the mass ejected from the dying star, and for the determination of the abundance of species there relative to hydrogen.

H II regions (such as e.g. the Orion Nebula) are giant clouds marking the birthplaces of stars; these nebulae are photoionized by hot, young massive stars (M $8\,M_\odot$) and their chemical analysis provides a snapshot of the constituents of the ISM in the latest episode of its evolution. Because of their brightness H II regions are observable over cosmological distances and so their study is one of the best established methods for deciphering the elemental composition of distant galaxies.

The most powerful technique for obtaining the composition of PNe or H II regions involves the observation (with long-slit, échelle or integral-field unit spectrographs) of

CP848, *Recent Advances in Astronomy and Astrophysics,*
7th *International Conference of the Hellenic Astronomical Society,* edited by N. H. Solomos

emission lines from He, C, N, O, Ne, Mg, S, Cl, Ar etc. ions in the UV, optical and IR spectral domains. The emission lines involved belong in two distinct categories: collisionally excited lines (CELs; e.g. [O III] $\lambda\lambda$4959, 5007) and optical recombination lines (ORLs) of hydrogen (e.g. the Balmer lines), helium and of heavy ions (e.g. C II λ4267, N II λ4041, O II λ4650). Abundances of heavy elements derived from ORLs scaled to a hydrogen recombination line should in principle be more accurate than those derived from CEL to hydrogen line ratios, due to the exponential dependence upon the adopted nebular temperature of the latter line ratios.

THE PROBLEM

Extensive surveys of a large number of PNe that were recently published (Tsamis et al. 2003b, 2004; Liu et al. 2004; Wesson et al. 2005) show that for most nebulae the abundances of C, N, O, and Ne derived from analyses of their faint ORLs (e.g. C II λ4267, O II λ4650) are systematically higher than those obtained from the classical method which utilizes the bright CELs, with abundance discrepancy factors (ADFs) of $\sim$2 obtained for most nebulae. For about 10% of nebulae however the ADFs have large values of 4–80. A similar situation prevails in H II regions where the ADFs are relatively low (2–5; Tsamis et al. 2003a).

A closely related problem (Liu et al. 2001; Tsamis et al. 2004) involves the fact that the hydrogen Balmer jump electron temperatures of PNe are lower, on average, than those derived from the [O III] λ4363/λ5007 line ratio (Peimbert 1971); the latter is the most widely used nebular thermometer. This temperature discrepancy, taken as a clue to the nebular thermal structure, has been widely used for empirical corrections to the CEL-based heavy element abundances via the introduction of the t^2 temperature fluctuation parameter which quantifies the excursions of the local electron temperature in the nebula from the mean temperature (Peimbert 1967). Similar corrections are also done by estimating t^2 from the observed ADFs of heavy ions, resulting in the adoption of final heavy-element abundances that are invariably very close to those obtained from ORL ratios for both PNe (e.g. Ruiz et al. 2003; Peimbert et al. 2004) and for H II regions (Peimbert 2003; Esteban et al. 2005). However, attempts by either high spatial resolution imaging or spectroscopy to detect significant local temperature fluctuations that would justify the empirical corrections to the derived abundances have so far been unsuccessful (e.g. Rubin et al. 2002; O'Dell et al. 2003; Wesson & Liu 2004).

This 'thermal and abundance discrepancy' problem is important since, if we were to casually adopt the high ORL abundances for CNONe, this would have profound implications for studies which rely on accurate PN and H II region abundances in the Galaxy and beyond in order to set constraints on cosmic chemical evolution models.

A NEAT SOLUTION: THE DUAL-ABUNDANCE PARADIGM

A proposition that aspires to explain both the temperature and abundance discrepancies is one according to which PNe are chemically inhomogeneous objects, incorporating two (or more) gas phases of very different elemental composition. This description assumes the existence of hydrogen-poor clumps or filaments within a PN, embedded in material of solar-like abundances. It predicts that the overwhelming fraction of heavy-element ORL emission arises from H-poor regions, which due to their enhanced metallicity have efficiently cooled down to ~1000 K by emitting IR fine-structure lines (Harrington & Feibelman 1984; Liu et al. 2000). Observational evidence in support of the dual-abundance model comes from direct measurements of the average temperatures in the inferred H-poor regions, by means of He I and O II ORL ratios, which yield very low electron temperatures indeed (Tsamis et al. 2004; Liu et al. 2004; Wesson et al. 2005). At the moment the exact nature and origins of the inferred H-poor regions in PNe remain unclear. Hypotheses include: (i) the evaporation of primitive material left over from the formation of the PN precursor star (e.g. planetesimals), (ii) incompletely mixed dredged-up material ejected along with the rest of the star's outer envelope during the PN formation phase. For H II regions one could appeal to incompletely mixed supernova ejecta which occur frequently in OB-type stellar associations (Tsamis & Péquignot 2005).

Testing the solution with VLT FLAMES observations

The proposed solution lacks further direct observational support. According to our analyses from self-consistent photoionization models, a few jovian masses of H-poor material immersed in gas of normal metallicity are enough to explain the CNONe ORL emission and resolve the discrepancy paradox. The small mass and low filling factor of the inferred inclusions means that their sizes will be similarly small and large telescopes providing high spatial resolution and excellent sensitivity are required to detect them – and the VLT 8.2-m/Kueyen is a leader in this class. We have thus acquired during Jun–Aug 2005, FLAMES integral-field unit (IFU) spectroscopy in Argus mode of 3 Galactic PNe (NGCs 5882, 6153, 7009) which show a wide range of ADF behaviour (~2–10).

A first glimpse of the secured raw spectra of NGC 7009 is provided in Fig. 1 taken with the 6.6″ × 4.2″ IFU at a spectral resolution, $\lambda/\delta\lambda$, of 32 500 (i.e. 9.2 km s^{-1}). The plotted spectra have been extracted over 5 and ~80 spatial pixels (top and bottom tracings in each of the two panels respectively) in order to highlight the change in the S/N ratio of the C II and O II ORLs. Note the split line profiles due to nebular expansion, and how the components of the C II λ4267 doublet are clearly resolved in each of the blue- and red-shifted nebular portions (Fig. 1, top panel). Several exposures will be combined so that the increased S/N per spatial pixel will allow us to map the fibre-to-fibre (spatial) variations of the velocity profiles of ORLs versus CELs (e.g. C II λ4267 vs. [O III] λ4363) and determine whether the two types of line arise from kinematically distinct regions. This might be one of the signatures of a discrete H-poor gas phase embedded within the nebulae.

In order to elucidate further the nature of the inferred H-poor regions we aim to: (i) construct temperature and density maps across the IFUs and determine the variations of chemical composition (and of the ADF) relative to the position of the PN nuclei; (ii) resolve or set strict upper limits to the sizes of the inferred H-poor clumps by comparing C II λ4267/Hγ vs. [O III] λ4363/Hγ intensity ratio maps at the highest (0.3″) spatial

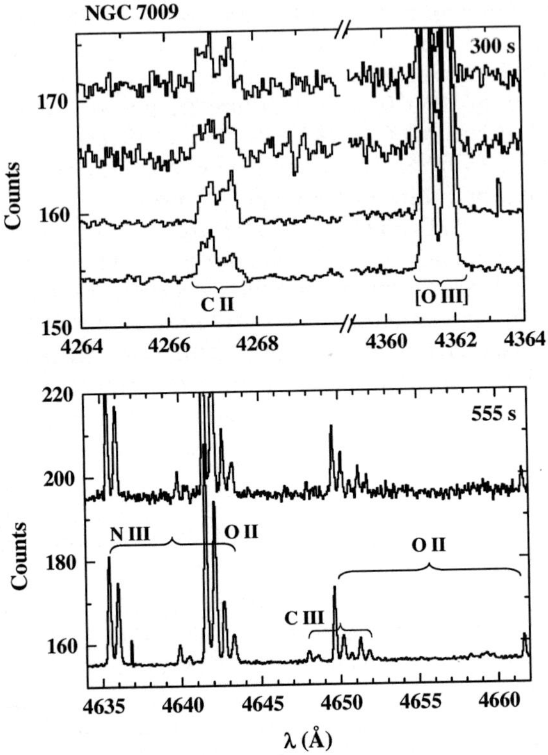

FIGURE 1. The emission line spectrum of the planetary nebula NGC 7009, taken with VLT FLAMES/Argus at a spatial resolution of 0.3″ per fibre, showing the C II λ4267 and O II V1 multiplet optical recombination lines at a spectral resolution of 0.05 Å. The lines are very well resolved. The seing was 0.5″.

resolution; (iii) create images of the nebulae in the light of heavy-element C II λ4267 and O II λ4649 ORLs. All this will help us to pinpoint the emitting regions of the elusive H-poor nebular component, constrain its physical properties and discriminate amongst the various hypotheses regarding its origin.

Small-scale chemical inhomogeneities within H II regions?

In order now to test the dual abundance paradigm on a template H II region, we constructed photoionization models of the giant H II region 30 Doradus in the Large Magellanic Cloud (LMC) which is the nearest extragalactic starburst (Tsamis & Péquignot 2005). The models were confronted to available UV, optical, IR (*ISO*) and radio spec-

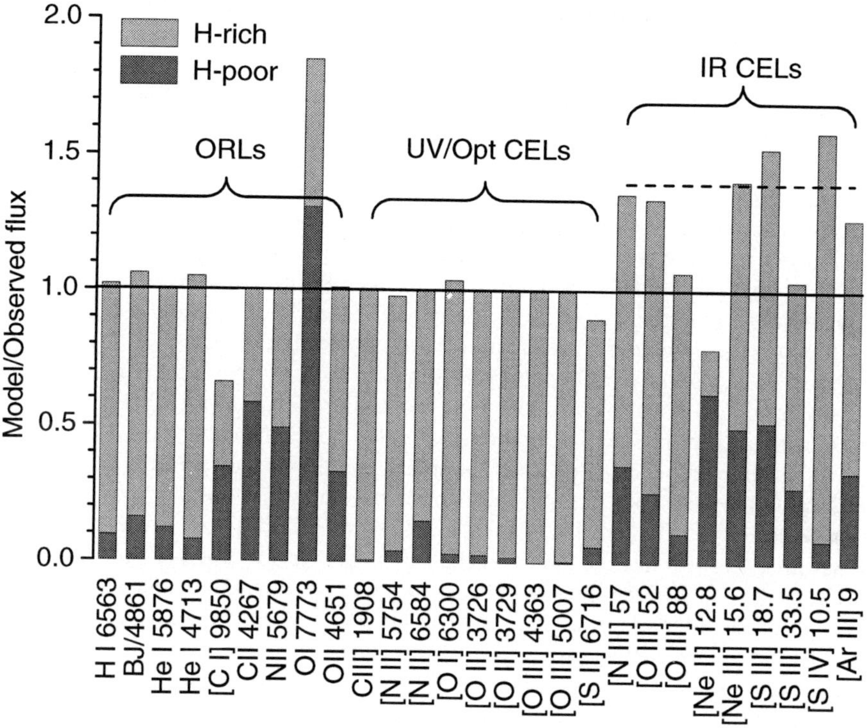

FIGURE 2. The model over observed flux ratios for the chemically inhomogeneous model of 30 Doradus: the relative contributions of each nebular component to the individual line fluxes are colour-coded. Three major groups of lines are identified: the optical recombination lines, the UV and optical collisionally excited lines, and the IR collisionally excited lines (from *ISO* observations). The dashed line shows the suggested re-calibration of the IR fluxes relative to the optical.

tra, under black-body or *CoStar* spectral energy distributions for the primary source and various density distributions for the nebular gas. We found that chemically homogeneous models of 30 Dor show very small *rms* electron temperature fluctuations [$t^2(H^+)$ $=0.005$] and fail to reproduce the heavy element ORL spectrum of the nebula. On the other hand, dual abundance models incorporating small-scale (<1 pc) chemical inhomogeneities in the form of hydrogen-deficient inclusions, which are in pressure balance with the normal composition ambient gas, provide a simultaneous fit to the observed heavy element ORLs *and* CEL-based diagnostics (see Fig. 2).

The inferred inclusions comprise ∼2 per cent of the total gaseous mass and are 2–3 times cooler (T_e = 4000–5000 K), and denser (N_e = 440–700 cm^{-3}) than the ambient nebula, emitting a substantial fraction of the total flux of metallic ORLs such as C II λ4267 and O II λ4650. The modelled density of the H-poor gas agrees very well with that inferred from the observed intensities of O II V1 multiplet components relative to the total multiplet intensity. The O/H abundance ratio of the H-poor gas is ∼0.9 dex larger than in the normal composition nebular component: helium is found to be about as deficient as hydrogen in the inclusions, while elements heavier than neon, such as sulfur and argon, are quite possibly enhanced in proportions similar to oxygen. This suggests that the posited H-deficient gas may have arisen from partial mixing of matter which was nucleosynthetically processed in a supernova event with gas of normal LMC composition.

For 30 Dor, the presence of H-deficient inclusions warrants an overall upward revision of the heavy element abundances by at most 0.1 dex, compared to empirical methods or models that do not take into account the heavy element ORLs, and indicates that the latter techniques provide only an upper limit to the helium abundance in H II regions; i.e. it is possible that the He/H ratios in extragalactic H II regions may have been systematically *overestimated* thus far. In 30 Dor however this effect is small ($\sim$3 per cent). Importantly, the mean C/H and Fe/H abundance ratios in the nebula are $\sim$0.3 dex *lower* than those advocated by workers who, following the t^2 empirical approach, apply upward corrections to the CEL-based heavy element abundances.

Even though the presence of chemical inhomogeneities is warranted in this particular case it should be pointed out that the chemical homogeneity of the ISM has not been seriously challenged thus far, except in a few cases. A recent review by Scalo & Elmegreen (2004) summarized the state of the art in the field: observationally, the Galactic ISM appears to be well mixed over scales of $\sim$1 kpc with oxygen abundance dispersions of only a few percent (e.g. Moos et al 2002; André et al. 2003). Abundance analyses of F-type stars in nearby young open clusters found no cluster-to-cluster variations implying a well-mixed progenitor ISM on scales $\sim$1–100 pc (Friel & Boesgaard 1992).

Importantly, Casassus et al. (2005) have reported *rms* variations of 16.2 per cent in the $^{13}C/^{12}C$ isotopic ratio on scales of $\sim$100 pc providing the first significant detection of chemical inhomogeneity in the local ISM. Moreover, theoretical studies suggest that the ISM *should* appear rather inhomogeneous given the initially localized introduction of heavy elements from supernovae and other sources: de Avillez & Mac Low (2002) note that in a supernova-driven ISM chemical inhomogeneities take a very long time to be erased (with 10^8 Myr being a lower limit since no sources of new metals were considered). Such long timescales may indeed indicate that it is quite plausible for 'oxygen-rich' material ejected by supernovae erupting within OB associations, and then raining down on to a galactic disk (the *galactic fountain model*, e.g. Tenorio-Tagle 1996), to disperse sufficiently slowly so that it can be 'detected' as a sub-pc sized, cool, and rather dense, ORL-emitting plasma upon its photoionization in the vicinity of a young H II region.

CONCLUSIONS

Emission-line nebulae and studies of their composition remain as important and exciting now as they were in the 1960's. In this contribution we dwelt on a specific problem, that of the 'temperature and abundance dichotomy', in order to emphasize how PNe were used as tools that fostered new thinking in the adjacent field of H II region research. We discussed how according to the dual abundance paradigm (i) self-consistent nebular models can be proposed that are free of spectroscopic 'anomalies' such as the ORL/CEL discrepancy, and that (ii) metallic ORLs do not provide an accurate estimate of the average heavy element to hydrogen ratios in either PNe or H II regions, but are useful indicators of small-scale chemical inhomogeneities.

Current efforts are focused in an attempt to isolate at high spatial/spectral resolution the inferred H-deficient inclusions in a sample of high-ADF PNe via means of IFU spectroscopy taken with VLT FLAMES (Tsamis et al. in preparation); they were briefly discussed in this report. Such spatially-resolved data would also be worthy of the new generation 3D plasma codes (e.g. Ercolano et al. 2003). Renewed emphasis should therefore be placed on investigating the origins, survivability, evolution, and detection of chemical inhomogeneities in diffuse and ionized interstellar media via a concerted observational and theoretical effort. It should be noted that on the observational front Greece is now favourably placed to foster this kind of research with the advent of the new Aristarchos telescope and its modern échelle spectrograph. An abundance of northern nebulae and their secrets, all the way out to Local Group galaxies, are within easy reach from the top of Helmos.

ACKNOWLEDGMENTS

In investigating the issues discussed in this report I have benefited from an ongoing collaboration with Drs. J. R. Walsh, D. Péquignot, M. J. Barlow and X.-W Liu.

REFERENCES

1. André M. K., Oliveira C. M., Howk J. C., et al. 2003, ApJ, 591, 1000
2. Barker, T. 1991, ApJ, 371, 217
3. de Avillez, M. A., & Mac Low, M. 2002, ApJ, 581, 1047
4. Casassus, S. et al. 2005, A&A, 441, 181
5. Ercolano, B., Barlow, M. J., Storey, P. J., Liu, X.-W., Rauch, T., & Werner, K. 2003, MNRAS, 344, 1145
6. Esteban, C., García-Rojas, J., Peimbert, M., Peimbert, A., Ruiz, M. T., Rodríguez, M., & Carigi, L. 2005, ApJL, 618, L95
7. Friel, E. D., & Boesgaard, A. M. 1992, ApJ, 387, 170
8. Garnett, D. R., & Dinerstein, H. L. 2001, ApJ, 558, 145
9. Harrington, J. P., & Feibelman, W. A. 1984, ApJ, 277, 716
10. Iben, I., Kaler, J. B., Truran, J. W., & Renzini, A. 1983, ApJ, 264, 605
11. Liu, X.-W. 2003, IAU Symposium, 209, 339
12. Liu, X.-W., & Danziger, J. 1993, MNRAS, 263, 256
13. Liu X.-W., Storey P. J., Barlow M. J., Danziger I. J., Cohen M., Bryce M., 2000, MNRAS, 312, 585
14. Liu X.-W., Luo S.-G., Barlow M. J., Danziger I. J., Storey P. J., 2001, MNRAS, 327, 141
15. Liu, Y., Liu, X.-W., Barlow, M. J., & Luo, S.-G. 2004, MNRAS, 353, 1251
16. Luo, S.-G., & Liu, X.-W. 2003, IAU Symposium, 209, 393
17. Moos H. W., Sembach K. R.,Vidal-Madjar, A., et al. 2002, ApJS, 140, 3
18. O'Dell, C. R., Peimbert, M., & Peimbert, A. 2003, AJ, 125, 2590
19. Peimbert, A. 2003, ApJ, 584, 735
20. Peimbert M., 1967, ApJ, 150, 825
21. Peimbert M. 1971, Bol. Inst. Tonantzintla, 6, 29
22. Peimbert, M., Peimbert, A., Ruiz, M. T., & Esteban, C. 2004, ApJS, 150, 431
23. Péquignot D., Amara M., Liu X.-W., Barlow M. J., Storey P. J., Morisset C., Torres-Peimbert S., & Peimbert M. 2002, RMxAA (Ser. de Conf.), 12, 142

24. Péquignot, D., Liu, X.-W., Barlow, M. J., Storey, P. J., & Morisset, C. 2003, in Planetary Nebulae: Their Evolution and Role in the Universe, IAU Symp. 209, Eds S. Kwok, M. Dopita, & R. Sutherland (ASP), p. 347
25. Rubin, R. H., Martin, P. G., Dufour, R. J., Ferland, G. J., Blagrave, K. P. M., Liu, X.-W., Nguyen, J. F., & Baldwin, J. A. 2003, MNRAS, 340, 362
26. Ruiz, M. T., Peimbert, A., Peimbert, M., & Esteban, C. 2003, ApJ, 595, 247
27. Scalo J., & Elmegreen B. G. 2004, ARA&A, 42, 275
28. Tenorio-Tagle G. 1996, AJ, 111, 1641
29. Tsamis Y. G., Barlow M. J., Liu X.-W., Danziger I. J., Storey P. J., 2003a, MNRAS, 338, 687
30. Tsamis Y. G., Barlow M. J., Liu X.-W., Danziger I. J., Storey P. J., 2003b, MNRAS, 345, 186
31. Tsamis, Y. G., Barlow, M. J., Liu, X.-W., Storey, P. J., & Danziger, I. J. 2004, MNRAS, 353, 953
32. Tsamis Y. G. & Péquignot, D., 2004, in press, in Proc. of ESO workshop 'Planetary Nebulae beyond the Milky Way', Springer-Verlag, `http://www.arxiv.org/abs/astro-ph/0407506`
33. Tsamis Y. G. & Péquignot D., 2005, MNRAS, 364, 687
34. Wesson, R., Liu, X.-W., & Barlow, M. J. 2003, MNRAS, 340, 253

Mapping and multi-wavelength properties of stellar complexes in the LMC

E. Livanou*, M. Kontizas*, I. Gonidakis*, E. Kontizas†, F. Maragoudaki*, S. Oliver**, A. Efstathiou‡ and U. Klein§

*Department of Astrophysics Astronomy & Mechanics, Faculty of Physics, University of Athens, GR-15783 Athens, Greece

†Institute for Astronomy and Astrophysics, National Observatory of Athens, P.O. Box 20048, GR-11810 Athens, Greece

**Astronomy Centre, Department of Physics and Astronomy, University of Sussex, Brighton, BN1 9QJ, UK

‡Department of Computer Science and Engineering, Cyprus College, 6 Diogenes Str, 1516 Nicosia, Cyprus

§Radioastronomisches Institut der Universitat, Bonn, Auf dem Hugel 71, 53121, Bonn, Germany

Abstract. Filamentary structures of early type stars are found to be a common feature of the Magellanic Clouds formed at an age of about $0.9-2\times10^{8}\,yr$. As we go to younger ages these large structures appear fragmented and sooner or later form young clusters and associations. In the optical domain we have detected 56 such large structures of young objects, known as stellar complexes in the LMC for which we give coordinates and dimensions. IR properties of these regions have been investigated using IRAS data. The colour-magnitude diagram (CMD) and two-colour diagrams from IRAS data of these regions have been compared with observations of starburst galaxies. It has been found that nearly 1/3 of the stellar complexes are extremely active resembling the IR behaviour of starburst galaxies and H II regions. These stellar complexes illustrating such properties are called here "starburst regions". They host an increased number of H II regions and SNRs. The main starburst tracers are their IR luminosity (F_{60} well above 5.4 Jy) and the 8.6-GHz radio emission. Finally the evolution of all stellar complexes is discussed based on the CO emission.

Keywords: LMC, star forming regions, stellar complexes, starburst galaxies
PACS: 01.30.Cc, 97.10.Bt, 98.56.Si

INTRODUCTION

Due to its proximity to our Galaxy and the SMC, the LMC offers two important advantages: i) the possibility to study its stellar content down to low stellar masses and ii) the possibility to follow the consequences of the close interaction, about $2-4\times10^{8}\,yr$ ago (Gardiner & Noguchi [1]; Kunkel et al. [2]), with the SMC. Stellar complexes are defined as large structures (clump like) in galaxies dominated by recently formed stellar component mixed with very young clusters, stellar associations and gas (Martin et al. [3]; van den Bergh [4]; Elmegreen B.G. et al. [5, 6, 7]; Feitzinger et al. [8]; Ivanov [9]; Larson et al. [10, 11]; Efremov et al. [12, 13, 14]; Elmegreen D.M. et al. [15]; Kontizas et al. [16, 17]; Battinelli et al. [18]). So far they have been detected as regions with young stellar component and enhanced stellar number density, compared to the surrounding region. In the LMC stellar complexes have been detected using UK Schmidt plates (Maragoudaki et al. [19]).

CP848, *Recent Advances in Astronomy and Astrophysics,*
7th *International Conference of the Hellenic Astronomical Society,* edited by N. H. Solomos

Concentrations of early type stars are found to form filamentary structures in both Magellanic Clouds (Maragoudaki et al. [20]) at an age of about $0.9-2\times10^{8}\,yr$. In the optical domain we have detected 56 such large structures of young objects. These stellar complexes are associated with the already known nine Shapley constellations (Shapley [21]), which are referred to as large stellar structures of early type supergiants.

In normal galaxies it has been assumed that the emission into the IRAS photometric bands is mostly due to thermal radiation from dust grains heated by the interstellar radiation field. So the IRAS emission is a function of the present stellar population, the spatial distributions of dust, the stars and the grain properties. Helou ([22]) found that the IRAS $log(F_{60}/F_{100})$ versus $log(F_{12}/F_{25})$ colour-colour diagram of "normal galaxies" shows a distribution that extends continuously from the cool ($\sim$ 20K) relatively constant "cirrus" emission from the neutral medium to the warmer ($\sim 40-50$K) emission from the active, starburst end.

IRAS far-IR fluxes for 6 H II regions in the LMC have been studied by DeGioia-Eastwood ([23]). These regions are the sites of massive star formation, where the radiative heating source is young stars rather than the general interstellar radiation field. Such regions are expected to lie in stellar complexes and/or to be the stage just before the stellar complex is formed.

The highly structured diffuse X-ray emission of the LMC has been imaged in detail by ROSAT. There is a strong correlation between diffuse features in the X-ray image and ESO colour images of the LMC in visible light. Bright knots in the X-ray map correspond to H II emission in the optical (Westerlund [24]). The distribution of the SNRs in the LMC shows (i) a clumping of objects in the 30 Doradus region, (ii) several remnants within the Bar of the LMC and (iii) that the remainder of the remnants are found in super-associations. Most SNRs in the Bar are in regions where many young clusters are located, clusters as young as $\leq 10\,Myr$ (Westerlund [24]).

The LMC has been well studied in the radio continuum, which is connected to the formation of new and massive stars. Fukui et al. ([25]) found that the molecular clouds in the LMC have a good correlation with the youngest ($\leq$10 Myr) stellar clusters, while there is little spatial correlation of these clouds with SNRs or with the older stellar clusters.

In this paper we describe how the stellar complexes were detected, using optical data and we give their location and dimensions. The total flux of each structure and IR two-colour diagrams from the IRAS images at 12, 25, 60, and 100 μm are presented. Their derived IR properties are compared to IR properties of galaxies in order to identify the distinction between ordinary stellar complexes and active star formation regions. Possible connection of the complexes determined here with known X-ray detected SNRs and candidates as well as H II regions is discussed. Finally, their correlation between high resolution radio data at 8.6-GHz and CO emission is examined, in order to determine the activity of the complexes and their evolution.

OPTICAL DATA

The large-scale structures in nearby galaxies require extended field observational material in order to be treated homogeneously. Therefore, we used direct photographic

TABLE 1. Characteristics of the UKST photographic plates.

Plate No	RA (1950)	DEC (1950)	Emulsion	Filter	exp time (min)	comments
U12346	05:26	-69:38	IIIaJ	UG1	150	
F56	05:12	-70:00	IIIaJ	RG630	150	
UJ5416P	05:11	-69:56	IIIaJ	none	60	obj. prism
He10562	05:40	-67:00	IIIaJ	SM468B	10	
He10625	05:10	-67:00	IIIaJ	SM468B	15	
He10633	05:10	-70:00	IIIaJ	SM468B	240	
He10562	05:40	-70:00	IIIaJ	SM468B	7.5	

plates taken with the UK 1.2m Schmidt Telescope, in various wavebands: U, R and HeII ($\lambda = 4686$ Å) down to a magnitude limit of $19 - 20\,mag$. The characteristics of these plates are summarised in Table 1. The plates were digitised using the fast measuring machines APM and SuperCosmos. The derived data were produced as catalogues of detected stellar images. Star counts were performed on the above catalogues in order to derive isopleth contour maps of the LMC in the various wavebands and magnitude slices, to trace the differences between faint and brighter stellar populations.

Star counts, in the optical wavelength range, of large areas in the LMC have revealed 56 groupings of young stars, known as stellar complexes (with dimensions from 150 pc to $\sim 1500\,pc$). They are found to be formed in an hierarchy of scales: the largest structures (super-complexes and complexes) contain smaller groupings (the aggregates) which again consist of even smaller stellar groups such as clusters and associations. These stellar complexes, selected to have stellar component densities greater than $3\,\sigma$ of the local background, have shapes usually irregular and are located along the filamentary density fluctuation, where they are formed (Maragoudaki et al. [19]). Their coordinates, dimensions and properties are given in Table 3, whereas their ages span from a few $\times 10^8\,yr$ to very young ($\sim 10^6\,yr$). The ages are derived from their stellar component, as described by Maragoudaki et al. ([19]), which consist of early type stars (O, B, A), dust, young clusters and associations, very often in groups. The parental density fluctuations, which were fragmented into stellar complexes can be explained as a result of the close encounter of the SMC with the LMC that happened a few $\times 10^8\,yr$ ago (Gardiner & Noguchi [1]; Kunkel et al. [2]).

IR DATA

IRAS all-sky survey Images taken at 12, 25, 60 and 100 μm have been obtained from SkyView (http://skyview.gsfc.nasa.gov/skyview.html) for the whole area of the LMC to study the IR properties of the stellar complexes detected in the optical domain. From the IRAS catalogues we calculated a total flux for each complex as the sum of the fluxes corresponding to the pixels, which cover the predetermined complex area. These total fluxes were corrected from the background. The background has been calculated twice under two assumptions. 1) The minimum flux value in the image was adopted as the background and 2) a background value was assumed as the average of four flux/pixel values at the four rather clear corners of each frame, for all IRAS wavelengths

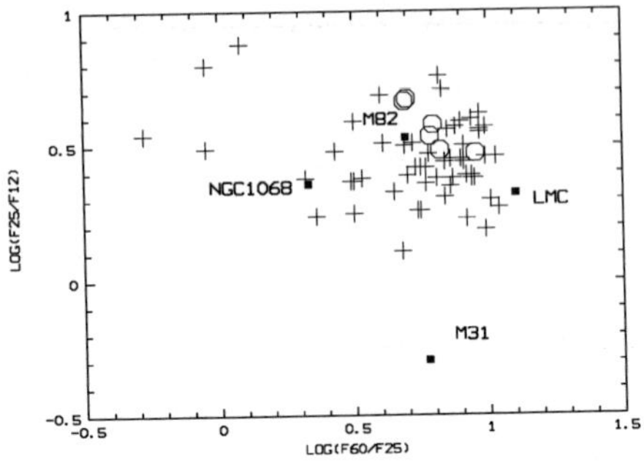

FIGURE 1. IRAS colour-colour diagram for the detected stellar complexes. The black-filled squares mark the observed colours of the LMC (irregular galaxy) and three typical galaxies: M31 (disk), M82 (starburst) and NGC1068 (AGN). The circles mark LMC H II regions.

respectively. The second case minimises better the effect of the background and thus is adopted in this investigation. The background corrected total fluxes were averaged to a mean value (flux/pixel in MJy/sr) for all complexes so that they can be more easily compared with the literature.

Fig. 1 and 2 show the IRAS colour-colour diagram of $log(F_{25}/F_{12})$ vs $log(F_{60}/F_{25})$ and $log(F_{100}/F_{60})$ vs $log(F_{60}/F_{25})$ for the detected stellar complexes. The black-filled squares represent the integrated values for NGC1068, M82, M31 and LMC respectively, indicating the loci of different types of galaxies on this diagram. NGC1068 is a prototype AGN (Rowan-Robinson & Crawford [26]), M82 is a prototype starburst galaxy, while M31 is a typical disk galaxy (Rice, W., et al. [27]). The LMC H II regions of DeGioia-Eastwood ([23]) are plotted with circles.

In Fig. 1 we notice that many stellar complexes lie around the location of M82, a prototype starburst, and the H II regions revealing enhanced star formation activity. In Fig. 2 we notice the presence of some complexes near H II regions, however, there is no significant concentration near the prototype starburst galaxy, as it happens in Fig. 1. This is probably an indication of the importance of F_{60} and F_{100} for tracing the starburst phenomenon. The mean colour for known main sequence stars detected by IRAS is plotted with a black triangle. The stellar complexes found around this spot, will later be proved not active considering star formation.

The IRAS flux densities in $logF_{60}$ versus the IRAS colour $log(F_{60}/F_{100})$ for all the large structures detected here are illustrated in Fig. 3. Lehnert and Heckman ([28]) seeking to improve our understanding of the range of physical processes occurring within IR selected galaxies, found that the starburst galaxies are IR "warm" if $F_{60}/F_{100} \geq 0.4$ ($log(F_{60}/F_{100}) \geq -0.39$) and IR "bright" if $F_{60} \geq 5.4$ Jy ($log(F_{60}) \geq 0.73$). From Fig. 3 it is obvious that nearly 1/3 of the detected stellar complexes fulfil the criteria defined by Lehnert & Heckman ([28]) for the starburst galaxies. The three upper points correspond to regions in 30 Doradus (complexes C17, A21, A24), where there is the highest concentration of SNRs. The 15 structures that have $log(F_{60}/F_{100}) \geq -0.39$ and $log(F_{60}) \geq 0.73$, resemble the behaviour of starburst galaxies, whereas the rest 41 do not

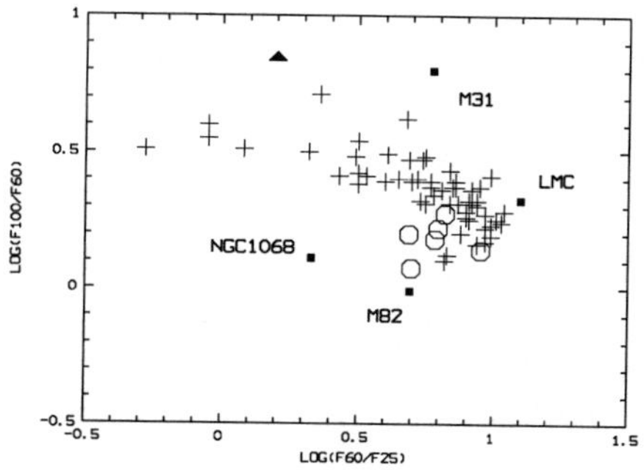

FIGURE 2. Same as Fig. 1 for $log(F_{100}/F_{60})$ vs $log(F_{60}/F_{25})$. The triangle marks the average IRAS colours of the main sequence stars.

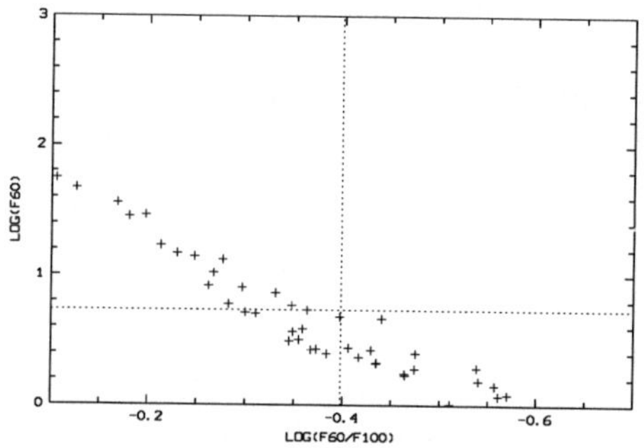

FIGURE 3. IRAS flux density in $logF_{60}$ versus IRAS colour $log(F_{60}/F_{100})$ for all the structures detected here. Two lines are plotted which represent $F_{60} = 5.4Jy$ and $F_{60}/F_{100} = 0.4$. The upper left area indicates the loci of very active star forming regions.

show enhanced star formation activity.

RADIO AND CO DATA

Gas surrounding newly formed stars is heated and ionised by UV light from massive young OB stars. This hot gas radiates thermal free-free emission over a wide range of wavelengths, and stands out clearly in the radio domain, tracing star formation. The molecular gas out of which stars form, can be traced by the CO line. We used the recently published 8.6-GHz map of Dickel et al. ([29]), with its angular resolution of 20 arcsec and linear resolution ~5 pc and the CO (1→0) line that was fully mapped by Fukui et al. ([25]), with angular resolution of 2.6 arcmin corresponding to a spatial resolution of ~40 pc. In Fig. 4 we have superimposed the stellar aggregates, complexes and super-complexes (dotted lines) onto the radio 8.6-GHz (grey-scale) and CO (contours)

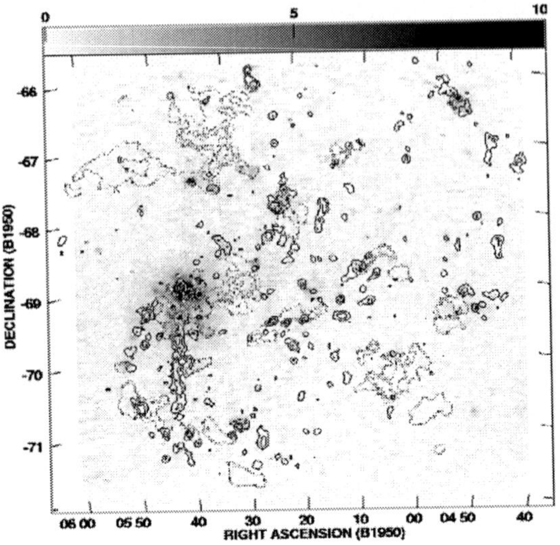

FIGURE 4. Maps of the LMC at 8.6-GHz (grey-scale) and in the CO(1→0) line (contours), with the stellar complexes superimposed with dotted lines.

map. The time sequence evident in this figure is such that the molecular gas (CO) traces regions of ongoing and (near-) future star formation, the ionised gas (8.6-GHz) locates ongoing and somewhat evolved star formation ($10^7 \sim 10^8$ yr). We examined the correlation between 8.6-GHz radio emission, the CO emission and the location of the 56 stellar complexes (Table 3, columns 8 and 9).

We notice that the majority (13 out of 15) of the regions with $F_{60} > 5.4Jy$, coincide well with the 8.6-GHz radio emission. The existence of CO emission in seven of them, reveals future star formation activity. The rest 2, that lack both 8.6-GHz and CO radiation, are actually found to have values of F_{60} very close to $5.4Jy$. Considering the 41 regions with $F_{60} < 5.4Jy$, the opposite behaviour is revealed. 36 out of 41 regions do not exhibit significant amount of 8.6-GHz radiation, while the remaining 5 that do exhibit 8.6-GHz radiation, are found to have F_{60} very close to $5.4Jy$.

The CO emission is indicative of on-going and possible future star formation activity and there is no pattern revealed regarding the F_{60} of the complexes. 16 out of the 56 complexes have significant CO emission, indicating possible future star formation activity and only 10 of them are correlated with 8.6-GHz emission as well.

DISCUSSION & CONCLUSIONS

From optical observations, 56 stellar complexes are revealed in an area of $6 \times 7\,\mathrm{deg}^2$ of the LMC. The IR properties are studied using the IRAS fluxes. The ratio of the flux densities F_{60}/F_{100}, the flux density F_{60} as discussed above, the 8.6-GHz and CO radio emission, will be used in order to classify these complexes.

As mentioned in section 3, Lehnert and Heckman ([28]) showed that starburst galaxies

TABLE 2. Complexes characterisation.

Complex Type	F_{60}	8.6-GHz	No. of Complexes
starburst	F_{60} well above 5.4 Jy	yes	13
starburst candidate	F_{60} grater but close to 5.4 Jy	yes	5
active complex candidate	F_{60} less but close to 5.4 Jy	no	2
active complex	F_{60} well bellow 5.4 Jy	no	36

are IR "warm" and "bright". In the same manner, we adopt the equivalent term **"starburst regions"** for complexes that are found to fulfil these criteria, namely $F_{60}/F_{100} \geq 0.4$ and $F_{60} \geq 5.4$ Jy. In addition, complexes that do not fulfil the previously stated criteria will be called **"active complexes"**.

A more detailed study of high resolution 8.6-GHz map (Fig. 4) revealed that all complexes with F_{60} well above the 5.4 Jy limit are very well correlated with the 8.6-GHz radio emission. In contrast, complexes with F_{60} well below the 5.4 Jy limit lack 8.6-GHz emission. However, 7 out of the 56 complexes (A5, A6, A7, C5, C7, C22 and A31) have F_{60} close to the classification limit and deviate from the above behaviour. Complex C22, for example, is well correlated with 8.6-GHz emission; its $F_{60}(5.07\pm0.20$ Jy) though is below the accepted limit. In contrast, A31 has F_{60}=*5.98*$\pm$0.20 *Jy*, which would classify it as a starburst region; however, there is no 8.6-GHz emission associated with it. Since the F_{60} criterion is not trustworthy for these regions, we additionally use the 8.6-GHz emission to classify the complexes as "starburst candidates" or "active complexes candidates" depending on its existence. Hence, using the 8.6-GHz radiation as an additional tracer in tandem with the F_{60}, we can classify the stellar complexes into 13 starbursts, 5 starburst candidates, 2 active complexes candidates and 36 active complexes as seen in Table 2. The characterisation of each complex is given in Table 3, column 10.

Moreover, a discrimination concerning the evolution of all the complexes was attempted based on the CO data. More than half of the starburst and starburst candidate regions show enhanced CO emission indicating ongoing and future evolution, while the rest of them are thought to be more or less evolved. Regarding the active stellar complexes (and the two candidates), we found only 5 of them with significant amount of CO emission, giving additional indication of current star formation and revealing potential future starbursts. The lack of CO in the rest of them could indicate that these complexes have never been starbursts or that their starburst activity previously exhausted the molecular gas.

In Fig. 5 we plotted the known SNRs and H II regions based on the literature (Williams et al. [30]; Haberl et al. [31]; Filipovic et al. [32] and Sasaki et al. [33]) over the stellar complexes with triangles and squares respectively. One would expect H II regions to be prominent in areas of high star formation activity since the very massive and bright stars forming in starbursts could ionise the hydrogen in their ambient. The overplotted H II regions show an increased concentration in and around the very active star forming regions, as expected. Moreover, SNRs should also be correlated with starbursts. Most of the defined starburst areas (and the candidates) are also associated with SNRs. The detected complexes are also cross-matched with the catalogue of stellar associations

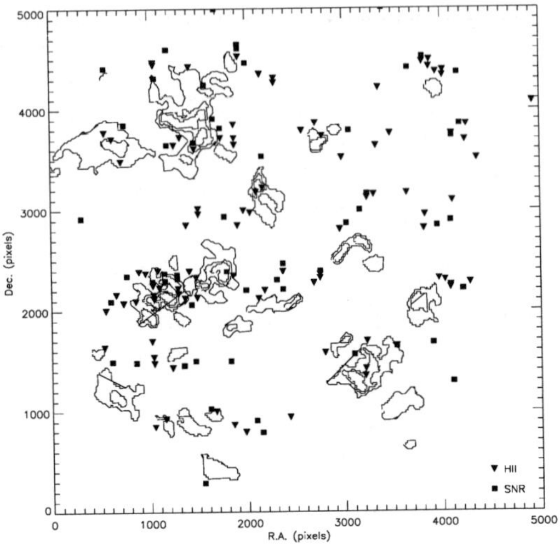

FIGURE 5. The H II regions (triangles) and SNRs (squares), plotted over the stellar complexes as defined here.

(Gouliermis et al. [34]) and that of nebulae in the LMC (Davies et al. [35]). The majority of the complexes are loci of stellar associations, whereas all of them are associated with a large number of nebulosities (Table 3).

ACKNOWLEDGMENTS

The authors wish to thank the UK Schmidt Plate Library for loan of the observational material as well as the staff of SuperCosmos at ROE for digitising the plates. They also thank the SkyView for the IRAS all-sky survey images. We are grateful to Y. Fukui for making the NANTEN CO map of the LMC available to us. We would also like to address our thanks to the British Council and the General Secretariat of Research and Technology for financial support. M. Kontizas would like to thank the University of Athens (ELKE) for partial financial support. Finally M. Kontizas, E. Livanou and I. Gonidakis are very much indebted to the Ministry of Education for the financial support through the "Pythagoras II" program.

REFERENCES

1. Gardiner, L.T. & Noguchi, M., 1996, MNRAS, 278, 191
2. Kunkel, W.E., Demers, S. & Irwin, M.J., 2000, AJ, 119, 2789
3. Martin, N., Prevot, L., Rebeirot, E., Rousseau, J., 1976, A&A, 51, 31
4. van den Bergh, S., 1981, A&AS, 46, 79
5. Elmegreen, B.G. & Elmegreen, D.M., 1983, MNRAS, 203, 31
6. Elmegreen, B.G. & Elmegreen, D.M., 1987, ApJ, 320, 182
7. Elmegreen, B.G. & Efremov, Yu.N., 1996, ApJ, 466, 802

TABLE 3. List of the LMC stellar complexes. Col. 1 indicates the adopted identification name. A denotes aggregate, C complex and SC super-complex. For a structure found inside a larger one, the name of the later is given in parenthesis. Col. 2, 3 and 4 give the RA, DEC of the "centre" of each stellar complex and its dimension. Col. 5,6, 7 show the number of SNRs and candidates, stellar associations and number of Nebulae and Henize objects found in each complex. Col. 8 and 9 indicate presence of 8.6-GHz and CO line emission respectively. Finally in Col. 10 the complexes are characterised. I indicates starbursts, II starburst candidates, III active complexes candidates and IV active complexes.

Name	RA(1950) h m s	DEC(1950) deg m s	Dim. (pc)	No SNRs	Association ID	No Nebulae	8.6 GHz	CO line	Type
SC1	05:44:39	-67:15:18	1421.5	1	173,257	46	-	-	IV
C1 (SC1)	05:43:30	-67:17:29	355.3	0	173,257	57	-	-	IV
A1	05:52:53	-67:29:46	205	0	173,257	22	-	-	IV
A2	05:43:44	-67:19:37	191	1	173,257	31	-	-	IV
SC2	05:31:22	-66:54:42	1526.5	7		24	-	-	IV
C2 (SC2)	05:32:45	-67:05:11	749	1	173	41	-	-	IV
C3 (C2)	05:32:47	-66:58:22	572	1		32	-	-	IV
A3 (SC2)	05:29:45	-66:37:17	204	0		31	-	-	IV
A4 (C2)	05:30:53	-66:39:15	245	0		26	-	-	IV
A5 (SC2)	05:32:55	-67:32:29	164	1	173,257	41	-	-	III
A6	05:25:58	-66:12:18	299	1		20	√	√	II
A7	05:27:01	-67:27:42	299	1	173,257	10	√	-	II
C4	05:13:28	-67:20:57	314	0	173,257	16	-	-	IV
A8 (C4)	05:13:10	-67:22:34	246	0	173,257	16	-	-	IV
A9	05:10:55	-67:11:35	192	0	173,257	26	-	√	IV
A10	04:57:01	-66:28:27	245	1		7	√	√	I
C5	05:20:58	-68:10:16	765	2		17	√	√	II
C6 (C5)	05:22:13	-67:58:26	342.6	1	173,257	36	√	√	I
A11 (C6)	05:22:13	-67:56:44	205	1	173,257	17	√	√	I
C7	05:07:30	-68:44:27	545	1		39	√	√	II
A12 (C7)	05:09:04	-68:49:37	287	1		17	√	√	I
A13 (C7)	05:05:38	-68:39:12	245	0		27	-	-	IV
A14	05:04:00	-68:57:45	230	0		11	-	-	IV
C8	04:55:33	-69:32:52	628	1		12	-	-	IV
A15 (C8)	04:56:55	-69:30:18	273	1		10	-	-	IV
C9	05:09:26	-70:08:58	328	0	150,153,151	19	-	-	IV
A16 (C9)	05:08:09	-70:05:05	246	0	150,151	10	-	-	IV
SC3	05:04:44	-70:29:23	1093.2	2	150,153,151	23	-	-	IV
C10 (SC3)	05:05:01	-70:32:56	478.3	1	150,153,151	13	-	-	IV
C11(C10)	05:06:16	-70:40:18	328	0	150,153,151	17	-	-	IV
C12	04:58:02	-70:53:27	574	0		2	-	-	IV
A17	04:56:28	-71:28:37	164	0		0	-	-	IV
C13	05:30:55	-71:50:37	574	0	76	6	-	-	IV
C14	05:39:55	-71:11:22	328	0	76	10	-	√	IV
C15	05:35:42	-71:12:22	437	0	76	17	-	√	IV
A18	05:31:26	-71:06:11	218.64	2	76	21	√	√	I
C16	05:47:50	-70:41:07	615	1 very close	150,151	15	-	√	IV
A19 (C16)	05:47:53	-70:44:31	210	0	150,153,151	14	-	√	IV
A20	05:49:53	-70:05:51	164	0	150,153,151	10	-	-	IV
C17	05:39:42	-69:21:58	724	10	172	23	√	√	I
C18 (C17)	05:38:17	-69:26:16	463	3	172	50	√	-	I
A21 (C17)	05:36:42	-69:10:51	272.3	5	172	29	√	-	I
A22 (C18)	05:39:34	-69:28:00	218	1	172	47	√	-	I
A23 (C18)	05:36:59	-69:24:30	245	1	172	32	√	-	I
A24 (C17)	05:40:52	-69:38:12	298	2	172	44	√	-	I
A25	05:35:48	-68:55:28	190.61	0		29	√	√	I
A26	05:35:18	-69:43:05	231.45	0	172	46	√	-	I
A27	05:36:33	-70:10:26	272.3	0	150,153,151	32	-	-	IV
C19	05:30:00	-69:11:52	792.57	2	172	10	-	-	IV
C20 (C19)	05:29:42	-69:08:07	517.37	2	172	46	-	-	IV
A28 (C20)	05:31:34	-69:15:19	163	0	172	63	-	-	IV
A29 (C20)	05:28:27	-69:04:25	285	2	172	71	-	-	IV
C21	05:26:03	-69:51:28	463	1 very close	172	36	-	-	IV
A30 (C21)	05:27:03	-69:50:03	163	0	172	28	-	-	IV
C22	05:19:27	-69:34:39	763	2	177	38	√	√	II
A31 (C22)	05:17:09	-69:31:39	300	0	772	20	-	-	III

8. Feitzinger, J.V. & Braunsfurth, E., 1984, A&A, 139, 104
9. Ivanov, G.R., 1987, Astrop.& Sp.Scien., 136, 113
10. Larson, R.B., 1988 In: Galactic and extragalactic star formation, eds. R. Pudritz & M. Fich, p. 459
11. Larson, R.B., 1992 In: Star formation in stellar systems, eds. G. Tenorio-Tagle et al. (Cambridge: Cambridge Univ. Press)
12. Efremov, Yu.N., 1989, Sov.Sci.Rev./Section E, 7, Part 2,
13. Efremov, Yu.N. & Chernin, A.D., 1994, Vistas in Astronomy, 38, 165
14. Efremov, Yu.N., 1995, AJ 110, 2757
15. Elmegreen, D.M., Elmegreen B.G., Lang, C. & Stephens, C., 1994, ApJ, 425, 57
16. Kontizas, M., Kontizas, E., Dapergolas, A., Argyropoulos S., Bellas-Velidis I., 1994, A&AS, 107, 77
17. Kontizas, M., Maravelias, S.E., Kontizas, E., Dapergolas, A., Bellas-Velidis, I., 1996, A&A, 308, 40
18. Battinelli, P., Efremov, Yu.N., Magnier, E., 1996, A&A, 314, 51

TABLE 4. List of the derived IRAS fluxes (MJy/sr per pixel), luminocities (in units of solar luminocities) of the identified stellar groupings at 12, 25, 60 and 100 μ, the IRAS ratio F_{60}/F_{100} and the flux at 60 μm in Jy per pixel.

Name	f12	L12	f25	L25	f60	L60	f100	L100	F_{60}/F_{100}	F_{60}
SC1	0.18	2.8E+06	0.50	3.7E+06	3.63	1.1E+07	8.87	1.7E+07	0.41	0.69 ± 0.25
C1 (SC1)	0.31	0.3E+06	1.02	0.5E+06	5.38	1.1E+06	13.38	1.6E+06	0.40	1.02 ± 0.23
A1	0.36	0.1E+06	1.28	0.2E+06	0.67	0.4E+06	2.16	0.8E+06	0.31	0.13 ± 0.49
A2	0.47	0.1E+06	1.54	0.2E+06	6.32	0.4E+06	19.39	0.7E+06	0.33	1.20 ± 0.23
SC2	1.14	2.0E+07	4.46	3.8E+07	13.96	5.0E+07	33.50	7.2E+07	0.42	2.65 ± 0.21
C2 (SC2)	0.16	0.7E+06	0.56	1.1E+06	4.03	3.5E+06	9.34	4.9E+06	0.43	0.77 ± 0.24
C3 (C2)	0.11	0.3E+06	0.52	0.6E+06	2.08	1.1E+06	5.13	1.5E+06	0.41	0.40 ± 0.29
A3 (SC2)	0.17	0.5E+06	1.10	0.2E+06	0.98	0.6E+05	3.51	0.1E+06	0.28	0.19 ± 0.40
A4 (C2)	0.12	0.6E+05	0.89	0.2E+06	1.06	0.1E+06	3.46	0.2E+06	0.31	0.20 ± 0.38
A5 (SC2)	1.39	0.3E+06	3.80	0.4E+06	30.79	1.3E+06	64.60	0.6E+06	0.48	5.85 ± 0.20
A6	1.16	0.8E+06	2.80	0.9E+06	25.17	3.4E+06	58.76	4.8E+06	0.43	4.78 ± 0.20
A7	0.82	0.6E+06	2.89	0.9E+06	27.08	3.7E+06	50.10	4.1E+06	0.54	5.15 ± 0.20
C4	0.68	0.5E+06	1.53	0.6E+06	11.15	1.7E+06	26.38	0.4E+06	0.42	2.12 ± 0.21
A8 (C4)	0.80	0.4E+06	1.90	0.4E+06	12.30	1.1E+06	28.14	1.5E+06	0.44	2.34 ± 0.21
A9	0.52	0.1E+06	1.66	0.2E+06	8.07	0.5E+06	23.91	0.8E+06	0.34	1.53 ± 0.22
A10	2.26	1.0E+06	8.83	1.9E+06	69.77	6.5E+06	127.91	7.1E+06	0.55	13.26 ± 0.20
C5	1.61	7.2E+06	4.71	1.0E+07	28.23	2.5E+07	61.20	3.3E+07	0.46	5.36 ± 0.20
C6 (C5)	0.10	0.9E+05	6.66	2.9E+06	54.65	9.9E+06	97.19	1.1E+07	0.56	10.38 ± 0.20
A11 (C6)	3.21	1.0E+06	11.89	1.8E+06	89.85	5.8E+06	142.78	5.6E+06	0.63	17.07 ± 0.20
C7	0.96	2.2E+06	2.31	2.5E+06	20.22	9.3E+06	42.25	1.2E+07	0.49	3.84 ± 0.21
A12 (C7)	1.89	1.2E+06	5.31	1.6E+06	42.20	5.4E+06	79.59	6.1E+06	0.53	8.02 ± 0.20
A13 (C7)	0.59	0.3E+06	1.56	0.4E+06	9.20	0.8E+06	22.76	1.3E+06	0.40	1.75 ± 0.22
A14	0.94	0.4E+06	1.90	0.4E+06	13.16	1.1E+06	35.27	1.7E+06	0.37	2.50 ± 0.21
C8	0.90	2.7E+06	2.44	3.6E+06	16.74	1.0E+07	33.96	1.2E+07	0.49	3.18 ± 0.21
A15 (C8)	0.81	0.4E+06	1.86	0.5E+06	11.02	1.3E+06	25.97	1.8E+06	0.42	2.09 ± 0.21
C9	1.20	1.0E+06	3.67	1.5E+06	9.98	1.7E+06	25.75	2.6E+06	0.39	1.90 ± 0.22
A16 (C9)	0.72	0.3E+06	1.77	0.4E+06	8.95	0.8E+06	22.08	1.2E+06	0.41	1.70 ± 0.22
SC3	0.40	3.7E+06	0.87	3.9E+06	3.87	7.1E+06	9.74	1.1E+07	0.40	0.74 ± 0.25
C10 (SC3)	0.46	0.8E+06	1.07	0.9E+06	3.38	1.2E+06	8.85	1.9E+06	0.38	0.64 ± 0.25
C11 (C10)	0.60	0.5E+06	1.44	0.6E+06	4.82	0.8E+06	12.47	1.2E+06	0.39	0.92 ± 0.24
C12	0.37	0.9E+06	0.87	1.1E+06	1.82	0.9E+06	5.75	1.8E+06	0.32	0.35 ± 0.30
A17	0.69	0.1E+06	2.14	0.2E+06	1.89	0.8E+05	7.51	0.2E+06	0.25	0.36 ± 0.30
C13	0.45	1.1E+06	0.79	9.6E+05	2.51	1.3E+06	8.61	2.6E+06	0.29	0.48 ± 0.27
C14	0.99	0.8E+06	1.81	0.7E+06	10.17	1.7E+06	31.07	3.1E+06	0.33	1.93 ± 0.22
C15	0.62	0.9E+06	1.13	0.8E+06	6.19	1.8E+06	18.46	3.3E+06	0.34	1.18 ± 0.23
A18	1.86	0.7E+06	4.59	0.8E+06	38.36	2.8E+06	78.27	3.5E+06	0.49	7.29 ± 0.20
C16	0.67	1.9E+06	0.87	1.2E+06	4.16	2.4E+06	17.37	6.1E+06	0.24	0.79 ± 0.24
A19 (C16)	1.08	0.4E+06	1.86	0.3E+06	4.24	0.3E+06	21.73	0.9E+06	0.20	0.81 ± 0.24
A20	0.05	0.1E+05	0.01	0.1E+04	13.07	0.5E+06	27.62	0.7E+06	0.47	2.48 ± 0.21
C17	7.21	2.9E+07	36.90	7.2E+07	248.00	2.0E+08	327.45	1.6E+08	0.76	47.12 ± 0.20
C18 (C17)	3.88	6.4E+06	16.18	1.3E+07	149.37	4.9E+07	222.33	4.4E+07	0.67	28.38 ± 0.20
A21 (C7)	7.62	4.3E+06	43.93	1.2E+07	293.57	3.4E+07	369.90	2.5E+07	0.79	55.78 ± 0.20
A22 (C18)	4.26	1.6E+06	15.66	2.8E+06	153.80	1.1E+07	238.40	1.0E+07	0.65	29.22 ± 0.20
A23 (C18)	2.30	1.1E+06	8.35	1.9E+06	77.75	7.2E+06	128.05	7.1E+06	0.61	14.77 ± 0.20
A24	5.50	3.8E+06	21.67	7.1E+06	190.57	2.6E+07	276.81	2.3E+07	0.69	36.20 ± 0.20
A25	1.42	0.4E+06	4.05	0.5E+06	43.42	2.4E+06	75.47	2.5E+06	0.57	8.25 ± 0.20
A26	2.62	1.1E+06	7.49	1.5E+06	73.60	6.1E+06	126.18	6.3E+06	0.58	13.98 ± 0.20
A27	1.64	0.9E+06	2.51	0.7E+06	24.35	2.8E+06	63.05	4.3E+06	0.39	4.63 ± 0.20
C19	0.77	3.7E+06	1.87	4.3E+06	16.38	1.6E+07	32.32	1.9E+07	0.51	3.11 ± 0.21
C20 (C19)	0.79	1.6E+06	1.90	1.9E+07	14.00	5.8E+06	28.65	7.1E+06	0.49	2.66 ± 0.21
A28 (C20)	1.30	0.3E+06	3.42	0.3E+06	19.26	0.8E+06	39.05	1.0E+06	0.49	3.66 ± 0.21
A29 (C20)	1.01	0.6E+06	2.66	0.8E+06	14.18	1.8E+06	29.49	2.2E+06	0.48	2.69 ± 0.21
C21	1.02	1.7E+06	1.74	1.4E+06	14.49	4.8E+06	32.92	6.6E+06	0.44	2.75 ± 0.21
A30 (C21)	1.01	0.2E+06	2.40	0.2E+06	7.42	0.3E+06	22.65	5.6E+06	0.33	1.41 ± 0.22
C22	1.28	5.7E+06	2.41	5.2E+06	26.67	2.4E+07	50.59	2.7E+07	0.53	5.07 ± 0.20
A31 (C22)	1.53	1.0E+06	3.05	1.0E+06	31.41	4.4E+06	56.30	4.7E+06	0.56	5.98 ± 0.20

19. Maragoudaki, F., Kontizas, M., Kontizas, E., et al., 1998, A&A, 338, L29, (paperI)
20. Maragoudaki, F., Kontizas, M., Morgan, D.H. et al., 2001, A&A, 379, 864.
21. Shapley, H., 1956, Amer. Scientist, 44, 73
22. Helou, G., 1986, ApJ, 311, L33
23. DeGioia-Eastwood, K., 1992, ApJ, 397, 542
24. Westerlund, B.E., 1997, The Magellanic Clouds, Camb. Astr.Ser., 29
25. Fukui, Y., Mizuno, N., Yamagushi, R. et al., 1999, PASJ, 51, 745
26. Rowan-Robinson, M. & Crawford, J., 1989, MNRAS, 238, 523
27. Rice, W., Lonsdale, C.J., Soifer, B.T. et al., 1988, ApJS, 68, 91
28. Lehnert, M.D. & Heckman, T.M., 1996, ApJ, 472, 546
29. Dickel, J.R., McIntire, V.J., Gruendl, R.A., Milne, D.K. 2005, ApJ, L29, 790
30. Williams R., Murphy, R., Chu, Y. et al., 1999, ApJS, 123, 467
31. Haberl, F. & Pietsch, W., 1999, A&AS, 139, 277
32. Filipovic, M.D., Pietsch, W., Haynes, R.F. et al., 1998, A&AS, 127, 119
33. Sasaki, M., Haberl, F., Pietsch, W., 2000, A&AS, 143, 391
34. Gouliermis, D., Kontizas, M., Kontizas, E., Korakitis, R., 2003 A&A, 405, 111
35. Davies, R.D., Elliot, K.H., Meaburn, J., 1976, MmRAS, 81, 89

AGB stars in nearby galaxies

P. Tsalmantza*, E. Kontizas†, L. Cambrésy**, F. Genova**, A. Dapergolas† and M. Kontizas*

*Department of Astrophysics Astronomy & Mechanics, Faculty of Physics, University of Athens, GR-15783 Athens, Greece
†Institute for Astronomy and Astrophysics, National Observatory of Athens, P.O. Box 20048, GR-118 10 Athens, Greece
**Observatoire Astronomique de Strasbourg, F-67000 Strasbourg, France

Abstract. This study focuses on very luminous ($M_{bol} < -6.0$ mag) AGB stars with $J - K_s > 1.5$ mag and $H - K_s > 0.4$ mag in the LMC, SMC, M31, and M33 from 2MASS data. A comparison of the most luminous AGB stars detected here with previous studies shows that our sample contains mainly OH/IR stars and carbon stars. The carbon star candidates detected here are not included in existing catalogues produced from data in the optical domain, where they are not visible since they are dust–enshrouded.

Keywords: galaxies:M31,M33,LMC,SMC, stars:AGB, photometry, infrared
PACS: 01.30Cc, 95.85.Hp, 97.30.Hk, 98.56.Ne, 98.56.Si

INTRODUCTION

The very red AGB stars, such as the various types of carbon and variable stars in galaxies, trace the intermediate–mass stellar population, thereby providing information about stellar evolution theories. Their spatial distribution is related to the star–formation history of the parent galaxy. These reddest and most extreme AGB stars cannot be detected easily at the conventional *B*, *R* wavelengths but can be observed in the near–infrared.

In the near–infrared domain carbon stars mainly populate the *red tail* (Nikolaev & Wenberg [20] and Cioni et al. [8]), which is easily detected in the color–magnitude diagram M_{K_s} vs $J - K_s$ as an inclined branch of stars departing from an almost vertical sequence of red giants. It is not yet clear whether all stars in that area of the color–magnitude diagram are carbon stars. Marigo et al. ([17]) point out that the majority of *red tail* stars are carbon, and Davidge ([10]) showed that, in NGC 205, carbon stars are those among the *red tail* with the additional constraints $J - K_s > 1.5$ mag and $H - K_s > 0.4$ mag.

Adopting the above criteria for carbon stars, we selected carbon star candidates in the Magellanic Clouds, M31 and M33 galaxies from the 2MASS All–Sky Survey restricted to objects for which the error is lower than 0.15 mag in all bands.The 2MASS detection limit allowed us to see only the brightest carbon star candidates in M31 and M33. On the other hand, the brightest carbon dust– enshrouded stars are not detected in the 4000-6000 Å range in which the published catalogues are based for both the SMC and LMC (Kontizas et al. [15]; Rebeirot et al. [21]; Morgan et al. [19]). Our search for the most luminous AGB stars will produce a homogeneous set of data for all four galaxies

CP848, *Recent Advances in Astronomy and Astrophysics,*
7th *International Conference of the Hellenic Astronomical Society,* edited by N. H. Solomos

as confined by the detection limit of 2MASS. Data retrieval and basic analysis were performed through the Virtual Observatory (VO) and relevant tools.

To compare the results found here with previous studies of the carbon star populations in galaxies, the K_s magnitudes of the stars were converted into bolometric magnitudes. The bolometric correction used is an approximation proposed by Bessel & Wood ([4]), $BC_K = 0.72 + 2.65 \times (J-K) - 0.67 \times (J-K)^2$, which was validated for the magnitude range $0.6 < J-K < 2.0$ mag.

In this study we detected the brightest AGB stars ($-8.4 < M_{\rm bol} < -6.0$ mag) in four galaxies: the LMC, the SMC, M31, and M33. This selection of the magnitude range is due to the limit of detection of 2MASS in M31 and M33. The spatial distribution and corresponding luminosity functions are derived and discussed for each galaxy.

COLOR–MAGNITUDE DIAGRAMS

The Large Magellanic Cloud

Modeling the color–magnitude diagram of the *red tail* of the LMC from the 2MASS data, Marigo et al. ([17]) found that field stars are not expected to reach colors that are redder than $J-K_s \approx 1.0$ mag, thus they hardly contaminate the features produced by the LMC population of AGB stars. We assumed the distance modulus $m-M = 18.55 \pm 0.04(\text{statistical}) \pm 0.08(\text{systematic})$ mag (Cioni et al. [8]) and the mean reddening $E_{B-V} = 0.13$ mag (Massey et al. [18]), which gives an absorption of $A_{K_s} = 0.04$ mag using $A_{K_S} = 0.34 \times E_{B-V}$ (Bessel & Brett [3]). Carbon star candidates were selected following Davidge criteria ([10]) in the 2MASS M_{K_s} vs $J-K_s$ diagram presented in Fig. 1. Keeping only objects with 2MASS photometric errors less than 0.15 mag, we found 7137 carbon star candidates among the 10055 stars that belong to the *red tail*, which is defined by $J-K_s > 1.3$ mag and $M_{K_s} < -6.5$ mag (Cioni et al. [8]). The cross–match within a 3 arcsec radius of the 7137 AGB stars with the 7716 spectroscopically confirmed carbon stars from Kontizas et al. ([15]) gives 3782 common objects. This confirms that *red tail* stars contain a large number of carbon stars.

A comparison with Kontizas et al. ([15]) data shows that the bright end of the optically detected carbon stars is two magnitudes fainter in the near–infrared than the bright end of the *red tail* stars detected in 2MASS. The bright AGB stars with $-8.4 < M_{\rm bol} < -6.0$ mag are found to be 216, whereas the total number of the bright *red tail* stars is 256. In addition, we searched the DSS catalogue in three colors $B(J)$, R, and I to confirm that these stars are not observable in the B, R wavelengths. We found that many stars (almost 1 out of 5) are not visible in B and R (traditionally accepted as the representative optical window), but they become detectable or appear very luminous in I, as shown in the example of Fig. 2. Deep surveys may reveal these stars, but available observations have not detected them yet.

The spatial distribution of the 7137 *red tail* AGB stars extracted from 2MASS is presented at the top of Fig. 3, and the bottom figure shows the distribution of the 216 most luminous of them. Figure 3 shows high density in the central regions for the *red tail* AGB stars, including the most luminous ones, in agreement with previous studies by Hughes & Wood ([14]) and Wood et al. ([24]).

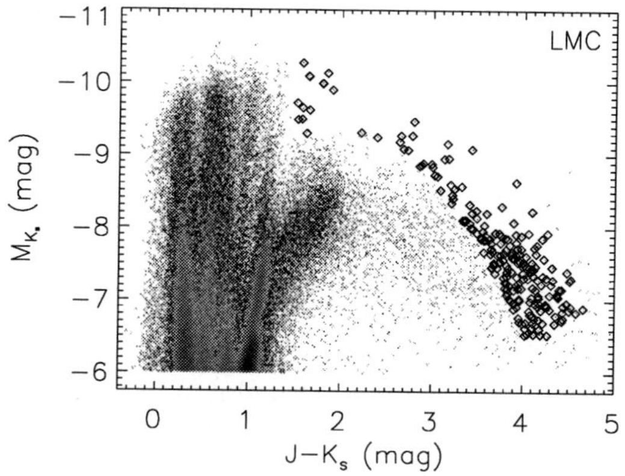

FIGURE 1. Color–magnitude diagram for the LMC stars with $M_{K_s} < -6.0$ mag. The *red tail* is defined by $J-K_s > 1.3$ mag and $M_{K_s} < -6.5$ mag, whereas the luminous AGB stars with $-8.4 < M_{\rm bol} < -6.0$ mag are shown with diamonds.

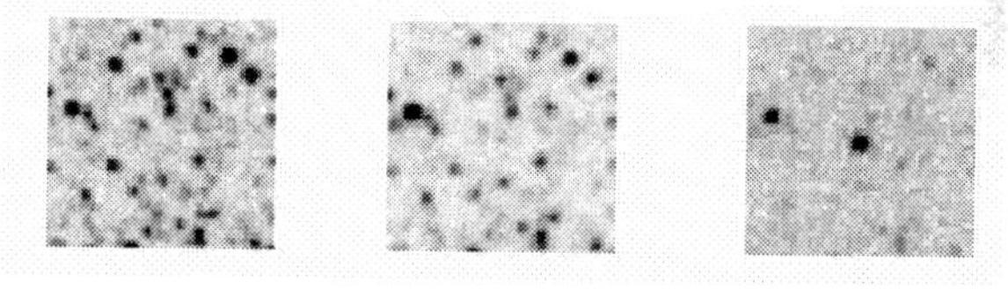

FIGURE 2. The $B(J)$, R, I images from the DSS survey of one LMC carbon star detected here (2MASS J04532183-7051449) show that it is only detected in the I band.

Egan et al. ([12]) combined LMC data from 2MASS and MSX to detect carbon stars. They defined carbon stars by using the criteria $1.5 < K_s - A < 3.75$ mag and $0.8 < H - K_s < 1.5$ mag and the OH/IR with $J - K_s > 3$ mag, $K_s - A > 3.75$ mag and $H - K_s > 1.5$ mag. Cross–matching the stars found here with those of Egan et al. ([12]) has shown that 20 of the luminous AGB stars are actually carbon stars, 109 of them are OH/IR stars, 5 are planetary nebulae, 1 is oxygen–rich AGB and 1 is carbon–rich AGB star, 24 are of unkown nature, and no match was found for the 56 other stars included in our sample.

The Small Magellanic Cloud

Like the LMC, no significant foreground star contamination is expected at $J - K_s >$ 1.0 mag (Marigo et al. [17]) toward the SMC. The distance modulus is $m - M = 18.99 \pm 0.03(\text{statistical}) \pm 0.08(\text{systematic})$ mag (Cioni et al. [8]). The mean reddening adopted here is the one given by Massey et al. ([18]) $E_{B-V} = 0.09$ mag, which corresponds to $A_{K_s} = 0.03$ mag. The *red tail* known to be located at $J - K_s > 1.2$ mag and $M_{K_s} <$

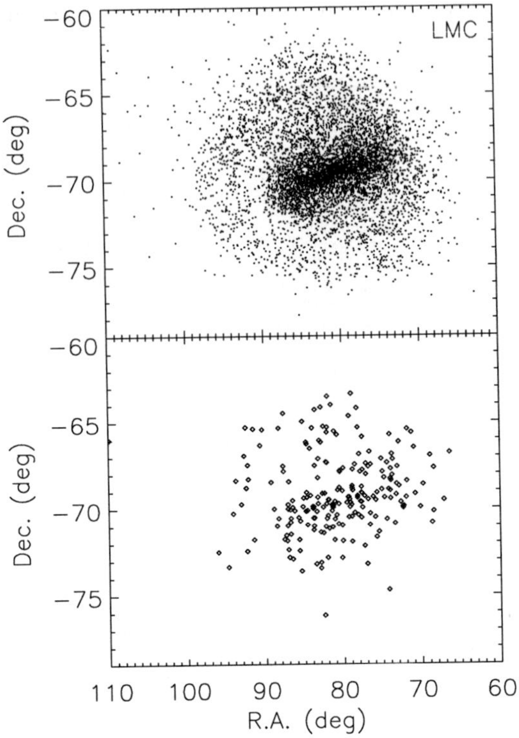

FIGURE 3. Spatial distribution of all LMC AGB stars selected in this paper (top), and the very bright ones with $-8.4 < M_{bol} < -6.0$ mag (bottom).

-7.0 mag can be seen in Fig. 4. It contains 1674 stars in the 2MASS catalogue when restricted to sources with photometric errors smaller than 0.15 mag. Among them, 911 are carbon star candidates if assuming the Davidge ([10]) criteria. There are 34 bright carbon stars with $-8.4 < M_{bol} < -6.0$ mag out of a total of 50 bright sources in the *red tail*. A 3 arcsec radius cross–match of the 911 carbon star candidates with the catalogues of SMC carbon stars confirmed by spectroscopy in the optical wavelengths (Rebeirot et al. [21]; Morgan et al. [19]) gives 676 sources in common. It is worth noting that only one cross–matched star belongs to the 34 very bright AGB stars of our sample.

The carbon star candidates selected here are distributed within an elliptical area, with no obvious central concentration (Fig. 5). The brightest sources with $-8.4 < M_{bol} < -6.0$ mag tend to be in the central area, although they are too few in number to draw a definite conclusion. Demers et al. ([11]) found that the carbon stars are located almost exclusively in and near the disk for the Magellanic type galaxy NGC 3109. This is consistent with our results in the SMC and the LMC.

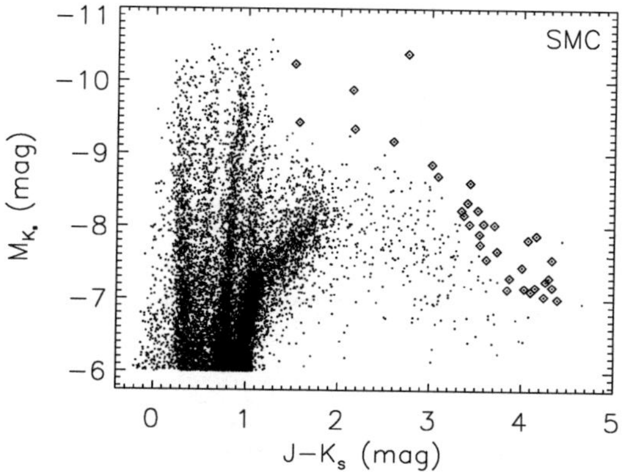

FIGURE 4. Color–magnitude diagram for all SMC stars with $M_{K_s} < -6.0$ mag. The *red tail* is defined by $J - K_s > 1.2$ mag and $M_{K_s} < -7.0$ mag, whereas the luminous ABG stars with $-8.4 < M_{\rm bol} < -6.0$ mag are shown with diamonds.

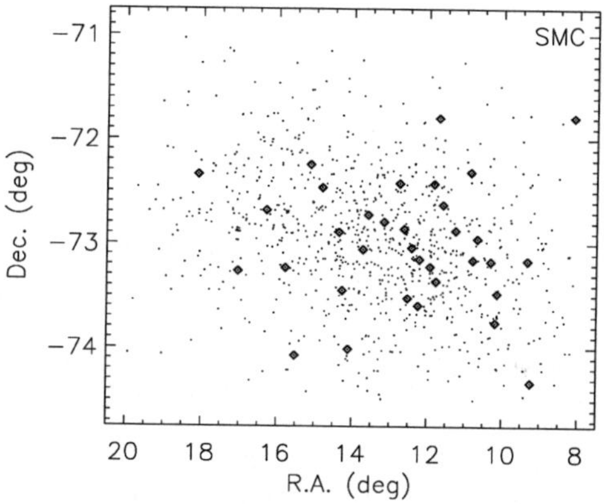

FIGURE 5. Spatial distribution of all SMC AGB stars selected in this paper (dots). The most luminous with $-8.4 < M_{\rm bol} < -6.0$ mag are represented with diamonds.

Messier 31

M31 is a more distant galaxy than the LMC and SMC, with a distance modulus of $m - M = 24.38 \pm 0.05$ mag (Brewer and Richer [7]). The color–magnitude diagram presented in Fig. 6 shows that only the upper part of the *red tail* is observable in M31 because of the 2MASS sensitivity limit. It was assumed that all $J - K_s \geq 1.1$ mag stars are *red tail* stars. In that area there are 959 stars with photometric errors less than 0.15 mag. Since M31 is located at an intermediate galactic latitude, the question of foreground star contamination needs to be properly addressed. We retrieved data from

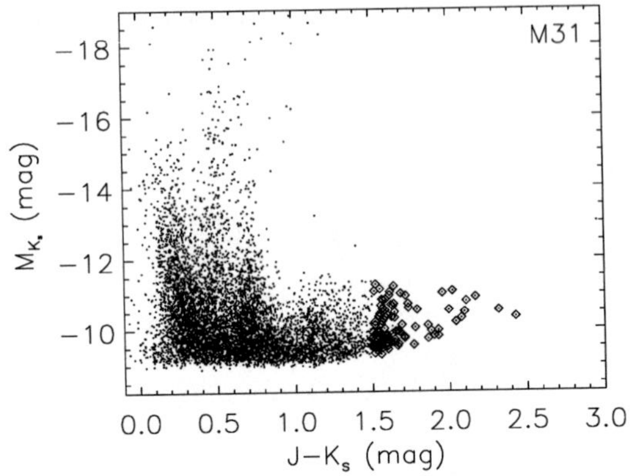

FIGURE 6. Color–magnitude diagram for all M31 stars detected by 2MASS. The upper part of the *red tail* is located at $J-K_s > 1.1$ mag, whereas the luminous AGB stars with $-8.4 < M_{\rm bol} < -6.0$ mag are shown with diamonds.

a region near M31, located at $16.0 < \mathrm{RA} < 17.5$ deg and $40.5 < \mathrm{Dec} < 42.0$ deg. The color–magnitude diagram for M31 is very populated at $J-K_s > 1.1$ mag (Fig. 6), whereas the neighboring field contains very few stars in this part of the diagram, indicating that the contamination from the Milky Way stellar population is very low.

For the total reddening correction near M31, $E_{B-V} \approx 0.23$ mag is adopted (Brewer & Richer [7]). Consequently the absorption at K_s is 0.08 mag, assuming again $A_{K_s} = 0.34 \times E_{B-V}$ (Bessel & Brett [3]). From these relatively low absorption values, we excluded the possibility that the 959 *red tail* stars are early type stars heavily obscured, although they are located in the spiral structure of M31.

Considering the criteria of Davidge ([10]), only 100 stars are carbon star candidates. In Fig. 7 we overplotted the carbon star candidates found in M31 on the GALEX image. These stars fall on the star forming regions detected at UV wavelengths, following the familiar M31 ring–structure, 10 kpc away from the galaxy center. The spatial distribution of the carbon star candidates suggests that the velocity dispersion and the differential galactic rotation have not had enough time to spread these bright AGB stars, which are among the youngest intermediate mass stars produced in spiral arms, like the current population of young stars. All the 100 bright AGB stars detected here are within the magnitude range of $-8.4 < M_{\rm bol} < -6.0$ mag.

In previous investigations of carbon stars in M31, either the studied areas are far from the ring of luminous AGB stars presented in this paper or candidate carbon stars are fainter (Brewer et al. [6]; Battinelli et al. [2]; Battinelli et al. [1]; Davidge et al. [9]).

Messier 33

To complete our survey of the brightest *red tail* stars in nearby galaxies, we took the 2MASS data for M33 and followed the same process as for the LMC, SMC, and M31.

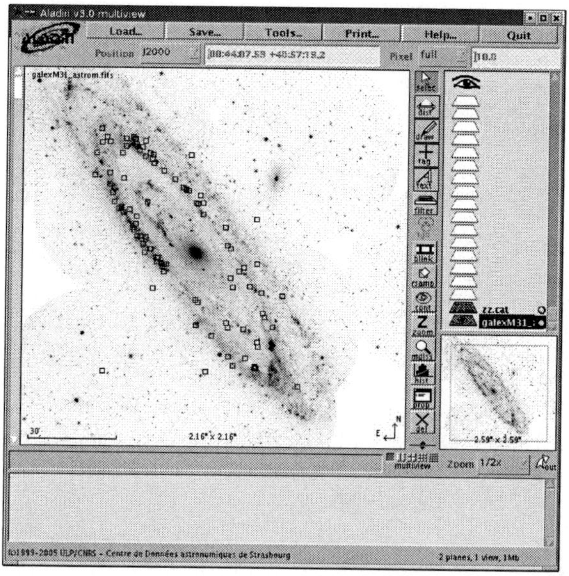

FIGURE 7. The brightest carbon star candidates found in M31 (squares) overplotted on the GALEX UV image, as processed by Aladin v3.0.

Wilson et al. ([23]) derive $E_{B-V} = 0.3 \pm 0.1$, including both the Milky Way foreground and M33 internal extinction. This color excess is translated into $A_{K_s} \approx 0.10$ mag. The distance modulus for M33 is $m - M = 24.52 \pm 0.14(\text{statistical}) \pm 0.13(\text{systematic})$ mag (Lee et al. [16]), and the *red tail* is located at $J - K_s > 1.0$ mag (Fig. 8). Block et al. ([5]) consider that stars detected by 2MASS with $J - K_s > 1.0$ mag cannot be M-stars in the low-metallicity regions, indicating the presence of *red tail* stars. In this area of the color–magnitude diagram, 916 stars have been found; and among them, 31 are supposed to be carbon stars, after we adopt the Davidge ([10]) criteria. We stress that all the 31 carbon star candidates are very bright with $-8.4 < M_{\text{bol}} < -6.0$ mag.

The spatial distribution of these stars (Fig.9) is not correlated with the arcs of red stars found by Block et al. ([5]) in the disk of M33. The cross–identification of these 31 carbon star candidates with those of Rowe et al. ([22]) gives no pair within a 1 arcsec search radius. Their study has not revealed the most luminous AGB star population detected in our paper here, most probably because their observations are in the optical domain where these stars are not detected.

LUMINOSITY FUNCTIONS

The luminosity function reflects the distribution of stellar masses formed in a given volume of space for a stellar system or a galaxy. The upper end of the luminosity function displays the most massive stars of any particular population. Carbon stars usually have intermediate mass and provide insight into the star–forming history of these masses.

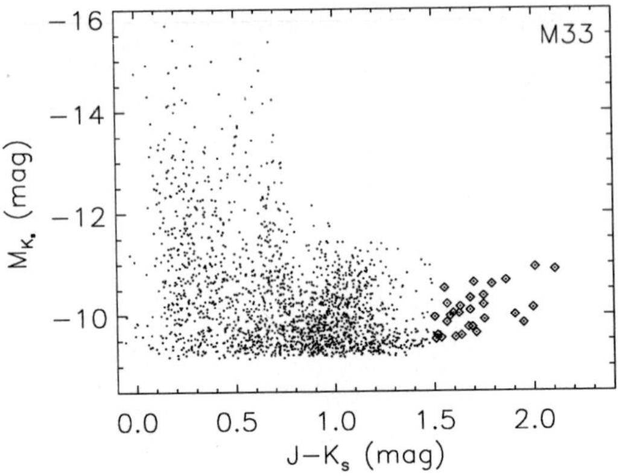

FIGURE 8. Color–magnitude diagram for the stars in M33. The upper part of the *red tail* is located at $J-K > 1.0$ mag, whereas the luminous AGB stars with $-8.4 < M_{\rm bol} < -6.0$ mag are shown with diamonds.

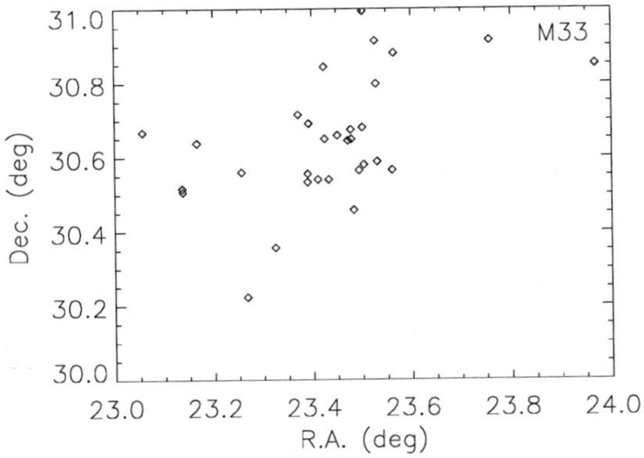

FIGURE 9. Spatial distribution of the most luminous AGB stars with $-8.4 < M_{\rm bol} < -6.0$ mag.

The luminosity function for the bright AGBs in each galaxy is presented in Fig. 10 (LMC, SMC, M31, M33). It illustrates that in all four galaxies, luminous carbon and OH/IR candidates (Egan et al. [12]) exhibit a similar luminosity function with slopes varying only slightly and within the errors of the fit. This result suggests that a similar mass distribution is expected in all four galaxies for the upper part of the mass function. Groenewegen ([13]) found that the faint part of luminosity functions are also similar. Luminosity functions for the bright *red tail* population are presented in Fig. 11. No significant difference is found between the population of bright ($-8.4 < M_{\rm bol} < -6.0$ mag) AGB *red tail* stars ($H-K_s > 0.4$ mag and $J-K_s > 1.5$ mag) and the whole population of *red tail* stars (LMC: $J-K_s > 1.3$ mag and $M_{K_s} < -6.5$ mag, SMC:

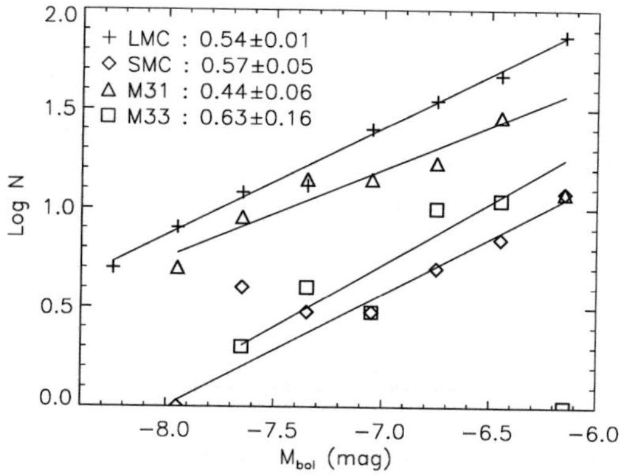

FIGURE 10. Luminosity functions for the luminous carbon and OH/IR candidates found in the LMC, the SMC, M31, and M33. Numbers in the upper–left corner are the slopes of the linear regression fits.

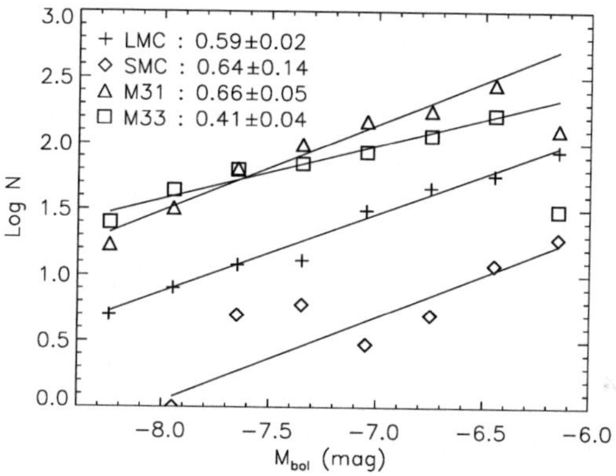

FIGURE 11. Luminosity function for the bright *red tail* stars found in the LMC, the SMC, M31, and M33. Numbers in the upper–left corner are the slopes of the linear regression fits.

$J-K_s>1.2$ mag and $M_{K_s}<-7.0$ mag, M31: $J-K_s>1.1$ mag, M33: $J-K_s>1.0$ mag).

CONCLUSION

We searched for the most luminous AGB stars in four nearby galaxies: the LMC, the SMC, M31, and M33. Stars were selected on the *red tail* of the color–magnitude diagram (M_{K_s}, $J-K_s$) with $J-K_s>1.5$ mag and $H-K_s>0.4$ mag. Known optical catalogues of carbon stars in the SMC and LMC do not include these luminous AGB stars, since they were constructed from observations at B and R wavelengths; therefore, the luminous

AGB stars detected in this work are a valuable complement to existing data.

Using the same range in $M_{\rm bol}$ ($-8.4 < M_{\rm bol} < -6.0$ mag) for the four galaxies, we analyzed the spatial distribution of these stars. We found that they follow the spiral ring of M31 and are mostly located in the central region of the LMC, where all the young population is concentrated. In the SMC and in M33, their number is too small to draw reliable conclusions about their location. Finally the luminosity functions of the *red tail* stars and the bright AGB stars, display similar slopes within the same bolometric magnitude range.

ACKNOWLEDGMENTS

The authors would like to thank the CDS and the VO for valuable support. P. Tsalmantza and M. Kontizas would like to thank the E.L.K.E. of the University of Athens and the Greek General Secretariat of Research & Technology. Special thanks go to the University of Strasbourg for financial support and the hospitality of E. & M. Kontizas.

REFERENCES

1. Battinelli, P., Demers, S., 2005, A&A, 430, 905
2. Battinelli, P., Demers, S., Letarte, B., 2003, AJ, 125, 1298
3. Bessel, M. S., Brett, J. M., 1988, PASP, 100, 1134
4. Bessel, M. S., Wood, P. R., 1984, PASP, 96, 247
5. Block, D. L., Freeman, K. C., Jarrett, T. H., Puerari, I., Worthey, G., Combes, F., Groess, R., 2004, A&A, 425, L37
6. Brewer, J. P., Richer, H. B., Crabtree, D. R., 1996, AJ, 112, 491
7. Brewer, J. P., Richer, H. B., Crabtree, D. R., 1995, AJ, 109, 2480
8. Cioni, M. R. L., Habing, H. J., Loup, C., Epchtein, N., Deul, E., 2001, The Messenger 115, 22
9. Davidge, T. J., Olsen, K. A. G., Blum, R., Stephens, A. W., Rigaut, F. 2005, AJ, 129, 201
10. Davidge, T. J., 2003, ApJ, 597, 289
11. Demers, S., Battinelli, P., Letarte, B., 2003, A&A, 410, 795
12. Egan, M. P., Van Dyk, S. D., Price, S. D., 2001, AJ, 122, 1844
13. Groenewegen, M. A. T., 2002, astro-ph/0208449
14. Hughes, S. M. G., Wood, P. R., 1990, AJ, 99, 784
15. Kontizas, E., Dapergolas, A., Morgan, D. H., Kontizas, M., 2001, A&A, 369, 932
16. Lee, M. G., Kim, M., Sarajedini, A., Geisler, D., Gieren, W., 2002, ApJ, 565, 959
17. Marigo, P., Girardi, L., Chiosi, C., 2003, A&A, 403, 225
18. Massey, P., Lang, C. C., DcGioia-Eastwood, K., Garmany, C. D., 1995, ApJ, 438, 188
19. Morgan, D. H., Hatzidimitriou, D., 1995, A&AS, 113, 539
20. Nikolaev S. and Wenberg M. D., 2000, ApJ, 542, 804
21. Rebeirot, E., Azzopardi, M., Westerlund, B. E., 1993, A&AS, 97, 603
22. Rowe, J. F., Richer, H. B., Brewer, J. B., Crabtree, D. R., 2005, AJ, 129, 729
23. Wilson, C. D., 1991, AJ, 101, 1663
24. Wood, P. R., Bessel, M. S., Paltoglou, G., 1985, ApJ, 290, 477

Mass Segregation in Star Clusters in the LMC HST/WFPC2 Observations

D. Gouliermis[1], A. Dapergolas[2], S. Lianou[3], E. Kontizas[2], & M. Kontizas[3]

1. *Max-Plank-Institut für Astronomie, Königstuhl 17, D-69117 Heidelberg, Germany*
2. *Institute of Astronomy and Astrophysics, National Observatory of Athens, P.O. Box 20048, Athens 118 10, Greece*
3. *Section of Astrophysics- Astronomy and Mechanics, Department of Physics, University of Athens, Greece*

Abstract. The Large Magellanic Cloud (LMC) provides a unique laboratory for the study of the star-formation process, because of its large sample of young stellar systems, such as stellar associations (e.g. Hodge [9]) and young compact star clusters (e.g. van den Bergh [20]). Mass segregation nowadays is a well known phenomenon, which manifests itself as a strong concentration of the heaviest stars towards the centre of the cluster, and recent investigations have shown that even young star clusters in the Magellanic Clouds (MCs) exhibit this phenomenon (e.g. Kontizas et al. [10]; Fischer et al. [6]; Kontizas et al. [11]; Santiago et al. [17]; de Grijs et al. [4]; Sirianni et al. [18]; Gouliermis et al. [7]). A large set of young and old clusters observed with HST has been selected for this study. Preliminary results are given for NGC 1916 an old cluster with age 2×10^8yr, in which mass segregation is detected from its Mass Function. The continuation of this study will determine in a definite manner which is the origin for this phenomenon (primordial or dynamical evolution).

Keywords: galaxies: star clusters–Magellanic Clouds–stars: luminosity function, mass function.
PACS: 01.30.Cc, 98.20.-d, 98.20.Jp, 98.35.Df, 98.56.-p

ORIGINS OF MASS SEGREGATION

Old globular clusters show mass segregation due to their dynamical evolution, through which equipartition of kinetic energy among different mass groups in the cluster can be achieved. According to Spitzer [19] this leads to stars of different masses being found in different "layers" with the heavier stars gravitating toward the centre, and thus mass stratification occurs. This phenomenon is mostly known as *Dynamical Mass Segregation* (see reviews by Lightman & Shapiro [12] and Meylan & Heggie [15]). The discovery of mass segregation in young clusters (Kontizas et al. [10]; Kontizas et al. [11], Clarke et al. [3]) in the MCs introduced new problems concerning its interpretation, since it can be either the signature of the initial conditions and loci, where massive stars are formed (e.g. Hillenbrand & Hartmann [8]; Bonnel & Davies [1]), or the result – at least partially – of early dynamics (e.g. Malumuth & Heap [14]; Brandl et al. [2]). This phenomenon is usually referred to as *Primordial Mass Segregation.*

CP848, *Recent Advances in Astronomy and Astrophysics,*
7th *International Conference of the Hellenic Astronomical Society,* edited by N. H. Solomos

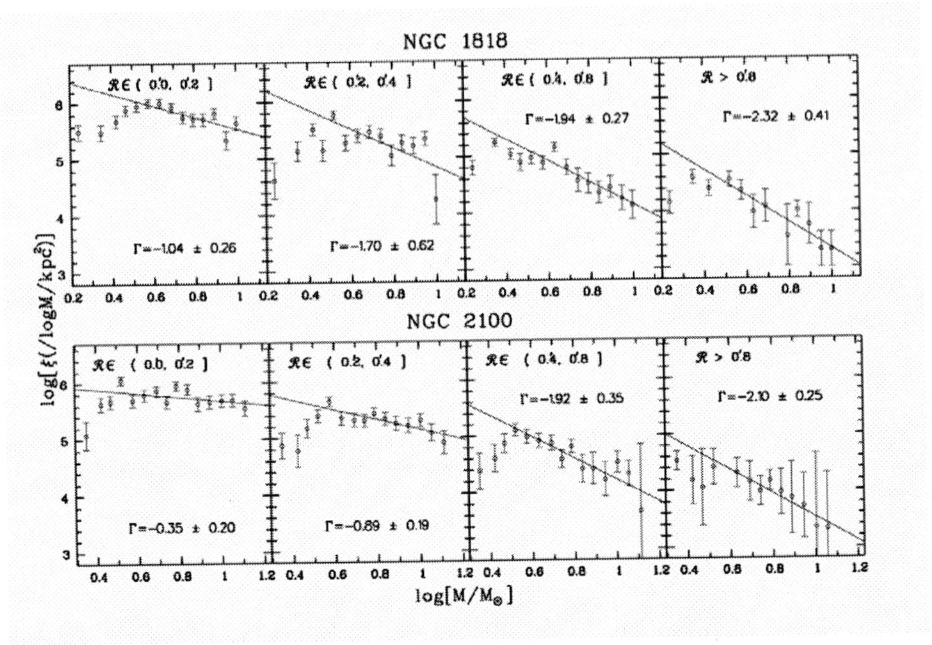

FIGURE 1. Main-sequence MFs of the clusters NGC 1818 and NGC 2100 for indicative selected radial distances. There is a dependence of the MF slope to the selected annuli. The drawn slopes are estimated for every cluster adopting a single-power law and they are given for the most complete mass ranges (4 – 11 $M_\odot$ for NGC 1818 and 4 – 14 $M_\odot$ for NGC 2100). Plot from Gouliermis et al. [7].

The analysis of the mass function (MF) of the cluster is an accurate diagnostic tool for the detection of mass segregation. If there is a higher concentration of massive stars toward the center of the cluster, then mass segregation should be exhibited from a radial variation of its MF slope, with the MF becoming steeper outwards, where less high-mass stars are expected to exist. The change of the slope of its MF becomes evident at a certain distance if mass segregation occurs. For this diagnosis imaging with high resolution, so that crowding effects to be eliminated, is very important. The best so far available camera for such a study is the Wide Field Planetary Camera (WFPC2) on-board the Hubble Space Telescope (HST). An example of detection of mass segregation is demonstrated for two LMC clusters in Figure 1.

Mass segregation in LMC clusters has been detected with various methods and from data of different quality, and thus it cannot be studied in a statistical manner, so that the global important parameters to be isolated. The study of the phenomenon in a consistent way and with the use of comparable observational material for all clusters is essential. We plan to make use of a large sample of data on LMC clusters observed with HST/WFPC2. These data are available in the *HST Data Archive* and will allow a search for Mass Segregation in LMC clusters to be made in a detailed manner, so that results from a consistent method to be extracted. We demonstrate the procedure to be applied on a template cluster, NGC 1916 (Figure 2).

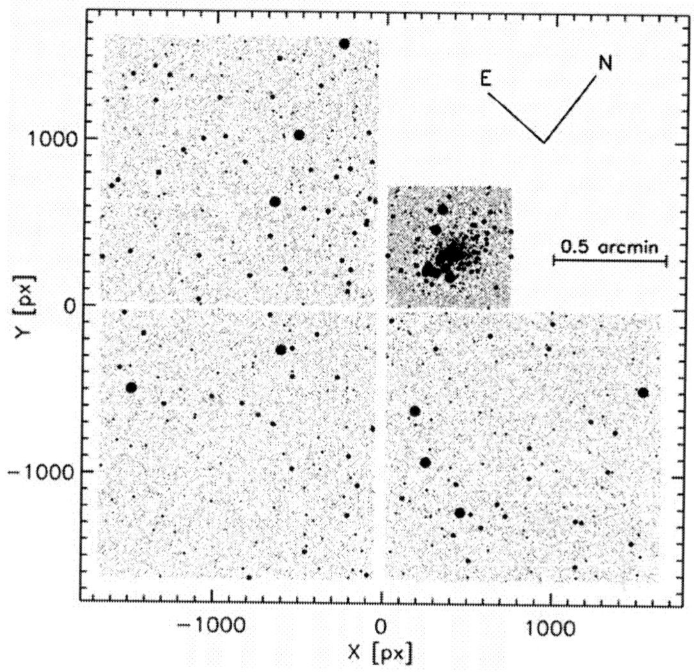

FIGURE 2. Star chart from HST/WFPC2 observations on the cluster NGC 1916. The main part of the cluster is covered by the PC Frame, while its periphery by the three WF Frames.

PHOTOMETRY ON THE LMC CLUSTER NGC 1916

The WFPC2 images on NGC 1916 were collected within the HST program GO-5916 and they were obtained from the data archive at the Space Telescope Science Institute (STScI) and European Southern Observatory (ESO). The observed field of view covers the main body of the cluster in the PC Frame of the camera (Figure 2). Data obtained with filters *F555W* (~*V*) and *F814W* (~*I*) are used. The photometry is performed with the package HSTphot as developed by Dolphin [5] (Figure 3).

The completeness of the data at different distances from the center of the cluster was evaluated by artificial star experiments. The completeness is found to change at radial distance of ≅ 0.4 arcmin. Isochrone models of different ages and for different Metallicities were used for the construction of the Mass-Luminosity Relation to be used for the transformation of the observed stellar magnitudes into masses. These isochrones were selected as the most representative for the age of the cluster and its field ($8.25 \leq \log \tau \leq 10.$).

The main-sequence (MS) population of the observed field is selected and the corresponding Mass Function (MF) is constructed by counting the stars in logarithmic (base ten) mass intervals for stars found in the PC and the WF Frames separately. The numbers have been corrected for incompleteness and normalized to a surface of $1 kpc^2$. The mass bins errors reflect the Poisson statistics.

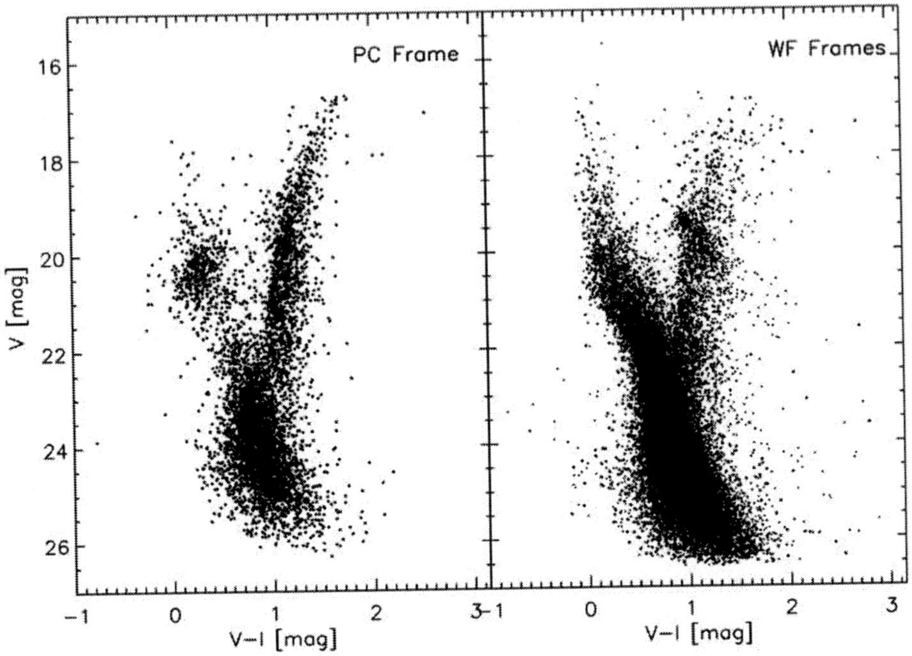

FIGURE 3. *V–I*, *V* Color-Magnitude Diagram of the detected stars in the observed WFPC2 field on NGC 1916. LEFT: Stars in the PC Frame of the Camera. RIGHT: Stars detected in the three WF Frames.

The difference found for the slope of the MF of NGC 1916 between the PC and the WF Frames, the second being steeper, shows that more massive stars are indeed concentrated toward the center of the cluster, making the MF shallower. This is also demonstrated by the larger number of stars with masses $2 \leq M/M_{\odot} \leq 4$ found in the PC Frame. A thorough investigation of the radial dependence of the MF in the PC Frame (Figure 4) shows that indeed the MF becomes steeper outwards, a clear indication that the cluster NGC 1916 is mass segregated. More detailed investigation of the MF within different mass limits will reveal the mass range of the segregated stars.

ACKNOWLEDGMENTS

The authors would like to thank the University of Athens and the Greek General Secretariat of Research & Technology for financial support.

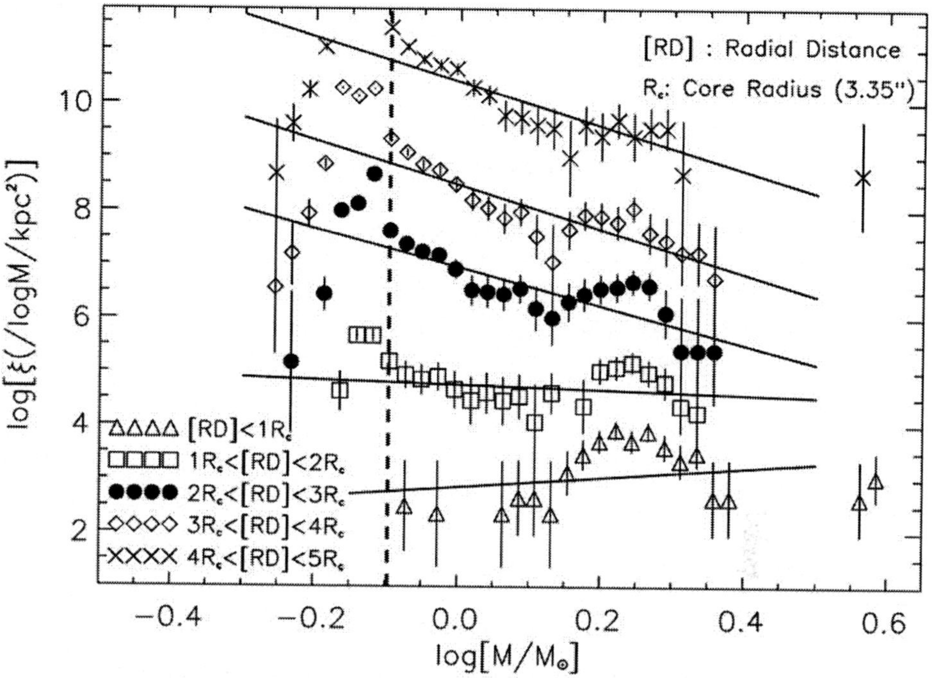

FIGURE 4. The Main-Sequence MF, down to ~ 0.8 $M_{\odot}$, of the central part of the cluster NGC 1916 covered by the PC Frame, for different radial distances. Five characteristic MFs, for distances from 1 up to 5 times the core radius of the cluster (3.35 ± 0.18 arcsec; Mackey & Gilmore [13]) have been overplotted. The differences in the slopes of these MFs is a direct proof that the central main body of NGC 1916 is mass segregated. The vertical dashed line indicates the limit of 50% of completeness.

REFERENCES

1. I. A. Bonnell and M. B. Davies 1998, MNRAS, 295, 691.
2. B. Brandl, B. J. Sams, F. Bertoldi, et al. 1996, ApJ, 466, 254.
3. C. J. Clarke, I. A. Bonnell, and L. A. Hillenbrand 2000, in Protostars and Planets IV, ed. V. Mannings, A. P. Boss, and S. S. Russell (University of Arizona Press), 151.
4. R. de Grijs, G. F. Gilmore, A. D. Mackey, et al 2002c, MNRAS, 337, 597.
5. A. E. Dolphin 2000, PASP, 112, 1383D.
6. P. Fischer, C. Pryor, S. Murray, M. Mateo and T. Richtler 1998, AJ, 115, 592.
7. D. Gouliermis, S. C. Keller, M. Kontizas, E. Kontizas, I. Bellas-Vellidis 2004, A&A, 416, 137G.
8. L. A. Hillenbrand and L. W. Hartmann 1998, ApJ, 492, 540.
9. P. Hodge 1986, PASP, 98, 1113H.
10. M. Kontizas, E. Theodossiou, E. Kontizas 1987, A&AS, 70, 15D.
11. M. Kontizas, D. Hatzidimitriou, I. Bellas-Vellidis, D. Gouliermis, E. Kontizas, R. D. Cannon 1998, A&A, 336, 503.
12. A. P. Lightman and S. L. Shapiro 1978, Rev. Mod. Phys., 50, 437.
13. A. D. Mackey and G. F. Gilmore 2003a, MNRAS, 338, 85.
14. E. M. Malumuth and S. R. Heap 1994, AJ, 107, 1054.
15. G. Meylan and D. C. Heggie 1997, A&ARv, 8, 1.
16. E. E. Salpeter 1955, ApJ, 121, 161.
17. B. Santiago, S. Beaulieu, R. Johnson and G. F. Gilmore 2001, A&A, 369, 74.
18. M. Sirianni, A. Nota, G. De Marchi, C. Leitherer and M. Clampin 2002, ApJ, 579, 275.
19. L. Jr. Spitzer 1969, ApJL, 158, 139.
20. S. van den Bergh 1990, BAAS, 22, 1226V.

The role of kink instability in Poynting-flux dominated jets

Dimitrios Giannios* and Henk Spruit*

*Max Planck Institute for Astrophysics, Box 1317, D-85741 Garching, Germany

Abstract.
The role of kink instability in magnetically driven jets is explored through numerical one-dimensional steady relativistic MHD calculations. The instability is shown to have enough time to grow and influence the dynamics of Poynting-flux dominated jets. In the case of AGN jets, the flow becomes kinetic flux dominated at distances $\gtrsim 1000 r_g$ because of the rapid dissipation of Poynting flux. When applied to GRB outflows, the model predicts more gradual Poynting dissipation and moderately magnetized flow at distances of $\sim 10^{16}$ cm where the deceleration of the ejecta due to its interaction with the external medium is expected. The energy released by the instability can power the compact "blazar zone" emission and the prompt emission of GRB outflows with high radiative efficiencies.

Keywords: MHD – Instabilities – Gamma rays: bursts – Quasars: general
PACS: 95.30.Qd, 98.62.Nx, 98.70.Rz, 98.54.Cm

INTRODUCTION

Relativistic collimated outflows have been extensively observed in active galactic nuclei (AGN) and X-ray binaries (XRB). Gamma-ray bursts (GRB) are also believed to be connected to ultrarelativistic and collimated outflows so that to overcome the "compactness problem" [1] and to explain the achromatic afterglow breaks [2]. It is also believed that all these sources are powered by accretion of matter by a compact object.

The widely accepted mechanism for jet acceleration and collimation in the context of AGN and XRB jets is that of magnetic driving. According to this paradigm, magnetic fields anchored to a rotating object can launch an outflow. The material is accelerated thermally up to the sonic point and centrifugally until the Alfvén point defined as the point where the flow speed equals the Alfvén speed. After the Alfvén point the inertia of mater does not allow corotation of the magnetic field. As a result the magnetic field lines bend developing a strong toroidal component.

Further out the flow passes through the fast magnetosonic point where most of the energy of the flow remains in the form of Poynting flux in the case of relativistic outflows [3, 4, 5, 6]. Further acceleration of the flow is not straightforward within ideal MHD. It can be shown, for example, that a radial flow is not accelerated after the fast point (e.g., see Ref. [6]). This is a result of the fact that the magnetic pressure and tension terms of the Lorentz force almost cancel each other [7].

Magnetized jets suffer from a number of instabilities. Interaction with an environment causes instabililty of the Kelvin-Helmholtz type, kink instability causes internal rearrangement of the field configuration. Here we focus on kink instability, since it internally dissipates magnetic energy associated with the Poynting flux. As demonstrated

CP848, *Recent Advances in Astronomy and Astrophysics,*
7th *International Conference of the Hellenic Astronomical Society,* edited by N. H. Solomos

elsewhere [8, 9, 10] such internal energy dissipation directly leads to acceleration of the flow. Dissipation steepens the radial decrease of magnetic pressure, thereby lifting the cancellation between outward pressure force and inward magnetic tension, and allowing the magnetic pressure gradient to accelerate the flow.

Using energy dissipation and field decay rates as indicated by linear stability analysis [11, 12] and numerical simulations [13, 14, 15] for the kink instability we find that the instability has plenty of time to grow in the AGN and XRB cases, dissipating energy in the toroidal component of the magnetic field while accelerating the flow at the same time. The dissipation of magnetic energy is almost complete and fast in the case of AGN jets, so that on parsec scales the flow has become kinetic energy dominated, in agreement with current interpretation of the observations (e.g. Ref. [16], where the possible effects of magnetic dissipation are also discussed briefly).

The structure of the paper is as follows. In the next section we discuss MHD instabilities in Jets and focus on the kink instability and its growth rate. The model is described including the assumptions, the dynamical equations and the parameters at the base of the flow. We apply the model to the case of AGN jets and GRBs, while the last Section includes the conclusions of the study. A more detailed description of this work can be found in Ref. [17]

THE KINK INSTABILITY

Magnetized outflows are subject to a variety of instabilities. These can be classified to pressure driven, Kelvin-Helmholtz, and current driven instabilities. Pressure driven instabilities are related to the interplay between the gas pressure and the curvature of magnetic field lines. Kelvin-Helmholtz (KH) instabilities arise from velocity gradients in the flow and may be important in the shearing layer between the outflow and the external medium. Current driven (CD; [18, 11, 12]) instabilities have received much less attention but are the most relevant ones for Poynting-flux dominated outflows, since they can convert the bulk Poynting flux into radiation and kinetic energy of the flow Among the CD instabilities the $m=1$ kink instability is generally the most effective. In this work, we focus on the effect of the kink instability on to the dynamics of these outflows.

The growth rate of the instability

While magnetized outflows can be accelerated 'centrifugally' by large scale poloidal fields [19, 5], already at the radius of the light-cylinder inertial forces become significant and the magnetic field cannot force corotation. At this radius the strength of the toroidal and the poloidal components are comparable. Further out, induction equation dictates that, within ideal MHD, the toroidal component dominates over the poloidal one since the strength of the former scales as $1/r$ while that of the latter as $1/r^2$. This magnetic configuration of a strongly wound-up magnetic field like this is, however, known to be highly unstable from tokamak experiments [20] to the kink $m=1$ mode. Linear stability analysis has shown that the growth time of the instability is given by the Alfvén crossing time across the outflow in a frame comoving with it [11, 12].

The study of the non-linear evolution of the instability demands three dimensional relativistic MHD simulations over many decades of radii and is, therefore, not surprising that the issue is not settled. In Refs. [13, 21], it is argued in favor of the dynamical importance of the instability in reorganizing the magnetic configuration inside the jet. It has been suggested, however, that the jet creates a "backbone" of strong poloidal field which slows down the development of instabilities [14, 15]. In view of these works and since the growth rate of the instability is important for this study, we consider two alternatives for the non-linear stage of the instability. One in which the instability proceeds at the Alfvén crossing time across the outflow (as suggested by linear stability analysis) and rearranges the magnetic field configuration to a more chaotic one. In this case the instability time scale is given by the expression (in the central engine frame)

$$t_{\rm k} = \frac{r\theta\gamma}{v_{{\rm A},\phi}}. \quad (1)$$

We will refer to this as the "fast kink" case.

For the second case, we reduce the dissipation rate by a suitable (but arbitrary) function of the poloidal-to-toroidal field ratio.

$$t_{\rm k} = \frac{r\theta\gamma}{v_{{\rm A},\phi}} e^{B_p^{\rm co}/B_\phi^{\rm co}}. \quad (2)$$

Where we will refer to this as the "slow kink" case. Numerical simulations of the instability would be needed to determine which of these prescriptions (if any) of the growth time scale is more close to describe its non-linear development. In the last expressions, $B_\phi^{\rm co}$, $B_p^{\rm co}$, are the toroidal and the poloidal components of the magnetic field as measured by an observer comoving with the flow, θ stands for the jet opening angle, γ is the bulk Lorentz factor of the flow and $v_{{\rm A},\phi}$ is the ϕ component of the Alfvén speed.

THE MODEL

A magnetically launched outflow passes through three characteristic points where the speed of the flow equals the speed of slow mode, the poloidal Alfvén wave and the fast mode and are called the slow magnetosonic, the Alfvén and the fast magnetosonic points respectively. For flows where the energy density of the magnetic field dominates that of matter, the Alfvén point lies very close to the light-cylinder

$$R_L = c/\Omega, \quad (3)$$

where Ω is the angular velocity of the foot point. At the Alfvén radius most of the centrifugal acceleration has already taken place and the magnetic field cannot force corotation of matter. At this location, the toroidal and the poloidal components of the magnetic field are comparable in magnitude. Further out, the flow passes through the fast magnetosonic point at a distance $\sim$ a few $R_{\rm L}$ (see, e.g. [5, 6]). At the location of the fast magnetosonic point the speed of the four-velocity of the flow equals $\sim \mu^{1/3}$, where μ is the Michel magnetization parameter (i.e., the energy flux per unit rest mass;

[3]). For Poynting-flux dominated flows (i.e., $\mu \gg 1$), most of the energy is still in magnetic form at this point since the ratio of magnetic to matter energy flux is $\sim \mu^{2/3}$. Further acceleration of the flow within ideal MHD appears to be hard to achieve.

We set the initial conditions of our calculation at the fast magnetosonic point r_0. To make the problem tractable we make a number of simplifying assumptions. First, We limit ourselves to a radial, static flow. Evidently, this approach does not allow us to explore the important issue of jet collimation. Furthermore, the flow is assumed one-dimensional by ignoring the structure of the jet is the θ direction. Finally, we ignore the azimuthal component of the velocity. This component is not dynamically important beyond the fast magnetosonic point and can be neglected from the dynamic equations [4]. These simplifying assumptions minimize the number of the free parameters of the model allowing us to study the effect of each on the jet dynamics, as will become clear in the next sections.

Dynamical equations

To determine the characteristics of the flow as a function of radius, one needs the conservation equations for mass, energy and angular momentum. These equations can be brought into the form (if, for the moment, we neglect radiative losses; e.g., see Ref. [9])

$$\partial_r r^2 \rho u = 0, \tag{4}$$

$$\partial_r r^2 \left(w\gamma u + \frac{\beta B_\phi^2}{4\pi} \right) = 0, \tag{5}$$

$$\partial_r r^2 \left(w u^2 + p + \frac{(1+\beta^2) B_\phi^2}{8\pi} \right) = 2rp, \tag{6}$$

where $w = \rho c^2 + e + p$ is the proper enthalpy density, e and p are the internal energy and pressure respectively and $u = \gamma\beta$ is the radial four-velocity. We still need to assume an equation of state that will provide a relation between the pressure p and the internal energy. Assuming ideal gas, we take $p = (\gamma_a - 1)e$, where γ_a is the adiabatic index.

Mass conservation (4) can be integrated to yield mass flux per sterad

$$\dot{M} = r^2 u \rho c, \tag{7}$$

while energy conservation gives the total luminosity per sterad

$$L = w r^2 \gamma u c + \frac{\beta c (r B_\phi)^2}{4\pi}. \tag{8}$$

The first term of the last expression corresponds to the kinetic energy flux and the second to the Poynting flux. A key quantity at this point is related to the "magnetic content" of the flow which we will refer to as magnetization parameter σ defined as the ratio of the radial Poynting to matter energy flux

$$\sigma = \frac{L_{\rm pf}}{L_{\rm kin}} = \frac{B_\phi^2}{4\pi\gamma^2 w}. \tag{9}$$

The dynamical equations can be easily modified to allow for radiative losses of the flow (see [17])

Modeling the kink instability, initial conditions

The set of equations presented in the previous section is not complete. There is one more equation needed to determine the problem at hand, which is the induction equation. For ideal MHD, the induction equation yields $\partial_r \beta r B_\phi = 0$ and can be integrated to give the scaling $B_\phi \propto 1/r$ for relativistic flows. One can immediately see that the Poynting-flux term in equation (8) is approximately constant and no further acceleration of the flow is possible within ideal MHD for a radial flow. This is a result of the fact that the magnetic pressure and tension terms of the Lorentz force almost cancel each other [7].

We argue, however, that when the toroidal component of the magnetic field becomes dynamically dominant the kink instability sets in. The instability drives its energy from B_ϕ^2 on the instability growth time scale. This effect can be primitively modeled by the addition of one sink term on the right hand side of the induction equation following Refs. [8, 9]

$$\partial_r \beta r B_\phi = -\frac{r B_\phi}{c t_{\rm k}}. \tag{10}$$

The kink instability time scale is given by the expressions (2) or (1) depending on whether a the poloidal component of the magnetic field is assumed to have a stabilizing effect or not. When the instability sets in B_ϕ drops faster than $1/r$ and the acceleration of the flow is possible at the expense of its magnetic energy.

The characteristics of the flow are determined when a number of quantities are specified at the fast magnetosonic point r_0 which is taken to be $\sim$ a few times the light cylinder radius [5, 7, 6], or expressed in terms of the gravitational radius $r_g = GM/c^2$ of the central engine $r_0 \sim 100 r_g$. These quantities are the initial magnetization σ_0, the luminosity L, the opening angle θ, the ratio $B_{r,0}/B_{\phi,0}$ and the cooling length scale κ (see Ref. [17] for details). The quantities one has to solve for so as to determine the characteristics of the flow are ρ, e, u and B_ϕ as functions or radius r. This is done by integrating numerically the mass, energy, momentum conservation equations and the modified induction equation. The parameters of the model determine the initial values of ρ, e, u and B_ϕ at r_0.

The initial four-velocity for our calculations is assumed to be

$$u_0 = \mu^{1/3} = \sqrt{\sigma_0}, \tag{11}$$

in accordance to previous studies [3, 4, 6] which show that at the fast point the ratio of Poynting to kinetic flux is $\mu^{2/3}$. The flow is assumed to be cold at r_0, i.e. $e = 0$ and using the previous expression with Eqs. (8), (9) one can solve for ρ_0 and $B_{\phi,0}$.

APPLICATIONS

Although at first sight different, jets in AGNs (and microquasars) and GRBs probably have central engines of similar characteristics. AGN jets are launched in the inner regions of magnetized accretion disks [19], or drive their power by magnetic fields that are threading the ergosphere of a rotating black hole [22]. In the case of GRBs, the same central engine may be at work, or the energy is tapped by a millisecond magnetar (e.g., [23]). In all of these situations, strong magnetic fields play an important role and most of the energy is released in the form of Poynting flux.

The above scenarios may give rise to magnetized outflows, which evolution depends, to a large extent, on the dominance of the magnetic energy, or in other words on the ratio of the Poynting-flux to matter energy flux at the base of the flow. By varying this ratio, one can apply the model to jets in both the cases of GRBs and AGNs.

AGN jets

Relativistic jets are commonly observed in AGNs to have bulk Lorentz factors in the range $\gamma \sim 10-20$. Such terminal Lorentz factors can be achieved for ratio σ_0 of Poynting to matter energy flux of the order of several at the fast magnetosonic Point r_0. The location of the fast point is most likely at a few light cylinder radii [5] and is taken to be $100r_g$.

Actually, σ_0 is very important parameter of the flow. Its effect on the acceleration of the flow is clearly seen in Fig. 1, where the bulk Lorentz factor is plotted as a function radius r for different σ_0. The rest of the parameters have the values $\theta = 10^o$, $B_{r,0}/B_{\phi,0} = 0.5$, while the energy released by the instability is assumed to be locally radiated away. The results do not depend on the luminosity L of the flow in the case of AGN jets, while r_0 sets the scale of the problem (since it is the only length scale) which means that the results can be trivially rescaled in case of a different choice of r_0.

From Fig. 1, one can see that the instability acts fast and accelerates the flow within 1-2 orders of magnitude in distance from the location of the fast magnetosonic point. The acceleration is faster in the "fast kink" case and much more gradual in the "slow kink" one. At larger distances, practically all the magnetic energy has been dissipated and the terminal Lorentz factors are very similar in the slow and fast kink cases.

Another quantity of special interest is the Poynting to matter energy flux ratio σ as a function of radius. While, the flow is initially moderately Poynting flux dominated, the σ drops rather fast as a function of distance and the flow is matter dominated at distances $r \gtrsim 10^3 r_g$ independently of the prescription of the instability or the cooling timescales. Far enough from the fast magnetosonic point, practically all the magnetic energy has been transfered to the matter, and the bulk Lorentz factor saturates. Also the calculations show that, despite the fact that the instability grows fast from the toroidal component of the magnetic field, this component still dominates over the radial one.

Having solved for the dynamics of the flow predicted by our modeling of the kink instability, we will turn to the implications of these findings with respect to observations of AGN jets. One highly debated issue is whether the AGN jets are Poynting-flux

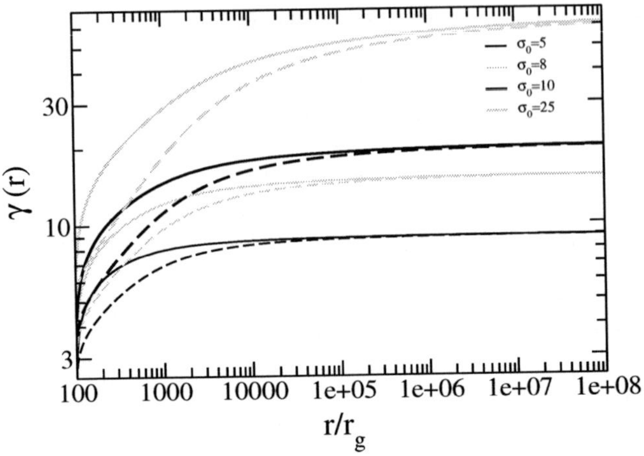

FIGURE 1. The bulk Lorentz factor of the flow as a function of radius for different values of σ_0. The solid curves correspond to the case where Eq. (1) is used for the instability growth time scale (fast kink) and the dashed to the one where Eq. (2) is used (slow kink case).

dominated on pc and kpc scales or not. The case-dependent arguments are reviewed in Ref. [16], where it is shown that there is no strong observational reason for assuming Poynting-flux dominated jets on scales larger than a few pc and that the observed emission on these scales can be understood as energy dissipated in shocks internally in the flow or due to interaction of the flow with the external medium.

Our model predicts that most of the energy is in the form of kinetic flux at distances say $\gtrsim 10^3 - 10^4 r_g \simeq 10^{17} - 10^{18} m_9$ cm, where m_9 stands for a black hole of 10^9 solar masses. So, on pc scales the magnetic fields are dynamically insignificant in agreement with observations. Further information on the dynamics of AGN jets comes from the shortest variability timescale in the optical and gamma-ray bands in blazars. This timescale can be as short as a few days indicating that most of the non-thermal radiation comes from a compact region of size $R \lesssim 10^{17}$ cm (the so-called blazar zone). Since most of the magnetic energy is dissipated on these scales, it is quite probable that the observed radiation is result of the instability-released energy provided that the dissipative processes lead to wide enough particle distributions (see also Ref. [16]). On the other hand, one cannot exclude the possibility that, within this model, the "blazar zone" emission is a result of internal shocks.

Gamma-ray bursts

The analysis we follow so as to apply the model to GRBs is very similar to this described in the previous sections. The only new ingredient that has to be added is related to the very high luminosities that characterize the GRB jets. As a result, the inner part

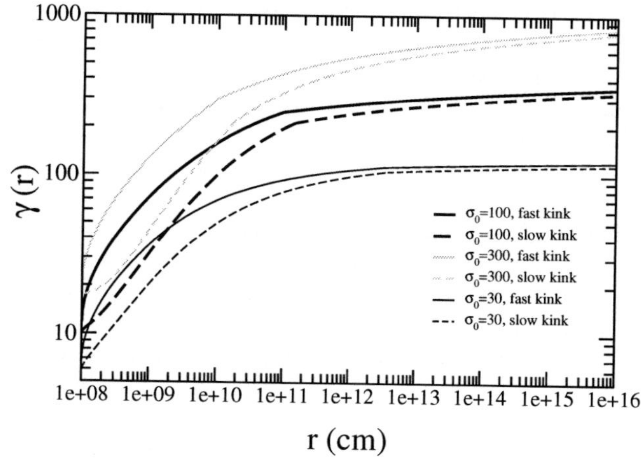

FIGURE 2. The bulk Lorentz factor of the flow for different σ_0 and for the fast and slow kink case. Notice that larger values for σ_0 result in faster outflows.

of the flow is optically thick to electron scattering and matter and radiation are closely coupled together. At the photospheric radius the optical depth drops to unity and further out the flow is optically thin. So, the high luminosity introduces a new length scale to the problem to be taken into account (for details see Ref. [17]).

Following the procedure described in the previous sections, we have calculated the bulk Lorentz factor of the flow for different values of σ_0 at the base of the flow and for the two prescriptions for the timescale of the kink instability [Eqs. (1), (2)]. The fast magnetosonic point is set to $R_0 = 10^8$ cm, the jet opening angle to $\theta = 10^o$, the initial ratio $B_{r,0}/B_{\phi,0} = 0.5$ and the luminosity of the flow $L = 10^{51}$ erg/sec · sterad.

The results are given in Fig. 2, where it is shown that the flow reaches terminal Lorentz factors $\gamma \gtrsim 100$ for $\sigma_0 \gtrsim 30$. The solid curves correspond to the case where the timescale for the kink instability is given by Eq. (1) (fast kink case) and the dashed to the case where Eq. (2) is used for the timescale of the growth of the instability (slow kink case). Notice that the initial acceleration of the flow differs in the two cases, being much faster in the fast kink case. This is expected since this case is characterized by rapid dissipation of magnetic energy and acceleration from the base of the flow, while in the slow kink case the non-negligible poloidal component of the magnetic field close to r_0 slows down the instability. The terminal Lorentz factors are, however, similar in the two cases.

A second key parameter of the model is the opening angle of the jet. For smaller opening angle, the instability timescale becomes shorter and the flow is accelerated faster and to higher terminal Lorentz factors. One important point is that, for $\sigma_0 \gtrsim 100$, the flow remains Poynting-flux dominated even at range distances away from the source where the deceleration of the outflow because of its interaction with the interstellar medium or the stellar wind is expected which means that the instability is not fast enough to convert most of the magnetic energy into bulk motion of matter. Afterglow observations can in principle probe into the magnetic content of the ejecta through early observations of the reverse shock emission [24]. Modeling of the forward and reverse shock emission in cases where quick follow ups were possible suggests the existence of frozen-in magnetic fields in the ejecta that are dynamically important with $\sigma \gtrsim 0.1$ [25]. Rapid follow-ups in the X-rays, UV and optical are now readily possible thanks to Swift satellite and ground based telescopes and can either give support or rule out our model which predicts magnetization parameter of the order of unity for the outflowing material at the afterglow region.

The strong magnetic fields in the flow are expected to reduce the radiative efficiency of internal shocks [26] in the flow. We, therfore, investigate the possibility that the energy released by the instability powers the prompt gamma-ray emission. We have calculated the radiative efficiency -defined as the radiated luminosity $L_{\rm rad}$ divided by the flow luminosity L- for different values of σ_0 and θ. For values of the parameters relevant for GRB outflows, the efficiency is shown to be quite high and of the order of 20% (for detailed plots see Ref. [17]).

CONCLUSIONS

The standard scenario for jet launching, acceleration and collimation involves large scale magnetic fields anchored to a rotating object (e.g., [19, 5]). The flow passes through three critical points, i.e. the slow, the Alfvén and the fast point. At the fast point the ratio of Poynting to matter energy flux is much larger than unity in the case of relativistic jets [3, 6] while further acceleration of the flow appears hard to achieve within ideal MHD.

In this work, we study how this picture is modified when one takes into account the fastest growing current driven instability, i.e. the $m = 1$ mode kink instability. We have modeled the instability by modifying the induction equation to account for non-ideal MHD processes and solving the relativistic MHD equations in the case of a radial flow. The instability is driven by B_ϕ, dissipates Poynting flux and has been shown to be an efficient mechanism to accelerate the flow.

The key parameter of the model is the ratio σ_0 of the Poynting to matter energy flux at the base of the flow. A large part of the AGN jet phenomenology can be understood in the context of this model for $\sigma_0 \sim$ several. On sub-pc scales the flow is Poynting-flux dominated with $B_\phi \gg B_r$. The flow is shown to be accelerated fast and to become matter dominated already at ~pc scales, while it reaches terminal bulk factors of a few tens. The emission at the blazar zone can be a result of either internal shocks or direct manifestation of the energy released by the instability.

Within the context of the same model, we propose that GRBs are a result of more Poynting flux dominated outflows with $\sigma_0 \sim 100$. For these values of σ_0 the flow reaches terminal bulk Lorentz factors of the order of a few to several hundreds, while it remains moderately magnetized (i.e. $\sigma_\infty \sim 1$) at the afterglow region region. Although, there is evidence for magnetized ejecta from afterglow modeling [24, 25], more results are anticipated from early afterglow follow-ups that can give support or rule out the model. The strong magnetic fields remaining in the flow at distances $10^{13} - 10^{15}$ cm are expected to reduce the efficiency of internal shocks. On the other hand, the instability-released energy can be radiated away converting $\sim 20\%$ of the luminosity of the flow into radiation and appears to be a more promising mechanism to produce the prompt emission.

ACKNOWLEDGMENTS

Giannios acknowledges support from the EU FP5 Research Training Network "Gamma Ray Bursts: An Enigma and a Tool."

REFERENCES

1. T. Piran, *Phys. Rep.*, **314**, 575 (1999)
2. J. E. Rhoads, *Astrophys. J.*, **487**, L1 (1997)
3. F. C. Michel, *Astrophys. J.*, **158**, 727 (1969)
4. P. Goldreich, and W. H. Julian, *Astrophys. J.*, **160**, 971 (1970)
5. T. Sakurai, *Astron. Astrophys.*, 152, 121 (1985)
6. V. S. Beskin, I. V. Kuznetova, and R. R. Rafikov, *Month. Not. Roy. Astron. Soc.*, **299**, 341 (1998)
7. M. C. Begelman, and Z.-Y. Li, *Astrophys. J.*, **426**, 269 (1994)
8. G. Drenkhahn, *Astron. Astroph.*, 387, 714 (2002)
9. G. Drenkhahn, and H. C. Spruit, *Astron. Astroph.* 391, 1141 (2002)
10. D. Giannios, and H. C. Spruit, *Astron. Astroph.*, **430**, 1 (2005)
11. M. C. Begelman, *Astrophys. J.*, **493**, 291 (1998)
12. S. Appl, T. Lery, and H. Baty, *Astron. Astroph.*, **355**, 818 (2000)
13. T. Lery, H. Baty, and S. Appl, *Astron. Astroph.*, **355**, 1201 (2000)
14. R. Ouyed, D. A. Clarke, and R. E. Pudritz, *Astrophys. J.*, **582**, 292
15. M. Nakamura, and D. L. Meier, *Astrophys. J.*, **617**, 123 (2004)
16. M. Sikora, M. C. Begelman, M. G. Madejski, and J.-P. Lasota, *Astrophys. J.*, **625**, 72 (2005)
17. D. Giannios, and H. C. Spruit, *Astron. Astroph.*, submitted (2005)
18. H. C. Spruit, T. Foglizzo, and R. Stehle, *Month. Not. Roy. Astron. Soc.*, **288**, 333 (1997)
19. R. D. Blandford, and Payne, D. G., *Month. Not. Roy. Astron. Soc.*, **199**, 883 (1982)
20. G. Bateman, *MHD Instabilities*, Cambridge: MIT Press (1978)
21. H. Baty, and R. Keppens, *Astrophys. J.*, **580**, 800 (2002)
22. R. D. Blandford, and R. L. Znajek, *Month. Not. Roy. Astron. Soc.*, **179**, 433 (1977)
23. V. V. Usov, *Nature*, **357**, 472 (1992)
24. P. Kumar, and A. Panaitescu, *Month. Not. Roy. Astron. Soc.*, **346**, 905 (2003)
25. B. Zhang, and S. Kobayashi, S. *Astrophys. J.*, **628**, 315 (2005)
26. M. J. Rees, and P. Mészáros, *Astrophys. J.*, **430**, L93 (1994)

Magnetic Driving of AGN Jets

Nektarios Vlahakis

Section of Astrophysics, Astronomy & Mechanics, Department of Physics, University of Athens, Panepistimiopolis, GR-15784 Zografos Athens, Greece; E-mail: vlahakis@phys.uoa.gr

Abstract. Jets in active galactic nuclei are collimated, relativistic flows that emanate from accretion disks around supermassive black holes. Electromagnetic stresses are the most plausible candidate for extracting energy at the source and converting it into outflow kinetic energy. Questions that need to be answered in order for these processes to be well understood are: Can we explain parsec-scale accelerations that the observations infer? How the conditions near the disk are related to the terminal Lorentz factor of the jet and what is the asymptotic value of the Poynting-to-matter energy flux ratio? Can we model the apparent kinematics of the observed jet components? I present solutions of the ideal magnetohydrodynamic equations that help to shed light on these questions.

Keywords: galaxies: active, galaxies: jets, ISM: jets and outflows, MHD
PACS: 95.30.Qd, 98.38.Fs, 98.54.Cm, 98.58.Fd

INTRODUCTION

The commonly accepted interpretation of AGN jets is that they represent relativistic flows ejected from the vicinity of supermassive black holes and powered by energy released by accretion in an underlying disk. The jets consist of various components that follow curved apparent trajectories and exhibit apparent superluminal motions, a manifestation of their relativistic velocities.

There is growing evidence that these components undergo extended (parsec-scale) acceleration. For example [1] deduced a change in the bulk Lorentz factor of the C7 component in 3C345, from $\gamma \sim 5$ to $\gamma \sim 10$ over a deprojected distance range of $\sim 3 - 20$pc (they also infer a decrease in the Doppler factor from 12 to 4 and an increase in the angle between the velocity of the component and the line of sight from 2 to 10° during the same period). [2] give a more general argument related to the extended acceleration: If the bulk flow near the disk is sufficiently fast ($\gamma \gtrsim 5$) it would Comptonize photons coming from the disk, producing bulk-Compton features. The absence of such features indicate that acceleration up to $\gamma \gtrsim 10$ must take at least 10^3 Schwarzschild radii. They also argue that the electron/positron kinetic energy (estimated from the emissivity of blazar events), is too small to support the energetics of blazars and of radio lobes in quasars. Thus, the dynamics of AGN jets is likely dominated by protons rather than leptons.

Extended acceleration is unlikely to be driven hydrodynamically in a proton-electron outflow. If T_i is the initial temperature, energy conservation implies that pure-hydrodynamic driving gives Lorentz factors $\gamma \sim k_B T_i / m_p c^2$, which is of order unity even if the temperature is as high as 10^{12}K. In addition, the distance on which the Lorentz factor attains its terminal value is of the order of the sonic distance, certainly much smaller than pc-scales where the acceleration is inferred from the observations,

CP848, *Recent Advances in Astronomy and Astrophysics,*
7th International Conference of the Hellenic Astronomical Society, edited by N. H. Solomos

since the sonic surface is located close to the event horizon of the central black-hole. Alternatively, a possible heating source makes $\gamma \gg 1$ possible (e.g., [3]). It is, however, unclear how such a heating source would be established in a natural way on pc-scales. The most likely alternative is magnetic driving, which is considered in the next section of this article.

AGN jets are also well collimated; e.g., in the galaxy M87, the jet is seen opening widely in its formation region, at an angle of about 60° nearest the black hole, but is squeezed down to only 6° at 100 Schwarzschild radii [4]. The jet opening angle continues to decrease at larger distances. Magnetic self-collimation has long been thought to be the underlying mechanism for the observed jet shape.

Summarizing, as far as the dynamics of AGN jets are concerned, we need to explain extended (pc-scale) acceleration, collimation and the observed jet kinematics.

MAGNETOHYDRODYNAMICS (MHD)

The dynamics of astrophysical outflows may be described to zeroth order by the set of steady, ideal magnetohydrodynamic equations. These consist of the Maxwell's equations $0 = \nabla \cdot \boldsymbol{B} = \nabla \times \boldsymbol{E} = \nabla \times \boldsymbol{B} - 4\pi \boldsymbol{J}/c = \nabla \cdot \boldsymbol{E} - 4\pi J^0/c$, Ohm's law $\boldsymbol{E} = \boldsymbol{B} \times \boldsymbol{V}/c$, mass conservation $\nabla \cdot (\rho_0 \gamma \boldsymbol{V}) = 0$, entropy equation, and the momentum equation

$$\gamma \rho_0 (\boldsymbol{V} \cdot \nabla)(\xi \gamma \boldsymbol{V}) = -\nabla P + \left(J^0 \boldsymbol{E} + \boldsymbol{J} \times \boldsymbol{B}\right)/c. \tag{1}$$

Here $\boldsymbol{V}$ is the velocity of the outflow, γ the associated Lorentz factor, $(\boldsymbol{E}, \boldsymbol{B})$ the electromagnetic field as measured in the central object's frame, $J^0/c, \boldsymbol{J}$ the charge and current density respectively, ρ_0 the gas density in the comoving frame, and $\xi c^2 = c^2 + (5/2)(P/\rho_0)$ the relativistic specific enthalpy.

Due to the axisymmetric nature of jets, which means that the ϕ cylindrical coordinate is ignorable, the above system can be partially integrated to yield several constants of motion. If $A = (1/2\pi) \int\int \boldsymbol{B}_p \cdot d\boldsymbol{S}$ is the poloidal magnetic flux function (with the inverse relation giving the poloidal magnetic field $\boldsymbol{B}_p = \nabla A \times \hat{\phi}/\varpi$), these are (e.g., [5])[1]: the field angular velocity

$$\frac{V_\phi}{\varpi} - \frac{V_p}{\varpi} \frac{B_\phi}{B_p} = \frac{c}{\varpi} \frac{E}{B_p} = \Omega(A), \tag{2}$$

the Michel's magnetization parameter $\sigma_{\rm M}(A)$ which is related with the magnetic-to-mass flux ratio $B_p/\gamma \rho_0 V_p = (4\pi c^3/A\Omega^2)\sigma_{\rm M}$, the entropy $P/\rho_0^{5/3} = Q(A)$ for adiabatic flows, the total energy-to-mass flux ratio

$$\xi \gamma c^2 - \frac{\sigma_{\rm M} c^3}{A\Omega} \varpi B_\phi = \mu(A) c^2, \tag{3}$$

and the total specific angular momentum $\xi \gamma \varpi V_\phi - (\sigma_{\rm M} c^3/A\Omega^2)\varpi B_\phi = \mu \Omega \varpi_{\rm A}^2(A)$. The left-hand side of equation (3) consists of the matter energy-to-mass flux ratio $\xi \gamma c^2$,

---

[1] The subscripts p/ϕ denote poloidal/azimuthal components.

and the Poynting-to-mass flux ratio $(\mu - \xi\gamma)c^2$. Note that since, for $B_p > 0$, B_ϕ is negative (required in order to have $V_\phi < c$ beyond the light cylinder; see equation [2]), the contribution of the electromagnetic field in equation (3) is positive.

Using the above expressions of the five integrals of motion we may express all the flow quantities as functions of γ and A. The equations that give the remaining unknowns γ and A are two components of the momentum equation (1): one along the flow (which gives the velocity for known shape of the flow) and one in the perpendicular with respect to the flow direction (the so-called transfield equation that determines the shape of the flow and the magnetic field distribution A). These two equations are coupled because not only the velocity depends on the shape of the flow and the magnetic field distribution, but also inertial forces that are present in the transfield equation affect the shape of the magnetic fieldlines. Despite the fact that the remaining equations are only two, of which one is algebraic, the system remains highly intractable. The reason is that the problem remains two-dimensional (with independent variables say the z and ϖ cylindrical coordinates which define the so-called poloidal plane) and the resulting partial differential equation is of mixed type, i.e., changes from elliptic to hyperbolic in unknown a-priori surfaces. Due to this fact, it is beyond the capability of existing numerical codes to solve this highly nonlinear problem, and no numerical solution has been obtained so far. An alternative numerical approach is to solve the time-dependent problem (which is hyperbolic in time) and expect to reach a steady-state. Although there has been significant progress over the last years in doing so for studying nonrelativistic flows, all existing codes fail to simulate relativistic magnetohydrodynamic disk-driven flows for more than a few rotational periods. On top of that, it is not always clear how the issue of the boundary conditions is handled.

Besides fully numerical methods, there have also been suggested semi-analytical ones. These are basically a way to separate the two independent variables in the poloidal plane (z, ϖ), and reduce the system to ordinary differential equations. There are two families of such models: the θ self-similar, appropriate for the description of coronal outflows near the rotation axis, and the r self-similar for outflows associated with disks [6]. [7] and [8] found independently an r self-similar relativistic model, which was further generalized including thermal effects and applied to γ-ray burst jets by [5], and to AGN jets by [9]. I'll present in the following a typical solution obtained using this model, showing that many of the observable characteristics of AGN jets can be explained.[2]

Acceleration

As discussed in the introduction, the most likely case for AGN outflows is to be driven magnetically. Thus, we are interested in examining Poynting flux-dominated flows at the top of the accretion disk, meaning that the electromagnetic part of the total energy-to-mass flux ratio is much larger that the matter part.

In general, there are three acceleration mechanisms.

[2] The presented solution is solution *a* from [9]; for a detailed description of the model see [5].

(i) Thermal acceleration: This is due to the expansion of the flow under its own pressure (or enthalpy) gradient. Remembering that the thermal acceleration is a result of a De-Laval nozzle coming from the interplay between gravity (that initially confines the flow) and pressure gradients (that expand the flow) we conclude that thermal acceleration is almost complete at distances where the sound speed is of the order of the escape speed from the central gravitational potential. This happens very close to the black hole and cannot be connected to pc-scale accelerations.

(ii) Magnetocentrifugal acceleration: This can be interpreted in the "bead on a rotating wire" picture [10]. The magnetic fieldlines act like rotating rigid wires and the plasma as beads moving along them. If the angle between the wires and the axis of rotation is sufficiently large, the centrifugal force overtakes the gravitational pull causing acceleration. The main ingredient of this mechanism is the rotation of the roots of the wires and the terminal speed that gives is of the order of the Keplerian velocities of the underlying disk. However, in relativistic flows this mechanism cannot play an important role. This is because the nonnegligible rotation that would be required in this case would constrain the maximum value of the poloidal speed (since their sum cannot exceed the speed of light). Note that this mechanism is directly connected to the magnetic field that causes corotation at the base of the jet up to the Alfvénic distance, and transforms Poynting flux to kinetic energy flux; hence the name "magnetocentrifugal".

(iii) Magnetic acceleration: This is the direct result of the $\boldsymbol{J} \times \boldsymbol{B}$ Lorentz force; its work describes the transformation from Poynting flux to kinetic energy flux. The maximum Lorentz factor that this mechanism can give is the initial ratio of the Poynting flux over the mass flux (over c^2). However, it is an important question if this mechanism continues to operate until the whole energy initially deposited in the magnetic field is transferred to the matter, or stops at some distance resulting in some finite Poynting-to-matter energy flux ratio. In other words, it is important to know what the efficiency of this acceleration mechanism is, and how the conditions near the disk affect this result. A general analysis concerning the efficiency of the magnetic acceleration is given in the last section of this article. For the moment I continue with the presentation of a typical solution of the self-similar model.

Figure 1a shows the various forces accelerating the jet as functions of the cylindrical distance. The thermal pressure gradient is negligible even for initial temperatures of the order of 10^{10}K (as in this particular solution; see figure 1b). The magnetocentrifugal mechanism is the dominant one very close to the disk and remains finite as long as the rotation speed is nonnegligible. However, the magnetic acceleration rapidly takes over.

The result of the acceleration is shown in figure 2a. It is seen that the Lorentz factor increases monotonically with distance. The upper curve in figure 2a indicates that the Lorentz factor increases due to a decreasing Poynting-to-mass flux ratio: the Poynting flux is converted into matter kinetic-energy flux. Note that the sum of these two quantities is a constant of motion (a result of the total energy-to-mass flux ratio conservation expressed by equation [3]).

The flow is initially Poynting-flux-dominated and its initial Poynting-to-mass flux ratio (over c^2) is $\approx \mu$, the total energy-to-mass flux ratio (over c^2). This number μ gives also the maximum Lorentz factor that the flow can have (in the case where the whole Poynting flux is transferred to the matter). Asymptotically, in the presented solution, the Lorentz factor is $\mu/2$, or equivalently the flow reaches a rough equipartition between

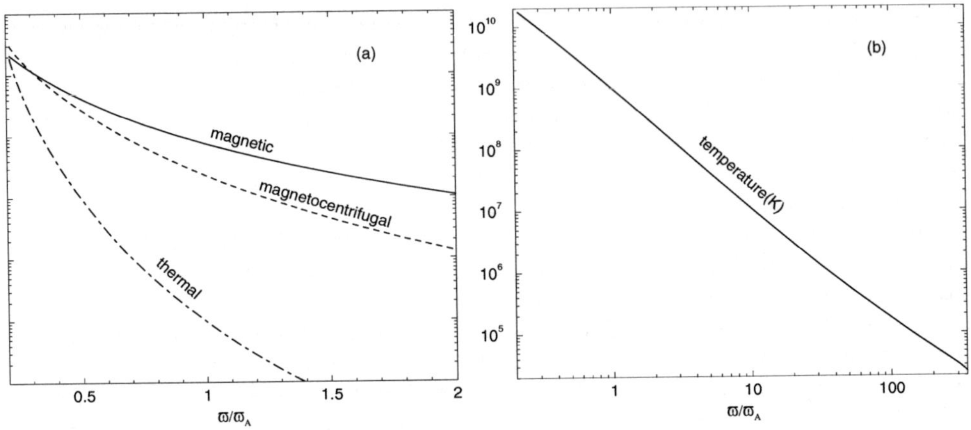

FIGURE 1. Left panel: Accelerating forces. Right panel: Temperature as function of the cylindrical distance from the rotation axis in units of the Alfvénic cylindrical distance.

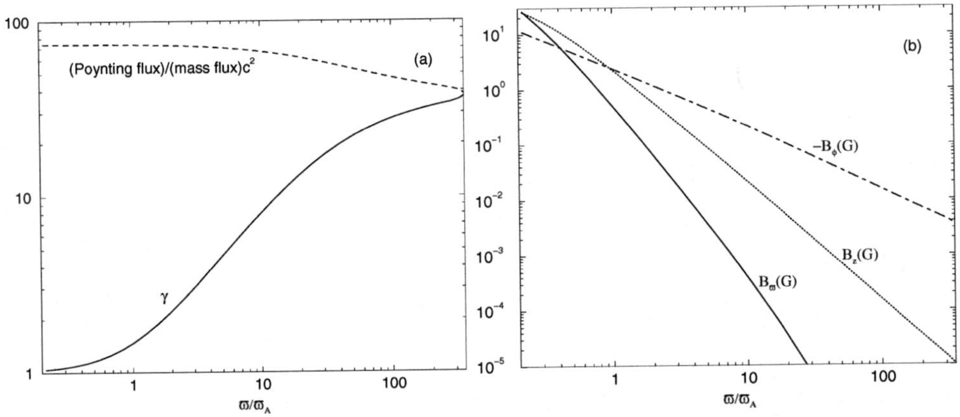

FIGURE 2. Left panel: Lorentz factor and Poynting-to-mass flux ratio (over c^2) as function of the dimensionless cylindrical distance. Right panel: Magnetic field components in cylindrical coordinates (ϖ, ϕ, z).

Poynting and kinetic energy fluxes. In other words, the efficiency of the magnetic acceleration in this particular solution is $\sim$50%.

Collimation

In order to examine the collimation process we need to think which are the forces acting on the plasma in the transfield direction (i.e., perpendicular to the flow). Besides the pressure gradient component (which is not expected to be important for Poynting-

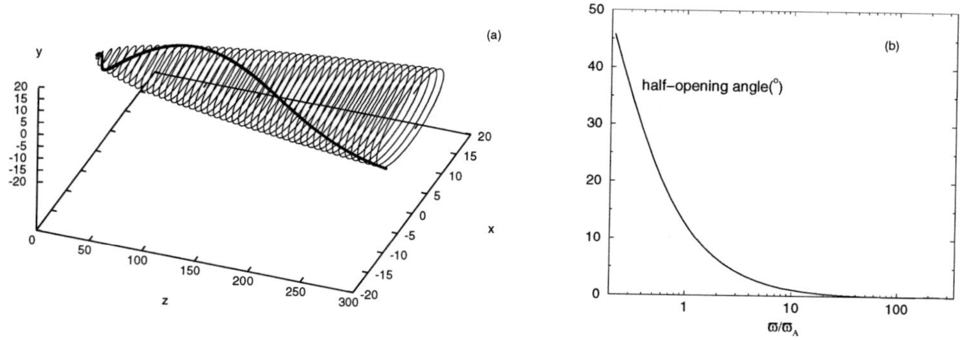

FIGURE 3. Left panel: The 3D field-line (thin) and stream-line (thick) shapes. The distances are in units of ϖ_A, the Alfvénic cylindrical distance. Right panel: Half-opening angle of the jet.

dominated flows), there are two forces of electromagnetic nature. The magnetic force (the component of the $\boldsymbol{J} \times \boldsymbol{B}$ force) is responsible for the well known and understood self-collimation process in nonrelativistic flows. The same force is of course present in relativistic flows as well. The important difference between these two categories is the appearance of the electric force. The electric field is comparable with the magnetic field for $V \approx c$, since from Ohm's law $\boldsymbol{E} = -(\boldsymbol{V}/c) \times \boldsymbol{B}$. Moreover, it is always acting against the collimating magnetic force. Thus, we expect the collimation to be much more difficult to achieve for extremely relativistic flows. The magnetic and electric forces almost cancel each other. Their difference, which is much smaller than each of the two terms, equals the inertia force in the transfield direction. Quantitatively, this can be written as $\gamma^2 \varpi / \mathscr{R} \leq 1$, where $\mathscr{R}$ is the curvature radius of the streamlines (e.g., [11]): it is much more difficult to change the shape of a relativistic flow because it has higher inertia. The above reasoning led [12] to suggest that highly relativistic jets could be collimated by an outer nonrelativistic magnetized wind. In this scenario, the self-collimation works in the outer wind, and the wind itself forces the inner highly relativistic flow to collimate (see also [13] for an application to the jet in M87). However, if the flow starts with $\gamma \approx 1$ there is a possibility that the main part of the collimation happens at relatively small values of γ. This is seen in the presented solution. Figure 3b shows the half-opening angle of the jet as function of the cylindrical distance. At the origin of the jet the angle between the flow motion and the axis of rotation is 45°. Following the Lorentz factor evolution from figure 2a we see that the collimation is indeed almost complete for relatively small values of γ. Note that the collimation and acceleration processes shown in figures 2a, 3b are found self-consistently by solving the two coupled components of the momentum equation, along and perpendicular to the flow direction.

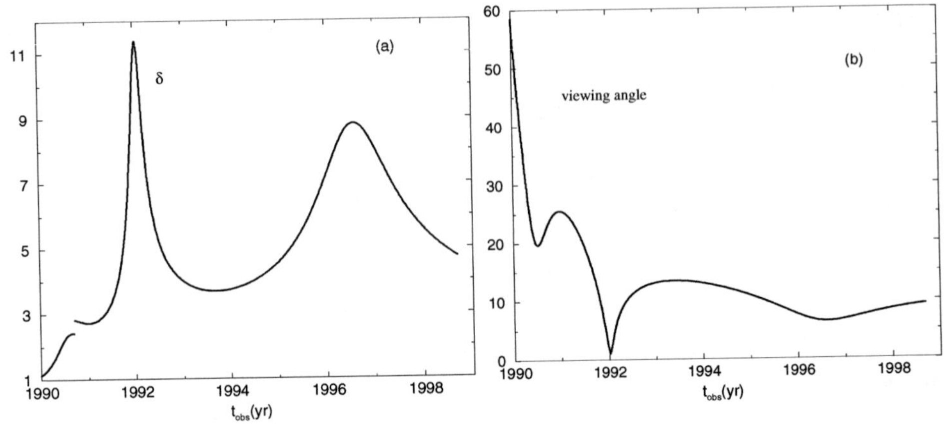

FIGURE 4. Doppler factor (left panel) and viewing angle (angle between the instant velocity of the jet component and the line of sight) as functions of the observer time for the C7 component in 3C345.

Apparent kinematics

The components of AGN jets follow curved (helical) trajectories over the years. Their apparent velocity, as well as their intensity, depends on two quantities: the Lorentz factor and the angle between the velocity of the component and the line of sight (the viewing angle). There have been various tries to explain the apparent jet kinematics: as Kelvin-Helmholtz instabilities [14], or due to source precession [15]. However, since a MHD outflow being rotating naturally follows a helical path (see figure 3a) it may explain (or at least play some partial role to) the observed kinematics. In order to explore this possibility I used the self-similar solution presented in the previous parts of the article.

Suppose that the angle between the line of sight and the axis of the jet is $\theta_{\rm obs}$. Assuming that each observed plasma component corresponds to ejection from a localized region of the disk (characterized by an angle ϕ_0 on the disk-plane and the distance from the center, or equivalently the Alfvénic distance of the field-line that is rooted at this region), one can easily project the plasma trajectory on the plane perpendicular to the line of sight, as well as to find Doppler factors and apparent speeds. As a test for the model we tried to fit the trajectory, Doppler factor, and viewing angle, as they were inferred from observations by [1].

For $\theta_{\rm obs} = 9^\circ$ and $\phi_0 = 180^\circ$ the model could explain the [1] results (Vlahakis, Marin, & Königl to be submitted). Between the years 1992-1993 the Doppler factor decreases from ~ 12 to ~ 4 and the viewing angle increases from ~ 2 to $\sim 10^\circ$ during the same period, see figures 4, 5.

Polarization maps

Another manifestation of magnetic driving is the observed polarization in radio maps of AGN jets. Fractional polarization of a few tens percent are typical in pc-scale jets, sup-

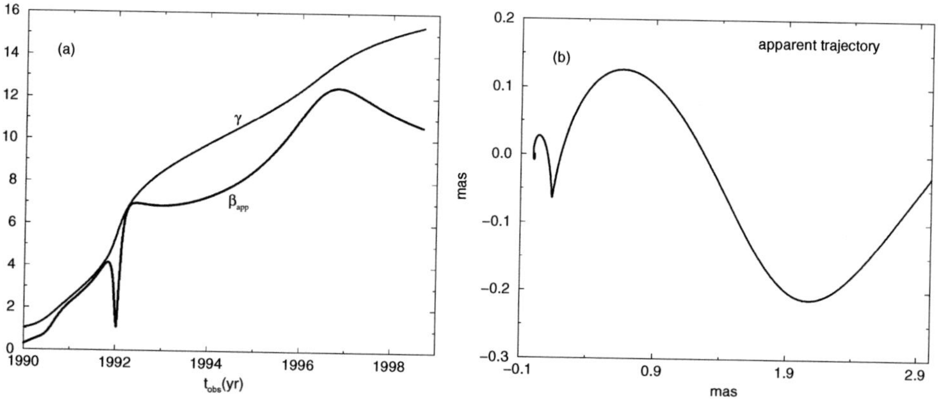

FIGURE 5. Left panel: Apparent velocity ($\beta_{\rm app}$) and Lorentz factor (γ). Right panel: Apparent trajectory of the C7 component in 3C345 (the distances are in mas).

porting the existence of a dominant large-scale magnetic field component. In addition, the observed electric field polarization vectors and the Faraday rotation measure gradients across the jet, support the existence of helical magnetic fields with strong transverse (azimuthal) component (e.g., [16]). This is consistent with the MHD picture, since the field is naturally helical (see figure 3a) and the azimuthal (B_ϕ) component is dominant beyond the Alfvénic distance (see figure 2b). This is another connection between models and observations that definitely need to be examined more closely (Vlahakis, Marin, & Königl in preparation).

THE EFFICIENCY OF THE MAGNETIC ACCELERATION

Equation (3) shows the two acceleration mechanisms causing an increase of γ: (1) thermal, corresponding to a decreasing specific enthalpy ξc^2, and (2) magnetic, corresponding to a decline in the poloidal current $I = (c/2)\varpi B_\phi$. The former works up to the point where ξ gets its minimum value ($= 1$, something that happens very close to the disk even for $T_i = 10^{12}$K), while the latter is the result of the Lorentz force and its final outcome depends on the value of the asymptotic current I_∞. (Note that the magnetocentrifugal acceleration also corresponds to a tiny decline in the poloidal current that happens at the corotating part of the outflow, from the base up to the Alfvénic distance.) The maximum value of the Lorentz factor is the constant of motion μ (corresponding to $\xi = 1$ and $I = 0$; see equation [3]).

We may re-write equation (3) as $\xi\gamma/\mu = 1 - \sigma_{\rm M} c^3 \varpi |B_\phi| / \mu A \Omega = 1 - 2\sigma_{\rm M} |I| / \mu A \Omega$. Downstream from the classical fast-magnetosonic surface (where the poloidal flow speed equals the phase speed of the fastest magnetosonic waves propagating along the flow) the magnetic field is mainly toroidal and, for a highly relativistic poloidal flow ($V_\phi \ll V_p$), equation (2) gives $B_\phi \approx -\varpi B_p \Omega / c$. As a result, the poloidal current is $|I| \approx B_p \varpi^2 \Omega / 2$ and the matter-to-total energy flux ratio depends on the key function

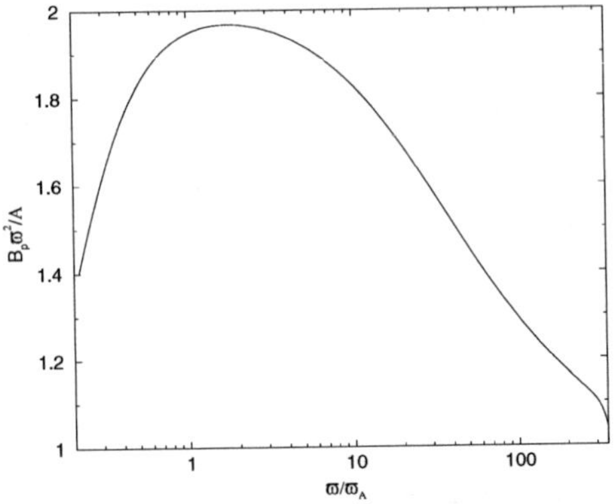

FIGURE 6. The function $B_p\varpi^2/A$ for the self-similar solution.

$B_p\varpi^2/A$:

$$\frac{\xi\gamma}{\mu} \approx 1 - \frac{\sigma_M}{\mu}\frac{B_p\varpi^2}{A}. \tag{4}$$

For an initially Poynting-dominated flow the matter part of the energy flux is negligible, $(\xi\gamma)_i/\mu \ll 1$[3]. This continues to be the case in the neighborhood of the classical fast magnetosonic surface, since at this point $\gamma \sim \mu^{1/3} \ll \mu$ (e.g., [17]). Suppose that the value of the function $B_p\varpi^2/A$ near the classical fast surface is $(B_p\varpi^2/A)_f$. Since $\xi\gamma \ll \mu$ at this point, equation (4) implies that $\sigma_M/\mu \approx 1/(B_p\varpi^2/A)_f$.

Denoting with $\delta\ell_\perp$ the distance between two neighboring poloidal field lines A and $A + \delta A$, magnetic flux conservation implies $2\pi\delta A = B_p(2\pi\varpi\delta\ell_\perp)$, or, $B_p\varpi^2 = (\varpi/\delta\ell_\perp)\delta A$. Thus, a decreasing $B_p\varpi^2$ – and hence, from equation (4), an accelerating flow – corresponds to poloidal field lines expanding in a way such that their distance $\delta\ell_\perp$ increases faster than ϖ. How fast the field lines expand is determined by the transfield force balance equation; thus, this equation indirectly determines the flow acceleration. Since the available solid angle for expansion of the field lines is finite, there is a minimum value of the $B_p\varpi^2/A$ function. The field lines asymptotically have a shape $z \approx z_0(A) + \varpi/\tan\vartheta(A)$, where $\vartheta(A)$ is their opening angle [18]. Differentiating the latter equation we get a decreasing function $B_p\varpi^2/A = \left(A\vartheta'/\sin\vartheta - Az_0'\sin\vartheta/\varpi\right)^{-1}$, reaching a minimum value $\sin\vartheta/A\vartheta'$ at $\varpi \gg z_0'\sin^2\vartheta/\vartheta'$. Since the factor $\sin\vartheta/A\vartheta'$ is ~ 1, the minimum value of the $B_p\varpi^2/A$ function is ~ 1, corresponding to

$$\frac{\gamma_\infty}{\mu} \approx 1 - \frac{\sigma_M}{\mu}\left(\frac{B_p\varpi^2}{A}\right)_\infty \approx 1 - \frac{(B_p\varpi^2)_\infty}{(B_p\varpi^2)_f} \sim 1 - \frac{\sigma_M}{\mu} \sim 1 - \frac{1}{(B_p\varpi^2/A)_f}. \tag{5}$$

Equivalently, the asymptotic Lorentz factor is $\gamma_\infty \sim \mu - \sigma_M$, and the asymptotic Poynting-to-mass flux ratio is $\sim \sigma_M c^2$.

---

[3] The subscript i denotes values at the top of the disk.

Another interesting connection with the boundary conditions on the disk can be found by using the relation between the poloidal current and the function $B_p\varpi^2/A$: $|I| = B_p\varpi^2\Omega/2$. Thus, $(B_p\varpi^2/A)_{\rm f} \approx 2|I|_{\rm f}/A\Omega$, and since $|I|$ remains practically a constant of motion inside the force-free subfast regime, $(B_p\varpi^2/A)_{\rm f} \approx 2|I|_{\rm i}/A\Omega$, and $\mu/\sigma_{\rm M} \approx 2|I|_{\rm i}/A\Omega$. Hence, equation (5) implies a direct connection of the acceleration efficiency and the asymptotic Lorentz factor with the current distribution at the disk

$$\frac{\gamma_\infty}{\mu} \approx 1 - \frac{A\Omega}{2|I|_{\rm i}}\left(\frac{B_p\varpi^2}{A}\right)_\infty \sim 1 - \frac{A\Omega}{2|I|_{\rm i}}. \tag{6}$$

The presented self-similar solution verifies the above analysis. Figure 6 shows that the value of the function $B_p\varpi^2/A$ near the fast point (which is located at $\varpi/\varpi_{\rm A} \sim$ a few) is ~ 2. According to the general analysis of this section the asymptotic Lorentz factor is expected to be $\sim \mu/2$ (see equation 5), meaning that there is an asymptotic equipartition between Poynting and kinetic energy fluxes. The solution indeed shows this behavior, see figure 2. In principle it is possible to get solutions with higher $(B_p\varpi^2/A)_{\rm f}$, and thus higher acceleration efficiencies as in [18].

Summarizing, acceleration of AGN jets on pc-scales is most plausibly explained in terms of magnetic driving. MHD outflows are also characterized by strong self-collimation and are thus consistent with the narrow opening angles inferred in AGN jets. Finally, the internal rotation in MHD flows could explain the observed jet kinematics.

ACKNOWLEDGMENTS

The author acknowledges partial support from the General Secretariat for Research and Technology of Greece under "PYTHAGORAS" grant No 70/3/7396.

REFERENCES

1. S. C. Unwin, A. E. Wehrle, A. P. Lobanov, Z. J. A., G. M. Madejski, and M. F. Aller, *ApJ* **480**, 596 (1997).
2. M. Sikora, M. C. Begelman, G. M. Madejski, and J.-P. Lasota, *ApJ* **625**, 72 (2005).
3. Z. Meliani, C. Sauty, K. Tsinganos, and N. Vlahakis, *A&A* **425**, 773 (2004).
4. J. A. Biretta, W. Junor, and M. Livio, *NewAR* **46**, 239 (2002).
5. N. Vlahakis, and A. Königl, *ApJ* **596**, 1080 (2003).
6. N. Vlahakis, and K. Tsinganos, *MNRAS* **298**, 777 (1998).
7. Z.-Y. Li, T. Chiueh, and M. C. Begelman, *ApJ* **394**, 459 (1992).
8. J. Contopoulos, *ApJ* **432**, 508 (1994).
9. N. Vlahakis, and A. Königl, *ApJ* **605**, 656 (2004).
10. R. D. Blandford, and D. G. Payne, *MNRAS* **199**, 883 (1982).
11. S. V. Bogovalov, *A&A* **371**, 1155 (2001).
12. S. Bogovalov, and K. Tsinganos, *MNRAS* **357**, 918 (2005).
13. J. Gracia, K. Tsinganos, and S. V. Bogovalov, *A&A* **442**, L7 (2005).
14. P. E. Hardee, and R. C. Walker, *ASP Conference Series* **340**, 92 (2005).
15. A. P. Lobanov, and J. Roland, *A&A* **431**, 831 (2005).
16. D. C. Gabuzda, E. Murray, and P. Cronin, *MNRAS* **351**, L89 (2004).
17. M. Camenzind, *A&A* **162**, 32–44 (1986).
18. N. Vlahakis, *ApJ* **600**, 324–337 (2004).

The Supercritical Pile Model for GRBs

A. Mastichiadis* and D. Kazanas†

*Department of Physics, University of Athens, Panepistimiopolis, GR 15783, Zografos, Greece
†Laboratory for High Energy Astrophysics, NASA/GSFC, Code 661, Greenbelt, MD 20771,USA

Abstract. We present the spectral and temporal radiative signatures expected within the "Supercritical Pile" model of Gamma Ray Bursts (GRBs). This model is motivated by the need for a process that provides the dissipation necessary in GRBs and presents a well defined scheme for converting the energy stored in the relativistic protons of the Relativistic Blast Waves (RBW) associated with GRB into radiation; at the same time it leads to spectra which exhibit a peak in the burst νF_ν distribution at an energy $E_p \simeq 1$ MeV in the observer's frame, in agreement with observation and largely independent of the Lorentz factor Γ of the associated relativistic outflow. Futhermore, this scheme does not require acceleration of particles at the shock other than that provided by the isotropization of the flow bulk kinetic energy on the RBW frame.

Keywords: Gamma Ray Bursts: Radiation Mechanisms; Plasmas: Relativistic
PACS: 98.70.Rz; 41.60.-m; 52.27.Ny

INTRODUCTION

The longstanding issue of the distance and absolute luminosity of GRBs has been settled in the past decade through the observational evidence collected by *BeppoSAX* [2] while the theoretical work of [13] provided the broader physical framework into which these events seem to generally fit. This framework associates GRBs with radiation emitted by relativistic blast waves (hereafter RBWs), produced by an unspecified todate agent, presumably associated with the formation of a neutron star or black hole. While the source of the energy associated with GRBs has remained uncertain, there remains little doubt about the presence of the RBWs, which power also the later time emissions at X-ray, optical and radio frequencies, known collectively as GRB afterglows (see review of [12]).

With the discovery of GRB afterglows [15], much of the theoretical activity has since shifted to the study of the physics of these later time emissions. Nonetheless, a number of issues associated with the prompt $\gamma-$ray emission, besides the nature of their energy source, still remain open. Chief among them are the conversion of the RBW kinetic energy into radiation and the fact that the photon energy at which the GRB luminosity peaks, $E_{\rm p}$, is narrowly distributed around a value intriguingly close to the electron rest mass energy. The purpose of the present note is to describe a process that provides a "natural" account of these generic, puzzling GRB features.

Following the work of [14] it has been generally accepted that a certain amount of baryons must be carried off with the blast waves responsible for the GRBs. This baryon contamination has even been deemed necessary, else the entire blast wave internal

CP848, *Recent Advances in Astronomy and Astrophysics,*
7th *International Conference of the Hellenic Astronomical Society,* edited by N. H. Solomos

energy would be converted into radiation on very short time scales, leading to events of very different temporal and spectral appearance (e.g. [11]) than observed in GRBs. While the low radiative efficiency of baryons is essential for the GRB energy transport to the requisite distances ($\gtrsim 10^{16}$ cm), it becomes problematic when demanded that their internal energy be radiated away on the short time scales associated with the GRB prompt γ–ray emission.

The narrow range of the GRB νF_ν spectral peak energy, E_p, is another well documented systematic feature of these events, a result of the extensive data base accumulated by *BATSE*. The compilation of [6] shows clearly a preference for an energy $E_\mathrm{p} \simeq 200$ keV at which the νF_ν GRB spectra exhibit a maximum. In fact, it is precisely this maximum in the spectral energy distribution that qualifies GRBs as such. Furthermore, when corrected for the redshift ($z_\mathrm{GRB} \sim 1$), E_p shifts close to the electron rest mass.

One of the few attempts to address these issues was that of [4], hereafter KGM: In search of a well defined and tractable dissipation mechansim, they proposed a process that utilises proton - photon collisions to convert the energy stored in protons behind the forward shock of the RBW to electrons (and then into radiation). KGM provided the kinematic and dynamic thresholds for this process to take place and showed that, in addition, it produced spectra that peaked at several well defined energies, namely at $\sim 2m_ec^2/\Gamma^2$, $2m_ec^2$, $m_ec^2\Gamma^2$, in the observer's frame. The interesting feature of this model is that the combination of the threshold for pair production energy and the final Lorentz boost to the observer's frame leads to a spectral peak at $\sim m_ec^2$ largely independent of the Lorentz factor Γ of the RBW, provided that the latter is larger than a threshold value which depends on the value of the magentic field of the RBW. Therefore, the process described by KGM provides a framework for resolving both the proton-to-electron-energy-transfer and the narrow E_p–range problems simultaneously. This picture was verified numerically by [7].

The purpose of the present paper is to present numerical calculations based on the model outlined in KGM, which for reasons explained below is referred to as the "Supercritical Pile" model for GRB. In §2 of this note we provide a qualitative description of the basic notions that underlie the "Supercritical Pile" model and how they apply to GRB. In §3 we describe the numerical method employed and show some basic results and in §4 we summarize our results.

DISSIPATION: THE PROTON-PAIR-SYNCHROTRON (PPS) NETWORK

The process we model in detail in the following has two distinct aspects, along the lines of the two GRB issues discussed above. The first one concerns the transfer of the energy stored in a population of relativistic hadrons (protons) into leptons. This has been discussed originally by [5] and modeled in more detail in [8] and [9]. This same process was then modified to include the effects of relativistic motion of the plasma in the possibility that some of its photons could be scattered by matter (a "mirror") located upstream along the direction of motion. This leads to a relaxation of the threshold conditions obtained in [5] and was discussed by [3]. The second aspect concerns the spectra that result from the combination of these processes as was originally discussed in KGM and their relation to the energetics of GRB.

To formulate the conversion of proton energy to leptons through the $p\gamma \to pe^+e^-$ process (known, also, as Bethe-Heitler pair production) one assumes the presence of a population of relativistic protons of the form $N(\gamma_p) = n_0 \gamma_p^{-\beta}$ in a volume which is taken for simplicity to be spherical with radius R. The requisite photons are provided by the synchrotron radiation of the pairs produced in the proton - photon intereactions, with these two processes combined into a reaction network. This network involves both kinematic and dynamic thresholds. The physical arguments that provide these thresholds are rather simple: There is a *kinematic* threshold for the reaction $p\gamma \to pe^+e^-$; for a photon of energy E_γ, there is a critical value $\gamma_c = 2m_ec^2/E_\gamma$ of the proton Lorentz factor for the reaction to take place with each lepton produced by this reaction with energy $\gamma_c\, m_ec^2$. The reaction network will be *self-contained* if these pairs can produce, through the synchrotron process, the photons needed to effect the pair production, i.e. if $E_\gamma = E_s = b\gamma_c^2 m_ec^2$ (b is the magnetic field B of the plasma in units of the critical magnetic field $B_c = m_e^2c^3/\hbar c$ = 4.4 10^{13} G). This leads to the following kinematic threshold for the reaction network $\gamma_c^3 b \simeq 2$ or $\gamma_c \simeq (2/b)^{1/3}$, as discussed in [5].

The combined network of the photon and pair producing reactions will also be *self - sustained* if *at least one* of the synchrotron photons produced by the e^+e^- pairs produces another pair before escaping the volume of the plasma in a reaction with a sufficiently energetic proton (i.e. one that fulfills the kinematic threshold). This results in a condition for the column density of the plasma which is identical to that of a critical nuclear chain reaction. Therefore, the plasma column density (at the proton critical energy), must be greater than the inverse of the number of synchrotron photons emitted by the electrons of Lorentz factor γ_c. The number of synchrotron photons just above the thershold energy is $\mathcal{N} = \gamma_c/b\gamma_c^2 = 1/b\gamma_c$ leading to the dynamic condition $\tau_{p\gamma} \simeq \sigma_{p\gamma}RN_p(\gamma_c)\,\gamma_c \gtrsim b\gamma_c$, where $\sigma_{p\gamma}$ is the photopair (Bethe-Heitler) cross section as given by [10]. Taking into account the kinematic threshold condition this reads $\sigma_{p\gamma}Rn_0 \gtrsim b^{(1-\beta/3)}$, in agreement with the result of [5].

The above processes have been explored numerically in [8] and [9]. It was found that the semianalytic estimates presented above are in excellent agreement with the more detailed numerical studies. The instability depends on both thresholds as described and converts the energy density stored in relativistic protons (those with energy above that of the threshold $\gamma_p \geq \gamma_c$) into pairs on the plasma crossing time scale R/c. During the initial (linear) regime of the instability the pair and photon numbers increase exponentially ($\propto e^{st}$) with the exponent s being greater the larger the value of the proton density n_0 (i.e. the farther the network is above threshold), while $s = 0$ at precisely the threshold value of this parameter.

The RBW associated with the GRB provide a natural setting for the production of relativistic proton populations, thus lending them as natural sites where the above considerations should apply. However, relativistic motion alone does not in general change the kinematic and dynamic threshold conditions obtained above. It is conceivable that the presence of a relativistic proton population that extends to sufficiently high energies could satisfy the kinematic threshold of the instability; however, the "standard" GRB model parameters yield column densities for the relativistic protons (these are just those in the matter "swept-up" by the RBW) far smaller than necessary to satisfy the dynamical threshold condition.

As pointed out in [3] the situation can change drastically in the presence of matter along the direction of relativistic motion which could isotropize through scattering the radiation emitted by the relativistically moving plasma. In this case, the RBW can "catch-up" with the scattered photons which, when re-intercepted by the RBW, appear on its rest frame blue-shifted by a factor $4\Gamma^2$. If those were the synchrotron photons of energy $\varepsilon_s \simeq b\gamma^2$ they will now have an energy $\varepsilon_s' \simeq 4b\gamma^2\Gamma^2$ (in units of $m_e c^2$). Therefore, the kinematic condition for the *reflected* photons to produce e^+e^--pairs gets modified to $\gamma\varepsilon_s' \gtrsim 2$ or $2b\gamma^3\Gamma^2 \gtrsim 1$. Making the further restrictive assumption that the only relativistic protons are those which are produced by the isotropization of the swept-up matter by the RBW, we can set $\gamma = \Gamma$ to obtain the kinematic threshold condition

$$\Gamma^5 b \gtrsim 2 \quad \text{or} \quad \Gamma \gtrsim \left(\frac{1}{2b}\right)^{1/5}. \tag{1}$$

One can employ the same considerations used above to obtain the dynamical threshold of the instability in this case too. The number of photons produced by an electron of Lorentz factor Γ on the RBW will now be $\mathcal{N} \simeq \Gamma/b\Gamma^2 = 1/b\Gamma$; assuming that a fraction $\tau_{\rm mir}$ of these photons are scattered by the mirror (and are therefore re-intercepted by the RBW) leads to

$$\sigma_{p\gamma} R n \gtrsim \frac{b\Gamma}{\tau_{\rm mir}} \quad \text{or} \quad 2\tau_{\rm mir}\,\sigma_{p\gamma} R n \Gamma^4 \gtrsim 1 \tag{2}$$

with the latter expression obtained from the former with the use of the kinematic threshold condition. The quantity n above now denotes the value of the ambient density (one should note that the column density is a Lorentz invariant so that its value as given above is identical to its value on the RBW rest frame). This condition can be satisfied for values of the GRB parameters $n \simeq 10^3 n_3$ cm^{-3} and $R = 10^{16} R_{16}$ cm for values of $\Gamma \gtrsim 100\,(n_3 R_{16})^{-1/4}$ (assuming $\tau_{\rm mir} \simeq 1$).

PARTICLE EVOLUTION AND RADIATION: THE KINETIC EQUATION APPROACH

Having outlined the qualitative aspects of the 'Supercritial Pile' model we proceed to describe the procedure used in simulating this model (for a more detailed description see [7]). We consider a Relativistic Blast Wave moving with speed $\upsilon_0 = \beta_\Gamma c$, where $\beta_\Gamma = (1-\Gamma^{-2})^{-1/2}$ and Γ the bulk Lorentz factor of the flow. The RBW has a radius $R_{\rm RBW}(t)$ as measured from the origin of our coordinate system (assumed to be the center of the original explosion) and it contains protons and electrons. These are assumed to have been picked up by the RBW and isotropized so that in the flow rest frame they have distributions with average Lorentz factors $\langle\gamma_p\rangle \simeq \langle\gamma_e\rangle \simeq \Gamma$. Under these conditions, protons carry a significant fraction of the blast wave energy. However, the electron and proton distributions will evolve in time on the RBW frame since both species will suffer from various types of energy losses, physical escape, etc and at the same time they will radiate by synchrotron, inverse Compton and possibly other processes. Our objective is to calculate in a time-dependent, self-consistent manner the particle evolution and the spectrum of radiation.

We also assume the presence of scattering material (which we will call the "mirror") ahead of the advancing shock. This is located between radii $R_{s,1}$ and $R_{s,2}$ and it is assumed to have a uniform electron density $n_{\rm e}^{\rm sc}$ and a corresponding Thomson optical depth $\tau_{\rm mir} = n_{\rm e}^{\rm sc}\sigma_{\rm T}(R_{s,2} - R_{s,1})$. Therefore part of the radiation emitted from the particles in the RBW will be intercepted by this mirror and scattered back, reentering the shock at a later time when this will have moved to a different location. The reflected (and amplified in energy) photons will contribute to the particles' energy losses.

To treat the radiative transfer we assume that the relativistic blast wave is expanding in a spherical fashion, however, due to relativistic beaming an observer receives the radiation coming mainly from a small section of it of lateral width $R_{\rm RBW}/\Gamma$ and longitudinal width $R_{\rm RBW}/\Gamma^2$ in the lab frame but $R_{\rm RBW}/\Gamma$ on the flow frame. Therefore, at the RBW frame, one can approximate the emitting region by a spherical 'blob' of radius $R_b = R_{\rm RBW}/\Gamma$ and solve the radiative transfer problem for that region.

The problem of particle evolution in a homogeneous source region containing protons, electrons and photons was formulated and solved numerically by [8] and more recently by [9] adopting the kinetic equation approach. Herein we follow the same method, making the changes which are appropriate for the present case. The equations to be solved can be written in generic form

$$\frac{\partial n_i}{\partial t} + L_i + Q_i = 0 \tag{3}$$

where the index i can be any one of the subscripts 'p', 'e' or 'γ' referring to protons, electrons or photons respectively. The operators L_i denote losses and escape from the system while Q_i denote injection and source terms. The unknown functions n_i are the differential number densities of the three species and the physical processes which are included in the kinetic equations are: (1) proton-photon (Bethe-Heitler) pair production; (2) photopion production; (3) electron synchrotron radiation and synchrotron self absorption; (4) inverse Compton scattering; (5) photon-photon pair production. One could add other processes in this list, however it turns out that the ones that provide the driving terms in the proton loop are the processes (1), (3) and (4). Note that processes (1) and (4) can take place either with photons produced internally in the RBW or with the photons which have been reflected by the mirror.

In order to calculate the reflected photon number density as seen in the rest frame of the blob we use Eqn. (2) of [1]. The causality condition implies that the blob does not receive any reflected photons for as long as it is at

$$z < z_{\rm crit} = R_{s,1}\frac{2\beta_\Gamma}{1+\beta_\Gamma}. \tag{4}$$

Therefore, if we assume for simplicity, that the RBW has reached a constant Lorentz factor Γ when it is some distance z_0 from the origin, then we can relate the current position of the RBW, z, with the time elapsed since then by the relation $z = z_0 + \beta_\Gamma ct$. With this hypothesis we can define a time

$$t_{\rm crit} = \frac{z_{\rm crit}}{\beta_\Gamma c} \tag{5}$$

with the property that only when $t > t_{\rm crit}$, photons produced at earlier times will be reentering, due to reflection, the RBW.

Thus we apply the numerical code in a continuous run covering two distinct time intervals: (1) when the blob is still out of the reflection region ($t \leq t_{\rm crit}$–Zone I) and (2) after it has entered this region ($t > t_{\rm crit}$–Zone II). Protons are assumed to have a Lorentz factor $\langle \gamma_p \rangle = \Gamma$ upon entering Zone II. This can be justified from the fact that protons do not suffer any substantial energy losses while in Zone I. However the situation is very different in the case of electrons which have much faster cooling timescales than protons. To treat this problem we assumed two extreme possibilities, i.e. that electrons either have the Lorentz factors of the flow upon entering the reflection zone (Energetic Electron case, EE) or that they have cooled before entering it (Cooled Electron case, CE). The cases where the electrons cool in various degrees between the site the blob first reaches the Lorentz factor Γ and the starting point of the reflection zone can be considered as intermediate cases.

The time evolution of the system shows that, depending on the main parameters of our problem which are the comoving magnetic field strength B and the number density of the protons, there are cases where the loop operates. The expected photon outgrowth depends sensitively on the assumptions about the initial electrons (CE or EE), however we can draw the following general results:

Overall Energetics

The runs that have initial conditions which satisfy the threshold criteria show an explosive energy transfer from the protons to the photons, in agreement with the theoretical predictions as these where outlined in §2. The loop becomes more efficient as we increase either the B-field (a fact that affects the kinematic threshold) or the proton number density n_p (which affects the dynamical threshold). In general EE runs are more efficient than CE, as the photons produced from the cooling of the initial energetic electrons faciliate the development of the instability.

Fig. 1 shows the evolution of the photon luminosity as a function of time after the RBW has entered the reflection zone. At low $B-$fields we get only the effects of the fast cooling electron population with the protons remaining practically unaffected. As the electrons cool very quickly on their own reflected radiation, the lightcurves peak equally fast (i.e. within one or two crossing times) after the entrance of the RBW inside the reflection zone. As the value of B increases the loop begins to operate and an increasing fraction of the proton energy is turned into radiation, as this is manifested by the higher peak luminosities and the longer decay times of the lightcurves.

These results are summarized in Fig. 2 that depicts the total photon energy radiated in each run as it is measured in the frame of the RBW. As we have already mentioned, the energy lost by the protons goes into electron-positron pairs and eventually escapes the system as radiation. Therefore, as the numerical code discussed above conserves energy we expect $\int dt L_\gamma(t) = \Delta U_p + \Delta U_{e,in}$, where ΔU_p and $\Delta U_{e,in}$ is the total energy lost by the protons and initial electrons respectively and L_γ is the radiated photon luminosity. The dashed and full lines in Fig. 2 represent the CE and EE cases respectively. At low B there is a marked difference between the two cases: In the EE case, a substantial fraction of the initial electron energy is radiated, despite the fact that the system is below the threshold. For this reason we get essentially no radiation in the CE case that lacks the energetic electrons. As B increases the loop begins to operate extracting energy from the protons. We note that in the EE case, the presence of the initial energetic electron population (and the photons resulting from their cooling) allows the extraction of energy from the protons for lower values of the magnetic field. At even higher values of B both cases saturate at the same value of luminosity as the loop is capable of cooling the protons completely irrespective of the presence of initial energetic electrons.

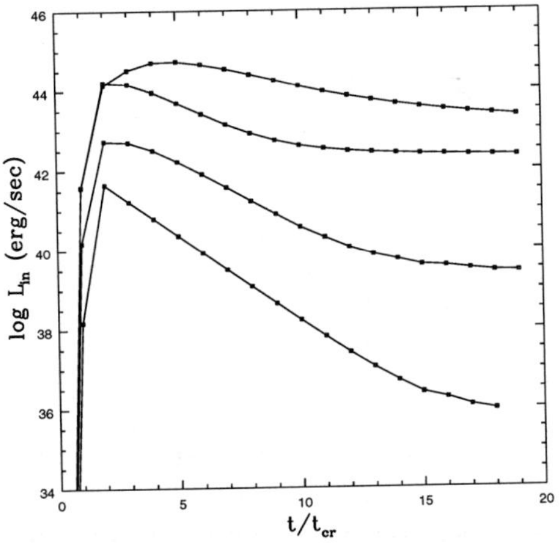

FIGURE 1. Plot of the photon luminosity evolution as this is measured in the comoving frame in the EE case where $n_p = n_e = 10^4\,\mathrm{cm}^{-3}$ while B varies (bottom to top) from 0.12 G to 120 G by increments of a factor of 10. The rest of the parameters are $R_{\mathrm{RBW}} = 3.10^{16}$ cm, $\Gamma = 10^{2.6}$ and $\tau_{\mathrm{mir}} = 1$. Time is measured in blob crossing times and the value $t = 0$ has been set at the instant when the RBW enters the reflection zone.

The Observed Spectra: Direct and Reflected Components

We find that in all cases the multiwavelength spectra produced on the RBW frame consist of a synchrotron component at low energies which peaks at $\varepsilon^{RBW}_{S,p} = b\Gamma^2$ and an inverse Compton component at high energies which peaks (due to Klein-Nishina effects) at $\varepsilon^{RBW}_{ICS,p} = \eta\Gamma$ with $0.1 < \eta < 1$ (all energies are expressed in units of $m_e c^2$). These are produced from the electron energy losses on the magnetic field and on the reflected photons respectively and the ratio of the luminosities of these two components

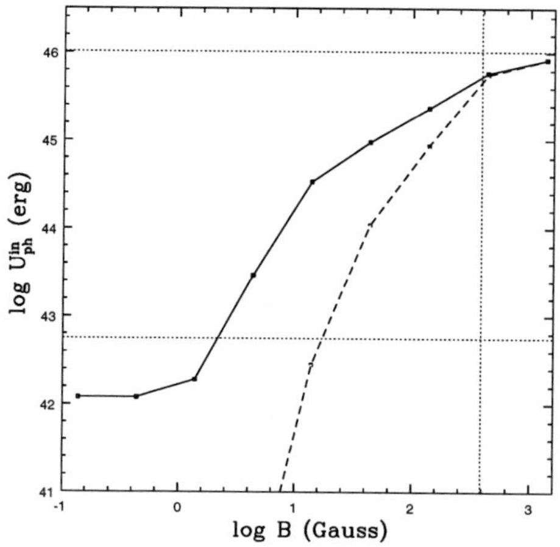

FIGURE 2. The total energy radiated by the protons in the case where $n_p = n_e = 10^4\,\mathrm{cm}^{-3}$ as a function of the magnetic field strength B - both quantities are measured in the comoving frame. The other parameters are the same as those of Fig.1 Full and dashed lines represent the EE and CE cases respectively. The lower horizontal dotted line is the total energy content stored in electrons while the upper one is the energy stored in protons. The vertical dotted line is the equipartition magnetic field.

in νF_ν will depend on the ratio of the corresponding energy densities. An observer at Earth will observe these two components shifted by Γ in energy and by Γ^3 in flux. This observer will not see the reflected component as this is directed away from him/her. However, any reflected photon which is scattered on the cooled pairs of the RBW will be isotropized in the frame of the flow and, upon transformation to the observer's frame, it will appear beamed within a cone of opening angle $\simeq 1/\Gamma$, with the same, as before, flux amplification ($\propto \Gamma^3$). Thus the observed spectrum will have a third component, which we shall refer to as the "doubly reflected" component and which is of special interest in our model as we propose that this is the observed prompt GRB emission.

The most conservative case concerning the flux of this "doubly reflected" component is that in which there are initially no "cold" pairs in the RBW; however, as the loop begins to operate, an increasing number of them accummulates near $\gamma \simeq 1$ by the cooling of those produced by the Bethe-Heitler process. Thus the relative importance of the doubly reflected component to the directly produced synchrotron one will depend solely on the Thomson optical depth of the RBW which is calculated self-consistently from the numerical code.

This component peaks around $\varepsilon_{R,p}^{obs} \simeq \Gamma^3 \varepsilon_{S,p}^{RBW}$ where one power in Γ comes from the Doppler blueshift while the other two come from the reflection on the mirror. Using the value of $\varepsilon_{S,p}^{RBW}$ given above, we arrive at $\varepsilon_{R,p}^{obs} = b\Gamma^5$ which is the relation derived in KGM. All three components can be seen in Fig.3 which shows a snapshot of the multiwavelength spectrum.

Finally Fig. 4 shows the peak energy of the three components as observed on Earth for various values of the comoving magnetic field strength. We notice that the synchrotron peak increases linearly with B, while the inverse Compton peak is largely independent of it and roughly equal to $\Gamma^2 m_e c^2$ i.e. the maximum electron energy as perceived in the lab frame. The doubly reflected peak increases with B (as the loop runs increasingly above its kinematic threshold) but the relation is slower than linear. Note that for combinations of Γ and B that are close to threshold, this peak lies between the 100 keV - 1 MeV range, and as we emphasized in §2 this is independent of the particular choice of Γ and B.

SUMMARY

In the present note we have presented a numerical approach to the "Supercritical Pile" model of GRBs. This model was conceived as a means of resolving two fundamental problems of GRBs, namely the dissipation and conversion into radiation of the energy stored in the protons of the RBW of a GRB and the narrow distribution in the energy of the GRB peak emission $E_{\rm p}$. We found that the presence of reflection reduces both the kinematic and dynamic thresholds of the loop discussed by [5] in accordance with the arguments put forward in KGM and that this process can indeed extract a substantial fraction of the proton energy within a few crossing times of the radiating segment of the RBW. If this picture is correct, then the prompt emission should be accompanied by emission at IR-optical-UV frequencies, the result of synchrotron radiation by the pairs produced by the loop, and by emission at high energies ($E_\gamma \simeq \Gamma^2 m_e c^2$), produced by the inverse Compton scattering of the reflected radiation on the relativistic electrons.

ACKNOWLEDGMENTS

AM would like to acknowledge partial support from the Special Funds for Research of the University of Athens and from the joint Greek-EU "Pythagoras" research grants. DK would like to acknowledge support by an INTEGRAL GO grant.

REFERENCES

1. Böttcher, M. & Dermer, C. D. 1998, ApJ, 501, L51
2. Costa, E. et al. 1997, Nature, 387, 783
3. Kazanas, D. & Mastichiadis, A., 1999, ApJ, 518, L17
4. Kazanas, D., Georganopoulos, M. & Mastichiadis, A. 2002, ApJ, 578, L15 (KGM)
5. Kirk, J. G. & Mastichiadis, A., 1992, Nature, 360, 135 (KM92)
6. Mallozzi, R. S. et al. 1995, ApJ, 454, 597
7. Mastichiadis, A. & Kazanas, D., to appear in ApJ (astro-ph/0512447)
8. Mastichiadis, A. & Kirk, J. G., A&A, 295, 1995, 613 (MK95)
9. Mastichiadis, A., Protheroe, R. J. & Kirk, J. G. 2005, A&A, 433, 765 (MPK05)
10. Motz, J.W., Olsen, H.A., & Koch, H.W. 1969, Rev.Mod.Phys., 41, 581
11. Paczyński, B. 1986, ApJ, 308, L43
12. Piran, T. 2004, Rev. Mod. Phys., 76, 1143
13. Rees, M. J. & Mészáros, P., 1992, MNRAS, 258, L41
14. Shemi, & Piran, T. 1990, ApJ, 365, L55
15. van Paradijs, J. et al. 1997, Nature, 386, 686

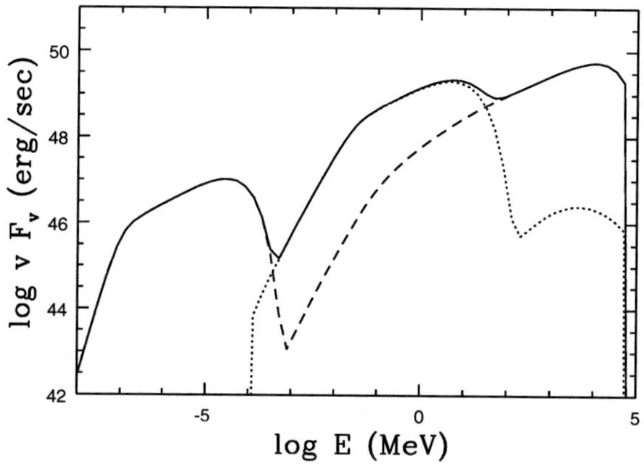

FIGURE 3. Snapshot of the multiwavelength spectrum at the peak of the direct emission as seen by an observer at Earth in the case where $n_p = n_e = 10^4$ cm^{-3} and $B = 44$ G. The dashed line corresponds to the directly produced component while the dotted line to the doubly reflected component. The full line is the composite spectrum.

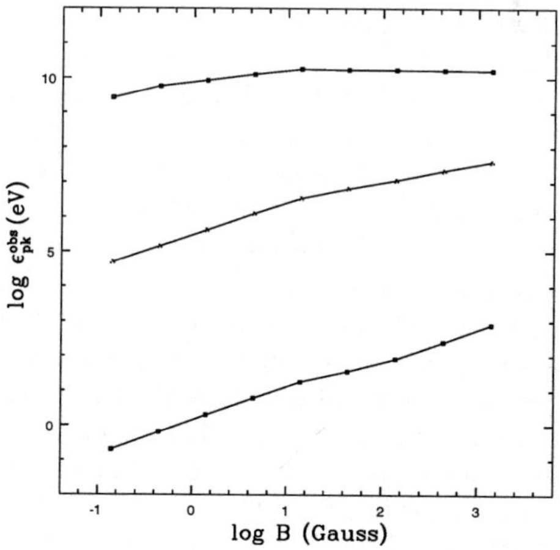

FIGURE 4. Plot of the peak of the observed synchrotron (lower curve), doubly reflected (middle) and inverse Compton (top) component versus the comoving magnetic field strength.

A GRMHD model for cosmical jets

K. Tsinganos*, Z. Meliani†, C. Sauty†, N. Vlahakis* and E. Trussoni**

*Department of Physics, University of Athens and IASA, 157 84, Greece
†Observatoire de Paris, LUTh, F-92190 Meudon, France
**INAF - Osservatorio Astr. di Torino, Str. Osservatorio 20, I-10025 Pino Torinese, Italy

Abstract. We present self-similar semi-analytical solutions obtained in the framework of general relativistic magnetohydrodynamics (GRMHD) which describe steady and axisymmetric outflows from the system of a hot coronal magnetosphere of a Schwarzschild black hole and its surrounding accretion disk. The model allows to extend previous non relativistic MHD studies for coronal winds from young stars to spine jets from Active Galactic Nuclei surrounded by disk-driven outflows. The collimation depends critically on an energetic integral measuring the efficiency of the magnetic rotator, similarly to the non relativistic case. The outflows are thermally driven and magnetically (thermally) collimated if the magnetic rotator is efficient (inefficient). It is also shown that relativistic effects affect quantitatively the depth of the gravitational well and the coronal temperature distribution in the launching region of the outflow. Similarly to previous analytical and numerical studies, relativistic effects tend to increase the efficiency of the thermal driving but reduce the effect of magnetic self-collimation.

INTRODUCTION

The formation of relativistic jets around compact objects and Active Galactic Nuclei (AGNs) with Lorentz factors ranging from values $\gamma \sim 2-10$ in Seyfert Galaxies and radio loud AGNs up to the inferred values $\gamma = 10^3$ in GRBs is one of the most intriguing and yet not fully understood astrophysical phenomena [1], [2]. The formation of collimated jets seems to be closely related to the presence of large scale magnetic fields and the existence of a gaseous disk and/or a hot corona around the central object. MHD models for coronal or disk-jets rely on the basic idea that the coronal and gravitational energies of the central object are transferred to the coronal and inner accretion disk plasma and a collimation mechanism then produces the jet. The energy released by accretion increases with the mass of the central object, a fact which may explain the wide variety of the powerful jets observed.

The energy source of the jet is usually assumed to be located at their base where they are powered either by a spinning black hole [3], [4], [5], or, a surrounding accretion disk [6]. The collimation of the outflow is likely to be due mainly to the hoop stress resulting from the toroidal magnetic field generated by the rotation of the source [7]. Recent VLBI observations suggest that the direction of the magnetic field vectors is transverse to the jet axis. This is the case in BL Lac objects, e.g., in Mrk 501 [8], or, in quasars where the central faster part of the jet is characterized by toroidal magnetic fields [9]. Magnetic self-collimation has been shown to be efficient in the non relativistic context [7], [10]. In the relativistic limit however it is slower due to the decollimating effect of the electric force and the higher inertia of the flow, but still possible [11], [12].

CP848, *Recent Advances in Astronomy and Astrophysics,*
7th International Conference of the Hellenic Astronomical Society, edited by N. H. Solomos

Alternatively, collimation in relativistic jets may be due to the external pressure of a surrounding slower and easily collimated disk wind, [13].

Analytically, radially self-similar disk-wind solutions [14] were extended to special relativistic cold winds by neglecting gravity to allow the separation of the variables [15]. Thermal effects were introduced into these relativistic models to analyze the formation of a relativistic flow from hot magnetized plasmas, showing that such solutions could be applied to Gamma Ray Bursts wherein the flow is thermally driven at the base [11]. However, most of the acceleration is of magnetic origin and there is an efficient conversion of Poynting to kinetic flux of the order of 50%. They also applied this disk wind solution to AGN jets [12] showing that they could trace the observed parsec scale expansion of the wind.

Radial self-similarity is usually used in disk-wind models due to the complexity of the non linear system of MHD equations. However, such solutions cannot describe the flow close to the rotational axis where they become singular. On the other hand, meridional self-similarity provides a better alternative to study the outflow close to the symmetry and rotation axis of the central corona. In the central part where the thermal energy is rather high, the wind may be thermally driven. Spherically symmetric relativistic hydrodynamical models have been proposed to study the formation of such outflows [16, 17]. Those models are restricted to the case where the magnetic effects in the acceleration are negligible. In these models, a wind forms in the hot corona because of the internal shock maintained by the centrifugal barrier [18, 19], or by the pressure induced via a first order Fermi mechanism [16].

An important alternative to analytical models are numerical simulations. In the special relativistic domain, simulations have been presented for coronal winds [7], and in the general relativistic domain for disk winds (e.g. [20, 21]) to model the formation and collimation of relativistic outflows in the vicinity of black holes. The difficulty for relativistic outflows in a single-component model to be collimated led [13] to propose a two-component model wherein a relativistic central wind is collimated by a surrounding non relativistic disk-wind. Shocks may also develop as the disk-wind collides and collimates the inner relativistic wind. Note that all such simulations are performed by using time-dependent codes. Analytical models conversely have presented more sophisticated steady solutions of outflows to be used as initial conditions in more complex simulations, albeit sacrifying freedom on the chosen boundary conditions.

In this article we present an extension of the non relativistic meridionally self-similar solutions ([22], hereafter ST94) to the case of relativistic jets emerging from a spherical corona surrounding the central part of a Schwarzschild black hole and its inner accretion disk. Attention is also given to the contribution of the different mechanisms, hydrodynamic and magnetic, to the acceleration and collimation of the outflow, as in our previous papers ([23], [24], [25], hereafter STT99, STT02 and STT04).

GRMHD EQUATIONS

In this paper we find convenient to use a 3+1 split of space-time, following the usual approach of MHD flow treatment in general relativity [26, 27]. The 3+1 approach allows to obtain equations similar to the familiar classical equations. We write all quantities in the FIDucial Observer frame of reference (FIDO), which corresponds to observers in free fall around the Schwarzschild black hole. For the FIDO, space time is locally flat. In Schwarzschild coordinates (ct, r, θ, φ) the background metric is :

$$ds^2 = -h^2c^2dt^2 + \frac{1}{h^2}dr^2 + r^2d^2\theta + r^2\sin^2\theta d\varphi^2\,, h = \sqrt{1-\frac{2\mathcal{G}\mathcal{M}_\bullet}{c^2r}} = \sqrt{1-\frac{r_G}{r}}. \quad (1)$$

Maxwell's equations written in the 3+1 formulation are [26] :

$$\nabla\cdot\vec{E} = 4\pi\rho_e\,, \quad (2)$$

$$\nabla\cdot\vec{B} = 0\,, \quad (3)$$

$$\nabla\times(h\vec{E}) = 0\,, \quad (4)$$

$$\nabla\times(h\vec{B}) = \frac{4\pi h}{c}\vec{J}_e\,, \quad (5)$$

$$\vec{E} = -\frac{\vec{V}\times\vec{B}}{c}. \quad (6)$$

Conservation of particles, $(nu^a)_{;a} = 0$, $u^a = (\gamma c, \gamma\vec{V})$ in the 3+1 formalism leads to:

$$\nabla\cdot(h\gamma n\vec{V}) = 0\,, \quad \gamma = \left(1-\vec{V}^2/c^2\right)^{-1/2}. \quad (7)$$

Euler's equation obtained by projecting the conservation of the energy-momentum tensor, $T^{ab}_{;b} = 0$, onto the spatial coordinates $(a = 1,2,3)$ and combined with Maxwell's equations is :

$$\gamma n(\vec{V}\cdot\nabla)\left(\frac{\gamma w\vec{V}}{c^2}\right) = -\gamma^2 nw\nabla\ln h - \nabla P + \rho_e\vec{E} + \frac{\vec{J}_e\times\vec{B}}{c}. \quad (8)$$

The first law of thermodynamics obtained by projecting the conservation of the energy-momentum tensor along the fluid four-velocity, $u_aT^{ab}_{;b} = 0$ is :

$$n\vec{V}\cdot\nabla w = \vec{V}\cdot\nabla P. \quad (9)$$

Assuming axisymmetry of the plasma flow allows us to reduce the number of differential equations by integrating some of them and thus obtaining conserved quantities along the streamlines (for notations see [28]). From Eqs. (3) and (7) we can introduce a magnetic flux function A and a stream function Ψ,

$$\vec{B}_p = \nabla\times\left(\frac{A}{r\sin\theta}\vec{e_\varphi}\right) = \frac{\nabla A}{\varpi}\times\vec{e_\varphi}\,, \quad 4\pi h\gamma n\vec{V}_p = \nabla\times\left(\frac{\Psi}{r\sin\theta}\vec{e_\varphi}\right) = \frac{\nabla\Psi}{\varpi}\times\vec{e_\varphi}\,, \quad (10)$$

With axisymmetry, $E_\varphi = 0$ implying that Ψ is constant on surfaces of constant A on which the corresponding streamlines and fieldlines are roped, $\vec{V}_p \parallel \vec{B}_p$,

$$\vec{V}_p = \frac{\Psi_A}{4\pi h\gamma n}\vec{B}_p. \quad (11)$$

From Eq. (6) the toroidal components V_φ and B_φ are related as

$$V_\varphi = \frac{\Psi_A}{4\pi h\gamma n} B_\varphi + \frac{\varpi\Omega}{h}, \tag{12}$$

where $\Omega(A)$ corresponds to the angular speed Ω of the footpoints of this poloidal stream/fieldline. The azimuthal component of the momentum equation yields the conservation of the total specific angular momentum,

$$L = \gamma \frac{w}{c^2} \varpi V_\varphi - h \frac{\varpi B_\varphi}{\Psi_A}. \tag{13}$$

Finally, the generalized Bernoulli integral (including rest mass)

$$\mathscr{E} = h\gamma w - h \frac{\varpi\Omega}{\Psi_A} B_\varphi, \tag{14}$$

may be obtained by integrating the equation of motion along each streamline. The first part of the r.h.s. represents the hydrodynamical energy flux transported by the fluid while the second part corresponds to the Poynting flux.

We have thus obtained the usual four integrals of motion Ψ_A, Ω, L and $\mathscr{E}$ that are constant along a fieldline for a stationary and axisymmetric plasma flow. They can be used to find algebraic relations between the Lorentz factor, the toroidal velocity and the toroidal magnetic field,

$$M^2 = \frac{4\pi h^2 n w \gamma^2 V_{\rm p}^2}{B_{\rm p}^2 c^2} = \frac{w\Psi_{\rm A}^2}{4\pi n c^2}, \quad x = \frac{\varpi\Omega}{ch}, \quad h\gamma w = \mathscr{E}\left[\frac{M^2 - h^2(1 - x_{\rm A}^2)}{M^2 - h^2(1 - x^2)}\right], \tag{15}$$

$$V_\varphi = \frac{c}{x}\left[\frac{M^2 x_{\rm A}^2 - x^2 h^2\left(1 - x_{\rm A}^2\right)}{M^2 - h^2\left(1 - x_{\rm A}^2\right)}\right], \quad B_\varphi = -\frac{\mathscr{E}\Psi_A}{cx}\left[\frac{x^2 - x_{\rm A}^2}{M^2 - h^2(1 - x^2)}\right], \tag{16}$$

The quantity $x_{\rm A}^2 = \Omega L/\mathscr{E}$ is a measure for the amount of energy carried by the electromagnetic field. It is a measure of the energy flux of the magnetic rotator in units of the total energy flux ([29]).

The MHD equations possess a well known singularity at the Alfvén surface where the denominator of Eqs. (16) and (15) vanishes. Then, the numerators should vanish simultaneously to ensure a regular behavior, implying at the Alfvén point,

$$x^2|_{\rm Alfvén} = \left(\frac{\varpi_{\rm A}\Omega}{h_\star c}\right)^2 = x_{\rm A}^2, \quad M^2|_{\rm Alfvén} = h_\star^2\left(1 - x_{\rm A}^2\right) = M_{\rm A}^2. \tag{17}$$

Using Eq. (15) we find that $V_\star^2 = B_\star^2 c^2 / 4\pi\gamma_\star^2 n_\star w_\star$ where the subscript $\star$ denotes quantities evaluated at the Alfvén point along the polar axis. Note that the Alfvén surface is shifted from the classical Alfvén surface because of the lapse function h and the existence of the light cylinder $x = 1$. In the Newtonian limit, Eq. (17) gives $L = m\varpi_{\rm A}^2\Omega$ where $m = w/c^2$ is the particle mass.

CONSTRUCTION OF THE MERIDIONALLY SELF-SIMILAR MODEL

We find semi-analytical solutions of the $r-$ and $\theta-$ components of the Euler Eq. (8), by separating the variables r and θ. First we use dimensionless quantities normalizing at the Alfvén radius along the polar axis,

$$R = \frac{r}{r_\star}\,, \; A = \frac{r_\star^2 B_\star}{2}\alpha\,, \nu = \frac{V_{\mathrm{esc},\star}}{V_\star} = \sqrt{\frac{2\mathcal{G}\mathcal{M}_\bullet}{r_\star V_\star^2}}\,, \; \mu = \frac{r_\mathrm{G}}{r_\star} = \frac{V^2_{\mathrm{esc},\star}}{c^2}\,, \; \frac{\mu}{\nu^2} = \frac{V_\star^2}{c^2} < 1\,. \quad (18)$$

As $\alpha = 0$ on the rotational axis and we are interested on the central component of the jet, we assume that the cross section area of a given magnetic flux tube can be written,

$$S(R,\alpha) = \pi\varpi^2 = \pi r_\star^2 G^2(R)\alpha\,, \mathrm{or}\,, \alpha = \frac{R^2}{G^2(R)}\sin^2\theta\,. \quad (19)$$

Normalizing at the Alfvén surface, we choose $G(R=1)=1$ such that G is the cylindrical radius in units of the Alfvénic cylindrical radius. Thus $G(R) = \varpi/\varpi_\mathrm{A}$ with $\varpi_\mathrm{A} = r_\star\sqrt{\alpha}$. We also introduce the expansion factor F

$$F = \left.\frac{\partial \ln\alpha}{\partial \ln R}\right|_\theta = 2 - \frac{\mathrm{d}\ln G^2}{\mathrm{d}\ln R}\,. \quad (20)$$

In addition, we assume that the surfaces of constant "Alfvénic" number are spheres, $M^2(R,\alpha) = M^2(R)$ with $M^2(R=1) = h_\star^2$, and the pressure dependence is,

$$P = P_0 + \frac{1}{2}\gamma_\star{}^2 n_\star \frac{w_\star}{c^2} V_\star^2 \Pi(R)(1+\kappa\alpha)\,, \quad (21)$$

with P_0, κ constants and $\Pi(R)$ a dimensionless function.
We have $4\pi nc^2M^2/w = \Psi_A^2$ and expanding the r.h.s. to first order in α, we find

$$\Psi_A^2 = 4\pi c^2 \frac{n_\star h_\star^2}{w_\star}(1+\delta\alpha)\,, \quad (22)$$

where δ is a free parameter describing the deviations from spherical symmetry of the ratio number density/enthalpy and not of the density itself as it was the case in ST94. Similarly we expand $L\Psi_A$, $L\Psi_A = h_\star \lambda B_\star r_\star \alpha$ where λ is a constant measuring rotation. Finally we choose for Ω a form similar to the one in ST94

$$\Omega = \lambda h_\star \frac{V_\star/r_\star}{\sqrt{1+\delta\alpha}}\,. \quad (23)$$

We can now express x^2 in terms of α and $G(R)$,

$$x^2 = \lambda^2 \frac{V_\star^2}{c^2}\frac{h_\star^2}{h^2}\frac{G^2\alpha}{1+\delta\alpha} = \frac{\mu\lambda^2}{\nu^2}\frac{h_\star^2}{h^2}\frac{G^2\alpha}{1+\delta\alpha}\,, \; x_\mathrm{A}^2 = \lambda^2\frac{V_\star^2}{c^2}\frac{\alpha}{1+\delta\alpha} = \frac{\mu\lambda^2}{\nu^2}\frac{\alpha}{1+\delta\alpha}\,, \quad (24)$$

which gives the form of the Bernoulli integral,

$$\mathscr{E} = h_\star \gamma_\star w_\star . \tag{25}$$

Conversely to the non relativistic limit, we cannot neglect the charge separation and the presence of the electric field. In order to separate the variables (R, θ) in the $r-$ and $\theta-$ components of the Euler Eq. (8) we need some further assumptions. Basically we expand these equations with respect to θ. We suppose that the rotational speed of the fluid remains always subrelativistic. In other words, we focus on streamlines that never cross the light cylinder such that the later does not affect the dynamics ($x \ll 1$ which implies $x_{\rm A} \ll 1$). Of course this reduces the domain of validity of the solutions to the vicinity of the rotational axis because $x_{\rm A}^2$ as it is given by Eq. (24) is sufficiently small only for relatively small α. The region of validity of our model depends on how small the parameter $\lambda^2 V_\star^2/c^2 = \lambda^2\mu/v^2$ is, though. The requirement that the light cylinder lies further away from the Alfvénic surface ($x_{\rm A} < 1$) constrains the parametric space to $\delta > \lambda^2\mu/v^2$.

The velocity and magnetic fields can now be written exclusively in terms of unknown functions of R. Ss in ST94, denoting by N_B, N_V and D the following quantities that appear in various components of the fields,

$$N_B = \frac{h^2}{h_\star^2} - G^2 , \quad N_V = \frac{M^2}{h_\star^2} - G^2 , \quad D = \frac{h^2}{h_\star^2} - \frac{M^2}{h_\star^2} , \tag{26}$$

we have

$$B_r = \frac{B_\star}{G^2}\cos\theta , \quad B_\theta = -\frac{B_\star}{G^2}\frac{hF}{2}\sin\theta , \quad B_\varphi = -\frac{\lambda B_\star}{G^2}\frac{h_\star}{h}\frac{N_B}{D} R\sin\theta , . \tag{27}$$

$$V_r = \frac{V_\star M^2}{h_\star^2 G^2}\frac{1}{\sqrt{1+\delta\alpha}}\left(\cos\theta + \frac{\mu\lambda^2}{v^2}\frac{N_B}{D}\alpha\right) , \tag{28}$$

$$V_\theta = -\frac{V_\star M^2}{h_\star^2 G^2}\frac{hF}{2}\frac{1}{\sqrt{1+\delta\alpha}}\sin\theta , V_\varphi = -\frac{h}{h_\star}\frac{\lambda V_\star}{G^2}\frac{N_V}{D}\frac{R\sin\theta}{\sqrt{1+\delta\alpha}} . \tag{29}$$

Similarly, the enthalpy and the particle number density are given by

$$h\gamma w = h_\star\gamma_\star w_\star\left(1 - \frac{\mu\lambda^2}{v^2}\frac{N_B}{D}\alpha\right) , \quad h\gamma n = h_\star\gamma_\star n_\star \frac{h_\star^2}{M^2}\left(1 + \delta\alpha - \frac{\mu\lambda^2}{v^2}\frac{N_B}{D}\alpha\right) , \tag{30}$$

where we used Eq. (22) to deduce Eq. (30), while the pressure is given by Eq. (21).

Finally, there are three equations given in [30] that determine the three unknown functions $\Pi(R)$, $F(R)$ and $M^2(R)$ which are integrated by taking care of all criticality and regularity conditions. It is worth noticing that the shape of the streamlines $F_\star$ at $R = 1$ is determined by the regular crossing of the slow magnetosonic surface. More details on the numerical technique can be found in ST94 and STT02.

As in ST94, it is possible to find a constant ε for all fieldlines. This parameter ε has been used in ST94 and the following papers to classify the various solutions. We shall use a similar technique to construct such a constant in the present model,

$$\varepsilon = \frac{M^4}{h_\star^4 R^2 G^2}\left(\frac{F^2}{4} - \frac{1}{h^2} - \kappa\frac{R^2}{h^2 G^2}\right) - \frac{(\delta-\kappa)\,\nu^2}{h^2 R} + \frac{\lambda^2}{G^2 h_\star^2}\left(\frac{N_V}{D}\right)^2 + \frac{2\lambda^2}{h^2}\frac{N_B}{D}, \quad (31)$$

where ε is a constant, the same for all fieldlines. Similarly to what was done in ST99, we can calculate this constant at the base of the flow R_o assuming the poloidal velocity is negligible there ($M(R_o) \approx 0$). Let's express $\varepsilon/2\lambda^2$ in terms of the conditions at the source boundary,

$$\frac{\varepsilon}{2\lambda^2} = \frac{\mathscr{E}_{\mathrm{R},o} + \mathscr{E}_{\mathrm{Poynt.},o} + \Delta\mathscr{E}_{\mathrm{G}}^*}{\mathscr{E}_{\mathrm{MR}}}, \quad (32)$$

where $\mathscr{E}_{\mathrm{MR}} = h^2 L\Omega$ is the energy of the magnetic rotator, $\mathscr{E}_{\mathrm{Poynt.}} = -h\varpi\Omega B_\varphi/\Psi_A$ is the Poynting flux and $\mathscr{E}_{\mathrm{R},o} = \mathscr{E}V_{\varphi,o}^2/2c^2$ is the rotational energy per particle. It is proportional to the specific rotational energy $V_{\varphi,o}^2/2$, with the factor $\mathscr{E}/c^2$ having the dimensions of a mass. Finally we have

$$\Delta\mathscr{E}_{\mathrm{G}}^* = -\frac{\mathscr{E}}{c^2}\frac{\mu c^2}{2R_o}\frac{(\delta-\kappa)\,\alpha}{1+\delta\alpha}, \quad (33)$$

a term similar to the nonrelativistic case where it measures the excess or the deficit on a non polar streamline compared to the polar one of the gravitational energy per unit mass which is not compensated by the thermal driving (STT99). As in the classical case, ε measures the efficiency of the magnetic rotator to collimate the flow. Thus if $\varepsilon > 0$ we have an Efficient Magnetic Rotator (EMR) where magnetic collimation may dominate, while if $\varepsilon < 0$ we have an Inefficient Magnetic Rotator (IMR) where collimation cannot be but of thermal origin.

SUMMARY OF RESULTS

In the following figures we show representative results for the efficient and inefficient magnetic rotator cases. In Fig. (1) is shown the morphology of the poloidal streamlines for two solutions corresponding to an Inefficient Magnetic Rotator (IMR, $\varepsilon = -1.747$) in (a), and an Efficient Magnetic Rotator (EMR, $\varepsilon = 1.128$) in (b). The solid lines correspond to the case where the rotational speed is low and the assumptions used valid. Note that in the EMR case a very well collimated jet is formed.

The solutions of this semi-analytical model are the first exact ones for relativistic jets which include general relativistic effects. We have made an extension of the non relativistic ST94 model for a black hole characterized by weak angular momentum, $a = J\mathscr{G}/(\mathscr{M}_\bullet^2/c) \ll 1$. Therefore, we treated the problem of GRMHD outflows in a Schwarzschild metric. Moreover, we concentrated our efforts on modelling the jet close to its polar axis. In the construction of this model, we were limited to describe jets possessing a weak rotation velocity compared to the light velocity and we neglected the

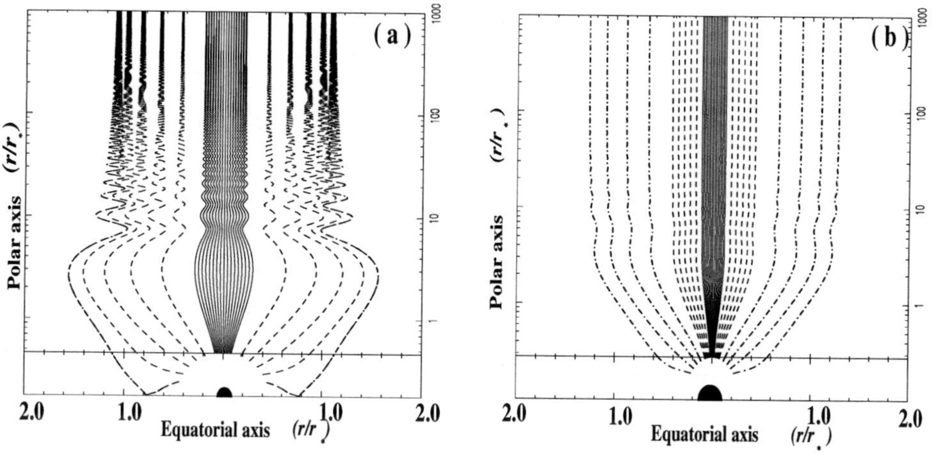

FIGURE 1. Morphology of the poloidal streamlines for two solutions corresponding to an Inefficient Magnetic Rotator (IMR, $\varepsilon = -1.747$) in (a), and an Efficient Magnetic Rotator (EMR, $\varepsilon = 1.128$) in (b). We chose $\mu = 0.1$ in both cases, while the other parameters are $\nu = 0.781$, $\kappa = 0.49$, $\delta = 2.613$, $\lambda = 0.880$ and $\Pi_\star = 1.4$ in (a), and $\nu = 0.541$, $\kappa = 0.39$, $\delta = 3.253$, $\lambda = 1.401$ and $\Pi_\star = 1.26$ in (b).

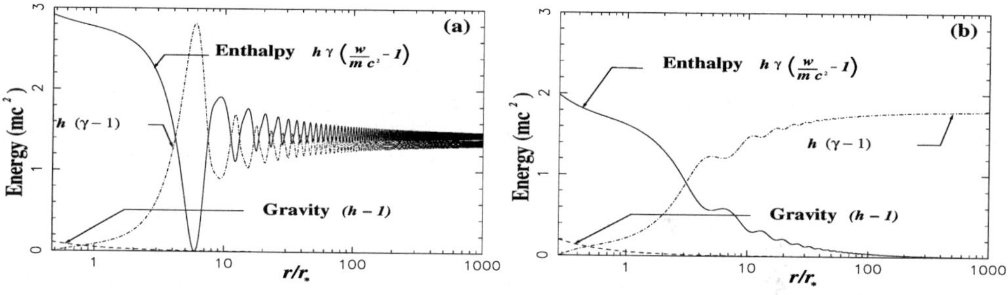

FIGURE 2. Variation of the energy flux normalized to the mass energy, along the external streamline for the IMR solution in (a) and the EMR solution in (b) of the previous figure. The Poynting energy is $\leq$ 1% of the enthalpy at the base of the flow and is not plotted. The parameters are the same as in Fig. 1.

effects of the light cylinder. Thus, we restricted our study to thermally accelerated jets with a weak contribution of the Poynting flux. However, in the collimation of the jet, both electric and magnetic terms are comparable to the inertial and pressure gradient ones. We have also studied the collimation effects by magnetic and thermal forces and the decollimation effect of centrifugal and electric forces.

We found that the influence of the electric force and the charge separation in the jet depends on the collimation regime. In the case of EMRs where jets are magnetically collimated, the electric decollimating force plays an important role. This force is of the order of the magnitude of the centrifugal force and of the magnetic pinching. Therefore, relativistic jets from EMRs are less collimated than their non relativistic counterparts.

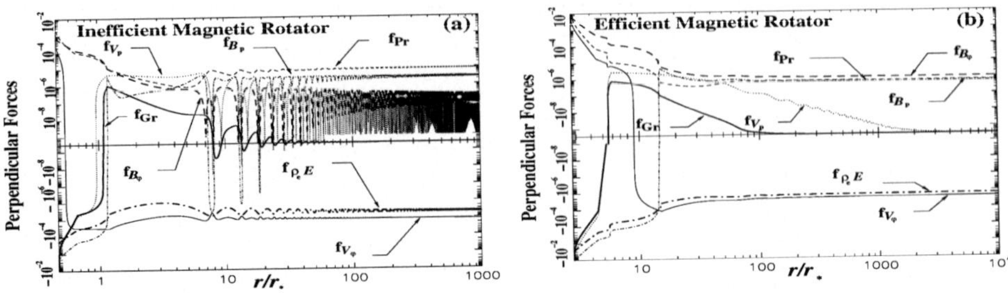

FIGURE 3. Plot of the forces acting perpendicular to a given streamline for the IMR solution in (a) and the EMR solution in (b). Forces are normalized by their maximum value, usually reached at the base of the flow. The parameters are the same as in Fig. 1.

Conversely, jets from IMRs, where collimation is mainly of thermal origin, are not very sensitive to electric decollimation. In this type of jets, the contribution of $\rho_e\vec{E}$ is balanced by the increase of the external pressure.

We have also used the model to compare classical jets to relativistic jets. We undertook this comparative study by changing the free parameter μ. In fact, this new parameter gives in the relativistic model the space curvature which is induced by gravity near the central object. We used typical values of μ from $\mu \sim 10^{-5}$ for Jets from Young Stellar Objects to $\mu \sim 0.1$ for jets from compact sources. We found that the difference between these two types of jets is only a scaling effect on μ for $\mu < 10^{-2}$. In this case, the spatial dimensions and energies are linear functions of μ. For $\mu > 10^{-2}$, the relativistic effects increase together with the thermal acceleration of the jet. Simultaneously, strongly relativistic effects tend to decollimate the jets because of the decrease of the electric current density.

To conclude we have seen that relativistic effects and particularly relativistic gravity tend to enhance the thermal acceleration (as in [17]) and reduce the magnetic collimation of the jet (as in [13]). Still collimation can be obtained either by thermal or magnetic means but relativistic effects lower the efficiency of the magnetic rotator. This means that despite quantitative changes, we can use this generalized model to verify, for the classification of AGN jets, the conjectures given in [31] using the non relativistic model. Mainly we see that the observed types of jets from radio loud galaxies can be connected to the efficiency of their magnetic rotator and to the environment of the host galaxy. By using the present model, a more precise and quantitative analysis of the observed jets will be presented in a following paper.

ACKNOWLEDGMENTS

This work was supported in part by EU Research Training Grant HPRN-CT-2002-00321.

REFERENCES

1. Ferrari, A. 1998, ARA&A, 36, 539
2. Mirabel I. F., 2003, NewAR, 47, 471
3. Blandford, R. D., & Znajek, R. L. 1977, MNRAS, 179, 433
4. Rees, M., Phinney, E. S., Begelman, M. C., & Blandford, R. D. 1982, Nature, 295,17
5. Belgelman, M. C., Blandford, R. D., & Rees, M. 1984, Rev. Mod. Phy., 56, 255
6. Miller, K. A., & Stone, J. M. 2000, ApJ, 534, 398
7. Tsinganos K., & Bogovalov S., 2002, MNRAS, 337, 553
8. Gabuzda, D. C. 2003, Ap&SS, 288, 39
9. Asada, K. et al., 2002, PASJ, 54, L39
10. Heyvaerts, J., & Norman, C. 2003, ApJ, 596, 1240
11. Vlahakis, N., & Königl, A. 2003, ApJ, 596, 1080
12. Vlahakis, N., & Königl, A. 2004, ApJ, 605, 656
13. Bogovalov, S., & Tsinganos, K. 2005, MNRAS, 357, 918
14. Blandford, R.D., & Payne, D. G. 1982, MNRAS, 199, 883
15. Contopoulos, J. 1994, ApJ, 432, 508
16. Das T. K., 1999, MNRAS, 308, 201
17. Meliani Z., Sauty C., Tsinganos K., & Vlahakis N., 2004, A&A, 425, 773
18. Chakrabarti S. K., 1989, MNRAS, 240, 7
19. Das T. K., 2002, APSS, 276, 267
20. Koide S., Shibata K., Kudoh T., ApJ, 1999, 522, 2, 727
21. Koide S., Shibata K., Kudoh T., Meier D. L., Journal of the Korean Astronomical Society, 2001, 34, 4, 215
22. Sauty C., & Tsinganos K., 1994, A&A, 287, 893 (ST94)
23. Sauty C., Tsinganos, K. & Trussoni E., 1999, A&A, 348, 327 (STT99)
24. Sauty C., Trussoni E., & Tsinganos K., 2002, A&A, 348, 327 (STT02)
25. Sauty C., Trussoni E., & Tsinganos K. 2004, A&A, 421, 797 (STT04)
26. Thorne K. S., & McDonald D. A., 1982, MNRAS, 198, 339
27. Thorne K. S., Price R. H., & Macdonald D. A. 1986, Black Holes: The Membrane Paradigm
28. Tsinganos K. 1982, ApJ, 252, 775
29. Breitmoser, E., & Camenzind, M. 2000, A&A, 361, 207
30. Meliani Z., Sauty C., Vlahakis N., Tsinganos K., & E. Trussoni, 2006, A&A, in press
31. Sauty, C., Tsinganos, K., & Trussoni,E. 2001, In: A. W. Guthmann et al (eds) Relativistic flow in astrophysics", Lecture Notes in Physics, Springer-Verlag, Heidlberg, 589, 41

TeV variability studies of PKS 2155-304 with H.E.S.S.

D.Emmanoulopoulos* and S.J.Wagner*

*Landessternwarte, Königstuhl 12, 69117 Heidelberg, Germany

Abstract. The High Energy Stereoscopic System (H.E.S.S.) has observed the high-frequency peaked BL Lac object PKS 2155−304 (z=0.117) in Very High Energy (VHE) γ-rays (E≥160 GeV) with high significance ∼34 sigma. The observations were carried out between October 19 and November 26, 2003 with a simultaneous coverage in X-ray, optical and radio bands, with the Proportional Counter Array (PCA) on board Rossi X-ray Timing Experiment (RXTE), the Robotic Optical Transient Search Experiment (ROTSE) and the Nançay Radio Telescope NRT respectively. Variability in the VHE regime is seen on intra-night timescales and in both the X-ray and the optical regime on ksec timescales. No evidence for correlation can be established between the X-ray and γ-ray fluxes or any of the other wave bands. Based on these variability studies we can estimate some source parameters such as the size of it and its Doppler boosting factor δ. Since the source has a non-negligible redshift the intrinsic VHE spectrum has been modified in shape and intensity due to absorption from Extragalactic Background Light (EBL). An estimation of the intrinsic VHE spectrum of the source is attempted based on three EBL models.

INTRODUCTION

Variability study of Active Galactic Nuclei (AGN) across the electromagnetic spectrum is the most efficient way to reveal the characteristics of the mechanism responsible for their luminosity. Correlations and/or time-lags between the different energy bands give us some hints about the location of the radiative component (jet or central black hole) as well as some source parameters (geometrical size and possible motion).

PKS 2155−304 was first claimed in VHE γ-rays by the Durham MK VI telescopes [7], situated in Narrabri-Australia and from then on it was classified as a Southern Hemisphere TeV blazar. Its redshift (z=0.117) makes it the second most distant TeV blazar after H 1426+428 (z=0.129) and therefore its emission is significantly modified from the EBL. Over the last decades PKS 2155−304 has been studied thoroughly (e.g. [10]) for correlations and variability patterns in different energy bands. However no simultaneous observations including the sampling and evolution of the VHE γ-ray component, had ever been conducted before. This presentation consists of a summary of the first simultaneous multiwavelength campaign conducted for PKS 2155−304 covering the energy bands from radio to VHE γ-rays observed by H.E.S.S.. For analytical information concerning the campaign see [4].

CP848, *Recent Advances in Astronomy and Astrophysics*,
7th *International Conference of the Hellenic Astronomical Society*, edited by N. H. Solomos

Observations and data analysis

H.E.S.S.

In its first phase, the H.E.S.S. array consists of four atmospheric Cherenkov telescopes operating in stereoscopic mode. However, the data shown here were taken during the construction of the system, initially with two telescopes, with one more as of mid-September, 2003. The fourth and final telescope was added to the array in December 2003, subsequent to the data presented here. Each telescope has a tessellated 13 m-diameter (107 m^2 surface area) mirror which focuses the Cherenkov light from the showers of secondary particles created by the interaction of γ-rays in the atmosphere onto a camera in the focal plane. This camera consists of 960 photomultipliers, each with a pixel size of 0.16°, giving a field of view of 5°. For the data sets presented here the γ-ray trigger threshold is $\approx$ 100 GeV and the spectral threshold is 160 GeV with an energy resolution $\simeq$ 15%. The experiment is located in the Khomas highlands in Namibia, (23° S, 15° E, 1800 m a.s.l.). A detailed description of the layout and the components of the telescope optical systems, including the segmented mirror with its support structure, the mirror facets for each telescope and the Winston cone light concentrators in front of the PMT camera can be found in [6], and a description of the mirror alignment is given in [8]. For details on the camera calibration see [2]. The trigger system is described in [12].

In the beginning of the observation period of October 2003, a single 5σ detection by H.E.S.S. achieved in 1h, triggered an approved RXTE ToO (Target of Opportunity) on this target.

The data were taken in the *Wobble* mode where the source direction is positioned $\pm 0.5°$ in declination relative to the centre of the field of view of the camera during observations. This allows for both on-source observations and simultaneous estimation of the background induced by charged cosmic rays by using all events passing cuts in a ring around the source (see section 4.3 in [3]). The data reported here are selected and analyzed with the "standard analysis" described in chapter 4 of [3]. The runs passing the quality selection criteria total 32.4 hours of livetime on the source. This yields a detection at the 34.3σ level, at the average rate of $0.91\gamma/\text{min}$ and a significance of $6.0\sigma/\sqrt{\text{h}}$.

The methods used here for reconstructing the energy of each event and for determining a spectrum are described in chapter 6 of [3]. The measured time-average spectrum is fitted by a power law of the form $\mathrm{d}N/\mathrm{d}E = I_0(E/1\ \text{TeV})^{-\Gamma}$ with I_0 the flux normalization at 1 TeV and Γ the photon index. The photon index obtained from the time-averaged spectrum is then used as a fixed parameter to estimate the integral flux above 300 GeV for each run. This integrated flux takes into account the effective area and the threshold variations due to the source moving through the sky, giving more reliable variability information than counting rates in units of γ-rays/minute. Overall systematic errors are estimated to be 20% for the integral flux and $\simeq 0.1$ for the photon index.

RXTE

The PCA [16] is an array of five Proportional Counter Units (PCUs), carried by RXTE, with a total collecting area of 6500 cm^2 sensitive to the nominal energy range 2-60 keV. Three units (PCU0/2/3) have observed PKS 2155−304 between October 22 and November 23 of 2003 with exposures of typically 10 ks in October and $\simeq$ 1 ks in November. The STANDARD2 data (i.e. time resolution 16 sec) were extracted using the ftools in the HEASOFT 5.3.1 analysis software package provided by NASA/GSFC and filtered using the RXTE Guest Observer Facility (GOF) recommended criteria. The changing configuration of these three PCUs throughout the observations was taken into account in the data reduction, and only the signals from the top layer (X1L and X1R) were used. When reducing the PCUs individually to establish the time-averaged spectrum, the average spectral fit parameters are similar within error bars but with a systematically higher χ^2 for spectral fits performed on PCU0 data alone. On shorter timescales this effect becomes negligible and PCU0 contributes to the statistical significance of the flux measurement. Therefore all PCUs were kept for the overall light curve which is binned in 400 s bins, but only PCU2 and PCU3 are used for the analysis of data segments that are simultaneous with H.E.S.S. runs and for the time-averaged spectrum.

To build a light curve in units of integrated flux in the 2–10 keV band, spectral data were derived from 400 s bins to probe short timescales with adequate statistical accuracy. These segments are then fitted by a power law in XSPEC with PCU configuration-dependent response matrices generated by the ftool `pcarsp v. 10.1` and a fixed column density of $N_{\mathrm{H}} = 1.7 \times 10^{20}\,\mathrm{cm}^{-2}$ obtained from PIMMS[1]. This yields the flux and the error (corresponding to the 1σ confidence level) on the flux reported in the light curves in Figure 1, in units of $10^{-11}\,\mathrm{erg\,cm^{-2}\,s^{-1}}$ in the 2–10 keV band. The fits did not improve by using a broken power-law for the 400 s binned observations.

ROTSE

The ROTSE-III array are a worldwide network of four 0.45 m robotic telescopes (described in detail in [5]) At the time of the observations of PKS 2155−304 in October and November 2003, two ROTSE-III telescopes were operational in the Southern hemisphere: ROTSE-IIIa located at the Siding Spring Observatory, Australia and ROTSE-IIIc at the H.E.S.S. site. The ROTSE-IIIc telescope is located in the centre of the H.E.S.S. telescope array. A 30% share of the total observation time is available to the H.E.S.S. collaboration, which has been used to perform an automated monitoring programme of blazars, including objects that are being observed with the H.E.S.S. telescopes. Both telescopes participated in the observation campaign on PKS 2155−304 in October and November 2003.

The telescopes observed PKS 2155−304 typically 10 times per night taking sequences of 2 frames with 60 s exposures. Overall, 323 bias-subtracted and flat-fielded

[1] See http://legacy.gsfc.nasa.gov/Tools/w3pimms.html

frames have passed visual inspection and are used to produce a light curve.

Using an overlay of 50 isolated comparison stars with similar brightness (12^{mag} – 14^{mag}) and co-located with PKS 2155−304 (< 15 arcmin), a two-dimensional Gaussian is fit to the intensity distribution characterising the point-spread function by σ_{psf}.

To estimate the local sky-background for the reference stars and the target object, an annulus with inner radius $2 \times \sigma_{psf}$ and an outer radius $6 \times \sigma_{psf}$ is chosen. Based upon a reference frame which is derived from co-adding 30 individual frames, a mask is calculated for each object excluding regions where faint objects coincide with the annulus; pixels exceeding 3 standard deviations of the local sky background are excluded. Using the local sky background, the intensity and error of each object is calculated. Using the 50 reference stars, a relative intensity and statistical error with respect to a reference frame is calculated.

NRT

The Nançay radiotelescope is a single-dish antenna with a collecting area of $200 \times 34.56\,m^2$ equivalent to that of a 94 m-diameter parabolic dish [25]. For the October and November campaign we have a total of 8 measurements at 11 cm taken every two or three days yielding an average flux of 0.30 ± 0.01 Jy.

Light Curves and Correlations

The October and November 2003 light curves of all the H.E.S.S., RXTE and ROTSE observations are shown in Figure 1. A χ^2 fit of the H.E.S.S. data to a constant yields a 3.4×10^{-10} χ^2 probability. The intra-night VHE flux on MJD52936 exhibits an increase of a factor of 1.9 ± 0.6 in 0.11 d. On MJD52932 the peak-to-peak flux shows an increase of a factor of 2.5 ± 0.9 within 0.09 d.

The 2–10 keV X-ray flux in this campaign ranges from $F_{2-10\,keV} = 2.0 \times 10^{-11}\,erg\,cm^{-2}\,s^{-1}$ to $4.4 \times 10^{-11}\,erg\,cm^{-2}\,s^{-1}$. The maximum is lower than the 20 November 1997 measurement of $2.3 \times 10^{-10}\,erg\,cm^{-2}\,s^{-1}$ [26] indicating that the X-ray state seen here is not exceptionally high. The minimum seen here is consistent with historically low fluxes [27]. The intra-night variability is also obvious here, but no flare was completely resolved.

A 60% flux variability in $t_{var} \approx 1.5$ ks on MJD52936 is the best marked transient episode in the observations reported here for which the H.E.S.S. observations were made at the end of the transit inducing a large associated error on the flux estimation due to the high zenith angle of the source. This timescale is comparable to those reported by [14] where doubling times as short as 15 min from Mkn 421 were observed in the VHE band. This flare is the fastest rise seen in this object to date since BeppoSAX saw a 5×10^4 s rise timescale [27] and [17] observed a doubling timescale of 3×10^4 s with the ASCA satellite.

In the optical regime the peak-to-peak amplitude of variability is moderate compared to the variability amplitude at shorter wavelengths and typically 0.1^{mag} peak-to-peak. The object has been monitored over longer time-scales with the ROTSE-IIIc telescopes showing variations with amplitudes close to 1^{mag}.

In order to quantitatively look for correlated variability between the VHE, X-ray and optical bands, the measured fluxes are plotted against each other in Figure 2 for all the observations carried out during this campaign. There is no evidence for correlation between VHE and X-ray since the correlation factor is $r = 0.37 \pm 0.13$. Also no correlation was traced between optical and VHE ($r = 0.24 \pm 0.27$) as well as between optical and X-rays ($r = -0.02 \pm 0.05$).

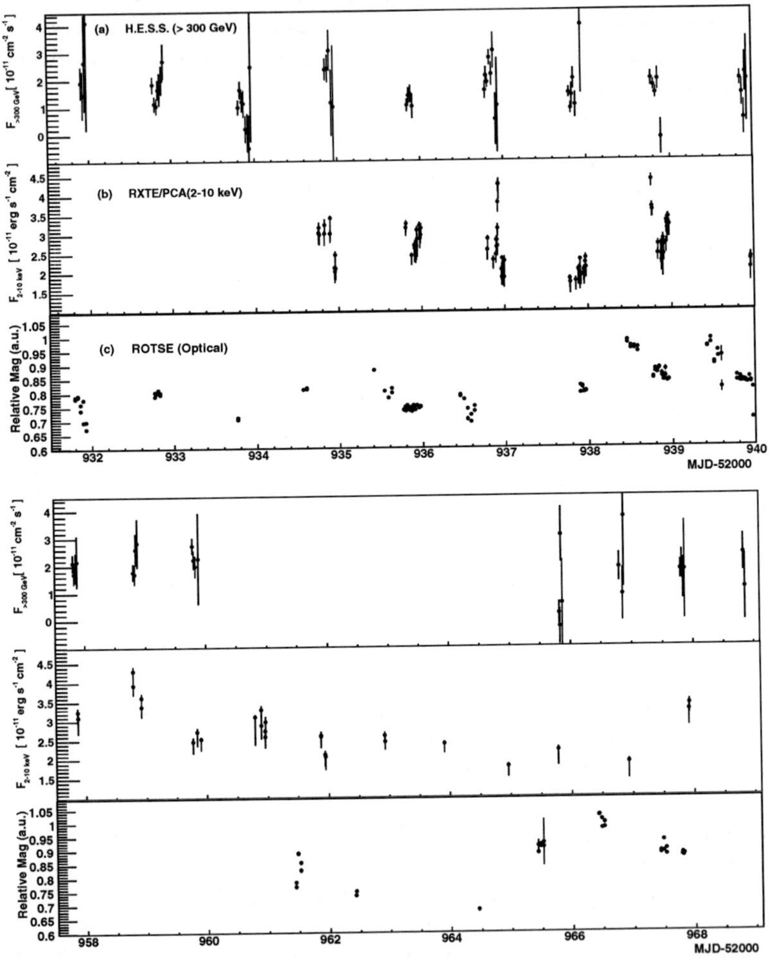

FIGURE 1. Light curves in three energy bands (Taken from [4]).
[Top] (a) October 2003 Light curve from H.E.S.S. binned in run lengths averaging 28 minutes each, in units of integral flux above 300 GeV derived using the average photon index 3.37 obtained in the next section. (b) Light curve expressed as flux in the 2–10 keV band for the RXTE observations, in 400 s bins. (c) Light curve derived from the ROTSE optical data.
[Bottom] Same as above but for the November 2003 data. Note that the RXTE observations were much shorter than in the previous month.

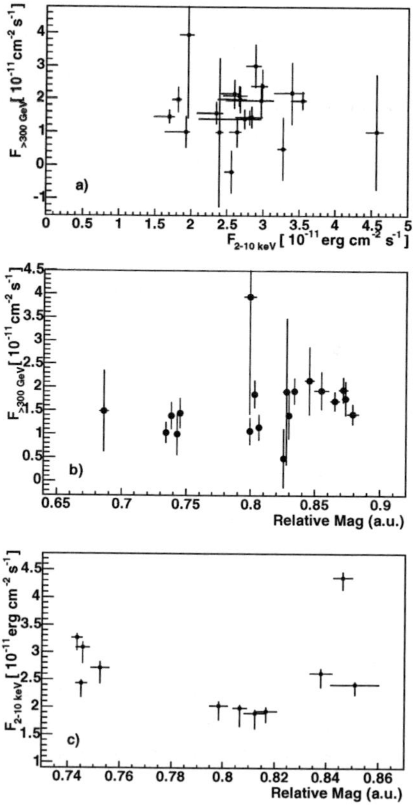

FIGURE 2. Correlation plots between different wavebands (Taken from [4]). No correlation is found between any of the simultaneous measurements.
[Top] Correlation plot for the 23 X-ray data segments coming from PCU2,PCU3 that overlapped exactly with a H.E.S.S. observation.
[Middle] Simultaneous H.E.S.S. and ROTSE observations. The ROTSE observations which are lying within a H.E.S.S. run are averaged and their errors are summed quadratically
[Bottom] Simultaneous RXTE and ROTSE observations. The optical data were binned to the 400 s long RXTE segments, which are overlapping on 6 different days.

TeV Spectrum and EBL Corrections

The October and November 2003 H.E.S.S. data were all combined in order to extract a common spectrum for the whole campaign. The best-fitting power law ($\chi^2 = 7$ for 8 degrees of freedom) is given by:

$$\frac{\mathrm{d}N}{\mathrm{d}E} = (2.73 \pm 0.17) \times 10^{-12} \left(\frac{E}{\mathrm{TeV}} \right)^{-3.37 \pm 0.07 \pm 0.10} \mathrm{cm}^{-2}\mathrm{s}^{-1}\mathrm{TeV}^{-1} \qquad (1)$$

which is comparable to $\Gamma = 3.32 \pm 0.06$ and $I_0 = (1.96 \pm 0.12) \times 10^{-12} \mathrm{cm}^{-2}\mathrm{s}^{-1}\mathrm{TeV}^{-1}$ previously reported in [3].

The EBL can significantly affect the intrinsic VHE spectrum of relatively high redshift sources such as PKS 2155−304 through $\gamma - \gamma$ interactions. Part of the TeV photons is absorbed from the soft background photons and the intrinsic spectrum changes considerably. Therefore in order to deconvolve the observed VHE spectrum (up to 3 TeV) from this absorption effect and reveal the intrinsic spectrum of the source we employ three EBL models Figure 3 (in order from higher to lower fluxes).

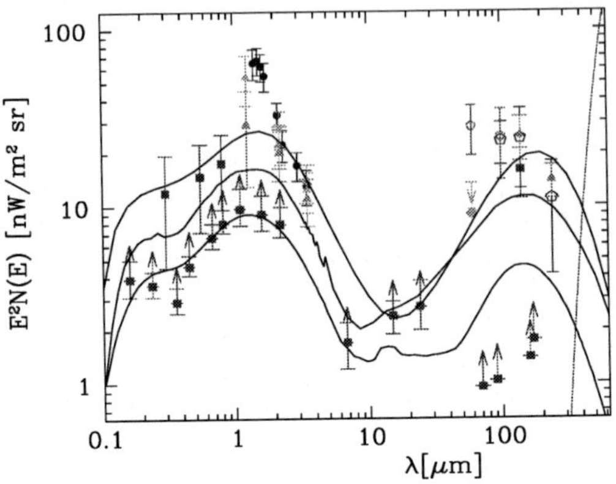

FIGURE 3. Spectral energy distribution of the EBL (Taken from [4]). It is thought to be characterized by two distinct bumps, around 1–2 and 100–200 μm, produced by the red-shifted stellar light and re-radiated thermal dust emission, respectively. The compilation of measurements have been taken from [15] and [1]. Lower limits are from HST and ISO source counts [18, 11, 13]. Above 400 μm, the contribution of the CMB starts to dominate. The three continuous lines represent the "Primack-type" EBL shapes: from higher to lower fluxes (at 1 μm), Phigh, Primack01 and Primack04 models (see text). The steep line starting at 300 μm is the onset of the CMBR.

The first model is the phenomenological shape used in [1], which is based on the original [22] calculation but smoothed and scaled up to match the data points below 1 μm and at 2–3.5 μm (hereafter *Phigh*). The second one, is the original [22] calculation for a Salpeter initial mass function (hereafter Primack01). Finally, the last EBL model is the new 2004 calculation ([23], hereafter Primack04), which takes advantage of the recent improvements in the knowledge of the cosmological parameters and of the local luminosity function of galaxies.

The opacities are calculated from the EBL SED shapes taking into account only the cosmology ($H_0 = 70\,\mathrm{km/s/Mpc}$, $\Omega_{\mathrm{Mat}} = 0.3$ and $\Omega_\Lambda = 0.7$). No evolution of the EBL has been introduced at this point.

The resulting absorption-corrected spectra are shown in Figure 4, together with the observed spectrum. The intrinsic spectra are all well fitted by a single power-law model, with a hard spectrum for Phigh ($\Gamma \approx 1.5$), and soft spectra for Primack01 and Primack04 models ($\Gamma \approx 2.3$ and $\Gamma \approx 2.8$, respectively). This effect is directly related to the different flux levels of the stellar peak component, which imprint a different amount of softening onto the original spectrum.

This direct link thus yields two simple scenarios for the location of the blazar's high energy peak, with the dividing line represented by the EBL flux which gives $\Gamma = 2.0$ (1.3 times the Primack01 model). Models with stellar peak fluxes above this (such as Phigh, and generally all those in agreement with the direct estimates of the fluxes between 2 and 3.5 μm) imply a hard intrinsic spectrum ($\Gamma < 2$), and thus an IC peak above 1–2 TeV. EBL models with lower fluxes (such as the Primack01 and Primack04) imply instead a soft spectrum ($\Gamma > 2$), locating the IC peak below the observed energy range (< 200 GeV).

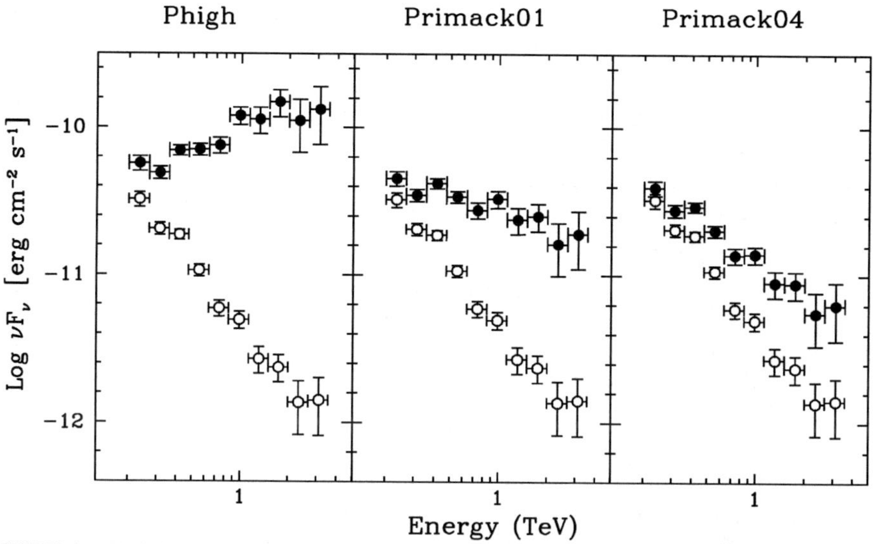

FIGURE 4. PKS 2155−304 absorption-corrected spectra, along with the observed spectrum, for each EBL model considered and in a νF_ν plot (i.e. a zoom in the blazar's SED) (Taken from [4]). Open dots: H.E.S.S. observed spectrum. Filled dots: reconstructed intrinsic spectrum. All spectra from "Primack-type" models are compatible with a power-law shape, with different slopes: hard (Phigh) and soft (Primack01, Primack04), locating the intrinsic SED peak for the high component above 1–2 TeV or below 200 GeV, respectively.

Source Parameters

The electromagnetic emission in blazars is very likely to be Doppler-boosted (or beamed) toward the observer. In the radio regime, the evidence for Doppler boosting in PKS 2155−304 comes from superluminal expansions observed with VLBI [21]. Relativistic beaming is also required in order to avoid absorption of GeV photons by X-ray photons via the e^+/e^- pair-production process (see e.g. [19]). It is thus possible to use the γ-ray variability to establish a limit for the Doppler factor δ, with δ defined in the standard way as $[\Gamma(1-\beta\cos\theta)]^{-1}$, where Γ is the bulk Lorentz factor of the plasma in the jet, $\beta = v/c$, and θ is the angle to the line of sight.

Following [9], the size of the γ-ray emission zone R is derived from the time-scale of variability $t_{\rm var}$ (supposing the timescale of intrinsic variability is negligible compared to the light crossing time) by $R \leq c t_{\rm var}\delta/(1+z)$. In this case (assuming a time scale $t_{\rm var} \approx 2$ ks from the "Light Curves and Correlations" section) the size of the emission region is

$$R\delta^{-1} \leq 5\times 10^{13}\,\text{cm}. \qquad (2)$$

Since for γ-rays in the 300 GeV–3 TeV range the opacity $\tau_{\gamma\gamma}$ cannot significantly exceed 1, independently of the emission mechanisms, another constraint on the minimum value of the Doppler factor can be derived by estimating how much Doppler boosting is necessary for photons with observed energy E_γ to escape from a source with radius R

and a flux density $F(E_t)$, where $E_t = (m_e c^2)^2 \delta^2 / E_\gamma (1+z)^2$. At this point one can follow [20] and especially Eq. 3.7 in [9], writing that the opacity is

$$\tau_{\gamma\gamma} \approx \frac{\sigma_T}{5} \frac{1}{hc} \frac{d_L^2}{R} \frac{1}{\delta^3 (1+z)} F(E_t) \,. \tag{3}$$

Imposing that $\tau_{\gamma\gamma} < 1$ for photons with observed energy $E_\gamma = 1\,\mathrm{TeV}$ yields a lower limit on δ for given R that can be derived numerically from the observed SED. For PKS 2155−304 this yields

$$R^{-1} \delta^{-6.4} \leq 5.6 \times 10^{-24}\,\mathrm{cm}^{-1}. \tag{4}$$

The minimum boosting factor δ together with an upper limit in the source size come from combining Eq. 2 and Eq. 4 and yields

$$\delta \geq 19 \ \text{ and } \ R \leq 10^{15}\mathrm{cm}. \tag{5}$$

The estimated Doppler factor δ is higher than the limit obtained in a similar way in [24], but at the lower end of the range usually obtained from SSC modelling as in [17]. It is useful to stress that this constraint is valid under the usual assumption that the region emitting the SED optical flux (i.e., the target photons) is cospatial with (or at least embeds) the high energy emitting region.

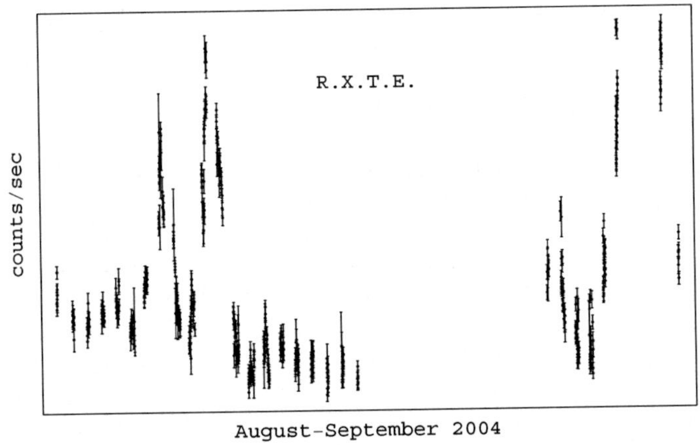

FIGURE 5. Preliminary results: X-ray observations (2-10 keV) of PKS 2155−304 during August-September 2004 with PCA (RXTE). The bin size is 5440 sec.

OUTLOOK

The H.E.S.S. collaboration has already performed observations of PKS 2155−304 during the period August-September 2004 (Figure 5), lasting 30 days in full stereoscopic mode using the four telescopes. Therefore we have ∼2.5 times better sensitivity (going from two to four telescopes) and longer time coverage consisting of about 110 twenty-minutes H.E.S.S. runs. The total duration of the X-ray observations is about 150 ksec.

ACKNOWLEDGMENTS

D.Emmanoulopoulos acknowledges support by the EU through HPRN-CT-2002-00321 and S.J.Wagner is grateful to the DFG for support through SFB 349.

REFERENCES

1. Aharonian F.A. et al. 2003, A&A, 403, 523
2. Aharonian, F.A., et al. 2004, APh, 22, 109
3. Aharonian, F.A., et al. 2005, A&A, 430, 865
4. Aharonian, F.A., et al. 2005, A&A, 442, 895
5. Akerlof, C. W., et al. 2003, PASP, 115, 132
6. Bernlöhr, K. et al. 2003, APh, 5 20, 111
7. Chadwick, P. M. et al. 1999,ApJ, 513, 161
8. Cornils, R. et al. 2003, APh, 20, 129
9. Dondi L. & Ghisellini G. 1995, MNRAS, 273, 583
10. Edelson R. et al. 1995, ApJ, 438, 120
11. Elbaz D. et al. 2002, A&A, 384, 848
12. Funk, S., Hermann, G., Hinton, J. et al. 2004, APh, 22, 285
13. Gardner, J.P. et al. 2000, ApJ, 542L, 79
14. Gaidos, J. A., Akerlof, C. W., Biller, S. D., et al. 1996, Nature, 383, 319
15. Hauser M.G. & Dwek E. 2001, ARAA, vol. 39
16. Jahoda, K. et al. 1996, in "EUV X-ray and Gamma-ray Instrumentation for Astronomy VII", SPIE Proceedings, 2808, 59
17. Kataoka, J. et al. 2000, ApJ, J., 528, 243
18. Madau P. & Pozzetti L. 2000, MNRAS, 312L, 9
19. Maraschi, L., Ghisellini, G. & Celotti, A. 1992, ApJ, 397, L5
20. Mattox, J. R., et al. 1993, ApJ, 410,609 Protheroe, R. J. 2001, APh, 15, 121
21. Piner, G. P. et al. 2004, ApJ, 600, 115
22. Primack, J. et al. 2001, AIP Conf. Proc. 558, 463
23. Primack, J. et al. 2004, AIP Conf. Proc. 745, 23
24. Tavecchio, F., et al. 1998, A&A, 509, 608 341, 733 P. 1995, PASP, 107, 803
25. van Driel, et al. 1996, in High sensitivity radio astronomy, ed. N. Jackson & R. J. Davis (Cambridge Univ. Press) 229.
26. Vestrand, W. T., & Sreekumar, P. 1999, APh, 11, 197
27. Zhang, Y. H., Treves, A. et al. 2002, ApJ, 572, 762

Identification of Gamma Ultra High Energy Cosmic Rays by Simulating Extensive Air Showers

A. Geranios*, E. Fokitis§, S. Maltezos§, A. Mastichiadis¶, O. E. Malandraki#, I. Antoniadou*

*Physics Department, Nuclear and Particle Physics Section, University of Athens, Panepistimioupoli, Athens Greece ZIP 15771

§Physics Department, Faculty of Applied Sciences, NTUA, 9 Heroon Polytechniou, Zografos, Athens, Greece ZIP15780

¶Physics Department, Section of Astrophysics, Astronomy and Mechanics, University of Athens, Panepistimioupoli, Athens Greece ZIP 15771

Research and Scientific Support Department of ESA, ESTEC, Noordwijk,The Netherlands & National Observatory of Athens, Institute for Space Applications and Remote Sensing, Pedeli, Athens, Greece

Abstract. Among the most interesting tasks for the identification of Ultra High Energy Cosmic Rays (origin, mechanisms of acceleration, top down model, direction, and maximum energy) is their isotopic composition. For example, if they are protons, heavier nuclei, gammas or even exotic. Air Shower Extended Simulation code (AIRES) is used to simulate the Extended Atmospheric Showers. This code gives, among others, the maximum number of secondary particles within the atmosphere along the longitudinal direction of the showers. From the distribution of the depths of maximum number of shower particles and their content in muons, we can distinguish between proton and gamma primary cosmic rays, both of energy of 100 EeV. The results show a maximum at lower atmospheric depths for protons and at higher for gammas, respectively. Due to the bremsstrahlung and pair production effects, the muon content of gamma showers is much poorer than of protons. Both facts enable the distinction between proton and gamma cosmic rays. For the simulation of the atmospheric EAS due to the passage of an Ultra High Energy Cosmic Ray, the Linsley standard atmospheric model is used [1].

INTRODUCTION

One of the primary goals of the Pierre Auger Observatory is to study the masses of ionizing radiation, cosmic rays, which are constantly striking the earth. Due to the high rate of low-energy, cosmic rays have become relatively well understood in the 80 years since their discovery. However, cosmic rays at higher energies are much rarer. For example, P. Auger Observatory will detect cosmic rays at 10^{19} eV with a rate of 1 per km^2 per 100 years. Primary cosmic rays at these energies can never be directly observed and must be studied through the properties of the Extensive Air Showers (EAS) which they cause by their nuclear interactions with the atmosphere.

CP848, *Recent Advances in Astronomy and Astrophysics,*
7th International Conference of the Hellenic Astronomical Society, edited by N. H. Solomos

The Pierre Auger Observatory project (PAO) is an international collaboration of 19 countries to study the highest energy cosmic rays and explore the upper energy region of their energy spectrum ($E > 5x10^{19}$ eV, Fig. 1) in which the presence of cosmic rays is prohibited according to the Greisen- Zatsepin -Kuzmin theory (Fig. 2) [2]. Due to the fact that in this energy region ten about cosmic ray particles are observed the last decade, the PAO with its high statistics will shed light in this controversy [3].

Two giant detector arrays, each covering 3000 square kilometres, are being constructed in the Northern and Southern Hemispheres. The PAO has been designed to work in a hybrid detection technique. Each will consist of 1600 water Cherenkov detectors and four atmospheric fluorescence detectors. The objective of the arrays is to measure the arrival direction, energy, and composition of cosmic rays above 10^{19} eV. The first detector soon reaches full integration in Mendoza province, Argentina. This site is especially interesting since it will make possible to explore the part of the sky not explored yet, with a preferential view to the Galactic Centre. A giant array of particle detectors will measure the lateral and temporal distribution of shower particles at ground level (Fig. 3). Air fluorescence detectors (FD) will measure the longitudinal development of the shower in the atmosphere above the surface array. The combination of both detectors will measure simultaneously the shower parameters of a subset of showers and will allow higher precision in the energy and arrival direction determination. This hybrid detector aims at a better separation of heavy nuclei, protons or gammas. In this way the anatomy of the showers will be derived.

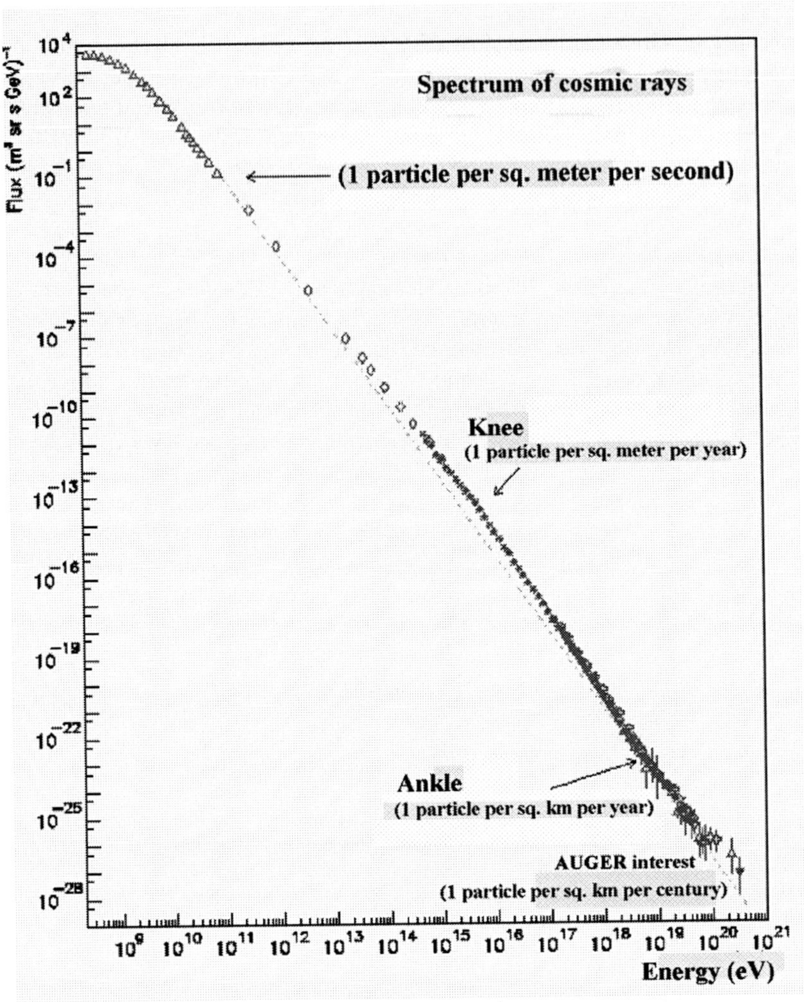

FIGURE 1. The cosmic ray energy spectrum from very low to very high energies. Energies above $5x10^{19}$ eV will explore the PAO [4].

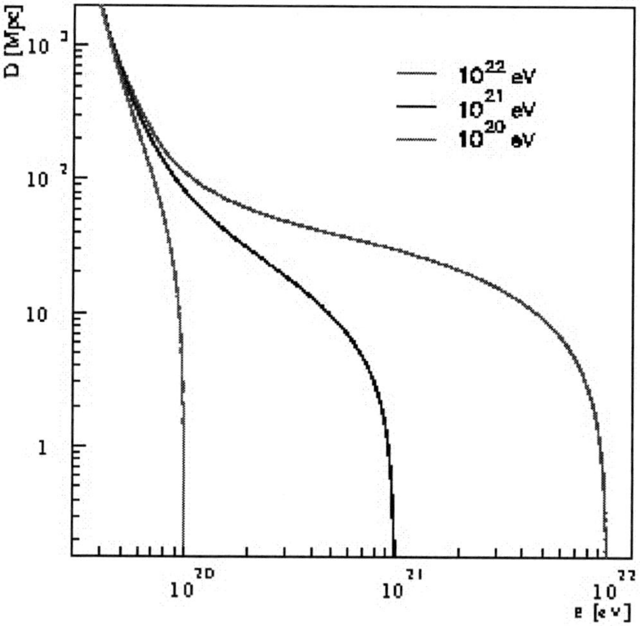

FIGURE 2. Due to the fact that very high energy cosmic protons and ions loose energy interacting with the microwave background radiation giving pions, there is an energy degradation as a function of distance to the observer for 3 different injection energies [5].

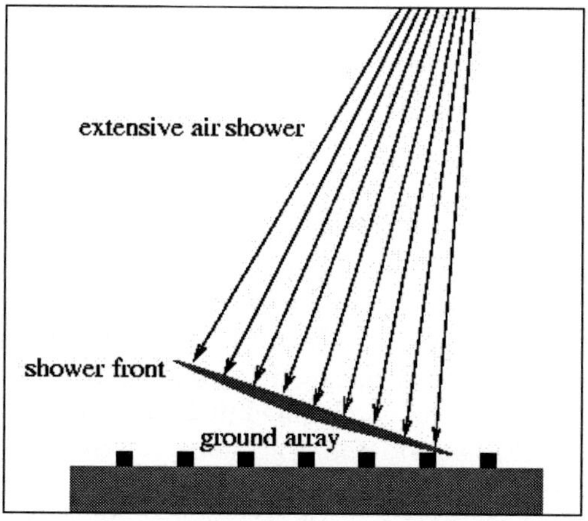

FIGURE 3. The shower front considered as spherical, triggers the ground array.

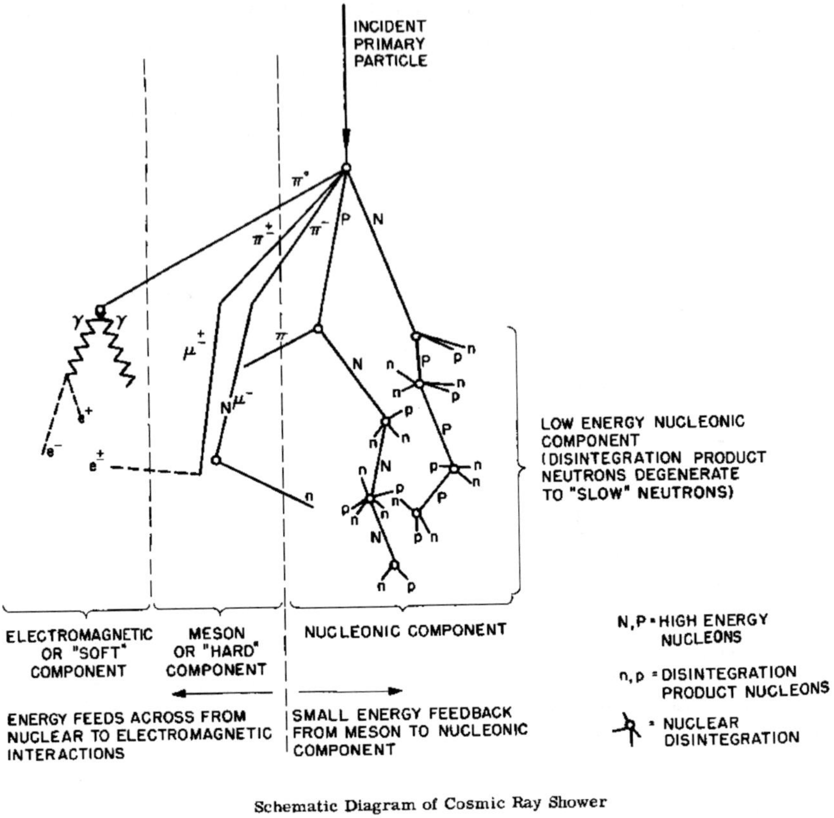

FIGURE 4. The EAS which is simulated by the AIRES technique.

THE AIRES SIMULATION TECHNIQUE

A practical method to determine the shower parameters is a fitting procedure; the experimental data for a specific shower will be fitted with Monte Carlo simulated event (for example, AIRES). The critical parameters (primary energy, direction and mass composition) of the simulated event will be modified until the fit gives best results. The whole evolution of an EAS due to the high energy primary interactions with the atmospheric air is here simulated by the AIRes technique [6]. The simulation takes into account among others, all possible nuclear interactions, decays and energy losses of particles evolved (Fig. 4). In particular, the AIRES program takes into account the pair production, electron-positron annihilation, bremsstrahlung, Compton and photoelectric effects, Landau-Pomeranchuk-Migdal effect, particle decays, such as pions, muons; hadronic processes, such as inelastic collisions hadron-nucleus and photon-nucleus, photonuclear reactions, nuclear fragmentation, elastic and inelastic propagation of particles, such as losses of energy in the medium (ionization), multiple Coulomb scattering and geomagnetic deflections. In the present analysis 250 EAS are simulated for primary protons and 500 for gammas. The energy of both cosmic particles was assumed 100 EeV, and their azimuth and elevation angle zero degrees. The simulations are shown in figure 5, which represent the atmospheric depth distribution of the maximum of showers.

HOW CAN WE DISTINGUISH A SHOWER CREATED BY A PROTON OR GAMMA?

Due to the larger interaction cross section of protons with the atmosphere, the maximum of the corresponding showers is at lower atmospheric depths for protons than for gammas. As the fluctuations of proton showers are less than for gammas, the latter corresponds to a wider distribution.

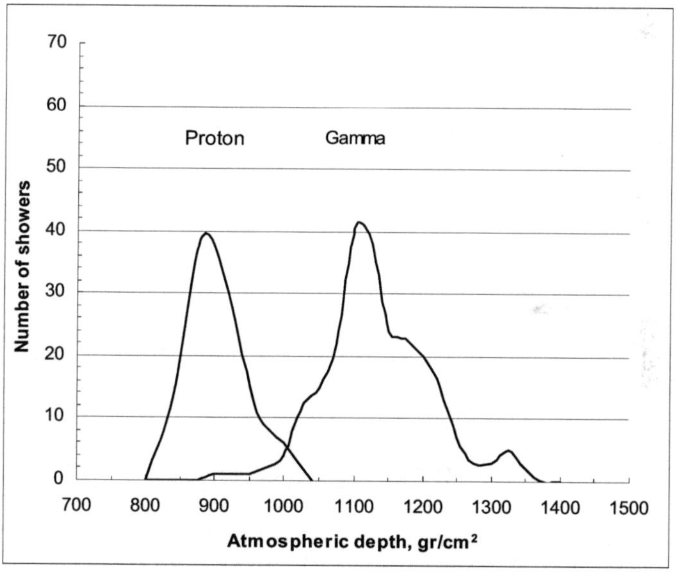

FIGURE 5. The mean maximum X_{max} of the proton showers is at lower atmospheric depth (900 gr/cm^2), while for gamma of the same energy is deeper (~1100 gr/cm^2).

A second characteristic of gamma showers is the fact that due to their pair production and Bremsstrahlung mechanism, their muon content is much lower than for protons. This fact can be shown by simulating the number of muons as a function of radial distance from the foot of the showers on the observing surface of the Auger Observatory (Fig. 6).

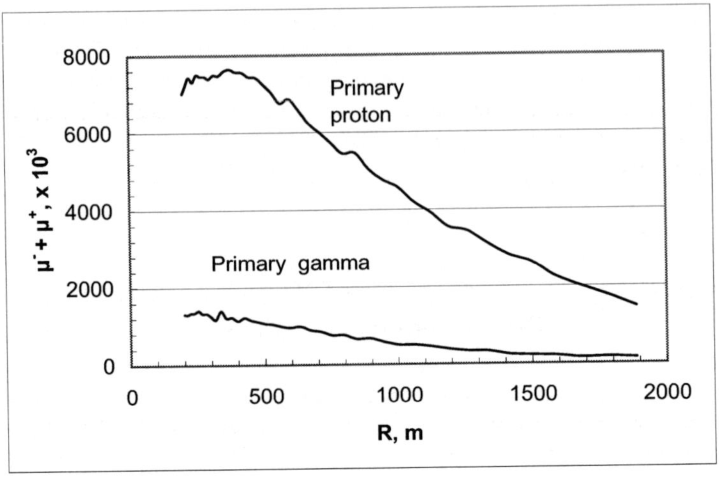

FIGURE 6. The lateral distribution of muons for vertical primaries of energy 10^{20} eV.

CONCLUSIONS

By estimating the atmospheric depths of the maximum of showers and the muon radial distributions, we will be able to distinguish gamma from proton cosmic rays when Pierre Auger Observatory will give its first measurements.

ACKNOWLEDGMENTS

We acknowledge S. Sciutto for the availability of the AIRES simulation program. This paper is in the frames of PYTHAGORAS II project.

REFERENCES

1. J. Linsley, *Proceedings of 15th ICRC*. Plovdiv, Vol. **12**, p. 89 (1977).
2. K. Greisen, *Phys. Rev. Lett.* **16**, 748 (1965).
3. M. Dova, *Proceedings of 27th ICRC*. Hamburg, Vol. **2**, p. 699 (2001).
4. A. Hillas, *Proc. 17th ICRC*, **8**, 193 (1981).
5. C..Hojvat, *GAP note* **97-056**, 1997 (http://www.auger.org/admin-cgi-bin/woda/gap_notes.pl).
6. S. Sciutto, AIRshower Extended Simulations, Department of Physics of the Universidad Nacional de La Plata, Argentina, (1999).

Jet-induced star formation in Seyfert galaxies unveiled through radio-UV correlations

E. Xanthopoulos*,†, P. Thomasson **, T. W. B. Muxlow ** and S. T. Garrington **

*University of California Davis, Department of Physics, Davis, CA 95616
†IGPP/Lawrence Livermore National Laboratory, Livermore, CA 94550
**University of Manchester, Jodrell Bank Observatory, Macclesfield, Cheshire SK11 9DL, England

Abstract. During the course of MERLIN L-band observations of the compact galaxy group, Stephan's Quintet, we found that the Seyfert 2 galaxy, NGC 7319, the brightest member of the compact group, has a triple radio structure typical of many extra-galactic radio sources which have a flat spectrum core and two steep spectrum lobes with hot spots. The two lobes are asymmetrically distributed on opposite sides of the core along the minor axis of the galaxy. Ultraviolet emission revealed in a high resolution HRC/ACS HST image is strongly aligned with the radio plasma and we interpret the intense star formation in the core and north lobe as an event induced by the collision of the north radio jet with over-dense ambient material. In addition, a re-mapping of archive VLA L-band observations reveals more extended emission along the major axis of the galaxy which is aligned with the optical axis. Images formed from the combined MERLIN and archive VLA data reveal more detailed structure of the two lobes and hot spots. NGC 7319 is the first example of jet induced star formation in a Seyfert galaxy unveiled through radio/UV correlations. Using high resolution HST ACS/HRC near-UV (F330W) images, and their corresponding published radio maps, we have identified a number of other possible jet-induced star formation Seyfert galaxies candidates.

Keywords: Group of galaxies, Radio observations, Stephan's Quintet, NGC 7319
PACS: 95.85.-e, 95.85.Bh, 98.65.-r, 98.54.Cm

INTRODUCTION

Stephan's Quintet (also known as Arp319 and VV228), which can be regarded as the prototype of the compact group class of galaxies (H92 in the Hickson 1982 catalog), was the first to be discovered 120 years ago by Stephan (1877). The members of the group (Fig. 1) were initially thought to be NGC 7317, NGC 7318A, NGC 7318B, NGC 7319, NGC 7320 and NGC 7320C. However, it was later found that NGC 7318B had a velocity almost 1000 km s^{-1} lower than the mean redshift of the group ($\approx$6600 km s^{-1}), while Moles et al. (1997) have argued that the discordant-redshift galaxy ($\approx$800 km s^{-1}), NGC 7320, is an unrelated foreground galaxy and have suggested that the kernel of three galaxies, NGC 7317, NGC 7318A and NGC 7319, has been visited by two "intruders" (NGC 7320C and NGC 7318B) at different times in the past few 10^8 years. VLA observations at 20 cm wavelength by van der Hulst & Rots (1981), with a resolution of 6″ (Fig. 1), have shown more clearly the main components of the radio emission associated with Stephan's Quintet. Their observations resolved the arc of emission (similar to a letter "W" on its side), between NGC 7319 and NGC 7318B into distinct sources embedded in more diffuse emission which is partly coincident with spiral-arm-like features in NGC 7318B. They also noted a discrete, unresolved source,

CP848, *Recent Advances in Astronomy and Astrophysics,*
7th International Conference of the Hellenic Astronomical Society, edited by N. H. Solomos

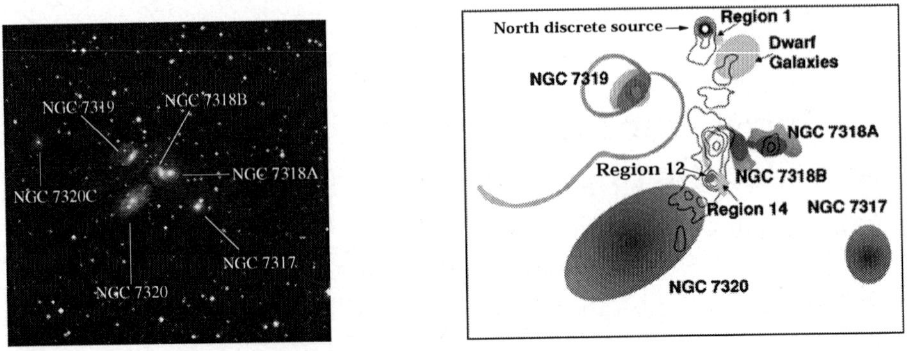

FIGURE 1. (Left:) Digital Sky Survey map covering the member galaxies of Stephan's Quintet. (Right:) Schematic showing the relative positions of the radio emission detected by van der Hulst & Rots (1981) and the optically visible galaxies. Also shown are the locations of a group of dwarf galaxies and regions of Hα emission, designated Regions 1, 12 and 14 by Mendes de Oliveira et al. (2001), all considered to be associated with NGC 7318B.

just to the north of the "W", which, with no optical counterpart, they considered to be an unrelated background object.

OBSERVATIONS AND RESULTS

The resulting naturally weighted VLA map of Stephan's Quintet overlaid on the optical Digital Sky Survey image of the same area is shown in Fig. 2. The lowest believable contour is at a level of approximately a factor of 2 lower than that in the published map of NGC 7319. As might be expected, in comparison with the MERLIN images, this shows the extended structure more clearly, though even at this resolution, much of the north-south arc of emission is considerably resolved and the emission that is visible in the region at $\alpha \approx 22^h35^m59.8^s$, $\delta \approx 33°57'\,34''$ (J2000) is at the limit of sensitivity. However, the northern unidentified source is clearly visible, though somewhat resolved. Also visible in the image is radio emission at a position corresponding to the core of NGC 7318A.

Aoki et al. (1999) have indicated that the previously known compact components (labelled A, B and C in Fig. 3 after Aoki et al. 1999) and diffuse emission of NGC 7319 all have steep spectra. However, our MERLIN only 20 cm image (Fig. 3), which resolves out much of the extended emission, indicates component B to be unresolved and to have a flux 1.07 mJy, which is very comparable with the value of 1 mJy quoted by Aoki et al. (1999) for its 6 cm flux. At the same time the flux densities of components A and C, although lower than the 20 cm VLA fluxes measured by Aoki et al. (1999), are more than double in flux than the 6 cm flux densities quoted by Aoki et al. (1999). Component B then would appear to have a flat spectrum, and as such is considered to be the core of the galaxy. We are also able to measure a size for component B. There is no evidence of a jet as is often the case. However, the size, luminosity and spiral host galaxy of NGC 7319 is consistent with it being a Seyfert 2 galaxy (Durret 1994). The

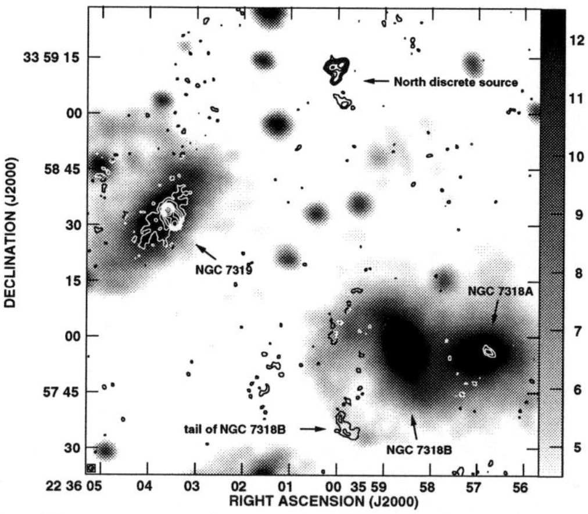

FIGURE 2. 20 cm naturally weighted VLA archive data image of Stephan's Quintet at a resolution of $1\overset{\prime\prime}{.}3$ overlaid on a Digital Sky Survey image. Contour levels are at $6.97\text{e-}5\times(1,2,4,8,16,32,64,128,256)$ Jy/beam. The peak flux is 10.3 mJy/beam and the rms noise level is $\sim$27 μJy/beam.

radio structure of NGC 7319 looks like a small, asymmetric FRII radio source, with a flat-spectrum core and two extended lobes, both containing compact hot-spots. This adds to the evidence which calls into question the usual assumption that the compact radio components of Seyfert galaxies represent the galaxy nucleus. The asymmetric arm ratio of 1:3.1 for the distances between the core (component B) and the two hotspots (components A and C respectively) possibly indicates that material to the north-east is blocking the outflow of plasma in that direction whereas plasma outflow to the south-west is much less inhibited. The new image formed from the VLA archive data also shows, for the first time, emission of extent $\approx 20''$ perfectly aligned with the strong bar at a PA = 142° (Durret 1994), which is also the position angle of the major axis of the 'optical' Seyfert galaxy.

The MERLIN L-band image is overlaid on the HST/ACS image in Fig. 3. Star formation is evident only in the bright core and the north lobe. There are only hints of UV emission from component C and three very compact regions of star formation, north of component C, trailing the region of the south lobe. The UV emission starts from component B, the core, and has an extent that matches the north lobe. The emission raps around the north lobe and strong star formation appears at the position of component A, the end hotspot of the north lobe. Almost midway between components A and B, an extended intense star formation region is almost perpendicular to the radio axis. The UV emission follows closely the elongated contours (jet-like extension) seen in the MERLIN image below component A. The SE-NW extent of this UV structure is $0\overset{\prime\prime}{.}85$ or 0.45 kpc. A $0\overset{\prime\prime}{.}13$ smoothed version of the HST/ACS UV image brings out an additional star formation region to the W of this region and just below to what seems now a more marked, pronounced and extended UV structure between components A and B, almost arclike in shape. A difference image (the smoothed image was subtracted from the original unsmoothed image) identifies the UV emission in the region of the core as

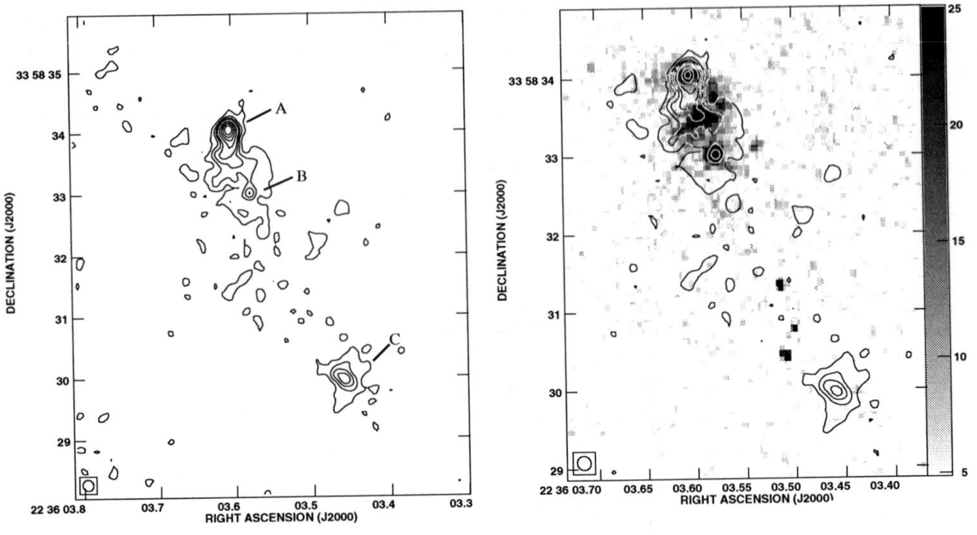

FIGURE 3. (Left:) 20 cm naturally weighted MERLIN image of NGC 7319 (resolution 0."15). The compact components are labelled A, B and C. Contour levels are set to 1.22e-4×(-1,1,2,4,8,16,32) Jy/beam. The peak flux is 1.84 mJy/beam and the rms noise level is ~47 μJy/beam. (Right:) The 20 cm naturally weighted MERLIN image of NGC 7319 (resolution 0."15) overlaid on the ACS/HRC UV image. North is up and East to the left. The sidebar shows the relative intensity of the greyscale image. See text for discussion of aligned features.

the brightest UV region with the most intense star formation.

DISCUSSION

Smith et al. (2003) have put forward an evolutionary scenario from spiral dominated groups to elliptical dominated groups. In this scheme Stephan's Quintet is probably in an intermediate transition state between HCG 16 (earlier stage), in which the HI mass is in both the inner region and tidal features, and NGC 4410 (later stage), with less HI mass than Stephan's Quintet distributed mostly in the tidal features. NGC 4410 is also the only group of the three that contains a radio galaxy with radio lobes extending beyond the optical galaxy. The existence of large radio lobes in the merging galaxy NGC 4410A shows that such structures can occur fairly early in the formation of an elliptical, when it is not yet an elliptical (Smith et al. 2003). Both HCG 16 and Stephan's Quintet contain Seyfert galaxies. However, in Stephan's Quintet, a triple radio structure is also present within the dominant galaxy NGC 7319. Is NGC 7319 an example of a spiral galaxy in which jets are present but smothered by dense interstellar clouds as suggested by Wang et al. (2000)? The evidence in this paper seems to point to this.

The jet-cloud interaction is evident in the arclike-structure seen in NGC 7319 midway between components B and A. We believe that incidence of the jet on the cloud/dense ambient gas at this region causes a strong bow-shock that is driven into the cloud and creates the bright star-formation perpendicular to the direction of the shock and aligned with the extended radio contours. The arclike shape of this structure, as revealed in the smoothed image, is further support of the presence of bow-shock (e.g. Bicknell et al. 2000). The jet deposits much of its momentum at this site and it continues upward to component A where the decelerated jet plasma accumulates as a radio "lobe". After the jet impacts the cloud, it makes slow headway and finally is effectively stalled by the cloud. A tilted interaction area forms in front of the jet nonetheless the vortical backflow exerts pressure on the jet from the top side, thus causing the penetrating part of the jet to bent slightly downward. The curved structure seen at the very end of the UV continuum can be a result of the bent of the jet as the UV emission trails the radio emission.

But is there evidence of such a cloud and dense material in or near NGC 7319? NGC 7319 has lost all detectable (upper limit $\sim 10^8 M\odot$) HI, the vast majority of HII regions (expected in a typical $\sim$SBb spiral) and an uncertain fraction of stellar mass. However, a very bright condensation of young blue stars is found in the northeast edge of the disk and both this Hα and CO emission (Yun et al. 1997) in NGC 7319 is $\sim$8 kpc north-northeast of the nucleus ($6.2\times 10^8 M\odot$). This region of material is most probably responsible for confining the North lobe within the spiral structure of NGC 7319. A second massive CO complex is found within 2 kpc of the nucleus (Yun et al. 1997), although Gao & Xu (2000) place this source on the nucleus. This material coincides with the UV bright core and is most possibly responsible for the birth of the bow-shock above the core.

Using high-resolution HST ACS HRC (F330W) near-UV archival images of a large sample of Seyfert galaxies as well as published radio maps and their FIR properties we identified new promising jet-induced star formation candidates following two methods: (1) Seyferts that in the FIR (60 micron) versus radio (1.4 GHz) plot fall in the $>$40 Jy FIR and $>$100 mJy radio area (2) objects that show the same UV and radio characteristics as NGC 7319.

REFERENCES

1. Aoki, K., Kosugi, G., Wilson, A. S., Yoshida, M., *Astrophys. J*, **521**, 565–571 (1999).
2. Bicknell, G. V., Sutherland, R. S., van Breugel, W. J. M., Dopita, M. A., Dey, A., Miley, G. K., *Astrophys. J*, **540**, 678–686 (2000).
3. Durret, F., *Astron. Astrophys. Suppl. Series*, **105**, 57–66 (1994).
4. Gao, Yu, Xu, C., *Astrophys. J*, **542**, L83–L87 (2000).
5. Hickson, P., *Astrophys. J*, **255**, 382–391 (1982).
6. Mendes de Oliveira, C., Plana, H., Amram, P., Balkowski, C., Bolte, M., *Astron. J*, **121**, 2524–2539, (2001)
7. Moles, M., Sulentic, J. W., Márquez, I., *Astrophys. J*, **485**, L69 (1997).
8. Smith, B., Nowak, M., Donahue, M., Stocke, J., *Astron. J*, **126**, 1763–1775 (2003).
9. Stephan, M. E., *Les Comptes rendus de l'Académie des sciences Paris,*, **84**, 641 (1877).
10. van der Hulst, J. M., Rots, A. H., *Astron. J*, **86**, 1775–1780 (1981).
11. Wang, Z., Wiita, P. J., Hooda, J. S., *Astrophys. J*, **534**, 201–212 (2000).
12. Yun, M. S., Verdes-Montenegro, L., Del Olmo, A., Perea, J., *Astrophys. J*, **475**, L21 (1997).

Linear radio structures in selected Seyfert and LINER galaxies

E. Xanthopoulos*,†, A. H. Thean **, A. Pedlar ** and A. M. S. Richards **

*University of California Davis, Department of Physics, Davis, CA 95616
†IGPP/Lawrence Livermore National Laboratory, Livermore, CA 94550
**University of Manchester, Jodrell Bank Observatory, Macclesfield, Cheshire SK11 9DL, England

Abstract. High resolution MERLIN 5 GHz observations (0.″04) of 7 Seyfert galaxies, selected as the ones previously showing evidence of collimated ejection, have been compared with high resolution archive HST data. The radio maps reveal rich structures in all the galaxies. NGC 2639 and TXFS 2226-184 have multiple knot parsec-scale extended structures, Mrk 1034, Mrk 1210, NGC 4922C and NGC 5506 reveal one-sided jets, while IC 1481 exhibits a jet-like extension. The close correlation between the radio-emitting relativistic plasma and the ionized gas in the inner regions of these galaxies allows us to study in detail the physics close to the center of low luminosity AGN.

Keywords: Seyfert and LINER galaxies, Radio observations, Jets
PACS: 98.54.Cm, 98.38.Fs

INTRODUCTION

High resolution radio observations have revealed double and triple radio sources in many Seyfert nuclei, suggesting that the collimated ejection of material is occurring in these objects in the same way as the radio jets in quasars and radio galaxies. But very few

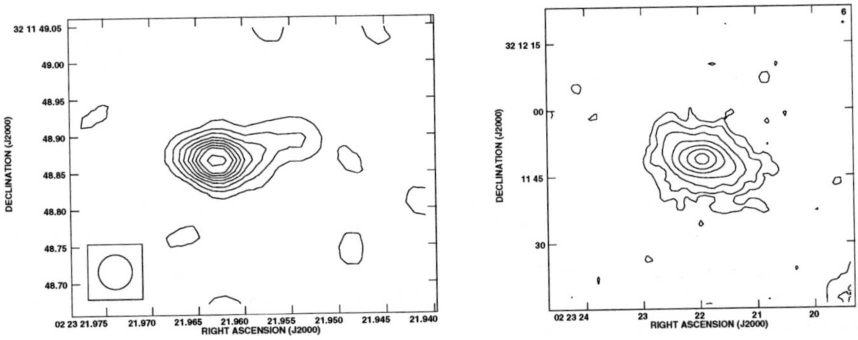

FIGURE 1. (Left:) 5 GHz MERLIN map of Mrk 1034. Contour levels are at 1.978e-4×(1,2,3,4,5,6,7,8,10,16,32,64) Jy/beam. The peak flux is 2.219 mJy/beam and the rms noise level is∼98 μJy/beam. (Right:) K_s (2.17 μm) Two Micron All Sky Survey (2MASS) image of Mrk 1034.

Seyfert galaxies have been found to have radio structures that can be described as jets (Kukula et al. 1993; Ghosh et al. 1994). The angular resolution of MERLIN at 5 GHz is equivalent to that of the *Hubble Space Telescope* (HST), making these radio images

CP848, *Recent Advances in Astronomy and Astrophysics*,
7th *International Conference of the Hellenic Astronomical Society*, edited by N. H. Solomos

ideal for comparison with the structure of the narrow-line region (NLR) and extended narrow line region (ENLR) and perfect cases to study in detail the individual small scale "jets" as revealed in e.g. Capetti et al. (1999). The purpose of the present study is to: **a)** Investigate the variety of collimated ejection in low luminosity AGN and increase the small number of multiple component "jets" known, **b)** Determine the alignment of the collimated ejection on scales of a few parsecs and compare this alignment with estimates made from extended NLR studies from HST and other optical observations, **c)** Involve hydrodynamical simulations and multi-wavelength data in order to constrain the bow shock and similar models and get a more complete picture of this class of AGN and, **d)** Compare the 5 GHz continuum observations with 22 GHz MERLIN observations of the same galaxies, since the coincidence of the position of the two emissions may mark the location of the obscured central engine.

MAPS AND RESULTS

Mrk 1034: The MERLIN 5 GHz map (Fig. 1) of Ark 81 - one of the Seyfert 1 galaxies of the pair Ark 80, Ark 81 that are interconnected- shows an E-W elongated radio structure (20 mas) with a 10 mas jet extension to the E consistent with a highly collimated radio jet in that direction. Mrk 1034 has not been observed yet with HST. However, the inner contours of a 2MASS near-IR image of Mrk 1034 (Fig. 1) follow the same elongation along the E-W direction.

Mrk 1210: The near-UV emission revealed in a high resolution ($0\farcs027$) ACS/HRC HST (F330W) image of Mrk 1210 follows closely the $\sim$15 mas one-sided jet unveiled in the MERLIN 5 GHz map (Fig. 2) of this amorphous Seyfert 2 galaxy.

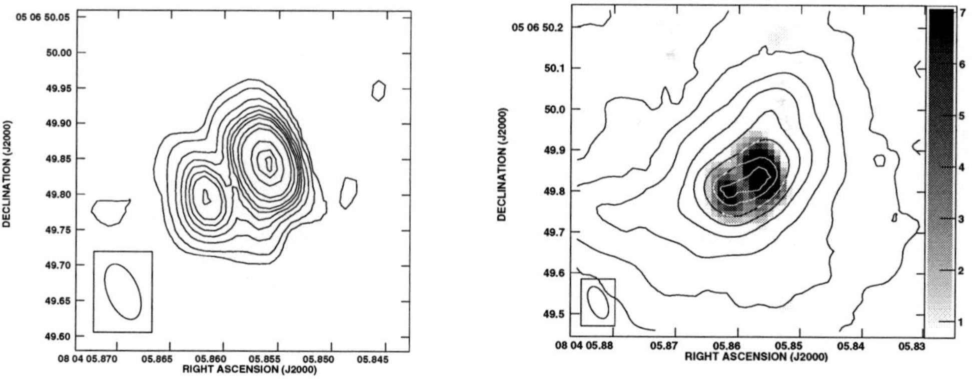

FIGURE 2. (Left:) 5 GHz MERLIN map of Mrk 1210. Contour levels are at 4.899e-4×(1,2,4,6,8,10,12,16,20,32,40,45) Jy/beam. The peak flux is 22.83 mJy/beam and the rms noise level is $\sim$ 163 μJy/beam. (Right:) HST ACS/HRC (F330W) contour map of Mrk 1210 overlaid on the MERLIN 5 GHz greyscale image.

NGC2639: The MERLIN 5 GHz map (Fig. 3) of this well-known LINER megamaser galaxy reveals a bright core and symmetrical E-W "wings" with rich structure (multiple knots), that is highly correlated with the UV star formation morphology of the HST ACS/HRC image (Fig. 3).

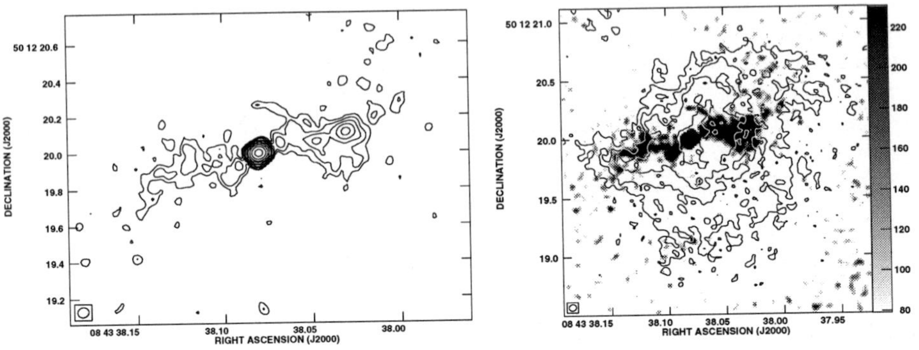

FIGURE 3. (Left:) 5 GHz MERLIN map of NGC 2639. Contour levels are at 1.297e-4×(1,2,4,8,16,32,64,128,256,512) Jy/beam. The peak flux is 93.5 mJy/beam and the rms noise level is ~48 μJy/beam. (Right:) HST ACS/HRC (F330W) contour map of NGC 2639 overlaid on the MERLIN 5 GHz greyscale image.

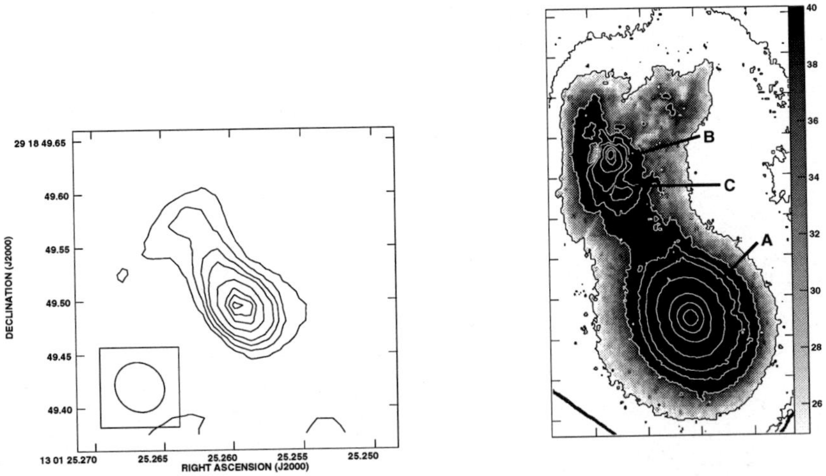

FIGURE 4. (Left:) 5 GHz MERLIN map of NGC 4922C (PGC 044896). Contour levels are at 3.803e-4×(1,2,4,6,8,9,10,16,32,64,128,256,512,1024) Jy/beam. The peak flux is 3.97 mJy/beam and the rms noise level is ~126 μJy/beam. (Right:) HST WFPC2 (F606W) image of the triple system NGC 4922.

NGC 4922: The MERLIN 5 GHz map (Fig. 4) detects the middle component C, PGC 044896 (FIRST J130125.2+291849 radio source), from the three (3) galaxy system

NGC 4922 (NGC 4922A/B are both Seyfert 2) as shown in HST WFPC2 (F606W) image (Fig. 4). The radio structure is elongated in the NE-SW direction, in good agreement with the orientation of the interacting galaxies.

NGC 5506: The MERLIN 5 GHz map shows a NW radio loop/jet (~5 mas in extent) that follows closely the rich near-UV emission (Fig. 5).

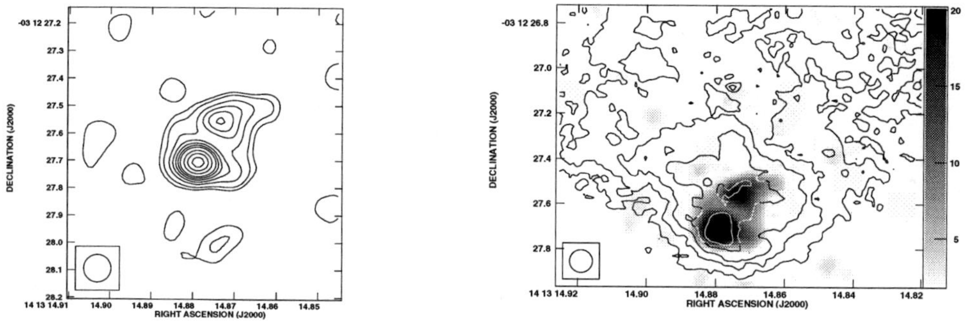

FIGURE 5. (Left:) 5 GHz MERLIN map of NGC 5506. Contour levels are at 2.185e-3×(1,2,4,6,10,12,16,20,24) Jy/beam. The peak flux is 57.29 mJy/beam and the rms noise level is ~ 546 μJy/beam. (Right:) HST ACS/HRC (F330W) contour map of NGC 5506 overlaid on the MERLIN 5 GHz greyscale image.

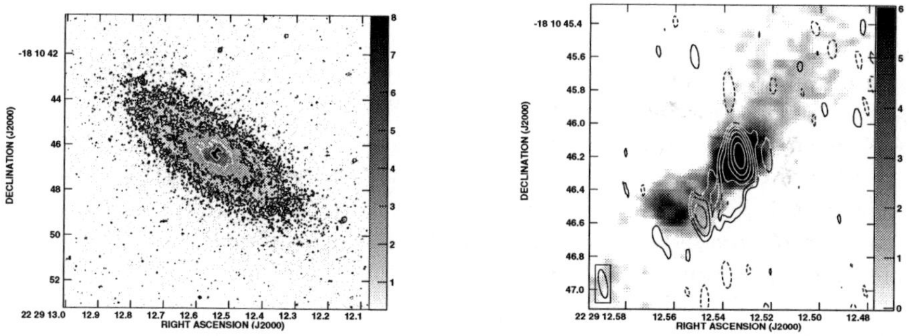

FIGURE 6. (Left:) The average red/green continuum HST WFPC2 image of the host galaxy of TXFS 2226-184. (Right:) 5 GHz MERLIN map of TXFS 2226-184 overlaid on the HST WFPC2 Hα image. Contour levels are at 1.750e-4×(1,2,4,8,16,32,64,128,256,512) Jy/beam. The peak flux is 14.25 mJy/beam and the rms noise level is ~ 65 μJy/beam.

TXFS 2226-184: The multiple knot radio structure (Fig. 6) coincides with the nucleus of the host galaxy and then is aligned with the Hα emission on the larger scale. This alignment reflects an interaction between the radio jet and the interstellar medium. The Hα and radio structures are perpendicular to the galaxy major axis (Fig. 6).

IC 1481: The radio map of this LINER water maser galaxy and a possible site of ongoing merger, shows elongated structure (Fig. 7) which agrees with VLBI observations that have shown that water maser emission is associated with thin, parsec-scale molecular disks located near the spout of radio continuum jets.

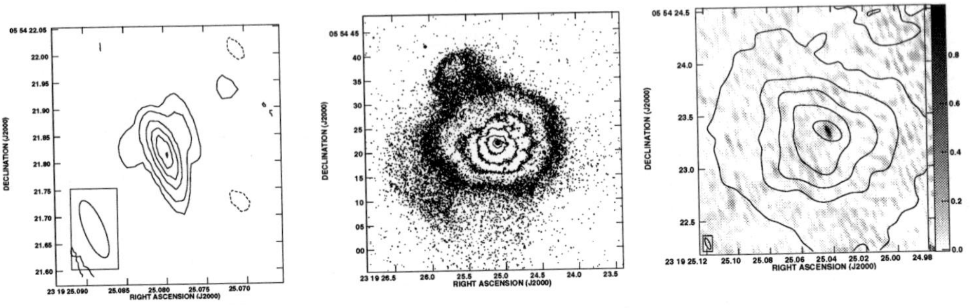

FIGURE 7. (Left:) 5 GHz MERLIN map of IC 1481. Contour levels are at 1.750e-4×(1,2,4,8,16,32,64,128,256,512) Jy/beam. The peak flux is 14.25 mJy/beam and the rms noise level is $\sim 65\ \mu$Jy/beam. (Middle:) The red continuum HST WFPC2 image of the host galaxy. (Right:) The Hα HST WFPC2 contour map overlaid on 5 GHz MERLIN greyscale image.

DISCUSSION

It is not widely appreciated that Seyferts, in common with radio galaxies, may show evidence for collimated ejection and/or radio jets. The conical or biconical morphology of the UV continuum in Seyferts has been found to be aligned with the radio axis, and we can postulate that the radio jets define the axis of the UV cone and also the central engine. In this area, high resolution images obtained by MERLIN at 5 GHz have provided the clearest evidence for linear radio jets in Seyfert galaxies that have been previously poorly resolved in VLA images. For this reason we observed with MERLIN at 5 GHz and 0″.04 resolution, 7 Seyfert galaxies that were found to show hints of such structures in previous lower resolution radio observations. The higher resolution observations allow us to identify the orientation and geometry of the nuclear disk and further examine, for the ones that was possible, the relation of the nuclear radio structure to the inner and large scale morphology, emission and dust lanes seen in optical images. This way we are able to build up a more complete picture of the physics of the low luminosity AGN.

REFERENCES

1. Capetti, A., Axon, D. J., Macchetto, F., Marconi, A., Winge, C., *Astrophys. J*, **516**, 187–194 (1999).
2. Ghosh, T., Schilizzi, R. T., Miley, G. K., de Bruyn, A. G., Kukula, M. J., Pedlar, A., Graham, D., Saikia, D. J., in "Multi-wavelength continuum emission of AGN", edited by Courvoisier T. J.-L., Blecha, A., Proceedings of IAU Symp. 159, Kluwer, Dordrecht, 1994, p. 426.
3. Kukula, M. J., Ghosh, T., Pedlar, A., Schilizzi, R. T., Miley, G. K., de Bruyn, A. G., Saikia, D. J., *Month. Not. Roy. Astron. Soc.*, **264**, 893–899 (1993).

Exploring Distant Galaxies with *Chandra*

P. Tzanavaris*, I. Georgantopoulos*, A. Georgakakis†, A. Akylas*, A. Zezas**, E. Koulouridis* and M. Plionis*

**Institute of Astronomy and Astrophysics, National Observatory Athens, I. Metaxa & V. Pavlou, 152 36 Penteli, Greece*
†Astrophysics Group, Blackett Laboratory, Imperial College London, Prince Consort Rd, London SW7 2BZ, United Kingdom
***Harvard-Smithsonian Center for Astrophysics, 60 Garden Street, Cambridge, MA 02138*

Abstract. We cross-correlate *Chandra* sources from the XAssist catalogue with the SDSS and identify galaxies which are dominated by stellar processes ('normal' galaxies). We obtain a further sample of such galaxies by also cross-correlating sources from the XBootes survey with the SDSS. We calculate photometric redshifts for this sample and combine both samples with 3 earlier samples of X-ray selected 'normal' galaxies to obtain the largest X-ray selected sample of such galaxies to date, consisting of 219 objects and spanning a range in X-ray flux f_X (0.5 – 2.0 keV) $\approx 10^{-13} - 10^{-16}$ erg cm^{-2} s^{-1}. We calculate the total sensitivity curve and number counts ('$\log N - \log S$') of 'normal' galaxies for our combined sample and compare our results with predictions from earlier work.

Keywords: X-rays:galaxies; X-rays:surveys; Galaxies:starburst; Galaxies:active
PACS: 98.52.-b; 95.85.Nv; 95.80.+p

INTRODUCTION

With the new generation of X-ray missions, *Chandra* and *XMM-Newton*, the study of galaxies which are not AGN dominated ('normal' galaxies, NGs) has progressed dramatically as systems at distances greater than ~ 100 Mpc became accessible to X-ray missions. For example, the *Chandra* Deep Fields North and South [CDF-N, CDF-S; 1, 2] reached fluxes $f(0.5-2.0\,\text{keV}) \sim 10^{-17}$ erg cm^{-2} s^{-1}. NGs are selected using a combination of criteria: Commonly used selection criteria in X-rays are low X-ray–to-optical flux ratios, $\log(f_X/f_O)$, and soft hardness ratios, $\text{HR} \equiv \frac{\text{H}-\text{S}}{\text{H}+\text{S}}$, where H (S) represents counts in a hard (soft) energy band. In the optical region extended image profiles and diagnostic spectral emission-line ratios [3, 4] are used. There is some uncertainty about the exact threshold for the $\log(f_X/f_O)$ criterion, but $\log(f_X/f_O) < -2$ appears to be appropriate. In particular, Tzanavaris et al. [5] used a new X-ray/optical cross-correlation sample as well as samples from the literature to show that the $\log(f_X/f_O) < -2$ criterion is adequate for separating star-forming galaxies from AGN, although it may be selecting against massive ellipticals. Extended work has been carried out on NG X-ray luminosity functions (XLFs) and number counts. Here we use a new NG sample to extend previous work and present new preliminary results on the statistics of the largest sample of X-ray selected NGs to date.

CP848, *Recent Advances in Astronomy and Astrophysics,*
7th International Conference of the Hellenic Astronomical Society, edited by N. H. Solomos

METHODOLOGY

In this paper we combine NG samples from independent wide-area shallow as well as pencil-beam surveys spanning a wide range of X-ray fluxes, f_X $(0.5-2.0$ keV$) \approx 10^{-13}-10^{-16}$ erg cm^{-2} s^{-1}. We now briefly describe how these samples were selected for each survey.

The Needles in the Haystack Survey Within the framework of the 'Needles in the Haystack Survey' (NHS) project [6, 7], 70 *XMM-Newton* fields were used in total, covering an area of 11 deg^2 and overlapping with the Sloan Digital Sky Survey Data Release-2 (SDSS DR-2). NGs were selected from optical counterparts found within $7''$ of X-ray sources, showing an extended optical light profile, having $\log(f_X/f_O) < -2$, soft HRs and optical spectra consistent with star-forming galaxies, as suggested by emission-line ratios and lack of broad emission lines. The final sample comprises 28 NGs to the $0.5-8.0$ keV band flux limit of $\approx 2\times 10{-}15$ erg cm^{-2} s^{-1}.

The First *XMM-Newton* Serendipitous Source Catalogue The NHS work was extended in Georgakakis et al. [8] were the First *XMM-Newton* Serendipitous Source Catalogue (1XMM) was used. The sample is taken from a total of 51 fields, covering an area of ≈ 6 deg^2 to a $0.5-2.0$ keV band flux limit of $\approx 10^{-15}$ erg cm^{-2} s^{-1}. In all 28 NGs were selected, based on the selection criteria mentioned above, with optical counterparts identified using the U.S. Naval Observatory Catalog (USNO), version A2.0.

The XAssist Catalogue Public *Chandra* observations, analysed for the XAssist catalogue [`http://www.xassist.org`, 9] and overlapping with the SDSS [10] were also used. Sources are detected in the $0.3-8.0$ keV band in a total of 913 fields. Full description of the NG galaxy selection will be presented in a forthcoming publication. This sample is available on-line at `http://www.astro.noa.gr/xray/XAssist-SDSS.txt`.

The total area of the 1XMM+NHS+XAssist sample is about 15deg^2 with a total of 68 NG candidates selected using the $\log(f_X/f_O) < -2$ criterion. The $\log(f_X/f_O)$ ratio was calculated by means of the Stocke et al. [11] definition, assuming $\Gamma = 1.8$ and $B-V = 0.8$. Spectroscopic identifications are available for 55 of the 68 galaxies.

The *Chandra* Deep Fields North and South We used the Bauer et al. [12] sample of NGs selected from the *Chandra* Deep Fields North and South [CDF-N, CDF-S; 1, 2]. The sample has a flux limit of 2.5×10^{-17} erg cm^{-2} s^{-1} in the $0.5-2.0$ keV band and galaxies were classified as NGs if classifying them as AGN, according to at least one of the following properties, was not possible: $f_{0.5-8.0\ \mathrm{keV}}/f_R > 0.1$, $N_H \geq 10^{22}$ cm^{-2}, HR > 0.8, $L_{0.5-8.0\ \mathrm{keV}} > 3\times 10^{42}$ erg s^{-1}, or broad/high-ionisation AGN emission lines.

The *Chandra* XBootes survey This is a *Chandra* survey of the 9 deg^2 Bootes field of the NOAO Deep Wide-Field Survey [NDWFS; 13], consisting of 126 contiguous ACIS-I observations of ~ 5 ks each and reaching a flux limit of $\sim 5 \times 10^{-15}$ erg cm^{-2} s^{-1} in the $0.5-2.0$ keV band. We cross-correlated the point source catalogue of Kenter et al. [13] for this survey with the SDSS (DR-3). Sources were classified as NGs if optical counterparts were within 3″ of an X-ray source, $\log(f_X/f_O) < -2$ and the optical SDSS image had a 'GALAXY' classification flag in the SDSS catalogue. Visual inspection was also performed. 47 NGs were identified in this way.

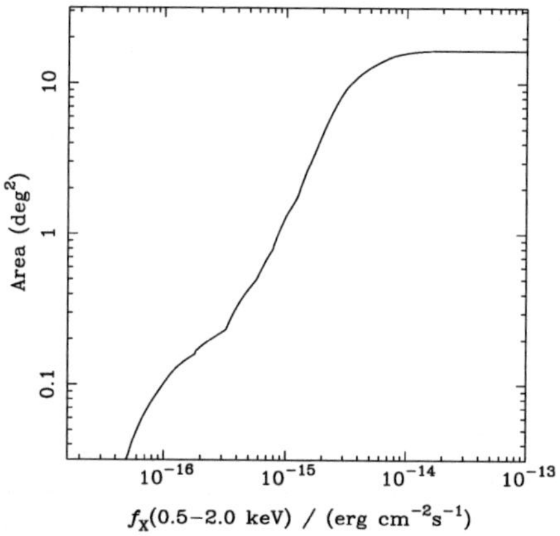

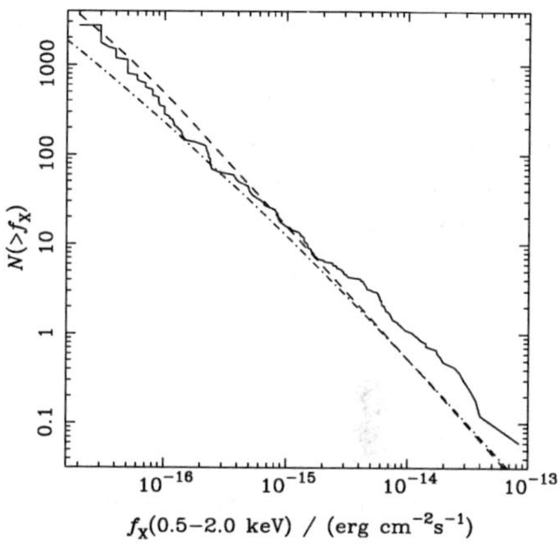

FIGURE 1. *Left*: Solid angle as a function of limiting X-ray flux in the $0.5-2.0$ keV energy band for the surveys used in this work. This is the total sensitivity curve for samples 1XMM+NHS+CDF+XAssist+XBootes. *Right*: Cumulative number counts of NGs in the $0.5-2.0$ keV energy band. The continuous line is the number counts for our combined 1XMM+NHS+CDF+XAssist+XBootes galaxy sample. The dashed and dot-dashed lines are XLF predictions as explained in the text.

RESULTS

Spectroscopic redshifts are available for some of the galaxies in the Xbootes/SDSS correlation sample. We also used the software *Hyperz* [`http://webast.ast.obs-mip.fr/hyperz/`, 14] to estimate photometric redshifts for this sample. We also calculated the X-ray luminosity and $\log(f_X/f_O)$ and give details for this sample in Table 1.

The total area curve for our extended sample is shown in Figure 1 (left). This was used to obtain the cumulative number counts of NGs per deg^2 as a function of X-ray flux in the $0.5-2.0$ keV energy band, shown by the continuous line in Figure 1 (right). The dashed line shows the prediction of the Georgakakis et al. [15] XLF for both early and late type galaxies, assuming the maximum likelihood evolution of the form $\propto (1+z)^p$ with $p = 1.92$. The dot-dashed line is the no evolution ($p = 0$) prediction of the same XLF. It is clear that for fluxes below $f_X \approx 2 \times 10^{-15}$ erg cm^{-2} s^{-1} the prediction which includes evolution is in good agreement with our results. The no-evolution prediction fails both at low and high fluxes. However, both predictions underestimate number counts at high fluxes.

The results shown here are thus based on the largest X-ray selected sample of NGs to date. We are currently analysing deeper fields from the Extended *Chandra* Deep Field South and the Extended Groth Strip in order to bridge the existing gap between shallow and deep observations.

ACKNOWLEDGMENTS

This work is funded in part by the Hellenic National Secretariat for Research and Technology within the framework of the Greece-USA collaboration programme *Study of Galaxies with the Chandra X-ray Satellite*. A. G. thanks PPARC (UK) for support. We acknowledge the use of data from the *XMM-Newton* Science Archive at VILSPA, the *Chandra*-XAssist archive, the High Energy Astrophysics Science Archive Research Center (HEASARC), and the NASA/IPAC Extragalactic Database (NED).

REFERENCES

1. D. M. Alexander, F. E. Bauer, W. N. Brandt, D. P. Schneider, A. E. Hornschemeier, C. Vignali, A. J. Barger, P. S. Broos, L. L. Cowie, G. P. Garmire, L. K. Townsley, M. W. Bautz, G. Chartas, and W. L. W. Sargent, *AJ* **126**, 539–574 (2003).
2. R. Giacconi, A. Zirm, J. Wang, P. Rosati, M. Nonino, P. Tozzi, R. Gilli, V. Mainieri, G. Hasinger, L. Kewley, J. Bergeron, S. Borgani, R. Gilmozzi, N. Grogin, A. Koekemoer, E. Schreier, W. Zheng, and C. Norman, *ApJS* **139**, 369–410 (2002).
3. J. A. Baldwin, M. M. Phillips, and R. Terlevich, *PASP* **93**, 5–19 (1981).
4. S. Veilleux, and D. E. Osterbrock, *ApJS* **63**, 295–310 (1987).
5. P. Tzanavaris, I. Georgantopoulos, and A. Georgakakis (2006), *A&A*, submitted.
6. A. E. Georgakakis, I. Georgantopoulos, S. Basilakos, M. Plionis, and V. Kolokotronis, *MNRAS* **354**, 123–126 (2004).
7. I. Georgantopoulos, A. Georgakakis, and E. Koulouridis, *MNRAS* **360**, 782–790 (2005).
8. A. E. Georgakakis, V. Chamvusyan, M. Plionis, I. Georgantopoulos, E. Koulouridis, I. Leonidaki, A. Mercado, and C. Goudis (2005), *MNRAS*, submitted.
9. A. Ptak, and R. Griffiths, "XAssist: A System for the Automation of X-ray Astrophysics Analysis," in *ASP Conf. Ser. 295: Astronomical Data Analysis Software and Systems XII*, 2003, p. 465.
10. D. P. Schneider, P. B. Hall, G. T. Richards, D. E. Vanden Berk, S. F. Anderson, X. Fan, S. Jester, C. Stoughton, M. A. Strauss, M. SubbaRao, W. N. Brandt, J. E. Gunn, B. Yanny, N. A. Bahcall, J. C. Barentine, M. R. Blanton, W. N. Boroski, H. J. Brewington, J. Brinkmann, R. Brunner, I. Csabai, M. Doi, D. J. Eisenstein, J. A. Frieman, M. Fukugita, J. Gray, M. Harvanek, T. M. Heckman, Ž. Ivezić, S. Kent, S. J. Kleinman, G. R. Knapp, R. G. Kron, J. Krzesinski, D. C. Long, J. Loveday, R. H. Lupton, B. Margon, J. A. Munn, E. H. Neilsen, H. J. Newberg, P. R. Newman, R. C. Nichol, A. Nitta, J. R. Pier, C. M. Rockosi, D. H. Saxe, D. J. Schlegel, S. A. Snedden, A. S. Szalay, A. R. Thakar, A. Uomoto, W. Voges, and D. G. York, *AJ* **130**, 367–380 (2005).
11. J. T. Stocke, S. L. Morris, I. M. Gioia, T. Maccacaro, R. Schild, A. Wolter, T. A. Fleming, and J. P. Henry, *ApJS* **76**, 813–874 (1991).
12. F. E. Bauer, D. M. Alexander, W. N. Brandt, D. P. Schneider, E. Treister, A. E. Hornschemeier, and G. P. Garmire, *AJ* **128**, 2048–2065 (2004).
13. A. Kenter, S. S. Murray, W. R. Forman, C. Jones, P. Green, C. S. Kochanek, A. Vikhlinin, D. Fabricant, G. Fazio, K. Brand, M. J. I. Brown, A. Dey, B. T. Jannuzi, J. Najita, B. McNamara, J. Shields, and M. Rieke, *ApJS* **161**, 9–20 (2005).
14. M. Bolzonella, J.-M. Miralles, and R. Pelló, *A&A* **363**, 476–492 (2000).
15. A. E. Georgakakis, V. Chamvusyan, M. Plionis, I. Georgantopoulos, E. Koulouridis, I. Leonidaki, and A. Mercado (2006), *MNRAS*, in press.

TABLE 1. The XBootes/SDSS correlation sample. From left to right columns give the source ID, right ascension in X-rays, declination in X-rays, SDSS Petrosian r magnitude, X-ray flux in units of 10^{-14} erg cm^{-2} s^{-1}, SDSS spectroscopic redshift (where available), photometric redshift, logarithmic X-ray luminosity, X-ray–to–optical flux ratio, hardness ratio (where available). Asterisks denote upper limit values.

ID	α_X	δ_X	r_{SDSS}	$f_{14,X}$	z_{spec}	z_{phot}	$\log L_X$	$\log(f_X/f_O)$	HR
1	216.332875	0.549562	17.712	0.2*		0.177	41.2217*	-2.1978	
2	216.334875	2.148504	18.355	0.58		0.010	39.1221	-2.3736	
3	216.384708	0.836659	17.668	0.2*		0.169	41.1689*	-2.4523	
4	216.386333	1.472453	17.794	0.14		0.160	40.9574	-2.4994	
5	216.572458	1.514476	18.611	0.51		0.248	41.9441	-2.1099	
6	216.649042	1.300395	17.134	0.57	0.077	0.031	40.8969	-2.7416	
7	216.684708	1.220260	16.96	1.76	0.173	0.139	42.1401	-2.0998	
8	216.748542	1.211570	15.508	2.46		0.010	39.7496	-2.7373	-1.00
9	216.786208	0.581452	12.392	1.97		0.019	40.1909	-4.0634	-0.78
10	216.797042	1.641792	15.973	0.50		0.097	41.0495	-3.2234	
11	216.844125	0.900225	18.753	0.26		0.252	41.6701	-2.0044	
12	216.953625	0.368677	17.575	0.14		0.167	41.0126	-2.5426	
13	217.057583	2.632075	17.477	0.44		0.159	41.454	-2.2938	
14	217.209458	0.310143	16.564	0.26		0.103	40.8325	-2.7739	
15	217.225875	0.786772	16.678	1.14		0.123	41.6358	-2.4391	-0.48
16	217.256542	0.284983	17.518	0.2*	0.129	0.159	40.9171*	-2.2877	
17	217.338583	1.093142	18.419	0.46		0.083	40.8675	-2.2239	
18	217.448417	0.564645	13.927	4.00	0.028	0.015	40.8313	-2.7935	0.20
19	217.460917	1.373583	16.292	0.61		0.108	41.2458	-3.0340	
20	217.513250	1.612435	17.051	0.15	0.126	0.139	40.7901	-2.5584	
21	217.683250	1.198647	18.425	0.64		0.217	41.9186	-2.1319	
22	217.748083	1.360613	18.501	0.15		0.219	41.2898	-2.1867	
23	217.765208	2.320921	17.119	0.61		0.154	41.566	-2.5316	
24	217.854500	1.682344	18.052	0.46		0.010	39.0214	-2.4401	
25	217.908375	1.171117	17.889	0.16		0.181	41.1502	-2.3721	
26	217.980333	2.402555	18.439	0.30		0.216	41.5889	-2.1347	
27	218.178417	2.776426	18.853	0.43		0.191	41.6297	-2.0962	
28	218.240042	0.862550	18.307	0.13		0.219	41.2277	-2.1629	
29	218.327125	0.580550	14.007	8.38	0.034	0.010	41.3346	-2.7157	-0.44
30	218.414875	0.928777	16.264	3.67		0.053	41.3716	-2.1050	-0.18
31	218.428875	2.788963	17.255	0.92	0.209	0.164	42.0328	-2.3439	
32	218.614958	0.704776	17.47	0.74	0.19	0.165	41.8433	-2.2092	
33	218.695000	1.116752	17.833	0.29		0.173	41.357	-2.1387	
34	218.770750	2.193170	17.503	0.73	0.187	0.163	41.8354	-2.2542	
35	218.789333	0.656417	18.277	0.13		0.217	41.2264	-2.2591	
36	218.876292	1.591918	16.955	0.71	0.123	0.112	41.4301	-2.6153	
37	218.934083	0.657431	14.443	1.73	0.029	0.036	40.516	-2.1997	0.87
38	219.005125	0.381335	16.764	0.43	0.127	0.104	41.2482	-2.8981	
39	219.097708	0.592493	17.041	0.82	0.129	0.128	41.5299	-2.5185	
40	219.118625	1.598737	17.289	0.61	0.159	0.143	41.5959	-2.5759	
41	219.163667	0.941165	16.77	0.57	0.128	0.118	41.3713	-2.6837	
42	219.178000	0.392132	15.528	0.96	0.044	0.047	40.613	-2.7546	0.22
43	219.291625	0.426897	17.801	0.76	0.127	0.118	41.4955	-2.3346	
44	219.319958	0.854484	16.901	0.72	0.086	0.059	41.0964	-2.7254	
45	219.428542	0.938571	17.974	0.59		0.180	41.6935	-2.1817	
46	219.449083	2.274069	16.766	0.59		0.145	41.4896	-2.6639	
47	219.597875	2.844940	17.656	0.88	0.204	0.169	41.9905	-2.1300	-0.21

Dust along a merging galaxy sequence

A. Georgakakis, *, E. M. Xilouris, *, A. Misiriotis, † and V. Charmandaris, **

*Institute of Astronomy & Astrophysics, National Observatory of Athens, I. Metaxa & Vas. Pavlou, Palaia Penteli, 15236, Athens, Greece
†University of Crete, Physics Department, PO Box 2208, 71003 Heraklion, Crete, Greece
**Astronomy Department, Cornell University, Ithaca, NY 14853, USA

Abstract. We investigate the cold and warm dust properties during galaxy interactions using a merging galaxy sample ordered into a chronological sequence from pre- to post-mergers. Our sample comprises a total of 29 merging systems selected to have far-infrared and sub-millimeter observations. We use the 100–to–850μm flux density ratio, f_{100}/f_{850}, as a proxy to the mass fraction of the warm and the cold dust in these systems. We find evidence for an increase in f_{100}/f_{850} along the merging sequence from early to advanced mergers and interpret this trend as an increase of the warm relative to the cold dust mass. We argue that the two key parameters affecting the f_{100}/f_{850} flux ratio is the star-formation rate and the dust content of individual systems relative to the stars. Using a sophisticated model for the absorption and re-emission of the stellar UV radiation by dust we show that these parameters can indeed explain both the increase and the observed scatter in the f_{100}/f_{850} along the merging galaxy sequence.

INTRODUCTION

One of the revolutionary results from the SCUBA bolometer array on the JCMT has been the direct detection at submillimeter (submm) wavelengths ($450, 850\mu m$) of cold dust (T=15-25 K) in isolated quiescent spiral galaxies that dominates the overall dust content. This component remained undetected by IRAS (60, 100 μm) that probes warmer dust (T> 30K) and is only detectable at longer submm wavelengths.

In more active interacting systems submm observations also suggest the presence of cold cirrus. Unlike quiescent spirals however, mergers also show a prominent warm dust component (T=30-60 K) most likely heated by the enhanced star-formation.

In this study we explore the role of galaxy interactions in heating the dust and hence, modifying the relative amounts of the cold and the warm dust component in galaxies. Our interacting galaxy sample comprises well separated spirals, systems close to nuclear coalescence and young merger remnants.

THE SAMPLE

The interacting galaxy sample used here is compiled from the Georgakakis et al. [1] study and comprises disk galaxy mergers of similar mass spanning a wide range of interactions stages: well separated spirals, systems close to nuclear coalescence and

CP848, *Recent Advances in Astronomy and Astrophysics,*
7th International Conference of the Hellenic Astronomical Society, edited by N. H. Solomos

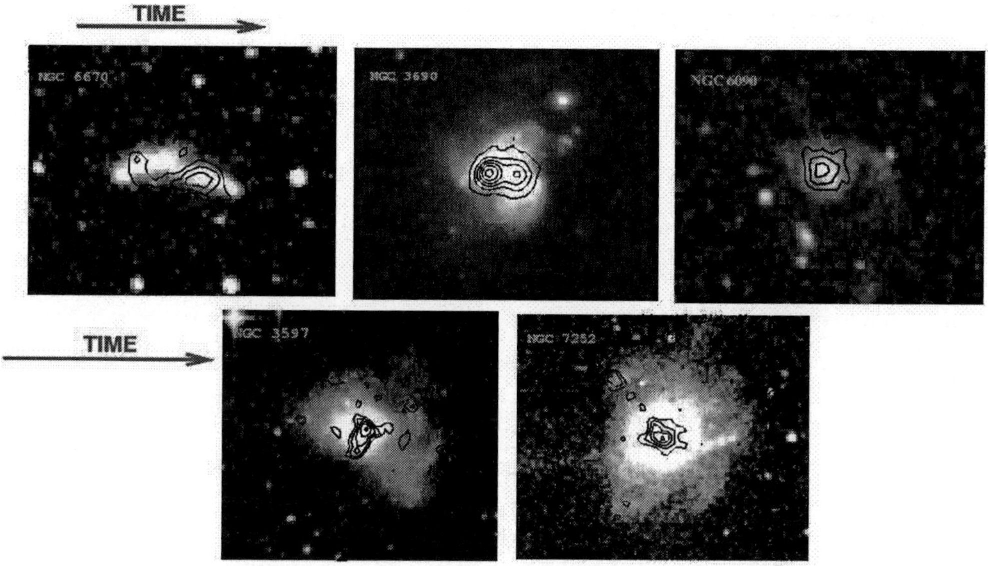

FIGURE 1. An example of merging galaxy sequence comprising well separated interacting galaxies on the left, progressively more advanced mergers (from left to right) and mergers remnants (bottom right). The merging evolves from left to right. The contours are submm 850 μm data plotted over the optical DSS images.

young merger remnants. All the systems employed here have submm data available from the SCUBA archive.

To explore the evolution of the galaxy dust properties at different times during the interaction we order our merging systems into a chronological sequence. For this we use the 'age' parameter providing an estimate of the evolutionary stage of each system relative to the time of the merging of the two nuclei. Negative 'ages' are for pre-mergers while positive 'ages' correspond to merger remnants. An example of a merging sequence is shown in Figure 1.

For pre-mergers the 'age' is estimated by dividing the projected separation of the two nuclei by an (arbitrary) orbital decay velocity v=30km/s. For post-mergers we adopt the evolutionary sequence defined by Keel & Wu [2] using dynamical and morphological criteria. It should be stressed that the 'age' parameter for both pre- and post-mergers does not represent an absolute galaxy age but is an indicator of the evolutionary stage of the interaction.

RESULTS

We quantify the dust properties of our sample using the 100–to–850 μm flux density ratio, f_{100}/f_{850}. This provides an estimate of the relative mass fraction of the warm and the cold dust component. The submm is sensitive to cold dust (T=15-25 K) with the

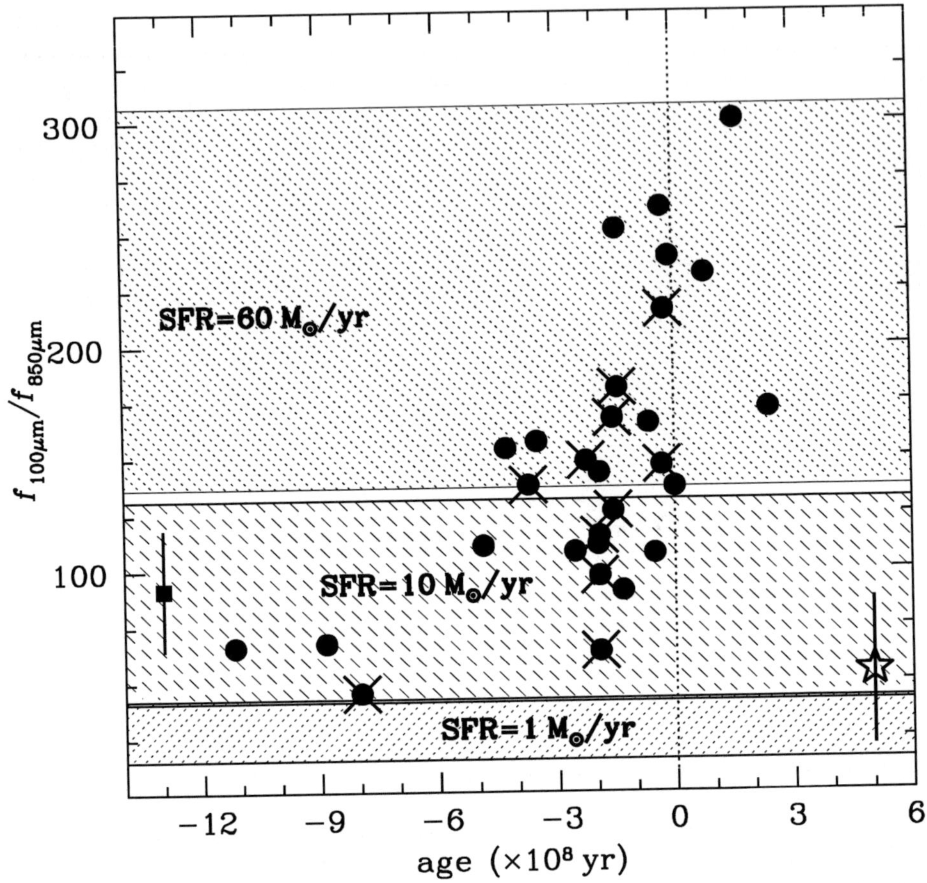

FIGURE 2. f_{100}/f_{850} flux ratio against the age parameter for our sample of interacting galaxies. The filled circles are the interacting galaxy sample in this study. A cross on top of symbol signifies an AGN in the catalogue of Veron-Cetty & Veron [4]. The square corresponds to the mean f_{100}/f_{850} of isolated spirals from the sample of Misiriotis et al. [3] arbitrarily plotted at an age parameter of –13. The star is the mean f_{100}/f_{850} of ellipticals from the sample of Temi et al. [5] arbitrarily plotted at an age parameter of +5.

far-infrared (FIR) probing the warmer dust (T> 30K).

Figure 2 plots the ratio f_{100}/f_{850} as a function of the age parameter for our interacting galaxy sample. There is evidence for an increase of f_{100}/f_{850} as the interaction progresses from well separated spirals toward the final stages of the merger. The Spearman rank test suggests that the correlation is significant at 3.5 σ confidence level. This statistically significant correlation is not due to f_{100} or f_{850} being individually correlated with the age parameter.

This trend can be interpreted as an increase of the warm relative to the cold dust mass from isolated spirals and early interacting systems to advanced mergers. Such a transition of the cold dust to warmer temperatures can be attributed to heating by the more intense UV radiation field of the enhanced star-formation triggered by the merging.

We attempt to interpret the trend between f_{100}/f_{850} and age using the model of Misiriotis et al. [3] that describes the absorption of UV/optical photons by dust and their re-emission at FIR and submm wavelengths. This model produces UV to submm SEDs for a given input system geometry, stellar population (e.g. young and old stars), dust content and total galaxy mass.

The dust mass of the fiducial galaxy is assumed to be in the range $10^7 - 10^8$ $M_\odot$, similar to the dust masses of our systems. The stellar content is directly associated with the density of the UV radiation responsible for the heating of the dust and is parameterized by the star-formation rate. Here we use three different values for the star-formation rate SFR=1, 10 and 60 $M_\odot$/yr. For a given SFR and dust mass we derive the f_{100}/f_{850} ratio predicted by the model. The model predictions for different SFRs are shown with the shaded regions in Figure 2. The lower and upper boundaries of these regions correspond to dust masses of 10^8 and 10^7 $M_\odot$ respectively.

The model predicts that higher SFRs heat the dust resulting in higher f_{100}/f_{850} ratios. Also, keeping the SFR fixed, more dusty systems have lower f_{100}/f_{850} which can be attributed to more efficient warming of the dust, i.e. higher fraction of UV/optical photons absorbed by the dust, heating it to temperatures T$\leq$ 30K and then being re-emitted primarily at submm wavelengths (850μm). This may also partly explain the scatter seen in the figure since the relative fraction of the warm–to–cold dust for a given SFR varies substantially with dust content.

CONCLUSIONS

Using a merging galaxy sequence comprising early interacting spirals, systems close to nuclear coalescence and merger remnants we explore the evolution of dust properties during galaxy mergers. We find evidence for dust heating from well separated young mergers to advanced interacting galaxies. Using a model that describes the absorption of the light by dust and its re-mission to FIR and submm wavelengths we argue that the observed trend can be attributed to the enhanced star-formation rate during interactions as well as differences in the dust masses of individual systems.

REFERENCES

1. Georgakakis, A., Forbes, D.A., Norris, R.P., 2000, MNRAS, 318, 124
2. Keel, W.C., Wu, W., 1995, AJ, 110, 129
3. Misiriotis A., Papadakis I.E., Kylafis N.D., Papamastorakis J., 2004, A&A, 417, 39
4. Veron-Cetty M.P., Veron P., 2003, A&A, 412, 399
5. Temi P., Brighenti F., Mathews W.G., Bregman J.D., 2004, ApJS, 151, 237

HII Region Detection in Local Group Census

E. Tremou*, D. Mislis* and J. H. Seiradakis*

*Aristotle University of Thessaloniki, Department of Physics, Section of Astrophysics, Astronomy and Mechanics, GR-541 24 Thessaloniki, Greece

Abstract. We have detected several new HII regions in two Local Group galaxies (IC10 and Sex B). The *hotstars* method, was used to define the HII regions. We also present preliminary results about some physical parameters of these galaxies.

Keywords: dwarf galaxies, HII regions, Local Group, IC 10, Sextan B
PACS: 98.58.Hf

INTRODUCTION

Most of the galaxies in the Local Group are dwarf irregulars and spheroidals. IC 10 is a highly obscured galaxy (E (B-V) $\cong 0.85$, Satai et.al 1999) located at a low galactic latitude $b = -3^o.3$. It is the only starburst galaxy in the Local Group and the presence of a large number of HII regions proves that it is undergoing massive star formation. Sextans B is a dwarf irregular galaxy (Irr IV-V morphological type). In spite of a probable common formation, Sextans B shows a different star formation history as indicated by the different amounts of stars in the various evolutionary phases and also reflected in the different number of PNe and HII regions observed.

FIGURE 1. The Wide Field Camera

OBSERVATIONS AND DATA REDUCTION

The observations were made with 2.5m Isaac Newton Telescope (INT) and the prime focus Wide Field camera (WFC - Figure 1). The pixel size of the WFC is 0.33 arc/pixel, the gain is 2.8 electrons / ADU and the read-out noise 6.2 electrons. We used data from

CP848, *Recent Advances in Astronomy and Astrophysics,*
7th International Conference of the Hellenic Astronomical Society, edited by N. H. Solomos

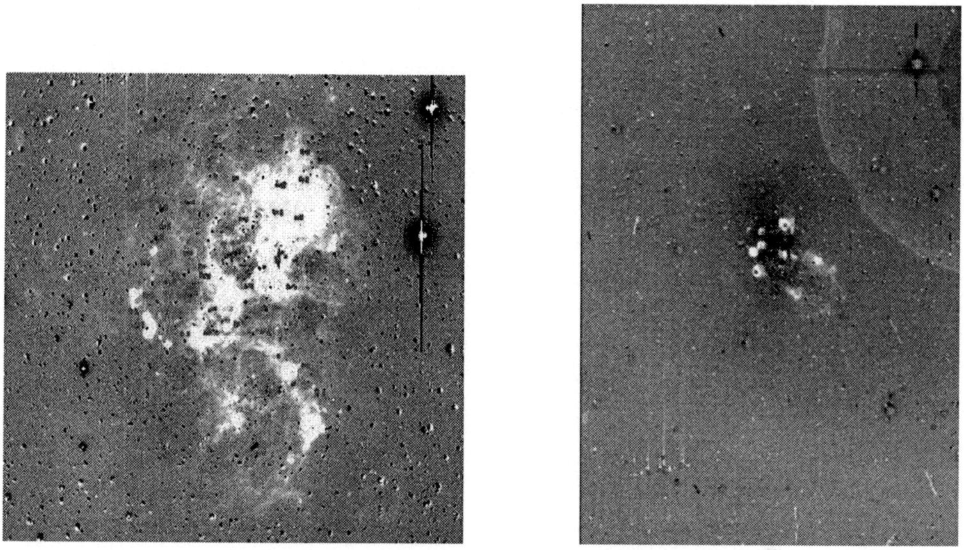

FIGURE 2. Blue Stars in IC10 and Sextan B

the INT observed in 2001 and 2002. The exposure time was about 1200 seconds per frame for the narrow band filters and 400 seconds for the broad band filters. We used the filters Ha, SII, stY, OIII, HeII, r, i and g at $\lambda 6568/95$Å,$\lambda 6725/80$Å, $\lambda 5505/240$Å, $\lambda 5008/100$Å, $\lambda 4686/100$Å, $\lambda 6380/1520$Å, $\lambda 6240/1347$Å, $\lambda 4846/1285$Å, respectively. The Ha filter includes the NII line and the OIII includes 52.9% of the OIII $\lambda 4959$Å. The data reduction was performed with the IRAF software package. The frames were debiased, flat-fielded and linearity corrected using the ING WFC data reduction pipeline (Irwinand Lewis 2001). Then we corrected for geometrical distortions and aligned all frames to the [Ha] one. The next step is continuum substraction. The aim is to substract the continue emission from the observations taken through the narrow band filters, in order to leave only the flux emitted by the spectra emission lines of our interest. Then we used the DAOPHOT package and some spectra-photometric stars for the flux callibration.

HOT STARS METHOD

The HII regions are clouds with ionized hydrogen. The clouds are usually ionized by the radiation of one (in small regions) or more (may be clusters, in big regions) hot stars. The luminosity of each region is a function of the number of hot stars. We detected the HII regions by plotting isophot contour lines at a pre-determined level, using the F.O.C.A.S. package of IRAF. Then, in order to separate the individual HII regions found in complex

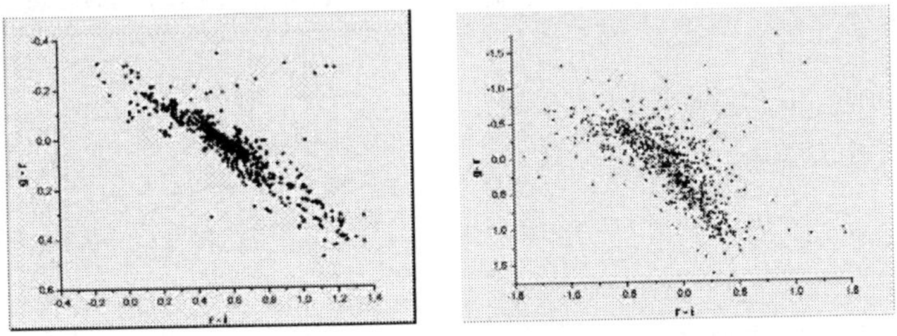

FIGURE 3. Color-Color Diagram of IC10 and Sextan B

and blended regions, we seeked and studied the distribution of hot (blue supergiant) stars in the vicinity, selected by the color-color diagram (Figure 3). In subsequent observations extinction was ignored, without any significant effect in our results.

RESULTS

We have detected 70 HII regions in IC 10 galaxy and 44 HII regions in Sex B galaxy. We estimated some characteristics of each region in all filters, shown in Figure 4.

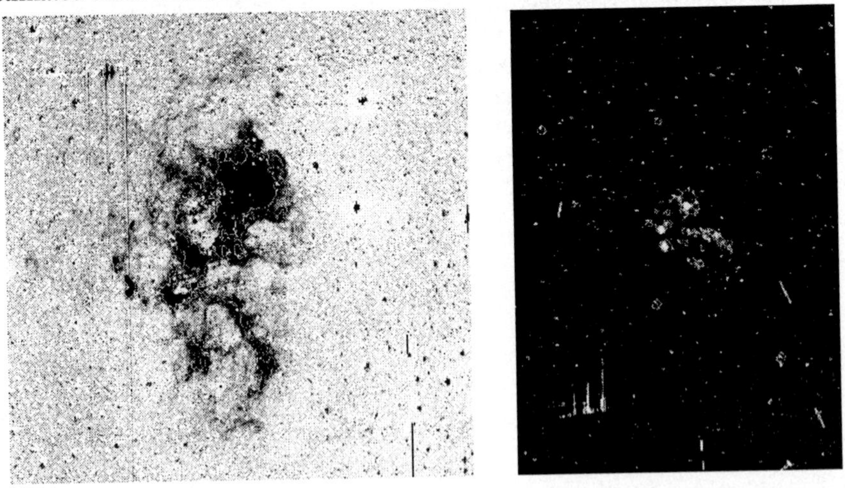

FIGURE 4. F.O.C.A.S results in IC10 and Sextan B

The space occupied by a region is a logarithmic function of the flux density, obeying the following law: $logS \sim log(r^3)$, where S stands for the flux density and r stands for

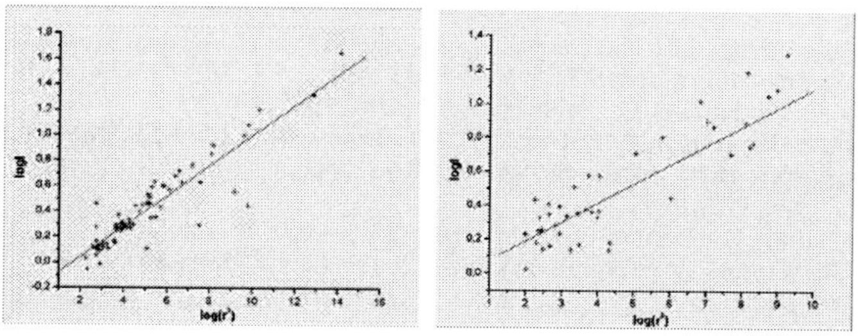

FIGURE 5. Luminosities vs Radius of HII regions in IC10 and Sextan B

the radius of the region (Figure 5). We plan to extend our studies in order to investigate the relation between the luminosity function of each region and the number of blue stars (Figure 6) in the vicinity of the region.

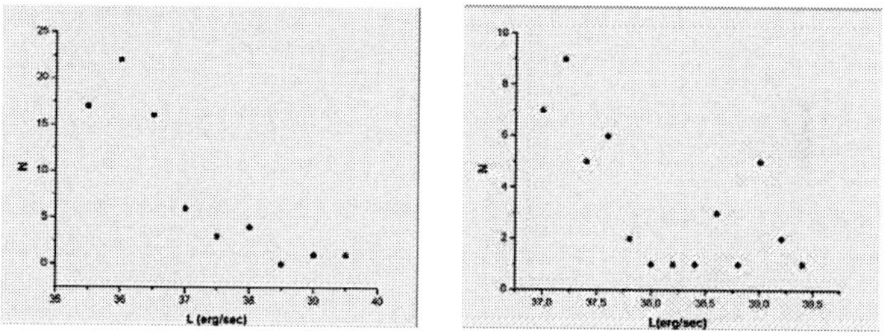

FIGURE 6. Luminosity Function of IC10 Sextan B

ACKNOWLEDGMENTS

We are grateful to the Isaac Newton Group (ING) for providing us the data.

REFERENCES

1. Vílchez, J. M., & Pagel, B. E. J. 1988, MNRAS, 231, 257
2. Paul Hodge, R.C. Kennicut, JR & Myung Cyoon Lee 1988, PASP, 100, 917-934
3. Paul Hodge, R.C. Kennicut, JR & Myung Cyoon Lee 1989, PASP, 101, 32-39
4. Paul Hodge, R.C. Kennicut, JR & Myung Cyoon Lee 1994, AJSS, 92, 119-123
5. R.J Dufour & D. L. Talent 1980, ASJ, 235, 22-29
6. R. Terlevich, M.S. Portal, A. I. Díaz & E. Terlevich, MNRAS, 249, 36-45
7. C.R O'Dell, P.W. Hodge & R.C Kennicut 1999 PASP, 111, 1382-1392

8. M.S. Portal, A.I. Díaz, R.Terlevich, E.Terlevich, M.A .Alvarez & I.Aretxaga 2000, MNRAS, 312, 2-32
9. C.W. McAlary, B.F. Madore, R. McGonegal, R.A. McLaren & D.L. Welch 1983, ASJ, 273, 539-543
10. G. Catanzaro, L. Bianchi, S. Scuderi & A. Manchado 2003, A&A, 403, 111-117
11. B.E.J. Pagel, M.G. Edmunds & G. Smith 1980, MNRAS, 193, 219-230
12. J.G. Nollenberg, E.D. Skillman, D.R. Garnett & H.L. Dinerstein 2002, ASJ, 581, 1002-2002
13. E.D. Skillman, R. Telverich, J. Melnick 1989, MNRAS, 240, 563-572
14. E.L. Fitzpatrick 1999, PASP, 111, 63-75
15. A.M. Hidalgo-Gamez, K. Olofsson & J. Masegosa 2001, A&A, 367, 388-404
16. R. Greimel, N.A. Walton, D.C Abrams, M.J Irwin, & J.R. Lewis, 2001, ASP Conf. Ser. 238
17. P. Massey, T.E. Armandroff, R. Pyke, K. Patel & C.D. Wilson, 1995, ASJ, 110, 2715-2738

PART 4

THEORETICAL INSIGHTS IN DYNAMICAL ASTRONOMY, RELATIVITY, AND COSMOLOGY

Celestial Mechanics
Galactic Dynamics
Relativistic Astrophysics
Cosmology

(Conveners: J. D. Hadjidemetriou, K. Kokkotas, and S. Persides)

Solar System dynamics, beyond the two-body-problem approach

Harry Varvoglis

Department of Physics, University of Thessaloniki, GR-541 24 Thessaloniki, Greece

Abstract. When one thinks of the solar system, he has usually in mind the picture based on the solution of the two-body problem approximation presented by Newton, namely the ordered clockwork motion of planets on fixed, non-intersecting orbits around the Sun. However, already by the end of the 18th century this picture was proven to be wrong. As discussed by Laplace and Lagrange (for a modern approach see [3] or [2]), the interaction between the various planets leads to secular changes in their orbits, which nevertheless were believed to be corrections of higher order to the Keplerian elliptical motion.

This idea has changed completely the last decades. Now it is well know that the solar system was created from a state of chaotic interactions of planetesimals, primordial bodies the size of a small asteroid, and that since this time many episodes of cataclysmic collisions have shaken all major planets, due to the pronounced chaotic motion of the minor bodies. A new discipline has emerged out of the above new ideas, which is based on the statistical approach to chaotic motion of bodies, in particular those in the asteroid belt. At the same time it has been understood that non-gravitational forces, in particular the Yarkovsky effect, may play an important role on the long-time evolution of the trajectories of kilometer-sized bodies.

Keywords: Secular theory, three-body problem, chaos, Fokker-Planck equation, Yarkovsky effect
PACS: 05.10.Gg, 95.10.Ce, 95.10.Fh, 96.30.Ys

1. KEPLERIAN MOTION (2 BODIES)

1.1. Newtonian approach

The equations of motion of one planet in orbit around the Sun are given by substituting the force of gravitational attraction between the two bodies

$$\vec{F} = \mathscr{G}\frac{\mathscr{M} m}{r^3}\vec{r} \tag{1}$$

into Newton's second law of motion

$$\vec{F} = m\vec{a} = m\frac{d^2\vec{r}}{dt^2} \tag{2}$$

where $\mathscr{M}$ and m define, respectively, the masses of the Sun and the planet, $\mathscr{G}$ the gravitational constant and r the distance between the bodies.

It was known since the time of Newton that the above differential equation can be solved exactly. The bounded solutions of a *test particle* moving around the Sun are *ellipses*, with semi-major axis

$$a = -\frac{\mathscr{G}\mathscr{M}}{2E} \tag{3}$$

CP848, *Recent Advances in Astronomy and Astrophysics,*
7th *International Conference of the Hellenic Astronomical Society,* edited by N. H. Solomos

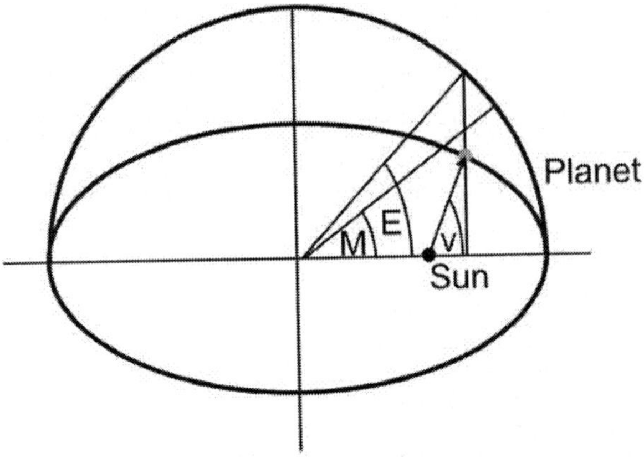

FIGURE 1. Definition of true (v), eccentric (E) and mean (M) anomaly.

and eccentricity

$$e = \sqrt{1 + \frac{2Eh^2}{\mathcal{G}^2 \mathcal{M}^2}} \tag{4}$$

where h denotes the angular momentum and E the energy (with $E < 0$). The orbit on the plane of the ellipse is given, in planar polar co-ordinates, by the function

$$r = \frac{a(1-e^2)}{1+e\cos(v)} \tag{5}$$

where the polar angle v is the *true anomaly*. The position of the planet on its orbit is given by the function $v(t)$, which is known in an *indirect way*. The true anomaly, v, is related to the *eccentric anomaly*, u, through the relation (see Fig. 1)

$$\cos(v) = \frac{\cos(u) - e}{1 - e\cos(u)} \tag{6}$$

The function $u(t)$, in turn, is found by solving *Kepler's equation*

$$u - e\sin(u) = n(t - t_0) \tag{7}$$

where t_0 is the *time of perihelion passage* and the *mean orbital frequency* (or *mean motion*) of the planet is given by Kepler's third law

$$n = \sqrt{\mathcal{G}\mathcal{M}}\, a^{-3/2}. \tag{8}$$

The study of an integrable conservative dynamical system is greatly facilitated if one goes to Hamiltonian formalism and writes the corresponding Hamiltonian in a special set of canonical co-ordinates (I_i, θ_i), known as *action-angle variables*. Then the "new"

Hamiltonian is in *normal form*, i.e. it depends only on the actions, I_i. In this case, as it is evident from Hamilton's equations of motion,

$$\frac{dI_i}{dt} = -\frac{\partial \mathcal{H}}{\partial \theta_i}, \frac{d\theta_i}{dt} = \frac{\partial \mathcal{H}}{\partial I_i} \tag{9}$$

the actions, I_i, are constants of the motion and the angles, θ_i, are linear functions of time. A simple inspection shows that neither v nor u are increasing linearly with time. The only angle with this property is the *mean anomaly*,

$$M = n(t - t_0) \tag{10}$$

and, as we will see in what follows, it is indeed the conjugate angle co-ordinate of an action-angle pair that transforms the Hamiltonian of the problem to a normal form. The "physical" interpretation of this angle is that it measures the percentage of the area of the ellipse spanned by the position vector within time $(t - t_0)$ (as the motion takes place at constant area velocity).

1.2. Hamiltonian approach

The Hamiltonian of the two-body problem in a heliocentric[1] co-ordinate system and in spherical co-ordinates (ρ, θ, ϕ) is:

$$\mathcal{H} = \frac{1}{2}\left(p_\rho^2 + \frac{p_\theta{}^2}{\rho^2} + \frac{p_\phi^2}{\rho^2 \sin^2\theta}\right) - \frac{\mathcal{G}\mathcal{M}}{\rho} \tag{11}$$

where, as in the previous section, we treat the planet as a test particle. The planet follows an elliptical orbit on the x-y plane (usually referred to as *invariant plane*), with the x-axis pointing from the Sun to the vernal equinox, γ, and the z-axis pointing to the north pole of the celestial sphere (see Fig. 2). The position of the planet, at each time, is fully defined by 6 constants, the so-called *orbit elements*: the *semi-major axis*, a, the *eccentricity*, e, the *inclination*, i, the *longitude of the ascending node*, Ω, the *argument of perihelion*, ω, and the *epoch of perihelion passage*, t_0. The first two constants give the shape of the ellipse, the next three are the Euler angles and give the *orientation* of the ellipse in space (Fig. 2) and the last one, in connection with eq.(8), gives the mean anomaly of the planet.

The complete set of canonical action-angle variables for the two-body problem are the Delaunay actions L, G, H and angles l, g, h, which are defined by the relations

$$L = \sqrt{\mathcal{G}\mathcal{M}a} \quad l = M \tag{12}$$

$$G = \sqrt{\mathcal{G}\mathcal{M}a(1-e^2)} \quad g = \omega \tag{13}$$

---

[1] Note that if we had selected a *barycentric* system instead, we should change its origin every time we would like to "add" a new body in our model!

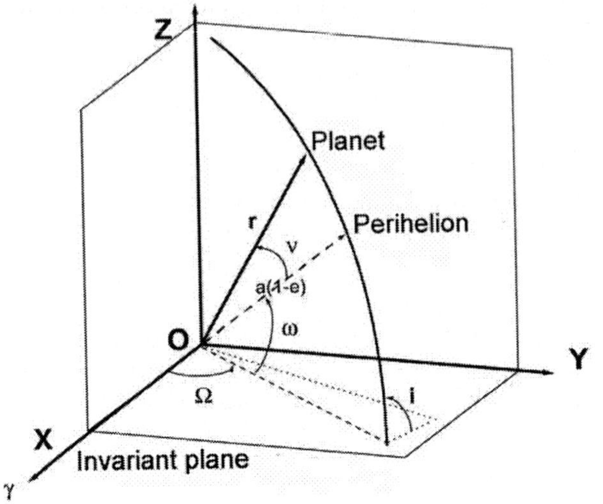

FIGURE 2. Definition of the six orbital elements.

$$H = \sqrt{\mathcal{G}\mathcal{M}a(1-e^2)}\cos i \quad h = \Omega \tag{14}$$

The Hamiltonian, in these variables, becomes simply

$$\mathcal{H}^* = -\frac{\mathcal{G}^2\mathcal{M}^2}{2L^2} \tag{15}$$

The angle $l = n(t - t_0)$, which is increasing linearly with time, gives the *position* of the body on the ellipse, while the other two angles are constants. The Delaunay variables, however, have a weak point, since the angle ω is not defined for $i = 0$ and the angle M is not defined for $e = 0$. This problem is solved if we go to *modified* Delaunay variables:

$$\Lambda = \sqrt{\mathcal{G}\mathcal{M}a} \quad l = \varpi + M \tag{16}$$

$$\Gamma = \Lambda(1 - \sqrt{1-e^2}) \quad \gamma = -\varpi \tag{17}$$

$$Z = \Lambda\sqrt{1-e^2}(1 - \cos i) \quad \zeta = -\Omega \tag{18}$$

which are, as well, action-angle variables for the two-body problem. The two new angles are the *longitude of perihelion*, $\varpi = \Omega + \omega$, and the *mean longitude*, $\lambda = \varpi + l$.

In these variables the Hamiltonian has the same functional form as eq. (15)

$$\mathcal{H}^* = -\frac{\mathcal{G}^2\mathcal{M}^2}{2\Lambda^2} \tag{19}$$

In this form becomes obvious the most important property of the two-body problem, the *degeneracy*, since the Hamiltonian depends only on one (out of three) action. The frequencies $\dot{\varpi}$ and $\dot{\Omega}$ are equal to 0 and, therefore, *besides* the three actions, there exist two more integrals of motion: the orientation angles ϖ and Ω. As a result of the degeneracy, there is only *one* non-zero frequency,

$$\dot{\lambda} = \frac{\partial \mathscr{H}^*}{\partial \Lambda} \tag{20}$$

which is independent of Γ and Z. That is why all bounded orbits of the two body problem are periodic and the period of the orbit does not depend on eccentricity or inclination.

2. THE RESTRICTED THREE-BODY PROBLEM

The next, just *slightly* more complicated than the two body problem, is the model of the *restricted three-body* problem. In this model our planetary system is approximated by three bodies: one large (the Sun), one small (Jupiter) in a *fixed* (circular or elliptic) orbit around the Sun and a massless body (a small planet or an asteroid)[2]. The problem posed is to find the trajectory of the third, massless body. The corresponding Hamiltonian (again in a heliocentric frame of reference) is:

$$\mathscr{H} = \left(\frac{\vec{p}^2}{2} - \frac{\mathscr{G}\mathscr{M}}{r}\right) - \mathscr{G}m'\left(\frac{1}{\| \vec{r} - \vec{r}' \|} - \frac{\vec{r}\cdot\vec{r}'}{r'^3}\right) = \mathscr{H}^* - \mathscr{G}m'\mathscr{R} \tag{21}$$

where p and r are the momentum and position of the third body, $\mathscr{M}$ is the mass of the Sun, m' and r' the mass and position of Jupiter and $\mathscr{R}$ is the *disturbing function*, which gives the "perturbation" of Jupiter to the orbit of the asteroid. We note that, apart from the disturbing function, the Hamiltonian of the three-body problem is identical to eq.(11).

Depending on whether Jupiter is moving on a circular or elliptic orbit and on whether the third body is restricted to move on the invariant plane or not, the restricted three-body problem may have from two degrees of freedom (circular-planar) to three plus time (elliptic-3D). In all cases we use a "particular" system of units, where $\mathscr{M}+m'=1$, $a'=1$, $\mathscr{G}=1$, $\mu = \frac{m'}{M+m'}$, $\mu_1 = 1-\mu$ and $T' = 2\pi$ or $n' = 1$, so that $\lambda' = n't + \lambda_0'$ (where by primes we denote Jupiter's elements). The modified Delaunay variables for the Hamiltonian (21) in these units become

$$\Lambda = \sqrt{\mu_1 a} \quad l = \varpi + M \tag{22}$$

$$\Gamma = \Lambda(1-\sqrt{1-e^2}) \quad \gamma = -\varpi \tag{23}$$

$$Z = \Lambda\sqrt{1-e^2}(1-\cos i) \quad \zeta = -\Omega \tag{24}$$

and the Hamiltonian in this co-ordinate system is written as

$$\mathscr{H} = -\frac{\mu_1^2}{2\Lambda^2} - \mu\mathscr{R}(\lambda,\gamma,\zeta,\Lambda,\Gamma,Z;\lambda'(t)) \tag{25}$$

[2] The massive bodies are usually named *primaries*.

The equations of motion of the third, massless, body are

$$\dot{\Lambda} = \mu \frac{\partial \mathcal{R}}{\partial \lambda} \quad \dot{\lambda} = \frac{\mu_1^2}{\Lambda^3} - \mu \frac{\partial \mathcal{R}}{\partial \Lambda} \tag{26}$$

$$\dot{\Gamma} = \mu \frac{\partial \mathcal{R}}{\partial \gamma} \quad \dot{\gamma} = -\mu \frac{\partial \mathcal{R}}{\partial \Gamma} \tag{27}$$

$$\dot{Z} = \mu \frac{\partial \mathcal{R}}{\partial \zeta} \quad \dot{\zeta} = -\mu \frac{\partial \mathcal{R}}{\partial Z} \tag{28}$$

First conclusion: We see that the "elements" of the trajectory of the third body are not constants any more, but their variations are of order $\mathcal{O}(\mu) \approx 0.001$. That is why we introduce the notion of *osculating* elements, i.e. elements of an ellipse which is *tangent* to the real trajectory of the third body and which would be followed by the body *if* the second primary would suddenly disappear.

Second conclusion: We see that in the three-body problem there exist two sets of frequencies, which differ by more than one order of magnitude. High frequency (*short period*)

$$\dot{\lambda}^{-1} = \mathcal{O}(1) \approx 3\text{-}8 \text{ yrs}$$

(as it can be calculated from Kepler's law in the region of the asteroid belt) and low frequency (*long period*)

$$g^{-1} = \dot{\gamma}^{-1} \approx s^{-1} = \dot{\zeta}^{-1} \approx \mathcal{O}(\mu) \approx 1{,}000 - 10{,}000 \text{ yrs.}$$

Therefore the perihelion of the osculating ellipse and the ascending node *are varying* with a period $\geq$ 1,000 years.

Third conclusion The problem possesses *resonances*. As we will see, the Fourier series expansion of $\mathcal{R}$ contains trigonometric terms, whose derivatives are of the form

$$\sin\Big(\sum_i k_i \dot{\theta}_i\Big) \tag{29}$$

where k_i are integers (positive, negative or zero) and $\dot{\theta}_i$ are frequencies. Whenever this sum is zero, a *resonance* exists. Resonances, which are *dense* in phase space, are the cause of the famous *small divisors problem*, which introduces *chaos* in the solutions of the problem, as it was shown by Poincaré. As a result, the solution of eqs. (27-29) is *non-analytic* with respect to the initial conditions and, therefore, it *cannot* be written as *power series* with respect to μ. But there was always the need to calculate the motion of the planets. That is why we still use the *secular theory* of the three-body problem developed by Laplace and Lagrange at the end of the eighteenth century, although we know that it leads to non-convergent series solutions, which are valid only for a limited time interval.

3. SECULAR THEORY

3.1. Secular Hamiltonian

The expansion of $\mathscr{R}$ is one of the older problems in Celestial Mechanics[3]. It has the general form:

$$\mathscr{R} = \sum_{p,q,k_i} A_{p,q,k_i}(a,e,i,a',e',i')\cos(p\lambda + q\lambda' + k_1\varpi + k_2\varpi' + k_3\Omega + k_4\Omega') \quad (30)$$

$$= \sum_{p,q,k_i} A_{p,q,k_i}(a,e,i,a',e',i')\cos\psi_{p,q,k_i}$$

where *each* coefficient A_{p,q,k_i} is a power series of $a/a', e, e', sn, sn'$, with $sn = \sin(i/2)$. For every combination of angles in eq.(30), the coefficient of the *leading* term of the corresponding Taylor series has the form

$$A^{\max}_{p,q,k_i} = f_{p,q,k_i}(a/a')e^{|k_1|}e'^{|k_2|}sn^{|k_3|}sn'^{|k_4|} \quad (31)$$

By keeping *only* the terms $A^{\max}_{p,q,k_i}$, the disturbing function, $\mathscr{R}$, takes the form

$$\mathscr{R} = \sum_{p,q,k_i} f_{p,q,k_i}(a/a')e^{|k_1|}e'^{|k_2|}sn^{|k_3|}sn'^{|k_4|}\cos\psi_{p,q,k_i} \quad (32)$$

The secular theory is a method of solving approximately the equations of motion originating from the above Hamiltonian. The first step is to average $\mathscr{H}$ over the "fast" angles λ' and λ. The "averaged" Hamiltonian is

$$\langle\mathscr{H}\rangle_{\lambda,\lambda'} = \mathscr{H}^* - \frac{\mu}{4\pi^2}\int_0^{2\pi}\int_0^{2\pi}\left(\sum_{p,q,k_i} A^{\max}_{p,q,k_i}\cos\psi_{p,q,k_i}\right)d\lambda d\lambda' = \quad (33)$$

$$-\frac{\mu_1}{2\Lambda^2} - \mu\sum_{0,0,k_i} A^{\max}_{0,0,k_i}\cos(k_1\varpi + k_2\varpi' + k_3\Omega + k_4\Omega').$$

Then λ is an ignorable co-ordinate, Λ becomes an integral of motion and, as a consequence, the osculating semi-major axis, a, of the orbit of the third body is *constant*. Therefore we may drop it from the Hamiltonian. Moreover, since the orbit of Jupiter is fixed, Ω' and ω' are constants and we may take them equal to zero as well. Then the Hamiltonian becomes

$$H_{\text{sec}} = \mu\sum_{k_i} A^{\max}_{k_i}\cos(k_1\varpi + k_3\Omega) \quad (34)$$

The appearance of μ as a factor in front of the Hamiltonian has the following consequence. If we *divide* the Hamiltonian by μ, in order to simplify it, we have to *multiply* time by μ, in order to keep the canonical form of the equations of motion. This implies that the "unit" of time is much longer in the secular solution, in agreement with the fact that the corresponding time scale, as we already showed, is of the order of μ^{-1}.

---

[3] J. Pierce, Astron. J. **1**, 1, 1849.

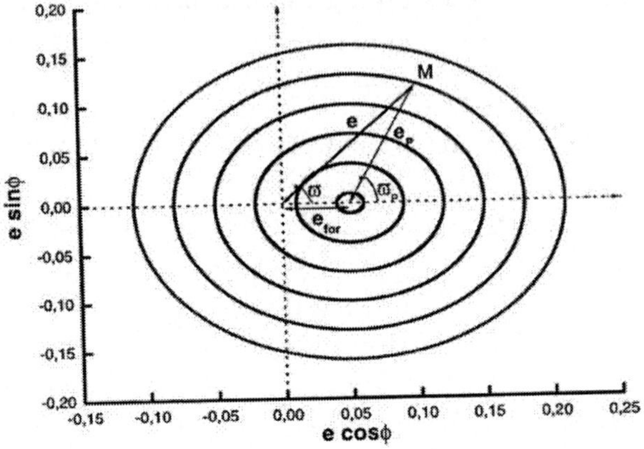

FIGURE 3. The linear solution of the secular theory for the osculating eccentricity

3.2. Linear Secular Theory

In the new Hamiltonian (35), μ is not anymore a "small" parameter. Therefore in order make expansions, we have to use other small parameters, such as the eccentricities and inclinations of the planets which, as a rule, are small. We expand H_{sec} and keep terms A_{ki} up to *second* order in the small quantities e, e', i, i'. This "truncated" Hamiltonian defines a *linear* system of differential equations of motion, whose solution is

$$e\cos\varpi = e_{for} + e_P\cos(gt+\beta) \tag{35}$$

$$e\sin\varpi = e_P\sin(gt+\beta) \tag{36}$$

$$i\cos\Omega = i_{for} + i_P\cos(st+\delta) \tag{37}$$

$$i\sin\Omega = i_P\sin(st+\delta) \tag{38}$$

The quantities e_{for} and i_{for} are called *forced elements* of the trajectory and

- e_f and i_f (named *forced eccentricity* and *inclination*, respectively) depend only (a) on the *motion of primaries* (mass - eccentricity) and (b) on the ratio, a/a' of the semi-major axis of the third body to that of Jupiter,
- e_P and i_P (named *proper eccentricity* and *inclination*, respectively) depend only on the motion of the third body.

The values of e_{for} and i_{for} are proportional, correspondingly, to e' and i'. I.e. in the *planar circular* three body problem we have $e_{for} = i_{for} = 0$.

The solution is better understood graphically (Fig. 3): the osculating eccentricity varies periodically with time and in the plane $(h = e\cos\varpi, k = e\sin\varpi)$ describes a cycle:

$$e^2 = e_f^2 + e_P^2 + 2e_f e_P\cos(gt+\beta) \tag{39}$$

A similar relation holds for the osculating inclination. In this way we see that, in this approximation, eccentricity and inclination are evolving independently. They are *decoupled*.

The quantities e_P and i_P are *integrals of motion* and are called *free* or *proper elements*. The variables Γ, Z are *not* actions for the *secular problem*. But we can define, through a series of canonical transformations, *new* actions

$$\Gamma = \Gamma(e_P), \qquad Z = Z(i_P) \tag{40}$$

which, together with the angles

$$\gamma_P = -\varpi_P = -(gt + \beta) \tag{41}$$

$$\zeta_P = -\Omega_P = -(st + \delta) \tag{42}$$

define a set of action-angle variables. The quantities (g, s) are the *proper frequencies* of the motion of the third body. The values of g and s depend *only* on a/a' and μ.

The method of *averaging* is the first step of an algorithm, which calculates the *normal form* of the Hamiltonian of the three-body problem. With this algorithm we construct, step by step, a canonical transformation, such that in the new variables the "new" Hamiltonian is a function of the "new" actions only. Therefore the new canonical variables are *action-angle variables* of the new Hamiltonian and the equations of motion are solved trivially. The "bad news" is that, in this transformation, which involves a power series in μ, e, e', i, i', appear *small divisors*, i.e. resonant combinations of the frequencies

$$\sum_i k_i \dot{\theta}_i = 0. \tag{43}$$

Therefore the transformation is only *formal*, in that the series does not converge!

4. THE MULTI-PLANETARY PROBLEM

In this model we have $N-2$ "planets" moving on *known* orbits around the Sun and we try to calculate the trajectory of a small "massless" body. By applying the same, as above, method, we find the solution for the massless body as

$$e\cos\varpi = e_P\cos(gt+\beta) + \sum_{j=1}^{N-2} B_j e_j cos(g_j t + \beta_j) \tag{44}$$

$$e\sin\varpi = e_P\sin(gt+\beta) + \sum_{j=1}^{N-2} B_j e_j sin(g_j t + \beta_j) \tag{45}$$

$$i\cos\Omega = i_P\cos(st+\delta) + \sum_{j=1}^{N-2} \Delta_j i_j cos(s_j t + \delta_j) \tag{46}$$

$$i\sin\Omega = i_P\sin(st+\delta) + \sum_{j=1}^{N-2} \Delta_j i_j sin(s_j t + \delta_j) \tag{47}$$

where the constants (B_j, Δ_j) depend on a/a' and the constants (e_j, i_j) are the components of the eccentricities and inclinations of the planets (which, in this case are considered as *osculating elements*, i.e. they are allowed to *vary* with time, but in a known way). The frequencies g_j and s_j are the *fundamental frequencies* of the planetary system and describe the "precession" of the planets and β_j, δ_j are the phases of the planets for the considered epoch. These $4j-2$ constants[4] are known to a high precision.

In sum, the results of the secular theory of our planetary system are as follows:

- The *semi-major axes* of the planets are *constants* of the motion, away from resonances and in the absence of non-gravitational forces
- The osculating eccentricities and the inclinations are varying with very long *secular periods* (of the order of $10^4 - 10^5$ years).

5. RESONANCES

5.1. Types of resonances

Due to the order of magnitude differences in the values of the frequencies involved, we distinguish two *basic* types of resonances:

- Orbital resonances appear in both the three-body and the planetary problem, between the *mean longitude* of a planet, λ', and that of the massless body, λ. We say that an asteroid is in the region of an orbital resonance $p/(p+q)$ with a planet if we have that

$$p\dot{\lambda} - (p+q)\dot{\lambda}' \approx 0. \tag{48}$$

  The location of orbital resonances, to zero-th order, is given by Kepler's law:

$$a_{res} = a'[p/(p+q)]^{2/3} \tag{49}$$

  and is independent of e and i (because of the degeneracy).
- Secular resonances appear only in the multi-planetary model, between the frequencies (g, s) and the proper frequencies of our planetary system

$$p\dot{\varpi}_P + q\dot{\Omega}_P + \sum_j (k_j \dot{\varpi}_j + l_j \dot{\Omega}_j) = 0 \tag{50}$$

  In particular the resonances $(g = g_k)$ or $(s = s_k)$ are symbolized as ν_k and ν_{1k}, respectively and are called *first-order apsidal* (ν_k) or *nodal* (ν_{1k}) resonances.

5.2. Resonant dynamics

The linear secular theory, presented so far, assumes that there exist no resonances, either orbital of secular. However we can find a similar solution if the motion takes place

---

[4] Because the angular momentum is a constant of the motion, one (s_i, δ_i) pair is equal to zero.

in a region where exists *just one* resonant condition, through the following algorithm.

- Make the resonant combination a new angle co-ordinate, say, θ_1, through a canonical transformation
- Perform a *normalization* for all the other actions, *except* the one whose conjugate angle is the *resonant angle*.

It is easy to show that the "new" Hamiltonian is integrable. It does not depend on the angles $\theta_2, \theta_3, ...$, so that the corresponding actions $I_2, I_3, ...$ are integrals of motion. Therefore they are constants and can be treated as parameters. The Hamiltonian contains *only* one pair of conjugate variables, θ_1, I_1, so that it has *only* one degree of freedom. Then it is integrable by definition and we can, *in principle*, write the corresponding last integral of motion. The topology of the phase space of a 1-D Hamiltonian is, locally, similar to that of the pendulum. Therefore we can use the analytical solution of the pendulum, to describe the evolution of (θ_1, I_1).

6. PROPER ELEMENTS AND ASTEROID FAMILIES

The secular Hamiltonian (35) is *integrable* and, therefore, is has as many first integrals of motion as degrees of freedom. Among all possible sets of integrals of motion the one of physical interest is the set $\{a_P, e_P, i_P\}$. These are the *proper elements* of the asteroid's orbit. As a result, we can classify asteroids through their proper elements, provided that their orbit is neither resonant nor chaotic. In practice these two restrictions are not really strict. In cases where the orbit is close to *a single* resonance, we can construct a secular integrable Hamiltonian as well, as discussed in the previous section, so that *resonant* proper elements may be defined. Finally in cases where the orbit is close to two or more resonances, it is chaotic. In this case the secular Hamiltonian gives a very bad approximation of the real motion, but proper elements may be of some use for time intervals less than the Lyapunov time (i.e. the time after which a chaotic trajectory "forgets" its initial conditions).

If the asteroids are plotted in proper elements space, we can distinguish that some of them are clearly "clustered" in groups, the *asteroid families*. Today we believe that asteroid families consist of the breakup pieces of catastrophic collisions between large parent bodies.

7. BEYOND "CLASSICAL" SOLAR SYSTEM DYNAMICS

7.1. Chaos

Many of the solar system bodies follow chaotic trajectories. This is due to the fact that they are under the influence of more than one resonances. The secular theory and the proper elements presented here hold for ordered trajectories. What about the chaotic ones? One possible solution of this problem, that appeared recently in the literature, is a *statistical approach*. In this method we assume that asteroids are diffusing in action (equivalently, proper element) space and we follow the evolution of an ensemble of

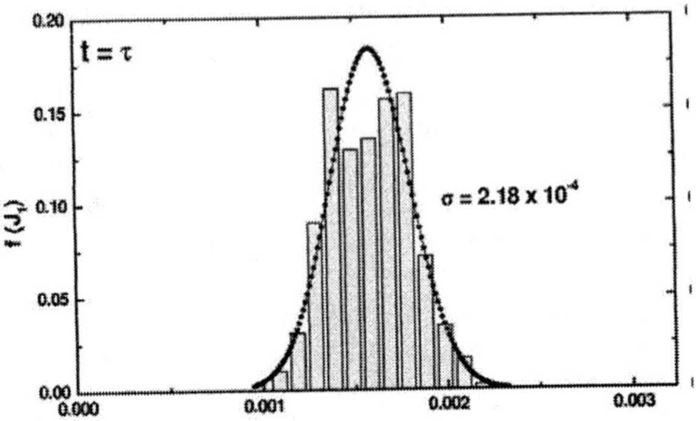

FIGURE 4. Chaotic chronology of the Veritas family. The actual distribution of asteroids is fitted to a Gaussian with a σ corresponding to an age of 8.9 Myrs.

asteroids in this space by using a diffusion equation of the form

$$\frac{\partial P}{\partial t} = \frac{\partial}{\partial I}\left(\frac{D}{2}\frac{\partial P}{\partial I}\right) \tag{51}$$

in which enters only one parameter, the *diffusion coefficient* $D(I)$. If we can calculate $D(I)$, either analytically [4] or numerically [9], we can then solve the equation and find the evolution of the initial distribution $P(I_0,0)$. In the simplest possible case, where the diffusion coefficient is constant and the space where the asteroids are diffusing is considered infinite, the solution is the well known *Fick's law*, a Gaussian centered at I_0 with a dispersion $\sigma^2 = Dt$.

In recent years the diffusion equation method has been used to attack some problems hard to solve through traditional approaches. One such example is the estimation of the age of the Veritas asteroid family through chaotic chronology, by assuming that the members of the family on chaotic orbits originated from the breakup of a parent body after a catastrophic collision [8]. The result, 8.9 Myrs, (Fig. 4) agrees very well with the estimate of 8.3 Myrs made by back integration of the members of the family on ordered orbits [5].

7.2. Non-gravitational forces

As mentioned above, in the case where the orbit of an asteroid is non-resonant or it is affected by a single resonance, the secular theory implies that the semi-major axis is a constant of the motion and the proper eccentricity and inclination remain constant. But even if the orbit is affected by two or more resonances, in which case it is chaotic, numerical integrations have shown that, although eccentricity and inclination are varying in proper elements space, the semi-major axis is hard to change.[5] The only way to "break" effectively the invariance of the semi-major axis is through non-gravitational forces. The most important mechanism for transport in semi-major axis space, for kilometer-size bodies, is the *Yarkovsky effect*, the recoil force acting on a spinning asteroid by the emission of infrared radiation.

The last years the Yarkovsky effect has been invoked in order to interpret otherwise difficult to understand phenomena. One such application was the accurate modelling of the motion of the asteroid Golevka [6] and the shape of the 7:3 Kirkwood gap [10]. In the first example it was shown that the very accurate measurements of the position and velocity of Golevka were not consistent with pure gravitational motion and could be understood only by assuming the action of the Yarkovsky effect. In the second example it was shown that the existence of asteroids with dynamical life-time of 10^7 years in the 7:3 Kirkwood gap can be understood only by assuming that what we observe is just a dynamical equilibrium between the transport of asteroids out of the resonant region and the injection of new asteroids from the adjacent Eos and Koronis asteroid families through the action of the Yarkovsky effect.

7.3. Simulations

The two approaches of studying the evolution of our planetary system under the influence of only gravitational forces (secular theory or statistical description of chaotic motion) are only the two "extreme" cases. For all "intermediate" cases a satisfactory appropriate theory does not exist today and, therefore, these cases have to be studied numerically. Indeed, the motion of a large number of bodies can be studied consistently, including the "hard to model" collisions and/or close encounters, provided that one has the appropriate computer power to follow their evolution for time intervals of the order of the age of the solar system. The last years this approach has gained momentum, through the continuous increase of speed and memory of "classical" computers or the use of dedicated machines. Interesting examples of this new approach are (a) the initial evolution of the outer solar system [7], (b) the simulation of the creation of the Moon through the impact of a Mars-sized body on Earth [1] and (c) the triggering of planetary formation, through the interaction of planetesimals of a protoplanetary disk with a close approaching star [11].

In the first example it has been shown, through a simulation on "traditional" computers, that the outer planets were created in a far more "compact" configuration than today's and that they were transported to their present places through a process called *migration*, caused by their interaction with a disk of planetesimals. In this way one may understand not only the present values of eccentricities and inclinations of the outer planets, but as well the phenomenon of the *Late Heavy Bombardment* of the inner planets by a swarm of Kuiper belt objects, $\sim$ 700 Myrs after the creation of the solar system

[5] This result stems from the degeneracy of the Kepler's problem: the resonances are at $a \simeq a_{res}$ for all values of e, i in the asteroid belt. The semi-major axis, a, undergoes non-periodic but zero-average variations around a_{res}, unless e increases so much, that overlapping with a nearby resonance is possible.

(i.e. 3.9 Gyrs ago).

The other two simulations were performed using a GRAPE machine, a dedicated computer accelerator which calculates the gravitational forces between N bodies. In the first of them it has been shown that most of the debris of such an impact are not ejected on hyperbolic orbits but remain in orbit around the Earth, forming a disk of "satellesimals". It was shown that the such a disk will end up in the creation of a satellite, in the same way that planetesimals create finally the planets, through successive disrupting and sticking collisions.

In the second we have shown that parabolic encounters between gas-free protoplanetary disks result into (i) the exchange of material between the disks, and (ii) the restructuring of the initial disk, into a "core" of nearly circular and co-planar orbits and an extended 3-D cloud of eccentric and inclined bodies. These processes may prove to be particularly important for the formation of planets and small-body belts.

ACKNOWLEDGMENTS

This work was supported by the research program EPEAEK II / Pythagoras I of the Greek Ministry of Education. I would like to thank Empeirikio Foundation of Greece and Aristotle University of Thessaloniki for funding the purchase of the Grape 6/Pro8 system and Dr. K. Tsiganis for reading the first draft of this article.

REFERENCES

1. S. Ida, R.M. Kanup, and G.R. Stewart, *Nature*, **389**, 353–357 (1997).
2. A. Morbidelli, *Modern Celestial Mechanics, Aspects of Solar System Dynamics*, Taylor & Francis, London, 2002, pp. 11–38, 127-230.
3. C.D. Murray and S.F. Dermott, *Solar System Dynamics*, Cambridge University Press, Cambridge, 1999, pp. 22–128, 225–318.
4. N. Murray and M. Holman, *Astron. J.*, **114**, 1246–1259 (1997).
5. D. Nesvorný, W.F. Bottke, H.F. Levison, and L. Dones, Astrophys. J., **591**, 486–497 (2003).
6. S.R. Chesley, S.J. Ostro, D. Vokrouhlický, D. Čapek, J.D. Giorgini, M.C. Nolan, J.L. Margot, A.A. Hine, L.A.M. Benner, and A.B. Chamberlin, *Science*, **302**, 1739–1742 (2003).
7. K. Tsiganis, R. Gomes, A. Morbidelli, and H.F. Levison, *Nature*, **435**, 459–461 (2005).
8. K. Tsiganis, Z. Knežević, and H. Varvoglis, *Icarus*, submitted (2006).
9. K. Tsiganis, H. Varvoglis, and A. Anastasiadis, "Diffusion of asteroids in the outer belt," in *Modern Celestial Mechanics: from Theory to Applications*, edited by A. Celletti, S. Ferraz-Mello, and J. Henrard D. F., Kluwer, Dordrecht, 2003, pp. 451–455.
10. K. Tsiganis, H. Varvoglis, and A. Morbidelli, *Icarus*, **166**, 131–140 (2003).
11. H. Varvoglis, K. Tsiganis, and Ch. Vozikis, "Large-scale simulations of gravitational Astrophysical systems," in *International Workshop: Rotation of Celestial Bodies*, edited by A. Lemaitre, in press, 2006.

Dynamics of KBOs near the 4:5 mean motion resonance with Neptune

Thomas Kotoulas, George Voyatzis

University of Thessaloniki, Department of Physics
Section of Astrophysics, Astronomy and Mechanics
GR-541 24 Thessaloniki, Greece
email=tkoto@skiathos.physics.auth.gr

Abstract. We study the dynamics at the 4:5 exterior mean motion resonance using the model of the restricted three-body problem with the Sun and Neptune as primaries. The position and the stability character of the periodic orbits determine critically the structure of the phase space and particularly the qualitative characteristics of the long-term evolution of trans-Neptunian objects (TNOs). Using the circular planar model as the basic model, we extend our study to other models by considering the ellipticity of Neptune's orbit and the inclination of TNO's orbit. Families of symmetric periodic orbits of the planar elliptic and the three-dimensional problem (circular or elliptic) are found. The stability of all orbits is also examined. Finally, we show the evolution of a real object, i.e. the 2003 FC128, which is located near the 4:5 resonance.

Keywords: Restricted 3-body problem, Periodic orbits, Kuiper Belt, methods:numerical
PACS: 45.50.Jf, 95.10.Ce

INTRODUCTION

The interest in studying exterior resonances is continuously growing after the discovery of many objects at the edge of the Solar System, beyond the orbit of Neptune, called Egdeworth-Kuiper (E-K) belt objects or trans-Neptunian objects (TNOs). Several studies of the dynamics of the E-K belt have been made up to now ([9], [8], [2], [11], [10]).

The restricted three-body problem (RTBP) is an efficient model for understanding the underlying dynamics of the resonant orbits in Kuiper Belt. In this model the primary bodies are the Sun and Neptune and the small body (of negligible mass) represents a potential TNO. As far as the Kuiper belt dynamics is concerned, the families of symmetric periodic orbits at the 1:2, 2:3 and 3:4 resonances have been studied previously ([5]). A complete study of periodic orbits for higher-order resonances was made in [14].

For the study of inclined orbits, the three-dimensional (3-D) RTBP should be considered as the basic model. Three-dimensional symmetric periodic orbits of the circular RTBP have been studied by several authors in the past ([1], [12]). These studies refer to families of 3D periodic orbits associated to the dynamics near the Lagrangian equilibrium points. As far as the dynamics in Kuiper belt is concerned, families of 3-D symmetric periodic orbits were computed in [5], [6], [7].

The study of the dynamics at the 4:5 mean motion resonance with Neptune will be done by computing the position and the stability of periodic orbits. Other known methods are: averaged Hamiltonians and mapping models. The periodic orbits are of particular interest since they define the structure of the associated resonance. In this

CP848, *Recent Advances in Astronomy and Astrophysics,*
7th International Conference of the Hellenic Astronomical Society, edited by N. H. Solomos

paper we shall consider a hierarchy of models: (i) planar circular, (ii) planar elliptic, (iii) 3-D circular and (iv) 3-D elliptic RTBP. We will compute families of symmetric periodic orbits, discuss about the topology of the phase-space and present the long-term evolution of real TNOs near the 4:5 resonance.

THE PLANAR CIRCULAR RTBP

We assume that the orbit of Neptune is circular and we consider a rotating frame of reference *Oxyz* whose *x*-axis is the line joining the Sun (*S*), with Neptune (*N*), the positive direction being from *S* to *N*, its origin is at their centre of mass, the *y*-axis is in the orbital plane of Neptune and the *z*-axis is perpendicular to the *xy* plane. In the usual normalized units where the radius $SN = 1$, the gravitational constant is $G = 1$ and the total mass $m_S + m_N = 1$, the differential equations of motion are the equations of the circular restricted 3-body problem (e.g. [13]).

The symmetric periodic orbits of the planar circular RTBP correspond to the initial conditions:

$$x(0) = x_0,\ y(0) = 0,\ \dot{x}(0) = 0,\ \dot{y}(0) = \dot{y}_0. \tag{1}$$

At half the period T of the orbit we must have:

$$y(x_0,\ \dot{y}_0;\ T/2) = 0,\ \dot{x}(x_0,\ \dot{y}_0;\ T/2) = 0. \tag{2}$$

Equations (2) give the periodicity conditions for a symmetric periodic orbit in the planar circular problem. A symmetric periodic orbit of multiplicity p starts perpendicularly from the x-axis and again crosses the x-axis perpendicularly after p intersections in the same direction.

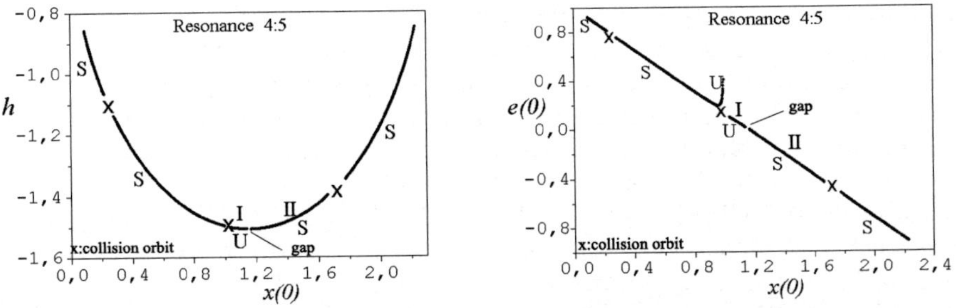

FIGURE 1. The families *I* and *II* of 4:5 resonant symmetric periodic orbits at the plane: (a) $x(0) - h$ and (b) $x(0) - e(0)$. The gap at the resonance is due to the non-continuation of the circular orbits. The symbols "*S*" and "*U*" denote stability and instability respectively. The symbol "*x*" stands for collision orbit. The small body is initially at perihelion ($e(0)>0$) or at aphelion ($e(0)<0$).

There are two families of periodic orbits for the 4:5 resonance, symmetric with respect to the *x*-axis, one of them corresponding to the small body at perihelion at $t = 0$ (Family *I*) and the other to the small body at aphelion (Family *II*). A monoparametric family

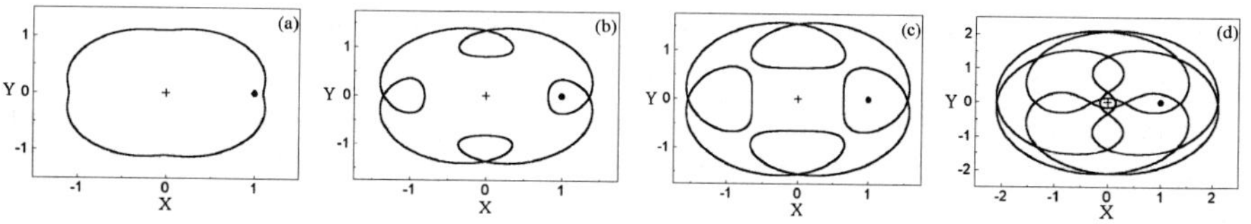

FIGURE 2. Periodic orbits along the family $I_{4:5}$: (a) e=0.05, U (b) e=0.30, S (c) e=0.44, S and (d) e=0.87, S. The development of a non-perpendicular collision orbit and the change of the multiplicity from 1 to 3 and from 3 to 7 is clearly seen. The symbols "+" and "•" denote the positions of Sun and Neptune respectively.

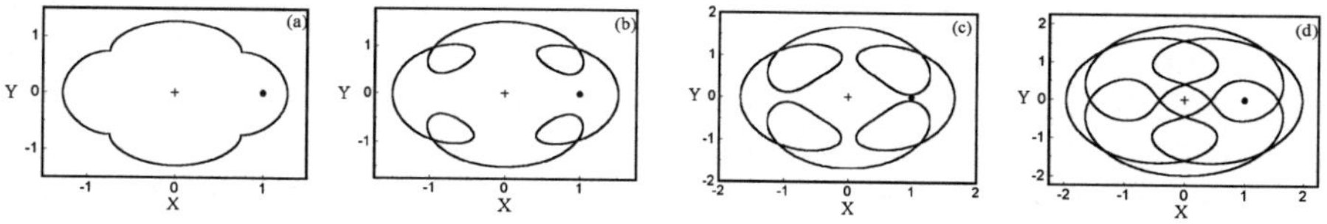

FIGURE 3. Periodic orbits along the family $II_{4:5}$: (a) e=0.10, S (b) e=0.30, S (c) e=0.44, U (close to collision) and (d) e=0.70, S. The development of a non-perpendicular collision orbit and the change of the multiplicity from 1 to 5 is shown.

of symmetric periodic orbits is defined by a smooth curve in the plane $x_0 - h$ of initial conditions, which is called the "*characteristic curve*". Along each family the eccentricity of the small body increases starting from zero.

In Fig. 1 we present the resonant families *I* and *II* of periodic orbits. The stability type is also indicated. In Figs. 2, 3 we show some samples of symmetric periodic orbits of family *I* and *II* respectively. The evolution of the orbits and the change of multiplicity is shown. In Fig. 4 we present two surfaces of section at the energy levels $h_1 = -1.50$ (Fig. 4a) and $h_2 = -1.46$ (Fig. 4b). We note that apart from the 4:5 resonance, several other first order resonances appear. As the h increases, the width of the resonant island decreases (see also [8]).

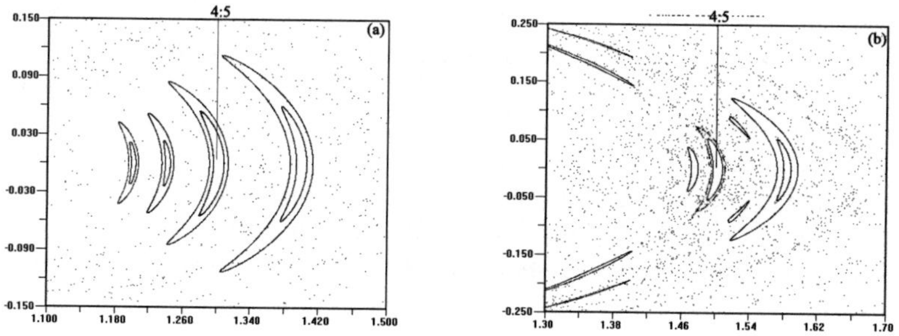

FIGURE 4. Poincaré maps taken at two characteristic energy levels: (a) $h_1 = -1.50$ and (b) $h_2 = -1.46$.

THE PLANAR ELLIPTIC RTBP

In this section we study the resonant dynamics of the elliptic restricted three body problem. We have two degrees of freedom, as in the circular problem, but now the system is *non*-autonomous. The distance between the Sun and Neptune is not constant and the xOy frame is not a uniformly rotating system. The initial conditions also involve the position and velocity of the Sun and Neptune at t=0. For a symmetric periodic orbit of the small body we get:

$$y(0) = 0,\ \dot{x}(0) = 0,\ \dot{r}(0) = 0,\ \theta(0) = 0,\ \pi, \tag{3}$$

with an indication that Neptune is at perihelion (p) or aphelion (a) at t=0. The nonzero initial conditions are then: $x(0)$=x_0, $\dot{y}(0) = \dot{y}_0$ and also $r(0)$, $\dot{\theta}(0)$, which, however, are determined from Keplerian theory in terms of the eccentricity of Neptune. Thus, the eccentricity of Neptune (e_N) appears as a parameter of the problem. By varying e_N we have a family of periodic orbits of the elliptic problem. For $\dot{r}(0)$=0 and $\theta(0)$=0 or π the periodicity conditions are now:

$$y(x_0,\ \dot{y}_0;\ e_N;\ T/2) = 0,\ \dot{x}(x_0,\ \dot{y}_0;\ e_N;\ T/2) = 0, \tag{4}$$

where T is the period of the periodic orbit in the rotating frame. Due to the fact that at $t = T/2$ we must have $\dot{r}$=0 and θ=0, π too, it arises that the period T must necessarily be a multiple of T_{Nep} where T_{Nep} is the period of Neptune. This means that the periodic orbits in the elliptic problem in the rotating frame are also periodic in the inertial frame.

There is one pair of families of periodic orbits bifurcating from the planar orbits of the family $II_{4:5}$, at e_1 =0.25, and another pair of such families bifurcating from the point e_2=0.87. The first pair is: $E_{1p}^{4/5}$, $E_{1a}^{4/5}$ and the second one is: $E_{2p}^{4/5}$, $E_{2a}^{4/5}$. These families are shown in Fig. 5. The curves on the x_0-$\dot{y}_0$ plane are the two families of the circular problem that are given in Fig. 1.

THE 3-D CIRCULAR RTBP

We present in this section the 4:5 resonant families of periodic orbits of 3-D circular RTBP. These families bifurcate from the families of type II of the planar circular problem at those orbits whose stability is critical with respect to perturbations along the z-axis. For the computation of the vertical stability we used the method developed by Hénon (1973). There are five vertical critical orbits (VCOs), which are starting points for the continuation of families of 3D periodic orbits. They belong to the family $II_{4:5}$ of the planar problem and are the following: $B_1(e_0$ =0.22), $B_2(e_0$ =0.23), $B_3(e_0$ =0.62), $B_4(e_0$ =0.73) and $B_5(e_0$ =0.83). No such VCOs exist on the family $I_{4:5}$.

The families of three-dimensional symmetric periodic orbits have the following symmetries:

Families F: These periodic orbits are symmetric with respect to xz-plane. In this case a periodic orbit can be defined by the initial conditions $x(0) = x_0$, $y(0) = 0$, $z(0) = z_0$, $\dot{x}(0) = 0$, $\dot{y}(0) = \dot{y}_0$, $\dot{z}(0) = 0$ and the periodicity conditions are:

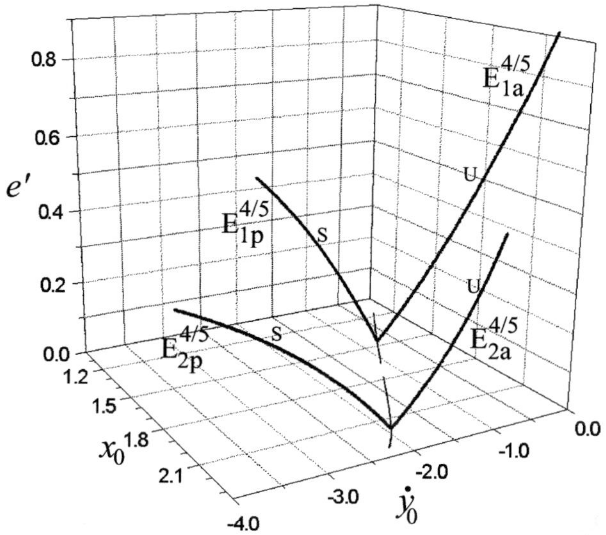

FIGURE 5. Families of periodic orbits of the elliptic restricted three-body problem at 4:5 resonance. Here e' is the eccentricity of Neptune (e_N).The symbols *S, U* denote for stability and instability respectively (from [14]).

$$y(0) = y(T/2) = 0, \ \dot{x}(0) = \dot{x}(T/2) = 0, \ \dot{z}(0) = \dot{z}(T/2) = 0, \tag{5}$$

where T is the period of the orbit. Such a periodic orbit can be represented as a point in 3-D space $x_0 - z_0 - h$, where h is the corresponding Jacobi constant. By varying the value of h we get a monoparametric family of periodic orbits.

Families G: These periodic orbits are symmetric with respect to x-axis. In this case a periodic orbit can be defined by the initial conditions $x(0) = x_0$, $y(0) = 0$, $z(0) = 0$, $\dot{x}(0) = 0$, $\dot{y}(0) = \dot{y}_0$, $\dot{z}(0) = \dot{z}_0$ and the periodicity conditions are:

$$y(0) = y(T/2) = 0, \ \dot{x}(0) = \dot{x}(T/2) = 0, \ z(0) = z(T/2) = 0, \tag{6}$$

where T is the period of the orbit. Such a periodic orbit can be represented as a point in 3-D space $x_0 - \dot{z}_0 - h$, where h is the corresponding Jacobi constant. By varying the value of h we get a monoparametric family of periodic orbits.

In Fig. 6 we present the families of 3-D resonant periodic orbits at the 4:5 exterior mean motion resonance with Neptune. The eccentricity and the inclination vary along a family, but the semimajor axis is almost constant, $a_{4:5} \approx 34.88$ A.U. (in non-dimensional units). Sample of periodic orbits are given in Fig.7. Finally, at Fig. 8, we show the projections of these families on the plane $e_0 - i_0$. The stability type is also indicated.

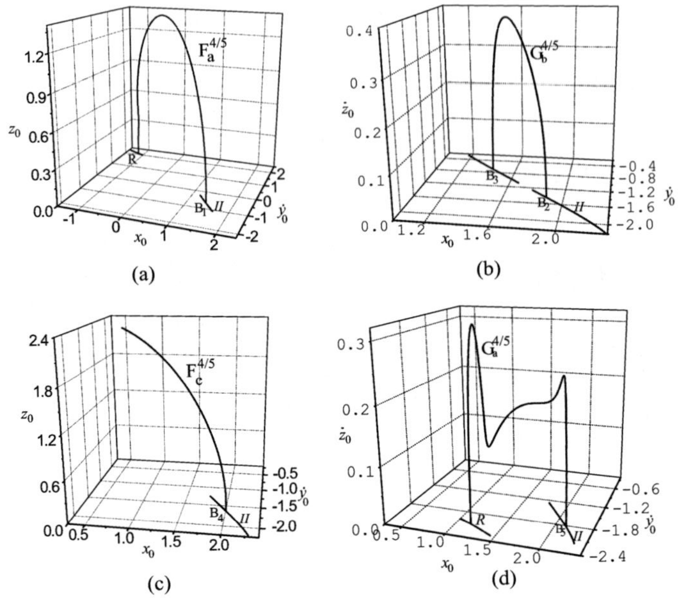

FIGURE 6. Families of periodic orbits of the 3-D circular restricted three-body problem at 4:5 resonance. The starting and the terminating points of each family are also indicated. The axes correspond to the initial conditions of periodic orbits. The symbol *R* stands for retrograde orbits.

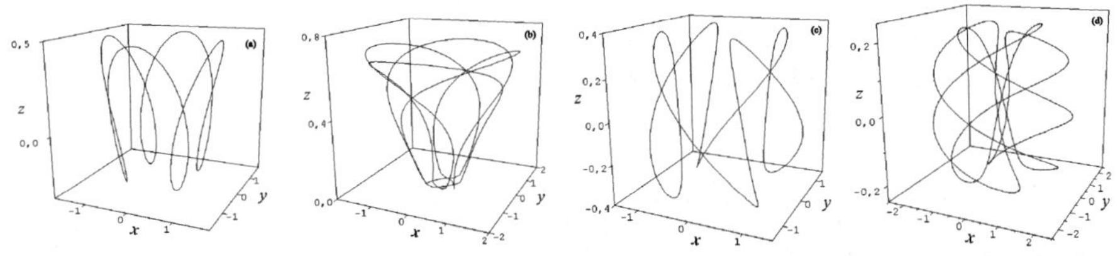

FIGURE 7. Samples of three-dimensional periodic orbits at 4:5 resonance. Each one of them belongs to the corresponding family presented in figure 6.

THE 3-D ELLIPTIC RTBP (ER3BP)

The 3-D elliptic RTBP is a non-autonomous Hamiltonian system with three degrees of freedom. This model is the most realistic one in the framework of the restricted three-body problem, since it describes the motion of the small body in space and the perturbing planet is on an elliptic orbit.

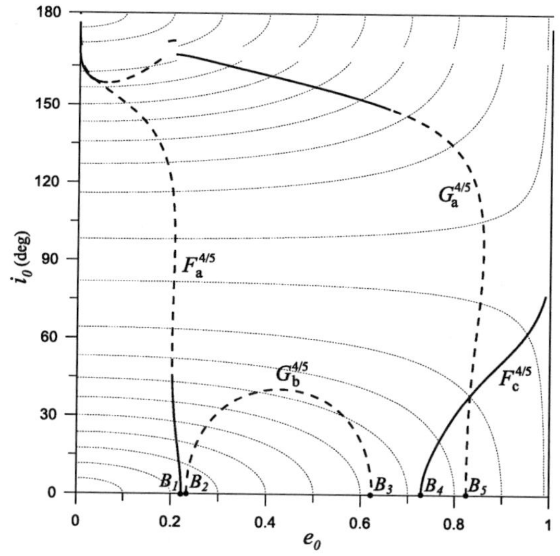

FIGURE 8. Projection of the families of periodic orbits of the 3-D circular restricted three-body problem at 4:5 resonance. Solid segments denote linear stability and dashed lines stand for linear instability. The thin dotted lines denote the curves of z-argument of the angular momentum related to the *Kozai* dynamics (from [7]).

There are two approximations for obtaining periodic orbits in the 3-D ER3BP:

(a) Families of periodic orbits of the 3-D ER3BP, with the eccentricity of Neptune (e_N) as a parameter, bifurcate from the families of the 3-D CR3BP mentioned above at those orbits whose period is a multiple of T_{Nep}. The results are summarized in Table 1.

TABLE 1
Bifurcation points from 3-D CR3BP to 3-D ER3BP

Resonance	Period	Family	Bifurcation Points
4 : 5	10π	F_a	e=0.204, i=80.9°
4 : 5	10π	G_a	e=0.865, i=86.8°

(b) The second way to search for periodic orbits that bifurcate to the 3-D ER3BP is to check the vertical stability of periodic orbits of the corresponding planar elliptic problem. From our results there are not any such periodic orbits which continue from the planar elliptic problem to the three-dimensional one.

A monoparametric family of symmetric periodic orbits is formed by analytic continuation varying the eccentricity of the planet (e') which is the parameter of the family in this case. Such a family can be represented by a curve in the 4D space $x_0 - \dot{y}_0 - z_0 - e'$ or $x_0 - \dot{y}_0 - \dot{z}_0 - e'$ accordingly to the symmetry of the periodic orbits. When $e' \neq 0$ we get two possible configurations of the system at $t = 0$. In the first configuration the planet is located at pericenter and in the second one at apocenter. Thus, from each generating orbit (or bifurcation point) two families bifurcate.

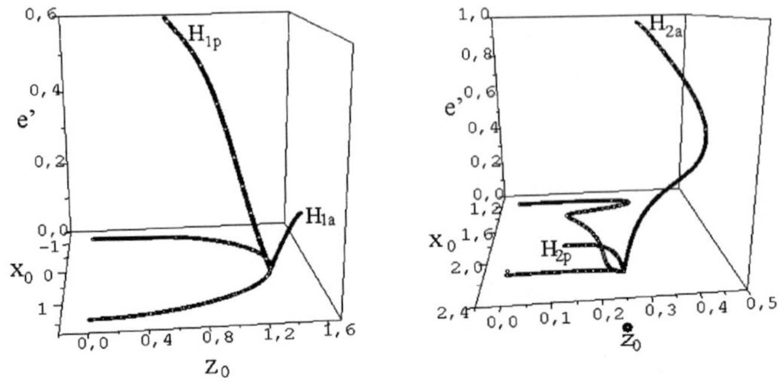

FIGURE 9. Projection of families of 3-D symmetric periodic orbits at the three-dimensional elliptic RTBP in (a) $x_0 - z_0 - e'$ and (b) $x_0 - \dot{z}_0 - e'$ 3-D space of initial conditions. The symbol e' stands for the eccentricity of the orbit of Neptune.

In Fig. 9 we present the families of periodic orbits in the 3-D space. From the first bifurcation point, given in Table 1, we obtain the families H_{1p} and H_{1a}, where p or a indicates the initial position of Neptune (pericenter or apocenter, respectively). Correspondingly, we obtain the families H_{2p} and H_{2a} bifurcating from the second bifurcation point. These families are linearly unstable.

DYNAMICAL EVOLUTION OF TNOS NEAR THE 4:5 NMMR

In this section we study the long-term evolution of the real object *2003 FC128* which is near the 4:5 mean motion resonance with Neptune. Its orbital elements are: a=35.127 A.U., e=0.088, i=2.4°, ω=41.0°, Ω=78.6°, M=41.4° (*Minor Planets Center*). This KBO is on a low eccentric orbit, almost circular. We have chosen a sufficient numerical method to study the long-term evolution of this KBO; the LIE integration method with an adaptive step size (Hanslmeier and Dvorak, 1984). This method uses direct integrations of the equations of motion. The accuracy was preserved at 10^{-11}. The time evolution of the orbital elements is shown in Fig. 10 and indicates a regular orbit. However, the evolution of the angle $\Omega - \Omega_8$ switches from libration to rotation which indicates that the orbit is rather weakly chaotic.

CONCLUSIONS

In this paper we presented families of periodic orbits at the 4:5 resonance in the model of the 3-D elliptic RTBP. The main results are summarized below:

1) There exist two families of elliptic orbits in the planar circular problem at the 4:5 resonance; families $I_{4:5}$ and $II_{4:5}$. The first one is unstable until $e \approx 0.10$ where a perpendicular collision with Neptune occurs; then it becomes stable. The other one ($II_{4:5}$) is all stable except for a small collision area around $e \approx 0.47$. The location of periodic orbits is important for obtaining the structure of the phase space.

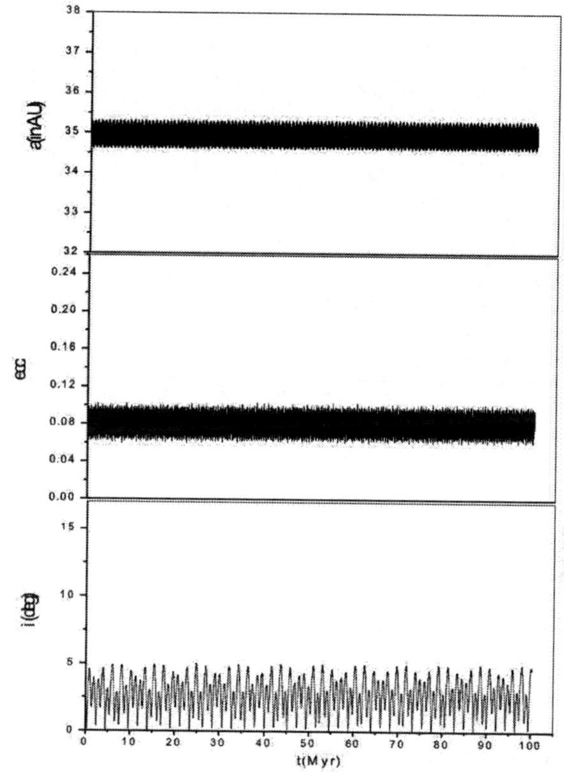

FIGURE 10. Time evolution of the orbital elements a, e i of the test body for 100 Myr. A quasi-periodically behaviour of the orbital elements is observed.

2) The families of periodic orbits of the planar elliptic RTBP bifurcate from the known families of the planar circular problem. We found totally four families of symmetric periodic orbits; two of them are linearly stable and the other ones are unstable.

3) As far as the 3-D circular RTBP is concerned, vertical critical orbits (VCOs) exist only in the family $II_{4:5}$. Families of three-dimensional periodic orbits which bifurcate from vertical critical orbits of the planar problem are symmetric with respect to the xz-plane (Type F) or to the x-axis (Type G). They continue up to high inclination values; stable and unstable orbits coexist in this model.

4) Families of periodic orbits of the 3-D ER3BP were also computed. These families bifurcate from the known families of the 3-D circular problem. They have the same symmetry as the families of periodic orbits in the 3-D circular model and are generated by varying the eccentricity of Neptune (e_N). All of them are linearly unstable.

5) The evolution of the test body (2003 FC128) seems to be weakly chaotic but practically stable.

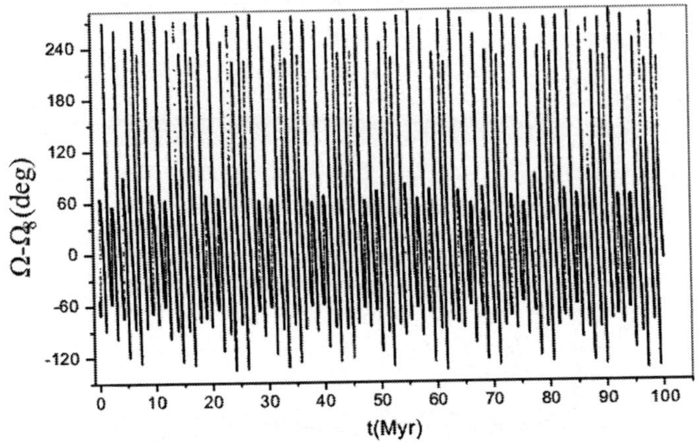

FIGURE 11. Time evolution of $\Omega - \Omega_8$ for 100 Myr. The orbit leaves the ν_{18} resonance; the argument $\Omega - \Omega_8$ switches between libration and circulation.

ACKNOWLEDGMENTS

The authors would like to thank Prof. J.D. Hadjidemetriou for useful discussions. This work has been supported by the program "*EPEAEK II, PYTHAGORAS*", No. 21878, of the Greek Ministry of Education and European Union.

REFERENCES

1. Broucke, R.A., 2001, *Cel. Mech. and Dyn. Ast.* **81**, 321
2. Gallardo T. and Ferraz-Mello. S:1998, *Planet Space Sci.*, Vol. 46, No.8, 945-965.
3. Hanslmeier A. and Dvorak R. 1984, A & A **132**, 203
4. Hénon, M., 1973, A & A, 28, 415
5. Kotoulas, T., & Hadjidemetriou, J.D., 2002, Earth, Moon and Planets, **91**, 63
6. Kotoulas, T., 2005, A & A, 429, 1107
7. Kotoulas, T., & Voyatzis, G., 2005, A & A 441, 807-814
8. Malhotra, R.:1996, *Astron. J.* **111 (1)**, 1996, 504-516
9. Morbidelli, A., Thomas, F. and Moons, M:1995, *Icarus* **118**, 322-340.
10. Morbidelli, A., Brown, M.E. & Levison H.F., 2003, Earth, Moon and Planets 92, 1-27
11. Nesvorný, D. & Roig, F.,II 2001, Icarus, 150, 104
12. Papadakis, K.E., 2004, A & A, 425, 1133
13. Roy, A.E.:1982, *Orbital Motion*, Adam Hilger, (2nd ed.).
14. Voyatzis, G., & Kotoulas T.A., 2005, Planetary and Space Science, **53**, 1189-1199

The Photo-Gravitational Restricted Four-Body Problem: An Exploration of Its Dynamical Properties

T.J. Kalvouridis[1], M. Arribas[2], A. Elipe[2]

[1] *Department of Mechanics, National Technical University of Athens, Greece*
[2] *Grupo de Mecánica Espacial, University of Zaragoza, Spain*

Abstract: We deal with the photo-gravitational version of the restricted four-body problem and we investigate the dynamical behaviour of a small particle, which is subjected both the gravitational attraction and the radiation pressure of three much bigger bodies, the primaries. These bodies are always in syzygy and two of them have equal masses and are located at equal distances from the central primary, which has a different mass. We study the effect of radiation on some dynamical characteristics of the system, such as the zero-velocity curves, the equilibrium positions and the periodic orbits.

Keywords: restricted four-body problem, radiation pressure, equilibrium points, periodic orbits

INTRODUCTION

The restricted four-body problem is a simple model and can be considered as an extension of the Copenhagen case, since a third primary with different mass is located at the center of mass of the previous configuration (Figure 1). This model was suggested by Marañhao in 1995. Here, we improve the original version by considering some or all the primaries as radiation sources. The treatment may be applicable to the investigation of phenomena such as the accumulation of matter around binaries or triple star systems, particularly where the stars are of luminous type and the study of the orbit perturbations of small space wanderers and drifters such as grains, meteors or other cosmic particles derived from a variety of sources; it can also prove useful for the exploration of some stellar wind problems where the wind forces have a similar modifying effect on the inverse square gravitational forces experienced by infinitesimal particles (Simmons et al., 1985; Wilck and Mann, 1996; Wada, et al. 2000).

PARAMETERS, ASSUMPTIONS AND EQUATIONS OF MOTION

The total radiation force exerted on a particle due to a radiating body, generally consists of three terms, namely the radiation pressure, the Doppler shift of the incident radiation and the Poynting drag (Poynting, 1903; Robertson, 1937). The first two act radially, while the third one is the transverse component and is opposite to the particle velocity vector. The latter two components are caused by the absorption and subsequent re-emission of radiation and constitute the Poynting-Robertson effect. Radzievski (1950, 1951) made the assumption that this effect is negligible and the only significant force is radiation pressure. This assumption is valid only when we study extremely small particles, such as interplanetary and interstellar dust particles, that move with velocities less than the speed of light and when the timespan is relatively short.

CP848, *Recent Advances in Astronomy and Astrophysics,*
7th *International Conference of the Hellenic Astronomical Society,* edited by N. H. Solomos

From this point of view, the effect of radiation pressure of a source on a small particle, is expressed by means of the reduction factor q=1-b where b, that is often mentioned as radiation coefficient (Ragos and Zagouras, 1991; Kazazakis and Kalvouridis, 2003), is the ratio of the force F_r which is caused by radiation to the force F_g which results from gravitation, that is $b = \frac{F_r}{F_g}$. We symbolize by q_0, q_1 and q_2, the reduction factors of the central body P_0 and the two peripheral bodies P_1 and P_2 respectively and by b_0, b_1 and b_2, their radiation coefficients. If we determine by $\beta = \frac{m_0}{m}$ the mass parameter of the system, then, the problem is characterized by four parameters.

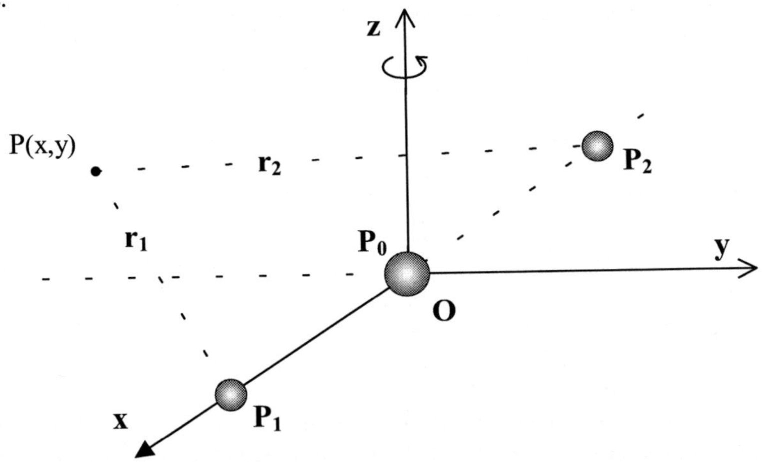

FIGURE 1. The photo-gravitational restricted four-body problem

Radiation coefficients depend on the physical properties of the radiating bodies, as well as those of the particle. If q=1, radiation has no effect. If q<0, then radiation surpasses gravity. Finally, if q>0, gravitational force exceeds radiation. The latter case is the most common one. Our analysis is based on some general hypotheses. We outline the essential points:

- The particle is assumed spherical and homogeneous with constant density. Its surface has a uniform reflection coefficient and its size must exceed the wavelength of the incident light.
- Light has no fluctuations, propagates in straight lines and screening or shadowing effects produced by other bodies of the system are negligible.

If we consider a synodic coordinate system Oxyz, where the line of syzygies of the primaries is the x-axis and the plane z=0 is the plane of the primaries' revolution, then the planar motion of the particle can be described by means of the following dimensionless second-order differential equations,

$$x - 2y = \frac{\partial U}{\partial x},$$
$$y + 2x = \frac{\partial U}{\partial y}$$

where

$$U(x,y)=\frac{1}{2}\left(x^2+y^2\right)+\frac{1}{2(1+4\beta)}\left[\frac{\beta q_0}{r_0}+\sum_{i=1}^{2}\frac{q_i}{r_i}\right]$$

and

$$r_0=\left[\left(x^2+y^2\right)\right]^{1/2}$$

$$r_1=\left[\left(x-\frac{1}{2}\right)^2+y^2\right]^{1/2},$$

$$r_2=\left[\left(x+\frac{1}{2}\right)^2+y^2\right]^{1/2}$$

are the distances of the particle from the primaries. The system is autonomous and there is a Jacobian-type integral of motion

$$C=2U(x,y)-\left(x^2+y^2\right)$$

where C is a constant.

The symmetry of the produced field is maintained only in case where the central body is a radiation source whatever the value of the radiation coefficient b_0 is, or the two peripheral bodies are radiation sources with the same radiation coefficients (b_1=b_2) no matter what the central primary is.

REGIONS OF PERMITTED MOTION AND EQUILIBRIUM POINTS

For a particular value of C, zero-velocity curves limit the regions where the particle motion is permitted. Here, they are drawn by means of the Jacobian integral of motion and for several values of C. In all the cases we have considered that β=2. Figure 2 shows these curves for three cases.

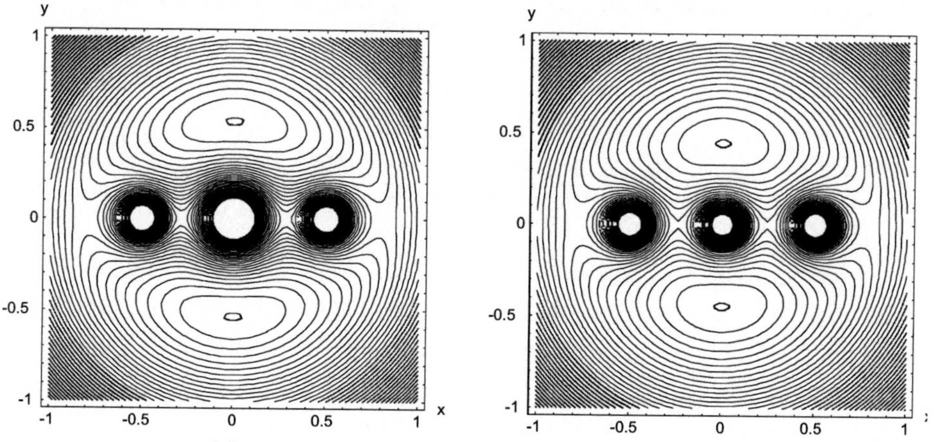

(a) (b)

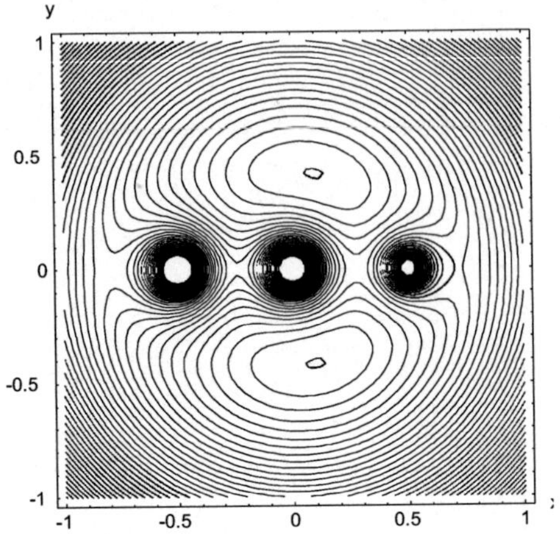

(c)

FIGURE 2. Zero-velocity curves for various values of the Jacobian constant C and for β=2.: (a) gravitational case, (b) b_0=0.5, (c) b_0=b_1=0.5

In the gravitational case there are six equilibrium positions. Four of them are collinear and they lie on the line of syzygies of the primaries (Figure 3). The other two lie on the y-axis. Two collinear points are disposed between the central primary and the peripheral ones in symmetric positions with respect to the origin O. The other two collinear points lie beyond the peripheral primaries in equal distances from the origin. We have computed the equilibrium positions in the photo-gravitational version, for $\beta \neq 0$ and for various values of the radiation coefficients. The results show that, as the radiation coefficient b_0 increases, all the equilibria are displaced towards the origin (Figure 3). When symmetry with respect to the y-axis is not preserved, the two triangular points are displaced outside this axis. In every case the number of the stationary solutions remains the same. By investigating the stability of the equilibria, we have found that all of them are unstable for every value of the parameters, in the considered range.

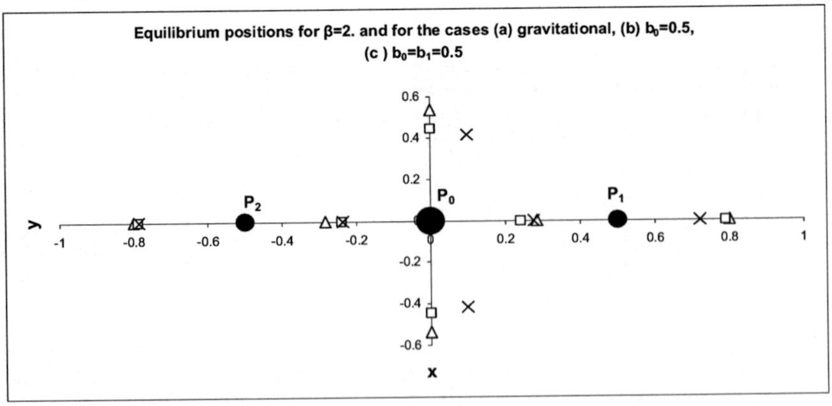

FIGURE 3. Equilibrium positions for β=2 in three cases: gravitational (Δ), b_0=0.5 ( ) and b_0=b_1=0.5 (×)

PERIODIC ORBITS

Apart of the gravitational case, we have treated two cases of the photo-gravitational version, where symmetry exists with respect either to both, or to a coordinate axis. We have investigated the symmetric periodic orbits in the x_0-C space of the initial conditions for various values of the parameters and we have found several families of simple periodic orbits, 9 of which we present here. Families SA, SB, SC, SD, SE, SH and SJ consist of retrograde orbits, and two, namely SG and SI consist of direct orbits. Families SB, SC and SD follow a spiral course during their evolution inside a very small area of the x_0-C diagram. A similar evolution is observed with families SG, SH and SJ, which also follow a spiral course in another small region. In these regions, we stress the creation of loops in the orbits of these families. Furthermore, a very fast increment of their periods is observed. Radiation produces a displacement of the zero-velocity curves toward lower values of C, as well as a simultaneous displacement of the characteristic curves toward the x_0 axis. This effect is more apparent, as the field produced by radiation is strong (Figure 14).

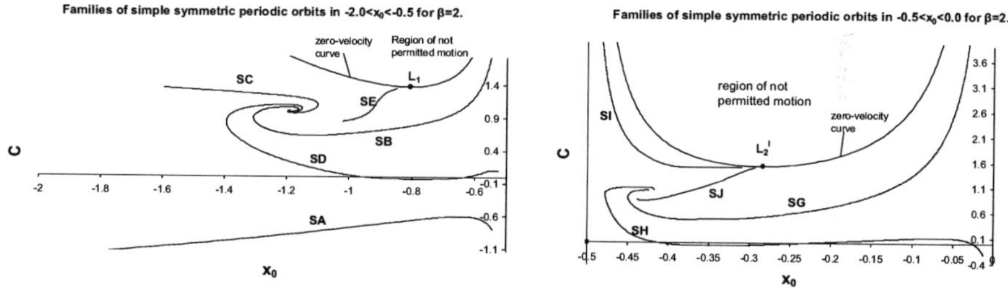

(a) (b)

FIGURE 4. Families of simple periodic orbits: (a) for $-2.5 < x_0 < -0.5$, (b) for $-0.5 < x_0 < 0$

Family SA. The characteristic curve bends near P_2 and continues its evolution close to the asymptote x_0=-0.5 (Figure 4a). The trajectories of the particle encircle all primaries and are symmetric with respect to both coordinate axes (Figure 5a). Family SA develops in the same manner when radiation is considered. However, as radiation increases, there is a slight displacement toward the x_0–axis (Figure 14a). Figure 5b depicts orbits of family SA, which have the same Jacobian constant, but belong to different cases. As it can be seen, orbits expand and their periods increase as radiation becomes strong.

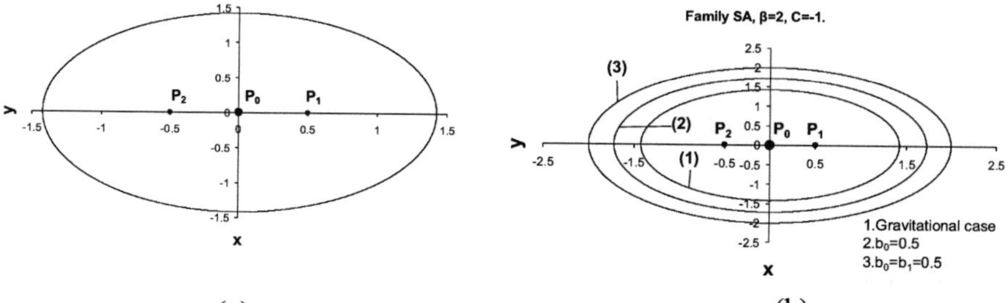

(a) (b)

FIGURE 5. Family SA. (a) a sample orbit (b) parametric evolution of orbits

Family SB. The particle traces paths around primary P_2 (Figure 6a). For high values of C the characteristic curve evolves inside the narrow channel that is formed between the zero-velocity curve and the asympote x_0=-0.5 (Figure 4a). In this part, the orbits are very small with very small periods. As C decreases, the orbits enlarge and their periods increase. Below the zero-velocity curve, the characteristic curve slightly bends upward and continues by spiraling inside the vortex area (Figure 4a). As the family evolves, the orbits become large and the particle approaches the central primary. Inside the vortex area two symmetric loops are created in each orbit. For the same variation of radiation, the characteristic curve of SB lowers and at the same time comes near the asympote that passes through primary P_2. When radiation increases further, family SB, terminates with a collision orbit, without reaching the vortex area (Figure 14b).

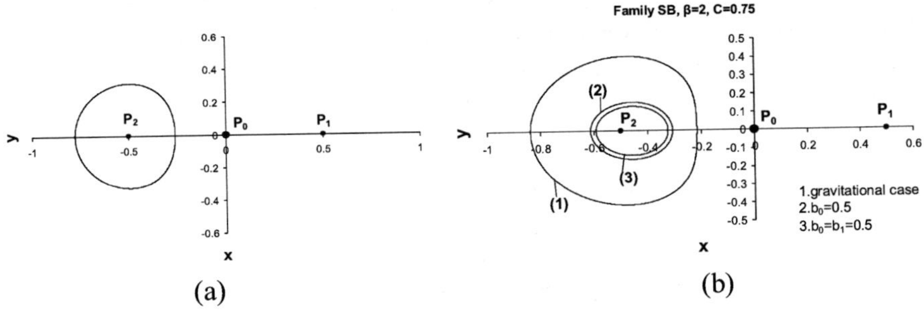

FIGURE 6. Family SB. (a) a sample orbit (c) parametric evolution of orbits

Family SC. It evolves smoothly and the orbits are similar to those of family SA for a large part of the characteristic curve (Figure 7a). The orbits are symmetric with respect to both coordinate axes. As C decreases, the trajectories shrink, but they do not approach the primaries. Near the vortex region, two loops symmetric with respect to the x-axis, are generated and they grow very fast. At the same time, the periods increase rapidly. When radiation is considered, the characteristic curve of the family evolves exactly in the same way as in the gravitational case, but the vortex area moves to small values of x_0 (Figure 14c).

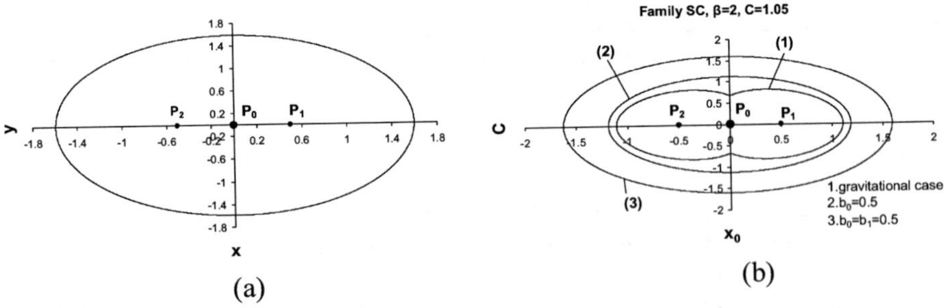

FIGURE 7. Family SC. (a) a sample orbit (b) parametric evolution of orbits

Family SD. The particle passes between P_1 and P_2 and traces trajectories around primaries P_2 and P_0 (Figure 8a). In the vortex area, two symmetric loops, close to the central primary, are generated in the orbits. As C decreases, the orbits become small and almost elliptic and they pass very close to these primaries. When radiation is considered, SD evolves exactly in the same way as in the gravitational case. However, there is no general rule governing the influence on the evolution of the orbits, since the orbits enlarge in some parts of the characteristic curves as radiation becomes stronger, while in some other parts, the orbits shrink (Figure 8b).

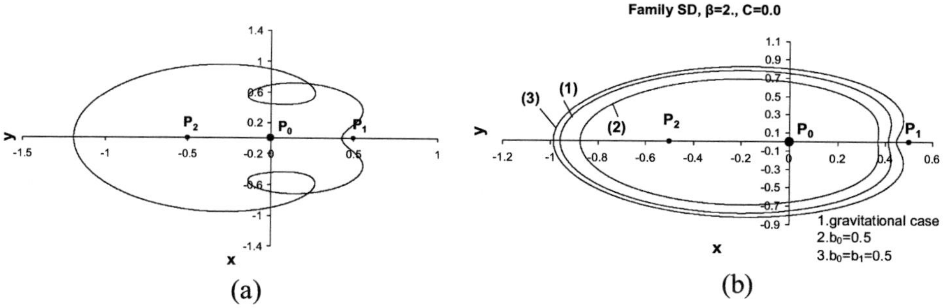

FIGURE 8. Family SD. (a) a sample orbit (b) parametric evolution of orbits

Family SE. It is a short family that emanates from the unstable collinear equilibrium position L_1 (Figure 4a). The orbits are described around this equilibrium point, and are heart-shaped (Figure 9a). As C decreases, the orbits become large and the period increases very fast. As radiation coefficients increase, the characteristic curve of SE shortens (Figure 14e) and the orbits shrink (Figure 9b).

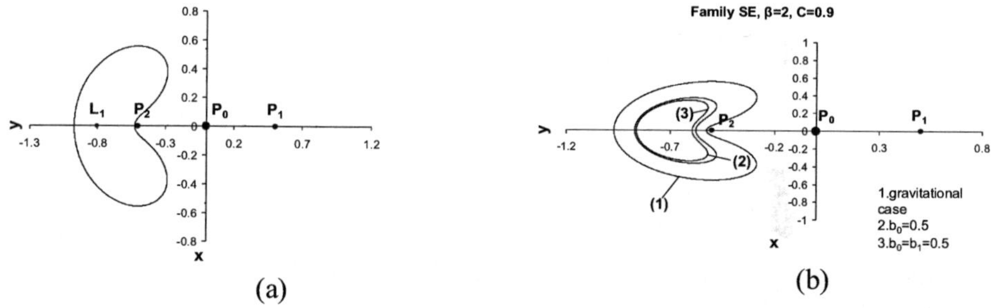

FIGURE 9. Family SE. (a) a sample orbit (b) parametric evolution of orbits

Family SG. For large values of C it evolves in the narrow channel created by the zero-velocity curve and the asymptote through P_0 (Figure 4b). The particle in this region, describes very small and almost circular satellite-type orbits around the central primary with very small period (Figure 10a). The orbits are symmetric with respect to both coordinate axes. As C decreases, the orbits enlarge. Near the vortex region the orbits present a symmetric bending near and opposite the two peripheral primaries, while at the same time a fast increment of the period occurs (Figure 10b). When radiation is considered, the vortex area is displaced to smaller values of C and to bigger values of x_0, than those of the gravitational case. A similar variation is observed on the characteristic curve of family SG, since a part of it evolves inside the referred vortex area (Figure 14f).

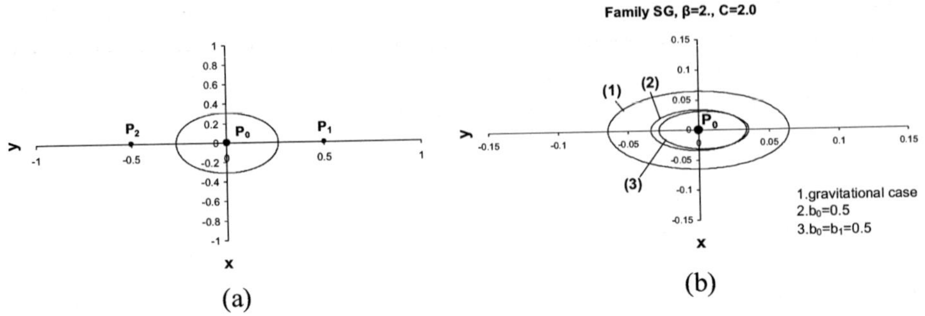

FIGURE 10. Family SG. (a) a sample orbit (b) parametric evolution of orbits

Family SH. The particle moves around P_0 and P_1 and passes between P_0 and P_2 (Figure 11a). Outside the vortex region the orbits enlarge and the particle approaches P_2 closely. Then, as C further decreases, the orbits shrink again, become oval and the particle comes very close to P_1. If one or more primaries are radiation sources, then the characteristic curve of the family approaches the x_0-axis.

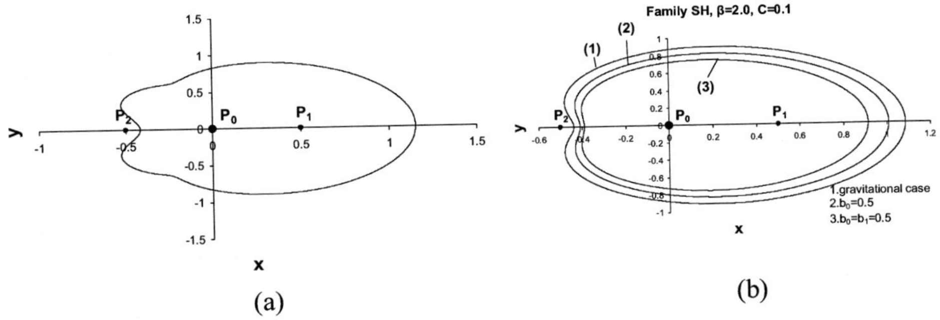

FIGURE 11. Family SH. (a) a sample orbit (b) parametric evolution of orbits

Family SI. The particle moves around P_2 (Figure 12a). For large values of C the family evolves inside the channel formed between the zero-velocity curve and the asymptote through P_2 (Figure 4b). In this part of the curve, the satellite-type orbits are very small and they are characterized by small periods. The orbits, as well as their periods slightly increase as C decreases (Figure 12b), and so do their periods.

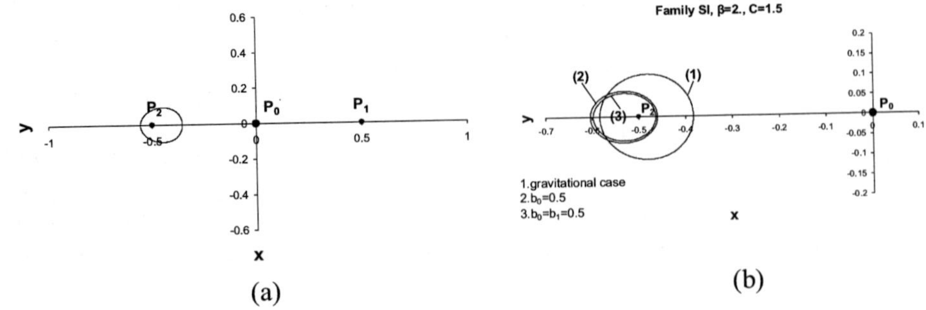

FIGURE 12. Family SI. (a) a sample orbit (b) parametric evolution of orbits

Family SJ. The family emanates from the unstable equilibrium position L_2 (Figure 4b). The orbits are described in the area between P_0 and P_1 and are very small near this point (Figure 13a). However, they soon become large. As C further decreases, the orbits start to distort symmetrically near primaries P_2 and P_0.

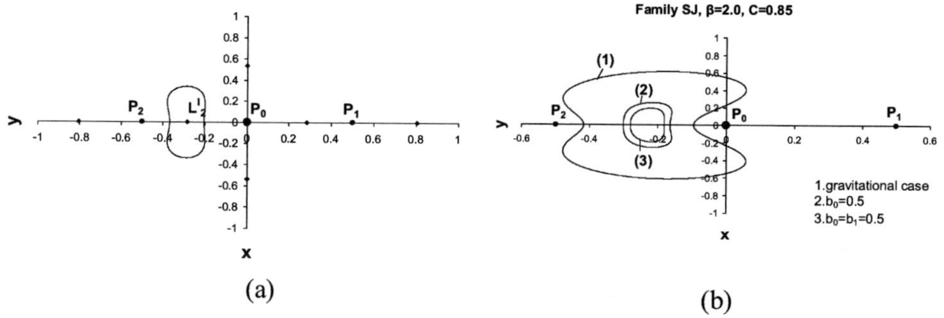

FIGURE 13. Family SJ. (a) a sample orbit (b) parametric evolution of orbits

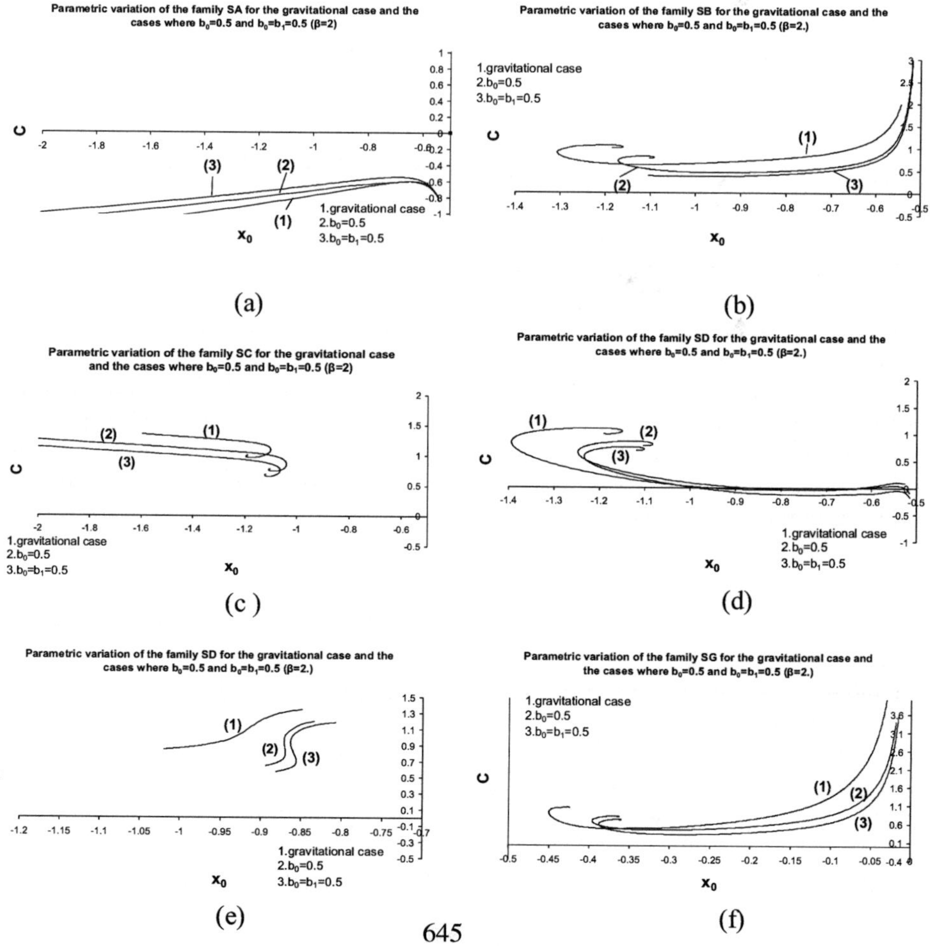

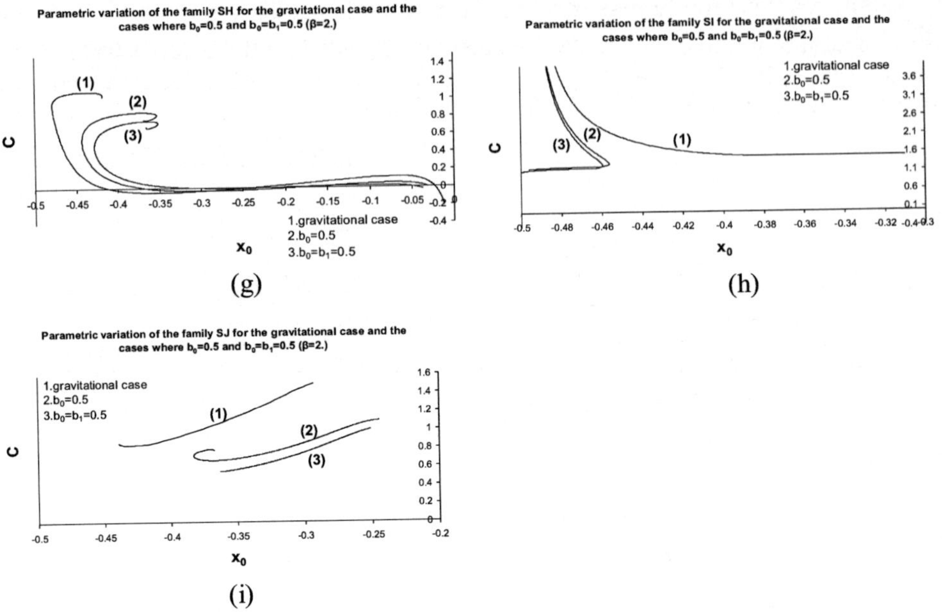

FIGURE 14. Parametric evolution of the families of simple periodic orbits (a) SA, (b) SB, (c) SC, (d) SD, (e) SE, (f) SG, (g) SH, (h) SI, (i) SJ

GENERAL CONCLUSIONS AND REMARKS

We have studied the photo-gravitational version of the restricted four-body problem and we have investigated the influence of the parameters on the main dynamic aspects of the periodic orbits. We have presented the parametric evolution of 9 families of simple periodic orbits. In general, when the Jacobian constant C decreases, the orbits enlarge and their periods increase. When radiation coefficients increase, the characteristic curves approach the x_0-axis of the (x_0, C) diagram. There are no concrete rules for the prediction of the perturbed particle motion under a certain radiation pressure. However, as a result, we can say that, for relatively small values of the coefficients, that is for $b_i<0.1$, i=0,1,2, the declinations from the purely gravitational orbits are almost negligible.

REFERENCES

1.Corben, H.C. and Stehle, P., *Classical Mechanics,* 2nd ed., Dover Publ., 1977.
2.Kazazakis, D. and Kalvouridis, T.J., *Earth, Moon and Planets* **93**, 75-95 (2003).
3.Marañhao, D.,L., *Estudi del flux d'un problema restringit de quatre cossos,*Ph.D. Thesis, Universitat Autonoma de Barcelona, (1995).
4.Poynting, J. H., *Phil. Trans. Roy. Soc. London* **202**, 525 (1903).
5.Radzievskii,V.V., *Astron.Zh.* **27**, 250-256 (1950) (in Russian).
6.Radzievskii,V.V., *Astronomiceskii Journal* **28** (5), 363-372 (1951) (in Russian).
7.Ragos, and Zagouras,Ch., *Celest.Mech. Dynam. Astron.* **50**, 325-347 (1991).
8.Robertson, H.P., *Monthly Not. Roy. Astr. Soc.* **97**, 423 (1937).
9.Wada, K., Sasaki, K., Masuhara, H., *Applied Physics Letters* **76**(20), 2815- 2817 (2000).

Can chaotic motion be responsible for the formation of spiral arms?

Nikos Voglis* and Ioannis Stavropoulos†,*

*Research Center for Astronomy, Academy of Athens, Soranou Efessiou 4, GR-11527, Athens, Greece

†Section of Astrophysics Astronomy and Mechanics, Department of Physics, University of Athens, Panepistimiopolis, GR 157 83, Athens, Greece

Abstract. We investigate a number of iso-energetic N-body models of rotating galaxies, i.e. models that have the same total mass and the same total binding energy, but different amount of total angular momentum. These models are also in the same scalar virial equilibrium.

We show that density waves grow so that a rotating strong bar is formed embedded in a rotating thick disk. The surface density of the disk has an exponential radial profile. On the disk, density waves form spiral arms. The longevity of the spiral arms increases with increasing angular momentum of the model.

Separating the particles moving in regular orbits from the particles moving in chaotic orbits, we find that chaotic orbits have a considerably different spatial distribution than the regular orbits. Regular orbits form mainly a part of the central bar that works as a ballast for the stability of the bar. The disk and the spiral arms are almost completely formed by chaotic orbits. We conclude therefore that chaotic orbits can form disks with exponential surface density profiles, endowed with spiral arms. This new result reveals that chaotic orbits could be important in understanding the structure and evolution of exponential disks with spiral arms, at least in the case of strong bars.

Keywords: Spiral arms and bars; galactic disks, Spiral galaxies; Spiral arms and galactic disk
PACS: 98.62.Hr 98.52.Nr , 98.35.Hj

INTRODUCTION

N-body simulations of galaxy formation from cosmological initial conditions show that dissipationless collapse and relaxation produces triaxial non-rotating (or very slowly rotating) equilibrium configurations. This happens either in monolithic collapses or in mergers of sub-clumps formed earlier in an hierarchical clustering scenario. Such triaxial non-rotating equilibrium configurations should be considered as the simplest possible models of galaxies and they resemble to elliptical galaxies.

Two models of this type with smooth centers have been examined in [16] as regards the fractions of mass in regular (ordered) motion and in chaotic motion. In these smooth center models the spatial density flattens inside a particular radius, which is about 1% of their half mass radius. The two models have been derived respectively from Clumpy (C model) and from Quiet (Q model) cosmological initial conditions.

The self-consistent run of these models is realized by a conservative technique code [1]. In this code the potential is smooth and the noise in running the orbits is low. Such codes are usually called also Smooth Field Codes (SFC). SFC codes give a solution of the collisionless Boltzmann equation

CP848, *Recent Advances in Astronomy and Astrophysics,*
7th *International Conference of the Hellenic Astronomical Society,* edited by N. H. Solomos

$$\frac{\partial f}{\partial t} + \dot{\vec{x}}\,\nabla_x f + \dot{\vec{v}}\,\nabla_v f = 0 \tag{1}$$

and the Poisson equation

$$\nabla^2 V = 4\pi G\rho \tag{2}$$

where the spatial density $\rho(\vec{x},t)$ at $\vec{x}$ is evaluated at the time t from the distribution function $f = f(\vec{x},\vec{v},t)$ through the integral

$$\rho = \int f(\vec{x},\vec{v},t)d^3\vec{v} \tag{3}$$

In this solution a smooth potential $V(\vec{x})$ is derived by the code, that is expressed as a series in terms of a particular basis of eigen-functions. The coefficient of the series are evaluated at every small time step Δt. In this way a self-consistent evolution of the system can be obtained.

The method applied for the distinction between regular and chaotic orbits is described in details in [16]. In brief the method is as follows: We fix the coefficients of the potential at a given snapshot of the self-consistent run, so that an autonomous Hamiltonian for the system can be written at this snapshot. From the equations of motion derived from this Hamiltonian we can also derive the variational equations along the orbit of every particle described in this potential. From the variational equations two indices can be found along an orbit, namely the Finite Time Lyapunov Characteristic Number (FT-LCN) and the Alignment Index (AI). We have shown that the combination of these two indices gives a sensitive reliable distinction between the regular and the chaotic character of the orbits. Thus, libraries of particles moving in one or the other type of motion can be constructed at any particular snapshot of the evolution of the N-body system.

In summary our results in [16] are a follows. The C model has a maximum ellipticity E4. The mass in chaotic motion in C is found to be about 24% of the total mass. The Lyapunov numbers of the orbits are found in general small, so that only 2% of the total mass is chaotic enough, able to develop chaotic diffusion in a Hubble time. The Q model has a high maximum ellipticity E7. The fraction of mass in chaotic motion in Q is found about 32% of the total mass and the values of the Lyapunov numbers are found higher than in C, so that a fraction of mass about 8% is able to develop chaotic diffusion in a Hubble time.

Similar results were found in [11], in a non-rotating triaxial smooth center model. An even larger fraction of mass in chaotic motion of about 53% was found. This larger fraction can be explained mainly by the fact that in this model the density flattens inside a radius smaller than in our cases, so the center is not equally smooth as in our models. Actually, the result of [11] verify our result that chaos in triaxial equilibrium models of non-rotating galaxies with smooth center is not negligible. The mass in chaotic motion is at least several tens per cent of the total mass.

Another interesting result found in [16] and in [11] is that the mass in chaotic motion has a different spatial distribution than the mass in regular motion. In particular, the total density profile can form in principle an observable hump, due to the difference of the two distributions.

However, the Lyapunov numbers of the chaotic orbits are in general small and the mass that can develop chaotic diffusion in a Hubble time is a small fraction of the total mass. This result is compatible with the simplicity and the stability of non-rotating equilibrium models of galaxies.

An important question that arises here is how rotation can affect the mass in chaotic motion and the mean level of the Lyapunov numbers as well as the stability of the pattern of the system.

In order to investigate this question, we use the Q model described above and we construct from it a series of so called iso-energetic rotating models of galaxies. Iso-energetic means that the models have the same mass and the same total binding energy, but diferent amount of total angular momentum. Also, they are in the same scalar virial equilibrium.

The scaling unit of length is equal to the half mass radius R_h in every model. Also, a scaling unit of the potential is defined, so that the deepest value of the potential well (at the center) of every model is V(0)=-100. The half mass crossing time

$$T_{hmct} = (2R_h^3/GM)^{1/2}$$

is defined as the unit of time. A Hubble time in these time units is $t_{Hub} \approx 300T_{hmct}$.

In the next section the initial conditions of the rotating models are described.

INITIAL CONDITIONS OF ISO-ENERGETIC ROTATING MODELS

The non-rotating Q model is in a well relaxed triaxial equilibrium configuration. The orbits have no preferable direction of rotation. The ratio of the direct to the retrograde orbits with respect to any axis is very close to $1:1$. The three components of the total angular momentum, J_x, J_y, J_z along the principal axes of the system are very close to zero.

Rotation can be inserted artificially by re-orienting the velocities of particles, so that all the orbits rotate initially around the same axis and along the same direction. (Of course, one can create rotating models by only a partial initial re-orientation of the velocities). This is a kind of "conspiracy" between orbits, introduced artificially, that increases the total amount of angular momentum in the system, but preserves the kinetic energy.

A measure of the angular momentum J, for a gravitational system of total mass M and total binding energy E, can be given in terms of the spin parameter λ introduced by [12]. This parameter is defined by the equation

$$\lambda = \frac{J|E|^{1/2}}{GM^{5/2}} \tag{4}$$

Notice that the value of λ of the disk of our Galaxy is $\lambda_{Galaxy} = 0.22$, as has been evaluated by [7] from the value of the angular momentum of our Galaxy given by [9].

The conspiracy of the directions of rotation of the orbits mentioned above is important for the formation of rotating patterns and gives to the spin parameter λ values comparable to the value of our Galaxy.

It should be stressed here that in a system with random directions of rotation of the orbits the total angular momentum of the system is very small, or zero. Even if the orbits rotate around the same axis, but those in direct rotation are comparable and mixed with the orbits in retrograde rotation, angular momentum of the system is again small, or zero. Pattern rotation in such models is negligible, or zero.

If galaxies are formed from small initial density perturbations in a hierarchical clustering cosmological scenario and they have acquired their angular momentum from tidal interactions with other density perturbations of their environment in the early Universe, a conspiracy between the directions of rotation of the orbits cannot be obtained in a dissipationless scenario either in monolithic collapses, or in mergers between sub-clumps formed hierarchically. Tidal interactions between density perturbations can only give structures with a mean value of the spin parameter of about $\lambda = 0.05$ (e.g. [7], [4], [18], [14], [15]). The main reason for such a small value of λ is the mixing between direct and retrograde orbits.

Various estimations show that in a galaxy formation process a value of λ near 0.2 can be achieved, if the collapse factors are remarkably larger than the value of 2, that results from the virial theorem when it is applied to a dissipationless collapse process. Such a large value of the collapse factor can be obtained by dissipative processes. Furthermore, dissipative processes absorb the retrograde motion (by collisions between gas clouds) and lead to systems where direct motion dominates.

Therefore, according to this scenario, only the dissipative mass, i.e. the baryonic component, of a galaxy can be characterized by large values of λ. This component is mostly related to the disk of galaxies. In principle, a dark matter halo should be characterized by small values of λ, compatible with those produced by cosmological tidal fields, as mentioned above.

In the present paper we compare iso-energetic rotating models of various angular momenta with a non-rotating model, in order to reveal the consequence of rotation as regards the level of chaos. For this reason, we examine the simple case where no distinction between disk and halo components is made. An extension of this work, combining a disky component of large λ with a halo component of small λ, will be presented elsewhere.

In order to create initial conditions for an iso-energetic rotating model based on the Q model, we apply the following velocity re-orientation process:

At a given snapshot of the self-consistent evolution of the Q model, at $t_Q = 100$, the component v_{yz} of the velocity of every particle (parallel to the plane of intermediate-longest axes) is re-oriented, so that it becomes perpendicular to the current radius r_{yz} of the particle on this plane, pointing along the same (clockwise) direction.

This velocity re-orientation process does not affects the total kinetic energy T and respects the scalar viral equation of the system

$$T = T_{rot,yz} + T_{rad,yz} + T_x = -E \tag{5}$$

where $T_{rot,yz}$ and $T_{rad,yz}$ are the rotational and the radial kinetic energies parallel to the $y-z$ plane. T_x is the kinetic energy along the (shortest) x axis.

The initial conditions created in this way form a rotating new model called QR1. At the time $t_{QR1} = 0$, when the running of this model starts, the radial components of the velocities parallel to the y-z plane are zero. All the kinetic energy parallel to this plane is rotational kinetic energy around the shortest axis (x), of the same direction. As a consequence, the total angular momentum of the system is equal to a non-zero component J_x. During the (self-consistent) evolution of this model, radial motion grows rapidly, but the total angular momentum J_x of the system remains constant.

In order to create initial conditions for the next model, QR2, we choose a snapshot of the self-consistent run of the QR1 model at $t_{QR1} \approx 20$. This period of time is long enough for some very fast transient effects of the QR1 model, due to the redistribution of the types of orbits, to decay. During this period considerable amount of radial motion has been grown. The initial conditions of the new model QR2 are created by applying again the velocity re-orientation process at this snapshot of the QR1 model. The total angular momentum J_x becomes larger in QR2 than in QR1.

Initial conditions of the model QR3 are created in a similar way by using a snapshot of the self-consistent run of the QR2 model at $t_{QR2} \approx 20$ on which the same velocity re-orientation process is applied. Angular momentum J_x is larger in QR3 than in QR2.

The initial conditions of the model QR4, with angular momentum J_x larger that in QR3, are created from the model QR3 in a similar way.

The values of λ of the above systems corresponding to the mass inside a cylinder with radius $r = 3$ on the equatorial plane and hight $x = \pm 1.5$ on either side of the equatorial plane, for the four models respectively, are $\lambda \simeq 0.20,\ 0.23,\ 0.27,\ 0.28$.

Thus, we have a sequence of five iso-energetic models, Q, QR1, QR2, QR3, QR4, with increasing amounts of their angular momentum J_x.

RESULTS

In Fig.1 projections of the four rotating models QR1, QR2, QR3, QR4 are displayed at a time $t_1 \approx 20$ of their self-consistent run. (The precise values of t_1 are shown in the figure). The egde-on projections (on the yx-plane) and the face-on projections (on the equatorial yz-plane) are shown respectively in the first and the second row of this figure.

During the self-consistent evolution of the models density waves grow to form a strong bar embedded in a thick disk. The bar is triaxial and rotates around its shortest axis x. The size of the bar decreases with increasing angular momentum along the sequence of the models.

In the models QR2, QR3, QR4 density waves form two symmetric spiral arms emanating from the ends of the bar, while in the model QR1 no such spiral arms appear. This means that spiral arms require angular momentum above a certain threshold in order to be present by the time of projection.

In fact spiral arms are spontaneously formed in all the models but in QR1, they only live for a time of about $t = 10$. In the models QR2, QR3, QR4 spiral arms live remarkably longer, i.e. for times about $t = 80, 90, 180$, respectively. During these times spiral arms temporarily disappear, or fade (for a short period of the order of T_{hmct}), but they grow

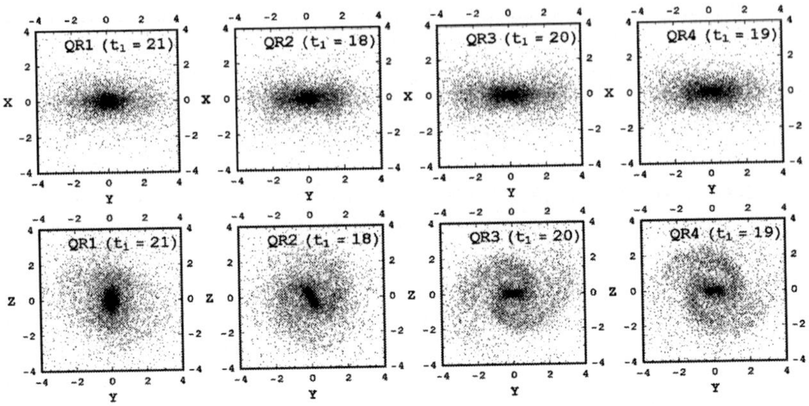

FIGURE 1. Edge on (first row) and face on projection (second row) of the four rotating models. Density waves form a bar embedded in a thick disk. At the time of projection spiral arms appear in QR2, QR3, QR4.

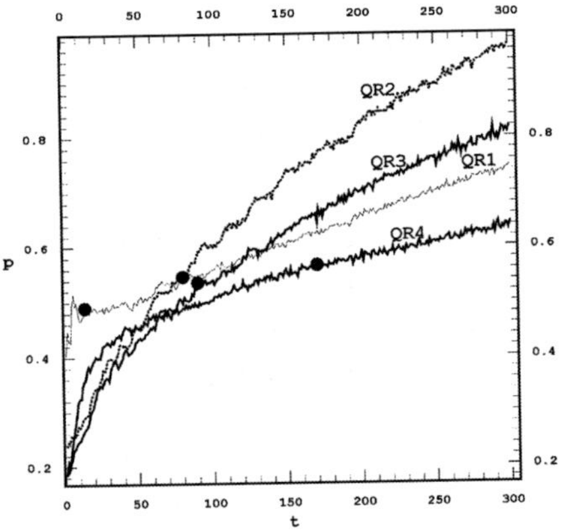

FIGURE 2. Evolution of the parameter p along the self-consistent run of the rotating models. Heavy dots indicate for how long spiral arms are discernible in each model.

again. This happens several times.

During their self-consistent runs for a Hubble time the rotating models develop considerable secular evolution, in contrast to the non-rotating model Q that remains almost unaltered.

We have found that the ratio of the kinetic energy in radial motion parallel to the equatorial plane with respect to the kinetic energy in rotational motion is a decisive parameter as regards the longevity of the spiral arms. In order to illustrate this effect we define the parameter

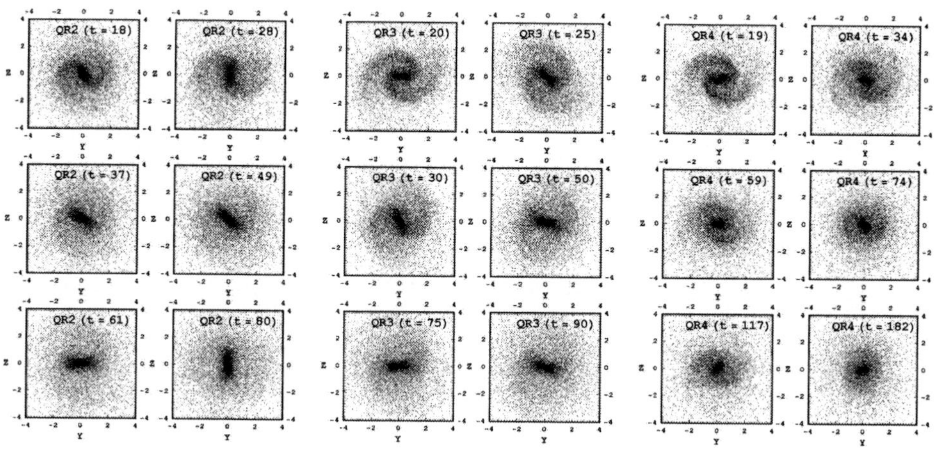

FIGURE 3. Projections on the equatorial plane of QR2 (first and second columns), QR3 (third and fourth columns) and QR4 (fifth and sixth columns) at various snapshot of their self-consistent evolution at which spiral arms are discernible.

$$p = \frac{< V^2_{rad,yz} >}{< V^2_{rot,yz} >} \qquad (6)$$

where $< V^2_{rad,yz} >$ is the mean square of the radial velocities parallel to the equatorial plane and $< V^2_{rot,yz} >$ is the mean square of the rotational components of the velocities.

In Fig.2 the parameter p is plotted as a function of time for all the rotating models. The heavy dots in this figure indicate for how long spiral arms can be discerned in each model. If p becomes greater than a value $p_c \approx 0.5$, spiral arms either disappear or become so faint that can hardly be discerned.

In Fig.3 a number of projections on the equatorial plane of the models QR2, QR3 and QR4 are shown at various snapshots as long as spiral arms are discernible.

The radial profiles of the mean surface density $\sigma_{yz}(r)$, as a function of the radius r, in log-linear scale, on the equatorial plane, of all the models are shown in the first column of Figs.4 . The time variations of these profiles during a Hubble time are small. For example, in the frames of QR1,QR2,QR3,QR4 in the first column of Fig.4 the surface density profiles are drawn at $t_1 \approx 20$ (solid line) and at $t_2 = t_{Hub} = 300$ (dashed line), showing no serious differences.

As the rotation increases along the sequence of the models, the value of the surface density at the center increases and the radial profiles near the center become steeper. The slope

$$k = -\frac{d \log \sigma_{yz}(r)}{dr} \qquad (7)$$

is shown as a function of the radius in the second column of Figs.4. In the rotating models this slope is quite constant beyond a radius, indicating an exponential profile of

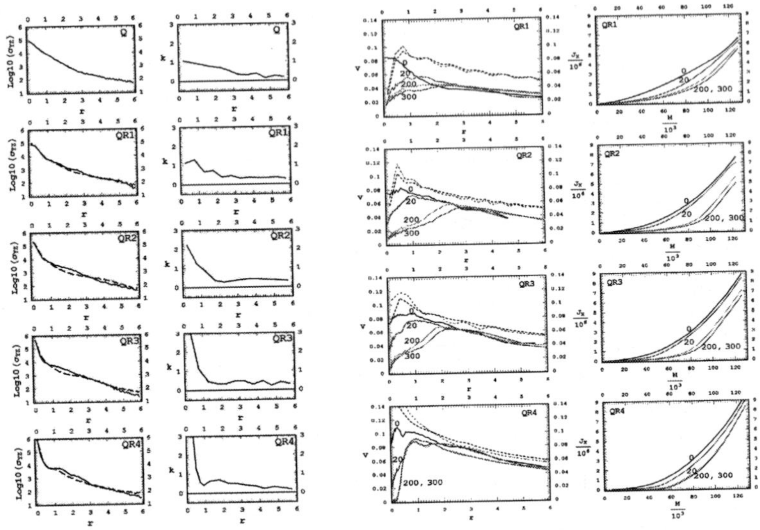

FIGURE 4. **First column:** Surface density profiles of all the models (Q, QR1, QR2, QR3, QR4) on the equatorial plane (yz). **Second column:** The slope of the profile in the first column indicates that beyond a radius the profiles are exponential. **Third column:** Dashed lines give the "dynamical" rotational velocity curve $v(r)$ at $t_1 \approx 20$ and $t_2 = 300$. Solid lines give the rotational velocity curves $v_i(r_i)$ from the distribution of the angular momentum along the radius at various snapshots. While $v(r)$ remains almost constant in time revealing a remarkably constant distribution of mass, $v_i(r_i)$ in the inner parts decreases in time due to the transference of angular momentum outwards. **Fourth column:** Cumulative angular momentum vs cumulative mass from the center at the same snapshots as the third column indicating the transference of angular momentum outwards.

$\sigma_{yz}(r)$ in this region.

It is worth noticing that the mean radial profiles of the surface density given by [8] for stellar disks of optically flocculent and grand design spirals are quite similar to the profiles of the first column of Figs.4.

Two types of rotational velocity curves can be evaluated for each model.

First, a "dynamical" rotational velocity $v(r) = \sqrt{rV'(r)}$ curve as evaluated from the centrifugal equilibrium under the forces $-V'(r)$ due to the monopole terms of the potential on the equatorial plane. This curve measures actually the distribution of mass along the radius. For each model two dynamical rotational velocity curves are plotted in the third column of Figs.4 by dashed lines corresponding to two different snapshots of the self-consistent run, at $t_1 \approx 20$ and at $t_2 \approx 300$, respectively. We see no significant variations between these two curves, indicating a remarkably constant distribution of the mass during this long period.

Second, rotational velocity curves can also be derived in terms of the distribution of the angular momentum. In order to find such a rotation curve at a given snapshot we define 50 successive cylindrical co-axial rings around the rotation axis (x), so that the mass inside every ring is the same for all the rings. The rotational velocity v_i at the mean radius r_i of the ring i is evaluated from the equation

$$v_i = \frac{J_i}{r_i} \tag{8}$$

where

$$J_i = \frac{1}{N_i} \sum_{j=1}^{N_i} (y_j \dot{z}_j - \dot{y}_j z_j) \tag{9}$$

is the mean angular momentum per particle along the x axis of the N_i particles in the ring i.

In the third column of Figs.4 the rotational velocity curves $v_i(r_i)$ of the four models are shown by solid lines corresponding to $t = 0,\ 20,\ 200,\ 300$, as indicated by the numbers next to the curves.

Comparing the two types of the rotational velocity curves shown in the third column of Fig.4 it becomes clear that although the dynamical rotational velocity curves remain almost constant, the rotational velocity curves $v_i(r_i)$, resulting from the distribution of the angular momentum evolve in time in all the rotating models. In the inner parts (e.g. for $r < 1$) the rotational velocity $v_i(r_i)$ becomes gradually smaller. This occurs in a decreasing rate and it is a result due to the transference of angular momentum outwards.

The transference of angular momentum is shown in the fourth column of Figs.4, where the cumulative angular momentum is plotted vs the cumulative mass from the center in each of the rotating models at the same snapshots as the rotation curves in the third column. It is clear that at later times more mass is required in order to collect the same amount of angular momentum.

In [10] it is shown that in a rotating disk angular momentum is transferred outwards. This transference is due to the interaction of a bar-like perturbation of the gravitational field with stars moving in resonant orbits particularly near ILR and CR. An extension of this analysis in [13] and [17] has shown that angular momentum transference can be interpreted as an effect caused by a particular type of dynamical friction.

N-body simulations indicate that angular momentum transference outwards is always present in rotating galactic models. This effect takes place not only between the bar and the disk, but it is extended to the halo particles (see for example, [5], [6], [2], [3]). Transference of angular momentum outwards in rotating galaxies produces secular dynamical evolution in which the bar grows and slows down. In the next section we show that chaotic orbits play an important role in this process.

Fractions of mass in regular and in chaotic motion

Following the method introduced in [16], we use the Finite Time Lyapunov Characterisitc Number (FT-LCN) and the Alignment Index (AI) to obtained a reliable distinction between regular and chaotic orbits at a particular snapshot of the self-consistent evolution of every system. This distinction refers to the type of the orbits described by the particles if their positions and velocities are used as initial conditions in the autonomous

Hamiltonian (in cylindrical coordinates (r,ϕ,x))

$$H=\frac{P_r^2}{2}+\frac{P_\phi^2}{2r^2}-\Omega_p P_\phi+\frac{\dot{x}^2}{2}+V_0(\sqrt{r^2+x^2})+V_1(r,\phi,x) \tag{10}$$

In this Hamiltonian $V_{tot}(r,\phi,x)=V(\sqrt{r^2+x^2})+V_1(r,\phi,x)$ is the potential given by the N-body code at a particular snapshot of the evolution. (The values of the coefficients of the expansion of the potential are fixed at this snapshot). $V_0(\sqrt{r^2+x^2})$ is the monopole component of the potential, while $V_1(r,\phi,x)$ is the triaxial component of the potential. The parameter Ω_p is the pattern speed estimated from the rotation of the bar. Namely, we find first, numerically, the function of time $\Phi(t)$, that gives the angle Φ of the orientation of the longest axis of the inertia ellipsoid of the mass inside a radius r (containing the bar), with respect to the initial orientation of this axis. The pattern speed of the bar is a function of time evaluated from the slope of $\Phi(t)$ at time t, i.e. $\Omega_p(t)=\frac{\Delta\Phi(t)}{\Delta t}$.

If the distinction between regular and chaotic orbits is applied at a snapshot $t_1\approx 20$ of the self-consistent run of the rotating models we find that a set R1 of particles move in regular orbits, while a set C1 move in chaotic orbits in every model. As in [16], the distinction between the two sets is obtained using a threshold of the Finite Time-Specific Lyapunov Characteristic Number (FT-SLCN), denoted by L_j, that measures the mean rate of the divergence between neighboring orbits in units of the inverse radial periods $1/T_{rj}$ of every orbit j. This threshold is $L_j > 10^{-2.8}$ and it is combined with a threshold in the values of the $AI < 10^{-3}$.

The set C1, as a fraction of the total mass, along the sequence of the models Q,QR1,QR2,QR3,QR4 is about 32%, 57%, 72%, 69%, 68%. Rotations remarkable increases the mass in chaotic motion.

If the same method of distinction between regular and chaotic orbits is applied at various snapshots of the self-consistent run of the models, we find that the large majority (more than 90%) of the particles in the set $R1$ (that are found in regular motion at t_1) maintain the character of their motion up to the end of the simulation $t_{Hub}=300$. The rest of the particles in R1 change their orbits to chaotic. Also, the majority of the particles in C1 (that are found in chaotic motion at t_1) maintain their chaotic character of motion.

In the models Q and QR1 the fraction of mass in regular or in chaotic motion are found roughly constant in time. However, in QR2, QR3 the fraction of mass in chaotic motion shows a tendency to decrease in time, so that at $t_{Hub}=300$ these fractions become 58% and 62%, respectively. This tendency for self-organization of the orbits is due to a secular evolution of the potential. In QR4 self-organization is more pronounced, so that the fraction of mass in chaotic motion at $t_{Hub}=300$ falls to the level of about 40%.

Fig.5 summarizes the results regarding the fractions of mass in chaotic motion found at various snapshots for each model.

The mass in regular and the mass in chaotic motion have different distributions along the energy axis. As an example we give the distributions of the regular and the chaotic orbits at the snapshot at $t_1\approx 20$, when the sets R1 and C1 are found.

In Figs.6a,b,c,d the effective potential in the rotating frame (right vertical axes V in each panel) as a function of the radius r (down horizontal axis r), when r is taken along the longest axis of the bar is shown by the dotted line. (The values of the potential are normalized so that the deepest value at the center of the every system is -100). A

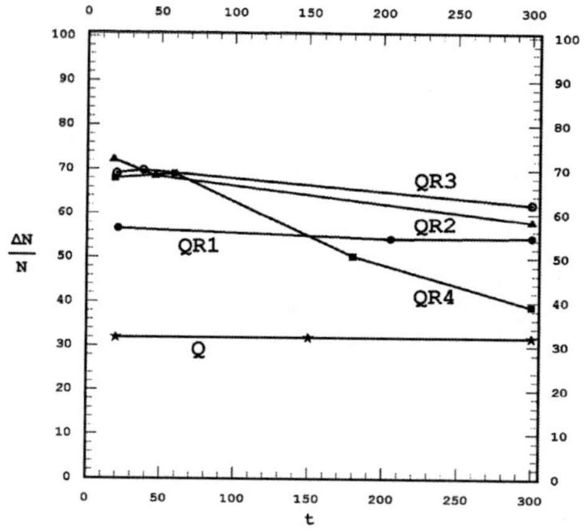

FIGURE 5. Fractions of mass in chaotic motion at various snapshots for all the models. Rotation produces a remarkable increase of the mass in chaotic motion. In Q and QR1 these fractions are roughly constant in time. In QR2 and QR3 some self-organization is observed. In QR4 self-organization is more pronounced.

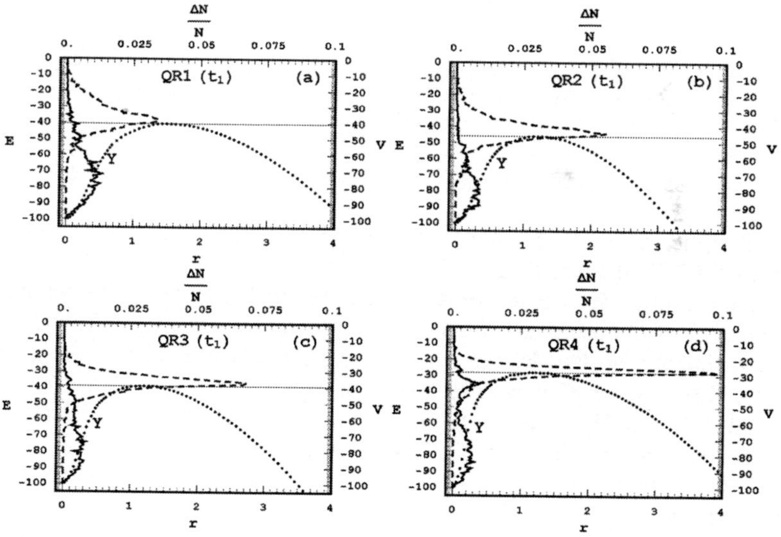

FIGURE 6. Distribution along the energy axis of the mass in chaotic orbits (dashed line) and the mass in regular orbits (solid line) in the rotating models. The dotted line gives the effective potential as a function of radius for comparison. Chaotic orbits are concentrated near the energy of corotation (horizontal straight line).

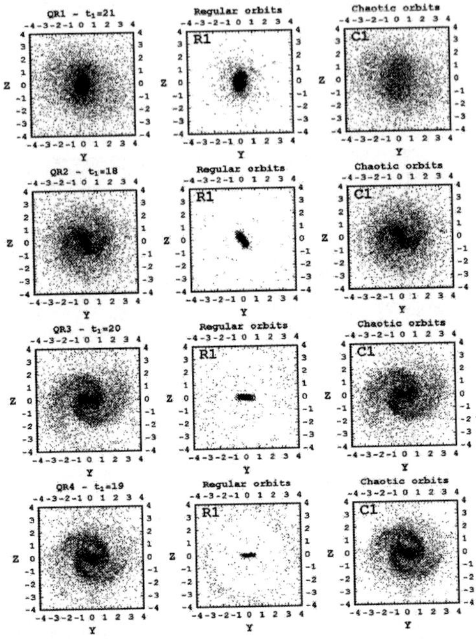

FIGURE 7. Projections on the equatorial plane of the rotating models. Every row belong to the same rotating model. **Left column:** a sample of particles uniformly distributed along the mass in each system, **Middle column:** regular orbits R1 of the sample, **Right column:** chaotic orbits C1 of the sample The exponential disk and the spiral arms are almost competely formed by chaotic.

horizontal straight line is also plotted at the maximum of this potential, i.e. the value of the Jacobi integral E_c at corotation.

In the same figures, for comparison, the left vertical axis gives the values of the Jacobi integral E of the orbits and the upper horizontal axis gives the fraction of mass $\Delta N_c(E)/N$ of chaotic orbits (dashed curve) and the fraction of mass in regular orbits $\Delta N_r(E)/N$ (solid curve) as they are distributed along the E axis.

As it can be seen in Figs.6a,b,c,d the maximum of the distribution of chaotic orbits occurs for energies around the corotation energy E_c.

Regular orbits dominate for values of the Jacobi integral $E < -60$, inside radii smaller than one half of the corotation radius.

The different distribution of regular and chaotic orbits along the energy axis implies an important difference in their spatial distribution.

In Figs.7 the projections of the rotating models on their equatorial plane are shown at $t_1 \approx 20$. The panels of the same row correspond to the same model. For a more clear illustration, a sample of 1 particle every 10 particles, uniformly distributed along the mass of the system, is plotted in the left panel for each model.

The set C1 of particles from this sample, that are detected in chaotic motion, is plotted separately in the right panel of the same row, while the set R1 of particles from the same sample, that are detected in regular motion, is plotted separately in the middle panel of the same row.

By comparing the projections of the sets R1 and C1 of the same model in these figures it becomes clear that the exponential disk with spiral arms appearing in QR2,QR3,QR4 are almost completely formed by chaotic orbits. This is a striking new result revealing that a mechanism able to produce such patterns should be sought in the properties of chaotic orbits rather than in the properties of regular orbits, at least in the case of large deviations from axial symmetry (strongly perturbed disk). Furthermore, chaotic orbits are probably responsible for the variability, the transference of angular momentum and the secular evolution of galactic models.

ACKNOWLEDGMENTS

We wish to thank Prof. G. Contopoulos and Drs P. Patsis and Ch. Efthymiopoulos for their useful comments and discussions. We wish to thank the Empeirikeion Foundation for financial support in the respective research program. I.S. wishes to thank the Greek State Scholarship Foundation for financial support.

REFERENCES

1. Allen A. J., Palmer P. L., Papaloizou J., 1990, MNRAS, 242,576.
2. Athanassoula E. 2002, ApJ, 569, L83.
3. Athanassoula E. 2003, MNRAS, 341, 1179.
4. Barnes J., Efstathiou G., 1987, ApJ, 319, 575.
5. Debattista V. P., Sellwood, J. A., 1998, ApJ, 493, L5.
6. Debattista V. P., Sellwood, J. A., 2000, ApJ, 543, 704.
7. Esstathiou G., Jones B.J.T., 1979, MNRAS, 186, 133.
8. Grosbol P.J., Patsis P., 1998, A&A, 336, 840.
9. Innanen K., 1966, ApJ, 143, 150.
10. Linden-Bell D., Kalnajs A.J., 1972, MNRAS, 157, 1.
11. Muzzio J. C., Carpintero D.D., Wachlin F.C., 2005, Cel. Mech. Dyn. Astr., 91, 173.
12. Peebles P.J.E., 1969, ApJ., 155, 393.
13. Tremaine S., Weinberg M., 1984, MNRAS, 209, 729.
14. Voglis N., Hiotelis, N., 1989, A&A, 218,1.
15. Voglis N., Hiotelis, N., Hoflich P. 1991, A&A, 249,5
16. Voglis N., Kalapotharakos C., Stavropoulos, 2002, MNRAS, 337, 619
17. Weinberg M., 1985, MNRAS, 213, 451.
18. Zurek W.H., Quinn P.J., Salmon J.K., 1988, ApJ, 330, 519

BARRED GALAXIES: SECULAR EVOLUTION AND THE ROLE OF THE HALO

E. Athanassoula

Laboratoire d'Astrophysique de Marseille (LAM), Observatoire Astronomique de Marseille-Provence (OAMP), 2 Place Le Verrier, 13248 Marseille cedex 04, FRANCE.

Abstract. In this talk I will summarize some of my recent results on secular evolution of barred galaxies, emphasizing the role of the halo. In particular, I will discuss the effect of a central mass concentration, the effect of velocity anisotropy in the halo velocity distribution, the orbital structure in the halo of strongly barred galaxies and how triaxiality can affect the expected annihilation signal from weakly interacting particles in the central regions of disc galaxies with bars.

Keywords: Stellar dynamics and kinematics; Galactic halos; Spiral arms and bars
PACS: 98.10.+z; 98.62.Gq; 98.62.Hr

Central mass concentrations (CMCs), be they massive black holes or compact massive central discs, affect the secular evolution of bars. If they are sufficiently massive or concentrated, they lead to a considerable decrease of the bar amplitude and some times to the total destruction of the bar. Orbital structure studies ([1], [2]) showed that with increasing CMC mass (or concentration) the fraction of stable bar-supporting orbits decreases, thus reducing the possibilities for barred equilibria. The effect of CMCs on bars was further elaborated by partly self-consistent *N*-body simulations ([3], [4], [5]) in which the disc is described by particles, but the halo is only considered as a rigid potential which does not evolve and does not interact with the other components of the galaxy. However, in order to study the effect of the halo on the evolution, we need to consider a live halo, composed of particles. This was done in [6], where a number of interesting results were shown. The devastating role of massive and/or compact CMCs was verified. It was further shown that the effect of the CMC depends also on the bar type and strength. Bars with an exponential radial density profile are more vulnerable than strong bars with a flat radial density profile. CMCs also affect the vertical profile of a bar, turning X-shaped profiles into peanuts and peanuts into boxy shapes. While its amplitude decreases, the pattern speed of the bar increases.

Very little is known on the velocity distribution in the halo. It is thus important to ask how an anisotropic halo velocity distribution can affect the interaction of the disc and the halo. This has been addressed analytically in [7], using the shearing sheet approximation to model a patch of a galactic disc. It was shown that velocity anisotropy increases considerably the maximum growth factor of perturbations in the disc. Although based on a number of approximations, this result can be extrapolated to predict that spirals and bars embedded in haloes with anisotropic velocity distributions should, on average, be stronger than those in haloes with isotropic distribution of velocities. *N*-body simulations to test this result are underway.

Secular evolution affects not only the disc, which forms the bar and also becomes

CP848, *Recent Advances in Astronomy and Astrophysics*,
7th *International Conference of the Hellenic Astronomical Society*, edited by N. H. Solomos

more centrally concentrated, but also the halo. The latter loses its spherical symmetry and also forms a bar, which is shorter and fatter than the disc bar ([8], [9], [10]). Preliminary results on the orbital structure in discs and haloes of strongly barred galaxies have been given in [9]. Halo near-resonant orbits are similar to the corresponding disc ones. Thus, they are trapped around the main families of periodic orbits such as the x1v1, x1v2, x1v4, long period banana orbits, or l_1 (for definitions see [11]). Particles which during the evolution get trapped at the corotation resonance are not a random sub-sample of the initial distribution function. Instead, they have preferentially smaller values of $|u_z|$ and larger values of J_z than the other particles.

The annihilation signal from weakly interacting massive particles (WIMPs) at the galactic center depends on their density distribution and could thus be influenced by the fact that this is not necessarily spherically symmetric ([8], [9], [10]). [12] showed that asphericity has a strong impact on the annihilation signal when the halo profile near the galactic center is flat (i.e. for profiles in agreement with observations [13]), but becomes less important as the profile becomes more cuspy (i.e. for profiles in agreement with cosmological N-body simulations [14], [15], [16]).

More information on these subjects can be found in the references cited below.

REFERENCES

1. H. Hasan, C. Norman, *Astrophys. J.*, **361**, 69-77, (1990).
2. H. Hasan, D. Pfenniger, C. Norman, *Astrophys. J.*, **409**, 91-109, (1993).
3. C. A. Norman, J. A. Sellwood, H. Hasan, *Astrophys. J.*, **462**, 114-124, (1996).
4. J. Shen, J. A. Sellwood, *Astrophys. J.*, **604**, 614-631, (2004).
5. S. Hozumi, L. Hernquist, *Publ. Astron. Soc. Japan*, **57**, 719-731, (2005).
6. E. Athanassoula, J. C. Lambert and W. Dehnen, *Mon. Not. R. Astron. Soc.*, **363**, 496-508, (2005).
7. B. Fuchs and E. Athanassoula, *Astron. Astrophys.*, **444**, 455-459, (2005).
8. E. Athanassoula, *Celest. Mech. and Dyn. Astron.* **91**, 9-31, (2005).
9. E. Athanassoula, *Nonlinear Dynamics in Astronomy and Physics (In memory of Henry E. Kandrup)*, eds. S. T. Gottesman, J. R. Buchler and M. E. Mahon, Annals of New York Academy of Sciences, **1045**, 168-192, (2005).
10. P. Colin, O. Valenzuela, A. Klypin, astro-ph/0506627, (2005).
11. Ch. Skokos, P. Patsis, E. Athanassoula, *Mon. Not. Roy. Astron. Soc.*, **333**, 847-860, (2002).
12. E. Athanassoula, F. S. Ling and E. Nezri, *Phys. Rev. D*, **72**, 3503-3508, (2005).
13. A. Bosma, *in* S. D. Ryder *et al.*, International Astronomical Union Symposium **220**, 39-48, (2004).
14. J. F. Navarro, C. S. Frenk, S. D. M. White, *Astrophys. J.* **490**, 493-508, (1997).
15. B. Moore *et al.*, *Mon. Not. Roy. Astron. Soc.*, **310**, 1147-1152, (1999).
16. J. Diemand *et al.*, *Mon. Not. Roy. Astron. Soc.*, **364**, 665-673, (2005).

BARRED GALAXIES: STUDYING THE CHAOTIC AND ORDERED NATURE OF ORBITS

T. Manos*,† and E. Athanassoula†

*Center for Research and Applications of Nonlinear Systems (CRANS), Department of Mathematics, University of Patras, GREECE.
†Observatoire Astronomique de Marseille-Provence (OAMP), 2 Place Le Verrier, 13248 Marseille cedex 04, FRANCE.

Abstract. The chaotic or ordered character of orbits in galactic models is an important issue, since it can influence dynamical evolution. This distinction can be achieved with the help of the *S*maller *AL*ingment *I*ndex - (*SALI*). We describe here briefly this method and its advantages. Then we apply it to that case of 2D and 3D barred galaxy potentials. In particular, we find the fraction of **chaotic and ordered orbits** in such potentials and present how this fraction changes when the main parameters of the model are varied. For this, we consider models with different bar mass, bar thickness or pattern speed. Varying only one parameter at a time, we find that bars that are more massive, or thinner, or faster, have a larger fraction of chaotic orbits.

Keywords: Barred galactic potentials, Chaotic motion, Ordered motion, Hamiltonian systems
PACS: 95.10.Ce, 95.10.Fh, 05.45.-a, 98.62.Hr, 98.62.Dm

INTRODUCTION

The chaotic and ordered motion in dynamical systems is a very important issue in non - linear sciences in general and in dynamical astronomy in particular. The qualitative distinction of the motion is, generally, not that trivial and it becomes harder in systems of many degrees of freedom. Thus, it is necessary to use fast and precise methods to indicate the nature of the orbits.

Many methods have been developed over the years trying to give an answer to this problem. The inspection of the successive intersections of an orbit with a Poincaré surface of section (PSS) [1] has been used mainly for 2 dimensional (2D) maps and 2 degrees of freedom Hamiltonian systems. One of the most common methods of chaos detection is the computation of the "Maximal Lyapunov Characteristic Number" (LCN) [2],[3], which can be applied for systems with many degrees of freedom. Another very efficient method is the "Frequency Map Analysis" [4],[5]. In recent years new methods have been introduced such as the study of spectra of "Short Time Lyapunov Characteristic Numbers" [6],[7] or "Stretching Numbers" [8],[9] and the "Spectral Distance" of such spectra [10], as well as the study of spectra of helicity and twist angles [11],[12],[13]. In addition, Froeschlé, Lega & Gonzi [14] introduced the "Fast Lyapunov Indicator" (FLI), while Vozikis, Varvoglis & Tsiganis [15] proposed a method based on the frequency analysis of "Stretching Numbers".

Recently, a new, fast and easy to compute indicator of the chaotic or ordered nature

CP848, *Recent Advances in Astronomy and Astrophysics,*
7th *International Conference of the Hellenic Astronomical Society,* edited by N. H. Solomos

of orbits, has been introduced [16] and was applied successfully in different dynamical systems [16],[17],[18],[19],[20],[21],[22],[23], the so-called "**S**maller **Al**ignment **I**ndex (**SALI**)", or, as elsewhere called **A**lignment **I**ndex **(AI)** [24],[25]. In the present paper, we first recall its definition and then we show its effectiveness in distinguishing between ordered and chaotic motion, by applying it to a barred potential of 2 and 3 degrees of freedom.

DEFINITION OF THE SMALLER ALIGNMENT INDEX

Let us consider the n - dimensional phase space of a conservative dynamical system, which could be a symplectic map or a Hamiltonian flow. We consider also an orbit in that space with initial condition $P(0) = (x_1(0), x_2(0), ..., x_n(0))$ and a deviation vector $v(0) = (dx_1(0), dx_2(0), ..., dx_n(0))$ from the initial point $P(0)$. In order to compute the SALI for a given orbit one has to follow the time evolution of the orbit with initial condition itself and also of two deviation vectors $v_1(t), v_2(t)$ which initially point in two different directions. The evolution of these deviation vectors is given by the variational equations for a flow and by the tangent map for a discrete-time system. At every time step the two deviation vectors $v_1(t)$ and $v_2(t)$ are normalized and the SALI is then computed as:

$$SALI(t) = min\left\{\left\|\frac{v_1(t)}{\|v_1(t)\|}+\frac{v_2(t)}{\|v_2(t)\|}\right\|, \left\|\frac{v_1(t)}{\|v_1(t)\|}-\frac{v_2(t)}{\|v_2(t)\|}\right\|\right\}. \qquad (1)$$

The properties of the time evolution of the SALI clearly distinguish between ordered and chaotic motion as follows: In the case of Hamiltonian flows or dimensional symplectic maps with $n \geq 2$, the SALI fluctuates around a non-zero value for ordered orbits, while it tends to zero for chaotic orbits [16],[18],[19]. In the case of 2D maps the SALI tends to zero both for ordered and chaotic orbits, following however completely different time rates, which again allows us to distinguish between the two cases. In 2 and 3 degrees of freedom Hamiltonian systems the distinction between ordered and chaotic motion is easy because the ordered motion occurs on a 2D or 4D torus respectively on which any initial deviation vector becomes almost tangent after a short transient period. In general, two different initial deviation vectors become tangent to different directions on the torus, producing different sequences of vectors, so that SALI does not tend to zero but fluctuates around positive values. On the other hand, for chaotic orbits, any two initially different deviation vectors tend to coincide in the direction defined by the nearby unstable manifold and hence either coincides with each other, or become opposite. This means that the SALI tends to zero when the orbit is chaotic and to a non-zero value when the orbit is ordered. Thus, the completely different behavior of the SALI helps us to distinguish between ordered and chaotic motion in Hamiltonian systems with 2 and 3 degrees of freedom and in general in dynamical systems of higher dimensionality.

THE MODEL

A 3D rotating bar model can be described by the Hamiltonian function:

$$H = \frac{1}{2}(p_x^2 + p_y^2 + p_z^2) + V(x,y,z) - \Omega_b(xp_y - ypx_x), \tag{2}$$

where x is the long axis, y the intermediate and z the short axis around which the bar rotates. The p_x, p_y and p_z are the canonically conjugate momenta. Finally, V is the potential, Ω_b represents the pattern speed of the bar and H is the total energy of the system.

The potential of the model V consists of three components:

a) A *disc*, represented by a Miyamoto disc [26]:

$$V_D = -\frac{GM_D}{\sqrt{x^2 + y^2 + (A + \sqrt{z^2 + B^2})^2}}. \tag{3}$$

where M_D is the total mass of the disc, A and B are the horizontal and vertical scale-lengths, and G is the gravitational constant.

b) The *bulge* is modeled by a Plummer sphere with the potential:

$$V_S = -\frac{GM_S}{\sqrt{x^2 + y^2 + z^2 + \varepsilon_s^2}} \tag{4}$$

where ε_s is the scalelength of the bulge and M_S is its total mass.

c) The triaxial Ferrers *bar*, the density $\rho(x)$ of which is:

$$\rho(x) = \begin{cases} \rho_c(1 - m^2)^2 & , m < 1 \\ 0 & , m \geq 1 \end{cases}, \tag{5}$$

where $\rho_c = \frac{105}{32\pi}\frac{GM_B}{abc}$ the central density, M_B is the total mass of the bar and

$$m^2 = \frac{x^2}{a^2} + \frac{y^2}{b^2} + \frac{z^2}{c^2}, \qquad a > b > c \geq 0, \tag{6}$$

with a, b and c are the semi-axes and M_B the mass of the bar component. The corresponding potential V_B and the forces are given in D. Pfenniger (1984) [27].

Applications in the 2D case

We first apply the SALI index to the 2 degrees of freedom barred potential. In this case, we can have a Poincaré Surface of Section (PSS) and can check the effectiveness of the SALI, by comparing the results.

In figure 1, we present two orbits of different kind: (i) $(x, y, p_x, p_y) = (1.5, 0, p_x(H), 0)$ and (ii) $(x, y, p_x, p_y) = (-0.9, 0, p_x(H), 0)$. For this application $H = -0.3$. In panels a) and b) of figure 1, we show the orbit projections in the (x, y)-plane and in panel c) we draw their corresponding PSS, where we can see that the chaotic orbit (i) tends to fill with scattered points the available part of the $(x, \dot{x})$ plane and that the ordered orbit (ii) creates a closed invariant curve. Finally, in panel d) we apply the SALI method for these

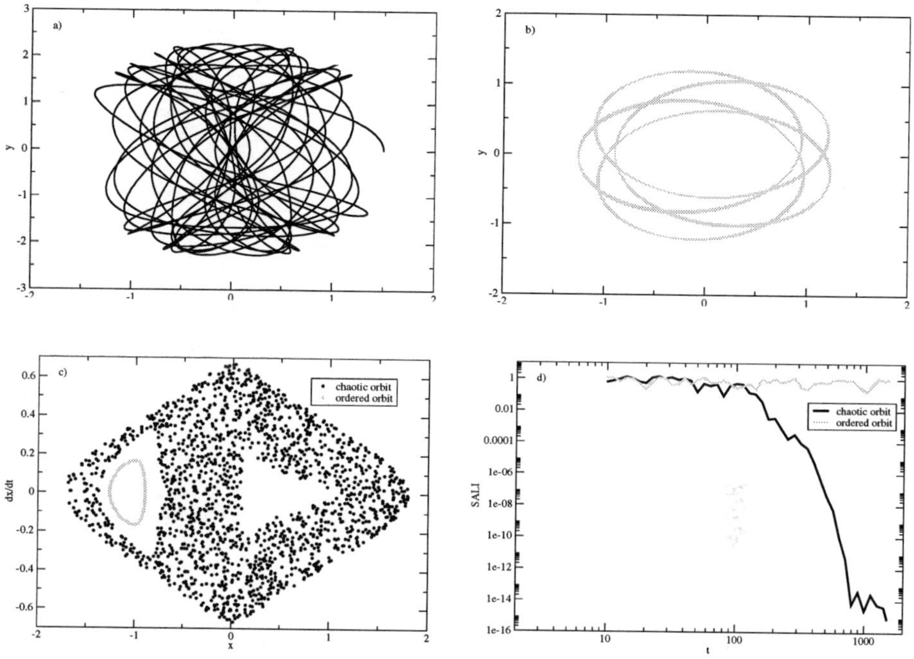

FIGURE 1. Projections of the orbits (i) and (ii) in the (x,y)-plane (panels a), b)). PSS of these two orbits in panel c). Behaviour of the SALI of the same orbits (panel d)).

two orbits. For the chaotic one, the SALI tends to zero (10^{-16}) exponentially after some time steps while for the regular orbit, it fluctuates around a positive number. By choosing initial conditions on the line $\dot{x} = 0$ of the PSS and calculating the values of the SALI, we can detect very small regions of stability that can not be visualized easily by the PSS method. Repeating this for many values of the energy, we are able to follow the change of the fraction of chaotic and ordered orbits in the phase space as the energy of the orbit varies.

Applications in the 3D case

In figure 2 we present percentages of chaotic, intermediate and regular trajectories, where we vary the mass of the bar component (panels A3-B3) and the length of the short z-axis (panels A2-B2) of the initial models A1 and B1. The two rows differ in the way we give the 27000 initial conditions. For the A1, we give initial conditions in the plane (x, p_y, z) with $(y, p_x, p_z) = (0,0,0)$ and for the B1, in the plane (x, p_y, p_z) with $(y, z, p_x) = (0,0,0)$. By comparing the results, we see that the increase of the bar mass causes more chaotic behavior in both cases (panels A3, B3). This confirms the results by Athanassoula et al. (1983) [28] in 2 degrees of freedom. On the other hand, it is obvious

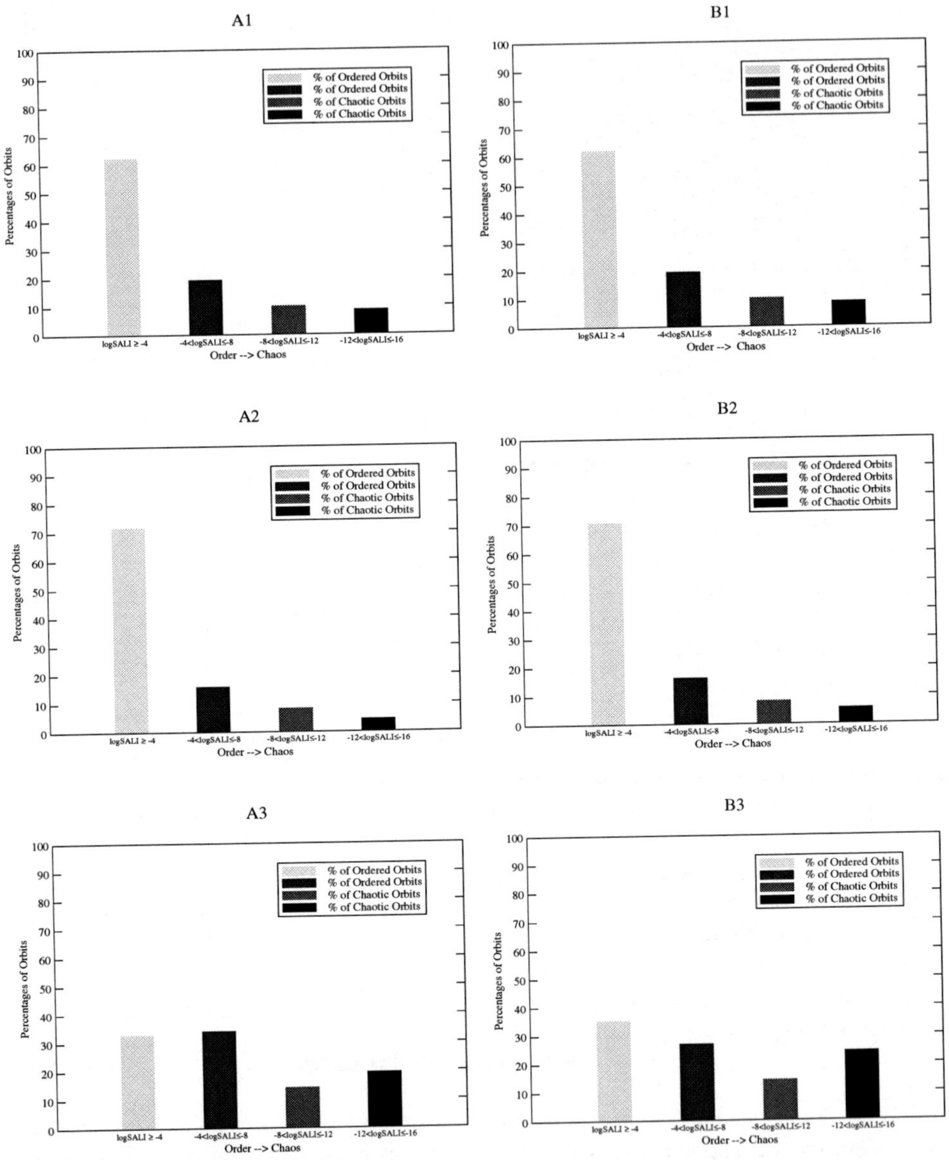

FIGURE 2. Percentages of regular (first and second bar) and chaotic (third and forth bar) orbits, for our standard model (first row of panels), a model with a thick bar (second row) and a model with a massive bar (third row). The two columns show two different ways of choosing the orbital population.

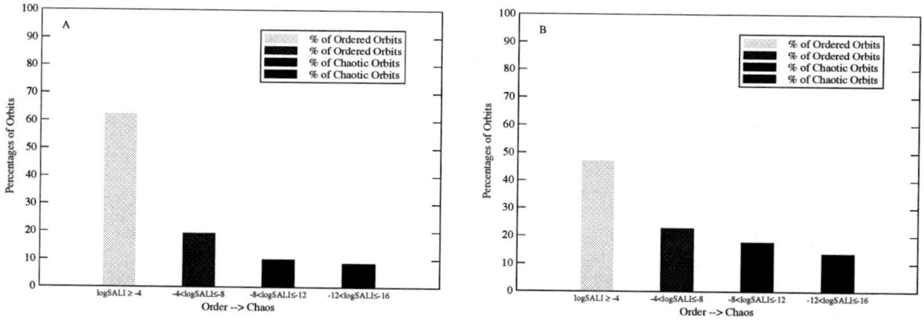

FIGURE 3. The percentages of the chaotic and ordered orbits for the two different values of the pattern speeds. Panel A corresponds to $\Omega_b = 0.0367032$ and panel B to $\Omega_b = 0.0554349$. The latter presents more chaotic behavior.

that when the bar is thicker, i.e. the length of the z-axis larger, the system gets more regular.

Finally, we calculated the same percentages for different pattern speeds Ω_b. From the orientation of periodic orbits, Contopoulos (1980) [29] showed that bars have to end before corotation, i.e. that $r_L > a$, where r_L the Lagrangian, or corotation, radius. Comparing the shape of the observed dust lanes along the leading edges of bars to that of the shock loci in hydrodynamic simulations of gas flow in barred galaxy potentials, Athanassoula (1992a,b) [30],[31] was able to set both a lower and an upper limit to corotation radius, namely $r_L = (1.2 \pm 0.2)a$. This restricts the range of possible values of the pattern speed, i.e. $\Omega_b = 0.0367032$, that corresponds to the Lagrangian radius $r_L = 1.4a$ and $\Omega_b = 0.0554349$, that corresponds to $r_L = 1.0a$. Using the extremes of this range, we investigated how the pattern speed of the bar affects the system and found that the percentage of the regular orbits is greater in slow bars.

CONCLUSIONS

In this paper, we applied the SALI method in the Ferrers barred galaxy models of 2 and 3 degrees of freedom. We presented and discussed our results comparing the SALI index with traditional methods, such as the PSS method for the 2 degrees of freedom and showed its effectiveness. We also, calculated percentages of chaotic and regular orbits and how they change with the main model parameters, for the 3 degrees of freedom case. In more detail, we calculated percentages of chaotic and ordered trajectories, as some important parameters, as the mass, the length of the short z-axis and the pattern speed of the bar vary in the initial basic model.

ACKNOWLEDGMENTS

Thanos Manos was partially supported by "Karatheodory" graduate student fellowship No B395 of the University of Patras and by "Marie - Curie" fellowship No HPMT-CT-

2001-00338.

REFERENCES

1. M. Lieberman and A. Lichtenberg, *Springer Verlag*, (1992).
2. C. Froeschlé, *Cel. Mech.*, **34**, 95, (1984).
3. G. Benettin, L. Galgani and J.M. Strelcyn, *Phys. Rev. A*, **14**, 2338, (1976).
4. J. Laskar, *Icarus*, **88**, 266, (1990).
5. J. Laskar, C. Froeschlé and A. Celletti, *Physica D*, **56**, 253, (1992).
6. C. Froeschlé, Ch. Froeschlé and E. Lohinger, *Celest. Mech. Dyn. Astron.*, **56**, 307, (1993).
7. E. Lohinger, C. Froeschlé and R. Dvorak, *Celest. Mech. Dyn. Astron.*, **56**, 315, (1993).
8. N. Voglis and G. Contopoulos *J. Phys. A*, **27**, 4899, (1994).
9. G. Contopoulos, E. Grousousakou and N. Voglis, *Astr. Astroph.*, **304**, 374, (1995).
10. N. Voglis, G. Contopoulos and C. Efthymiopoylos, *Celest. Mech. Dyn. Astron.*, **73**, 211, (1999).
11. G. Contopoulos and N. Voglis, *Celest. Mech. Dyn. Astron.*, **64**, 1, (1996).
12. G. Contopoulos and N. Voglis, *Astr. Astroph.*, **317**, 73, (1997).
13. C. Froeschlé and E. Lega, *Astr. Astroph.*, **334**, 355, (1998).
14. C. Froeschlé, E. Lega and R. Gonzi, *Celest. Mech. Dyn. Astron.*, **67**, 41, (1997).
15. Ch. L. Vozikis, H. Varvoglis and K. Tsiganis, *Astr. Astroph.*, **359**, 386, (2000).
16. Ch. Skokos, *J. Phys. A: Math. Gen.*, **34**, 10029, (2001).
17. Ch. Skokos, C. Antonopoulos, T. Bountis and M. Vrahatis, in *Proceedings of the 4th GRACM*, edited by D. T. Tsahalis, **IV**, 1496. (2002).
18. Ch. Skokos, C. Antonopoulos, T. Bountis and M. Vrahatis, *Prog. Theor. Phys. Suppl.*, **150**, 439, (2003).
19. Ch. Skokos, C. Antonopoulos, T. Bountis and M. Vrahatis, in *Proceedings of the Conference Libration Point Orbits and Applications*, edited by G. Gomez, M. W. Lo and J. J. Masdemont , World Scientific, 653, (2003).
20. Ch. Skokos, C. Antonopoulos, T. Bountis and M. Vrahatis, *J. Phys. A*, **37**, 6269, (2004).
21. P. Panagopoulos, T. Bountis and Ch. Skokos, *J. Vib. & Acoust.*, **126**, 520, (2004).
22. C. Antonopoulos, T. Manos and Ch. Skokos, in *Proceedings of the 1st IC-SCCE*, edited by D. T. Tsahalis, Patras Univ. Press, **III**, 1082, (2005).
23. A. Széll, B. Érdi, Zs. Sándor and B. Steves, *Mon. Not. R. Astron. Soc.*, **347**, 380, (2004).
24. N. Voglis , C. Kalapotharakos and I. Stavropoulos, *Mon. Not. R. Astron. Soc.*, **337**, 619, (2002).
25. C. Kalapotharakos , N. Voglis and G. Contopoulos, *Mon. Not. R. Astron. Soc.*, **428**, 905, (2004).
26. M. Miyamoto and R. Nagai, *PASJ*, **27**, 533, (1975).
27. D. Pfenniger, *Astr. Astroph.*, **134**, 373, (1984).
28. E. Athanassoula , O. Bienayme, L. Martinet and D. Pfenniger, *Astr. Astroph*, **127**, 349, (1983).
29. G. Contopoulos, *Astr. Astroph*, **81**, 198, (1980).
30. E. Athanassoula, *Mon. Not. R. Astron. Soc.*, **259**, 328, (1992a).
31. E. Athanassoula, *Mon. Not. R. Astron. Soc.*, **259**, 354, (1992b).

Mining for Observables: A New Challenge in Numerical Relativity

Deirdre Shoemaker*, Harald Pfeiffer†, Larry Kidder**, Pablo Laguna‡, Lee Lindblom†, Mark Scheel† and Saul Teukolsky**

*Department of Physics, Institute for Gravitational Physics and Geometry, Penn State, University Park PA 16802

†Theoretical Astrophysics 130-33, California Institute of Technology, Pasadena CA 91125

**Center for Radiophysics and Space Research, Cornell University, Ithaca NY 14853

‡Department of Astronomy and Astrophysics, Institute for Gravitational Physics and Geometry, Penn State, University Park PA 16802

Abstract. One of the motivations behind numerical relativity is to provide gravitational wave signals of compact objects to observers using the new gravitational wave detectors. Yet, because of the complexities involved, no dependable signals of binary-black hole coalescences have been established. The work in this proceedings is motivated by how numerical relativity can be used today to predict robust features in gravitational wave signals of binary black-hole coalescence by making approximations to the full problem. To illustrate this, we present results from evolving a Klein-Gordon equation on a frozen background. The background is set by a sequence of initial data in which the binary is in quasi-equilibrium. We probe the data resulting from the evolution for the transition between the linear and non-linear regimes using oscillations of the black holes as our guide. This information is used to motivate a qualitative picture of the gravitational signal of a black-hole coalescence.

Keywords: numerical relativity, black holes, gravitational waves
PACS: 04.25.Dm, 04.30.Db, 04.70.Bw, 95.30.Sf, 97.60.Lf

INTRODUCTION

Binary black holes are excellent candidates of detection by ground and space-based gravitational-wave observatories. The success of ground-based detectors may depend on the quality of modeled waveforms due to the low signal-to-noise ratios. Numerical relativity's goals are to model general relativistic sources that are strongly non-linear, of which the binary black hole coalescence is an excellent example. These models will provide information for gravitational wave detection and characterization. In the absence of an exhaustive set of waveforms, numerical relativity needs to provide information abbout the robust features of binary black hole coalescence to aid in the detection of the gravitational waves. The binary black hole coalescence problem is traditionally broken down into three epochs including inspiral, merger and ringdown. The inspiral phase is a slow adiabatic inspiral of the two black holes, and has been successfully modeled with post-Newtonian methods [1]. The merger phase involves the highly non-linear dynamics of curved spacetime and necessitates a heavily computational approach. The ringdown phase is dominated by an exponentially damped sinusoid and includes information about the mass and spin of the final black hole.

Recently, the numerical relativity community has made great strides in producing

CP848, *Recent Advances in Astronomy and Astrophysics*, 7th *International Conference of the Hellenic Astronomical Society*, edited by N. H. Solomos reliable simulations of the binary black hole coalescence [2, 3, 4, 5, 6, 7, 8, 9]. The startling success of the past few months has seen the first orbits of binary black holes (BBH), orbits through to merger and the generation of a series of waveforms involving equal mass and unequal mass [8] black holes. In the next few years, we will see the range of parameter space studied increase to include spinning black holes, more astrophysically relevant mass ratios and increasing black-hole separations. Although not-exhaustive, already a picture of the key features of the binary black hole coalescence is emerging, a picture of the BBH gravitational wave that is strikingly simple.

There have been hints at the simplicity of the waveforms produced by an equal-mass binary black hole orbit and collision from both the close-limit [10, 11] and Lazarus [12] approaches. One of the remarkable outcomes of the close-limit approach to BBH collisions was the range of agreement that these results had with non-linear numerical calculations [13]. Close-limit results closely followed the non-linear simulations beyond the point where the close-limit approximation was in principle valid. The reason for this agreement is not well understood. One possible explanation, which is the main motivation of this work, is that perhaps a potential barrier emerges hiding the non-linear dynamics taking place. The Lazarus approach took the close-limit work further by combining perturbative calculations with the then short-lived, fully non-linear computations to produce the first waveforms from plunge-like BBH dynamics. As the ability to evolve the final orbit and plunge has emerged, the Lazarus work has been surprisingly correct in its prediction of the bulk features of the gravitational waves from a binary black hole coalescence.

A comparison between the results of close-limit, Lazarus and full numerical relativity suggests that one of the key features leaving an imprint on the emitted gravitational radiation is the ringing of the potential. For a single black hole, a potential is clearly identifiable. The well-known quasi-normal ringing and power-law tail behavior is associated with the presence of this potential [14]. Because quasi-normal modes (QNMs) provide unique complex frequencies depending only on the black-hole mass and spin (we will assume neutral black holes for this paper), the QNMs can act as probes of the black hole spacetime.

One can then speculate that, if a potential barrier is formed before the BBH system reaches ISCO, this barrier could unfortunately hide rich non-linear signatures in the emitted gravitational radiation. The signal would be dominated by the ringing modes of this potential. A starting point to address this issue is to study the dynamics of a scalar field on the "frozen" space-time of a BBH system. The scalar field acts as a tracer to detect and map the potential.

KLEIN-GORDON EQUATION

We solve the Klein-Gordon equation, $\Box\psi = 0$, on a curved background using the Cornell/Caltech spectral code [15]. Following Ref. [16], we rewrite this equation in a first-order symmetric hyperbolic form. This is accomplished by introducing the following variables:

$$\Pi \equiv -\frac{1}{\alpha}\left(\partial_t\psi - \beta^i\partial_i\psi\right) \tag{1}$$

$$\Phi_i \equiv \partial_i \psi . \tag{2}$$

In terms of these variables, the Klein-Gordon equation $\Box\psi = 0$ can be written as

$$\begin{aligned}
(\partial_t - \beta^j \partial_j)\,\psi &= -\alpha\Pi & (3)\\
(\partial_t - \beta^j \partial_j)\,\Pi &= -\alpha g^{ij}\partial_j\Phi_i + \alpha\Gamma^i\Phi_i \\
&\quad - g^{ij}\partial_j\alpha\,\Phi_i + \alpha\Pi K & (4)\\
(\partial_t - \beta^j \partial_j)\,\Phi_i &= -\alpha\partial_i\Pi + \Phi_j\,\partial_i\beta^j - \Pi\,\partial_i\alpha , & (5)
\end{aligned}$$

where

$$\begin{aligned}
\Gamma^i &\equiv g^{jk}\Gamma^i_{jk} \\
&= -\partial_j g^{ij} - g^{ij}\,\partial_j \ln\sqrt{g} & (6)
\end{aligned}$$

Eqs. (3–5) can be rewritten as

$$\partial_t u + A^i \partial_i u = F , \tag{7}$$

where

$$\phi = \begin{pmatrix} \psi \\ \Pi \\ \Phi_j \end{pmatrix} \tag{8}$$

$$A^i = \begin{pmatrix} -\beta^i & 0 & 0 \\ 0 & -\beta^i & \alpha g^{ij} \\ 0 & \alpha\delta^i_j & -\beta^i \end{pmatrix} \tag{9}$$

and

$$F = \begin{pmatrix} -\alpha\Pi \\ \alpha\Gamma^i\Phi_i - g^{ij}\partial_j\alpha\,\Phi_i + \alpha\Pi K \\ \Phi_i\,\partial_j\beta^i - \Pi\,\partial_j\alpha \end{pmatrix} . \tag{10}$$

If ξ^i is a unit normal to any two-dimensional surface, the characteristic speeds and fields can be found from the eigenvalues and eigenfields of the matrix $A^i\,\xi_i$. The characteristic fields are:

$$\begin{aligned}
u^{\pm} &= \Pi \pm \xi^i\,\psi_i & (11)\\
u_i &= \psi_i - \xi_i\,\xi^j\,\psi_j & (12)\\
u^{\psi} &= \psi . & (13)
\end{aligned}$$

The fields $u^{\pm}$ have characteristic speeds $-\beta^i \pm \alpha\,\xi^i$ (speed of light). On the other hand, the fields u_i and u^{ψ} have characteristic speed $-\beta^i$ (zero velocity, normal to the spatial hypersurface). One can see from these equations that the information from the background that must be supplied is g_{ij}, K, α and β^i. The next section discusses our choice for the background quantities.

BINARY BLACK HOLE SPACETIME AS A CURVED BACKGROUND

We need to specify the background for the Klein-Gordon equation given by g_{ij}, K, α and β^i. To study the formation of a potential during the BBH coalescence, ideally we would use the geometry given by a numerical calculation of the binary system. In this proceedings, we use an approximation to the spacetime of two black holes in orbit given by the quasi-circular sequence of initial data generated by the thin-sandwich [17, 18, 19]. The thin-sandwich method of initial data provides us with g_{ij}, K, α and β^i for the background. The coordinates are co-rotating with the black holes such that the black holes remain stationary for timescales $T_{orbit} << T_{radiationreaction}$ Table (1) lists the details about wach initial data set in the sequence of quasi-circular equilibrium. The data is constructed using Harald Pfeiffer's multigrid, spectral code for solving elliptic equations [20].

TABLE 1. Parameters for each initial data set in a quasi-circular sequence. The first column is d, separation, the second is the orbital frequency, ΩM_{irr}, the third is the irreducible mass, $M_{irr} = \sqrt{\frac{A_{AH}}{16\pi}}$, the fourth is the infinite norm of the Hamiltonian constraint, the fifth is the angular momentum of the spacetime, J_{ADM}, the sixth is the mass of the spacetime, M_{ADM}, and the last column is the apparent horizon mass, $M_{AH/M}$. The black holes have equal mass.

d	ΩM_{irr}	M_{irr}	Ham-Linf	J_{ADM}	M_{ADM}/M	M_{AH}/M
6	0.168	2.63	3.33e-06	1.75e-07	2.59	1.31
7	0.135	2.48	2.08e-06	1.24e-07	2.44	1.24
8	0.112	2.39	1.40e-06	9.83e-08	2.35	1.20
9	0.0949	2.33	9.97e-07	8.16e-08	2.29	1.17
10	0.0817	2.29	7.38e-07	6.98e-08	2.25	1.14
14	0.0505	2.19	1.33e-08	1.83e-09	2.17	1.09

The background is frozen in time for each solution to the Klein-Gordon equation. We need to specify the initial data for ψ to represent an incoming wave of gravitational radiation:

$$\psi = \psi_0 e^{-(r-r_0)^2/\sigma^2} Y_{l=1,m=0} \tag{14}$$

where r_0 is set to be $12M$ away from the center of the black holes and $\sigma = 2$.

PROCEDURE

This is a model problem intended to get at a feature of the BBH merger, the early formation of the BH potential. As the scalar field is scattered off of the BBH background, what do we expect it to tell us about the spacetime? There are two general cases to consider. When the black holes are far apart, we expect each individual black hole to act as an isolated black hole; exhibiting quasi-normal ringing. Thus, at large separations for equal-mass black holes, there are two characteristic frequencies, the frequency associated with the individual black holes' dominant QNM and the orbital frequency of the binary system itself. The second case is when the black holes have coalescenced and formed a single, more massive black hole. In this case there is a single frequency, that

TABLE 2. Parameters for each run, where ω_{qnm} is the quasi-normal mode frequency found for each run.

d	ω_{qnm}	ΩM_{irr}	M_{irr}	M_{ADM}/M
6	0.130	0.168	2.63	2.59
7	0.150	0.135	2.48	2.44
8	0.202	0.112	2.39	2.35
9	0.212	0.0949	2.33	2.29
10	0.265	0.0817	2.29	2.25
14	0.268	0.0505	2.19	2.17

of the QNM associated with the black hole. We use the ringing frequencies to predict how long we maintain individual, isolated black holes and then how quickly we see a single black hole form. However, the approximation we are making in using a sequence of initial data breaks down the closer the black holes get in the sequence.

The procedure to find the frequency of the dominant QNM is to fit the scattered pulse, ψ to the expected functional form of quasi-normal ringdown, namely

$$\psi \approx e^{-\tau+i\omega t} \tag{15}$$

where ω is the real part of the frequency and τ is imaginary part associated with the decay time of the mode. We make the assumption that the fitted frequency is that of the dominant QNM, in this case the $l = 1$, $m = 0$ mode which has $\omega = 0.293$. We then compute the mass of the black hole based on this fitted frequency such that

$$M_{bh} = \frac{0.293}{w_{qnm}} \tag{16}$$

Table (2) lists some of the same parameters as in Table (1) but now with the QNM-frequency, ω_{qnm} also given.

RESULTS AND CONCLUSIONS

Fig. (1) plots a series of waveforms from a sample of the different initial data sets parameterized by the separation, d. For reference, ISCO in this sequence of initial data is estimated to be located at $d = 8M$. The separations of the black holes are $d = 14$ in the uppermost plot, $d = 8$ in the middle and $d = 6$ at the bottom. These separations represent well outside of the ISCO of this sequence of initial data at $d = 14$, near ISCO at $d = 8$ and within the predicted ISCO at $d = 6$.

Fig.(2) is a plot of the fitted masses versus the orbital frequency of each initial data set. This fit is done assuming the frequency of the black hole is $\omega_{l=1,m=0} = 0.293/M_{bh}$ which is the frequency for a scalar perturbation. The plotted masses indicate that 90% of the mass of the spacetime is acting like a single black hole by the end of the experiment. More interestingly, before the predicted ISCO has been reached, the black holes loose their individual quasi-normal ringing which would indicate that some non-linear process was present, perhaps even an early potential formation.

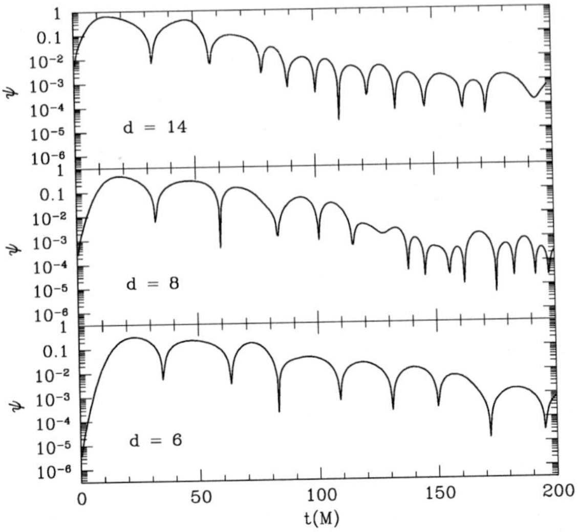

FIGURE 1. This plot is of $log\psi$ versus time for three separations from Table (1). The uppermost plot is of the largest separation, the middle is at approximately ISCO and the bottommost is the closest binary separation, $d = 6M$.

As with photons and test particles, one can gain physical insight of numerically generated space times by studying the dynamics of scalar fields. Our numerical experiment supports the view that QNM ringing potential forms early on by the pre-ISCO loss of a fundamental frequency associated with the two, individual black holes. Secondly, this experiment also shows that an early, single black hole potential may form prior to an enveloping apparent horizon. This is seen by the 90% of the total mass already "ringing" before the single apparent horizon forms. No apparent horizon was found enveloping both black holes in any of the initial data sets.

The main limitation of the work is the application of the frozen background approximation. This approximation may be valid at the pre-ISCO part of the simulations, and we did not find a common apparent horizon in the initial data even at the closest separation. We will not know how accurate the approximation is until we use a scalar field to map the spacetime during a real evolution of the binary black holes. To do this correctly, we would use the time-evolution of the BBH problem such as the head-on, plunge or orbit collisions as the background taking one time step for each solution of the Klein-Gordon equation. In future work, we will return to this issue. We can use the type of experiment described in this proceeding to compare initial data sets and to probe how dependent the early formation of a potential might be on the choice of initial data, most importantly the freely specifiable parts and assumptions such as conformal flatness. Finally, this ex-

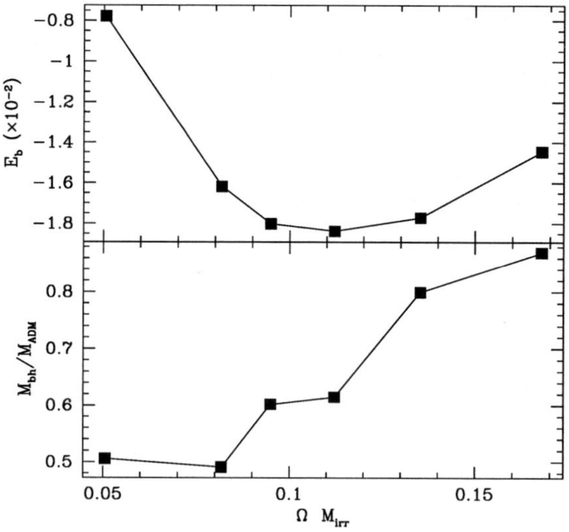

FIGURE 2. In this plot, the uppermost plot is the binding Energy, E_b scaled by 10^3 versus the orbital frequency, $\Omega M)irr$ for the quasi-circular sequence of initial data. The bottom plot is our results showing the fitted black hole mass M_{bh}/M_{ADM} versus the orbital frequency. The black holes are furthest apart to the left ($d = 14$) and closest to the right ($d = 6$).

periment may give us insight into how early during the coalescence non-linear effects may become important.

ACKNOWLEDGMENTS

This work was supported by NSF grants PHY-9800973 and PHY-0114375. and PHY-0354821. Work supported in part by the Center for Gravitational Wave Physics funded by the National Science Foundation under Cooperative Agreement PHY-0114375.

REFERENCES

1. L. Blanchet, *Living Rev. Rel.* **5**, 2 (2002).
2. B. Brügmann, W. Tichy, and N. Jansen, *Phys. Rev. Lett.* **92**, 211101 (2004), `gr-qc/0312112`.
3. M. Alcubierre, B. Brügmann, P. Diener, F. S. Guzmán, I. Hawke, S. Hawley, F. Herrmann, M. Koppitz, D. Pollney, E. Seidel, and J. Thornburg, *Phys. Rev. D* **72**, 044004 (2005), URL `http://link.aps.org/abstract/PRD/v72/e044004`, `gr-qc/0411149`.
4. F. Pretorius, *Phys. Rev. Lett.* **95**, 121101 (2005), `gr-qc/0507014`.

5. J. G. Baker, J. Centrella, D.-I. Choi, M. Koppitz, and J. van Meter, Gravitational wave extraction from an inspiraling configuration of merging black holes (2005), gr-qc/0511103.
6. M. Campanelli, C. O. Lousto, P. Marronetti, and Y. Zlochower, Accurate evolutions of orbiting black-hole binaries without excision (2005), gr-qc/0511048.
7. P. Diener, F. Herrmann, D. Pollney, E. Schnetter, E. Seidel, R. Takahashi, J. Thornburg, and J. Ventrella, Accurate evolution of orbiting binary black holes (2005), gr-qc/0512108.
8. F. Herrmann, D. Shoemaker, and P. Laguna (2006), gr-qc/0601026.
9. M. Campanelli, C. O. Lousto, and Y. Zlochower, The last orbit of binary black holes (2006), gr-qc/0601091.
10. R. Price, and J. Pullin, *Phys. Rev. Lett.* **72**, 3297 (1994).
11. G. Khanna, R. Gleiser, P. Laguna, O. Nicasio, H.-P. H.-P. Nollert, R. Price, and J. Pullin, *Phys. Rev. Lett.* **83**, 3585 (1999).
12. J. Baker, M. Campanelli, C. Lousto, and R. Takahashi, *Phys. Rev. D* **65**, 124012 (2002).
13. R. J. Gleiser, C. O. Nicasio, R. H. Price, and J. Pullin, *Physical Review Letters* **77**, 4483–4486 (1996).
14. K. Kokkotas, and B. Schmidt, *Living Rev. Rel.* **2**, 2 (1999).
15. L. Kidder, M. Scheel, S. Teukolsky, E. Carlson, and G. Cook, *Phys. Rev. D* **62**, 084032 (2000).
16. M. A. Scheel, A. L. Erickcek, L. M. Burko, L. E. Kidder, H. P. Pfeiffer, and S. A. Teukolsky, *Phys. Rev. D* **69**, 104006 (2004).
17. E. Gourgoulhon, P. Grandclément, and S. Bonazzola **65**, 044020 (2002).
18. P. Grandclément, E. Gourgoulhon, and S. Bonazzola, *Phys. Rev. D* **65**, 044021 (2002).
19. H. Pfeiffer, G. Cook, and S. Teukolsky, *Phys. Rev. D* **66**, 024047 (2002).
20. H. Pfeiffer, L. Kidder, M. Scheel, and S. Teukolsky, *Comput. Phys. Commun.* **52**, 253–274 (2003).

Measuring mass moments and electromagnetic moments of a massive, axisymmetric body, through gravitational waves.

Theocharis A. Apostolatos* and Thomas P. Sotiriou†

*Section of Astrophysics, Astronomy, and Mechanics, Department of Physics, National and Kapodistrian University of Athens Panepistimiopolis, Zografos GR-15783, Athens, Greece.
†SISSA-International School for Advanced Studies, Via Beirut 2-4, 34014 Trieste, Italy, and INFN, Sezione di Trieste, via Valerio 2, 34014, Trieste, Italy

Abstract. The electrovacuum around a rotating massive body with electric charge density is described by its multipole moments (mass moments, mass-current moments, electric moments, and magnetic moments). A small uncharged test particle orbiting around such a body moves on geodesics if gravitational radiation is ignored. The waves emitted by the small body carry information about the geometry of the central object, and hence, in principle, we can infer all its multipole moments. Due to its axisymmetry the source is characterized now by four families of scalar multipole moments: its mass moments M_l, its mass-current moments S_l, its electrical moments E_l and its magnetic moments H_l, where $l \quad 0 \; 1 \; 2, \ldots$. Four measurable quantities, the energy emitted by gravitational waves per logarithmic interval of frequency, the precession of the periastron (assuming almost circular orbits), the precession of the orbital plane (assuming almost equatorial orbits), and the number of cycles emitted per logarithmic interval of frequency, are presented as power series of the newtonian orbital velocity of the test body. The power series coefficients are simple polynomials of the various moments.

Keywords: Gravitational waves, relativistic multipole moments
PACS: 04.25.Nx, 04.30.Db

INTRODUCTION

The geometry around a massive object is uniquely characterized byo the multipole moments [1, 2, 3, 4] of the central object, as in newtonian gravity. Ryan [5] showed how we could reveal the spacetime geometry of a central axisymmetric body, by measuring a few physical quantities that are related to the kinematics of a hypothetical test body that is orbiting around the central object while it is emitting gravitational waves. This geometry-mapping could be achieved in principle by an advanced network of gravitational wave detectors with high sensitivity. Of course, to determine all moments we need infinitely sensitive detectors. However, knowing a few moments of the central body, we could put restrictions on the various models that are assumed to describe the interior of the central body.

Since the electromagnetic moments of the central object have their own contribution on the geometry around the central object, we could generalize the work of Ryan to yield the relation of all moments to the physical quantities used by Ryan.

In our paper we have considered binaries consisting of (i) a central object is assumed to be stationary and axisymmetric, and characterized by reflection symmetry with re-

CP848, *Recent Advances in Astronomy and Astrophysics, 7th International Conference of the Hellenic Astronomical Society,* edited by N. H. Solomos

spect to its equatorial plane, (ii) a test particle with no charge that is orbiting around the first one on a nearly circular and equatorial geodesic orbit. The axisymmetry of the central object ensures that the spacetime geometry could be described by scalar moments. Since we are dealing with electrovaccuum solutions 4 sets of scalar moments are needed: Its mass moments $M_0, M_2, M_4, \ldots$, its mass-current moments $S_1, S_3, \ldots$, its electric charge moments $E_0, E_1, E_2, \ldots$, and its magnetic moments $H_0, H_1, H_2, \ldots$, where e.g., the $M_0 \equiv M$ moment is the mass of the object, S_1 is its angular momentum, $E_0 \equiv E$ is its charge, and H_1 is its magnetic dipole moment. In every set of moments, each moment appears in steps of 2, and this holds good for the electromagnetic moments as well, due to the reflection symmetry of the metric itself (cf., [2]). The fact that all electromagnetic moments are written above, instead of every second one, is related to the fact that two kind of objects could have axisymmetric metrics, either the ones with even electric moments and odd magnetic moments, or the other way around.

The two precessing frequencies of the orbits (the periastron precession and the orbital precession) are written as functions of the moments of the metric on which the particle is moving. Moreover, although the orbit of the test body is assumed to be a geodesic one, we are going to use the quadrupole formula to infer the energy emitted by the binary. The evolution of the orbit will then be used to obtain alternative information about the moments of the geometry.

Throughout the paper units are chosen so that $G = c = 1$.

THE KINEMATICS OF THE TEST PARTICLE

If we perturb a circular equatorial orbit on a fixed axisymmetric background then two new frequencies arise, the one connected to the periastron precession and the one connected to the orbital precession, that are related to the circular orbit frequency and the metric itself by [5]

$$
\begin{aligned}
\Omega_\alpha &= \Omega - \left(-\frac{g^{\alpha\alpha}}{2} \left[(g_{tt} + g_{t\phi}\Omega)^2 \left(\frac{g_{\phi\phi}}{\rho^2} \right)_{,\alpha\alpha} \right.\right. \\
&- 2(g_{tt} + g_{t\phi}\Omega)(g_{t\phi} + g_{\phi\phi}\Omega) \left(\frac{g_{t\phi}}{\rho^2} \right)_{,\alpha\alpha} \\
&+ \left.\left. (g_{t\phi} + g_{\phi\phi}\Omega)^2 \left(\frac{g_{tt}}{\rho^2} \right)_{,\alpha\alpha} \right] \right)^{1/2},
\end{aligned} \tag{1}
$$

where α stands for ρ (periastron precession), or z (orbital precession). Actually, the frequencies written above correspond to the difference between the orbital frequency and the frequency of perturbations in ρ, or z, since these differences will show up in gravitational radiation emitted by the binary as a modulating frequency.

The energy per unit test-body mass for an equatorial circular orbit in an axially symmetric spacetime is

$$\frac{E}{\mu}=\frac{-g_{tt}-g_{t\phi}\Omega}{\sqrt{-g_{tt}-2g_{t\phi}\Omega-g_{\phi\phi}\Omega^2}}, \tag{2}$$

and the specific energy released as gravitational radiation per logarithmic interval of frequency is

$$\frac{\Delta E}{\mu}=-\Omega\frac{d(E/\mu)}{d\Omega}. \tag{3}$$

The expression above assumes that all the energy lost from the test body has been emitted at infinity as radiation.

The number of gravitational-wave cycles spent in a logarithmic interval of frequency, the quantity that could be most accurately measured by a gravitational wave detector, is

$$\Delta N=\frac{f\Delta E(f)}{dE_{\text{wave}}/dt}, \tag{4}$$

where dE_{wave}/dt is the gravitational-wave luminosity, which is assumed to be exactly the rate of energy loss of the orbiting test body. The main contribution of dE_{wave}/dt comes from the mass quadrupole radiative moment of the binary. Therefore, when we give the connection between the various orders of $v \equiv (M\Omega)^{1/3}$ (the orbital velocity) in the power series expansion of ΔN, and the moments of spacetime, only the highest order moments enter our expressions, since lower order moments contribute through higher than quadrupole radiative moments as well in a rather complicated and unclear way. However, at least the first 4 power parameters of ΔN could be written explicitly in terms of the multipole moments.

THE METRIC AS A POWER SERIES OF MOMENTS

To get expressions for these 3 measurable quantities, that are straightforwardly connected to the moments of the central object, one has to reexpress the metric functions in terms of all moments, as well as the radius ρ and the orbital frequency of the test body Ω.

In our paper, we consider only stationary axisymmetric objects that are reflection symmetric with respect to their equatorial plane. The metric of such a central object alone could be written in (t,ρ,z,ϕ) coordinates, in the form of Papapetrou metric [6].

$$ds^2=-F(dt-\omega\, d\phi)^2+\frac{1}{F}\left[e^{2\gamma}(d\rho^2+dz^2)+\rho^2 d\phi^2\right], \tag{5}$$

where F,ω, and γ are the three functions that fully determine a specific metric. These are functions of ρ and $|z|$ only, due to axisymmetry and reflection symmetry. Einstein's equations in vacuum guarantee that once F and ω are given, γ can be easily computed (see [7]). Once we incorporate an electromagnetic field in the vacuum around the compact object, the source of which is the compact object itself, which allows spacetime to have the same symmetries, the metric above still describes the electrovacuum spacetime, but now the metric and the electromagnetic field should satisfy the Einstein-Maxwell equations. In order to fully compute the metric functions one more complex function, Φ, which is related to the electromagnetic field, is necessary. F, ω, and Φ themselves can be determined by solving the so-called Ernst equations [8, 9], which are the Einstein-Maxwell equations written in a different form. It is a system of non-linear complex differential equations of second order:

$$(\Re(\ \ ) + |\Phi|^2)\nabla^2 \ \ = (\nabla \ \ + 2\Phi^* \nabla\Phi)\cdot\nabla \ \ , \tag{6}$$

$$(\Re(\ \ ) + |\Phi|^2)\nabla^2\Phi = (\nabla \ \ + 2\Phi^* \nabla\Phi)\cdot\nabla\Phi, \tag{7}$$

where ∇ denotes the gradient in a cartesian 3D space (ρ, z, ϕ) and $\Re(\ldots), \Im(\ldots)$, here and henceforth, denote the real and imaginary part, respectively, of the complex function in parentheses, while an asterisc * denotes complex conjugate. The third metric function, γ, is then easily computed by integrating the partial derivatives $\partial\gamma/\partial\rho$, $\partial\gamma/\partial z$, which are given as functions of derivatives of all other functions (see [10]).

From the two complex functions, and Φ, we could construct the metric functions as

$$\ \ = (F - |\Phi|^2) + i\varphi, \tag{8}$$

where φ is related with $g_{t\phi}$ through

$$g_{t\phi} = F\omega = F\int_\rho^\infty d\rho' \frac{\rho'}{F^2}\left(\frac{\partial\varphi}{\partial z} + 2\Re(\Phi)\frac{\partial\Im(\Phi)}{\partial z} - \right. \\ \left. - 2\Im(\Phi)\frac{\partial\Re(\Phi)}{\partial z}\right)\Bigg|_{z=\mathrm{const}}. \tag{9}$$

Note that there is a sign difference in Eq. (22) of [5], which has been corrected in a later paper of Ryan [11], and comes from an odd convention of ω used by Ernst (see relevant comment of [12]).

Now, instead of and Φ, one could use two new complex functions $\tilde{\xi}$ and $\tilde{q}$, that play the role of gravitational potential and Coulomb potential respectively, and are more directly connected to the mass and electromagnetic moments of the central body. These potentials are related to the Ernst functions by

$$\ \ = \frac{\sqrt{\rho^2+z^2} - \tilde{\xi}}{\sqrt{\rho^2+z^2} + \tilde{\xi}} \tag{10}$$

$$\Phi = \frac{\tilde{q}}{\sqrt{\rho^2+z^2} + \tilde{\xi}}, \tag{11}$$

and can be written as power series expansions at infinity

$$\tilde{\xi} = \sum_{i,j=0}^{\infty} a_{ij}\bar{\rho}^i\bar{z}^j, \qquad \tilde{q} = \sum_{i,j=0}^{\infty} b_{ij}\bar{\rho}^i\bar{z}^j, \tag{12}$$

where

$$\bar{\rho} \equiv \frac{\rho}{\rho^2+z^2}, \qquad \bar{z} \equiv \frac{z}{\rho^2+z^2}, \tag{13}$$

and a_{ij}, b_{ij} are coefficients that vanish when i is odd. This reflects the analyticity of the potentials on the z-axis. The tilded quantities, here and henceforth, are the conformally transformed ones, which are essential for calculating the moments (see [1]).

Due to Ernst equations (6,7) the above power expansion coefficients, a_{ij} and b_{ij}, are interrelated through complicated recursive relations (see [10]). Essentially, these relations are simply an algebraic version of Einstein-Maxwell equations for the coefficients of the power expansion of the metric and the electromagnetic field tensor. The recursive relations could be used to build the whole power series of $\tilde{\xi}$ and $\tilde{q}$ from a full knowledge of the metric on the axis of symmetry

$$\tilde{\xi}(\bar{\rho}=0)=\sum_{i=0}^{\infty} m_i \bar{z}^i, \qquad \tilde{q}(\bar{\rho}=0)=\sum_{i=0}^{\infty} q_i \bar{z}^i. \tag{14}$$

In [13] a method of calculating the complex multipole moments of the central object in terms of the m_i's and q_i's is presented. In brief, the gravitational moments are given by

$$P_n = \frac{1}{(2n-1)!!} S_0^{(n)}, \tag{15}$$

where $S_a^{(n)}$ are given by recursive formulas analyzed in [10].

The mass moments M_n and the mass-current moments S_n are related to P_n by

$$P_n = M_n + iS_n, \tag{16}$$

whereas the electric moments E_n and the magnetic moments H_n are related to Q_n by

$$Q_n = E_n + iH_n. \tag{17}$$

Since this algorithm can be used to evaluate the moments in terms of the m_i's and q_i's, one can invert these relations and express the m_i's and q_i's in terms of the moments:

$$\begin{aligned} m_n &= a_{0n} = M_n + iS_n + \text{LOM}, \\ q_n &= b_{0n} = E_n + iH_n + \text{LOM}, \end{aligned} \tag{18}$$

where "LOM" stands for lower order multipole moments of any type. Thus, we can use the recursive relations mentioned above to evaluate the a_{ij} and b_{ij} coefficients in terms of the moments.

Finally, following the procedure presented in the beginning of this section we can express the metric functions, and their first and second derivatives as power series of ρ and z with coefficients that are simple algebraic functions of the moments of the massive body. Since in our study we have confined the motion of the test particle on the equatorial plane, we actually need to compute everything at $z=0$ which makes calculations far simpler than what they seem.

One could argument (see Sec. IIC of [10]) that the reflection symmetry assumed for the metric is consistent with a set of even electric and odd magnetic moments or odd electric and even magnetic moments.

EXPRESSIONS RELATING THE MEASURABLE QUANTITIES WITH MOMENTS

Combining the formulae that are given in Sec. 2 we can express the three measurable quantities as power series of $v \equiv (M\Omega)^{1/3}$ with coefficients that have explicit dependence on all four types of moments. The choice of v as a dimensionless parameter to expand all physical quantities is warranted from the fact that the inspiral phase of a binary, the best exploitable part in gravitational-wave analysis [14], involves comparatively low magnitudes of v. All measurable quantities have been transformed to a dimensionless form as well, for example by dividing the two frequencies Ω_ρ, Ω_z, by the orbital frequency Ω.

We need also a power series expansion of ρ with respect to v, or equivalently Ω. Thus we have to invert the function $\Omega(\rho)$, at least as a power expansion. From an elementary analysis of circular geodesics on the equatorial plane (see [5]) we know that

$$\Omega = \frac{-g_{t\phi,\rho} + \sqrt{(g_{t\phi,\rho})^2 - (g_{tt,\rho})(g_{\phi\phi,\rho})}}{g_{\phi\phi,\rho}}. \tag{19}$$

In the following part of this section we explain the algorithm that one should follow, in order to obtain the power series for Ω_ρ/Ω, Ω_z/Ω, and ΔN. One starts with a power series of $\tilde{\xi}$ and $\tilde{q}$ of the form given by Eq. (12). Since no higher than second derivatives of the metric functions with respect to z are necessary, one should keep a_{ij}'s and b_{ij}'s with $0 \leq j \leq 2$, and as many values of i as one needs to carry the power series expansion of the measurable quantities at a desirable order. All quantities that are expressed as power series of $\bar{\rho}$ and $\bar{z}$, are evaluated at $\bar{z} = z = 0$ at the end, and thus, all expressions are finally power series of $\bar{\rho} = 1/\rho$, due to Eq. (13). Although the a_{ij} and b_{ij} are polynomials of various moments, from the practical point of view it is preferable to keep them as they are, and replace them by their moments dependence only at the final expressions. Then from $\tilde{\xi}$ and $\tilde{q}$ we construct , Φ, and F, φ (cf., Eqs. (8,10)). These are sufficient to build all metric functions. Next, following the procedure described above, we expand Ω as a power series of $1/\rho$, by virtue of Eq. (19). This series is inverted and in this way we obtain $1/\rho$ as a power series of Ω, which then can easily be turned into a power series of the dimensionless parameter v.

Now, the power series representing $1/\rho$ will replace all $1/\rho$ terms appearing at the expansions of the metric, its derivatives, and all other physical quantities depending on them. Finally, one has to rewrite the a_{ij} and b_{ij} terms appearing at the coefficients of all these power series as polynomials of the various moments. The recursive relations that relate all a_{ij} and b_{ij} with $m_k \equiv a_{0k}$ and $q_k \equiv b_{0k}$, which are directly related to the scalar moments of spacetime through Eqs. (24,25) of Ref. [13].

The algorithm described in the previous two paragraphs has been carried out with Mathematica, and has been checked for the following two subcases: (i) When all electromagnetic fields are turned off, by erasing all electromagnetic moments ($E_l = H_l = 0$), our expressions for Ω_ρ, Ω_z, ΔN are identical to the ones computed by Ryan [5]. (ii) For the Kerr-Newman metric it is quite easy to compute Ω, Ω_ρ, Ω_z, and $\Delta E/\mu$ for a quasi-equatorial, quasi-circular orbit. The expressions we obtain are identical to the ones obtained directly from the Kerr-Newman metric.

The power series expansion for Ω_ρ, Ω_z, take the following form:

$$\frac{\Omega_\rho}{\Omega} = \sum_{n=2}^{\infty} R_n v^n, \tag{20}$$

$$\frac{\Omega_z}{\Omega} = \sum_{n=3}^{\infty} Z_n v^n, \tag{21}$$

while the corresponding coefficients, up to 5th order for the two frequencies, for the two distinct electromagnetic cases (es) and (ms) are

$$R_2^{(es)} = \left(3 - \frac{1}{2}\frac{E^2}{M^2}\right), \quad R_3^{(es)} = -4\frac{S_1}{M^2},$$

$$R_4^{(es)} = \frac{9}{2} - \frac{3}{2}\frac{M_2}{M^3} - 2\frac{E^2}{M^2} - \frac{13}{24}\frac{E^4}{M^4}, \quad R_5^{(es)} = -10\frac{S_1}{M^2} - \frac{10}{3}\frac{S_1 E^2}{M^4} + 5\frac{H_1 E}{M^3}, \tag{22}$$

$$R_2^{(ms)} = \left(3 - \frac{1}{2}\frac{H^2}{M^2}\right), \quad R_3^{(ms)} = -4\frac{S_1}{M^2},$$

$$R_4^{(ms)} = \frac{9}{2} - \frac{3}{2}\frac{M_2}{M^3} - 2\frac{H^2}{M^2} - \frac{13}{24}\frac{H^4}{M^4}, \quad R_5^{(ms)} = -10\frac{S_1}{M^2} - \frac{10}{3}\frac{S_1 H^2}{M^4} - 5\frac{E_1 H}{M^3}, \tag{23}$$

$$Z_3^{(es)} = 2\frac{S_1}{M^2}, \quad Z_4^{(es)} = \frac{3}{2}\frac{M_2}{M^3}, \quad Z_5^{(es)} = -\frac{H_1 E}{M^3} + 2\frac{S_1 E^2}{M^4}, \tag{24}$$

$$Z_3^{(ms)} = 2\frac{S_1}{M^2}, \quad Z_4^{(ms)} = \frac{3}{2}\frac{M_2}{M^3}, \quad Z_5^{(ms)} = \frac{E_1 H}{M^3} + 2\frac{S_1 H^2}{M^4}, \tag{25}$$

The fact that we have two new sets of moments (the electromagnetic ones) with respect to Ryan allows many more combinations of moments in high order terms. Actually, from a practical point of view these expansions are far more advanced than what will be used in gravitational wave data analysis in the near future. On the other hand the expressions above present an important feature: in every new order term a new moment shows up. This suggests that a very accurate observational estimation of the series could in principle reveal any moment.

A glance at the corresponding terms of the two electromagnetic cases shows that each combination of moments for the (es) case is numerically equal to the corresponding combination for the (ms) case, if the electric and magnetic moments are interchanged. The sign though is the same for combinations of pure electric, or pure magnetic moments, but opposite for combinations of electric and magnetic moments.

Finally, in order to express ΔN also as a power series of v, we need to expand dE_{wave}/dt as power series of v. As was explained in Sec. 2 we cannot work out the perturbative analysis of gravitational wave emission at a generic spacetime background; we can only obtain accurate expressions for $dE_{\rm wave}/dt$ up to v^4 after the leading order. Therefore, in the following formulae for ρ and $dE_{\rm wave}/dt$, we write simply the power series coefficients explicitly up to the fourth order. More specifically, in order to compute the power expansion of $dE_{\rm wave}/dt$ we add up all power series contributions (quadrupole radiative moment, current quadrupole radiative moment, and Post-Newtonian corrections) which are the only ones that show up up to the 4th order. Thus, we yield

$$\Delta N = \frac{5}{96\pi}\left(\frac{M}{\mu}\right)v^{-5}\left(1+\sum_{n=2}^{\infty}N_n v^n\right), \tag{26}$$

where the N_n coefficients for the two electromagnetic cases are given by the following polynomials of the moments

$$\begin{aligned}
N_2^{(es)} &= \frac{743}{336}+\frac{14}{3}\frac{E^2}{M^2}\\
N_3^{(es)} &= -4\pi+\frac{113}{12}\frac{S_1}{M^2}\\
N_4^{(es)} &= \frac{3058673}{1016064}-\frac{1}{16}\frac{S_1^2}{M^4}+5\frac{M_2}{M^3}+\\
&\quad+\frac{12431}{504}\frac{E^2}{M^2}+\frac{179}{9}\frac{E^4}{M^4}
\end{aligned} \tag{27}$$

$$\begin{aligned}
N_2^{(ms)} &= \frac{743}{336}+\frac{14}{3}\frac{H^2}{M^2}\\
N_3^{(ms)} &= -4\pi+\frac{113}{12}\frac{S_1}{M^2}\\
N_4^{(ms)} &= \frac{3058673}{1016064}-\frac{1}{16}\frac{S_1^2}{M^4}+5\frac{M_2}{M^3}+\\
&\quad+\frac{12431}{504}\frac{H^2}{M^2}+\frac{179}{9}\frac{H^4}{M^4}.
\end{aligned} \tag{28}$$

As in the rest three measurable quantities, the power expansion of ΔN is such that in every order term a new moment, which was not present in any lower order term, occurs. This is an indication that all moments can in principle be unambiguously extracted from accurate measurements of ΔN.

COMMENTS

As is shown in [15] the first generation of LIGO is not expected to be able to extract the first two moments (S_1 and M_2) with high accuracy (~ 0.05 for the former and ~ 0.5 for the latter one), by analyzing the phase of the waves. If we allow for electromagnetic fields as well, the corresponding monopole (which classically is expected to be very close to zero) will be measured with even higher accuracy than the other two mass moments, since the charge of the source (or the magnetic monopole in case of some exotic body) is present at even lower order, namely in the v^2 term, while the electric dipole, or the magnetic dipole, that first show up at the v^5 term will be measured with rather disappointing accuracy. On the other hand, analyzing the data of LISA leads to accuracies almost two orders of magnitude higher than the corresponding for LIGO. Thus, it seems quite promising that LISA will give us the opportunity to measure the first few moments, including the electromagnetic dipole moments, quite accurately. Also, the fact that in every new order in the power series of ΔN a new moment appears is significant, since this means that in principle a unique set of moments arises from an accurate estimation of all power series terms. Actually, there are two possible sets of moments; one for each electromagnetic case, since we cannot *a priori* exclude one of them. We can only exclude one of the two sets on physical grounds, if only one of them leads to a physically reasonable classical object (for example a highly magnetized compact object is physically preferable to a compact object with a huge electric dipole). If we manage to measure a few lower moments, we can check if they are interrelated as in a Kerr-Newman metric [13]. A positive outcome of such a test will be of support to the black-hole no-hair theorem in the case that the central object is a black-hole. The case of observational violation of the black-hole no-hair theorem could either mean that the central object is not a black-hole, or that the theorem does not hold. Of course to assume the latter an extra verification that the central object is indeed a black hole is necessary.

While the phase of a gravitational wave is the quantity that can be most accurately measured, since a large number of cycles (a few thousand for LIGO and a few hundred thousand for LISA in case of binaries with high-ratio of masses) is sweeping up the sensitive part of the detectors, the two precession frequencies Ω_ρ and Ω_z, can in principle be measured if the detectors become more sensitive and templates that describe modulating waves are used [16]. If this ever become possible one could use any of them to test the no-hair theorem. This would demand no more than the four lower order terms, since according to this theorem all moments depend on only three quantities (mass, angular momentum, and total charge). Actually, from measurements of modulating frequencies we could not at first determine which frequency corresponds to each precession. However, the power expansions of the two frequencies begin at a different order, and thus we could discern them. Unfortunately, there are terms in these expansions that contain more than one first occurring moment (for example the first term of Ω_ρ, R_2, is a function of E or H and M). However, expansions of Ω_ρ and Ω_z, if used simultaneously, along with the intrinsic dependence of v on M, could finally lead to the full determination of the moments. Notice though that some corresponding terms in the series expansions of the two frequencies depend on the same set of multipole moments, like R_3 and Z_3. Assuming that the observed system is sufficiently well described by our model for the binary, this multipole information could serve to determine these moments with better precision.

If we manage to obtain information about the precessing frequencies or the evolution of the orbits of small bodies around compact stars by other observational means, we could measure the multipole moments (mass moments, mass-current moments, and electromagnetic moments) of the compact central object and thus put strict restrictions on the models that are used to describe the interior of these objects.

Our analysis demonstrates that it will be possible to determine all types of multipole moments of the central object, from future gravitational wave measurements. Thus, apart from spacetime geometry, we could also determine the central body's electromagnetic fields. Although the data of LISA should be suitable for extracting such information with high accuracy, the assumptions of circular and equatorial orbit are not that realistic. From this point of view we consider our work as a step towards a more detailed analysis with not so restrictive assumptions.

ACKNOWLEDGMENTS

This research was supported in part by Grant No 70/4/4056 of the Special Account for Research Grants of the University of Athens, and in part by Grant No 70/3/7396 of the "PYTHAGORAS" research funding program.

REFERENCES

1. R. Geroch, J. Math. Phys. **11**, 2580 (1970).
2. R. O. Hansen, J. Math. Phys. **15**, 46 (1974).
3. W. Simon and R. Beig, J. Math. Phys. **24**, 1163 (1983).
4. W. Simon, J. Math. Phys. **25**, 1035 (1984).
5. F. D. Ryan, Phys. Rev. D **52**, 5707 (1995).
6. A. Papapetrou, Ann. Phys. **12**, 309 (1953).
7. R. M. Wald, *General Relativity* (The University of Chicago Press, Chicago, 1984).
8. F. J. Ernst, Phys. Rev. **167**, 1175 (1968).
9. F. J. Ernst, Phys. Rev. **168**, 1415 (1968).
10. T. P. Sotiriou and T. A. Apostolatos, Phys. Rev. D **71**, 044005 (2005).
11. See Ref. [16] of F. D. Ryan, Phys. Rev. D **56**, 7732 (1997) and Eq. (2.10) of M. Shibata and M. Sasaki, Phys. Rev. D **58**, 104011 (1998).
12. W. Israel, Phys. Rev. D **2**, 641 (1970).
13. T. P. Sotiriou and T. A. Apostolatos, Class. Quantum Grav. **21**, 5727 (2004).
14. C. Cutler and É. E. Flanagan, Phys. Rev. D **49**, 2658 (1994).
15. F. D. Ryan, Phys. Rev. D **56**, 1845 (1997).
16. T. A. Apostolatos Phys. Rev. D **54**, 2421 (1996).

Considerations on the excitation of black hole quasinormal modes

Emanuele Berti*, Vitor Cardoso† and Clifford M. Will*

*McDonnell Center for the Space Sciences, Department of Physics, Washington University, St. Louis, Missouri 63130, USA

†Department of Physics and Astronomy, The University of Mississippi, University, MS 38677-1848, USA

Abstract. We provide some considerations on the excitation of black hole quasinormal modes (QNMs) in different physical scenarios. Considering a simple model in which a stream of particles accretes onto a black hole, we show that resonant QNM excitation by hyperaccretion requires a significant amount of fine-tuning, and is quite unlikely to occur in nature. Then we summarize and discuss present estimates of black hole QNM excitation from gravitational collapse, distorted black holes and head-on black hole collisions. We emphasize the areas that, in our opinion, are in urgent need of further investigation from the point of view of gravitational wave source modeling.

Keywords: black holes, gravitational waves, quasinormal modes
PACS: 04.70.-s, 04.30.Db, 04.80.Cc

A leading candidate source of detectable waves for Earth-based interferometers such as LIGO, Virgo, GEO600 and TAMA, as well as for the space-based interferometer LISA, is the inspiral and merger of binary black holes. The waveform should comprise three parts, usually referred to as inspiral, merger and ringdown. The inspiral waveform, originating from that part of the decaying orbit leading up to the innermost stable orbit, can be analyzed using post-Newtonian theory. Extensive studies of the detectability of this phase of the signal have been carried out, both for Earth-based [1] and for space-based interferometers [2]. The nature of the merger waveform is largely unknown at present, and is the subject of work in numerical relativity.

The ringdown waveform originates from the distorted final black hole, and consists of a superposition of quasi-normal modes (QNMs). Each mode has an oscillation frequency and a damping time that are uniquely determined by the mass M and specific angular momentum $j \equiv J/M^2$ of the black hole [3]. The amplitudes and phases of the various modes are determined by the specific process that formed the final hole.

The uniqueness of the modes' frequencies and damping times is directly related to the "no hair" theorem of general relativistic black holes. Therefore a reliable detection and accurate identification of QNMs could provide the "smoking gun" for black holes and an important test of general relativity in the strong-field regime [4].

The detectability of ringdown waves, and the accuracy with which we can measure their frequencies and damping times to test the no-hair theorem, depend mainly on the energy carried by each mode [5]. In turn, the energy distribution depends on the details of the merger process. Given our poor understanding of the merger phase, we have at best sketchy information concerning the energy distribution between different modes.

In the first part of this paper we show by a simple toy model that QNM excitation

CP848, *Recent Advances in Astronomy and Astrophysics,*
7th International Conference of the Hellenic Astronomical Society, edited by N. H. Solomos

by infalling matter requires a significant amount of fine-tuning, and is quite unlikely in astrophysically realistic scenarios. In the second part we briefly review present estimates of QNM excitation in various physical situations, including gravitational collapse, simulations of single distorted black holes and head-on collisions of two black holes. This second part is an extended version of Sec. VB in [5]. Most of our considerations can be applied to the solar-mass black holes detectable by Earth-based interferometers, but the main motivation for this short review comes from our study of the massive black holes detectable by LISA [5]. Throughout this paper we use geometrical units ($G = c = 1$).

EXCITATION BY INFALLING MATTER

Various authors [6, 7] suggested that quasinormal ringing could be resonantly excited by clumps of matter falling into the black hole at some appropriate rate. This rate should be such that the spatial separation of the clumps equals a multiple of the typical QNM wavelength. Here we show by a simple model that, in principle, the modes of a black hole can indeed be excited in this way. We also show that a simple addition of damped sinusoids provides a good fit of the resulting gravitational waveforms: in other words, we can interpret the resonant excitation of the modes in terms of interference between gravitational waves. For simplicity we consider clumps of mass μ much smaller than the black hole mass, $\mu \ll M$, so that we can apply a perturbative analysis.

Consider first one single clump falling from infinity. This process was first analyzed in a classical paper by Davis, Ruffini, Press and Price [8] (hereafter DRPP). They found that the total radiated energy is $\simeq 0.0104\mu^2/M$, most of it $(0.0091\mu^2/M)$ being emitted as quadrupolar ($\ell = 2$) waves, and that the total energy spectrum is peaked at $\omega \sim 0.32/M$, very close to the fundamental $\ell = 2$ QNM frequency $\omega \simeq 0.3737/M$.

We can now superpose the waveform and its Fourier transform for the one-particle case to represent two or more particles falling into a black hole and to study interference phenomena (we are assuming that the clumps are non-interacting, so that there is no extra contribution to the total energy-momentum tensor). A similar sum-over-point-particles approach has been used many times in the past [9], but the particular process we consider here has not been studied before.

Let us first suppose that we drop two particles with a temporal separation of T (which, bearing in mind that the retarded time is the measured quantity, also means a spatial separation of T between the two bodies). If $\Psi_1(t,r)$ represents the waveform for the first body (normalized by μ, where μ is the mass of the infalling body), then the total waveform will be

$$\Psi(t,r) = \mu\Psi_1(t,r) + \mu\Psi_1(t-T,r)\,. \tag{1}$$

Likewise the Fourier transform $\tilde{\Psi}(\omega,r)$ will be

$$\tilde{\Psi} = \mu\tilde{\Psi}_1 + \mu e^{i\omega T}\tilde{\Psi}_1\,. \tag{2}$$

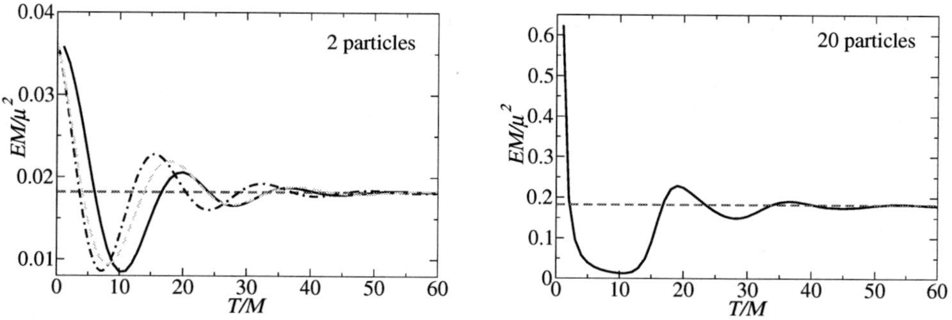

FIGURE 1. Characteristic interference pattern in the total energy radiated by two (left) and twenty (right) particles falling into a Schwarzschild black hole, as a function of the particles' initial separation. As $T \to 0$ the particles behave as a single particle of mass $(N\mu)$, and the total energy scales like $(N\mu)^2$. The red, dashed horizontal line corresponds to the sum of the single-particle energies ($E \sim N \times 0.0091\mu^2/M$ for N particles). In the two-particle case we also overplot the prediction of the two-mode ringdown waveform, Eq. (8). The blue, dot-dashed line corresponds to $\omega_r = 0.3737/M$ (the fundamental QNM frequency with $\ell = 2$). To obtain the green, dashed line we choose the "phenomenological" value $\omega_r = 0.32/M$ (corresponding to the peak of the one-particle energy spectrum).

For N bodies dropped regularly, such that the temporal difference between the dropping of the first and of the last is T, we have

$$\Psi(t,r) = \sum_{j=0}^{N-1} \mu \Psi_1 \left(t - \frac{iT}{N-1} j, r \right), \quad \tilde{\Psi}(\omega, r) = \sum_{j=0}^{N-1} \mu e^{\frac{i\omega T}{N-1} j} \tilde{\Psi}_1 . \tag{3}$$

The infall of a finite-sized body can also be studied with this formalism [1]. Incidentally, one can prove that $|\mu \sum_{j=0}^{N-1} e^{\frac{i\omega T}{N-1} j}|^2 \leq \mu^2 N^2$, so the total energy $E_{\rm rad} \sim \omega^2 |\tilde{\Psi}|^2$ radiated by a stream of plunging particles is always less than or equal to the energy radiated by a single point particle with total mass μN.

The total energy radiated in $\ell = 2$ as a function of separation for two and twenty particles is shown in Fig. 1. Let us first consider the two-particle case (left panel). As we increase the distance between the two bodies the total energy decreases. At the first minimum ($T \sim 9.9M$) the total energy radiated is $\sim 0.0085\mu^2/M$. The next maximum is attained at $T \sim 19.8M$, the total radiated energy being $\simeq 0.0206\mu^2/M$. Since the one-particle spectrum is highly peaked at $\omega \sim 0.32/M$, it is tempting to explain

[1] We can think of a finite body as being composed of many particles, the separation between the particles being infinitesimal. The total waveform Ψ will be a superposition of single-particle waveforms Ψ_1, of the form $\Psi(t,r) = \lim_{N\to\infty} \sum_{j=0}^{N-1} \frac{\mu}{N} \Psi_1 \left(t - \frac{iT}{N-1} j, r \right)$. The Fourier transform of this waveform is $\tilde{\Psi} = \lim_{N\to\infty} \sum_{j=0}^{N-1} \frac{\mu}{N} e^{\frac{i\omega T}{N-1} j} \tilde{\Psi}_1$. By taking this limit we obtain the gravitational radiation generated by a massive thin body of length T and total mass μ. Summation of the series yields $\lim_{N\to\infty} \sum_{j=0}^{N-1} \frac{\mu}{N} e^{\frac{i\omega T}{N-1} j} = \mu \frac{-1+e^{i\omega T}}{i\omega T}$, so that $\tilde{\Psi} = \mu \frac{-1+e^{i\omega T}}{i\omega T} \tilde{\Psi}_1$.

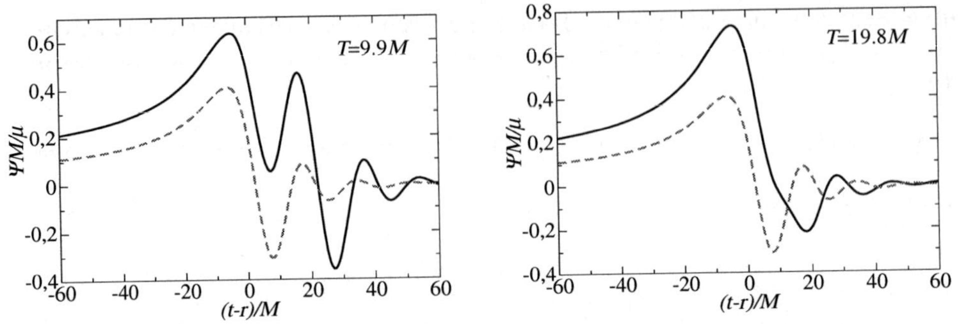

FIGURE 2. Two-particle waveforms (solid black lines) are compared with the single-particle waveforms (dashed red lines). In the left panel $T = 9.9M$, and the ringdown is suppressed by destructive interference. In the right panel $T = 19.8M$, and the ringdown is enhanced by constructive interference.

this behavior as an interference phenomenon. The typical dominant wavelength of the fundamental QNM is $\lambda = 2\pi/\omega \sim 19.6M$: this means that, for two particles, destructive interference should occur at $T \sim 9.8M$ and constructive interference at $T \sim 19.6M$, in good agreement with our numerical results. In the limit of large distances, interference becomes negligible and the total radiated energy tends to the sum of the individual particle energies ($\simeq 0.0182\mu^2/M$ for our two particles). This interpretation of the result is confirmed by an explicit calculation of the gravitational waveforms (Fig. 2).

The situation for more than two particles is similar. Notice that in the two-particle case, for $T < 6M$ there is always an enhancement on the total radiated energy (as compared to the sum of the single-particle energies). However, as we increase the number of particles, this enhancement interval decreases. So we should expect it will be hard to resonantly excite the black hole QNMs using a large number of particles.

Interpreting the results in terms of ringdown waveforms - The DRPP signal is dominated by quasinormal ringing, so it makes sense to try and interpret the preceding results in terms of ringdown waveforms. To start with, we assume our signal has the form

$$\psi = \Theta(t) A e^{-\omega_i t} \sin \omega_r t\,, \tag{4}$$

where $\Theta(t)$ is the Heaviside function. For the fundamental mode of a Schwarzschild black hole $\omega_r \simeq 0.3737/M$ and $\omega_i \simeq 0.0890/M$. The amplitude A can be obtained imposing that the total radiated energy in $\ell = 2$ is equal to $0.0091\mu^2/M$. The total energy radiated in the ℓ-th multipole is

$$E_{\rm rad} = \frac{1}{64\pi}\frac{(l+2)!}{(l-2)!}\int_{-\infty}^{-\infty}\left(\frac{d\psi}{dt}\right)^2 dt\,, \tag{5}$$

which for the waveform (4) gives

$$E_{\rm rad} = \frac{A^2 (l+2)!\,\omega_r^2}{256\pi (l-2)!\,\omega_i}\,. \tag{6}$$

Setting $\ell = 2$ and equating this to $0.0091\mu^2/M$ gives $A \simeq 0.441\mu$, in qualitative agreement with the DRPP waveform. Let us now consider two particles by superposing two ringdown waveforms:

$$\psi = \Theta(t)Ae^{-\omega_i t}\sin\omega_r t + \Theta(t-T)Ae^{-\omega_i(t-T)}\sin\omega_r(t-T). \qquad (7)$$

The total energy radiated would be

$$E_{\rm rad}^{\rm 2ptcles} = \frac{A^2 e^{-T\omega_i}\omega_r(l+2)!}{128\pi(l-2)!\omega_i}\left(e^{\omega_i T}\omega_r + \omega_r\cos\omega_r T - \omega_i\sin\omega_r T\right). \qquad (8)$$

A plot of this function is shown in the left panel of Fig. 1. It is in good qualitative agreement with the full numerical calculation. If we use a "phenomenological" QNM frequency $\omega_r = 0.32/M$ (corresponding to the peak of the one-particle energy spectrum, and yielding an amplitude $A \simeq 0.515\mu$) instead of the "true" fundamental $\ell = 2$ QNM frequency $\omega_r = 0.3737/M$, the agreement between Eq. (8) and the numerical results becomes even better.

This agreement suggests the possibility to extend our study to Kerr black holes, by simply assuming that ω_r and ω_i are functions of the black hole's angular momentum. Using (6) and (8) we get

$$\frac{E_{\rm rad}^{\rm 2ptcles}}{E_{\rm rad}} = 2e^{-T\omega_i}\left(e^{\omega_i T} + \cos\omega_r T - \frac{\omega_i}{\omega_r}\sin\omega_r T\right) \sim 2(1+\cos\omega_r T). \qquad (9)$$

In the last step we considered rapidly rotating (near-extremal) black holes and modes with $\ell = m$, for which $\omega_i \to 0$ and $\omega_r \to m/(2M)$. Therefore, for corotating modes of rapidly rotating black holes the resonance is almost perfect if we choose $T = 2\pi M/m$ (the ratio is then 4, the maximum allowed) while destructive interference results if $T = \pi M/m$.

In summary: constructive resonance, and consequently QNM excitation, is possible (in principle) when a stream of particles accretes onto a black hole. Unfortunately, the separation between each particle required for an enhancement in the total energy becomes very small as the number of particles increases. Unless some delicate fine tuning is at work [7], the separation between any two infalling particles or clumps should be driven by a random process, which means that on average there should be no resonance at all. This is confirmed by simulations of generic configurations of infalling matter, where quasinormal ringing is usually absent from the waveforms [10].

ENERGY DISTRIBUTION IN DIFFERENT SCENARIOS

In this Section we present a brief review of results on QNM excitation in different physical scenarios, including gravitational collapse, evolutions of single distorted black holes and head-on black hole collisions. Far from being complete or comprehensive, our review aims to emphasize areas that require further investigation from the point of view of gravitational wave source modeling.

Gravitational collapse - There are just a few calculations of gravitational wave emission from rotating gravitational collapse to a black hole. Perturbative calculations were first carried out by Cunningham, Price and Moncrief [11, 12, 13], and later improved upon by Seidel and collaborators [14]. These studies suggest that gravitational waves are mainly generated in the region where the Zerilli (or Regge-Wheeler) potential is large, and that the signal is always dominated by quasinormal ringing of the finally formed black hole. Simplified simulations based on a free-fall (Oppenheimer-Snyder) collapse model yield a rather pessimistic energy output, with a typical core-collapse radiating up to $E \simeq 10^{-7}M$ in gravitational wave energy, and most of the radiation being emitted in modes with $\ell = 2$. Radiation in $\ell = 3$ is typically two to three orders of magnitude smaller than the $\ell = 2$ radiation (see eg. Fig. 9 in [11]). Cunningham *et al.* also provide analytical expressions for the energy radiated in different multipoles as a function of the initial quadrupole deformation of the star and of the initial collapse radius [11, 12].

For a long time the only *nonperturbative*, axisymmetric calculation of gravitational wave emission from collapse has been provided by the seminal work of Stark and Piran [15]. Surprisingly, the waveform resembles that emitted by a point particle falling into a black hole, but with a reduced amplitude (we will see below that a similar result emerges from head-on black hole collisions in full general relativity). As expected for axisymmetric configurations, the amplitude of the cross polarization mode is always $<$ 0.2 that of the even mode, and the emission in the plus mode is dominantly quadrupolar, with a $\sin^2\theta$ angular dependence (the cross polarization has a dependence of the form $\cos\theta \sin^2\theta$). The total energy emitted is quite low; it increases with rotation rate, ranging from $\sim 10^{-8}M$ for $j = 0$ to $\sim 7\times 10^{-4}M$ as $j \to 1$. Rotational effects halt the collapse for some critical value of $j = j_{\rm crit}$ which is very close to unity, and depends on the amount of artificial pressure reduction used to trigger the collapse. Stark and Piran find that the energy emitted scales as j^4 for low j: $E/M \simeq \min\{1.4\times 10^{-3} j^4, \varepsilon_{\rm max}\}$ for $0 < j < j_{\rm crit}$, where $\varepsilon_{\rm max} \sim 10^{-4}$ again depends on the amount of artificial pressure reduction used to trigger the collapse. The h_+ waveform is very well fitted by a combination of the two lowest QNMs with $\ell = 2$ and it is only weakly sensitive to j, reflecting the weak dependence of the slowly-damped Kerr QNM frequencies on j.

This two-dimensional calculation has recently been improved upon using a three-dimensional code [16], but still keeping a high degree of axisymmetry (so only modes with $m = 0$ are excited). Ref. [16] picks as the initial configuration the most rapidly rotating, dynamically unstable model described by a polytropic EOS with $\Gamma = 2$ and $K = 100$, having a dimensionless rotation rate $\simeq 0.54$, and triggers collapse reducing the pressure by less than about 2%. Gravitational waves are extracted by a gauge-invariant approach in which the spacetime is matched with nonspherical perturbations of a Schwarzschild black hole. The "plus" polarization is essentially a superposition of modes with $\ell = 2$ and $\ell = 4$, and the "cross" polarization is a superposition of modes with $\ell = 3$ and $\ell = 5$. Fig. 1 in [16] shows that the amplitude of the $\ell = 2$ mode is about an order of magnitude larger than the amplitude of the $\ell = 4$ mode, and Fig. 3 shows that the cross polarization is roughly suppressed by one order of magnitude with respect to the plus polarization, with maximum amplitudes in a ratio $|(r/M)h_\times|_{\rm max}/|(r/M)h_+|_{\rm max} \simeq 0.06$ (here r denotes the distance between source and detector). Odd-parity perturbations should be zero in spacetimes with axial and equatorial symmetries: they are nonzero because of the rotationally induced coupling with even-parity perturbations [11, 12, 13, 14]. The maximum amplitude from these three-dimensional simulations is $|(r/M)h_+|_{\max} \simeq 0.00225$, about one order of magnitude smaller than the amplitude found by Stark and Piran. Correspondingly, the total energy lost to gravitational waves is $E \simeq 1.45 \times 10^{-6}(M/M_\odot)$. This is about two orders of magnitude smaller than the estimate by Stark and Piran for the same value of the angular momentum, but larger than the energy losses found in recent calculations of rotating stellar core collapse to protoneutron stars [17].

Distorted black holes - Distorted black hole simulations shed some light on the (crucial) issue of the dependence of QNM excitation on the initial data: that is, on the "shape" and "size" of the black hole distortion.

Ref. [18] carries out a pioneering study of linearized gravitational perturbations of the Kerr spacetime. Given initial data with a certain angular dependence - that is, assigned values of (l, m) - their study finds that modes with positive and negative m are always excited together. The simultaneous excitation of corotating and counterrotating modes depends on a certain reflection-symmetry property of the Kerr QNM spectrum [5], and is confirmed by more recent time-evolutions of scalar perturbations of the Kerr spacetime [19]. In addition, these new simulations show that the relative excitation of corotating and counterrotating modes depends on the radial profile of the initial data, confirming expectations based on analytic investigations of the same problem [20]. Unfortunately, detailed studies of the excitation of gravitational perturbations of Kerr black holes in the perturbative regime are still lacking.

Ref. [21] provides one of the first attempts to extract gravitational waveforms from *non-linear* simulations of distorted black holes. They use a two-dimensional code to evolve initial data corresponding to a single, nonrotating black hole superimposed with time-symmetric gravitational waves (Brill waves), extract gravitational waves using gauge-invariant perturbation theory and show that, for low-amplitude Brill waves, non-perturbative and perturbative waveforms with $\ell = 2$ and $\ell = 4$ are in good agreement.

A more complete analysis is due to Anninos *et al.* [22]. They use three different initial data sets, corresponding to very different physical situations: non-rotating black holes distorted by time-symmetric Brill waves, distorted rotating black holes, and the time-symmetric "two black hole" Misner data. In all three cases the system quickly settles down to a nearly spherical (or oblate, for rotating holes) configuration. They study the dynamics of the black hole horizon monitoring the ratio of polar and equatorial circumference $C_r = C_p/C_e$, and find that the horizon oscillates at the QNM frequencies of the final black hole. Oscillations with $\ell = 2$ and $\ell = 4$ are well fitted by a superposition of the fundamental QNM and the first overtone. In the "black hole plus (small) Brill wave" case, even when the initial data contain a significant component with $\ell = 4$ (that is, when the index n in the analytic expression of the Brill wave is equal to 4) the $\ell = 4$ component of the horizon distortion is a factor $\sim 10^{-4}$ smaller than the $\ell = 2$ component (Figs. 2 and 3 in [22]). For Misner initial data, corresponding to colliding black holes, $\ell = 2$ waves are also strongly dominant. An important conclusion of Ref. [22] is that the dynamics of the apparent horizon geometry can be used not only to find the fundamental QNM frequency of the hole, but also its mass and angular momentum.

Brandt and Seidel [23] present an extensive study of distorted, rotating "Kerr-like" black holes with a wide range of rotation parameters (as high as $j \simeq 0.87$). Quasi-normal ringing is observed both in the even and in the odd components of the emitted radiation. The angular momentum parameter j is well approximated by the simple fit $j = \sqrt{1-(-1.55+2.55C_r)^2}$ (the fit being generally accurate within $\sim 2.5\%$): this means that a measurement of the horizon geometry provides a value of the rotation parameter. For black holes with $j > \sqrt{3}/2$, a Kerr black hole cannot be embedded in flat space: an alternative method to obtain j consists in measuring the angle at which the embedding ceases to exist.

A crucial result of Ref. [23] is that the black hole dynamics depend critically on the properties of the initial data sets. For example, initial data corresponding to nonlinear, odd perturbations produce oscillations of the Gaussian curvature at frequencies given by linear combinations of QNMs with $\ell = 3$ and $\ell = 5$, because of beating phenomena. In Sec. IVB, they present a method to "dig out" frequencies with stronger damping from the slowly-damped modes in a Fourier transform, which can find useful applications in gravitational-wave data analysis. Sec. V of [23] presents waveforms corresponding to different initial distortions of the hole. Odd-parity distortions of Schwarzschild black holes (run "o1" in their terminology) produce a waveform in which 99.9% of the energy is radiated in $\ell = 3$, 0.05% of the radiation goes into $\ell = 2$, 0.02% into $\ell = 4$ and 0.0001% into $\ell = 5$; but for different "shapes" of the distortion (run "o2") , the energy radiated in $\ell = 5$ can be as large as 21%. Similarly, runs corresponding to distorted Kerr black holes ("r0" to "r5") contain different mixtures of the different multipoles depending on the shape of the initial distortion and on the rotation rate of the black hole. As a trend, large rotation seems to yield a large $\ell = 3$ component: the maximum corresponds to run "r4", for which the $\ell = 2$ component carries 92.1% of the energy and $\ell = 3$ carries $\sim 7\%$ of the energy.

Three-dimensional, non-axisymmetric evolutions are urgently needed. A handful of non-axisymmetric simulations can be found in Ref. [24]. They find that the total energy radiated in a given multipole, $E_{\ell m}$, ranges from $E_{20} \sim 3\times10^{-4}M$ to $E_{42} \sim 3\times10^{-7}M$. Only modes with $\ell = 2$, $m = 0$, 2 and $\ell = 4$, $m = 0$, 2 should occur at linear order in the amplitude, all other modes being quadratic in the amplitude (for one of these "nonlinear" modes they find a completely negligible energy, $E_{62} \sim 10^{-10}M$).

In fact, one of the most interesting features introduced by nonlinearities is mode-mode coupling. Ref. [25] studies this coupling in detail for nonrotating (Schwarzschild) black holes. Large-amplitude incident waves effectively add mass to the black hole, producing a small but visible "redshift" in the QNM frequencies. The waveform is also slightly shifted by nonlinearities. This nonlinear dephasing might be an important issue when applying matched-filtering techniques. On the other hand, when the amplitude A of incident waves exceeds a certain threshold, nonlinearities *amplify* the outgoing wave, which of course is good news for detection. Beyond the "nonlinearity threshold" the outgoing amplitude of the ℓ-th multipole scales as $A^{\ell/2}$. The reason is that, for ordinary spherical harmonics, $(Y_{\ell m})^k \sim Y_{(k\ell)m}$+terms with smaller ℓ. Given a linear incident wave with $\ell = 2$, an $\ell = 2k$ mode arises from order k (and higher) nonlinear terms; this mode's amplitude scales as $A^k + \mathcal{O}(A^{k+1})$. So, the quadratic coupling of an ($\ell = 2$, $m = 0$) mode with itself produces an ($\ell = 4$, $m = 0$) mode of order $A^2 + \mathcal{O}(A^3)$ (and higher), in addition to an ($\ell = 2$, $m = 0$) mode of order $A + \mathcal{O}(A^2)$ (notice that $\ell = 0$ is ruled out for gravitational perturbations, being non-radiative). In addition, nonlinearities produce beating phenomena. In general, quadratic terms of an harmonic with frequency ω contain the frequencies $(0,\ 2\omega)$, cubic terms contain $(\omega,\ 3\omega)$, and so on. The selection rules for mode coupling in Kerr backgrounds should be more interesting: even in the linear regime, the Lense-Thirring effect couples axial and polar perturbations of a slowly rotating star according to Laporte's rule [26]. Mode couplings in the Kerr background can be very important for gravitational wave data analysis, and deserve further study.

In conclusion, simulations suggest that (at least in situations with a certain degree of symmetry) high overtones are heavily suppressed. Investigating an extensive catalogue of initial data, Ref. [23] shows that we just *don't know* how energy is going to be partitioned between different modes in a realistic merger event, unless we can predict with accuracy the shape of the initial data. Nonetheless, for data-analysis purposes it should be reasonable to assume that modes with $\ell = 2$, $m = 0$, ± 2 have the largest amplitude (the components with $|m| = 2$ probably being dominant). The contribution from $\ell = 4$, $m = 0$, ± 2 could be smaller by about two orders of magnitude in energy. Modes with $\ell = 3$ may also be relevant, especially at large rotation rates. Nonlinearities could produce a cascade of energy into higher-order modes, but we need more simulations with realistic initial data to assess the relevance of this effect for QNM detection.

Head-on black hole collisions -

In four dimensions, the total energy radiated by a freely-falling particle released from rest at infinity in the Schwarzschild spacetime [8, 27] is in surprising agreement with numerical simulations of head-on black hole collisions. An extensive comparison of numerical results with perturbation theory can be found in Refs. [28, 29, 30, 31, 32].

Fig. 14 and Sec. IV of Ref. [30] show that the perturbation-theory prediction for the radiated energy, $E = 0.0104\mu^2/M$ [8], is in excellent agreement with numerical results when we replace the particle's mass by the system's reduced mass. In fact, perturbation theory slightly *overestimates* the energy output predicted by numerical simulations, which (according to state-of-the-art simulations) is $E \simeq 0.0013M$ [33]. In the so-called "particle-membrane" picture, this disagreement is compensated for multiplying the perturbative prediction by three fudge factors [30]: i) a factor F_{r_0}, accounting for the fact that in numerical simulations the infall starts at finite distance r_0; ii) a factor F_h coming in because the black hole (unlike the falling particle) has finite size, and tidal deformations heat up the horizon; iii) a factor $F_{\rm abs}$ to account for reabsorption of the emitted gravitational waves by each black hole. If the collision starts at infinite separation, $F_{r_0} = 1$; F_h has a weak dependence on r_0, and tends to $F_h \simeq 0.86$ as $r_0 \to \infty$; $F_{\rm abs} \simeq 0.99$ is even less relevant.

If the total energy output from head-on black hole collisions is well understood, unfortunately the relative excitation of higher multipoles with respect to $\ell = 2$ is not. Ref. [32] compares numerical calculations with the close-limit approximation (valid when the black holes are close enough to be considered as perturbations of a single Schwarzschild black hole) and with the particle-membrane approach. Fig. 12 of Ref. [32] compares the radiated energy according to the close-limit approximation with results from the numerical simulations. Extraction of the $\ell = 4$ component from the simulations is very sensitive to numerical noise, especially when the black holes are very close and the $\ell = 4$ component is more heavily suppressed.

Numerical simulations have been extended to include unequal-mass black holes [34] and boosted black holes [35, 36]. More recently, mesh refinement techniques were used to obtain more reliable estimates of the energy radiated in $\ell = 2$ [33]. Even for highly boosted black holes, the agreement between perturbation theory (in the close-limit approximation) and numerical simulations is very good. Numerical studies of boosted black holes show that, fixing the initial separation between the holes, the energy radiated saturates at $E \sim 0.01M$ for very large values of the initial momentum of the holes (see eg. Fig. 2 of [36]). Ref. [35] combines Newtonian dynamics and numerical simulations of boosted black holes to suggest that the maximum energy emission could actually be much lower than this, $E < 0.0016M$. In general, these studies suggest that the role of the weak-field phase of the evolution is only to determine the momentum of the black holes when they start to interact nonlinearly, emitting most of the radiation in the ringdown phase. This expectation is confirmed by the comparison of perturbation theory and Post-Newtonian calculations [37]: the bremsstrahlung radiation emitted by a particle at separations larger than $4M$ (in Schwarzschild coordinates) contributes only $\sim 3\%$ of the total energy, most of the radiation coming from the final (ringdown) phase.

CONCLUSIONS

Most numerical simulations so far consider nearly-axisymmetric situations, so they don't have much to say about the distribution of energy between modes with different values of (l,m) in a realistic black hole merger. This is a crucial issue for data analysis from ringdown waveforms [5]. Recently there has been some remarkable progress in non-axisymmetric simulations of black hole mergers (see eg. [38] and references therein), but a discussion of these developments is beyond the scope of this paper.

Some estimates of the energy radiated in the plunge phase by solar-mass binaries of rotating black holes are provided by the effective-one-body approach [39]. The error introduced by extrapolating to supermassive black holes should not be too large, since the energy balance of black hole binaries involves a single fundamental scale (their total mass M). It is useful to compare the effective-one-body estimate with earlier estimates of the energy radiated in the plunge *plus ringdown* phases, as provided by the Lazarus approach [40]. For aligned (anti-aligned) spins and specific angular momenta $j_1 = j_2 = 0.17$ ($j_1 = j_2 = 0.25$) the effective-one-body approach predicts an energy release $\sim (0.6-0.9)\%M$ [$\sim (1-3)\%M$]. The Lazarus approach finds $\sim (1.7-1.9)\%M$ [$\sim (1.9-2.1)\%M$], respectively. In principle, the difference of these two estimates should bracket the most reasonable range of ringdown efficiencies. However, the overall error on both estimates is still too large to draw any definite conclusion. Numerical relativity should be able to provide a reliable answer to this problem within a very short time.

ACKNOWLEDGMENTS

We thank Leonardo Gualtieri, Luis Lehner and Alessandro Nagar for useful discussions. This work was supported in part by the National Science Foundation under grant PHY 03-53180.

REFERENCES

1. A. Buonanno, Y. Chen and M. Vallisneri, Phys. Rev. D **67**, 024016 (2003).
2. E. Berti, A. Buonanno and C. M. Will, Phys. Rev. D **71**, 084025 (2005).
3. K. D. Kokkotas and B. G. Schmidt, Living Rev. Rel. **2**, 2 (1999); H.-P. Nollert, Class. Quantum Grav. **16**, R159 (1999).
4. O. Dreyer, B. Kelly, B. Krishnan, L. S. Finn, D. Garrison and R. Lopez-Aleman, Class. Quantum Grav. **21**, 787 (2004).
5. E. Berti, V. Cardoso and C. M. Will, gr-qc/0512160.
6. C. L. Fryer, D. E. Holz and S. A. Hughes, Astrophys. J. **565**, 430 (2002).
7. R. A. Araya-Góchez, MNRAS **355**, 336 (2004).
8. M. Davis, R. Ruffini, W. H. Press and R. H. Price, Phys. Rev. Lett. **27**, 1466 (1971).
9. M. P. Haugan, S. L. Shapiro and I. Wasserman , Astrophys. J. **257**, 283 (1982); S. L. Shapiro and I. Wasserman, Astrophys. J. **260**, 838 (1982); L. I. Petrich, S. L. Shapiro and I. Wasserman , Astrophys. J. Suppl. Ser. **58**, 297 (1985); T. Nakamura and K. Oohara, Phys. Lett. **98A**, 403, (1983); T. Nakamura, K. Oohara and Y. Kojima, Prog. Theor. Phys. Suppl. **90** (1987).
10. P. Papadopoulos and J. A. Font, Phys. Rev. D **63**, 044016 (2001); A. Nagar, G. Diaz, J. A. Pons and J. A. Font, Phys. Rev. D **69**, 124028 (2004); A. Nagar, J. A. Font, O. Zanotti and R. De Pietri, Phys. Rev. D **72**, 024007 (2005).
11. C. T. Cunningham, R. Price and V. Moncrief, Astrophys. J. **224**, 643 (1978).
12. C. T. Cunningham, R. Price and V. Moncrief, Astrophys. J. **230**, 870 (1979).
13. C. T. Cunningham, R. Price and V. Moncrief, Astrophys. J. **236**, 674 (1980).
14. E. Seidel and T. Moore, Phys. Rev. D **35**, 2287 (1987); E. Seidel, E. S. Myra and T. Moore, Phys. Rev. D **38**, 2349 (1988); E. Seidel, Phys. Rev. D **42**, 1884 (1990); E. Seidel, Phys. Rev. D **44**, 950 (1991).
15. R. F. Stark and T. Piran, Phys. Rev. Lett. **55**, 891 (1985); in *Proceedings of the Workshop Dynamical Spacetimes and Numerical Relativity, Philadelphia* (Cambridge University Press, Cambridge, U.K., 1986), pp. 40-73.
16. L. Baiotti, I. Hawke, L. Rezzolla and E. Schnetter, Phys. Rev. Lett. **94**, 131101 (2005).
17. E. Müller *et al.*, Astrophys. J. **603**, 221 (2004).
18. W. Krivan, P. Laguna, P. Papadopoulos and N. Andersson, Phys. Rev. D **56**, 3395 (1997).
19. N. Dorband, P. Diener, E. Schnetter, M. Tiglio and E. Berti, in preparation.
20. K. Glampedakis and N. Andersson, Phys. Rev. D **64**, 104021 (2001). K. Glampedakis and N. Andersson, Class. Quantum Grav. **20**, 3441 (2003).
21. A. Abrahams, D. Bernstein, D. Hobill, E. Seidel and L. Smarr, Phys. Rev. D **45**, 3544 (1992).
22. P. Anninos, D. Bernstein, S. R. Brandt, D. Hobill, E. Seidel and L. Smarr, Phys. Rev. D **50**, 3801 (1994).
23. S. R. Brandt and E. Seidel, Phys. Rev. D **52**, 870 (1995).
24. G. Allen, K. Camarda and E. Seidel, gr-qc/9806036.
25. Y. Zlochower, R. Gomez, S. Husa, L. Lehner and J. Winicour, Phys. Rev. D **68**, 084014 (2003).
26. S. Chandrasekhar and V. Ferrari, Proc. R. Soc. Lond. A **433**, 423 (1991).
27. V. Ferrari and R. Ruffini, Phys. Lett. B **98**, 381 (1981).
28. L. L. Smarr, in *Sources of Gravitational Radiation*, edited by L. L. Smarr (Cambridge Univ. Press, Cambridge, 1979).
29. P. Anninos, D. Hobill, E. Seidel, L. Smarr and W. M. Suen, Phys. Rev. Lett. **71**, 2851 (1993).
30. P. Anninos, D. Hobill, E. Seidel, L. Smarr and W. M. Suen, Phys. Rev. D **52**, 2044 (1995).
31. R. J. Gleiser, C. O. Nicasio, R. H. Price, and J. Pullin, Phys. Rev. Lett. **77**, 4483 (1996);
32. P. Anninos, R. H. Price, J. Pullin, E. Seidel and W. M. Suen, Phys. Rev. D **52**, 4462 (1995).
33. U. Sperhake, B. Kelly, P. Laguna, K. L. Smith, E. Schnetter, Phys. Rev. D **71**, 124042 (2005).
34. P. Anninos and S. Brandt, Phys. Rev. Lett. **81**, 508 (1998).
35. A. M. Abrahams and G. B. Cook, Phys. Rev. D **50**, R2364 (1994).
36. J. Baker, A. M. Abrahams, P. Anninos, S. Brandt, R. Price, J. Pullin and E. Seidel, Phys. Rev. D **55**, 829 (1997).
37. L. E. Simone, E. Poisson and C. M. Will, Phys. Rev. D **52**, 4481 (1995).
38. F. Herrmann, D. Shoemaker and P. Laguna, gr-qc/0601026.
39. A. Buonanno, Y. Chen and T. Damour, gr-qc/0508067.
40. J. Baker, M. Campanelli, C. O. Lousto and R. Takahashi, Phys. Rev. D **69**, 027505 (2004).

Accelerating Universe:Observational Status and Theoretical Implications

L. Perivolaropoulos

Department of Physics, University of Ioannina
e-mail: leandros@cc.uoi.gr

Abstract. This is a pedagogical review of the recent observational data obtained from type Ia supernova (SnIa) surveys that support the accelerating expansion of the universe. The methods for the analysis of the data are reviewed and some of the theoretical implications obtained from their analysis are discussed. In particular, the existence of dark energy with negative pressure required by the data is demonstrated and the simplest type of dark energy (the cosmological constant) is shown to fit well the SnIa data. Dynamical forms of evolving dark energy are also shown at their best fit form to the data and their common features are discussed.

Keywords: Cosmological Expansion,Dark Energy,Cosmological Constant
PACS: 98.80.Es,98.65.Dx,98.62.Sb

INTRODUCTION

Recent distance-redshift surveys [1, 2, 3, 4, 5, 6] of cosmologically distant Type Ia supernovae (SnIa) have indicated that the universe has recently (at redshift $z \simeq 0.5$) entered a phase of accelerating expansion. This expansion has been attributed to a dark energy [7] component with negative pressure which can induce repulsive gravity and thus cause accelerated expansion. The evidence for dark energy has been indirectly verified by Cosmic Microwave Background (CMB) [8] and large scale structure [9] observations.

The simplest and most obvious candidate for this dark energy is the cosmological constant [10] with equation of state $w = \frac{p}{\rho} = -1$. The extremely fine tuned value of the cosmological constant required to induce the observed accelerated expansion has led to a variety of alternative models where the dark energy component varies with time. Many of these models make use of a homogeneous, time dependent minimally coupled scalar field ϕ (quintessence[11, 12]) whose dynamics is determined by a specially designed potential $V(\phi)$ inducing the appropriate time dependence of the field equation of state $w(z) = \frac{p(\phi)}{\rho(\phi)}$. Given the observed $w(z)$, the quintessence potential can in principle be determined. Other physically motivated models predicting late accelerated expansion include modified gravity[13, 14, 15], Chaplygin gas[16], Cardassian cosmology[17], theories with compactified extra dimensions[18, 19], braneworld models[20] etc. Such cosmological models predict specific forms of the Hubble parameter $H(z)$ as a function of redshift z. The observational determination of the recent expansion history $H(z)$ is therefore important for the identification of the viable cosmological models.

The most direct and reliable method to observationally determine the recent expansion history of the universe $H(z)$ is to measure the redshift z and the apparent luminosity of

CP848, *Recent Advances in Astronomy and Astrophysics,*
7th *International Conference of the Hellenic Astronomical Society,* edited by N. H. Solomos

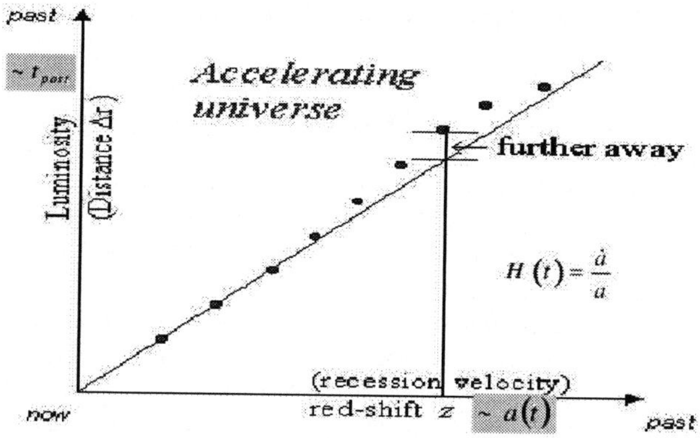

FIGURE 1. The Hubble diagram. In an accelerating universe luminous objects at a given redshift appear to be dimmer.

cosmological distant indicators (standard candles) whose absolute luminosity is known. The luminosity distance vs. redshift is thus obtained which in turn leads to the Hubble expansion history $H(z)$.

The goal of this review is to present the methods used to construct the recent expansion history $H(z)$ from SnIa data and discuss the most recent observational results and their theoretical implications. In the next section I summarize the method used to determine $H(z)$ from cosmological distance indicators and discuss SnIa as the most suitable cosmological standard candles. The three cosmological puzzles that emerge from the observed expansion history $H(z)$ and other cosmological observations are also discussed. In section 3 I show that the three puzzles may be resolved simultaneously by assuming that 70% of the energy content of the universe consists of a substance having negative pressure, the *dark energy*. The simplest form of dark energy, the *cosmological constant* is also discussed and its good fit to the SnIa data is demonstrated. Finally, in section 4 I discuss more general forms of dark energy whose energy density evolves dynamically with time. It is shown that these forms of dark energy can provide better fits to the observed expansion history $H(z)$.

EXPANSION HISTORY AND COSMOLOGICAL PUZZLES

A particularly useful diagram which illustrates the expansion history of the Universe is the *Hubble diagram*. The x-axis of a Hubble diagram (see Fig. 1) shows the redshift z of cosmological luminous objects while the y-axis shows the physical distance Δr to these objects.

In the context of a cosmological setup the redshift z is connected to the scale factor $a(t)$ at the time of emission of radiation by $1+z = \frac{a(t_0)}{a(t)}$ where t_0 is the present time. On the other hand, the distance to the luminous object is related to the time in the past t_{past} when the radiation emission was made. Therefore, the Hubble diagram contains information about the time dependence of the scale factor $a(t)$. The slope of this diagram at a given redshift denotes the inverse of the expansion rate $\frac{\dot{a}}{a}(z) \equiv H(z)$ ie

$$\Delta r = \frac{1}{H(z)} c\, z \tag{1}$$

In an accelerating universe the expansion rate $H(z)$ was smaller in the past (high redshift) and therefore the slope H^{-1} of the Hubble diagram is larger at high redshift. Thus, at a given redshift, luminous objects appear to be further away (dimmer) compared to an empty universe expanding with a constant rate (see Fig. 1).

The luminous objects used in the construction of the Hubble diagram are objects whose absolute luminosity is known and therefore their distance can be evaluated from their apparent luminosity along the lines discussed above. Such objects are known as *distance indicators* or *standard candles*. The best choice distance indicators for cosmology are SnIa not only because they are extremely luminous (at their peak they are as luminous as a bright galaxy) but also because their absolute magnitude can be determined at a high accuracy.

SnIa emerge in binary star systems where one of the companion stars has a mass below the Chandrasekhar limit $1.4 M_{\odot}$ and therefore ends up (after hydrogen and helium burning) as white dwarf supported by degeneracy pressure. Once the other companion reaches its red giant phase the white dwarf begins gravitational striping of the outer envelop of the red giant thus accreting matter from the companion star. Once the white dwarf reaches a mass equal to the Chandrasekhar limit, the degeneracy pressure is unable to support the gravitational pressure, the white dwarf shrinks and increases its temperature igniting carbon fussion. This leads to violent explosion which is detected by a light curve which rapidly increases luminosity in a time scale of less than a month, reaches a maximum and disappears in a timescale of 1-2 months.

SnIa are the preferred distance indicators for cosmology for several reasons:

1. They are exceedingly luminous. At their peak luminosity they reach an absolute magnitude of $M \simeq -19$ which corresponds to about $10^{10} M_{\odot}$.
2. They have a relatively small dispersion of peak absolute magnitude.
3. Their explosion mechanism is fairly uniform and well understood.
4. There is no cosmic evolution of their explosion mechanism according to known physics.
5. There are several local SnIa to be used for testing SnIa physics and for calibrating the absolute magnitude of distant SnIa.

On the other hand, the main problem for using SnIa as standard candles is that they are not easy to detect and it is impossible to predict a SnIa explosion. In fact the expected number of SnIa exploding per galaxy is 1-2 per millenium.

Our current knowledge of the expansion history of the universe can be summarized as follows: The universe originated at an initial state that was very close to a density singularity known as the Big Bang. Soon after that it entered a phase of superluminal accelerating expansion known as inflation. During inflation causally connected regions of the universe exited out of the horizon, the universe approached spatial flatness and the primordial fluctuations that gave rise to structure were generated. At the end of inflation the universe was initially dominated by radiation and later by matter whose attractive gravitational properties induced a decelerating expansion.

The SnIa data discussed in more detail below, strongly suggest that the universe has recently entered a phase of accelerating expansion at a redshift $z \simeq 0.5$. This accelerating expansion can not be supported by the attractive gravitational properties of regular matter. The obvious question to address is therefore 'What are the properties of the additional component required to support this acceleration?'.

This is one of the three important cosmological puzzles that emerged during the past decade in the context of a matter dominated cosmology. The second puzzle can be summarized as follows: The location of the first peak of the Cosmic Microwave Background (CMB) angular spectrum corresponds to the size of the sound horizon at the last scattering surface and can therefore probe the geometry of the universe. Indeed, the angular scale at which we observe this peak is tied to the geometry of the universe since in a negatively (positively) curved universe, photon paths diverge (converge), leading to a larger (smaller) apparent angular size as compared to a flat universe. We can therefore relate the spatial curvature (determined by the total energy density of the universe ρ_{tot} to the observed peak in the CMB spectrum [8]. Using this method and the latest CMB data from WMAP it was found that the geometry of the universe is very nearly flat and therefore, in the context of General Relativity (GR) the total energy density ρ_{tot} is equal to the critical energy density ρ_{crit} required for flatness ie

$$\rho_{tot} = \rho_{crit} \tag{2}$$

On the other hand large scale structure observations based on dynamics of galaxies and clusters as well as on gravitational lensing of background galaxies [21] and temperature profiles of the X-ray gas in clusters [22], indicate a smaller present energy density ρ_{0m} for matter. Observations converse towards

$$\rho_{0m} \simeq 0.3\rho_{crit} \tag{3}$$

instead of the expected value $\rho_{0m} \simeq \rho_{crit}$. This discrepancy raises the question: 'Where (and what) is the missing energy density which is not in the form of matter (or radiation)?'

The third cosmological puzzle is obtained by comparing the age of globular clusters (low metallicity old star populations) with the age of a flat matter dominated universe. By comparing a lower limit on the age of the oldest globular clusters in our galaxy estimated to be between $11 Gyrs$ and $21 Gyrs$ [23] with the expansion age of a matter dominated universe($\sim 10 Gyrs$), determined by measurements of the Hubble constant $H_o \geq 60 km/(sec \cdot Mpc)$, an apparent inconsistency arose: globular clusters appeared to be older than the Universe. This is known as 'the globular cluster age problem' or the 'age crisis of cosmology'.

DARK ENERGY AND THE COSMOLOGICAL CONSTANT

The simultaneous resolution to the above described three cosmological puzzles can come by assuming that 70% of the universe energy density consists of a substance called 'dark energy' which is characterized by *'negative pressure'*. This means that if a piston capping a cylinder with dark energy is pulled out, the negative pressure of dark energy will resist and will try reduce its volume.

As it is shown below, dark energy with negative pressure can induce repulsive gravity and thus cause accelerating expansion, it has positive energy and can therefore provide the missing mass required for flatness and can also lead to a larger age of the universe (with the same Hubble constant) thus resolving the cosmic age crisis. It can therefore resolve simultaneously the three cosmological puzzles described above.

To understand this resolution, we must consider the dynamical equation that determines the evolution of the scale factor $a(t)$. This equation is the Friedman equation which is obtained by combining General Relativity with the cosmological principle of homogeneity and isotropy of the universe. It may be written as

$$\frac{\ddot{a}}{a} = -\frac{4\pi G}{3}\sum_i(\rho_i + 3p_i) = -\frac{4\pi G}{3}[\rho_m + (\rho_X + 3p_X)] \tag{4}$$

where ρ_i and p_i are the densities and pressures of the contents of the universe assumed to behave as ideal fluids. The only directly detected fluids in the universe are matter ($\rho_m, p_m = 0$) and the subdominant radiation ($\rho_r, p_r = \rho_r/3$). Both of these fluids are unable to cancel the minus sign on the rhs of the Friedman equation and can therefore only lead to decelerating expansion. Accelerating expansion in the context of general relativity can only be obtained by assuming the existence of an additional component ($\rho_X, p_X = w\rho_X$) termed 'dark energy' which could potentially change the minus sign of eq. (4) and thus lead to accelerating expansion. Assuming a positive energy density for dark energy it becomes clear that negative pressure is required for accelerating expansion. In fact, writing the Friedman eq. (4) in terms of the dark energy equation of state parameter w as

$$\frac{\ddot{a}}{a} = -\frac{4\pi G}{3}[\rho_m + \rho_X(1+3w)] \tag{5}$$

it becomes clear that a $w < -\frac{1}{3}$ is required for accelerating expansion implying repulsive gravitational properties for dark energy for the resolution of the first cosmological puzzle. The second puzzle can be simply resolved by assuming that

$$\rho_{0m} + \rho_{0X} = \rho_{crit} \tag{6}$$

The third puzzle (the cosmic age crisis) is also resolved because a flat accelerating universe with the same present Hubble parameter as a flat decelerating matter dominated universe has a larger age (time since the Big Bang when $a(t) = 0$). This is shown schematically in Fig. 2 which shows the scale factor evolution for a closed matter dominated universe, a flat universe without dark energy and a flat accelerating universe with dark energy. All three curves are normalized with a common Hubble parameter at present (common tangent). Clearly, the age of the accelerating universe with dark energy is larger than the other cases and may be shown to be consistent with globular cluster constraints.

The redshift dependence of the dark energy can be easily connected to the equation of state parameter w by combining the energy conservation $d(\rho_X a^3) = -p_x d(a^3)$ with

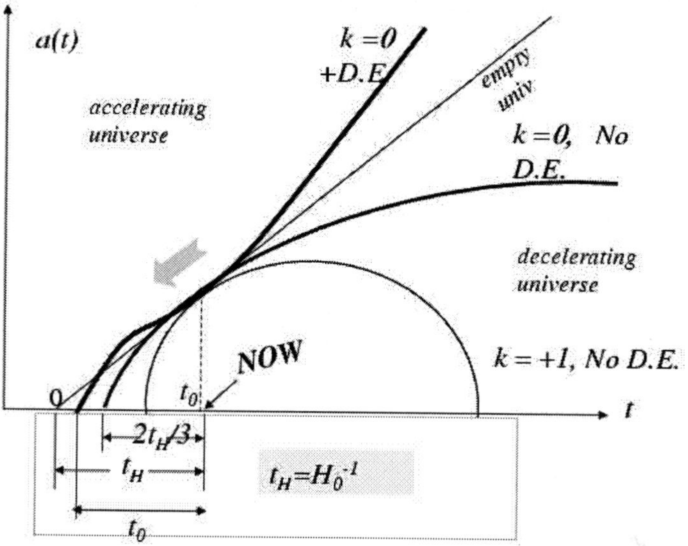

FIGURE 2. The age of the universe is larger in an accelerating universe than in a flat universe.

the equation of state $p_X = w\rho_X$ as

$$\rho_X \sim a^{-3(1+w)} = (1+z)^{3(1+w)} \tag{7}$$

This redshift dependence is related to the observable expansion history $H(z)$ through the Friedman equation

$$H(z)^2 = \frac{\dot{a}^2}{a^2} = \frac{8\pi G}{3}[\rho_{0m}(\frac{a_0}{a})^3 + \rho_X(a)] = H_0^2[\Omega_{0m}(1+z)^3 + \Omega_X(z)] \tag{8}$$

where the density parameter $\Omega \equiv \frac{\rho}{\rho_{0crit}}$ for matter is constrained by large scale structure observations to a value (prior) $\Omega_{0m} \simeq 0.3$. Using this prior, the dark energy density parameter $\Omega_X(z) \equiv \frac{\rho_X(z)}{\rho_{0crit}}$ and the corresponding equation of state parameter w may be constrained from the observed $H(z)$.

In addition to $\Omega_X(z)$, the luminosity distance-redshift relation $d_L(z)$ obtained from SnIa observations can constrain other cosmological parameters. The only parameter however obtained directly from $d_L(z)$ is the Hubble parameter $H(z)$ which is related to $d_L(z)$ by

$$H(z) = c[\frac{d}{dz}(\frac{d_L(z)}{1+z})]^{-1} \tag{9}$$

as indicated by the Hubble diagram. Other cosmological parameters can be obtained from $H(z)$ as follows:

- The age of the universe t_0 is obtained as:

$$t_0 = \int_0^\infty \frac{dz}{(1+z)H(z)} \tag{10}$$

- The present Hubble parameter $H_0 = H(z=0)$.
- The deceleration parameter $q(z) \equiv \frac{\ddot{a}}{a}\dot{a}^2$

$$q(z) = (1+z)\frac{dlnH}{dz} - 1 \tag{11}$$

  and its present value $q_0 \equiv q(z=0)$.
- The density parameters for matter and dark energy are related to $H(z)$ through the Friedman equation (8).
- The equation of state parameter $w(z)$ obtained as [24, 25]

$$w(z) = \frac{p_X(z)}{\rho_X(z)} = \frac{\frac{2}{3}(1+z)\frac{d\ln H}{dz} - 1}{1 - (\frac{H_0}{H})^2\Omega_{0m}(1+z)^3} \tag{12}$$

  obtained using the Friedman equations (5) and (8).

The most interesting parameter from the theoretical point of view (apart from $H(z)$ itself) is the dark energy equation of state parameter $w(z)$. This parameter probes directly the gravitational properties of dark energy which are predicted by theoretical models. The downside of it is that it requires two differentiations of the observable $d_L(z)$ to be obtained and is therefore very sensitive to observational errors.

The simplest form of dark energy corresponds to a time independent energy density obtained when $w = -1$ (see eq. (7)). This is the well known cosmological constant which was first introduced by Einstein in 1917 two years after the publication of the General Relativity (GR) equation

$$G_{\mu\nu} = \kappa T_{\mu\nu} \tag{13}$$

where $\kappa = 8\pi G/c^2$. At the time the 'standard' cosmological model was a static universe because the observed stars of the Milky Way were found to have negligible velocities. The goal of Einstein was to apply GR in cosmology and obtain a static universe using matter only. It became clear that the attractive gravitational properties of matter made it impossible to obtain a static cosmology from (13). A repulsive component was required and at the time of major revolutions in the forms of physical laws it seemed more natural to obtain it by modifying the gravitational law than by adding new forms of energy density. The simplest generalization of eq. (13) involves the introduction of a term proportional to the metric $g_{\mu\nu}$. The GR equation becomes

$$G_{\mu\nu} - \Lambda g_{\mu\nu} = \kappa T_{\mu\nu} \tag{14}$$

where Λ is the cosmological constant. The repulsive nature of the cosmological constant becomes clear by the metric of a point mass (Schwarschild-de Sitter metric) which, in the newtonian limit leads to a gravitational potential

$$V(r) = -\frac{GM}{r} - \frac{\Lambda r^2}{6} \tag{15}$$

which in addition to the usual attractive gravitational term has a repulsive term proportional to the cosmological constant Λ. This repulsive graviatational force can lead to a static (but unstable) universe in a cosmological setup and in the presence of a matter fluid. A few years after the introduction of the cosmological constant by Einstein came Hubble's discovery that the universe is expanding and it became clear that the cosmological constant was an unecessary complication of GR. It was then that Einstein (according to Gamow's autobiography) called the introduction of the cosmological constant *'the biggest blunder of my life'*. In a letter to Lemaitre in 1947 Einstein wrote: *'Since I introduced this term I had always had a bad conscience. I am unable to believe that such an ugly thing is actually realized in nature'*. As discussed below, there is better reason than ever before to believe that the cosmological constant may be non-zero, and Einstein may not have blundered after all.

If the cosmological constant is moved to the right hand side of eq. (14) it may be incorporated in the energy momentum tensor as an ideal fluid with $\rho_\Lambda = \frac{\Lambda}{8\pi G}$ and $w = -1$. In the context of field theory such an energy momentum tensor is obtained by a scalar field ϕ with potential $V(\phi)$ at its vacuum state ϕ_0 ie $\partial_\mu \phi = 0$ and $T_{\mu\nu} = -V(\phi_0) g_{\mu\nu}$.

The negative pressure that corresponds to the cosmological constant may be physically understood by considering a piston capping a cylinder filled with a 'cosmological constant' substance. Since the energy density of this substance is by definition independent of the volume V, an increase of the volume by ΔV leads to an increase of the internal energy of the system by

$$\Delta U = \rho_\Lambda \Delta V > 0 \tag{16}$$

By conservation of energy, the increase of the internal energy is equal to the work done by the external cause that increases the volume ie

$$\Delta U = -p_\Lambda \Delta V \tag{17}$$

Since now $\Delta U > 0$ and $\Delta V > 0$ eq. (17) implies that $p_\Lambda < 0$ and therefore the cosmological constant 'pulls' the piston instead of pushing it like a normal gas.

Even though the cosmological constant may be physically motivated in the context of field theory and consistent with cosmological observation there are two important problems associated with it:

- *Why is it so incredibly small?* Observationally, the cosmological constant density is 120 orders of magnitude smaller than the energy density associated with the Planck scale - the obvious cut off. Furthermore, the standard model of cosmology posits that very early on the universe experienced a period of inflation: A brief period of very rapid acceleration, during which the Hubble constant was about 52 orders of magnitude larger than the value observed today. How could the cosmological constant have been so large then, and so small now? This is sometimes called *the cosmological constant problem.*
- *The 'coincidence problem':* Why is the energy density of matter nearly equal to the dark energy density today?

FITTING TO THE SNIA DATA

Despite the above problems and given that the cosmological constant is the simplest dark energy model, it is important to investigate the degree to which it is consistent with the SnIa data. I will now describe the main steps involved in this analysis. According to the Friedman equation the predicted Hubble expansion in a flat universe and in the presence of matter and a cosmological constant is

$$H(z)^2 = \frac{\dot{a}^2}{a^2} = \frac{8\pi G}{3}\rho_{0m}(\frac{a_0}{a})^3 + \frac{\Lambda}{3} = H_0^2[\Omega_{0m}(1+z)^3 + \Omega_\Lambda] \tag{18}$$

where $\Omega_\Lambda = \frac{\rho_\Lambda}{\rho_{0crit}}$ and

$$\Omega_{0m} + \Omega_\Lambda = 1 \tag{19}$$

This is the LCDM (Λ+Cold Dark Matter) model which is currently the minimal standard model of cosmology. The predicted $H(z)$ has a single free parameter which we wish to constrain by fitting to the SnIa luminosity distance-redshift data.

Observations measure the apparent luminosity vs redshift ($l(z)$) or equivalently the apparent magnitude vs redshift ($m(z)$) which are related to the luminosity distance by

$$2.5log_{10}(\frac{L}{l(z)}) = m(z) - M - 25 = 5log_{10}(\frac{d_L(z)_{obs}}{Mpc}) \tag{20}$$

where M is the absolute magnitude which is assumed to be constant for standard candles like SnIa. From the theory point of view the predicted observable is the Hubble parameter (18) which is related to the theoretically predicted luminosity distance $d_L(z)$ by (see eg [24, 26])

$$d_L(z;\Omega_{0m})_{th} = c\,(1+z)\int_0^z \frac{dz'}{H(z';\Omega_{0m})} \tag{21}$$

Constraints on the parameter Ω_{0m} are obtained by the maximum likelihood method [27] which involves the minimization of the $\chi^2(\Omega_{0m})$ defined as

$$\chi^2(\Omega_{0m}) = \sum_{i=1}^{N} \frac{[d_L(z)_{obs} - d_L(z;\Omega_{0m})_{th}]^2}{\sigma_i^2} \tag{22}$$

where N is the number of the observed SnIa luminosity distances and σ_i are the corresponding 1σ errors which include errors due to flux uncertainties, internal dispersion of SnIa absolute magnitude and peculiar velocity dispersion. If flatness is not imposed as a prior through eq. (19) then $d_L(z)_{th}$ depends on two parameters (Ω_{0m} and Ω_Λ) and the relation between $d_L(z;\Omega_{0m},\Omega_\Lambda)_{th}$ and $H(z;\Omega_{0m},\Omega_\Lambda)$ takes the form

$$d_L(z)_{th} = \frac{c(1+z)}{\sqrt{\Omega_{0m}+\Omega_\Lambda-1}} Sin[\sqrt{\Omega_{0m}+\Omega_\Lambda-1}\int_0^z dz' \frac{1}{H(z)}] \tag{23}$$

In this case the minimization of eq. (22) leads to constraints on both Ω_{0m} and Ω_Λ. This is the only direct and precise observational probe that can place constraints directly on

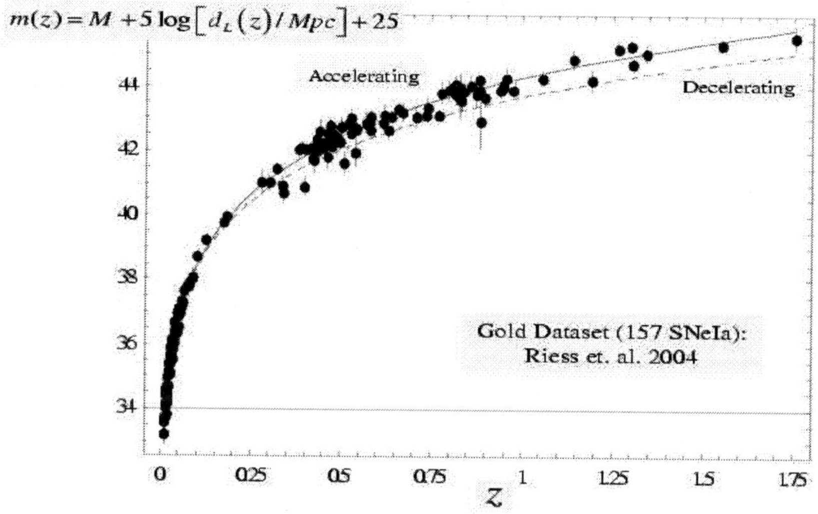

FIGURE 3. The Hubble diagram in the form of $m(z)$ as obtained from the Gold dataset.

Ω_Λ. Most other observational probes based on large scale structure observations place constraints on Ω_{0m} which are indirectly related to Ω_Λ in the context of a flatness prior.

The acceleration of the universe has been confirmed using the above maximum likelihood method since 1998 [1, 2]. Even the early datasets of 1998 [1, 2] were able to rule out the flat matter dominated universe (SCDM: $\Omega_{0m} = 1$, $\Omega_\Lambda = 0$) at 99% confidence level. The latest datasets are the Gold dataset[5] ($N = 157$ in the redshift range $0 < z < 1.75$) and the first year SNLS[6] (Supernova Legacy Survey) dataset which consists of 71 datapoints in the range $0 < z < 1$ plus 44 previously published closeby SnIa.

The Hubble diagram obtained from the Gold dataset is shown in Fig. 3 where the corrected apparent magnitude $m(z)$ of the 157 SnIa is plotted versus the redshift z. The apparent magnitude $m(z)$ is related to the corresponding luminosity distance d_L of the SnIa by eq. (20). The 68% and 95% χ^2 contours in the (Ω_{0m} and Ω_Λ) parameter space obtained using the maximum likelihood method are shown in Fig. 4 for the SNLS dataset, a truncated version of the Gold dataset (TG) with $0 < z < 1$ and the Full Gold (FG) dataset[28]. The following comments can be made on these plots:

- The two versions of the Gold dataset favor a closed universe instead of a flat universe ($\Omega_{tot}^{TG} = 2.16 \pm 0.59$, $\Omega_{tot}^{FG} = 1.44 \pm 0.44$). This trend is not realized by the SNLS dataset which gives $\Omega_{tot}^{SNLS} = 1.07 \pm 0.52$.
- The point corresponding to SCDM $(\Omega_{0m}, \Omega_\Lambda) = (1, 0)$ is ruled out by all datasets at a confidence level more than 10σ.
- If we use a prior constraint of flatness $\Omega_{0m} + \Omega_\Lambda = 1$ thus restricting on the corresponding dotted line of Fig. 4 and use the parametrization

$$H(z)^2 = H_0^2[\Omega_{0m}(1+z)^2 + (1 - \Omega_{0m})] \tag{24}$$

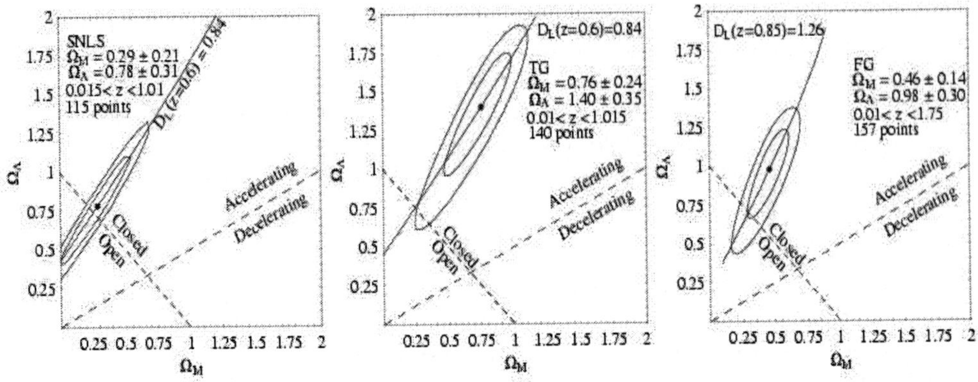

FIGURE 4. The 68% and 95% χ^2 contours in the (Ω_{0m} and Ω_Λ) parameter space obtained using the SNLS, TG and FG datasets (from Ref. [28]).

we find minimizing $\chi^2(\Omega_{0m})$ of eq (22)

$$\Omega_{0m}^{SNLS} = 0.26 \pm 0.04 \tag{25}$$

$$\Omega_{0m}^{TG} = 0.30 \pm 0.05 \tag{26}$$

$$\Omega_{0m}^{FG} = 0.31 \pm 0.04 \tag{27}$$

These values of Ω_{0m} are consistent with corresponding constraints from the CMB[8] and large scale structure observations[9].

GENERALIZED DARK ENERGY MODELS

Even though LCDM is the simplest dark energy model and is currently consistent with all cosmological observations (especially with the SNLS dataset) the question that may still be addressed is the following: 'Is it possible to get better fits (lowering χ^2 further) with different $H(z)$ parametrizations and if yes what are the common features of there better fits?' The strategy towards addressing this question involves the following steps:

- Consider a physical model and extract the predicted recent expansion history $H(z;a_1,a_2,...,a_n)$ as a function of the model parameters $a_1,a_2,...,a_n$. Alternatively a model independent parametrization for $H(z;a_1,a_2,...,a_n)$ (or equivalently $w(z;a_1,a_2,...,a_n)$) may be constructed aiming at the best possible fit to the data with a small number of parameters (usually 3 or less).
- Use eq. (21) to obtain the theoretically predicted luminosity distance $d_L(z;a_1,a_2,...,a_n)_{th}$ as a function of z.
- Use the observed luminosity distances $d_L(z_i)_{obs}$ to construct χ^2 along the lines of eq. (22) and minimize it with respect to the parameters $a_1,a_2,...,a_n$.
- From the resulting best fit parameter values $\bar{a}_1,\bar{a}_2,...,\bar{a}_n$ (and their error bars) construct the best fit $H(z;\bar{a}_1,\bar{a}_2,...,\bar{a}_n)$, $d_L(z;\bar{a}_1,\bar{a}_2,...,\bar{a}_n)$ and $w(z;\bar{a}_1,\bar{a}_2,...,\bar{a}_n)$. The quality of fit is measured by the depth of the minimum of χ^2 ie $\chi^2_{min}(\bar{a}_1,\bar{a}_2,...,\bar{a}_n)$.

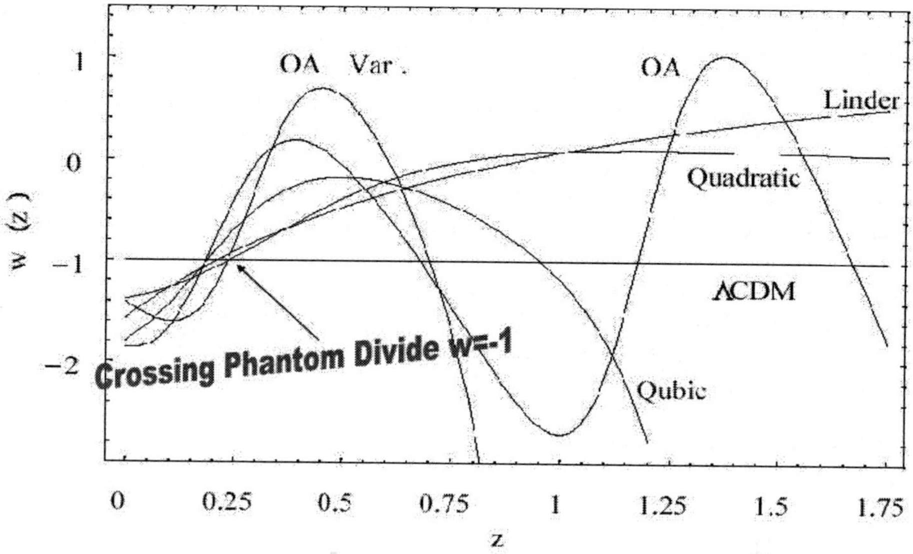

FIGURE 5. The best fit forms of $w(z)$ obtained from a variety of parametrizations [26] in the context of the Full Gold dataset. Notice that they all cross the line $w = -1$ also known as the Phantom Divide Line (PDL).

Most useful parametrizations reduce to LCDM of eq. (18) for specific parameter values giving a χ^2_{LCDM} for these parameter values. Let

$$\Delta\chi^2_{LCDM} \equiv \chi^2_{min}(\bar{a}_1, \bar{a}_2, ..., \bar{a}_n) - \chi^2_{LCDM} \tag{28}$$

The value of $\Delta\chi^2_{LCDM}$ is usually negative since χ^2 is usually further reduced due to the larger number of parameters compared to LCDM. For a given number of parameters the value of $\Delta\chi^2_{LCDM}$ gives a measure of the probability of having LCDM physically realized in the context of a given parametrization [26]. The smaller this probability is, the more 'superior' this parametrization is compared to LCDM. For example for a two parameter parametrization and $|\Delta\chi^2_{LCDM}| > 2.3$ the parameters of LCDM are more than 1σ away from the best fit parameter values of the given parametrization. This statistical test has been quantified in Ref. [26] and applied to several $H(z)$ parametrizations.

As an example let us consider the two parameter polynomial parametrization allowing for dark energy evolution

$$H(z)^2 = H_0^2[\Omega_{0m}(1+z)^3 + a_2(1+z)^2 + a_1(1+z) + (1 - a_2 - a_1 - \Omega_{0m})] \tag{29}$$

in the context of the Full Gold dataset. Applying the above described χ^2 minimization leads to the best fit parameter values $a_1 = 1.67 \pm 1.03$ and $a_2 = -4.16 \pm 2.53$. The corresponding $|\Delta\chi^2_{LCDM}|$ is found to be 2.9 which implies that the LCDM parameters values ($a_1 = a_2 = 0$) are in the range of $1\sigma - 2\sigma$ away from the best fit values.

The best fit forms of $w(z)$ obtained from a variety of parametrizations [26] in the context of the Full Gold dataset are shown in Fig. 5. Even though these best fit forms appear very different at redshifts $z > 0.5$ (mainly due to the two derivatives involved in obtaining $w(z)$ from $d_L(z)$), in the range $0 < z < 0.5$ they appear to have an interesting common feature: they all cross the line $w = -1$ also known as the Phantom Divide Line (PDL). This feature is difficult to reproduce[29, 30] in most theoretical models based on minimally coupled scalar fields and therefore if it persisted in other independent datasets it could be a very useful tool in discriminating among theoretical models.

For example the PDL can not be crossed in any model where the role of dark energy is played by a minimally coupled to gravity scalar field. Quintessence scalar fields[11] with small positive kinetic term ($-1 < w < -\frac{1}{3}$) violate the strong energy condition ($\rho + 3p > 0$) but not the dominant energy condition $\rho + p > 0$. Their energy density scales down with the cosmic expansion and so does the cosmic acceleration rate. Phantom fields[31] with negative kinetic term ($w < -1$) violate the strong energy condition, the dominant energy condition and maybe physically unstable. However, they are also consistent with current cosmological data and according to recent studies[32, 25, 26] they maybe favored over their quintessence counterparts.

Homogeneous quintessence or phantom scalar fields are described by Lagrangians of the form

$$\mathtt{L} = \pm\frac{1}{2}\dot{\phi}^2 - V(\phi) \tag{30}$$

where the upper (lower) sign corresponds to a quintessence (phantom) field in equation (30) and in what follows. The corresponding equation of state parameter is

$$w = \frac{p}{\rho} = \frac{\pm\frac{1}{2}\dot{\phi}^2 - V(\phi)}{\pm\frac{1}{2}\dot{\phi}^2 + V(\phi)} \tag{31}$$

For quintessence (phantom) models with $V(\phi) > 0$ ($V(\phi) < 0$) the parameter w remains in the range $-1 < w < 1$. For an arbitrary sign of $V(\phi)$ the above restriction does not apply but it is still impossible for w to cross the PDL $w = -1$ in a continous manner. The reason is that for $w = -1$ a zero kinetic term $\pm\dot{\phi}^2$ is required and the continous transition from $w < -1$ to $w > -1$ (or vice versa) would require a change of sign of the kinetic term. The sign of this term however is fixed in both quintessence and phantom models. This difficulty in crossing the PDL $w = -1$ could play an important role in identifying the correct model for dark energy in view of the fact that data favor $w \simeq -1$ and furthermore parametrizations of $w(z)$ where the PDL is crossed appear to be favored over the cosmological constant $w = -1$ according to the Gold dataset as shown in Fig. 5.

CONCLUSION

The recent SnIa data have opened new directions in cosmological research and have raised important questions related to the physical origin and dynamical properties of dark energy. In particular these questions can be structured as follows:

- Can the accelerating expansion be attributed to a dark energy ideal fluid with negative pressure or is it necessary to implement extensions of GR to understand the origin of the accelerating expansion?

- Is w evolving with redshift and crossing the PDL? If the crossing of the PDL by $w(z)$ is confirmed then it is quite likely that extensions of GR will be required to explain observations.
- Is the cosmological constant consistent with data? If it remains consistent with future more detailed data then the theoretical efforts should be focused on resolving the coincidence and the cosmological constant problems which may require anthropic principle arguments.

The main points of this brief review may be summarized as follows:

- *Dark energy* with *negative pressure* can explain SnIa cosmological data indicating accelerating expansion of the universe.
- The existence of a *cosmological constant* is consistent with SnIa data but other *evolving* forms of dark energy *crossing the* $w = -1$ line may provide better fits to some of the recent data (Gold dataset).
- New *observational projects* are underway and are expected to lead to significant progress in the understanding of the properties of *dark energy*.

REFERENCES

1. S. Perlmutter et al, Nature (London) **391**, 51(1998);
2. Riess A *et al.*, Astron. J. **116** 1009 (1998);
3. Tonry, J L *et al.*, 2003 Astroph. J. **594** 1;
4. Perlmutter S J *et al.*, Astroph. J. **517** 565 (1999); Barris, B *et al.*, 2004 Astroph. J. **602** 571; Knop R *et al.*, 2003 Astroph. J. **598** 102;
5. A. G. Riess *et al.* [Supernova Search Team Collaboration], Astrophys. J. **607**, 665 (2004).
6. P. Astier *et al.*, arXiv:astro-ph/0510447.
7. V. Sahni, arXiv:astro-ph/0403324.
8. D. Spergel et al., Astrophys.J.Suppl. 148 (2003) 175, [arXiv:astro-ph/0302209]
9. M. Tegmark et al, Phys.Rev. D69 (2004) 103501,astro-ph/0310723
10. V. Sahni and A. A. Starobinsky, Int. J. Mod. Phys. D **9**, 373 (2000) [arXiv:astro-ph/9904398].
11. P. J. Peebles and B. Ratra, Astrophys. J. **325**, L17 (1988); R. R. Caldwell, R. Dave and P. J. Steinhardt, Phys. Rev. Lett. **80**, 1582 (1998) [arXiv:astro-ph/9708069];I. Zlatev, L. M. Wang and P. J. Steinhardt, Phys. Rev. Lett. **82**, 896 (1999).
12. V. Sahni and L. M. Wang, Phys. Rev. D **62**, 103517 (2000) [arXiv:astro-ph/9910097].
13. F. Perrotta, C. Baccigalupi and S. Matarrese, Phys. Rev. D **61**, 023507 (2000) [arXiv:astro-ph/9906066];A. Riazuelo and J. P. Uzan, Phys. Rev. D **66**, 023525 (2002) [arXiv:astro-ph/0107386];G. Esposito-Farese and D. Polarski, Phys. Rev. D **63**, 063504 (2001) [arXiv:gr-qc/0009034].
14. D. F. Torres, Phys. Rev. D **66**, 043522 (2002) [arXiv:astro-ph/0204504].
15. S. Nojiri and S. D. Odintsov, arXiv:hep-th/0308176.
16. A. Y. Kamenshchik, U. Moschella and V. Pasquier, Phys. Lett. B **511**, 265 (2001) [arXiv:gr-qc/0103004].
17. K. Freese and M. Lewis, Phys. Lett. B **540**, 1 (2002) [arXiv:astro-ph/0201229].
18. L. Perivolaropoulos and C. Sourdis, Phys. Rev. D **66**, 084018 (2002) [arXiv:hep-ph/0204155].
19. L. Perivolaropoulos, Phys. Rev. D **67**, 123516 (2003) [arXiv:hep-ph/0301237].
20. V. Sahni and Y. Shtanov, JCAP **0311**, 014 (2003) [arXiv:astro-ph/0202346].
21. I. Smail, R. S. Ellis and M. J. Fitchett, arXiv:astro-ph/9402048.
22. A. D. Lewis, E. Ellingson, S. L. Morris and R. G. Carlberg, Astrophys. J. **517**, 587 (1999) [arXiv:astro-ph/9901062].
23. B. Chaboyer, Astrophys. J. **444**, L9 (1995) [arXiv:astro-ph/9412015].
24. D. Huterer and M. S. Turner, Phys. Rev. D **64**, 123527 (2001) [arXiv:astro-ph/0012510].

25. S. Nesseris and L. Perivolaropoulos, Phys. Rev. D **70**, 043531 (2004) [arXiv:astro-ph/0401556].
26. R. Lazkoz, S. Nesseris and L. Perivolaropoulos, arXiv:astro-ph/0503230.
27. W. H. Press *et. al.*, 'Numerical Recipes', Cambridge University Press (1994).
28. S. Nesseris and L. Perivolaropoulos, arXiv:astro-ph/0511040.
29. A. Vikman, Phys. Rev. D **71**, 023515 (2005) [arXiv:astro-ph/0407107].
30. L. Perivolaropoulos, JCAP **0510**, 001 (2005) [arXiv:astro-ph/0504582].
31. R. R. Caldwell, Phys. Lett. B **545**, 23 (2002) [arXiv:astro-ph/9908168]; P. Singh, M. Sami and N. Dadhich, Phys. Rev. D **68**, 023522 (2003) [arXiv:hep-th/0305110]; V. B. Johri, Phys. Rev. D **70**, 041303 (2004) [arXiv:astro-ph/0311293].
32. U. Alam, V. Sahni, T. D. Saini and A. A. Starobinsky, Mon. Not. Roy. Astron. Soc. **354**, 275 (2004) [arXiv:astro-ph/0311364]; Y. Wang and P. Mukherjee, Astrophys. J. **606**, 654 (2004) [arXiv:astro-ph/0312192].

A non-singular black hole model as a possible end-product of gravitational collapse

Demosthenes Kazanas* and Manasse R. Mbonye†

*NASA/Goddard Space Flight Center, Greenbelt, MD 20771
†Rochester Institute of Technology,84 Lomb Drive, Rochester, NY 14623

Abstract. We present a non-singular black hole model as a possible end-product of gravitational collapse and whose metric is an exact solution of the Einstein field equations. In this model, the matter fields of the final equilibrium configuration are concentrated in the vicinity of the origin of the radial coordinate r. This configuration comprises a core with an equation of state that approaches $p = -\rho$ as $r \rightarrow 0$, surrounded by matter fields with sound speed bounded by c. In the direction of decreasing r, the pressure $p(\rho)$ is a point-wise well-behaved functional of a monotonically increasing density function $\rho(r)$ and obeys an equation of state that smoothly matches the vacuum-like behavior near $r = 0$ to a polytrope at larger r (lower ρ) values. The space outside the matter fields is that of the Schwarzschild vacuum, while it becomes de Sitter as $r \rightarrow 0$. In the intervening space matter transits smoothly between normal matter (i.e with $P(r) > 0$) and a "quintessence-like" fluid state (i.e one with $-1 < w(r) = P(r)/\rho(r) < 0$). We estimate the fraction of matter with a conventional equation of state to be $\simeq 0.34$.

INTRODUCTION

It is generally accepted that the formation of a black hole proceeds with the formation of a marginally trapped surface (the apparent horizon) into which matter is collapsing and one that encloses a region of space interior to which the light cones are tipped [1] so much so that even outgoing null geodesics proceed toward ever decreasing values of the radial coordinate. The famous singularity theorems [2, 3, 4], then posit that under certain conditions on the properties matter, the eventual state of the collapse is an infinite density singularity surrounded by an event horizon defined by the outermost trapped surface. It is, however, a fact that the existence of such a singularity is neither easy to verify nor refute observationally because of the limits set by the black hole event horizon. In the absence of such observational evidence, the validity of the arguments leading to the singularity theorems has generally shaped the widespread conviction that indeed a black hole must contain a physical singularity.

The issue of the presence of a singularity, even one shrouded by a horizon, leave one with unsettling feelings and it has been generally argued on general grounds that these singularities are the result of the classical (i.e. non-quantum) character of our gravitational theories and that quantum gravitational effects will "smooth" these singularities out over a region of density comparable to the Planck density. An altogether different view is that the singularities implied by the aforementioned theorems simply reflect our lack of knowledge of the properties of matter under extreme conditions. Thus it has been suggested that the singularity theorems would be circumvented if (some of) the condi-

CP848, *Recent Advances in Astronomy and Astrophysics,*
7th *International Conference of the Hellenic Astronomical Society,* edited by N. H. Solomos

tions imposed on the properties of collapsing matter were relaxed. Simple considerations indicate that the collapse of a sufficiently dense star becomes inevitable because of the gravitating effects of the pressure that it is invoked to provide its hydrostatic support; as a result a higher pressure only serves to hasten (rather than arrest) the collapse. It is therefore not surprising that avoidance of singularities maybe possible for equations of matter for which pressure becomes *negative*.

Indeed such a concept of non-singular collapse dates back to Sakharov's consideration of the equation of state $p = -\rho$ for a superdense fluid and Gliner's view [7] that such a fluid could be the final state of *(non-singular)* gravitational collapse. Such ideas are now attracting a growing amount of attention. For example, Dymnikova [8] has constructed an exact solution of the Einstein field equations for a non-singular black hole *in which the core consists of a fluid with vacuum-like equation of state and anisotropic pressure*. This solution is Petrov Type [(II),(II)] by classification[1]. An alternative type of investigation along the same lines is that initiated by [9], based on a limiting curvature approach. This idea has been explored further by several authors, see for example [10, 11, 12].

A common feature in all these treatments is that the geometry of the spacetime in question is Schwarschild at large r as expected by Birkoff's theorem, and de Sitter-like at small r values as implied by the $p = -\rho$ equation of state assumed to prevail near $r = 0$. These approaches can be put in two broad classes. (a) Those for which the transition to this "exotic" state of matter take place well inside the Schwarzschild horizon [8]. (b) Those which replace the entire volume of the Schwarschild metric interior to $r = 2M$ by a substance with the $p = -\rho$ equation of state [6], dispensing in this way with the presence of the horizon altogether and the "information paradox "problems it engenders.

The problem of transition from the usual matter equation of state to that appropriate for these non-singular solutions is usually not addressed at all. This problem is particularly acute for models of the second class discussed above, since the density at horizon formation scales like $\rho \simeq 10^{16}\,(M/M_{\odot})^{-2}$ g cm^{-3} and for objects of mass $M \geq 10^8\,M_{\odot}$, is not greater than that of water, which is known to have an equation of state very different from $p = -\rho$. Moreover, issues to do with direct matching of an external Schwarschild vacuum to an interior de Sitter have previously been discussed based on junction conditions [13]. In fact, in terms of gravitational collapse the associated difficulties have been used to suggest modifications in some of these treatments [14]. The present work is motivated by these questions and is undertaken as a first effort to provide some answers by constructing models using an explicit equation of state with the desired properties. It would appear that the junction constraints [13] when applied to the end-product of non-singular gravitational collapse imply the matter fields across the Schwarschild/de Sitter boundary should have a radially dependent equation of state of the form $1 \lesssim p/\rho = w(r) \lesssim -1$ that smoothly changes the matter-energy from a stiff fluid to a cosmological constant.

The model sketched in this paper describes the geometry of a body that passed during its collapse through a stiff fluid state to settle into the final state of a non-singular black

---

[1] We shall present a solution with the same asymptotic behavior as in [8] which however is different in both classification (it is Petrov Type [II,(II)] and physical interpretation.

hole. The solution allows for a matter fluid region (with a family of equations of state $0 \leq w(r) \leq 1$ which envelopes an intermediate region with a family of equations of state $(-1 < w(r) < 0)$ which in turn envelopes a de Sitter like region with $w = -1$ at the center. Our model differs both from the traditional singular black hole solutions which contain the usual matter at the singularity and from the relatively new non-singular solutions which introduced a $p = -\rho$ equation of state at some finite volume near $r = 0$ without any consideration of how this state is achieved by the collapsing matter of the usual equation of state. The main features of our model is that it

(1) addresses the issue of matter inside a non-singular black hole and

(2) offers a reasonable explanation of how part of this matter can evolve towards a de Sitter-like vacuum to provide the radial tension or negative pressure that supports the remaining matter fields against the formation of a singularity.

The rest of the paper is organized as follows. In section 2 we summarize the working equations to be solved. We then construct the working equation of state and discuss some of its desirable features. In section 3 we solve the Einstein Field Equations for the desired geometry and obtain an exact solution. Section 4 concludes the discussion.

PROBLEM FORMULATION

The field equations

It is assumed, for simplicity, that the collapsed object can be reasonably described by a spherical, static geometry. The desired line element then takes the general form

$$ds^2 = -e^{\nu(r)}dt^2 + e^{\lambda(r)}dr^2 + r^2\left(d\theta^2 + \sin^2\theta d\varphi^2\right), \tag{1}$$

where $\lambda(r)$ and $\nu(r)$ are to be determined from the Einstein Field Equations

$$G_{\mu\nu} = -8\pi G T_{\mu\nu}. \tag{2}$$

Under the assumed spacetime symmetry, Eqs. (2) reduce to

$$e^{-\lambda}\left(\frac{\lambda'}{r} - \frac{1}{r^2}\right) + \frac{1}{r^2} = 8\pi G\rho(r), \tag{3}$$

$$e^{-\lambda}\left(\frac{\nu'}{r} + \frac{1}{r^2}\right) - \frac{1}{r^2} = 8\pi G p_r(r), \tag{4}$$

and

$$e^{-\lambda}\left(\frac{\nu''}{2} - \frac{\lambda'\nu'}{4} + \frac{\nu'^2}{4} + \frac{\nu' - \lambda'}{2r}\right) = 8\pi p_\perp(r). \tag{5}$$

Here $\rho(r) = T_0^0$ is the energy density, $p_r = T_1^1$ is the radial pressure and $\left(p_\theta = T_2^2\right) = \left(p_\varphi = T_3^3\right) = p_\perp$ is the tangential pressure. In this treatment the fluid is anisotropic with $T_2^2 = T_3^3 \neq T_1^1$ and $T_1^1 \neq T_0^0 \neq T_2^2$. The spacetime is thus Petrov Type [II, (II)], where ( ) implies a degeneracy in the eigenvalues of the Weyl tensor.

Equation of state

In the remaining part of this section we construct the working equation of state $p_r(\rho)$ in the radial direction which reproduces the characteristics of the collapsed body as highlighted in the previous section. We justify the choice by pointing out its desired features. On the other hand, the tangential pressure equation of state $p_\perp(\rho)$ is a consequence of the Einstein Field Equations (Eq. 5) and the radial pressure $p_r(\rho)$ obtained from the solution of the remaining equations.

The main desired features are that: (1) The equation of state $p_r(\rho)$ change from the one associated with the usual matter, at low densities, to that associated with the vacuum at sufficiently high densities near the center of the configuration and; (2) That this transitin be well-behaved. To this end we consider an equation of state that takes the general form[2]

$$p_r(\rho) = \left[\alpha - (\alpha+1)\left(\frac{\rho}{\rho_{\max}}\right)^m\right]\left(\frac{\rho}{\rho_{\max}}\right)^{1/n}\rho, \tag{6}$$

where m and $1/n$ are, are positive real numbers and α is a free parameter constrained by the properties demanded of the equation of state. The above expression implies presence of a maximum limiting density $\rho_{\max}$ concentrated in a region of size $r_0 = 1/\sqrt{G\rho_{\max}}$, the core region of the configuration. One notes that the main desirable features are readily apparent in the functional form of Eq. (6). At low densities, $(\rho/\rho_{\max})^m < \alpha/(\alpha+1)$, the equation of state reduces to that of a polytrope of index n ($p_r \propto \rho^{1+1/n}$) while for $(\rho/\rho_{\max})^m > \alpha/(\alpha+1)$ the pressure decreases to eventually approach that of the vacuum $p_r = -\rho_{\max}$ as $\rho \to \rho_{\max}$. We look for the simplest form of Eq. (6) by judiciously constraining both the indices m and n and the parameter α. While no explicit microscopic prescription leading to its form is presently given, we consider that its softening and its eventual conversion to an equation of state appropriate to that of the vacuum ($p = -\rho$), is effected by the coupling of matter to a scalar field akin to that of Higgs that gives rise to particle masses.

Since Eq. (6) already satisfies the general features desired by the model, its specific form, based on an appropriate choice of the indices m and n and the parameter α is made through the demand that it must satisfy the following basic conditions.

1. It must have no pathologies, i.e.
   (a) the sound speed $c_s^2 = dp/d\rho$ can not be a maximum at $\rho = 0$
   (b) in order to rule out superluminal behavior, $dp/d\rho \leq 1$, the maximum sound speed, i.e. the value of $dp/d\rho$ at the point where $d^2p/d\rho^2 = 0$ must be given by $dp/d\rho\,|_{\rho_{stiff}} = c_s^2 = c = 1$.
2. It must satisfy minimum acceptable energy conditions, namely the Weak Energy Condition, $\rho \geq 0$, $\rho + p_r \geq 0$ and the Dominant energy Condition, $\rho \geq 0$, $p_r \in [-\rho, {}^+\rho]$.

---

[2] This choice may not be unique.

Based on these conditions we select the quartic function corresponding to $m = 2, n = 1$. Thus Eq. (6) now takes the form

$$p_r(\rho) = \left[\alpha - (\alpha+1)\left(\frac{\rho}{\rho_{\max}}\right)^2\right]\left(\frac{\rho}{\rho_{\max}}\right)\rho, \tag{7}$$

with the constraint that $\alpha = 2.2135$. It is the equation of state we adopt throughout the remaining part of the paper.

THE SOLUTION

We now solve the Einstein equations (3)-(5) for the spacetime of the model. Integration of Eq.(3) gives

$$e^{-\lambda} = 1 - 8\pi\frac{1}{r}\int_0^r \rho\left(r'\right) r'^2 dr' = 1 - \frac{2m(r)}{r}, \tag{8}$$

where, as stated before, $m(r)$ is the mass enclosed by the a 2-sphere of radius r. Further, using Eqs. (7) and (8), integration of T_1^1 (Eq. 4), gives

$$e^{\nu} = \left(1 - \frac{2m(r)}{r}\right) e^{\int 8\pi\left[\alpha-(\alpha+1)\left(\frac{\rho}{\rho_{\max}}\right)^2\right]\left(\frac{\rho}{\rho_{\max}}\right)\rho\frac{dr}{1-\frac{2m(r)}{r}}}, \tag{9}$$

where $\alpha = 2.2135$.

Substituting Eqs.(8) and (9) into Eq.(1) gives

$$ds^2 = -\left(1 - \frac{2m(r)}{r}\right) e^{\Gamma(r)} dt^2 + \frac{1}{1 - \frac{2m(r)}{r}} dr^2 + r^2\left(d\theta^2 + \sin^2\theta d\varphi^2\right), \tag{10}$$

where

$$\Gamma(r) = \int 8\pi\left[\alpha - (\alpha+1)\left(\frac{\rho}{\rho_{\max}}\right)^2\right]\left(\frac{\rho}{\rho_{\max}}\right)\left(\frac{r}{r-2m(r)}\right)\rho dr \tag{11}$$

Eq. (10) is the general solution for the geometry of our model. It describes the spacetime of a non-singular black hole with both matter and a de Sitter core. We shall describe below a particular solution resulting from the choice of a prescribed density function $\rho(r)$.

To formally complete the solution one must also solve equation 5 for the tangential pressure $p_\perp = \{p_\theta, p_\varphi\}$. On doing this one finds (see also [15, 16]) that

$$p_\perp = p_r + \frac{r}{2}p_r' + \frac{1}{2}(p_r + \rho)\left[\frac{Gm(r) + 4\pi G r^3 p_r}{r - 2Gm(r)}\right], \tag{12}$$

where p_r is given by Eq. (7). Note that in this model the last term in Eq. (12) is not vanishing, in general, as is the case considered in some previous treatments where it was assumed that $p_r = -\rho$ for all ρ (see for example [17]). Eq.12) is a generalization of the Tolman-Oppenheimer-Volkoff equation [18].

One can now discuss the results of Eqs. (8) - (11). The mass $m(r)$ enclosed a 2-surface at any radial coordinate $0 < r < R$ is given by $m(r) = 4\pi \int_0^r \rho(r') r'^2 dr'$. The entire mass of the body M is given by $M = \int_0^\infty m(r) dr = \int_0^\infty \rho(r) r^2 dr$. It is then clear that outside the mass of the body $T_0^0 = 0$. Further, as ρ vanishes outside the body, so does the pressure (see Eqs.7 and (10)). In particular, from Eq. (10) one observes that in the limit $\rho \to 0$ and $m(r) \to M$ the lapse function and the shift vector give $e^\nu = e^{-\lambda} = 1 - \frac{2GM}{r}$. Thus outside the matter fields (i.e. for $r \gg R$) the spacetime becomes asymptotically Schwarzschild. Further, in the limit $r \to r_0$, $\rho \to \rho_{\max}$ (and as both Eqs. (7) and (11) show) $p_r \to -\rho_{\max}$ and $p_\perp \to -\rho_{\max}$. It follows, therefore that for $0 \le r < r_0$ the fluid has constant positive density $\rho_{\max}$ and constant negative pressure $-\rho_{\max}$ and takes on the character of a de Sitter spacetime, through the equation of state $p = -\rho_{\max}$.

In order to establish the actual asymptotic behavior of the spacetime at small r one has to choose a density profile for the matter. We choose to use the one employed by Dymnikova [8] because of its simplicity and convenient form for integration over the source volume $\rho = \rho_{\max} \exp\left[-\frac{r^3}{r_g (r_0)^2}\right]$, where r_g is a length scale of order of the Schwarzschild radius. The mass $m(r)$ enclosed by a 2-sphere at the radial coordinate r is then given by $m(r) = \int_0^r \rho_{\max} \{\exp\left[-\frac{r^3}{r_g (r_0)^2}\right]\} r'^2 dr' = M\left[1 - \exp\left(-\frac{r^3}{r_g (r_0)^2}\right)\right]$, where M is the total mass $M = \int_0^\infty \rho_{\max} \{\exp\left[-\frac{r^3}{r_g (r_0)^2}\right]\} r^2 dr = (4\pi/3) r_0^2 r_g \rho_{\max}$. The entire mass is therefore concentrated in a region of size $R \simeq (r_0^2 r_g)^{1/3}$, which in general is close to $r = 0$, with its precise value depending on the value of the maximum density assumed $\rho_{\max}$.

The introduction of the density function $\rho = \rho_{\max} \exp\left[-\frac{r^3}{r_g (r_0)^2}\right]$ in Eq. (10) provides a particular solution for the model. In the limit $r \to 0$, it is seen that the metric of Eq. (10) asymptotically leads to

$$ds^2 = -\left(1 - \frac{r^2}{r_0^2}\right) dt^2 + \frac{1}{\left(1 - \frac{r^2}{r_0^2}\right)} dr^2 + r^2 \left(d\theta + \sin^2\theta d\varphi^2\right). \tag{13}$$

indicating that the spacetime becomes asymptotically de Sitter.

Finally, one can estimate the fraction of the mass of this configuration that resides in the form of "normal" matter, $\mathscr{M}_{\text{matter}}$, i.e. one with $P(r) > 0$ (or equivalently $w(r) > 0$). For that we need to integrate the given density profile from $\rho = 0$ to $\rho_\varepsilon = [\alpha/(\alpha + 1)]^{1/2} \rho_{\max}$, i.e. to the density for which the pressure of the fluid considered vanishes. In this respect it is of help to note that $r^2 dr = (1/3) r_0^2 r_g \, d(\rho/\rho_{\max})$. Direct integration then yields $\mathscr{M}_{\text{matter}} = (4\pi/3) r_0^2 r_g (1/\rho_{\max})(\rho_\varepsilon^2/2) = (4\pi/3) r_0^2 r_g \rho_{\max} (1/2)[\alpha/(\alpha+1)]$. From the definition of the total mass, $M = (4\pi/3) r_0^2 r_g \rho_{\max}$, one then obtains that $\mathscr{M}_{\text{matter}}/M = (1/2)[\alpha/(\alpha+1)]$, which for the given value of $\alpha \simeq 2.213$ yields $\mathscr{M}_{\text{matter}}/M \simeq 0.34$, suggesting that the majority of mass resides in the fluid with the negative pressure. This is not terribly surprising, since the negative pressure is called to "undo" the positive pull of the rest mass density. The singularity is avoided by putting most of the rest mass into the fraction of the fluid with the negative pressure.

CONCLUSION

In this paper we have proposed a possible model for a non-singular black hole as a product of gravitational collapse. It is an exact solution of the Einstein Equations and it is of Type [II, (II)] by the Petrov classification. This model is based on a judicious choice we have made of the equation of state of the collapsed matter. A non-singular final state is permitted because at high densities this equation of state violates some of the energy conditions [2, 3] originally used to prove the existence of black hole singularities. In our model this violation leads to a well defined non-singular state. On the other hand, the equation of state in our approach still satisfies the Weak Energy Condition $\rho \geq 0$, $\rho + p_r \geq 0$, for the entire range of densities considered, a basic requirement for physical fields.

The solution we provide depicts a spacetime which is of the Schwarzschild form at distances greater than $r = (r_0^2 r_g)^{1/3}$, it becomes dominated by matter-like fluid near this distance and by vacuum-like fluid for $r \lesssim r_0$), where the metric looks very similar to de Sitter *after passing through an inner horizon at* $r \simeq r_0$.

Finally, one should note that our model does not provide an estimate of the values of the constants r_0, or $\rho_{\max}$, which can, but are not obliged to, have the Planck values. On the other hand the value of $\rho_{\max}$ is arguably as least as large as m_H^4, where m_H is the mass of the Higgs ield, as this is the value at which we expect the contribution of scalar fields to the energy momentum tensor to become comparable to that of usual matter, a contribution needed to produce the negative pressure necessary for our model.

REFERENCES

1. S. Hawking and R. Penrose, *The Nature of Spacetime*, Princeton University Press (1995).
2. R. Penrose, *Phys. Rev. Lett.*, **14**, 57 (1965).
3. S. W. Hawking and R. Penrose, *Proc. Roy. Soc. Lond.*, **A 314**, 529 (1970).
4. S. W. Hawking and G. F. R. Ellis, *ApJ*, **152**, 25 (1968).
5. J. D. Bekenstein, *Phys. Rev. D*, **7** 2333 (1973).
6. P.O.Mazur and E. Mottola, gr-qc/010903.
7. E. B. Gliner, gr-qc/9808042.
8. I. Dymnikova, *Int. J. Mod. Phys. D* **5**, 529 (1996). See also I. Dymnikova, gr-qc/0304110.
9. M. A. Markov *JEPT Lett.*, **36**, 265 (1982); *Ann. Phys.*, **155**, 333 (1984)
10. V. P. Frolov, M. A. Markov, V. F. Mukhanov, *Phys. rev D*, **41**, 383 (1990).
11. V. F. Mukhanov, R. Brandenberger, *Phys. Rev. Lett.*, **68**, 1969 (1992).
12. M. Trodden, V. F. Mukhanov, Brandenberger, *Phys.Lett. B*, **316**, 483 (1993).
13. O. Grøn, *Lett.Nuovo Cimento*,**44**, 177 (1985)
14. M. Visser and D. Wiltshire, *Class. Quant. Grav.*, **21** 1135 (2004); also gr-qc/0310107.
15. R. L. Flowers, E.P.T. Liang, ApJ 188 657(1974).
16. I. Dymnikova, gr-qc/0112052.
17. I. Dymnikova,*Class. Quant. Grav.*, **19**, 725 (2002).
18. R. M. Wald, *General Relativity*, Univ. of Chicago Press (1984).

Spherically symmetric, static spacetimes in a tensor-vector-scalar theory

Dimitrios Giannios

Max Planck Institute for Astrophysics, Box 1317, D-85741 Garching, Germany

Abstract.
Recently, a relativistic gravitation theory has been proposed by Bekenstein [1] that gives the Modified Newtonian Dynamics (or MOND) in the weak acceleration regime. The theory is based on three dynamic gravitational fields and succeeds in explaining a large part of extragalactic and gravitational lensing phenomenology without invoking dark matter. In this work we consider the strong gravity regime of TeVeS. We study spherically symmetric, static and vacuum spacetimes relevant for a non-rotating black hole or the exterior of a star. Two branches of solutions are identified: in the first the vector field is aligned with the time direction while in the second the vector field has a non-vanishing radial component. We show that in the first branch of solutions the β and γ PPN coefficients in TeVeS are identical to these of general relativity (GR) while in the second the β PPN coefficient differs from unity violating observational determinations of it (for the choice of the free function F of the theory made in Bekenstein's paper). For the first branch of solutions, we derive analytic expressions for the physical metric and discuss their implications. Applying these solutions to the case of black holes, it is shown that they violate causality (since they allow for superluminal propagation of metric, vector and scalar waves) in the vicinity of the event horizon and/or that they are characterized by negative energy density carried by the fields.

Keywords: Dark Matter – General Relativity – Black Holes
PACS: 95.35.+d, 97.60.Lf, 04.20.Jb

INTRODUCTION

On cosmological scales, Newtonian gravitational theory under-predicts the acceleration of stars and gas. Furthermore, galaxies and clusters of galaxies show anomalously large gravitational lensing when only baryonic matter is taken into account. A natural "cure" for these discrepancies is to assume the existence of dark matter, which dominates over visible matter [2, 3, 4]. Such dark matter might solve the "missing mass" problem within the standard theory of gravity (i.e., general relativity or GR). For this picture to be complete, however, the origin of dark matter needs to be identified.

A second approach to the acceleration discrepancy and lensing anomaly is to look for alternative theories of gravity, that modify GR on large scales. Among many other attempts, the MOND paradigm has been proposed [5, 6]. It is characterized by an acceleration scale a_0 so that

$$\tilde{\mu}(|\vec{a}|/a_0)\vec{a} = -\nabla\Phi_{\rm N}, \tag{1}$$

where $\Phi_{\rm N}$ is to be understood as the Newtonian potential, $\tilde{\mu}(x) = x$ for $x \ll 1$, while $\tilde{\mu}(x) = 1$ for $x \gg 1$.

This empirical law has been very successful in explaining the rotation curves in a large number of (spiral, low surface brightness and elliptical) galaxies using the

CP848, *Recent Advances in Astronomy and Astrophysics,*
7th *International Conference of the Hellenic Astronomical Society,* edited by N. H. Solomos

observed distributions of gas and stars as input (see, e.g., [7]). MOND can also explain the observed correlation between the infrared luminosity of a disk galaxy L_K and the asymptotic rotational velocity v_a (i.e. the the Tully-Fisher law: $L_K \propto v_a^4$ [8]), on the (suggested by population synthesis models) assumption that the M/L_K ratio is constant.

However, MOND is merely a prescription for gravity and not a self consistent theory. It violates, for example, conservation of momentum and angular momentum and does not provide the formulation to describe light deflection or to build a cosmological model. A theory of gravity is needed that has the MOND characteristics in the weak acceleration limit but also has full predictive power.

In this context, a new relativistic theory of gravity has been introduced by Bekenstein [1]. It consists of three dynamical gravitation fields: a tensor field ($g_{\alpha\beta}$), a vector field (U_α) and a scalar field (ϕ) leading to the acronym TeVeS. The theory involves a free function F, a length scale ℓ (that can be related to a_0) and two positive dimensionless constants κ, K.

TeVeS gives MOND in the weak acceleration limit (and therefore inherits the successes of MOND on large scale), makes similar prediction on gravitational lensing as GR (with dark matter) and provides a formulation for constructing cosmological models (see Ref. [9] for structure formation in TeVeS).

The predictions of the theory have thus been explored to some extent for a range of strengths of the gravitational field. The strong gravity regime (e.g. in the vicinity of a black hole or a neutron star) of TeVeS has been studied in Ref. [10]. The basic findings of that study are summarized here. The motivation for this work is two-fold. First, it is interesting to see how the physics of compact objects (i.e. black holes, neutron stars) differ in TeVeS with respect to GR and what constraints (if any) observations can put on its free parameters. Second, one can check the consistency of the theory (e.g., its causality, positivity of energy carried by the fields) in these extreme conditions.

In the next section, we summarize the fundamentals of TeVeS and then we consider its strong gravity limit. We limit ourselves to static, spherically symmetric and vacuum spacetimes relevant for a non-rotating black hole or the exterior of a star. Two branches of solutions are identified: in the first the vector field is aligned with the time direction while in the (not previously explored) second branch the vector field has a non-vanishing radial component. We show that the β and γ PPN coefficients in TeVeS are identical to these of GR in the first branch of solutions while the β PPN coefficient differs in the second. For the choice of the free function F made in Ref. [1], we find that TeVeS predicts a value for β that is in conflict with recent observational determinations of it. Then we focus on the first branch of solutions and derive exact solutions for the metric for arbitrary values of the parameters of the theory. The observational properties of the black holes in TeVeS are discussed along with the issue of superluminal propagation of waves in the black hole vicinity. The last section of this manuscript contains our conclusions.

THE BASIC EQUATIONS OF TEVES

TeVeS is based on three dynamical gravitational fields: a tensor field (the Einstein metric $g_{\alpha\beta}$), a 4-vector field U_α and a scalar field ϕ. The physical metric $\tilde{g}_{\alpha\beta}$ in TeVeS is connected to these fields through the expression

$$\tilde{g}_{\alpha\beta} \equiv e^{-2\phi} g_{\alpha\beta} - 2U_\alpha U_\beta \sinh(2\phi). \tag{2}$$

The total action in TeVeS is the sum of 4 terms S_g, S_s, S_U and S_m (see Ref. [1]), where S_g is identical to the Hilbert-Einstein action and is the part that corresponds to the tensor field, while S_s, S_U, S_m are the actions of the scalar fields, the vector field and the matter respectively. The basic equations of TeVeS are derived by varying the total action S with respect to the fields.

Doing so for $g^{\alpha\beta}$, one arrives to the metric equations

$$G_{\alpha\beta} = 8\pi G[\tilde{T}_{\alpha\beta} + (1-e^{4\phi})U^\mu \tilde{T}_{\mu(\alpha}U_{\beta)} + \tau_{\alpha\beta}] + \Theta_{\alpha\beta}, \tag{3}$$

where a pair of indices surrounded by parenthesis stands for symmetrization, i.e. $A_{(\alpha}B_{\beta)} = A_\alpha B_\beta + A_\beta B_\alpha$, the $G_{\alpha\beta}$ denotes the Einstein tensor for $g_{\alpha\beta}$, $\tilde{T}_{\alpha\beta}$ is the energy momentum tensor and

$$\tau_{\alpha\beta} \equiv \sigma^2 \left[\phi_{,\alpha}\phi_{,\beta} - \frac{1}{2} g^{\mu\nu}\phi_{,\mu}\phi_{,\nu} g_{\alpha\beta} - U^\mu \phi_{,\mu}\left(U_{(\alpha}\phi_{,\beta)} - \frac{1}{2}U^\nu \phi_{,\nu} g_{\alpha\beta}\right)\right] - \frac{1}{4} G\ell^{-2}\sigma^4 F(kG\sigma^2) g_{\alpha\beta}, \tag{4}$$

$$\Theta_{\alpha\beta} \equiv K\left(g^{\mu\nu}U_{[\mu,\alpha]}U_{[\nu,\beta]} - \frac{1}{4} g^{\sigma\tau} g^{\mu\nu} U_{[\sigma,\mu]} U_{[\tau,\nu]} g_{\alpha\beta}\right) - \lambda U_\alpha U_\beta, \tag{5}$$

where a pair of indices surrounded by brackets stands for antisymmetrization, i.e. $A_{[\alpha}B_{\beta]} = A_\alpha B_\beta - A_\beta B_\alpha$.

Similarly one derives a scalar equation which depends on a free function F of the theory. For the choice of F made by Bekenstein, one can put a lower limit on the value of the κ parameter of the theory so that it is not in conflict with the measured motions of planets of the outer solar system (see Ref. [1]; Sec. IV). On the other hand, small values of κ are relevant for cosmological models. Together, these constraints indicate a value of κ around ~ 0.03. As it will turn out, however, the results derived here are essentially independent of the exact choice of F and quite general.

Finally, the vector equation is derived through variation of S with respect to U_α. The vector equation includes a Lagrange multiplier λ that enforces the normalization condition $g^{\alpha\beta}U_\alpha U_\beta = -1$.

SPHERICAL SYMMETRIC, STATIC SPACETIMES IN TEVES

From this point on we focus on the strong gravity limit of TeVeS and explore the spacetime in the vicinity of a spherically symmetric mass. The isotropic form of a spherical symmetric, static metric is

$$g_{\alpha\beta}dx^\alpha dx^\beta = -e^\nu dt^2 + e^\zeta(dr^2 + r^2 d\theta^2 + r^2\sin^2\theta d\varphi^2), \tag{6}$$

where both ν, ζ are only functions of r. For the static system of our case the vector field has two non-vanishing components

$$U^\alpha = (U^t, U^r, 0, 0), \tag{7}$$

where U^t and U^r are functions of the radial coordinate.

Taking $\phi = \phi(r)$, the scalar-field is given by

$$\phi' = \kappa \mathrm{Gm_s}/(4\pi)\frac{\mathrm{e}^{-(\nu+3\zeta)/2}}{\mathrm{r}^2[\mathrm{e}^{-\zeta} - (\mathrm{U^r})^2]}, \tag{8}$$

where one can define the "scalar" mass m_s as a (non-negative) particular integral over $\tilde{\rho}$ and $\tilde{P}$ (defined as the proper energy density and pressure expressed in the physical metric) of the star's matter.

The r component of the vector equation can be brought into the form

$$U^r\left(\lambda + \frac{\kappa(Gm_s)^2}{2\pi}\frac{e^{-(\nu+4\zeta)}}{r^4[e^{-\zeta} - (U^r)^2]^2}\right) = 0. \tag{9}$$

This equation shows that there are two cases: either U^r vanishes or one has a constraint on the Lagrange multiplier λ. In the former case the vector field is aligned with the time direction, while in the latter there is a non-vanishing radial component of the vector field. Since the mathematical analysis of the two cases is rather different, we examine them separately.

Case I: the vector field is aligned with the time direction

When U^r vanishes, the vector field is determined by the normalization expression which yields

$$U^\alpha = (e^{-\nu/2}, 0, 0, 0). \tag{10}$$

The physical metric is then given by Eq. (2) which reduces to $\tilde{g}_{tt} = g_{tt}e^{2\phi}$, $\tilde{g}_{ii} = g_{ii}e^{-2\phi}$. To determine $\tilde{g}^{\alpha\beta}$, one needs to solve for ν and ζ and, through them, for the metric $g_{\alpha\beta}$ and the scalar ϕ. To this end, the differential equations resulting from the tt, rr and components of the metric equation (3) must be solved. This is equivalent to the procedure one follows to arrive at the GR solutions.

Since we are looking for vacuum spacetimes, the terms that include the matter energy density in Eq. (3) are zero. So, we are left with the $\tau_{\alpha\beta}$ and $\Theta_{\alpha\beta}$ terms. The $\tau_{\alpha\beta}$ [see Eq. (4)] contains $\phi_{,\alpha}$ terms while the last term depends on the function F. In the strong acceleration limit, it is possible to show (see Sec. V in Ref. [1]) that the last term is completely negligible in comparison to the other terms. This is exactly the limit we are interested in here and so we will neglect the F term. Using Eqs. (4), (8) and (10), we then compute τ_{tt} and τ_{rr}. Similarly one computes the relevant $\Theta_{\alpha\beta}$ terms [defined by Eq. (5)] (see [10] for details).

We can now use the tt and rr components of the metric equation [Eq. (3)] to derive a system of ordinary differential equations for ζ and ν. After some rearrangement of terms, one finds that

$$\zeta'' + \frac{(\zeta')^2}{4} + \frac{2\zeta'}{r} = -\frac{\kappa(Gm_s)^2}{4\pi}\frac{e^{-(\zeta+\nu)}}{r^4} - K\left(\frac{(\nu')^2}{8} + \frac{\nu''}{2} + \frac{\nu'\zeta'}{4} + \frac{\nu'}{r}\right) \tag{11}$$

and

$$\frac{(\zeta')^2}{4} + \frac{\zeta'\nu'}{2} + \frac{\zeta'+\nu'}{r} = \frac{\kappa(Gm_s)^2}{4\pi}\frac{e^{-(\zeta+\nu)}}{r^4} - K\frac{(\nu')^2}{8}. \tag{12}$$

The study of the properties of these equations for the appropriate boundary conditions constitutes most of the rest of this work.

Case II: the vector field has a non-vanishing r component

In the non-aligned case, $U^r \neq 0$, the Lagrange multiplier λ is given by [see Eq. (9)]

$$\lambda = -\frac{\kappa(Gm_s)^2}{2\pi}\frac{e^{-(\nu+4\zeta)}}{r^4[e^{-\zeta}-(U^r)^2]^2}. \tag{13}$$

The components of the vector field are connected to the functions ν and ζ through the normalization condition for the vector $g^{\alpha\beta}U_\alpha U_\beta = -1$

$$e^{\nu}(U^t)^2 - e^{\zeta}(U^r)^2 = 1 \tag{14}$$

and the t component of the vector equation (given in a compact form) yields

$$KU^{[t;\beta]}{}_{;\beta} - \frac{\kappa(Gm_s)^2}{2\pi}\frac{e^{-(\nu+4\zeta)}}{r^4[e^{-\zeta}-(U^r)^2]^2}U^t = 0, \tag{15}$$

where the $U^{[t;\beta]}{}_{;\beta}$ term involves derivatives of the four unknown functions ν, ζ, U^r, U^t.

The last two expressions can be combined with the tt and rr components of the metric equation (3) to arrive to a closed system of four differential equations with four unknown functions. To this end, one calculates the relevant $\tau_{\alpha\beta}$ and $\Theta_{\alpha\beta}$ terms.

The task in the next subsection is to study the asymptotic behavior of the physical metric far from the source in the two cases I and II and derive the Post-Newtonian corrections predicted by TeVeS.

Asymptotic behavior of the metric far from the source

Far from the source (but not too far, so that the MOND corrections can be safely neglected), the metric can be taken to be asymptotically flat. Expanding the e^{ζ}, e^{ν} to powers of r/r_g (where r_g is a length scale to be determined) we have

$$e^{\nu} = 1 - r_g/r + a_2(r_g/r)^2 + \cdots \tag{16}$$

and

$$e^{\zeta} = 1 + b_1 r_g/r + b_2(r_g/r)^2 + \cdots, \tag{17}$$

where the proportionality constant of the second term in the expansion (16) has been absorbed by r_g. We now proceed to calculate the coefficients a_i and b_i of the metric and equivalent coefficients of the physical metric $\tilde{g}_{\alpha\beta}$ for the two cases I, II (defined in the previous Section).

Case I: $U^r = 0$

If the vector field is aligned with the time direction one can substitute the expansions (16), (17) into the metric equations (11) and (12), match coefficients of like powers if $1/r$ and solve for the coefficients a_i, b_i. The physical metric $\tilde{g}_{\alpha\beta}$ is given by the expressions $\tilde{g}_{tt} = g_{tt}e^{2\phi}$, $\tilde{g}_{ii} = g_{ii}e^{-2\phi}$; so we also calculate the asymptotic behavior of ϕ.

The detailed calculation shows that the physical metric is *identical* to GR up to post-Newtonian order [1, 10]. No constrains can be set to the parameters of TeVeS κ and K from measurements of the standard Post-Newtonian coefficients, if the radial component of the vector field vanishes.

Case II: $U^r \neq 0$

In the non-aligned case one needs to consider the asymptotic expansion of the vector field components. For $r_g/r \ll 1$ the vector field relaxes to its cosmological value, i.e., $U^t \to 1$ and $U^r \to 0$ (since there is no preferred spatial direction). So, expanding to powers of r_g/r, we have

$$U^t = 1 + c_1 r_g/r + c_2(r_g/r)^2 + \cdots \tag{18}$$

and

$$U^r = d_1 r_g/r + d_2(r_g/r)^2 + \cdots. \tag{19}$$

From this point on the method we follow to calculate the Post-Newtonian corrections is similar to this of the previous subsection. Substituting these expansions and Eqs. (16), (17) into Eqs. (14), (15) and the tt and rr components of the metric equations (3) and matching coefficients of like powers if $1/r$, we derive the coefficients a_i, b_i, c_i and d_i. This analysis is carried out down to the order necessary to calculate the post-Newtonian coefficients for $K \ll 1$, $\kappa \ll 1$.

The physical metric is given by Eq. (2) and an appropriate rescaling of the r, t coordinates is needed so that it asymptotes to the Minkowskian form. Notice, however, that ϕ_c is not absorbed by this rescaling unlike the $U^r = 0$ case, because of the more complicated connection of the physical metric with the fields of TeVeS in this case. Assuming again $K \ll 1$, $\kappa \ll 1$ and furthermore that $\phi_c \ll 1$, we have for the standard β, γ Post-Newtonian coefficients, as predicted by TeVeS for the case that $U^r \neq 0$

$$\beta = 1 + \frac{\kappa}{8\pi} + \frac{K}{4} + \phi_c\left(3 + \frac{\kappa}{\pi K} \pm \sqrt{\frac{2\kappa}{\pi K} + 5}\right) \tag{20}$$

and $\gamma = 1$.

The $\pm$ sign comes from the fact that the normalization expression (14) contains squares of the vector components. The $+$ sign corresponds to $U^r > 0$ and vice versa.

While the γ coefficient coincides with the GR prediction, the β differs from unity. The best determination of β comes from lunar laser ranging tests (see for example Ref. [13]) which yields $\beta - 1 \lesssim 10^{-4}$ [14]. How does this result compare with the prediction of Eq. (20)? It is important to note that the term multiplied with the (positive) ϕ_c in Eq. (20) is positive for any value of $\kappa/K > 0$ and choice of the $\pm$ sign, so one has the inequality $\beta - 1 \geq \kappa/(8\pi) + K/4$. For the choice of the function F made in Ref. [1], κ is constrained to be $\simeq 0.03$ which results in $\beta - 1 \gtrsim 2.5 \cdot 10^{-3}$ (taking the K term much smaller). This is in conflict with observations.

ANALYTIC SOLUTIONS WHEN U^R VANISHES

Until now, we have kept the study of spherical symmetric spacetimes in TeVeS quite general. From this point on, we focus on the branch of solutions for which $U^r = 0$, i.e., the vector field is aligned with the time direction. As it turns out, exact analytic solutions are possible in this case.

Solutions in the $K \to 0$ limit

The system of Eqs. (11) and (12) is rather complicated. Here, we first consider some special cases and then use the intuition we gain to derive the general solution. In the simplest case where both $\kappa = 0$ and $K = 0$, the metric equations in TeVeS coincide with these in GR and their write hand side is zero (i.e. no source terms appear). In this limit the integration of Eqs. (11), (12) is straightforward leading to the familiar GR solution.

As a next step toward the most general solution, we take the limit $K \to 0$ but allow κ to be arbitrary. In this limit we essentially decouple the theory from the vector field and it turns out that that Eqs. (11) and (12) are equivalent to spherical symmetric spacetimes in some metric-massless scalar theories of gravity. The exact solution was originally written down by Buchdahl in Ref. [11] (see also Ref. [12])

$$e^{\nu} = \left(\frac{r - r_c}{r + r_c}\right)^{r_g/2r_c} \tag{21}$$

and

$$e^{\zeta} = \frac{(r^2 - r_c^2)^2}{r^4} \left(\frac{r - r_c}{r + r_c}\right)^{-r_g/2r_c}, \tag{22}$$

where

$$r_c \equiv \frac{r_g}{4} \sqrt{1 + \frac{\kappa}{\pi} \left(\frac{Gm_s}{r_g}\right)^2}. \tag{23}$$

It is straightforward to check that in the limit where $\kappa m_s \to 0$, one derives the well known general relativistic expressions.

Having solved for the metric components, one can integrate Eq. (8) to derive the r dependence of the scalar field and then the physical metric through Eq. (2). However, the results derived in this Section are of limited generality since they correspond to the $K = 0$ case, where the effect of the vector field to the metric equations is ignored. The generalization of the solutions to the case where $K \neq 0$ is the task of the next Section.

Spherically symmetric, vacuum solution for the metric for arbitrary κ, K

We turn to the general case where both κ and K are non-zero. While at first sight the metric equations look quite complicated in this case, it turns out that it is possible to derive quite general *exact* spherical symmetric, vacuum solutions for the metric which are identical to (21), (22) provided that one redefines r_c [in Eq. (23)] as [10]

$$r_c \equiv \frac{r_g}{4}\sqrt{1+\frac{\kappa}{\pi}\left(\frac{Gm_s}{r_g}\right)^2-\frac{K}{2}}. \tag{24}$$

One can check that, for the derived solution, the energy density contributed by the vector field is negative. The fact that $\Theta_{\rm tt} < 0$ can have important consequences for the theory since it may lead to instability of the vacuum from the quantum point of view.

The behavior of the scalar field can be followed by integrating Eq. (8)

$$\phi(r) = \phi_c + \frac{\kappa G m_s}{8\pi r_c}\ln\left(\frac{r-r_c}{r+r_c}\right), \tag{25}$$

where ϕ_c stands for the cosmological value of the scalar field at a specific epoch. Just as in metric-scalar theories, one can see that, unless $\kappa m_s = 0$ the scalar field diverges logarithmically at $r = r_c$ and that there is always a radius $r_1 > r_c$ where $\phi(r_1) = 0$ and becomes negative further in. We will return to this point in the next Section where we discuss how black holes look like in TeVeS, and how much they differ from the ones predicted by GR.

The components of physical metric is related to ν, ζ and ϕ through the expressions (2), (21), and (22) that yield

$$\tilde{g}_{tt} = -\left(\frac{r-r_c}{r+r_c}\right)^a \tag{26}$$

and

$$\tilde{g}_{rr} = \frac{(r^2-r_c^2)^2}{r^4}\left(\frac{r-r_c}{r+r_c}\right)^{-a}, \tag{27}$$

where $a \equiv \frac{r_g}{2r_c} + \frac{\kappa G m_s}{4\pi r_c}$.

The expressions (26) and (27) describe spherically symmetric, vacuum spacetimes, i.e., they describe the spacetime down to the surface of a star. The two dimensionless parameters κ and K of the theory provide the parameter space that is to be explored. In addition to these, we have kept the scalar mass m_s and the gravitational radius r_g as free parameters so that the derived results to be quite general and applicable to both the case of a black hole and the exterior of a star. In appendix D of Ref. [1] a detailed description of the procedure to calculate m_s and r_g in terms of its gravitational mass m_g of the star is given. This method is not applicable to the case of a black hole and a different approach is needed to determine m_s and r_g.

HOW DO BLACK HOLES LOOK LIKE IN TEVES?

The characteristic radius of the physical metric described by Eqs. (26) and (27) is r_c. At r_c the tt component of the metric vanishes and the question that rises is whether and under which conditions r_c is the location of the horizon of a black hole. A first step toward answering this question is to calculate the surface area at this radius, which turns out to be proportional to $\tilde{g}_{rr}(r_c)$. A black hole must have a finite surface area at r_c which [in view of Eq. (27)] constrains a to be ≤ 2.

A second constraint on a (Bekenstein private communication) comes from the demand that there is no essential singularity at r_c. For our solution, the Ricci scalar R is

$$R = \frac{2(a^2-4)r_c^2 r^4 (r-r_c)^{a-4}}{(r+r_c)^{a+4}}. \tag{28}$$

From this expression one can see that the Ricci scalar is finite when $a = 2$ or $a > 4$. Considered together, the two constraints (i.e. of finite surface area and Ricci scalar at r_c) imply that only the value $a = 2$ describes a black hole. Using the definition of a we have that for $a = 2$

$$r_c = \frac{r_g}{4} + \frac{\kappa G m_s}{8\pi} \tag{29}$$

and the physical metric has a form identical to that predicted by GR. So, the physical metric in TeVeS is identical to this of GR in the case of a non-rotating black hole.

Furthermore, one can use the definition of r_c [see Eq. (24)] in Eq. (29) to solve for m_s which is non-zero. We have already shown in the previous Section that when $m_s \neq 0$, there is always a region close to r_c where the scalar field turns negative. Bekenstein in ref. [1], on the other hand, has shown that TeVeS becomes acausal (i.e. it suffers from superluminal propagation of metric, vector and scalar field disturbances) when $\phi < 0$. As a result, the theory appears to behave in an unphysical way in the vicinity of our black hole solution. On the other hand, our solutions have been derived under the assumption that $U^r = 0$. Perhaps the causality problem can be overcome by allowing for $U^r \neq 0$.

CONCLUSIONS

Bekenstein's recent relativistic gravitational theory (TeVeS) that leads to MOND in the relevant limit has been proposed as a modification to GR. TeVeS has several attractive features, for example it predicts the right amount of gravitational lensing when only the observed mass is used and provides a covariant formulation to construct cosmological models.

The free parameters in TeVeS can be constrained by the large extra-galactic phenomenology. In this work, instead, we have looked at TeVeS in the strong gravity limit. Two branches of solutions are identified: the first is characterized by the vector field being aligned with the time direction while in the (not previously explored) second branch the vector filed has a non-vanishing radial component. We have shown that the β and γ PPN coefficients in TeVeS are identical to these of GR in the first branch of solutions while the β PPN coefficient differs in the two theories in the second. Despite the fact that the results derived here are essentially independent of the exact choice of the free function F of the theory, our final conclusions do depend on it, since the choice of F influences the way in which observations put constraints on the parameters of the theory. For the second branch of solutions and for the choice of the free function F made in Ref. [1], TeVeS predicts β that is in conflict with recent observational determinations of it.

For the first branch of solutions, we derive exact analytic expression for the physical metric depends on the values of the two dimensionless parameters κ, K of TeVeS. One of the findings of this work is that the energy density contributed by the vector field is negative possibly turning the vacuum unstable from the quantum point of view.

In the case of a black hole, our solutions for the metric are identical to the Schwarzschild solution in GR. On the other hand, these solutions are shown to be acausal in the vicinity of the black hole. Possibly, the issues of the negative energy density contributed by the vector field, and of causality close to a black hole do not appear in the case where $U^r \neq 0$. In this case, however, a different choice of the free function F will be needed so that TeVeS is not in conflict with solar system phenomenology.

ACKNOWLEDGMENTS

I thank Demos Kazanas for bringing exact solutions of metric-scalar theories of gravity to my attention and Jacob Bekenstein and Henk Spruit for valuable suggestions and comments on a previous version of the manuscript. This work was supported by the EU FP5 Research Training Network "Gamma Ray Bursts: An Enigma and a Tool."

REFERENCES

1. J. D. Bekenstein, *Phys. Rev. D*, **70**, 083509 (2004)
2. F. Zwicky, *Helv. Phys. Acta*, **6**, 110 (1933)
3. J. P. Ostriker, P. J. E. Peebles, and A. Yahil, *Astrophys. J.*, **193**, 10 (1974)
4. V. Trimble, *Astron. Astrophys.*, **25**, 425 (1987)
5. M. Milgrom, *Astrophys. J.*, **270**, 365 (1983)
6. M. Milgrom, *Astrophys. J.*, **270**, 371 (1983)
7. K. G. Begeman, A. H. Broeils, and R. H. Sanders, *Month. Not. Roy. Astron. Soc.*, **249**, 523 (1991)
8. R. P. Tully, and J. R. Fisher, *Astron. Astrophys.*, **54**, 661 (1977)
9. C. Skordis, D. F. Mota, P. G. Ferreira, and C. Boehm, *Phys. Rev. Lett.*, in press, (2005; astro-ph/0505519)
10. D. Giannios, *Phys. Rev. D*, **71**, 103511 (2005)
11. H. A. Buchdahl, *Phys. Rev.*, **115**, 1325 (1959)
12. P. H. Mannheim, and D. Kazanas, *Astrophys. Sp. Science*, **185**, 167 (1991)
13. J. D. Anderson, and J. G. Williams, *Classical Quantum Gravity*, **18**, 2447 (2001)
14. J. G. Williams, S. G. Turyshev, and D. H. Boggs, *Phys. Rev. Lett*, **93**, 261101 (2004)

W-Modes in Rotating Relativistic Stars

Nikolaos Stergioulas*, Kostas D. Kokkotas* and Ian Hawke†

*Department of Physics, Aristotle University of Thessaloniki, Thessaloniki 54124, Greece
†School of Mathematics, University of Southampton, Southampton SO17 1BJ, UK

Abstract. We obtain w-mode frequencies of rapidly rotating, relativistic stars, using a 3D fully nonlinear numerical code. The so-called "inverse-Cowling approximation" is used by fixing the evolution of the fluid and allowing only for spacetime dynamics. We demonstrate that in the nonrotating limit the code reproduces known w-mode frequencies obtained through perturbative methods. In the rotating case, we focus on a dynamically unstable model and show that its w-mode frequencies agree well with the gravitational-wave spectrum produced during the initial phase of the dynamical collapse to a Kerr black hole.

INTRODUCTION

The dynamics of compact stars is closely related to the emission of gravitational waves. For example, a prime source of gravitational waves is the coalescence of two neutron stars, while the formation and subsequent dynamical evolution of proto-neutron stars, which involves oscillations and/or rotational instabilities, is also a phase during which strong gravitational waves signals are emitted [1]. The oscillations and instabilities of newly born compact stars have typically been studied using perturbation theory i.e. using the linearized version of the Einstein equations. Through this analysis many oscillation modes have been identified, most of them being related to the oscillations of the fluid (f-, p-, g-modes, etc.) However, during the past decade the study of oscillations modes characterizing the perturbations of the spacetime itself (spacetime modes) also become possible. This study showed the existence of three families of spacetime modes, the so called w-modes [2], the w_{II}-modes or *scattering modes* [3] and (for extremely compact stars) the *trapped modes* [4, 5]. Some of these families of modes could in principle become unstable due to the Chandrasekhar-Friedman-Schutz (CFS) instability [6, 7] if the star is rotating very rapidly and is very compact [8].

So far, the spectrum of spacetime oscillations is well studied and understood for non-rotating or slowly rotating stars, but for the most interesting case, which is the case of rapid rotation, theoretical and numerical difficulties have prevented the computation of pulsation modes through perturbative techniques (for zero-frequency modes or when the Cowling approximation is used, see [9] for a review). Recently, a different approach in studying pulsations of rapidly rotating compact stars (relying on the time-evolution of the fully nonlinear equations) has been employed in obtaining frequencies of axisymmetric fluid pulsations in the Cowling approximation [10, 11] as well as for dynamical spacetimes [12, 13]. What is still missing is the oscillation spectrum of the *spacetime modes*. Ideally, both the fluid and spacetime must be evolved in order to obtain an accurate description of the spacetime modes of compact stars. However, the coupling of

CP848, *Recent Advances in Astronomy and Astrophysics,*
7th *International Conference of the Hellenic Astronomical Society,* edited by N. H. Solomos

these modes to the fluid is very weak, so that in practice one could neglect the evolution of the fluid and only consider the dynamics of the spacetime. This approximation, called *Inverse Cowling Approximation* (ICA) [14] has been shown to yield quite accurate spacetime mode frequencies for nonrotating stars. In the slow rotation approximation the spectrum of w-modes has been studied only recently, in the ICA approximation [15]. Here, we extend this approach to rapid rotation, using fully nonlinear dynamical evolutions.

NUMERICAL SETUP

For our simulations, we use a 3-dimensional, nonlinear, general-relativistic evolution code that was presented in detail in [18, 19, 20]. To summarize, the code has been constructed within the framework of the `Cactus` Computational Toolkit [21], which provides the basic grid structure, parallelization and spacetime evolution routines. The code can solve the hydrodynamical equations using High-Resolution Shock Capturing (HRSC) methods, based on the Valencia formulation (see [22, 23]) while the spacetime is evolved by finite differencing the so-called BSSN formulation (see [24, 25]) of the Einstein equations. The spacetime differencing is fourth order accurate in both space and time, while the method of lines is used.

Due to the resources required, the numerical accuracy of such 3-dimensional simulations can be rather low (when a uniform grid is employed), due to the small grid sizes used or due to the close location at which artificial outer boundaries must be placed. To avoid these problems, we use *structured mesh refinement* to increase the resolution in the regions of interest (in this case in the vicinity of the compact star) and to place the outer boundaries in the wave zone. The mesh refinement code used is detailed in [26] and was also required to extract gravitational wave signals in [20]. The mesh refinement code enables us to causally disconnect the artificial outer boundary from the surface where the gravitational waves are detected, ensuring that the errors made in imposing an artificial boundary condition have no effect on our results. By comparing against results obtained with uniform grids we see that our results are robust against changes in resolution, but the use of mesh refinement assures us that we are in the convergent regime. However, due to restrictions caused by time interpolation (see [26]), the use of the mesh refinement code limits the accuracy to second order. The is still a small price to pay, compared to the huge savings in computational costs.

The initial data are constructed using stationary equilibrium configurations of rotating relativistic polytropes [17, 9]. The initial data are then perturbed by adding a small quadrupolar disturbance in the density (of the order of 1%). This disturbance appears as a nonstationary source term in the Einstein equations and triggers the excitation of spacetime modes. Far from the source, we extracted the emitted gravitational waves by matching the dynamical spacetime to non-spherical perturbations of a Schwarzschild background, using the gauge-invariant Moncrief variables [16] (see [27] for a recent review). Since our background spacetime deviates from Schwarzshild's spacetime (due to rotational effects) the gravitational waveforms contain initial off-sets that need to be subtracted. Fourier transforms for the resulting waveforms reveal the frequencies of the excited modes.

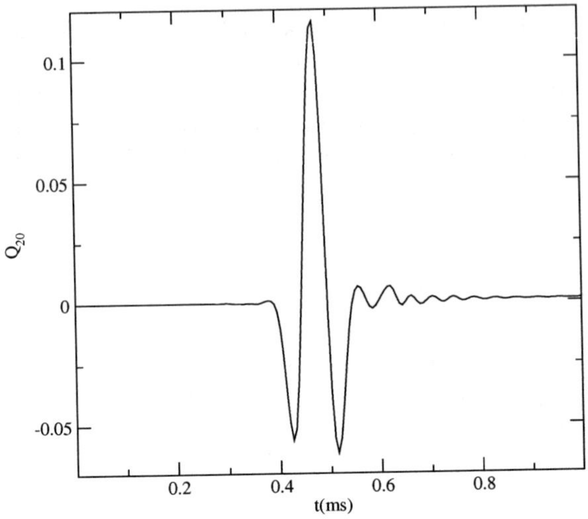

FIGURE 1. Gauge-invariant Moncrief variable for $l = 2, m = 0$ gravitational radiation, extracted at $r = 60M$ after adding a quadrupolar perturbation to a nonrotating polytropic star. The signal is dominated by a rapidly damped spacetime oscillation.

We use nondimensional quantities by requiring that $c = G = M_\odot = 1$.

COMPARISON TO PERTURBATIVE RESULTS

In order to test our numerical method, we first use a spherical polytropic model with $N = 1$ and $K = 100$ and central density $\rho_c = 1.28 \times 10^{-3}$. The gravitational mass of the star is $M = 1.4$ while its equatorial circumferential and coordinate radii are $R = 9.59$ and $r_e = 8.13$, respectively. We add a quadrupolar density perturbation of central amplitude 0.5% in order to excite $l = 2, m = 0$ spacetime modes. For our test runs, we use a low-resolution uniform grid with $dx = dy = dz = 1.6$ in octant symmetry, placing the outer boundaries at $x_{max} = y_{max} = z_{max} = 91.2$. The gauge-invariant Moncrief variable Q_{20}, corresponding to $l = 2, m = 0$ even parity radiation is extracted at various radii.

Figure 1 shows the time-evolution of Q_{20}, as extracted at a coordinate radius of $r = 84.0$, or $r = 60M$. The signal shows the existence of a dominant oscillation frequency which is rapidly damped. On a logarithmic scale (Figure 2) one can also observe the characteristic exponential damping of the oscillations, as expected for black hole modes or spacetime modes of compact stars. A Fourier transform of the signal in Figure 1 is shown in Figure 3. In addition, the frequencies of the lowest-order even-parity $w-$modes are displayed as dashed vertical lines. These frequencies were obtained with an independent perturbative code and have values of 10.09, 17.84, 25.13, 32.48 and 39.41. Figure 3 shows excellent agreement between the spacetime oscillation frequencies obtained with our nonlinear evolution method in the ICA approximation and the expected linear w-mode frequencies, for several modes. Due to the low resolution used in these test runs and/or the simultaneous excitation of additional modes in our simula-

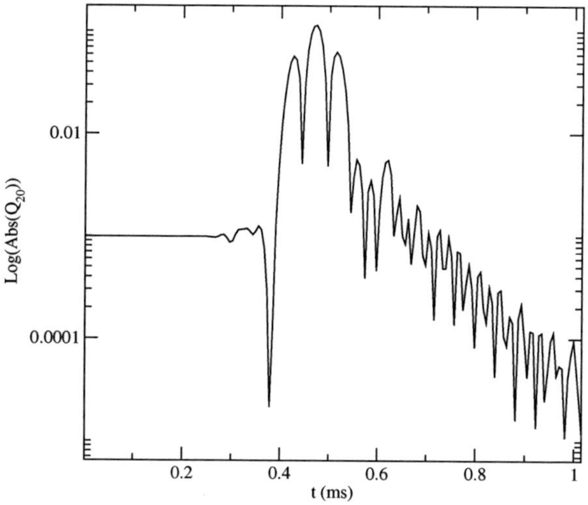

FIGURE 2. Same as Figure 1, but in logarithmic scale, where one can observe the characteristic exponential damping of the oscillations expected for black hole modes or spacetime modes of compact stars.

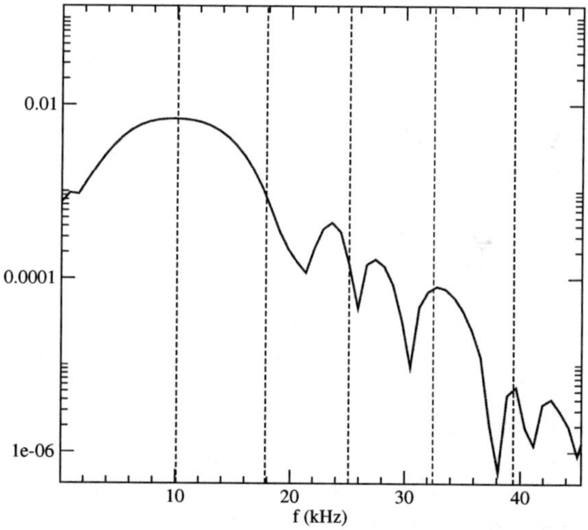

FIGURE 3. Fourier transform of the gravitational wave signal shown in Figure 1. The dashed vertical lines correspond to the frequencies of the even-parity $l = 2, m = 0$ w-modes, as obtained by an independent perturbative code. There is excellent agreement for several mode frequencies.

tions there is a difficulty identifying the expected mode at 25.13 kHz, but this difficulty should be removed once better resolution and initial excitation are used.

In order to test the accuracy of our low-resolution test run in obtaining the fundamental w−mode, i.e. the dominant frequency in Figure 1, we compared the evolution

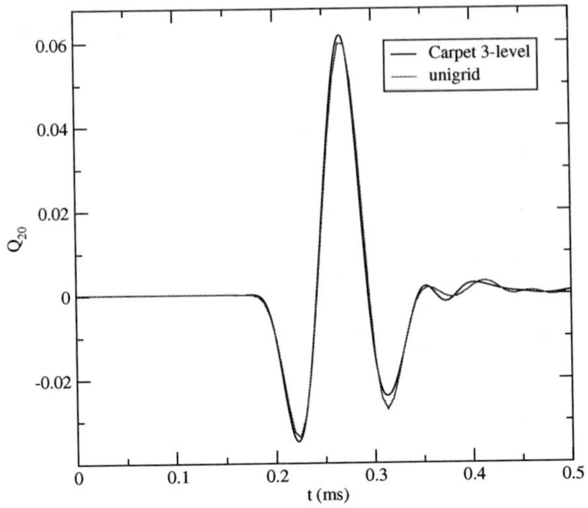

FIGURE 4. Comparison of the gravitational waveform extracted at a coordinate radius of $r = 44.0$ for a low-resolution run of uniform spacing and for a run with 3-level mesh refinement. There is good agreement between the two runs, sufficient to yield the frequency of the fundamental w-mode with good accuracy.

of Q_{20} to a run in which structured mesh refinement was used to increase the effective resolution. Specifically, we used three refinement levels, starting from a grid-spacing of $dx = 1.6$ and refining the region with $x < 24$ to $dx = 0.8$ and the region with $x < 5.6$ to $dx = 0.4$ (similarly for the y and z directions). In this way, the region where the w-modes are created, roughly the interior of the star, is resolved with much better accuracy than in the original test run with uniform spacing. The generated gravitational waves show good agreement between the two runs, as can be seen in Figure 4.

$W-$MODES OF ROTATING STARS

We have applied the numerical method discussed in the previous section to a sequence of rotating stars that have the same baryonic mass as the nonrotating test model (models AU1 to AU4 of the uniformly rotating, constant baryonic mass sequence AU of [11]). The dependence of the *axisymmetric* w-mode frequencies on rotation is rather weak along this sequence. This is expected, as the axisymmetric w-mode eigenfunctions depend on the global properties of the spacetime and are not very sensitive to the rotational distortions induced by the rotating fluid on the various metric components (these distortions are of second order or higher in the rotation). Even for fluid modes, the frequency dependence on rotation of the fundamental polar mode frequencies is weak [11, 13]. Based on the results of [15], (obtained in the slow rotation approximation) we expect a much stronger dependence on rotation for *nonaxisymmetric* $w-$modes. Similarly, we expect a much stronger rotational dependence for *trapped modes* [8].

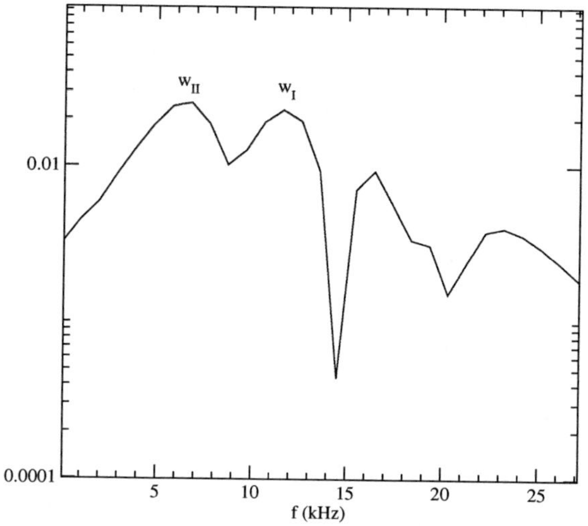

FIGURE 5. Fourier transform of the polar $l = 2, m = 0$ gravitational waveform produced after a disturbance was introduced in a dynamically unstable rapidly rotating model (see text for details). The spectrum is dominated by two spacetime modes which we tentatively identify as the lowest-order $w-$mode and $w_{II}-$mode.

COMPARISON TO COLLAPSE WAVEFORM

In [20] the gravitational waves produced by the collapse of a rapidly rotating neutron star to a Kerr black hole were computed. The neutron star was modelled by a $\Gamma = 2$ polytrope that was initially uniformly rotating near the mass-shedding limit; see [19] for details of the initial data used. As the initial data is close to axisymmetric the vast majority of the power contained in the gravitational wave signal is contained in the leading $\ell = 2$ and $\ell = 4$ multipoles.

The essential conclusion of the calculations performed in [20] was that the gravitational waveform produced closely mirrored the properties of the near-stationary spacetimes at the beginning and end of the simulations. That is, the collapse can be viewed as the rapid transition between the spacetime of the initial equilibrium star and the spacetime of a rotating black hole. Thus the power spectral density of the gravitational waveforms is expected to show imprints of the fundamental modes of these near-stationary spacetimes: that is, the quasi-normal modes of the black hole and the w-modes of the initial star.

In [20] it was confirmed that there was a close relation between the peaks in the PSD, the QNMs of the black hole, and the w-modes of a "typical" star of the same mass. At the time the accurate computation of the w-mode frequencies produced here were not available. Using the w-mode frequencies computed here for the initial data used in the collapse calculation of [20] we see that the correspondence between the peaks in the PSD of the simulated gravitational wave signal, the QNMs of the black hole, and the w-modes, is greatly improved. In some cases, particularly the $\ell = 2$ multipole, there is

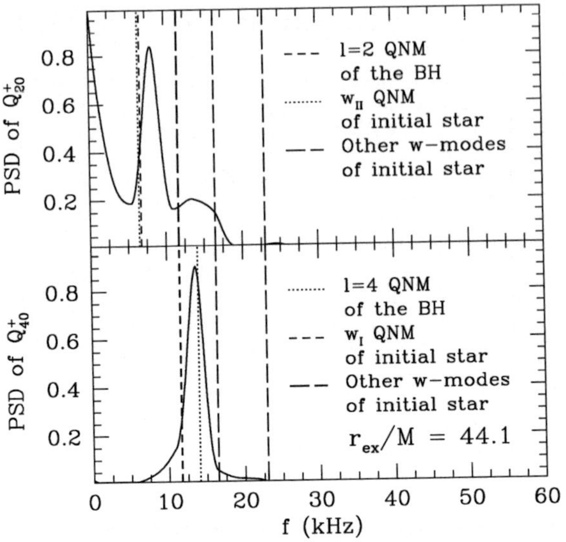

FIGURE 6. The two leading spherical harmonics of the gravitational wave signal PSD from [20]. By overlaying the QNM of a black hole with the same mass and spin, and *w*-modes for a typical star of equivalent mass, [20] noted the close similarity between the GW modes excited during the collapse of the neutron star to a BH with these standard oscillation modes. By overlaying the *w*-modes computed in this work using the inverse Cowling approximation, we see that the BH and NS modes are practically identical, and match well the modes computed during the collapse.

practically no difference in the computed frequencies.

DISCUSSION

The results presented here represent a first step towards the computation of spacetime mode frequencies in rapidly rotating compact stars. In a subsequent publication we will present complete results for a large number of rotational sequences and for a larger number of modes, including nonaxisymmetric modes, obtained with higher resolution and with a more accurate excitation of specific *w*-modes.

ACKNOWLEDGMENTS

Thanks to Luca Baiotti, Luciano Rezzolla and Erik Schnetter for comments and collaborative work. This work has been supported by the Heracleitus and Pythagoras II grants of the Greek Ministry of Education and by the European Network of Theoretical Astroparticle Physics (ENTApP) as part of the ILIAS Network (RII3-CT-2004-506222).

REFERENCES

1. K. D. Kokkotas, and N. Stergioulas, *Proceedings of 5th International Workshop "New Worlds in Astroparticle Physics"* World Scientific (2005), gr-qc/0506083
2. K. D. Kokkotas, and B. F. Schutz *M.N.R.A.S.* **255**, 119-128 (1992)
3. M. Leins, H.-P Nollert, and M. H. Soffel, *Phys.Rev. D* **48**, 3467–3472 (1993)
4. S. Chandrasekhar and V. Ferarri *Proc.R.Soc. London* **A434**, 449 (1991).
5. K. D. Kokkotas *M.N.R.A.S.* **268**, 1015-1018 (1994)
6. S. Chandrasekhar *Phys. Rev. Lett.* **24** 611 (1970)
7. J. L. Friedman and B. F. Schutz *Astrophys. J.* **221** 937 (1978)
8. K. D. Kokkotas, J. Ruoff and N. Andersson, *Phys. Rev. D* **70** 043003 (2004)
9. N. Stergioulas, *Living Rev. Relativity*, **6** 3 (2003)
10. J. A. Font, H. Dimmelmeier, A. Gupta and N. Stergioulas *M.N.R.A.S.* **325** 1463 (2001)
11. N. Stergioulas, Th. Apostolatos and J. A. Font *M.N.R.A.S.* **352** 1089 (2004)
12. J. A. Font et al., *Phys.Rev. D* **65** 084024 (2002)
13. H. Dimmelmeier, N. Stergioulas and J. A. Font *Preprint* astro-ph/0511394
14. N. Andersson, K. D. Kokkotas and B. F. Schutz *M.N.R.A.S.* **280**, 1230–1234 (1996)
15. A. Stavridis and K. D. Kokkotas *Int.J.Mod.Phys.* **D14** 543 (2005)
16. V. Moncrief, *Annals of Physics* **88** 323 (1974)
17. N. Stergioulas and J. L. Friedman, *Astrophys. J.* **444** 306 (1995)
18. L. Baiotti, I. Hawke, P. Montero, and L. Rezzolla, in *Computational Astrophysics in Italy: Methods and Tools*, edited by R. Capuzzo-Dolcetta, Mem. Soc. Astron. It. Suppl., Trieste, 2003, vol. 1, p. 327.
19. L. Baiotti, I. Hawke, P. J. Montero, F. Löffler, L. Rezzolla, N. Stergioulas, J. A. Font, and E. Seidel, *Phys. Rev. D* **71**, 024035 (2005).
20. L. Baiotti, I. Hawke, L. Rezzolla, and E. Schnetter, *Phys. Rev. Lett.* **94**, 131101 (2005).
21. Cactus Computational Toolkit `www.cactuscode.org`
22. F. Banyuls, J. A. Font, J. M. Ibánez, J. M. Martí, and J. A. Miralles, *Astrophys. J.* **476**, 221 (1997).
23. J. A. Font, *Living Rev. Rel.* **2000-2** (2000).
24. M. Shibata, and T. Nakamura, *Phys. Rev. D* **52**, 5428 (1995).
25. T. W. Baumgarte, and S. L. Shapiro, *Phys. Rev. D* **59**, 024007 (1999).
26. E. Schnetter, S. H. Hawley, and I. Hawke, *Class. Quantum Grav.* **21**, 1465–1488 (2004).
27. A. Nagar, and L. Rezzolla, *Classical and Quantum Gravity* **22**, R167–R192 (2005).

High velocity chaotic motion in an active galaxy model

N.J. Papadopoulos* and N. D. Caranicolas*

*Department of Physics, Section of Astrophysics, Astronomy and Mechanics, University of Thessaloniki, 541 24, Thessaloniki, Greece

Abstract. We present a dynamical model for an active galaxy with a disk, nucleus and halo components. Numerical calculations and theoretical evidence show that, for a fixed value of mass of the disk and the nucleus, the stellar velocities in the central region decrease as the mass of halo increases. With a massive halo, the motion tends to be regular while, when the halo component is absent, the majority of orbits are chaotic. The dynamical evolution of the system is also studied when mass is transported from the halo to the disk and the nucleus simultaneously. Our results are compared to the recently obtained observation data for active galaxies.

Keywords: active galaxies, kinematics and dynamics.
PACS: 98.54.Cm, 98.62.Ck, 98.62.Dm, 98.62.Gq, 98.62.Js

INTRODUCTION

Observations in the central regions of the Galaxy show high velocities and lead to the identification of a massive and dense nucleus (see [1],[2]). High rotational velocities were also observed in the central regions of Virgo spirals using newly-develop methods. This indicates the presence of massive cores, of order of $10^9 M_\odot$ or more, in the centre of those systems ([3],[4], [5]).

In the present article we shall adopt a rather simple dynamical model given by the potential

$$V(r,\phi) = -\frac{M_n}{(r^2+c_n^2)^{1/2}} - \frac{M_d}{\{r^2[1+(b^2-1)sin\phi^2+c_d^2]\}^{1/2}} - \frac{M_n}{(r^2+c_h^2)^{1/2}}, \quad (1)$$

where r,ϕ are polar coordinates. Equation (1) describes plane motion in a disk galaxy with a nucleus and a halo component. Here M_n, M_d, M_h is the mass of the nucleus, disk and halo, c_n, c_d, c_h represent the scale length of the nucleus, disk and halo respectively while $1 \leq b \leq 2$ gives the deviation of the disk from axial symmetry.

The aim of the present article is to investigate and measure the magnitude of the velocities in the central regions of the model and to see how this magnitude is affected by the mass of disk nucleus and halo. Particularly, we shall try to find how the magnitude of velocities is affected by the mass of halo. Furthermore, we shall try to see, if high or low velocities are connected with the nature of motion (regular or chaotic). First we shall investigate the time independent model and then we shall come to the evolving model, where mass is transported from the halo to the disk and nucleus simultaneously.

CP848, *Recent Advances in Astronomy and Astrophysics*,
7th *International Conference of the Hellenic Astronomical Society*, edited by N. H. Solomos

In all numerical and analytical calculations the well known system of galactic units is used, where the unit of length is $1kpc$, the unit of time is $0.97746 \times 10^8 yr$ and the unit of mass is $2.325 \times 10^7 M_\odot$. The velocity and energy unit (per unit mass) is $10km/s$ and $100(km/s)^2$ respectively, while G is equal to unity. In the above units we use the values $c_n = 0.25kpc, c_d = 1.5kpc, c_h = 5kpc, b = 2$. We consider a system with a fixed value of total mass thus we take $M_n + M_d + M_h = 10000$.

We consider the case where the disk rotates clockwise at a constant angular velocity Ω. Then the Hamiltonian, which is known as the Jacobi integral, in rectangular cartessian coordinates x, y, writes

$$H_J = \frac{1}{2}(p_x^2 + p_y^2) + V_{eff}(x,y) = E_J \quad , \tag{2}$$

where p_x, p_y are the momenta, per unit mass, conjugate to x and y

$$V_{eff}(x,y) = -\frac{M_n}{\sqrt{x^2+y^2+c_n^2}} - \frac{M_d}{\sqrt{x^2+b^2y^2+c_d^2}} - \frac{M_h}{\sqrt{x^2+y^2+c_h^2}} - \frac{1}{2}\Omega^2(x^2+y^2) \quad ,(3)$$

is the effective potential and E_J is the numerical value of the Jacobi integral.

The results for the time dependent potential are given in Section 2, while Section 3 contains the discussion.

MOTION IN THE TIME DEPENDENT POTENTIAL

Numerical work, not given here, suggest that in a time independent potential, where the mass of nucleus and disk is kept constant, the stellar velocities near the centre decrease as the mass of halo increases. When a massive halo is present,the chaotic regions ,if any, are negligible while most orbits are chaotic, when the halo is absent. These results show similarities to the outcomes derives in previous papers (see [6], [7], [8])

Let us now come to study the properties of orbits when the galaxy evolves with the time. It is assumed that mass is transported from the halo to the disk and the nucleus simultaneously. The equations describing the mass transportation are

$$M_h(t) = M_{hi} - m_1[1 - \exp(-kt)] \quad , \tag{4}$$

$$M_d(t) = M_{di} + (m_1 - m_2)[1 - \exp(-kt)] \quad , \tag{5}$$

$$M_n(t) = M_{ni} + m_2[1 - \exp(-kt)] \quad , \tag{6}$$

where M_{hi}, M_{di}, M_{ni} are the initial values of mass of halo, disk and nucleus while $M_h(t), M_d(t), M_n(t)$, are the values of mass of halo, disk and nucleus at time t respectively and m_1 is the portion of the halo mass that is transported. The mass m_2 goes to the nucleus, the mass $m_1 - m_2$ goes to disk and $k > 0$ is a parameter. Remember that the total mass of the galaxy remains constant so that we always have $M_n(t) + M_d(t) + M_h(t) = 10000$.

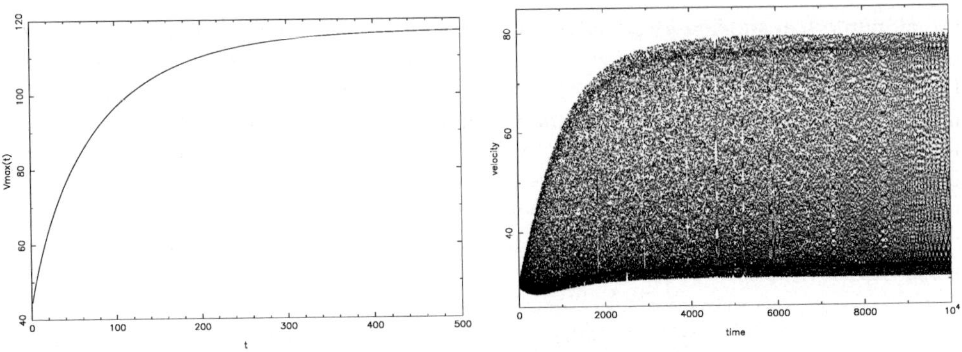

FIGURE 1. (a) left: Plot of v_{max} vs time, as the system evolves following Equations (4) - (6). We take $M_{hi} = 9900, M_{di} = 95, M_{ni} = 5, k = 0.01$.
(b) right: The evolution of the total velocity a regular orbit. See text for details.

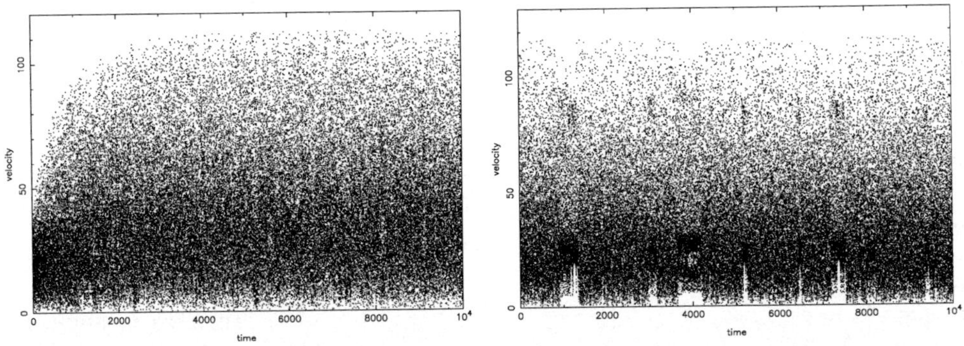

FIGURE 2. (a) left: Plot of the total velocity vs time as the system evolves following Equations (4) - (6) obtained using numerical integration. The initial values of the mass are as in Fig. 5 while $k = 0.001$.
(b) right: Same as (a) for the time independent system. The values of the parameters are as in (a).

It is interesting to follow the evolution of the maximum velocity as mass is transported from the halo to the disk and nucleus as described by Equations (4) - (6). Using Equation (2), we express the maximum velocities in the form

$$v_{max}(t) = [2(E_J + \frac{M_n(t)}{c_n} + \frac{10000}{c_d} - \frac{(c_h - c_d)M_h(t)}{c_h c_d})]^{1/2} \quad . \tag{7}$$

Equation (7) clearly shows that, as the mass of halo decreases, the maximum velocity increases. A plot of v_{max} vs time, as the system evolves following Equations (4) - (7), is shown in Fig. 1a. Here we take $M_{hi} = 9900, M_{di} = 95, M_{ni} = 5, k = 0.01$. Note that after the transportation of the total mass of halo v_{max} remains constant.

Figure 2a shows a plot of the total velocity vs time as the system evolves following Equations (4) - (6) obtained using numerical integration. The initial values of the mass are as above and $k = 0.001$. Here we can make the following interesting comments: (i) The evolution of the maximum velocity is same as that obtained analytically (see Fig.

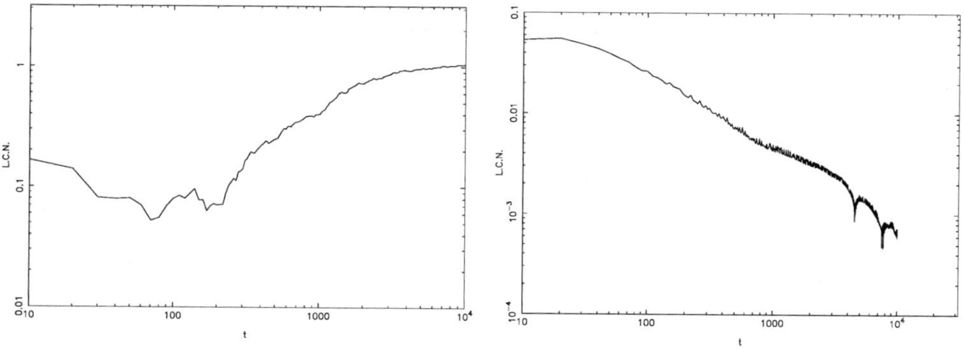

FIGURE 3. (a) left: The LCN of a chaotic orbit vs time.
(b) right: The L.C.N. of a regular orbit vs time.

1a). (ii) Although very high velocities, more than $1000km/sec$, are observed, the test particle travels most of its time at velocities below $500km/sec$. For better understanding, the diagram shown in Fig.2a we give the same diagram, for the time independent system in Fig.2b. Here, we do not have any mass transportation and star's maximum velocity is constant. Here again, we see that the test particle travels most of its time at velocities below $500km/sec$.

It is interesting to follow the evolution of orbits as the galaxy evolves following Equations (4) - (6). As the stars energy changes it is not possible to use the phase $x - p_x$ plane. In order to find the regular or chaotic character of orbits we use the Lyapunov Characteristic Number (L.C.N.) (see [9]). We start when the galaxy has $M_n = 5, M_d = 95, M_h = 9900, k = 0.001, m_2 = 395, m_1 = 9505$. Then, as mentioned above, all orbits are regular and no chaotic motion is observed. We shall follow an orbit starting at $x = -3, y = 0$, with $p_x = 0$ while the value of p_y is found from the initial value of energy $E_J(t = 0) = -1150$. Fig. 3a shows the L.C.N. of this orbit vs time. At $t = 10000$ all the mass of halo has been transported to disk and the nucleus and the orbit has become chaotic as the L.C.N. is close to unity. The evolution of the total velocity of this orbit is that given in Fig. 2a. On the contrary, Fig. 3b shows the L.C.N. of an orbit having the same mass and evolution parameters as that of Fig. 3a but starting at $x = 4, y = 0$. It is evident that this orbit starts as regular and remains regular during the galactic evolution.

The evolution of the total velocity of this regular orbit is that given in Fig. 1b. Comparing this diagram with that shown in Fig. 2a, we observe similarities and significant differencers as well: (i) The maximum velocity is smalller from that obtained by the chaotic orbit. (ii) The test particle travels for significant time periods at velocities below $300km/sec$ as the dark part in the bottom of the Fig. 1b indicates. (iii) The velocity range is smaller than that observed in chaotic orbits.

DISCUSSION

In this article we have adopted a rather simple dynamical model in order to study the motion in an active galaxy. Interesting phenomena are observed in the evolving model. In this model we have an exponential mass transportation from the halo to disk and nucleus simultaneously, while the total galactic mass is kept constant. As this happens, one observes that the velocity in the centre of galaxy increases with time until finally reaches a maximum value. Using the L.C.N. we come to the conclusion that low velocities support regular motion, while high velocities are connected to chaos. Thus, it is evident, that in time dependent (and time independent) potentials, the high velocity motion observed in active galaxies is chaotic. An additional feature, that is strongly suggested by this investigation, is that chaotic motion is observed in active galaxies with a fixed total mass value and, particularly, in galaxies with weak halo concentrations, dense disks and massive nuclei.

Finally, we must note that recent observation data support our model, as similar models with a central mass point and an extended rotating stellar disk were used to reproduce spectroscopic observations ([4]). On the other hand, recent studies in active galaxies strongly indicate that matter falling from the halo is collected from the outer disk and by complex mechanisms part of this matter is driven in the central region ([3]). We believe that the main contribution of this article will be that our results indicate that the motion of this matter, is up to a significant degree, very fast and chaotic.

REFERENCES

1. Genzel, R., Townes, C.H., 1987, An.Rev.Astr. Ap 25, 377
2. Serabyn, E., Lacy, J.H., Townes, C.H., Bharat, R. 1988, Ap.J. 326, 171
3. Rubin, V., Kenney, J.D.P., Young, J.S., 1997, AJ 113, 1250
4. Bertola, F., Cappellari, M., Funes, J. G. et al., 1998, Ap. J 509, L93
5. Sofue, Y. Koda, J., Naganishi, H., Odonera, S., Takamiya, T., Kohno, Okumura, S., Japan - German Seminar, 2001, Sendai
6. Caranicolas, N.D., Innanen, K. A., 1991, A.J. 102, 1343
7. Caranicolas, N.D., 1997, A&SS, 246, 15
8. Caranicolas, N.D., Papadopoulos, N. J., 2003, A&A 399, 957
9. Lichtenber, A.J., Lieberman, M. A, 1983, Regular and stochastic motion (Berlin; Heidelberg; New York)

Sitnikov Motions in the Photogravitational Restricted Three–Body Problem

E.A. Perdios and V.S. Kalantonis

Department of Engineering Sciences, University of Patras, Greece
e-mail: {e.perdios,v.kalantonis}@des.upatras.gr

Abstract. We consider the family of Sitnikov motions in the photogravitational circular restricted three-body problem. The linear stability of the family of periodic orbits of this problem is computed and critical orbits are determined at which families of three-dimensional periodic orbits bifurcate. Some of the bifurcating families are presented.

Keywords: Sitnikov motions, restricted three–body problem, bifurcations, stability
PACS: 45.10.-b, 45.50.Pk, 95.10.Ce

INTRODUCTION

The restricted three–body problem is the most celebrated problem in dynamics and much attention has been paid to its solutions. An interesting case of periodic solutions of this problem is the family of rectilinear motions of the infinitesimal moving body perpendicular to the plane of motion of the primaries. This particular case of motion exists only for $\mu = 1/2$ and may be considered as an equilibrium solution in the sense that it appears as an equilibrium point in the orbital plane but not in three–dimensions. These solutions were first studied by Sitnikov [1] and are called Sitnikov motions.

Recently a certain amount of work has been done on the elliptic Sitnikov problem (see e.g. [2,3]). In the case of the classical circular Sitnikov problem Perdios and Markellos [4] studied numerically the stability and bifurcations of Sitnikov motions. They also presented some results on the continuation of these bifurcations into the case of non–equal masses of the primaries. Belbruno et al. [5] extended these numerical results and studied this problem analytically using elliptic functions.

In this paper we consider Sitnikov motions in the photogravitational circular restricted three–body problem. The photogravitational problem is an interesting variant of the classical one and has received considerable attention in recent years (see e.g. [6,7]). In the photogravitational circular Sitnikov problem we consider that the two primaries are of equal mass and radiation factors, i.e. $m_1 = m_2 = 1/2$ and $q_1 = q_2 = q$. The primaries perform circular motion of constant angular velocity 1 and the massless body is moving in the z direction perpendicular to the plane of the primaries. In the next section, we introduce the equations of motion of the problem. Then, using Floquet theory, the linear stability of the family of straight line oscillations is studied and critical orbits are determined at which families of three-dimensional periodic orbits bifurcate. Some of the bifurcating families are presented.

CP848, *Recent Advances in Astronomy and Astrophysics,*
7th International Conference of the Hellenic Astronomical Society, edited by N. H. Solomos

EQUATIONS OF MOTION

We consider a barycentric, rotating and dimensionless coordinate system $Oxyz$, where the Ox axis always contains the two main bodies having masses μ and $1-\mu$, $\mu \leqslant 1/2$, and radiation pressure parameters q_1 and q_2, $q_i \leqslant 1$, $i=1,2$, respectively. The equations of motion of a third body of a negligible mass moving under the gravitational attraction of the two primaries are [8]:

$$\ddot{x}-2\dot{y}=x-\frac{q_1(1-\mu)}{r_1^3}(x+\mu)-\frac{q_2\mu}{r_2^3}(x+\mu-1),$$
$$\ddot{y}+2\dot{x}=y\left[1-\frac{q_1(1-\mu)}{r_1^3}-\frac{q_2\mu}{r_2^3}\right], \qquad \ddot{z}=z\left[-\frac{q_1(1-\mu)}{r_1^3}-\frac{q_2\mu}{r_2^3}\right], \tag{1}$$

where $\mu=m_2/(m_1+m_2)$ and $r_1=\sqrt{(x+\mu)^2+y^2+z^2}$, $r_2=\sqrt{(x+\mu-1)^2+y^2+z^2}$, are the distances of the moving body from the two primaries. System (1) admits the following integral:

$$C=\frac{1}{2}(\dot{x}^2+\dot{y}^2+\dot{z}^2)-\frac{1}{2}(x^2+y^2)-\frac{q_1(1-\mu)}{r_1}-\frac{q_2\mu}{r_2}, \tag{2}$$

where C is the Jacobi constant. The equation governing the Sitnikov motions $z(t)$ is:

$$\ddot{z}=-\frac{zq}{(z^2+\frac{1}{4})^{3/2}}, \tag{3}$$

and can be easily obtained from System (1) for $\mu=0.5$, $q_1=q_2=q$ and $x(t)=y(t)=0$, while the equation of the Jacobi integral becomes:

$$C=\frac{1}{2}\left[\dot{z}^2-\frac{2q}{(z^2+\frac{1}{4})^{1/2}}\right]. \tag{4}$$

STABILITY OF SITNIKOV MOTIONS

We now consider small perturbations $x=\xi$ and $y=\eta$ of the horizontal components of the rectilinear motion described by (3). For $\mu=1/2$ and equal radiation factors the linearized equations of the perturbed motion are:

$$\ddot{\xi}-2\dot{\eta}=[F_1(z)+F_2(z)]\xi, \qquad \ddot{\eta}+2\dot{\xi}=F_1(z)\eta, \tag{5}$$

where we have abbreviated:

$$F_1(z)=1-\frac{q}{(z^2+\frac{1}{4})^{3/2}}, \qquad F_2(z)=\frac{3q}{4(z^2+\frac{1}{4})^{5/2}}. \tag{6}$$

System (5) can be written in the form $\dot{\Xi}=A\Xi$, $\Xi=(\xi,\eta,\dot{\xi},\dot{\eta})$. These are the variational equations of the rectilinear motion and describe the evolution of the planar perturbations

ξ and η of the Sitnikov motion in the photogravitational problem. The characteristic roots for the equations of variation will determine the stability of the Sitnikov motions and will be approximated by a numerical analysis based on the classical Floquet theory. In this paper we use the method of Perdios and Markellos [4].

The characteristic roots s_k, $k = 1,2,3,4$, are the solutions of the characteristic polynomial $\det(B - Is) = 0$, where I is the 4×4 identity matrix and $B = X^{-1}(t)X(t+T)$, where $X(t)$ is a fundamental solution of the variational equations and T is the period of a particular solution of System (3). Without loss of generality we can set $X(0) = I$, so $B = X(T)$. In case where the roots of the characteristic equation are distinct there are four independent solutions x_k satisfying the property $x_k(t+T) = s_k x_k(t)$, $k = 1,2,3,4$. Thus, a solution is periodic if s_k is unity, while in case of $|s_k| < 1$ ($|s_k| > 1$) the motion is bounded (unbounded). The characteristic equation is quartic and can be written as the product of two quadratic factors $(s^2 + a_1 s + 1)(s^2 + a_2 s + 1) = 0$, with:

$$a_1 = \frac{1}{2}(p_1 + \sqrt{\Delta}), \quad a_2 = \frac{1}{2}(p_1 - \sqrt{\Delta}), \quad \Delta = p_1^2 - 4(p_2 - 2), \tag{7}$$

where we have set for abbreviation:

$$p_1 = -\mathrm{Tr}B, \qquad p_2 = \sum_{j=i+1}^{4} \sum_{i=1}^{4} (b_{ii}b_{jj} - b_{ij}b_{ji}), \tag{8}$$

and b_{ij}, $i, j = 1,2,3,4$ are the elements of the matrix B. The stability conditions are:

$$\Delta > 0, \quad |a_1| \leqslant 2, \quad |a_2| \leqslant 2. \tag{9}$$

For economy in computing time the matrix B can be determined by integrating numerically the fundamental solution from $t = 0$ to $t = T/4$ and applying the transformation $X(T) = [MX^{-1}(T/4)MX(T/4)]^2$, where M is the constant symmetry matrix

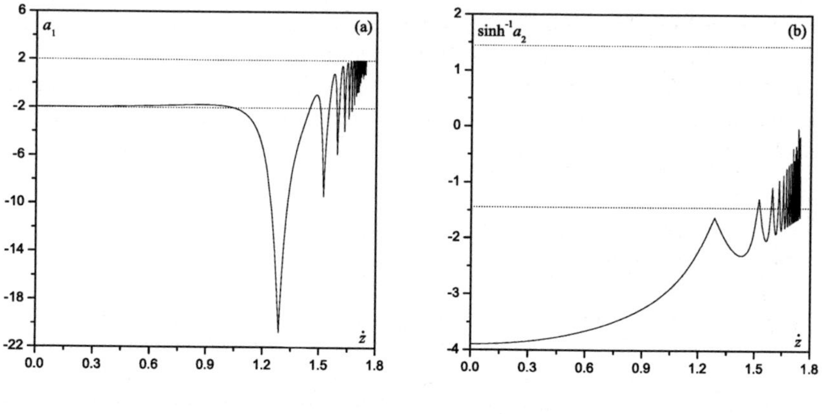

FIGURE 1. The evolution of the stability parameters (a) a_1 and (b) a_2.

TABLE 1. Critical orbits of the Sitnikov family of the photogravitational problem for which $a_1 = -2$. The value of the radiation factor is $q = 0.8$.

	$\dot{z}_0$	$z(T/4)$	a_2	$T/4$	C
B_1	1.04044786	0.56653386	−421.104	1.01283056	−1.05873413
B_2	1.44724480	1.35822292	−94.3494	2.37809708	−0.55274124
B_3	1.50290936	1.62464329	−25.5764	2.97155364	−0.47063173
B_4	1.54880720	1.93340740	−43.8052	3.72714776	−0.40059813
B_5	1.59008038	2.32914982	−9.48627	4.79165060	−0.33582219
B_6	1.60168564	2.47118521	−21.2040	5.19810612	−0.31730156
B_7	1.62949193	2.89422257	−5.57409	6.47969364	−0.27237802
B_8	1.63347515	2.96704776	−11.1642	6.71059256	−0.26587946
B_9	1.65338003	3.39438594	−3.69107	8.12302732	−0.23316723
B_{10}	1.65478597	3.42931960	−6.09030	8.24270880	−0.23084170
B_{11}	1.66980114	3.85341606	−2.55225	9.74425200	−0.20588208
B_{12}	1.67015801	3.86479106	−3.25018	9.78573524	−0.20528611

$M = \text{diag}[1,-1,-1,1]$. In Figure 1 the evolution of the stability parameters a_1 and a_2 along the family of the Sitnikov motions is shown. The value of the radiation factor is arbitrary chosen to be $q = 0.8$.

CRITICAL ORBITS AND BIFURCATIONS

The orbits of the Sitnikov family at which $|a_i| = 2$, $i = 1,2$, are called critical orbits and are of special interest since at these points families of three–dimensional periodic orbits bifurcate. In Figure 1 we see that the stability parameters a_i, $i = 1,2$, take the value -2 many times indicating the existence of a large number of critical orbits. In this study we compute only critical orbits for which $a_1(\dot{z}_0) = -2$. To compute these orbits accurately we seek proper correction $\delta\dot{z}_0$ which can be obtained by linearizing the criticality condition. We have computed critical orbits for which $z(T/4) \leqslant 4$ where T is the period of such an orbit. Up to this value of z we found twelve critical orbits for which $a_1 = -2$. These orbits, named B_i, $i = 1,\ldots,12$, are presented in Table 1. For $z(T/4) \leqslant 4$ all members of the family are unstable because always $|a_2| > 2$.

We have computed all the families of three–dimensional periodic orbits, for $\mu = 1/2$ and $q = 0.8$, branching from the Sitnikov motion at the bifurcation points of Table 1. According to the names of the critical orbits we denote these families by B_i, $i = 1,\ldots,12$. Each one of these families begins at its critical orbit of the rectilinear motion and ends in a planar periodic orbit of the photogravitational circular restricted three–body problem. To compute a family of three–dimensional periodic orbits we have used known

TABLE 2. Members of families of three–dimensional periodic orbits of double symmetry branching from the critical orbits $B_2, B_3, B_6, B_7, B_{10}$ and B_{11}.

	x_0	z_0	$\dot{y}_0$	$T/4$	C	p	q
B_2	0.57399941	1.26903224	−0.60869170	2.37939728	1.06950435	−3.50916	−41.0096
B_3	0.42077096	1.58075640	−0.40451617	2.96555783	0.95617524	−4.01875	−11.7229
B_6	−0.24184003	2.46072859	0.25051937	5.19857442	0.63032873	−3.02219	−18.9070
B_7	−0.14174887	2.89098411	0.13789321	6.47917915	0.54584654	−3.00554	−5.20300
B_{10}	0.37006205	3.41058878	−0.37815081	8.24339206	0.45562587	−5.09054	−6.60876
B_{11}	0.15069156	3.85057379	−0.14801256	9.74386716	0.41256871	−2.50013	−3.68936

TABLE 3. Members of families of three–dimensional periodic orbits of simple symmetry branching from the critical orbits B_1, B_4, B_5, B_8, B_9 and B_{12}.

	y_0	z_0	$\dot{x}_0$	$T/2$	C	p	q
B_1	−0.42672994	0.39920000	−0.28102167	1.99826885	−2.18357692	−2.14992	−360.020
B_4	−0.29722446	1.91430000	−0.30917256	7.45600667	−0.79246618	−3.00536	−36.6380
B_5	−0.23556753	2.31840000	−0.22639037	9.57971854	−0.67555674	−4.00521	−7.54063
B_8	−0.34129162	2.94900000	−0.35024356	13.42263997	−0.52528345	−5.01861	−9.10467
B_9	−0.09113021	3.39322000	−0.08910059	16.24571544	−0.46669253	−2.50012	−3.61838
B_{12}	−0.17352721	3.86109000	−0.17659702	19.57171762	−0.40947690	−3.00286	−3.19443

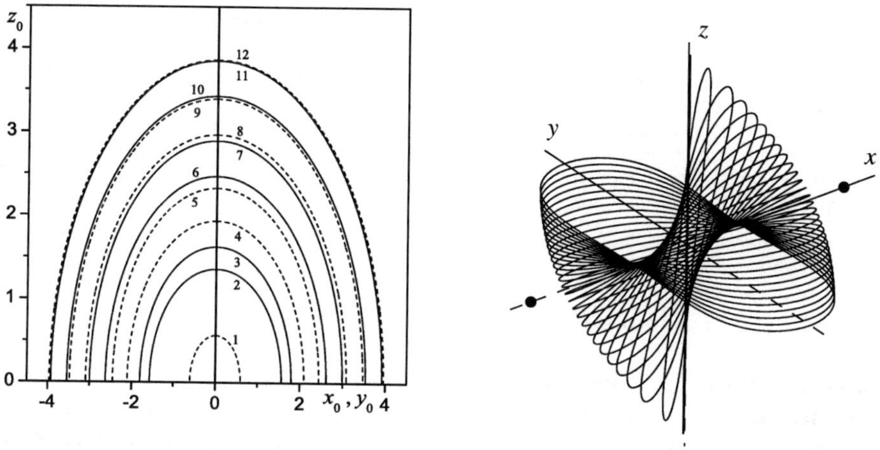

FIGURE 2. (Left frame) Projections of the characteristics of the families B_i, $i = 1, \ldots, 12$ (marked by the value of i). Continuous lines: families of double symmetry (horizontal axis x_0). Dashed lines: families of simple symmetry (horizontal axis y_0). (Right frame) Typical orbits of family B_1.

techniques. Families $B_2, B_3, B_6, B_7, B_{10}$ and B_{11} are of double symmetry with respect to the Ox axis and the Oxz plane and their initial conditions are given in Table 2. Families B_1, B_4, B_5, B_8, B_9 and B_{12} are of simple symmetry with respect to the Oyz plane and their initial conditions are given in Table 3. In these tables we also include the values of the stability parameters p, q of the three–dimensional orbits as well as the Jacobi constant C. Projections of the characteristics of the families B_i, $i = 1, \ldots, 12$ are shown in Figure 2(a). In Figure 2(b) we show some members of family B_1.

REFERENCES

1. K. Sitnikov, *Soviet Phys.*, **5,** 647 (1961).
2. R. Dvorak, *Celest. Mech. Dyn. Astr.*, **56,** 71 (1993).
3. S.B. Faruque, *Celest. Mech. Dyn. Astr.*, **87,** 353 (2003).
4. E.A. Perdios, and V.V. Markellos, *Celest. Mech.*, **42,** 187 (1988).
5. E. Belbruno, J. Llibre, and M. Ollé, *Celest. Mech. Dyn. Astr.*, **60,** 99 (1994).
6. Z. Niedzielska, *Celest. Mech. Dyn. Astr.*, **58,** 203 (1994).
7. E.A. Perdios, *Astr. Space Sci.*, **286,** 501 (2003)
8. J.F.L. Simmons, A.J.C. McDonald, and J.C. Brown, *Celest. Mech.*, **35,** 145 (1985).

On the RTBP with oblate primary and radiating secondary

C.N. Douskos, E.A. Perdios and V.V. Markellos

Department of Engineering Sciences, University of Patras, Greece
e-mail: {c.douskos, e.perdios, markellos}@des.upatras.gr

Abstract. We consider a version of the restricted three–body problem which includes the effects of oblateness of the primary body and radiation of the secondary body, determine the equilibrium points and their stability and discuss the zero–velocity curves. We find that both oblateness of the primary and radiation of the secondary reduce the stability region of the isosceles triangle equilibrium points in the parameter space. These effects also reduce the Roche lobe, and thus the sphere of influence, of the secondary.

Keywords: Restricted problem, oblate primary, radiating secondary, equilibrium points, stability
PACS: 45.50.Jf, 45.50.Pk, 95.10.Ce

INTRODUCTION – EQUATIONS

Versions of the restricted three–body problem with added effects of radiation or oblateness of primaries have attracted considerable interest (e.g. [1,2]). Here we consider a mixed effect restricted three–body problem including oblateness of the larger primary body of mass m_1 (hereafter called "the primary") as well as radiation of the smaller primary ("the secondary") of mass m_2. In the following sections we determine the positions and stability of the triangular and collinear equilibrium points and discuss the zero–velocity curves. The restricted three–body problem considered here is described by the equations of motion [3]:

$$\ddot{x} - 2n\dot{y} = \frac{\partial \Omega}{\partial x}, \quad \ddot{y} + 2n\dot{x} = \frac{\partial \Omega}{\partial y}, \quad \Omega = n^2 \frac{x^2 + y^2}{2} + (1-\mu)\left(\frac{1}{r_1} + \frac{A_1}{2r_1^3}\right) + \frac{q_2 \mu}{r_2}, \tag{1}$$

where $0 \leqslant A_1 << 1$ is the oblateness coefficient of the primary, an oblate spheroid in this case, $0 < q_2 \leqslant 1$ is the radiation coefficient of the secondary, and:

$$n = \sqrt{1 + \frac{3A_1}{2}}, \quad r_1 = \sqrt{(x-\mu)^2 + y^2}, \quad r_2 = \sqrt{(x - \mu + 1)^2 + y^2}. \tag{2}$$

The coordinate system is rotating with angular velocity n. The problem admits the Jacobi integral:

$$2\Omega - (\dot{x}^2 + \dot{y}^2) = C, \tag{3}$$

where C is the Jacobi constant.

CP848, *Recent Advances in Astronomy and Astrophysics,*
7th International Conference of the Hellenic Astronomical Society, edited by N. H. Solomos

ISOSCELES TRIANGLE EQUILIBRIA AND THEIR STABILITY

It is easy to see that for $y \neq 0$ and zero velocity components $\dot{x} = \dot{y} = 0$ the equations of motion are satisfied at the "triangular" equilibrium points L_4, L_5 with coordinates:

$$(x_0, y_0) = (-1 + \mu + \frac{R^2}{2}, \pm\frac{R}{2}\sqrt{4 - R^2}), \qquad R = \left(\frac{2q_2}{2 + 3A_1}\right)^{1/3} \leqslant q_2^{1/3} \leqslant 1, \quad (4)$$

where R is the distance of these points from the secondary. Their distance from the primary, placed at $(\mu, 0)$, is 1 for all values of the oblateness and radiation coefficients A_1 and q_2. Thus, for all values of the parameters these equilibrium points form isosceles triangles with the two primary bodies (the distance of the primaries from each other also being 1). In the classical restricted problem $(q_2 = 1, A_1 = 0)$ R reduces to 1 and the triangles become equilateral.

To determine the linear stability of the equilibrium point (x_0, y_0) we transfer the origin to this point and expand the equations of motion to first order terms arriving at the variational equations. The characteristic equation of these is found to be a quartic:

$$\lambda^4 + b\lambda^2 + c = 0, \quad (5)$$

$$\text{with :} \quad \mathrm{b} = \frac{1}{2}[2 - 3\mathrm{A}_1(1 - 2\mu)], \quad \mathrm{c} = \frac{9\mu(1-\mu)}{16}(2 + 3\mathrm{A}_1)(2 + 5\mathrm{A}_1)(4 - \mathrm{R}^2). \quad (6)$$

Note that in the classical restricted problem (5) reduces to $\lambda^4 + \lambda^2 + 27\mu(1-\mu)/4 = 0$ (see e.g. [4]). For all $\mu \in (0, 1/2]$ we have $b > 0$ and $c > 0$, since $A_1 < 2/3$ and $R < 1$. It follows that the characteristic equation has four imaginary roots, and the triangular equilibrium points are stable, only when the discriminant is positive. Substituting R from (4) into the discriminant we obtain the condition for stability in the full three–dimensional space of the parameters μ, A_1, q_2:

$$q_2 > \frac{2 + 3A_1}{2}\left\{4 - \frac{[2 - 3A_1(1 - 2\mu)]^2}{9\mu(1-\mu)(2 + 3A_1)(2 + 5A_1)}\right\}^{3/2}. \quad (7)$$

For small $(1 - q_2)$ and A_1 we can write this to first order terms as:

$$\mu < \mu_1 = \mu_1^{(0)} - \frac{2}{27\sqrt{69}}(1 - q_2) - \frac{1}{9}\left(1 + \frac{13}{\sqrt{69}}\right)A_1, \quad (8)$$

where $\mu_1^{(0)} = 1/2 - \sqrt{69}/18 \cong 0.03852$ is Routh's critical value for stability in the classical restricted problem. Both radiation of the secondary and oblateness of the primary clearly reduce the stability interval on the μ–axis. For $A_1 = 0$ we obtain from (7) the simplified condition:

$$\mu < \frac{1}{2} - \frac{1}{6}\sqrt{(32 - 9q_2^{2/3})/(4 - q_2^{2/3})}, \quad (9)$$

giving the full separate effect of the secondary's radiation in reducing the stability interval on the μ–axis. We note that if μ and A_1 are sufficiently small, then the roots of the characteristic equation are given approximately by:

$$\lambda = \pm\frac{i}{2}\sqrt{2-3A_1 \pm [9\mu(4-q_2^{2/3})+3A_1-2]}. \tag{10}$$

With regard to the nonlinear stability of the triangular points, we note that within the linear stability region we have four imaginary roots $\lambda_{1,2} = \pm i w_1$, $\lambda_{3,4} = \pm i w_2$, and there are three main cases of resonance: $w_1 - K w_2 = 0$, $K = 1,2,3$, which leads to a quadratic equation for μ, and working as in [5], we obtain:

$$\mu_K = \frac{1}{2} + \frac{1-n^2-\sqrt{g[g+(1-n^2)^2-1]}}{2[g+(1-n^2)^2]}, \qquad g = \frac{c}{\mu(1-\mu)}\left(\frac{K^2+1}{2K}\right)^2, \tag{11}$$

where g contains oblateness and radiation effects through c. Equation (11) generalises in the present case Equation (24) of [6], and for $K = 1$ gives the boundary of the linear stability region (7), while for $K = 2,3$ it gives inner resonance surfaces.

Nonlinear stability zones for the triangular points were determined numerically, in the sense described by McKenzie and Szebehely [7], by Markellos *et al.* [8] for the classical restricted problem. Similar computations were carried out in [5] where the effect of oblateness of the primary was considered. In those computations resonances were seen to play an important role in reducing the nonlinear stability zones in the configuration space. In the present case of combined oblateness and radiation effects, similar reductions of the nonlinear stability zones due to resonances may be expected, though we have not carried out such computations as this would be outside the scope of the present paper. For small values of $1-q_2$ and A_1 the main resonances (11) are given to first order terms by (8) for $K = 1$ ($\mu = \mu_1$), and by:

$$\mu_2 = \mu_2^{(0)} - \frac{32}{135\sqrt{1833}}(1-q_2) - \frac{16(141+5\sqrt{1833})}{31725}A_1, \tag{12}$$

$$\mu_3 = \mu_3^{(0)} - \frac{2}{45\sqrt{213}}(1-q_2) - \left(\frac{1}{25}+\frac{13}{15\sqrt{13}}\right)A_1, \tag{13}$$

for $K = 2$ and 3, respectively. In these expressions the values: $\mu_2^{(0)} = 1/2-\sqrt{1833}/90 \cong 0.024294$, $\mu_3^{(0)} = 1/2-\sqrt{213}/30 \cong 0.013516$, together with Rouths's value $\mu_1^{(0)}$ given earlier, are the critical values of the classical restricted problem [9].

COLLINEAR EQUILIBRIA AND THEIR STABILITY

For $y = 0$ and zero velocity components the equations of motion are satisfied at the collinear equilibrium points $(x,0)$ obtained by putting the right hand–side of the first of the equations of motion (1) equal to zero and solving for x. We distinguish among the three collinear equilibrium points L_1, L_2 and L_3 by setting: $x_{L_1} < -1+\mu < x_{L_2} < \mu < x_{L_3}$. The corresponding roots of the equation are expanded as follows:

$$x_{L_1} = -1 - M - \left(\frac{1}{3}+\frac{5A_1}{6}\right)M^2 + \left[\frac{28}{9}+3\varepsilon_2+\left(\frac{145}{18}+\frac{15\varepsilon_2}{2}\right)A_1\right]M^3 -$$

$$-\left[\frac{50}{81}+\frac{2\varepsilon_2}{3}+\left(\frac{103}{54}+2\varepsilon_2\right)A_1\right]M^4+O(M^5), \tag{14}$$

$$x_{L_2}=-1+M-\left(\frac{1}{3}+\frac{5A_1}{6}\right)M^2+\left[\frac{26}{9}+3\varepsilon_2+\left(\frac{125}{18}+\frac{15\varepsilon_2}{2}\right)A_1\right]M^3+$$

$$+\left[\frac{58}{81}+\frac{2\varepsilon_2}{3}+\left(\frac{113}{54}+2\varepsilon_2\right)A_1\right]M^4+O(M^5), \tag{15}$$

$$x_{L_3}=1+\left[\frac{4+q_2}{5}+\left(\frac{28}{5}+\frac{q_2}{2}\right)A_1\right]N+\frac{1}{50}(128-56q_2+5q_2^2)A_1N^2+O(N^3), \tag{16}$$

where : $$\mathrm{M}=\left[\frac{2q_2\mu}{3(2+5A_1)}\right]^{1/3}, \quad \varepsilon_2=\frac{1-q_2}{q_2}, \quad \mathrm{N}=\frac{5\mu}{3(2+5A_1)^2}, \tag{17}$$

and the coefficients of the powers of M and N are given to first order terms in A_1.

To determine the stability of the collinear points we proceed as in the previous section and find that for all three collinear points the characteristic equation is of the form (5) with the coefficients b and c now being complicated expressions of the parameters. For all three points we have established numerically that c is negative for all values of $\mu \in (0,1/2]$, $A_1 \in [0,1/2]$ and $q_2 \in (0,1]$. It follows that we have two pure imaginary and two real roots of the characteristic equation. Due to the real roots the collinear points are unstable as in the classical restricted problem.

THE ZERO–VELOCITY CURVES

In [10,11] the effects of radiation and oblateness on the topology of the zero–velocity curves were treated separately. Here both effects considered reduce the influence of the secondary and we present in Figure 1 typical stages of the evolution of the curves under the simultaneous effects of radiation of the secondary and oblateness of the primary using arbitrary ratio of the two effects: $A_1=(1-q_2)/10$. We note in particular that as these effects increase, the Roche lobe of the secondary shrinks. Only the curves corresponding to the Jacobi constant values $C(L_1)$, $C(L_2)$, $C(L_3)$ are shown.

The topology of the curves changes when the ordering of the Jacobi constant values for the collinear equilibrium points changes. For $\mu=0.3$, for example, at $q_2=1$ (and $A_1=0$) we have $C(L_2)>C(L_1)>C(L_3)$ and the curves behave as in the top left frame of Figure 1. This arrangement of the curves remains unaltered for smaller values of q_2 (and larger A_1) until the critical value $q_2\cong 0.70453$ (and $A_1=(1-0.70453)/10\cong 0.02955$) is reached. Below this critical value we have $C(L_2)>C(L_3)>C(L_1)$ and the curves behave as in the top middle frame of Figure 1, until a second critical value $q_2\cong 0.51415$ (and $A_1=(1-0.51415)/10\cong 0.04858$) is reached. Below this second critical value we have $C(L_3)>C(L_2)>C(L_1)$ and the curves behave as in the top right frame.

The two critical values discussed above depend on the value of μ used. Their behaviour as functions of μ has been described by Schuerman [12] for the photogravitational restricted three–body problem (without oblateness effects) and is not reproduced here (see also [13]). The combined total effect explored here (Figure 1) is qualitatively the same as if only radiation of the secondary were present (Figure 3 of [12]).

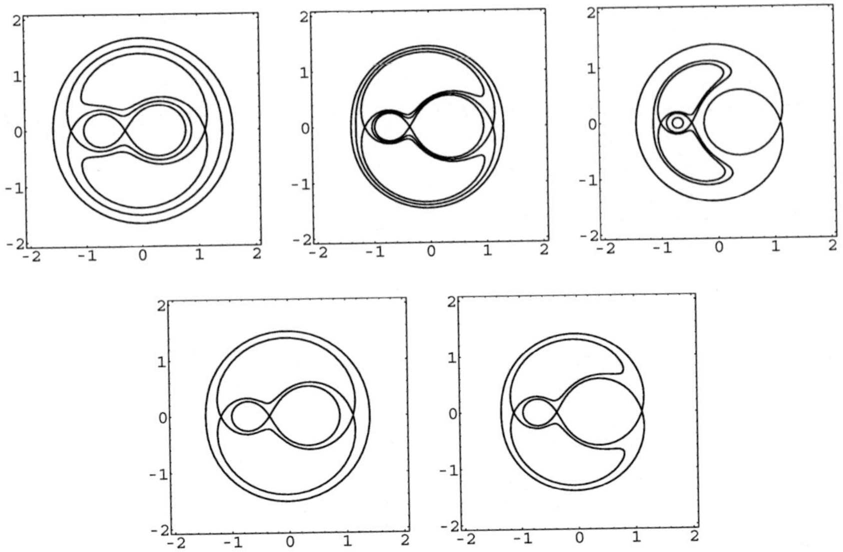

FIGURE 1. Top row from left to right: zero–velocity curves for $\mu = 0.3$, $q_2 = 1, 0.6, 0.2$ and $A_1 = (1 - q_2)/10$. Bottom row, the curves for the transition values: from left to right: $q_2 = 0.70453$ and $q_2 = 0.51415$.

ACKNOWLEDGMENTS

During this work one of the authors (V.V.M.) was on leave at the Department of Physics and Astronomy, University of Glasgow.

REFERENCES

1. J.F.L. Simmons, A.J.C. McDonald, and J.C. Brown, *Celes. Mech.*, **35,** 145 (1985).
2. R.K. Sharma, and P.V. Subba Rao, *Celes. Mech.*, **13,** 137 (1976).
3. R.K. Sharma, *Astrophys. Space Sci.*, **135,** 271 (1987).
4. V. Szebehely, *Theory of Orbits*, Academic Press (1967).
5. V.V. Markellos, K.E. Papadakis, and E.A. Perdios, *Astrophys. Space Sci.*, **245,** 157 (1996).
6. D.G. Tuckness, *Celes. Mech.*, **61,** 1 (1995).
7. R. McKenzie, and V. Szebehely, *Celes. Mech.*, **23,** 223 (1981).
8. V.V. Markellos, K.E. Papadakis, and E.A. Perdios, in *From Newton to chaos*, edited by A.E. Roy, Plenum Press, New York, 1995, p. 371.
9. A. Deprit, and A. Deprit–Bartholome, *AJ*, **72,** 173 (1967).
10. V.V. Markellos, A.E. Roy, M.J. Velgakis, and S.S. Kanavos, *Astrophys. Space Sci.*, **271,** 293 (2000).
11. V.V. Markellos, A.E. Roy, E.A. Perdios, and C.N. Douskos, *Astrophys. Space Sci.*, **278,** 295 (2001).
12. D.W. Shuerman, *Astrophys. Space Sci.*, **19,** 351 (1972).
13. H.–N. Zhou, and K.–C. Leung, *Astrophys. Space Sci.*, **141,** 257 (1988).

A Criterion On The Nonexistence Of Exact Invariants In Adiabatic Systems

Voyatzi Despina and Meletlidou Efi

Department of Physics
University of Thessaloniki
Thessaloniki 54124, Greece
e-mail: dvoya@skiathos.physics.auth.gr

Abstract. In the present paper we investigate the nonintegrability of adiabatic one degree of freedom Hamiltonian systems, with the additional assumption that the frozen system possesses an unstable fixed point with two asymmetric homoclinic loops. We prove a criterion for the nonexistence of an integral for such systems, and therefore we prove the nonexistence of a quantity which is conserved in an arbitrarily high order on ε. A specific application is given in the asymmetric quartic oscillator with adiabatic time dependence.

Keywords: Adiabatic systems, Nonintegrability.
PACS: 05.45.-a, 45.20.Jj

A NONINTEGRABILITY CRITERION

We consider Hamiltonians systems of the form

$$H = H(q, p, z)$$

where $z = \varepsilon t + z_0$, is the slow time and ε is a small parameter. The equations of motion for this system are

$$\dot{q} = \frac{\partial H}{\partial p}, \quad \dot{p} = -\frac{\partial H}{\partial q}, \quad \dot{z} = \varepsilon. \tag{1}$$

When $\varepsilon = 0$, (1) becomes a one-parameter family of integrable planar autonomous Hamiltonian systems, with respect to z_0,

$$\dot{q} = \frac{\partial H}{\partial p}(p,q,z_0), \quad \dot{p} = -\frac{\partial H}{\partial q}(p,q,z_0), \quad \dot{z} = 0,$$

which will be referred to as the unperturbed system, with z_0 as the parameter, also known as the "frozen system" (see [1]-[5]).

Proposition: Assume that in the frozen system there exist two asymmetric homoclinic loops, encircling two separate monoparametric families of periodic orbits of the unperturbed system with the constant value h of the Hamiltonian as a parameter, at

CP848, *Recent Advances in Astronomy and Astrophysics,*
7th *International Conference of the Hellenic Astronomical Society,* edited by N. H. Solomos

least for one value of z_0. If the average value $\langle \partial H / \partial z \rangle_{z_0}$, evaluated along two periodic orbits, each one belonging to a different family and corresponding to the same value of h, attains different values, then system does not possess an integral, analytic in h in an open interval around $\varepsilon = 0$ (see also [6], [7]).

Suppose that there exists a nontrivial integral $\Phi = \Phi(q, p, z, \varepsilon)$ of (1) which can be written as a power series with respect to ε

$$\Phi = \Phi_0(q, p, z) + \varepsilon \Phi_1(q, p, z) + O(\varepsilon^2),$$

i.e. we suppose that Φ is analytic with respect to ε. By identity it must hold that

$$\frac{d\Phi}{dt} = [\Phi, H] + \frac{\partial \Phi}{\partial t} = 0.$$

In zeroth order terms with respect to ε it holds that $[\Phi_0, H] = 0$ therefore Φ_0 is an integral of autonomous Hamiltonian H for any z_0. Consequently, Φ_0 has the form

$$\Phi_0 = \Phi_0(H(q, p, z_0), z_0). \tag{2}$$

Lemma 1: The integral Φ can always be selected, such that Φ_0 is a nontrivial constant ([8]).

In first order terms with respect to ε we have

$$[\Phi_1, H] + \frac{\partial \Phi_0}{\partial H}\frac{\partial H}{\partial z} + \frac{\partial \Phi_0}{\partial z} = 0. \tag{3}$$

Now we integrate identity (3) along a periodic orbit of the unperturbed system for one period T,

$$\int_0^T \frac{d\Phi_1}{dt} dt + \frac{\partial \Phi_0}{\partial H} \int_0^T \frac{\partial H}{\partial z} dt + \frac{\partial \Phi_0}{\partial z} = 0. \tag{4}$$

However,

$$\int_0^T \frac{d\Phi_1}{dt} dt = 0,$$

since, along the periodic orbit, Φ_1 is a single-valued periodic function of time. So we obtain the equation

$$\frac{\partial \Phi_0}{\partial H} \frac{1}{T} \int_0^T \frac{\partial H}{\partial z} dt + \frac{\partial \Phi_0}{\partial z} = 0. \tag{5}$$

Note that the above integrals are evaluated along the unperturbed periodic solution $q(t - t_0, q_0, p_0)$, $p(t - t_0, q_0, p_0)$, and $z = z_0$.

Lemma 2: The zeroth order term Φ_0 of the expansion of Φ with respect to ε is such that $\partial \Phi_0 / \partial H \neq 0$ in an open domain of values of H and z.

Indeed, from (5) we have that if $\partial \Phi_0 / \partial H = 0$ holds, then it must also hold that $\partial \Phi_0 / \partial z = 0$, i.e. Φ_0 mast be a trivial constant, in contradiction to Lemma 1.

Consider that, fixing z_0, we have two periodic orbits (q_+, p_+) and (q_-, p_-) each one inside a different loop of the unperturbed system, which correspond to the same constant value of the unperturbed Hamiltonian H. In both orbits, the quantity $\frac{\partial \Phi_0}{\partial z}(H, z_0)$ is the same, so we conclude that it must hold

$$\frac{\partial \Phi_0}{\partial H}(H, z_0)\left\{\frac{1}{T}\int_0^T \frac{\partial H}{\partial z}(q_+(t), p_+(t), z_0)dt \ -\frac{1}{T'}\int_0^{T'} \frac{\partial H}{\partial z}(q_-(t), p_-(t), z_0)dt\right\} = 0. \quad (6)$$

If the average value is different along the two periodic orbits for a particular value of H and z_0, then it will also be true for values of H, z in an open domain, due to continuity with respect to H and z. So by (6) it must hold that

$$\frac{\partial \Phi_0}{\partial H}(H, z) = 0$$

in this open domain. Equation (5) also gives $\partial \Phi_0 / \partial z = 0$ on the periodic orbits and therefore Φ_0 is a trivial constant in contradiction to Lemma 1.

APPLICATION TO THE GENERALIZED QUARTIC OSCILLATOR

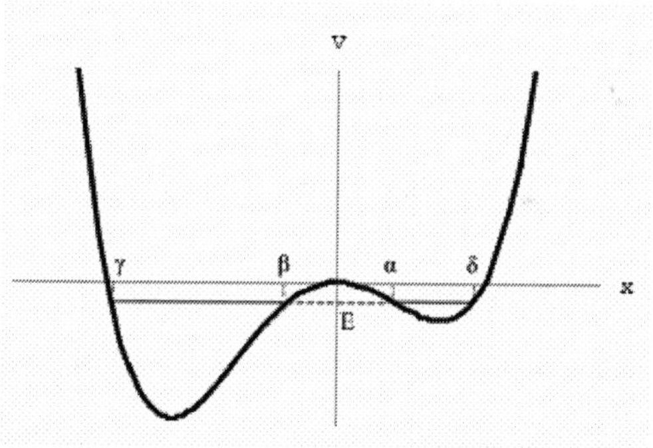

FIGURE 1. A generic potential of a forth order polynomial with two homoclinic loops.

We consider a "natural" Hamiltonian of the form

$$H = \frac{1}{2}p^2 + V(x, z)$$

with the potential

$$V(x, z) = \frac{1}{4}x^4 + \frac{1}{3}x^3 - \frac{\mathrm{a}(z)}{2}x^2. \quad (7)$$

For different values of z we acquire different forms of the potential V and consequently different forms of H through the function $\mathrm{a}(z)$. In order to find the periodic solutions of the frozen system (i.e. for $\mathrm{a}(z)=\mathrm{a}(z_0)=\sigma\tau\alpha\theta.$), we will apply the method in Davis [9] [1962, p. 209].

From the integral of energy we have

$$\dot{x}^2 = -\frac{1}{2}\left(x^4+\frac{4}{3}x^3-2\mathrm{a}(z)x^2-4E\right). \tag{8}$$

Let $\alpha,\beta,\gamma,\delta$ be the roots of the polynomial equation

$$x^4+\frac{4}{3}x^3-2\mathrm{a}(z)x^2-4E=0,$$

with $\delta>\alpha>0>\beta>\gamma$. In the open domain of z_0, a value of E can be selected such that the difference $(\delta-\alpha)$ can be made arbitrarily small

The solution is

$$x_+=\beta+\frac{\alpha-\beta}{1-\dfrac{\alpha-\delta}{\beta-\delta}y^2}\quad \text{for } x>0 \quad \text{and} \quad x_-=a+\frac{\alpha-\beta}{1-\dfrac{\beta-\gamma}{\alpha-\gamma}y^2}\quad \text{for } x<0$$

where

$$y=\mathrm{sn}(hM(t-t_0),k)$$

with

$$h^2=-\frac{1}{2},\qquad M^2=\frac{(\beta-\delta)(a-\gamma)}{4},\qquad k^2=\frac{(\beta-\gamma)(\alpha-\delta)}{(\alpha-\gamma)(\beta-\delta)}.$$

In order to prove nonintegrability of the complete system, we must show that, along a pair of solutions $x_\pm$ for the same value of E, the average value

$$\left\langle\frac{\partial H}{\partial z}\right\rangle_{z_0}=\frac{da}{dz}(z_0)\left\langle x_\pm^2\right\rangle$$

attains different values and, since both solution correspond to the same z_0, this must be shown only for the term $\left\langle x_\pm^2\right\rangle$.. We select a suitable value of E such that $\delta=a+\mu$, with μ arbitrarily small, we expand the above solutions with respect to μ and take the average values of $x_\pm^2$ along both solutions which correspond to the same value of E, i.e. to the same values of α,β,γ and δ. We finally obtain (see [10] for details)

$$\left\langle x_-^2\right\rangle-\left\langle x_+^2\right\rangle=-\frac{2}{3}\frac{(\alpha-\beta)}{\sqrt{1+\alpha_1^2}}+O(\mu)\neq 0$$

where $\alpha_1^2=(\beta-\gamma)/(\alpha-\gamma)$ and this proves nonintegrability of the system for $\varepsilon\neq 0$ in an open interval around zero.

CONCLUSIONS

We prove a nonintegrability criterion for systems of one degree of freedom with adiabatic time dependence, i.e. a system of the form $H = H(q,p,z)$ where $z = z_0 + \varepsilon t$. Namely, if for a particular constant value h of the Hamiltonian, the frozen system, for at least one z_0, possesses two different periodic orbits, such that the average value $\langle \partial H / \partial z \rangle_{z_0}$, evaluated along these orbits, attains different values, then the full system cannot possess an integral of motion for values of $\varepsilon \neq 0$ in an open interval around zero. The criterion has been applied in the generalized quartic oscillator with a slowly varying quadratic term in the potential and nonintegrability of the system has been shown. Note that if one additionally assumes that the zeroth order term of Φ depends on z only through the Hamiltonian, then the present criterion would become merely $\langle \partial H / \partial z \rangle_{z_0} \neq 0$ in only one periodic cycle of the frozen system. Therefore, if one tries to construct an adiabatic invariant ([11], [12]) by defining a succession of canonical transformations pushing the dependence of the new action variable on ε to higher orders, it must necessarily hold that $\langle \partial H / \partial z \rangle_{z_0} = 0$. In the general case, in order to construct an adiabatic invariant of a general form (i.e. that may depend also explicitly on the slow time) that does not undergoes a jump through the separatrix, one needs to ensure that the above average value is the same for every pair of periodic cycles of the frozen system with the same constant energy value. Finally, if the dependence of the Hamiltonian on z is periodic (or at least quasiperiodic) and the frozen system possesses a homoclinic loop one may use in order to prove nonintegrability the Melnikov-like results proven in [4]. Our criterion, however, is applicable for arbitrary dependence of H on the slow time z.

ACKNOWLEDGEMENTS

This work is supported by Research Programme "Pythagoras" Nr. 21879 of the Greek Ministry of Education and the European Union.

REFERENCES

1. C. J. Cotter and S. Reich, *BIT* **44**, 439-455 (2004).
2. C. S. Gardner, *Phys. Rev.* **115**, 791-794 (1959).
3. J. Henrard, "The adiabatic invariant in Classical Mechanics", in *Dynamics Reported*, New Series 2, edited by Jones, Kirchgraber & Walther, Berlin: Springer-Verlag, 1993, pp. 117-235.
4. T. J. Kaper and G. Kovacic, *J. Math. Phys.* **35**, 1202-1218 (1994).
5. M. Kruskal, *J. Math Phys.* **3**, 806–828 (1962).
6. E. Meletlidou and S. Ichtiaroglou, *Physica* **D71**, 261-268 (1994).
7. E. Meletlidou, and S. Ichtiaroglou, *J. Phys.* **A27**, 3919-3926 (1994).
8. H. Poincarè, *Les Mèthodes Nouvelles de la Mèchanique Cèleste*, Vol I, Paris: Gauthier-Villars, 1892. English translation: , *New Methods in Celestial Mechanics,* edited by D. L. Goroff, American Institute of Physics, 1993.
9. H. T. Davis, *Introduction to Nonlinear Differential and Integral Equations*, New York: Dover, 1962.
10. D. Voyatzi and E. Meletlidou, *Int. J. of Bifurcation and Chaos*, 2006 (to appear).
11. A. I. Neishtadt, *J. Appl. Math. Mech.* **45**, 58–63 (1981).
12. A. I. Neishtadt, *J. Appl. Math. Mech.* 48, 133-139 (1984).

Classical Relativistic Tests in Zatrikean Pregeometry

T.G. Dallas* and V.S. Geroyannis†

*Department of History, Archaeology and Social Anthropology, University of Thessaly, GR-38221 Volos, Greece

†Department of Physics, University of Patras, GR-26500 Patras, Greece

Abstract. Three well-known problems (perihelion advance, photon deflection, radar echo delay) are studied in the framework of Zatrikean Pregeometry. The results differ less than 0.1% from those of General Relativity.

Keywords: General Relativity, Zatrikean Pregeometry
PACS: 04.20, 95.30.Sf, 98.80.Jk

PERIHELION ADVANCE

In classical mechanics, the differential equation governing the orbit of a planet around the Sun has the form

$$\frac{\mathrm{d}^2 u_\mathrm{T}}{\mathrm{d}\theta^2} + u_\mathrm{T} = \frac{GM}{h^2}, \tag{1}$$

where $u_\mathrm{T} = 1/r_\mathrm{T}$, r_T is the radius of the planet's orbit, h is the planet's orbital angular momentum, and M is the mass of the Sun.

Applying the fundamental pregeometric transformation ([1], §5, Eq. [5.16])

$$r = r_\mathrm{T}\Phi, \qquad \text{where } \Phi = 1 + \Psi_0 \gamma \frac{GM}{c^2 r} \tag{2}$$

to Eq. (1), yields the differential equation

$$\frac{\mathrm{d}^2 u}{\mathrm{d}\theta^2}\left(1 + 2\Psi_0\gamma\frac{GM}{c^2}u\right) + 2\gamma\frac{GM}{c^2}\frac{\mathrm{d}u}{\mathrm{d}\theta}\left(2\frac{\mathrm{d}\Psi_0}{\mathrm{d}\theta}u + \Psi_0\frac{\mathrm{d}u}{\mathrm{d}\theta}\right) + u\left(1 + \Psi_0\gamma\frac{GM}{c^2}u + \gamma\frac{GM}{c^2}\frac{\mathrm{d}^2\Psi_0}{\mathrm{d}\theta^2}u\right) = \frac{GM}{h^2}. \tag{3}$$

A solution of Eq. (1) is

$$u_\mathrm{T} = \frac{GM}{h^2}(1 + e\cos\theta). \tag{4}$$

Substituting the above in Eq. (3), keeping terms up to the second order, eliminating terms containing the small quantity $\mathrm{d}\Psi_0/\mathrm{d}\theta$, and using the formula $\Psi_0 + \mathrm{d}^2\Psi_0/\mathrm{d}\theta^2 = -1$ ([1], §7), we get

$$\frac{\mathrm{d}^2 u}{\mathrm{d}\theta^2} + u \cong 2\left(\frac{GM}{h^2}\right)^2 \Psi_0\gamma\frac{GM}{c^2}\,u\,e\cos\theta, \tag{5}$$

CP848, *Recent Advances in Astronomy and Astrophysics,*
7th *International Conference of the Hellenic Astronomical Society,* edited by N. H. Solomos

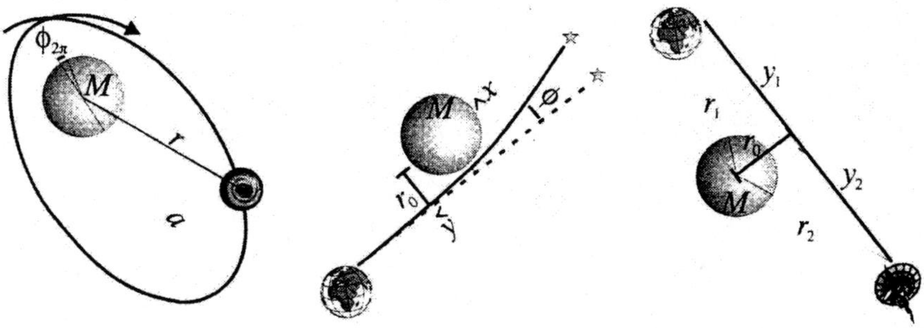

FIGURE 1. The three classical problems: perihelion advance, photon deflection, radar echo delay

which has the solution

$$u \cong \frac{GM}{c^2}\left(1 + e\cos\left(\theta - (1+\Psi_0)\,\gamma\left(\frac{GM}{ch}\right)^2\theta\right)\right). \tag{6}$$

After one revolution the perihelion advances by

$$\phi_{2\pi} \cong 2\pi\,(1+\Psi_0)\,\gamma\,\frac{GM}{c^2\alpha\,(1-e^2)}, \tag{7}$$

where we have used the well-known relation $h = \sqrt{GM\alpha(1-e^2)}$ (α is the semimajor axis). On the other hand, the relativistic perihelion advance is

$$\phi_{2\pi} \cong 6\pi\,\frac{GM}{c^2\alpha\,(1-e^2)}. \tag{8}$$

Taking into account the mean value $(1+\overline{\Psi}_0)\,\gamma \cong 2.99738$, we find that the difference between the estimates of Eqs. (7) and (8) is -0.087%.

Using the values $\overline{\Phi}_0$ for the planets of the solar system in Eq. (7), we can calculate improved values for the perihelion shift of the planets; the results are presented in Table 1. The quantity $1 - \Phi\gamma/3$ shows the difference between such improved values and the corresponding relativistic results of Eq. (8).

TABLE 1. Perihelion advance for the planets (in arcsecons per century)

Planet Name	Mean Value of $1 - \Phi\gamma/3$	Perihelion Advance
Mercury	$+2.339 \cdot 10^{-4}$	42.969
Venus	$-8.589 \cdot 10^{-3}$	8.696
Earth	$-1.580 \cdot 10^{-2}$	3.898
Mars	$-1.200 \cdot 10^{-2}$	1.367
Jupiter	$-1.314 \cdot 10^{-2}$	0.063

In case of a binary pulsar, we face the reverse proble, namely, calculating the total mass of the system from the periastron advance. To estimate the effect of the missing terms in Eq. (7), we solve numerically Eq. (3). Our results are given in Table 2. We remark that the numerical solution results in an increased periastron shift.

TABLE 2. Total mass of the binary pulsars (in solar units)

Pulsar Name	Numerical Result based on Eq. (3)	Approximation based on Eq. (7)	Relativity based on Eq. (8)
PSR 1534+12	2.680	2.696	2.694
PSR 1913+16	2.829	2.836	2.834
PSR 2127+11C	2.709	2.714	2.712

PHOTON DEFLECTION

The photon deflection is governed by the same newtonian equation as the perihelion problem, but, since the photon is massless, Eq. (1) reduces to

$$\frac{d^2u_{\mathrm{T}}}{d\theta^2}+u_{\mathrm{T}}=0. \tag{9}$$

Applying Eq. (2) to Eq. (9), we find

$$\frac{\mathrm{d}^2u}{\mathrm{d}\theta^2}\left(1+2\Psi_0\gamma\frac{GM}{c^2}u\right) + 2\gamma\frac{GM}{c^2}\frac{\mathrm{d}u}{\mathrm{d}\theta}\left(2\frac{\mathrm{d}\Psi_0}{\mathrm{d}\theta}u+\Psi_0\frac{\mathrm{d}u}{\mathrm{d}\theta}\right)+ + u\left(1+\Psi_0\gamma\frac{GM}{c^2}u+\gamma\frac{GM}{c^2}\frac{\mathrm{d}^2\Psi_0}{\mathrm{d}\theta^2}u\right)=0. \tag{10}$$

A solution of Eq. (9) is

$$u_{\mathrm{T}}=u_{\mathrm{T}0}\cos\theta. \tag{11}$$

Using the same technique as in the previous section, we get the equation

$$\frac{\mathrm{d}^2u}{\mathrm{d}\theta^2}+u\cong(1+\Psi_0)\,\gamma\frac{GM}{c^2}\left(u_0\cos\theta\right)^2, \tag{12}$$

which has the solution

$$u=\frac{1}{3}\left(1+\Psi_0\right)\gamma\frac{GM}{c^2}u_0^2\left(\cos^2\theta+2\sin^2\theta\right)+u_0\cos\theta. \tag{13}$$

Multiplying the above with $r_0 r$ and transforming to cartesian coordinates ($x=r\cos\theta$, $y=r\sin\theta$), we find

$$x\cong r_0-\frac{1}{3}(1+\Psi_0)\,\gamma\frac{GM}{c^2r_0}\frac{x^2+2y^2}{\sqrt{x^2+y^2}}. \tag{14}$$

The above for $y>>x$ gives

$$x\cong r_0-\frac{1}{3}(1+\Psi_0)\,\gamma\frac{GM}{c^2r_0}2y \tag{15}$$

and the angle between the asymptotes is

$$\phi \cong \frac{(1+\Psi_0)\gamma}{3} \frac{4GM}{c^2 r_0}. \tag{16}$$

The corresponding relativistic angle between the asymptotes is

$$\phi \cong \frac{4GM}{c^2 r_0}. \tag{17}$$

The difference between the values of Eqs. (16) and (17) is -0.087%.

To estimate the effect of the missing term in the approximate solution, we solve numerically Eq. (10). The results are listed in Table 3. We remark that the numerical solution results in a slightly stronger deflection.

TABLE 3. Photon deflection due to Sun's mass (in arcsec)

Approach of Photon Ray	Numerical Result based on Eq. (10)	Approximation based on Eq. (16)	Relativity based on Eq. (17)
Grazing Sun's Limb	1.7509	1.7488	1.7504
Perpendicular to Sun-Earth Line	0.00409	0.00406	0.0040

RADAR ECHO DELAY

To calculate the radar echo delay, we have to examine the Sun-photon system from the photon's viewpoint. The miniscule turn $\delta\phi$ that the photon feels at every point of its orbit is, according to Eq. (16),

$$\delta\phi = \frac{2(1+\Psi_0)\gamma GM}{3c^2 r}. \tag{18}$$

The above relation shows that the photon should feel the Sun's attraction perpendicular to its path; this is in agreement with the principle of the least pregeometric action ([1], §6). Integrating Eq. (18) from $-\infty$ to $+\infty$ via r_0, we return to Eq. (16). Therefore, the additional road during this turn is approximately equal to $\delta\phi$.

The photon moves on a curve $d\vec{s} = d\vec{x} + d\vec{y}$. Since $y >> x$, we get

$$c\,\mathrm{d}t = \mathrm{d}s = \left(1 + \frac{2(1+\Psi_0)\gamma GM}{3c^2\sqrt{y^2+r_0^2}}\right)\mathrm{d}y. \tag{19}$$

Let r_1 and r_2 be the distances of the transmitter and receiver from the Sun, respectively. Integrating the above from r_1 to r_2, we find

$$t \cong \frac{y_1+y_2}{c} + \frac{2(1+\Psi_0)\gamma GM}{3c^3} \cdot \ln\left(\frac{\left(y_1+\sqrt{y_1^2+r_0^2}\right)\left(y_2+\sqrt{y_2^2+r_0^2}\right)}{r_0^2}\right). \tag{20}$$

Since the first part of the right term is the classical time need for the photon to go from r_1 to r_2 via r_0, the time delay is

$$\Delta t \cong \frac{2(1+\Psi_0)\gamma GM}{3c^3} \cdot \ln\left(\frac{\left(r_1+\sqrt{r_1^2-r_0^2}\right)\left(r_2+\sqrt{r_2^2-r_0^2}\right)}{r_0^2}\right). \tag{21}$$

For a round trip the time delay is twice that given by Eq. (21).

Compare the above with the relativistic formula

$$\Delta t \cong \frac{2GM}{c^3} \cdot \ln\left(\frac{\left(r_1+\sqrt{r_1^2-r_0^2}\right)\left(r_2+\sqrt{r_2^2-r_0^2}\right)}{r_0^2}\right). \tag{22}$$

The difference between the estimates of Eqs. (21) and (22) is -0.087%. The results are listed in Table 6.

TABLE 4. Radar echo delay due to Sun's mass (in μsec)

Return trip from Earth to	Approximation based on Eq. (21)	Relativity based on Eq. (22)
Mercury	219.63	219.82
Venus	231.91	232.11
Mars	246.54	246.75

REFERENCES

1. V. S. Geroyannis, *Astrophysics and Space Science*, **199**, 53–76 (1993).

A Three-Dimensional Extension to Zatrikean Pregeometry

V.S. Geroyannis* and T.G. Dallas†

*Department of Physics, University of Patras, GR-26500 Patras, Greece
†Department of History, Archaeology and Social Anthropology, University of Thessaly, GR-38221 Volos, Greece

Abstract. The zatrikean abacus was originally defined as a two-dimensional chessboard-like lattice with square geobits. In this paper we generalize the zatrikean abacus in three dimensions by using a three-dimensional lattice with cubic geobits. We then calculate the values of certain interesting pregeometric quantities for the solar system.

Keywords: Zatrikean Pregeometry
PACS: 04.20, 95.30.Sf, 98.80.Jk

INTRODUCTION

According to Zatrikean Pregeomentry [1], i.e., a chess-like pregeometry, space is represented by a chessboard-like lattice, $A_{\rm N}$, called abacus and consisting of a sufficiently large number of square plaquettes called geobits. The particles move on the abacus from one geobit to a next following certain rules that resemble the game of chess. Such motions are represented by chain-like subabaces, $R_{\rm i} \subset A_{\rm N}$, called alyssoids and determined by certain alternative sets of rules called premetrics. The overall effect is a kind of transformation to classical mechanics that causes mass dependent dilations of distances. Detailed calculations of the number of alyssoids and the dilation factors ([1], §5, Eq. [5.14]) in the so-called Q and R premetrics ([1], §§3–4) have been performed in [2] (§§2–3).

Since in this paper we deal with Q premetrics only, we drop for convenience the Q subscript from all quantities; e.g., distances are denoted by r instead of $r_{\rm Q}$. Subscripts are used to denote a particular premetric if necessary, especially the so-called T premetric ([1], §§2,5). For instance, the fundamental pregeometric transformation ([1], §5, Eq. [5.16]) is

$$r = r_{\rm T}\Phi, \quad \text{where } \Phi = 1 + \Psi_0\,\gamma\,\frac{GM}{c^2 r}. \tag{1}$$

The quantity Ψ_0 in the above equation can represent either $\Psi^*_{\rm Q0}(\theta)$, the angle dependent pure pregeometric component ([1], §5, Eq. [5.18]) or its mean value over 2π, $\widetilde{\Psi}^*_{\rm Q0}$ ([2], §3). This is a minor point and it should not concern us, because the problems currently studied are examined over large distances and/or over a complete period of rotation. Generally speaking, we use the pure pregeometric component for our numerical results and its mean value for our analytical solutions.

CP848, *Recent Advances in Astronomy and Astrophysics,*
7th *International Conference of the Hellenic Astronomical Society,* edited by N. H. Solomos

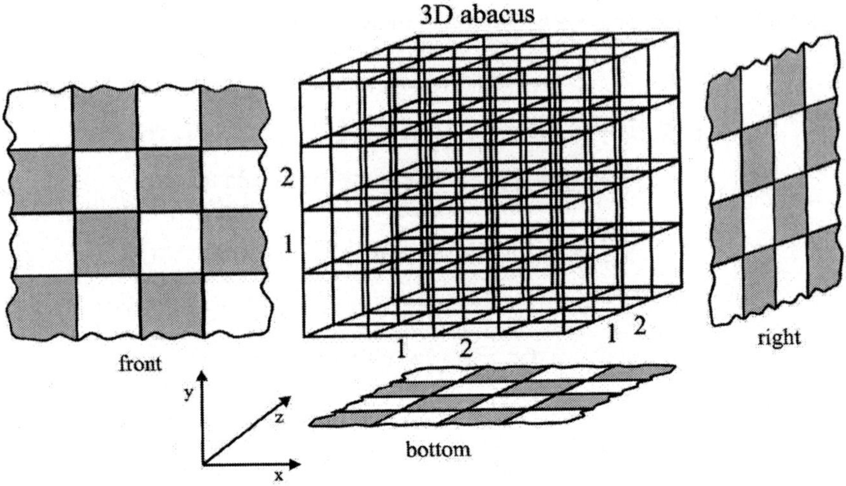

FIGURE 1. Schematic of a three-dimensional abacus

THREE-DIMENSIONAL EXTENSION

In [1] the abacus is considered as a two-dimensional chessboard-like lattice. In this case the geobits are squares of side α and diagonal β. An immediate three-dimensional extension is to consider a cubic geobit of edge α. Then the abacus becomes a three-dimensional structure made up from such cubic geobits (see Fig. 1).

A three-dimensional geobit has a total of 26 neighbours instead of 8 neighbours in the case of a two-dimensional geobit. The distance between two neighbouring geobits depends on the contact: side (surface), edge (straight line) or corner (point). Notice, in particular, that the mean distance δ between one geobit and its neighbour has increased from the value 1.20711α to the value 1.41642α (see Table 1).

TABLE 1. Distances between neighbouring geobits in a three-dimensional abacus

Contact	Number of Neighbours	Distance Between Geobits
side (surface)	6	α
edge (straight line)	12	$\alpha\sqrt{2}$
corner (point)	8	$\alpha\sqrt{3}$
total and mean	26	$(6+12\sqrt{2}+8\sqrt{3})\alpha/26$

The pregeometric dilation constant γ ([1], §7, Eq. [7.12]) was originally set equal to $\gamma = (\beta/\alpha)^3 \cong 2.82843$, resulting as a characteristic quantity of the microstructure of the two-dimensional abacus ([1], §9, Eq. [9.8]). In three dimensions, a fundamental quantity is the mean distance between adjoining geobits, which, as said above, is equal to $\delta \cong 1.41642\alpha$. In this case γ can be defined as

$$\gamma \equiv \left(\frac{\delta}{\alpha}\right)^3 \cong 2.84169. \tag{2}$$

Next we turn our attention to the three-dimensional calculation of Φ. Let x, y, z be the predistances of the alyssoids joining the geobits g_{000} and g_{ijk} along the row-axis, column-axis, and depth-axis, respectively, of the abacus A_{N}. Let x_{k}, y_{k}, z_{k} be the corresponding predistances of the part of the geodesic $G(g_{000}, g_{\mathrm{ijk}})$ joining g_{000} with the ν-th geobit of the alyssoid, counting from its starting geobit. Then the distance is

$$r = \sum_{\nu=1}^{d} \alpha\sqrt{(x_\nu - x_{\nu-1})^2 + (y_\nu - y_{\nu-1})^2 + (z_\nu - z_{\nu-1})^2}, \tag{3}$$

where d is the total number of geobits in the alyssoid. In T premetrics, this equation reduces to

$$r_{\mathrm{T}} = \alpha\sqrt{(x^2 + y^2 + z^2)}. \tag{4}$$

In Q premetrics, at least one of $(x_\nu - x_{\nu-1})$, $(y_\nu - y_{\nu-1})$, $(z_\nu - z_{\nu-1})$ equals unity. We have $\min(x,y,z)$ steps on the shortest direction, $\max(x,y,z)$ steps on the longest direction and $x+y+z-\max(x,y,z)-\min(x,y,z)$ steps on the intermediate direction.

In polar co-ordinates, we can write

$$x = \frac{r_{\mathrm{T}}}{\alpha}\sin\theta\cos\phi, \qquad y = \frac{r_{\mathrm{T}}}{\alpha}\sin\theta\sin\phi, \qquad z = \frac{r_{\mathrm{T}}}{\alpha}\cos\theta. \tag{5}$$

We then define the quantities

$$X = \min(|\sin\theta\cos\phi|, |\sin\theta\cos\theta|, |\cos\theta|), \tag{6}$$

$$Y = \max(|\sin\theta\cos\phi|, |\sin\theta\cos\theta|, |\cos\theta|), \tag{7}$$

$$Z = |\sin\theta\cos\phi| + |\sin\theta\cos\theta| + |\cos\theta| - (X+Y). \tag{8}$$

Substituting these quantities in the equation $\Phi = r/r_{\mathrm{T}}$, we find

$$\Phi(\theta,\phi) = X\sqrt{3} + (Z-X)\sqrt{2} + (Y-X). \tag{9}$$

Integrating the above, we calculate the mean value $\overline{\Phi}$ over a whole sphere. It is $\overline{\Phi} \cong 1.05478$, the same as that calculated in two dimensions ([2], §3). Finally, the interesting quantity $\overline{\Phi}\gamma$ becomes equal to $\overline{\Phi}\gamma \cong 2.99736$.

SOLAR SYSTEM CALCULATIONS

We shall now apply the three-dimensional calculation of Φ to the solar system. Based on the ideas of pregeometry and effective rotation ([1], §§6–7), we take into account the fact that this system is not isotropic. In particular, it possesses a main and significant

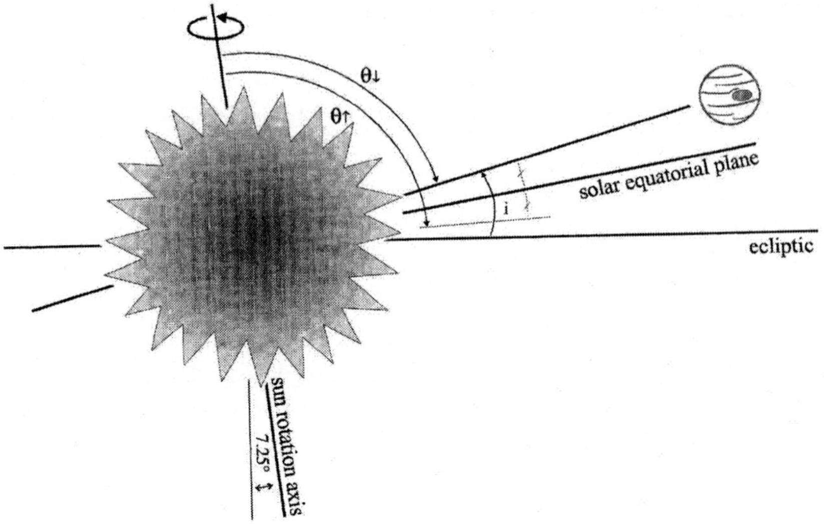

FIGURE 2. Definition of angles $\theta\downarrow$ and $\theta\uparrow$ used to calculate the mean value $\overline{\Phi}$ for a planet

axis, namely, the axis of rotation of the Sun. The planets rotate, in turn, on a disk almost perpendicular to this axis. Therefore, the planets do not scan all the values of θ, but restrict themselves within a narrow range of angles (see Fig. 2). The mean value $\overline{\Phi}$ for each planet (see Table 2) is calculated by the integral

$$\overline{\Phi} = \frac{1}{2\pi\,(\theta\uparrow - \theta\downarrow)} \int_0^{2\pi} \int_{\theta\downarrow}^{\theta\uparrow} \Phi(\theta,\phi)\mathrm{d}\theta\mathrm{d}\phi. \tag{10}$$

TABLE 2. Values of Φ for the planets in the solar system

Planet Name	Inclination of Planetary Orbit to the Solar Equator (°)	Mean Value $\overline{\Phi}$
Mercury	0.246	1.05546
Venus	3.856	1.06477
Earth	7.250	1.07238
Mars	5.400	1.06837
Jupiter	5.942	1.06958
Saturn	4.762	1.06691
Uranus	6.476	1.07075
Neptune	5.476	1.06858
Pluto	9.900	1.07755

REFERENCES

1. V. S. Geroyannis, *Astrophysics and Space Science*, **199**, 53–76 (1993).
2. V. S. Geroyannis, and T. G. Dallas,*Astrophysics and Space Science*, **200**, 117–127 (1993).

Stellar Dynamics in Scalar-Tensor Gravity

K. D. Kokkotas* and H. Sotani*

*Department of Physics, Aristotle University of Thessaloniki, Thessaloniki 54124, Greece

Abstract. We study perturbations of relativistic stars in scalar-tensor theory of gravity and examine the effects of the scalar field on the corresponding oscillation spectrum. Scalar waves which might be produced from such oscillations can be a unique probe for the theory, but their detectability is questionable if the radiated energy is small. However we show that there is no need for a direct observation of scalar waves: the shift in the gravitational wave spectrum could unambiguously signal the presence of a scalar field.

Keywords: Neutron Stars, Gravitational Waves, Scalar tensor Gravity
PACS: 04.40Nr,04.30Db,04.50.+h,04.80.Cc

INTRODUCTION

Scalar-tensor theories of gravity are an alternative or generalization of Einstein's theory of gravity, where in addition to the tensor field a scalar field is present. The theory has been proposed in its earlier form about half century ago [1, 2, 3], and it is a viable theory of gravity for a specific range of the coupling strength of the scalar field to gravity [4, 5]. Actually, the existence of scalar fields is crucial (e.g. in inflationary and quintessence scenarios) to explain the accelerated expansion phases of the universe. Experimentally, the existence of a scalar field has not yet been probed, but a number of experiments in the weak field limit of general relativity set severe limits on the existence and coupling strengths of scalar fields [6].

A basic assumption is that the scalar and gravitational fields φ and $g_{\mu\nu}$ are coupled to matter via an "effective metric" $\tilde{g}_{\mu\nu} = A^2(\varphi)g_{\mu\nu}$. The Fierz-Jordan-Brans-Dicke [1, 2, 3] theory assumes that the "coupling function" has the form $A(\varphi) = \alpha_0\varphi$, i.e., it is characterized by a unique free parameter $\alpha_0^2 = (2\omega_{\rm BD}+3)^{-1}$, and all its predictions differ from those of general relativity by quantities of order α_0^2 [7]. Solar system experiments set strict limits in the value of the Brans-Dicke parameter $\omega_{\rm BD}$, i.e., $\omega_{\rm BD} > 40000$, which suggests a very small $\alpha_0^2 < 10^{-5}$ (see [6]).

In the early 1990s, based on a simplified version of scalar tensor theory where $A(\varphi) = \alpha_0\varphi + \beta\varphi^2/2$, Damour and Esposito-Farese [7, 8] found that for certain values of the coupling parameter β the stellar models develop some strong field effects which induce significant deviations from general relativity. This sudden deviation from general relativity for specific values of the coupling constants has been named "spontaneous scalarization". Harada [9] studied in more detail models of non-rotating neutron stars in the framework of the scalar-tensor theory and he reported that "spontaneous scalarization" is possible for $\beta < -4.35$.

CP848, *Recent Advances in Astronomy and Astrophysics,*
7th *International Conference of the Hellenic Astronomical Society,* edited by N. H. Solomos

Recently, we have examined the possibility to obtain the information for the presence of the scalar field via gravitational wave observations of oscillating neutron stars [10, 11]. The effect of the scalar field on the fluid perturbations, i.e., f and p modes has been examined in [10] (in the Cowling approximation, i.e., no spacetime perturbations). In [11] the effect of the scalar field on the spacetime perturbations (w modes) has been studied.

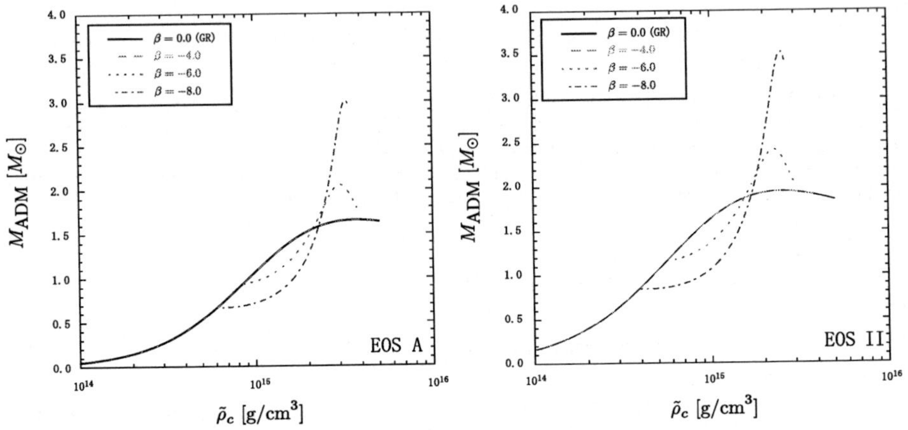

FIGURE 1. Stellar models with β=0, −4, −6 and −8 are shown. The left column corresponds to models for the EOS A while in the right to models for EOS II. The effect of the scalar field is apparent for $\rho_c \geq 5 \times 10^{14}$gr/cm^3. In both panels, the general relativistic stellar models are shown by a solid line (GR)

STELLAR MODELS IN SCALAR-TENSOR THEORY

In this section we study neutron star models in scalar-tensor theory of gravity with one scalar field. This is a natural extensions of Einstein's theory, in which gravity is mediated not only by a second rank tensor (the metric tensor $g_{\mu\nu}$) but also by a massless long-range scalar field φ. The action is given by [4]

$$S = \frac{1}{16\pi G_*} \int \sqrt{-g_*} \left(R_* - 2g_*^{\mu\nu} \varphi_{,\mu}\varphi_{,\nu}\right) d^4x + S_m \left[\Psi_m, A^2(\varphi) g_{*\mu\nu}\right], \qquad (1)$$

where all quantities with asterisks are related to the "Einstein metric" $g_{*\mu\nu}$, then R_* is the curvature scalar for this metric and G_* is the bare gravitational coupling constant. Ψ_m represents collectively all matter fields, and S_m denotes the action of the matter represented by Ψ_m, which is coupled to the "Jordan-Fierz metric tensor" $\tilde{g}_{\mu\nu}$. The field equations are usually written in terms of the "Einstein metric", but all non-gravitational experiments measure the "Jordan-Fierz" or "physical metric". The "Jordan-Fierz metric" is related to the "Einstein metric" via the conformal transformation, $\tilde{g}_{\mu\nu} = A^2(\varphi) g_{*\mu\nu}$. Hereafter, we denote by a tilde quantities in the "physical frame" and by an asterisk those in the "Einstein frame". From the variation of the action S we get the field equations in the Einstein frame

$$G_{*\mu\nu} = 8\pi G_* T_{*\mu\nu} + T^{(\varphi)}_{*\mu\nu} \quad \text{and} \quad \nabla^\mu_* \nabla_{*\mu}\varphi = -4\pi G_* \alpha(\varphi) T_*, \tag{2}$$

where $T^{(\varphi)}_{*\mu\nu}$ is the energy-momentum of the massless scalar field and $T^{\mu\nu}_*$ is the energy-momentum tensor in the Einstein frame. It is apparent that $\alpha(\varphi)$ is the only field-dependent function which couples the scalar field with matter, for $\alpha(\varphi) = 0$ the theory reduces to general relativity. Finally, the law of energy-momentum conservation $\tilde{\nabla}_\nu \tilde{T}_\mu{}^\nu = 0$ is transformed into

$$\nabla_{*\nu} T_{*\mu}{}^\nu = \alpha(\varphi) T_* \nabla_{*\mu}\varphi, \tag{3}$$

In this paper, we adopt the same form of conformal factor $A(\varphi)$ as in Damour and Esposito-Farese [7], which is

$$A(\varphi) = e^{\frac{1}{2}\beta\varphi^2}, \tag{4}$$

i.e., $\alpha(\varphi) = \beta\varphi$ where β is a real number. In the case $\beta = 0$ this scalar-tensor theory reduces to general relativity.

Here we model the neutron stars as self-gravitating perfect fluid of cold degenerate matter in equilibrium and the stellar matter is assumed to be a perfect fluid.

BASIC PERTURBATION EQUATIONS

In this section we present the equations describing perturbations of the spacetime, scalar field, and fluid in a spherically symmetric background. The equations we provide describe the non-radial oscillations of spherically symmetric neutron stars in scalar-tensor theories. We assume, in the physical frame, using the Regge-Wheeler gauge [16], the following form of the perturbed metric tensor

$$\tilde{h}_{\mu\nu} = \tilde{h}^{(-)}_{\mu\nu} + \tilde{h}^{(+)}_{\mu\nu}, \tag{5}$$

where $\tilde{h}^{(-)}_{\mu\nu}$ denotes the *axial* (or *odd parity*) part of metric perturbations and $\tilde{h}^{(+)}_{\mu\nu}$ denotes the *polar* (or *even parity*) part of metric perturbations.

Following the previous definitions the perturbed metric tensor $h_{*\mu\nu}$ in the Einstein frame has the form:

$$h_{*\mu\nu} = \frac{1}{A^2}\tilde{h}_{\mu\nu} - \frac{2}{A} g_{*\mu\nu}\delta A \tag{6}$$

where $\delta A \equiv A\beta\varphi\delta\varphi$ is the perturbation of the conformal factor A and $\delta\varphi$ is the perturbation of the scalar field, where $\delta\varphi$ is a function of t and r only. The above definition of $h_{*\mu\nu}$ will be used to derive the perturbation equations: in other words, we will work out the perturbations in the Einstein frame and we will transform back to the physical frame whenever we need it.

The perturbation equations will be derived by taking the variation of equations (2)

$$\delta G_{*\mu\nu} = 8\pi G_* \delta T_{*\mu\nu} + \delta T^{(\varphi)}_{*\mu\nu} \quad \text{and} \quad \delta\left(\nabla^\mu_* \nabla_{*\mu}\varphi\right) = -4\pi G_* \delta\left[\alpha(\varphi) T_*\right]. \tag{7}$$

The various components of $\delta T^{(\varphi)}_{*\mu\nu}$ are expressed as linear combinations of $\delta\varphi$ and $\tilde{h}_{\mu\nu}$.

RESULTS AND DISCUSSION

The spectrum of an oscillating neutron star is directly related to its parameters, mass, radius and EOS [13]. As we have seen in Section II the presence of the scalar field on the background star is influencing both the mass and the radius, while in the way that it enters in the equilibrium equations it has a role of an extra pressure term, i.e., it seems to alter the actual EOS. As before we will restrict our study only two equations of state, i.e., EOS A and EOS II.

The oscillation frequencies of the various types of modes have been derived using two complementary techniques. One way is to assume a harmonic time dependence of the perturbation functions, e.g., $W(t,r) = e^{i\omega t}W(r)$ and transform the problem into a boundary value one. Alternatively, we can just use time evolution of the perturbation equations and by Fourier transformations of the derived time-series one can get the characteristic oscillation frequencies [10, 11].

First, we will examine the effect of the scalar factor β on the frequency and especially whether the "spontaneous scalarization" can be traced in the spectrum. A discontinuous change in a system, as one varies its parameters, signals a catastrophic behavior and the so called "spontaneous scalarization"[7] is related to it. In order to study this effect we constructed a sequence of stellar models with $M_{ADM} = 1.4M_{\odot}$ by varying β, where M_{ADM} is ADM mass.

The idea of a gravitational wave asteroseismology [13, 15] was based on empirical relations that can be drawn for the relation of the stellar parameters to the eigenfrequencies of an oscillating neutron star. These empirical relations were derived by taking into account data for a dozen or more EOS, and it was shown that through these relations one can extract the stellar parameters by analyzing the gravitational wave signal of an oscillating neutron star. The easiest relation to be understood on intuitive physical grounds is the one between the fundamental oscillation mode the f mode and the average density. This relation emerges naturally by combining the time that a perturbation needs to propagate across the star and the sound speed, this boils down to a linear relation between the period of oscillation and the average density or the star or better $\omega^2 \sim M/R^3$. According to [13] the empirical relation between the f mode frequency and the average density of typical neutron stars is

$$f_{\text{f-mode}}(\text{kHz}) \approx 0.78 + 1.63 \left(\frac{M}{1.4M_{\odot}}\right)^{1/2} \left(\frac{R}{10\text{km}}\right)^{-3/2} . \tag{8}$$

Almost, all EOS follow this empirical relation (including the two EOS used in this article) for the case of general relativity. This observation suggests a unique way in estimating the average density of a star via its f mode frequency, and it can be a very good observational test for the neutron stars in scalar-tensor theories. In Figure 2 we draw the frequency as function of the averaged density. The introduction of a scalar field even in moderate central densities alters completely the behavior of the f mode for both EOS. The frequencies grow considerably faster as functions of the average density, for $\beta < -4.35$, and the three examples that we have chosen ($\beta = -6, -7$, and -8) show exactly this behavior. The change is quite dramatic even for typical neutron stars with average density. Depending on the value of the parameter β they become $30-50\%$

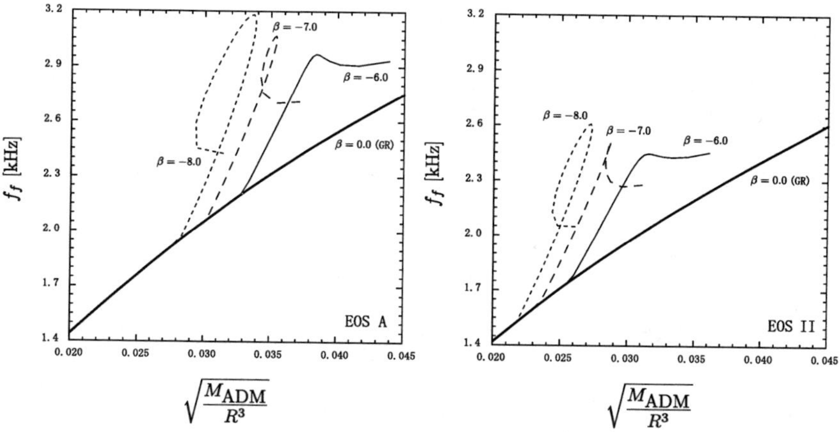

FIGURE 2. The frequency of the f mode as function of the averaged density $(M_{\mathrm{ADM}}/R^3)^{1/2}$ of the star (notice that $f_f = \omega_f/2\pi$). The thick solid line corresponds to the values of the mode for $\beta = 0.0$ (GR) while we also show the effect of the scalar field for three values of the scalar parameter β (-6, -7, and -8). The left panel corresponds to EOS A and the right panel to EOS II.

larger than those of a general relativistic neutron star. This can be an observable effect, since the detection of frequencies are higher than those expected from a typical neutron star, would signal in a unique way the presence of a scalar field.

The results of the two methods we described briefly earlier agree very well, providing a good consistency check on our calculations. In Figure 3 we present the eigenvalues for different stellar models derived for EOS A and II. Our results suggest that the presence of a spontaneous scalarization can be inferred from the w modes emitted by a newly born, oscillating neutron star. The modes that might be relevant for gravitational wave detectors are the lowest w modes [17]. The w_{II} modes [18] damp out roughly twice as fast as the w modes, but having lower frequencies they could also be relevant for detection by Earth-based interferometers. The higher-frequency w modes (w_2, w_3, w_4, ...)are difficult, if not impossible to detect. In the study of w modes as a tool for asteroseismology [12, 13, 14, 15] it has been suggested that a proper normalization for $Re(\omega)$ is to multiply it with the radius R of the star and to scale it as a function of the compactness M/R. This phenomenological argument has been recently verified analytically by Tsui and Leung [19]. Introducing $f = Re(\omega)/2\pi$, it is clear that Rf scales linearly as function of the compactness M/R. This applies both to w_{II} and w_{I} modes (and even to the higher overtones). The linear relations that can be derived from Figure 3 are

$$f_{\omega_{\mathrm{I}}-\mathrm{mode}}(\mathrm{kHz}) = \frac{1}{\bar{R}}\left(\alpha_{\mathrm{I}} - \beta_{\mathrm{I}}\frac{\bar{M}}{\bar{R}}\right) \quad \text{and} \quad f_{\omega_{\mathrm{II}}-\mathrm{mode}}(\mathrm{kHz}) = \frac{1}{\bar{R}}\left(\alpha_{\mathrm{II}} + \beta_{\mathrm{II}}\frac{\bar{M}}{\bar{R}}\right), \tag{9}$$

where the constants α_{I}, β_{I}, α_{II} and β_{II} are fitting factors given [11].

Another reason why it is harder to detect high-damped quasinormal modes such as the w_{II} modes for compact stars is that the effective amplitude scales as the square root

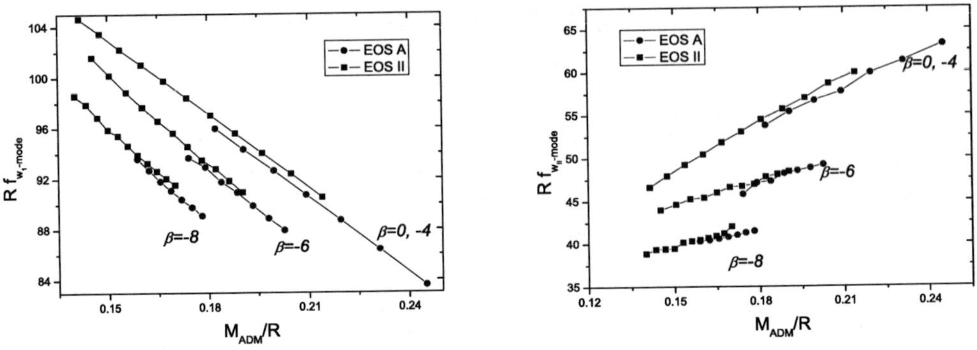

FIGURE 3. The w_1 modes (left panel) and the $w_{\rm II}$ modes (right panel) for EOS A and II and for values of $\beta = 0$, -4, -6, and -8.

of the number of oscillations [15]. Typically we can hardly observe more than $2-3$ cycles for highly damped quasinormal modes of black-holes and for the $w_{\rm II}$ modes of compact stars. Spontaneous scalarization might help in this direction. The reason is that the damping time of the $w_{\rm II}$ mode for stars with $\beta < -4.35$ is significantly longer than for typical stars in general relativity. The reason is that the presence of a scalar field increases the maximum mass of the stars and their compactness. Since the damping scales with compactness, the $w_{\rm II}$ modes live considerably longer. On the contrary, the damping times of the w_1 modes become shorter as the compactness increases [11].

For the study of the stellar oscillations in scalar theory theory of gravity we conclude that the presence of a scalar field affects the equilibrium model, and consequently the oscillation spectrum. The scalar field perturbations couple with the polar perturbations of the spacetime and fluid, but they don't couple with the axial perturbations. Since the spacetime modes of polar and axial perturbations have the same qualitative behavior, we have chosen to study the effect of the scalar field on the axial perturbations.

The results show that in the presence of spontaneous scalarization, a scalar field reduces the oscillation frequency of the w_1 modes by about 10% (i.e. by about 1kHz). The decrease in frequency for the $w_{\rm II}$ modes is about 25% the frequency of (i.e., about 1.5 kHz). The effect on the damping time is even more pronounced. The damping rate of $w_{\rm II}$ modes can decrease by as much as 30%, while it can increase by as much as 50% for the w_1 modes. Detectors operating at these high frequencies are under development. Through a detection of the w mode spectrum, they could provide a unique proof for the existence of scalar fields with $\beta < -4.35$.

Acknowledgement. This work is supported by the Pythagoras I program, the European Program ILIAS and the Marie-Curie grant FP6-021979.

REFERENCES

1. M.Fierz Helv. Phys. Acta **29**, 128 (1956).
2. P. Jordan Z.Phys. **157**, 112 (1959).
3. C. Brans and R. H. Dicke, Phys. Rev. **124**, 925 (1961).
4. T. Damour and G. Esposito-Farèse, Class. Quantum Grav. **9**, 2093 (1992).
5. C. M. Will, *Theory and Experiment in Gravitational Physics* (Cambridge University Press, Cambridge, England 1993).
6. G.Esposito-Farese gr-qc/0402007.
7. T. Damour and G. Esposito-Farèse, Phys. Rev. Lett. **70**, 2220 (1993).
8. T. Damour and G. Esposito-Farèse, Phys. Rev. D **54**, 1474 (1996).
9. T. Harada, Phys. Rev. D **57**, 4802 (1998).
10. H. Sotani and K. D. Kokkotas, Phys. Rev. D, **70**, 084026 (2004).
11. H. Sotani and K. D. Kokkotas, Phys. Rev. D, **71**, 124038 (2005).
12. N. Andersson, K.D. Kokkotas, Phys. Rev. Letters, **77**, 4134 (1996).
13. N. Andersson, K.D. Kokkotas, MNRAS, **299**, 1059 (1998).
14. H. Sotani, K. Kohri and T. Harada, Phys. Rev. D, **69**, 084008 (2004).
15. K. D. Kokkotas, T. A. Apostolatos and N. Andersson, MNRAS, **320** , 307 (2001).
16. T. Regge and J. A. Wheeler Phys. Rev. **108**, 1063 (1957)
17. K.D.Kokkotas and B.F.Schutz MNRAS **255**, 119 (1992).
18. M. Leins, H-P. Nollert and M. H. Soffel Phys.Rev. D, **48**, 3467 (1993)
19. L. K. Tsui and O. T. Leung MNRAS **357** 1029 (2005)

Predicting and analyzing topological correlations in the CMB

A. Niarchou* and A. H. Jaffe*

*Astrophysics Group, Blackett Laboratory, Imperial College London, Prince Consort Road, London SW7 2AZ, UK

Abstract. The Cosmic Microwave Background (CMB) power spectrum derived from the first year WMAP data demonstrates an intriguing lack of power at large scales that cannot be accounted for within the framework of the standard cosmological model. We explore the possibility that this anomaly could be explained by waiving the implicit assumption of a simply connected topology, rather than modifying the physics of the standard model. In particular, we assume that the Universe is slightly closed and its spatial section can be described by one of the simplest spherical multi-connected manifolds (the quaternionic, the octahedral, the truncated cube and the Poincare space). We discuss the implications for the CMB in each case.

Keywords: cosmology, cosmic microwave background, topology
PACS: 98.80.-k

INTRODUCTION

The CMB has proved to be an invaluable tool in our quest to unveil the mysteries of the cosmos. It dates back to the epoch of matter-radiation decoupling and the expansion of the Universe has caused its mean temperature to drop to a mere 2.726 K today. It exhibits minute temperature fluctuations whose properties depend on fundamental cosmological parameters and initial conditions. It is customary to expand the temperature fluctuation in a given direction $\hat{n}$ on the sky in a series of spherical harmonics:

$$\frac{\Delta T}{T}(\hat{n}) = \sum_{\ell=0}^{\infty} \sum_{m=-\ell}^{\ell} a_{\ell m} Y_{\ell m}(\hat{n}) \tag{1}$$

and use the coefficients $a_{\ell m}$ to infer the parameters of interest. To this end, we usually employ their correlation, $\langle a_{\ell m} a^*_{\ell' m'} \rangle$. In the case of a simply connected universe all off-diagonal terms are zero:

$$\langle a_{\ell m} a^*_{\ell' m'} \rangle = C_\ell \delta_{\ell \ell'} \delta_{m m'} \tag{2}$$

and we define the diagonal (rotationally invariant) part of the correlation matrix, C_ℓ, to be the temperature anisotropy power spectrum. Another derived quantity often used is the two-point correlation function, which can either be calculated directly from the CMB map, or constructed from the power spectrum:

$$C(\theta) \equiv \left\langle \frac{\Delta T}{T}(\hat{m}) \frac{\Delta T}{T}(\hat{n}) \right\rangle_{\hat{m}\cdot\hat{n}=\cos\theta} = \frac{1}{4\pi} \sum_\ell C_\ell P_\ell(\cos\theta) \tag{3}$$

CP848, *Recent Advances in Astronomy and Astrophysics*,
7th *International Conference of the Hellenic Astronomical Society*, edited by N. H. Solomos

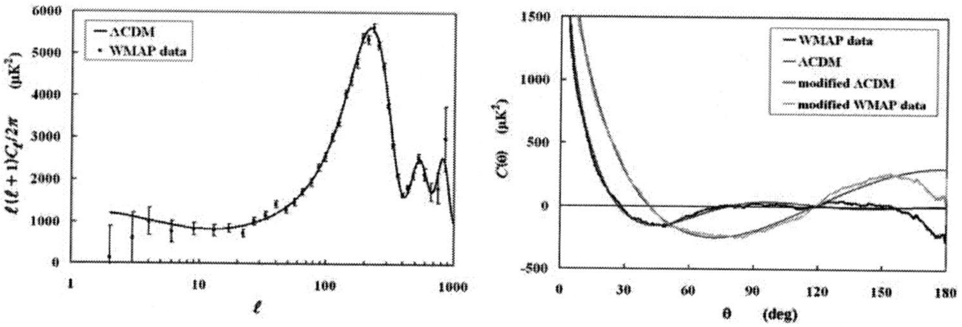

FIGURE 1. Left: The CMB power spectrum for the WMAP best-fit model. There is a deficit of power at large scales (specifically $\ell = 2,3$). **Right:** The temperature anisotropy correlation function from the WMAP data, as predicted from the WMAP best-fit ΛCDM, for the best-fit ΛCDM with C_2 and C_3 changed to the observed values and from the WMAP data with C_2 and C_3 matching the predictions of the best-fit ΛCDM. It is clear that the disagreement between theory and observation is due to the lack of power at $\ell = 2,3$.

where P_ℓ are the Legendre polynomials. Since the aforementioned anisotropies are very small (of the order 10^{-5}), extremely accurate measurements of the CMB temperature are of the essence. The WMAP satellite, launched in 2001, has been the most accurate CMB experiment to date in the range of large to medium scales. Its first year data seem to be overall consistent with the standard cosmological model (ΛCDM), a flat cold dark matter universe, dominated by dark energy and whose primordial perturbations are adiabatic and nearly scale-invariant [1]. However, there is a striking disagreement between the predictions of that model and the actual data at large scales, which is prominent in the $\ell \leq 3$ power spectrum and the correlation function (Fig. 1). There have been some attempts to remove this discrepancy by invoking new physics, but most of them result in contrived models with low statistical significance [2]. However, this phenomenon arises naturally in the context of non-trivial topologies, a possibility we will consider next.

SPHERICAL MULTI-CONNECTED MANIFOLDS

A slightly closed universe is marginally consistent with the WMAP data ($\Omega_0 = 1.02 \pm 0.02$). If it has a non-trivial topology, its spatial section could be described by one of the spherical multi-connected manifolds. Closed multi-connected manifolds are the quotient spaces $\mathbf{S}^3/\Gamma$, where $\mathbf{S}^3$ is the 3-sphere and Γ is the symmetry group particular to a given manifold. According to the way it acts on $\mathbf{S}^3$, we obtain different categories of manifolds. We are only interested in some of the simplest manifolds (single action manifolds) [3]:

- **the quaternionic space**, where $\Gamma = D_2^*$, the binary dihedral group of order 4. The fundamental domain is a 4-sided prism.
- **the octahedral space**, where $\Gamma = T^*$, the binary tetrahedral group of order 24. The fundamental domain is a regular octahedron.

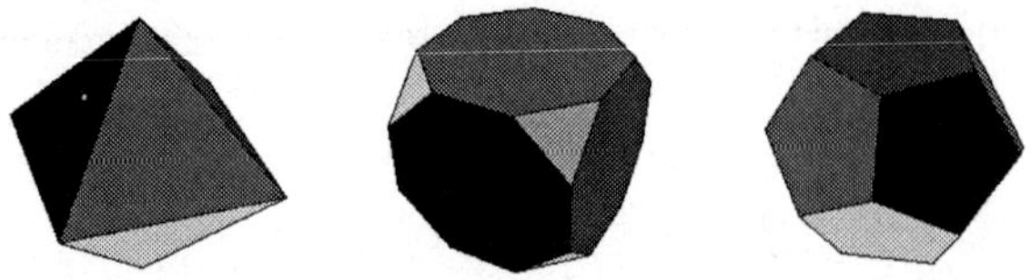

FIGURE 2. Fundamental domains of some spherical multi-connected manifolds. From left to right depicted are the regular octahedron, the truncated cube and the regular dodecahedron.

- **the truncated cube space**, where $\Gamma = O^*$, the binary octahedral group of order 48. The fundamental domain is a truncated cube.
- **the Poincaré space**, where $\Gamma = I^*$, the binary icosahedral group of order 120. The fundamental domain is a regular dodecahedron.

In each case, the number of fundamental domains that tile the 3-sphere is equal to the order of the group Γ. Figure 2 shows fundamental domains of some spherical multi-connected manifolds.

IMPRINTS OF A NON-TRIVIAL TOPOLOGY

The most obvious effect of a non-trivial topology would be a suppression of power at large scales, due to the finiteness of the fundamental domain: evidently, no wavelengths larger than the fundamental domain can exist, therefore the primordial power spectrum should be truncated at scales close to the size of the fundamental domain. The latter depends on the curvature radius, since the number of fundamental domains that tile the 3-sphere is constrained. Another crucial effect would be the breakdown of global isotropy, as all the manifolds we consider are orientable. This would manifest as the emergence of non-diagonal terms in the temperature anisotropy correlation matrix, appearing as non-Gaussianity. Table 1 summarizes the differences between simply connected and multi-connected spaces. We observe that the causal micro-physics governing the evolution of CMB anisotropies remain unchanged; what changes are the eigenmodes of the Laplacian. This affects CMB observables in two ways: firstly, in multi-connected spaces not all wavenumbers $\beta \geq 3$ correspond to Laplacian eigenmodes (in contrast to simply connected spaces) so that the β-contribution to each multipole is not the same; secondly, since the eigenmodes affect the evolution of fluctuations along the line of sight, we expect the latter to evolve differently in spaces of different topology. The Laplacian eigenmodes of a multi-connected space are the Γ-invariant eigenmodes of its universal covering space and they can be expressed as a series of the actual eigenmodes of the latter. Thus, in spherical multi-connected manifolds, they can be expressed on a basis of the eigenmodes of $\mathbf{S}^3$ (e.g. as a product of a hyperspherical Bessel function and a spherical harmonic). The expansion coefficients $\xi_{\beta\ell m}$ are particular to a given topology. The effect of that topology on the CMB is expressed through the presence of the $\xi_{\beta\ell m}$ in the correlation matrix.

Figures 3 and 4 show CMB sky realizations and power spectra for a closed universe with simply/multi-connected topology and with $\Omega_b = 0.057$, $\Omega_{CDM} = 0.275$, $\Omega_\Lambda =$

TABLE 1. Effects of a multi-connected topology. The eigenmodes of the Laplacian can be decomposed on a basis of the eigenmodes in the universal covering space, with the expansion coefficients encoding the topological properties of the given space (first row) [4, 5]. We see that the correlation matrix is no longer diagonal in the multi-connected case, however a rotationally invariant power spectrum can still be defined [5]. Note that the physics remain unchanged and all effects of a non-trivial topology are expressed through the coefficients $\xi_{\beta\ell m}$. Here $\mathscr{Y}$ and Ψ are the eigenmodes of the universal covering space and the multi-connected space respectively, $P(\beta)$ is the primordial power spectrum and $\Delta_{T\ell}$ are the transfer functions. s is the multiplicity of a given eigenmode, V is the volume of the fundamental domain and the e's are uncorrelated, appropriately normalized Gaussian random variables.

simply connected	**multi-connected**
$\mathscr{Y}_\beta(\chi,\theta,\phi)$	$\Psi^s_\beta(\chi,\theta,\phi)=\Sigma_{\ell=0}^{\beta-1}\Sigma_{m=-\ell}^{\ell}\xi^s_{\beta\ell m}\mathscr{Y}_\beta(\chi,\theta,\phi)$
$a_{\ell m}=i^\ell\int\beta^2\sqrt{P(\beta)}\Delta_{T\ell}(\beta)e^s_{\beta\ell m}d\beta$	$a_{\ell m}=\frac{(2\pi)^3}{V}\Sigma_\beta\sqrt{P(\beta)}\Delta_{T\ell}(\beta)\Sigma_s\xi^s_{\beta\ell m}e_{\beta s}$
$\langle a_{\ell m}a^*_{\ell' m'}\rangle=C_\ell\delta_{\ell\ell'}\delta_{mm'}$	$\langle a_{\ell m}a^*_{\ell' m'}\rangle=C^{\ell' m'}_{\ell m}\quad C_\ell=\frac{1}{2\ell+1}\Sigma_{m=-\ell}^{\ell}C^{\ell m}_{\ell m}$

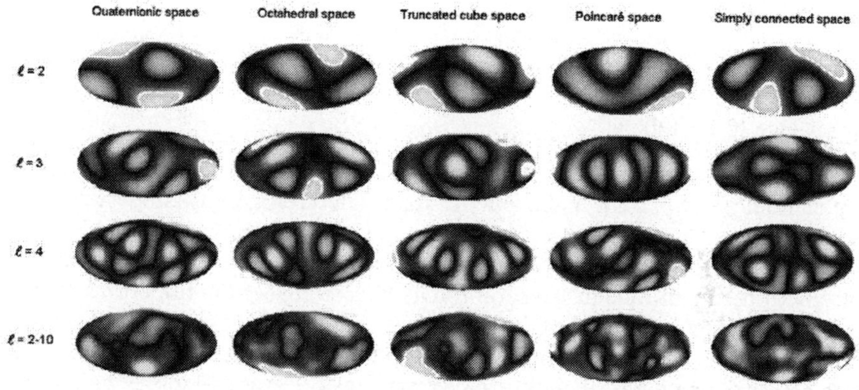

FIGURE 3. CMB sky realizations for low multipoles for a closed ΛCDM universe with simply/multi-connected topology.

0.683, $\Omega_k=-0.015$ and H_0=65 km/sec/Mpc. Note that the drop at higher ℓ in Fig. 4 is partly artificial and due to the fact that only wavenumbers $\beta<33$(~0.88/kpc) were used in the calculation. The simply connected universe has the same power spectrum as the WMAP best-fit at large scales and since it matches those of the multi-connected spaces from $\ell\simeq 9$ onwards, we expect the latter to coincide with the power spectrum of the WMAP best-fit at smaller scales. Finally, we observe that spaces with smaller fundamental domains exhibit greater suppression of power at low multipoles, while non-trivial topologies can also induce features such as moment alignment in CMB maps.

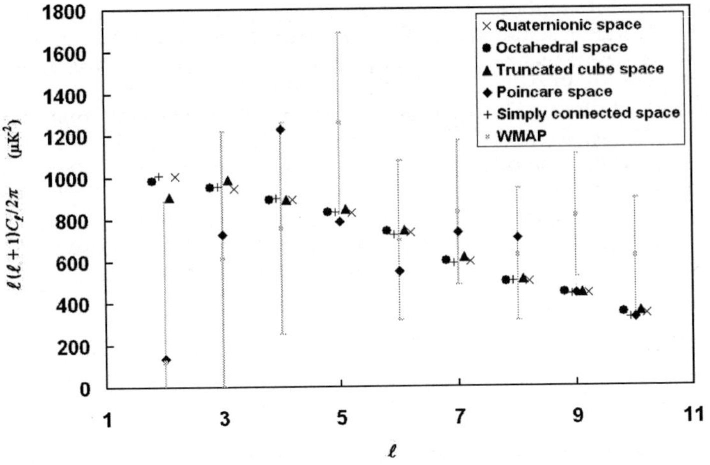

FIGURE 4. CMB power spectra for closed ΛCDM universes. The smaller the fundamental domain the larger the suppression at low ℓ: the power spectra of the simply connected, the quaternionic and the octahedral spaces almost coincide, while that of the truncated cube space has visibly less power. The power deficit is most prominent in the Poincaré space. Note that some points have been slightly displaced from their original x-position for reasons of clarity.

FUTURE PROSPECTS

Our calculations show that spherical non-trivial topologies reproduce the observed low power at large scales by construction. In addition, they can lead to moment alignment, a feature claimed to be present in the WMAP data. Therefore, multi-connected topologies seem a viable scenario that could explain the intriguing findings of WMAP, especially since there are no known constraints on the topology of the Universe. The intrinsic anisotropy of the manifolds in question forces us to consider the full correlation matrix and not just its diagonal part, enabling us to gain better insight into the physics of large scales. We are now calculating the likelihood functions for these topological spaces, in order to assess their statistical significance by means of Bayesian model comparison.

REFERENCES

1. D. N. Spergel, L. Verde, H. V. Peiris, E. Komatsu, M. R. Nolta, C. L. Bennett, M. Halpern, G. Hinshaw, N. Jarosik, A. Kogut, M. Limon, S. S. Meyer, L. Page, G. S. Tucker, J. L. Weiland, E. Wollack, and E. L. Wright, *The Astrophysical Journal Supplement Series* **148**, 175–194 (2003).
2. A. Niarchou, A. H. Jaffe, and L. Pogosian, *Phys.Rev.D* **69**, 063515 (2004).
3. E. Gausmann, R. Lehoucq, J.-P. Luminet, J.-P. Uzan, and J. Weeks, *Classical and Quantum Gravity* **18**, 5155–5186 (2001).
4. R. Lehoucq, J. Weeks, J.-P. Uzan, E. Gausmann, and J.-P. Luminet, *Classical and Quantum Gravity* **19**, 4683–4708 (2002).
5. A. Riazuelo, J.-P. Uzan, R. Lehoucq, and J. Weeks, *Phys.Rev.D* **69**, 103514 (2004).

A locally anisotropic geometrical model of space–time based on CMBR

Panagiotis G. Labropoulos* and Panayiotis C. Stavrinos†

*Physics Department, University of Athens, Panepistimiopolis Zografou 15784, Athens, Greece, plabrop@phys.uoa.gr
†Mathematics Department, University of Athens, Panepistimiopolis Zografou 15784, Athens, Greece, pstavrin@math.uoa.gr

Abstract. We study a locally anisotropic model of space–time induced by the existence of inflationary scalar fields in the framework of a more general geometrical structure than the Riemannian one. In this model the observable anisotropy of the CMBR (WMAP) is represented by a tensor of anisotropy and it is included in the metric structure of space-time. As well, some interesting special cases of spaces are considered.

Keywords: Finsler geometry, inflation, locally anisotropic spaces
PACS: 98.80.Jk, 04.20.-q, 95.30.Sf, 02.40.-k

INTRODUCTION

In 1994, Bekenstein, during his study on the applicability of Finsler geometry to physics [3, 2], considered the following generalization of a conformal transformation which respects the weak equivalence principle and causality, but is not, though, angle-preserving:

$$\tilde{g}_{\alpha\beta} = e^{2\psi}\left[A(I)\, g_{\alpha\beta} + L^2 B(I)\, \psi_{,\alpha}\psi_{,\beta}\right] \tag{1}$$

with the following properties:

1. $B(I) \leq 0$
2. $\underset{ref}{A(I) > 0}, \quad C(I) = A(I) + IB(I) > 0$
3. $\frac{d}{dI}\left[\frac{I}{C(I)}\right] > 0$

where I is a invariant constructed by a scalar field ψ and its first derivative ψ_μ, A, B, C are functions of I and L is a length scale. C can be interpreted as a modulus of anisotropy. The meaning of the transformation (1) is clear if it is examined under the prism of the Finslerian angular metric tensor: this transformation has an extra contribution which reaches its maximum along the gradient of the field. Later, in 2003, Kaloper [4] studied an inflationary model with the gravitational coupling to matter being introduced by

$$\tilde{g}_{\mu\nu}\left(x^\mu, \psi_\mu(x)\right) = g_{\mu\nu} - \frac{1}{m^4}\partial_\mu\psi\partial_\nu\psi. \tag{2}$$

which is a special case of Bekenstein' s disformal transformation. The pseudo-scalar field ψ here plays the role of the inflaton, which must describe a spin–0 particle to

CP848, *Recent Advances in Astronomy and Astrophysics*, 7th *International Conference of the Hellenic Astronomical Society*, edited by N. H. Solomos guarantee that inflation' s negative pressure condition is meaningful and has shift $\psi \leftrightarrow \psi + C$ and reflection $\psi \leftrightarrow -\psi$ symmetries to protect the inflaton from matter loop corrections.

In this work we investigate some aspects induced by this model, by exploiting the rich geometrical structure of Finsler geometry and the non–linear connection on N–anholonomic manifolds in the framework of a vector bundle.

MODIFIED CONNECTION STRUCTURE

Let us consider the tangent bundle TE of the total space E of a fiber bundle. The study of this space is justified because if we would like to introduce a coordinate system on E we need to define vectors over E, which belong to its tangent space. The fact that E is constituted of fibers suggests a physical way to analyze a vector X at a point p: X must have a component tangent with respect to the fiber corresponding to p. So we can split T_pE into two subspaces, one constructed of vectors tangent to the corresponding fiber called vertical V_pE, and its complimentary constructed of all the other vectors which belong to T_pE, called the horizontal subspace H_pE. This global splitting is represented by the Whitney sum of the non–intersecting subspaces:

$$\mathrm{T}_pE = V_pE \oplus H_pE \tag{3}$$

A base of the vertical subspace will be of the form $\left\{\frac{\partial}{\partial \varphi_i}\right\}$, as these vectors are tangent to the fiber. The definition of the horizontal subspace is more complex as the demand of being complimentary to V_pE is not sufficient. The reason beyond this is that the construction of the bundle does not suggest the points were neighboring fibers are glued, although we know the neighborhoods of these points. Generally, the vectors of H_pE are linear combinations of $\frac{\partial}{\partial x^i}$ and $\frac{\partial}{\partial \varphi_i}$, so we can define the following base:

$$\left\{\frac{\partial}{\partial x^i} + N_{ij}(x,\varphi)\frac{\partial}{\partial \varphi_j}\right\} \tag{4}$$

We can equip the bundle with a structure directly related to the parallel displacement of the elements of the fiber in order to compute the functions N_{ij}. This structure will provide us a rule of connecting fibers. This supplementary structure must be a connection structure.

We first consider a pseudo-Riemannian space–time as the base manifold and the space spanned by $\partial_\mu \varphi$ as the typical fiber. Next we examine the case where the previous bundle is the base manifold and the fiber is spanned by a second field $\partial_\mu \sigma$. The existence of a scalar field in this framework entering the metric by means of the so-called disformal transformation changes the way vectors are parallelly transported.

First we define for simplicity the covector $\varphi_i = \frac{\partial \psi}{\partial x^i}$. We notice immediately that our independent variables will be (x^μ, φ_i). The adapted frame is set as follows:

$$\left\{\begin{array}{l} dx^A \equiv \left(dx^\mu, \delta\varphi_i = d\varphi_i - M_{\nu i}dx^\nu\right) \\ \frac{\partial}{\partial x^A} \equiv \left(\frac{\delta}{\delta x^\nu} = \frac{\partial}{\partial x^\nu} + M_{\nu i}\frac{\partial}{\partial \varphi_i}, \frac{\partial}{\partial \varphi_i}\right) \end{array}\right. \tag{5}$$

Then the connection structure is given by:

$$\Gamma^{A}_{BC} = \left\{ L^{\mu}_{\nu\lambda}, L^{i}_{j\lambda}, h_{kl}C^{\mu l}_{\nu}, h_{kl}C^{il}_{j} \right\} \tag{6}$$

The following covariant derivatives can be defined:

$$\begin{aligned} V^{\mu}|_{\lambda} &= \frac{\delta V^{\mu}}{\delta x^{\lambda}} + L^{\mu}_{\nu\lambda} V^{\nu} \\ V^{\mu}||_{i} &= \frac{\partial V^{\mu}}{\partial \varphi_i} + C^{\mu i}_{\nu} V^{\nu} \end{aligned} \tag{7}$$

The metrical structure is introduced by:

$$\tilde{g}\left(x^{\mu}, \varphi_i\right) = g_{\mu\nu} dx^{\mu} \otimes dx^{\nu} + h^{ij} \delta\varphi_i \otimes \delta\varphi_j \tag{8}$$

where h^{ij} is the metric of the vertical space V_pE [9].

Had we started with a non–linear connection, instead of defining a priori the adapted frame, and determining the adapted frame we would have ended up with an off–diagonal metric of the following form:

$$\tilde{g} = \begin{bmatrix} g_{\mu\nu} + M_{\mu i} M_{\nu j} h^{ij} & M_{jk} h^{ik} \\ M_{ik} h^{jk} & h^{ij} \end{bmatrix} \tag{9}$$

The previous canonical metric structure is constructed by a Sasaki type lift from the base manifold. A number of ansatzs with off–diagonal metrics were investigated in higher dimensional gravity (see, for instance, the Salam and Strathee and the Percacci and Randjbar–Daemi works [5]) which showed that off–diagonal components in higher dimensional metrics are equivalent to including $U(1), SU(2)$ and $SU(3)$ gauge fields. We can write the metric in the form (8) only if there exists a coordinate transformation under which it can be diagonalized.

We shall remark that the non–linear connection can be physically interpreted as the interaction between the inflaton and the matter sector. The metric is written in block form which expresses the independence of the inflaton' s motion in its "internal" space from the motion in space–time.

No one restricts us from considering more light scalar fields. In the curvaton scenario a primordial fluctuation of a second scalar field σ generates the density fluctuations needed to to explain the hierarchical large scale structure of the Universe. Several fields such as the Affleck-Dine field, pseudo-Nambu-Goldstone bosons, cosmological moduli fields, flat direction in the minimally supersymmetric model (MSSM) etc. are candidates for generating curvature perturbations. In this case we also define $\xi_a = \frac{\partial \sigma}{\partial x^a}$ and the independent variables become $(x^{\mu}, \varphi_i, \xi_\alpha)$. The spatial structure becomes second order vector bundle–like and the adapted frame is set as follows:

$$\begin{cases} dx^A \equiv \left(dx^{\mu}, \delta\varphi_i = d\varphi_i - M_{\nu i} dx^{\nu}, \delta\xi_\alpha = d\xi_\alpha - P_{\nu\alpha} dx^{\nu} + Q^{i}_{\alpha} d\varphi_i\right) \\ \frac{\partial}{\partial x^A} \equiv \left(\frac{\delta}{\delta x^{\nu}} = \frac{\partial}{\partial x^{\nu}} - M_{\nu i}\frac{\partial}{\partial \varphi_i} + P_{\nu\alpha}\frac{\partial}{\partial \xi_\alpha}, \frac{\delta}{\delta \varphi_i} = \frac{\partial}{\partial \varphi_i} - Q^{i}_{\alpha}\frac{\partial}{\partial \xi_\alpha}, \frac{\partial}{\partial \xi_\alpha}\right) \end{cases} \tag{10}$$

where we introduce three kinds of non-linear connections $\left\{M_{\nu i}, P_{\nu\alpha}, Q^{i}_{\alpha}\right\}$ representing inflaton–matter, curvaton–matter and inflaton–curvaton interactions respectively. Therefore, the connection and metrical structures are given by:

$$\Gamma^A_{BC} = \left(L^\mu_{\nu\lambda}, L^i_{j\lambda}, L^a_{b\lambda}, C^\mu_{k\nu}, C^i_{kj}, C^a_{kb}, H^\mu_{c\nu}, H^i_{cj}, H^a_{cb}\right) \tag{11}$$

and the metric sstructure of this space–time becomes

$$\tilde{g} = g_{\mu\nu} dx^\mu \otimes dx^\nu + h^{ij} \delta\varphi_i \otimes \delta\varphi_j + k^{ab} \delta\xi_a \otimes \delta\xi_b \tag{12}$$

where k^{ab} is the metric of a Hamilton space of order 2 [9].

In the single scalar field case case the curvture tensors are

$$R^\mu_{MN\mu} = \frac{\delta L^\mu_{MN}}{\delta x^\mu} - \frac{\delta L^\mu_{M\mu}}{\delta x^N} + L^\mu_{K\mu} L^K_{MN} - L^\mu_{KN} L^K_{M\mu}$$
$$R^i_{MNi} = \frac{\delta C_{iMN}}{\delta\varphi_i} - \frac{\delta C^i_{iM}}{\delta\varphi_N} + C^i_{Ki} C^K_{MN} - C^i_{KN} C^K_{Mi}$$

and the Ricci tensors and curvature scalar are respectively

$$\mathscr{R}_{MN} = \left\{R^\mu_{MN\mu}, R^i_{MNi}\right\} \tag{13}$$

$$\mathscr{R} = \tilde{g}^{MN} \mathscr{R}_{MN} = g_{\mu\nu} R^{\mu\nu} + h_{ij} R^{ij} = R + R^{(\mathrm{inf}\,lat\,on)} \tag{14}$$

Although R is the usual Riemannian curvature scalar constructed from $g_{\mu\nu}$, $\mathscr{R}$ constructed from $\tilde{g}$ will differ because of the internal variables h_{ij}. This could indicate a way to solve the flatness problem through its contribution to $\Omega_{\,\kappa}$ [10].

DEVIATION OF GEODESICS

The relation that characterizes isotropic points in a Finslerian space–time of constant curvature is analogous to the Riemannian one and has the form

$$L_{\mu\nu\kappa\lambda} = K\left(\tilde{g}_{\mu\kappa}\tilde{g}_{\nu\lambda} - \tilde{g}_{\mu\lambda}\tilde{g}_{\nu\kappa}\right) \tag{15}$$

where K is the constant curvature. If we set into correspondence the events on two null geodesics C and C' in a space of constant curvature their deviation reduces to:

$$\frac{d}{ds}\left(n^i\varphi_i\right) + \varepsilon K n^i \varphi_i = 0 \tag{16}$$

where $\varepsilon = \pm 1$, s is the conventional affine parameter and n^i the deviation vector. This equation describes the rate of change of the cross–section of a beam of light rays passing through an inhomogenous mass distribution. In this space the equation of geodesic deviation may be integrated and the deviations of space–like and time–like geodesics are given by:

$$\begin{gathered} \mathrm{K} > 0, \quad n^i\varphi_i = c_1 \sin s\,(\varepsilon K)^{1/2} + c_2 \cos s\,(\varepsilon K)^{1/2} \\ \mathrm{K} = 0, \quad n^i\varphi_i = c_1 s + c_2 \\ \mathrm{K} < 0, \quad n^i\varphi_i = c_1 \sinh s\,(-\varepsilon K)^{1/2} + c_2 \cosh s\,(-\varepsilon K)^{1/2} \end{gathered} \tag{17}$$

Null geodesics are of great importance in relativity because nearly all astronomical information cames to us optically. The deviation vector in this case takes the form $n^2 = as^2 + bs + c$

We can define the Hubble parameter as $\tilde{H} = \frac{\dot{\tilde{a}}}{\tilde{a}}$ where $\tilde{a} = \tilde{a}(v(s))$ and $\dot{\tilde{a}} = \frac{d\tilde{a}}{ds} = \frac{\partial \tilde{a}}{\partial v^i}\frac{dv^i}{ds}$ with $v^i(s)$ the unit vector tangent to the flow lines. The Hubble parameter is defined as [8] $\tilde{H} = \frac{1}{3}\tilde{\Theta} = v^{\mu}_{|\mu} - C^{\mu}_{\mu\lambda}\dot{v}^{\lambda}$ and the anisotropic character of the Hubble parameter is revealed by the Cartan torsion tensor. The Cartan torsion tensor is defined as $C_{\mu\nu\lambda} = \frac{1}{2}\frac{\partial \tilde{g}_{\mu\nu}(x,\varphi)}{\partial \varphi_{\lambda}}$. This tensor which depends only on the coupling constant and the field' s derivative is very important because its vanishing is a criterion that the space is Riemannian. The Cartan torsion tensor plays the role of the Christoffel symbols with respect to φ. We must also note that in this form Hubble' s parameter accounts for a locally corrected version due the contribution from Cartan' s tensor.

For a scalar field ψ the energy density ρ_{ψ} and pressure p_{ψ} are $\rho_{\psi} = \frac{1}{2}\left(\frac{d\psi}{ds}\right)^2 + V(\psi)$ and $p_{\psi} = \frac{1}{2}\left(\frac{d\psi}{ds}\right)^2 - V(\psi)$ where $V(\psi)$, so that the Einstein equations can be written in the form:

$$L^{\mu\nu}\varphi_{\mu}\varphi_{\nu} = \mathrm{K}\left(\left(\frac{d\psi}{ds}\right)^2 - V(\psi)\right) \tag{18}$$

We also get the following form for the Raychauduri equation in the spitit of [8] $\frac{d\tilde{\Theta}^2}{ds} = -\frac{1}{3}\tilde{\Theta} - \tilde{\sigma}_{ik}\tilde{\sigma}^{ik} + \tilde{\omega}_{ik}\tilde{\omega}^{ik} - \mathrm{K}\left(\left(\frac{d\psi}{ds}\right)^2 - V(\psi)\right)$. The continuity equation becomes $\dot{\rho}_{\psi} = -3\tilde{H}\left(\left(\frac{d\psi}{ds}\right)^2\right)$, yielding the inflaton field equation $\frac{d^2\psi}{ds^2} + 3\tilde{H}\frac{d\psi}{ds} = -\frac{dV}{d\psi}$. The potential V remains to be taken from a more fundamental theory. Once we have inserted the potential we can calculate the slow roll parameters $\varepsilon(\psi) = \frac{m^2_{Planck}}{2}\left(\frac{1}{V}\frac{dV}{d\psi}\right)$ and $\eta(\psi) = m^2_{Planck}\left(\frac{1}{V}\frac{d^2V}{d\psi^2}\right)$ and calculate the logarithmic derivative $\frac{d\ln P_{\psi}}{d\ln(\tilde{a}\tilde{H})}$ in order to calculate the scalar spectral index.

REFERENCES

1. H. Rund, *The Differential Geometry of Finsler Spaces* (1959).
2. J. D. Bekenstein, *The Sixth Marcel Grossman meeting on general relativity*, ed. H. Sato, World Publishing, (1992)
3. J. D. Bekenstein, Phys. Rev. D., **48** (1993) 3641.
4. N. Kaloper, Phys. Lett. B **583**, 1 (2004) [arXiv:hep-ph/0312002].
5. A. Salam and J. Strathdee, Ann. Phys. (NY) **141**, (1982) 316; R. Percacci and S. Randjbar–Daemi, J. Math. Phys. **24** (1983) 807.
6. P. Stavrinos and S. Ikeda, *Variational Principle to the Generalized Scalar Ű Tensor Theory of Gravitation II* .Rev. Bulletin of Calcutta Mathematical Society **8** (2000) , no.1-2 pp.11-20.
7. S. Vacaru and P. Stavrinos, *Spinors and Space time Anisotropy*, University of Athens, (2003)
8. P. Stavrinos 2005 J. Phys.: Conf. Ser. **8** 49-57
9. R. Miron, *The geometry of higher-order hamilton spaces*, Kluwer Academic Publishers, 2003
10. G. Efstathiou , Mon. Not. Roy. Astron. Soc. 343 (2003) L95

PART 5

EMILIOS HARTLAFTIS' LECTURE

INSTRUMENTATION AND METHODS OF ASTRONOMICAL OBSERVATIONS

(Conveners: J. Seiradakis and N. H. Solomos)

The Quest for Extra-Solar Planets

Keith Horne

SUPA Physics & Astronomy, St Andrews University, North Haugh, St.Andrews, Scotland, UK.
(kdh1@st-and.ac.uk)

Abstract. The quest to discover planets beyond the solar system is in full swing. Over 170 exoplanets are now known, from three successful discovery methods: Doppler wobbles, transits, and microlensing. We have information on the abundance pattern of planets vs mass, orbit size and eccentricity, radius, density and on exoplanet atmospheres, challenging current ideas of how planetary systems form and evolve, and placing our own solar system in a broader context. With microlensing and space transits missions, we will discover hot and cool Earths. If habitable planets found to be common, and hence relatively nearby, we will know within this generation about extra-solar life.

Keywords: Planets:Extra-Solar
PACS: 97.82.-j

EMILIOS HARLAFTIS

This talk is dedicated to Emilios Harlaftis, who died tragically in an avalanche on 14 Feb 2005. Emilios was a dear friend and colleague. His loss shook me deeply. There were so many experiments that we were planning to do together, using the new Aristarchus telescope at Chelmos. My prevailing memory of Emilios is of his seemingly boundless energy, and enthusiasm, and of his laughter, for he brought real joy to the enterprise of exploring ideas and doing the science that he loved. His loss leaves a sore hole in many hearts.

THE EXTRA-SOLAR PLANET DISCOVERY ERA

We are a priviledged generation to be living in the era of exoplanet discovery. Astronomical technology enables us for the first time to detect planets orbiting nearby and distant stars. Prior to 1995, human knowledge of planetary systems was confined to our own solar system – with four small rocky inner planets, four gas giants, Pluto and similar Kujper Belt objects, a variety of moons, asteroids, comets, and smaller debris. The 1995 discovery of 51 Peg b, the first "Hot Jupiter", launched the exo-planet revolution. We now know of $\sim$ 170 exoplanets, ranging from Hot to Cool Jupiters, which are driving a new understanding of how planets form and migrate within their protoplanetary disks. In the next decade, by 2010-2015, we should know the abundance of "habitable" planets, by which we mean small rocky planets with surface temperatures suitable for oceans of liquid water. And then, from 2015 to 2025, we should have the technology to directly image and take spectra of the nearest habitable planets, to search for dis-equilibrium atmospheres, the signature of life. Thus within this generation we should have direct evidence, or otherwise, of life on planets among the stars beyond the solar system.

CP848, *Recent Advances in Astronomy and Astrophysics,*
7th International Conference of the Hellenic Astronomical Society, edited by N. H. Solomos

Direct Imaging

Direct detection of light from extra-solar planets is very difficult. An enormous contrast, $\sim 10^9$ in reflected optical starlight and $\sim 10^4$ in the thermal infrared, burries faint planets in the stellar glare. Special techniques – coronography, pupil shaping, extreme adaptive optics, nulling interferometry – can supress the starlight enough to reveal brown dwarf companions to main-sequence stars. We can also target dimmer stars – white and brown dwarfs – in search of faint common-proper motion companions. Recent successes (Neuhaeuser et al. 2005; Chauvin et al. 2005a,b) include GQ Lup (22 m_J, 100 AU), AB Pic (14 m_J, 280 AU) and 2M1207 (5 m_J, 55 AU). But for the planets of main-sequence stars, direct detection remains a major technological challenge.

Debris Disks

Planets affect the appearance of debris disks. These arise from the small dust grains generated in comet-comet collisions, and can be directly imaged at sub-mm wavelengths, and in scattered starlight. SCUBA images indicate that debris disks are lumpy more often than they are smooth (Holland et al. 1998, Greaves et al. 1998, 2004, 2005). Gravitational tugs shape the orbiting debris, imposing asymmetric patterns co-rotating with interior planets. Paul Kalas gave a nice talk on this (Kalas, Graham, Clampin 2005).

Doppler Star Wobbles

Fig. 1 shows the exoplanet discovery space, planet mass vs orbit size, comparing the solar system planets with 172 known extra-solar planets. Doppler, transits, astrometry, and microlensing have different selection biases, as indicated.

The first and by far most successful planet hunting method uses high-precision radial velocity measurements to detect the planet-induced reflex orbit of the host star. A key technology is the iodine gas cell, giving a precise reference frame for differential Doppler measurements, and of course careful data analysis is essential. The Doppler detection zone indicated in Fig. 1 is limited on the low-mass edge by the precision of Doppler measurements, and ultimately by turbulence in stellar atmospheres, and at long orbit periods by the survey duration, now ~ 10 years at 3 m s^{-1} and a few years at 1 m s^{-1}. Doppler surveys should reach Jupiter analogs at 5 AU in a few years, but it will not be easy to press down to much lower masses.

The first exoplanet discovery occurred by accident while searching for short-period brown dwarf companions (Mayor & Queloz 1995). The velocity curve of 51 Peg exhibits very short-period sinusoidal wobble with $P = 4.2$ day and $K = 55$m s^{-1}. The planet follows a circular orbit with semi-major axis $a = 0.05$ AU, and a lower limit to its mass is $m \sin i = 0.56\ m_J$. This established a new and unexpected class of planets, the "Hot Jupiters", with surface temperatures $T \sim 2000$ K, launching the exoplanet revolution.

Doppler surveys are now monitoring ~ 2000 nearby F,G,K and ~ 200 M dwarfs. In a spectacular decade, with 1-2 new planets per month, some 160 Doppler wobble planets

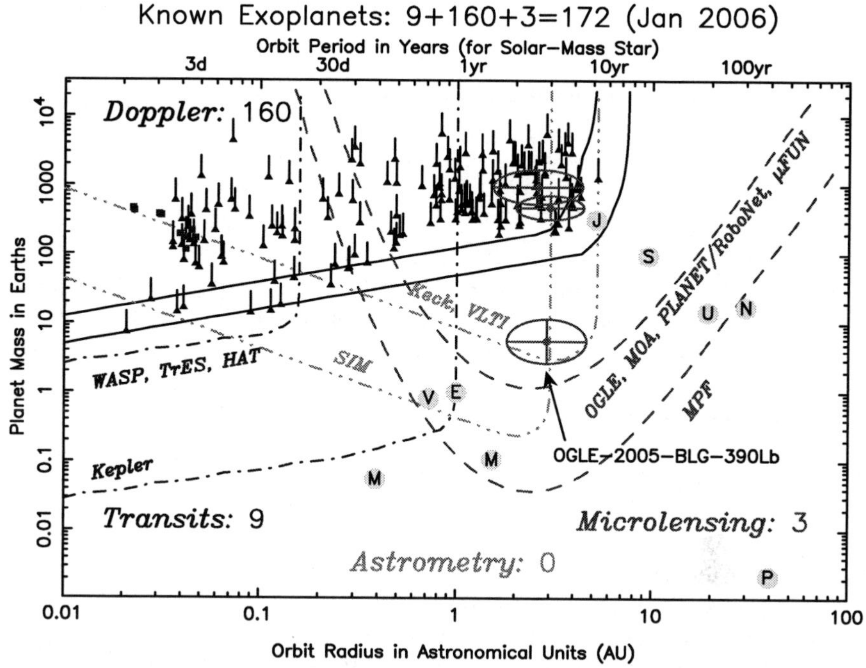

FIGURE 1. Exoplanet Discovery Space – planet mass vs orbit size – showing the solar system planets for comparison with the detection zones for and exoplanets discovered by different methods. Parameters of the extra-solar planets are from `http://vo.obspm.fr/exoplanetes/encyclo/encycl.html`

have now been found [1], many in multi-planet systems. These fill the Doppler detection zone (Fig. 1), but with $m \lesssim 10\ m_J$ and $P \gtrsim 3d$, revealing interesting limits to planet grown and inward orbital migration.

Tidal interaction with the star circularises the short-period Hot Jupiter orbits, but at longer periods the orbital eccentricities are roughly uniformly distributed from 0 to 0.8. This large range suggests that some mechanism, perhaps planet-planet interactions, pumps up the eccentricity of initially circular orbits, probably also ejecting smaller planets. Thus the planet formation history may be more violent in these systems tham it was in our own system, where the planets retain co-planar circular orbits from their origin in the proto-solar accretion disk.

Another important result is that metallicity clearly matters. Roughly $\sim 5\%$ of sun-like stars exhibit detectable wobbles, and this increases to $\sim 25\%$ for stars with three times solar metalicity (Santos et al. 2003, Fischer & Valenti 2005).

A most interesting case (Marcy et al 2001, Rivera et al. 2005) is the M4 star GJ 876, which has 2 giant planets locked in resonant orbits ($m \sin i = 0.56, 1.9\ m_J$; $P = 30, 60$d)

[1] `http://exoplanets.org/`

plus the smallest hot planet to date ($m \sin i \sim 8\ m_J$; $P = 1.8$d).

PLANETARY TRANSITS

Planets with nearly edge-on orbits can transit their host star, dimming the starlight for a few hours during each orbit. The transit depth and duration measure the planet radius r and orbit inclination i. Combining this information with $m \sin i$ from the Doppler data, the planet mass m and density follow. Since $r_J \approx 0.1\ R_\odot$, the transit depth is typically $\Delta f/f \approx 1\%\,(r/r_J)^2\,(R_\star/R_\odot)^{-2}$. The duration is $\Delta t \approx 3\ \mathrm{h}\,(M/M_\odot)^{2/3}\,(P/4\ \mathrm{d})^{1/3}$. The probability of transits, for a random orbit orientation, is $P_t \approx 10\%\,(R/R_\odot)\,(M/M_\odot)^{-1/3}\,(P/4\ \mathrm{d})^{-2/3}$. The discovery of Hot Jupiters with such short orbital periods makes the search for planetary transits of other stars a viable proposition.

The First Transits: HD209458

Planetary transits were first detected with the remarkably small 10 cm aperture STARE[2] telescope (Charbonneau & Brown 2000). The planet orbiting HD209458 was already known from Doppler wobble studies, and the transits were found at predicted times when the starlight dims by $\approx 1.6\%$ for $\Delta t \approx 3$ hours every $P \approx 3.5$ days. Fits to the observed Doppler amplitude plus the STARE lightcurve yield $m \approx 0.63\ m_J$, $r \approx 1.3\ r_J$, $i \approx 87°$. This planet is a "bloated" gas giant with a density less than $1/3$ that of Jupiter.

The HST lightcurve (Brown et al. 2001) is one of the most spectacular results in the field. The observed lightcurve is beautifully consistent with a model limb-darkened star traversed by a spherical planet. The fit leaves photometric residuals of order 10^{-4}, sufficient to place strong upper limits on small ($r < 2r_\oplus$) moons and rings. It is not surprising that a 2.5m telescope in space achieves significantly better photometric accuracy than the ground-bound 10 cm STARE telescope, but remarkable that such small scopes can detect planets. The wide field is in fact essential to discover which stars are 'winking' at us, so that we can tell the HST where and when to look.

Probing Hot Jupiter Atmospheres

Once a transiting planet has been identified, transmission spectroscopy (Brown 2001) offers an exciting technique for probing the exoplanetary atmosphere. A transiting planet's silhouette is slightly larger at wavelengths of high opacity in the planetary atmosphere, causing a slightly deeper transit in the lightcurve. By observing transits with a spectrophotometer, and measuring the transit depth vs wavelength, opacity features should be visible at the 0.1% level, including molecular bands (e.g. methane, ammonia,

---

[2] `http://www.hao.ucar.edu/public/research/stare/stare.html`

water, carbon monoxide) in the infrared, and resonance lines (e.g. Na I D) in the optical. These absorption features are weaker (stronger) for high (deep) reflective cloud decks, expected to form at various depths in Hot Jupiter atmospheres from droplets of molten Fe, corrundum, and enstatite.

The first success of transmission spectroscopy (Charbonneau et al. 2002) was the detection of Na I D absorption using the HST/STIS data on HD 209458. This is a delicate experiment requiring careful differential spectrophotometry. Na I absorption lines, present in both the planetary and stellar atmospheres, have transits deeper by $\sim 10^{-4}$ compared with those of numerous Fe transitions that are present in the stellar atmosphere only.

A more extensive evaporating 'exosphere' was also found (Vidal-Madjar et al. 2003) by examining ultraviolet emission lines from the stellar chromosphere. These become weaker during transits, indicating a fractional coverage of 5-10% of the stellar disk by HI, OI, CII wafting away from the irradiated planetary atmosphere, perhaps trailing behind the planet like the dust and ions a comet's tail.

Eclipse of the planet by the star enables direct detection of thermal infrared emission from the heated faces of Hot Jupiters. This has been accomplished for HD209458 (Deming et al. 2005) and TrES-I (Charbonneau et al. 2005) using the *Spitzer* infrared space telescope to observe a subtle $\approx 0.2\%$ dimming as the planet passes behind the stellar limb.

Transit Surveys: Deep

Some two dozen photometric transit surveys are currently underway [3] to discover new transiting planets. By obtaining high-precision CCD photometry of tens of thousands of stars in parallel, rather than high-precision Doppler measurements of one star at a time, these photometric transit surveys have the potential to reach much fainter and more distant stars than the Doppler surveys. Two approaches – Deep and Wide – are being pursued.

Deep surveys employ wide-field CCD cameras on existing 1-4 m telescopes, typicaly with $\Delta\theta \lesssim 1$ arcsec pixels and $\theta \lesssim 1°$ fields of view. OGLE III[4] is the most successful deep transit survey, using the 1.3 m telescope at Las Campanas Observatory with its 8-CCD mosaic camera covering an 0.6° field of view. This system is capable of detecting transits on 18th mag stars out to about 3 kpc. Survey fields are in the Galactic plane, to keep the survey volume filled with stars, and as a result there are many false alarms. Of the several hundred candidates found (Udalski et al. 2002a,b,c,2003,2004), subsequent follow-up indicates that most are impostors, e.g. grazing or blended eclipsing binaries, or eclipses by brown dwarfs.

The spectroscopic follow-up of OGLE III candidates (e.g. Konacki et al. 2004,2005; Bouchy et al. 2004, Pont et al. 2004, Moutou et al 2004) has confirmed 5 new transiting planets (OGLE-TR-10,56,111,113,132). The orbital periods are shorter than those

---

[3] `http://star-www.st-and.ac.uk/ kdh1/transits/table.html`
[4] `http://www.astrouw.edu.pl/ ogle/`

found in Doppler surveys (Fig. 1), due to the strong selection effects favouring short-period planets. OGLE-TR-10 is a bloated giant, with mass, radius and density similar to HD209458, but the others are smaller, close to Saturn and Jupiter. The mass-radius distribution is clearly not a 1-dimensional sequence, indicating that at least 2 parameters are required. One possibility is that planets that migrate inward rapidly remain bloated because they are unable to radiate their internal heat, while those migrating more slowly are able to cool and shrink. Another is that some of these have larger rocky cores than others. The smallest, a Hot Saturn orbiting HD 149026 (Sato et al. 2005) seems to have a large dense core.

Transit Surveys: Wide

Wide surveys employ purpose-built wide-field CCD cameras with smaller 2-20 cm apertures. These surveys sacrifice angular resolution to expand the survey field of view. Typical parameters, e.g. HATNet [5] and SuperWASP[6], are $D = 11$ cm F/1.8 lenses imaging a $\theta \approx 8°$ starfield onto a $2k \times 2k$ CCD with $\Delta\theta \approx 15$ arcsec pixels.

The discovery potential for small wide-angle surveys is good, and these will find the brightest stars transited by Hot Jupiters, for which the most interesting follow-up studies are possible. A simple calculation assumes that HD209458, the brightest star with a transiting planet, is typical. We then expect 1 transiting planet in the whole sky down to 8th mag, and 4 times more for each magnitude – 4 brighter than 9th mag, 16 to 10th, ... 500 to 12th and 1000 to 13th mag. The 10cm wide-angle instruments are expected to reach 12th to 13th mag, and thus may find 500 to 1000 transiting Hot Jupiters. Given the large number of cameras currently on sky, it may take 5-6 years to find them all.

So far TrES is the first and so far only successful wide survey, (TrES-1, Alonso et al. 2004). TReS employs 3 cameras at different longitudes in the northern hemisphere, including the original STARE telescope relocated to Tenerife.

The SuperWASP experiment employs 8 CCD cameras with 11cm F/1.8 lenses each covering an 8° field of view, on a robotic mount, with one on La Palma for the northern sky, and one at SAAO for the southern sky. Candidates identified in the 2004 dataset are currently undergoing spectroscopic follow-up, and we are hopeful that confirmed planets will emerge in 2006.

OTHER EARTHS

An important goal for the next 5-10 years is to discover other Earth-like planets in sufficient numbers to determine whether planets like our own are common or rare.

---

[5] `http://www.cfa.harvard.edu/ gbakos/HAT/`
[6] `http://www.superwasp.org/index.html`

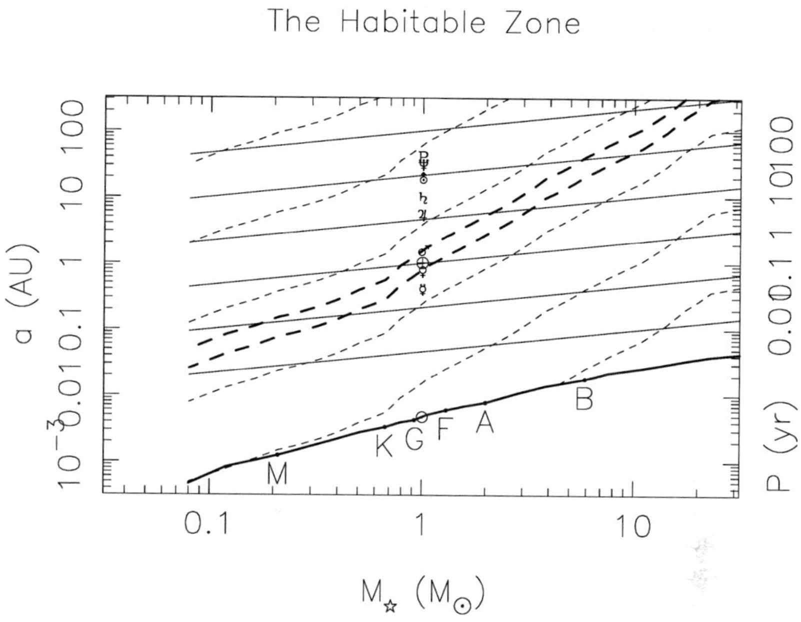

FIGURE 2. The "Habitable Zone", where the temperature is suitable for liquid water to persist on the surface of a small rocky planet, is closer to cooler low-mass stars and farther from hotter high-mass stars.

The Habitable Zone

Because liquid water is thought to be essential for life-forms like our own, attention focusses on planets in the so-called "Habitable Zone", where surface temperatures near 300 K are suitable for liquid water to persist on the planetary surface. As shown in Fig. 2, this zone is relatively narrow, and is closer in for planets orbiting a cool low-mass star, and farther out for those orbiting a hot high-mass star.

It turns out that detection of Earth-mass planets in the Habitable Zone is quite difficult. Fortunately, we have at present two complementary methods that are capable of discovering planets as small as the Earth both inside and outside the Habitable Zone. Transit experiments from space can discover Hot Earths, while microlensing experiments can discover Cool Earths. By doing both of these experiments, we can measure the abundance of Earth-mass planets both inside and outside the Habitable Zone, and thereby interpolate to find the abundance of Habitable Earths.

Hot Earths: Space Transits

Two missions, Corot[7] and Kepler[8], with launch dates in 2006 and 2008 respectively, will exploit the stable environment of space to perform high-precision photometry of tens to hundreds of thousands of stars in search of transiting planets. The accuracy achieved by the HST/STIS spectrophotometry of HD209458 transits suggests that, with careful design to achieve even higher accuracy, transits of much smaller planets, down to the size of Earth, may be detectable by these missions. The chief concern is in separating the transit signal from low-level stellar variability, about which we have limited knowledge at present.

Corot's mission plan includes 6-month continuous observation of several fields. This should pick up transiting planets with orbit periods up to a few months. Corot will certainly identify many Hot Jupiters, and may also pick up secondary eclipses and orbital modulations from starlight reflected from their atmospheres. Corot may be the first experiment to detect transits by Hot Earths, if they are relatively abundant and if low-amplitude stellar variability is not too severe.

Kepler's mission plan extends for 4 years in order to be able to detect transits by Earth analogs. The Earth transiting the Sun, if viewed from a distant star, would produce a transit with a fractional depth $\Delta f/f \approx 10^{-4}$. The transit would last up to 13 hours, and would recur once per year. Observing a single transit is insufficient because of possible confusion with low-amplitude stellar variability. Thus Kepler's 4-year mission aims to pick up 3 or 4 equally-spaced transits of the same duration and depth. If successful, Kepler may detect Earth-sized planets in the Habitable Zone by about 2010.

Cool Earths: Microlensing

Microlensing is the fastest way to find small cool planets. To date 3 cool planets have been discovered by the gravitational lensing technque (Fig. 1). Two are giant planets with $m \sim m_J$, and the third is a small one only 5 $m_\oplus$. The microlensing detection zone extends down to Earth mass, with greatest sensitivity for orbit sizes $\sim$ 3 AU, thus cool planets.

The microlensing technique involves watching a background source star in the Galactic Bulge whose light is being focused on the Earth by the gravitational field of an intervening lens star. The lens star magnifies the light for a few weeks, and planets near the lens star can act as smaller lenses, briefly altering the magnification for a few hours to days.

The star's Einstein Ring size is $R_E \sim 3$ AU, or $\sim 10^{-3}$ arcsec, in a typical Galactic Bulge microlensing event. With perfect alignment the source star would appear as a ring with this radius around the lens star. With imperfect alignment, there are 2 distorted images of the source star, on opposite sides of the lens, one outside and one inside the lens star's Einstein Ring. These two images cannot be seen directly – it is the increase

[7] http://corot.oamp.fr/

[8] http://www.kepler.arc.nasa.gov/

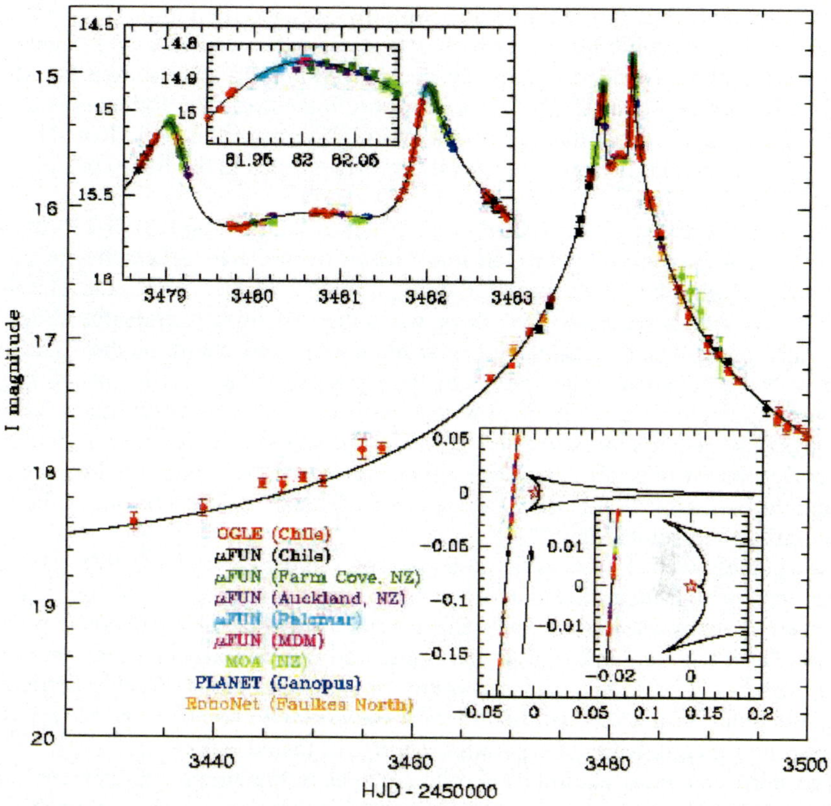

FIGURE 3. The lightcurve of microlensing event OGLE-2005-BLG-071 exhibits a 4-day 'divot' near the peak, caused by lensing of a 3 m_J planet. Many telescopes observed the event in order to obtain nearly continuous coverage during the anomaly.

in brightness due to the magnification that is observed. When one of the images passes close to a planet, that planet's gravity further deflects the light ray, causing a brief but easily observable flash or dip in the lightcurve. The planet signal can be large, even for a small planet. But the signal is brief, the duration scaling as $\sqrt{m}$, a few days for a Jupiter and a few hours for an Earth. The observed duration determines the planet/star mass ratio, and the time of the signal determines the planet-star separation, and hence the orbit size, in units of R_E. Finite size of the source star dilutes the signal and reduces the sensitivity to planets smaller than Earth.

Discovery of cool planets by microlensing requires international collaboration of several teams because observations are needed from telescopes at different longitudes in order to acquire quasi-continuous coverage of enough microlens events to discover the brief flashes and dimmings produced by the cool planets. By observing rich starfields in the Galactic Bulge, through low-dust windows, OGLE III[9] employs a 1.3m telescope at Las Campanas in Chile to survey $\sim 10^8$ stars with a cadence of 1-4 days, discovering over 600 microlensing events every year. Follow-up observations by MOA[10], by PLANET[11], and by microFUN[12], monitor more intensively the lightcurves of the most promising events – those that are bright and highly magnified. The RoboNet[13] experiment adds observations from three 2m robotic telescopes to the microlens planet search effort.

The first microlens planet (Bond et al. 2004) is OGLE-2003-BLG-235/MOA-2003-BLG-053, co-discovered by the OGLE and MOA teams. In this event the planet's gravity bent the starlight enough to form 2 additional images of the source star. The lightcurve exhibits two spikes separated by 7 days, when the new images first appeared, and when they later merged and disappeared. Fits to the detailed shape of the lightcurve give accurately the planet/star mass ratio, and the projected separation in units of the radius of the star's Einstein Ring. Combined with an estimate for the lens star mass and distance, this gives the planet's mass $m \sim 1.3\ m_J$ and orbit size $a \sim 3$ AU. Uncertainties here are currently a factor of order 2, but this will decrease to $\sim 30\%$ once the two stars separate enough to be resolved, allowing the mass and distance to the lens star to be measured from multi-colour imaging.

Two planets were found in the 2005 season. OGLE-2005-BLG-071 (Udalski et al. 2005) was a high-magnification event exhibiting a 4-day dip near the peak (Fig. 3) observed by many telescopes, including several from amateur astronomers participating in the microFUN network. Fits to the lightcurve indicate a planet mass $m \sim 3\ m_J$ and orbit size $a \sim 2$ or 4 AU. The ambiguity in a arises because in this particular high-magnification event a very similar dip near the peak can be produced by a planet either outside the Einstein Ring, at separation $1.6\ R_E$, or inside at $R_E/1.6$.

The third and most exciting microlens planet is the small one, detected as a brief 12-hour 20% brightening on the declining part of the lightcurve (Fig. 4) of OGLE-2005-BLG-390 (Beaulieu et al. 2006). In this case the planet mass is $m \sim 6\ m_\oplus$ with an orbit size $a \sim 2.9$ AU, again with a factor 2 uncertainty until the two stars separate.

The main significance of the first 3 microlens events is the inferrence that small cool planets are more abundant than large ones. Microlensing is more sensitive to giant planets, by a factor of perhaps 20 to 50, and thus we would expect to find 20 to 50 giant planets before the first small one. In fact the third one is a small one, so the abundance is likely to be a factor 10-20 higher for the small ones.

Future microlens studies will let us measure the abundance of cool planets as a function of planet mass, orbit size, and star mass. The pattern that emerges will test predictions of planet formation and migration. In particular, recent models (Ida & Lin 2004,2005) predict a large number of small rock/ice cores forming in the outer regions of proto-planetary systems where it is cold enough for dust grains to retain ice mantles.

---

[9] http://www.astrouw.edu.pl/ ogle/

[10] http://www.physics.auckland.ac.nz/moa/

[11] http://planet.iap.fr/

[12] http://www.astronomy.ohio-state.edu/ microfun/

[13] http://www.astro.livjm.ac.uk/RoboNet/

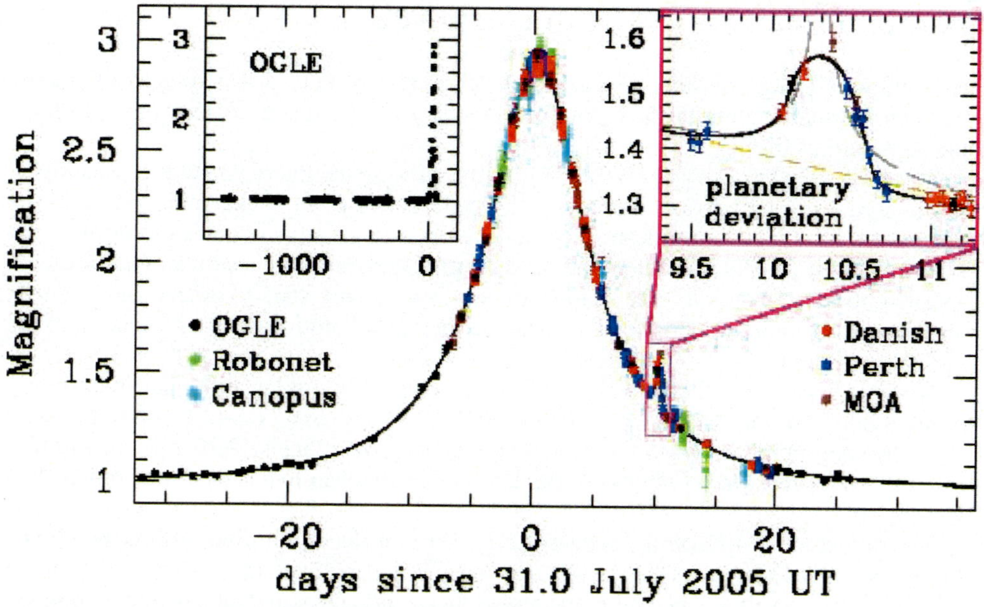

FIGURE 4. The lightcurve of microlensing event OGLE-2005-BLG-390 shows the source star magnified by a factor up to 3 over a few weeks due to passage of an intervening M dwarf lens star. The brief flash, lasting 12 hours is caused by a small cool planet with $m \sim 5m_{\oplus}$ and $a \sim 3$AU.

The dust grains accumulate rapidly into rock/ice cores, and then accrete gas slowly until the gas envelope mass approaches the core mass. At this point gas accretion runs away, growing the core into a gas gaint. The largest cores grow rapidly into gas giants, open a gap in the gas disk, and then migrate in toward the star, forming the planets we see in the Doppler surveys. The detection of the small planet in OGLE-2005-BLG-390 most likely represents the first discovery of one of these small rock/ice cores.

Habitable Earths: Common or Rare?

With microlensing sensing small cool planets, and upcoming space missions finding small hot ones, we will soon have good measurements of the abundance of small planets on both sides of the Habitable Zone. The abundance of habitable planets will then be securely established, by interpolating between these two results. If the abundance of habitable planets is relatively high, then the nearest ones will be relatively close by, and it will then be feasisible to take the next major step – a search for life on those planets.

Extra-Solar Life

Ambitious space missions are under development to take direct images of nearby planetary systems, identify the habitable planets, and take spectra of them to search for the signature of life.

The ESA mission *Darwin* ($\sim$ 2015) is an infrared space interferometer consisting of 5 free-flying infrared space telescopes. The telescopes point at the same star, and the beams are combined using destructive interference to greatly suppress the light of the star, making it possible to detect the light from planets orbiting that star. One possible *Darwin* mission plan is to take snapshots of $\sim$ 500 nearby stars to detect their planets, and then study $\sim$ 50 in more detail to observe the orbital motions and take spectra of the habitable planets.

NASA's *Terrestrial Planet Finder* (*TPF*) mission has similar aims, but is currently divided into two stages *TPF-C* ($\sim$ 2014) and *TPF-I* ($\sim$ 2020). *TPF-C*, is a 4-6 m visible light coronagraph, aiming for direct imaging of giant planets. *TPF-I* is an infrared interferometer like *Darwin* but with the telescopes attached to a rigid frame rather than free-flying.

The signature of life on an extra-solar planet is a dis-equilibrium atmosphere. The high oxygen abundance of the Earth's atmosphere is generated and maintained by plant life. If life were extinguished here, the atmospheric oxygen abundance would be quickly fall, the oxygen rusting into the rocks. From the infrared spectra of Venus, Earth, and Mars, it is easy to see which one has oceans and life. *Darwin* and *TPF-I* will take infrared spectra of nearby habitable planets to determine how many of them display dis-equilibrium atmospheres revealing the presence of life.

Thus we are making steady progress along the exoplanet roadmap, leading up to a search for extra-solar life. If we look ahead perhaps 10-20 years, we should know whether or not there are other habitable worlds like our own, how far away they are, and what fraction of these potential habitats already supports life.

ACKNOWLEDGMENTS

Thanks for a great meeting and warm hospitality on a beautiful island!

REFERENCES

1. Alonso, R., et al. *ApJ* **613**, 153 (2004).
2. Beaulieu, J-P., et al. *Nature* **439**, 437 (2006).
3. Bond, I.A., et al. *ApJ* **606**, 155 (2004).
4. Bouchy, F., et al. *A&A* **421**, 13 (2004).
5. Brown, T.M. *ApJ* **553**, 1006 (2001).
6. Brown, T.M., et al. *ApJ* **552**, 699 (2001).
7. Charbonneau, D., Brown, T.M., Latham, D.W., Mayor, M. *ApJ* **529**, 45 (2000).
8. Charbonneau, D., Brown, T.M., Noyes, R.W., Gilliland, R.L., *ApJ* **568**, 377 (2002).
9. Charbonneau, D., et al. *ApJ* **626**, 523 (2005).
10. Chauvin, G., et al. *A&A* **438**, 25 (2005a).
11. Chauvin, G., et al. *A&A* **438**, 29 (2005b).

12. Deming, D., Seager, S., Richardson, L.J., Harrington, J. *Nature* **434**, 740 (2005).
13. Fischer, D.A., Valenti, J. *ApJ* **622**, 1102 (2005).
14. Greaves, J.S. et al. *ApJ* **619**, 187 (2005).
15. Greaves, J.S. et al. *MNRAS* **351**, 54 (2004).
16. Greaves, J.S. et al. *ApJ* **506**, 133 (1998).
17. Holland, W.S., et al. *Nature* **392**, 788 (1998).
18. Ida, S., Lin, D.N.C. *ApJ* **616**, 567 (2004).
19. Ida, S., Lin, D.N.C. *ApJ* **626**, 1045 (2005).
20. Konacki, M., et al. *ApJ* **609**, 37 (2004).
21. Konacki, M., et al. *ApJ* **624**, 372 (2005).
22. Kalas, P., Graham, J.R., Champin, M., *Nature* **435**, 1067 (2005).
23. Marcy, G.W., et al. *ApJ* **556**, 296 (1995).
24. Mayor, M., Queloz, D. *Nature* **378**, 355 (1995).
25. Moutou, C., Pont, F., Bouchy, F., Mayor, M. *A&A* **424**, 31 (2004).
26. Neuhaeuser, R., et al. *AN* **426**, 630 (2005).
27. Pont, F., et al. *A&A* **426**, 15 (2004).
28. Rivera, E.J., et al. *ApJ* **634**, 625 (2005).
29. Santos, N.C., Israelian, G., Mayor, M. *A&A* **415**, 391 (2004).
30. Santos, N.C., et al. *A&A* **398**, 363 (2003).
31. Sato, B., et al. *ApJ* **633**, 465 (2005).
32. Udalski, A., et al. *ApJL* **628**, 109 (2005).
33. Udalski, A., et al. *Acta Astron.* **52**, 1 (2002a).
34. Udalski, A., et al. *Acta Astron.* **52**, 11 (2002b).
35. Udalski, A., et al. *Acta Astron.* **52**, 317 (2002c).
36. Udalski, A., et al. *Acta Astron.* **53**, 133 (2003).
37. Udalski, A., et al. *Acta Astron.* **54**, 313 (2004).
38. Vidal-Madjar, A. et al. *Nature* **422**, 143 (2003).

ARISTARCHOS: The New 2.3 m Greek Telescope

C.D. Goudis, P. Hantzios, P. Boumis and E. M. Xilouris

Institute of Astronomy & Astrophysics, National Observatory of Athens, Greece

Abstract. The state of the telescope, now fully assembled and installed at the Helmos Observatory is presented. The completion of our new Transient Aristarchos Spectrometer (ATS), the arrival of the refurbished high resolution Manchester Echelle Spectrometer (MES-AT) and the progress made towards the construction of the Vernikos-Eugenides CCD Camera (VEC), in combination with our presence in OPTICON, maintain alive all our hopes for an eventual successful and fruitful international operation of the ARISTARCHOS telescope.

Keywords: Telescopes, Astronomical instruments.
PACS: 95.45.+i

GENERAL OUTLINE

The Helmos Observatory is equipped with the 2.3 m ARISTARCHOS optical telescope built by Carl Zeiss Jena. The telescope is located on mountain Helmos in North-Central Peloponnese at an altitude of 2340 m and it is one of the darkest locations in Europe (see Fig. 1). The telescope is now in the final testing phase and its operation is anticipated in summer 2006. The telescope will be equipped with two CCD cameras and two spectrometers.

The 2.3 m ARISTARCHOS telescope is part of the OPTICON consortium of medium size (2-4 m class) telescopes and is expected (after its successful beginning of operation; summer 2006) to be used as an international facility, through OPTICON and through collaborative programs, providing state-of-the-art instrumentation and facilities.

The HELMOS Observatory

The National Observatory of Athens completed the construction of the new astronomical station on the top of mount Helmos (Longitude = 22° 13' E, Latitude = 37° 59' N) at an altitude of 2340 m (nearby the city of Kalavryta). Road access, infrastructure, meteorological conditions, darkness of the night sky (see Fig. 1) and image quality were the criteria for the selection of the Helmos mountaintop.

CP848, *Recent Advances in Astronomy and Astrophysics,*
7th *International Conference of the Hellenic Astronomical Society,* edited by N. H. Solomos

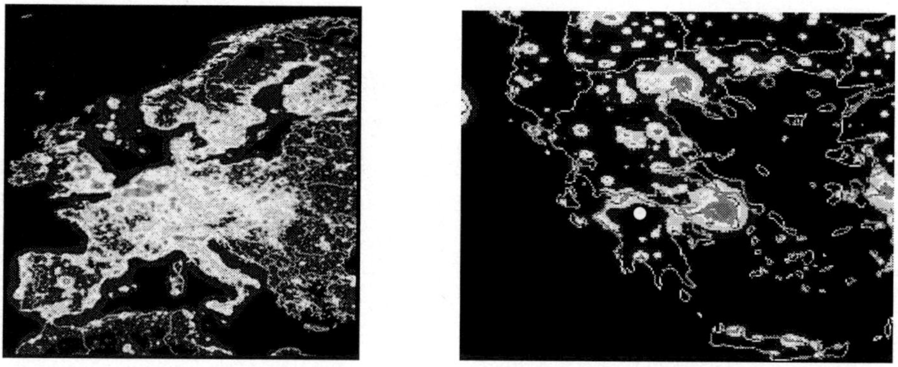

Figure 1.: The maps of light pollution in Europe (left panel) and in Greece (right panel) taken from Cinzano et al [1]. The white circle in North-Central Peloponnese indicates the location of the telescope site.

The astronomical station consists of two main compartments: the telescope tower and the control building, with a distance of 35 meters between them (see Fig. 2). This way, human activity and the heat radiated from engines, computers, etc will not affect the performance of the telescope and particularly the image quality.

Figure 2.: The Helmos observing site (left panel) and the enclosure with the telescope (right panel).

The Helmos Astronomical Station disposes the necessary infrastructure for the support of the observing activities and operation. The site (250 km from Athens) is accessible by a four-wheel-drive car. The road is asphalt paved except the last 8 km which are through a mountain road which was vastly improved recently (see Fig. 3).

Figure 3.: Part of the road through the woods (left panel) and the mountaintop (right panel).

The telescope is, for the moment, running on electrical power from two generators (of 75 kVA and 12 kVA). Connection with the Public Electric Company network is anticipated from summer 2006. The connection of the astronomical station with the headquarters of our Institute in Athens will be effected through a microwave link. The capacity of the link is 2 Mb/s with the ability to be extended to 11 Mb/s. Finally, the astronomical station is equipped with a computer network (server, work stations, PCs and related items), which will support the operation of the telescope, the astronomical instruments, communications, etc.

The 2.3m ARISTARCHOS Telescope

The 2.3 m ARISTARCHOS telescope (Fig. 4) utilizes new technologies and techniques arising mainly as a spin-off from the development of the new generation of 8-metre optical/IR telescopes. The modern design of the telescope makes it unique in this class of telescopes, since technologies developed for the new, largest 8-10 m telescopes are now applied to medium-size telescopes (2.3 m ARISTARCHOS, 2 m Liverpool telescope, 2.5 m VST/ESO). A back-illuminated 1024 x 1024 pixel CCD camera, as well as a filter-wheel, are already attached to the telescope.

Figure 4.: The ARISTARCHOS telescope.

Mechanical Structure

The alt-azimuth design makes the telescope lighter and easier to construct, whereas at the same time offers a number of instrument positions which are easily accessible by a simple movement of a small flat mirror. It has four cassegrain side ports able to bear instruments up to a total of 300 kgr, while the weight of one instrument can be up to 100 kgr, and one main cassegrain port sustaining an instrument up to 300 kgr (by distance of the center of gravity of the instrument less or equal to 300 mm from the instrument flange). The total weight of the moving parts is 29.000 kgr rising up to 34.000 kgr if the rest of the infrastructure is included.

Optics

The 2.3 m ARISTARCHOS telescope is a Ritchey Chretien telescope, consisting of a Sitall primary mirror of 2.3 meters in diameter with a focal ratio f/2.4, and a Sitall secondary mirror of 74 cm in diameter with focal ratio f/3.1. The final focal ratio of the telescope is f/8.

The optical design is such as to offer an uncorrected field of view of ~10 arcmin (in the four side ports) and a corrected field of view of ~1 degree (in the main cassegrain port) with an image scale of 1.17 arcsec/mm. The telescope has been designed to provide image quality of less than 0.35 arcsec spreading for 80% of the encircled energy.

Performance

Some useful parameters characterizing the performance of the telescope are:

Range of movement in azimuth	:	+/- 200 degrees
Range of movement in altitude	:	+/- 90 degrees
Rate of movement in azimuth	:	max 2 degrees per second
Rate of movement in altitude	:	max 2 degrees per second
Accuracy of pointing:		< 4 arcsec up to zenith distances of 70 degrees
Accuracy of tracking in open loop	:	< 0.5 arcsec in 10 minutes
		< 2 arcsec in 1 hour
Accuracy of tracking in closed loop	:	< 0.25 arcsec in 10 minutes
		< 0.5 arcsec in 1 hour
Accuracy of rotator tracking	:	0.25 arcsec in 10 minutes
		0.5 arcsec in 1 hour
Radius of Zenith blind spot	:	2 degrees
Operating temperature range	:	from – 10 C to + 35 C
Operating humidity range	:	up to 80% relative humidity
Wind tolerance	:	up to 15 m/s (in operation)
		>15 m/s (parked)
Earthquake resistance:		accelerations up to 2m/ square sec in any direction

As has already pointed out, the telescope building design is such as to minimize any heat trapped within the structure which could degrade the ambient 'seeing' (control room outside telescope building, oil cooling system, heat extractor, advanced ventilation of building, advanced shielding and insulation of building). The control system is designed to support remote-control operation of the telescope.

INTSRUMENTATION

The modern design of the 2.3 m ARISTARCHOS telescope, similar to that of the design of 8-10 m class telescopes, along with the excellent dark-sky site on mount Helmos that this telescope is installed at, makes it a very unique telescope in its class with which a plethora of research projects can be performed. Several instruments have already been developed and are ready to be tested and installed to the telescope [the "ARISTARCHOS Transient Spectrometer" (ATS) and the "Manchester Echelle Spectrometer-ARISTARCHOS Telescope" (MES-AT)] and others are currently being developed [the "Vernikos-Eugenedis Camera" (VEC)].

(a) "First light" CCD camera This CCD camera is a back-illuminated 1024 x 1024 pixel CCD camera (liquid nitrogen cooled). The pixel size is 24 microns and the camera covers a 5 arcminute field-of-view on the sky. This camera is already installed on the telescope and the "first light" images have already been taken (Fig. 5).

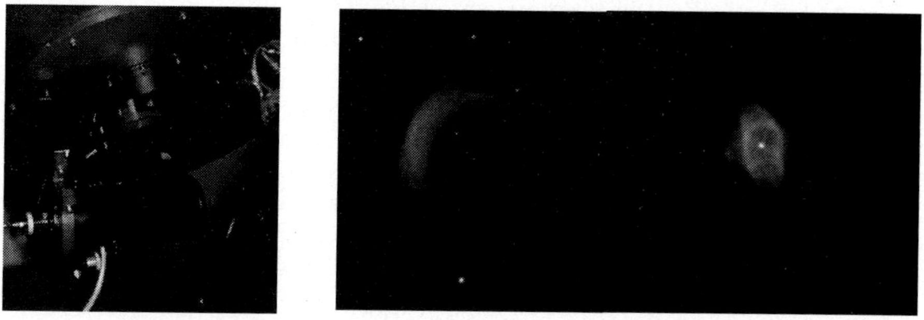

Figure 5.: The CCD camera attached to the telescope's main port (left panel) and the "First light" images, of the Ring and the Saturn planetary nebulae, taken with this camera (right panel).

(b) Aristarchos Transient Spectrometer (ATS)

The Aristarchos Transient Spectrometer (ATS; see Fig. 6), is a low/medium dispersion spectrometer with a Red 1200 groove/mm grating centred on 6000 Å with a 2.5 Å resolution and 103 Å/mm, a Green 1200 groove/mm grating centred on 5000 Å with a 2.5 Å resolution and 95 Å/mm as well as a 600 groove/mm grating centred on 6000 Å with a resolution of 6 Å and 245 Å/mm. The wavelength range of this spectrometer is between 4270 Å and 7730 Å. The spectrometer is designed to work with a dedicated CCD camera (AP7 back-illuminated 500 x 500 chip, thermo-cooled). The spectrometer's long slit input is fed by 50 fibers (each 50 micron diameter) in a 10 arcsec diameter, bundle in the telescope's focal plane. It is designed and fabricated in the University of Manchester, and is similar in design with the spectrometer of the 2 m Liverpool Telescope in La Palma but more evolved with 3 gratings available instead of one. The spectrometer can obtain sufficient quality spectra (signal-to-noise of 10) down to B=19th magnitude for point sources. This instrument is already assembled and it is immediately, after some final tests, available to be adapted to the telescope.

Figure 6: The ATS spectrometer.

(c) Manchester Echelle Spectrometer – Aristarchos Telescope (MES-AT)

The Manchester Echelle Spectrometer (MES-AT; see Fig. 7) is a high resolution spectrometer which has been successfully used in the past on the 3.9 m Anglo-Australian Telescope (AAT - Meaburn et al. [2]), on the 2.5 m Isaac Newton Telescope (INT) and 4.2 m William Herschel Telescope (WHT - Meaburn & Bryce [3]). A duplicate of this spectrometer is currently used with the 2.1 m San Pedro Martin telescope (SPM - Meaburn et al. [4]). In its primary mode a single order of its echelle grating (nominally δ=63.54° with 31.6 grooves/mm) is isolated by a broad, efficient interference filter eliminating the need for cross-dispersion. Consequently, its primary use is to obtain spatially-resolved profiles of individual emission lines from faint extended sources emitting in the range 3900-9000 Å with a spectral resolving power of $\lambda/\Delta\lambda \leq 10^5$. Several secondary modes are available, a direct image of the field can be obtained by both the insertion of a clear area to replace the slit and of a mirror before the grating. Slit positions against images of extended sources can be obtained using this mode. Furthermore, insertion of a grism along with the plane mirror permits long-slit, low-dispersion (76.3 Å/mm) spectra to be obtained. This spectrometer will be equipped with the 1000 x 1000 (24 micron) pixel nitrogen-cooled CCD camera which is currently used as a testing instrument on the telescope.

Figure 7: The MES-AT spectrometer.

(d) Vernikos-Eugenides Camera (VEC).

The Vernikos-Eugenedis Camera (VEC; see Fig. 8) is a back-illuminated 4000 x 4000 pixel CCD camera (Fairchild CCD486, liquid nitrogen cooled).

Figure 8: The VEC CCD camera.

The pixel size is 15 microns and the camera covers a 10 arcminute field-of-view on the sky.

REFERENCES

1. Cinzano et al., 2001, MNRAS, 328, 689
2. Meaburn J., Blundell B., Carling R., Gregory D.E., Keir D.F. & Wynne C.G., 1984, MNRAS, 210, 463
3. Meaburn J. & Bryce M., 1993, Opt. Astr., 32nd Herst.Conf., p.9
4. Meaburn J., Lopez J.A., Gutierrez L., Quiroz F., Murillo J.M., Valdez J. & Petrayez M., 2003, RMAA, 39, 185.

The high-speed camera ULTRACAM

T.R. Marsh* and V.S. Dhillon†

*Department of Physics, University of Warwick, Coventry CV4 7AL, United Kingdom
†Department of Physics and Astronomy, University of Sheffield, Sheffield S3 7RH, United Kingdom

Abstract. ULTRACAM is a high-speed, tri-band CCD camera designed for observations of time variable celestial objects. Commissioned on the 4.2m WHT in La Palma, it has now been used for observations of many types of phenomena and objects including stellar occultations, accreting black-holes, neutron stars and white dwarfs, pulsars, eclipsing binaries and pulsating stars. In this paper we describe the salient features of ULTRACAM and discuss some of the results of its use.

Keywords: High-speed photometry
PACS: 95.55.Aq, 95.75.De, 97.30.Qt

INTRODUCTION

The high quantum efficiency of CCDs and their 2D imaging properties have lead to their displacing photomultipliers for most astronomical observations. However, in the area of high-speed observations these advantages are negated by slow readout times, with several 10s of seconds not uncommon. This started to change with the use of frame-transfer CCDs in which exposures are rapidly shifted into a masked area which is read out while the next exposure is taking place [1, 2]. Still, these single-CCD photometers did not match the most advanced photoelectric instruments which feature multiple colour channels [3]. Thus in the late 1990s we set out to build a high-speed CCD camera which could observe simultaneously in more than one band. We call this camera ULTRACAM, and it was first commissioned on the 4.2m WHT in May 2002 and in May 2005 on the visitor focus of the 8m UT3 of the VLT. This paper briefly describes ULTRACAM and some of the science it has been used to address.

TECHNICAL SUMMARY

The salient features of ULTRACAM are as follows[1]. *3 channels:* dichroics are used to separate the light in ultraviolet (SDSS u'), green (g') and red (r', i' or z') channels. Several narrow-band filters are also available. *Frame-transfer CCDs:* three E2V 47-20 frame-transfer CCDs are used to detect the light. These have imaging areas of 1024x1024 pixels and fields of 5 and 2.5 arcminutes on a side on the 4.2 WHT and 8m VLT respectively. They allow sampling rates from 0.01 to several 100 Hz with minimal deadtime. *GPS timing:* times are taken from the GPS satellites and are accurate to 50 μs. *Real-time reduction:* the software allows one to plot light-curves and indicators of the observing

[1] Full details are available at `http://www.shef.ac.uk/people/vdhillon/ultracam`

CP848, *Recent Advances in Astronomy and Astrophysics,*
7th International Conference of the Hellenic Astronomical Society, edited by N. H. Solomos

conditions as the data come in. *Portable:* the optics start with a collimator which can be changed allowing ULTRACAM to be mounted on different telescopes.

SCIENTIFIC RESULTS FROM ULTRACAM

ULTRACAM has been used in a wide variety of different applications. Observations of pulsating sdB stars to measure the wavelength variations of the pulsation amplitudes have aided the identification of the normal modes [4, 5]. It allows complete solutions of the physical parameters of faint eclipsing cataclysmic variable stars [6, 7], while in a challenging application, it was used to detect pulsations from the anomolous X-ray pulsar 4U 0142+61 ($I \sim 24$, $P = 8\,$s) [8]. ULTRACAM has been used to observe X-ray binaries [9] including a novel application in which a narrow-band filter in the green arm was used to detect flourescent emission from the face of the donor star [10]. Most recently the first results of a program to measure very precise eclipse timings of white dwarf/M dwarf binary stars were published [11]. Times of individual eclipses of the $V = 16.5$ NN Ser could be measured to a precision of ~ 0.05 seconds, allowing a measurement of a period decrease from observations covering only a 2 year baseline. The most novel application of all perhaps has been an attempt to detect small Kuiper belt objects from occultations of stars in the ecliptic plane [12].

ACKNOWLEDGMENTS

TRM acknowledges the support of a PPARC Senior Fellowship. We thank the staff of both the La Palma and Paranal observatories for their great help during observing runs.

REFERENCES

1. R. J. Stover, and S. L. Allen, *PASP* **99**, 877–886 (1987).
2. D. O'Donoghue, *Baltic Astronomy* **4**, 519–526 (1995).
3. H. Barwig, R. Schoembs, and C. Buckenmayer, *A&A* **175**, 327–344 (1987).
4. C. S. Jeffery, V. S. Dhillon, T. R. Marsh, and B. Ramachandran, *MNRAS* **352**, 699–707 (2004).
5. C. S. Jeffery, C. Aerts, V. S. Dhillon, T. R. Marsh, and B. T. Gänsicke, *MNRAS* **362**, 66–78 (2005).
6. W. J. Feline, V. S. Dhillon, T. R. Marsh, M. J. Stevenson, C. A. Watson, and C. S. Brinkworth, *MNRAS* **347**, 1173–1179 (2004).
7. W. J. Feline, V. S. Dhillon, T. R. Marsh, and C. S. Brinkworth, *MNRAS* **355**, 1–10 (2004).
8. V. S. Dhillon, T. R. Marsh, F. Hulleman, M. H. van Kerkwijk, A. Shearer, S. P. Littlefair, F. P. Gavriil, and V. M. Kaspi, *MNRAS* **363**, 609–614 (2005).
9. T. Shahbaz, V. S. Dhillon, T. R. Marsh, J. Casares, C. Zurita, P. A. Charles, C. A. Haswell, and R. I. Hynes, *MNRAS* **362**, 975–982 (2005).
10. J. Casares, T. Muñoz-Darias, I. G. Martínez-Pais, R. Cornelisse, P. A. Charles, T. R. Marsh, V. S. Dhillon, and D. Steeghs, "Echo Tomography of Sco X-1 using Bowen Fluorescence Lines," in *AIP Conf. Proc. 797: Interacting Binaries: Accretion, Evolution, and Outcomes*, 2005, pp. 365–370.
11. C. S. Brinkworth, T. R. Marsh, V. S. Dhillon, and C. Knigge, *MNRAS* **365**, 287–295 (2006).
12. F. Roques, A. Doressoundiram, V. Dhillon, T. Marsh, J. Kavellars, S. Bickerton, N. Peixinho, A. Fitzsimmons, F. Colas, J. Lecacheux, S. Pau, I. Belskaya, and O. Mousis, *AAS/Division for Planetary Sciences Meeting Abstracts* **36** (2004).

WASP0: A wide field CCD search for extra solar planets. Instrumentation, data analysis and preliminary results

D. Mislis*, S. Pyrzas†, E.T. Harlaftis**, R. A. Street‡, K. Horne§, D. Pollacco‡, J. H. Seiradakis†, S. R. Kane§, T. Lister§, Collier Cameron§ and W.I. Clarkson§

*Aristotle University of Thessaloniki, Department of Physics, Section of Astrophysics, Astronomy and Mechanics, GR-541 24 Thessaloniki, Greece
†University of Thessaloniki, Department of Physics, Section of Astrophysics, Astronomy and Mechanics, GR-541 24 Thessaloniki, Greece
**National Observatory of Athens, Institute of Astronomy and Astrophysics, Lofos Koufou, GR-152 36 Athens, Greece
‡Queen's University of Belfast, School of Physics, APS Division, Belfast, BT7 1NN, UK
§University of St Andrews, School of Physics Astronomy, North Haugh, St Andrews, Fife, KY16 9SS, UK

Abstract. We present the observing strategy, the method of analysis and first results from the WASP0-project, with data taken at the Kryonerion Astronomical Station of the National Observatory of Athens, in 2002. WASP0 (Wide Angle Search for Planets) is a project aiming to discover extrasolar planets using the transit method. Besides extrasolar planets, the project is suitable for detecting new variable stars, novae, asteroids, etc. For the reduction we used the 'SuperWASP pipeline' developed for this specific purpose by the University of St. Andrews and Queens University, UK, with in situ modifications for our computing facilities.

Keywords: Stars: variables, extra solar planets
PACS: 97.82.Cp,97.82.Fs

OBSERVATIONS AND FIRST STAGE ANALYSIS

The observations were undertaken at Kryonerion Astronomical Station, Greece, with the WASP0 Instrument. WASP0 is a wide-field (9°), 6.3 cm aperture F/2.8 Nikon camera lens, equipped with an Apogee-10 CCD detector. The detector is a $2K \times 2K$ chip with 16" pixels, built by Don Pollacco at Queen's University, Belfast (Figure 1). The Hyades Open Cluster was observed for 20 nights over a period of three months in winter 2002. The observed field contained about 25 000 stars to the limiting magnitude of our equipment. Between 80 and 120 photographs were taken every night, depending on the weather conditions. We used the Super Wasp pipeline to reduce the data. The pipeline estimates the fluxes, coordinates, standard deviation, RMS errors and air-mass of each star in the field.

CP848, *Recent Advances in Astronomy and Astrophysics,*
7th *International Conference of the Hellenic Astronomical Society,* edited by N. H. Solomos

FIGURE 1. Apogee 10 CCD from Kryonerion Observatory (Greece)

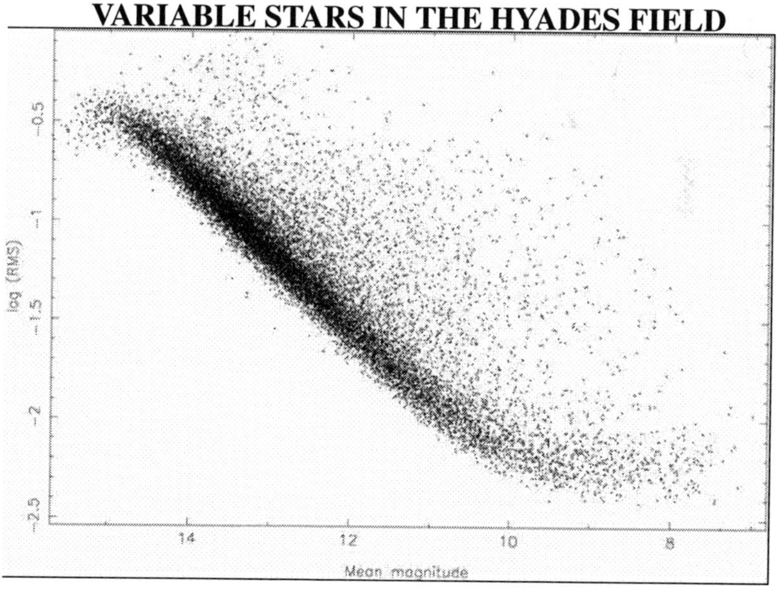

FIGURE 2. RMS vs Mean Magnitude from Hyades field. The variable stars are those with the highest distance from average linear plot

The variable stars exhibit the highest standard deviation among all stars. Therefore, they are easily detected by studying the stars with the highest standard deviation (Figure 2). We have discovered about 10 new variable stars up to now, none of which appear in

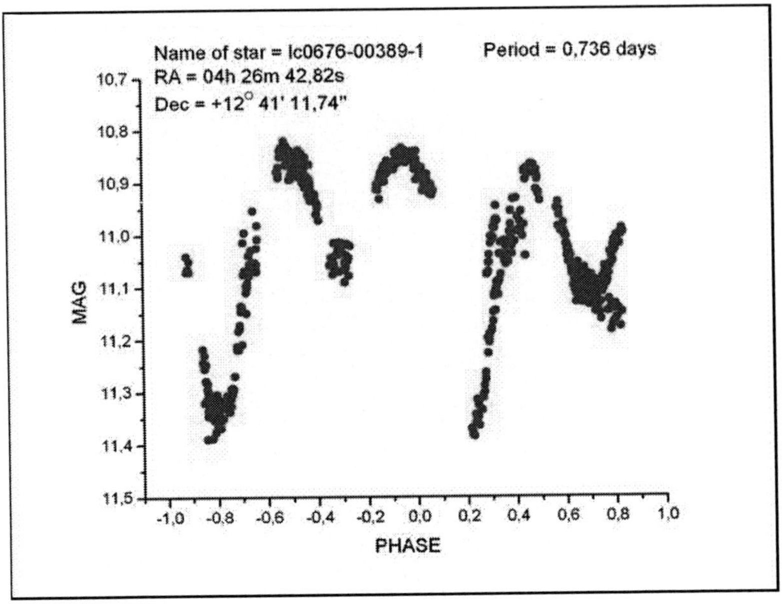

FIGURE 3. The First Algol type Variable Star from Kryonerion data

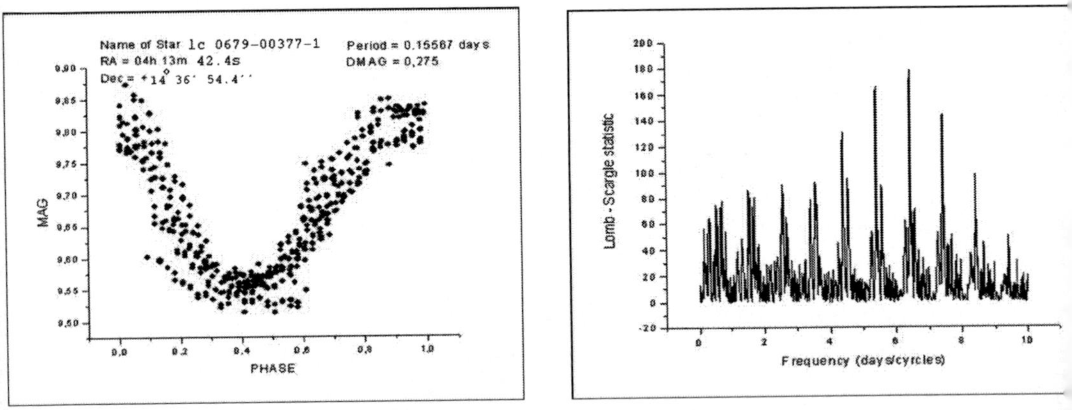

FIGURE 4. δ Scuti New Variable Star and the Periogram

the SIMBAD database. We have studied both the light curves and the periodograms of the new stars, some of which are shown below. The analysis is still underway. We present the light-curves of 3 new variable stars in the Hyades open cluster and the periodograms of 2 of them. The light curve of a new Algol-type (binary) system is presented (Figure 3) and the light curve and the periodogram of one δ–Scuti type star is depicted (Figure

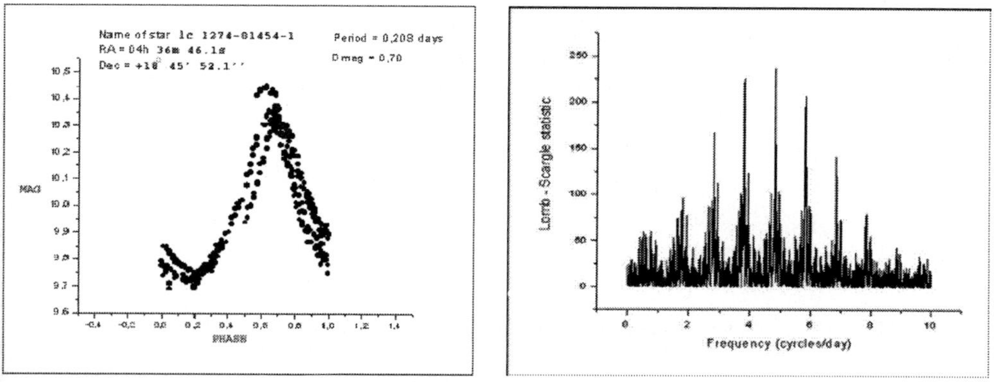

FIGURE 5. W UMa Variable Star and the Periogram

4). Also a WUma variable star was found (Figure 5).

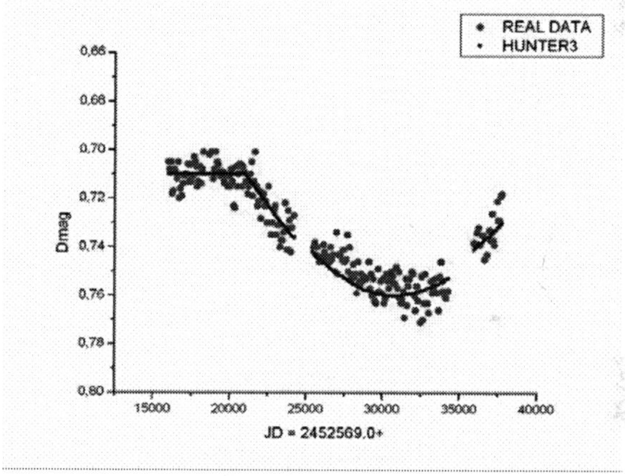

FIGURE 6. Theoretical Fitting from Hunter3 code in HD209458 data

PLANET HUNTING

We search for planets using the *Hunter3* code developed for this project. The code fits a number of models to the data., searching a four dimensional space, parametrized by the position of the event, the duration, the depth of the transit and the slope of the transit curve. We fit all possible theoretical transit models in the light curves of all 25000 stars in the field and compare the standard deviation of the data before and after subtracting the fitted curve. If the standard deviation is below a certain threshold, we call it a suspect-transit. Then we fit the same model to the light curves from the same star in other nights. If in one more night we find a suspect-transit with the same model, we call it real transit. In this way, we perform 7.1 trillion different combinations in the light curves of the

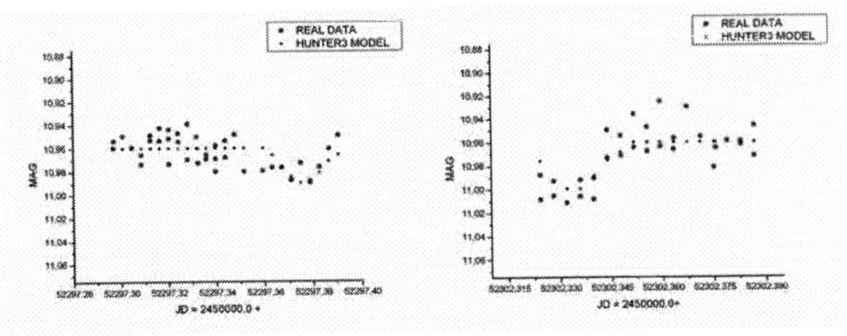

FIGURE 7. This two transits found in two seperate nights 22 and 27 January 2002

stars in our field. In order to test our technique and the *Hunter3* code, we observed and detected the well known planet, HD209458, under full moon conditions (Figure 6).

A NEW EXTRA SOLAR PLANET?

This graphs above (Figure 7) show the first suspect planets discovered by the transit method from observations with the WASP0 camera from Greece. The transit was found in two nights. The time between the two transits define the period (or multiples of the period). It was not detected in the data of other nights. This sets rigid constraints to the period of the transit. If it is a real extra solar planetary system, we can derive some useful information about the system.

Name : lc0679-00223-1
RA: $04^h14^m58.50^s$
Dec: $+14^o42'55.80$"
Radius = 3.21 RJ
Period: 4.956 days
Depth of transit: 0.021 mag
Duration of Transit: 32 min
Inclination of orbit, i : 82.35^o
Possibility of transit detection: 6.7%

FUTURE WORK

We plan to investigate the suspect extra solar planet, using small or medium size telescopes, equipped with more sensitive detectors and, in particular, with spectroscopic capabilities. Following these first results from the WASP0 prototype, we plan to continue working towards the detection of new variable stars and extra solar planets. We shall try to make the *Hunter*3 code more intelligent to fit planet transits with higher probability to detect physical planetary systems. We shall apply it in data that we already have from the SUPER WASP and WASP0 archive.

ACKNOWLEDGMENTS

The authors would like to thanks D.J. Christian (Belfast Open University), N.A. Evans (Keele), A. Fitzsimmons (Belfast), C.A. Haswell (Open University), C. Hellier (Keele), S.T. Hodgkin (Cambridge), F.P. Keenan (Belfast), A.J. Norton (Open University), R. Ryans (Belfast), I. Skillen (ING) , R.G. West (Leicester), P.J. Wheatley (Leicester), V. Manimanis and K. Gazeas (University of Athens), K. Stathoulis (Financial Technologies SA , Athens), G. Dimou (National Observatory of Athens), E. & H. Haralabidi (Thessaloniki)

Dedicated in the memory of Emilios Harlaftis, who passed away on Sunday February the 13th 2005, leaving a huge gap in the Greek astronomical community. We shall miss him. May he rest in peace.

REFERENCES

1. Street R.A., Lister T.A., Cameron A.C., Horne K., "Shadow of the Planet Orbiting HD 209458", *ASPC*, Vol. 219, 2000, pp. 621
2. Kane S.R., Cameron C.A., Horne K., James D., Lister T.A., Pollacco D.L., Street R.A., Tsiapas Y., "Results from the Wide Angle Search for Planets Prototype (WASP0) - I. Analysis of the Pegasus Field", *MNRAS*, Vol. 353, 2004, pp. 689-696
3. Kane S.R., Lister T.A., Cameron C.A., Horne K., James D., Pollacco D.L., Street R.A., Tsiapas Y., "Results from the Wide Angle Search for Planets Prototype (WASP0) - II. Stellar Variability in the Pegasus Field", *MNRAS*, Vol. 353, 2004, pp. 689-696
4. Kane S.R., Cameron C.A., Horne K., James D., Lister T.A., Pollacco D.L., Street R.A., Tsiapas Y., "Results from the Wide Angle Search for Planets Prototype (WASP0) - III. Planet Hunting in the Draco Field", *MNRAS*, Vol. 364, 2004, pp. 1091-1103
5. Hubbard W.B., Fortey J.J., Lunine J.I., "Theory of Extrasolar Giant Planet Transits", *SALT technical report, SAAO*, 2000

In Search Of Axions: The CAST Experiment

George K. Fanourakis
(For the CAST Collaboration[1])

Institute of Nuclear Physics, NCSR "Demokritos"
Neapoleos & Patr. Grigoriou, 15310 Ag. Paraskevi Attikis, Greece

Abstract. The CERN Axion Solar Telescope (CAST) experiment uses a decommissioned LHC test magnet in conjunction with three different technology X-ray photon detectors and tracks the sun to look for the elusive axion. Axions or axion-like particles with a two-photon interaction can be produced in the Sun by the Primakoff process, according to the theory. In a laboratory magnetic field they can be transformed into X-rays with energies of a few keV. CAST has been running for about 6 months during 2003 and most of 2004. The first results from the analysis of the 2003 data are presented here. No signal above background was observed, implying an upper limit to the axion to photon coupling $g_{\alpha\gamma\gamma} < 1.16 \times 10^{-10}$ GeV^{-1} at 95% CL for $m_\alpha \leq 0.02$ eV. This limit is comparable to the limit from stellar energy loss arguments and considerably more restrictive than any previous experiment in this axion mass range.

Keywords: axions, dark matter, X-ray detection, Micromegas.
PACS: 95.35.+d; 14.80.Mz; 96.60.Vg; 07.85Nc; 84.71Ba.

INTRODUCTION

The Lagrangian of Quantum Chromodynamics (QCD), the gauge theory of the strong interactions of quarks and gluons, is constructed with the requirement that it is Lorentz as well as local gauge invariant. Denoting the fermion matter fields (quarks) by ψ_n, N the number of quark flavors, the gauge fields (gluons) by A^α, g the strong

---

[1] The CAST Collaboration:
S. Andriamonje[2], V. Arsov[13], S. Aune[2], D. Autiero,[1] F. Avignone[3], K. Barth[1], A. Belov[11], B. Beltran[6], H. Brauninger[5], J. M. Carmona[6], S. Cebrian[6], E. Chesi[1], J. I. Collar[7], R. Creswick[3], T. Dafni[4], M. Davenport[1], T.Decker[16], L. Di Lella[1], C. Eleftheriadis[8], J. Englhauser[5], G. Fanourakis[9], H. Farach[3], E. Ferrer[2], H. Fischer[10], J. Franz[10], P. Friedrich[5], T. Geralis[9], I. Giomataris[2], S. Gninenko[11], N. Goloubev[11], M. D. Hasinoff[12], F. H. Heinsius[10], R. Hill[16], D.H.H. Hoffmann[4 1], I.G. Irastorza[2], J. Jacoby[13], D. Kang[10], K. Koenigsmann[10], R. Kotthaus[14], M. Krcmar[15], K. Kousouris[9], M. Kuster[4,5], B. Lakic[15], C. Lasseur[1], A. Liolios[8], A. Ljubicic[15], G. Lutz[14], G. Luzon[6], D. W. Miller[7], A. Morales[6], J. Morales[6], M. Mutterer[4], A. Nikolaidis[8], A. Ortiz[6], T. Papaevangelou[1], M. Pivovaroff[16], A. Placci[1], G. Raffelt[14], J. Ruz[6], H. Riege[4], M. L. Sarsa[6], I. Savvidis[8], W. Serber[14], P. Serpico[14], Y. Semertzidis[4], R. Soufli[16], L. Stewart[1], K. Van Bibber[16], J. D. Vieira[7], J. Villar[6], L. Walckiers[1], K. Zachariadou[9], and K. Zioutas[8].
[1]European Organization for Nuclear Research (CERN), Geneve, Switzerland. [2]DAPNIA, Centre d' Etudes Nucleaires de Saclay (CEA-Saclay), Gif-sur-Yvette, France, [3]Department of Physics and Astronomy, University of South Carolina, Columbia, SC, USA, [4]GSI-Darmstadt and Institut fur Kernphysik, TU Darmstadt, Darmstadt, Germany, [5]Max-Planck-Institut fur Extraterrestrische Physik, Garching, Germany, [6]Instituto de Fisica Nuclear y Altas Energıas, Universidad de Zaragoza, Zaragoza, Spain, [7]Enrico Fermi Institute and KICP, University of Chicago, Chicago, IL, USA, [8]Aristotle University of Thessaloniki, Thessaloniki, Greece, [9]National Center for Scientific Research "Demokritos", Athens, Greece, [10]Albert-Ludwigs-Universitaet Freiburg, Freiburg, Germany, [11]Institute for Nuclear Research (INR), Russian Academy of Sciences, Moscow, Russia, [12]Department of Physics and Astronomy, University of British Columbia, Vancouver, Canada, [13]Johann Wolfgang Goethe-Universitaet, Institut fur Angewandte Physik, Frankfurt am Main, Germany, [14]Max-Planck-Institut fur Physik (Werner-Heisenberg-Institut), Munich, Germany, [15]Rudjer Boskovic Institute, Zagreb, Croatia, [16]Lawrence Livermore National Laboratory, 7000 East Avenue, Livermore, CA94550, USA.

CP848, *Recent Advances in Astronomy and Astrophysics,*
7th International Conference of the Hellenic Astronomical Society, edited by N. H. Solomos

coupling constant and by t^{α} the generators of the $SU(3)_c$ color algebra of the rotations of the quark fields in the color space, $\psi_n \rightarrow \exp\{i\alpha^a(x)\, t^a\}\psi_n$ with $\alpha(x)$ an x-dependent parameter, which satisfy the relation: $[t^a, t^b] = i f^{ab}{}_c\, t^c$, with a, b, c = 1,…,8, and $f_{\alpha bc}$ the structure constants of this algebra, the Lagrangian can be written as:

$$L = \sum_{n=1}^{N} \overline{\psi}_n \left(i\gamma^{\mu}\partial_{\mu} - g\gamma^{\mu} A_{\mu}^{a} t^{a} - m_n \right) \psi_n - \frac{1}{4} G^{a\mu\nu} G_{\mu\nu}^{a} \tag{1}$$

With $G_{\mu\nu}^{a}$ the kinetic term of the gluons, $G_{\mu\nu}^{a} = \partial_{\mu} A_{\nu}^{a} - \partial_{\nu} A_{\mu}^{a} - g f_{abc} A_{\mu}^{b} A_{\nu}^{c}$ (2)

The inherent local gauge symmetry of QCD is not spontaneously broken so that the gluons remain massless at all scales. The non Abelian character of this theory introduces non linear phenomena which result in a non trivial topological structure of the vacuum. In fact the true vacuum is actually a linear combination of an infinite number of degenerate vacua, denoted by |n>, the so-called θ vacuum:

$$|\theta\rangle = \sum_{n=-\infty}^{\infty} e^{in\theta} |n\rangle \tag{3}$$

The parameter θ is an observable and as a consequence it must be included in the QCD Lagrangian in an additional effective term which is CP violating:

$$L_{\theta} = \frac{g^2 \theta}{32\pi^2} G_{\mu\nu}^{a} \widetilde{G}^{a\mu\nu} \tag{4}$$

The QCD Lagrangian, in the limit of zero up and down quark masses, has a large global symmetry described by the group $U(2)_V$**X**$U(2)_A$. For small masses this is approximately true. Indeed experimentally we know that $U(2)_V = SU(2)_V$**X**$U_V(1)$ which is the Isospin**X**Baryon symmetry is true and results in the (p,n) and (π^{+-}, π^0) multiplets in the hadron spectrum. However, the axial symmetry $U(2)_A$ is spontaneously broken and one would expect four light mesons. Only three pions are observed though and this is known as the $U(1)_A$ problem. The resolution of this problem comes again from the complexity of the QCD vacuum which makes $U(1)_A$ not a quantum symmetry of QCD even though it is a classical symmetry in the limit of zero quark masses.

Finally in the process of spontaneously symmetry breaking of the local gauge symmetry in the strong and electroweak interactions, the quarks acquire in general complex masses and in order to obtain real values one has to diagonalize the quark mass matrix M, a fact which inserts another CP violating term in the QCD Lagrangian:

$$L_m = \frac{g^2 Arg(\det M)}{32\pi^2} G_{\mu\nu}^{a} \widetilde{G}^{a\mu\nu} \tag{5}$$

The total θ parameter is then: $\overline{\theta} = \theta + Arg(\det M)$

If based on this information we calculate the neutron electron dipole moment, we get $d_n \approx 10^{-16}\ \bar{\theta}$ e·cm. Comparing with the experimental limit $d_n < 0.63 \times 10^{-25}$ e·cm, we deduce that θ has the very small value of: $\bar{\theta} \approx 10^{-9}$[1]. Thus the strong CP problem appears as the question of why this parameter coming from the strong and weak sector is so small.

The most elegant solution of the strong CP problem comes from the work of Peccei and Quinn [2] who introduced an additional axial global $U(1)_{PQ}$ symmetry, which is spontaneously broken and it replaces $\bar{\theta} = \theta + \mathrm{Arg}(\det M)$ with $a(x)/f_a$, where a(x) is a new Nambu-Goldstone (NG) field and f_a is the symmetry breaking scale. This axion field has non zero vacuum expectation value <a> and the Lagrangian written in terms of the physical axion $a_{ph}(x)=a(x)-\langle a\rangle$ has no longer the θ-term.

The original model of Peccei and Quinn as supplemented by Weinberg [3] and Wilczek [4] introduced the standard axion, an axion tied with the electroweak symmetry breaking scale. This axion, with near MeV mass and appreciable couplings to matter and radiation, would be easily detectable. However, the PQWW axion was subsequently ruled out by experiment. Later models such as the DFSZ model [5] and the KSVZ model [6] introduced the invisible axions, thus called because they are very weakly interacting since $f_a >> f_{ew}$. The detection of this type of axion is based on its coupling to two photons and many experiments have been and are being conducted to search for its existence. Two detailed reviews on the evolution of the theory and the axion searches are written by Kim [7] and Cheng [8].

AXION PROPERTIES AND ASTROPHYSICAL ISSUES

The axion, as every NG boson, couples derivatively to matter:

$$L_{a\psi\psi} = \sum_i c_i \frac{1}{f_a} \left(\bar{\psi}_i \gamma^{\mu} \gamma_5 \psi_i\right)\left(\partial_{\mu} a\right) \tag{6}$$

The parameters c_i are model dependent parameters, and i stand for the fermion matter fields. The axion is special because it couples to two gluons with a non-derivative term:

$$L_{agg} = \frac{1}{f_a} \frac{g}{8\pi} G^{a\mu\nu} \tilde{G}^{a}_{\mu\nu} a \tag{7}$$

At low energies this term generates the axion mass:

$$m_a = \frac{f_{\pi} m_{\pi}}{f_a} \frac{\sqrt{m_u m_d}}{m_u + m_d} = 0.6\, eV \left(\frac{10^7\, GeV}{f_a}\right) \tag{8}$$

All effective coupling constants of axions with matter and radiation depend on the inverse of the symmetry breaking scale f_a, and are therefore linear in the axion mass. The axion also couples to photons:

$$L_{a\gamma\gamma} = c_\gamma \frac{g}{\pi f_a} F^{\mu\nu} \widetilde{F}_{\mu\nu} a = -g_{a\gamma\gamma} \vec{E} \cdot \vec{B} a \qquad (9)$$

For the DFSZ axion we have c_γ=0.36 and for the KSVZ c_γ=-0.97. This coupling is very important for the detection of the axions.

One of the main axion production mechanisms, stemming from the two photon interaction thus common to all axion models, is the Primakoff process:

$$Z + \gamma(k) \rightarrow Z + a(k_a) \quad or \quad e(p_1) + \gamma(k) \rightarrow e(p_2) + a(k_a) \qquad (10)$$

Here Z is the nucleus, e=electron, γ=photon, a=axion and the symbols in parentheses represent the initial and final momenta of these particles.

Any light and weakly interacting scalar or pseudoscalar can be produced in hot plasmas and has the potential to cause energy loss in the stellar evolution [9]. Axion emission from stellar objects (red giants, stars in globular clusters, supernova 1987A, etc…) must be weak enough not to change their known evolutionary process. This puts a limit on the scale which, although model dependent, is roughly $f_a > 6*10^9$GeV, corresponding to a mass limit of m_a<0.01eV. On the other hand masses less than about 10 μeV cannot be allowed because so light axions would have abundantly produced in the early universe so that to overcome the closure density. Assuming that the main source of axion radiation is string loops rather than long strings the axions produced abundantly in the Big Bang would be the main candidate for cold dark matter if its mass is roughly in the range of 1μeV - 1 meV [10]. Axions with mass in the sub-eV to eV region could on the other hand be good candidates for the hot dark matter [11].

EXPERIMENTAL TECHNIQUES

Experimental efforts to discover the axion particle have been active since the proposal of the standard axion and continued all along the subsequent theoretical models [12,13]. The new generation axion search includes experiments looking for halo axions of very low masses and experiments looking for solar axions of masses in the allowed region:

The US Large Scale Dark Matter Axion Search experiment (ADMX), employing a large scale microwave cavity at LLNL to search for relic axions in our galactic halo [14]. In the presence of a magnetic field, an axion can decay into a single photon of the same energy, and the conversion probability is resonantly enhanced in a high-Q microwave cavity [15,16]. The technique involves a tunable helium-cooled high-Q cavity placed in the bore of a superconducting solenoid, and the resonant frequency of its lowest TM mode is slowly swept while the cavity output is monitored for excess power from resonant axion conversions. The expected signal is very weak, even with their expected local density of axions, 10^{14} cm^{-3} for a 10 μeV axion, the expected conversion power is extremely weak, ~10^{-23} W, so the physical temperature of the cavity and the intrinsic noise of the electronics must be reduced to minimal values. For the detection of the signal, Heterostructure Field Effect Transistor (HFET) microwave amplifiers are being used achieving total system noise temperatures of ~3 K and representing the world's quietest spectral radio receivers. The experiment is looking for axions with masses between 1.3 μeV and 13 μeV and has already reported null results in the range of 2.1 - 3.3 μeV [13]. An upgrade to ADMX experiment is being pursued in order to improve the signal to noise using SQUID amplifiers.

The University of Kioto experiment (CARRACK), seeks to exploit the extremely low-noise photon counting capability of Rydberg atoms in a Sikivie-type microwave cavity experiment [17]. Rydberg atoms are atomic systems that have been excited to near the ionization limit and in which an electron has been promoted to a level with large principal quantum number n (n~100). Rydberg atoms are quite sensitive as microwave photon detectors because the energy difference between adjacent levels is in the microwave region, they are highly efficient in absorbing these photons and the lifetime of the excited states is appropriate (1ms). This detector would scan the region of 2 to 30 μeV axion masses and results are reported for the 10 μeV [18].

The SOLAX [19] in Sierra Grande, COSME [20] in the Canfrac laboratory and DAMA in Gran Sasso [21], were designed as solar axion experiments which exploit the coherent conversion of axions on the crystalline planes in germanium detectors [22]. Indeed, solar axions could transform efficiently in electric crystal fields when the Bragg condition is satisfied. The obtained limits are not as restrictive as the solar axion experiments based on strong magnets described below.

The recent CERN Axion Solar Telescope (CAST) [23] and the earlier Japanese experiment in Tokyo [24] represent a more sensitive technique to detect solar axions with mass up to ~1 eV. The axion production in the sun's core is obtained by transforming thermal photons in the plasma to axions through the Primakoff process (figure 1a). The expected solar axion flux can be described by the following analytical fit [25,26,27]:

$$\frac{d\Phi_{\alpha}}{dE} = 6.02\times10^{10}\left(\frac{g_{\alpha\gamma}}{10^{-10}GeV^{-1}}\right)^{2}\left(\frac{E}{keV}\right)^{2.481} e^{\left(-\frac{E}{1.205keV}\right)} axions/cm^{2}/\sec/keV \qquad (11)$$

This expression resembles a Maxwell-Boltzmann with a mean value of ~ 4.2 keV, peaking at about 3.2 keV.

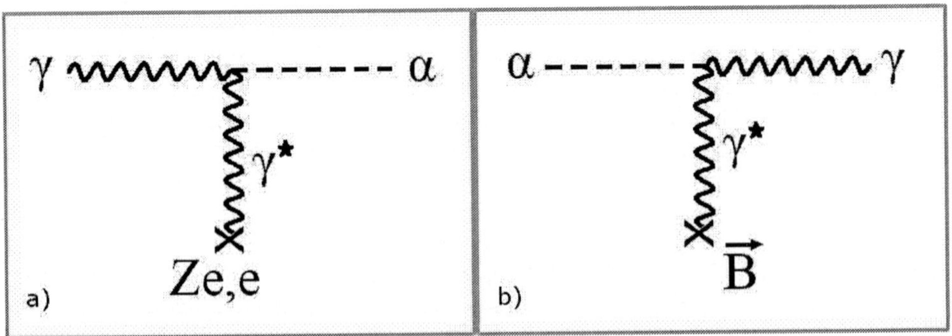

FIGURE 1. a) The axion production through the Primakoff process. b) The axion detection in a magnetic field through the inverse Primakoff process.

If the axions are allowed to enter a transverse magnetic field region, they can convert back to X-ray photons, which retain the original direction of axions, via the inverse Primakoff process (figure 1b) with a probability of:

$$P_{a\to\gamma} = \left(\frac{g_{\alpha\gamma}B}{q}\right)^2 \sin^2\left(\frac{qL}{2}\right) \tag{12}$$

With B the magnetic field strength, L the path length and q the axion-photon momentum difference which in the vacuum is: $q = m_\alpha^2/2E$. For $qL<<1$ this becomes $P_{\alpha\to\gamma} = (g_{\alpha\gamma}BL/2)^2$, so the X-ray flux does not explicitly depend on the axion mass:

$$\Phi_\gamma = 0.51 cm^{-2} day^{-1} \left(\frac{g_{\alpha\gamma}}{10^{-10}\,GeV^{-1}}\right)^4 \left(\frac{L}{9.26m}\right)^2 \left(\frac{B}{9.0\,T}\right)^2 \tag{13}$$

If, however, $qL\geq 1$ the axion-photon oscillation length is less than L the rate is reduced because of the axion-photon momentum mismatch. In this case one can use a low-Z gas in order to provide a refractive photon mass m_γ so that $q=|m_\gamma^2-m_\alpha^2|/2E$. If $m_\gamma \approx m_\alpha$ (as accomplished by adjusting the gas pressure), then the X-ray flux is restored to the above rate.

THE CAST EXPERIMENT

Experimental Details

The CAST experiment was designed to dramatically improve the sensitivity to the axion detection [23]. A 9.26m long decommissioned superconducting test LHC magnet was made operational again providing a 9 Tesla field and cross sectional area of 2x14.5 cm^2. The magnet is mounted on a rotating platform and is able to track the sun for about 1.5 hours during sunrise and 1.5 hours during sunset with an accuracy of better than 0.01°. In order to control the systematics, the experiment uses three different technology detectors to search for the X-rays expected by the axion conversions inside the two magnet bores. A conventional Time Projection Chamber (TPC) is mounted on one side of the magnet covering both bore openings and looks for sunset axions. At the other end two more X-ray detectors are looking for the sunrise axions. One novel Micromegas (Micromesh Gaseous) detector with X-Y readout strip structure [28,29] covers one bore and a Charge Coupled Device (CCD) detector [30], used in conjunction with an X-ray mirror telescope [31], covers the second bore. The X-ray telescope can produce an 'axion' image of ~ 6 mm^2 on the CCD, greatly reducing the effective background of the device. All detectors are made of low background materials and external shielding is used, when required, to keep the overall background rate to a minimum.

All three CAST detectors are recording 'axion' data during their corresponding tracking time while the rest of the time they are devoted to background (mostly) and calibration data taking. Figure 2 shows the CAST experiment in action.

FIGURE 2. View of the CAST experiment. The TPC, mounted at the left end of the magnet, searches for sunset axions whereas the Micromegas and CCD detectors, at the right end, look for sunrise axions.

Data Analysis And Results

CAST operated for a total of about 6 months in 2003 and the data taken are already analyzed and published [32]. The data from the experiment running in 2004 are at the final analysis stages. An important feature of the CAST data treatment, for the TPC and Micromegas data, is that the detector backgrounds are measured with about ten times longer exposure during the non tracking periods. The use of these data to estimate and subtract the true experimental background during sun tracking data is the most sensitive step in the CAST analysis. A more sensitive analysis is obtained with the CCD. The CCD measures the background and potential signal at the same time, since the signal is expected at the focusing area of the telescope, so an axion signal will be seen as excess of events in this area. It is worth mentioning the low background levels which can be reached by the CAST detectors after the filtering of the events. As an example the Micromegas detector, during 2003, reached a level of 2.0×10^{-5} events/keV/cm^2/s with better than 85% software efficiency [29].

The three detectors are analyzing their data independently, so in order for an observed excess of events over the background to be considered as a credible signal of axions it must be seen by all detectors. Each detector after filtering the tracking and the background data performs a comparison of the two sets. In the no-excess, no-signal case each analysis comes up with an individual result on the axion-photon coupling by either chi-square minimization (TPC and Micromegas) or the use of Likelihood method (CCD). Input for this analysis is the knowledge of the axion induced photon spectrum for fixed m_α, multiplied with the detector efficiency curves. The final answer is produced by combining the individual results.

As no axion signal was observed, the conclusion from the 2003 data analysis was a greatly improved limit for the axion-photon coupling $g_{\alpha\gamma}$. In the mass range of $m_\alpha \leq$ 0.02eV and with a confidence level of 95%, the CAST experiment obtained [32]:

$$g_{a\gamma\gamma} < 1.16 \times 10^{-10} GeV^{-1} \tag{14}$$

Figure 3 shows the obtained limit in comparison with the limits obtained from other experiments in the same axion mass range and also shows the limits imposed by astrophysical considerations. The results from the 2004 data analysis are expected to reach the planned sensitivity of the experiment.

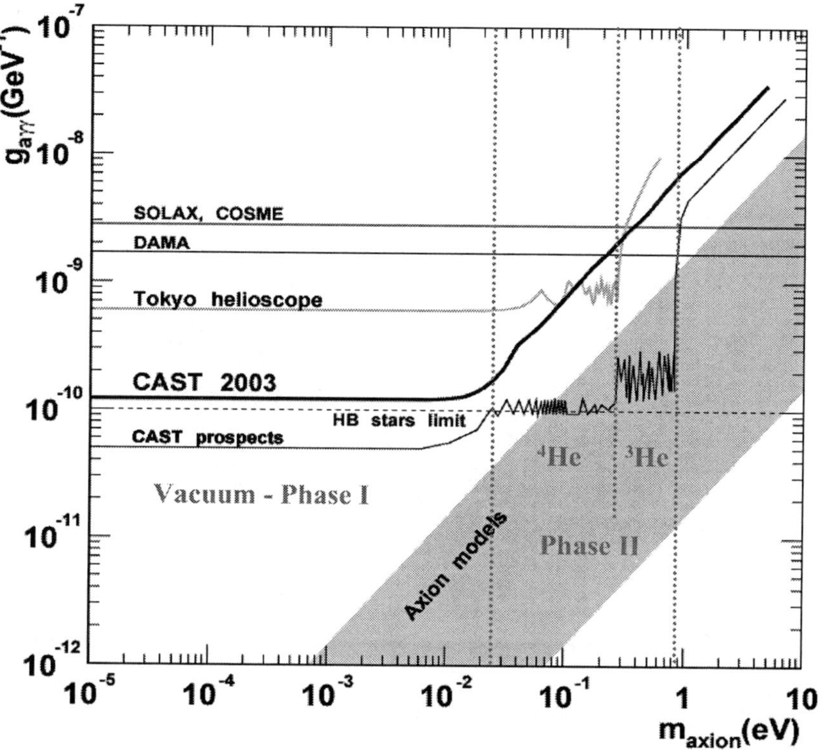

FIGURE 3. Exclusion limit (95% CL) from the CAST 2003 data compared with other constraints discussed in the text. The shaded band is the best motivated by typical theoretical models [33] to search for axions. Also shown is the future CAST sensitivity as foreseen in the experiment proposal.

Future Prospects

CAST, as seen in figure 3, has not reached the axion model band as yet. It is expected to enter this region in the already approved second phase of the experiment, in which search for higher mass axions is to be conducted. This phase is scheduled to start in October 2005 and involves filling the magnet bores with helium gas so that to obtain zero momentum difference between axion and photon. The vapor pressure of He^4 at 1.8^{o} K, is such that a maximum axion mass of 0.26 eV can be reached. The pressure settings will be incremented in steps to achieve maximum sensitivity for different mass values. Using He^3 that has a higher vapor pressure, one can reach a higher axion mass of up to 0.8 eV. For even higher masses an isolating gas cell in the bore will be required where He^3 at 5.4° K would allow reaching a mass of 1.4 eV.

An important upgrade of the Micromegas detector is planned, for phase II. The contribution of the Livermore group, a member of the CAST collaboration, involves the design and construction of an X-ray focusing telescope for use with a scaled down version of the Micromegas detector. This development combined with the addition of shielding around the Micromegas detector is expected to improve the background level of this detector by two orders of magnitude. Figure 4 shows the layout of the new Micromegas line.

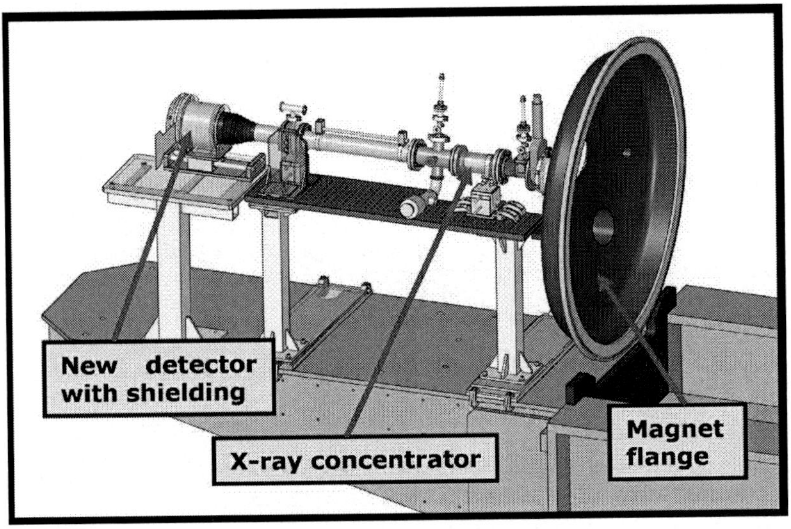

FIGURE 4. The schematic of the new Micromegas detector line for phase II of the CAST experiment is shown. The CCD line (next to Micromegas) is not shown for clarity.

CONCLUSIONS

A new generation of experiments is active and its detector arsenal continuously improving with the purpose to become more efficient for the search of the elusive axions. Axions constitute the most elegant solution to the strong CP problem and the theory will greatly benefit in predictive power and self consistency when this kind of particle is discovered.

The CAST experiment has been looking for axions of mass under 0.02eV during the 2003 and 2004 data taking periods. The result is an enormous improvement of the previous axion-photon coupling limit. During the second phase of the CAST experiment the axion mass region of at least up to 0.8 eV will be explored, by filling with appropriate helium gas the magnet bores. This way CAST enters the axion model region of figure 3, a region which is favored by theoretical modes.

ACKNOWLEDGMENTS

We thank CERN for hosting the experiment and the enormous help of CERN's technical and scientific personnel on practical matters. We acknowledge support from NSERC (Canada), MSES (Croatia), CEA (France), BMBF (Germany), GSRT (Greece), RFFR (Russia), CICyT (Spain), and NSF (USA). Work at LLNL was performed under the auspices of the U.S. Department of Energy by the University of California Lawrence Livermore National Laboratory under Contract No. W-7405-ENG-48. We also acknowledge support from the European RTN under contract No. HPRN-CT-2002-0029.

REFERENCES

1. V. Baluni, *Phys. Rev.* **D19**, 2227 (1979); R. Crewther, P. di Vecchia, G. Veneziano and E. Witten, *Phys. Lett.* **88B**, 123 (1979); **91B**, 487 (1980).
2. R.D. Peccei and H.R. Quinn, *Phys. Rev. Lett.* **38**, 1440 (1977); R.D. Peccei and H.R. Quinn, *Phys. Rev.* **D16**, 1791 (1977).
3. S. Weinberg, *Phys. Rev. Lett.* **40**, 223 (1978).
4. F. Wilczek, *Phys. Rev. Lett.* **40**, 279 (1978).
5. M. Dine, W. Fischler, and M. Srednicki, *Phys. Lett.* **B104**, 199 (1981); A.R. Zhitnitsky, *Yad. Fiz.* **31**,497 (1980) [*Sov. J. Nucl. Phys.* **31**, 260 (1980)].
6. J.E. Kim, *Phys. Rev. Lett.* **43**, 103 (1979); M.A. Shifman, A.I. Vainshtein and V.I. Zakharov, *Nucl. Phys.* **B166**, 493 (1980).
7. J.E. Kim, *Phys. Rep.* **150**, 1 (1987).
8. H.-Y. Cheng, *Phys. Rep.* **158**, 1 (1988).
9. G. G. Raffelt, "Axions and other very light bosons: astrophysical constraints," in Review of particle properties, K. Hagiwara et al., *Phys. Rev.* **D66**, 010001(2002).
10. S. Eidelman et al. (Particle Data Group), "Review of particle physics," *Phys. Lett. B* **592**, 1 (2004).
11. S. Hannestad, A. Mirrizi, G. Raffelt, "New cosmological mass limit on thermal relic axions", hep-ph/0504059 (2005).
12. C. Hagmann, K. Van Bibber, L.J. Rosenberg, "Axions and other very light bosons: experimental limits," in Review of particle properties, K. Hagiwara et al., *Phys. Rev.* **D66**, 010001(2002).
13. L.J Rosenberg, K.A. van Bibber, *Physics Reports* **325**, 1-39 (2000).
14. C. Hagmann et al., *Phys. Rev. Lett.* **80**, 2043 (1998).
15. P. Sikivie, *Phys. Rev. Lett.* **51**, 1415 (1983); **52**, 695 (1984)(E); *Phys. Rev.* **D 32**, 2988 (1985).
16. R. Bradley et al, *Rev. Mod. Phys.*, **Vol. 75, No. 3**, 777-817 (2003).
17. Matsuki, S., and K. Yamamoto, 1991, *Phys. Lett. B* **263**, 523 (1991); Ogawa, I., S. Matsuki, and K. Yamamoto, *Phys. Rev. D* **53**, R1740 (1996).
18. Yamamoto, K., *et al.*, *e-print hep-ph*/0101200 (2001); Tada, M., et al., *Phys. Lett. A* **303**, 285 (2002); Tada, M., *et al.*, *Nucl. Phys. B* **72**, 164 (1999).
19. A.O. Gattone et al., *Nucl. Phys. B (Proc. Suppl.)* **70**, 59 (1999).
20. A. Morales et al., *Astropart. Phys.* **16**, 325 (2002).
21. R. Bernabei et al., *Phys. Lett. B* **515**, 6 (2001).
22. E. A. Paschos and K. Zioutas, *Phys. Lett. B* **323**, 367 (1994).
23. K. Zioutas et al., *Nucl. Instrum. Methods A* **425**, 480 (1999).
24. S. Moriyama, et al., *Phys. Lett. B* **434** 147 (1998).
25. G. G. Raffelt, *Phys. Rev. D* **33**, 897 (1986); T. Altherr et al., *Astropart. Phys.* **2**, 175 (1994) [hep-ph/9310304].
26. J. N. Bahcall and M. H. Pinsonneault, *Phys. Rev. Lett.* **92**, 121301 (2004) [astro-ph/0402114].
27. G.G. Raffelt, hep-ph/0504152 V2, 2 May 2005.
28. Y. Giomataris et al., *Nucl. Instrum. Meth. A* **376**, 29(1996).
29. S. Andriamonje et al., *Nucl. Instrum. Meth. A* **535**, 309-313 (2004).
30. L.Strueder et al., 'The European Photon Imaging Camera on XMM-Newton: The pn-CCD camera", *Astronomy & Astrophysics*, **A 365**, L18-L26 (2001).
31. R. Kotthaus et al., astro-ph/0511390 (2005).
32. K. Zioutas et al., *Phys. Rev. Letters.* **94**, 121301 (2005).
33. D. B. Kaplan, *Nucl. Phys. B* **260**, 215 (1985).

Quantum Optics in Astrophysics: The Potential of a New Window

Nikolaos H. Solomos

Hellenic Naval Academy, Physics Sector, Hatzikyriakeion Piraeus, 18503, Greece,
National Astronomy Center "EUDOXOS", Mount Ainos, Kefallinia, Greece

Abstract.
The entire optical astronomy relies upon the detection of light. In this contribution, we put emphasis to the fact that a new window to the universe could be opened with the obvious idea of applying the ...Quantum Theory to describe incoming light Quanta (!). It is clearly the appropriate approach but, nevertheless, it never deemed necessary in main stream astrophysics: Customarily, traditional astronomy not only prefers time-averaged quantities, (although fluctuations in time of a measurement can be a source of information which is getting entirely lost in any time-averaged value) but misses much more information content by continuing to use old semi-classical approaches to treat photon detection processes. Thus, we fail to describe and appreciate in full very important properties of cosmic light, like spatiotemporal coherence. Nevertheless, 45 years of knowledge accumulation in Quantum Optics and technology can now result to the development of instruments capable to extract intimate quantum information scrambled in the incoming optical light fields from celestial sources, provided their ability to detect light emission alterations in the <ns timescales, can be demonstrated. In the emerging field of *Quantum or "Photonic" Astrophysics (QAp)* random fluctuations of light are not considered undesirable as unavoidable photon-noise that hampers the photometrical precision, but, instead, as a subject of investigation per se. Such light fluctuations can frequently be valuable as a rich source of information. Statistical fluctuations, are able to provide radically new diagnostic methods of physical conditions in astrophysical plasmas by directly identifying characteristic quantum signatures of light-generating or light-scattering processes. However, it has to be noted that, close enough to the quantum limit, the bosonic character of photon cannot be ignored and, therefore, a detailed Quantum Electrodynamical treatment of the interaction of the incoming photon stream with every component comprising the chain of physical systems from the source to the detecting instrument, has to be elaborated. To propagate the main ideas regarding the framework of QAp is not an easy task. Here, summarize the conceptual foundations and experimental aspects of such a new information channel and then describe the main strategy aspects, to develop the subject. We discuss also, briefly, the scientific rationale, prospects and difficulties of the above mentioned instruments and describe the aims of our key-program to contribute in their realization.

Keywords: photonic astronomy – quantum optics –new methods –non-thermal astrophysical processes
PACS: 42.50Ar, 95.75kk, 95.75Fg, 95.30

INTRODUCTION: THE NOISE IS THE SIGNAL!

In what follows, I review the physical basis and current status of the application of quantum optical approaches in time-domain astrophysical investigations. The idea is the following: We can extract energy from the EM fields generated by celestial sources by insertion of aerials or in the form of light quanta -using classical photon detectors-, but, from the point of view of theory we keep describing the light fields semi-classically:

CP848, *Recent Advances in Astronomy and Astrophysics,*
7th *International Conference of the Hellenic Astronomical Society,* edited by N. H. Solomos

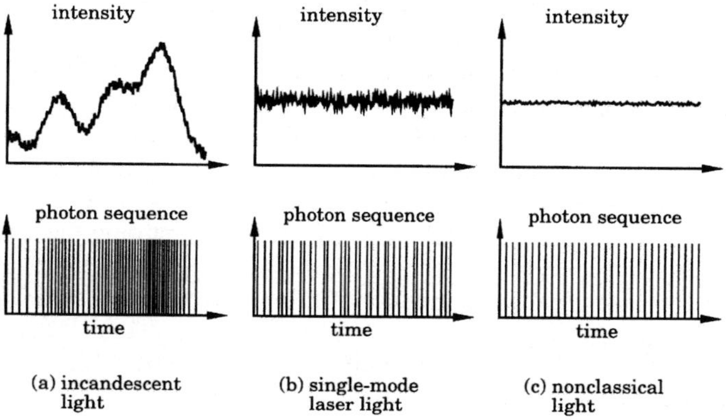

FIGURE 1. Classical and non-classical states of a light field distinguished from their fluctuations which are invisible if seeking for time averages (adopted from Lauternborn, Kurz, Wiesenfeldt: *Coherent Optics*, Springer-Verlag, 1995, with kind permission from Springer Science & Business Media)

i.e. applying EM-Wave Theory for the field and Quantum Theory for the detection. As a matter of fact,though, all types of light contain some feature which is demonstrably quantum-mechanical in origin which derives from the discreteness and bosonic character of light quanta which make up the field. In practice these non-classical features are ignored in the way we perform our measurements since for reasonably good models of large scale objects, it suffices to treat astrophysical light-fields as classical fluctuating sources, hence their predicted measurable properties are hardly influenced by their underlying quantum nature. Instead, if we do focus on fluctuation properties of light putting the photon noise as the subject of our investigation, any semi classical approach fails to reproduce statistical moments higher than the first-order mutual coherence function. Thus, the traditionally measurable properties of energy, momentum, polarization, position, time and coherence, being essentially "single photon" -classical- properties of light are predicted effectively, but the description is incomplete: Coherent and thermal fields are, for example, two distinct cases whose quantum characteristics are, for most practical purposes, indistinguishably scrambled. Other cases include light-field states which are totally non-classical having gross properties (i.e. photon number) below that of a perfectly coherent field like the sub-poissonian and squeezed states. Nevertheless the quantum properties of these inherently quantum mechanical fields are recoverable, at least in principle, through the careful scrutiny of their photon arrival temporal statistics.

Moreover, purely quantum mechanically predicted "multi-photonic" (collective) properties of the photon gas, associated with the actual "quantum states" of the light field established between the object and the detecting instruments, are not yet being investigated in astrophysics, although they could offer a potentially new astrophysical condition diagnostic tool (essentially a new window in astronomy), on the ground that the photon arrival times (or the number of photons emitted by a light source in a given time interval) in certain spatial points, are almost always random processes with *statistical properties depending on the very nature of the source under study.*

In fact, different physical processes in the generation of light or subsequent conditions in the course of its propagation, may imprint inherent quantum-statistical signatures (like different degrees of photon bunching in time) in light fields with otherwise identical spectrum, polarization, intensity etc. Despite the potential for a direct diagnostic tool of the physical condition within celestial sources, the field is not yet mature. We have to pursue a difficult in-depth theoretical examination of the expected *physical properties* of the incoming light and in parallel the development (and physical investigation) of new instrumentation lying on the technological frontier regarding the speed of detector response.

A theoretical appreciation of the time resolutions involved comes from Heisenberg's uncertainty principle $\sigma_E \sigma_t \geq \frac{h}{4\pi}$. Quantum correlation effects are fully developed and anticipated over time scales of the order of the inverse bandwidth of light, thus for 1 nm passband optical filter ($10^{12}Hz$), of $10^{-12}s$ duration. It thus appears well justified that simply working with *very short time scales* we could bring into emergence the above mentioned inherent quantum properties in optical light field which are scrambled by the classical instruments. Our whole knowledge-space on traditional astrophysics (except neutrino, gravitational, cosmic ray, in situ studies) is shown in Fig-2 where the unexplored domain targeted by the quantum optical initiatives mentioned, is depicted.

However, QAp. is *terra incognita*: There do not yet appear to exist established predictions for macroscopic processes in astrophysical sources that would be observable in the *ns* domain. We also have not yet confirm that such properties might ultimately become experimentally measurable for such sources. At present, even if we achieve such quantum-optical time resolutions needed to probe microscopic processes we face a gap in theoretical modelling due to the large scale complexity in the astrophysical sources, although the existence of regions of very low temperatures may partially alleviate the poor visibility of 'clean' quantum conditions. High speed astrophysics is in its infancy, thus much preparatory experimental and theoretical work is needed. In the last section, we describe indicative aspects of our program (1999-)to advance the subject in Greece [15].

THE INTERNAL WORKINGS OF QUANTUM ASTROPHYSICS

R. Glauber [9],[10] developed generic quantum mechanical methods to describe the results of photon correlation experiments. His construction has been and remains the only self-consistent theory able not only to explain the Hanboury-Brown & Twiss effect, but to retain its predictive power far beyond such a phenomenon. Although investigations on High Time Resolution Astrophysics involving mere classical observables (i.e. intensity) exist long ago, Dravins [4] was the first to have started in depth exploratory studies on the actual possibility to open the quantum optical information channel in astronomy, an attempt culminated to the conceptual design of the QuantEye instrument for ESO/OWL [6].

Glauber [10], was in fact able to devise a QED theory extension for light, combining both, a high explanatory and predicting efficiency (after the introduction of the new basis of "coherent-states" as well as the elaboration of the proper expressions describing the various degrees of coherence) and a preservation of the classical formalism of

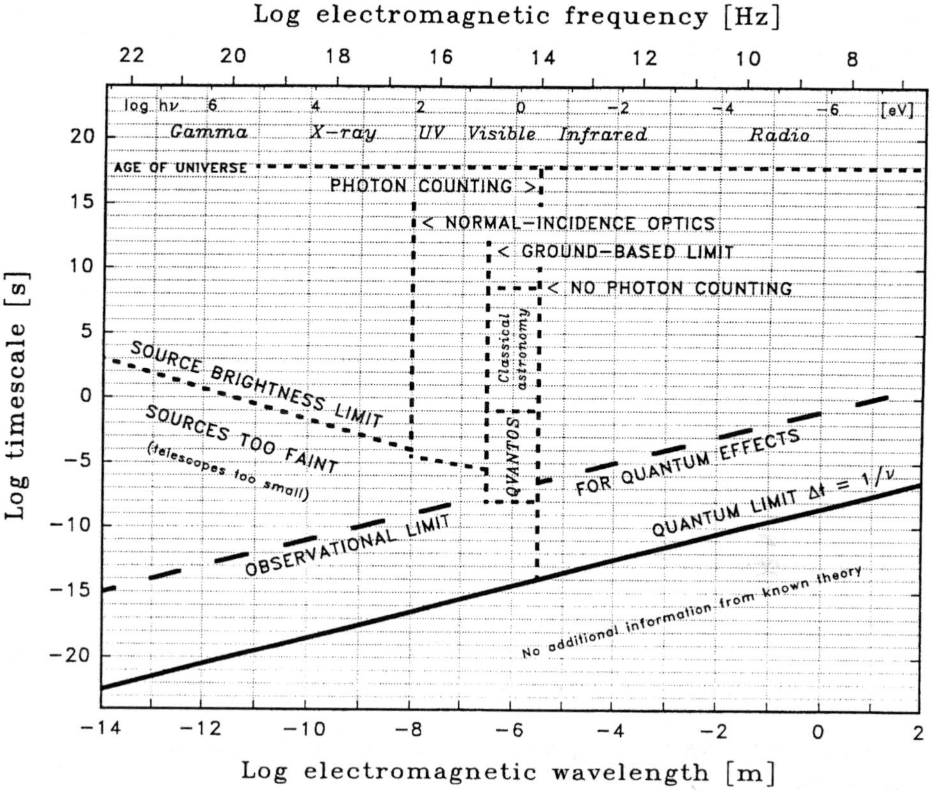

FIGURE 2. A panorama of our knowledge-space of astrophysics in parametric form -not all regions are accessible today (courtesy D.Dravins, [4])

electromagnetic Optics. Thus, he not only retained the legacy of the efficient vectorial formalism of Quantum Mechanics but, also, the familliar formalism of the radio- engineering description of light. His formalism on coherence properties evolves as follows: in single-photon experiments we describe light as an EM wave of a single linear polarization component whose electric field E contains terms of the type $e^{-i\omega t}$ for angular frequencies ω. By quantizing the electromagnetic field and taking into account that each single-photon detection alters the photon number states by removing one quantum from the field, he is able to formulate the proper expressions for the "quantum" degree of optical coherence of any order. The following expression, for example, is the form of second order coherence (g^2) in Glauber's formalism:

$$g^{(2)}(r_1t_1, r_2t_2; r_2t_2, r_1t_1) = \frac{\left\langle \widehat{E^-}(r_1,t_1)\cdot\widehat{E^-}(r_2,t_2)\cdot\widehat{E^+}(r_2,t_2)\cdot\widehat{E^+}(r_1,t_1)\right\rangle}{\left\langle \widehat{E^-}(r_1,t_1)\cdot\widehat{E^+}(r_1,t_1)\right\rangle\cdot\left\langle \widehat{E^-}(r_2,t_2)\cdot\widehat{E^+}(r_2,t_2)\right\rangle}$$

where $\widehat{E^-}(r,t)$ and $\widehat{E^+}(r,t)$are the photon-creation and photon-annihilation operators respectively (acting on quantum state vectors) as a function of position in space and time, and angle-bracket expressions denote ensemble averages evaluated in accordance with $\left\langle \widehat{E^-}(r_1,t_1)\cdot\widehat{E^+}(r_2,t_2)\right\rangle = Tr\left(\hat{\rho}\left\langle \widehat{E^-}(r_1,t_1)\cdot\widehat{E^+}(r_2,t_2)\right\rangle\right)$ with Tr denoting Trace and $\hat{\rho}$ the density-matrix operator. Glauber showed that arbitrary states of light can be decomposed in a series of 1st, 2nd, 3rd ... order coherence functions corresponding to single-, two-, three-...etc. photon-correlations. $G^{(2)}$ for example (which is often normalized to the 2nd -order coherence function of light ($g^{(2)}$)) describes two-photon properties of light since a single-photon detection (measurement of I) enters twice. This powerful formalism leads to calculations of final states from initial ones, as well as the expected photocount distributions on the presumption that the light field under consideration is known.

A good theory, though, is significant not only for its power to predict outcomes of future experiments but, -equally important- for changing the way we look at the old ones:

With all our classical optical instruments we are in fact performing "one-photon experiments" i.e. measure light properties that can be deduced from the first-order correlation function of light $G^{(1)}(r,t)$, for two coordinates in space r and time t. What our existing instruments actually measure in "classical astronomy" can be appreciated if we express the 1st-order coherence function as $G^{(1)}(r_1,t_1 : r_2,t_2) = \overline{E^*(\mathbf{r}_1,t_1).E(\mathbf{r}_2,t_2)}$ as is summarized in Table-1. Conventional photometry measures $\overline{E^*(\mathbf{0},0).E(\mathbf{0},0)}$ ($^-$denoting time average) yielding the well known classical quantity Intensity, irrespective of the geometry or the spectrum of the celestial source. The "spectrum" of light measured by conventional spectrometers is the "power spectral density PSD" (Power Density vs EM frequency) derived from the Fourier Transform (FT) of $G^{(1)}(r_1 = r_2, t_1 \neq t_2)$ (i.e. the time autocorrelation function). A Fourier Transform Spectrometer samples the PSD directly, while a grating spectrograph performs the transform to it through constructive interference.

Also in the case of distant light collectors ($r_1 \neq r_2$) but $t_1 = t_2$ (provided we can deploy accurate time keeping for delay equalization along optical paths) we get the spatial autocorrelation function whose FT -being measured by classical (phase) interferometers- yields the angular distribution of the source's power density.

Thus, all such measurements can be ascribed to quantities of the type E^*E corresponding to intensity I, which, in the quantum limit, means observations (detections) of individual photons or of statistical one-photon properties. Table-1 shows that what is actually measured in our "observations" carried out using traditional astronomical instruments corresponds to special cases of the r,t combinations. Moreover, it is very important to realize that theoretically allowed multi-photon phenomena in the incoming photon gas reaching the observer can escape identification, as classical astronomical measurements do not even distinguish among light sources with identical $G^{(1)}$.

On the other hand, a successful formalism contains, also, the power of generalization: in our case it allows to go beyond photometry, imaging and spectroscopy, if we ask for descriptions of collective multi-photon properties of incoming light (which cannot be given without a proper quantum mechanical treatment of the boson fluid consisting of

TABLE 1. $G^{(1)}((\mathbf{r}_1,t_1:\mathbf{r}_2,t_2)$ in classical astronomy

Instrument - type	*space − time coordinates*	form of $G^{(1)}$
Bolometric Photometer	$r_1=r_2,\ t_1=t_2$	$\overline{E^*(\mathbf{0},0).E(\mathbf{0},0)}$
Spectrometer	$r_1=r_2,\ t_1\neq t_2$	$\overline{E^*(\mathbf{0},0).E(\mathbf{0},t)}$
Interferometer (phase)	$r_1\neq r_2,\ t_1=t_2$	$\overline{E^*(\mathbf{0},0).E(\mathbf{r},0)}$

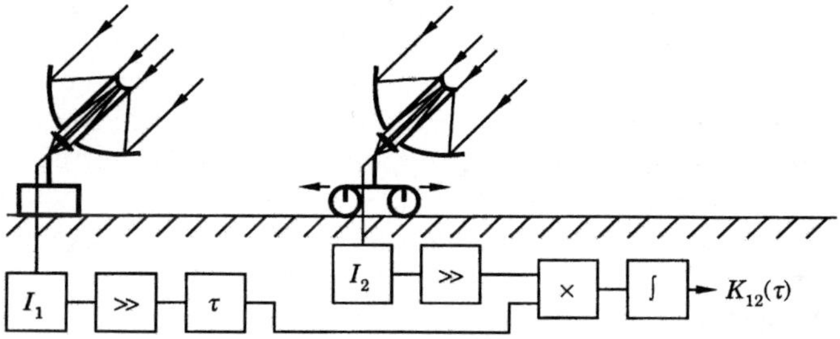

FIGURE 3. layout of the Intensity Interferometer principle of H-Brown & Twiss. Its high angular resolution power relies upon the ability to measure $G^{(2)}$ via the multiplication of the two photocurrents.(adopted from Lauternborn, Kurz, Wiesenfeldt: *Coherent Optics*, Springer-Verlag, 1995, with kind permission from Springer Science & Business Media)

the indistinguishable photons arriving on the detector). In several special cases, single and two-photon properties are interrelated and we can deduce one from the other. For a perfectly chaotic distribution of photons for example (photon gas in a maximum entropy state) it holds $g^{(2)}=[g^{(1)}]^2+1$.

In "Quantum-Astrophysics", we focus on $G^{(2)}(\mathbf{r}_1,t_1:\mathbf{r}_2,t_2)=\overline{I(\mathbf{r}_1,t_1).I(\mathbf{r}_2,t_2)}$. The anticipated observables associated with new emerging observational techniques are given in Table-2. It is stressed that the semi-classical theory fails to predict the correct forms for g_2 and higher-order correlations.

The original concept of the intensity interferometer (I-I) of Hanbury-Brown and Twiss (HBT) (1974) shown in Fig.3, is the archetypal example of a quantum-optical instrument which had been materialized. HBT interferometry employs two photodetectors at equal distance r from a source of diameter a and measures a correlated current if their separation d is smaller than the transverse coherence length $d_c\sim\lambda r/a$ of the source's radiation at wavelength λ. The result holds true if, the source is a perfect black-body (so that at each frequency any incident radiation is either fully absorbed of fully reflected). Within the framework of the new formalism HBT measures $g^{(2)}$ for $r_1\neq r_2,\ \ t_1=t_2$ hence deduces angular sizes of stars and, as stated, works correctly only for sources in LTE state (and of shape known a-priori). HBT, gives us a characteristic example of the predicting efficiency of the QED approach: It can be theoretically shown [1] that in cases of more realistic systems where there is a certain frequency range over which only partial absorption occurs, the correlation function does not decay completely to zero, but to a non-zero value independent of d. Thus this long range correlation conveys information on deviations of the thermal radiation from the black-body limit. On the extreme,

TABLE 2. $G^{(2)}(\mathbf{r}_1,t_1:\mathbf{r}_2,t_2)$ in Quantum-Astrophysics

Instrument - type	space-time coordinates	form of $G^{(2)}$
quantum spectrometer	$r_1 = r_2,\ t_1 = t_2$	$I(\mathbf{0},0).I(\mathbf{0},0)$
correlation spectrometer	$r_1 = r_2,\ t_1 \neq t_2$	$I(\mathbf{0},0).I(\mathbf{0},t)$
intensity interferometer	$r_1 \neq r_2,\ t_1 = t_2$	$I(\mathbf{0},0).I(\mathbf{r},0)$

even systems which are chaotic in the sense that scatter radiation in an irregular random way so that they feature very weak absorption, can be theoretically studied employing efficient techniques like random-matrix theory [2] to deduce predictions on the degree of deviations from local thermodynamic equilibrium (LTE). In the "random laser" cases for example, likely to occur in astrophysical context, statistical properties of spontaneous emission would be distinctly different from the typical case of laser with feedback [12]. Such theoretic predictions can be very specific and testable like calculated photocount distributions reflecting anticipated excess noise of amplified spontaneous emission in a random laser below the laser threshold [1].

Crucial theoretical problems that need to be solved now in astrophysical context are, the cases of coherent radiation propagating through amplifying random media, propagation of non-classical radiation through absorbing or amplifying random media, correlation properties of thermal radiation in conditions deviating from the ideal thermal equilibrium (for example, much larger effects of photon bunching are predicted for amplifying media). Techniques to tackle the problems of interest do exist today and can be developed even more. A particularly promising direction of work for a theorist, is the applicability of the random-matrix theory of Beenaker [2] as a method to solve the quantum radiation-transport problem in various astrophysical environment cases.

Under the generic scheme of the new quantum electrodynamical formalism for the coherence function, Table-2 summarizes what we could measure with "second generation" non-classical astronomical instruments. Such instruments can of course be subjected to further development but what is needed to emerge within the coming decade, is more theoretical predictions of diagnostic value.

What all non-classical instruments share in common, is the necessity to achieve extremely high temporal resolution in their photon registration. This brings into emergence very interesting potential applications of the second generation instrumentation. Thus, possibilities for digital intensity interferometry (as opposed to the analog correlator used by Hanbury-Brown and Twiss (1974)) as well as Very Long Baseline Optical Interferometry (Optical VLA) that is essentially post-detection "Software" Interferometry) (giving, in principle, angular resolution of the order needed for observing fine structure on stellar surfaces !) could be realized, provided a powerful technique to distribute extremely precise time reference signals -in a synchronous manner- to several observatories, is available. The HBT interferometer in particular, can serve not only as a mere LTE-deviation seeker, as is theoretically predicted and was mentioned above, but to even partially recover phase information (to allow image recovery in very high spatial resolution). Recently Ofir & Ribak [13] building upon the formalism of Fontana [7] have proposed a scheme of a very redundant intensity interferometer which, in theory, allows to reconstruct the missing phase of the *complex degree of coherence*. In conclusion, the

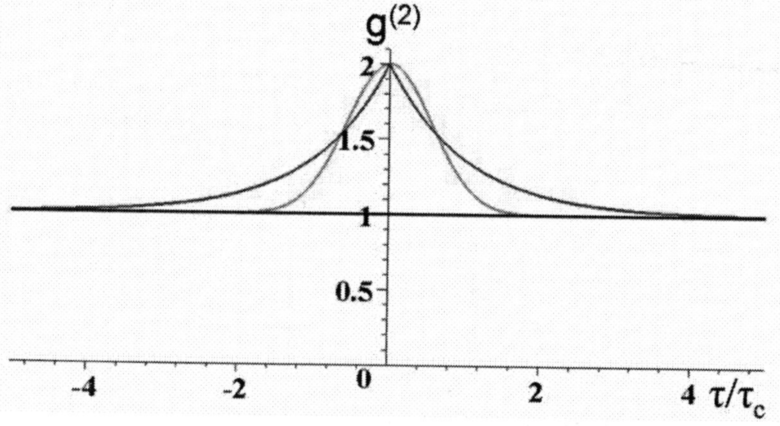

FIGURE 4. plot of g(2) as a function of time-delay (normalized to the coherence length t/tc). The horizontal curve is for a pure coherent state (monochromatic laser above threshold). The wider curve is for collision-broadened (Lorentzian) chaotic light . The narrow peaked curve corresponds to Doppler broadened (Gaussian) chaotic light.

future seems to point to the direction of obtaining access to $\geq$ hectometric-class giant optical telescopes not only on the obvious grounds of high resolution and faint source coverage but for prospects offered by materializing the quantum channel as well.

NEW GENERATION ASTROPHYSICAL EXPERIMENTS: NON-CLASSICAL INSTRUMENTATION AND THE KB.X INITIATIVE

In the pursuit of experimentation with an instrument capable to detect photon-noise, we launched (1999) a program for the development of a Greek experimental 2nd generation instrument for QAp studies in the 3D photon gas stream [15]. The device, to be called "$quantum-chrono-phasmatometer$"$(KB.X)$, is of the quantum - spectrometer type of Table-2. Initial contemplation of the conceptual study resulted in a few basic guidelines on its internal workings and on the type of detector to be chosen: Photon-counting detectors (unavailable beyond the near-IR) is obviously a prerequisite that excludes exploitation of the radio domain (although the quantum limit corresponds to lower (more feasible)- time resolutions). Similarly, X-rays, although of closer origin to compact objects, having both, much higher ν and too much lower photon fluxes than the optical, are excluded too. Therefore, ultrarapid phenomena remain, practically, within the reach of ground or space-based optical astronomy in *ms*, μs, and perhaps *ns* timescales.

For reference, the main properties of existing Fast/Ultra fast instruments, (among them the Swedish QVANTOS (Dravins et. al. [4]), pioneering the field of Q.Ap., which followed work on the MANIA photometric complex (Beskin et.al [3]) and references therein) are summarized in Table-4 below.

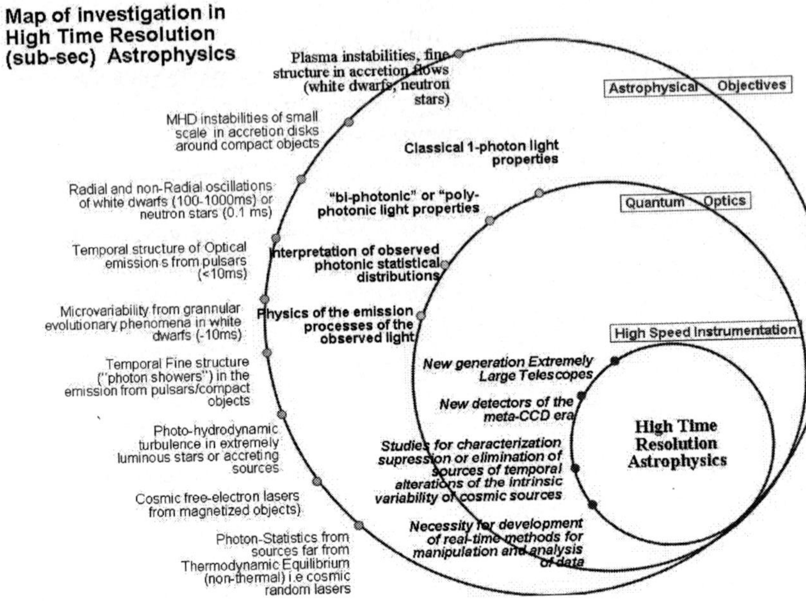

FIGURE 5. The development space for HTR astrophysics. Indicative science cases placed in the outer periphery are adopted from Dravins, ESO-Messenger, No78 (1994)

TABLE 3. Speed and characteristics of photonic detectors available today

Type of Detector	*Speed (approx.)*	*Spectral Rng*	*QE*
PMT (PhotoMultiplier Tube)	5-10 ns	360 -1000 nm	10-25 %
APD (Avalance Photodiode)	10 ns	opt -UV	60-80 %
MCP (microchannel Plate)	0.1 -2 μs	360 - 800 *nm*	4-8 %
STJ (Superconducting Tunnel Junction)	$\pm 5\mu$s [tag acc/cy]	310 - 720 *nm*	70 %
μM (MicroMegas)	1-5ns	$\leq$hard UV	10-25%
CCD- Keck (new readout mode of LRIS)	72 *ms*	350 - 850 *nm*	$\leq$90%
CCD ESO-VLT(new mode of FORS2)	2.3 *ms*		
VLPC (Visible Light Photon Counters)	8 ns	350 - 850 *nm*	80 %

TABLE 4. Existing and forthcoming instruments for Time Domain Astrophysics

Instrument	Detector Employed	Time Resolution.	Reference
MANIA	PMT/MCP	0.2 x10^{-7} sec -3.3 μs	Beskin (1982)
QVANTOS	PMT/APD	100 ns -100 ms	Dravins (1994)
OPTIMA	APD	1-3 μs	Kanbach et.al.(14*)
TRIFFID	MCP	1 μs	Butler-Golden (14*)
ULTRACAM	CCD	$\geq$0.237 s	Dhillon et.al.(14*)
QuantEYE	SPAD	$\sim$ *tens* ps -100 ns	Dravins (2005)
AquEYE	SPAD	$\sim$ *tens* ps - $\sim$ *ns*	Barbieri (2006)(14*)
KB.X	HPD/μMegas	1-5 ns	'Eudoxos' group

Of course, traditional *light curves*, are of questionable use and even impossible to be obtained in such ultra-high time resolutions. The enormous data throughput (far beyond the storage capacity of current technologies) demands a real-time reduction pipeline with 'pre-filtering' on the arriving photon stream, by architectures employing fast DSPs. Under this perspective it is noted that quantum astrophysics lends itself even to quantum computation. Furthermore, the role of very large telescopes (8-50m class) is critical: not only they offer $\geq x100$ more photon gathering, but, since the photometric measurements are forced to rely upon computation of quantities of statistical nature (like autocorrelations, power spectra, and other collective properties of the light quanta) which depend on a power of the intensity >1 (e.g. autocorrelation (ACF $\sim I^2$)) they are *far more sensitive for the detection of rapid variability* than the "large" ones available today.

As a matter of fact, in order to reveal ultrarapid intrinsic fluctuations in astronomical objects, we do need to segregate high speed phenomena of atmospheric origin like sky scintillation. The statistical description and characterization of scintillation noise has been initiated in a series of three papers by D. Dravins et.al. [5].They found persistent 2D structure from turbulent eddies and discover existence of measurable power in the scintillation for timescales $< 1ms$. Besides the need to characterize or monitor the temporal structure of the sky background and the that of the stray light in general as well as that of scattered light inpinging in the structural parts of the telescope, one has to deconvolve time variability due to telescope vibrations (seismic or wind driven).

Moreover, the actual physics of light detection needs to be rigorously understood: we need to understand the working physical principles in the most fundamental level possible, concerning the *physical observables* we measure during interaction of the KBX instrument with the incoming light field. In pursuing the *astronomical quantum optics* studies of this stage, we required that the optical scheme should be 2-arms of simple imaging of the entrance pupil onto 2 detectors. This is so because only such internal constituents of the KBX ought to be used, which have their effect upon the statistical behaviour of the input photon gas, been adequately described by QED in experiments of the past. This implies intention to build upon existing laboratory experience on absorbing elements (lenses), beam splitters, and *photocathode conversion* so that the requirement of the existence of an experimentally confirmed theory connecting the photocount statistics with that of light field falling on the detector in terms of measurable probabilities, is fulfilled. A beam splitter for instance does maintain the Poissonian distribution as it preserves the properties of coherent states. Other design criteria set, have been: highest possible time resolution (Visual: $\frac{1}{\Delta\nu} 10^{-14} s$), highest possible S/N ratio (implying really large telescopes + efficient detectors), ultrahigh data throughput/ handling capacity [1], ability to study faint sources (proper statistical analysis for variability in timescales shorter than typical successive photon arrival intervals). Obviously the development of the whole field is presently detector-driven. In Table-3 we summarize critical parameters of the spectrum of detectors available today. Our *KB.X* will employ HPDs or gaseous 'microMegas'detectors- (Giomataris [8]) which with a proper gas filling (*He* and *Isobutane*) can easily reach a gain 10^6, permitting single photo-e^- detection. A joint R&D

[1] (1 *ms* results in 30 Gb/night !)

program plan for extending their response in the optical through deployment of photocathodes is being finalized among NAC "Eudoxos" Kefalinia and NRCPS "Demokritos" Athens.

In addition, a dual exploratory experiment is currently carried out using the 0.62m fully-robotic telescope "Andreas Michalitsianos" (of the *"Eudoxos" observatories* on mount Ainos, Greece), as a test-bed. Firstly, we configured the optical setup in such a way as to measure the real time 2D brightness distribution over the pupil and thus characterize the scintillation. Secondly, we continue collecting and analyzing measurements on the effect of propagation through the atmosphere on photon statistical distributions. We use inclined optical paths connecting two telescope stations separated by 17 Km at an altitude difference of 600m over the Livadion Bay in Kefallinia.

REFERENCES

1. Beenakker, C.W.J.: (1998) Phys.Rev.Lett. 81, 1829-1832
2. Beenakker, C.W.J.: (1997) Rev.Mod.Phys. 69, 731-808
3. Beskin, G., Neizvestnyi, S., Pimonov, A,. Plakhotnichenko,V., Pustilnik, L, Shvartsman, V: (1982), *Astronomicheskij Tsirkulyar*, 1245, 1+
4. Dravins, D., Hagerbo, H.O, Lindegren, L., Mezey, E., Nilsson, B.: (1994), *Instrumentation in Astronomy VIII* SPIE Proc. Vol-2198, 289-301
5. Dravins, D.,Lindegren, L., Mezey, E., Young, A.T.,: (1997), PASP, 109, 173.
6. D.Dravins, C.Barbieri, R.A.E.Fosbury, G.Naletto, R.Nilsson, T.Occhipinti, F.Tamburini, H.Uthas, L.Zampieri (2006), *"QuantEYE : The Quantum Optics Instrument for OWL"*, in T.Herbst, ed. *"Instrumentation for Extremely Large Telescopes"*, MPIA Heidelberg special publication 0106, pp.85-91
7. Fontana, P. R,: (1983), J. Appl. Phys. 54, 473.
8. Giomataris, I.: (1998), *Nucl. Instr. and Meth.* A419, 239
9. Glauber, R.: (1970), in *Quantum Optics*, Academic, pp.53+
10. Glauber, R.: (1963), in *Phys. Rev.*,Vol. 131 (6), 2766-2788
11. Hanbury-Brown, R.: (1974), *The Intensity Interferometer*, Taylor & Francis
12. Letokhov, V.S.: (1968) Sov. Phys. JETP, 26, 835-840.
13. Ofir, A., Ribak, E. N., (2006): MNRAS, Vol. 368, 4 p.1652
14. (*)*2nd Galway Workshop on High Time Res. Ap.*: (2006) National University of Ireland, Galway.
15. Solomos, N.: (2001) in *4th Conf. of Hellenic Astron. Soc* Ed. J. Seimenis, Samos

ACKNOWLEDGMENTS

I have benefited a lot from discussions with D. Dravins, C. Barbieri and G. Fanourakis. The NAC "Eudoxos" and the Hellenic Naval Academy are thanked for generously providing partial support funds.

The new Athens Center applied to Space Weather Forecasting

H. Mavromichalaki*, C. Sarlanis*, G. Souvatzoglou*, G. Mariatos*, M. Gerontidou*, C. Plainaki*, A. Papaioannou*, S. Tatsis*

* *University of Athens, Physics Department, Section of Nuclear & Particle Physics, Zografos 15771 Athens Greece emavromi@cc.uoa.gr*

A. Belov†, E. Eroshenko†, V. Yanke†

† *IZMIRAN, Russian Academy of Science,1420092 Moscow, Russia abelov@izmiran.rssi.ru*

Abstract. The Sun provides most of the initial energy driving space weather and modulates the energy input from sources outside the solar system, but this energy undergoes many transformations within the various components of the solar-terrestrial system, which is comprised of the solar wind, magnetosphere and radiation belts, the ionosphere, and the upper and lower atmospheres of Earth. This is the reason why an Earth's based neutron monitor network can be used in order to produce a real time forecasting of space weather phenomena.

Since 2004 a fully functioned new data analysis Center in real-time is in operation in Neutron Monitor Station of Athens University, (ANMODAP Center) suitable for research applications. It provides a multi sided use of twenty three neutron monitor stations distributing in all world and operating in real-time given crucial information on space weather phenomena. In particular, the ANMODAP Center can give a preliminary alert of ground level enhancements (GLEs) of solar cosmic rays which can be registered around 20 to 30 minutes before the main part of lower energy particles. Therefore these energetic solar cosmic rays provide the advantage of forth warning. Moreover, the monitoring of the precursors of cosmic rays gives a forehand estimate on that kind of events should be expected (geomagnetic storms and/or Forbush decreases).

Keywords: Cosmic rays, Neutron Monitors, solar energetic particles, Space weather
PACS: 94.20 wq

INTRODUCTION

Since 1960 the nucleonic component of the galactic cosmic rays (CR) is recorded in Neutron Monitor (NM) stations [1]. These stations are essential to the detection of low and mid energy particles that arrive at the Earth from solar or/and galactic sources because of the significant monitoring window that these stations open in the atmosphere. The cut-off rigidity of the measured particles varies from zero (0) to fifteen (15) GV, while the energy cut-off applies from 0 value at polar stations, to the value of 20 GeV at the equatorial stations. The technological process made possible

CP848, *Recent Advances in Astronomy and Astrophysics,*
7th *International Conference of the Hellenic Astronomical Society,* edited by N. H. Solomos

the construction of very effective and low cost recording systems that could record and provide data of the cosmic ray intensity in real time. The first real time NM station was established in 1997 at Moscow. Nowadays twenty five NM stations provide their data in real or quasi real time and their distribution is presented in Figure 1.

Athens NM station (Super 6NM-64) is operating in the Physics Department at the University of Athens (37.58°N 23.47°E, cut-off rigidity 8.53 GV)) since November 2000, and is the 6th station to operate in real time mode. The measurements of the station are being elaborated automatically in order to be compatible to other stations data because of the necessity to compare a number of high rigidity stations in a good quality data that is required for a detailed study of CR variations and space weather conditions. The resolution of the measurements reaches as far as one second – which is uniquely worldwide. Specifically a modern type of atmospheric pressure sensor and registration system which was developed by the Athens Cosmic Ray Group helped a number of NM stations that were out of service to operate again [2].

A number of scientific groups working on the field try to establish a reliable network of neutron monitors operating in real time, because of the necessity to have a large number of recorded activity from as many stations as possible through out the globe. For instance: 'Spaceship Earth' is an effort that combines eleven neutron monitor stations strategically located around the world in order to provide 3-dimensional measurements of the cosmic ray angular distribution [3]. Another operating network of neutron monitors is Aragats Space Environmental Center (ASEC), located in Armenia which consists of two high altitude stations [4]. Those

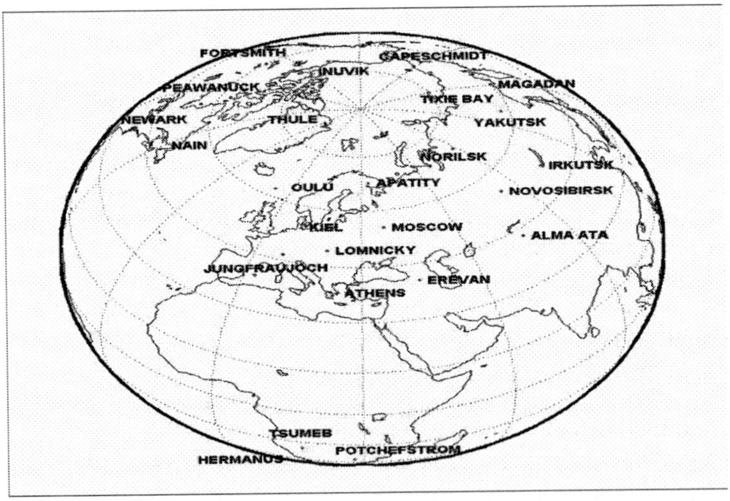

FIGURE 1. A global distribution of NM stations provided data in real time

efforts provide the direction of Space weather forecasting research by cosmic rays [5]. It is of great importance to establish a network of as many neutron monitors as possible combined with satellite data in order to be led into successful prediction.

During the last two years the Athens Neutron Monitor Data processing (ANMODAP CENTER) is being created at Athens University on the basis of the cosmic ray group activity.

DATA COLLECTION SYSTEM

In order to perform a real time prediction of space weather phenomena, only real time data from a neutron monitor network should be employed. The system of the ANMODAP Center collects data in real time from twenty- three widely distributed real time cosmic ray stations using the Internet [6]. The characteristics of these stations (geographic coordinates, altitude, standard pressure, cut-off rigidity) are listed in Table I. The main server in Athens station collects 1-minute, 5-minute and hourly cosmic ray data. The station measurements are processed automatically and converted to a suitable format for forecasting purposes. All programs have been written in an expandable form for easy upgrade of the NM network. Programs that make use of these data for forecasting purposes are already running in experimental mode. The increased number of NM stations operating in real-time gives a good basis for using NM network as a tool of forecasting the arrival of the interplanetary disturbances at the Earth [7].

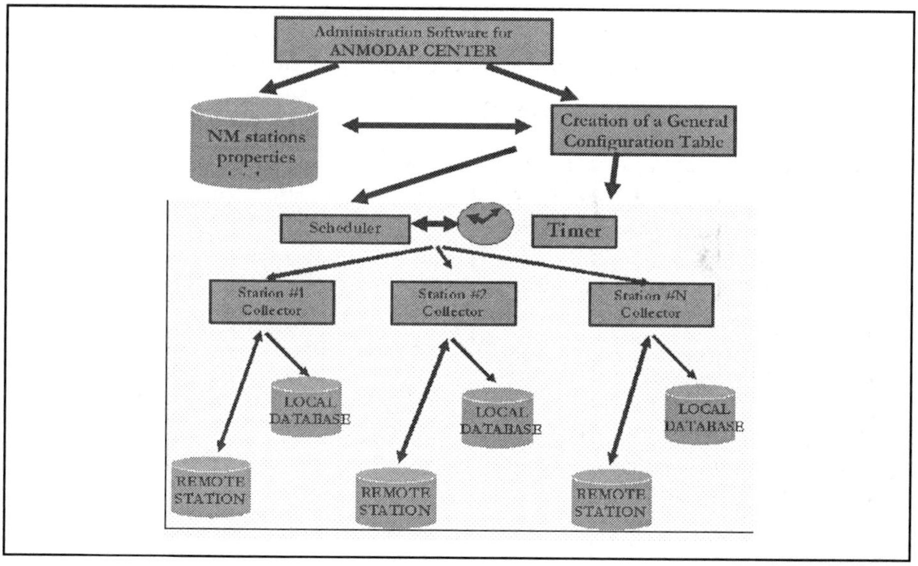

FIGURE 2. A schematic diagram of ANMODAP Center in real time

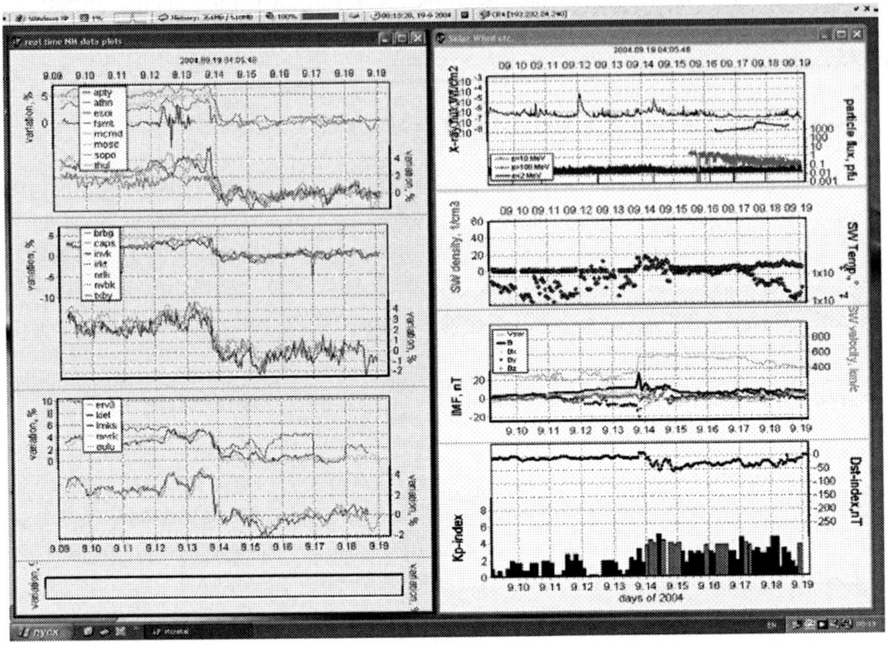

FIGURE 3. An example of the graphical presentation of ANMODAP CENTER data

Stations	Abbrev.	Lat (^{0})	Long(^{0})	Alt (m)	H_0 (mb)	R_C (GV)
APATITY	APTY	67.55	33.33	177	977.80	0.55
ATHENS	ATHN	37.97	23.72	260	974.70	8.53
BAKSAN	BKSN	43.28	42.69	0	818.50	6.91
BARENTSBURG	BRBG	78.12	14.42	0	964.70	0.20
CAPESCHMIDT	CAPS	68.92	180.53	0	1021.30	0.52
EREVAN						
EREVAN-3						
ESOI	ESOI	33.30	35.79	2025	800	10.41
FORTSMITH	FSMT	60.00	-112.00	0	996.10	0.30
INUVIK	INVK	68.35	-133.72	21	1019.10	0.14
IRKUTSK	IRKT	52.47	104.03	433	965	3.49
KIEL	KIEL	54.34	10.13	54	981.40	2.36
LOMNICKY STIT	LMKS	49.20	20.22	2634	733	3.88
McMURDO	MCMD	-77.85	166.72	48	985.10	0.00
MOSCOW	MOSC	55.47	37.32	200	991.90	2.30
NORILSK	NRLK	69.26	88.05	0	1015.30	0.53
NOVOSIBIRSK	NVBK	54.80	83.00	163	999.20	2.69
NEWARK	NWRK	39.68	-75.75	50	1008.60	2.21
OULU	OULU	65.05	25.47	0	990.00	0.77
SOUTH POLE	SOPO	-88.00	210.00	2820	687.70	0.05
THULE	THUL	76.50	-68.70	260	1011.50	0.00
TIXIE BAY						
YAUTSK						

TABLE I. List of real-time Neutron Monitor stations connected to the Athens Center

The properties of each station in the network are input to a database program, the 'Properties Database' (Figure 2). This program includes the characteristics that are described in Table I, the ftp directory of each station, the station operating mode (active or not) and the data update period for each station. A 'Scheduler' program reads the properties of each station and decides whether or not to make a data collection call. Each station has a separate data collection program (triggered by the scheduler) that uploads data from the remote station to the database in Athens. Every station has a recording system and a local database. The data collection program of each station interfaces with the recording system of the remote station and transfers one minute, 5 minute or hourly data to the main server. The main database in Athens can be refreshed with customizable rate (usually every 5min or every hour). For our study not only collecting data but also data presentation is very important. A special program that interfaces with the 'scheduler' program creates a graphical file once per hour which is displayed on the web page of the station presented at the server of Athens Center. An advanced and quick data processing system of 1-, 5-, 15-, and 60-minute refreshes the ANMODAP database providing both graphical and digital form of the measurements (http://cosray.phys.uoa.gr /ANMODAP). On the same page interplanetary space parameters data from GOES and ACE satellite in real time are also presented (Figure 3).

METHOD OF PREDICTION

The early detection of an Earth-directed solar proton event by NMs gives a good chance of preventive prediction of dangerous particle flux and can provide an alert with a very low probability of false alarm [8], [9]. The flux cannot be recorded on satellites with enough accuracy because of their small detecting area. However, it can be measured by ground-based neutron monitors with high statistical accuracy (in average, 0.5% for 5 min) as Ground level enhancement (GLE). The first step towards collecting data from a number of stations and analyzing them in real time have been made by the Bartol cosmic ray group (Bartol Research Institute - BRI) in the frame of the Space Ship Earth project (http://neutronm.bartol.udel.edu//). Subsequently a new real time data collection system was developed by the IZMIRAN cosmic ray group (http://Helios.izmiran.rssi.ru/ cosray/main.htm) using the latest networking methods in order to get data from the maximum possible number of stations to provide sufficiently reliable data of the further analysis. The use of all stations as a unified multidirectional detector substantially improves the measurement accuracy (<0.1% for hourly data).

The Athens Neutron Monitor Data Processing (ANMODAP) Center is being developed by the cosmic ray group of Athens University during the last two years. The prediction is mainly based on a feasible and statistically proven method [10] using total counts from several stations in real time. It consists of the following steps:

GLE onset: Data from at least three NM stations at Earth (two high latitudinal and one/two low latitudinal) and two independent satellite channels, for example X-ray on

GOES10 and GOES12, are processed to search for the start of ground level enhancement. The initiation of a GLE is identified as simultaneous detection of

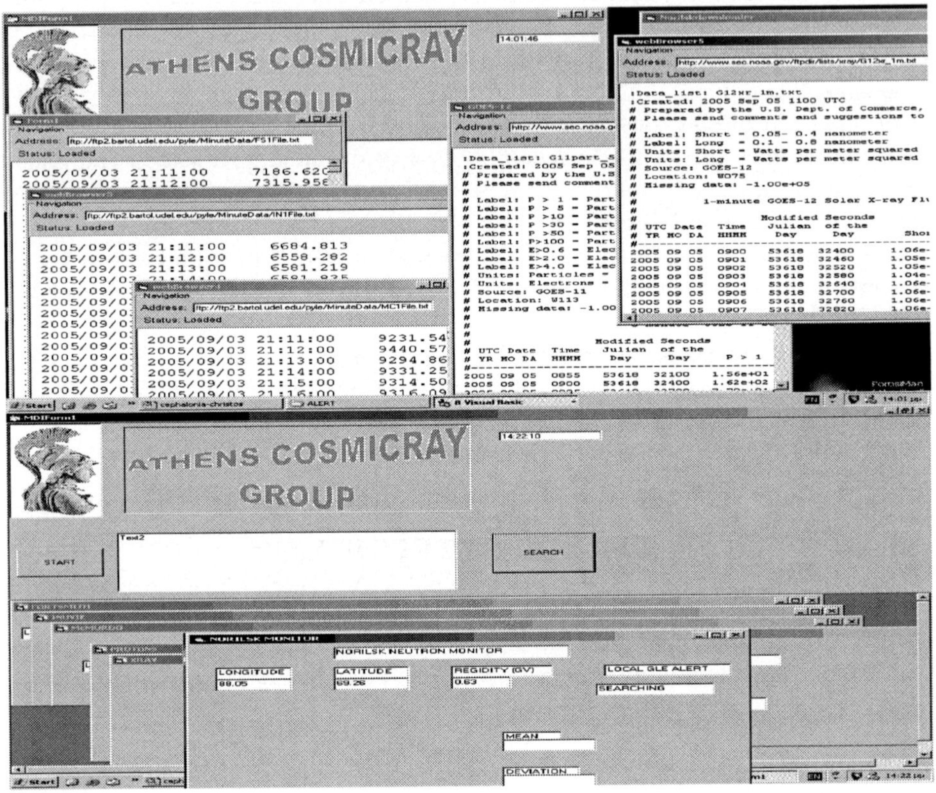

FIGURE 4. The operation of the Alert program at ANMODAP Center

enhancement in at least two neutron monitors and in an X-ray channel. If these conditions are met, data is collected from all other NMs. The real time algorithm in order to be accurate takes different kind of inputs from all the available sources. For large events the Alert stage is envisaged to be of about 99 % accuracy. Our statistic is still poor because of the continuous upgrades of the input data source code in the program. It is possible that very soon we will present a full analysis of all the old available GLE data. It is well known that GLEs accompany only small fractions of the strong solar proton events and also that not all proton events result in ground enhancements. However, a technique of calculating the relative flux increases using the data of high latitude stations eliminates the possibility from loosing an event.

An improved version of the ALERT program [11] will soon be available via a link from the main page of the Athens NM station visualizing all the forecasted events in real time (Figure 4). Also, it calculates spectra and other parameters for the estimation of the expected CR profiles for lower energies at different altitudes, several hours ahead. After the alert is obtained by the procedure described in the previous paragraph, the forecasting is sent by e-mail to the whole NM network in order to activate a system of minute data collection. During the last two years GLEs Alerts were also estimated using this program. Taking into consideration that in the analysis the data from high latitude stations are included and also the fact that our program uses 5min-averages and X-ray data from GOES for a comparison, the probability that the observed increase is not associated with GLE is very low. Preliminary results of the Alert program for the events of October 28, 2003, August 24, 2002 and January 20, 2005 are presented in Tables IIa, IIb and IIc.

Geomagnetic storms: According to the proposed method [8] , [9] , [11] for the analysis of geomagnetic storms the system collects hourly data from at least fifteen stations. An analysis based on the global survey method (GSM) results in the derivation of certain parameters such as CR density, the spectral parameters of density, and the three components of CR anisotropy vector (in x, y, and z axes), the amplitude of isotropic variation and the characteristics of the CR anisotropy at this current hour. In summary, a first analysis of the current heliospheric situation is being performed. To carry out more reliable analysis it would be desirable to use these results together with other data on solar and solar wind measurements.

TABLE IIa. Event of the 20th January 2005

NM Stations	Event started at (UT)	Established Alert signal at (UT)	Maximum of the event (UT)
Moscow	6:51	6:55	7:05
Kiel	6:51	6:55	7:05
Oulu	6:52	6:56	7:00
Norilsik	6:31	6:35	6:41

TABLE IIb. Event of the 24th August 2002

NM Stations	Event started at (UT)	Established Alert signal at (UT)	Maximum of the event (UT)
Norslik	1:29	1:33	1:44
Oulu	1:28	1:32	1:49
Cape Smidt	1:27	1:31	1:41

TABLE IIc. Event of the 28th October 2003

NM Stations	Event started at (UT)	Established Alert signal at (UT)	Maximum of the event (UT)
Cape Smidt	11:46	11:50	12:07
Irkutsk	11:40	11:44	11:54
Oulu	11:44	11:48	11:51

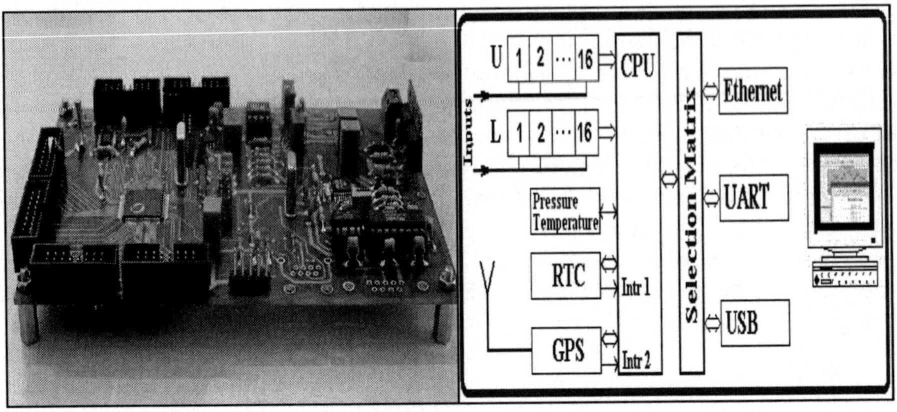

FIGURE 5. A new electronic pad is in development at the ANMODAP CENTER

DISCUSSION- CONCLUSIONS

The new Athens Neutron Monitor Data processing Center r based on the activity of the cosmic ray group of Athens University provides real time monitoring of cosmic ray variations. It is created also with the aim to make feasible the use of the neutron monitor network data in real time for the space weather tasks. The Athens Center at the same time with some other Centers (IZMIRAN, BARTOL University) gathers data to detect possible abrupt changes in the cosmic rays associated with the real solar wind and geomagnetic disturbances. This Center put the basis of a world wide data collection system in a real time from the network and elaborates adjusted software capable of the real time data processing and forecasting.

It is interesting to note that the Athens Center except of the possibility of successful prediction of the behaviour of the low energy part of the solar energetic events near Earth, it is proposed to be suitable also for other applications. A system of data accumulation for multidirectional muon telescopes and neutron monitor stations is in development (Figure 5) [12]. This electronic pad will provide hard real time implementation, GPS synchronization and time stamping, Ethernet connectivity, accuracy of μsec of event recording, low power consumption, portability and analog parameters measurement and recording. Such construction is very important due to the fact that a complete monitoring in mid, high and low energies in synchronization will be in place to indicate extreme phenomena well in advance providing an alert signal.

Moreover a study of a light unmanned aerial vehicle(UAV) carrying a linear energy transfer (let) radiation spectrometer and networked to the new Athens Neutron

Monitor data Processing Center applied to space weather events forecasting is carried out (Figure 6) the ANMODAP CENTER capability will be increased from high

FIGURE 6. The Unmanned Aerial Vehicle (UAV)

altitudes of Greek area with the LET radiation spectrometer carried by a special designed UAV [13].

In the near future this ANMODAP center is managing to improve the data collecting process, the resolution of the measurements and the data timestamp drift. Furthermore more measurement parameters like multiplicity and more sensor like multidirectional muon telescopes In order to succeed these the ANMODAP CENTER needs a implementation of a new data acquisition system that meet the needs of compact design, high performance, low cost and easy setup in any Real or non Real time NM station. The new type of this registration system is under development and it fulfill hard real time processing and acquisition needs. Typical board characteristics are GPS synchronization and time stamping, Ethernet connectivity, μsec accuracy of event recording, low power consumption, portability and analog parameters measurement and recording. The new type of registration system for NM and muon telescopes is very important due to the fact that a complete monitoring in mid, high and low energies in synchronization will be in place to indicate extreme phenomena well in advance providing an alert signal and best data for research combine measurements that recorded first time with so high resolution.

Concluding we can say that the joint multidimensional analysis of the relevant information from space borne and ground based detectors and from other sources as well, will minimize the number of false alarms and will maximize the reliability and the timely forecasting of the arrival of dangerous fluxes and disturbances from the space.

ACKNOWLEDGMENTS

Acknowledgments: Thanks are due to all NM groups for posting the data in the Internet and kindly provided these to the NM Network : Bartol Research Institut NSF grant ATM-oooo315; IEP SAS in Kosice by VEGA grant 1147; Oulu ; all Russian stations etc. This work is partly supported by PYTHAGOPAS II Project and by IZMIRAN RFFR grants 03-07-90389 and 04-02-16763.

REFERENCES

1. J. A. Simpson, *Space Science Reviews* **93**, 1-20 (2000)

2. H. Mavromichalaki, C. Sarlanis, G. Souvatzoglou, S. Tatsis, A. Belov, E. Eroshenko, V. Yanke and A. Pchelkin: Proc. 27th ICRC 2001 **10**, 4099-5002 (2001)

3. J. W. Bieber, P. Evenson, W. Dröge, R. Pyle, D. Ruffolo, M. Rujiwarodom, P. Tooprakai, and T. Khumlumlert, *Astrophys. J. (Lett.)*, **601**, L103-L106 (2004).

4. A. Chiligarian, V. Badayan, N. Bostanjyan, A. Hovanissyan, G. Hovsepyan, G. Gharagyozyan, N. Gevorgyan, S. Kazaryan, L. Melkumyan, S. Zarunyan, *Adv. Space Res*. **31**, 861 – 865 (2003)

5. K. Kudela., M. Storini, M. Hoffer, and A. Belov, *Space Sci. Rev*., **93**, 153-174 (2000)

6. H. Mavromichalaki, G. Souvatzoglou, C. Sarlanis, G. Mariatos, M. Gerontidou, A. Papaioannou, C. Plainaki, S. Tatsis, A. Belov, E. Eroshenko, V. Yanke, *Annales Geophys*, **28**, 1-8 (2005)

7. H. Mavromichalaki, Yanke, V., Dorman, L., Iucci, N., Chillingaryan, A. and Kryakunova, O., *Monograph NATO series, Effects of Space weather on Technology Infrastructure*, 301-317 (2004)

8. L.I. Dorman, L.A. Pustilnik, A. Sternlieb, I.G. Zukerman, A.V. Belov, E.A. Eroshenko, V.G. Yanke, H. Mavromichalaki, C. Sarlanis, G. Souvatzoglou, S. Tatsis, N. Iucci, G. Villoresi, Yu. Fedorov, B. A. Shakhov, M. Murat, *IEEE for Plasma Science*, **32,** 1478-1488 (2004)

9. G. Villoresi, L.I. Dorman, N. Iucci, and N.G. Ptitsyna, *J. Geophys. Res*. **105,** 21025-21034 (2000)

10. H. Mavromichalaki, M. Gerontidou, G. Mariatos, C. Plainaki, A. Papaioannou, C. Sarlanis, G. Souvatzoglou, A. Belov, E. Eroshenko, V. Yanke and S. Tsitomeneas , IEEE NSREC No TNS-00273 (2005)

11. Mariatos G., Mavromichalaki H., Sarlanis C., Souvatzoglou G, *Modern Journal of Physics A,* **20**, 6711-6714 (2005)

12. .C. Sarlanis, G. Souvatzoglou, H. Mavromichalaki, V. Kartyshov, E. Klepach, V. Yanke, K. Yudachin, *Proc. 29th ICRC* (2005)

13. E. G. Stassinopoulos, H. Mavromichalaki, C. Sarlanis, G. Souvatzoglou and S. Tsitomeneas *RADECS 2005 Conference* LN-10 (2005)

SWAP: An EUV imager for solar monitoring on board of PROBA2

A.C. Katsiyannis*, D. Berghmans*, B. Nicula*, J.-M. Defise†, G. Lawrence*, J.-H. Lecat†, J.-F. Hochedez* and V. Slemzin**

*Royal Observatory of Belgium, Ave. Circulaire -3- Ringlaan, B-1180, Brussels, Belgium.
†Centre Spatial de Liège, Liège Science Park, Ave. Du Pre-Aily, B-4031, Belgium.
**P. N. Lebedev Physical Institute, Leninsky pr. 53, Moscow, R-119991, Russia.

Abstract. PROBA2 is an ESA technology demonstration mission to be launched in 2007. The prime instrument on board of Proba2 is SWAP (Sun Watcher using Active Pixel System detector and Image Processing), a full disk solar imager with a bandpass filter centred at 17.5 nm (Fe IX-XI) and a fast cadence of ≈ 1 min. The telescope is based on an off-axis Ritchey Chretien design while an extreme ultraviolet (EUV) enhanced APS CMOS will be used as a detector. As the prime goal of the SWAP is solar monitoring and advance warning of Coronal Mass Ejections (CME), on-board intelligence will be implemented. Image recognition software using experimental algorithms will be used to detect CMEs during the first phase of eruption so the event can be tracked by the spacecraft without human intervention.

PACS: 96.60.-j, 95.40.+s, 07.87.+v, 95.55.Ev, 95.55.Fw, 95.85.Mt

INTRODUCTION

ESA's Project for On-board Autonomy (PROBA) mission line serves as a test-bed of new technologies for satellites and on-board systems. As a part of this program, PROBA2 will provide a platform for testing a number of instruments and techniques relevant to solar physics, space weather, aeronomy, avionics, spacecraft attitude control, power system and propulsion. The two main instruments, SWAP and LYRA are expected to reveal a wealth of information on the physics of the solar corona (coronal seismology, EIT waves, etc), space weather (CME acceleration and solar wind, etc) and aeronomy (ozone balance between the stratosphere and mesosphere, etc).

PROBA2 is a micro-satellite of ≈120 kg to be developed under the ESA General Support Technology Program (GSTP) by a consortium led by Verhaert Design & Development (Kruibeke, Belgium). Development of SWAP is led by Centre Spatial de Liège (CSL), Belgium. As a follow on of the PROBA1 Teston et al. [1] micro-satellite, PROBA2 is expected to match or even exceed the autonomous capabilities of it's predecessor, which operates successfully since October 2001.

The satellite is scheduled to be launched in February 2007 as secondary payload on a Rockot (former SS-19 Inter-Continental Ballistic Missile) rocket from the Plesetsk Cosmodrome in Northern Russia. The main load of the launcher will be ESA's Soil Moisture and Ocean Salinity Explorer (SMOS) and both satellites are expected to be placed into a dawn-dusk orbit at an altitude of ≈800 km with ascending node at 6am (exact orbit TBD). Such orbits are subject to eclipse periods between May and July that can last up to 18 min. During those eclipses, SWAP will be utilised in calibration

CP848, *Recent Advances in Astronomy and Astrophysics,*
7th *International Conference of the Hellenic Astronomical Society,* edited by N. H. Solomos

TABLE 1. Orbit Parameters for PROBA2

Semi-major axis (km)	7128
Local time of the Descending node	06:00
Orbital period (min)	95.65
Eccentricity	0
Inclination (deg)	97.598
Argument of perigee (deg)	0
R.A.A.N. (deg)	40.98

programs.

THE SWAP INSTRUMENT

SWAP has been designed as the successor of the Extreme ultraviolet Imaging Telescope (EIT) (Delaboudinière et al. [2], Dere et al. [3]) on board the SOlar and Heliospheric Observatory (SOHO). SWAP will continue EIT's systematic 'CME watch' program that continuously monitors solar activity for space weather-related events such as Coronal Mass Ejections (CME). It offers significant advantages over EIT as the field of view will be ≈ 20 % larger and it has a much higher cadence of 1 image per min (i.e. 12 times higher than EIT's 1 image per 12 min). The only disadvantage of PROBA2's sun-synchronous orbit is the interruption of sun visibility due to eclipses in every orbit for a few months every year.

Optical Design

PROBA2's strict power consumption and mass restrictions have led to the application of many innovative technologies. The overall mass of the instrument is limited to 10 kg while the power consumption must not exceed 5 W at any given time. To address those issues SWAP is based on a novel off-axis Ritchey-Chrétien design (Defise et al. [4]) that offers several important advantages (Fig. 1):

- Smaller (and therefore lighter) primary mirror for a given pupil area (as there is no central obstruction).
- Smaller optical filter. Apart from offering the obvious advantage of lighter weight it also minimises the risk of mechanical damage to the thin aluminium foil filter used in SWAP.
- A simple, efficient baffling system without protruding parts.

The INVAR optical bench used in SWAP will provide mechanical stability and insensitivity to thermal variations. The distance between the two mirrors of SWAP is not

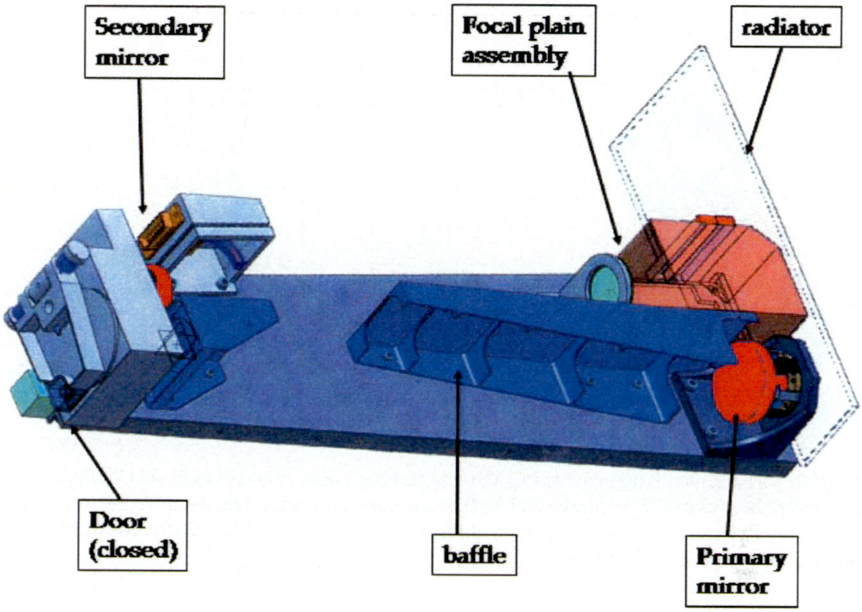

FIGURE 1. Schematic view of the interior of the SWAP instrument.

expected to vary more than 50 μm throughout it's operation range of temperatures.

The aspheric mirrors used on the instrument will be polished from zerodum with a micro-roughness below 0.5 nm. Multilayer coating of 30 layers of 2 different materials (Mo/Si) deposited on the mirrors will provide the EUV reflectivity and spectral selection. Institut d'Optique Théoretique et Appliquée (IOTA, France) will apply the coating and optimise it for near normal reflection in the 17 – 18 nm band. Fig. 2 contains efficiency measurements made on a sample. The bandpass provided by the mirror-multicoating combination is well suited to the needs of the SWAP experiment as it is centred almost exactly at the desired wavelength and the peak reflectivity is measured at 40.5%. The higher effective area of SWAP comparing to EIT's 171 Åfilter will allow a much higher cadence. Fig. 3 contains a comparison of the effective area of SWAP comparing to those of EIT 171 & 195 Å.

The CMOS APS detector

A complementary metal oxide semiconductor (CMOS) active pixel sensor (APS) will be used as the detector of SWAP. It is manufactured by Fillfactory N.V. as an improved version of the STAR1000 detector (see Dierickx et al. [5]) used for high-accuracy start tracking. This makes SWAP the first in-orbit solar physics instrument that will utilise a CMOS APS detector for scientific research. The most immediate implication of this is that a physical shutter could be replaced by an electronic which leads to a simpler, more

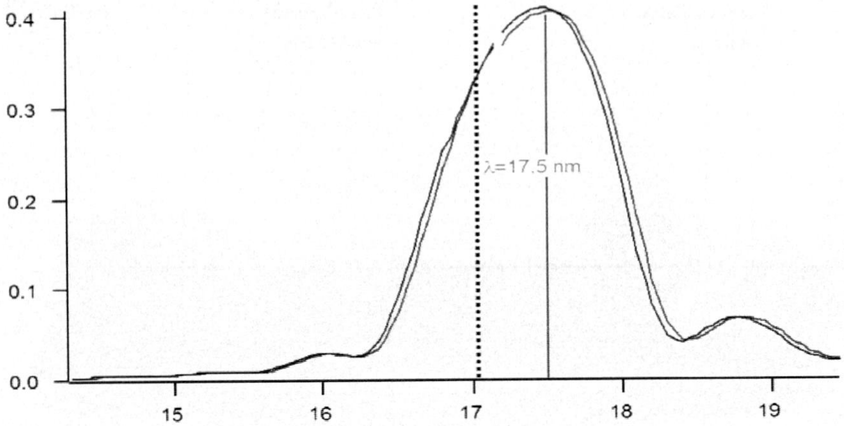

FIGURE 2. The Reflectivity provided by the multilayer coating (first bounce). The bandpass is 1.3 nm and the peak reflectivity 40.5% (source IOTA). The two lines correspond to two different incidence angles (1 and 4 deg). The dashed line indicates the aluminium cut-off of the filter. All wavelengths below this line will not be detected by the sensor.

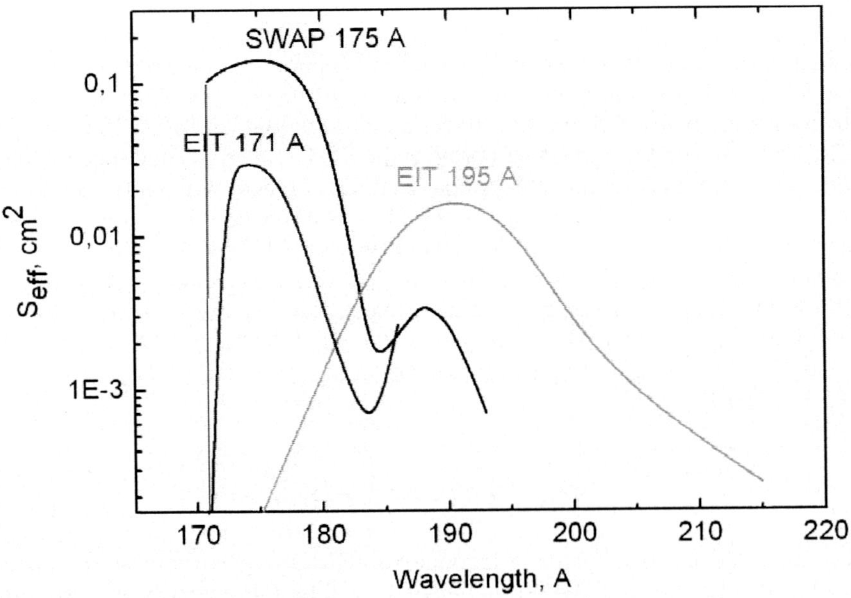

FIGURE 3. A comparison between the effective areas (S_{eff}) of SWAP and EIT's 171 & 195 Å. SWAP is estimated to have a similar bandpass to EIT's 171 Åand larger effective area. The later will allow a higher cadence of 1 image per minute.

reliable mechanical design with only a single-shot door on the sun-facing side of the experiment. This door is needed for protection of the instrument (especially the optical filters) during launch and the first phases of commissioning of the instrument.

Improved Signal-to-noise ratio (S/N) will be achieved via the use of the non-destructive readout functionality of the detector. This will be achieved by applying the 'Correlated Double sampling' that reduces static fixed pattern noise and the KTC noise. EUV sensitivity will be achieved by applying scintillator coating on the sensor. Passively cooling the detector by attaching it to a radiator constantly viewing cold space will further reduce the detector noise. Any condensation problems that may appear will be dealt by a decontamination heater that will be attached to the instrument and periodically used to outgas any condensation on the CMOS.

The proximity electronics (PE) that drive the detector and the calibration diodes will be mounted on the focal plane. A compact PCI (cPCI) card, the Memory, Compression and Packetisation Module (MCPM), plugged in the satellite's central computer connects to the PE and provides 512 MB storage capabilities for the instrument. This configuration has the advantage of utilising the satellite's spare CPU power of the PROBA2's 100 MHz AT697 LEON II microprocessor for the running of the instrument. PROBA2 will be the first satellite mission to fly ESA's new high performance, radiation hardened processor which is expected to be used for many other ESA missions. The extra computing power delivered by the new microprocessor will make possible to implement some basic on-board image processing capabilities that will lead to some spacecraft autonomy (e.g. CME watch mode below).

Software Innovations

The PROBA mission line is developed by ESA as a series of technology demonstrations platforms that provide autonomous operations in space. A number of software algorithms will be implemented on PROBA2's on-board computer that will allow the platform to perform image processing and some autonomous operations.

1. The acquired images can be stored in the MCPM until transmission to the ground either raw, or compressed with the lossless LZW algorithm, or compressed by the lossy JPEG algorithm.
2. When operated in the default cadence ($\approx$1 min) and the whole detector is read (i.e. CMOS windowing is disabled), SWAP will produce a significantly larger amount of data than can be sent to the ground station. To make the best use of the limited telemetry bandwidth, the storage capabilities of the MCPM will be utilised for buffering the extra observations and a prioritization algorithm will be implemented. The purpose of this algorithm is to identify potentially interesting events and transfer the associated images to ground first.
3. When the JPEG routine is used, discontinuities (for example cosmic rays) will be removed and replaced by the local average to avoid artifacts of the JPEG compression.

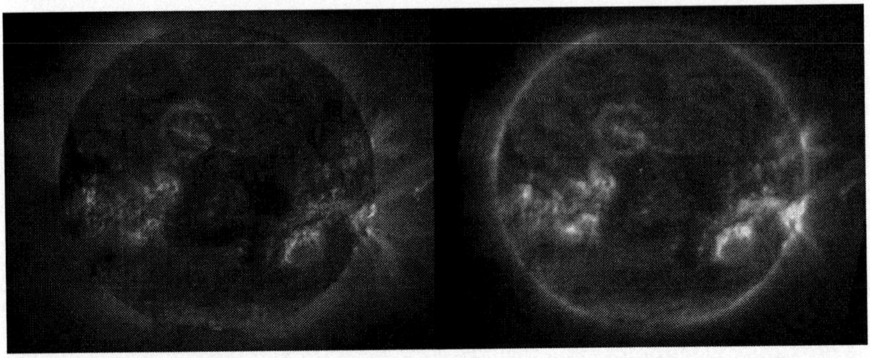

FIGURE 4. An EIT (19.5 nm) image on the left and a SPIRIT (17.5 nm) image on the right taken almost simultaneously. The contrast of the EIT image is slightly larger due to the active regions being brighter in 19.5 nm than in 17.5 nm. The smaller contrast of the signal on the 17.5 nm bandpass suits well the smaller pixel depth of SWAP compared to EIT.

4. Weak pixels will be compensated by autonomous on-board routines based on flat-field information uploaded to the satellite from the ground.
5. For a more efficient compression of the data, when the JPEG algorithm is used, the 12 bit pixel values will be re-coded to 8 bit. A sophisticated re-coding algorithm will be used for this purpose, to ensure that the "real" loss of information for the re-coding is minimal. This is made possible as the data alteration by the change to the pixel depth will be smaller than the expected noise produced by the photon shot noise. For exact details of the re-coding algorithm and the comparisons between information lost from this algorithm and the photon shot noise see Nicula et al. [6].
6. Another innovation that adds to SWAP's autonomous capabilities is the CME watch mode. Under this mode, eruptions of material from the Sun's surface will be detected by analysing the observations using on-board software and notifying the spacecraft's positioning system. The satellite then will autonomously off-point to the direction of the potential CME and a configurable number of observations will then be taken. This mode has two distinct goals: a) to test an automated algorithm for early detection of such events and b) to provide important, much needed observations of CMEs while in the first stages of acceleration.

SCIENTIFIC APPLICATIONS FOR SWAP

Although SWAP's main goal is to demonstrate and test a number of technologies first used on a satellite, SWAP is also expected to produce important scientific results and breakthroughs in solar physics. A number of the instrument's characteristics together with proven instrumental concepts, should became very useful on providing important information on the understanding of space weather. Below is a presentation of those characteristics and how combined together they can potentially provide a deep inside into the physics of the solar corona.

TABLE 2. Preliminary estimation of the relative contribution of various spectral lines based on CHIANTI simulations.

Ion	log T_{max}	Quite Sun (%)	Active Region (%)	Coronal hole (%)	Flare (%)
O_{VI}	5.5	0.7	0.5	4.7	2.6
Fe_{IIX}	5.8	0.5	0.3	1.6	0.5
Fe_{IX}	5.9	38.2	31.6	57.5	26.0
Fe_{X}	6.0	51.4	54.6	31.8	35.1
Fe_{XI}	6.1	8.9	11.5	2.3	7.0
Fe_{XII}	6.2	0.3	0.4	0.0	0.3
Fe_{XX}	7.0	0.0	0.0	0.0	14.3

Choice of Wavelength for SWAP

A bandpass cantered at 17.5 nm was chosen as it covers some of the brightest spectral lines of iron ions emitted by the solar corona and contains only one strong non-coronal line ([O_{VI}]). SPIRIT, on-board the Russian satellite CORONAS (see Zhitnik et al. [7]), has successfully used the same bandpass over the last few years, allowing us to make detailed comparisons between their instrument and EIT. Fig. 4 contains EIT (19.5 nm) and SPIRIT (17.5 nm) observations taken simultaneously. The smaller contrast of the SPIRIT (17.5 nm) image is apparent. This difference in contrast is not due to instrumental factors but rather in differences in the EUV emitting mechanisms of the two wavelengths. The smaller contrast of the images of the 17.5 nm bandpass fits well with the smaller dynamical range of SWAP comparing to EIT.

Estimations of the relative contribution of a number of spectral lines was achieved by applying the CHIANTI code by Dere et al. [8], Young et al. [9] and assuming ionisation equilibrium, convolved with mirror reflectivity estimations based on test mirrors. Table 2 contains the results of those calculations for different types of activities (quite Sun, active region, corona hole and solar flare) for all the emission lines within the SWAP's filter bandpass. It is apparent that most of the signal will be contributed from Fe ions with maximum excitation temperatures (T_{max}) of 800.000-1.300.000 °K ($log\ T_{max} = 5.9$-6.1). Based on previous experience with SPIRIT (Slemzin et al. [10]), it is expected that SWAP's wavelength band will be more effective for observations of CME-associated phenomena such as EIT waves, dimming, off-limb ejections etc (for example Fig. 4).

Choice of Field Of View (FOV) for SWAP

The FOV covered by SWAP will be 54 arcmin, ie 20% larger than that of EIT (45 arcmin) which also means a 20% larger pixel size than EIT since both instruments have the same number of pixels. An even more extended coverage of Sun's corona is expected to be achieved by utilising the spacecraft's ability off-point from the sun-centred direction. This off-pointing, together with the spacecraft's autonomous abilities (see earlier section), will allow the tracking of CME activity during the first few, critical

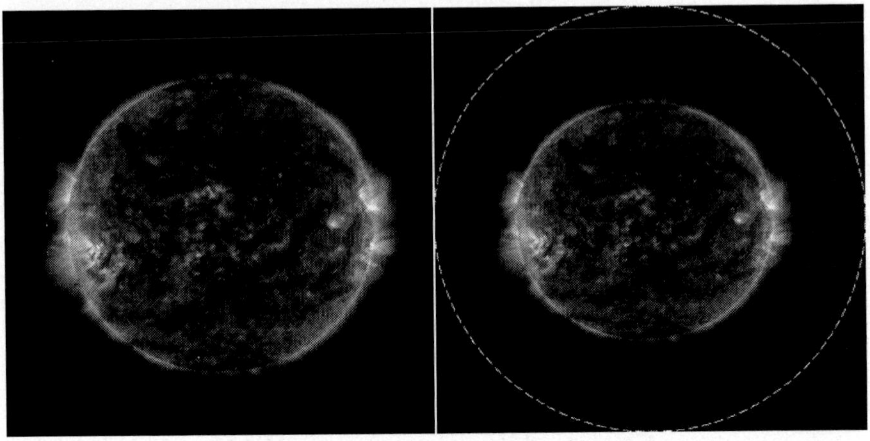

FIGURE 5. A comparison of EIT (on the left) and SWAP (on the right) field of views. The dashed circle corresponds to 1.67 solar radii while the corners of SWAP's FOV are at a distance of 2.36 solar radii. The same EIT image of the Sun shown on the left was used as the solar image on the right for comparison.

hours of creation and acceleration of a CME. Fig. 5 contains a comparison of the FOV of the two instruments (EIT and SWAP). It is apparent that SWAP's FOV is more suited for the study of Solar weather phenomena as it allows the imaging of a more extended area of the solar corona.

THE FUTURE

Although PROBA2 is a low-budget micro-satellite, it is expected to provide valuable data for a number of scientific purposes. It is also expected to be an important test platform for technologies that will be deployed in the more ambitious ESA mission, the Solar Orbiter (SolO). New detector technology, large on-board memory capacity, software design that will allow autonomous pointing decisions and preferential download of observations to the ground are only a few of the innovations deployed in PROBA2 that are expected to make their way to the SolO mission scheduled for launch in 2015.

The SWAP project is committed to an open data policy. All data and analysis software will be available through the Royal Observatory of Belgium web site (http://swap.oma.be) and interested parties are invited to analyse data and to plan their own special observation campaigns.

ACKNOWLEDGEMENTS

SWAP is funded by the Belgian Federal Science Policy Office (BELSPO) through the ESA/PRODEX program. The SWAP project is also supported by the Max-Plank-Institut für Sonnensystemforschung, Germany.

REFERENCES

1. F. Teston, D. Bernaerts, and K. Gantois, “Proba, an ESA technology demonstration mission, results after 3 years in orbit,” in *Small Satellites, Systems and Services*, edited by B. Warmbein, 2004, vol. 571 of *ESA SP*, p. 7.1.
2. J.-P. Delaboudinière, G. E. Artzner, J. Brunaud, A. H. Gabriel, J. F. Hochedez, F. Millier, X. Y. Song, B. Au, K. P. Dere, R. A. Howard, R. Kreplin, D. J. Michels, J. D. Moses, J. M. Defise, C. Jamar, P. Rochus, J. P. Chauvineau, J. P. Marioge, R. C. Catura, J. R. Lemen, L. Shing, R. A. Stern, J. B. Gurman, W. M. Neupert, A. Maucherat, F. Clette, P. Cugnon, and E. L. van Dessel, *Sol. Phys.* **162**, 291–312 (1995).
3. K. Dere, J.-P. Moses J.D., Delaboudinière, J. Brunaid, C. Carabetian, J.-F. Hochedez, X. Song, R. Catura, F. Clete, and J.-M. Defise, *Sol. Phys.* (2000).
4. J. Defise, D. Berghmans, J. E. Hochedez, J. M. Lecat, E. Mazy, P. L. Rochus, T. Thibert, P. Nicolosi, M. G. Pelizzo, U. H. Schühle, R. A. M. Van der Linden, and A. N. Zhukov, “SWAP: Sun watcher using APS detector on-board PROBA2, a new EUV off-axis telescope on a technology demonstration platform,” in *Telescopes and Instrumentation for Solar Astrophysics*, edited by M. A. Fineschi, Silvano; Gummin, SPIE, 2004, vol. 5171 of *Proceedings of the SPIE*, pp. 143–154.
5. B. Dierickx, G. Meynants, and D. Scheffer, “Near 100% fill factor CMOS active pixels",” in *IEEE CCD & AIS workshop proceedings*, 1997, p. 1.
6. B. Nicula, D. Berghmans, and H. J.-F., *Solar Physics* **228**, 255–266 (2005).
7. I. A. Zhitnik, O. I. Bugaenko, J.-P. Delaboudinière, A. P. Ignat'ev, V. Korneev, V. Krutov, A. Mitrofanov, S. Oparin, V. Oraevsky, A. Pertsov, V. Slemzin, S. I.I., S. A.I., and S. J., “SPIRIT X-ray telescope/spectroheliometer results.,” in *Solar Variability: From Core to Outer Frontiers*, ESA, 2002, vol. SP-506, pp. 915–918.
8. K. P. Dere, E. Landi, H. E. Mason, B. C. Monsignori Fossi, and P. R. Young, *A&A* **125**, 149–173 (1997).
9. P. R. Young, G. Del Zanna, E. Landi, K. P. Dere, H. E. Mason, and M. Landini, *ApJS* **144**, 135–152 (2003).
10. V. Slemzin, V. Grechnev, I. Zhitnik, S. Kuzin, I. Chertok, S. Bogachev, A. Ignatiev, A. Pertsov, and D. Lisin, “EUV-observations of CME-associated eruptive phenomena with the CORONAS-F/SPIRIT telescope/spectroheliograph,” in *Proc. of the IAU Symposium 226*, edited by K. Dere, J. Wang, and Y. Yan, IAU, 2004, vol. 2004, pp. 21–26.

VLBI: Networks and Developments (Abstract)

Anastasios Tzioumis

Australia Telescope National Facility (CSIRO)

Very Long Baseline Interferometry (VLBI) achieves resolutions of milliarcseconds, the highest by any astronomical instrument. It is used to study small scales structures, from the inner regions of Active Galactic Nuclei (AGN) in distant galaxies and quasars, to stellar explosions such as X-ray binaries in our own galaxy. It also has helped to define and continues to improve the International Celestial reference Frame (ICRF).

A review was given of current VLBI networks around the world and their capabilities. A lot of development is under way to upgrade these arrays to higher sensitivities and wide-field imaging. The fast networks that are becoming available are also facilitating e-VLBI, with the potential of world-wide real-time networks. A progress report of current developments and plans was presented.

CP848, *Recent Advances in Astronomy and Astrophysics,*
7th *International Conference of the Hellenic Astronomical Society,* edited by N. H. Solomos

Adapting and expanding the Plateau de Bure interferometer

A. Karastergiou & R. Neri

IRAM, 300 rue de la Piscine, Domaine Universitaire, Saint Martin d'Hères, France

Abstract. We were recently faced with the following problem: The *T*-shaped Plateau de Bure Interferometer is expanding its tracks to achieve higher spatial resolution in astronomical images at mm wavelengths. Two more stations for positioning the antennas during observations are being built at the ends of the extended tracks. Which of the given stations should the remaining four antennas occupy to accommodate the new stations? What is the optimal set of antenna configurations, given the new extended one, to achieve necessary coverage of the *uv*-plane at a variety of spatial resolutions? We present in this paper the solutions to the above questions, resulting from a novel method we have recently developed. The method is based on identifying which placement of elements provides the most appropriate *uv*-plane sampling.

Keywords: instrumentation: interferometers – methods: numerical
PACS: 95.75.Kk

INTRODUCTION TO THE PROBLEM

The IRAM interferometer is located in the South of the French Alps, at an altitude of 2560 m. It currently comprises $N_A = 6$ identical antennas of 15-m diameter, that can be positioned on $N_p = 32$ pads along a *T*-shaped track. The north-south arm is 232 m long, and the almost east-west oriented arm extends 216 m west and 192 m east of the intersection. The angle between the arms is 75°. In order to better sample the *uv*-plane, the IRAM interferometer operates in four different configurations (specific sets of antenna positions), ranging between compact and extended baselines. The choice of the total number of configurations and the design of each of them has been heavily influenced in the past by the very demanding requirements of antenna movement. This currently takes place two to three times in each of the 6-month observing terms. Astronomical observations are carried out within two frequency bands around 100 and 230 GHz.

At the end of the summer of 2005, the northern track has been extended to 368 m and one new pad was constructed at the end of this track. Also, the east-west track has been extended from 408 m to 760 m, with one new pad at the far eastern end of the extension. A diagram showing the positions of all the current stations is shown in Figure 1. The new stations can be clearly identified at the easternmost and northernmost positions of the tracks. Despite the track extensions, the total number of configurations will remain constant, due to the practical difficulties of moving antennas on a plateau at the top of a snowy mountain. However, the 4 new configurations and combinations of consecutive configurations have been designed with an emphasis on the quality of the synthesized beams. The first constraint for designing the new configurations, therefore, is to preserve the total number of 4, and optimize consecutive configurations to match each other well.

CP848, *Recent Advances in Astronomy and Astrophysics,*
7th *International Conference of the Hellenic Astronomical Society,* edited by N. H. Solomos

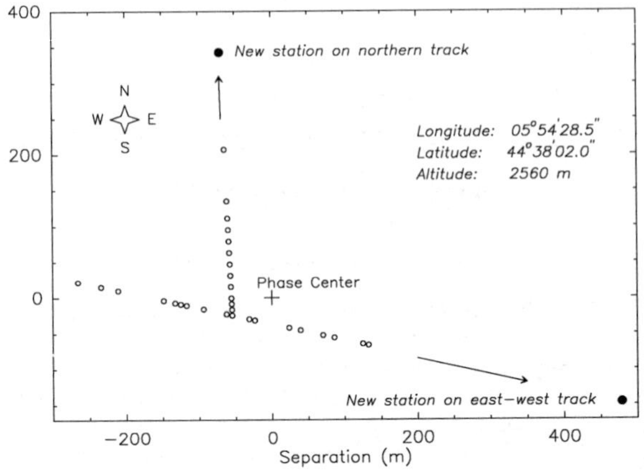

FIGURE 1. The station positions of the IRAM interferometer. The two new stations at the far north and far east of the tracks are emphasized. The latitude and longitude refer to the phase center of the array

TABLE 1. Synthesized beam sizes and position angles for the new and previous configurations of the IRAM interferometer array, computed for an 8-hour observation of a source at +45° declination at a frequency of 300 GHz.

Configuration name	beam size at 300 GHz New	Old
A	0.30″ × 0.24″ @ 38°	0.45″ × 0.43″ @ 22°
B	0.42″ × 0.35″ @ 68°	0.67″ × 0.54″ @ 63°
C	0.95″ × 0.85″ @ 61°	0.86″ × 0.71″ @ 48°
D	1.83″ × 1.54″ @ 103°	2.03″ × 1.37″ @ 90°
AB	0.35″ × 0.30″ @ 45°	0.54″ × 0.47″ @ 41°
BC	0.62″ × 0.54″ @ 59°	0.76″ × 0.62″ @ 55°
CD	1.26″ × 1.12″ @ 82°	1.25″ × 0.99″ @ 68°

A SUMMARY OF THE METHOD

In order to identify the most appropriate positioning of the antennas, we have followed a method described in detail in Karastergiou et al. [1]. In brief, we compute the *uv*-samples of a typical observation in all possible configurations, and identify the most appropriate coverage. In doing so, we assume that for a sparse interferometer such as the one in question, where the baselines are extended and the number of antennas small, the most uniform *uv*-coverage is most appropriate for astronomical imaging. The benefits of uniform *uv*-sampling are clearly pointed out in Keto [2]. On the other hand, Boone [3] correctly points out that the Fourier components with a Gaussian radial distribution

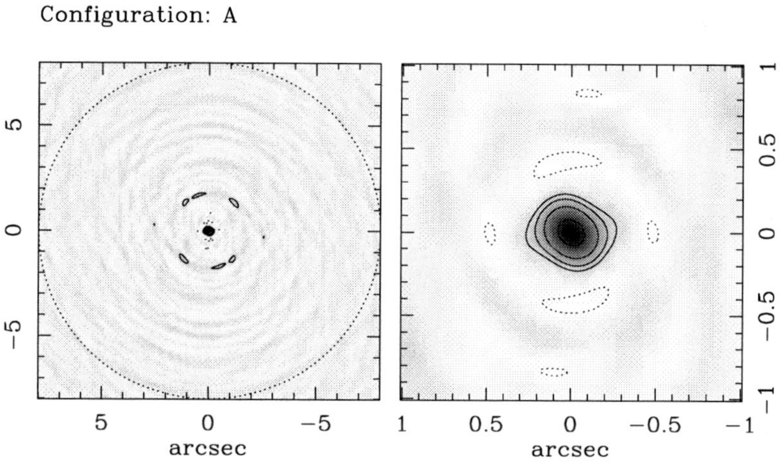

FIGURE 2. The synthesized beam obtained by a 9-hour observation of a source at +45° declination, using the new stations of the Plateau de Bure interferometer, in the most extended configuration (A). The image on the left has approximately the size of the primary beam at 300 GHz, whereas on the right we have zoomed in on the central region. The contour levels are at ±10%, ±20%, ±40% and ±80% help in evaluating the sidelobe levels.

achieve a more desirable Gaussian beam. However, in the case of a sparse interferometer, the Gaussian radial distribution of Fourier samples is so broad, that it is essentially equivalent to uniform.

The advantages of working in the Fourier domain rather than the image domain permit this automated procedure. Identifying the configuration with the most uniform *uv*-coverage is based on the analogy between the Fourier domain and samples, and a closed 2-d surface populated by charged particles; the most uniform distribution of such particles results in the lowest energy of the system. In Karastergiou et al. [1] we show how this "energy" is computed for each configuration. The execution time of the computer software that performs this search ranges between less than one minute and a few days, depending on the number of possibilities it is required to loop through.

THE NEW CONFIGURATIONS

The first step in designing the 4 new configurations of the IRAM interferometer was to design the new, most extended array. With an antenna placed on each of the two new stations, a third antenna is placed on the far western station of the east-west track, to provide the most extended baselines possible. The positioning of the remaining three antennas in the most extended configuration is achieved by using the method described above. Firstly, we choose a particular declination for which we would like to optimize the configurations. In doing so, we consider the average declination of the sources observed at the telescope. Also, and quite importantly, due to the asymmetric shape of the array, configurations that are optimized for low declination sources (which do not rise beyond a certain elevation) are better for sources of all declinations than configurations which

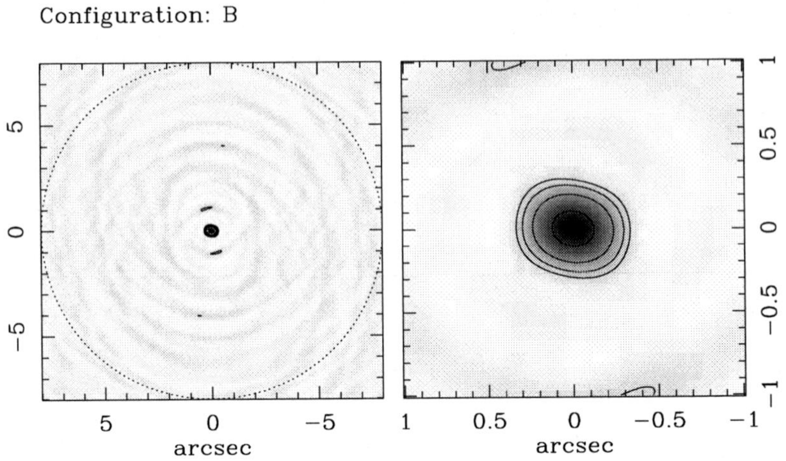

FIGURE 3. Same as Figure 2, but showing the second most extended configuration (B).

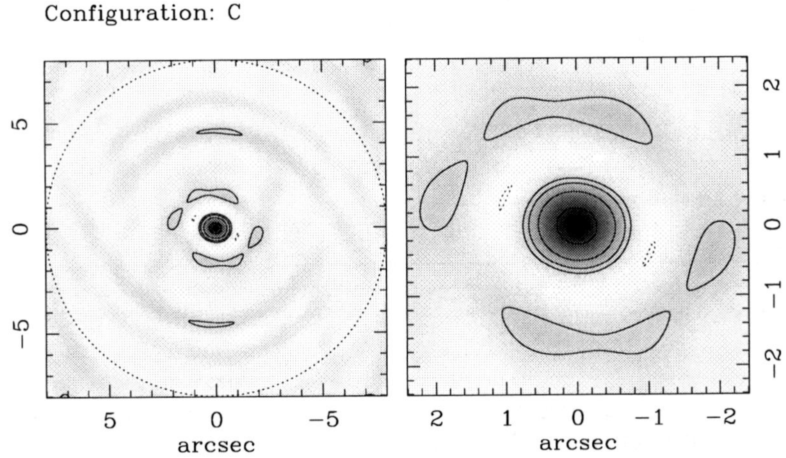

FIGURE 4. Same as Figure 2, but showing the second most compact configuration (C).

are optimized for high declinations. We have found that optimizing for a declination of $+20°$ proves a good compromise for sources of various declinations. Also, the tracking time of the Earth rotation synthesis for the optimization process is 9 hours, symmetric around transit.

After optimizing the most extended configuration, we proceed to find a second extended one, which only includes one of the two new stations. The aim is to optimize the second configuration in such way, that the *uv*-samples complement the sampling of the first, most extended configuration. We do this by adding the *uv*-samples resulting from that configuration into the optimization procedure, taking them into account while finding the most uniform distribution of samples. Observing in these two configurations,

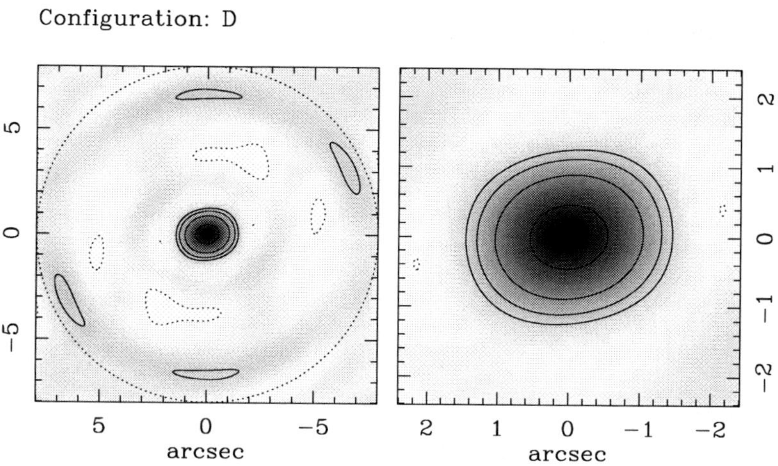

FIGURE 5. Same as Figure 2, but showing the most compact configuration (D).

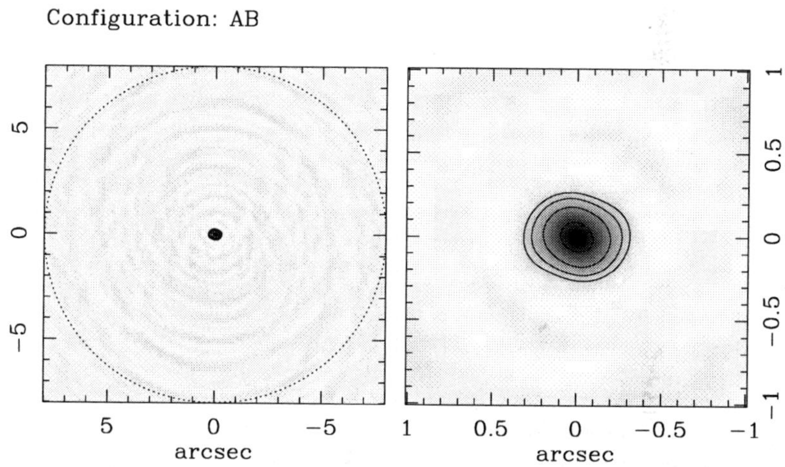

FIGURE 6. Same as Figure 2, but showing the combination of configurations A and B, each for an observation of 9 hours.

one after the other, results in a better beam shape and an intermediate spatial resolution between the two. The next step we take is to optimize the most compact configuration, by limiting the maximum baseline of the array. The compact configuration is then scrutinized in terms of shadowing effects, which result in a loss of sensitivity. Having found three out of the desired four optimal configurations, the last one is optimized to match both the most compact and the second most extended in terms of *uv*-sampling.

Figures 2 to 5 show the beams achieved by the new, optimized configurations of the IRAM interferometer. The names **A**, **B**, **C**, **D**, are used to describe the configurations in descending order of the size of the baselines, i.e. **A** is the most extended configuration and **D** the most compact. All beams shown are results of a 9-hour observation of a +45°

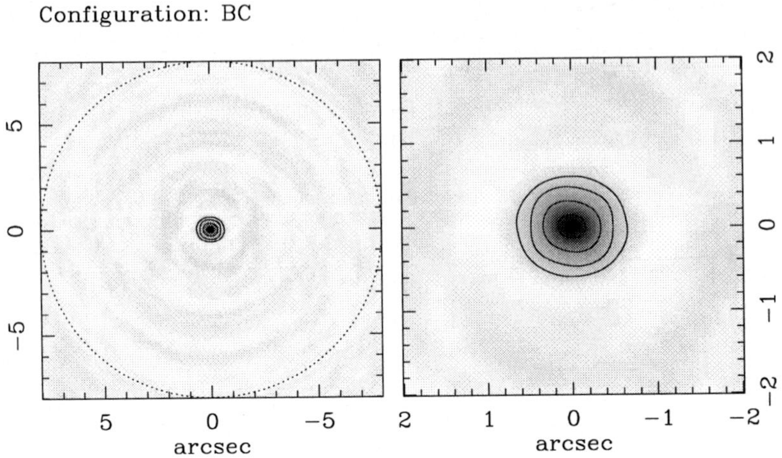

FIGURE 7. Same as Figure 2, but showing the combination of configurations B and C, each for an observation of 9 hours.

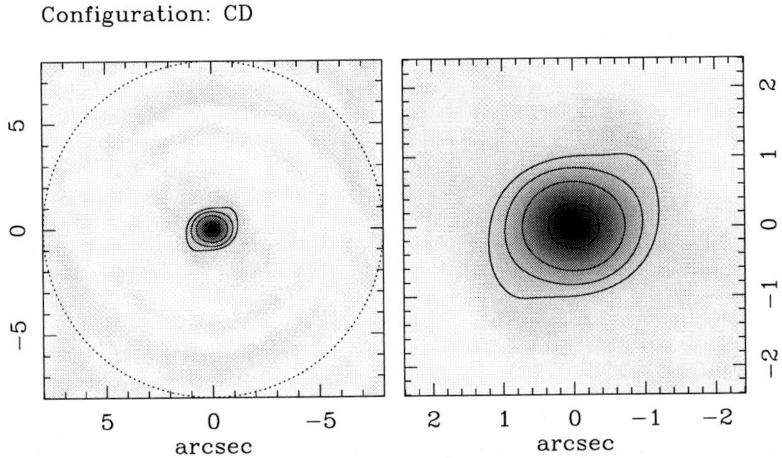

FIGURE 8. Same as Figure 2, but showing the combination of configurations C and D, each for an observation of 9 hours.

declination source. The two plots on the top row of each figure show the beam of the new configuration at a scale of the primary and synthesized beams (left/right). The beams in Figures 6, 7 and 8 are the results of adding the *uv* data from 9-hour observations in each of two configurations denoted by the letters **AB**, **BC** and **CD**.

Table 1 shows the characteristic beam sizes of the new and old antenna configurations of the IRAM interferometer, computed at the observing frequency of 300 GHz. The various beam sizes shown in Table 1, are represented in terms of the beam surface in Figure 9. Squares denote the current beams, circles the new ones. Above the symbols, the configuration name is given, along with the ratio of the surface size of the old versus

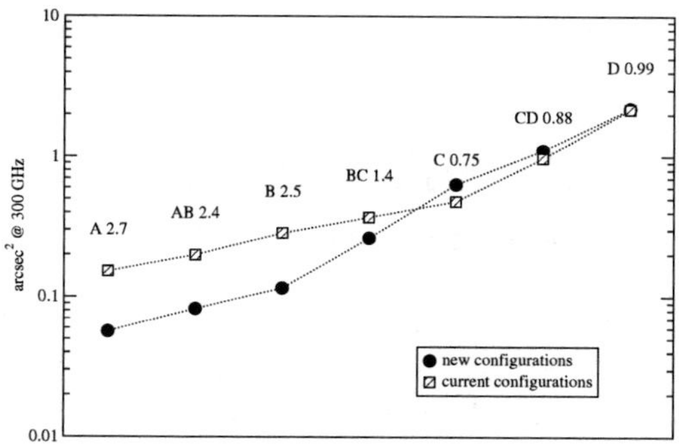

FIGURE 9. The size of the beam surface at 300 GHz as given in Table 1 for the previous (squares) and new (circles) IRAM interferometer configurations. The labels denote the configuration name, as explained in the text. The appended numbers are the ratio between the surface size of the current versus the new beams.

the new beams. In the most extended configurations, the current beams are ≈ 2.5 times the surface of the new beams. For obvious reasons related to the physical limits in the most compact configuration, the current and new beams are of the same size, although the new beam is significantly rounder.

CONCLUDING REMARKS

We present here the possibilities of the new configurations of the IRAM interferometer on the Plateau de Bure. We have applied the method described in Karastergiou et al. [1] to obtain four new configurations. Simulated observations with the new array yield circular beams with low sidelobes. The new configurations are optimised in such a way that combinations also yield good quality beams and allow more possible spatial resolutions for astronomical observations.

REFERENCES

1. A. Karastergiou, R. Neri, and M. Gurwell, ApJS (2006) submitted.
2. E. Keto, ApJ **475**, 843–+ (1997).
3. F. Boone, A&A **386**, 1160–1171 (2002).

Ten Years of the Solar Radiospectrograph ARTEMIS–IV

C. Caroubalos*, C. E. Alissandrakis†, A. Hillaris*,**, P. Preka–Papadema*, J. Polygiannakis*, X. Moussas*, P. Tsitsipis‡, A. Kontogeorgos‡, V. Petoussis‡, C. Bouratzis*, J.-L. Bougeret§, G. Dumas§ and A. Nindos†

**University of Athens, 15783 Athens, Greece*
†University of Ioannina, 45110 Ioannina, Greece
***Hellenic Naval Academy, Piraeus 18503, GREECE*
‡Technological Education Institute of Lamia, Lamia, Greece
§Observatoire de Paris, CNRS UA 264, 92195 Meudon Cedex, France

Abstract. The Solar Radiospectrograph of the University of Athens (ARTEMIS-IV[1]) is in operation at the Thermopylae Satellite Communication Station since 1996. The observations extend from the base of the Solar Corona (650 MHz) to about 2 Solar Radii (20 MHz) with time resolution 1/10-1/100 sec. The instruments recordings, being in the form of dynamic spectra, measure radio flux as a function of height in the corona; our observations are combined with spatial data from the Nancay Radioheliograph whenever the need for 3D positional information arises. The ARTEMIS-IV contribution in the study of solar radio bursts is two fold– Firstly, in investigating new spectral characteristics since its high sampling rate facilitates the study of fine structures in radio events. On the other hand it is used in studying the association of solar bursts with interplanetary phenomena because of its extended frequency range which is, furthermore, complementary to the range of the WIND/WAVES receivers and the observations may be readily combined. This reports serves as a brief account of this operation. Joint observations with STEREO/WAVES and LOFAR low frequency receivers are envisaged in the future.

Keywords: Solar activity; Corona; Radio; Radiation and spectra; Solar electromagnetic emission
PACS: 96.60.Q; 96.60.P; 95.85.Bh; 96.25.Tg; 96.60.Tf

INTRODUCTION

The radio emission of solar flares and transients includes a plethora of emission processes which serve as diagnostic tools for the analysis of non-thermal electron distributions, plasma waves and turbulence (cf. [3] for a review). They are caused by plasma instabilities driving various wave modes that in turn may emit observable radiation. Some of the processes are reasonably well understood others are open research issues.

Solar radio spectrographs have revealed a large variety of Solar radio–bursts taking place in the Solar Corona. As observations at different wavelengths sample different heights and physical conditions in the solar atmosphere, with longer wavelengths referring to higher heights above the photosphere, these phenomena can in principle be probed from the bottom of the corona to large distances in the interplanetary medium. This report is a brief account of a decade of the ARTEMIS–IV Solar Radiospectrograph

---

[1] Appareil de Routine pour le Traitement et l' Enregistrement Magnetique de l' Information Spectral

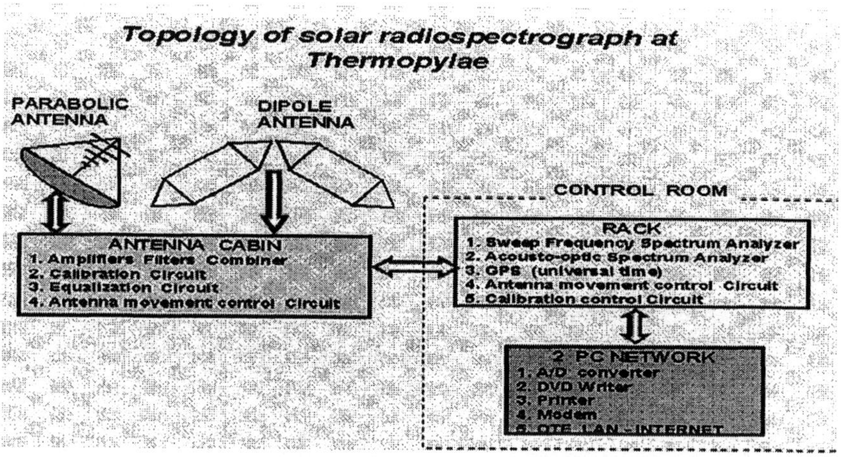

FIGURE 1. Block Diagram of the The Artemis–IV radiospectrograph architecture

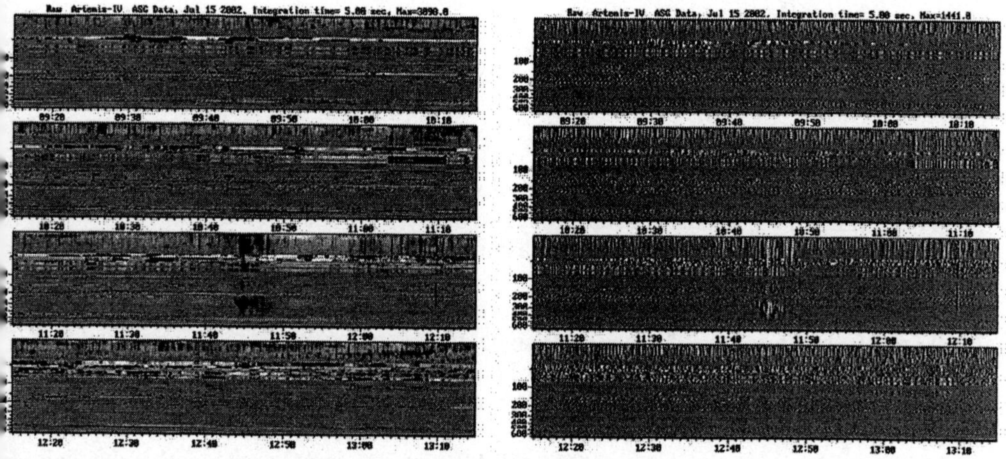

FIGURE 2. An ASG QUICKLOOK Intensity Spectrum July 15 2001. The parallel lines are terrestrial interference. LEFT: Intensity Spectrum, RIGHT: Differential Spectrum.

operation in the field of observational analysis of solar radio bursts.

OPERATION AND SYSTEM ARCHITECTURE OF THE ARTEMIS–IV SOLAR RADIOSPECTROGRAPH

The Solar Radiospectrograph of the University of Athens (ARTEMIS-IV) is in operation at the Thermopylae Satellite Communication Station since 1996. The observations extend from the base of the Solar Corona (650 MHz) to about 2 Solar Radii (20 MHz)

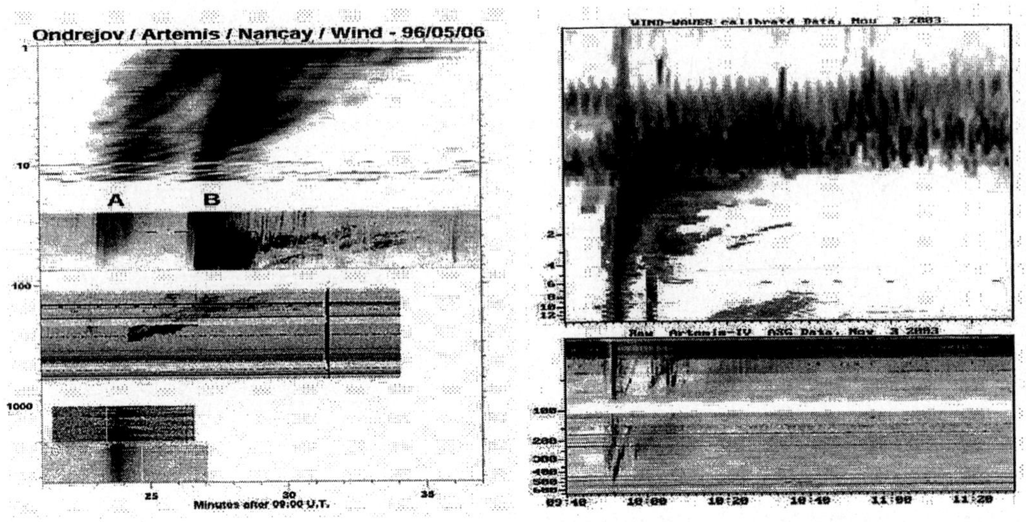

FIGURE 3. Global Dynamic Spectra. LEFT: The May 6, 1996 event from 4.2GHz to 1 MHz. From Bottom to Top: Ondrejov Radiospectrogram; ARTEMIS-IV data; Nançay Decametric Array; Wind/Waves/RAD2 ([4]) RIGHT: Combined ARTEMIS–IV and WIND–WAVES spectra of the November 3, 2003 event.

It offers high quality recordings with time resolution 1/10-1/100 sec, in the form of dynamic spectra; which is radio flux as a function of height in the corona.

The radio emission is collected by two antennas. Firstly, a 7-meter parabolic antenna with equatorial mounting covering the frequency range 100–650 MHz; it tracks the Sun in Declination and Hour Angle. Secondly, a stationary *inverted V* antenna is used for the 20 to 110 MHz range. The incoming signal from the two antennas is combined and then shared between two receivers (Figure 1). The sweep frequency receiver (Analyseur de Spectre Globalor ASG), covers the range of 20-650 MHz at 10 samples/sec, with a dynamic range of 70 db. The Acousto-Optical receiver (Spectrographe Acousto Optique or SAO), which covers the range of 270-450 MHz at 100 samples/sec, with a dynamic range of 25 db. A detailed description of the receivers can be found in [5].

The data acquisition is performed by two PCs (equipped with 12 bit, 225 ksamples/sec DAC, one for every receiver), connected through Ethernet. The daily operation is fully automated.

The recordings are in the form of dynamic spectra from the ASG and the SAO. The ASG gives good observations of strong, broad band events, such as type III groups and type II/type IV bursts. The SAO, on the other hand, has a higher sensitivity and lower noise compared to classic sweep–frequency receivers, although its dynamic range is rather limited; there is a wealth of fine structure comprising of weak, small time scale as well as narrow band features observable at this resolution.

In order to enhance small fluctuations and features which are usually undetected as they are superimposed on high level activity the time derivative of the intensity for each frequency channel is often used. Positive derivatives are coded in black, negative in white

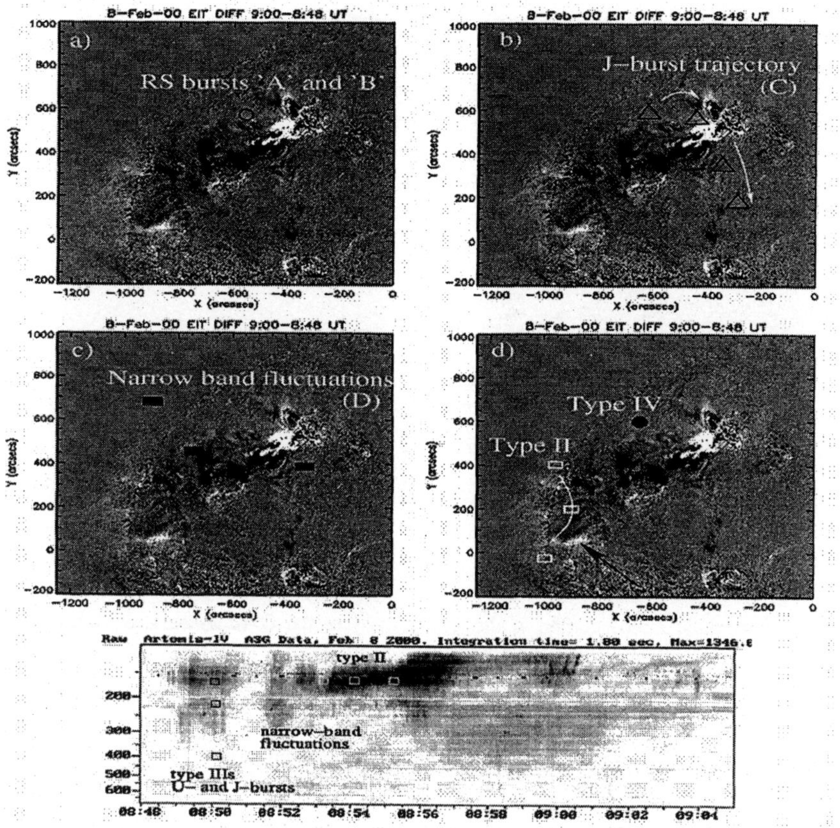

FIGURE 4. The February 08, 2000 Event. UPPER FOUR PANELS: NRH Positions of reverse slope (RS) and J-bursts and narrowband fluctuations overlayed on an EIT difference image. BOTTOM PANEL: Artemis–IV Dynamic radio spectrum, showing J-bursts, narrowband fluctuations,and the 2nd harmonic lane of the type II burst. Black and white circles indicate the radio sources that are imaged.

and all intermediate grey levels have been adjusted to obtain an optimal presentation of the events.

The daily spectra of the ASG with a reduced time resolution of 5 seconds (Quick Looks) are available on line from the ARTEMIS-IV web page[2] (cf figure 2 for an example). Higher resolution data are available on request.

---

[2] http://www.cc.uoa.gr/artemis

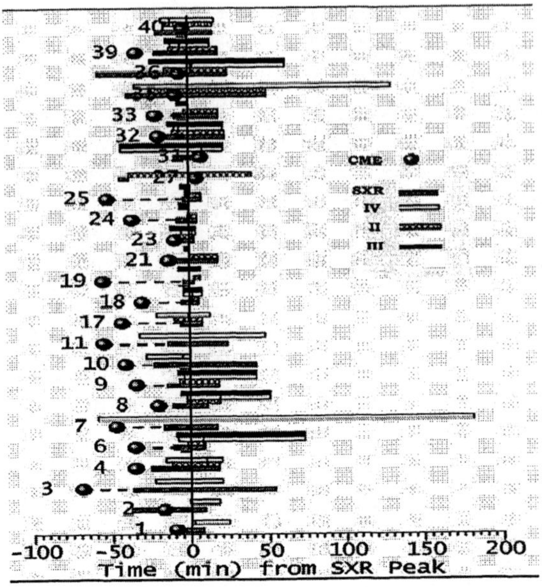

FIGURE 5. Schematic representation of the total activity of the CME associated events. It depicts durations of SXR flares, Type III, II, IV activity and extrapolated CME launch time; all times are measured from the SXR flare peak [7].

OBSERVATIONAL RESULTS

In the following subsections, we review in brief, observational results of solar radio bursts, recorded by ARTEMIS-IV and, mostly combined with data from other sources:

Joint Observations from the base of the Solar Corona to Earths Orbit: The SA Electrons

The SA Electrons constitute a class of intense interplanetary type III bursts closely associated to meter-wave (coronal) type II bursts. The acronym SA is interpreted either as *Shock Accelerated* or as *Shock Associated*; thus expressing an uncertainty on the origin of the electrons responsible for the SA events. They are thought to be produced by electrons accelerated by shock waves in the solar corona as an extension of *herringbone structures* or by electrons accelerated in the chromosphere in the same energy release that has launched the shock.

An almost complete frequency coverage of an SA radio event and related phenomena observed on May 6, 1996 at 9:27 UT was presented for the first time ([4], cf. also Figure 3). It was observed from the base of the solar corona up to almost 1 AU from the Sun by

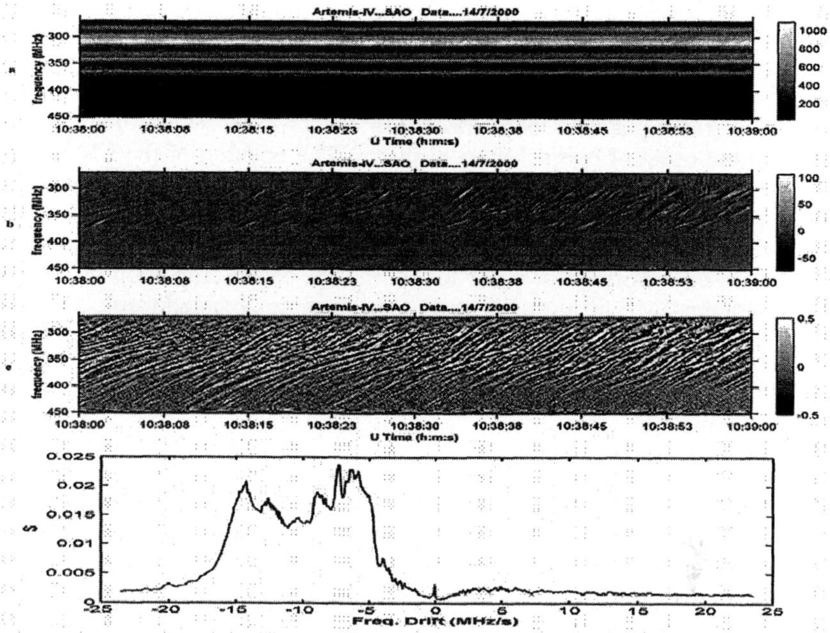

FIGURE 6. Detection of fiber bursts embedded in a type IV continuum on a dynamic spectrum recorded by ARTEMIS-IV on the 14 July 2000 [6]. Top: Dynamic Spectrum of a solar continuum; the lines parallel to the time axis represent fixed frequency terrestrial interference. Second panel from the top: Same images after a high pass filtering along the time axis; this has eliminated the interference and has suppressed the continuum enhancing the embedded fine structure. The latter consists of pulsations & fibers.Third panel from top: High pass filtering along the frequency axis has suppressed the pulsations enhancing the *fibers*). Bottom: The application of the *Slope Detection Algorithm*, [13], provides a distribution of fiber *Slopes* for the filtered dynamic spectrum; these correspond to the distribution of frequency drift rates for the fibers.

the following radio astronomical instruments:

- the Ondrejov spectrometer operating in 4.5–1 GHz (radiation produced near the chromosphere)
- The Artemis-IV spectrograph operating in 600–110MHz (1.1–1.4R from sun center)
- The Nançay Decametric Array Decameter Array in 75–25 MHz (1.4–2 R)
- The RAD2 and RAD1 radio receivers on the WIND spacecraft covering the range from 14 MHz–20 kHz (3R–1AU).

The observations show that the SA event starts from a coronal type II radio burst which traces the progression of a shock wave through the corona above $1.8R - 2R$ from the sun center. This SA event had no associated radio emission in the decimetric-metric range, thus there is no evidence for electron injection in the low/middle corona and, as a case study, corroborates the Shock Acceleration.

CME Events Associated with Large Scale Field Line Openings and Disappearing Transequatorial Loops

Disappearing transequatorial loops have been observed on several occasions in association with the onset of coronal mass ejections. The opening of the CME magnetic field can be revealed by radio observations since the radio emission sources trace the propagation paths of electrons (along the loops before they open up completely).

A multi–wavelength analysis of two events (February 8, 2000, cf. [11] & May 2, 1998 cf. [10]) was performed in order to test the transequatorial loop disappearance association with the CME lift off. Observational data were obtained from:

- SOHO LASCO, in white light).
- SOHO EIT, in EUV
- YOHKOH, in SXR
- The ARTEMIS–IV
- The Nançay Radioheliograph (NRH).

This study provided combined observations of large–scale loop–like SXR & EUV dimmings coincident, on disk, with type III (J) family bursts tracing the abovementioned loops and type IV emission marking the CME lift–off. This evidence in support of large scale field line opening preceding the lift–off of material corroborates the *breakout* model of CME launch formulated by [2].

CME–Flare Events Associated with Coronal Waves

The question whether a given coronal wave was driven by a CME or it was ignited by a flare is an open research issue; this issue was addressed in a *case study*([14]) of the large flare/CME event that occurred close to the west solar limb on 3 November 2003. The comprehensive multi–wavelength coverage of this event, from radio to HXR, marked this event as an appropriate candidate for this type of analysis. The data included:

- Ha Filtergrams from the Kanzelhohe Solar Observatory (KSO)
- X–ray Images in SXR from GOES-12 and in HXR from the Reuven Ramaty High Energy Solar Spectroscopic Imager
- EUV images from SOHO EIT
- Dynamic spectra from: The ARTEMIS–IV, the Radiospectrograph of the Solar Radio Laboratory of Izmiran, the Potsdam–Tremsdorf Radiospectrograph, the RT4/5 spectrograph of Ondrejov, the PHOENIX-spectroghraph of ETH Zurich & the RAD1 radio receiver on the WIND spacecraft.
- The Nançay RADIOHELIOGRAPH.
- Magnetograms from the Michelson Doppler Imager onboard SoHO
- White light observations from SOHO LASCO.

The analysis indicated that the physical relationship between CME/flare and the coronal wave (shock) may be quite complex. In this case it was found that the coronal shock that produced the Moreton wave and the associated type II burst was generated by the energy release in the flare rather than by the CME expansion; this evidence in support of the flare initiated shock however does not preclude the CME initiation scenario.

Association of type II and type IV Burst with other type of Solar Active Phenomena

A catalogue of the type II and type IV solar radio bursts in the 110–687 MHz range, observed with ARTEMIS-IV in 1998–2000 was presented ([7]); these were compared with the LASCO archives of Coronal Mass Ejections and the Solar Geophysical Reports of solar flares (Ha & SXR) and examined for possible associations. The main results of this work corroborating previous works ([1], [8], [15], [12]) are:

- 68% of the catalogue events were associated with CMEs.
- 67% of the type II events were associated with CMEs, in accordance with previous results. This percentage rises to 79% in the case of composite type II/IV events.
- 77% of the type IV continua were associated with CMEs, which is higher that the CME–type II association probability.
- The type II associated CMEs had an average velocity of $(835 \pm 380) km/s$, while the CMEs without type IIs had an average velocity of $(500 \pm 150) km/s$.
- All events, but one, were well associated with Ha and/or SXR flares.
- Most of the CME launch times precede by 5–60 min (30 min on average) the associated SXR flare peak; an important fraction (72%) precede the flare onset as well.
- Most of the type II associated CMEs have velocities greater than 400 km/sec.

In figure 5, we present a *comprehensive* schematic of the CME, flare and radio burst sequence of events for the total of our data set.

A new Algorithm for Fine Structure Detection & Analysis in Dymamic Spectra

An efficient algorithm for the detection of linear and quasi–linear structures in grey-scale images was developed ([13]). This algorithm introduces the *Angular Energy Density* and detects angular distribution of these structures in images; this statistical approach proves to be of value in case of a great number of line segments. As is based on Fast Fourier Transform methods, the computational efficiency and, in turn, the rate of estimation are high.

As regards solar radio burst fine structure the method proposed appears quite sensitive in slope detection. Due to the correspondence between frequency and height in the Corona the slopes of bursts on the time-frequency dynamic spectra provide direct estimation of exciter speed. This algorithm is a useful tool in the study of solar radio bursts wih a drift, such as type III, fibers etc; an example is demonstrated in figure 6 .

Super Short Structures *SSS*

A new type of super–short structures (SSSs) recorded during metric type IV bursts is reported in [9]. Their duration, at half power ranges from 4-50 ms, almost 10 times shorter than spikes at corresponding frequencies. The solar origin of the SSSs was confirmed by one-to-one correspondence between spectral recordings of Artemis–IV and high time resolution single frequency measurements (Trieste–Italy). The recorded of *SSS* were divided in three categories:

- *Simple broad-band SSSs*; characterized by a bandwidth $\Delta f \geq 100\ MHz$. They are further subdivided in:
  - *SSS-pulses* have duration 10–20 ms, and bandwidth $\approx\ 100MHz$. They appear in groups and, occasionally, exhibit quasi-periodic behaviour.
  - *Drifting SSSs* have duration 30-70 ms. Their bandwidth in general exceeds the 100 MHz, and measured frequency drift rates ($|\Delta f/\Delta t| \approx$ 400-1000 MHz/sec).
- *Simple narrow-band SSSs* are distinguished by their narrow bandwidth $\Delta f \leq\ 20MHz$. They are, also, subdivided into:
  - *Spike-like SSSs* are the shortest *SSSs* with duration 4-30 ms. Their bandwidth is mostly $\Delta f\ \leq\ 20$ MHz. If measurable, frequency drifts are $|\Delta f/\Delta t| \geq\ 800\ MHz/sec$.
  - *Patch–like SSSs* exhibit, due to their morphological diversity, a rather broad range of duration varying between 4 and 50 ms. The frequency bandwidth is $\Delta f \leq$ 15 MHz, and it can be as low as a few MHz. This qualifies *patch–like SSS* as the *SSSs* of the narrowest bandwidth. Their spectral appearance varies; they can resemble dots, sails or flags and were further subdivided accordingly to: *dot–like SSS*, *sail-like SSS* and *flag–like SSS*.
- *Complex Super Short Structures* are characterised by an emission and an absorption element. Two subcategories could be distinguished:
  - *Rain-drop bursts* consist of a narrow-band emission *HEAD* ($\Delta f \approx 5MHz$) and a broad-band absorption *TAIL* ($\Delta f\ \approx\ 40MHz$). The durations are approximately 50 ms for the *HEAD* and 30 ms for the *TAIL*. Both the *HEAD* and the *TAIL* exhibit frequency drift which is $|\Delta f/\Delta t|\ \approx\ 60 \pm 10$ MHz/sec and $|\Delta f/\Delta t| \approx 1000 \pm 400$ MHz/sec, respectively.
  - *Blinkers* are drifting bursts ($|\Delta f/\Delta t| \approx$ 650 MHz/sec) consisting of absorption element that is switching abruptly to emission element. Opposite cases have also been found (emission in the high-frequency part, and absorption in low-frequency part of the burst). The duration $d_{1/2} = 30 - 40\ ms$. They are the *SSSs* with the largest bandwidth $\Delta f > 150\ MHz$.

CONCLUSIONS & FUTURE PERSPECTIVES

ARTEMIS–IV is characterized by high time and frequency resolution as well as autonomous operation. We have obtained high quality data which are used in improving our understanding of solar radio bursts and of the underlying physics of the corona and the interplanetary medium. The particular areas of interest have been, transient activity and flares, energy dissipation, electron acceleration and transport during flares, radio signatures of CMEs and coronal radio radiation amongst others.

The most comprehensive results have been obtained from coordinated studies of radio observations and data recorded in other spectral ranges. Further comparisons with observations of HXR, energetic electrons, EUV and SXR imaging can put coherent emissions into context and may open exciting new possibilities for radio diagnostics unfolding their full potential as a tool for understanding plasma processes and energy release in the solar corona.

The use of industry standard hardware and the modular architecture of the system provide a significant potential for further expansion. In our future plans three Acusto–Optic receivers are envisaged for full spectral coverage. We also intend to develop and integrate processing algorithms for supervised learning and pattern recognition, in an attempt to minimize recording of inactive periods and to automate classification of observational data.

REFERENCES

1. M. D. Andrews, *AGU Fall Meeting*, Abstract **SH42A–0766**, (2001).
2. S. K. Antiochos, C. R. DeVore, and J. A. Klimchuk, *Astrophys.J.*, **510**, 485–493, (1999).
3. A. O. Benz, *Lecture Notes in Physics*, **612**, 80–95, (2003)
4. J.–L. Bougeret,P. Zarka, C. Caroubalos et al, *GeoRL*, **25**, 2513–2516, (1998).
5. C., Caroubalos, D., Maroulis, N., Patavalis, et al, *Experimental Astronomy*, **11**,23–32, (2001).
6. C., Caroubalos, C. E., Alissandrakis, A. Hillaris, et al, *Sol. Phys*, **204**, 167–179, (2001).
7. C., Caroubalos, A., Hillaris, C., Bouratzis, C. E., Alissandrakis et al, *A&A* **413** 1125-1133 (2004).
8. H. T., Classen, & H., Aurass, *A&A*, **384**, 1098–1106, (2002)
9. J., Magdalenic, A., Hillaris, P., Zlobec, B., Vrsnak, *These Proceedings*
10. S., Pohjolainen, D., Maia, M., Pick, et al, *ApJ* **556**, 421–431, (2001).
11. S. Pohjolainen, N., Vilmer, J. I., Khan, and A. E. Hillaris, *A&A* **434**, 329-341, (2005).
12. N., Sheeley, R. T., Stewart, R. D., et al, *ApJ*, **279**, 839–847,(1984)
13. P., Tsitsipis, A., Kontogeorgos, A., Hillaris, X., Moussas, C., Caroubalos, P., Preka–Papadema, *Pattern Recognition* (submitted)(2005).
14. B., Vrsnak, A., Warmuth, M., Temmer, A., Veronig, J., Magdalenic, A., Hillaris, M., Karlicky., *A&A*, (accepted) (2005).
15. J., Zhang, K. P., Dere, R. A., Howard, M. R., Kundu and S. M., White, *ApJ*, **559**, 452–462 (2001).

Detection of slopes of linear and quasi-linear structures in noisy background, using 2D–FFT

P. Tsitsipis*, A. Kontogeorgos*, X. Moussas†, P. Preka–Papadema†, A. Hillaris†,**, V. Petoussis*, C. Caroubalos†, C. E. Alissandrakis‡, J.-L. Bougeret§ and G. Dumas§

*Technological Education Institute of Lamia, Lamia, Greece
†University of Athens, 15783 Athens, Greece
**Hellenic Naval Academy, Piraeus 18503, GREECE
‡University of Ioannina, 45110 Ioannina, Greece
§Observatoire de Paris, CNRS UA 264, 92195 Meudon Cedex, France

Abstract. We present a fast algorithm for slope detection on grey scale images, based on 2D–FFT, which may be used for line or edge detection. Our approach is based on the calculation of *"energy"* per direction of the image, thus obtaining the Energy Spectrum on Slope (α). This exhibits local maxima at the points where α equals the slopes of linear or quasi–linear segments within the image, yet it is not affected by their position within it. The process thus outlined was motivated by the study of astrophysical images (Solar Dynamic Radio Spectra) which necessitated the introduction of a method for fast extraction of *drifting structures*, since they appear as linear or quasi linear segments on these spectra.

Keywords: Solar activity; Corona; Radio; Radiation and spectra; Solar electromagnetic emission
PACS: 96.60.Q; 96.60.P; 95.85.Bh; 96.25.Tg; 96.60.Tf

INTRODUCTION

The Solar Radiospectrograph (ARTEMIS-IV[1]) is in operation at the Thermopylae Satellite Communication Station since 1996 (cf. [1], [4]). The observations extend from the base of the Solar Corona (650 MHz) to about 2.7 Solar Radii (20 MHz) with time resolution 1/10-1/100 sec. The instruments recordings, being in the form of *"Dynamic Spectra"*, depict the variations in the bursts radiation intensity, represented by pixel brightness, in terms of time and logarithm of frequency which form, respectively, the x and y axis of an image. The frequency is characteristic of a certain layer of the solar atmosphere and proportional to the square of the electron number density at that level; thus the logarithm of the frequency is proportional to the layer's height. This is due to the almost barometric (exponential) scaling of the coronal density near the solar surface. The solar radio bursts are broadly classified according to the form of their dynamic spectrum to *"drift bursts"* and *"continua"*; the former being narrow band structures drifting in frequency with time, the latter broadband forms covering most of the spectral range simultaneously.

The *"continua"* family of bursts often excibits *"fine structure"*, which may become ac-

[1] Appareil de Routine pour le Traitement et l' Enregistrement Magnetique de l' Information Spectral

CP848, *Recent Advances in Astronomy and Astrophysics, 7th International Conference of the Hellenic Astronomical Society,* edited by N. H. Solomos

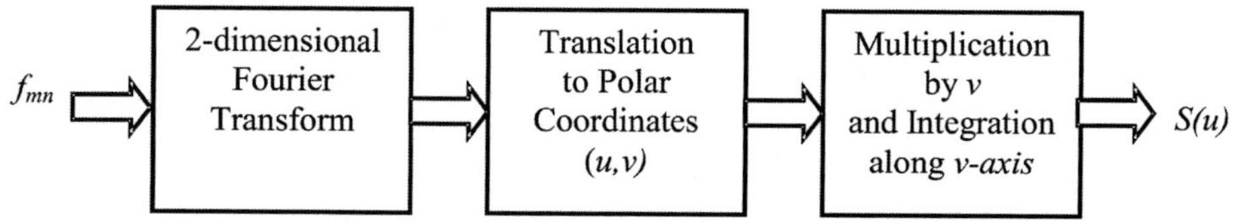

FIGURE 1. Representation of the algorithm for derivation of *angular spectra*.

centuated once the continuum background is suppresed. Certain classes of fine structure features appear in form of groups of drifting elements.

This outlines the importance of the detection of linear segments on dynamic spectra and the necessity of the introduction of the appropriate algorithms, since drifting structures trace the path of an exciter ascending at an almost constant velocity within the solar atmosphere, they form linear segments on the dynamic spectra.

In [5] an efficient method based on Fast Fourier Transform was proposed. The basic idea was the calculation of energy density as a function of angle, with the peaks of *"energy"* per direction on the image indicating, in a statistical sense, the *"dominant"* slopes of linear or quasi linear segments within the image. Certain applications of this algorithm on ARTEMIS–IV dynamic spectra are outlined in this report.

DEFINITION OF THE ANGULAR ENERGY DENSITY–OUTLINE OF THE ALGORITHM

Definition of the Angular Energy Density based on 2D–FFT

An image, or a dynamic spectrum in this case, may be represented by the function $f(x,y)$ of pixel brightness in terms of image coordinates; we may calculate the total *energy* of the image integrating over x and y. Furthermore, using the Parsevals Theorem we can rewrite the resulting integral in Fourier space:

$$\int\int \|f(x,y)\|^2 dxdy = \int_{-\infty}^{\infty}\int_{-\infty}^{\infty} \|F(\xi_1,\xi_2)\|^2 d\xi_1 d\xi_2 = 2\int_0^{\pi} d\theta \int_0^{\infty} \|F(\xi,\theta)\|^2 \xi d\xi \qquad (1)$$
$$\xi_1 = \xi\cos(\theta),\ \xi_2 = \xi\sin(\theta)$$

where, in the last term of equation 1 we have changed from cartesian (ξ_1,ξ_2) into polar (ξ,θ) coordinates in Fourier space. We may define the *"Angular Energy Density"* ($S(\theta)$) as follows:

$$S(\theta) = 2\int_0^{\infty} \|F(\xi,\theta)\|^2 \xi d\xi = \frac{dE}{d\theta} \qquad (2)$$
$$-\pi/2 \leq \theta \leq \pi/2$$

The *"Angular Energy Density"* represents the amount of *"Energy"* (E) in the (θ,θ+dθ) range and peaks at $\theta=\phi_i+\pi/2$ where ϕ_i is the slope of the ith linear segment(cf. [5]). This characteristic is exploited in the proposed process of *slope detection*.

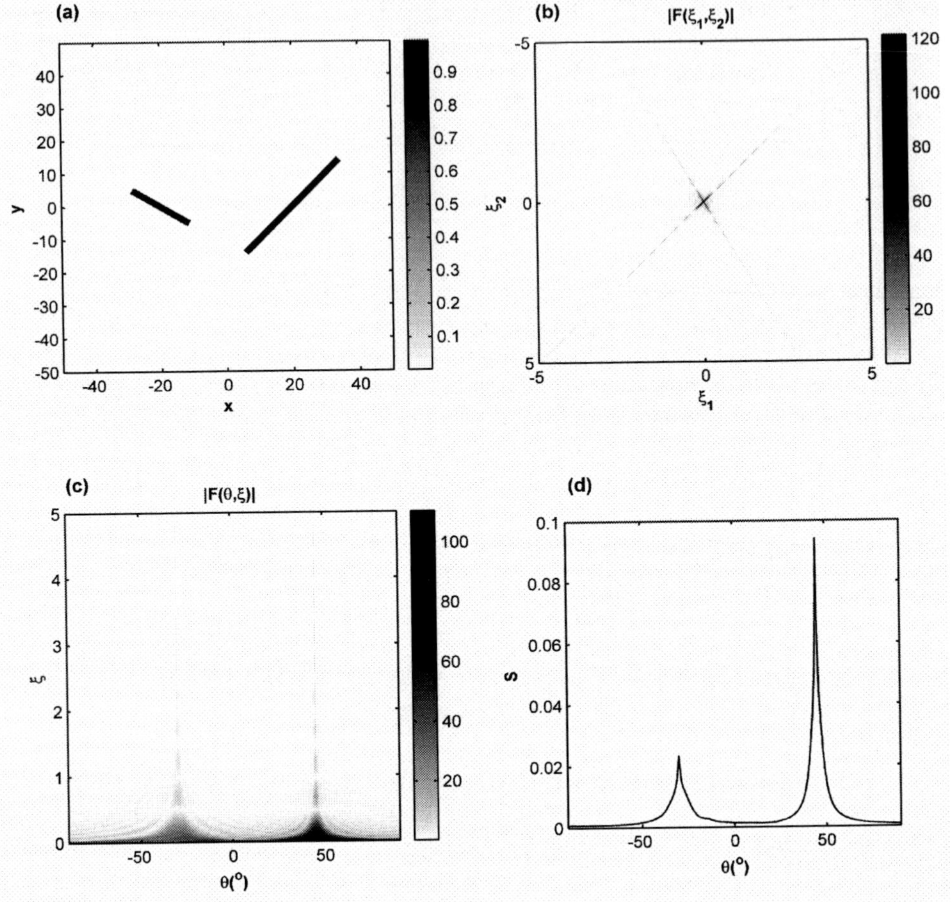

FIGURE 2. (a) Original image representing two line segments in -45° and 60°. (b) Two-dimensional Fourier transform of original image (image is rotated by 90°). Each segment is transformed to a line segment in Fourier space at right angle to the original. (c) The Fourier transform in polar coordinates. (d) The angular spectrum of original image

Discrete Form of the Angular Energy Density

Upon discretization the variables (x,y) may be replased by (m,n), (ξ_1,ξ_2) by (k,l) and (ξ,θ) by (v,u). We have

$$\begin{aligned} F_{k,l} &= \sum_{m=0}^{M-1}\sum_{n=0}^{N-1} f_{m,n}\left(e^{-i\frac{2\pi}{M}}\right)^{mk}\left(e^{-i\frac{2\pi}{N}}\right)^{nl} \\ E_{k,l} &= \frac{1}{M\cdot N}\left\|F_{k,l}\right\|^2 \Leftrightarrow \sum_{k=1}^{M}\sum_{l=1}^{N} E_{k,l} = E \end{aligned} \tag{3}$$

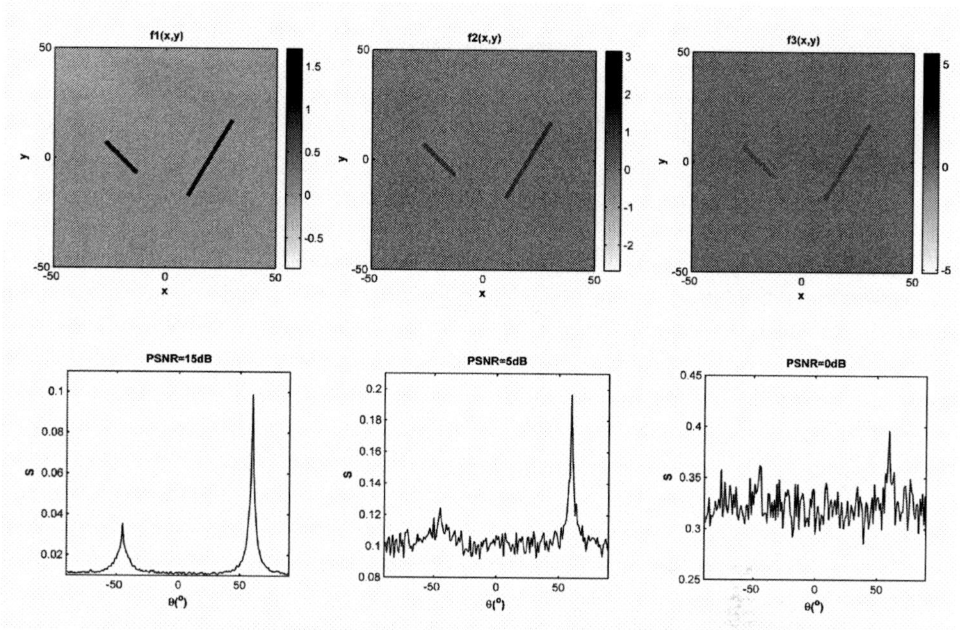

FIGURE 3. Angular spectra of images with noisy backgrounds. Top: Original images with PSNR (Peak Signal to Noise Ratio) of 15, 5 and 0 db respectively. Bottom: The corresponding angular spectra.

where we have assumed that the original function and the corresponding 2D Fourier transform were sampled on a $M \times N$ grid. $F_{k,l}$ stands for the two dimensional discrete Fourier Transform of $f(x,y)$ and $E_{k,l}$ for the *Energy* on pixel (k,l), in Fourier space. In discrete polar coordinates equation 3 assumes the form:

$$E = \sum_u S_u = \sum_u \frac{2}{\pi N_B^2} \sum_{\upsilon=1}^{N_B} \upsilon \cdot E_{u,\upsilon} \tag{4}$$

where we have taken $N_B = \frac{N}{2} - 1$, under the assumptions $M = N$ and $N_B >> 1$ which do not result in any loss of generality, and S_u stands for the energy along the, discrete, direction u. We note that for the Fourier Transform in polar form the sampling is not uniform as it is the case for the original function and its transform in Cartesian Coordinates, therefore the multiplier υ is used in the summation. Equation 4 constitutes the definition of the *"Angular Energy Density"* in discrete form, which is used henceforward throughout the text. Examples of the computation of the *"Angular Energy Density"* of linear segments on image are presented in Figure 2.

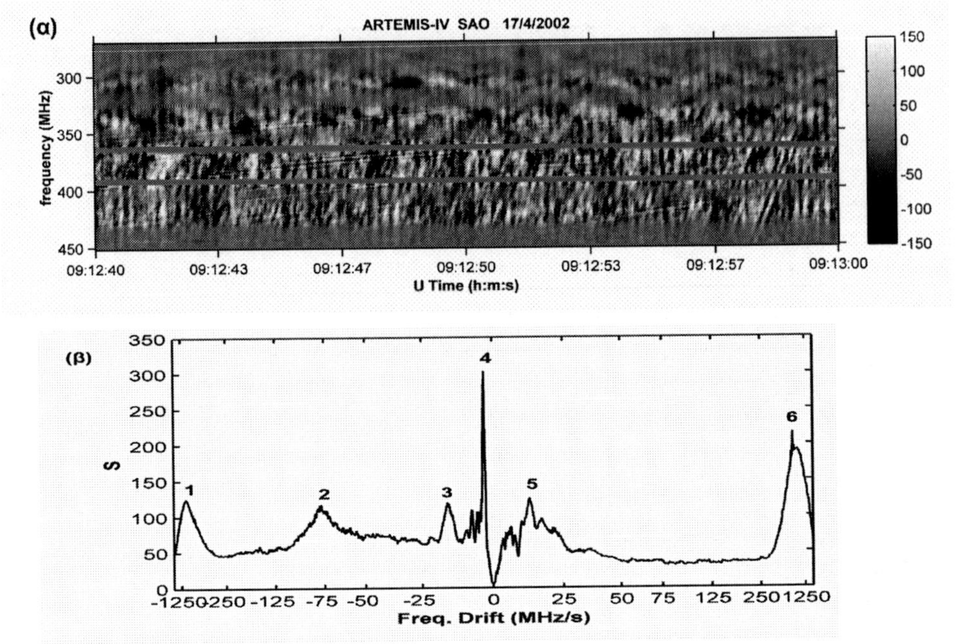

FIGURE 4. Detection of linear segments corresponding to drifting structures on a dynamic spectrum recorded by the ARTEMIS-IV radiospectrograph on the 17 April 2002. Top: Dynamic Spectrum (after an application of high pass filter along the x–axis which has suppressed the background) of a solar continuum; the lines parallel to the time axis represent fixed frequency terrestrial interference; the fine structure consists of pulsations & fibers. Bottom: The application of the *Slope Detection Algorithm*, [5], provides a distribution of pulsation & fiber *Slopes* for the filtered dynamic spectrum; these correspond to the distribution of frequency drift rates for each.

Robustness of the Method in the Presence of Additive White Gaussian Noise

The presence of additive white gaussian noise (AWGN) in the image will increase the mean and variance of backgound level of the *"Angular Energy Density"*. In particular, additive white noise with zero mean and variance equal to σ^2, may raise the *"Mean Square Error"* (henceforward MSE) of a noiseless image with maximum equal to unity to MSE=σ^2.

$$\mathrm{var}[S_u] = \frac{2\sigma^4}{3\pi^2}\cdot\frac{(N_B+1)(2N_B+1)}{N_B^3} \approx \frac{4\sigma^4}{3N_B\pi^2},\, N_B \gg 1 \tag{5}$$

This result suggests the use of a reasonable threshold (such as 5σ) in order to discriminate the spectral peaks representing line segments, from the noise spectrum. Hence we have $S_u \geq 10\sigma^2/\pi\sqrt{3N_B}$.

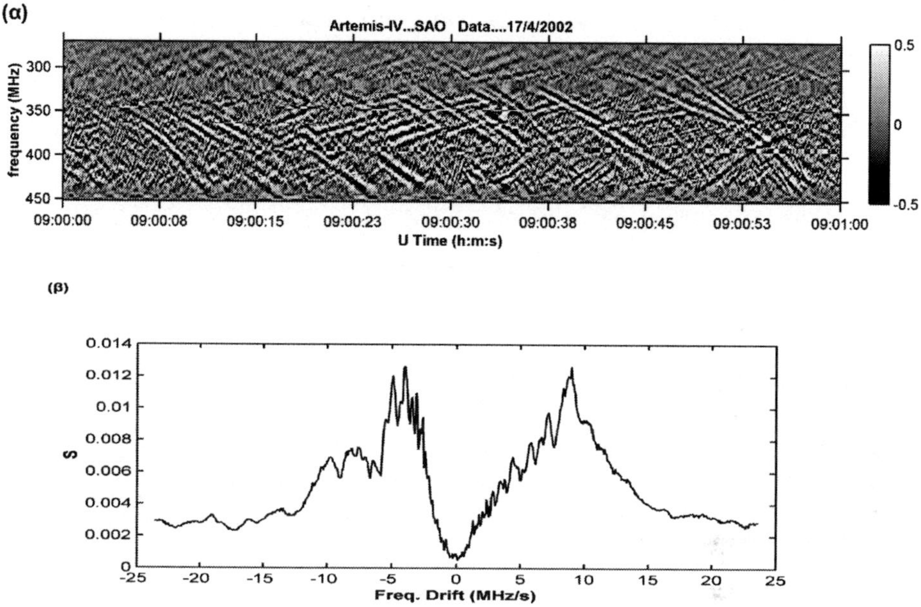

FIGURE 5. Angular energy distribution of fibers on a dynamic spectrum recorded by the ARTEMIS-IV radiospectrograph [2] on the 14 July 2000. Top: Dynamic Spectrum (Continuum background has been suppressed by high pass filtering and terrestrial interference has been eliminated by directional filtering. Bottom: Angular Energy Density of the filtered image.

An example of the angular energy density of two linear segments embedded in a noisy backgound for different signal to noise ratios is excibited in Figure 3.

A more detailed analysis of the edge–detection algorithm, with the appropriate derivation of the above results is presented in [5].

DETECTION OF FINE STRUCTURE APPEARING AS LINEAR SEGMENTS ON SOLAR DYNAMIC SPECTRA

General Description–Application of the Algorithm

Often some part of the image background needs be suppressed in order to enhance the linear structures before their detection, hence an initial stage of high pass filtering is, usually, included. As regards the dynamic spectra of solar radio bursts in particular, the detection of fine structure is customarily obtained by differentiation on time; this however, may introduce loss of information or distortion of the original signal. The use of directional filters to suppress interference at specific angles, such as θ=0 corresponding to fixed frequency terrestrial signals seems promising. Gaussian filters seem also appropriate to this task as they may also suppress slowly varying background

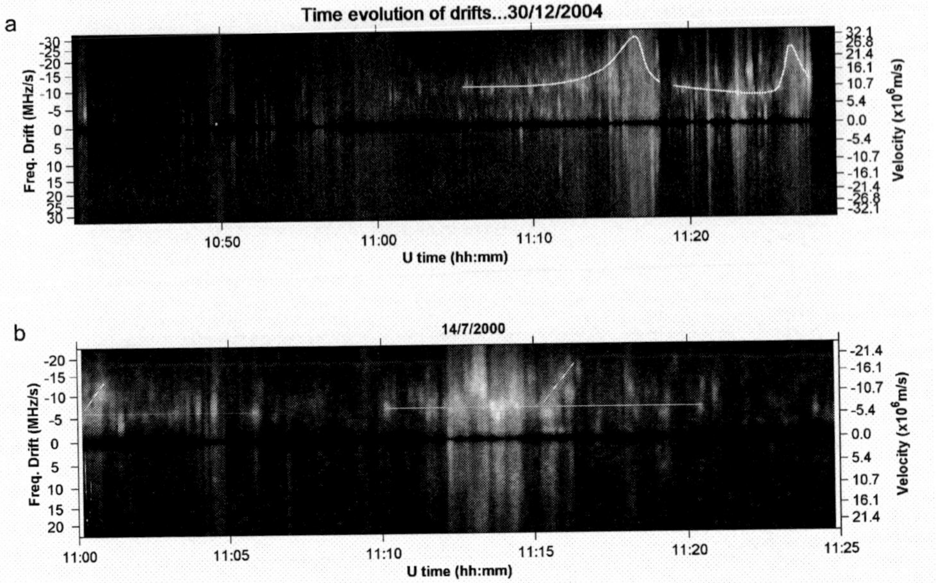

FIGURE 6. Top panel: Evolution of frequency drift rate in a period of 30 minutes during the great solar event of 30–12–2004 in the range 270–450 MHz. The overlayed curve represents the variation of peak frequency drift rate of fibers during this time period. Bottom panel: Same as top panel but for the event of 14–7–2000.

components.

Further more, since the angle θ on a dynamic spectrum does not carry any physical meaning, we can readily replace it by the *slope* (α) of the angular spectrum which is expressed in MHz/sec, if the axis of frequencies[2] is linear or in sec^{-1} if it is logarithmic; the slope represents the radial velocity of the drifting structure exciter as it traverses the solar corona. Suffice it to note that $\alpha \propto \tan(\theta)$, hence the transition is straightforward.

An application of detection and measurement of frequency drifts is represented in Figures 4 and 5; after the removal of the continuous background a composite fine structure arises including intermediate drift bursts (fibers) and pulsations in the former and complex fiber groups in the latter.

---

[2] The y–axis usually

The identification of each element within a composite spectral structure and the measurement of individual drift rates is quite difficult, due to the complexity of the recorded spectrum. Yet, the proposed method can detect slopes, corresponding to frequency drift rate, in a statistical sense; the peaks of the angular energy function, in particular, mark the dominant drift rates at each period of observation. In the example presented in Figure 4 significant drift rates of -920, -77, -15.5, -2.9, +12.7 and +700 MHz/sec appear in the time interval 09:12:40–09:13:00 UT on 2002 April 17. The same method is used to detect the dominant frequency drift rates of fibers in a rather complicated spectrum with intertwined fiber groups (Figure 5); in this example the dominant drift rates are -5 and 10 MHz/sec.

The slope detection algorithm presented, may be applied on a, relatively, short window of time *sliding* on a much larger period of activity. The result of this *windowed* algorithm is an *evolution of slope spectra* (cf. Figure 6); the y–axis represents frequency drift and the x–axis time, while the pixel brightness represents the corresponding *Angular Energy Density*. This representation facilitates a comprehensive analysis of the evolution of frequency drift rate, which can lead to estimation and representation of exciter velocity. These graphs could be used to study correlation of exciter velocity with other time varying quantities as total radio flux etc.

CONCLUSIONS

In this report we present a fast algorithm for the detection of linear and quasi–linear structures in gray–scale images. This algorithm introduces the *Angular Energy Density*and detects angular distribution of these structures in images; this statistical approach proves to be very useful in case of a great number of line segments. As is based on Fast Fourier Transform methods, the computational efficiency and, in turn, the rate of estimation are high.

The Noise immunity of the method is satisfactory since PSNR[3] along a given direction angle is greater than a reasonably low threshold.

As regards the detection of solar radio burst fine structure the method proposed appears quite sensitive in slope detection. Since there is a direct correspondence between frequency and height in the solar atmosphere (*Corona*) the slopes of bursts on the time-frequency plane can be, almost immediately, indicate the corresponding velocities of their exciters. This suggests this method as a promising tool in the study of solar radio bursts.

---

[3] Peak Signal to Noise Ratio

ACKNOWLEDGMENTS

This work was partly supported by the Greek Ministry of Education with and *Archimedes II* fund. The ARTEMIS–IV group thanks the Hellenic Telecommunication Organization for hosting the ARTEMIS–IV in the the Thermopylae Satellite Communication Station

REFERENCES

1. C. Caroubalos, D. Maroulis, N. Patavalis, J. L. Bougeret, G. Dumas, C. Perche, C. Alissandrakis, A. Hillaris, X. Moussas, P. Preka–Papadema, A. Kontogeorgos, P. Tsitsipis and G. Kanellakis, *Experimental Astronomy* **11**, 23–32, (2001).
2. C. Caroubalos, C. E. Alissandrakis, A. Hillaris, A. Nindos, P. Tsitsipis, X. Moussas, J. L. Bougeret, C. Bouratzis, G. Dumas, G. Kanellakis, A. Kontogeorgos, D. Maroulis, N. Patavalis, C. Perche, J. Polygiannakis and P. Preka–Papadema, *Solar Physics* **204**, 167–179, (2001).
3. C. Caroubalos, A. Hillaris, C. Bouratzis, C. E. Alissandrakis, P. Preka-Papadema, J. Polygiannakis, P. Tsitsipis, A. Kontogeorgos, X. Moussas, J. L. Bougeret, G. Dumas, C. Perche, *A&A* **413** 1125-1133 (2004).
4. C. Caroubalos, C. E. Alissandrakis, A. Hillaris, P. Preka–Papadema, J. Polygiannakis, X. Moussas, P. Tsitsipis, A. Kontogeorgos, V. Petoussis, C. Bouratzis, J. L. Bougeret, G. Dumas and A. Nindos, *These Proceedings* (2006).
5. P. Tsitsipis, A. Kontogeorgos, A. Hillaris, X. Moussas, C. Caroubalos, P. Preka–Papadema, *Pattern Recognition* (submitted) (2005).

The EUDOXOS National Observatory from 1996 to 2005

Nikolaos H. Solomos

Physical Sciences Sector, Applied Physics Laboratory, Hellenic Naval Academy, Piraeus
EUDOXOS National Astronomy Center, Mount Ainos, Kefalinia Island, Greece

Abstract. An account of the status and prospects of the National Robotic Observatories Complex '*EUDOXOS*' is given. 'EUDOXOS' is the oldest Greek center for robotic astronomy with astronomical and educational activity that started effectively in 1999, as the official observatory of the Greek secondary education system. EUDOXOS hosts a variety of remotely controlled telescopes and implementations of their associated diverse technologies and has adopted novel network-centric approaches reflecting in its distributed telescope architectures. In parallel, it aims in studying and testing those software hardware and middleware implementations which facilitate the development of such an observational facility as one that features the high availability needed to achieve continuous operation, inherent capacity for effective multi-user support, fully robotic unattended and redundant operation, fast response to targets of opportunity and accommodation of tele-operating instruments.

Keywords: Astronomical Observatories, Robotic Astronomy, Astronomy Education
PACS: 95.55.Cs, 95.55.-n, 1.40

INTRODUCTION

'EUDOXOS' as a new governmental astronomical institution consists of two constituent divisions bearing the names '*National Observatory of Education*' and '*National Astronomy Center*'. It was conceived in 1993, started development in 1996, and operates since 1999. Its headquarters are located within its own 48.000 m^2 base at a plateau 600m below the peak of mount Ainos (1628m altitude) in Kefallinia Island Greece (Fig-1).

Scientifically, EUDOXOS invests on the strength of the Robotic Astronomy which, in fact, is based on the inherent advantages of the small (≤2.0m) telescopes emanating from:

- the continuing existence of astrophysical problems where large collecting power and extreme sky quality are not critical or constraining factors for the success of the relevant observations.

CP848, *Recent Advances in Astronomy and Astrophysics,*
7th *International Conference of the Hellenic Astronomical Society,* edited by N. H. Solomos

- the realization that there are still extremely significant research opportunities in astronomical issues the investigation of which is intimately associated with the factor of time. Such projects could be the long term monitoring of celestial objects the latter being stars or galaxies.

Largely based on these grounds we essentially introduced Robotic Astronomy in Greece, on the grounds of the following line of reasoning: Traditional "hands on" observing modes where the astronomer works next to the telescope and controls the instruments monitoring weather and data quality during the acquisition, leads to limited productivity and do not satisfy the demands posed by the nature of the astrophysical problems, appropriate for Advanced Small Telescopes (ASTs) (Solomos, 1995), in the most optimal way [1]. The disadvantages of the traditional observing mode are serious: Time awarded is always the minimum needed for a programme completion, therefore weather-prone observations and easy failures often occur, resulting in precision, accuracy and time coverage considerations to be compromised, as vital observations and calibrations are not always performed in the proper way, yet experimental self-consistency is not easily maintained.

On the contrary, automated telescope operation provide optimum utilization of time resources, facilitates execution of intensive repetitive tasks, supports long-term photometric/spectroscopic work, increase number of objects observable in a clear night, minimize failures during vital observations and if a "flexible scheduler" is implemented, bad weather-prone observations are also minimized. Automated telescopes under unattended operation are characterized by at least 1 order of magnitude higher throughput than ordinary ones[1]. These considerations combined with inherent advantages of small telescopes such as: less demand for time allocation, unique performance on repetitive tasks (low-inertia structures, rapid switching among astronomical objects), efficient and economical dedication to special purposes and experiments, photometry and spectroscopy from larger field of views (FoV) and attractive capacity for co-ordinated observations, make AST's the ideal platforms of long-term continuous operation of astronomical instruments. This, in turn, implies dedicated small telescopes pre-designed and preconfigured so as to maximize their capacity of autonomously executing repetitive tasks.

NATIONAL OBSERVATORY OF EDUCATION 'EUDOXOS'

'Eudoxos' is the National Observatory for Education. The stimulus for its creation has been the recognition that the significance of the role of Astronomical Instrumentation as a major astronomical branch, (despite its criticality as the lever for the sustained development of observational astronomy), is still underestimated in Greece and for a prolonged period was reduced to mere management of foreign instrument imports. By breaking links in the interplay among theory and observation, a stagnation point was reached long ago by most of the existing astronomical facilities of the country which reflects in the fields of competitiveness, technological evolution

[1] For example a typical UBV photometric sequence of 10sec integrations executed in -7min with an AST, leads in -6000sets/year (assuming 100 nights/year usage) as opposed to the 600 sets/year hardly reached by a dedicated observer

and efficiency in satisfying current astrophysical needs. The impact in observational astronomy was a distinct lag behind mainstream astrophysics, which can only be compensated by the anticipated scientific input of a new generation of skilled experimental astrophysicists. NOE- 'Eudoxos' main aim is to contribute to their emergence through better education, offering of new opportunities and enhancement of their experiences. Eudoxos puts the emphasis on developing discovery attitudes by personal experimentation, *in conjunction with early acquaintance with open astrophysical problems.* Its vision and major teaching goal is to provide pupils and undergraduate students hands-on access to a variety of modern scientific instruments which allow them to acquire and analyze their own observations to test astronomical hypotheses and theories.

In addition to making astronomical infrastructure, several interdisciplinary laboratory curricula have been developed in order to attract student's interest, provide theoretical guidance and create motivation in using the new facilities. The emphasis is again put on building upon the students own curiosity to guide them to adopt a reductionist approach of understanding the Universe using the scientific method of hypothesis, observation, and analysis. Moreover, EUDOXOS develops astronomy curricula and distant learning activities tailored to the equipment and capabilities of its observatories (Solomos et.al., [2],[3]). At first, we released the Eudoxos' Modern Telelaboratory Curriculum (EMTC) tailored to the characteristics and special advantages of the TAM. 13 subjects are currently available to the participating schools and can be also available on demand to other interested Greek Universities. At second a further extension of the EMTC to the field of radioastronomy was also designed and is currently in preparation. At third, a major extension on astronomy-driven Interdisciplinary targets (data mining, informatics, radio science, astrobiology) is being planned (following suggestions by Kafatos [5]). Within their framework, European high school students are carrying out observations and analyses through the Internet. EUDOXOS has also established working collaborations with similar initiatives and major international projects.

'EUDOXOS' NATIONAL ASTRONOMY CENTER

'EUDOXOS' operates a complex of research grade networked telescopes: These are the 0.62m *'Andrew Michalitsianos'* telescope (TAM), the 0.27m diameter *'Ouranos'* (TO), the *'Apollon'* Solar Telescope (HTA) -a custom designed 0.15m diameter heliostat on top of a 4.5m high solar tower-, and the "Vikendios Damodos" Telescope (TVD), a custom-made Ritchey-Chrétien 0.51m equatorial reflector to be installed in September 2006. The observatory has developed scalable optical designs for a novel 3-mirror robotic telescope (1.1 m dia. - TKK) and plans to create an experimental robotic station in Australia. Besides, a new project aiming to the release of a quartet of 20m diameter radiotelescopes for educational use is ongoing.

(TAM) (Fig-2) consists of a fully autonomous optical system, automated enclosure, meteorological sensors, a large format imaging CCD camera and photometric filters as well as a fleet of peripheral spectrometric instruments currently under design. A Web-based interface allows any observer with internet access to select, observe, and

retrieve their own astronomical images. Additional data on the specifications of the telescopes and their instrumentation can be found in [2].

Work is ongoing on collaboration plans with other institutes abroad on GRBs, Reverberation mapping of AGN, Symbiotic stars, AGN monitoring, time-series studies. Also, since 1997 we had conceived and push forward preparatory work on a visionary program aiming at the advancement of time-domain *'Ultra High Speed Astrophysics'* in Greece. A brief presentation of such a promising initiative is given by the author (2006, this volume).

FIGURE 1. Location of Kefallinia Island in the Ionian Sea (left panel) and an aerial view of the EUDOXOS base (right) in Mount Ainos. The TKK, TAM, TO domes and Radio Dishes can be seen from top to bottom.

FIGURE 2. A closer view of the TKK, TO and TAM domes (left) and a photo of the 'Andrew Michalitsianos' Robotic Telescope (TAM) from inside its shelter (right).

In anticipation of the critical role that the automated systems and facilities will play in astronomy during forthcoming years, a reorganization project to enhance its current capabilities has been established. Eudoxos' Observing Complex is thus undertaking a major transformation in order to integrate all its telescopic nodes into a *Heterogeneous Telescope Network* [4].

The Eudoxos' heterogeneous telescope network is controlled by dedicated automated telescope control and analysis software, it conforms to international standards and is in working partnership with other similar facilities worldwide, in order to implement radically new modes of observing.
Eudoxos main achievement so far has been the realization and gaining of experience on the "tele-laboratory" concept, the latter be applied for the first time in Greece under the auspices of the Ministry of Education and the EU. The facility is the first robotic observatory in Greece. It allocates 65% of telescope time in education and the remaining 35% is available to research.
More information can be found in: http://www.astro-eudoxos.org.

In conclusion, the "EUDOXOS" center, aims not only to fulfill its mission as the official National Observatory of Education accessible through the World Wide Web, but, also, to become and remain one of the country's rapidly evolving and competitive observational establishments, capable to contribute original high quality data for the study of a wide spectrum of well defined and timely astrophysical problems.

ACKNOWLEDGMENTS

We thank the Hellenic Air Force for the generous transfer of the Ainos mountain military base to NOE *'Eudoxos'* and to the Ministry of Education and Prefecture of Kefallinia & Ithaki for their legal support. The contribution of the Ministry of Defence to *Eudoxos'* creation, is gratefully acknowledged as a symbolic contribution to the progress of the Youth.

REFERENCES

1. Solomos, N.: (1995), *'The research potential of small telescopes'*, in Proc. 2nd Conf. of Hellenic Astronomical Society. (Editors, Condadakis, M., Hadjidemetriou, J., Mavridis, L., Seiradakis, J.), Thessaloniki , pp. 643-648.
2. Solomos N., Fanourakis G., Hatzilau I., Hatzilau I., Zachariadou K., et al.: (2002), in Proc. 5th Conf. of Hellenic Astronomical Society, Fodele, Crete. http://astrophysics.physics.uoc.gr/conf/sessions/Session08/S806/solomos29.ps
3. Solomos, N. et. al: (2004) in in Proc. 6th Conf. of Hellenic Astronomical Society, (Ed. P. Laskarides) Athens, pp. 375-378
4. Solomos N., Georgatos, F.: (2006) Astron. Nachr. **327**, No.8, (in press)
5. Kafatos, M. : (2002), Abstract in Proc. 7th Conf. of Hellenic Astronomical Society, Fodele Crete. http://www.astro.auth.gr/elaset/helasmtg/2001/sessions/Session08/S825/absses8pap25.htm

On the Optical Prototype of a New Scalable 3-Mirror Telescope Design

N. H. Solomos [1,2] & I. Belyankin [2]

1. *Physical Sciences Sector, Hellenic Naval Academy, Piraeus*
2. *EUDOXOS National Astronomy Center, Mount Ainos, Kefalinia Island, Greece*

Abstract. We have studied, designed built and demonstrated a 270mm diameter scaled-down prototype of a larger 3-mirror telescope (3M-T) of low cost/performance ratio offering an anastigmatic field of 1-deg. A higher performance/cost variant could reach 5-deg field of view, and can be scaled-up to very high diameters. The original concept, the particular optical characteristics and functional aspects are briefly explained. The design has already been materialized in the Applied Optics Laboratory of the Hellenic Naval Academy where the instrument is assembled. The experimental working-model prototype of such an unusual telescope is based on a modified Paul/Baker/Willstrop design and was studied from the beginning as an inseparable entity with its dedicated single-beam photon counting High Speed Photometer. The instrument is being used in our laboratory to characterize its imaging performance, to feed light at the Nasmyth focus for various atmospheric studies, and direct laser beams to the sky for experiments on laser guide-star excitation. The 3MT prototype has also been used to devise and test perplex methods of aligning its mirrors in order to obtain the sharpest possible images for future research on faint extended objects.

Keywords: Astronomical instrumentation, 3-mirror telescopes, optical design
PACS: 95.55.Cs, 95.55.-n, 95.55.Qf, 42.15 Eq.

TELESCOPE DESIGN

Background on the Evolution History of Telescope Designs

The concept of the two mirror telescope (2MT) began to emerge in the 17th century with the conception of the Gregorian and Cassegrain optical schemes, differentiated primarily by their secondary mirrors[1]. Optimization had been developed in the early 1800s by Josef Fraunhofer, and advanced even further by Steinheil and Abbe (for aplanatic refractor objective lenses) [1], but, it had not been well understood when applied to reflecting systems until Martin Schwartzchild[2] (1906) [9], published his celebrated set of equations. These formulae describe in mathematical language, the principal characteristics of 2M-Telescopes giving predictive clues about their effective performance. This was the first time in history that the idea of combined optimization in a reflecting telescope was conceived and subsequently scientifically understood. In

[1] [Gregorian utilizes a concave mirror placed beyond the focal point of the primary mirror. Cassegrain employs a convex secondary placed somewhat inside of the focus of the primary].
[2] responding to increased interest in wide-field photography which takes advantage of 2M-T design)

CP848, *Recent Advances in Astronomy and Astrophysics,*
7th *International Conference of the Hellenic Astronomical Society,* edited by N. H. Solomos

particular, applied to the existing Cassegrain design these relatively simple formulae began to reveal the important fact *that by adjusting the asphericity between the two mirrors off-axis aberrations could be appropriately manipulated.*

Constantine Caratheodory [2],[3],[4],[5],[6] with his monumental contribution of transforming the Calculus of Variations into a synectic and systematic mathematical theory, had set the intellectual framework from which the field of multivariable optimization theory (and therefore the kernel of the modern ray-tracing software packages ZEMAX-EE, Code-V etc.) has been emerged. Exploiting the capabilities of optimixation tools we, at the Electrooptics Laboratory of the Hellenic Naval Academy, have used specific packages to experiment with possible new designs to be adopted for the forthcoming 1.1m "Constantine Caratheodory Telescope" (TKK) of the EUDOXOS Observatory in Mount Ainos Kefalinia.

The following historic account and critical appraisal of existing 2M-T designs for astronomy is given, because it had set the scene for the consideration of the actual 3-Mirror variant (3-MT) presented in this paper, which was designed in 2002 and experimentally implemented in 2003:

Major advancements were made after the conception and employment of *asphericity-testing procedures* for Cassegrain systems, which require the use of an auto-collimating flat mirror equal to the size of the primary mirror[3]. Alternative tests have been devised to test the secondary mirrors separately, although is always much better to employ star-testing of the telescope in its entirety. Notably, combination of aspheric curves yields excellent correction over moderate fields for image-critical astronomical applications. The geometry of the typical Cassegrain[4] , however, not only results in a relatively small field view, but, two kinds of aberrations are still present. These are *astigmatism* and *coma,* the first being quite small, while the latter comparatively large.

On the other hand, the fundamental breakthrough of the Ritchey-Chrétien scheme, is the reduction to a minimum of the off-axis aberrations (appearing in wide-field applications), through deployment of more extensive aspheric optimization. It turned out that, by making the primary mirror slightly more prolate, (i.e. going beyond a parabola into a hyperbola), and compensating by making the secondary mirror even more hyperbolic, we can correct coma at the expense of astigmatism. But, since astigmatism is so small it can be allowed to raise considerably before it becomes larger than the coma resulting in a two mirror system optimized over a specific field of view. Because the Ritchey-Chrétien (R-C) camera is tailored to the specific purpose of wide field photography and not high resolution imaging over a small field, R-C systems, are usually made faster that a 2Mirror Cassegrains[5]. However, extreme aspherization and the desire for a high quality wavefront are mutually incompatible, so that the final optical performance (wave front) of such fast highly-aspherized systems is usually inferior to the one achieved in slower less-aspherized Cassegrain systems. As to the FoV, what a R-C camera can cover it is relatively small (-10' to 60' max.)

---

[3] Consequently, the Cassegrain telescope became popular after 1850s, when sophisticated mirror-tests became available to the optics shop.

[4] All these systems are designed to work at focal ratios ranging from f/12 to f/30

[5] (for R-C systems to work with f/8 or f/6 they have primary mirrors in f/2.5 or f/2 and thus require massive aspherization of their strongly curved primary and secondary mirrors)

compared to other wide-angle systems[6] or the ultra-wide-field (~*3-4 deg* FoV, but very enlongated) Schmidt cameras.
Some additional considerations are the following: rather than achieving optimized asymmetric comatic images, the original R-C design forms circular (round) stellar images somewhat aberrated (out of focus)[7]. Besides, even a slightly aberrated R-C telescope exhibits considerable focal plane curvature -like a Schmidt-, a fact which is of a serious concern if extended focal plane exploitation is intended to take place.

As far as the other type of commonly encountered 2-M Cassegrain i.e. the Dall-Kirkham[8] is concerned, we note that D-K uses a spherical (easiest to figure) secondary mirror combined with a primary mirror (the latter aspherized slightly less than an actual paraboloid). However, as a matter of fact, figuring a convex prolate (ellipsoidal, paraboloidal or hyperboloidal) surface is one of the more difficult tasks in optics. The central problem with the Dall-Kirkham design is that it does not correct for comatic off-axis images. In practice, [7] the Dall-Kirkham performs only moderately well with primaries as fast as f/4 and better with primaries of f/5 and higher. Overall focal ratios are typically not lower than 15 with the best systems at f/18 or above. However, a carefully designed a Dall-Kirkham can deliver the finest images of any Cassegrain type (2-Mirror) telescope [7].

Design and construction of the model 27cm 'ΝΗΟΠΤΗΡ' Telescope

Outline of the Optical Architecture

The present 3M-T design is based on a simpler variant discovered in 1935 by the French optician Maurice Paul [8]. It was rediscovered independently in 1945 by the American James Baker and had been optimized by Roderick Willstrop [10]. The original Paul-Baker design had a paraboloidal primary mirror, a convex spherical secondary mirror and a concave spherical tertiary mirror. If the aperture of the primary mirror was f/4, it would have given acceptable images, and a field of view of about 1 degree. Its optical design is unique because it is the only form of telescope which combines three advantages, namely: a wide field of view, very small sharp images, and all-reflection optics.
In our investigation we adopted a particular strategy: using routines included in ZEMAX-EE ray-tracing package of the Electrooptics Laboratory of the Hellenic Naval Academy, we searched for possible alterations of the fundamental parameters of the 3MT design of Paul-Baker (paraboloidal primary, convex spherical secondary and concave spherical tertiary), which allow the addition of a retractable diagonal in such a way, so that the flexibility of Nasmyth and Goude foci could be added to the system. We then have determined the optimum aspheric surfaces and have found that a primary mirror of prolate ellipsoid shape, a spherical secondary and an ellipsoidal tertiary, are required to retain anastigmatism and a very good wide field performance.

---

[6] (e.g. Wright, Houghton or Maksutov, Pimenov, Klevtsov)
[7] to meet increased centroiding requirements, the early R-C systems were optimized even further than today, to the extent of forming perfectly circular stellar images.
[8] developed in the 1930s by the English H. Dall

On the other hand we can retain moderate and logical demands of cost in actual glass processing and testing.
In (Fig-1), for clarification, the optical scheme has been inserted into the mechanical outline of the telescope.

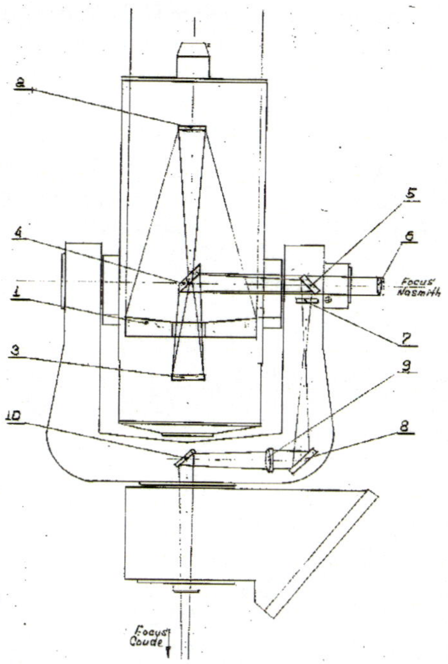

FIGURE 1 & 2: Mechanical layout of the model 3M-Telescope (left) and its laboratory realization (right) (Photo: Electrooptics Laboratory, Hellenic Naval Academy).

Explanation of operation: Light rays impinging on the primary mirror (1) are reflected towards the secondary (2) where being subjected to another reflection, go through the hole of the auxiliary flat diagonal mirror (4) in the vicinity of which they form an intermediate focus. Their trajectory passes that focus and crosses the tertiary mirror (3) producing another reflection towards the backside of the flat diagonal mirror the latter being the last reflection making a final deviation of the rays parallel to the DEC axis of the telescope onto the Nasmyth Focus. The Goude focus is materialized after the reflections following the insertion of the second (5) third (8), and forth (10) diagonals and the passage of the beam through a number of correcting lens combinations. The focus is brought to a static point in the interior of the Blue "Goude compartment" underneath the mount (see photo), where heavy instruments (e.g the Nd:YAG laser or bulky receivers can be attached)

Paul-Baker vs. NHOΠTHP-27cm: Comparative Description

Our optical scheme (see Fig.3) stemmed from the above mentioned principles of the 3M-Ts, can be described as a non-astigmatic type of pre-focal and post-focal system. It is much closer to Maksutov, but more rational and feasible with its realization[9].

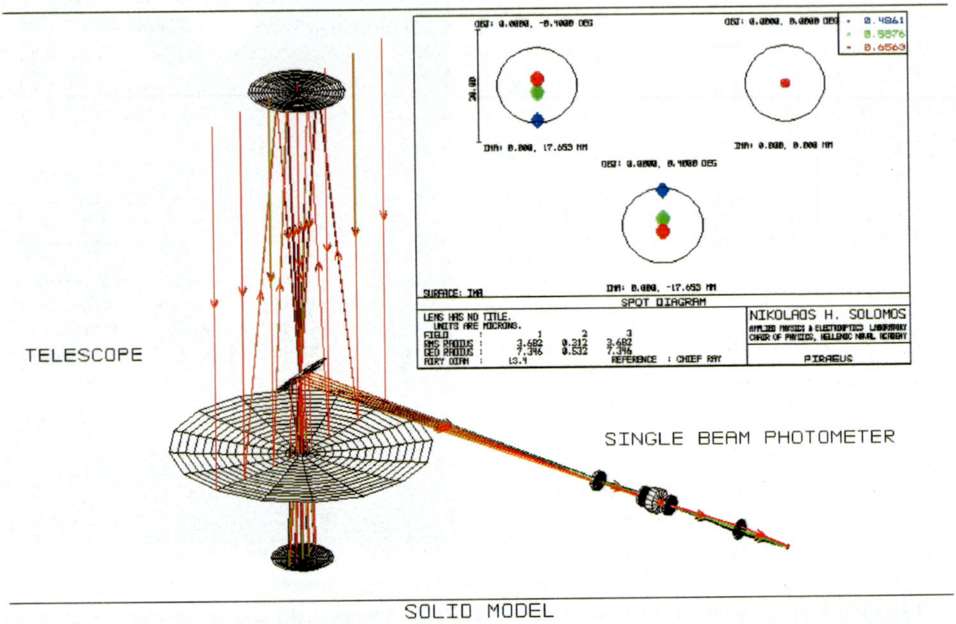

FIGURE 3: Optical layout of the model 3M telescope in conjunction with its High Speed Photometer (@ Nasmyth). The inserted upper right panel depicts the ray-tracing results (spot diagram) of the design at that focus.

- **Image quality**: The main aberrations (spherical, astigmatism) are corrected, the curvature of the field and distortion are uncorrected. The curvature of the field is correctable with a positive lens, placed near the focal plane (the lens introduces little chromatism, but it is within limits; better results could be achieved using two-lens corrector). In the optical scheme, a flat diagonal mirror (providing Nasmyth focus) is inserted (justified after instrumentation considerations), which turns the beam by 90 degrees to the axis of declination, but does not influence the image quality in any way.
- **Immunity:** This the big advantage due to minimal exposition of the detector to stray (not focused) direct light, when comparing with a usual R-C telescope: light-protecting baffles which enlarge the central screening, and light losses, are absent.

---

[9] (We keep interested in this very aspect of optical systems, as it is, in principle, possible to design a perfect theoretical system, but there are serious problems with its construction, e.g. difficulty in production of complex aspheric surfaces, mechanical & inertial stability, convenience of observation, setting aside financial restrictions which can degenerate the merits of it).

- **Compactness:** At similar light power (f#) of the primary mirror in R-C telescopes, the tube ought to be much longer as the distance between mirrors is larger.

The main advantage, though, is technological: In R-C telescopes primary and secondary mirrors have hyperbolic surfaces, in our telescope, they have ellipsoid form, where eccentricities of mirrors is about 0.6, so that the asphericity of the surface is approximately smaller by 1.8 times that of the R-C ones. The big advantage of this, is the high exactness and simplicity of control of ellipsoidal mirrors in conjugate focuses. At the same time, to control R-C hyperbolic mirrors, complex compensators are needed. As to the complicated question in exactness of the system alignment, it is mostly dependent on the f# of the primary mirror.

TABLE 1. Optical Characteristics – specifications of the model 3M-Telescope.

Optical Architecture	3-mirror Anastigmatic with auxiliary flat diagonal mirror
Foci	Simultaneous availability of Nasmyth & Goude
Primary	concave ellipsoid
Secondary	convex spheroid
Tertiary	concave ellipsoid
Mirror Substrate	ASTRO-SITALL zero expansion glass ceramic(CO115M)
Aperture Diameter	270mm
Equivalent focal Length	2670mm
Field of View	0.75°
Image scale (Nasmyth)	0.017mm/arcsec
Focal Length at Goude plane	6750 mm
F# @ Nasmyth	9.34
F# @ Goude	30.2
Object finder diameter	50 mm

TABLE 2. Electromechanical Characteristics of the model 3M-Telescope.

Mechanical Layout	Enclosed Telescope Tube
Mounting	Forked Equatorial or Altazimuth with the addition of a wedge.
Electronic Design	Stepper motors for slewing, DC analogue servomotor for focusing, Remote control unit
Computer control	Embedded controller based on the 89C51 microprocessor, computer control via serial interface (RS232),
Control software	Written in MATLAB. LabView drivers and user interface is in preparation. RTML compatibility issues currently under consideration

The disadvantages of the system are as follows:
Light losses from the larger number of mirrors, a fact that can be (partially) compensated by introducing multi-layer dielectric coatings onto the reflective

surfaces. On the other hand, mirror alignment demands are more complicated in our case, due to the number of mirrors and new customized alignment procedures should be devised. Also, the curvature of the field is approximately twice as big compared with the R-C ones (which is not severe problem for modern dedicated CCD mosaic detectors)

DESIGN OF THE HIGH-SPEED PHOTOMETER 'POLIFIMOS'

The new compact telescope is designed to work in conjunction with its own tailored instrumentation i.e. the very compact & lightweight double detector CCD/PMT photometer 'Πολύφημος'. It allows both Photon-Counting or Imaging CCD photometry. The main features of 'Πολύφημος' are as follows:

Optomechanics: The photometer provides a temperature stabilized collimator section forming a parallel beam in order to accommodate very narrow band (10-100 Å) interference filters attached in a stepper motor controlled rotating filter wheel (12 positions 1" diameter.). A viewing port for intensified CCD camera is provided by switching among two positions of an electronically controlled rotating diagonal mirror. An motorized iris wheel with 8 apertures to match different seeing conditions is located in the focal plane following a field lens. By removing the Fabry lens compartment which images the pupil onto the photocathode and replacing it with an imaging objective we can re-image the field on a relayed focal plane onto which a larger photometric CCD camera (Apogee Ap7b) head can be placed. The unit has an excellent overall performance and can easily be scaled-up for deployment in much larger telescopes.

Detectors: The main detector is the compact self-cooled PMT photon counting head HP7421 (HAMAMATSU) for ultrafast studies and LIDAR experiments in the atmosphere. The alternative CCD head is the Apogee Ap7b working under Maxim-DL software.

CONCLUDING REMARKS

Reflecting telescopes have been built with apertures up to 10 metres, but they have fields of view limited to 40 arc minutes (Ritchey- Chretien two-mirror design) or 1 or 2 degrees with a 3 or 4 lens Wynne-correctors near to the focus of the main mirror. Schmidt cameras can give fields of view of 7 or 8 degrees, but their apertures are limited to about 1.3 metres because they use a thin glass lens large enough to cover the whole aperture. The 3M-T designs like the one presented here, can reach a foV of 5^0 in diameter, while retaining a ray-tracing theoretical image size less than 0.33 arcsec everywhere, (and less than 0.1 arcsec over the central 10' of the field. Because no lenses are needed the images are perfectly achromatic. Furthermore it is possible, in principle, to build this telescope with a larger aperture than any Schmidt camera.

In conclusion, we have built and demonstrated an experimental prototype version of a 3MT. A working model of 270 mm aperture telescope named NHOΠTHP ('Nioptir') (completed in 2003) is being used in our laboratory to measure imaging performance, to feed light to the single-beam High Speed Photometer POLIFIMOS at the Nasmyth focus, or even to work in its active mode where it is able to direct powerful laser beams to the sky (after the placement of the bulky laser source in its Goude focus compartment). The justification for building that prototype in Piraeus was to demonstrate that the design works as effective as a Schmidt camera does, and pave the way for the construction and deployment of larger systems of this type on the much darker sites of Eudoxos National Astronomy Centre.

NHOΠTHP optical design could be slightly modified to allow the use of many CCDs in the focal plane. The prototype of 0.27m aperture (2003) has been used both for CCD imaging, and to test perplex methods of aligning the mirrors in order to exploit its theoretical potential in obtaining sharp images, for research on faint extended objects. Such studies latter are particularly favored by such a fast instrument as this. It is thus expected to help making realistic predictions on the anticipated performance of the EUDOXOS National Observatory new instruments, namely the forthcoming 1.1m C. Caratheodory Telescope and its associated High Speed Three Beam Photometer. Future prospects of applicability in astrophysical context will be discussed elsewhere.

The design concept under consideration, exhibits the necessary flexibility and characteristic features of a new generation of *'special-purpose telescopes'* which have not the performance sacrifices of the general purpose ones. It also provides us with an excellent model to educate the younger generation of observers with instruments of another philosophy: Those belonging to a class designed to play a part in dedicated experiments to answer very specific, open and timely astrophysical questions.

ACKNOWLEDGMENTS

N.H.S is indebted to the Hellenic Naval Academy for funding his research program on Applied Optics.I.B is thankful to NAC- EUDOXOS for his engineering fellowship.

REFERENCES

1. Volkmann, J.: (1966), *Ernst Abbe and his work,* Appl. Optics, **5,** 1720-1731.
2. Caratheodory, C: (1926) *"Uber den Zusammenhang der Theorie der absoluten optischen Instrumente mit einem Satze der Variationsrechnung"*, in Munchener Sitzungsb. Math. Natur.
3. Caratheodory, C: (1959) *"Geometrische Optic",* Birkhäuser Verlag
4. Caratheodory, C: (1930) "*Les transformations canoniques de glissement et leur application a l' optique geometrique,* in Rend. Lincei, Munchen
5. Caratheodory, C: (1940) *"Das parabolische Spiegeltelescop,* Munchen
6. Caratheodory, C: (1942) *"Die Fehler hoherer Ordnung der optischen Instrumente*" Munchen
7. Schroeder, D.: (2000) "*Astronomical Optics*", Academic Press
8. Paul, M. (1935): Rev. Opt. **14**, 169.
9. Schwartzchild. M. (1905):
10. Willstrop, R. (1984) *MNRAS*, **210**, 597

Increasing the Field of View for the Cassegrain telescope at the Kryoneri Astronomical Station

Kosmas D. Gazeas

Department of Astrophysics, Astronomy and Mechanics, Faculty of Physics, University of Athens

Abstract. During recent observations it was possible to increase the field of view for the Cassegrain telescope at the Kryoneri Astronomical Station of the National Observatory of Athens. The very small area of sky observable with the existing CCD, is a problem when larger fields need to be observed. A focal reducer was installed on the telescope thus increasing the observed area by a factor of about 4. The advantage is faster photometry and detection of very faint stars (>20 mag). A permanent focal reducer should be mounted on the telescope for the benefit of all observers.

Keywords: instrumentation, CCD camera, focal reducer
PACS: 95.45.+I, 95.55.-n, 95.55.Aq, 95.55.Cs

INTRODUCTION

The Cassegrain telescope of the Kryoneri Astronomical Station of the National Observatory of Athens was built in 1975 with a large focal ratio (f/13) ideal for planetary observations and high resolution cluster studies. It is a "slow" instrument for stellar photometry and the small field of view on the CCD chip (2.5×2.5 arcmin) is a problem when a comparison star is needed in the same field for differential photometry.

The idea to decrease the effective focal length of the telescope by installing a focal reducer before the prime focus of the telescope was thought some years ago after making a similar installation in the Cassegrain telescope of the University of Athens Observatory.

The goal was to install a focal reducer f/6.3 to decrease the effective focal ratio of the telescope thus making the instrument "faster" for stellar photometry.

OBSERVATIONS

Test observations with the focal reducer as well as photometric observations of eclipsing variable stars were made in June, August and September 2005. The instruments used were the 1.22m f/13 Cassegrain reflector at the Kryonerion Astronomical Station of the National Observatory of Athens, Greece, and a Photometrics CH250 CCD camera. A detailed description of the CCD chip is given by Sinachopoulos et al. (1998).

CP848, *Recent Advances in Astronomy and Astrophysics,*
7th *International Conference of the Hellenic Astronomical Society,* edited by N. H. Solomos

The original scale on the chip is 0.3 arcsec/pixel (without binning), giving a field of view of 2.5×2.5 arcmin. The image scale changed after the installation, decreasing the resolution of the image but offering a wider field of view. The image scales calculated for the mostly used binned modes are shown in Table 1.

Table 1. Image scale for various binned modes before and after the reducer's installation.

Binned mode	Without focal reducer (field 2.5×2.5 arcmin)	With focal reducer (field 5×5 arcmin)
1×1	0.3 arcsec	0.6 arcsec
2×2	0.6 arcsec	1.2 arcsec
3×3	0.9 arcsec	1.8 arcsec

In June 2005 the first focal reducer f/6.3 with an adaptor was attached to the telescope. The adaptor was installed between the filter wheel and the CCD camera, since there was no space between the telescope's rear mount and the filter wheel. The adaptor "pushed" the camera back a distance of 8 cm (the length of the adaptor). The focal reducer could be moved into the adaptor by sliding the lens inside the cylindrical adaptor.

The achieved field of view with the first adaptor was 3.3×3.3 arcmin for the closest distance of the focal reducer to the CCD camera and 4.1×4.1 arcmin for the farthest. A second focal reducer (f/3.3) was also installed, but strong vignetting in all possible distances from the CCD chip was the reason to discard this option.

The field of view obviously depends on the distance of the focal reducer to the CCD chip. The closer the lens, the smaller the field.

In August 2005 a longer adaptor was attached, in order to study the correlation between the distance and the field of view. With the new adaptor it was possible to further increase the field of view up to 7×7 arcmin. With this adaptor it was also possible to measure the distance at which the vignetting was apparent.

After these tests, a third adaptor was constructed and installed on September 2005. This adaptor was constructed according to the measurements made before, in order to check how the final version of the adaptor will behave under real conditions.

DATA REDUCTION AND TESTING

The first task was to measure the ideal distance of the focal reducer from the CCD camera, in order to have the desired field of view. Various observations were made, measuring the respective distance of the CCD camera and the focal reducer, in comparison to the observed field of view (Figure 1). The field was increased from 2.5×2.5 arcmin gradually to 3.3×3.3, 4.1×4.1, 4.7×4.7, 5×5 and 5.7×5.7 arcmin (Figures 2a and 2b).

All images wider than 5×5 arcmin were suffering from vignetting, which was stronger as the field increased. Images with field of view narrower that 4.5×4.5 arcmin were homogeneously illuminated, without any vignetting at all. The critical area between 4.5 and 5 arcmin shows very small vignetting, which can be easily corrected by flat fielding.

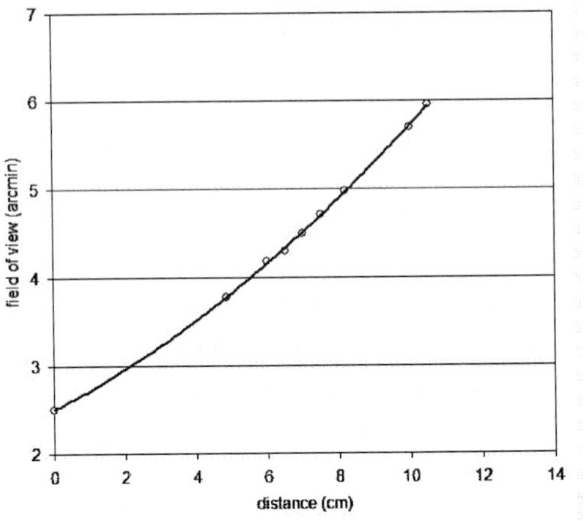

FIGURE 1. The respective distance of the focal reducer from the CCD chip, in comparison to the observed field of view.

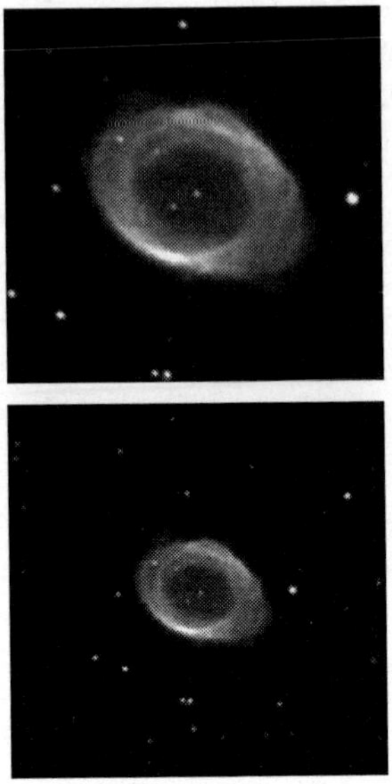

FIGURES 2a & 2b. The 2.5×2.5 arcmin (above) and 4.7×4.7 arcmin (below) fields around the planetary nebula M57 (Ring Nebula)

The focal reducer is constructed in such a way to reduce the focal length by a factor of 63% for a telescope with a focal ratio of f/10 (i.e. f/6.3). The focal ratio for the Cassegrain telescope at Kryoneri after the installation of the focal reducer should be f/8 and the resulted field of view 4×4 arcmin. After all tests, it was noticed that we can shift the focal reducer away from the CCD chip slightly more, so that the focal ratio will become even smaller (f/6.5) and the field of view 5×5 arcmin without any coma effect, spherical aberration or strong vignetting. The dimensions of the CCD chip are small enough (13×13 mm) for such errors to appear.

The next task was to test the effect of flat fielding and vignetting through the linearity of the CCD camera. For this test, images of the open cluster NGC 7790 were obtained in order to measure the brightness of the standard stars of the field and check if the vignetting and the flat fielding affect the photometry and the measurements.

The open cluster NGC 7790 contains standard stars with magnitudes that vary from 13 mag to 18 mag (Figure 3). Aperture photometry was applied on many stars of the field, close to the center of the images as well as away from it. The data reduction was made with the package AIP4WIN (Berry & Burnell 2000). A diagram of instrumental magnitude vs. the standard magnitude (Romeo et al. 1989) is shown in Figure 4. This graph shows a very good linearity over the field, no matter if the measured stars were close to the edges of the field or not. The same procedure was repeated for the fields around the open cluster NGC 7654 and the BL Lac field with similar results. Indeed, the small vignetting that appears in fields around 4.5 and 5 arcmin can be perfectly corrected and thus the use of such a wide field is absolutely safe.

The linearity test gave the opportunity to check the ability of faint star photometry. It was easy to measure stars as faint as the 18th magnitude with S/N approximately equal to 5. Thus, photometry of faint stars could be done down to the 18th magnitude with exposures down to 60 seconds. The S/N ratio was high enough for the faint stars to be measured with an acceptable accuracy.

FIGURE 3. The 5×5 arcmin field around the open cluster NGC 7790.

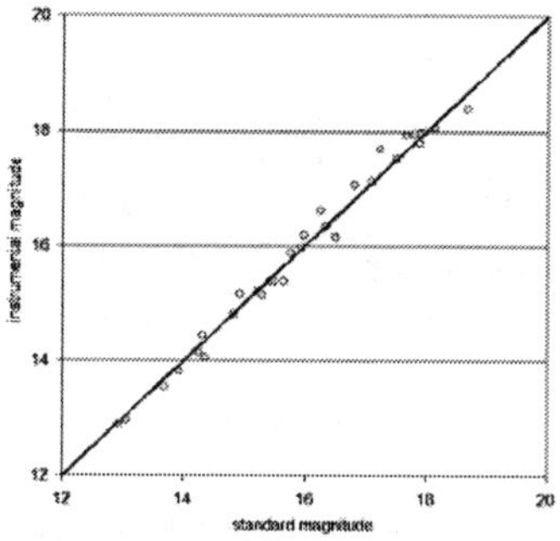

FIGURE 4. Standard magnitude vs instr. magnitude for selected standard stars in the field of NGC 7790.

ADVANTAGES OF THE FOCAL REDUCER'S INSTALLATION

- **The optical quality of images is increased**. Since the scale is different, the stars spread their light in fewer pixels. The signal is higher and the S/N ratio is 4 times bigger, allowing 4 times less exposure times. A sample of the achieved image quality is shown on Figures 5a and 5b, in comparison with the quality of the DSS-2 scanned plates.
- **The wider field of view shows more stars in each frame**, which is very useful for the observers who use the differential photometry technique to observe variable stars.
- **The poor tracking of the telescope at Kryoneri can also be overcome**, since any possible random movements of the telescope or bad tracking will still keep the object inside the frame, without losing useful information.
- **A 20-magnitude star can be easily observed with a 60-sec exposure** and a good signal in comparison to the sky background. The limiting magnitude in a 60 sec exposure was measured to about 20.5 mag, which is probably a result (limitation) from the light pollution of the observing site from the nearby cities (Kane et al. 2004).
- **The sky background is lower**, since shorter exposure times are needed to achieve the desired S/N ratio.
- **Spatial information is not lost even with the resulted lower resolution**, after the focal reducer's installation, since the seeing at Kryoneri is never below 1 arcsec.
- **Fast photometry** can be achieved since the exposures can me smaller.

FIGURES 5a & 5b. The 3.3×3.3 arcmin field around the variable V401 Cyg, as seen from the DSS-2 scanned plates (left) and Kryoneri (right).

SUMMARY AND CONCLUSIONS

The installation of the focal reducer on the telescope of the Kryoneri Astronomical Station offers many advantages to the observers. The observed area increased by a factor of about 4. Faster photometry and detection of very faint stars (>20 mag) are some of the advantages.

Up to date, time-series photometry of variable stars was made with a wider field of view with remarkable results. Photometry of faint stars was not conducted since the installation of the focal reducer is under test so far. The telescope at Kryoneri after the installation can be "faster" and more powerful with a focal ratio of f/6.5 (decreased from the initial value of f/13). A permanent focal reducer should be mounted on the telescope for the benefit of all the observers.

ACKNOWLEDGEMENTS

Dr. Anastasios Dapergolas is gratefully acknowledged for the advice and discussions during the reduction of images. The author would also like to thank the National Observatory of Athens for the observational time at the Kryoneri Astronomical Station. Constantinos Stathoulis is also acknowledged for his useful comments of the present work.

REFERENCES

1. Berry, R., Burnell, J., *The Handbook of Astronomical Image Processing*, Willmann-Bell, Richmond (Virginia) (2000)
2. Kane, S.R, Cameron, A.C., Horne, K., James, D., Lister, T.A., Pollaco, D.L., Street, R.A., Tsapras, Y., *MNRAS*, **353**, 689 (2004)
3. Romeo, G., Bonifazi, A., Fusi Pecci, F., Tosi, M., *MNRAS*, **240**, 459 (1989)
4. Sinachopoulos, D., Dapergolas, A., Kontzias, E., *Memoirs of the National Observatory of Athens*, Ser. I, No 36 (1998)

A test of the Kryonerion "seeing" with DIMM equipment

D. Mislis*, E.T. Harlaftis†, D. Buckley**, K. Gazeas‡, K. Stathoulis§, G. Dimou† and J. H. Seiradakis*

*Aristotle University of Thessaloniki, Department of Physics, Section of Astrophysics, Astronomy and Mechanics, GR-541 24 Thessaloniki, Greece
†National Observatory of Athens, Institute of Astronomy and Astrophysics, Lofos Koufou, GR-15236 Athens, Greece
**South African Astronomical Observatory, PO Box 9, Observatory, 7935 Cape Town, South Africa
‡University of Athens, Department of Physics, Section of Astrophysics, Astronomy and Mechanics, GR-15783 Zografos, Athens, Greece
§Financial Technologies SA, Xenias 24, GR-11528 Athens, Greece

Abstract. We present a test-study of the "seeing" at the Kryonerion Station of the National Observatory of Athens. We used a Differential Image Motion Monitor (DIMM) during July and October 2002 and found a median "seeing" of 0.68 arcseconds and 1.42 arcseconds, respectively. We also compared the technique against the Hartmann-version of the ESO-type DIMM method and found similar results within 0.05 arcseconds. For some reason, which is under investigation by international teams, the measurements of our ESO-type DIMM are underestimated by $\approx 10\%$ compared to ESO-standard DIMM measurements. Simultaneous ESO-type DIMM measurements and standard gauss-fitting- technique measurements, taken with the Kryonerion 1.2m telescope (inside the dome) indicate that the "seeing", as measured by the telescope is by 1 arcsecond larger. The above tests were performed in order to calibrate, test and make a brief sampling of the "seeing" at the Kryonerion Station with the aim to move eventually the equipment to the *Emilios Harlaftis Astronomical Station* the site of the new 2.3m ARISTARCHOS telescope - for a long-term "seeing" monitoring. As a by-product of this project, a new DIMM Station became operational at Mt. Holomon, in the premises of the University of Thessaloniki, in 2004.

Keywords: atmospheric effects, site testing
PACS: 95.45.+i

INTRODUCTION

The new station for the 2.3m telescope of the National Observatory of Athens lies at the peak of the 2.3km Helmos mountain, close to the historic-town of Kalavryta in North Peloponese (longitude $22^o13'E$ and latitude $37^o59'N$). The site characterization of the Station, which has been named *"Emilios Harlaftis Astronomical Station"* was immediately one of the concerns for developing further international collaboration and exploiting in full the scientific potential of the new site. In particular, addressing concerns within the OPTICON network regarding the unknown properties of the Helmos night sky, we took an initiative to put together a package with the aim to address at least the "seeing" characteristics by setting-up a portable differential image motion monitor (DIMM: Sarazin & Roddier 1990 - Tokovinik 2002). Previous measurements of the site "seeing", using the stellar trail technique, were indicating a median seeing as good as

CP848, *Recent Advances in Astronomy and Astrophysics,*
7th *International Conference of the Hellenic Astronomical Society,* edited by N. H. Solomos

FIGURE 1. The CELESTRON telescope and the PC driving the "seeing" measurements, next to the small enclosure built to protect the telescope from weather. At the background, the 1.2m telescope dome at the Kryonerion Astronomical Station

0.7 arcseconds from a sample of 30 winter nights (Sinachopoulos et al. 2000), though this technique is being replaced by the DIMM method, considered to be instrument-dependent. As a matter of interest, the stellar trail technique gave similar relative "seeing" measurements between Calar Alto and Mount Parnon (Pyrgaki, 1.8 km) at south-east Peloponese (longitude $22^o.6E$ and latitude $37^o.2N$) back in the 70's (Birkle et al. 1976). The DIMM method removes every error due to bad tracking, wind or mirror blotches by using a differential measurement of the stellar image motion (through two apertures). Given the minimal resources available, we adopted a portable-version of DIMM, leaving for the next stage the construction of a fixed tower (Figure 1).

SEEING METHODS AND AIRMASS CALIBRATION: RESULTS

In addition to the mask/prism DIMM, there is a simpler method just using the Hartmann mask (H-DIMM; Bally et al. 1996) which we also used for comparison. This method uses the same equipment as the SAAO DIMM, but instead of using a prism we just defocus the telescope by a few millimetres to create two image spots. The distance to defocus is given by $X = \frac{F}{D} \cdot P$, where X is distance from focus in mm, F is the telescope's focal length in mm, P is the pixel size in mm and D is the distance between the two apertures. We typically found that only 1.3 mm was needed away from focus. Our purpose was to check if the H-DIMM method gives the same results as the SAAO DIMM method. We made two masks, one with a prism (prism-DIMM method) and one without (H-DIMM method). During the test observations, the masks were changed every ten minutes. The result of observations on a photometric night (29 May 2003), using the star α Boo, is shown in the figure (Figure 2) beneath proving that the two methods give identical

results within a 0.05 arcseconds accuracy (prism-DIMM Seeing Average = 0.98 arcseconds and H-DIMM Seeing Average = 1.00 arcseconds. Every time that we measure the seeing from a star, we need to calibrate the data for the changing airmass. This is

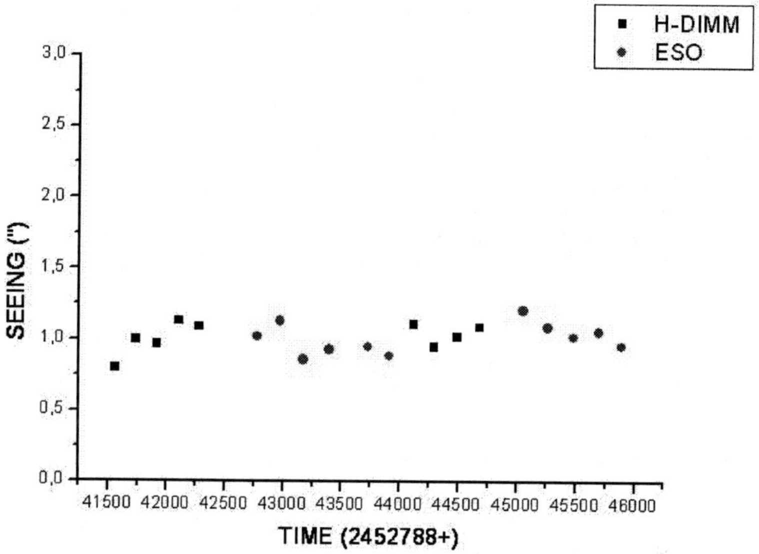

FIGURE 2. Comparison between the H-DIMM method and he ESO-DIMM method

necessary because "seeing" is a ofunction of the air mass, therefore it is also a function of the star's altitude. For this reason, we reduce all measurements at the air mass minimum zenith. To do that we use equation (Humphries et al. 1984), $S \propto S_o \cdot \alpha^{-0.6}$ where S is the calibrated value at zenith and S_o the "seeing" at airmass α. In the figures following we give the corrected "seeing" measurements during our test DIMM observations in July and October 2002 (Figure 3 & Table 1).

TABLE 1. Seeing Statistics

Month (2002)	**Median Seeing** [*arcsec*]	**Average Seeing** [*arcsec*]	**Standard Deviation**
July	0.68"	0.87"	0.15"
October	1.42"	1.49"	0.21"

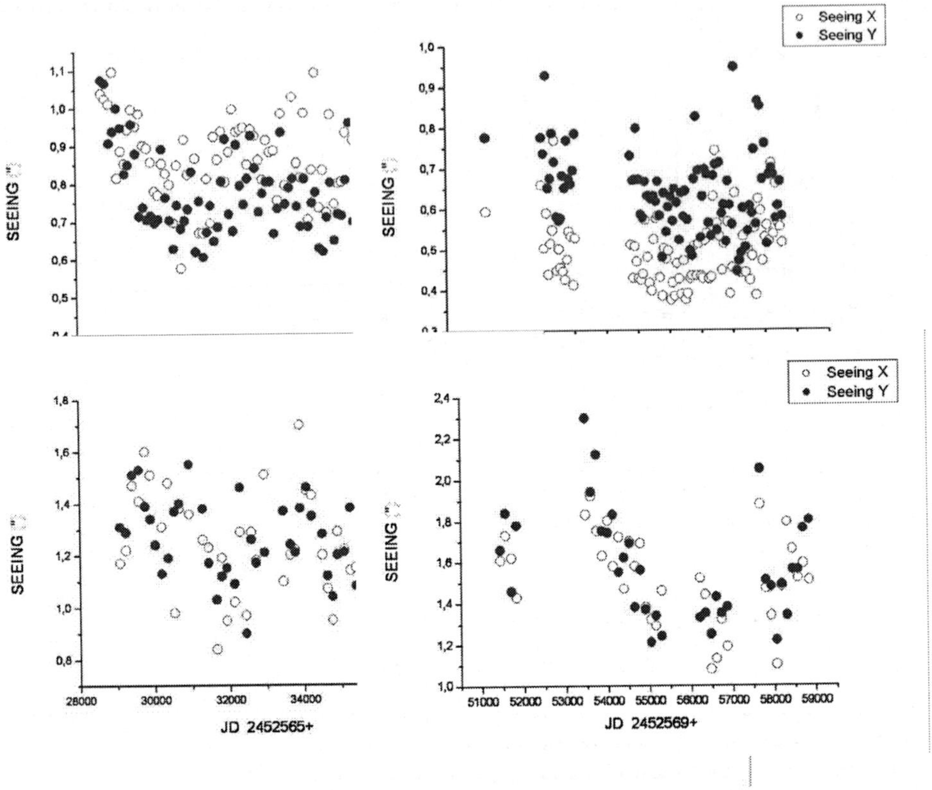

FIGURE 3. A Sample from DIMM measurments in Kryonerion Observatory July & Octomber 2002

ACKNOWLEDGMENTS

We would like to thank Neil O'Mahony (ING group) and David M. Russel (Southampton), for useful discussion on DIMM aspects

REFERENCES

1. Bally J., Theil D., Billawala Y., Potter D., Loewnstein R.F., Mrozek F., Lloyd J.P., "A Hartmann Differential Image Motion Monitor (H-DIMM) for Atmos Turbulence Characterisation", *A&A*, Vol. 13, 1995, pp. 22-27

2. Birkle K., Elsasser H., Neckel T., Schnur G., "Seeing Measrments in Greece, Spain, South West Africa and Chile", *A&A*, Vol. 46, 1976, pp. 397
3. Erasmus A., "Meteorological Characterzation of the Sutherland Astronomical Observatory and Evaluation of Candidate Sites for the Southern African Large Telescope", *SALT technical report, SAAO*, 2000
4. Fried D., "Optical Resolution Through a Randomly Inhomogeneous Medium for Very Long and Very Short Exposure", *J. Opt. Soc. Am.*, Vol. 55, 1965, pp. 1427
5. Humphries C.M., Reddish V.C., Walshaw D.J., "Cost scaling laws and their origin - desing strategy for on optical array telescopes", *Very Large Telescopes, their Instrumentations and Programs - IAU Colloquium*, Vol. 79, 1984, pp. 379
6. Sarazin M., Roddier F., "The ESO Differential Image Motion Monitor", *A&A*, Vol. 227, 1990, pp. 294-300
7. Tokovinik A., "From Differential Image Motion Monitor to Seeing", *PASP*, Vol. 104, 2002, pp.760
8. Vernin J., Munoz-Tunon C., "Measuring Astroomical Seeing. The DA/IAC DIMM", *PASP*, Vol. 107, 1995, pp. 760

SEEING MEASURMENTS FROM Mt. HOLOMON

J.S.Nestoras*, D. Mislis*, S. Pyrzas*, E. Tremou*, J.H.Seiradakis* and S.I. Avgoloupis*

*Aristotle University of Thessaloniki, Department of Physics, Section of Astrophysics, Astronomy and Mechanics, GR-541 24 Thessaloniki, Greece

Abstract. We present seeing measurements from Mt. Holomon (Chalkidiki, Greece), a possible site for the establishment of a new University of Thessaloniki Observatory, using the Hartman - DIMM method. After two years of systematic observations we present the first robust results and give a first evaluation of the astronomical seeing characteristics, which are very promising. In order to check our system and method of analysis we have also taken similar seeing measurements at the Kryonerion Astronomical Station of the National Observatory of Athens. The Kryonerion measurements fully agree with previous, established, seeing results taken there. This agreement makes us feel comfortable of the credibility of our system.

Keywords: Seeing, Site testing, DIMM
PACS: 97.80.Jp

INTRODUCTION

DIMM is an abbreviation of the term *Differential Image Motion Monitor*, a method used for aquiring very accurate measurements of the astronomical seeing. *Seeing* is the astronomical jargon for the turbulent flows of the atmosphere, responsible for example for the twinkling of the stars. Earth's atmosphere causes random motion of the diffraction limited image formed by a telescope. Random fluctuations in the index of refraction throughout the Earth's atmosphere result in image degradation, thus limiting the telescope's angular resolution. Figure 1 shows how the DIMM method works.

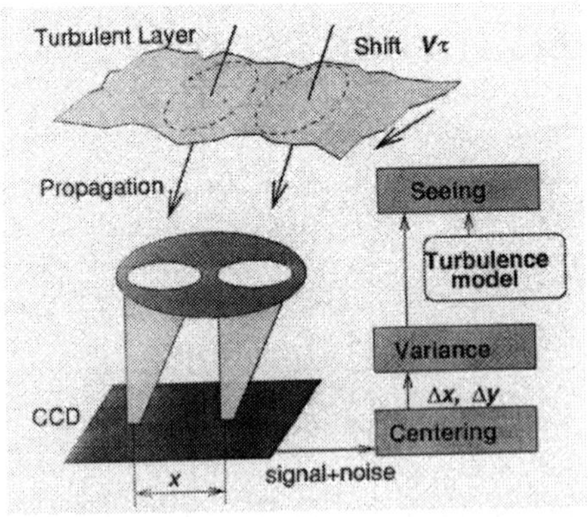

FIGURE 1. The DIMM method and how it works

CP848, *Recent Advances in Astronomy and Astrophysics*,
7th *International Conference of the Hellenic Astronomical Society*, edited by N. H. Solomos

The Fried parameter r_0 (Fried 1965) describes all seeing effects and it can be measured from the image motion in a small telescope. The value of the astronomical seeing is actually the FWHM of the star image in the CCD sensor. From Fried's theory we know that:

$$FWHM = 0.98\frac{\lambda}{r_0}$$

where λ is the wavelength and thus we measure the seeing in two axis x and y respectively.

There are two distinct DIMM methods: In the so called *ESO-DIMM*, proposed by M. Sarazin (1990), light that passes through two separated small apertures of a "mask" - a solid circular cover with two holes, called *Hartmann mask* - is separated by means of an optical wedge in one of the apertures before falling on a detector. This produces two well separated images of the same star on the detector, eg a CCD chip. Thus, telescope vibrations affect each sub-image in the same way, while turbulence induces differential motions in the sub-images. Variations in the image separations can be used to obtain a quantitative estimate of the Fried parameter. The second method is called the *Hartmann-DIMM*. Here we use the same mask but without the optical wedge. In order to produce the two well seperated images one has to place the light sensitive detector (CCD) slightly out of focus. The DIMM method is now a globally accepted method for accurate estimation of seeing values.

OBSERVATIONS AND DATA REDUCTION

Using the H-DIMM method we started observations on Mt. Holomon at an altitude of 900m. We used an 8-inch telescope with an SBIG ST-4 CCD camera, connected to a PC. The observations on Mt. Holomon date from 27/03/2003 until 19/02/2005. To compare our method we also took measurments at the 1.2m Kryonerion Astronomical Station from 23/06/2003 until 24/02/2005. DIMM data also need to undergo an airmasss correction. This took place in the University of Thessaloniki. Below we present some characteristic examples of the seeing on Mt. Holomon:

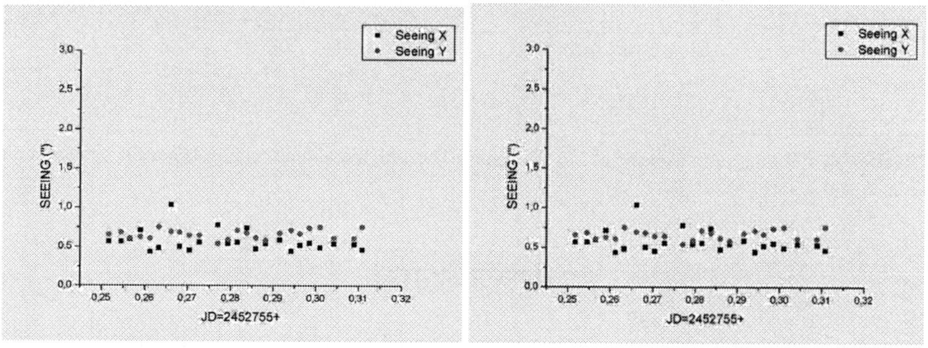

FIGURE 2. The nights of 24/04/2003 (left) and 25/04/2003 (right). The seeing average is 0.62" and 0.61" respectively.

Some other single night measurments:

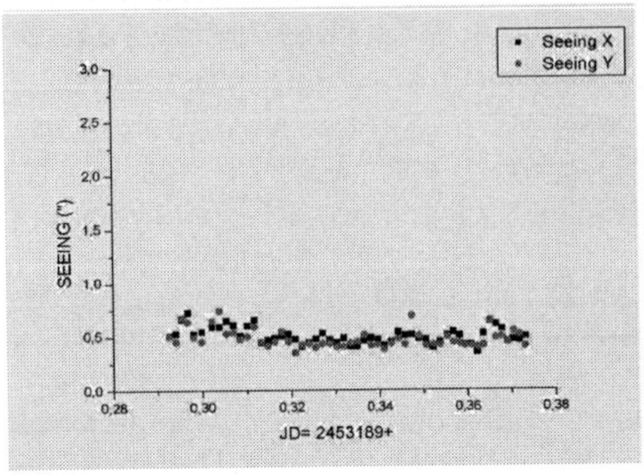

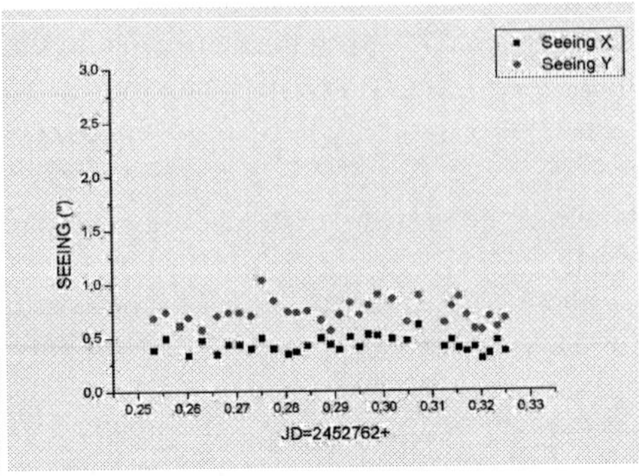

FIGURE 3. The nights of 19/02/2005 (left) and 01/05/2003 (right). The seeing average is 0.55" and 0.58" respectively.

The figure below is a plot generated by all the seeing measurments from Holomon:

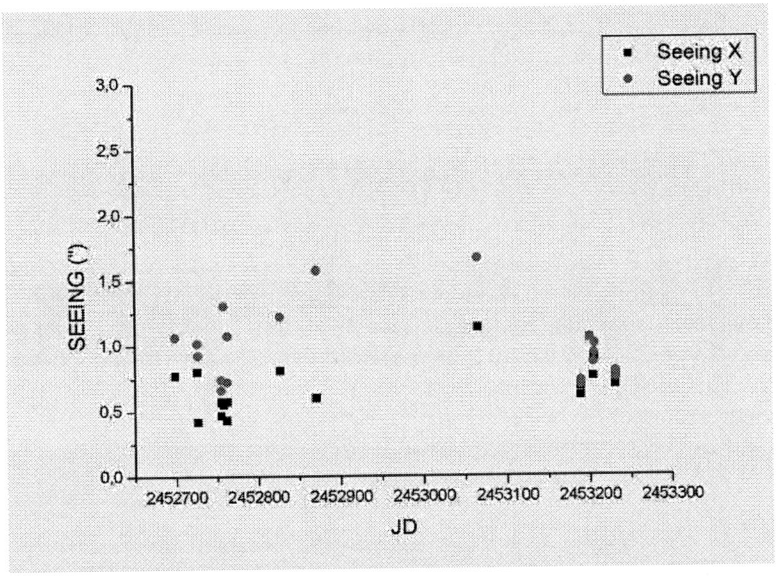

FIGURE 4. Total seeing average at Mt. Holomon. The squared and circular dots are the seeing values in the X and Y direction respectively

DISCUSSION

A good - ie low - seeing value is something very important for astronomical observations and the quality of the aquired data. After two years of observations at Mt. Holomon we have a firm certainty regarding the good quality of the data that will be obtained in future work there. We have measure an average of astronomical seeing equal to 0.82" over 40 nights. In all above graphs it is clear that only south winds (parallel with y axis) can sometimes intervene and worsen the good seeing average. Also it's clear that the seeing is very steady through the night. It is established that the difference between the seeing inside and outside a dome is approximately 1.00-1.50". In the Kryonerion Astronomical Station this difference is 0.93" (namely 2.17" inside the dome and 1.25" outside the dome). As a result we can estimate that if an Astronomical Station is build on Mt. Holomon, the seeing inside the dome is expected not to be more than 1.80", if we take the Kryoneri example into account.

ACKNOWLEDGMENTS

The authors would like to thank E. Tsorlinis (University of Thessaloniki, Department of Physics) for his valuable assistance. We would also like to acknowledge the Department of Forestry and Natural Environment, Aristotle University of Thessaloniki and the regional forest office, for allowing us to use the existing faciities of the Department on the Holomon site.

REFERENCES

1. Bally J., Theil D., Billawala Y., Potter D., Loewnstein R.F., Mrozek F., Lloyd J.P., "A Hartmann Differential Image Motion Monitor (H-DIMM) for Atmos Turbulence Characterisation", *A&A*, Vol. 13, 1995, pp. 22-27
2. Birkle K., Elsässer H., Neckel T., Schnur G., "Seeing Measrments in Greece, Spain, South West Africa and Chile", *A&A*, Vol. 46, 1976, pp. 397
3. Erasmus A., "Meteorological Characterzation of the Sutherland Astronomical Observatory and Evaluation of Candidate Sites for the Southern African Large Telescope", *SALT technical report, SAAO*, 2000
4. Fried D., "Optical Resolution Through a Randomly Inhomogeneous Medium for Very Long and Very Short Exposure", *J. Opt. Soc. Am.*, Vol. 55, 1965, pp. 1427
5. Humphries C.M., Reddish V.C., Walshaw D.J., "Cost scaling laws and their origin - desing strategy for on optical array telescopes", *Very Large Telescopes, their Instrumentations and Programs - IAU Colloquium*, Vol. 79, 1984, pp. 379
6. Sarazin M., Roddier F., "The ESO Differential Image Motion Monitor", *A&A*, Vol. 227, 1990, pp. 294-300
7. Tokovinik A., "From Differential Image Motion Monitor to Seeing", *PASP*, Vol. 104, 2002, pp.760
8. Vernin J., Munoz-Tunon C., "Measuring Astroomical Seeing. The DA/IAC DIMM", *PASP*, Vol. 107, 1995, pp. 760

PART 6

HISTORY AND EDUCATION IN ASTRONOMY

(Conveners: E. Theodossiou and H. Tomboulides)

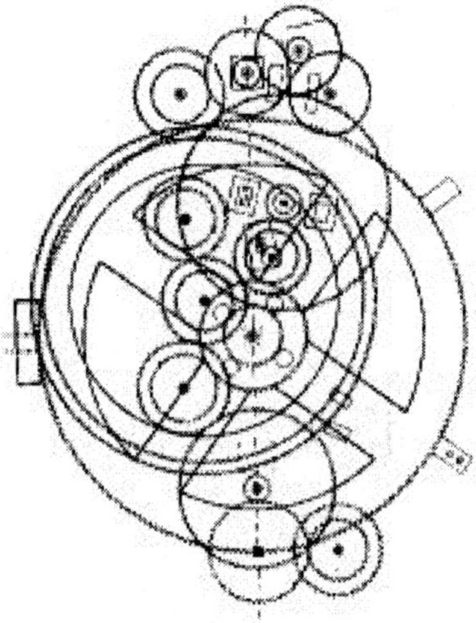

The Antikythera Mechanism gives us a unique and quite extraordinary glimpse into the science and technology of the Classical World. Prof. Edmunds summarizes what is currently known about its structure and origin, and places it in the context of the development of Greek tronomy during the last two centuries B.C. This background will be developed into some fascinating speculation about the purpose and function of the mechanism. It is explained how the radically new techniques of an international research programme will at last allow resolution of its mysteries, and create a definitive database of x-ray and imaging data for future generations.

The Antikythera Mechanism – Real Progress Through Greek/UK/US Research

M.G.Edmunds for the Antikythera Research Project

M.G. Edmunds & A. Freeth, School of Physics & Astronomy, Cardiff University, Cardiff CF24 3YB, UK; Y. Bitsakis &X. Moussas, University of Athens; J. Seiradakis, Aristotle Uinversity, Thessaloniki; E. Magkou & M. Zafeiropoulou, National Archaeological Museum, Athens; P.Haycock, Keele University, U.K.; P.Hadland & A. Ramsay, X-Tek, Tring, UK; D.Gelb & T. Malzbender, Hewlett-Packard, Palo Alto, USA.

Abstract. We give a brief report on the aims and objectives of a new international research effort to investigate the Antikythera Mechanism. The extraordinary nature and importance of the mechanism is emphasized, and the new techniques employed are outlined. As well as addressing the question of the functions of the mechanism, the project will provide a generally available web-based experimental database.

Keywords: Antikythera Mechanism, History of Astronomy, History of Technology
PACS: 01

INTRODUCTION

"“History with its flickering lamp stumbles along the trail of the past, trying to reconstruct its scenes, to revive its echoes, and kindle with pale gleams the passions of former days”

This quotation, from Sir Winston Churchill – himself an accomplished historian - mirrors our own reaction to the unique Antikythera Mechanism. The Mechanism is a physical artifact of almost tangible power, more complicated than any known mechanism for a millennium after its construction, forcing us to confront its implications for the development of human thought and technology. Past attempts to understand its purpose are by now quite well documented, yet it seems as though its profound implications are still not widely recognised among astronomers, classicists and historians of ideas and technology.

The mechanism was found in 1900 as a single corroded lump, about the size of a shoe box, by sponge divers near the Mediterranean island Antikythera – hence its name – in the sea-floor remains of a shipwreck. Through some of the very first underwater archaeology, the wreck subsequently yielded a fabulous collection of bronze and stone statues, glass and other artifacts, many of which are displayed in the rich collection of the National Archaeological Museum in Athens. (some illustrations can be found at http://www.culture.gr/2/21/214/21405m/e21405m4.html). The shipwreck may have been of a trading or plunder vessel on its way to Rome around 80-60 BC.

CP848, *Recent Advances in Astronomy and Astrophysics*,
7th *International Conference of the Hellenic Astronomical Society*, edited by N. H. Solomos

Investigation of the mechanism began with the discovery of gearwheels when the "lump" split open. Once radiography was carried out by Ch. Karakalos and Derek de Solar Price the true extent and complication of the mechanism became apparent – over 30 gear wheels, with multiple display scales of uncertain function. Price published a fine account of his pioneering researches in his book "Gears from the Greeks" in 1975. But one has the impression that he had rather fixed ideas, and it has become apparent since that time that his reconstructions are probably not correct in detail. Much subsequent interesting and imaginative work has been done by Michael Wright and his collaborators. A major problem that faces any new potential investigator is the lack of access to real data. Illustrations in books and journals are often poor, and "image enhancement" on the few published X-rays has tended to mean an ink line drawn around gear teeth! One of the major goals of our new investigation is to provide a web-based data archive which will allow much of future research without the need for direct access to the original mechanism, except of course where radically new techniques may be developed. Our campaign, described below, is designed to gather and make available full surface imaging of the extant fragments of the mechanism, and full 3-dimensional X-ray tomography at sufficient resolution to extract all possible physical information on structure, construction and inscription.

A short basic general illustrated review of the mechanism was given by one of us a few years ago (Edmunds & Morgan 2000), to which the reader is referred for source references where they are not given below.

THE MYSTERY OF THE MECHANISM

To establish a date for the mechanism within a window of 90 years is not too difficult. The shipwreck contained coins dating to 86-60 BC, so the mechanism must be at least as old as this. The epigraphy of inscriptions is characteristic of the beginning of the first century BCE. The use of the older form "XYΛAI" in a zodiac dial suggests an earlier origin than Geminus' *Isagogue* which was written around 55 BC and used a later form. Circumstantial evidence comes from Cicero's account of orrerys existing in Rhodes around 78-44 BC. Taken together, a date within the period 150-60 BC seems appropriate. This is a potentially interesting period, since it falls within a rather "dark age" in our knowledge of advances in Greek astronomy. As Dreyer (1906) put it "From Hipparchus [190-120 BC] we have to pass over two centuries and a half……[to Ptolomy, 85-165 AD]". A brief reading of Ptolomy's Almegeist will convince anyone of the intellectual brilliance of both Hipparchus and Ptolomy. Hipparchus himself was dead by 120 BC, but as realised by Price, Hipparchus had worked in his later years on Rhodes. From that account by Cicero of an orrery on Rhodes, and the evidence we have of Rhodes as a major naval (and hence probably technological) centre, certainly suggests Rhodes as the most likely site for the construction of the Anikythera Mechanism. It would perhaps be wrong to completely rule out other possible sites (Pergamon?) - but a tradition or school of mechanical representations of the Universe originating from Hipparchus on the island of Rhodes is rather attractive. Archimedes, working some one hundred years before Hipparchus is reported to have made mechanical devices to display astronomical phenomena, but his book on the subject, *de Spherae*, is unfortunately lost. What seems certain is that the Antikythera Mechanism could *not* have been unique. It is surely inconceivable that so complicated a device was the single example – it must have been the result of considerable development. No similar device is known for a thousand years afterwards, until the coming of the mediaeval cathedral clocks. There is an interesting sundial-calendar from 400-600 A.D. Byzantium (Field & Wright 1985), which suggests some continuation of the geared mechanism tradition, but it is much simpler with only eight gears.

The Antikythera gears are hand-cut or filed from 2mm bronze sheet. We wait for definitive gear teeth counts from our X-ray images, and the unpublished images of other investigations. It is interesting to speculate on the implications of what is so far known, and memory of school geometry with a straight edge and pair of compasses might suggest that powers of two, and three times powers of two, would be easy to lay out. Indeed it seems that many of the gears conform to this pattern. But for astronomical cycle ratios, some more difficult gears are needed – it seems likely, for example, that there are gears with 38 and 127 teeth. Rough numbers may be about 21 easy gears, 9 not difficult, and 2 rather challenging! – but present day metalworkers point out that there are various uncomplicated schemes for setting out *any* number of gear teeth. It is obviously crucial in determining the astronomical function that the teeth counts are known accurately, but this may not be easy where only partial evidence remains, and we are developing simple statistical methods to help understand the confidence limits. Price was greatly impressed that the mechanism appeared to contain a differential gear mechanism – something of which did not appear again until modern times, and not widely known before the 19th century. It will be important to establish whether the gearing actually is a differential – the doubt comes because there appear to be much easier straightforward gearing trains that would carry out the function proposed for the differential by Price. Why go to the complication of a differential when simple gearing would suffice? And if the structure is not a differential gear, then what is it for?

What astronomical cycles should we be looking for? Price's reconstruction and the numbers clearly visible in the inscriptions - 19, 76 and 223 - are already very suggestive. The inscription fragment with these numbers on (in Greek notation of course) can be viewed at the Hewlett-Packard website mentioned below – image AK01a-512. There are scales with zodiac signs and months, and a parapegma inscription – calendrical data. The obvious calendrical cycles are Metonic (*19* years of 235 lunar months) and the Callipic (*76* years of 940 lunar months, and a day less than four Metonic cycles). Hipparchus had formally improved the cycle with 304 years, and a day less than four Callipic cycles, but this period might well be too long for useful representation in a mechanical mechanism. The *223* suggests the Saros eclipse cycle. Another possibility that might be worth looking for is the approximately 19 year cycle of lunar standstills – the times when the moon moves its at furthest North and South of the ecliptic, and a phenomenon which it is believed was followed by megalithic astronomers in Northern Europe (Ruggles 1999). That the device showed lunar phase seems very likely, with both Price and Wright noting that the circular structure on the reverse of Fragment C might well have been associated with a mechanical moon phase display.

Did the mechanism show planetary positions? This is a crucial question. The only real evidence that it might have done is rather circumstantial. The main clue is the single tantalising word "Aphrodite" (Venus) on the inverse inscription on Fragment B, together with rather half-hearted early reports by Price (1959) that he had read the word "stationary" in the inscriptions, implying planetary retrograde motion. Secondary support is the statement by Cicero that devices existed in Rhodes that *did* show movement of the planets. Much enjoyable speculation has been made by ourselves (Edmunds & Morgan 2000, Freeth 2002,) and Wright (2002) in suggesting how simple epicyclic planetary mechanisms might have fitted in to the device. But as yet we have to admit that the evidence is very thin – and will obviously be a major target for our new investigations. If evidence were to be found it might shed some light on the development of planetary theories. By about 55 BC Geminus was supposing that the sun, moon and planets move so that "their circles are eccentric or that they move on epicycles..", but the earlier views of Hipparchus are not known. Ptolomy comments that Hipparchus "being a great lover of truth...did not even make a beginning in establishing theories for the five planets, *not at least in his writings which have come down to us"*. There are hints (A. Jones 2005, private communication) that the inscriptions on the tablet from Keskinto in Rhodes, dated to around 100 BC – and hence perhaps contemporary with the Antikythera Mechanism, and within 20 years of Hipparchus' death – deal with a planetary model rather different from that eventually adopted by Ptolomy.

So back to the obvious question; just what was the mechanism for? Six possibilities suggest themselves, although there could of course be a combination of functions:

1) It was a device for performing calendrical calculations
2) It was a device for performing calendrical and astronomical calculations
3) It was a tellurium – a device for representing (for display or educational purposes) the relation of the Earth, Moon and Sun.
4) It was an orrery – displaying the relation of the Earth, Sun, Moon and planets
5) It was a navigational instrument
6) It was a device for performing astrological calculations

The fifth of these can probably be dismissed straight away. Although it is true that the mechanism was found in a shipwreck, so were many other high status artifacts like the sculptures and glass, and there is no obvious way (at least to a present-day sailor) in which the mechanism could have performed a navigational function. The last suggestion, an astrological purpose, is harder to dismiss. But for horoscopic calculation one would probably have to have a display of planetary positions and one would expect astrological references in the inscriptions – again not yet found. Real evidence in favour of the idea would be the discovery that the mechanism was able to calculate and display the "lot of fortune" – a relatively easy and characteristic calculation off the sun, moon and horoscopic point position, although historically there were some variants in the calculation (Bouche-Leclerc 1899).

Who made the mechanism? The traditional view that Greek philosophers would not stoop to mechanical work might suggest that it had to be a partnership between and able astronomer and a master craftsman. Neither would have the skills or astronomical/mathematical knowledge on their own. If, following Price, Rhodes is indeed the site of construction, then a school perhaps 140 - 60 BC based around Hipparchus and later Posidonius (died 51 BC) seems a reasonable choice. Exactly when this particular mechanism was designed and constructed within that period remains an open question, as does the actual role of these two prominent astronomers.

THE NEW INVESTIGATIONS

Our new investigations will primarily concentrate on the obvious need for better and readily available data on inscriptions, gears, axles, scales, constructional details and overall structure. We also hope that further consideration of the historical, astronomical, economic and technological *context* in which the mechanism was made may help illuminate its purpose and implications. It will also be important to be able to trace the sequence of photographic material showing how the many fragments have originated from the original single find, in order to understand the overall mechanism structure. In direct investigation technique we are using two very powerful new tools, through excellent collaboration with, and support from, industry. The first is to use for surface studies a reflectance imaging technique pioneered by Tom Malzbender at Hewlett- Packard in California (http://www.hpl.hp.com/research/ptm/ri.html). A small lightweight hemispherical dome contains a carefully-spaced array of electronic flashbulbs and a digital camera. From a controlled sequence of exposures, computer images are produced in which the research worker can subsequently greatly alter the lighting conditions and angle – allowing subtle surface detail to be seen, particularly so with image enhancement processing. The technique has already proved very successful in investigating palimpsests and archaeological tablets. The excellent Hewlett-Packard Website explains and illustrates the technique. It may be noted that subsequent to the conference talk in September, the imaging of the fragments of the mechanism in the National Archaeological Museum in Athens was very successful. Examples of the images obtained, which allow image enhancement and variation of lighting are displayed at:
http://www.hpl.hp.com/research/ptm/antikythera_mechanism/index.html

The other technique is X-ray tomography, offering full 3-dimensional internal reconstruction at sub-mm resolution. Here we befitted enormously from the collaboration with X-Tek, a company who make rugged specialist X-ray instrumentation typically used in the detailed examination of components such as turbine blades and printed circuits. We will also undertake some "element specific" tomography to differentiate iron and bronze details. X-Tek's web site is at: http://www.xtekxray.com/systems.htm. It is very pleasing to report that after the conference we were able to use a new version of their "Bladerunner" machine, which was transported to the Athens museum from the UK especially for the investigation.

We have much data processing to do, with some of the data sets stretching to gigabytes, but we can say that the initial results are very exciting – the technique is beginning to allow us to read inscriptions *inside* the fragments which have not been seen for over two thousand years. We hope that initial publication of the major results will begin in the summer of 2006, and that the detailed database of experimental material will be established by sometime in 2007.

ACKNOWLEDGMENTS

MGE is very conscious of the honour in being invited to present this progress report to the Helenic Astronomical Society, and thanks them very much.
We are particularly grateful to the relevant Ministry in Athens, and our colleagues at the National Museum – particularly the Director Dr Kaltzas – for their cooperation in enabling this extraordinary project. Financial support is from the Leverhulme Foundation, the Walter Hudson Bequest at Cardiff University, and the very generous donation of time and loan of equipment by our industrial partners, Hewlett-Packard and X-Tek.
We hope that the structure, purpose and significance of the Antikythera Mechanism will soon be available for all to share. The technical achievement represented by this artifact can only cause a sense of increasing wonder at the extraordinary achievements of the ancient Greek world, and engender a resolve and cooperation among us all that "dark ages" should never again be allowed to close in and destroy the beneficial fruits of civilization.

REFERENCES

1. A. Bouche-Leclerc, *L'Astronomie Grecque, Paris* (1899).
2. W.S. Churchill, *Parliamentary tribute to Neville Chamberlain November 12th* (1940).
3. J.L.E. Dreyer, *A History of Astronomy from Thales to Kepler, Cambridge Univ. Press* (1906).
4. M.G. Edmunds & P.Morgan, *The Antikythera Mechanism: still a Mystery of Greek Astronomy? Astronomy & Geophysics* **41**, 6.10 (2000).
5. J.V. Field & M.T. Wright, *Annals of Science* **42**(2): 87-138 (1985): A picture of the sundial is at: http://www.scienceandsociety.co.uk/results.asp?image=10308322&wwwflag=2&imagepos=8
6. A. Freeth, *Mediterranean Arch.* **2**, 21 & 45 (2002).
7. D. de Solar Price, *Scientific American* **200**, 66 (1959).
8. D. de Solar Price, *Gears from the Greeks, Science History Publications, New York* (1975).
9. M.T. Wright, *Horological Journal*, **144**, No. 5, 169-173 (2002); see also Bull.Scientific Instrument Society **85**, and **87**, (2005).
10. C. Ruggles, *Astronomy in Prehistoric Britain and Northern Ireland, Yale University Press*, (1999).

ANCIENT CELESTIAL SPHERES FROM GREECE

S. Dimitrakoudis, P. Papaspyrou, V. Petoussis, X. Moussas

Section of Astrophysics, Astronomy and Mechanics, Department of Physics, University of Athens, Panepistimopolis, GR 15783 Zographos, Greece

Abstract: We present several ancient celestial spheres from the 8th century B.C. found throughout Greece, mainly in Thessaly, at the temple of Itonia Athena, but also in Olympia and other places. These celestial spheres have an axis, equator and several meridians and they have several markings with the symbol of stars (today's symbol for the Sun) $\odot$. Such instruments could have been used to measure the time, the latitude of a location, or the coordinates of stars.

By the 8th century BC, the transition from the Geometric to the Archaic Age was producing the first signs of a renaissance of Greek civilization, after the collapse of the Mycenaean order and the subsequent wars that left Greece depopulated. The Olympic Games were revived in 776 BC, the Homeric epics were becoming popular throughout the Hellenic world, and Hesiod produced the first detailed Theogony, as well as his *Works and Days*, that contained a repository of the cosmological (and meteorological) knowledge of that era. The accepted wisdom of those days only dimly foreshadowed the amazing series of scientific and philosophical revolutions that were to occur only a couple of centuries later.

At that time, as we know from Hesiod, the earth was considered to be flat, possibly disk shaped, while the heavens were spherical, endlessly revolving around the still Earth. Many of the constellations catalogued in antiquity had been named by the early Archaic age, as had many of the brightest stars (Orion, Ursa Major, Ursa Minor, Bootes, Sirius, Arcturus, the Pleiades and Hyades). The colonization of the Mediterranean and the Black Sea had boosted the mercantile and navigational experience of those early Greeks, who, venturing far from their native ports, had almost certainly used the position of fixed stars for navigational purposes. Aratus mentions (although some centuries later) that Greek seamen used the constellation of Ursa Major for navigational purposes, while the Phoenecians relied on Ursa Minor to chart their courses on the sea.

The sphere of the heavens had already been incorporated in mythology, with the titan Atlas holding it on his shoulders. In later ages, a variety of devices were invented to calculate the time or latitude of a place, by comparing the position of the Sun or another star with respect to the horizon. Those devises ranged from the simple casting of the Sun's shadow on a graduated quadrant pitch circle, to elaborate astrolabes and celestial spheres crafted during the Hellenistic Age. In earlier ages, sun-dials (dating from the 15th century BC in Egypt, and possibly the Aegean region as well) had provided the first small scale application of astronomical observations for the practical needs of people (large scale applications consisted of the study of yearly and monthly variations of the Sun, Moon and stars, for the creation of farming calendars, and are evidenced by the astronomical orientation of Neolithic monuments in Europe and elsewhere). However, of that particular age little is known concerning possible

CP848, *Recent Advances in Astronomy and Astrophysics,*
7th *International Conference of the Hellenic Astronomical Society,* edited by N. H. Solomos

advancements in astronomical devices, or the possible existence of astronomical relics from the Mycenean Age. An archeological hint in that direction may be inferred from certain pendant offerings discovered in Thessaly, Macedonia and the rest of the Greek world, corresponding to that period.

Archaeological excavations at the temple of Itonia Athena in Philia, a religious and political center of the Thessalian League, by D.R. Theocharis in 1962-67 [14-17], Angeliki Pilali-Papasteriou and Aekaterina Papastamou-Papaevthymiou in 1980 [13], and Imma Kilian-Dirlmeier in 1997-2001 [18], have unearthed a number of bronze votive artifacts, perhaps pendants, whose form resembles a representation of the celestial sphere, and may be indicative of a larger, more functional device. These artifacts correspond to a period in time when such offerings to the gods were very common.

Of the numerous bronze artifacts discovered in that site, each unique in its details, those that are sphere-shaped can be classified into three broad categories. The first category consists of objects whose main feature is a hollow bronze sphere, decorated with symbols traditionally associated with stars, and an upper extension in the form of one or two birds. Three artifacts of this type have been unearthed at the temple of Itonia Athena, with heights ranging from 5 cm to 7 cm. On the first one, the solid bronze sphere is decorated with three 'parallel circles' (Fig. 1) formed by star symbols (the astronomical symbol we still commonly use for the Sun). The second is characterized by seven meridian incisions (Fig. 2), in between of which the star symbols are repeatedly crafted. The third (Fig. 3) is similar to the second, in that the same pattern is seen, only on a larger scale. This one has ten meridian incisions, and two bird-shaped protrusions, as opposed to only one in the previous two artifacts.

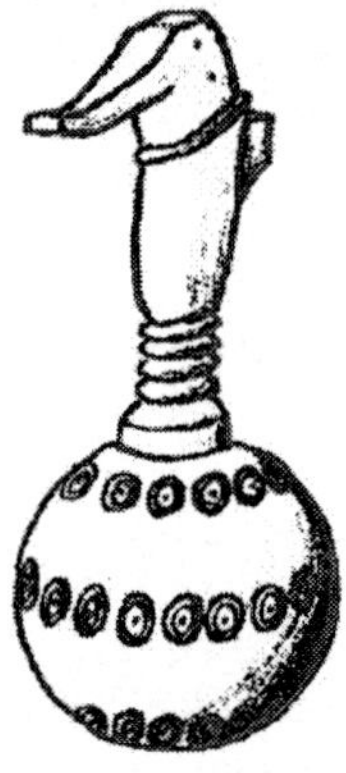

Fig. 1

Fig. 2

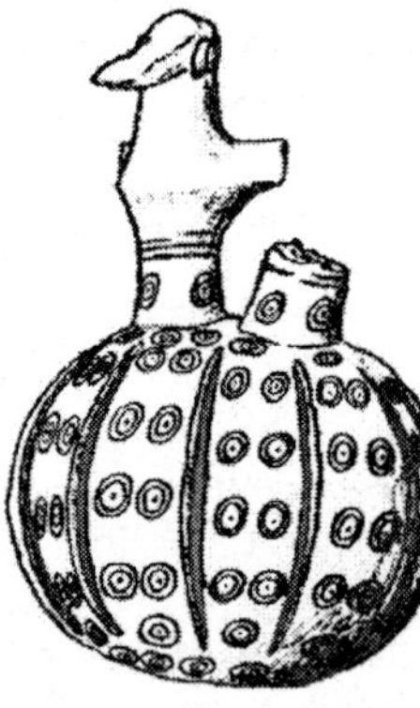

Fig. 3

The second category can be broken down to two sub-categories. The first one is represented by two artifacts. Their characteristics are a spherical main body, made up of bronze foils that form an equator and four or six meridians (respectively), an upper part in the form of one or two birds (respectively), and a lower part, in the form of a hollow cylindrical extension, that perhaps fitted onto a pole (reminiscent of a 14th-13th c. BC Mycenaean scepter discovered in Cyprus, with the same decorative motive of two birds on a sphere, Fig.6). The foils are all curved (where not damaged), and they are decorated with the same star (sun) symbols found on the previous artifacts (Figs. 4, 5).

Fig. 4

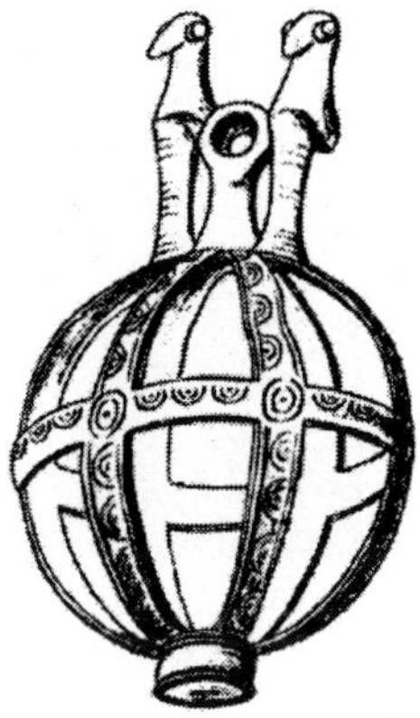

Fig. 5

Fig. 6

These objects are the ones that bear the closest resemblance to the armillary spheres that were to become very common in subsequent centuries.

The second sub-category is the most numerous, with representative artifacts found not only at the temple of Itonia Athena, but also at temples of Zeus and Artemis in Thessaly, in Olympia, and Macedonia. They appear to be a simplified sub-category of the second category objects, having the same basic form as the first sub-category, but with rougher, undecorated straight foils. The number of meridians ranges from four to nine, the lower extension is rarely made to accommodate a pole (it is mostly flat, curved, or in some cases conical), and the top part resembles a variety of forms; birds (overwhelmingly), abstract geometrical forms, animals (artifact from Olympia), and humans (artifact from Macedonia). Some examples from this category can be seen in figures 7-12.

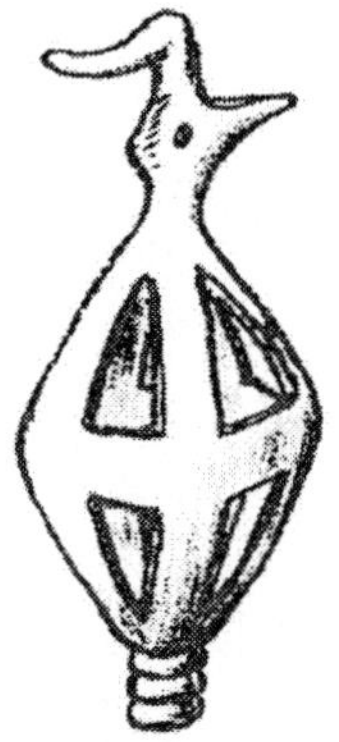

Fig. 7

Fig. 8

Fig. 9

Fig. 10

Fig. 11

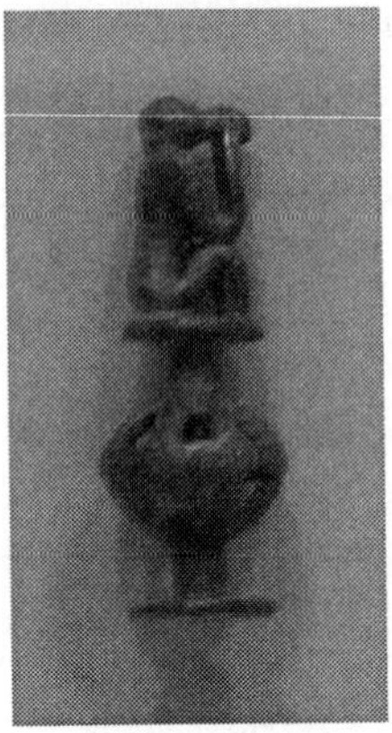

Fig. 12

The third category consists of artifacts that incorporate more than one sphere into their design. A representative artifact from Philia is a bronze horse-shaped object, whose legs are made of two identical spheres, with four meridians each (Fig. 13). Another example comes from Olympia, where a hollow bronze pole incorporates two spheres (of four meridians), before culminating in a bird-shaped protrusion (Fig. 14). In both cases the spheres are undecorated (as in category 2b). In these cases the hollow sphere is degraded from being the core of the artifact to being a simple decorative element. If there is indeed some sequence of imitations originating from an astronomically useful original, this category represents the furthest deviation from the initial votive intent.

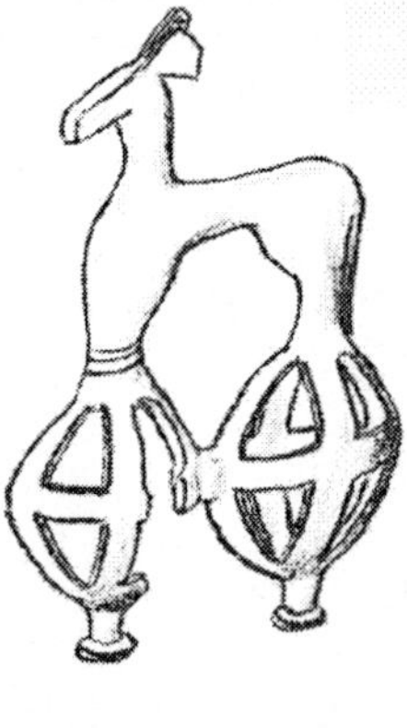

Fig. 13

Fig. 14

Another group of bronze offerings, similar to the sphere-shaped ones and with possible astronomical allusions, are the seal-shaped or disk-shaped artifacts. As with the spherical variety, there is a bird-shaped upper part. In these cases, however, it is attached to a flat lower segment, at an angle that varies from zero to ninety degrees.

The artifacts from the temple of Itonia Athena with the upper part perpendicular to the lower part (seal-shaped) are displayed in figures 15-19. In all cases the bottom of the lower segment has a relief, perhaps used as an actual seal. In the first artifact, the relief consists of three concentric circles, the outer one delineating the seal's border. Between that and the two inner circles there are seven circular holes, one of them not completely punched through. The seven holes are not crafted in regular intervals, but the composition as a whole is not without a sense of symmetry. It is possible that the circles may represent the known planets around the earth. The shallow seventh circle may or may not have some special significance. It is worth noting that during that era the planet Venus was considered to be two distinct planets, the Morning Star (Augerinos) and the Evening Star (Hesperinos). The same applied to Mercury, who was called Mercury (Hermes) when he appeared after sunset, and Apollo when he was seen before dawn. Therefore the known planets were seven, plus the Sun and Moon, bringing the total number up to nine. It is unclear whether the holes in the first seal represent some of the known planets, the Pleiades, or nothing of astronomical context. The second seal has an elaborate geometrical relief of semicircles forming what appears to be a double axe symbol. The double axe had been a popular symbol of power among the Minoans and the Mycenaeans, and continued to be so, throughout the next millennium, all over Greece. The third seal has a simple 'X' pattern, while the top of the lower segment is engraved with zigzag lines. The fourth seal is a curious artifact, in that the seal relief is an empty circle. The fifth seal is shaped like a Boeotean shield (also reminiscent of a double axe, or an abstract rendering of the waxing and waning of the moon).

A more unusual seal-shaped artifact is the one in figure 20. In this case, the upper part is tilted at an angle of approximately 45°, a unique feature among the artifacts studied. The seal relief is a clockwise swastika (depicted counter-clockwise if used as a seal), a common Indo-European symbol that has been theorized (by Carl Sagan and X. Moussas independently) to be associated with comets (in accordance with ancient depictions of comets in the form of a swastika).

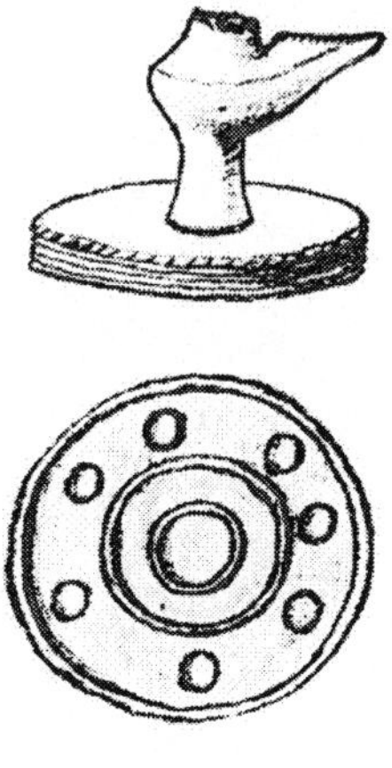

Fig. 15

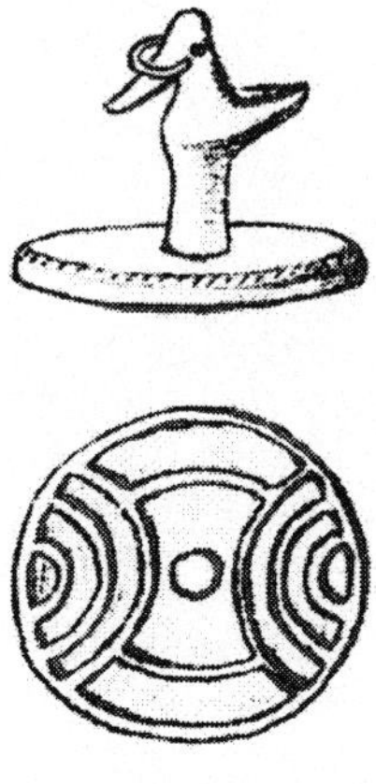

Fig. 16

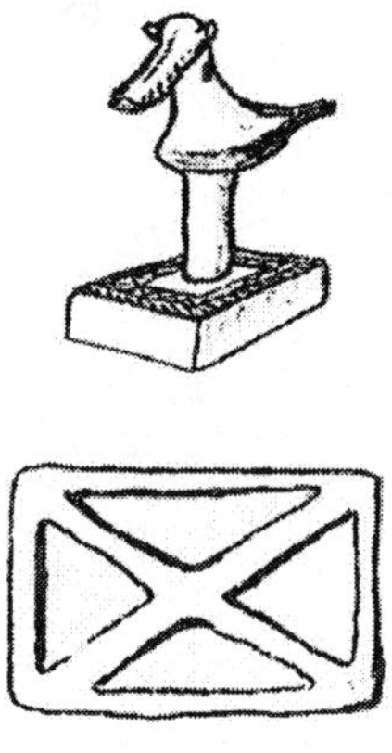

Fig. 17

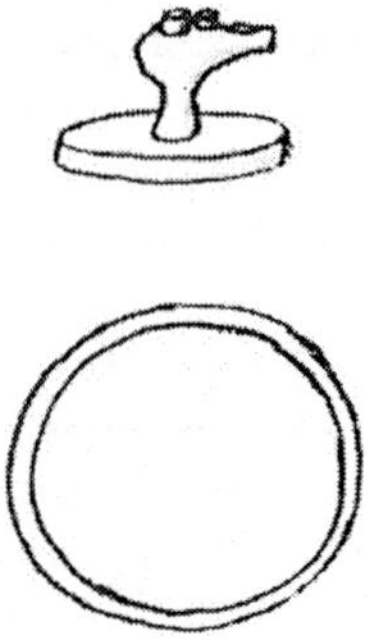
Fig. 18

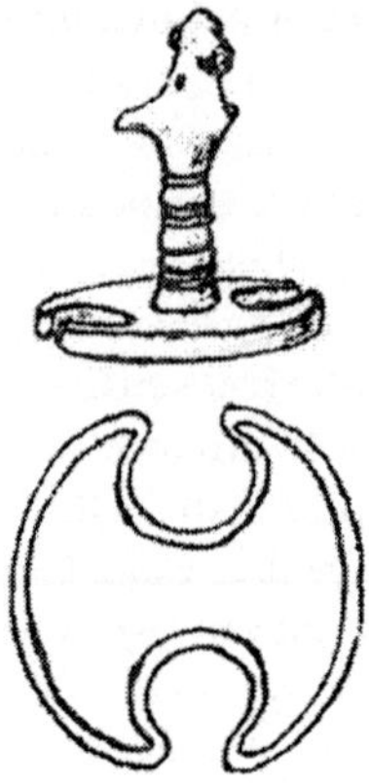
Fig. 19

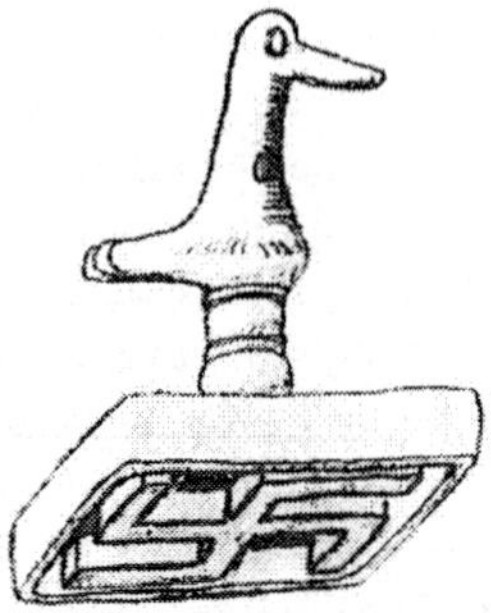
Fig. 20

A number of disk-shaped bronze artifacts from Thessalian temples are on display at the Archeological Museum in Athens. In most cases the disk is undecorated, but in one artifact (Fig. 21) it is adorned with 17-18 star symbols, identical to the ones on the spherical objects already mentioned. The density and regular positioning of the symbols precludes the representation of some star map. They are, most likely, just a star-themed decoration.

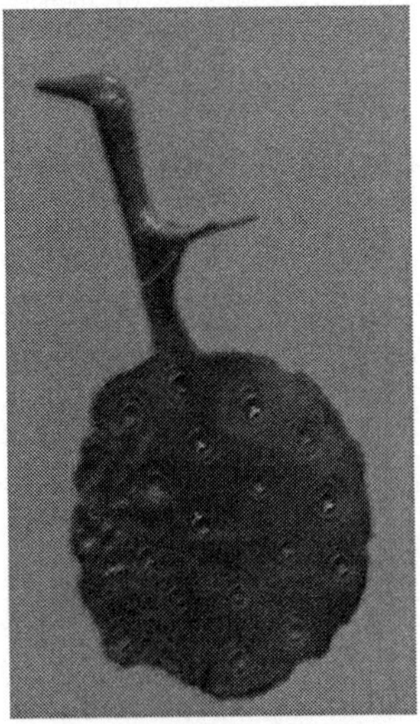
Fig. 21

The great diversity within the not so great quantity of astronomically linked votive artifacts is a testament more to our ignorance of the particulars of that time period, than to any new insight obtained through parallelism and conjecture. Written documents in that era were scarce, and only the more popular of the first epics have survived to this age. Through these epics, fragments of astronomical wisdom stare back at us in artistic purity, detached from the prosaic details of their discoveries and the means used by those inclined to study them. The birth of natural philosophy overshadowed the more empiric achievements of the multitude of star-gazers of aeons past, to the effect that Eudemus (a pupil of Aristotle), in his work 'the History of Astronomy' wrote that Thales was the first man who undertook the study of astronomy. In the Geometric age, the study of the stars was probably conducted by priests, men whose inclination was the drafting of practical calendars and not scientific curiosity, and whose names were not recorded in history books. The artifacts presented here may shed some dim light into that dark period (made all the darker in contrast to the more documented centuries that followed), if the assumptions made about their decorative origin are correct.

The main source of our investigations remains the excellent monograph of Kilian-Dirlmeier [18]. It is an excellent piece of work that embraces in an exhaustive and systematic way all the current archeological knowledge we possess at the moment about the numerous findings of the sanctuary, and as far as we know, it serves this purpose uniquely till today. This is followed with some useful extracts from the bibliography of the above cited work, that are general in character, thus serving as an introduction, or of specific nature, dealing with the archeological research in the sanctuary. The history of the archeological investigations in Philia are presented in [7], [9], [10], [13], [14], [15], [16] and [17]. One could mention among them the works of K. Kilian in [5], [6], [8], and these of I. Kilian-Dirlmeier in [11] and [12]. Some archeological remarks about Thessaly are given in [1] and [2]. Two collections of miniatures are mentioned in [3] and [4].

REFERENCES

1. Y. Béquignon, *Recherches archéologiques à Phères de Thessalie* (Paris 1937).
2. H. Biesantz, *Die thessalischen Grabreliefs* (Mainz 1965).
3. Collection Hélène Stathatos IV. *Bijoux et petits objets* (Athens 1971).
4. H. Donder, Die Fibeln. Katalog der Sammlung antiker Kleinkunst des Archäologischen Instituts der Universität Heidelberg Bd. 3, Teil 2 (Mainz 1994).
5. K. Kilian, *Zum italischen und griechischen Fibelhandwerk des 8. und 7.* Jahrhunderts. Hamburger Beiträge zur Archäologie 3/1, 1973, 1-40.
6. K. Kilian, *Zur eisenzeitlichen Transhumanz in Nordgriechenland. Arch. Korrbl.* 3,1973, 431-435.
7. K. Kilian, *Fibeln in Thessalien von der mykenischen bis zur archaischen* Zeit. Prächistorische Bronzefunde XIV, 2 (München 1975).
8. K. Kilian, *Trachtzubehör der Eisenzeit zwischen Ägäis und Adria.* Praehist. Zeitschr. 50, 1975, 9-140.
9. K. Kilian, Weihungen aus Eisen und Eisenverarbeitung im Heiligtum zu Philia (Thessalien). In: R. Hägg (Ed.), *The Greek Renaissance of the Eighth Century B.C.: tradition and innovation.* Proceedings of the Second International

Symposium at the Swedish Institute in Athens, 1-5 June, 1981 (Stockholm 1983) 131-146.
10. I. Kilian-Dirlmeier, *Anhänger in Griechenland von der mykenischen bis zur spätgeometrischen* Zeit. Prähistorische Bronzefunde XI,2 (München 1979).
11. I. Kilian-Dirlmeier, Bemerkungen zu den Fingerringen mit Spiralenden. Jahrb. RGZM 27, 1980 (1982) 249-269.
12. I. Kilian-Dirlmeier, *Nadeln der frühhelladischen bis archaischen* Zeit von der Peloponnes. Prähistorische Bronzefunde XIII, 8 (München 1984).
13. A. Pilali-Papasteriou/A. Papaevthymiou-Papastamou, *Νέα ανασκαφική έρευνα στο ιερό της Φίλιας*. Ανθρωπολογικά 4, 1983, 49-68.
14. D. Theoharis, *Φίλια*. Arch. Deltion 18, 1963 Chronica 135-139.
15. D. Theoharis, *Ιερόν Αθηνάς εν Φίλια Καρδίτσης*. Arch. Deltion 19, 1964, Chronika 244-251.
16. D. Theoharis, *Ιερόν Αθηνάς Ιτονίας (Φίλια-Καρδίτσης)*. Arch. Deltion 20, 1965 Chronika 311-313.
17. D. Theoharis, *Ανασκαφή ιερού Αθηνάς παρά την Φίλιαν*. Arch. Deltion 22, 1967, Chronika 295-296.
18. Imma Kilian-Dirlmeier *Kleinfunde aus dem Athena Itonia-Heiligtum bei Philia (Thessalien)*, Mainz 2002, Verlag des Römisch-Germanischen Zentralmuseums, Bonn.

The Measurement of Small Time Intervals By the Ancient Astronomers

E. Theodossiou, D. Papathanassoglou and V.N. Manimanis

Dept. of Astrophysics-Astronomy and Mechanics, School of Physics, National & Kapodistrian University of Athens, Panepistimiopoli Zographou, Athens 15784, GREECE

Abstract. The use of divisions of the day by the Chaldaeans marked the beginning of time measurements using sundials, astronomical instruments and water clocks. Ptolemy introduced the "first small parts" and "second small parts", i.e. the minutes and seconds we use today; however, we argue that back then they represented purely theoretical units since no time measurement had an accuracy higher than 1 minute.

Keywords: Antiquity, time measurement, Ptolemy, ancient astronomical instruments.
PACS: 95.90

DAY, WEEK, YEAR, MONTH

Since the prehistoric times, the alternation between day and night, the phases of the Moon and the regular return of the year's seasons were the phenomena that instilled the notion of time into humans and were studied by them in order to measure this entity. As a natural consequence, the first time unit established was the day-and-night period. From that point on, all civilizations that created time measurement systems defined the longer time units as integer multiples of the solar day-and-night period, even when these large units were not based on the solar motion, but rather on the motions of other celestial bodies, e.g. the Moon.

The existence of the seasons and their rotation had been realized from the prehistoric times, due to the food-gatherers' struggle for food and protection from the weather. Later on, when they started their occupation with agriculture, the need to predict the coming of the seasons led humans to the notion of the tropical year. Thus they had the ability to adequately prepare themselves against natural phenomena such as the rain period, cold and drought, and for the hunting, sowing and harvest seasons.

These "climatologic seasons" of the year were a basic calendrical fact. However, it gradually became evident that the precise starting time of each season was impossible to be determined from some vague weather phenomena. Then the ancient sages turned to the Moon for a solution. They noticed that the time necessary for a complete lunar cycle was 29.5 solar days-and-nights, thus discovering a new time unit, the lunar or synodic month. A division of this unit was the week of 7 solar days-and-nights and 9 hours, the time corresponding to one quarter of the lunar phases. This division presupposed the equal subdivisions of the day, i.e. the hours.

CP848, *Recent Advances in Astronomy and Astrophysics,*
7th *International Conference of the Hellenic Astronomical Society,* edited by N. H. Solomos

It must be acknowledged that even before the dominance of the lunar month, all eastern civilizations had divided their year into smaller periods based on the rising or setting time of very bright stars, such as Sirius. The Chaldaeans and the Egyptians were using 36 bright stars, the rise of each signaling the start of a ten-day period, one of the 36 of their year; these stars were accordingly called *decanoi.* The cuneiform writing tablets Mul-apin of the Chaldaeans (7th Century B.C.) there are catalogs of 36 stars known as Ea's, Anou's and Enlil's stars. This means that the months of the year were 12, since three stars corresponded to a month. These texts define a calendrical system similar to the ancient Egyptian one: 12 months of 3 ten-day weeks or 6 five-day intervals (*hamustu*) forming a month (72 *hamustu* were making a 360-day year). In Egypt the decanoi were headed by the brightest star, Sirius, the deity Isis-Sothis and the dog-headed god Anubis (*Anpu*). Sirius (*Thoth*) was defining with its "heliacal rising"[1] the beginning of the Egyptian year and its first month, also called Thoth. This month coincided with the annual floods of the Nile.

Similar to their dividing of the year into 12 months, was the division by the eastern people of both day and night into 12 hours (*haru* for the Egyptians). The beginning of both the day and the 24-hour period was the moment of the sunrise. The time from the sunrise to the sunset was initially divided into 10 hours and they were adding 2 hours for the morning and evening twilight. Similar was the division of the night. They regulated the two 12-hour spans so that both noon and midnight were coinciding with the sixth hour. These *seasonal hours* were of variable duration according to the season of the year, because the length of the natural day also varies from one solstice to the next. For instance, the hours of the night were longer in the winter and shorter in the summer. Therefore, the hour was not the 1/24 of the whole cycle. The daily hour was the 1/12 of the time between sunrise and sunset, while the nightly hour was accordingly the 1/12 of the time between sunset and sunrise. The daily hour was equal to the nightly hour only during the equinoxes. So the duration of the Egyptian hour was changing continuously.

The Egyptians were measuring time with the sundials and the water clocks or hourglasses, made especially to show the hours according to the season of the year. **In any case, the fixed-duration hour, as we know it today, was never used by the ancient Egyptians.**

In the ancient Greek world, Homer divided the day into three parts (*eos, meson hemar, deile*) and the night into other three (*espera, mese nykta, amphilyke*). Later on, with the wide use of gnomons, sundials and hourglasses, the division into 12+12 unequal hours was established. Nevertheless, the ancient Athenians were using the *nychthemeron*, i.e. they were starting the 24-hour period at the sunset, as do the Jews and the Muslims, or the Orthodox Church in its traditional worship. The ancient Persians were starting the cycle at midnight, as we do. The Romans were initially using sunrise as the start, as most eastern civilizations, but later they also transferred the beginning at midnight.

[1] The **heliacal rising** of a star is the time and date of its rising when it precedes sunrise by the minimum time interval necessary for that star to be visible.

MINUTES, SECONDS AND MEASUREMENT INSTRUMENTS

The Chaldaeans were probably the first to define sunrise as the starting point of the 24-hour cycle. It is possible that they divided each hour into 60 equal parts, since the sexagesimal system was the basis of their numbering. However, it is obvious that even an elementary study of the length of the subdivisions of the basic time units required the invention and use of some instruments.

The first instrument invented, **gnomon**, was measuring the horizontal coordinates of the Sun, altitude and azimuth, and it formed the basis for all sundials.

Unfortunately, the sundials have serious disadvantages. They cannot be used on cloudy days or during the night, and the position of the gnomon's shadow cannot be measured with accuracy. Therefore, it is impossible to measure small time intervals using them. When sunlight was not present, the ancients were using another natural phenomenon, the flow of water in the form of drops. Thus the first *mechanical* mean to measure time was born, the water clock.

However, none of these instruments was of any accuracy. Much more accurately they could measure the instant of the upper culmination of a given star, i.e. its transit from the meridian of the place, due to its negligible dimensions. Therefore, the ancient astronomers defined as "day" the interval between two successive upper culminations of a star; that is, they were measuring the hour angle H. The local sidereal time (t) is the sum of the right ascension (α) of the star and the hour angle: $t = \alpha + H$. At the moment of culmination the hour angle is zero, and so $t = \alpha$. Measuring the interval between two successive upper culminations of the same star, they found it approximately equal to 24 solar hours, since we know it is exactly 23h 56min 4.1sec or the length of a "sidereal day".

We do not know what other instruments they had besides the gnomon and the water clock, but we cannot rule out more complicated constructions, especially during the Hellenistic period, since we know of the famed Antikythera Mechanism finding.

We should note the stableness of the sidereal day relative to the variable length of the solar day, which increases and decreases substantially during the course of a year.

Instruments

The Gnomon

The gnomon is the oldest and simplest astronomical instrument. It is a vertical stick of known length, fixed with respect to horizontal ground. It was known since very ancient times to the eastern and Mediterranean people, and to the Chinese since 2000 B.C., and the ancient astronomers were able to obtain many interesting facts and data with this elementary instrument, namely:

1. The direction of the meridian line, i.e. the North-South direction, and through this of all cardinal points.
2. The equinoxes, the solstices and subsequently the length of the tropical year, from the changing length of the stick's shadow. At solstices, the shadow of

the gnomon at local noon becomes longest or shortest, depending on which solstice is.

3. The geographical latitudes of various places.
4. The obliquity of the ecliptic.
5. The declination (δ) of the Sun.
6. The true solar time, since with the sundial they were measuring hours of unequal length.
7. The apparent diameter of the Sun and of the Moon with good approximation.
8. The altitude and the azimuth of the Sun. The ratio of the gnomon's length (M) to the shadow's length (S) gives the tangent of the altitude (a) of the Sun: $\tan a$ = M/S, while the azimuth is calculated from the angle between the shadow and the direction of the meridian line.

One of the most ancient branches of astronomy is *gnomonics*, which studies the gnomon and the sundials. Sundials show the true solar time and usually are accompanied by tables as an aid to find the mean solar and civilian time of the location.

The Parallactic Ruler

The ancient astronomer Hipparchus (190–120 B.C.) managed to calculate the time at any moment by measuring the altitude of bright stars whose position was known. The altitude of the stars was measured with the so-called "parallactic ruler", an instrument with the shape of a gun-sight, which is described in every detail by Claudius Ptolemy in his *Great Mathematical Syntaxis* (the *"Almagest"*).

At about the same period the first *spherical astrolabe* or *organon* should have been constructed by either Apollonius of Perga (262/1–190/179 B.C.) or Hipparchus. This instrument combined the representation of cosmos in a stereographic projection. It was bearing calibrated circles that were accurate replicas of the main great circles of the celestial sphere (the meridian, the celestial equator, the horizon, the ecliptic, etc.). With the help of certain bright stars, the position of which was known, the circles of the spherical astrolabe were positioned to coincide with the respective circles of the celestial sphere. In this way, the equatorial coordinates (R.A., dec.) of the stars were being found.

Other Instruments

Some other instruments were:

- The *plinthis* of Ptolemy, in fact most probably an invention of Hipparchus. It was used to calculate the angular distance of the Sun from zenith at noon.
- The *copper* or *equatorial armilla*. It consisted of a copper concave hemisphere and a copper great circle attached in its interior in such a way that it coincided with the celestial equator. This astronomical instrument was used by the ancient Greek astronomers to find the dates and hours of the solstices and of the equinoxes.

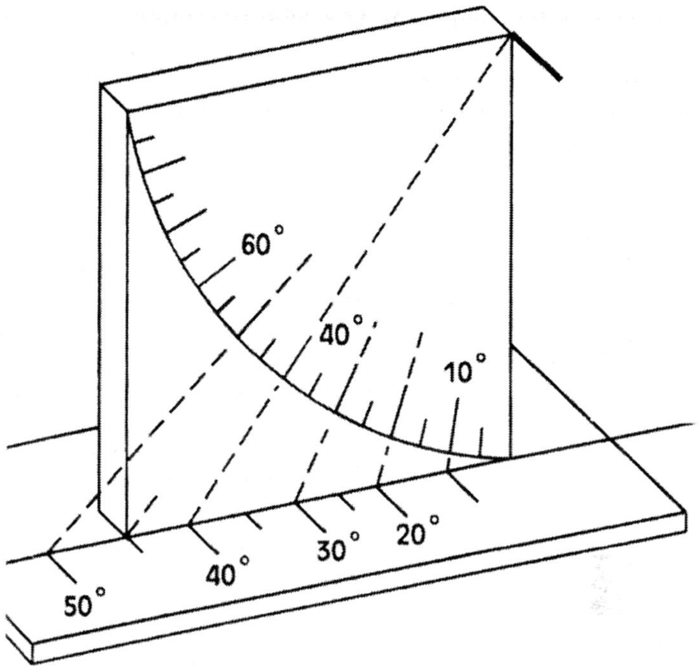

FIGURE 1. The ancient astronomical instrument plinthis.

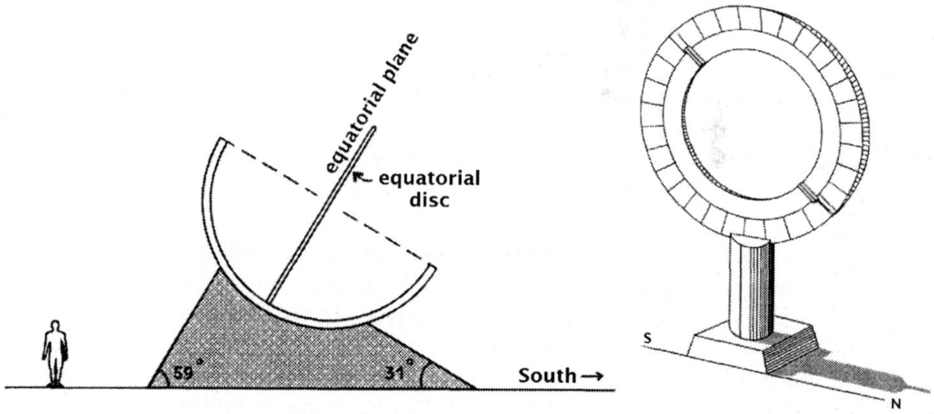

FIGURE 2. The copper or equatorial armilla (left) and the tropical armilla or ring instrument (right).

- The *tropical armilla* or *ring instrument*. This instrument was used by the ancient Greek astronomers to calculate the altitude of the Sun at its upper culmination. This *armilla* is described by Ptolemy, but also by Proclus (411–485 A.D.) in his work *Hypotyposis of the Astronomical Hypotheses*.

However, the question remains: what was the accuracy of these instruments?

The Precision Of The Measurements

Could the ancient astronomers measure the second of time, or they were just referring to it theoretically? The latter is the most probable; however, we cannot exclude the measurements of such short time intervals, since the water clock can be fine-tuned to one drop per second. The fact is that Claudius Ptolemy was the first to define and name the minutes and seconds, during the 2nd Century A.D.. He subdivided both the hour and the degree into 60 "first minute parts" (minutes) and each "first minute part" into 60 "second minute parts" (seconds)[2], according to the sexagesimal system that was traditionally in use during that period, a legacy from the Mesopotamian numerical system.

The use of the sexagesimal system in the measurement of circular arcs (degrees, minutes) and time (hours, minutes) was very convenient for the Greek astronomers, because in this way they could readily use the data of the Babylonian astronomical observations without any reductions, in order to check their own theoretical models of planetary motions. Ptolemy in the *Almagest* (Book I, 10) divides the circle into 360 degrees. In Book III, he presents essentially unaltered the theory of Hipparchus on the solar motion, adopting the same value for the length of the tropical year, 365 and ¼ minus 1/300 day-and-night periods. Ptolemy writes that the length of the year is $365 + 14/60 + 48/60^2 = 365.24666...$ days and nights, which is exactly the same with Hipparchus' value. Indeed, $365 + 1/4 - 1/300 = (365 + 15/60) - 12/3600 = 365.24666...$

However, Ptolemy himself did not write "minutes" or "seconds", but "parts of the day", while he used the former terms in the case of the circle. In the First Book of the *Almagest* he subdivides the circle into 360 "degrees" and its diameter into 120 parts, each of which forms the unit to express the length of a chord. Each degree of the circumference is divided into 60 parts (first minutes) and each minute is further subdivided into 60 parts (second minutes).

In support of the view that the time second was a purely theoretical and not a measurable unit is the fact that through the end of Antiquity the so-called *seasonal hours* (mentioned already) were being used. Only after 1300 A.D., with the invention and use of mechanical clocks, the unequal seasonal hours were abolished as inadequate.

The arc second was also a theoretical unit; this can be demonstrated by the accuracy of the observations made by the great observer of the skies Tycho Brahe (1546–1601), who made the most accurate observations before the invention of the telescope. The error of these observations ranges from about 1 to 2 arc minutes as compared with the 10 to 15 arc minutes of the Ptolemaic era or of the Arab astronomers. This comes as a natural consequence of the physical fact that the observational limit of the human eye is approximately 1 arc minute.

For historical reasons we mention that it was with the method of the respective altitudes of the astronomer Jean Picard (1620–1682) and with the use of a telescope that the time measurement was achieved for the first time with an approximation of 1 second. Besides, the precise duration of the time second was defined after the French

---

[2] The English terms "minute" and "second" originate from the Latin translations "partes minutae primae" and "partes minutae secundae" of the respective Greek expressions.

Revolution, when a committee of French scientists proposed in 1820 the mean second as the 1/86,400 of the mean solar day. Subsequently, the second was defined based on the tropical year 1900 and, by the International Convention for Weights and Measures in 1964, as the duration of 9,192,631,770 periods of the electromagnetic radiation corresponding to a specific transition between two energy levels of the ground state of the atom of the only stable isotope of caesium (^{133}Cs).

CONCLUSIONS

After they defined the day-and-night interval, the ancient peoples subdivided both day and night into 12 hours. However, their next step was not the subdivision of the hour, but the classification of the days into groups, a step that created the week, the lunar month and the year itself. Ptolemy knew that the year needs subdivisions of the hour to accurately describe its duration, since he himself mentions that it is equal to $365 + 14/60 + 48/60^2$ days-and-nights.

Nevertheless, it seems that the measurement accuracy of the ancient astronomers was no higher than about a few minutes and therefore, although both time minutes and seconds were defined by Ptolemy they were never used in their measurements. The ancient knowledge of at least time seconds was rather theoretical and conceptual.

ACKNOWLEDGMENTS

We would like to express our thanks to the Special Account of Research Funds (ELKE) of the National and Kapodistrian University of Athens for its founding of this research.

REFERENCES

1. C. Ptolemy, *Ptolemy's Almagest*, translated & annotated by G. J. Toomer, Princeton: Princeton University Press, 1998.
2. E. Th. Theodossiou and E. Danezis, *Measuring the Timeless Time – Time in Astronomy*, Athens: Diavlos Publ., 1994 [in Greek].
3. E. Th. Theodossiou and E. Danezis, *The Odyssey of the Calendars – Searching For the Roots of Knowledge*, Athens: Diavlos Publ., 1995 [in Greek].
4. D. Tsimbourakis, *Mathematical Measurements in Ancient Greece*, Athens: Aiolos Publ., 2002.

The Vertical Sundial of the Monastery of Panaghia Scripou

E. Theodossiou and V.N. Manimanis

Dept. of Astrophysics-Astronomy and Mechanics, School of Physics, National & Kapodistrian University of Athens, Panepistimiopoli Zographou, Athens 15784, GREECE

Abstract. As part of our ongoing project to correctly identify and describe all the Byzantine sundials in Greece, the vertical marble sundial in the main church of the "Koimesis tis Theotokou" (Dormition of the Virgin) or "Panaghia Scripou" monastery, in Orchomenos, Boeotia, central Greece, is presented. It is almost certainly the oldest (and arguably the most beautiful) Byzantine sundial in the whole Greece, being contemporary to the church (874 A.D.).

Keywords: sundials: Byzantine, Greece, Boeotia, Orchomenos, monasteries.
PACS: 95.90

THE MONASTERY OF PANAGHIA SCRIPOU

The Monastery of "Koimesis tis Theotokou" (Dormition of the Virgin) or "Panaghia Scripou" monastery, in Boeotia, central Greece, is located opposite from the archaeological site of Orchomenos, a most ancient Greek city, and its Citadel. Pausanias mentions (9, 38, 1) the existence of two sanctuaries, of Charites (the Graces) and of Dionysus, in the area. These have not been identified up to this day, and may very well lie at the exact position where, in 874 A.D., the Monastery of Panaghia Scripou was erected; a strong indication is that excavations revealed the remains of a Mycenaean building next to the Catholicon (the main church). The same excavations revealed a proto-Christian mosaic inside the church.

The monastery itself is, as we mentioned, a 9th Century building and from the original complex of its buildings only the main church survives today, which is dedicated to the Dormition of the Virgin Mary, but also to the Apostles Peter and Paul, as indicated by the triple sanctum; the two eastern partitions serve, according to the inscriptions, as chapels of St. Peter (Petros) and Paul (Pavlos) "the sacred dust of whose is covered by the soil of Rome".

The main church of the convent is probably the most important example of the church type "transitional domed cross-in-square" in Greece, and also the largest and most sumptuous known Byzantine temple of its century outside Constantinople. This convent church stands apart for its size (22.3 m × 18.6 m without its narthex), for its rich marble decoration and finally for the historical information given by four monument-style inscriptions. From these we learn that the founder of the church was Leon, not the Byzantine Emperor, but the "Royal Protospatharius", i.e. the leader of

CP848, *Recent Advances in Astronomy and Astrophysics,*
7th *International Conference of the Hellenic Astronomical Society,* edited by N. H. Solomos

spatharii, the guards of the Palace in Constantinople. Leon was the owner of the area and, according to the inscription incorporated on outer side of the sanctum's arc, he built the church in 873-874 A.D.. This makes us certain on the major issue of the temple's chronology. Incidentally, the peculiar name "Scripou" itself is probably derived from these wall in-scriptions (Latin: scriptus).

The walls of the church contain a lot of ancient building blocks (as is the case with many other Christian temples of the Byzantine era), which originate from the nearby archaeological site of the historical city of Orchomenos: column sections, carved corner-stones, even sepulchral columns were freely used, creating unexpected decorations and hues on the large walls.

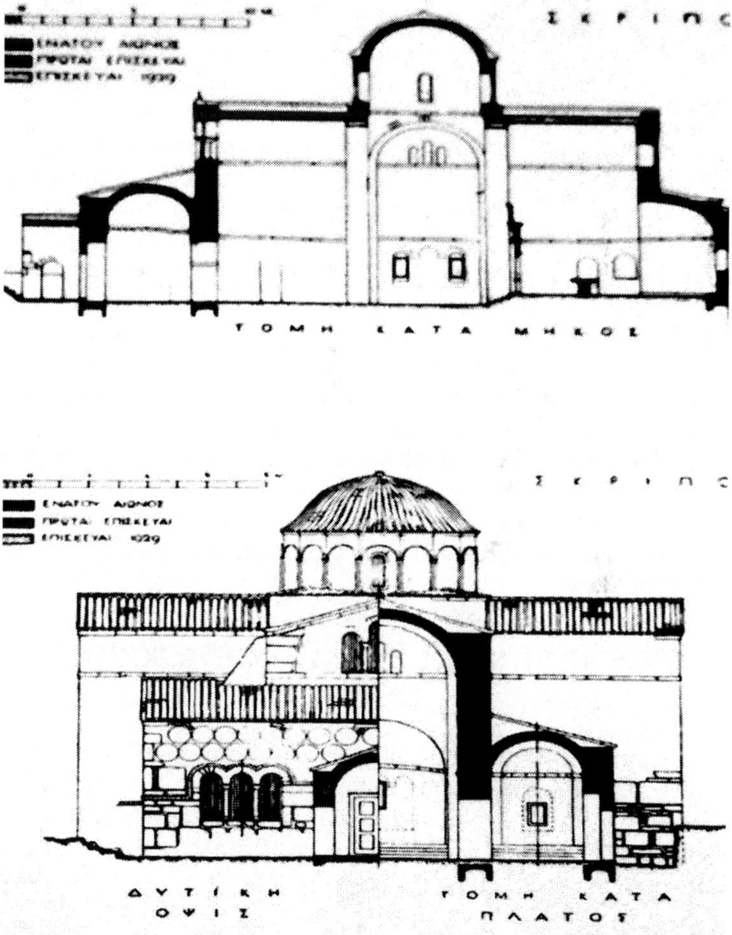

FIGURE 1. The main church ("catholicon") of the Panaghia Scripou monastery in its views from the South (upper) and from the West (lower), showing the successive restoration levels; the black color denotes the oldest (9th Century A.D.) existing parts, among which is the southern wall, the one that bears the sundial.

It is probable that the church, a remarkable architectural example of the transition from the palaeo-Christian basilica to the Byzantine rhythm, was designed as a burial monument of its founder, although this was uncommon practice in Byzantine Greece. In that case, the vertical sundial on its wall, being decorated with two peacocks, could be a kind of reference to the eternal life. Anyway, the founder of the temple wanted it to have *"a joyful glint of splendour, beautiful from all directions"*.

The sculptural decoration of the Catholicon is rich on both the internal and external surfaces; moreover, several carved pieces (originating from ancient buildings or from a proto-Christian cemetery) were incorporated during the following centuries in the monk cells to the south, in the western gate and in the storehouses. The older cells are to the west of the newer ones, in a building consisting of five successive tile-covered spaces.

The archaeological research discovered that the oldest murals decorating the church date from the 12th Century. Since 1930 restoration activities are under way, as well as the cleaning of the murals, while in 1939 a belfry for the church was built, designed by the Ministry of Culture, in the northwestern area of the monastery. The preservation works in the monastery were included for financing in the EU's Community Support Framework of 1994-1999. And while the treatment of the murals and the stabilisation and cleaning of the relief decoration of the narthex were being completed, arson was committed against the temple. The fire caused some serious damage, nevertheless the interior was cleaned from the soot and currently a second phase of restoration is about to start with the replacement of the burned windows and the cleaning of the murals of the narthex.

The monastery of "Panaghia Scripou" operated for many years as such, but not anymore; the Catholicon is currently in use as a parish church. Its Saint's Day comes on the 15th of August each year (celebration of the Dormition of Mary); celebrations also occur on the 23rd of August and on the 10th of September. The September celebration honours a 20th -Century miracle: On that day in 1943 Orchomenos and its inhabitants were spared from retaliation by the German occupation forces thanks to a miracle by Virgin Mary.

THE VERTICAL SUNDIAL OF PANAGHIA SCRIPOU'S MAIN CHURCH

A handsome vertical sundial, probably the most beautiful of all Byzantine sundials of Greece, decorates the Catholicon (main church) of the Scripou monastery. Since there is no evidence suggesting otherwise, this sundial was placed on the wall during the construction of the church, in 874 A.D., a fact that makes it the most ancient of all Byzantine/Mediaeval sundials of Greece! This was supported by the late archaeologist, Byzantinologist and architect Professor Anastassios Orlandos (1935-1936).

This vertical sundial is carved on the exterior surface of a marble rectangular building stone incorporated in the southern wall of the Catholicon. The dimensions of this parallelepiped stone are length 125 cm, width 35 cm and height 66 cm. The plate surface (125 cm × 66 cm) bears only ten hour lines, which are numbered with the ancient Greek number symbols – letters of the Greek alphabet: A, B, Γ, Δ, E, F, Z, H,

Θ and I; a horizon line is also present. The lower side of the plate is at a height of approximately 2 meters from the ground.

The hour lines are 50 cm long and they end on a carved semicircle, which has a radius of 50 cm. The simply carved capital letters of the Greek alphabet that symbolise the hour numbers are just beyond the end of the hour lines, inside a semicircular ring section formed by the first 50-cm semicircle and a second semicircle which has a radius of 58 cm. The lines of the semicircles are considerably thicker than the hour lines.

The marble plate surface outside the semicircles is decorated at the left-hand and the right-hand side with two carved peacocks facing away from the centre of the plate (the one at left faces towards the West and the one at right faces towards the East). The peacock at left has a total length of 46 cm, while the one at right, being more elongated, has a total length of 42 cm. The depiction of these birds is most probably a kind of reference to the eternal life. Finally, below the hours E and F there is some other carved decoration: Two flower-like leaves connected with a hook-like form as the leaf stalk.

FIGURE 2. The vertical sundial of the main church of the monastery.

We note here that the decorative elements of the palaeo-Christian art come from the Greek gentile world. In the oldest Christian art appears a large set of sculptured motifs, e.g. flowers, birds, dolphins, etc.. Later on, symbolic representations start to appear, such as the fish, a palm branch, the sheep, the pigeon and the peacocks. The fish symbolizes Jesus Christ, since the letters of the Greek word for it form the famous acrostic Jesus Christ, Son of God, Savior. The palm branch symbolizes the victory of the Church. The pigeon stands for the Holy Ghost and purity. Finally, the peacocks symbolize the eternal life in Paradise. Indeed, in the Byzantine and Christian Museum of Athens there is a funerary inscription from Athens (BMX 400, 5th – 6th Cent. A.D.) with facing peacocks as "symbols of paradisiacal happiness, being the birds of paradise" (as the relative museum explanation states). Even today, peacocks are a very common motif in Greek churches; they are carved on the wooden icon-stand (reredos) of the temples, on the epitaph and at the external decoration of several modern temples.

Unfortunately, while the housing of the Scripou sundial's gnomon is clearly visible, the gnomon itself is not there; it was stolen, lost or destructed in the distant past. This is also the case with the Byzantine sundial in St. Lavrentios convent, in Mount Pelion of Central Greece.

The sundial of the Panaghia Scripou monastery, in spite of the ten numbers written on it, divides the day not into ten but into eleven hours, since on its semicircle there are eleven sectors, exactly as on the St. Lavrentios sundial. The horizontal line at the upper part of the sundial's plate represents the horizon. At true solar noon the shadow would follow the meridian; on this 10-hour-line dial there is no such meridian line, as it should on any sundial with an odd number of hour lines. For example, on an 11-hour-line dial the sixth hour line is vertical and in the middle, having symmetrically at its sides the rest ten hour-lines.

REFERENCES

1. A. Orlandos, *Ergasiai Anastiloseos Vyzantinon Mnemeion tis Ellados, Archeion ton Vyzantinon Mnemeion tis Ellados*, vols. A and B, Athens, 1935/36 [in Greek].
2. Pausanias, *Description of Greece* (with an English translation by W.H.S. Jones), The Loeb classical library (9, 38, 1), Cambridge, Mass.: Harvard University Press, 1934.
3. E. Theodossiou, J. Kouris and V.N. Manimanis, *British Sundial Society Bulletin* **16** (iii), 101-103 (2004).

The Colour Of The Stars And The Pupils In The Kindergarten And The Primary School

Hariton Tomboulides

1st Lyceum of Naousa, Aimil. Veaki 9, 592 00 Naousa, Greece, haring5@oneway.gr

Abstract. A collaboration between the author of this article and the 7th Primary school of Naousa has been established eight years ago. It started with the kindergarten of the school and has been extended to the other classes of the school, as well as to other primary schools in Naousa, including the school with physically and mentally handicapped children. The collaboration with the kindergarten has as a result the production of some handicrafts, paintings or even theatrical plays. Last school year's project was the colour of the stars and the product was a year calendar with paintings of planets, star clusters, nebulae and stars.

Keywords: Primary education, astronomy, collaboration.
PACS: 01.40.E-; 01.40.-d

PRIMARY SCHOOL IN GREECE

Education in Greece is compulsory for all children 6-15 years old; namely it includes Primary (Dimotiko) and Lower Secondary (Gymnasio) education. The school life of the students, however, can start from the age of 2,5 years (pre-school education) in institutions (private and public) called "Vrefonipiakoi Paidikoi Stathmoi" (creches). In some Vrefonipiakoi stathmoi there are also Nipiaka Tmimata (nursery classes) which operate along with the Nipiagogia (kindergartens). Usually the kindergarten is located in the same building as the Primary school it belongs to. There are ten Primary schools in Naousa and eight kindergartens.

Attendance at Primary education (Dimotiko) lasts for six years, and children are admitted at the age of six. Along with the regular kindergartens and the dimotika, All-day Primary schools are in operation, with an extended time table and an enriched Curriculum.

COLLABORATION OF AN ASTROPHYSICIST WITH A KINDERGARTEN

A good collaboration between the teacher of the kindergarten of the 7th Primary School of Naousa and me was established eight years ago, regarding the study of nature and the universe.

A subject is chosen every year in the beginning of the school year and I participate with some lectures or starwatchings with telescopes. A working project is

CP848, *Recent Advances in Astronomy and Astrophysics,*
7th *International Conference of the Hellenic Astronomical Society,* edited by N. H. Solomos

planned in different levels and with different means every year. The project is finished with productions of paintings, handicrafts, theatrical performances or even public performances.

"Light pollution", "The air we breathe", "My role in nature, "The planets", "The colour of stars", "The Christmas star", "The food chain", "The bee society" were some of the subjects in the last years.

The teacher – or sometimes two teachers – are preparing the children two months earlier with stencils, paintings, video films or books. In November or December we are gathered in the main concert hall of the Municipal Music School where I have one or two lectures mainly with pictures in form of slides, videos or computer projections.

The subjects I show them are chosen by the children and each of them is working on its own subject for the winter months.

In the spring time sometimes the children produce handicrafts or paintings on the subject they choose, sometimes they use the telescope together with me to observe the stars, the moon or the planets and sometimes they themselves participate in a play with nature as a main subject and their own role in it.

Last year's project was the colour of the stars. I showed them pictures of stars, star clusters and planets in different spectal classes and in different colours. This has been connected with colours in nature and in the moon. I presented this with my astronaut uniform I received by NASA in the Space Camp of Huntsville four years ago and it was an impulse giving event for the children. The children's product was 12 paintings corresponding to the 12 months of the year and the year calendar was produced with each month in its own colour. A copy of it has been given to me as a thanks giving.

The collaboration of so many years with the infants is a very positive factor in the children's growth. The children get the feeling of creating and at the same time the participation in the creation. Their imagination and fantasy is expanding in a cosmic scale. This good collaboration with the infants of the 7th Primary school has been expanded to one more kindergarten in Naousa (the kindergarten of the 6th Primary School), where a lecture on the sun and the planets has been hold by me. This presentation led the children to the study of the solar system. Lectures have been given even to the pupils of the other classes in three Primary schools, mainly in their lectures on natural sciences.

THE COLLABORATION WITH THE "SPECIAL SCHOOL" IN NAOUSA

In the town of Naousa there exists a "Special School" (Eidiko Sxoleio), as it is called, for children with mental or physical problems. This school has its own teachers and it has been existing for many years in Naousa with very good results. The children produce every year some handicrafts which they expose to the public every year.

The teachers of this school asked for my help six years ago in order for the pupils to "fly to the moon" and to the stars. I accepted their invitation and I had a day's presentation of the solar system and the Milky Way, as well as the space exploration of the solar system.

It was a very special experience for me, since the imaginary travel we participated in, with the Space Shuttle, was so strong for two of the pupils, that they did not allow me to leave their school without taking them with me. I can not forget this event and this gave me a very strong feeling for the handicapped persons. A tender and intensive engagement of their teachers was needed in order to leave the school without hurting the pupils feelings.

CONCLUSION

This kind of collaboration I presented here between a scientist and the children of the kindergarten and the pupils of the first grades of the Primary school, and the pupils of the Special school, is the best way for the teaching of natural sciences. The infants have the feeling of experience and adventure from their birth and we have to make it stronger and not disappear.

This feeling is unfortunately disappearing in the upper grades of the Secondary school. There are many reasons for this, and some of them I present in my papers on the teaching of Astronomy in the 1st Lyceum of Naousa. (Tomboulides 1996, 2005).

REFERENCES

1. H. Tomboulides, "The teaching of Astronomy in the 1st Lyceum of Naousa, Greece" in *Proceedings of the 2nd Hellenic Astronomical Conference*, 1996.
2. H. Tomboulides, "*Sunspots and Venus transit from the school yard, factors for Astronomy teaching in the Upper High School in Greece*." (unpublished).

The Teaching Of Astronomy In The Lyceums Of Naousa Eight Years After The School Reform Of 1997

Hariton Tomboulides

1st Lyceum of Naousa, Aimil. Veaki 9, 59200 Naousa, Greece, haring5@oneway.gr

Abstract. The teaching of Astronomy as a separate subject in the second grade of the upper high school in Greece, has more or less disappeared since the school reform of 1997. This applies even in the two Lyceums in Naousa, which have a good tradition in Astronomy teaching, where the number of pupils attending the subject has been diminished to 1/5 of the total amount of pupils. The main reason for this is the national character of the examinations a pupil has to take, in order to pass the grade. Astronomy has changed to an optional subject since the reform 1997, and the pupil chooses mainly a foreign language, computers or sketching in order to get higher grades to help him to advance. In order to keep the subject of Astronomy as a separate subject in both Lyceums of Naousa many activities take place, both in the schoolyard and in the open areas in the city. The use of the school telescope for observing the solar sunspots, the Venus transit and the planets was applied last school year for the pupils to choose Astronomy as a separate subject.

Keywords: School reform, astronomy, pupils.
PACS: 01.40.E-; 01.40.-d

ASTRONOMY IN THE UPPER HIGH SCHOOL IN GREECE AFTER THE SCHOOL REFORM OF 1997

The child in the Greek school normally comes in contact with Astronomy in the first year of the Primary school but it is not until the second grade of the Upper High school (Lyceum) that Astronomy is taught as a separate subject. The subject was a compulsory subject for every pupil before the school reform of 1997 and lots of working projects in different areas of Astronomy were produced in the two Lyceums of Naousa (Tomboulides, 1996). The school reform of 1997 has changed the subject from compulsory to an optional one. Fifteen subjects in the second grade of the Lyceum are compulsory and one is optional. Nine of the compulsory subjects are examined in a national level and the rest in local school examinations. For a pupil to proceed in the next grade an average degree of 9,5 (20 is the maximum) is required (for details see Nomos Plaisio, XX 2525/1997). The optional subjects in the second grade are:

1) Principles of Environmental Sciences, 2) Modern European Literature, 3) History Subjects, 4) Sketching (Linear or free), 5) Second Foreign Language, 6) Computer Applications, 7) Astronomy and 8) Biology. For a subject to be taught a

CP848, *Recent Advances in Astronomy and Astrophysics,*
7th *International Conference of the Hellenic Astronomical Society,* edited by N. H. Solomos

number of at least fifteen pupils have to choose it. Every subject is examined in a written way except Computer Application and Sketching where a group project is required.

CONSEQUENCES FOR THE ASTRONOMY TEACHING AFTER THE SCHOOL REFORM OF 1997

The school reform of 1997 has had a disastrous effect on the teaching of Astronomy in the Lyceum. The number of pupils who choose the subject is very low and there are prefectures (as for example the prefectures of Pieria, Imathia and Pella) which have only a couple of high schools (or not even one) where the subject of Astronomy is taught. There are many reasons for this but the main reasons are the following:

I) Many pupils choose subjects as Computer Applications, Sketching or a Second Foreign Language mainly because they don't have to take a written exam or because they know the subjects from earlier private school teaching they have attended. In their choice the local school rector's contribution takes also place. They manipulate the pupils and channel the pupils up to the above named subjects.

II) By the time the school reform of 1997 was applied, new books had to be written. The earlier book of "Elements of Astronomy and Space Research" has been changed to a new one with an astrophysical main character than an astronomical one with great difficulties for a mathematician to teach the subject. (The mathematician has the first choice to teach the subject according to the Greek law) (Tomboulides, 2000). So, the lack of a capable teacher contributes to the lack of teaching of Astronomy.

The stress for a pupil to manage the advancement in the high school because of the national examinations he or she has to take in nine subjects is also an important factor. The marks are usually much higher in a Foreign Language, the Computer Applications and the Sketching than in Astronomy. And these marks participate in the average degree which is required for a pupil to pass the class, so they choose an "easier" subject than a "difficult" one, as Astronomy is regarded.

THE CASE OF NAOUSA

The case of Naousa is an exception to this general rule. There is at least one group in each of the two Lyceums every year that attends the subject of Astronomy. Of course the number of pupils is changing every year but there is an average of 30-50 pupils who attend the subject of Astronomy.

The main reasons for this are the following:

I) There is an astrophysicist as a teacher who teaches the subject.

II) There are lots of astronomy activities both in the schoolyard and in the town. Last year, for example, the Venus transition in June, the large solar sunspots in November and the conjunctions of the planets Saturn and Jupiter with the moon were observed by the pupils with the school telescope from the schoolyard.

III) There is a good collaboration of the Lyceums with the Amateur Astronomical Society of Naousa. The pupils and their teacher were the founders of this society in 1995 and lots of activities contribute to the spreading of Astronomy in Naousa and the central Macedonia.

IV) There is a series of lectures every last Saturday of each month in cooperation with the Amateur Society and The Municipal Cultural Society under the name "Man and stars in the new Millennium" for the last five years. A university teacher holds a lecture and an average of 30-70 pupils participate in the discussion which follows the lecture.

V) The radio program in a local radio station (Ergatikos FM, 101,5) "Astro...logic", I held every week for one hour for three years with astronomical subjects and the participation of pupils too.

VI) The good library and the optical and audio materials (CDs, Videos, telescopes) that exist in the two Lyceums.

VII) The participation of the pupils every year to the National Astronomical Examination which is held by the Astronomical Society of Volos in collaboration with the local authorities and the Ministry of Education and Religious Affair. We can mention that pupils from the Lyceums of Naousa have been honoured seven of the ten years the examination takes place.

VIII) And last, but not least, the existence of earlier Astronomy projects that pupils have produced in the subject of Astronomy. Their booklets and stencils are exposed many times in the school library for both Lyceums as paradigms for imitation. Part of these have been exposed in the 2nd Hellenic Conference in Chalkidiki in 1995, (Tomboulides, 1996).

Of course the number of the booklets and the CDs that are produced now, after the school reform 1997, has been dramatically diminished. Only 20 works have been produced the last seven years, while almost 60 were produced every year earlier, when the subject of Astronomy was compulsory for the pupils of the second grade.

A good sign of renaissance for the Astronomy subject has been the last year's reform. The national character of the examination in the nine subjects for the pupils of the second grade has changed. A consequence of this change has been the production of sixteen Astronomy projects in different subjects last year in Naousa and the Astronomy has also been taught in two more high schools in Veria, capital of Imathia.

CONCLUSION

The school reform of 1997 has changed the number of pupils attending the Astronomy course in the Naousa's Upper High schools – Lyceums – to 1/5, since the subject has been changed from an obligatory to an optional one. The number of Astronomy working projects has also been affected, but not the quality of them. The main reasons for this are: the existence of an astrophysicist as a teacher for both Lyceums, the regular uses of the telescopes by the pupils for night observations and solar ones, the good collaboration of the schools both with the Amateur Astronomical Society in Naousa and the Municipal Cultural Society (ΔΗΠΟΝ) and the monthly lectures and discussions that are held by scientists and university professors.

These activities make the Astronomy bloom in the town of Naousa, a town of 23 000 inhabitants, and this is reflected in the success the high school pupils have in the National Astronomical Examinations organized by the Volos Amateur Astronomical Society in collaboration with the National Ministry of Education the last ten years, since pupils from Naousa have been honoured seven times.

REFERENCES

1. Nomos Plaisio XX2525/1997, "*Eniaio Lykeio. To nomothetiko plaisio*" ΥΠ.Ε.Π.Θ. (Ministry of Education and Religious Affairs), Pedagogical Institute, Athens, 1997.
2. H. Tomboulides, "The Teaching of Astronomy in the 1st Lyceum of Naousa, Greece" in *Proceedings of the 2nd Hellenic Astronomical Conference*, 1996.
3. H. Tomboulides, "The new Astronomy book in the Second Grade of Lyceum and the teaching of Astronomy in the Lyceums of the Imathia prefecture" in *Proceedings of the 1st Hellenic Amateur Conference* , *OURANOS* 35, Volos, 2000.

The ESA Hubble 15th Anniversary Campaign: A Trans-European collaboration project

Manolis Zoulias*, Lars Lindberg Christensen† and Martin Kornmesser†

*36, Argonafton St., Ilioupolis, Athens, Greece, mzoulias@yahoo.com
†ESO/ESA/ST-ECF Karl-Schwarzschild-Strasse 2 D-85748 Garching bei Munchen Germany

Abstract. On April 24th 2005, NASA/ESA Hubble Space Telescope had been in orbit for 15 years. The anniversary was celebrated by ESA with the production of an 83 min. scientific movie and a 120 pages book, both titled "Hubble, 15 years of discovery". In order to cross language and distribution barriers a network of 16 translators and 22 partners from more than 10 countries was established. The DVD was distributed in approximately 700,000 copies throughout Europe. The project was amongst the largest of its kind with respect to collaboration, distribution and audience impact. It clearly demonstrated how international collaboration can produce effective cross-cultural educational and outreach products for astronomy.

Keywords: Astronomy, Education, Public Outreach, Media
PACS: 01.20.+x, 01.50.-i, 01.75.+m

INTRODUCTION

Science communication and Public Outreach is becoming a vital part of the scientific process. Communicating astronomy and science in general has changed during the last decade and continues to change constantly. The creation of an Education and Public Outreach (EPO) product has become competitive work - competing for getting public attention. Impressive New Media products are made widely available, raising the requirement for high quality content in order to get the message through to the public and especially the younger generation. Furthermore, cultural and linguistic diversity across Europe demands the development of multilingual products.

On the other hand evolution of computing and media technologies provides us with new instructional and informational tools that should be incorporated in the communication strategy of a modern EPO office or collaborative scheme. The cost for reproduction of digital media has diminished and production has never been easier. A high quality DVD movie can now be authored using a moderate personal computer. The DVD technology allows a movie with multiple language audio and subtitles to be compiled in a single disk.

Finally there are new distribution channels that must be taken into account. Low reproduction cost has facilitated the use of digital products as insets to newspapers and magazines, hence increasing demand for such productions by mass media organizations.

CP848, *Recent Advances in Astronomy and Astrophysics,*
7th International Conference of the Hellenic Astronomical Society, edited by N. H. Solomos

THE PROJECT

The 15th anniversary of Hubble's launch presented the ideal opportunity for a dynamic collaborative project, aimed to attract the attention of the public with a special emphasis on the younger generation, promoting the advancement of knowledge of astronomy and science in general.

In this project Hubble, a well known space based astronomical instrument, was used as a centerpiece and a guide through our universe.

The project consisted of a number of activities:

- Production of a full-length documentary movie "Hubble - 15 Years of Discovery" issued:
  - on DVD for wide distribution.
  - for TV Broadcast
  - as a Show Package for planetarium show production
- Production of a full-color 120 page book
- Downloadable educational Material
- Promotional movie Poster
- CD production of the Movie soundtrack.

The Movie

The movie has a duration of 83 minutes, it is in other words a full time movie. An extra 60 minutes of contributed material from the partner institutions is available on the DVD. Many new, high quality animations were produced specifically for this film. Dr. Bob Fosbury, a scientist from the European Space Agency who has frequently used Hubble for his own research, presents the movie. As an active, but approachable scientist himself, he brings an added depth and insight to the material. This helps to demystify the image of the scientist and associates the man with the research instruments and the knowledge that derives from them. Bob can perhaps serve as a role model for the younger generation.

The movie's script takes us to a visual journey through Space and Time. Starting from Hubble's place in space around the Earth, a journey begins taking us through our Solar System, our Milky Way Galaxy's stars, nebulae and clusters to other distant celestial bodies, neighboring and distant galaxies, cosmic "monsters" such as black holes, supernovae and quasars. A cosmic voyage from the birth of the Universe, to the birth, evolution and death of stars and galaxies through Hubble's spectacular images.

More than 22 national partners collaborated to prepare translations and narrations, to sponsor and promote distribution of the movie all over Europe. The DVD contained subtitles in 15 languages: Bulgarian, Danish, French, German, Greek, Italian, Dutch, Norwegian, Polish, Portuguese, Russian, Spanish, Finnish and Swedish (Romanian subtitles were added lately). Narration was available in three languages: English, German and Greek (Italian language was added in the second version of the DVD).

FIGURE 1. Greek edition DVD cover and label.

The Book

The book consists of 120 pages in full color. It's cover is matte finished with partial varnish and includes a full copy of the DVD. The light, image-rich design aims to attract the attention of the less science-attentive part of the population. The first edition was printed in 5,000 copies in English. The book has already been translated to Finnish, German and Portuguese while translations are pending to other languages (Danish, Dutch, Greek, Italian and Spanish).

The Images

In order to attract more interest, two new Hubble images were produced in collaboration with NASA. The images were embargoed until the 24th of April and their publication drew additional public attention. They were simultaneously released all over Europe and throughout the world. The first was a composite of M51 revisited, with enhanced resolution and more detailed features shown, while the second one depicted an impressive detail from the Eagle nebula region.

THE RESULTS

More than 60 events took place all over Europe, in Austria, Belgium, Bulgaria, Czech Republic, Denmark, France, Germany, Greece, Hungary, Italy, Lithuania, Poland, Portugal, Slovak Republic, Spain, Sweden, The Netherlands, UK and Ukraine.

518,307 DVD copies of the first edition of the movie was distributed through more than 80 delivery points all over Europe (magazines, newspapers, science centers etc.). Currently more than 700.000 copies have been distributed. This makes the Hubble Anniversary project's DVD probably the most widely distributed science documentary ever. An estimated 10 million viewers will watch the movie through various TV channels.

To match the individual needs and achieve maximum impact to different states and cultures a record number of 23 different DVD labels and 17 different cover versions were produced.

FIGURE 2. The two embargoed images released on April 24th: Eagle nebula (left) and M51 (right). *Credits: NASA, ESA, and The Hubble Heritage Team (STScI/AURA)* and *NASA, ESA, S. Beckwith (STScI), and The Hubble Heritage Team (STScI/AURA)* respectively.

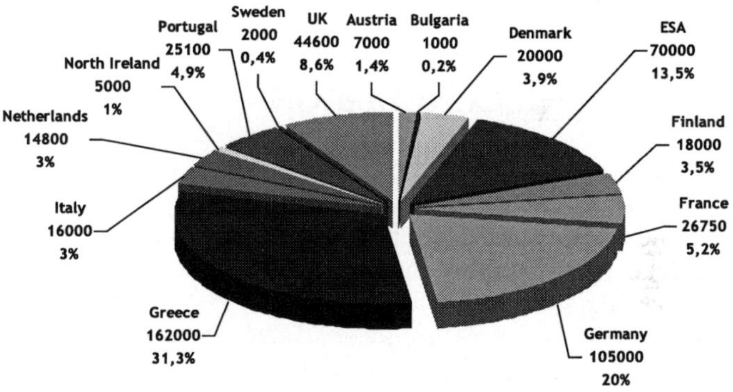

FIGURE 3. DVD distribution statistics.

Around 20,000 DVD/CD Soundtrack bundles are sold inexpensively via the company SPV. This DVD/CD Soundtrack bundle even managed to make it into Amazon.de top 100 CD Chart at number 83. The bundle managed to stay in top 3 of Amazon.de Top10 Chart for Documentaries for months.

The production budget was kept under 20.000 Euros. This figure does not include the production costs of the physical copies of the products (which were paid by partners in each country).

TABLE 1. European partners that collaborated in the Hubble Anniversary project

Partner	Country
Academy of Athens	Greece
Eugenides Foundation Planetarium	Greece
Eleftherotypia Newspaper	Greece
Planetarium Hamburg	Germany
Sterne und Weltraum	Germany
Astronomie Heute	Germany
New Scientist	UK
ESPACE magazine	France
Expresso	Portugal
IMAX Tycho Brahe Planetarium	Denmark
Danish National Space Center	Denmark
Politiken	Denmark
Ursa Astronomical Association	Finland
The Finnish National Technology Agency	Finland
Centro Multimeios de Espinho	Portugal
Fundação Navegar	Portugal
Le Stelle	Italy
Zeiss Planetarium Vienna	Austria
SDC	Germany
Armagh Planetarium	UK
DeKoepel	The Netherlands
Veen Magazines	The Netherlands

CONCLUSION

Exploiting the trans-national dimension was seen as a natural and mandatory component of this project. Collaborators and partners from more than 20 EU member states and third countries joined in the collaboration. This truly unique multinational initiative created a multiple win-win situation for everyone involved from the participants in the production to the national partners and the end-users.

This project will hopefully act as a pilot for future international collaboration in astronomy communication and public outreach activities throughout Europe and the informal network created will be exploited for many years to come.

More information regarding the project details and products can be found at: `http://www.spacetelescope.org`.

ACKNOWLEDGMENTS

We would like again to thank everyone involved in this massive project! The list of people to thank is unfortunately too large for this article, but most are listed in the end-titles of the DVD and in the website. This project would never have happened without them!

AUTHOR INDEX

A

B

C

D

E

F

G

H

J

K

L

M

N

O

P

R

S

Town of Lixourion, (ancient Palli), Kefallinia: Reception in the gardens of the "Iakovateios" Library and National Museum (Photographs Charalambos Solomos).

100 km
Lixourion

HELLENIC ASTRONOMICAL SOCIETY
7th Hellenic Astronomical Conference
Recent Advances in Astronomy & Astrophysics
Lixourion, Kefallinia Island, Greece 8-11 September 2005
Special Meetings:
Future of Greek Astronomy: Ideas, Plans and Policies
Greece and the European Space Agency
http://comas.interzone.gr/conf7/
Supported by:
Co-organized with:

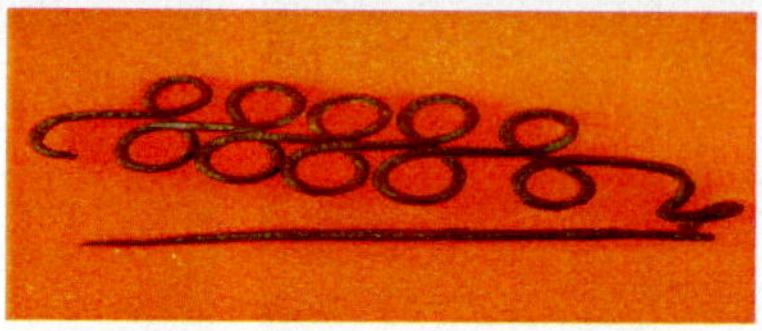

7TH ASTRONOMY CONFERENCE OF THE
HELLENIC ASTRONOMICAL SOCIETY
RECENT ADVANCES IN ASTRONOMY & ASTROPHYSICS
LIXOURION, KEFALLINIA ISLAND, GREECE
SEPTEMBER 8-11, 2005
ORGANIZED BY
THE HELLENIC ASTRONOMICAL SOCIETY (HEL.A.S)